Interfacial Transport Phenomena

John C. Slattery

Interfacial Transport Phenomena

With 185 Illustrations

Springer-Verlag
New York Berlin Heidelberg
London Paris Tokyo Hong Kong

John C. Slattery
Department of Chemical Engineering
Texas A&M University
College Station, Texas 77843-3122
USA

Library of Congress Cataloging-in-Publication Data
Slattery, John Charles, 1932–
Interfacial transport phenomena / John C. Slatery.
p. cm.
Includes bibliographical references and index.
ISBN 0-387-97387-7 (alk. paper). — ISBN 3-540-97387-7 (alk. paper)
1. Transport theory. 2. Surfaces (Physics) 3. Energy transport. 4. Mass transfer. 5. Heat—Transmission. 6. Entropy. I. Title.
QC175.2.S59 1990
530.1'38—dc20 90-10123
CIP

Printed on acid-free paper.

Camera-ready copy prepared by the author using Egg Bookmaker Interface.
Printed and bound by: Edwards Brothers, Inc., Ann Arbor, Michigan.
Printed in the United States of America.

9 8 7 6 5 4 3 2 1

ISBN 0-387-97387-7 Springer-Verlag New York Berlin Heidelberg
ISBN 3-540-97387-7 Springer-Verlag Berlin Heidelberg New York

Preface

Transport phenomena is used here to describe momentum, energy, mass, and entropy transfer (Bird *et al.* 1960, 1980). It includes thermodynamics, a special case of which is thermo*statics*. *Interfacial transport phenomena* refers to momentum, energy, mass, and entropy transfer within the immediate neighborhood of a phase interface, including the thermodynamics of the interface.

In terms of qualitative physical observations, this is a very old field. Pliny the Elder (Gaius Plinius Secundus, 23-79 A.D.; Pliny 1938) described divers who released small quantities of oil from their mouths, in order to damp capillary ripples on the ocean surface and in this way provide more uniform lighting for their work. Similar stories were retold by Benjamin Franklin, who conducted experiments of his own in England (Van Doren 1938).

In terms of analysis, this is a generally young field. Surface thermostatics developed relatively early, starting with Gibbs (1948) and continuing with important contributions by many others (see Chapter 5). Derjaguin and Landau (1941) and Verwey and Overbeek (1948) indicated how London-van der Waals and electrostatic double-layer forces were to be incorporated in continuum mechanics, now often referred to as DLVO theory. But prior to 1960, there were relatively few notable papers concerned with the analysis of dynamic systems. Two stand out in my mind. Boussinesq (1913) recognized the surface stress tensor and proposed the constitutive equation that we now refer to as the Boussinesq surface fluid model (Sec. 2.2.2). Unfortunately, he did not carry out an experiment in which the effects of the interfacial viscosities could be clearly recognized. While many studies of the surface viscosities followed, the corresponding data analyses were not convincing. Brown et al. (1953) appear to have been the first to demonstrate how the interfacial shear viscosity could be measured in a limit where the viscous effects in the adjacent phases could be neglected with respect to those in the interface (Sec. 3.4.1).

More recently, interest in analysis has begun to flourish within this area. Since many people have made important contributions, the best that I can do briefly is to indicate a few papers that have had particular meaning for me. Scriven (1960) restated the Boussinesq surface fluid model in a form more convenient for analysis. Burton and Mannheimer (1967; Osborne 1968; Mannheimer and Schechter 1968, 1970; Pintar *et al.* 1971) analyzed and demonstrated the deep channel surface viscometer, which is still the recommended technique for measuring relatively small surface shear viscosities (Exercise 3.4.1-3 and Sec. 3.5.1). Dussan V. and Davis (1974), through both analysis and experiment, pointed out with unusual clarity the contradictions to be reconciled in describing a moving common line (Secs. 1.2.9 through 1.2.11 and 1.3.9). By analyzing a thin film, Israelachvili (1985) derived an expression for interfacial tension that is in excellent agreement with experimental measurements, demonstrating that continuum mechanics can be usefully extended to regions having molecular dimensions (Exercise 4.1.4-3).

With the appearance of these papers, there were also questions. Were the surface viscosities real physical parameters or were they artifacts of the manner in which the surface viscometer was analyzed? Was the measured value of the surface shear viscosity consequently dependent upon the viscometer used to measure it? Was the introduction of the surface stress tensor consistent with some general view of continuum mechanics? Could the effects of the surface viscosities be observed in any situations judged to be of practical importance? Was there really slip in the neighborhood of a moving common line? Was it possible to successfully apply continuum mechanics to the very thin films within the neighborhood of a common line? In trying to answer questions like these for my students, I decided to prepare this book.

This book is written both as a guide for those preparing for active research in transport phenomena and as a reference for those currently working in the area. The emphasis is upon achieving understanding starting from the fundamental postulates. The dominant theme is the translation of physical problems into mathematical terms.

I normally introduce my students to this book after they have completed the first semester of lectures from my first book (Slattery 1981). The text is self-contained, but I would prefer to see the reader already conversant with analogous discussions for single phases. Although I have lectured from this text here at Texas A & M, it is written with the intention of being sufficiently complete to be used for self-study. This is the manner in which most of my students have employed the text as it was being written. All of the exercises have answers. Where appropriate, the reader is led through an exercise, since the objective is not to test his comprehension of the preceding text. The exercises are used as a literary device to transmit information relevant to the text without overwhelming the reader with additional details.

In many respects this book was a group effort. Many colleagues have influenced and directed my thinking through conversations, by listening to their talks at meetings, and by reading their papers. While I have not been able to provide complete answers to all of their questions, I have been able to finish this book only through the continued probing, encouragement, and active help of my students. Jing-Den Chen and M. Sami Selim offered comments on portions of the final manuscript. My wife Bea and Brenda Wilson cheerfully typed and retyped through many revisions over many years, never questioning whether the book would finally be completed. The final manuscript was prepared by Cheri Sandlin, with assistance from Ruth Heeremans and Izora Brown. Alfred Li provided invaluable help and support through the long months of proof reading, correcting the final manuscript, and preparing indices. The Peregrine Falcon Company made available a test copy of *THE EGG BOOKMAKER INTERFACE* (The Peregrine Falcon Co., P. O. Box 8155, Newport Beach, CA 92658-8155), in which the camera-ready copy was typed. David Adelson further modified this test copy, permitting me to use boldface greek, boldface script, boldface brackets (for jumps at interfaces), and boldface parentheses (for jumps at common lines). Joel Meyer and Peter Weiss prepared the final forms of the figures. Stephen H. Davis shared with me the original photographs from his work with Elizabeth B. Dussan in Sec. 1.2.9. Richard Williams and the David Sarnoff Research Center provided both the previously published and the previously unpublished photographs from his work that also appear in Sec. 1.2.9. My friends and colleagues at Northwestern University, where most of this book was written between 1972 and 1989, gave me their patience and encouragement. Thanks to you all.

College Station, Texas
July 10, 1990

References

Bird, R. B., W. E. Stewart, and E. N. Lightfoot, "Transport Phenomena," John Wiley, New York (1960).

Bird, R. B., W. E. Stewart, and E. N. Lightfoot, *Advances in Chemistry Series No. 190*, p. 153, edited by W. F. Furter, American Chemical Society, Washington, D.C. (1980).

Boussinesq, J., *Comptes Rendus des Seances de l' Acade'mie des Sciences* **256**, 983, 1035, 1124 (1913).

Brown, A. G., W. C. Thuman, and J. W. McBain, *J. Colloid Sci.* **8**, 491 (1953).

Burton, R. A., and R. J. Mannheimer, "Ordered Fluids and Liquid Crystals," Advances in Chemistry Series No. 63, p. 315, American Chemical Society, Washington, D.C. (1967).

Derjaguin, B. V., and L. D. Landau, *Acta physicochim. URSS* **14**, 633 (1941).

Dussan V., E. B., and S. H. Davis, *J. Fluid Mech.* **65**, 71 (1974).

Gibbs, J. W., "The Collected Works," vol. 1, Yale University Press, New Haven, Conn. (1948).

Israelachvili, J. N., "Intermolecular and Surface Forces," Academic Press, London (1985).

Mannheimer, R. J., and R. S. Schechter, *J. Colloid Interface Sci.* **27**, 324 (1968).

Mannheimer, R. J., and R. S. Schechter, *J. Colloid Interface Sci.* **32**, 195 (1970).

Osborne, M. F. M., *Kolloid-Z. Z. Polym.* **224**, 150 (1968).

Pintar, A. J., A. B. Israel, and D. T. Wasan, *J. Colloid Intefface Sci.* **37**, 52 (1971).

Pliny, "Natural History," vol. 1, p. 361 (book II, 234 in original), Harvard University Press, Cambridge, MA (1938).

Scriven, L. E., *Chem. Eng. Sci.* **12**, 98 (1960).

Slattery, J. C., "Momentum, Energy, and Mass Transfer in Continua," McGraw-Hill, New York (1972); second edition, Robert E. Krieger, Malabar, FL 32950 (1981).

Van Doren, C., "Benjamin Franklin," p. 433, Viking Press, New York (1938).

Verwey, E. J. W., and J. Th. G. Overbeek, "Theory of the Stability of Lyophobic Colloids," Elsevier, Amsterdam (1948).

Contents

1

Kinematics and conservation of mass

This chapter as well as appendix A may be thought of as introductory for the main story that I have to tell. In appendix A, I introduce the mathematical language that we shall be using in describing phenomena at phase interfaces. In this chapter, I describe how the motions of real multiphase materials can be represented using the continuum point of view. To bring out the principal ideas as clearly as possible, I have chosen to confine my attention in these first chapters either to a material composed of a single species or to a material in which there are no concentration gradients. The conditions under which these results are applicable to multicomponent materials will be clear later.

There are two basic models for real materials: the particulate or molecular model and the continuum model. We all agree that the most realistically detailed picture of the world around us requires that materials be composed of atoms and molecules. In this picture, mass is distributed discontinuously throughout space; mass is associated with protons, neutrons, and electrons, which are separated by relatively large voids. In contrast, the continuum model requires that mass be distributed continuously through space.

The continuum model is less realistic than the particulate model, but far simpler. Experience has shown that for many purposes the more accurate details of the particulate model are not necessary. To our sight and touch, mass appears to be continuously distributed throughout the water which we drink and the air which we breathe. Our senses suggest that there is a large discontinuity in density across the static surface defined by our desk top or the moving and deforming surface of the ocean. The problem may be analogous in some ways to the study of traffic patterns on an expressway: the speed and spacing of the automobiles are important, but we probably should not worry about their details of construction or the clothing worn by the drivers.

The distinction between the particulate and continuum models should be maintained. In the context of a continuum representation, one sometimes hears a statement to the effect that a region is large enough to contain many molecules ... but small enough to represent a point in space ...

or small enough to be used as an element for integration. This makes little sense for two reasons. In continuum mechanics, mass is continuously distributed through space and molecules are not defined. When we talk about *material particles*, we are not using another word for *molecule*. The material particle is a primitive concept (primitive in the sense that it is not defined) that allows us to attach a convenient name to the material at a particular location in a reference configuration.

But in reading this and the succeeding chapters, I am not asking you to forget that real materials are actually composed of atoms and molecules. The molecular picture may help us to decide what to say in terms of a continuum model, as when we assign a mass density to the dividing surface. It is only through the use of a molecular model and statistical mechanics that a complete a priori prediction about material behavior can be made. Continuum mechanics alone can never yield the density or viscosity of carbon dioxide at room temperature and pressure.

For an interesting discussion of the role of continuum mechanics in physics, I recommend the introduction to Truesdell and Toupin's (1960, pp. 226-235) review of the foundations of continuum mechanics.

Reference

Truesdell, C., and R. A. Toupin, "Handbuch der Physik," vol. 3/1, edited by S. Flügge, Springer-Verlag, Berlin (1960).

1.1 Motion

1.1.1 Body, motion, and material coordinates We are concerned in this text with moving and deforming bodies of material undergoing momentum, energy, and mass transfer. Let us begin with the model for a material body that will be the basis for our discussion. In order not to introduce too many ideas at once, let us confine our attention for the moment to a body consisting of a single phase.

There are four equivalent methods for describing the motion of a body: the material, the spatial, the referential, and the relative (Truesdell 1966, p. 9). All four are important to us.

Material description A **body** is a set; any element ζ of the set is called a particle or a **material particle**. A one-to-one continuous mapping of this set onto a region of the three-dimensional euclidean space (E^3, V^3) (see Sec. A.1.1) exists and is called a **configuration** of the body:

$$\mathbf{z} = \boldsymbol{\chi}(\zeta) \tag{1-1}$$

$$\zeta = \chi^{-1}(\mathbf{z}) \tag{1-2}$$

The point $\mathbf{z} = \boldsymbol{\chi}(\zeta)$ of (E^3, V^3) is called the place occupied by the particle ζ; ζ is the particle at the place $\mathbf{z}$ in (E^3, V^3).

A **motion** of a body is a one-parameter family of configurations; the real parameter t is time. We write

$$\mathbf{z} = \boldsymbol{\chi}(\zeta, t) \tag{1-3}$$

and

$$\zeta = \chi^{-1}(\mathbf{z}, t) \tag{1-4}$$

Let A be any quantity: scalar, vector, or tensor. We shall have occasion to talk about the time derivative of A following the motion of a particle:

$$\frac{d_{(m)}A}{dt} \equiv \left(\frac{\partial A}{\partial t}\right)_{\zeta} \tag{1-5}$$

We shall refer to this as the material derivative of A. For example, the **velocity** vector $\mathbf{v}$ represents the time rate of change of position of a material particle,

$$\mathbf{v} \equiv \frac{d_{(m)}\mathbf{z}}{dt} \equiv \left(\frac{\partial \boldsymbol{\chi}}{\partial t}\right)_{\zeta} \tag{1-6}$$

In the **material description**, we deal directly with the abstract particles in terms of which the body is defined.

Spatial description Of necessity, experimental measurements are made on a body in its current configuration, the region of space currently occupied by the body. In the **spatial description**, we focus our attention upon the body in its present configuration.

Using (1-4), we may replace any function $g(\zeta, t)$ by a function of $\mathbf{z}$ and t.

$$g(\zeta, t) = G(\mathbf{z}, t) \equiv g(\chi^{-1}(\mathbf{z}, t), t) \tag{1-7}$$

In particular, this is how we obtain from (1-6) an expression for velocity as a function of position in space and time,

$$\mathbf{v} = \mathbf{v}(\mathbf{z}, t) \tag{1-8}$$

Granted that the spatial description is convenient for experimentalists. But it is awkward for any discussions of principles that may be more naturally expressed in terms of the body itself rather than its present configuration.

Referential description The concept of a material particle is abstract. We have no way of directly following the material particles of a body. We are able to observe only spatial descriptions of a body. This suggests that we identify the material particles of a body by their positions in some particular configuration. This **reference configuration** may be, but need not be, one actually occupied by the body in the course of its motion. We might choose a fluid to have been in its reference configuration while sitting in a beaker on the lab bench before the experiment of interest was begun. The place of a material particle in the reference configuration $\boldsymbol{\kappa}$ will be denoted by

$$\mathbf{z}_{\kappa} = \boldsymbol{\kappa}(\zeta) \tag{1-9}$$

The particle at the place $\mathbf{z}_{\kappa}$ in the configuration $\boldsymbol{\kappa}$ may be expressed as

$$\zeta = \kappa^{-1}(\mathbf{z}_{\kappa}) \tag{1-10}$$

If $\boldsymbol{\chi}$ is a motion of the body in (1-3), then

$$\mathbf{z} = \boldsymbol{\chi}_{\kappa}(\mathbf{z}_{\kappa}, t) \equiv \boldsymbol{\chi}(\kappa^{-1}(\mathbf{z}_{\kappa}), t) \tag{1-11}$$

and from (1-4)

$$\mathbf{z}_{\kappa} = \boldsymbol{\chi}_{\kappa}^{-1}(\mathbf{z}, t) \equiv \boldsymbol{\kappa}(\chi^{-1}(\mathbf{z}, t)) \tag{1-12}$$

These expressions describe the motion in terms of the reference configuration κ. We say that they define a family of **deformations** from κ. The subscript is to remind you that the form of $\boldsymbol{\chi}_\kappa$ depends upon the choice of reference configuration.

The **deformation gradient** is defined in terms of (1-11):

$$\mathbf{F} \equiv \mathbf{F}_\kappa(\mathbf{z}_\kappa, t) \equiv \text{grad}\, \boldsymbol{\chi}_\kappa(\mathbf{z}_\kappa, t) \tag{1-13}$$

It tells how position in the current configuration is changed as the result of a small change of location in the reference configuration. It transforms a vector described with respect to the reference configuration into one expressed in terms of the current configuration. When we speak about the fixed reference configuration of a body, let us use as rectangular cartesian coordinates ($z_{\kappa 1}$, $z_{\kappa 2}$, $z_{\kappa 3}$); the corresponding basis fields are ($\mathbf{e}_{\kappa 1}$, $\mathbf{e}_{\kappa 2}$, $\mathbf{e}_{\kappa 3}$). In talking about the current configuration of the body, we use as rectangular cartesian coordinates (z_1, z_2, z_3); the corresponding basis fields are ($\mathbf{e}_1$, $\mathbf{e}_2$, $\mathbf{e}_3$). For clarity, we will use grad when taking a gradient with respect to the reference configuration and ∇ when taking a gradient with respect to the current configuration. In terms of these coordinates, the deformation gradient may be expressed as

$$\mathbf{F} = \frac{\partial \boldsymbol{\chi}_\kappa}{\partial z_{\kappa M}} \mathbf{e}_{\kappa M}$$

$$= \frac{\partial z_m}{\partial z_{\kappa M}} \mathbf{e}_m \mathbf{e}_{\kappa M} \tag{1-14}$$

The inverse

$$\mathbf{F}^{-1} \equiv \nabla \boldsymbol{\chi}_\kappa^{-1}$$

$$= \frac{\partial \boldsymbol{\chi}_\kappa^{-1}}{\partial z_n} \mathbf{e}_n$$

$$= \frac{\partial z_{\kappa N}}{\partial z_n} \mathbf{e}_{\kappa N} \mathbf{e}_n \tag{1-15}$$

tells how position in the reference configuration is changed as the result of an incremental move in the current configuration. The inverse has the usual properties:

$$\mathbf{F} \cdot \mathbf{F}^{-1} = \frac{\partial z_m}{\partial z_{\kappa M}} \frac{\partial z_{\kappa N}}{\partial z_n} \delta_{MN}\, \mathbf{e}_m \mathbf{e}_n$$

$$= \delta_{mn}\, \mathbf{e}_m \mathbf{e}_n$$

$$= \mathbf{I} \tag{1-16}$$

and

$$\mathbf{F}^{-1} \cdot \mathbf{F} = \frac{\partial z_{\kappa N}}{\partial z_n} \frac{\partial z_m}{\partial z_{\kappa M}} \delta_{mn} \, \mathbf{e}_{\kappa N} \mathbf{e}_{\kappa M}$$

$$= \delta_{MN} \, \mathbf{e}_{\kappa N} \mathbf{e}_{\kappa M}$$

$$= \mathbf{I} \tag{1-17}$$

The physical meaning of the material derivative introduced in (1-5) may now be clarified somewhat, if we think of it as a derivative with respect to time holding position in the reference configuration fixed:

$$\frac{d_{(m)} \mathbf{A}}{d t} = \left(\frac{\partial \mathrm{A}}{\partial t} \right)_{\mathbf{z}_\kappa} = \left(\frac{\partial \mathrm{A}}{\partial t} \right)_{z_{\kappa 1}, z_{\kappa 2}, z_{\kappa 3}} \tag{1-18}$$

In particular, the velocity vector becomes in terms of (1-11)

$$\mathbf{v} \equiv \frac{d_{(m)} \mathbf{z}}{d t} = \left(\frac{\partial \boldsymbol{\chi}_\kappa (\mathbf{z}_\kappa, t)}{\partial t} \right)_{\mathbf{z}_\kappa} = \left(\frac{\partial \boldsymbol{\chi}_\kappa}{\partial t} \right)_{z_{\kappa 1}, z_{\kappa 2}, z_{\kappa 3}} \tag{1-19}$$

Relative description When we think of a reference configuration, we almost invariably visualize one that is fixed with respect to time. But this is not necessary. There is no reason why a reference configuration can not be a function of time. The choice of the current configuration as the reference is particularly appropriate when we wish to compare the past with the present. The corresponding description of the motion is called **relative**.

Let $\bar{z}$ be the position occupied by the particle ζ at some time $\bar{t}$:

$$\bar{\mathbf{z}} = \boldsymbol{\chi}(\zeta, \bar{t}) \tag{1-20}$$

At the present time t, this same particle is at the position z:

$$\mathbf{z} = \boldsymbol{\chi}(\zeta, t) \tag{1-21}$$

It follows immediately that the motion is described by

$$\bar{\mathbf{z}} = \boldsymbol{\chi}_t (\mathbf{z}, \bar{t})$$

$$\equiv \boldsymbol{\chi}(\chi^{-1} (\mathbf{z}, t), \bar{t}) \tag{1-22}$$

which tells us the position at time $\bar{t}$ of the particle that is currently at $\mathbf{z}$. The function $\boldsymbol{\chi}_t$ might be called the relative motion, but is usually referred to as the **relative deformation.**

For this choice of reference configuration, the deformation gradient becomes

$$\mathbf{F}_t \equiv \mathbf{F}_t(\bar{t}) = \nabla\boldsymbol{\chi}_t(\mathbf{z}, \bar{t}) \tag{1-23}$$

This is known as the **relative deformation gradient**. Note

$$\mathbf{F}_t(t) = \mathbf{I} \tag{1-24}$$

From (1-13), it follows that

$$\mathbf{F}(\bar{t}) = \mathbf{F}_t(\bar{t}) \cdot \mathbf{F}(t) \tag{1-25}$$

Reference

Truesdell, C., "The Elements of Continuum Mechanics," Springer-Verlag, New York (1966).

1.1.2 Stretch and rotation (Truesdell 1966, p. 17) Because the deformation gradient $\mathbf{F}$ is non-singular, the polar decomposition theorem (Ericksen 1960, p. 841; Halmos 1958, p. 169; Sec. A.3.8) allows us to express it in two different forms:

$$\mathbf{F} = \mathbf{R} \cdot \mathbf{U} = \mathbf{V} \cdot \mathbf{R} \tag{2-1}$$

Here $\mathbf{R}$ is a unique orthogonal tensor; $\mathbf{U}$ and $\mathbf{V}$ are unique, positive-definite symmetric tensors.

Physically, (2-1) tells us that $\mathbf{F}$ may be obtained by imposing pure stretches of amounts $u_{(j)}$ along three mutually orthogonal directions $\mathbf{x}_{(j)}$

$$\mathbf{U} \cdot \mathbf{x}_{(j)} = u_{(j)}\mathbf{x}_{(j)} \tag{2-2}$$

followed by a rigid rotation of those directions. Alternatively, the same rotation might be effected first, followed by the same stretches in the resulting directions:

$$\begin{aligned} \mathbf{V} \cdot (\mathbf{R} \cdot \mathbf{x}_{(j)}) &= (\mathbf{R} \cdot \mathbf{U} \cdot \mathbf{R}^T) \cdot (\mathbf{R} \cdot \mathbf{x}_{(j)}) \\ &= u_{(j)}\mathbf{R} \cdot \mathbf{x}_{(j)} \end{aligned} \tag{2-3}$$

The tensors **U** and **V** have the same principal values, but different principal axes: **R** is the rotation that transforms the principal axes of **U** into the principal axes of **V**. We refer to **R** as the **rotation** tensor; **U** is the **right stretch** tensor and **V** the **left stretch**.

The orthogonal tensor **R** need not be proper orthogonal: det $\mathbf{R} = \pm 1$. However, continuity requires that det **R** keep either one value or the other for all time t (Slattery 1981, Exercise 1.3.3-3). As a result

$$J \equiv | \det \mathbf{F} | = \det \mathbf{U} = \det \mathbf{V} \tag{2-4}$$

Instead of **U** and **V**, it is often more convenient to use the **right** and **left Cauchy-Green** tensors

$$\mathbf{C} \equiv \mathbf{U}^2 = \mathbf{F}^T \cdot \mathbf{F} \tag{2-5}$$

$$\mathbf{B} \equiv \mathbf{V}^2 = \mathbf{F} \cdot \mathbf{F}^T \tag{2-6}$$

since they can be rather simply calculated from the deformation gradient.

If we start with the relative deformation gradient $\mathbf{F}_t$, we can introduce in the same manner the **relative rotation** $\mathbf{R}_t$, the **relative stretch** tensors $\mathbf{U}_t$ and $\mathbf{V}_t$, and the **relative Cauchy-Green** tensors $\mathbf{C}_t$ and $\mathbf{B}_t$:

$$\mathbf{F}_t = \mathbf{R}_t \cdot \mathbf{U}_t = \mathbf{V}_t \cdot \mathbf{R}_t \tag{2-7}$$

$$\mathbf{C}_t \equiv \mathbf{U}_t^2 = \mathbf{F}_t^T \cdot \mathbf{F}_t \tag{2-8}$$

$$\mathbf{B}_t \equiv \mathbf{V}_t^2 = \mathbf{F}_t \cdot \mathbf{F}_t^T \tag{2-9}$$

References

Ericksen, J. L., "Handbuch der Physik," vol. 3/1, edited by S. Flügge, Springer-Verlag, Berlin (1960).

Halmos, P. R., "Finite-Dimensional Vector Spaces," Van Nostrand, Princeton, NJ (1958).

Slattery, J. C., "Momentum, Energy, and Mass Transfer in Continua," McGraw-Hill, New York (1972); second edition, Robert E. Krieger, Malabar, FL 32950 (1981).

Truesdell, C., "The Elements of Continuum Mechanics," Springer-Verlag, New York (1966).

1.2 Motion of multiphase bodies

1.2.1 What are phase interfaces? Let us define a **phase interface** to be that region separating two phases in which the properties or behavior of the material differ from those of the adjoining phases.

There is considerable evidence that density and the concentrations of the various species present are appreciably different in the neighborhood of an interface (Defay *et al.* 1966, p. 29). As the critical point is approached, density is observed to be a continuous function of position in the direction normal to the interface (Hein 1914; Winkler and Maass 1933; Maass 1938; McIntosh *et al.* 1939; Palmer 1952). This suggests that the phase interface might be best regarded as a three-dimensional region, the thickness of which may be several molecular diameters or more.

Molecular models for the interfacial region are a separate subject that we will not discuss here. It is perhaps sufficient to mention that the three-dimensional character of the phase interface is explicitly recognized in statistical mechanical calculations (Ono and Kondo 1960).

There are two continuum models for the phase interface. One model represents it as a three-dimensional region; the other as a two-dimensional surface. They are considered in the next two sections.

References

Defay, R., I. Prigogine, A. Bellemans, and D. H. Everett, "Surface Tension and Absorption," John Wiley, New York (1966).

Hein, P., *Z. Phys. Chem.* **86**, 385 (1914).

Maass, O., *Chem. Rev.* **23**, 17 (1938).

McIntosh, R. L., J. R. Dacey, and O. Maass, *Can. J. Res. Sect. B* **17**, 241 (1939).

Ono, S. and S. Kondo, "Handbuch der Physik," vol. 10, edited by S. Flügge, Springer-Verlag, Berlin (1960).

Palmer, H. B., Doctoral Dissertation, University of Wisconsin (1952), as reported by J. O. Hirschfelder, C. F. Curtiss, and R. B. Bird, "Molecular Theory of Gases and Liquids," second corrected printing with notes added, John Wiley, New York (1954), p. 373-374.

Winkler, C. A., and O. Maass, *Can. J. Res.* **9**, 613 (1933).

1.2.2 Three-dimensional interfacial region Perhaps the most obvious continuum model to propose for the interface is a three-dimensional region of finite thickness.

Korteweg (1901) suggested that the stress-deformation behavior in such a region could be described by saying that the stress tensor is a function of the rate of deformation tensor, the gradient of density and the second gradient of density. He used a linear form of this relationship to analyze the stresses in a spherical shell that represented the interface of a static spherical bubble. In the limit as the thickness of the shell was allowed to approach zero, the result took the same form as that obtained by assuming a uniform tension acts in a two-dimensional spherical surface separating the two phases.

While Korteweg's approach is appealing, there are inherent difficulties. We have no way of studying experimentally the stress distribution and velocity distribution in the very thin interfacial region. His model for interfacial stress-deformation behavior can be tested only by observing the effect of the interfacial region upon the adjoining phases. Such observations are complicated by the fact that apparently no dynamic problems have been solved using Korteweg's model for interfacial behavior.

In a variation on the approach of Korteweg, Deemer and Slattery (1978; see Sec. 2.3.1) have modeled a dilute solution of surfactant molecules in the interfacial region by a dilute suspension of rigid bodies.

References

Deemer, A. R., and J. C. Slattery, *Int. J. Multiphase Flow* **4**, 171 (1978).

Korteweg, D. J., *Arch. Neerl. Sci. Exactes Nat.* (2) **6**, 1 (1901), as reported by C. Truesdell and W. Noll, “Handbuch der Physik,” vol. 3/3, edited by S. Flügge, Springer-Verlag, Berlin (1965), p. 513.

1.2.3 Dividing surface Gibbs (1928, p. 219) proposed the following model for a phase interface in a body at rest or at equilibrium: a hypothetical two-dimensional **dividing surface** that lies within or near the interfacial region and separates two homogeneous phases. By a homogeneous phase, he meant one in which all variables, such as mass density and stress, assume uniform values. He suggested that the cumulative effects of the interface upon the adjoining phases be taken into account by the assignment to the dividing surface of any excess mass or energy not accounted for by the adjoining homogeneous phases.

Gibbs' approach may be extended to include dynamic phenomena, if we define a homogeneous phase to be one throughout which each constitutive equation or description of material behavior applies uniformly.

As in the static case, the cumulative effects of the interface upon the adjoining phases can be described by associating densities and fluxes with the dividing surface.

Our primary model for an interface in this text will be the dividing surface.

The natural question is whether we have compromised the theory by choosing the simplest of the three models available: the particulate model of statistical mechanics, the three-dimensional interfacial region, and the dividing surface. The advantage of the particulate model of statistical mechanics is realism and the ability to predict macroscopic material behavior, given a description of intermolecular forces. One chooses to use continuum mechanics for simplicity in solving problems and for ease in analysis of experimental data. In much the same way, the potential advantage of the three-dimensional interfacial region is greater realism and the ability to predict excess properties, given a description of the material behavior within the interfacial region. The advantage of the dividing surface is simplicity in solving problems and ease in analysis of experimental data. Ultimately, the real test of a continuum theory is whether or not it is in agreement with experimental observations (Truesdell and Noll 1965, p. 5).

The dividing surface is normally said to be sensibly coincident with the phase interface. More precise definitions for the location of the dividing surface are introduced in Secs. 1.3.6 and 5.2.3.

References

Gibbs, J. W., "The Collected Works of J. Willard Gibbs," vol. 1, Yale University Press, New Haven, CT (1928).

Truesdell, C., and W. Noll, "Handbuch der Physik," vol. 3/3, edited by S. Flügge, Springer-Verlag, Berlin (1965).

1.2.4 Dividing surface as model for three-dimensional interfacial region If both a three-dimensional region of finite thickness and a dividing surface can be used as models for a phase interface, then we should be able to express one model in terms of the other. The result is a more detailed interpretation for the quantities to be associated with the dividing surface.

In Figure 1.2.4-1, we show a dividing surface superimposed upon a three-dimensional interfacial region. For any variable such as mass density, there are two distributions to be considered on either side of the dividing surface: the interfacial distribution corresponding to the three-dimensional interfacial region of finite thickness and the bulk distribution appropriate to

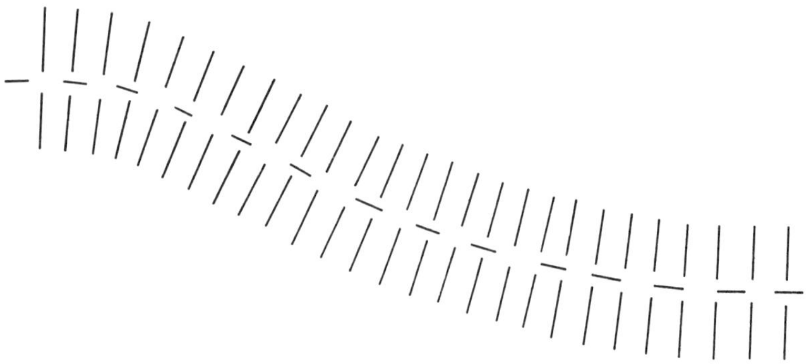

Figure 1.2.4-1: Dividing surface superimposed upon three-dimensional interfacial region of finite thickness

the dividing surface model. The surface mass density[a] can be interpreted in terms of the difference between these interfacial and bulk distributions (see Sec. 1.3.2).

Buff (1956) and Buff and Saltsburg (1957) assumed a simple expression for the actual stress tensor in the interfacial region at equilibrium and studied surface tension.

Deemer and Slattery (1978) have developed expressions for the densities, fluxes and source terms appropriate to the dividing surface model in terms of the densities, fluxes and source terms arising in the context of the three-dimensional model for the interfacial region. These will be developed later as I introduce each of the primary postulates (conservation of mass, Euler's first law, ...).

References

Buff, F. P., *J. Chem. Phys.* **25**, 146 (1956).

Buff, F. P., and H. Saltsburg, *J. Chem. Phys.* **26**, 23 (1957).

Deemer, A. R., and J. C. Slattery, *Int. J. Multiphase Flow* **4**, 171 (1978).

---

a) Densities and fluxes associated with a dividing surface have often been referred to as **excess.**

1.2.5 Motion of dividing surface In Sec. 1.1.1, we considered bodies consisting of a single phase. Now we turn our attention to our primary concern: multiphase bodies.

It may help to start with a physical picture in mind. A tray of water, initially consisting of a single liquid phase, is slipped into the freezer and begins to turn into ice. Sometime later there are two phases in the tray. A portion of the water can be readily assigned to the liquid phase, a portion to the solid phase, and (possibly) a portion to the interface.

We said in Sec. 1.1.1 that a body is a set, any element ζ of which is called a particle or a material particle. A subset is known as a portion of the body. There exists a one-to-one continuous mapping onto a region of the three-dimensional euclidean space (E^3, V^3) for any portion of a body whose current configuration is a single phase, and the discussion given in Sec. 1.1.1 applies without change. This means that we must give special attention only to the dividing surface.

We will require all four methods for describing the motion of a dividing surface: the material, the spatial, the referential, and the relative.

Material description A dividing surface Σ in (E^3, V^3) is the locus of a point whose position is a function of two parameters y^1 and y^2 (see Sec. A.1.3):

$$\mathbf{z} = \mathbf{p}^{(\sigma)}(y^1 , y^2) \tag{5-1}$$

The two **surface coordinates** y^1 and y^2 uniquely determine a point on the surface.

In order to allow the fraction of a body to be associated with a dividing surface to vary, we will say that there exists a many-to-one mapping of a portion of a multiphase body onto Σ,[a]

$$y^\alpha = X^\alpha(\zeta^{(\sigma)}) \qquad \alpha = 1, 2 \tag{5-2}$$

This mapping is the **intrinsic configuration** for the surface particles on Σ. Not only may there be many material particles occupying any given point on Σ but also, because of mass transfer across the dividing surface, the members of the set of material particles occupying any particular point on the dividing surface may vary with time. The set of all material particles at any point on Σ will be denoted by $\zeta^{(\sigma)}$ and will be referred to as a **surface particle**. Consequently, (5-2) and

---

a) In what follows a superscript $...^{(\sigma)}$ is used to emphasize that the quantity is associated with the dividing surface Σ. It will also distinguish surface entities from similar ones describing the bulk phases that are separated by Σ: compare (5-4) with (1.1.1-1).

$$\zeta^{(\sigma)} = X^{-1}(y^1, y^2) \tag{5-3}$$

may also be thought of as a one-to-one mapping of the set of surface particles onto Σ. From (5-2), the point (y^1, y^2) on Σ is the place occupied by $\zeta^{(\sigma)}$; (5-3) tells us which surface particle is at the place (y^1, y^2).

Equations (5-1) and (5-2) together give us an expression for the **configuration** of the surface particles in (E^3, V^3):

$$\begin{aligned} \mathbf{z} &= \boldsymbol{\chi}^{(\sigma)}(\zeta^{(\sigma)}) \\ &\equiv \mathbf{p}^{(\sigma)}(X^1(\zeta^{(\sigma)}), X^2(\zeta^{(\sigma)})) \end{aligned} \tag{5-4}$$

The mapping $\boldsymbol{\chi}^{(\sigma)}$ does not have an inverse; although there is a position in space corresponding to every surface particle, the converse is not true.

A moving and deforming dividing surface Σ in (E^3, V^3) is the locus of a point whose position is a function of two surface coordinates and time t,

$$\mathbf{z} = \mathbf{p}^{(\sigma)}(y^1, y^2, t) \tag{5-5}$$

An **intrinsic motion** of the surface particles on Σ is a one-parameter family of intrinsic configurations:

$$y^\alpha = X^\alpha(\zeta^{(\sigma)}, t) \qquad \alpha = 1, 2 \tag{5-6}$$

$$\zeta^{(\sigma)} = X^{-1}(y^1, y^2, t) \tag{5-7}$$

Equations (5-6) and (5-7) tell us how the surface particles move from point to point on the surface independently of how the surface itself is moving. Equations (5-5) and (5-6) together give us an expression for the **motion** of the surface particles in (E^3, V^3),

$$\begin{aligned} \mathbf{z} &= \boldsymbol{\chi}^{(\sigma)}(\zeta^{(\sigma)}, t) \\ &\equiv \mathbf{p}^{(\sigma)}(X^1(\zeta^{(\sigma)}, t), X^2(\zeta^{(\sigma)}, t), t) \end{aligned} \tag{5-8}$$

The motion of the surface particles in (E^3, V^3) is a one-parameter family of configurations. It is the result both of the movement of the surface Σ itself (5-5) and of the intrinsic motion of the surface particles within Σ (5-6 and 5-7).

Let A be any quantity: scalar, vector, or tensor. We shall have occasion to talk about the time derivative of A following the motion of a surface particle,

$$\frac{d_{(s)}A}{dt} \equiv \left(\frac{\partial A}{\partial t} \right)_{\zeta^{(\sigma)}} \tag{5-9}$$

We shall refer to this as the **surface material derivative** of A. For example, the **surface velocity** vector $\mathbf{v}^{(\sigma)}$ represents the time rate of change of position of a surface particle,

$$\mathbf{v}^{(\sigma)} \equiv \frac{d_{(s)}\mathbf{z}}{dt} \equiv \left(\frac{\partial \boldsymbol{\chi}^{(\sigma)}}{\partial t} \right)_{\zeta^{(\sigma)}} \tag{5-10}$$

Since the surface particles are confined to the dividing surface, they must at all times move with a normal component of velocity equal to the normal component of velocity of the surface (the speed of displacement of the surface as explained in Sec. 1.2.7). The tangential components of the velocity of a particular surface particle can be visualized as equal to the tangential components of the velocity of the collection of material particles represented by the surface particle. This suggests that we adopt in Chapter 3 the usual assumption of fluid mechanics: continuity of the tangential components of velocity across a dividing surface.

In the **material description**, we deal directly with the abstract particles in terms of which the body is defined.

Spatial description Experimental measurements are made on a body in its current configuration, the region of space currently occupied by the body. In the **spatial description**, we fix our attention upon the surface particles in their present configuration in (E^3, V^3).

Using (5-7), we may replace any function g($\zeta^{(\sigma)}$, t) by a function of y^1, y^2, and t:

$$g(\zeta^{(\sigma)}, t) = G(y^1, y^2, t)$$

$$\equiv g(\mathcal{X}^{-1} (y^1, y^2, t), t) \tag{5-11}$$

In particular, this is how we obtain from (5-10) an expression for surface velocity as a function of position on the surface and time,

$$\mathbf{v}^{(\sigma)} = \mathbf{v}^{(\sigma)}(y^1, y^2, t) \tag{5-12}$$

The spatial description is convenient for experimentalists. However, it is awkward in any discussions of principles that may be more naturally expressed in terms of the body itself rather than its present configuration.

Referential description The concept of a surface particle is abstract. We have no way of following the surface particles in a dividing surface. We are able to observe only spatial descriptions of a dividing surface. For example, at some **reference time** t_κ the dividing surface (5-5) takes the form

$$\mathbf{z}_\kappa = \boldsymbol{\kappa}^{(\sigma)}(y_\kappa^1, y_\kappa^2)$$

$$\equiv p^{(\sigma)}(y^1_\kappa, y^2_\kappa, t_\kappa) \tag{5-13}$$

which we can call the **reference dividing surface**. This together with (5-6) and (5-7) suggests that we identify surface particles by their **reference intrinsic configuration** or their position in this reference dividing surface:

$$y^A_\kappa = K^A(\zeta^{(\sigma)})$$

$$\equiv X^A(\zeta^{(\sigma)}, t_\kappa) \qquad A = 1, 2 \tag{5-14}$$

$$\zeta^{(\sigma)} = K^{-1}(y^1_\kappa, y^2_\kappa)$$

$$\equiv X^{-1}(y^1_\kappa, y^2_\kappa, t_\kappa) \tag{5-15}$$

Equations (5-13) and (5-14) describe the **reference configuration** of the surface particles in (E^3, V^3):

$$\mathbf{z}_\kappa = \boldsymbol{\kappa}^{(\sigma)}_K(\zeta^{(\sigma)})$$

$$\equiv \boldsymbol{\kappa}^{(\sigma)}(K^1(\zeta^{(\sigma)}), K^2(\zeta^{(\sigma)})) \tag{5-16}$$

In choosing a reference configuration and an intrinsic reference configuration, we have done more than say at what time we are looking at the dividing surface[b]. We have done more than say where the surface particles are at the reference time. Equation (5-13) having been given, the two linearly independent vector fields (A = 1, 2)

$$\mathbf{a}_{\kappa A} \equiv \frac{\partial \boldsymbol{\kappa}^{(\sigma)}}{\partial y^A_\kappa} \tag{5-17}$$

provide an orientation with respect to this reference dividing surface. Any set of surface coordinates could be used in describing the reference dividing surface. I have identified the (y^1_κ, y^2_κ) with y^1, y^2 only for ease in explanation.

---

b) The reference dividing surface could be described in the form

$$f(\mathbf{z}_\kappa) = 0$$

without ever mentioning the surface coordinates. See Sec. A.1.3.

We can now use (5-14) and (5-15) together with (5-6) and (5-7) to describe the **intrinsic deformation** from the reference intrinsic configuration:

$$\begin{aligned} y^{\alpha} &= X_{K}^{\alpha}(\, y_{\kappa}^{1}, y_{\kappa}^{2}, t\,) \\ &\equiv X^{\alpha}(\, K^{-1}(\, y_{\kappa}^{1}, y_{\kappa}^{2}\,), t\,) \end{aligned} \tag{5-18}$$

$$\begin{aligned} y_{\kappa}^{A} &= K_{X}^{A}(\, y^{1}, y^{2}, t\,) \\ &\equiv K^{A}(\, X^{-1}(\, y^{1}, y^{2}, t\,)) \end{aligned} \tag{5-19}$$

These are really alternative expressions for the intrinsic motion. Equation (5-18) gives the position on the dividing surface at time t of the surface particle that was at (y_{κ}^{1}, y_{κ}^{2}) at the reference time t_{κ}. The surface particle that currently is at (y^{1}, y^{2}) was at (y_{κ}^{1}, y_{κ}^{2}) according to (5-19). Using (5-7), (5-8), (5-15), and (5-16), we may express the **deformation** in (E^{3}, V^{3}) by

$$\begin{aligned} \mathbf{z} &= \boldsymbol{\chi}_{K}^{(\sigma)}(\, y_{\kappa}^{1}, y_{\kappa}^{2}, t\,) \\ &\equiv \boldsymbol{\chi}^{(\sigma)}(\, K^{-1}(\, y_{\kappa}^{1}, y_{\kappa}^{2}\,), t\,) \end{aligned} \tag{5-20}$$

$$\begin{aligned} \mathbf{z}_{\kappa} &= \boldsymbol{\kappa}_{K\,X}^{(\sigma)}(\, y^{1}, y^{2}, t\,) \\ &\equiv \boldsymbol{\kappa}_{K}^{(\sigma)}(\, X^{-1}(\, y^{1}, y^{1}, t\,)) \end{aligned} \tag{5-21}$$

The **surface deformation gradient**

$$\begin{aligned} \boldsymbol{\mathcal{F}} &\equiv \boldsymbol{\mathcal{F}}_{\kappa}(\, y_{\kappa}^{1}, y_{\kappa}^{2}, t\,) \\ &\equiv \text{grad}_{(\kappa)}\boldsymbol{\chi}_{K}^{(\sigma)}(\, y_{\kappa}^{1}, y_{\kappa}^{2}, t\,) \end{aligned} \tag{5-22}$$

is defined in terms of (5-20) and tells us how position in the current dividing surface is changed as the result of a small change of location on the reference dividing surface. It is a tangential transformation from the space of tangential vector fields on the reference dividing surface to the space of tangential vector fields on the dividing surface in its current form. The operation $\text{grad}_{(\kappa)}$ in (5-22) denotes the surface gradient on the reference dividing surface. When we use (y_{κ}^{1}, y_{κ}^{2}) as surface coordinates in speaking about the fixed reference dividing surface, the corresponding basis fields ($\mathbf{a}_{\kappa 1}$, $\mathbf{a}_{\kappa 2}$) are defined by (5-17). In talking about the dividing surface in its current form, we use as surface coordinates (y^{1}, y^{2}); the corresponding basis fields are ($\mathbf{a}_{1}$, $\mathbf{a}_{2}$). In terms of these surface coordinate systems, the surface deformation gradient may be expressed as

$$\mathcal{F} = \frac{\partial \chi_{\kappa}^{(\sigma)}}{\partial y_{\kappa}^{A}} \mathbf{a}_{\kappa}^{A}$$

$$= \frac{\partial \mathbf{p}^{(\sigma)}}{\partial y^{\alpha}} \frac{\partial X_{\kappa}^{\alpha}}{\partial y_{\kappa}^{A}} \mathbf{a}_{\kappa}^{A}$$

$$= \frac{\partial X_{\kappa}^{\alpha}}{\partial y_{\kappa}^{A}} \mathbf{a}_{\alpha} \mathbf{a}_{\kappa}^{A} \tag{5-23}$$

In the second line, I have used (5-5) and (5-18).

The **inverse** of the **surface deformation gradient**

$$\mathcal{F}^{-1} \equiv \mathcal{F}_{\kappa}^{-1} (y^1, y^2, t)$$

$$\equiv \nabla_{(\sigma)} \kappa_{\kappa}^{(\sigma)}(y^1, y^2, t) \tag{5-24}$$

is defined in terms of (5-21). It tells how position on the reference dividing surface is changed as the result of an incremental move on the current dividing surface. It may be expressed in terms of its components as

$$\mathcal{F}^{-1} = \frac{\partial \kappa_{\kappa}^{(\sigma)}}{\partial y^{\beta}} \mathbf{a}^{\beta}$$

$$= \frac{\partial \kappa^{(\sigma)}}{\partial y_{\kappa}^{B}} \frac{\partial K_{X}^{B}}{\partial y^{\beta}} \mathbf{a}^{\beta}$$

$$= \frac{\partial K_{X}^{B}}{\partial y^{\beta}} \mathbf{a}_{\kappa B} \; \mathbf{a}^{\beta} \tag{5-25}$$

In the second line, I have employed (5-13) and (5-19). The inverse has the properties (Sec. A.3.5)

$$\mathcal{F} \cdot \mathcal{F}^{-1} = \frac{\partial X_{\kappa}^{\alpha}}{\partial y_{\kappa}^{A}} \mathbf{a}_{\alpha} \frac{\partial K_{X}^{B}}{\partial y^{\beta}} \delta_{B}^{A} \; \mathbf{a}^{\beta}$$

$$= \mathbf{a}_{\alpha} \; \mathbf{a}^{\alpha}$$

$$= \mathbf{P} \tag{5-26}$$

$$\mathcal{F}^{-1} \cdot \mathcal{F} = \frac{\partial K^{\beta}_{X}}{\partial y^{\beta}} \mathbf{a}_{\kappa B} \frac{\partial X^{\alpha}_{K}}{\partial y^{A}_{\kappa}} \delta^{\beta}_{\alpha} \mathbf{a}^{A}_{\kappa}$$

$$= \mathbf{a}_{\kappa A} \mathbf{a}^{A}_{\kappa}$$

$$= \mathbf{P}_{\kappa} \tag{5-27}$$

By $\mathbf{P}_{\kappa}$, I mean the projection tensor for the tangential vector fields on the reference dividing surface.

The physical meaning of the surface derivative introduced in (5-9) may now be clarified, if we think of it as a derivative with respect to time holding position in the reference intrinsic configuration fixed:

$$\frac{d_{(s)}A}{dt} = \left(\frac{\partial A}{\partial t}\right)_{y^1_{\kappa}, y^2_{\kappa}} \tag{5-28}$$

In particular, the surface velocity becomes in view of (5-20)

$$\mathbf{v}^{(\sigma)} \equiv \frac{d_{(s)}\mathbf{z}}{dt} = \left(\frac{\partial \chi^{(\sigma)}_{\kappa}}{\partial t}\right)_{y^1_{\kappa}, y^2_{\kappa}} \tag{5-29}$$

Relative description We usually think of a reference configuration as being one that is fixed with respect to time. But there is no reason why a reference configuration cannot be a function of time. The current configuration is convenient, when we wish to compare the past with the present. The description of the motion with respect to the present is called **relative**.

From (5-5), the moving and deforming dividing surface at any prior time $\bar{t}$ takes the form

$$\bar{\mathbf{z}} = \mathbf{p}^{(\sigma)}(\bar{y}^1, \bar{y}^2, \bar{t}) \tag{5-30}$$

The intrinsic motion of the surface particles with respect to this dividing surface is described by (5-6) and (5-7)

$$\bar{y}^{\alpha} = X^{\alpha}(\zeta^{(\sigma)}, \bar{t}) \tag{5-31}$$

$$\zeta^{(\sigma)} = X^{-1}(\bar{y}^1, \bar{y}^2, \bar{t}) \tag{5-32}$$

At the current time t, the dividing surface takes the form (5-5), which becomes the reference dividing surface. We will identify surface particles by their intrinsic configuration (5-6) and (5-7) with respect to this reference dividing surface.

The **relative intrinsic deformation** follows from (5-6), (5-7), (5-31), and (5-32):

$$\bar{y}^\alpha = X^\alpha_t (y^1, y^2, \bar{t})$$
$$\equiv X^\alpha(X^{-1}(y^1, y^2, t), \bar{t}) \tag{5-33}$$

$$y^\alpha = X^\alpha_{\bar{t}} (\bar{y}^1, \bar{y}^2, t)$$
$$\equiv X^\alpha(X^{-1}(\bar{y}^1, \bar{y}^2, \bar{t}), t) \tag{5-34}$$

It describes the intrinsic motion relative to the current configuration. Equation (5-33) gives the position on the dividing surface at time $\bar{t}$ of the surface particle that currently is located at (y^1, y^2). The surface particle whose position was ($\bar{y}^1$, $\bar{y}^2$) at time $\bar{t}$ currently is at (y^1, y^2) according to (5-34). The **relative deformation** from the current configuration in (E^3, V^3) is determined by (5-7), (5-8), and (5-32):

$$\bar{\mathbf{z}} = \boldsymbol{\chi}^{(\sigma)}_t(y^1, y^2, \bar{t})$$
$$\equiv \boldsymbol{\chi}^{(\sigma)}(X^{-1}(y^1, y^2, t), \bar{t}) \tag{5-35}$$

$$\mathbf{z} = \boldsymbol{\chi}^{(\sigma)}_{\bar{t}}(\bar{y}^1, \bar{y}^2, t)$$
$$\equiv \boldsymbol{\chi}^{(\sigma)}(X^{-1}(\bar{y}^1, \bar{y}^2, \bar{t}), t) \tag{5-36}$$

The **relative surface deformation gradient** is defined in terms of (5-35):

$$\boldsymbol{\mathcal{F}}_t \equiv \boldsymbol{\mathcal{F}}_t(\bar{t})$$
$$\equiv \boldsymbol{\mathcal{F}}_t(y^1, y^2, \bar{t})$$
$$= \nabla_{(\sigma)}\boldsymbol{\chi}^{(\sigma)}_t(y^1, y^2, \bar{t}) \tag{5-37}$$

It tells how position on some past dividing surface is changed as the result of a small change of position on the current dividing surface. It is a tangential transformation from the space of tangential vector yields on the current dividing surface to the space of tangential vector fields on the dividing surface at some past time $\bar{t}$. In referring to a past configuration of the dividing surface, we denote the surface coordinates as ($\bar{y}^1$, $\bar{y}^2$); the corresponding basis fields are ($\bar{\mathbf{a}}_1$, $\bar{\mathbf{a}}_2$). The relative surface deformation gradient becomes

$$\boldsymbol{\mathcal{F}}_t = \frac{\partial \boldsymbol{\chi}^{(\sigma)}_t}{\partial y^\alpha}\mathbf{a}^\alpha$$

$$= \frac{\partial \mathbf{p}^{(\sigma)}}{\partial \overline{y}^{\beta}} \frac{\partial X^{\beta}_{t}}{\partial y^{\alpha}} \mathbf{a}^{\alpha}$$

$$= \frac{\partial X^{\beta}_{t}}{\partial y^{\alpha}} \overline{\mathbf{a}}_{\beta} \mathbf{a}^{\alpha} \tag{5-38}$$

Note that this reduces to the projection tensor at the current time t:

$$\mathcal{F}_{t}(t) = \mathbf{P}(t) = \mathbf{a}_{\alpha} \mathbf{a}^{\alpha} \tag{5-39}$$

We will later find it useful to write the surface deformation gradient (5-23) in terms of this relative surface deformation gradient:

$$\mathcal{F}(\overline{t}) = \left(\frac{\partial X^{\alpha}_{\kappa}}{\partial y^{A}_{\kappa}} \right)_{\overline{t}} \overline{\mathbf{a}}_{\alpha} \mathbf{a}^{A}_{\kappa}$$

$$= \left(\frac{\partial X^{\alpha}_{t}}{\partial y^{\beta}} \right)_{\overline{t}} \left(\frac{\partial X^{\beta}_{\kappa}}{\partial y^{A}_{\kappa}} \right)_{t} \overline{\mathbf{a}}_{\alpha} \mathbf{a}^{A}_{\kappa}$$

$$= \mathcal{F}_{t}(\overline{t}) \cdot \mathcal{F}(t) \tag{5-40}$$

In the second line, I have employed (5-18) and (5-33).

1.2.6 Stretch and rotation within dividing surfaces The discussion in Sec. 1.1.2 suggests how we might talk about stretch and rotation within dividing surfaces.

The surface deformation gradient $\mathcal{F}$ is a non-singular tangential transformation from the space of tangential vector fields on Σ_{κ}, the reference dividing surface, to the space of tangential vector fields on Σ, the current dividing surface. The polar decomposition theorem (Sec. A.3.8) allows us to express it in two different forms:

$$\mathcal{F} = \mathcal{R} \cdot \mathbf{U}^{(\sigma)} = \mathbf{V}^{(\sigma)} \cdot \mathcal{R} \tag{6-1}$$

Here $\mathcal{R}$ is a unique, orthogonal, tangential transformation from the space of tangential vector fields on Σ_{κ} to the space of tangential vector fields on Σ; $\mathbf{U}^{(\sigma)}$ is a unique, positive-definite, symmetric, tangential tensor for the space of tangential vector fields on Σ_{κ}; $\mathbf{V}^{(\sigma)}$ is a unique, positive-definite, symmetric, tangential tensor for the space of tangential vector fields on Σ.

Physically, (6-1) says that $\mathcal{F}$ may be obtained by imposing pure stretches of amounts $u_{(j)}^{(\sigma)}$ along two mutually orthogonal directions $\mathbf{x}_{(j)}^{(\sigma)}$ that are tangent to the dividing surface in its reference configuration

$$\mathbf{U}^{(\sigma)} \cdot \mathbf{x}_{(j)}^{(\sigma)} = u_{(j)}^{(\sigma)} \, \mathbf{x}_{(j)}^{(\sigma)} \tag{6-2}$$

followed by a rigid rotation of those directions. Alternatively, the same rotation might be effected first, followed by the same stretches in the resulting directions:

$$\begin{aligned} \mathbf{V}^{(\sigma)} \cdot (\mathcal{R} \cdot \mathbf{x}_{(j)}^{(\sigma)}) &= (\mathcal{R} \cdot \mathbf{U}^{(\sigma)} \cdot \mathcal{R}^T) \cdot (\mathcal{R} \cdot \mathbf{x}_{(j)}^{(\sigma)}) \\ &= u_{(j)}^{(\sigma)} \, \mathcal{R} \cdot \mathbf{x}_{(j)}^{(\sigma)} \end{aligned} \tag{6-3}$$

The tensors $\mathbf{U}^{(\sigma)}$ and $\mathbf{V}^{(\sigma)}$ have the same principal values, but different principal axes: $\mathcal{R}$ is the rotation that transforms the principal axes of $\mathbf{U}^{(\sigma)}$, which are tangential vector fields on Σ_κ, into the principal axes of $\mathbf{V}^{(\sigma)}$, which are tangential vector fields on Σ. We can refer to $\mathcal{R}$ as the **surface rotation**; $\mathbf{U}^{(\sigma)}$ is the **right surface stretch** tensor and $\mathbf{V}^{(\sigma)}$ is the **left surface stretch** tensor.

The orthogonal tangential transformation $\mathcal{R}$ need not be proper orthogonal: $\det_{(\sigma)}\mathcal{R} = \pm 1$. However, continuity requires that $\det_{(\sigma)}\mathcal{R}$ keep either one value or the other for all time t (see Exercise 1.3.5-4). As a result

$$J^{(\sigma)} \equiv | \det_{(\sigma)}\mathcal{F} | = \det_{(\sigma)}\mathbf{U}^{(\sigma)} = \det_{(\sigma)}\mathbf{V}^{(\sigma)} \tag{6-4}$$

Instead of $\mathbf{U}^{(\sigma)}$ and $\mathbf{V}^{(\sigma)}$, it is often easier to use the **right** and **left surface Cauchy-Green tensors**

$$\mathbf{C}^{(\sigma)} \equiv \mathbf{U}^{(\sigma)2} = \mathcal{F}^T \cdot \mathcal{F} \tag{6-5}$$

$$\mathbf{B}^{(\sigma)} \equiv \mathbf{V}^{(\sigma)2} = \mathcal{F} \cdot \mathcal{F}^T \tag{6-6}$$

since they can be simply calculated from the surface deformation gradient.

If we start with the relative surface deformation gradient $\mathcal{F}_t$, we can introduce in the same way the **relative surface rotation** $\mathcal{R}_t$, the **relative surface stretch** tensors $\mathbf{U}_t^{(\sigma)}$ and $\mathbf{V}_t^{(\sigma)}$, and the **relative surface Cauchy-Green tensors** $\mathbf{C}_t^{(\sigma)}$ and $\mathbf{B}_t^{(\sigma)}$:

$$\mathcal{F}_t = \mathbf{R}_t \cdot \mathbf{U}_t^{(\sigma)} = \mathbf{V}_t^{(\sigma)} \cdot \mathcal{R}_t \tag{6-7}$$

$$\mathbf{C}_t^{(\sigma)} \equiv \mathbf{U}_t^{(\sigma)2} = \mathcal{F}_t^T \cdot \mathcal{F}_t \tag{6-8}$$

$$\mathbf{B}_t^{(\sigma)} \equiv \mathbf{V}_t^{(\sigma)2} = \mathcal{F}_t \cdot \mathcal{F}_t^T \tag{6-9}$$

I think that you will find it helpful to compare the ideas introduced here with those in Sec. 1.1.2. There we discuss the analogous concepts that are standard in treatments of the deformation of materials outside the immediate neighborhood of the phase interface.

1.2.7 More about surface velocity Let A be any quantity: scalar, vector, or tensor. Let us assume that A is an explicit function of position (y^1, y^2) on the dividing surface and time.

$$A = A(y^1, y^2, t) \tag{7-1}$$

The surface material derivative of A, introduced in Sec. 1.2.5, becomes

$$\begin{aligned}\frac{d_{(s)}A}{dt} &\equiv \left(\frac{\partial A}{\partial t}\right)_{\zeta^{(\sigma)}} \equiv \left(\frac{\partial A}{\partial t}\right)_{y^1_\kappa, y^2_\kappa} \\ &= \frac{\partial A(y^1, y^2, t)}{\partial t} + \frac{\partial A(y^1, y^2, t)}{\partial y^\alpha}\frac{\partial X^\alpha_\kappa(y^1_\kappa, y^1_\kappa, t)}{\partial t} \\ &= \left(\frac{\partial A}{\partial t}\right)_{y^1, y^2} + \nabla_{(\sigma)}A \cdot \dot{\mathbf{y}}\end{aligned} \tag{7-2}$$

where we have used (5-18) in defining the **intrinsic surface velocity**

$$\begin{aligned}\dot{\mathbf{y}} &= \dot{y}^\alpha \mathbf{a}_\alpha \\ &\equiv \frac{\partial X^\alpha_K(y^1_\kappa, y^2_\kappa, t)}{\partial t}\mathbf{a}_\alpha \\ &\equiv \frac{d_{(s)}y^\alpha}{dt}\mathbf{a}_\alpha\end{aligned} \tag{7-3}$$

We will have particular interest in the **surface velocity**, the time rate of change of spatial position following a surface particle:

$$\begin{aligned}\mathbf{v}^{(\sigma)} &\equiv \frac{\partial \boldsymbol{\chi}^{(\sigma)}_K(y^1_\kappa, y^2_\kappa, t)}{\partial t} \\ &\equiv \frac{d_{(s)}\mathbf{z}}{dt}\end{aligned} \tag{7-4}$$

From Sec. 1.2.5, any dividing surface in (E^3, V^3) is the locus of a point whose position may be a function of the two surface coordinates and time:

$$\mathbf{z} = \mathbf{p}^{(\sigma)}(y^1, y^2, t) \tag{7-5}$$

Applying (7-2) to (7-5), we find

$$\mathbf{v}^{(\sigma)} = \frac{\partial \mathbf{p}^{(\sigma)}(y^1, y^2, t)}{\partial t} + \nabla_{(\sigma)}\mathbf{p}^{(\sigma)}(y^1, y^2, t) \cdot \dot{\mathbf{y}}$$

$$= \mathbf{u} + \dot{\mathbf{y}} \tag{7-6}$$

Here we have defined $\mathbf{u}$ to be the time rate of change of spatial position following a surface point (y^1, y^2)

$$\mathbf{u} \equiv \frac{\partial \mathbf{p}^{(\sigma)}(y^1, y^2, t)}{\partial t} \tag{7-7}$$

and we have noted that the tangential gradient of the position vector is the projection tensor $\mathbf{P}$ (see Sec. A.5.1)

$$\nabla_{(\sigma)}\mathbf{p}^{(\sigma)}(y^1, y^2, t) = \mathbf{P} \tag{7-8}$$

Note that $\dot{\mathbf{y}}$ does not represent the tangential component of $\mathbf{v}^{(\sigma)}$. It is perhaps easiest to see that in general $\mathbf{u}$ has both normal and tangential components. From Sec. A.1.3, a moving and deforming dividing surface may be described by the single scalar equation

$$f(\mathbf{z}, t) = 0 \tag{7-9}$$

Upon differentiating this with respect to time following a particular point (y^1, y^2) on the surface, we find

$$\frac{\partial f}{\partial t} + \nabla f \cdot \mathbf{u} = 0 \tag{7-10}$$

But we know that the unit normal $\boldsymbol{\xi}$ to the surface can be defined by (Slattery 1981, p. 619; see also Sec. A.2.6)

$$\boldsymbol{\xi} \equiv \frac{\nabla f}{|\nabla f|} \tag{7-11}$$

From (7-10), this means that

$$
\begin{aligned}
v_{\{\xi\}} &\equiv \mathbf{v}^{(\sigma)} \cdot \boldsymbol{\xi} \\
&= \mathbf{u} \cdot \boldsymbol{\xi} \\
&= -\frac{\partial f/\partial t}{|\nabla f|}
\end{aligned}
\tag{7-12}
$$

We shall refer to $v_{\{\xi\}}$ as the **speed of displacement** of the surface (Truesdell and Toupin 1960, p. 499). All possible velocities $\mathbf{u}$ (corresponding to different choices for the surface coordinates y^1 and y^2) have the same normal component $v_{\{\xi\}}$. But in general there is only one choice of space coordinates for which

$$
\mathbf{u} = v_{\{\xi\}}\, \boldsymbol{\xi} \tag{7-13}
$$

References

Slattery, J. C., "Momentum, Energy, and Mass Transfer in Continua," McGraw-Hill, New York (1972); second edition, Robert E. Krieger, Malabar, FL 32950 (1981).

Truesdell, C., and R. A. Toupin, "Handbuch der Physik," vol. 3/1, edited by S. Flügge, Springer-Verlag, Berlin (1960).

Exercise

1.2.7-1 *surface divergence of surface velocity*

i) Prove

$$
\text{div}_{(\sigma)}\mathbf{u} = \frac{\partial \mathbf{a}_\alpha}{\partial t} \cdot \mathbf{a}^\alpha
$$

ii) Determine

$$
\frac{\partial \mathbf{a}_\alpha}{\partial t} \cdot \mathbf{a}^\alpha = -\frac{1}{2}\frac{\partial a^{\alpha\beta}}{\partial t}\, a_{\alpha\beta}
$$

iii) Show

$$\text{div}_{(\sigma)}\mathbf{u} = \frac{1}{2}\, a^{\alpha\beta}\, \frac{\partial a_{\alpha\beta}}{\partial t}$$

iv) Find

$$\text{div}_{(\sigma)}\mathbf{u} = \frac{1}{2a}\frac{\partial a}{\partial t}$$

v) Conclude

$$\text{div}_{(\sigma)}\mathbf{v}^{(\sigma)} = \frac{1}{2a}\frac{\partial a}{\partial t} + \text{div}_{(\sigma)}\dot{\mathbf{y}}$$

1.2.8 Rate of deformation I would like to characterize the rate of deformation of a dividing surface.

Let us take a dividing surface in its reference configuration and lay out two intersecting curves on it. We might visualize doing this either with a dye or with a set of very small floating particles. We will say that these are surface material curves, in the sense that we will think of them as being attached to the surface particles. As the dividing surface moves and deforms, they move and deform with it. In the current configuration of the surface, they appear as shown in Figure 8-1.

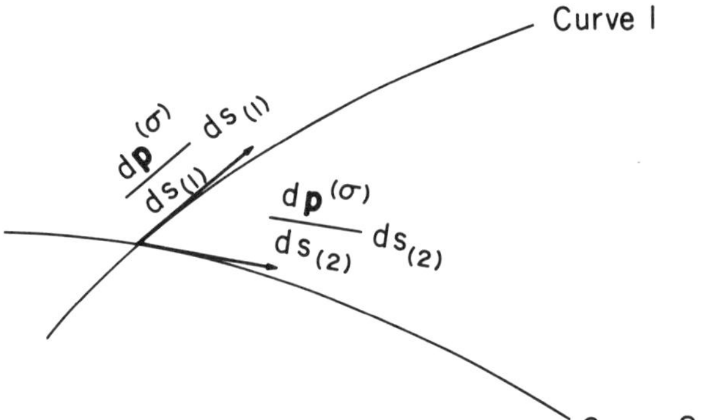

Figure 1.2.8-1: Intersecting surface material curves

At the point of intersection of these curves, the rate of deformation of the surface is described by the instantaneous rates at which the lengths of these curves change and by the rate of change of the angle between them. This suggests that we follow these curves as the surface is deformed in some arbitrary fashion.

Let $s_{(1)}$ and $s_{(2)}$ be the arc lengths measured along curves 1 and 2 in the present configuration. The arc lengths along these curves in the reference configuration will be denoted as $s_{\kappa(1)}$ and $s_{\kappa(2)}$. We should be able to learn about the local rate of deformation of the surface by examining

$$\frac{d_{(s)}}{dt}\left(\frac{d\mathbf{p}^{(\sigma)}}{ds_{(1)}}\cdot\frac{d\mathbf{p}^{(\sigma)}}{ds_{(2)}}ds_{(1)}ds_{(2)}\right)$$

$$=\frac{d_{(s)}}{dt}\left(\frac{d\boldsymbol{\chi}_{\kappa}^{(\sigma)}}{ds_{\kappa(1)}}\cdot\frac{d\boldsymbol{\chi}_{\kappa}^{(\sigma)}}{ds_{\kappa(2)}}ds_{\kappa(1)}ds_{\kappa(2)}\right)$$

$$=\frac{d\mathbf{v}^{(\sigma)}}{ds_{\kappa(1)}}\cdot\frac{d\boldsymbol{\chi}_{\kappa}^{(\sigma)}}{ds_{\kappa(2)}}ds_{\kappa(1)}ds_{\kappa(2)}+\frac{d\boldsymbol{\chi}_{\kappa}^{(\sigma)}}{ds_{\kappa(1)}}\cdot\frac{d\mathbf{v}^{(\sigma)}}{ds_{\kappa(2)}}ds_{\kappa(1)}ds_{\kappa(2)}$$

$$=\frac{d\mathbf{v}^{(\sigma)}}{ds_{(1)}}\cdot\frac{d\mathbf{p}^{(\sigma)}}{ds_{(2)}}ds_{(1)}ds_{(2)}+\frac{d\mathbf{p}^{(\sigma)}}{ds_{(1)}}\cdot\frac{d\mathbf{v}^{(\sigma)}}{ds_{(2)}}ds_{(1)}ds_{(2)} \tag{8-1}$$

This can be simplified by noting that

$$\frac{d\mathbf{v}^{(\sigma)}}{ds_{(1)}}\cdot\frac{d\mathbf{p}^{(\sigma)}}{ds_{(2)}}=\left(\frac{\partial\mathbf{v}^{(\sigma)}}{\partial y^{\alpha}}\frac{dy^{\alpha}}{ds_{(1)}}\right)\cdot\frac{d\mathbf{p}^{(\sigma)}}{ds_{(2)}}$$

$$=\left(\nabla_{(\sigma)}\mathbf{v}^{(\sigma)}\cdot\frac{d\mathbf{p}^{(\sigma)}}{ds_{(1)}}\right)\cdot\frac{d\mathbf{p}^{(\sigma)}}{ds_{(2)}} \tag{8-2}$$

and similarly

$$\frac{d\mathbf{p}^{(\sigma)}}{ds_{(1)}}\cdot\frac{d\mathbf{v}^{(\sigma)}}{ds_{(2)}}=\frac{d\mathbf{p}^{(\sigma)}}{ds_{(1)}}\cdot\left(\nabla_{(\sigma)}\mathbf{v}^{(\sigma)}\cdot\frac{d\mathbf{p}^{(\sigma)}}{ds_{(2)}}\right) \tag{8-3}$$

This allows us to write (8-1) as

$$\frac{d_{(s)}}{dt}\left(\frac{d\mathbf{p}^{(\sigma)}}{ds_{(1)}}\cdot\frac{d\mathbf{p}^{(\sigma)}}{ds_{(2)}}ds_{(1)}ds_{(2)}\right)$$

$$= \left(\nabla_{(\sigma)} \mathbf{v}^{(\sigma)} \cdot \frac{d\mathbf{p}^{(\sigma)}}{ds_{(1)}} \right) \cdot \frac{d\mathbf{p}^{(\sigma)}}{ds_{(2)}} ds_{(1)} ds_{(2)}$$

$$+ \frac{d\mathbf{p}^{(\sigma)}}{ds_{(1)}} \cdot \left(\nabla_{(\sigma)} \mathbf{v}^{(\sigma)} \cdot \frac{d\mathbf{p}^{(\sigma)}}{ds_{(2)}} \right) ds_{(1)} ds_{(2)}$$

$$= 2 \frac{d\mathbf{p}^{(\sigma)}}{ds_{(2)}} \cdot \left(\mathbf{D}^{(\sigma)} \cdot \frac{d\mathbf{p}^{(\sigma)}}{ds_{(1)}} \right) ds_{(1)} ds_{(2)} \tag{8-4}$$

in which[a]

$$\mathbf{D}^{(\sigma)} \equiv \frac{1}{2} \left[\mathbf{P} \cdot \nabla_{(\sigma)} \mathbf{v}^{(\sigma)} + (\nabla_{(\sigma)} \mathbf{v}^{(\sigma)})^T \cdot \mathbf{P} \right] \tag{8-5}$$

We call $\mathbf{D}^{(\sigma)}$ the **surface rate of deformation** tensor. Using (8-4), we can measure in terms of this tensor the instantaneous rates of change of length and angle of material elements in a deforming surface. Arc length is that parameter such that (Willmore 1959, p. 6)

$$\frac{d\mathbf{p}}{ds} \cdot \frac{d\mathbf{p}}{ds} = 1 \tag{8-6}$$

When curve 1 instantaneously coincides with curve 2 in Figure 8-1, (8-4) and (8-6) give [Oldroyd (1955) gave this relationship in defining the surface rate of deformation tensor for a stationary dividing surface.]

---

a) We could just as well have written (8-4) in terms of

$$\mathbf{D}^{(\sigma)\prime} \equiv \frac{1}{2} \left[\nabla_{(\sigma)} \mathbf{v}^{(\sigma)} + \left(\nabla_{(\sigma)} \mathbf{v}^{(\sigma)} \right)^T \right] \tag{8-5a}$$

Instead, we have immediately recognized that only the components of the corresponding tangential tensor

$$\mathbf{P} \cdot \mathbf{D}^{(\sigma)\prime} \cdot \mathbf{P} = \mathbf{D}^{(\sigma)} \tag{8-5b}$$

play a role.

$$\frac{d_{(s)}}{dt}\left(\frac{d\mathbf{p}^{(\sigma)}}{ds}\cdot\frac{d\mathbf{p}^{(\sigma)}}{ds}(\,ds\,)^2\right)$$

$$=\frac{d_{(s)}}{dt}(\,ds\,)^2$$

$$=2\,\frac{d\mathbf{p}^{(\sigma)}}{ds}\cdot\left(\mathbf{D}^{(\sigma)}\cdot\frac{d\mathbf{p}^{(\sigma)}}{ds}\right)(\,ds\,)^2 \tag{8-7}$$

This may be rewritten as

$$\frac{d_{(s)}}{dt}(\,\ln ds\,)=\frac{d\mathbf{p}^{(\sigma)}}{ds}\cdot\left(\mathbf{D}^{(\sigma)}\cdot\frac{d\mathbf{p}^{(\sigma)}}{ds}\right) \tag{8-8}$$

where the derivative on the left is known as the **rate of stretching** in the direction $d\mathbf{p}^{(\sigma)}/ds$. If we assume, for example, that $d\mathbf{p}^{(\sigma)}/ds$ instantaneously is tangent to the y^1 coordinate curve, we have for the stretching in the y^1 direction

$$\frac{d_{(s)}}{dt}(\,\ln ds_{(1)}\,)=D_{11}^{(\sigma)}\left(\frac{dy^1}{ds_{(1)}}\right)^2$$

$$=\frac{D_{11}^{(\sigma)}}{a_{11}} \tag{8-9}$$

For an orthogonal surface coordinate system (y^1, y^2), (8-9) may be expressed in terms of the physical component of $\mathbf{D}^{(\sigma)}$:

$$\frac{d_{(s)}}{dt}(\,\ln ds_{(1)}\,)=D_{<11>}^{(\sigma)} \tag{8-10}$$

In this way we obtain a physical interpretation for the diagonal components of the surface rate of deformation tensor.

Let us denote by β_{12} the angle between the tangents to the two material curves at their intersection:

$$\cos\beta_{12}=\frac{d\mathbf{p}^{(\sigma)}}{ds_{(1)}}\cdot\frac{d\mathbf{p}^{(\sigma)}}{ds_{(2)}} \tag{8-11}$$

Equation (8-4) becomes

$$\frac{d_{(s)}}{dt}\left(\cos \beta_{12}\, ds_{(1)} ds_{(2)} \right)$$

$$= - \sin \beta_{12} \frac{d_{(s)}\beta_{12}}{dt} ds_{(1)} ds_{(2)} + \cos \beta_{12} \frac{d_{(s)}}{dt}\left(ds_{(1)} \right) ds_{(2)}$$

$$+ \cos \beta_{12}\, ds_{(1)} \frac{d_{(s)}}{dt}\left(ds_{(2)} \right)$$

$$= 2 \frac{d\mathbf{p}^{(\sigma)}}{ds_{(2)}} \cdot \left(\mathbf{D}^{(\sigma)} \cdot \frac{d\mathbf{p}^{(\sigma)}}{ds_{(1)}} \right) ds_{(1)} ds_{(2)} \qquad (8\text{-}12)$$

which may be rearranged to read

$$- \sin \beta_{12} \frac{d_{(s)}\beta_{12}}{dt}$$

$$= 2 \frac{d\mathbf{p}^{(\sigma)}}{ds_{(2)}} \cdot \left(\mathbf{D}^{(\sigma)} \cdot \frac{d\mathbf{p}^{(\sigma)}}{ds_{(1)}} \right)$$

$$- \cos \beta_{12} \left[\frac{d_{(s)}}{dt}\left(\ln ds_{(1)} \right) + \frac{d_{(s)}}{dt}\left(\ln ds_{(2)} \right) \right] \qquad (8\text{-}13)$$

The rate of decrease in the angle β_{12} is called the **rate of shear** of the directions $d\mathbf{p}^{(\sigma)}/ds_{(1)}$ and $d\mathbf{p}^{(\sigma)}/ds_{(2)}$. If curves 1 and 2 are instantaneously orthogonal in Figure 8-1, we find that

$$- \frac{d_{(s)}\beta_{12}}{dt} = 2 \frac{d\mathbf{p}^{(\sigma)}}{ds_{(2)}} \cdot \left(\mathbf{D}^{(\sigma)} \cdot \frac{d\mathbf{p}^{(\sigma)}}{ds_{(1)}} \right) \qquad (8\text{-}14)$$

If $d\mathbf{p}^{(\sigma)}/ds_{(1)}$ and $d\mathbf{p}^{(\sigma)}/ds_{(2)}$ are instantaneously tangent to the orthogonal surface coordinates y^1 and y^2, this reduces to

$$- \frac{d_{(s)}\beta_{12}}{dt} = 2\, D_{21}^{(\sigma)} \frac{dy^1}{ds_{(1)}} \frac{dy^2}{ds_{(2)}}$$

$$= \frac{2}{\sqrt{a_{11}}\sqrt{a_{22}}} D_{12}^{(\sigma)}$$

$$= 2\, D^{(\sigma)}_{<12>} \tag{8-15}$$

To summarize, (8-10) and (8-15) show us that, with respect to an orthogonal surface coordinate system, the physical components of the surface rate of deformation tensor may be interpreted as equal to the rates of stretching and the halves of the rates of shearing in the coordinate directions:

$$[\, D^{(\sigma)}_{<\alpha\beta>} \,] = \begin{bmatrix} \frac{d_{(s)}}{dt}(\ln ds_{(1)}) & -\frac{1}{2}\frac{d_{(s)}\beta_{12}}{dt} \\ -\frac{1}{2}\frac{d_{(s)}\beta_{12}}{dt} & \frac{d_{(s)}}{dt}(\ln ds_{(2)}) \end{bmatrix} \tag{8-16}$$

References

Oldroyd, J. G., *Proc. R. Soc. London A* **232**, 567 (1955).

Scriven, L. E., *Chem. Eng. Sci.* **12**, 98 (1960).

Slattery, J. C., *Chem. Eng. Sci.* **19**, 379 (1964).

Willmore, T. J., "An Introduction to Differential Geometry," Oxford University Press, London (1959).

Exercises

1.2.8-1 *alternative expression for* $\mathbf{D}^{(\sigma)}$

i) Observe that

$$\mathbf{P} \cdot \nabla_{(\sigma)} \mathbf{v}^{(\sigma)} = \mathbf{P} \cdot \nabla_{(\sigma)} \mathbf{u} + \mathbf{P} \cdot \nabla_{(\sigma)} \dot{\mathbf{y}}$$

ii) Reason that

$$\mathbf{P} \cdot \nabla_{(\sigma)} \mathbf{u} = \mathbf{a}^{\alpha}\, \mathbf{a}_{\alpha} \cdot \frac{\partial \mathbf{a}_{\beta}}{\partial t}\, \mathbf{a}^{\beta}$$

iii) Conclude

$$\mathbf{D}^{(\sigma)} = \frac{1}{2}\left(\dot{y}_{\alpha,\beta} + \dot{y}_{\beta,\alpha} + \frac{\partial a_{\alpha\beta}}{\partial t} \right) \mathbf{a}^{\alpha} \mathbf{a}^{\beta}$$

This agrees with Scriven (1960, eq. 5) and Slattery (1964, eq. 4.1).

1.2.8-2 *another alternative expression for* $\mathbf{D}^{(\sigma)}$ Prove that

$$\mathbf{D}^{(\sigma)} = \frac{1}{2}\left(v^{(\sigma)}_{i,\beta} \frac{\partial x^i}{\partial y^{\alpha}} + v^{(\sigma)}_{i,\alpha} \frac{\partial x^i}{\partial y^{\beta}} \right) \mathbf{a}^{\alpha} \mathbf{a}^{\beta}$$

I have found this to be the most useful expression for the components of $\mathbf{D}^{(\sigma)}$.

1.2.8-3 *still another alternative expression for* $\mathbf{D}^{(\sigma)}$ We are often willing to neglect mass transfer between the dividing surface and the adjoining phases, in which case

at the dividing surface: $\mathbf{v}^{(\sigma)} = \mathbf{v}$

i) Prove that under these conditions (see Sec. A.5.2)

$$\mathbf{D}^{(\sigma)} = \mathbf{P} \cdot \mathbf{D} \cdot \mathbf{P}$$

$$= D_{ij} \frac{\partial x^i}{\partial y^{\alpha}} \frac{\partial x^j}{\partial y^{\beta}} \mathbf{a}^{\alpha} \mathbf{a}^{\beta}$$

ii) If we confine our attention to orthogonal spatial coordinate systems, reason that

$$\mathbf{D}^{(\sigma)} = \left[\sum_{i=1}^{3} \sum_{j=1}^{3} \sqrt{g_{ii}} \sqrt{g_{jj}}\, D_{<ij>} \frac{\partial x^i}{\partial y^{\alpha}} \frac{\partial x^j}{\partial y^{\beta}} \right] \mathbf{a}^{\alpha} \mathbf{a}^{\beta}$$

When applicable, these expressions are more convenient than the one recommended in Exercise 1.2.8-2.

1.2.8-4 *still another alternative expression* for $\mathbf{D}^{(\sigma)}$ Prove that

$$\mathbf{D}^{(\sigma)} = \frac{1}{2}\left(v^{(\sigma)}_{\alpha,\beta} + v^{(\sigma)}_{\beta,\alpha} - 2\, v^{(\sigma)}_{(\xi)} B_{\alpha\beta} \right) \mathbf{a}^{\alpha} \mathbf{a}^{\beta}$$

1.2.9 Moving common lines: qualitative description In our discussion of the motion of multiphase bodies, we have not as yet mentioned their common lines.

A **common line** (contact line or three-phase line of contact) is the curve formed by the intersection of two dividing surfaces. When a drop of water sits on a china plate, the water-air dividing surface intersects the solid in a common line. We are primarily concerned here with the motion of common lines. For example, when the plate is tipped, the drop begins to flow resulting in the displacement of the common line.

If we dip a glass capillary tube in a pan of water, the water-air dividing surface rises in the tube to some equilibrium level. When it has come to rest, the dividing surface intersects the wall of the tube in an easily observed common line. This experiment suggests some questions. Is the air displaced by the water at the wall of the tube or is the water separated from the tube wall by a thin film of air? How does the common line move with respect to the adjoining air, water, and glass?

In attempting to answer these questions, let us begin with a qualitative picture of the phenomena involved. Some simple experiments are helpful.

Initial formation of common line Dussan V. and Davis (1974) released a small drop of water in a tank of silicone oil. It appeared to be spherical as it fell and remained spherical for 5-10 seconds after it came to rest on the bottom. Suddenly, as shown in Figure 9-1, the water drop *popped* onto the Plexiglas base with a markedly different configuration. Apparently what happened is that, when the drop initially came to rest, it was separated from the Plexiglas by a thin film of silicone oil. When the spherical drop appeared to be resting on the bottom of the tank, it was actually squeezing the silicone oil film, forcing it to become thinner. After 5-10 seconds the film was so thin that it became unstable. It ruptured, rolled back, and the water drop came into contact with the Plexiglas base.

Dussan V. and Davis (1974) obtained the more detailed view of this popping phenomena shown in Figure 9-2 using a glycerol drop (whose viscosity is about 1500 times that for water). Following its rupture, the thin film of silicone oil rolled back: a curve dividing two distinct types of reflections appeared after the initial rupture of the film of silicone oil and swept across the lower surface of the drop.

After the thin film of silicone oil ruptured and rolled back allowing the glycerol drop to come into direct contact with the Plexiglas surface, some silicone oil was almost certainly left behind, adsorbed in the glycerol-Plexiglas interface. As time passed and the multiphase system approached a new equilibrium, some of this adsorbed silicone oil was then desorbed into the glycerol bulk phase. For a more detailed view of the multicomponent dividing surface, see Chapter 5.

In both of these experiments, there is the strong implication that, following its rupture, the thin film of bulk silicone oil was displaced without violating the requirement that the tangential components of velocity be continuous across a phase interface. The silicone oil film was simply rolled to one side in much the same manner as we might handle a rug.

Figure 1.2.9-1: A drop of water containing food coloring is released from a hypodermic needle within a pool of silicone oil, the kinematic viscosity of which is 10 centistokes. The bottom surface is Plexiglas. After resting on the *bottom* of the tank for 5-10 seconds, it *pops* onto the Plexiglas surface in less than 0.75 seconds. The motion picture was taken at 18 frames per second (Dussan V. and Davis 1974, Figure 2).

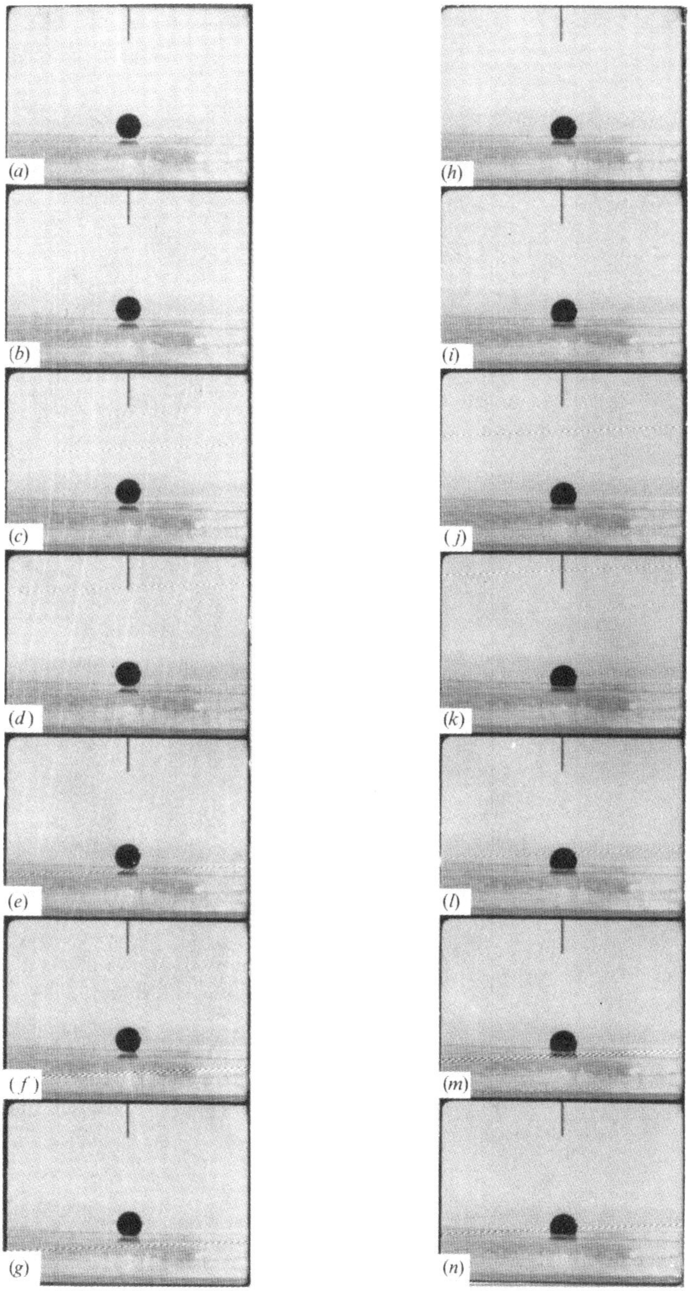
(a)
(b)
(c)
(d)
(e)
(f)
(g)
(h)
(i)
(j)
(k)
(l)
(m)
(n)

Figure 1.2.9-2: Glycerol drop surrounded by silicone oil *pops* onto a Plexiglas surface (Dussan V. and Davis 1974, Figure 3).

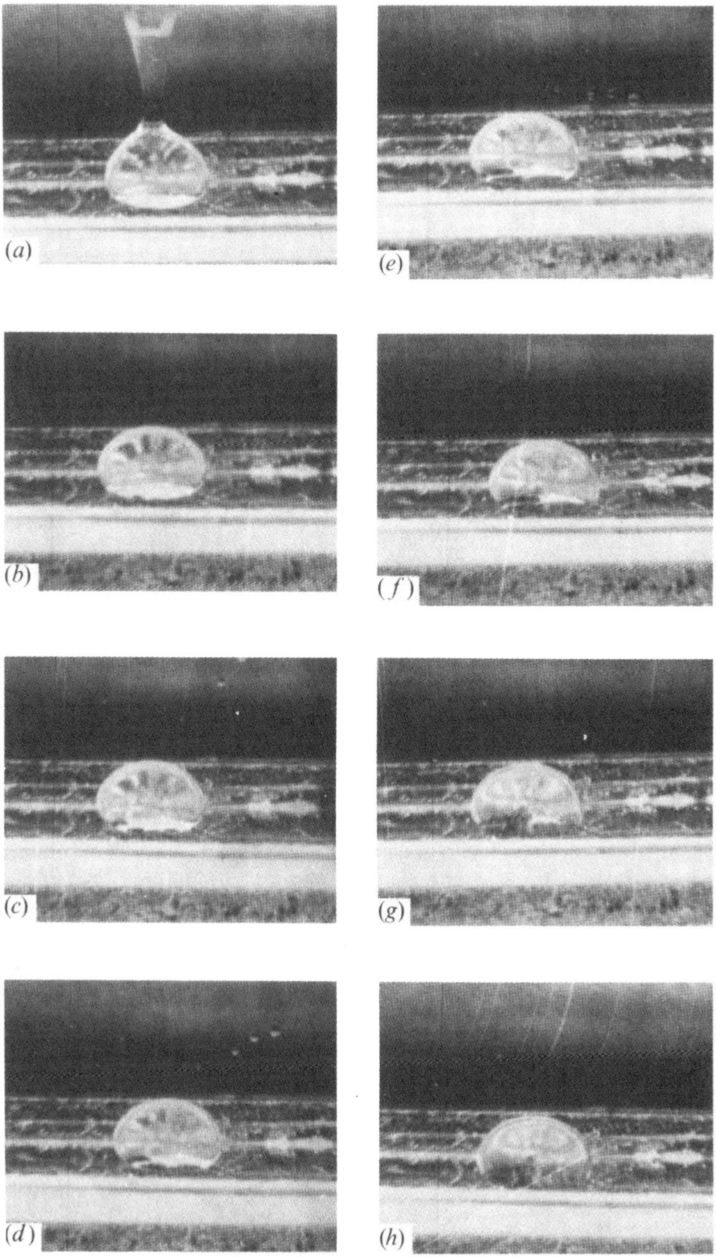
(a)
(e)
(b)
(f)
(c)
(g)
(d)
(h)

Precursor film Sometimes a very thin film or precursor film is observed to precede the advance of a macroscopic film.

Bascom *et al.* (1964) studied the spontaneous spreading behavior of hydrocarbon liquids on both horizontal and vertical surfaces using interference microscopy and ellipsometry, which enabled them to study the thickness of the spreading films in detail. All of the hydrocarbons showed zero contact angles on the metal surfaces employed. In all cases, breath patterns formed by breathing over the spreading liquids showed that the outer edge of the film was considerably beyond the edge of the macroscopic film as detected by the first-order interference band: a very thin precursor film preceded the advance of the macroscopic film. The presence of this precursor film was also demonstrated by placing ahead of the macroscopic film minute drops of another liquid having a higher surface tension against air. Placed on the precursor film, these drops would immediately retract from the spreading liquid; placed on a clean metal surface, they would spread uniformly in all directions. For the first few hours, they were not able to detect the precursor film using ellipsometry. After a relatively long period of time (18 hours in the case of squalane), they were able to determine that the precursor film was as much as several millimeters long although less than 50 Å thick. Considering the relatively low volatility of the liquids employed, they attributed the formation and movement of the precursor film to capillary flow in microscratches on the surface and to surface diffusion (Sec. 5.9.8) rather than to evaporation from the macroscopic film and subsequent condensation upon the plate. Teletzke *et al.* (1987), who have re-examined these experiments, offer new evidence supporting the importance of surface diffusion.

Radigan *et al.* (1974) used scanning electron microscopy to detect during the spreading of drops of glass on Fernico metal at 1000°C a precursor film whose height was of the order of 1 μm.

In studies described by Williams (1977, 1988), the leading edge of this precursor film has a periodic structure. Depending upon perhaps both the liquid and the solid, either it assumes a scalloped periodic structure or it moves by a series of random advances with some approximate periodicity.

Figure 9-3 shows the advancing front of absolute ethyl alcohol spreading over aluminum that had been evaporated onto a glass substrate. A precursor film of alcohol 1,000 - 2,000 Å thick moved ahead of the drop as much as 1 mm. In this case, the leading edge of the precursor film maintained an early sinusoidal form as it advanced. Williams (1977) found similar results with a variety of liquids spreading on aluminum: isopropyl alcohol, n-heptane, toluene, n-octane, and dimethyl silicone oil (molecular weight 340).

Figure 9-4 shows the leading edge of an advancing precursor film of a non-volatile, hydrocarbon-substituted silicone oil (General Electric SF-1147, mean molecular weight 2,000) spreading over a surface of evaporated gold. Although the configuration of the leading edge of the precursor film is more random than that of the alcohol precursor film shown in Figure 9-3, it still displays an approximate periodicity. As in the case of alcohol,

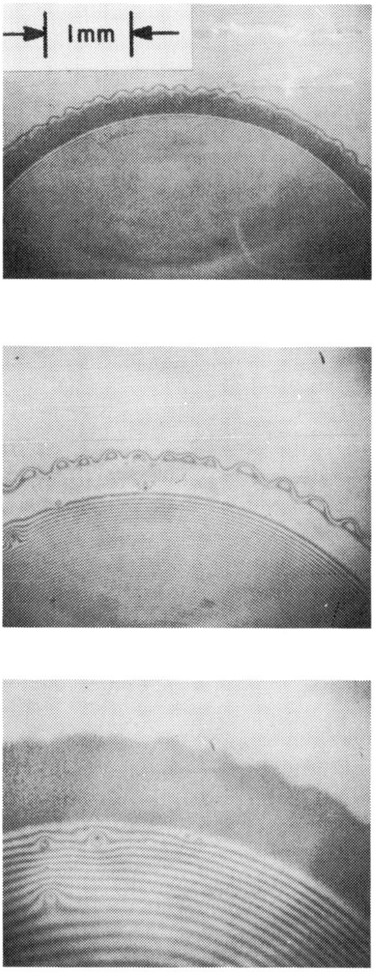

Figure 1.2.9-3: Advancing front of a drop of ethyl alcohol spreading on evaporated aluminum as photographed by Williams (1977). Successive fringes correspond to a thickness difference of 2,100 Å. The three pictures correspond to different times after application of drop: (a) 30 s, (b) 60 s, and (c) 200 s. At (c), the precursor film is about 1,000 Å thick.

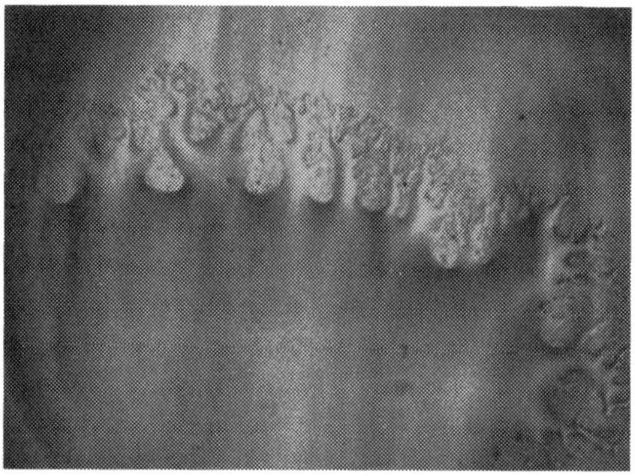

Figure 1.2.9-4: Advancing front of a drop of silicone oil spreading on evaporated gold 5 minutes after application (Williams 1977).

Williams (1977) found that the precursor film advanced far ahead of the bulk liquid, and, after several hours, covered nearly all of the solid surface.

Figures 9-5 and 9-6 show the same silicone oil used in Figure 9-4, now spread as a stripe on a surface of SiO_2. As time passed, the leading edge of the precursor film grew progressively more convoluted.

Teletzke *et al.* (1987) suggest that these precursor films are probably the result of what they refer to as *primary* and *secondary spreading* mechanisms.

By primary spreading, they refer to the manner in which the first one or two molecular layers are deposited on the solid. For a nonvolatile liquid, primary spreading is by surface diffusion (Bascom et al. 1964; Cherry and Holmes 1969; Blake and Haynes 1969; Ruckenstein and Dunn 1977). For a volatile liquid, primary spreading may be by the condensation of an adsorbed film (Hardy 1919).

By secondary spreading, they refer to the motion of films whose thickness varies from several molecules to roughly a micron. These films spread as the result of either a positive disjoining pressure or a surface tension gradient. Surface tension gradients can be the result of either temperature gradients (Ludviksson and Lightfoot 1971) or composition gradients, perhaps resulting from the evaporation of a more volatile component (Bascom *et al.* 1964).

For more on wetting and spreading including precursor films, see Teletzke *et al.* (1987), Sec. 1.3.10, and an excellent review by de Gennes (1985).

In the displacement of one fluid by another, will a common line move? Not necessarily.

Dip a clean piece of metal partially into a can of oil-based paint and

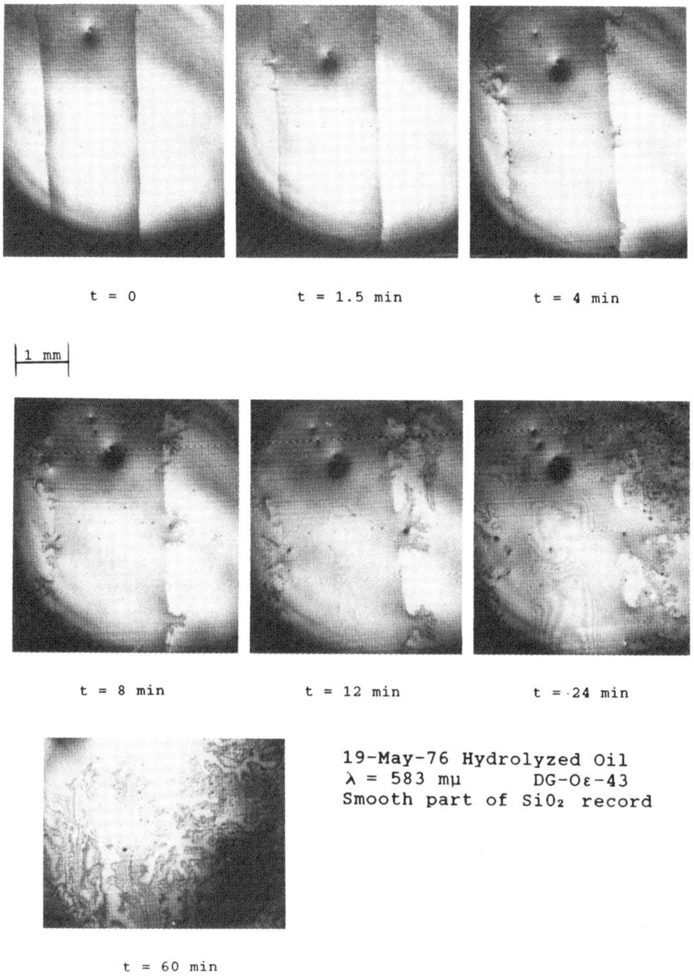

Figure 1.2.9-5: The same silicone oil used in Figure 1.2.9-4 was spread as a stripe on a surface of SiO_2 using an artist's brush. As time passed, the stripe widened, and projections began to appear and to become successively more convoluted (Richard Williams, May 19, 1976; photograph provided by Dr. Williams and the David Sarnoff Research Center, Princeton, NJ 08543-5300).

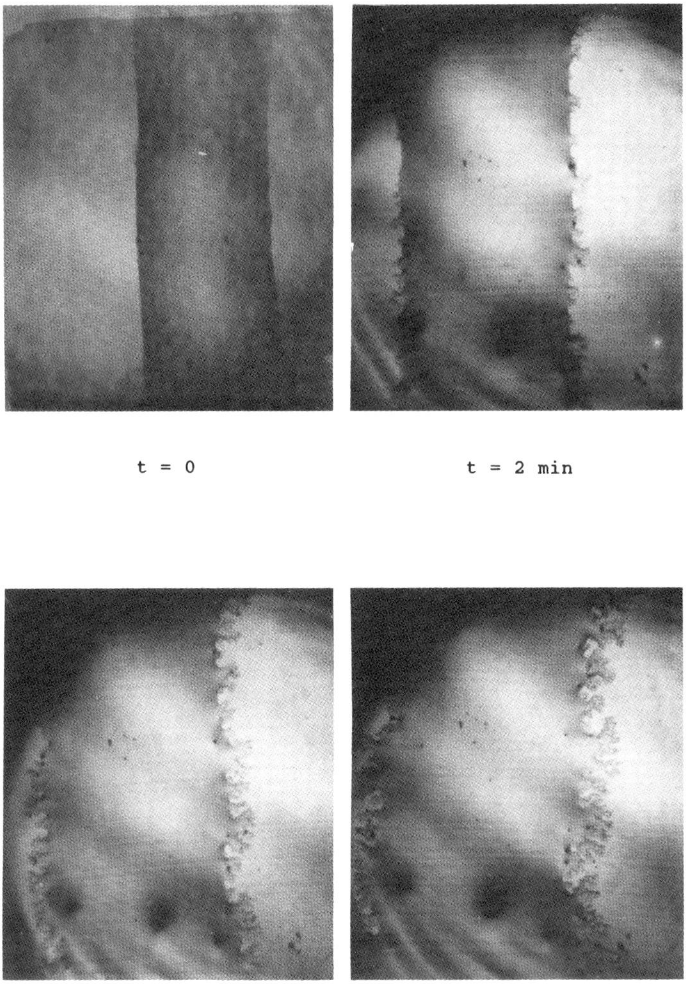

Figure 1.2.9-6: Another experiment similar to the one shown in Figure 1.2.9-5 at the same magnification (Richard Williams, May 19, 1976; photograph provided by Dr. Williams and the David Sarnoff Research Center, Princeton, NJ 08543-5300).

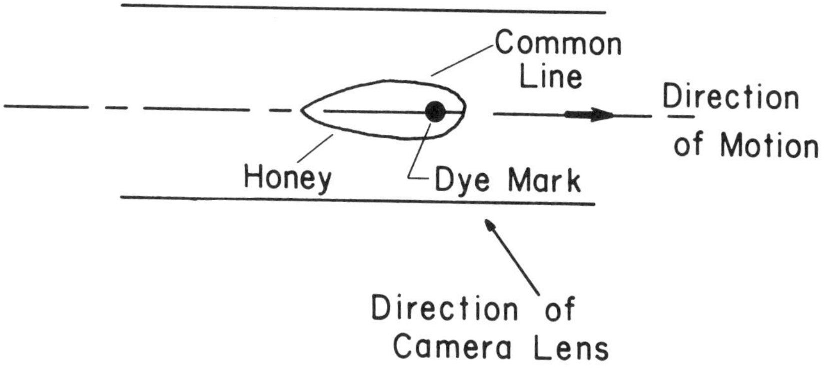

Figure 1.2.9-7: Plane view of drop of honey on a Plexiglas surface.

slowly withdraw it. The apparent common line does not move, although the film of paint begins to thin and drain. Before the film of paint can thin appreciably by drainage, loss of solvent through the paint-air interface causes the film to solidify.

More generally, thin films of liquid left behind on a solid as a liquid phase retreats can be stabilized by electrostatic forces, by reaction with the solid phase, or in the case of paint by the loss of solvent and resultant solidification.

If a common line moves, will it move smoothly? Generally not.

In discussing the precursor film that sometimes precedes an advancing macroscopic film, we noted Williams' (1977, 1988) observations illustrated by Figures 9-3 through 9-6. The configuration of the leading edge of the precursor film is often markedly different from that of the apparent leading edge of the macroscopic film. It may have a periodic or even random structure. Here I would like to consider the apparent leading edge of the macroscopic film.

Take a glass from the back of your cupboard, preferably one that has not been used for a long time, and add a little water to it, being careful not to wet the sides. Now hold the glass up to the light, tip it to one side for a moment, and return it to the vertical. The macroscopic film of water left behind on the side retreats very rapidly and irregularly, sometimes leaving behind drops that have been cut off and isolated. Now take a clean glass and repeat the same experiment. The macroscopic film of water now retreats much more slowly and regularly. But if you look closely, the apparent common line still moves irregularly, although the irregularities are on a much smaller scale than with the dirty glass. Very small drops that have been cut off and isolated in the irregular retreat are still left behind.

Poynting and Thomson (1902) forced mercury up a capillary tube and then gradually reduced the pressure. Instead of falling, the mercury at first adjusted itself to the reduced pressure by altering the curvature of the air-

Figure 1.2.9-8 A drop of honey flows over a Plexiglas surface (Dussan V. and Davis 1974, Figure 5).

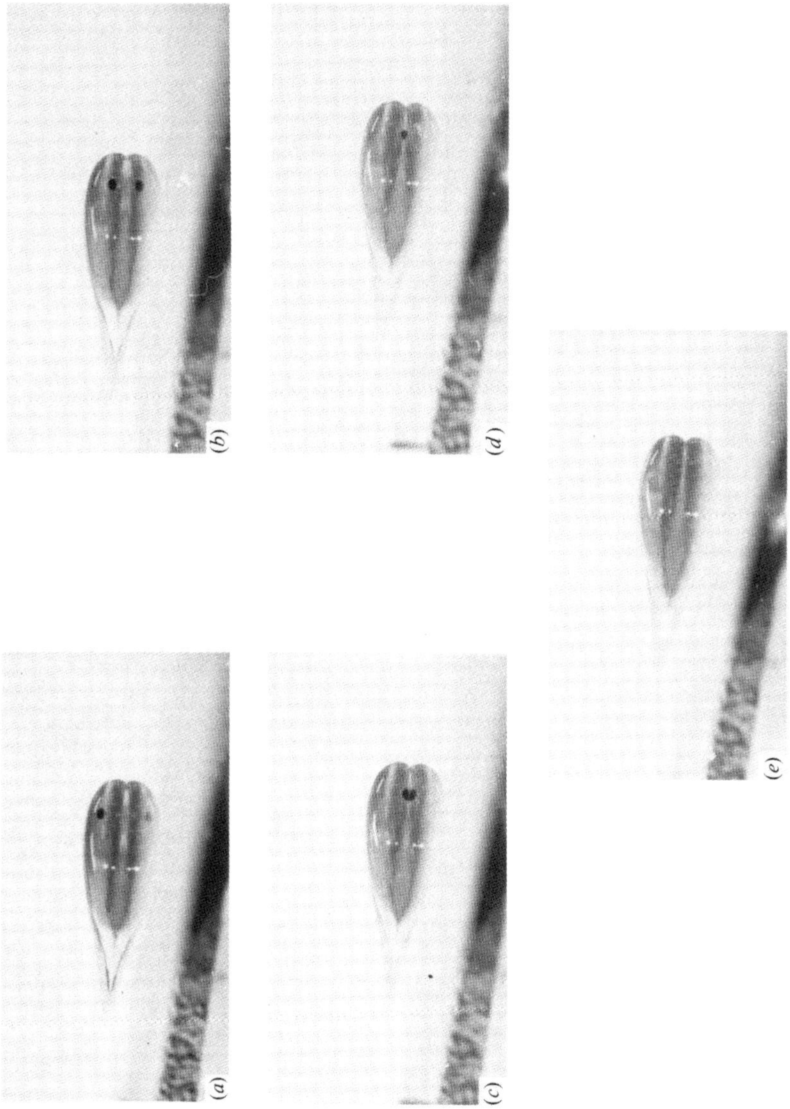
(a)
(b)
(c)
(d)
(e)

mercury interface. When the pressure gradient finally grew too large, the configuration of the meniscus became unstable and the mercury fell a short distance in the tube before stopping, repeating the deformation of the interface, and falling again when a new instability developed.

Yarnold (1938) saw this same sticking phenomena as a liquid index moved slowly through a glass capillary tube.

Elliott and Riddiford (1967) described a technique for measuring dynamic contact angles. In their experiment, one fluid was displaced by the other in the gap between two narrowly spaced horizontal parallel plates made of or coated with the solid material under investigation. The interface moved as fluid was injected or withdrawn at the center of the circle by means of a syringe controlled by a cam. For proper interpretation of the data, the projection of the interface between the fluids should have been a circle. Elliott and Riddiford (1967) reported that, when fluid was withdrawn through the syringe, the interface often did not move concentrically, irregular jerking or sticking being observed instead. Wilson (1975) pointed out that the flow which results when fluid is withdrawn through the syringe must always be unstable. The flow is more likely to be stable when fluid is injected through the syringe, even if the less viscous fluid is required to displace the more viscous one.

In some experiments, observed irregularities and episodic movements probably may be attributable to variations in the contact angle along the common line, which in turn may be due to a non-uniform distribution of contaminants. In other cases, the multiphase flow may be hydrodynamically unstable and no amount of cleaning can alter the situation.

In a displacement, one of the phases exhibits a rolling motion. In an effort to understand how material adjacent to a phase interface moves with a displacement of the common line, Dussan V. and Davis (1974) carried out the following four experiments.

i) Approximately one cm^3 of honey was placed on a horizontal Plexiglas surface. A small dye mark, consisting of honey and food coloring, was inserted by means of a needle in the honey-air interface at the plane of symmetry of the honey. The Plexiglas was tilted, the honey began to flow, and the trajectory of the dye mark was photographed from the direction shown in Figure 9-7. Because of the camera angle, a mirror image of the drop on the Plexiglas surface was seen in Figure 9-8. The common line could be distinguished as the intersection of the honey-air interface with its reflected image. As the honey flowed down the plane, the dye mark approached the common line. Finally, in Figure 9-8 (c), it made contact with the Plexiglas and in Figure 9-8 (d) appeared to be in contact with the Plexiglas. Although Figure 9-8 (e) showed no dye mark, it could be seen from above still in contact with the Plexiglas.

ii) Two drops of glycerol, one transparent and the other dyed with food coloring, were placed side by side on a solid bee's wax surface. The

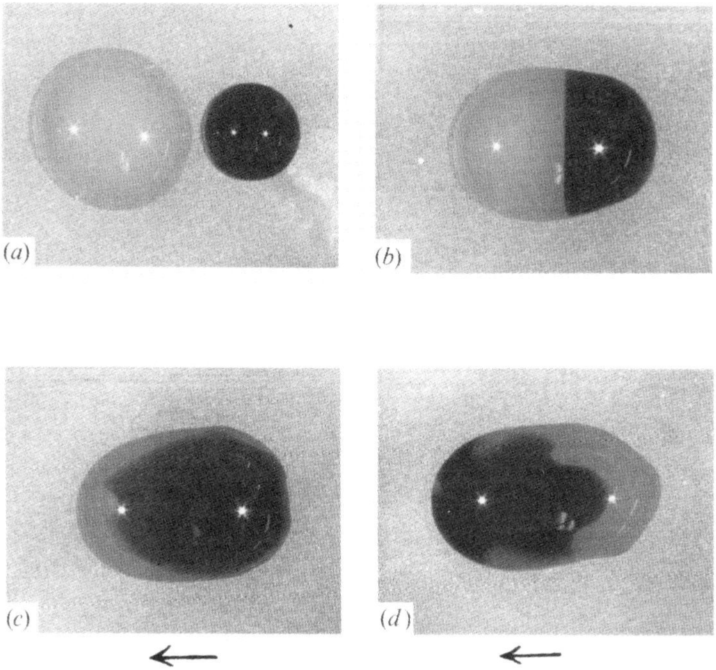

Figure 1.2.9-9: (a) Two drops of glycerol on a bee's wax surface. (b) After merging, there is only one drop, part of which is dyed. (c) The left end of the bee's wax plate is lowered. (d) After a finite length of time, the entire common line is composed of clear glycerol. (Dussan V. and Davis 1974, Figure 7)

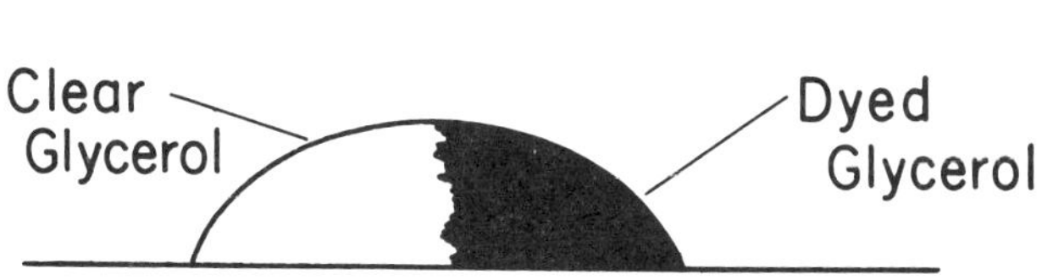

Figure 1.2.9-10: Cross-sectional view of a drop of glycerol that is partially dyed.

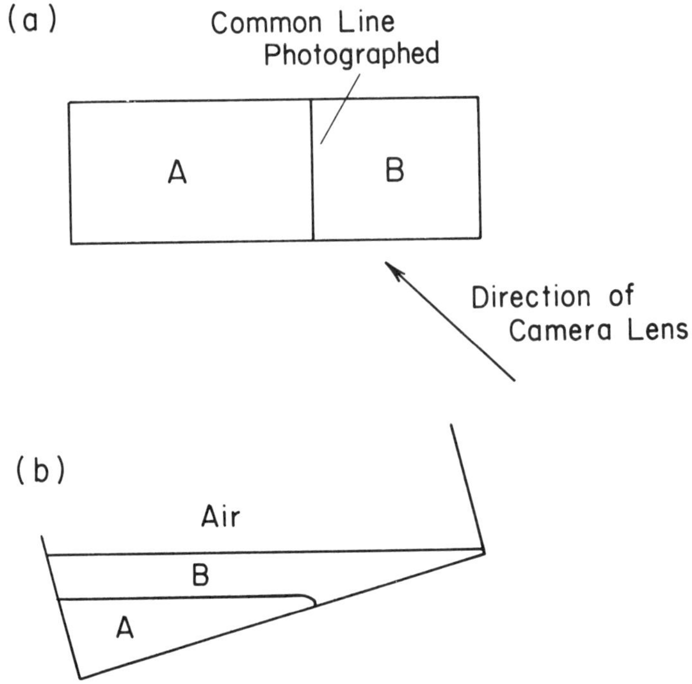

Figure 1.2.9-11: (a) Plan view of the bottom surface of the container. (b) Side view of the container. In experiments (iii) and (iv), fluid A is glycerol and fluid B is silicone oil.

surface was tilted, causing the two drops to merge, and it was then returned to its original horizontal position; see Figures 9-9 (a) and (b). In cross section, the resulting single drop probably looked like that shown in Figure 9-10. The left end was then lowered in Figure 9-9 (c). After a finite length of time, the entire common line was composed of clear glycerol in Figure 9-9 (d). The dyed glycerol adjacent to the right end portion of the common line moved to assume a position adjacent to the glycerol-air interface.

iii) A rectangular Plexiglas container was tilted with respect to the horizontal, partially filled with glycerol, and then with silicone oil. The result is sketched in Figure 9-11. A small drop of glycerol mixed with food coloring was added to the glycerol-silicone oil interface near the common line as shown in Figure 9-12 (a). The right end of the container was slowly lowered, the common line moved to the right, and the dye mark approached the common line; see Figures 9-12 (b) and (c). It finally became part of the common line in Figure 9-12 (d) and disappeared from sight; it was no longer adjacent to the glycerol-oil interface, but remained adhered to the Plexiglas.

iv) With the same system as in (iii), the common line was forced to move left by slowly raising the right end of the container. First a drop of dyed glycerol was placed on the Plexiglas initially covered with silicone oil. After a few moments, the drop *popped* onto the Plexiglas. The right end of the container was lowered, the clear glycerol moved forward and eventually merged with the dyed glycerol drop [Figure 9-13 (b)]. The right end of the container was then gradually raised, the common line moved to the left, and the dyed mark came off the bottom surface in Figure 9-13 (e).

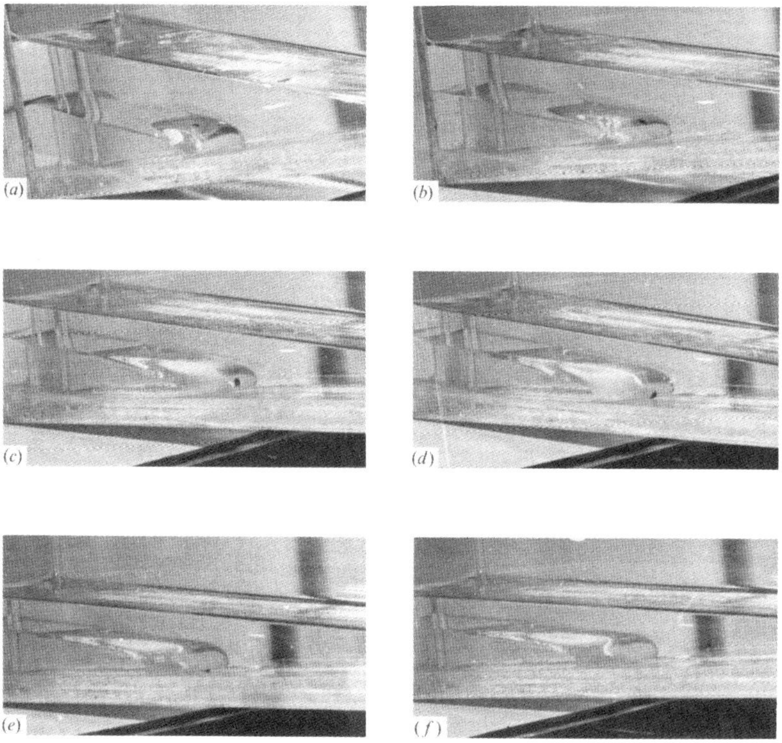

Figure 1.2.9-12: A dyed portion of glycerol adjacent to the glycerol-silicone oil interface is shown moving as the common line advances to the right (Dussan V. and Davis 1974, Figure 10). The lower fluid is glycerol; the upper fluid is silicone oil.

Figure 1.2.9-13 The lower fluid and the dyed fluid are glycerol; the upper fluid is silicone oil (Dussan V. and Davis 1974, Figure 11). The arrows indicate the direction of motion of the common line.

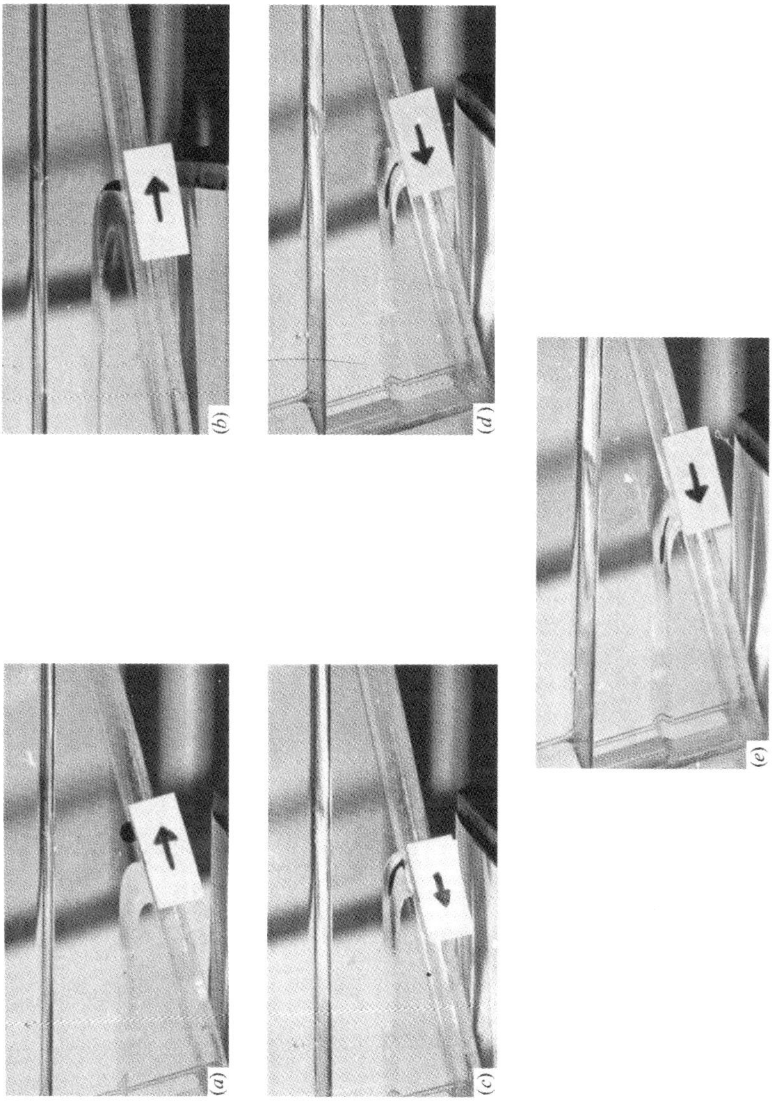
(a)
(b)
(c)
(d)
(e)

All four of these experiments illustrate a *rolling* motion exhibited by one of the phases[a] in a displacement. In a **forward rolling motion**, the material in this phase adjacent to the fluid-fluid interface moves down to the common line and is left adjacent to the fluid-solid interface. In a **backward rolling motion**, material in this phase originally adjacent to the fluid-solid interface is lifted off as the common line passes, and it is transported up along the fluid-fluid interface.

Allen and Benson (1975) have closely observed the motion in a drop moving down an inclined plane. Although this three-dimensional motion is more complicated than the primarily two-dimensional flows that we have just finished discussing, the motion in the neighborhood of the common line continues to exhibit a rolling character.

References

Allen, R. F., and P. R. Benson, *J. Colloid Interface Sci.* **50**, 250 (1975).

Bascom, W. D., R. D. Cottington, and C. R. Singleterry, in "Contact Angles, Wettability and Adhesion," *Advances in Chemistry Series* No. 43, p. 355, American Chemical Society, Washington, D.C. (1964).

Blake, T. D., and J. M. Haynes, *J. Colloid Interface Sci.* **30**, 421 (1969).

Cherry, B. W., and C. M. Holmes, *J. Colloid Interface Sci.* **29**, 174 (1969).

de Gennes, P. G., *Rev. Mod. Phys.* **57**, 827 (1985).

Dussan V., E. B., and S. H. Davis, *J. Fluid Mech.* **65**, 71 (1974).

Elliott, G. E. P., and A. C. Riddiford, *J. Colloid Interface Sci.* **23**, 389 (1967).

Hardy, W. B., *Phil. Mag. (6)* **38**, 49 (1919).

Ludviksson, V., and E. N. Lightfoot, *AIChE J.* **17**, 1166 (1971).

Poynting, J. H., and J. J. Thomson, "A Text-Book of Physics - Properties of Matter," p. 142, Charles Griffin, London (1902).

---

a) Dussan V. and Davis (1974) do not report either static or apparent contact angles (see Sec. 2.1.9).

Radigan, W., H. Ghiradella, H. L. Frisch, H. Schonhorn, and T. K. Kwei, *J. Colloid Interface Sci.* **49**, 241 (1974).

Ruckenstein, E., and C. S. Dunn, *J. Colloid Interface Sci.* **59**, 135 (1977).

Teletzke, G. F., H. T. Davis, and L. E. Scriven, *Chem. Eng. Commun.* **55**, 41 (1987).

Williams, R., *Nature London* **266**, 153 (1977).

Williams, R., *personal communication* (1988).

Wilson, S. D. R., *J. Colloid Interface Sci.* **51**, 532 (1975).

Yarnold, G. D., *Proc. Phys. Soc. London* **50**, 540 (1938).

1.2.10 Moving common lines: emission of material surfaces (Dussan V. and Davis 1974) If, as suggested by the experiments described in the last section, one phase exhibits a rolling motion as a common line moves, what is the character of the motion of the other phase?

In Figure 10-1, fluid A is being displaced by fluid B as the common line moves to the left over the solid S. I suggest we make the following assumptions.

i) There is a forward rolling motion in phase B in the sense defined in the previous section.

ii) The tangential components of velocity are continuous across all phase interfaces.

iii) The A-S dividing surface is in chemical equilibrium (see Sec. 5.10.3) with phase A prior to displacement. There is no mass transfer between the bulk phase A and the solid surface.

iv) The mass density of phase A (see Sec. 1.3.1) must be finite everywhere. A finite mass of phase A must occupy a finite volume.

Let us use referential descriptions for the motion in phase A (see Sec. 1.1.1)

$$\mathbf{z}^{(A)} = \boldsymbol{\chi}_{\kappa}^{(A)}(\,\mathbf{z}_{\kappa}^{(A)}, t\,) \tag{10-1}$$

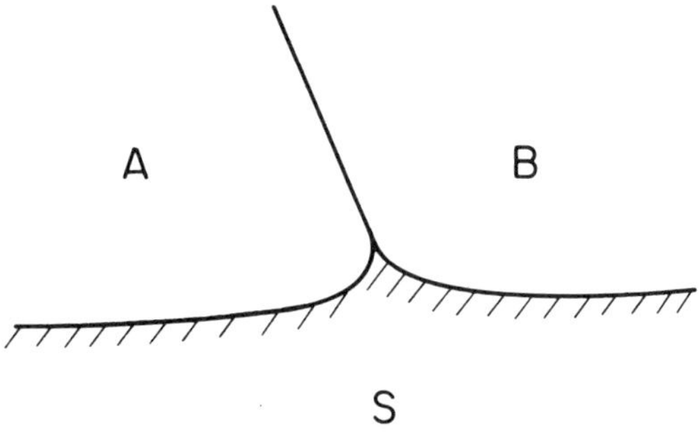

Figure 1.2.10-1: The common line formed by fluids A and B on the solid S. The solid is shown with a dimple at the common line, although the analysis described in this section applies equally well to both flat and deformed solids.

and for the motion in the solid S

$$\mathbf{z}^{(S)} = \boldsymbol{\chi}_{\kappa}^{(S)}(\mathbf{z}_{\kappa}^{(S)}, t) \tag{10-2}$$

To be specific, let us focus our attention on the position $\mathbf{z}_0$, which in the reference configurations for phases A and S denotes a position on the A-S dividing surface. Our particular concerns are the trajectory of the material particle of phase A that is adjacent to $\mathbf{z}_0$ in the reference configuration

$$\mathbf{z}_0^{(A)}(t) \equiv \text{limit } \mathbf{z}_{\kappa}^{(A)} \to \mathbf{z}_0 : \quad \boldsymbol{\chi}_{\kappa}^{(A)}(\mathbf{z}_{\kappa}^{(A)}, t) \tag{10-3}$$

and the trajectory of the material particle of the solid that is adjacent to $\mathbf{z}_0$ in the reference configuration

$$\mathbf{z}_0^{(S)}(t) \equiv \text{limit } \mathbf{z}_{\kappa}^{(S)} \to \mathbf{z}_0 : \quad \boldsymbol{\chi}_{\kappa}^{(S)}(\mathbf{z}_{\kappa}^{(S)}, t) \tag{10-4}$$

Since the common line is moving, we know that at some particular time t_1 the solid point $\mathbf{z}_0^{(S)}(t_1)$ will be adjacent to the common line. We know that $\mathbf{z}_0^{(S)}(t)$ for $t > t_1$ must be adjacent to the B-S dividing surface.

Where is $\mathbf{z}_0^{(A)}(t)$ for $t > t_1$? There are only four possible locations.

1) It remains adjacent to the same solid material point

$$\text{for } t > t_1 : \quad \mathbf{z}_0^{(A)}(t) = \mathbf{z}_0^{(S)}(t) \tag{10-5}$$

2) It is mapped adjacent to the A-B dividing surface.

3) It remains adjacent to the common line.

4) It is mapped into the interior of phase A.

Alternative (1) is not possible. Let $\mathbf{z}^{(cl)}(s, t)$ denote position on the common line as a function of arc length s measured along the common line from some convenient point. A moving common line implies that there exists a time $t_2 > t_1$ such that for some D:

$$\text{for all } s: \quad | \mathbf{z}_0^{(S)}(t_2) - \mathbf{z}^{(cl)}(s, t_2) | > D \tag{10-6}$$

Equation (10-3) implies that for any $\varepsilon > 0$ there exists an $\eta > 0$ such that

$$| \mathbf{z}_0^{(A)}(t_2) - \boldsymbol{\chi}_\kappa^{(A)}(\mathbf{z}_\kappa^{(A)}, t_2) | < \varepsilon \tag{10-7}$$

for all material points within phase A satisfying the relation

$$| \mathbf{z}_\kappa^{(A)} - \mathbf{z}_0 | < \eta \tag{10-8}$$

Physically this means that all of the material of phase A located within a distance η of $\mathbf{z}_0$ in the reference configuration is mapped to within a distance ε of $\mathbf{z}_0^{(A)}(t_2)$ at time t_2. Since ε is arbitrary, we are free to choose $\varepsilon < D$. As a result of the motion of the common line, $\mathbf{z}_0^{(S)}(t_2)$ is located adjacent to the B-S dividing surface, $\mathbf{z}_0^{(A)}(t_2)$ is also adjacent to the B-S dividing surface by (10-5), and all of phase A within an η neighborhood of $\mathbf{z}_0$ in the reference configuration must also be located on the B-S dividing surface at time t_2. In view of assumption (iii), this implies that a finite quantity of the bulk phase A has a zero volume at time t_2 which contradicts assumption (iv).

Alternative (2) contradicts assumption (i).

Alternative (3) is also impossible, since it violates assumption (ii).

The only possibility is alternative (4): material adjacent to the A-S dividing surface is mapped into the interior of phase A as phase A is displaced by phase B.

If the motion of the common line in Figure 10-1 were reversed, we would have a backward rolling motion in phase B. All of the arguments above would apply with relatively little change, but the final conclusion would be that material from the interior of phase A would be mapped into positions adjacent to the A-S dividing phase as phase B is displaced by phase A.

Motions of this character have been observed in three experiments carried out by Dussan V. and Davis (1974).

a) The rectangular Plexiglas container shown in Figure 9-10 was again tilted with respect to the horizontal, partially filled with a mixture of 60% water and 40% methyl alcohol (fluid A), and then silicone oil (fluid B). A drop of the alcohol-water mixture containing food coloring was injected onto the interface between the Plexiglas and the aqueous solution of alcohol. The dye remained undisturbed for a time to allow it to diffuse very close to the wall. The right end of the tank containing the oil was then slowly raised; the common line moved to the left in Figures 9-10 and 10-1 and

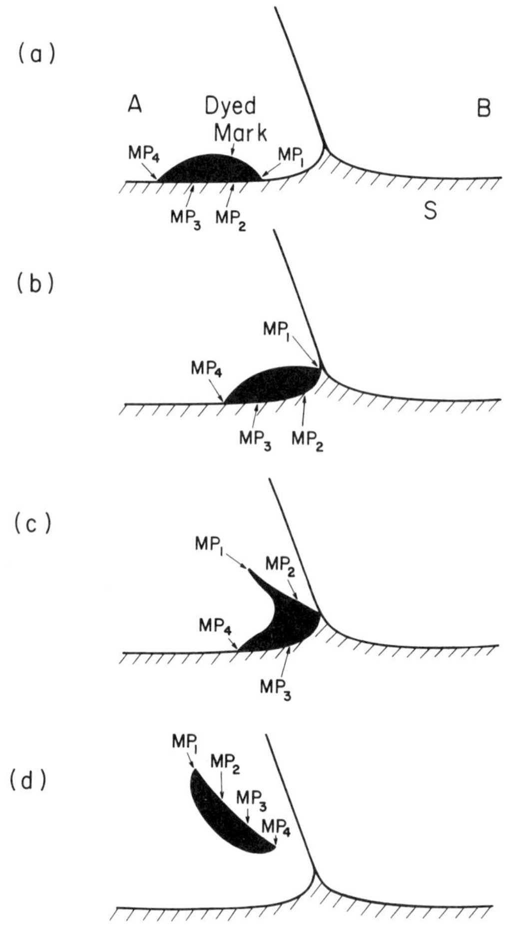

Figure 1.2.10-2: The trajectory of a dyed mark in the aqueous alcohol solution (A), initially adjacent to the aqueous alcohol solution-Plexiglas (S) phase interface, as the aqueous alcohol solution-silicone oil (B) phase interface moves to the left.

approached the dye mark. Their photographs for this experiment are not clear, but I believe that Figure 10-2 illustrates what they saw. The dyed mark was lifted off the Plexiglas surface and *ejected* to the interior of the aqueous alcohol solution along an otherwise invisible material surface in the flow field. (A *material* surface is one that is everywhere tangent to the local material velocity.)

b) This same experiment was performed again, but this time the dye mark was adjacent to the aqueous alcohol solution-silicone oil interface. Since

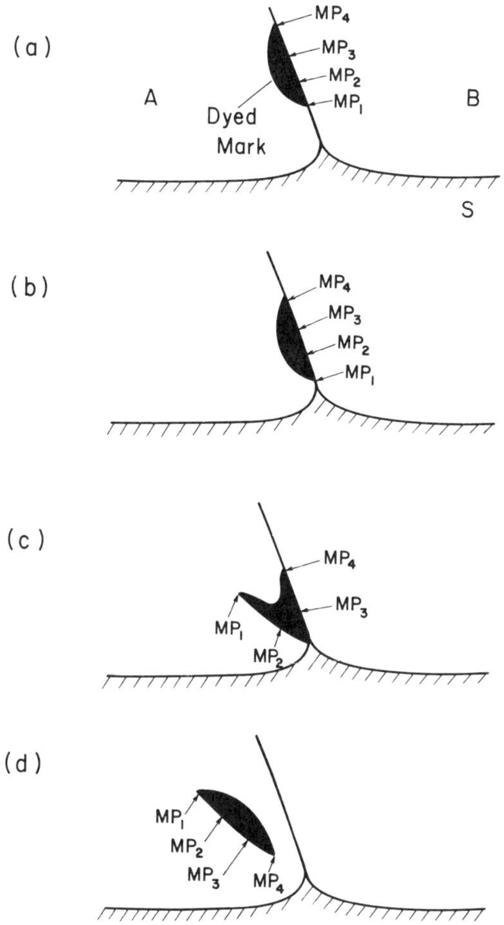

Figure 1.2.10-3: The trajectory of a dyed mark in the aqueous alcohol solution (A), initially adjacent to the aqueous alcohol-silicone oil (B) interface, as this interface moves to the left.

their photographs are not clear, I have drawn Figure 10-3 from their description. As the right hand end of the container was raised, the common line moved to the left in Figures 9-10 and 10-1, the dyed mark moved down the fluid-fluid interface to the common line, and it was finally *ejected* to the interior of the aqueous alcohol solution along a material surface. This material surface within the aqueous alcohol solution appeared to coincide with that observed in the previous experiment.

c) The backward version of this same system was also examined. In this case, the common line moves to the right in Figures 9-10 and 10-1 towards

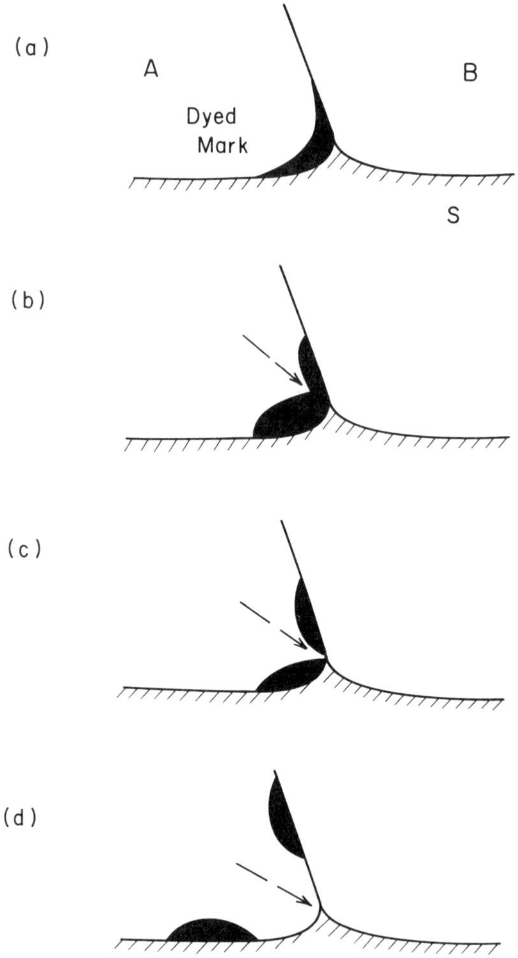

Figure 1.2.10-4: The trajectory of a dyed mark in the aqueous alcohol solution (A), initially adjacent to the common line, as the aqueous alcohol solution-silicone oil (B) phase interface moves to the right.

the oil. Figure 10-4 illustrates the evolution of a dyed spot placed at the common line within the aqueous alcohol phase. The dyed mark distinctly divides into two parts: one is transported up along the aqueous alcohol solution-silicone oil interface and the other is left behind on the aqueous alcohol solution-Plexiglas interface. In this case, there appears to have been *injection* of aqueous alcohol solution into the immediate neighborhood of the common line along both sides of a material surface within the aqueous alcohol phase.

The preceding discussion and experiments suggest the following generalization concerning the character of the motions involved when one phase displaces another over a solid surface. If one phase exhibits a *forward* rolling motion (see Sec. 1.2.9, footnote a), there will be *ejection* from the neighborhood of the common line into the interior of the other phase along both sides of a material surface originating at the common line and dividing the flow field within the phase. If one phase exhibits a *backward* rolling motion, there will be injection from the interior of the other phase to the immediate neighborhood of the common line along both sides of such a surface.

Reference

Dussan V., E. B., and S. H. Davis, *J. Fluid Mech.* **65**, 71 (1974).

1.2.11 Moving common lines: velocity is multivalued on a rigid solid Under certain conditions, the velocity field will be multivalued at a common line moving over a rigid solid. The proof does not require the surface of the solid to be a smooth plane. For this reason, in the displacement of A by B over S in Figure 11-1, I have pictured the surface of the solid phase S as an arbitrary curve.

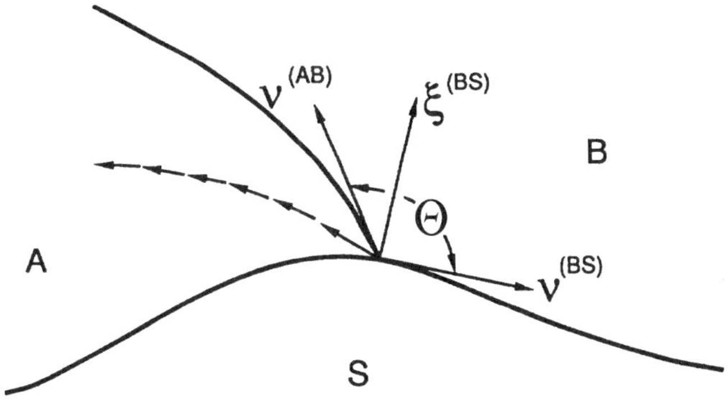

Figure 1.2.11-1: In the displacement of A by B over S, the solid phase S is assumed not to deform. The material surface ejected from the common line is drawn with the assumption that there is a forward rolling motion within phase B. I denote by $\mathbf{v}^{(BS)}$ the unit vector normal to the common line and tangent to the B-S dividing surface, by $\mathbf{v}^{(AB)}$ the unit vector normal to the common line and tangent to the A-B dividing surface, and by $\boldsymbol{\xi}^{(BS)}$ the unit vector that is normal to both the common line and the B-S dividing surface.

In a frame of reference that is fixed with respect to the common line, the velocity distribution in the solid

$$\mathbf{v}^{(S)} = \mathbf{v}^{(S)}(\mathbf{z}, t) \tag{11-1}$$

and the velocity distribution in the A-B dividing surface

$$\mathbf{v}^{(\sigma)} = \mathbf{v}^{(\sigma)}(\mathbf{z}, t) \tag{11-2}$$

are both functions of position and time. Choosing the origin of our coordinate system to coincide with whatever point is of interest on the common line, we wish to compare

$$\mathbf{v}^{(S)}(\mathbf{0}, t) \equiv \text{limit } \mathbf{z} \to \mathbf{0},\ \mathbf{z} \in \text{ solid: } \mathbf{v}^{(S)}(\mathbf{z}, t) \tag{11-3}$$

$$\mathbf{v}^{(\sigma)}(\mathbf{0}, t) \equiv \text{limit } \mathbf{z} \to \mathbf{0},\ \mathbf{z} \in \text{ dividing surface: } \mathbf{v}^{(\sigma)}(\mathbf{z}, t) \tag{11-4}$$

In this frame of reference, $\mathbf{v}^{(S)}(\mathbf{0}, t)$ is tangent to the solid surface

$$\boldsymbol{\xi}^{(BS)} \cdot \mathbf{v}^{(S)}(\mathbf{0}, t) = 0 \tag{11-5}$$

$$\boldsymbol{\nu}^{(BS)} \cdot \mathbf{v}^{(S)}(\mathbf{0}, t) > 0 \tag{11-6}$$

and $\mathbf{v}^{(\sigma)}(\mathbf{0}, t)$ does not have a component normal to the A-B dividing surface, which means that it can be written in the form

$$\mathbf{v}^{(\sigma)}(\mathbf{0}, t) = C\,\boldsymbol{\nu}^{(AB)} + D(\boldsymbol{\xi}^{(BS)} \wedge \boldsymbol{\nu}^{(BS)}) \tag{11-7}$$

Equation (11-7) implies that

$$\boldsymbol{\xi}^{(BS)} \cdot \mathbf{v}^{(\sigma)}(\mathbf{0}, t) = C(\boldsymbol{\xi}^{(BS)} \cdot \boldsymbol{\nu}^{(AB)}) \tag{11-8}$$

and

$$\boldsymbol{\nu}^{(BS)} \cdot \mathbf{v}^{(\sigma)}(\mathbf{0}, t) = C(\boldsymbol{\nu}^{(BS)} \cdot \boldsymbol{\nu}^{(AB)}) \tag{11-9}$$

Our objective is to compare these expressions with (11-5) and (11-6). Four cases can be identified.

Case 1 Let us assume that the contact angle Θ in Figure 11-1 is not 0, $\pi/2$ or π. This means that

$$\boldsymbol{\xi}^{(BS)} \cdot \boldsymbol{\nu}^{(AB)} \neq 0 \tag{11-10}$$

and

$$\boldsymbol{\nu}^{(BS)} \cdot \boldsymbol{\nu}^{(AB)} \neq 0 \tag{11-11}$$

If C ≠ 0, (11-5) and (11-8) imply

$$\boldsymbol{\xi}^{(BS)} \cdot \mathbf{v}^{(S)}(\mathbf{0}, t) \neq \boldsymbol{\xi}^{(BS)} \cdot \mathbf{v}^{(\sigma)}(\mathbf{0}, t) \tag{11-12}$$

If C = 0, (11-6) and (11-9) tell us that

$$\boldsymbol{\nu}^{(BS)} \cdot \mathbf{v}^{(S)}(\mathbf{0}, t) \neq \boldsymbol{\nu}^{(BS)} \cdot \mathbf{v}^{(\sigma)}(\mathbf{0}, t) \tag{11-13}$$

Our conclusion is that

$$\mathbf{v}^{(\sigma)}(\mathbf{0}, t) \neq \mathbf{v}^{(S)}(\mathbf{0}, t) \tag{11-14}$$

and the velocity vector must have multiple values at the common line.

Case 2 If the contact angle Θ is $\pi/2$

$$\boldsymbol{\nu}^{(BS)} \cdot \boldsymbol{\nu}^{(AB)} = 0 \tag{11-15}$$

For all values of C, we find (11-13) and (11-14). Again the velocity vector must have multiple values at the common lines.

Case 3 If the contact angle Θ is π

$$\boldsymbol{\xi}^{(BS)} \cdot \boldsymbol{\nu}^{(AB)} = 0 \tag{11-16}$$

There is a nonzero value of C for which

$$\boldsymbol{\nu}^{(BS)} \cdot \mathbf{v}^{(S)}(\mathbf{0}, t) = \boldsymbol{\nu}^{(BS)} \cdot \mathbf{v}^{(\sigma)}(\mathbf{0}, t) > 0 \tag{11-17}$$

and

$$\mathbf{v}^{(\sigma)}(\mathbf{0}, t) = \mathbf{v}^{(S)}(\mathbf{0}, t) \tag{11-18}$$

However, if we examine the velocity distribution in the ejected surface within phase A

$$\mathbf{v}^{(e)} = \mathbf{v}^{(e)}(\mathbf{z}, t) \tag{11-19}$$

assuming[a] forward rolling motion in phase B, we see that

---

a) If there is a backward rolling motion in Phase A, the arguments presented in cases 3 and 4 are reversed.

$$\mathbf{v}^{(BS)} \cdot [\text{ limit } \mathbf{z} \to 0, \mathbf{z}\ \varepsilon \text{ ejected surface: } \mathbf{v}^{(e)}(\mathbf{z}, t)] \leq 0 \tag{11-20}$$

Our conclusion is that the velocity vector will assume multiple values at the common line, if material is ejected directly from the common line. In this case, our conclusion would be based upon the assumptions predicated in Sec. 1.2.10, particularly assumptions ii and iii.

Case 4 If the contact angle Θ is 0,

$$\mathbf{v}^{(BS)} \cdot \mathbf{v}^{(\sigma)}(\mathbf{0}, t) < 0 \tag{11-21}$$

assuming[a] that there is a forward rolling motion in phase B. This implies (11-13) and (11-14): the velocity vector is multiple-valued at the common line.

To summarize, we have shown here that under certain conditions the velocity vector will be multiple-valued at a fixed common line on a moving rigid solid. For more on this velocity singularity, see Secs. 1.3.7 through 1.3.10.

The discussion given in this section was inspired by that of Dussan V. and Davis (1974).

Reference

Dussan V., E. B., and S. H. Davis, *J. Fluid Mech.* **65**, 71 (1974).

1.3 Mass

1.3.1 Conservation of mass This discussion of continuum mechanics is based upon several axioms, the first and most familiar of which is **conservation of mass:**

The mass of a body is independent of time.

Physically, this means that, if we follow a portion of a material body through any number of translations, rotations, and deformations, the mass associated with it will not vary as a function of time.

We can express this idea mathematically by saying that mass is a non-negative, time-independent, scalar measure m defined for every body in the universe. The mass $\mathcal{M}$ of a body B may be written in terms of this measure as

$$\mathcal{M} \equiv \int_{B} dm \tag{1-1}$$

Because this measure is independent of time, it follows immediately that

$$\frac{d\mathcal{M}}{dt} = \frac{d}{dt}\int_{B} dm = 0 \tag{1-2}$$

At first thought, it appears that mass should be an absolutely continuous function of volume. But this can not be true in general, when a dividing surface is used to represent the phase interface. We must allow for concentrated masses in the dividing surface.

In order to keep our discussion as simple as possible, let us restrict our attention for the moment to a body consisting of two phases occupying regions $R^{(1)}$ and $R^{(2)}$ respectively. The mass density of each phase i (i = 1, 2) will be denoted by $\rho^{(i)}$, a continuous function of position within the phase. No matter what equations of state are used to describe the thermodynamic behavior of each phase, they will generally not be able to account for the mass distribution in the interfacial region. To compensate, we assign to the dividing surface $\Sigma^{(1,2)}$ a mass density $\rho^{(\sigma:1,2)}$, a continuous function of position on $\Sigma^{(1,2)}$ having the units of mass per unit area. With these definitions, the mass $\mathcal{M}$ of the body can be described by

$$\mathcal{M} = \int_{R^{(1)}} \rho^{(1)}\, dV + \int_{R^{(2)}} \rho^{(2)}\, dV + \int_{\Sigma^{(1,2)}} \rho^{(\sigma:1,2)}\, dA \tag{1-3}$$

and the postulate of mass conservation takes the form

$$\frac{d\mathcal{M}}{dt} = \frac{d}{dt}\int_{R^{(1)}} \rho^{(1)}\, dV + \frac{d}{dt}\int_{R^{(2)}} \rho^{(2)}\, dV$$

$$+ \frac{d}{dt}\int_{\Sigma^{(1,2)}} \rho^{(\sigma:1,2)}\, dA = 0 \tag{1-4}$$

In writing (1-3) and (1-4), we have introduced as a shorthand notation a single integral to represent what must be a triple or volume integration in space when expressed in terms of rectangular cartesian coordinates z_1, z_2, and z_3:

$$\int_{R^{(1)}} \rho^{(1)}\, dV = \iiint_{R^{(1)}} \rho^{(1)}\, dz_1 dz_2 dz_3 \tag{1-5}$$

Here we denote a volume integration by dV; $R^{(1)}$ indicates the region in space over which the integration is to be performed. Similarly, a single integral is also used to represent an area integration, which is a double integration when expressed in terms of the surface coordinates (see Sec. A.6.2):

$$\int_{\Sigma^{(1,2)}} \rho^{(\sigma:1,2)}\, dA \equiv \iint_{\Sigma^{(1,2)}} \rho^{(\sigma:1,2)} \sqrt{a}\; dy^1 dy^2 \tag{1-6}$$

We intend an area integration when we write dA; $\Sigma^{(1,2)}$ tells us the particular surface over which the integration is to be carried out.

Equations (1-3) and (1-4) are easily generalized for a body consisting of M phases. Let $\Sigma^{(i,j)}$ be the dividing surface separating phases i and j and let $\rho^{(\sigma:i,j)}$ be the mass density assigned to this dividing surface. The mass of the body becomes

$$\mathcal{M} = \sum_{i=1}^{M} \int_{R^{(i)}} \rho^{(i)}\, dV + \sum_{i=1}^{M-1} \sum_{j=i+1}^{M} \int_{\Sigma^{(i,j)}} \rho^{(\sigma:i,j)}\, dA \tag{1-7}$$

and conservation of mass requires

$$\frac{d\mathcal{M}}{dt} = \frac{d}{dt}\sum_{i=1}^{M} \int_{R^{(i)}} \rho^{(i)}\, dV + \frac{d}{dt}\sum_{i=1}^{M-1} \sum_{j=i+1}^{M} \int_{\Sigma^{(i,j)}} \rho^{(\sigma:i,j)}\, dA$$

$$= 0 \tag{1-8}$$

Later derivations will be simplified, if we introduce a less formidable notation. If ρ is a piecewise continuous function defined by $\rho^{(i)}$ in phase i, then we may write

$$\int_R \rho \, dV \equiv \sum_{i=1}^{M} \int_{R^{(i)}} \rho^{(i)} \, dV \tag{1-9}$$

where R is the region occupied by the body:

$$R \equiv \sum_{i=1}^{M} R^{(i)} \tag{1-10}$$

Similarly, let the surface mass density $\rho^{(\sigma)}$ be a piecewise continuous function defined by $\rho^{(\sigma:i,j)}$ on the dividing surface $\Sigma^{(i,j)}$. The mass of the body to be associated with the dividing surface may be denoted more compactly as

$$\int_\Sigma \rho^{(\sigma)} \, dA \equiv \sum_{i=1}^{M-1} \sum_{j=i+1}^{M} \int_{\Sigma^{(i,j)}} \rho^{(\sigma:i,j)} \, dA \tag{1-11}$$

where Σ represents the union of all the dividing surfaces:

$$\Sigma \equiv \sum_{i=1}^{M-1} \sum_{j=i+1}^{M} \Sigma^{(i,j)} \tag{1-12}$$

In this notation, conservation of mass requires

$$\begin{aligned} \frac{d\mathcal{M}}{dt} &= \frac{d}{dt}\left(\int_R \rho \, dV + \int_\Sigma \rho^{(\sigma)} \, dA \right) \\ &= 0 \end{aligned} \tag{1-13}$$

Our next objective will be to determine the relationships that express the idea of conservation of mass at every point in a material. In order to do this, we will find it necessary to interchange the operations of differentiation with respect to time and integration over space in (1-4), (1-8) or (1-13). Yet the limits on these integrals describe the boundaries of the various phases in their current configurations. Because they generally are functions of time, a simple interchange of these operations is not possible. In Sec. 1.3.3, we explore this problem in more detail.

1.3.2 Surface mass density But before we go on, let us stop and ask how the surface mass density is to be interpreted, when the dividing surface is regarded as a model for a three-dimensional interfacial region (see Sec. 1.2.4).

In Figure 2-1, we see a material body that occupies a region R and consists of two phases. In addition to the dividing surface Σ, we also show two surfaces Σ^+ and Σ^-

i) that are parallel[a] to Σ,

ii) that move with the speed of displacement of Σ (see Sec. 1.2.7), and

iii) that enclose all of the material in R whose behavior is not described by the constitutive equations appropriate to either of the neighboring phases during the time of observation.

We will refer to the region enclosed by Σ^+ and Σ^- as $R^{(I)}$. Notice that $R^{(I)}$ always includes the interfacial region, but at any particular time may include a portion of the neighboring phases as well.

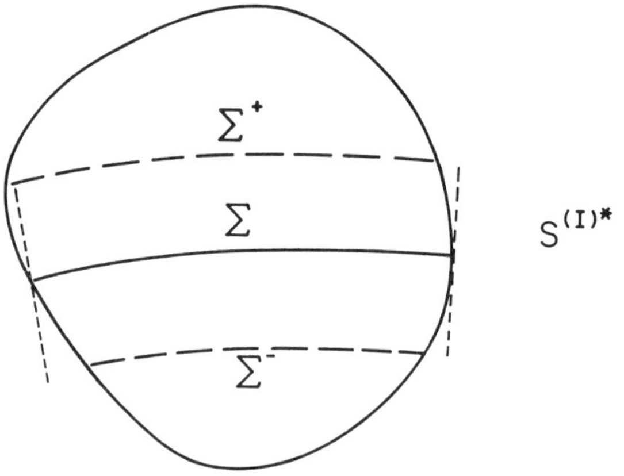

Figure 1.3.2-1: Region R occupied by the material body consisting of two phases: S is the closed surface bounding R; $R^{(I)}$ and $S^{(I)}$ are those portions of R and S enclosed by Σ^+ and Σ^-; $S^{(I)*}$ is the locus of all the straight lines stretching from Σ^+ to Σ^- that are normal to Σ and that pass through C, the closed curve bounding Σ; $R^{(I)*}$ is bounded by the surfaces Σ^+ , Σ^- , and $S^{(I)*}$.

---

a) A surface whose points lie at a constant distance along the normal $\boldsymbol{\xi}$ from a surface Σ is said to be **parallel** to Σ (Willmore 1959, p. 116).

Conservation of mass in the previous section is described in terms of the *bulk* mass density ρ and the *surface* mass density $\rho^{(\sigma)}$. But if we think of the interface as a three-dimensional region of finite thickness, then we would have to say that conservation of mass requires

$$\frac{d}{dt}\int_{R} \rho^{(I)}\, dV = 0 \tag{2-1}$$

Here $\rho^{(I)}$ is the *interfacial* mass density distribution such that

$$\text{outside } R^{(I)}: \quad \rho^{(I)} = \rho \tag{2-2}$$

From (2-1) and (2-2), this means

$$\frac{d}{dt}\left(\int_{R} \rho\, dV + \int_{R^{(I)}} (\, \rho^{(I)} - \rho \,)\, dV \right) = 0 \tag{2-3}$$

and we can identify

$$\int_{\Sigma} \rho^{(\sigma)}\, dA = \int_{R^{(I)}} (\, \rho^{(I)} - \rho \,)\, dV \tag{2-4}$$

Unfortunately, this is as far as we can go rigorously.

The surface $S^{(I)*}$ in Figure 2-1 is the locus of all the straight lines stretching from Σ^- to Σ^+ that are normal to Σ and that pass through C, the closed bounding curve of Σ. Considerably more progress can be made towards understanding the surface mass density, if we approximate

$$\int_{R^{(I)}} (\, \rho^{(I)} - \rho \,)\, dV \doteq \int_{R^{(I)*}} (\, \rho^{(I)} - \rho \,)\, dV \tag{2-5}$$

where $R^{(I)*}$ is the region bounded by the surface Σ^-, Σ^+, and $S^{(I)*}$.

Using the theorem developed in Exercise 1.3.2-1, we find

$$\int_{R^{(I)*}} (\, \rho^{(I)} - \rho \,)\, dV$$

$$= \int_{\Sigma} \int_{\lambda^-}^{\lambda^+} (\, \rho^{(I)} - \rho \,)(\, 1 - \kappa_1 \lambda \,)(\, 1 - \kappa_2 \lambda \,)\, d\lambda dA \tag{2-6}$$

By λ, we mean the distance measured along the normal to the dividing surface Σ; λ^+ and λ^- are the values of λ at Σ^+ and Σ^- respectively; κ_1 and κ_2 are the principal curvatures of the dividing surface (see Exercise A.5.3-

3). Equations (2-4), (2-5), and (2-6) suggest that we take

$$\rho^{(\sigma)} \equiv \int_{\lambda^-}^{\lambda^+} (\rho^{(I)} - \rho)(1 - \kappa_1 \lambda)(1 - \kappa_2 \lambda)\, d\lambda \tag{2-7}$$

Normally we expect $\kappa_1 \lambda^+ << 1$, $\kappa_1 \lambda^- << 1$, $\kappa_2 \lambda^+ << 1$, and $\kappa_2 \lambda^- << 1$, in which case this can be further simplified to

$$\rho^{(\sigma)} \doteq \int_{\lambda^-}^{\lambda^+} (\rho^{(I)} - \rho)\, d\lambda \tag{2-8}$$

Equations (2-7) and (2-8) are interesting conceptually. But with the exception of carefully specified special situations, they have not proved to be useful in estimating $\rho^{(\sigma)}$. The problem is that we rarely have an expression for $\rho^{(I)}$. However, it does tell us that $\rho^{(\sigma)}$ may be either positive or negative, depending upon the location of Σ. For more on this point, see Sec. 1.3.6.

References

Buff, F. P., *J. Chem. Phys.* **25**, 146 (1956).

McConnell, A. J., "Applications of Tensor Analysis," Dover, New York (1957).

Slattery, J. C., *Ind. Eng. Chem. Fundam.* **6**, 108 (1967); *ibid.* **7**, 672 (1968).

Willmore, T. J., "Introduction to Differential Geometry," Oxford, London (1959).

Exercises

1.3.2-1 *integral over $R^{(I)}$ as an integral over* Σ (Slattery 1967) Referring to Figure 2-1, let $\boldsymbol{\xi}$ be the unit normal to the dividing surface Σ. A surface $\Sigma^{(\lambda)}$ is said to be **parallel** to Σ, if its points are at a constant distance λ from Σ measured in the direction $\boldsymbol{\xi}$ (Willmore 1959, p. 116). Let the y^α ($\alpha = 1, 2$) be an arbitrary coordinate system defined on Σ and imposed on all surfaces that are parallel to it by projecting along the normal $\boldsymbol{\xi}$. Then the position vector $\mathbf{z}^{(\lambda)}$ of a point on $\Sigma^{(\lambda)}$ may be expressed simply in terms of the position vector $\mathbf{z}$ of the point on Σ with the same surface coordinates:

$$\mathbf{z}^{(\lambda)} = \mathbf{z} + \lambda\boldsymbol{\xi} \tag{1}$$

From this relationship, we see that position in $R^{(I)*}$ is fixed once we have specified the surface coordinates and the distance from Σ; (y^1, y^2, λ) form a new curvilinear coordinate system for the region $R^{(I)*}$.

i) Prove that

$$\int_{R^{(I)*}} \psi \, dV = \int_{\lambda^-}^{\lambda^+} \int_{\Sigma^{(\lambda)}} \psi \, dA d\lambda$$

$$= \int_{\lambda^-}^{\lambda^+} \int_{\Sigma} \psi \, (a^{(\lambda)}/a)^{1/2} \, dA d\lambda$$

$$= \int_{\Sigma} \int_{\lambda^-}^{\lambda^+} \psi \, (a^{(\lambda)}/a)^{1/2} \, dA d\lambda \tag{2}$$

where

$$\text{for } \Sigma^{(\lambda)} : \quad a^{(\lambda)} \equiv \det(a^{(\lambda)}_{\alpha\beta}) \tag{3}$$

$$\text{for } \Sigma : \quad a \equiv \det(a_{\alpha\beta}) \tag{4}$$

and ψ is any scalar, vector or tensor.

ii) From (1), we have

$$\mathbf{a}^{(\lambda)}_{\alpha} = \mathbf{a}_{\alpha} + \lambda \frac{\partial \boldsymbol{\xi}}{\partial y^{\alpha}} \tag{5}$$

Starting with this relation, find

$$a^{(\lambda)}_{\alpha\beta} = a_{\alpha\beta} - 2\lambda B_{\alpha\beta} + \lambda^2 B_{\alpha\mu} \, a^{\mu\nu} \, B_{\nu\beta} \tag{6}$$

In arriving at this result, you will find it helpful to use

$$\frac{\partial \boldsymbol{\xi}}{\partial y^{\alpha}} = - B_{\gamma\alpha} \, \mathbf{a}^{\gamma} \tag{7}$$

which is derived in Sec. A.5.3.

iii) Noting (McConnell 1957, p. 205)

$$B_{\alpha\mu} \, a^{\mu\nu} \, B_{\nu\beta} = 2HB_{\alpha\beta} - Ka_{\alpha\beta} \tag{8}$$

show

$$a_{\alpha\beta}^{(\lambda)} = (1 - \lambda^2 K) a_{\alpha\beta} + 2\lambda (H\lambda - 1) B_{\alpha\beta} \tag{9}$$

Here H is the mean curvature of Σ and K is the total curvature of Σ, both of which are defined in Exercise A.5.3-3.

iv) With the help of Exercise A.3.7-4, determine that

$$\frac{a^{(\lambda)}}{a} = (1 - \lambda^2 K)^2 + 4\lambda^2 (H\lambda - 1)^2 \det_{(\sigma)}\mathbf{B}$$

$$+ 2\lambda(1 - \lambda^2 K)(H\lambda - 1) \operatorname{tr}\mathbf{B} \tag{10}$$

or

$$\frac{a^{(\lambda)}}{a} = 1 - 4H\lambda + 2(2H^2 + K)\lambda^2 - 4HK\lambda^3 + K^2\lambda^4 \tag{11}$$

v) Show

$$\frac{a^{(\lambda)}}{a} = (1 - \kappa_1 \lambda)^2 (1 - \kappa_2 \lambda)^2 \tag{12}$$

where κ_1 and κ_2 are the principal curvatures of Σ introduced in Exercise A.5.3-3, and conclude

$$\int_{R^{(I)*}} \psi \, dV = \int_{\Sigma} \int_{\lambda^-}^{\lambda^+} \psi (1 - \kappa_1 \lambda)(1 - \kappa_2 \lambda) \, d\lambda dA \tag{13}$$

The derivation of (13) given here is an alternative to that presented by Buff (1956).

1.3.2-2 *general balance equation when interface is represented by a three-dimensional region* With the assumption that the effect of the interface may be attributed to a dividing surface, the general balance or general conservation law for some quantity associated with a multiphase material body takes the form

$$\frac{d}{dt}\left(\int_R \psi \, dV + \int_\Sigma \psi^{(\sigma)} \, dA\right) = -\int_S \boldsymbol{\phi} \cdot \mathbf{n} \, dA - \int_C \boldsymbol{\phi}^{(\sigma)} \cdot \boldsymbol{\mu} \, ds$$

$$+ \int_R \rho\zeta \, dV + \int_\Sigma \rho^{(\sigma)}\zeta^{(\sigma)} \, dA \qquad (1)$$

Here S is the closed surface bounding the body, C the lines formed by the intersection of Σ with S, ψ the density of the quantity per unit volume within the bulk phases, $\psi^{(\sigma)}$ the density of the quantity per unit area on Σ, $\boldsymbol{\phi}$ the flux of the quantity (per unit area) through S, $\mathbf{n}$ the unit vector normal and outwardly directed with respect to the closed surface S, $\boldsymbol{\phi}^{(\sigma)}$ the flux of the quantity (per unit length of line) through C, $\boldsymbol{\mu}$ the unit vector normal to C that is both tangent to and outwardly directed with respect to Σ, ζ the rate of production of the quantity per unit mass at each point within the bulk phases, and $\zeta^{(\sigma)}$ the rate of production of the quantity per unit mass at each point on Σ.

Let us now assume that the effect of the interface may be attributed to a three-dimensional region of finite thickness. The text suggests that with this point of view, the general balance equation may be written as

$$\frac{d}{dt}\int_R \psi^{(I)} \, dV = -\int_S \boldsymbol{\phi}^{(I)} \cdot \mathbf{n} \, dA + \int_R \rho^{(I)}\zeta^{(I)} \, dV \qquad (2)$$

or as

$$\frac{d}{dt}\left(\int_R \psi \, dV + \int_{R^{(I)}} (\psi^{(I)} - \psi) \, dV\right)$$

$$= -\int_S \boldsymbol{\phi} \cdot \mathbf{n} \, dA - \int_{S^{(I)}} (\boldsymbol{\phi}^{(I)} - \boldsymbol{\phi}) \cdot \mathbf{n} \, dA$$

$$+ \int_R \rho\zeta \, dV + \int_{R^{(I)}} (\rho^{(I)}\zeta^{(I)} - \rho\zeta) \, dV \qquad (3)$$

with the understanding that outside the interfacial region

$$\psi^{(I)} = \psi \qquad \boldsymbol{\phi}^{(I)} = \boldsymbol{\phi}$$

$$\rho^{(I)} = \rho \qquad \zeta^{(I)} = \zeta \qquad (4)$$

By $S^{(I)}$ I refer to that portion of S bounding $R^{(I)}$.

Referring to Figure 2-1, argue that

$$\int_{R^{(I)}} (\psi^{(I)} - \psi)\, dV$$

$$\doteq \int_{R^{(I)*}} (\psi^{(I)} - \psi)\, dV$$

$$= \int_{\Sigma} \int_{\lambda^-}^{\lambda^+} (\psi^{(I)} - \psi)(1 - \kappa_1 \lambda)(1 - \kappa_2 \lambda)\, d\lambda dA \tag{5}$$

$$\int_{R^{(I)}} (\rho^{(I)}\zeta^{(I)} - \rho\zeta)\, dV$$

$$\doteq \int_{R^{(I)*}} (\rho^{(I)}\zeta^{(I)} - \rho\zeta)\, dV$$

$$= \int_{\Sigma} \int_{\lambda^-}^{\lambda^+} (\rho^{(I)}\zeta^{(I)} - \rho\zeta)(1 - \kappa_1 \lambda)(1 - \kappa_2 \lambda)\, d\lambda dA \tag{6}$$

$$\int_{S^{(I)}} (\boldsymbol{\phi}^{(I)} - \boldsymbol{\phi}) \cdot \mathbf{n}\, dA \doteq \int_{S^{(I)*}} (\boldsymbol{\phi}^{(I)} - \boldsymbol{\phi}) \cdot \mathbf{n}^*\, dA$$

$$= \int_{C} \left(\int_{\lambda^-}^{\lambda^+} (\boldsymbol{\phi}^{(I)} - \boldsymbol{\phi})\, d\lambda \right) \cdot \boldsymbol{\mu}\, ds \tag{7}$$

and that (3) may be written alternatively as

$$\frac{d}{dt}\left(\int_{R} \psi\, dV + \int_{\Sigma} \int_{\lambda^-}^{\lambda^+} (\psi^{(I)} - \psi)(1 - \kappa_1 \lambda)(1 - \kappa_2 \lambda)\, d\lambda dA \right)$$

$$= - \int_{S} \boldsymbol{\phi} \cdot \mathbf{n}\, dA - \int_{C} \left(\int_{\lambda^-}^{\lambda^+} (\boldsymbol{\phi}^{(I)} - \boldsymbol{\phi})\, d\lambda \right) \cdot \boldsymbol{\mu}\, ds + \int_{R} \rho\zeta\, dV$$

$$+ \int_{\Sigma} \int_{\lambda^-}^{\lambda^+} (\rho^{(I)}\zeta^{(I)} - \rho\zeta)(1 - \kappa_1 \lambda)(1 - \kappa_2 \lambda)\, d\lambda dA \tag{8}$$

Upon comparing (1) and (8), we can conclude that

$$\psi^{(\sigma)} \equiv \int_{\lambda^-}^{\lambda^+} (\psi^{(I)} - \psi)(1 - \kappa_1 \lambda)(1 - \kappa_2 \lambda)\, d\lambda \tag{9}$$

$$\boldsymbol{\phi}^{(\sigma)} \equiv \left(\int_{\lambda^-}^{\lambda^+} (\boldsymbol{\phi}^{(I)} - \boldsymbol{\phi})\, d\lambda \right) \cdot \mathbf{P} \tag{10}$$

$$\rho^{(\sigma)}\zeta^{(\sigma)} \equiv \int_{\lambda^-}^{\lambda^+} (\rho^{(I)}\zeta^{(I)} - \rho\zeta)(1 - \kappa_1 \lambda)(1 - \kappa_2 \lambda)\, d\lambda \tag{11}$$

As a particular example, consider conservation of mass for which

$$\psi \equiv \rho \qquad \psi^{(\sigma)} \equiv \rho^{(\sigma)}$$

$$\boldsymbol{\phi}^{(I)} = \boldsymbol{\phi} = \boldsymbol{\phi}^{(\sigma)} = 0 \qquad \zeta^{(I)} = \zeta = \zeta^{(\sigma)} = 0 \tag{12}$$

Equation (2-7) follows immediately from (9).

1.3.3 Surface transport theorem Let us consider the operation

$$\frac{d}{dt} \int_{\Sigma} \psi^{(\sigma)}\, dA$$

Here $\psi^{(\sigma)}$ is any scalar-, vector-, or tensor-valued function of time and position on the dividing surface. The indicated integration is to be performed over the dividing surface in its current configuration Σ. We should expect that Σ, or the limits on this integration, is a function of time.

Generally, the dividing surface will not be composed of a fixed set of material particles; there will be mass transfer between the dividing surface and the two adjoining phases. But we can always confine our attention to a portion of a dividing surface composed of a fixed set of surface particles. Either the dividing surface is closed or no surface particles cross the closed curve C bounding Σ. In this case, we can express the integration over the dividing surface in terms of its fixed reference configuration Σ_κ (see Exercise 1.3.3-1).

$$\int_{\Sigma} \psi^{(\sigma)}\, dA = \int_{\Sigma_\kappa} \psi^{(\sigma)}\, J^{(\sigma)}\, dA \tag{3-1}$$

Here

$$J^{(\sigma)} \equiv | \det_{(\sigma)}\mathcal{F} | = \sqrt{(\det_{(\sigma)}\mathcal{F})^2} \tag{3-2}$$

is area in the current configuration per unit area in the reference configuration, and $\mathcal{F}$ is the deformation gradient for the dividing surface introduced in Sec. 1.2.5. The advantage is that the integration limits on the right side of (3-1) are independent of time:

$$\begin{aligned} \frac{d}{dt}\int_{\Sigma} \psi^{(\sigma)}\, dA &= \frac{d}{dt}\int_{\Sigma_\kappa} \psi^{(\sigma)}\, J^{(\sigma)}\, dA \\ &= \int_{\Sigma_\kappa} \left(\frac{d_{(s)}\psi^{(\sigma)}}{dt} J^{(\sigma)} + \psi^{(\sigma)} \frac{d_{(s)} J^{(\sigma)}}{dt} \right) dA \\ &= \int_{\Sigma} \left(\frac{d_{(s)}\psi^{(\sigma)}}{dt} + \frac{\psi^{(\sigma)}}{J^{(\sigma)}} \frac{d_{(s)} J^{(\sigma)}}{dt} \right) dA \end{aligned} \tag{3-3}$$

In the second line of (3-3) we have used the surface derivative, because at this point we are thinking of $\psi^{(\sigma)}$ and $J^{(\sigma)}$ as explicit functions of time and position in the reference configuration of the surface. From (3-2) we have

$$\frac{1}{J^{(\sigma)}} \frac{d_{(s)} J^{(\sigma)}}{dt} = \frac{1}{\det_{(\sigma)}\mathcal{F}} \frac{d_{(s)}}{dt} (\det_{(s)}\mathcal{F}) \tag{3-4}$$

In Exercises 1.3.3-3 through 1.3.3-5, we outline proofs for

$$\frac{1}{\det_{(\sigma)}\mathcal{F}} \frac{d_{(s)}}{dt} (\det_{(\sigma)}\mathcal{F}) = \text{div}_{(\sigma)}\mathbf{v}^{(\sigma)} \tag{3-5}$$

We conclude as a result

$$\frac{d}{dt}\int_{\Sigma} \psi^{(\sigma)}\, dA = \int_{\Sigma} \left(\frac{d_{(s)}\psi^{(\sigma)}}{dt} + \psi^{(\sigma)}\, \text{div}_{(\sigma)}\mathbf{v}^{(\sigma)} \right) dA \tag{3-6}$$

The surface divergence theorem (Sec. A.6.3) may be used to write this last as

$$\frac{d}{dt}\int_{\Sigma} \psi^{(\sigma)}\, dA$$

$$= \int_{\Sigma} \left(\frac{\partial \psi^{(\sigma)}}{\partial t} - \nabla_{(\sigma)} \psi^{(\sigma)} \cdot \mathbf{u} + \mathrm{div}_{(\sigma)} (\psi^{(\sigma)} \mathbf{v}^{(\sigma)}) \right) dA$$

$$= \int_{\Sigma} \left(\frac{\partial \psi^{(\sigma)}}{\partial t} - \nabla_{(\sigma)} \psi^{(\sigma)} \cdot \mathbf{u} - 2H \psi^{(\sigma)} v_{\{\xi\}} \right) dA$$

$$+ \int_{C} \psi^{(\sigma)} \mathbf{v}^{(\sigma)} \cdot \boldsymbol{\mu} \, ds \tag{3-7}$$

Equations (3-6) and (3-7) are alternative forms of the **surface transport theorem.**

In deriving the surface transport theorem, we have followed that portion of a dividing surface associated with a set of surface particles. Sometimes we will wish to follow as a function of time a **surface system** $\Sigma_{(sys)}$ in the dividing surface that does not contain a fixed set of surface particles. For example, a liquid-air dividing surface rising or falling in a glass capillary tube is a surface system that may not be composed of a fixed set of surface particles: a rolling motion in one of the phases (Sec. 1.2.9) may force surface material particles to cross the common line. The surface transport theorem just derived does not apply to such a surface system. On the other hand, there is nothing to prevent us from associating a set of fictitious **surface system particles** with this system. It is not necessary for us to fully define these surface system particles. Let $\boldsymbol{\mu}$ be the unit tangent vector that is normal to the curve $C_{(sys)}$ bounding the surface system and that is outwardly directed with respect to the system. Let $\mathbf{v}_{(sys)} \cdot \boldsymbol{\mu}$ be the component of the velocity of $C_{(sys)}$ in the direction $\boldsymbol{\mu}$. We will require only that at $C_{(sys)}$ the $\boldsymbol{\mu}$ component of the velocity of the surface system particles be equal to $(\mathbf{v}_{(sys)} \cdot \boldsymbol{\mu})$. If in the derivation of (3-7) we replace the set of surface particles with this set of surface system particles, the result is

$$\frac{d}{dt} \int_{\Sigma_{(sys)}} \psi^{(\sigma)} \, dA$$

$$= \int_{\Sigma_{(sys)}} \left(\frac{\partial \psi^{(\sigma)}}{\partial t} - \nabla_{(\sigma)} \psi^{(\sigma)} \cdot \mathbf{u} - 2H \psi^{(\sigma)} v_{\{\xi\}} \right) dA$$

$$+ \int_{C_{(sys)}} \psi^{(\sigma)} (\mathbf{v}_{(sys)} \cdot \boldsymbol{\mu}) \, ds \tag{3-8}$$

We can refer to this as the **generalized surface transport theorem.**

Exercises

1.3.3-1 *meaning of* $J^{(\sigma)}$ A physically motivated definition for $J^{(\sigma)}$ might be the ratio of the area ΔA of a set of surface particles in the current configuration to the area ΔA_κ occupied by the same set of surface particles in the reference configuration in the limit as $\Delta A_\kappa \to 0$:

$$J^{(\sigma)} \equiv \text{limit } \Delta A_\kappa \to 0: \ \frac{\Delta A}{\Delta A_\kappa} \tag{1}$$

Now let us be more specific. In the limit as $\Delta y^1_\kappa \to 0$, a change in position $\Delta y^1_\kappa \, \mathbf{a}_{\kappa 1}$ in the reference configuration corresponds to a position change $\Delta y^1_\kappa \, \mathcal{F} \cdot \mathbf{a}_{\kappa 1}$ in the current configuration. In the limit as $\Delta y^2_\kappa \to 0$, a change in position $\Delta y^2_\kappa \, \mathbf{a}_{\kappa 2}$ in the reference configuration means a shift in location $\Delta y^2_\kappa \, \mathcal{F} \cdot \mathbf{a}_{\kappa 2}$ in the current configuration. Equation (1) implies that $J^{(\sigma)}$ is the ratio of the area of the parallelogram spanned by $\Delta y^1_\kappa \, \mathcal{F} \cdot \mathbf{a}_{\kappa 2}$ and $\Delta y^2_\kappa \, \mathcal{F} \cdot \mathbf{a}_{\kappa 2}$ in the current configuration to the area of the parallelogram spanned by $\Delta y^1_\kappa \, \mathbf{a}_{\kappa 1}$ and $\Delta y^2_\kappa \, \mathbf{a}_{\kappa 2}$ in the reference configuration in the limits $\Delta y^1_\kappa \to 0$ and $\Delta y^2_\kappa \to 0$. Prove as a result

$$J^{(\sigma)} = | \det{}_{(\sigma)} \mathcal{F} | \tag{2}$$

Because (2) expresses $J^{(\sigma)}$ in a form better suited for direct computation than (1), I have chosen it as our definition for $J^{(\sigma)}$ in the text.

1.3.3-2 *alternative expression for* $J^{(\sigma)}$ Starting with (3-2), derive

$$J^{(\sigma)} = \frac{\sqrt{a}}{\sqrt{a_\kappa}} \left| \det\left(\frac{\partial X^\alpha_K}{\partial y^A_\kappa} \right) \right|$$

Here $\det(\partial X^\alpha_K / \partial y^A_\kappa)$ means the determinant of the matrix whose typical element is $\partial X^\alpha_K / \partial y^A_\kappa$.

This expression for $J^{(\sigma)}$ is consistent with more general formulas often derived in calculus textbooks for changing the set of integration variables in a multidimensional integral.

1.3.3-3 *Proof of (3-5)* We will break the proof of (3-5) into several steps, beginning with (see Sec. A.3.7)

$$2 \det{}_{(\sigma)} \mathcal{F} = (\mathcal{F} \cdot \mathbf{a}_{\kappa M}) \cdot \left(\boldsymbol{\varepsilon}^{(\sigma)} \cdot [\mathcal{F} \cdot \mathbf{a}_{\kappa N}] \right) \varepsilon^{MN}_\kappa$$

i) Determine that

$$2 \frac{d_{(s)}}{dt} (\det_{(\sigma)} \boldsymbol{\mathcal{F}}) = \left(\frac{d_{(s)} \boldsymbol{\mathcal{F}}}{dt} \cdot \mathbf{a}_{\kappa M} \right) \cdot \left(\boldsymbol{\varepsilon}^{(\sigma)} \cdot [\boldsymbol{\mathcal{F}} \cdot \mathbf{a}_{\kappa N}] \right) \varepsilon_{\kappa}^{MN}$$

$$+ (\boldsymbol{\mathcal{F}} \cdot \mathbf{a}_{\kappa M}) \cdot \left(\frac{d_{(s)} \boldsymbol{\varepsilon}^{(\sigma)}}{dt} \cdot [\boldsymbol{\mathcal{F}} \cdot \mathbf{a}_{\kappa N}] \right) \varepsilon_{\kappa}^{MN}$$

$$+ (\boldsymbol{\mathcal{F}} \cdot \mathbf{a}_{\kappa M}) \cdot \left(\boldsymbol{\varepsilon}^{(\sigma)} \cdot \left[\frac{d_{(s)} \boldsymbol{\mathcal{F}}}{dt} \cdot \mathbf{a}_{\kappa N} \right] \right) \varepsilon_{\kappa}^{MN}$$

ii) Calculate with the help of Exercise A.3.3-5

$$\left(\frac{d_{(s)} \boldsymbol{\mathcal{F}}}{dt} \cdot \mathbf{a}_{\kappa M} \right) \cdot \left(\boldsymbol{\varepsilon}^{(\sigma)} \cdot [\boldsymbol{\mathcal{F}} \cdot \mathbf{a}_{\kappa N}] \right) \varepsilon_{\kappa}^{MN}$$

$$= (\boldsymbol{\mathcal{F}} \cdot \mathbf{a}_{\kappa M}) \cdot \left(\boldsymbol{\varepsilon}^{(\sigma)} \cdot \left[\frac{d_{(s)} \boldsymbol{\mathcal{F}}}{dt} \cdot \mathbf{a}_{\kappa N} \right] \right) \varepsilon_{\kappa}^{MN}$$

$$= \operatorname{tr} \left(\frac{d_{(s)} \boldsymbol{\mathcal{F}}}{dt} \cdot \boldsymbol{\mathcal{F}}^{-1} \right) \det_{(\sigma)} \boldsymbol{\mathcal{F}}$$

iii) Show

$$(\boldsymbol{\mathcal{F}} \cdot \mathbf{a}_{\kappa M}) \cdot \left(\frac{d_{(s)} \boldsymbol{\varepsilon}^{(\sigma)}}{dt} \cdot [\boldsymbol{\mathcal{F}} \cdot \mathbf{a}_{\kappa N}] \right) \varepsilon_{\kappa}^{MN}$$

$$= - \operatorname{tr} \left(\frac{d_{(s)} \boldsymbol{\varepsilon}^{(\sigma)}}{dt} \cdot \boldsymbol{\varepsilon}^{(\sigma)} \right) \det_{(\sigma)} \boldsymbol{\mathcal{F}}$$

iv) Find

$$- \operatorname{tr} \left(\frac{d_{(s)} \boldsymbol{\varepsilon}^{(\sigma)}}{dt} \cdot \boldsymbol{\varepsilon}^{(\sigma)} \right) = 0$$

v) Conclude from the results of (i) through (iv) that

$$\frac{1}{\det_{(\sigma)}\boldsymbol{\mathcal{F}}}\frac{d_{(s)}}{dt}(\det_{(\sigma)}\boldsymbol{\mathcal{F}}) = \operatorname{tr}\left(\frac{d_{(s)}\boldsymbol{\mathcal{F}}}{dt}\cdot\boldsymbol{\mathcal{F}}^{-1}\right)$$

vi) Calculate

$$\frac{d_{(s)}\boldsymbol{\mathcal{F}}}{dt} = \frac{\partial X^{\alpha}_{K}}{\partial y^{A}_{\kappa}}\frac{\partial \mathbf{u}}{\partial y^{\alpha}}\,\mathbf{a}^{A}_{\kappa} + \frac{\partial X^{\alpha}_{K}}{\partial y^{A}_{\kappa}}\,B_{\beta\alpha}\,\dot{y}^{\beta}\xi\,\mathbf{a}^{A}_{\kappa} + \dot{y}^{\alpha}_{,\beta}\frac{\partial X^{\beta}_{K}}{\partial y^{A}_{\kappa}}\,\mathbf{a}_{\alpha}\,\mathbf{a}^{A}_{\kappa}$$

vii) Show

$$\boldsymbol{\mathcal{F}}^{-1}\cdot\frac{d_{(s)}\boldsymbol{\mathcal{F}}}{dt} = \frac{\partial K^{\beta}_{Y}}{\partial y^{\mu}}\,\mathbf{a}_{\kappa B}\frac{\partial X^{\alpha}_{K}}{\partial y^{A}_{\kappa}}\,\mathbf{a}^{\mu}\cdot\frac{\partial \mathbf{u}}{\partial y^{\alpha}}\,\mathbf{a}^{A}_{\kappa} + \frac{\partial K^{\beta}_{X}}{\partial y^{\alpha}}\,\mathbf{a}_{\kappa B}\;\dot{y}^{\alpha}_{,\beta}\frac{\partial X^{\beta}_{K}}{\partial y^{A}_{\kappa}}\,\mathbf{a}^{A}_{\kappa}$$

viii) Conclude

$$\operatorname{tr}\left(\boldsymbol{\mathcal{F}}^{-1}\cdot\frac{d_{(s)}\boldsymbol{\mathcal{F}}}{dt}\right) = \operatorname{div}_{(\sigma)}\mathbf{v}^{(\sigma)}$$

and

$$\frac{1}{\det_{(\sigma)}\boldsymbol{\mathcal{F}}}\frac{d_{(s)}}{dt}(\det_{(\sigma)}\boldsymbol{\mathcal{F}}) = \operatorname{div}_{(\sigma)}\mathbf{v}^{(\sigma)}$$

1.3.3-4 *Alternative proof of (3-5)* It is helpful to test your understanding by giving an alternative proof of (3-5), starting with

$$\det_{(\sigma)}\boldsymbol{\mathcal{F}} = \frac{1}{2}\,\varepsilon^{MN}_{\kappa}\,\varepsilon_{\alpha\beta}\,\mathcal{F}^{\alpha}_{.M}\,\mathcal{F}^{\beta}_{.N}$$

i) Calculate

$$\frac{d_{(s)}}{dt}(\det_{(\sigma)}\boldsymbol{\mathcal{F}}) = \frac{1}{2}\,\varepsilon^{MN}_{\kappa}\frac{\partial\varepsilon_{\alpha\beta}}{\partial t}\,\mathcal{F}^{\alpha}_{.M}\,\mathcal{F}^{\beta}_{.N} + \varepsilon^{MN}_{\kappa}\,\varepsilon_{\alpha\beta}\frac{\delta_{(s)}\mathcal{F}^{\alpha}_{.m}}{\delta t}\,\mathcal{F}^{\beta}_{.N}$$

where

$$\frac{\delta_{(s)} \mathcal{F}^{\alpha}_{.M}}{\delta t} \equiv \frac{d_{(s)} \mathcal{F}^{\alpha}_{.M}}{dt} + \left\{ {\alpha \atop \gamma \ \mu} \right\}_a \mathcal{F}^{\mu}_{.M} \dot{y}^{\gamma}$$

ii) Determine

$$\frac{\partial \varepsilon_{\alpha\beta}}{\partial t} = \frac{1}{2a} \frac{\partial a}{\partial t} \varepsilon_{\alpha\beta}$$

iii) Find

$$\varepsilon^{MN}_{\kappa} \varepsilon_{\alpha\beta} \frac{\delta_{(s)} \mathcal{F}^{\alpha}_{.M}}{\delta t} \mathcal{F}^{\beta}_{.N} = \frac{\delta_{(s)} \mathcal{F}^{\gamma}_{.M}}{\delta t} \mathcal{F}^{-1M}_{\ \ \ .\gamma} \det_{(\sigma)} \boldsymbol{\mathcal{F}}$$

iv) Show

$$\frac{\delta_{(s)} \mathcal{F}^{\gamma}_{.M}}{\delta t} \mathcal{F}^{-1M}_{\ \ \ .\gamma} = \dot{y}^{\gamma}_{,\gamma}$$

v) Use Exercise 1.2.7-1, to conclude

$$\frac{1}{\det_{(\sigma)} \boldsymbol{\mathcal{F}}} \frac{d_{(s)}}{dt} \left(\det_{(\sigma)} \boldsymbol{\mathcal{F}} \right) = \frac{1}{2a} \frac{\partial a}{\partial t} + \dot{y}^{\gamma}_{,\gamma} = \text{div}_{(\sigma)} \mathbf{v}^{(\sigma)}$$

1.3.3-5 *Alternative proof of (3-5)* Let us begin with the result of Exercise 1.3.3-3(i) in constructing another proof of (3-5)

i) Calculate

$$\frac{d_{(s)} \boldsymbol{\mathcal{F}}}{dt} = \frac{d_{(s)}}{dt} \left(\frac{\partial X^{\alpha}_{K}}{\partial y^{A}_{\kappa}} \mathbf{a}_{\alpha} \mathbf{a}^{A}_{\kappa} \right)$$

$$= \frac{\partial X^{\alpha}_{K}}{\partial y^{A}_{\kappa}} \frac{\partial \mathbf{u}}{\partial y^{\alpha}} \mathbf{a}^{A}_{\kappa} + \frac{\partial X^{\alpha}_{K}}{\partial y^{A}_{\kappa}} B_{\beta\alpha} \dot{y}^{\beta} \boldsymbol{\xi} \mathbf{a}^{A}_{\kappa} + \dot{y}^{\alpha}_{,\beta} \frac{\partial X^{\beta}_{K}}{\partial y^{A}_{\kappa}} \mathbf{a}_{\alpha} \mathbf{a}^{A}_{\kappa}$$

ii) Determine

$$\frac{d_{(s)}\,\boldsymbol{\varepsilon}^{(\sigma)}}{dt} = \frac{\partial \varepsilon_{\alpha\beta}}{dt}\,\mathbf{a}^{\alpha}\mathbf{a}^{\beta} + \varepsilon_{\alpha\beta}\left(\frac{\partial \mathbf{a}^{\alpha}}{\partial t}\,\mathbf{a}^{\beta} + \mathbf{a}^{\alpha}\,\frac{\partial \mathbf{a}^{\beta}}{\partial t}\right)$$

$$+ \varepsilon_{\alpha\beta}\,a^{\alpha\mu}\,B_{\gamma\mu}\,\dot{y}^{\gamma}(\,\boldsymbol{\xi}\mathbf{a}^{\beta} - \mathbf{a}^{\beta}\boldsymbol{\xi}\,)$$

iii) Starting with the result of (i), show that

$$\left(\frac{d_{(s)}\mathcal{F}}{dt}\cdot \mathbf{a}_{\kappa M}\right)\cdot\left[\boldsymbol{\varepsilon}^{(\sigma)}\cdot(\,\mathcal{F}\cdot \mathbf{a}_{\kappa N}\,)\right]\varepsilon_{\kappa}^{MN}$$

$$= (\,\mathcal{F}\cdot \mathbf{a}_{\kappa M}\,)\cdot\left[\boldsymbol{\varepsilon}^{(\sigma)}\cdot\left(\frac{d_{(s)}\mathcal{F}}{\partial t}\cdot \mathbf{a}_{\kappa N}\right)\right]\varepsilon_{\kappa}^{MN}$$

$$= \text{div}_{(\sigma)}\mathbf{v}^{(\sigma)}\ \det{}_{(\sigma)}\mathcal{F}$$

iv) Beginning with (ii), show

$$(\,\mathcal{F}\cdot \mathbf{a}_{\kappa M}\,)\cdot\left[\frac{d_{(s)}\,\boldsymbol{\varepsilon}^{(\sigma)}}{dt}\cdot(\,\mathcal{F}\cdot \mathbf{a}_{\kappa N}\,)\right]\varepsilon_{\kappa}^{MN} = 0$$

Exercise 1.2.7-1 may be helpful.

v) Use these results to conclude that (3-5) follows from Exercise 1.3.3-3(i).

1.3.3-6 *intuitive derivation of surface transport theorem* As we have seen in the preceding exercises, the proof of (3-5) is not trivial. For this reason, I think it is helpful to have an intuitive derivation for comparison.

Consider a polyhedron

$$\sum_{z=1}^{N} \Sigma_k$$

each planar element Σ_k (k = 1, ..., N) of which is tangent to Σ. Identify the maximum diagonal D_N of any element of this polyhedron. Form a sequence of such polyhedra, ordered with decreasing maximum diagonals. The areas of these polyhedra approach the area of Σ as $N \to \infty$ and $D_N \to 0$ (see Sec. A.6.2).

Let C_k denote the closed curve bounding Σ_k. If $\mathcal{A}$ is the area of Σ, we can say *intuitively* that

$$\frac{d\mathcal{A}}{dt} = \text{limit } N \to \infty,\ D_N \to 0: \sum_{k=1}^{N} \int_{C_k} \mathbf{v}^{(\sigma)} \cdot \boldsymbol{\mu}\, ds \tag{1}$$

An application of the surface divergence theorem (Sec. A.6.3) permits this to be written as

$$\frac{d\mathcal{A}}{dt} = \text{limit } N \to \infty,\ D_N \to 0: \sum_{k=1}^{N} \int_{\Sigma_k} \text{div}_{(\sigma)}\mathbf{v}^{(\sigma)}\, dA$$

$$= \int_{\Sigma} \text{div}_{(\sigma)}\mathbf{v}^{(\sigma)}\, dA \tag{2}$$

From (3-3), observe that

$$\frac{d\mathcal{A}}{dt} = \int_{\Sigma} \frac{1}{J^{(\sigma)}} \frac{d_{(s)} J^{(\sigma)}}{dt}\, dA \tag{3}$$

Compare (2) and (3) to conclude that

$$\int_{\Sigma} \left(\frac{1}{J^{(\sigma)}} \frac{d_{(s)} J^{(\sigma)}}{dt} - \text{div}_{(\sigma)}\mathbf{v}^{(\sigma)} \right) dA = 0 \tag{4}$$

or that

$$\frac{1}{J^{(\sigma)}} \frac{d_{(s)} J^{(\sigma)}}{dt} = \text{div}_{(\sigma)}\mathbf{v}^{(\sigma)} \tag{5}$$

since (4) is valid for every dividing surface and for every portion of a dividing surface.

The surface transport theorem (3-6) follows immediately from (3-3) and (5).

1.3.4 Transport theorem for body containing dividing surface Using the notation introduced in Sec. 1.3.1, let us now consider the operation

$$\frac{d}{dt}\left(\int_R \psi \, dV + \int_\Sigma \psi^{(\sigma)} \, dA \right)$$

In the first integral, ψ is any scalar-, vector-, or tensor-valued function of time and position in the region R occupied by a material body. The quantity ψ may be discontinuous at any dividing surface $\Sigma^{(i,j)}$; $\psi^{(i)}$ is continuous within $R^{(i)}$, the region occupied by phase i. We have already shown in Sec. 1.3.3 how the second term may be transformed using the surface transport theorem. So let us begin by considering the time derivative of an integral over R:

$$\frac{d}{dt}\int_R \psi \, dV \equiv \frac{d}{dt}\sum_{i=1}^{M} \int_{R^{(i)}} \psi^{(i)} \, dV \tag{4-1}$$

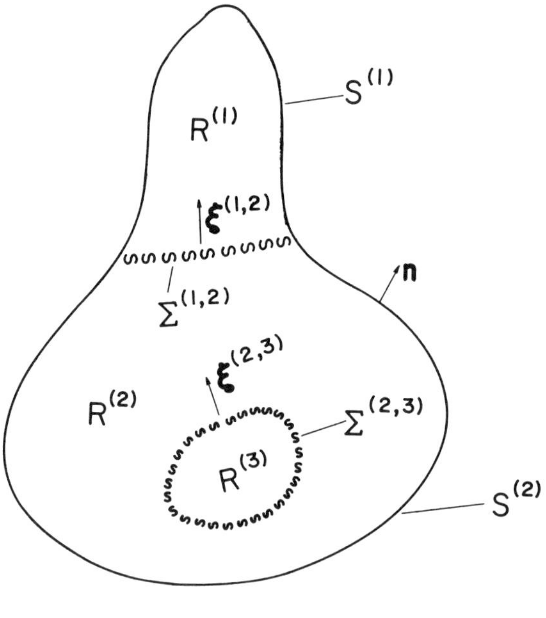

Figure 1.3.4-1: A body that is composed of three phases occupies the region $R = R^{(1)} + R^{(2)} + R^{(3)}$. Its closed bounding surface is $S = S^{(1)} + S^{(2)}$.

If S is the closed bounding surface of R, we define $S^{(i)}$ to be that portion of S bounding phase $R^{(i)}$ as illustrated in Figure 4-1. Phase i is not a material body, because the phase boundaries may be moving across the material. Think of a cube of ice melting on the table: water that was initially solid becomes liquid at some later time as the result of the movement of the phase boundary relative to the material. The velocities of the boundaries of phase i are

$$\text{on } S^{(i)} : \quad \mathbf{v}^{(i)}$$

$$\text{on } \Sigma^{(i,j)} : \quad \mathbf{u}^{(i,j)} \tag{4-2}$$

The generalized transport theorem (see Exercise 1.3.4-1) requires for each phase i

$$\frac{d}{dt}\int_{R^{(i)}} \psi^{(i)}\, dV = \int_{R^{(i)}} \frac{\partial \psi^{(i)}}{\partial t}\, dV + \int_{S^{(i)}} \psi^{(i)}\, \mathbf{v}^{(i)} \cdot \mathbf{n}\, dA - \sum_{\substack{j=1\\ j\neq i}}^{M} \int_{\Sigma^{(i,j)}} \psi^{(i)}\, \mathbf{u}^{(i,j)} \cdot \boldsymbol{\xi}^{(i,j)}\, dA \tag{4-3}$$

with the understanding that $\boldsymbol{\xi}^{(i,j)}$ is the unit normal to $\Sigma^{(i,j)}$ pointing into $R^{(i)}$.

It follows that

$$\frac{d}{dt}\int_{R} \psi\, dV \equiv \frac{d}{dt}\sum_{i=1}^{M}\int_{R^{(i)}} \psi^{(i)}\, dV$$

$$= \sum_{i=1}^{M}\int_{R^{(i)}} \frac{\partial \psi^{(i)}}{\partial t}\, dV + \sum_{i=1}^{M}\int_{S^{(i)}} \psi^{(i)}\, \mathbf{v}^{(i)} \cdot \mathbf{n}\, dA - \sum_{i=1}^{M}\sum_{\substack{j=1\\ j\neq i}}^{M} \int_{\Sigma^{(i,j)}} \psi^{(i)}\, \mathbf{u}^{(i,j)} \cdot \boldsymbol{\xi}^{(i,j)}\, dA \tag{4-4}$$

If we understand that ψ and $\mathbf{v}$ denote piecewise continuous functions defined by $\psi^{(i)}$ and $\mathbf{v}^{(i)}$ in $R^{(i)}$ and that $\mathbf{u}$ is a piecewise continuous function defined by $\mathbf{u}^{(i,j)}$ at $\Sigma^{(i,j)}$, we can write this as (Slattery 1981, p. 22; Truesdell and Toupin 1960, p. 525)

$$\frac{d}{dt}\int_{R} \psi\, dV = \int_{R} \frac{\partial \psi}{\partial t}\, dV + \int_{S} \psi\, \mathbf{v} \cdot \mathbf{n}\, dA - \int_{\Sigma} [\,\psi \mathbf{u} \cdot \boldsymbol{\xi}\,]\, dA \tag{4-5}$$

where

$$[\,\psi\boldsymbol{\xi}\,] \equiv \psi^{(i)}\boldsymbol{\xi}^{(i,j)} + \psi^{(j)}\,\boldsymbol{\xi}^{(j,i)}$$
$$= (\,\psi^{(i)} - \psi^{(j)}\,)\,\xi^{(i,j)} \tag{4-6}$$

and $\boldsymbol{\xi}$ is understood to be the unit normal to Σ that points into the phase concerned. If we observe that by the divergence theorem

$$\int_S \psi\,\mathbf{v}\cdot\mathbf{n}\,dA = \int_R \text{div}(\,\psi\,\mathbf{v}\,)\,dV + \int_\Sigma [\,\psi\,\mathbf{v}\cdot\boldsymbol{\xi}\,]\,dA \tag{4-7}$$

and that

$$\frac{\partial\psi}{\partial t} + \text{div}(\,\psi\mathbf{v}\,) = \frac{d_{(m)}\psi}{dt} + \psi\,\text{div}\,\mathbf{v} \tag{4-8}$$

we can finally express (4-5) as

$$\frac{d}{dt}\int_R \psi\,dV = \int_R \left(\frac{d_{(m)}\psi}{dt} + \psi\,\text{div}\,\mathbf{v}\right) dV$$
$$+ \int_\Sigma [\,\psi\,(\,\mathbf{v}\cdot\boldsymbol{\xi} - v_{(\xi)}\,)]\,dA \tag{4-9}$$

Here

$$v_{(\xi)} \equiv \mathbf{v}^{(\sigma)}\cdot\boldsymbol{\xi}$$
$$= \mathbf{u}\cdot\boldsymbol{\xi} \tag{4-10}$$

is the **speed of displacement** of the dividing surface (see Sec. 1.2.7).

We can now use (4-9) together with the surface transport theorem derived in Sec. 1.3.3 to conclude

$$\frac{d}{dt}\left(\int_R \psi\,dV + \int_\Sigma \psi^{(\sigma)}\,dA\right) = \int_R \left(\frac{d_{(m)}\psi}{d\,t} + \psi\,\text{div}\,\mathbf{v}\right) dV$$
$$+ \int_\Sigma \left(\frac{d_{(s)}\,\psi^{(\sigma)}}{dt} + \psi^{(\sigma)}\,\text{div}_{(\sigma)}\mathbf{v}^{(\sigma)} + [\,\psi(\,\mathbf{v}\cdot\boldsymbol{\xi} - v_{(\xi)}\,)]\right) dA \tag{4-11}$$

We shall refer to (4-11) as the **transport theorem for a body containing a**

dividing surface.

References

Slattery, J. C., "Momentum, Energy, and Mass Transfer in Continua," McGraw-Hill, New York (1972); second edition, Robert E. Krieger, Malabar, FL 32950 (1981).

Truesdell, C., and R. A. Toupin, "Handbuch der Physik," vol. 3/1, edited by S. Flügge, Springer-Verlag, Berlin (1960).

Exercise

1.3.4-1 *generalized transport theorem* For a body consisting of a single phase, the **transport theorem** (Slattery 1981, p. 18; Truesdell and Toupin 1960, p. 347) says that

$$\frac{d}{dt}\int_R \psi \, dV = \int_R \left(\frac{d_{(m)}\psi}{dt} + \psi \operatorname{div} \mathbf{v} \right) dV \tag{1}$$

or

$$\frac{d}{dt}\int_R \psi \, dV = \int_R \frac{\partial \psi}{\partial t} dV + \int_S \psi \mathbf{v} \cdot \mathbf{n} \, dA \tag{2}$$

where ψ is a scalar-, vector-, or tensor-valued function of time and position in the region R occupied by the body and **n** is the outwardly directed unit normal vector to the closed bounding surface of the body.

We have occasion in the text to ask about the derivative with respect to time of a quantity while following a system that is not necessarily a material body. For illustrative purposes, let us take as our system the air in a child's balloon and ask for the derivative with respect to time of the volume associated with the air as the balloon is inflated. Since material (air) is being continuously added to the balloon, we are not following a set of material particles as a function of time. On the other hand, there is nothing to prevent us from introducing a particular set of fictitious system particles to be associated with our system. It will not be necessary for us to give a complete description of these system particles. We shall say only that the normal component of the velocity of any system particle at the boundary of the system must be equal to the normal component of velocity of the boundary.

Prove that the transport theorem stated above can be extended to an arbitrary system by replacing (i) derivatives with respect to time while following material particles, $d_{(m)}/dt$, by derivatives with respect to time while following fictitious system particles, $d_{(sys)}/dt$, and (ii) the velocity of a material particle, $\mathbf{v}$, by the velocity of a fictitious system particle, $\mathbf{v}_{(sys)}$. Conclude that

$$\frac{d}{dt}\int_{R_{(sys)}} \psi \, dV = \int_{R_{(sys)}} \frac{\partial \psi}{\partial t} dV + \int_{S_{(sys)}} \psi \, \mathbf{v}_{(sys)} \cdot \mathbf{n} \, dA \tag{3}$$

Here $R_{(sys)}$ signifies that region of space currently occupied by the system; $S_{(sys)}$ is the closed bounding surface of the system. We will refer to (3) as the **generalized transport theorem** (Slattery 1981, p. 20; Truesdell and Toupin 1960, p. 347).

For an extension of the generalized transport theorem to a system containing intersecting dividing surfaces, see Exercise 1.3.7-1.

1.3.5 Jump mass balance Our objective here is to learn what conservation of mass requires at each point on a dividing surface. After an application of the transport theorem for a material region containing a dividing surface, the postulate of mass conservation stated in Sec. 1.3.1 takes the form

$$\begin{aligned}
\frac{d}{dt}\left(\int_R \rho \, dV + \int_\Sigma \rho^{(\sigma)} \, dA\right)
&= \int_R \left(\frac{d_{(m)}\rho}{dt} + \rho \operatorname{div} \mathbf{v}\right) dV \\
&\quad + \int_\Sigma \left\{\frac{d_{(s)}\rho^{(\sigma)}}{dt} + \rho^{(\sigma)} \operatorname{div}_{(\sigma)} \mathbf{v}^{(\sigma)} + [\rho(\mathbf{v} \cdot \boldsymbol{\xi} - v_{(\xi)})]\right\} dA \\
&= 0
\end{aligned} \tag{5-1}$$

Equation (5-1) must be true for every body that does not include a common line. In particular, for a body consisting of a single phase, the integral over the region R is zero, which implies (Slattery 1981, p. 20)

$$\frac{d_{(m)}\rho}{dt} + \rho \operatorname{div} \mathbf{v} = 0 \tag{5-2}$$

This is the **equation of continuity**. It expresses the constraint of conservation of mass at each point within a phase.

In view of the equation of continuity, (5-1) reduces to

$$\int_{\Sigma} \left\{ \frac{d_{(s)} \rho^{(\sigma)}}{dt} + \rho^{(\sigma)} \operatorname{div}_{(\sigma)} \mathbf{v}^{(\sigma)} + [\, \rho(\mathbf{v} \cdot \boldsymbol{\xi} - v_{\{\xi\}})] \right\} A = 0 \qquad (5\text{-}3)$$

Equation (5-1) is valid for every material body and every portion of a material body, so long as there are no intersecting dividing surfaces. This implies that (5-3) must be true for every portion of a dividing surface excluding a common line. We conclude (see Exercises 1.3.5-1 and 1.3.5-2)

$$\frac{d_{(s)} \rho^{(\sigma)}}{dt} + \rho^{(\sigma)} \operatorname{div}_{(\sigma)} \mathbf{v}^{(\sigma)} + [\, \rho(\mathbf{v} \cdot \boldsymbol{\xi} - v_{\{\xi\}})] = 0 \qquad (5\text{-}4)$$

This is called the **jump mass balance**. It expresses the requirement that mass be conserved at every point on a dividing surface.

There are two special cases of the jump mass balance to which I would like to call your attention. If there is no mass transfer to or from the dividing surface, then

$$\text{at } \Sigma : \quad \mathbf{v} \cdot \boldsymbol{\xi} = v_{\{\xi\}} \qquad (5\text{-}5)$$

and (5-4) reduces to

$$\frac{d_{(s)} \rho^{(\sigma)}}{dt} + \rho^{(\sigma)} \operatorname{div}_{(\sigma)} \mathbf{v}^{(\sigma)} = 0 \qquad (5\text{-}6)$$

If in addition the surface mass density $\rho^{(\sigma)}$ is a constant on the dividing surface, (5-6) further simplifies to

$$\operatorname{div}_{(\sigma)} \mathbf{v}^{(\sigma)} = 0 \qquad (5\text{-}7)$$

Physically this means that there is no local dilation of the phase interface.

References

Kaplan, Wilfred, "Advanced Calculus," Addison-Wesley, Cambridge, MA (1952).

Slattery, J. C., "Momentum, Energy, and Mass Transfer in Continua," McGraw-Hill, New York (1972); second edition, Robert E. Krieger, Malabar, FL 32950 (1981).

Exercises

1.3.5-1 *The integrand must be zero, if an integral over an arbitrary portion of dividing surface is zero.* Let us examine the argument that must be supplied in going from (5-3) to (5-4).

We can begin by considering the analogous problem in one dimension. It is clear that

$$\int_0^{2\pi} \sin \theta \, d\theta = 0$$

does not imply that $\sin \theta$ is identically zero. But

$$\int_{\xi_1}^{\xi_2} f(y) \, dy = 0$$

does imply that

$$f(y) = 0$$

so long as it is understood that ξ_2 is arbitrary.

Proof: The Leibnitz rule for the derivative of an integral states that (Kaplan 1952, p. 220)

$$\frac{d}{dx} \int_{a(x)}^{b(x)} g(x, y) \, dy = g\Big(x, b(x) \Big) \frac{db}{dx} - g\Big(x, a(x) \Big) \frac{da}{dx}$$

$$+ \int_{a(x)}^{b(x)} \frac{\partial g}{\partial x} \, dy$$

If we apply the Leibnitz rule to

$$\frac{d}{d\xi_2} \int_{\xi_1}^{\xi_2} f(y) \, dy = 0$$

(1) follows immediately.

Let us now consider the analogous problem for

$$\int_{\xi_1}^{\xi_2} \int_{\eta_1 (y^2)}^{\eta_2 (y^2)} g(y^1, y^2) \sqrt{a} \, dy^1 \, dy^2 = 0$$

where $\eta_1 (y^2)$, $\eta_2 (y^2)$, ξ_1, and ξ_2 are completely arbitrary. Prove that this implies

$$g(y^1 , y^2) = 0$$

1.3.5-2 *another argument why the integrand must be zero, if an integral over an arbitrary portion of dividing surface is zero* Let an integral over an arbitrary portion of a dividing surface be zero.

i) Construct a sequence of portions of this surface with monotonically decreasing areas, all the members of which include the same arbitrary point.

ii) Normalize the sequence by dividing each member by its surface area.

iii) Take the limit of this normalized sequence to conclude that the integrand is zero at any arbitrary point on the surface.

1.3.5-3 *alternative form of transport theorem for material region containing dividing surface* Assuming mass is conserved, prove that

$$\frac{d}{dt}\left(\int_R \rho\hat{\psi}\, dV + \int_\Sigma \rho^{(\sigma)}\, \hat{\psi}^{(\sigma)}\, dA \right) = \int_R \rho\, \frac{d_{(m)}\hat{\psi}}{dt}\, dV$$

$$+ \int_\Sigma \left(\rho^{(\sigma)}\, \frac{d_{(s)}\hat{\psi}^{(\sigma)}}{dt} + [\, \rho(\hat{\psi} - \hat{\psi}^{(\sigma)})(\mathbf{v} \cdot \boldsymbol{\xi} - v_{\{\xi\}})] \right) dA$$

1.3.5-4 *physical meaning of* $J^{(\sigma)}$ From the jump mass balance as well as (3-4) and (3-5), determine that under conditions of no mass transfer to or from the phase interface

$$\frac{d_{(s)}}{dt}[\, \log(\rho^{(\sigma)}\, J^{(\sigma)})] = 0$$

Integrate this equation to learn

$$J^{(\sigma)} = \frac{\rho_\kappa^{(\sigma)}}{\rho^{(\sigma)}}$$

where $\rho_\kappa^{(\sigma)}$ denotes the surface density distribution in the reference configuration.

1.3.5-5 *alternative form of jump mass balance* Rearrange the jump mass balance (5-4) in the form

$$\frac{\partial \rho^{(\sigma)}}{\partial t} - \nabla_{(\sigma)} \rho^{(\sigma)} \cdot \mathbf{u} + \text{div}_{(\sigma)}(\rho^{(\sigma)} \mathbf{v}^{(\sigma)}) + [\rho(\mathbf{v} \cdot \boldsymbol{\xi} - \mathrm{v}\left\{\begin{matrix} g \\ (\xi) \end{matrix}\right\})] = 0$$

1.3.6 Location of dividing surface At every point in time, the dividing surface should be sensibly coincident with the phase interface. This is a sufficient description for most experimental measurements, but it is not sufficiently precise to define the location of the dividing surface. Small variations in position may lead even to a change in the sign assigned to a surface mass density (Defay *et al.* 1966, p. 25).

For single-component bodies, it will usually prove convenient to define the location of the dividing surface at each point in time to be such that

$$\rho^{(\sigma)} = 0 \tag{6-1}$$

Note that this is consistent with the conception of surface mass density developed in Section 1.3.2.

For a discussion of the Gibbs surface of tension, see Sec. 5.8.6, footnote a. Still other definitions suggest themselves for multicomponent bodies. These are discussed in Sec. 5.2.3.

Reference

Defay, R., I. Prigogine, A. Bellemans, and D. H. Everett, "Surface Tension and Adsorption," John Wiley, New York (1966).

1.3.7 Transport theorem for body containing intersecting dividing surfaces In Sec. 1.3.4 we derived the transport theorem for a body containing any number of non-intersecting dividing surfaces. In what follows I extend the transport theorem to multiphase bodies with intersecting dividing surfaces.

Consider for example a three phase body consisting of air, oil, and water. During the period of observation, the oil spreads over the air-water interface forming three interfaces: air-water, air-oil and oil-water. These three interfaces intersect in a moving common line.

As in Sec. 1.3.4, we are ultimately interested in the operation

$$\frac{d}{dt}\left(\int_R \psi \, dV + \int_\Sigma \psi^{(\sigma)} \, dA\right)$$

We have already considered the first term of this expression in Sec. 1.3.4. So let us begin by considering the time derivative of an integral over Σ:

$$\frac{d}{dt}\int_\Sigma \psi^{(\sigma)} \, dA \equiv \frac{d}{dt}\sum_{i=1}^{P} \int_{\Sigma^{(i)}} \psi^{(\sigma,i)} \, dA \tag{7-1}$$

It will be understood here that Σ represents the union of the P dividing surfaces in the body,

$$\Sigma \equiv \sum_{i=1}^{P} \Sigma^{(i)} \tag{7-2}$$

This discussion is in the context of an arbitrary multiphase body. However, it may be helpful to think in terms of the simpler body illustrated in Figure 7-1. It consists of three phases separated by three dividing surfaces: $\Sigma^{(i)}$, $\Sigma^{(j)}$, and $\Sigma^{(k)}$. The three dividing surfaces intersect in the common line $C^{(cl,ijk)}$. The three curves $C^{(i)}$, $C^{(j)}$, and $C^{(k)}$ represent respectively the intersections of the three dividing surfaces $\Sigma^{(i)}$, $\Sigma^{(j)}$, and $\Sigma^{(k)}$ with the boundary of the body.

Now let us focus our attention on any one of the dividing surfaces in the body, say $\Sigma^{(i)}$. The velocities of the boundaries of $\Sigma^{(i)}$ are

$$\text{on } C^{(i)}: \quad \mathbf{v}^{(\sigma,i)}$$

$$\text{on } C^{(cl,ijk)}: \quad \mathbf{u}^{(cl,ijk)} \tag{7-3}$$

The generalized surface transport theorem (Sec. 1.3.3) requires for each dividing surface i

$$\frac{d}{dt}\int_{\Sigma^{(i)}} \psi^{(\sigma,i)} \, dA$$

$$= \int_{\Sigma^{(i)}} \left(\frac{\partial \psi^{(\sigma,i)}}{\partial t} - \nabla_{(\sigma)}\psi^{(\sigma,i)} \cdot \mathbf{u}^{(i)} - 2H^{(i)} \, \psi^{(\sigma,i)} \, v_{(\xi)}^{(\xi,i)}\right) dA$$

$$+ \int_{C^{(i)}} \psi^{(\sigma,i)} \, \mathbf{v}^{(\sigma,i)} \cdot \boldsymbol{\mu}^{(i)} \, ds$$

$$- \sum_{\substack{j=1 \\ j \neq i}}^{P-1} \sum_{\substack{k=j+1 \\ k \neq i}}^{P} \int_{C^{(cl,ijk)}} \psi^{(\sigma,i)} \, \mathbf{u}^{(cl,ijk)} \cdot \boldsymbol{\nu}^{(i,ijk)} \, ds \qquad (7\text{-}4)$$

with the understanding that $\boldsymbol{\nu}^{(i,ijk)}$ is the unit vector normal to $C^{(cl,ijk)}$ tangent to and pointing into $\Sigma^{(i)}$.

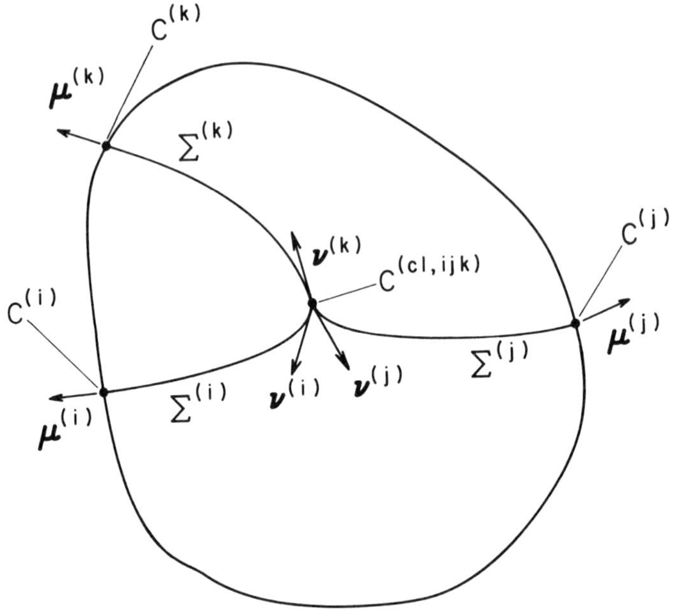

Figure 1.3.7-1: This is a body consisting of three phases separated by three dividing surfaces: $\Sigma^{(i)}$, $\Sigma^{(j)}$, and $\Sigma^{(k)}$. The three dividing surfaces intersect in the common line $C^{(cl,ijk)}$. The three curves $C^{(i)}$, $C^{(j)}$, and $C^{(k)}$ represent respectively the intersections of the three dividing surfaces $\Sigma^{(i)}$, $\Sigma^{(j)}$, and $\Sigma^{(k)}$ with the boundary of the body.

It follows that

$$\frac{d}{dt} \sum_{i=1}^{P} \int_{\Sigma^{(i)}} \psi^{(\sigma,i)} \, dA$$

$$= \sum_{i=1}^{P} \int_{\Sigma^{(i)}} \left(\frac{\partial \psi^{(\sigma,i)}}{\partial t} - \nabla_{(\sigma)} \psi^{(\sigma,i)} \cdot \mathbf{u}^{(i)} - 2H^{(i)} \, \psi^{(\sigma,i)} \, v_{(\xi)}^{(i)} \right) dA$$

$$+\sum_{i=1}^{P} \int_{C^{(i)}} \psi^{(\sigma,i)}\, \mathbf{v}^{(\sigma,i)} \cdot \boldsymbol{\mu}^{(i)}\, ds$$

$$-\sum_{i=1}^{P} \sum_{\substack{j=1\\ j\neq i}}^{P-1} \sum_{\substack{k=j+1\\ k\neq i}}^{P} \int_{C^{(cl,ijk)}} \psi^{(\sigma,i)}\, \mathbf{u}^{(cl,ijk)} \cdot \boldsymbol{\nu}^{(i,ijk)}\, ds \qquad (7\text{-}5)$$

If we understand that $\psi^{(\sigma)}$, $\mathbf{u}$, H, and $\mathbf{v}^{(\sigma)}$ are piecewise continuous functions defined by $\psi^{(\sigma,i)}$, $\mathbf{u}^{(i)}$, $H^{(i)}$, and $\mathbf{v}^{(\sigma,i)}$ on the dividing surface $\Sigma^{(i)}$ and that $\mathbf{u}^{(c\ell)}$ is a piecewise continuous function defined by $\mathbf{u}^{(cl,ijk)}$ on the common line $C^{(cl,ijk)}$, we can write (7-5) as

$$\frac{d}{dt}\int_{\Sigma} \psi^{(\sigma)}\, dA$$

$$= \int_{\Sigma} \left(\frac{\partial \psi^{(\sigma)}}{\partial t} - \nabla_{(\sigma)}\psi^{(\sigma)} \cdot \mathbf{u} - 2H\, \psi^{(\sigma)}\, v_{\{\xi\}} \right) dA$$

$$+ \int_{C} \psi^{(\sigma)}\, \mathbf{v}^{(\sigma)} \cdot \boldsymbol{\mu}\, ds - \int_{C^{(cl)}} (\, \psi^{(\sigma)}\, \boldsymbol{\nu}\,) \cdot \mathbf{u}^{(cl)}\, ds \qquad (7\text{-}6)$$

where

$$(\, \psi^{(\sigma)}\, \boldsymbol{\nu}\,) \equiv \psi^{(\sigma,i)}\, \boldsymbol{\nu}^{(i)} + \psi^{(\sigma,j)}\, \boldsymbol{\nu}^{(j)} + \psi^{(\sigma,k)}\, \boldsymbol{\nu}^{(k)} \qquad (7\text{-}7)$$

and $\boldsymbol{\nu}$ is understood to be the unit normal to $C^{(cl)}$ that points into the dividing surface concerned. If we observe that by the surface divergence theorem

$$\int_{C} \psi^{(\sigma)}\, \mathbf{v}^{(\sigma)} \cdot \boldsymbol{\mu}\, ds - \int_{\Sigma} 2H\, \psi^{(\sigma)}\, v_{\{\xi\}}\, dA$$

$$= \int_{C^{(cl)}} (\, \psi^{(\sigma)}\, \mathbf{v}^{(\sigma)} \cdot \boldsymbol{\nu}\,)\, ds + \int_{\Sigma} \mathrm{div}_{(\sigma)}(\, \psi^{(\sigma)}\, \mathbf{v}^{(\sigma)}\,)\, dA \qquad (7\text{-}8)$$

and that

$$\frac{\partial \psi^{(\sigma)}}{\partial t} - \nabla_{(\sigma)}\psi^{(\sigma)} \cdot \mathbf{u} + \mathrm{div}_{(\sigma)}(\, \psi^{(\sigma)}\, \mathbf{v}^{(\sigma)}\,)$$

$$= \frac{d_{(s)} \psi^{(\sigma)}}{dt} + \psi^{(\sigma)} \operatorname{div}_{(\sigma)} \mathbf{v}^{(\sigma)} \tag{7-9}$$

we can finally express (7-6) as

$$\frac{d}{dt} \int \psi^{(\sigma)} \, dA = \int_{\Sigma} \left(\frac{d_{(s)} \psi^{(\sigma)}}{dt} + \psi^{(\sigma)} \operatorname{div}_{(\sigma)} \mathbf{v}^{(\sigma)} \right) dA$$

$$+ \int_{C^{(cl)}} \left(\psi^{(\sigma)} [\, \mathbf{v}^{(\sigma)} \cdot \mathbf{v} - u_{(\nu)}^{(cl)} \,] \right) ds \tag{7-10}$$

Here

$$u_{(\nu)}^{(cl)} \equiv \mathbf{u}^{(cl)} \cdot \mathbf{v} \tag{7-11}$$

is the **speed of displacement** of the common line.

We can now use (7-10) together with

$$\frac{d}{dt} \int_{R} \psi \, dV$$

$$= \int_{R} \left(\frac{d_{(m)} \psi}{dt} + \psi \operatorname{div} \mathbf{v} \right) dV + \int_{\Sigma} [\, \psi (\, \mathbf{v} \cdot \boldsymbol{\xi} - v_{(\xi)} \,)] \, dA \tag{7-12}$$

which was derived in Sec. 1.3.4 to conclude

$$\frac{d}{dt} \left(\int_{R} \psi \, dV + \int_{\Sigma} \psi^{(\sigma)} \, dA \right) = \int_{R} \left(\frac{d_{(m)} \psi}{dt} + \psi \operatorname{div} \mathbf{v} \right) dV$$

$$+ \int_{\Sigma} \left(\frac{d_{(s)} \psi^{(\sigma)}}{dt} + \psi^{(\sigma)} \operatorname{div}_{(\sigma)} \mathbf{v}^{(\sigma)} + [\, \psi (\, \mathbf{v} \cdot \boldsymbol{\xi} - v_{(\xi)} \,)] \right) dA$$

$$+ \int_{C^{(cl)}} \left(\psi^{(\sigma)} [\, \mathbf{v}^{(\sigma)} \cdot \mathbf{v} - u_{(\nu)}^{(cl)} \,] \right) ds \tag{7-13}$$

We shall refer to (7-13) as the **transport theorem for a body containing intersecting dividing surfaces.**

In deriving (7-13), an important assumption has been made: ψ and $\mathbf{v}$ are piecewise continuous with piecewise continuous first derivatives. This must be resolved with our observation in Sec. 1.2.11 that velocity may be multivalued at a common line moving over a rigid solid. Alternatives are discussed in Sec. 1.3.10.

Exercise

1.3.7-1 *generalized transport theorem for system containing intersecting dividing surfaces* Reasoning as we did in Exercise 1.3.4-1, extend the transport theorem for a body containing intersecting dividing surfaces (7-13) to an arbitrary system containing intersecting dividing surfaces. If ψ and $\psi^{(\sigma)}$ are scalars,

$$\frac{d}{dt}\left(\int_{R_{(sys)}} \psi \, dV + \int_{\Sigma_{(sys)}} \psi^{(\sigma)} \, dA\right)$$

$$= \int_{R_{(sys)}} \frac{\partial \psi}{\partial t} dV + \int_{S_{(sys)}} \psi \, \mathbf{v}_{(sys)} \cdot \mathbf{n} \, dA + \int_{C_{(sys)}} \psi^{(\sigma)} \, \mathbf{v}_{(sys)} \cdot \boldsymbol{\mu} \, ds$$

$$+ \int_{\Sigma_{(sys)}} \left(\frac{\partial \psi^{(\sigma)}}{\partial t} - \nabla_{(\sigma)} \psi^{(\sigma)} \cdot \mathbf{u} - 2H\psi^{(\sigma)} \, v_{\{\xi\}}^{\{g\}} - [\, \psi \, v_{\{\xi\}}^{\{g\}} \,]\right) dA$$

$$- \int_{C_{\{sys\}}^{(cl)}} (\, \psi^{(\sigma)} \, u_{\{v\}}^{(cl)} \,) \, ds \tag{1}$$

If $\boldsymbol{\psi}$ and $\boldsymbol{\psi}^{(\sigma)}$ are vectors,

$$\frac{d}{dt}\left(\int_{R_{(sys)}} \boldsymbol{\psi} \, dV + \int_{\Sigma_{(sys)}} \boldsymbol{\psi}^{(\sigma)} \, dA\right)$$

$$= \int_{R_{(sys)}} \frac{\partial \boldsymbol{\psi}}{\partial t} dV + \int_{S_{(sys)}} \boldsymbol{\psi} \, \mathbf{v}_{(sys)} \cdot \mathbf{n} \, dA + \int_{C_{(sys)}} \boldsymbol{\psi}^{(\sigma)} \, \mathbf{v}_{(sys)} \cdot \boldsymbol{\mu} \, ds$$

$$+ \int_{\Sigma_{(sys)}} \left(\frac{\partial \boldsymbol{\psi}^{(\sigma)}}{\partial t} - \nabla_{(\sigma)} \boldsymbol{\psi}^{(\sigma)} \cdot \mathbf{u} - 2H \, \boldsymbol{\psi}^{(\sigma)} \, v_{\{\xi\}}^{\{g\}} - [\, \boldsymbol{\psi} \, v_{\{\xi\}}^{\{g\}} \,]\right) dA$$

$$- \int_{C_{\{sys\}}^{(cl)}} (\, \boldsymbol{\psi}^{(\sigma)} \, u_{\{v\}}^{(cl)} \,) \, ds \tag{2}$$

Here $\Sigma_{(sys)}$ are the dividing surfaces contained within the arbitrary system,

$C_{(sys)}$ are the lines formed by the intersection of these dividing surfaces with $S_{(sys)}$, $C^{(cl)}_{(sys)}$ are the common lines contained within the system, and $\boldsymbol{\mu}$ is the unit vector normal to $C_{(sys)}$ that is both tangent to and outwardly directed with respect to the dividing surface. On $S_{(sys)}$, we define $\mathbf{v}_{(sys)} \cdot \mathbf{n}$ to be the normal component of the velocity of this boundary; on $C_{(sys)}$, $\mathbf{v}_{(sys)} \cdot \boldsymbol{\mu}$ is the component of the velocity of this curve in the direction $\boldsymbol{\mu}$. We can refer to (1) and (2) as forms of the **generalized transport theorem for a system containing intersecting dividing surfaces.**

These forms of the generalized transport theorem assume that $S_{(sys)}$ and $\Sigma_{(sys)}$ do not overlap. Equations (1) and (2) may be extended to the case where $S_{(sys)}$ and $\Sigma_{(sys)}$ coincide by first considering a new system in which $S'_{(sys)}$ and $\Sigma'_{(sys)}$ do not overlap and then letting $S'_{(sys)}$ shrink to $S_{(sys)}$.

1.3.8 Mass balance at common line What does conservation of mass require at each point on a common line?

After an application of the transport theorem for a body containing intersecting dividing surfaces, the postulate of mass conservation stated in Sec. 1.3.1 takes the form

$$\begin{aligned}
&\frac{d}{dt}\left(\int_R \rho\, dV + \int_\Sigma \rho^{(\sigma)}\, dA\right) \\
&\quad = \int_R \left(\frac{d_{(m)}\rho}{dt} + \rho\, \operatorname{div} \mathbf{v}\right) dV \\
&\qquad + \int_\Sigma \left\{\frac{d_{(s)}\rho^{(\sigma)}}{dt} + \rho^{(\sigma)} \operatorname{div}_{(\sigma)} \mathbf{v}^{(\sigma)} + [\rho(\mathbf{v}\cdot\boldsymbol{\xi} - v_{(\xi)})]\right\} dA \\
&\qquad + \int_{C^{(cl)}} \left(\rho^{(\sigma)}[\mathbf{v}^{(\sigma)}\cdot\boldsymbol{\nu} - u^{(cl)}_{(\nu)}]\right) ds \\
&\quad = 0 \qquad (8\text{-}1)
\end{aligned}$$

In view of the equation of continuity (Sec. 1.3.5)

$$\frac{d_{(m)}\rho}{dt} + \rho \operatorname{div} \mathbf{v} = 0 \tag{8-2}$$

and the jump mass balance (Sec. 1.3.5)

$$\frac{d_{(s)}\rho^{(\sigma)}}{dt} + \rho^{(\sigma)}\operatorname{div}_{(\sigma)}\mathbf{v}^{(\sigma)} + [\rho(\mathbf{v}\cdot\boldsymbol{\xi} - v_{(\xi)})] = 0 \tag{8-3}$$

(8-1) reduces to

$$\int_{C^{(cl)}} (\rho^{(\sigma)}[\mathbf{v}^{(\sigma)}\cdot\boldsymbol{\nu} - u_{(\nu)}^{(cl)}])\, ds = 0 \tag{8-4}$$

Since (8-1) is valid for every material body and every portion of a material body, (8-4) must be true for every portion of a common line. We conclude (see Exercises 1.3.8-1 and 1.3.8-2)

$$(\rho^{(\sigma)}[\mathbf{v}^{(\sigma)}\cdot\boldsymbol{\nu} - u_{(\nu)}^{(cl)}]) = 0 \tag{8-5}$$

This can be called the **mass balance at the common line**. It expresses the requirement that mass can be conserved at every point on a common line.

Let me repeat my caution of Sec. 1.3.7. In deriving the transport theorem for a body containing intersecting dividing surfaces, an important assumption has been made: ρ and $\mathbf{v}$ are piecewise continuous with piecewise continuous first derivatives. This must be resolved with our observation in Sec. 1.2.11 that velocity may be multivalued at a common line moving over a rigid solid. Alternatives are discussed in Sec. 1.3.10.

Equation (8-5) tells us that, if there is mass transfer at the common line, $\mathbf{v}^{(\sigma)}\cdot\boldsymbol{\nu} - u_{(\nu)}^{(cl)}$ may assume different values in each of the dividing surfaces at the common line. This is compatible with the conclusion of Sec. 1.2.11 that velocity may be multivalued at a moving common line on a rigid solid.

If there is no mass transfer across the common line, we would expect the velocity field to be single-valued, since the velocity field would be continuous everywhere as the common line was approached. (I assume here that the tangential component of velocity that is also tangent to the common line will be continuous across the common line.) In a frame of reference such that the common line is fixed in space, the only motion would be along the common line in the limit as the common line was approached. The common line would be stationary relative to the intersecting dividing surfaces, although it might be in motion relative to the material outside its immediate neighborhood.

If two adjoining phases are saturated with one another, we can with reasonable confidence assume that there is no mass transfer across the phase interface separating them. Under what conditions are we free to say that there is no mass transfer across the common line?

a) If we are concerned with a stationary common line on a solid phase interface, it should be possible after a period of time to assume that chemical equilibrium has been established on the phase interfaces in the neighborhood of the common line and that there is no mass transfer across the common line.

b) If the various species present are assumed not to adsorb in any of the phase interfaces and if the adjoining bulk phases have been pre-equilibrated, it may be reasonable to eliminate mass transfer at the common line.

c) When slip is assumed in the neighborhood of a moving common line on a rigid solid, it is usual to assume that velocity is single-valued at the common line and as a result that there is no mass transfer at the common line (see Secs. 1.3.10 and 3.1.1). With this assumption, no portion of an adsorbed film in one phase interface can be rolled directly into an adsorbed film in another phase interface. All mass transfer must take place through one of the adjoining phases.

In stating (8-1) and in deriving (8-5) we have assumed there is no mass in the common line. I am not aware of any experimental evidence that suggests mass should be associated with the common line. For related references, see the concluding remarks of Sec. 2.1.9.

Exercises

1.3.8-1 *The integrand must be zero, if an integral over an arbitrary portion of common line is zero.* Construct a proof along the lines of the initial argument offered in Exercise 1.3.5-1.

1.3.8-2 *another argument why the integrand must be zero, if an integral over an arbitrary portion of common line is zero* Construct a proof along the line of that suggested in Exercise 1.3.5-2, replacing "dividing surface" by "common line" and "area" by "arc length".

1.3.8-3 *alternative form of transport theorem for body containing intersecting dividing surfaces* Assuming mass is conserved, prove that

$$\frac{d}{dt}\left(\int_R \rho \, \hat{\psi} \, dV + \int_\Sigma \rho^{(\sigma)} \, \hat{\psi}^{(\sigma)} \, dA \right)$$

$$= \int_R \rho \frac{d_{(m)}\hat{\psi}}{dt} \, dV$$

$$+ \int_\Sigma \left\{ \rho^{(\sigma)} \frac{d_{(s)}\hat{\psi}^{(\sigma)}}{dt} + [\, \rho(\hat{\psi} - \hat{\psi}^{(\sigma)})(\mathbf{v} \cdot \boldsymbol{\xi} - v_{\{\xi\}})] \right\} dA$$

$$+ \int_{C^{(cl)}} (\, \rho^{(\sigma)} \hat{\psi}^{(\sigma)} [\, \mathbf{v}^{(\sigma)} \cdot \boldsymbol{\nu} - u^{(cl)}_{(\nu)} \,]) \, ds$$

1.3.8-4 *general balance equations when interface is represented by a dividing surface* In Exercise 1.3.2-2, I wrote the general balance or general conservation law for some quantity associated with a multiphase material body as

$$\frac{d}{dt}\left(\int_R \psi \, dV + \int_\Sigma \psi^{(\sigma)} \, dA \right)$$

$$= - \int_S \boldsymbol{\phi} \cdot \mathbf{n} \, dA - \int_C \boldsymbol{\phi}^{(\sigma)} \cdot \boldsymbol{\mu} \, ds$$

$$+ \int_R \rho\zeta \, dV + \int_\Sigma \rho^{(\sigma)} \zeta^{(\sigma)} \, dA \qquad (1)$$

Use the transport theorem for a body containing intersecting dividing surfaces to rearrange (1) in the form

$$\int_R \left(\frac{d_{(m)}\psi}{dt} + \psi \operatorname{div} \mathbf{v} + \operatorname{div} \boldsymbol{\phi} - \rho\zeta \right) dV$$

$$+ \int_\Sigma \left\{ \frac{d_{(s)}\psi^{(\sigma)}}{dt} + \psi^{(\sigma)} \operatorname{div}_{(\sigma)}\mathbf{v}^{(\sigma)} + \operatorname{div}_{(\sigma)}\boldsymbol{\phi}^{(\sigma)} - \rho^{(\sigma)} \zeta^{(\sigma)} \right.$$

$$\left. + [\, \psi(\mathbf{v} \cdot \boldsymbol{\xi} - v_{\{\xi\}}) + \boldsymbol{\phi} \cdot \boldsymbol{\xi} \,] \right\} dA$$

$$+ \int_{C^{(cl)}} (\, \psi^{(\sigma)}[\, \mathbf{v}^{(\sigma)} \cdot \boldsymbol{\nu} - u^{(cl)}_{(\nu)} \,] + \boldsymbol{\phi}^{(\sigma)} \cdot \boldsymbol{\nu} \,) \, ds$$

$$= 0 \tag{2}$$

Conclude that at each point within a phase the **general balance** equation is

$$\frac{d_{(m)}\psi}{dt} + \psi \operatorname{div} \mathbf{v} + \operatorname{div} \boldsymbol{\phi} - \rho\zeta = 0 \tag{3}$$

At each point on a dividing surface, the **general jump balance** is

$$\frac{d_{(s)}\psi^{(\sigma)}}{dt} + \psi^{(\sigma)} \operatorname{div}_{(\sigma)}\mathbf{v}^{(\sigma)} + \operatorname{div}_{(\sigma)}\boldsymbol{\phi}^{(\sigma)} - \rho^{(\sigma)}\zeta^{(\sigma)}$$

$$+ [\, \psi(\mathbf{v} \cdot \boldsymbol{\xi} - v_{(\xi)}^{\{g\}}) + \boldsymbol{\phi} \cdot \boldsymbol{\xi} \,] = 0 \tag{4}$$

Finally, the **general balance equation at a common line** is

$$\left(\psi^{(\sigma)}[\, \mathbf{v}^{(\sigma)} \cdot \boldsymbol{\nu} - u_{(\nu)}^{(cl)} \,] + \boldsymbol{\phi}^{(\sigma)} \cdot \boldsymbol{\nu} \right) = 0 \tag{5}$$

1.3.9 Comment on velocity distribution in neighborhood of moving common line on rigid solid In Sec. 1.2.11, we found that under certain conditions velocity will be multivalued at a moving common line on a rigid solid. This suggests that, at least in some cases, one or more derivatives of the velocity component could be unbounded within the adjoining phases in the limit as the common line is approached. What are the physical implications?

If the rate of deformation tensor

$$\mathbf{D} \equiv \frac{1}{2}\left(\nabla\mathbf{v} + (\nabla\mathbf{v})^{T} \right) \tag{9-1}$$

is unbounded as the common line is approached, the stress tensor in the adjoining phases may be unbounded as well. For example, we know that for a Newtonian fluid (Slattery 1981, p. 49)

$$\mathbf{T} = (- P + \lambda \operatorname{div} \mathbf{v}) \mathbf{I} + 2\mu \mathbf{D} \tag{9-2}$$

From a formal point of view, the stress tensor should be bounded everywhere, since every second-order tensor in three space is bounded (Stakgold 1967, p. 140). From a more physical point of view, we should require at least the forces in the neighborhood of the common line be finite.

In order to better appreciate the issue, we extend the argument of Dussan V. and Davis (1974) to consider a three-dimensional flow of an incompressible, Newtonian fluid within one of the phases in the neighborhood of a common line moving over a *rigid*[a] solid. We will assume that there is no mass transfer between the dividing surfaces and the adjoining bulk phases in the limit as the common line is approached and we will adopt the no-slip boundary condition requiring the tangential components of velocity to be continuous across the dividing surface.

Let us choose a frame of reference in which the moving common line shown in Figure 9-1 is stationary. In terms of a cylindrical coordinate system centered upon the common line, assume that the flow is fully three-dimensional but independent of time in phase B:

$$\begin{aligned} v_r &= v_r(\, r, \theta, z\,) \\ &= f(\, \theta, z\,) + F(\, r, \theta, z\,) \end{aligned} \tag{9-3}$$

$$\begin{aligned} v_\theta &= v_\theta(\, r, \theta, z\,) \\ &= g(\, \theta, z\,) + G(\, r, \theta, z\,) \end{aligned} \tag{9-4}$$

$$v_z = v_z(\, r, \theta, z\,) \tag{9-5}$$

The form of this velocity distribution is restricted by the equation of continuity for an incompressible fluid

$$\text{div } \mathbf{v} = 0$$

$$r \frac{\partial v_r}{\partial r} + v_r + \frac{\partial v_\theta}{\partial \theta} + r \frac{\partial v_z}{\partial z} = 0$$

---

a) With two restrictions, a two-dimensional plane flow in the neighborhood of a moving common line formed by three deformable phases is not possible (Exercise 1.3.9-1): in the limit as the common line is approached, there is no interphase mass transfer (Exercise 1.3.9-2) and all of the surface mass densities must be non-negative with at least one different from zero.

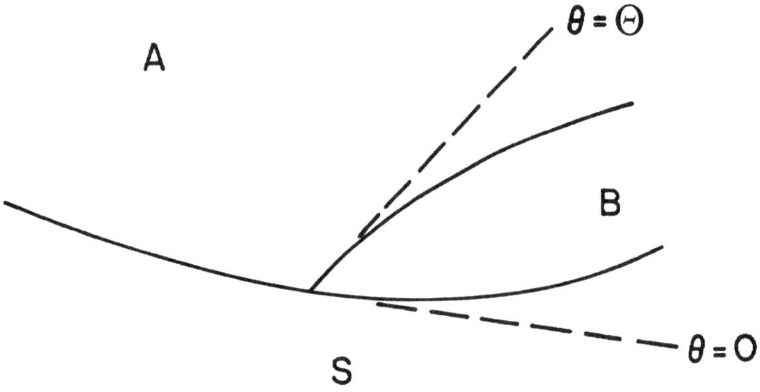

Figure 1.3.9-1 A frame of reference is chosen in which the moving common line is stationary.

$$r \frac{\partial F}{\partial r} + f + F + \frac{\partial g}{\partial \theta} + \frac{\partial G}{\partial \theta} + r \frac{\partial v_z}{\partial z} = 0 \qquad (9\text{-}6)$$

In order to identify $f(\theta, z)$ and $g(\theta, z)$ as the r- and θ- components of velocity at the common line, let us require

$$\text{limit } r \to 0,\ \theta \text{ and } z \text{ fixed:} \quad F(\, r, \theta, z\,) = 0 \qquad (9\text{-}7)$$

$$\text{limit } r \to 0,\ \theta \text{ and } z \text{ fixed:} \quad G(\, r, \theta, z\,) = 0 \qquad (9\text{-}8)$$

Following the discussion in Sec. 1.2.11, we will assume

$$f(\,0, z\,) \neq f(\,\Theta, z\,) \qquad (9\text{-}9)$$

If this fails in phase B, the comparable relation would necessarily be true in phase A. (The comparable relation would also be true in phase B, if the ejected or emitted material surface described in Sec. 1.2.10 is in phase B and if we identify Θ with this surface.)

Since our concern here is the possibility of an unbounded force developed within phase B in the neighborhood of the common line attributable to (9-9), we need to consider only the r-component of force (per unit width) exerted upon one side of plane θ = a constant within phase B:

$$\mathcal{F}_r = \int_0^r T_{r\theta}\, dr$$

$$= 2\mu \int_0^r D_{r\theta}\, dr$$

$$= \mu \int_0^r \left(\frac{1}{r}\frac{\partial v_r}{\partial \theta} + \frac{\partial v_\theta}{\partial r} - \frac{v_\theta}{r} \right) dr$$

$$= \mu \int_0^r \left[\frac{1}{r} \left(\frac{\partial f}{\partial \theta} + \frac{\partial F}{\partial \theta} - g - G \right) + \frac{\partial G}{\partial r} \right] dr \qquad (9\text{-}10)$$

So long as G is assumed to be absolutely continuous in r for fixed θ,

$$\int_0^r \frac{\partial G}{\partial r}\, dr = G \qquad (9\text{-}11)$$

In order that $\mathcal{F}_r$ be finite, we must require

$$\text{limit } r \to 0: \quad \frac{\partial f}{\partial \theta} + \frac{\partial F}{\partial \theta} - g = 0 \qquad (9\text{-}12)$$

Differentiating (9-6) with respect to θ, we have

$$r\frac{\partial^2 F}{\partial r \partial \theta} + \frac{\partial f}{\partial \theta} + \frac{\partial F}{\partial \theta} + \frac{\partial^2 g}{\partial \theta^2} + \frac{\partial^2 G}{\partial \theta^2} + r\frac{\partial^2 v_z}{\partial \theta \partial z} = 0 \qquad (9\text{-}13)$$

Equation (9-13) can be used to eliminate $\partial f/\partial\theta + \partial F/\partial\theta$ from (9-12):

$$\text{limit } r \to 0: \quad \frac{\partial^2 g}{\partial \theta^2} + g + r\frac{\partial^2 F}{\partial r \partial \theta} + \frac{\partial^2 G}{\partial \theta^2} + r\frac{\partial^2 v_z}{\partial \theta \partial z} = 0 \qquad (9\text{-}14)$$

Let us assume that the velocity distribution in the neighborhood of the common line is such that

$$\text{limit } r \to 0: \quad \frac{\partial F}{\partial \theta} = 0 \qquad (9\text{-}15)$$

$$\text{limit } r \to 0: \quad r\frac{\partial^2 F}{\partial r \partial \theta} = 0 \qquad (9\text{-}16)$$

$$\text{limit } r \to 0: \quad \frac{\partial^2 G}{\partial \theta^2} = 0 \tag{9-17}$$

Because the solid is rigid[a] and because we are assuming that the tangential components of velocity are continuous across the dividing surfaces,

$$\text{limit } r \to 0: \quad r \frac{\partial^2 v_z}{\partial \theta \partial z} = \frac{\partial v_z}{\partial z} = 0 \tag{9-18}$$

With the restrictions (9-16) through (9-18), (9-14) reduces to

$$\frac{\partial^2 g}{\partial \theta^2} + g = 0 \tag{9-19}$$

Since there is no interphase mass transfer in the limit as the common line is approached,

$$g(0, z) = g(\Theta, z) = 0 \tag{9-20}$$

If $\Theta \neq 0$ or π, the only solution is

$$g = 0 \tag{9-21}$$

But (9-12) and (9-15) imply

$$\frac{\partial f}{\partial \theta} - g = 0 \tag{9-22}$$

or

$$\frac{\partial f}{\partial \theta} = 0 \tag{9-23}$$

which contradicts our assumption (9-9) that the r-component of velocity within the A-B dividing surface differs from that within the B-S dividing surface at the common line.

If $\Theta = \pi$ and the emitted or ejected material surface described in Sec. 1.2.10 is in phase B, the same argument as above applies. It is only necessary to identify Θ ($\neq \pi$) with the location of emitted or ejected surface.

Finally, consider $\Theta = 0$ and the emitted or ejected material surface in phase B. In this case, the solution to (9-19) consistent with boundary conditions (9-20) is

$$g = B \sin \theta \tag{9-24}$$

Equation (9-22) consequently requires

$$f = - B \cos \theta + C \tag{9-25}$$

The constants B and C are not sufficient to satisfy the constraints of an emitted or ejected surface originating at the common line.

If $\mathcal{F}_r$ is to be finite, one of the other assumptions made in this argument must be incorrect. Notice in particular that this discussion is unaffected by the existence of body forces such as gravity or of mutual forces such as London-van der Waals forces or electrostatic double-layer forces (see Exercises 2.1.3-1 and 2.1.3-2). Possibilities are considered in the next section.

References

Dussan V., E. B., and S. H. Davis, *J. Fluid Mech.* **65**, 71 (1974).

Giordano, R. M., personal communication, April 25, 1979.

Slattery, J. C., "Momentum, Energy, and Mass Transfer in Continua," McGraw-Hill, New York (1972); second edition, Robert E. Krieger, Malabar, FL 32950 (1981).

Stakgold, I., "Boundary Value Problems of Mathematical Physics," vol. 1, MacMillan, New York (1967).

Exercises

1.3.9-1 *velocity distribution in neighborhood of moving common line formed by three deformable phases* Repeat this same argument for a two-dimensional plane flow in the neighborhood of a common line formed by the intersection of three deformable phases:

$$\begin{aligned} v_r &= v_r(r, \theta) \\ &= f(\theta) + F(r, \theta) \end{aligned}$$

$$\begin{aligned} v_\theta &= v_\theta(r, \theta) \\ &= g(\theta) + G(r, \theta) \end{aligned}$$

$$v_z = 0$$

Justify (9-9) on the basis of the mass balance at the common line, so long as all of the surface mass densities are non-negative and at least one is different from zero. Conclude that two-dimensional plane flows are incompatible with finite forces within the neighborhood of a moving common line (Dussan V. and Davis 1974), unless interphase mass transfer is allowed in the limit as the common line is approached (Exercise 1.3.9-2).

1.3.9-2 *more on velocity distribution in neighborhood of moving common line formed by three deformable phases* (Giordano 1979) Two-dimensional motions are consistent with finite forces in the neighborhood of a moving common line formed by three deformable phases, when interphase mass transfer is allowed in the limit as the common line is approached.

Instead of (9-20), let us allow mass transfer between the solid S and phase B in the limit as the common line is approached:

$$g(0) = -C$$

$$g(\Theta) = 0 \tag{1}$$

For the moment, let us leave C unspecified.

i) Integrate (9-19) consistent with (1) to find

$$g = C\,(\cot\Theta \sin\theta - \cos\theta) \tag{2}$$

Let us visualize that

$$f(\Theta) = -A \tag{3}$$

where A is specified by the mass balance at the common line.

ii) Integrate (9-22) consistent with (2) and (3) to discover

$$f = -A + C\left(-\cot\Theta\cos\theta - \sin\theta + \frac{1}{\sin\Theta}\right) \tag{4}$$

Normally the speed of displacement of the solid wall would be given

$$f(0) = V$$

in which case

$$C = \frac{V + A}{\left(-\cot\Theta + \dfrac{1}{\sin\Theta} \right)}$$

1.3.10 More comments on velocity distribution in neighborhood of moving common line on rigid solid In Sec. 1.2.11, we found that under certain conditions velocity may be multivalued at a common line moving over a rigid solid. Although we noted that this is consistent with mass transfer at the common line in Sec. 1.3.8, it may be inconsistent with the development of the transport theorem for a body containing intersecting dividing surfaces (Sec. 1.3.7), and it may lead to unbounded forces at the common line (Sec. 1.3.9). At least one of the assumptions underlying these inconsistencies must be incorrect. Let us consider the possibilities.

i) A solid is not rigid.

For most solids, the deformation caused by the forces acting at a common line is very small (Lester 1961; Wickham and Wilson 1975). The deformation of metals and other materials with high tensile strength will be of the same order or smaller than imperfections we will wish to ignore in real surfaces. But this deformation is non-zero. An unbounded force at the common line might be interpreted as a natural response to our refusal to allow the surface to deform.

On the other hand, it can not be denied that a rigid solid is a convenient idealization.

ii) The tangential components of velocity are not continuous at phase interfaces in the limit as a moving common line is approached.

Outside the immediate neighborhood of the common line, there is no debate. Goldstein (1938, p. 676) presents a sound case favoring the no-slip boundary condition, except in the limit of a rarefied gas where it is known to fail. Richardson (1973) comes to a similar conclusion in considering single phase flow past a wavy wall. He concludes that, even if there were no resistance to relative motion between a fluid and a solid in contact, roughness alone would ensure that the boundary condition observed on a macroscopic scale would be one of no slip. Still another more restricted argument supporting the continuity of the tangential components of velocity at a dividing surface is presented in Exercise 2.1.6-3.

Hocking (1976) finds that, in the displacement of one fluid by another over a wavy surface, a portion of the displaced phase may be left behind, trapped in the valleys. As a result, the displacing fluid moves over a composite surface, particularly within the immediate neighborhood of the moving common line. Hocking (1976) concludes that it is appropriate to describe the movement of the displacing phase over this composite surface though a slip boundary condition. On the other hand, his argument could not be used to justify the use of a slip boundary condition on a geometrically smooth surface such as we have considered here.

Huh and Scriven (1971) as well as Dussan V. and Davis (1974) suggest that slip might be allowed within the immediate neighborhood of the common line. The introduction of slip within the immediate neighborhood of the common line eliminates the appearance of an unbounded force (Dussan V. 1976; Hocking 1977; Huh and Mason 1977; Greenspan 1978; Hocking and Rivers 1982; Cox 1986).

It is reassuring to note that computations outside the immediate neighborhood of the common line appear to be insensitive to the details of the slip model used (Dussan V. 1976; Kafka and Dussan V. 1979).

iii) There is interphase mass transfer in the limit as the common line is approached.

Interphase mass transfer could eliminate these inconsistencies. This possibility should be investigated further.

iv) A liquid does not spread over a rigid solid, but rather over a precursor film on the rigid solid.

Teletzke *et al.* (1987) suggest that there are three classes of spreading: primary, secondary, and bulk.

By primary spreading, they refer to the manner in which the first one or two molecular layers are deposited on the solid. For a nonvolatile liquid, primary spreading is by surface diffusion (Bascom *et al.* 1964; Cherry and Holmes 1969; Blake and Haynes 1969; Ruckenstein and Dunn 1977). For a volatile liquid, primary spreading may be by the condensation of an adsorbed film (Hardy 1919).

By secondary spreading, they refer to the motion of films whose thickness varies from several molecules to roughly a micron. These films spread as the result of either a positive disjoining pressure or a surface tension gradient. Surface tension gradients can be the result of either temperature gradients (Ludviksson and Lightfoot 1971) or composition gradients, perhaps resulting from the evaporation of a more volatile component (Bascom *et al.* 1964).

By bulk spreading, they refer to the motion of films thicker than a micron. In these films, the flow is driven by interfacial tension forces, gravity, and possibly a forced convection.

From this point of view, the precursor films discussed in Sec. 1.2.9 are probably the result of both primary and secondary spreading. There may be no moving common line, but simply a smooth transition from film flow to surface diffusion.

This conception of the spreading process certainly seems to explain some classes of observations. For example, Teletzke *et al.* (1987) argue effectively that the experiments of Bascom *et al.* (1964) can be understood in this manner.

But it does not explain how common lines move more generally. For example, consider a common line driven by forced convection where a thin film of the liquid exhibits a negative disjoining pressure; a common line driven by forced convection where the time required for displacement is much smaller than the time required the primary and secondary spreading phenomena described above; a common line that recedes as the result of a draining liquid film.

v) The common line does not actually move.

This proposal, perhaps initially surprising, will be investigated further in Secs. 2.1.13 and 3.3.3.

No significance should be attached to my developing only proposal v in the remainder of the text, other than it is simpler to employ than i, iii or iv and it seems to retain more of the underlying chemistry than ii. I see merit in all of these proposals. The last three in particular are not mutually exclusive. I don't expect to see one mechanism developed to describe all common line movements, since different situations may involve different physical phenomena. It certainly is too early to begin ruling out any ideas.

References

Bascom, W. D., R. L. Cottington, and C. R. Singleterry, in "Contact Angle, Wettability and Adhesion," Advances in Chemistry Series No. 43, p. 355, American Chemical Society, Washington, D.C. (1964).

Blake, T. D., and J. M. Haynes, *J. Colloid Interface Sci.* **30**, 421 (1969).

Cherry, B. W., and C. M. Holmes, *J. Colloid Interface Sci.* **29**, 174 (1969).

Cox, B. G., *J. Fluid Mech.* **168**, 169 (1986).

Dussan V., E. B., and S. H. Davis, *J. Fluid Mech.* **65**, 71 (1974).

Dussan V., E. B., *J. Fluid Mech.* **77**, 665 (1976).

Goldstein, S., “Modern Developments in Fluid Dynamics,” Oxford University Press, London (1938).

Greenspan, H. P., *J. Fluid Mech.* **84**, 125 (1978).

Hardy, W. B., *Phil. Mag. (6)* **38**, 49 (1919).

Hocking, L. M., *J. Fluid Mech.* **76**, 801 (1976).

Hocking, L. M., *J. Fluid Mech.* **79**, 209 (1977).

Hocking, L. M., and A. D. Rivers, *J. Fluid Mech.* **121**, 425 (1982).

Huh, C., and S. G. Mason, *J. Fluid Mech.* **81,** 401 (1977).

Huh, C., and L. E. Scriven, *J. Colloid Interface Sci.* **35**, 85 (1971).

Kafka, F. Y., and E. B. Dussan V., *J. Fluid Mech.* **95**, 539 (1979).

Lester, G. R., *J. Colloid Sci.* **16**, 315 (1961).

Ludviksson, V., and E. N. Lightfoot, *AIChE J.* **17**, 1166 (1971).

Richardson, S., *J. Fluid Mech.* **59**, 707 (1973).

Ruckenstein, E., and C. S. Dunn, *J. Colloid Interface Sci.* **59**, 135 (1977).

Teletzke, G. F., H. T. Davis, and L. E. Scriven, *Chem. Eng. Commun.* **55**, 41 (1987).

Wickham, G. R., and S. D. R. Wilson *J. Colloid Interface Sci.* **51**, 189 (1975).

1.4 Frame

1.4.1 Changes of frame The Chief of the United States Weather Bureau in Milwaukee announces that a tornado was sighted in Chicago at 3 PM (Central Standard Time). In Chicago, Harry tells that he saw a black funnel cloud about two hours ago at approximately 800 North and 2400 West. Both men described the same event with respect to their own particular frame of reference.

The time of some occurrence may be specified only with respect to the time of some other event, the **frame of reference for time**. This might be the time at which a stopwatch was started or an electric circuit was closed. The Chief reported the time at which the tornado was sighted relative to the mean time at which the sun appeared overhead on the Greenwich meridian. Harry gave the time relative to his conversation.

A **frame of reference for position** might be the walls of a laboratory, the fixed stars, or the shell of a space capsule that is following an arbitrary trajectory. When the Chief specified Chicago, he meant the city at 41° north and 87° west measured relative to the equator and the Greenwich meridian. Harry thought in terms of eight blocks north of Madison Avenue and 24 blocks west of State Street. More generally, a frame of reference for position is a set of objects whose mutual distances remain unchanged during the period of observation and which do not all lie in the same plane.

To help you get a better physical feel for these ideas, let us consider two more examples.

Extend your right arm and take as your frame of reference for position the direction of your right arm, the direction of your eyes, and the direction of your spine. Stand out at the street with your eyes fixed straight ahead. A car passes in the direction of your right arm. If you were standing facing the street on the opposite side, the automobile would appear to pass in the opposite direction from your right arm.

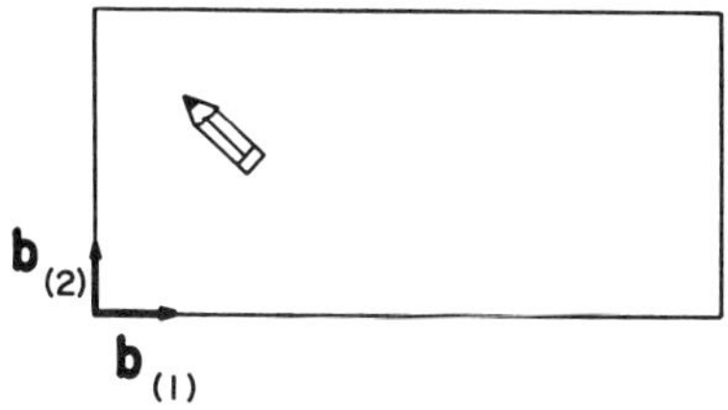

Figure 1.4.1-1: Pencil points away from the direction of $\mathbf{b}_{(1)}$ and towards the direction of $\mathbf{b}_{(2)}$.

Lay a pencil on your desk as shown in Figure 1-1 and take the edges of the desk that meet in the left-hand front corner as your frame of reference for position. The pencil points away from $\mathbf{b}_{(1)}$ and towards $\mathbf{b}_{(2)}$. Without moving the pencil, walk around to the left-hand side and take as your new frame of reference for position the edges of the desk that meet at the left-hand rear corner. The pencil now appears to point towards the intersection of $\mathbf{b}^*_{(1)}$ and $\mathbf{b}^*_{(2)}$ in Figure 1-2.

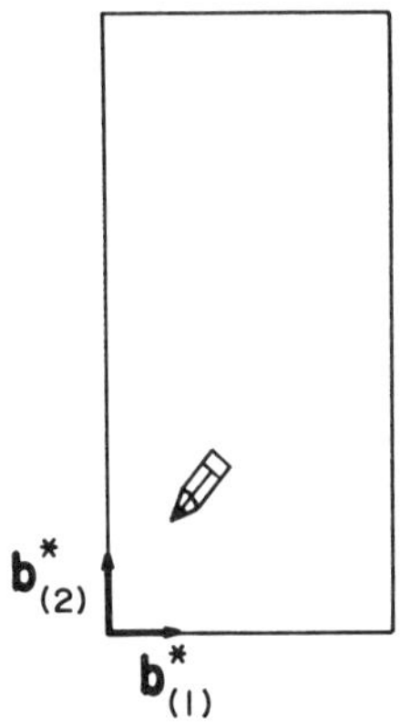

Figure 1.4.1-2: Pencil points towards the intersection of $\mathbf{b}^*_{(1)}$ and $\mathbf{b}^*_{(2)}$.

Since all of the objects defining a frame of reference do not lie in the same plane, we may visualize replacing them by three mutually orthogonal unit vectors. Let (E^3, V^3) be the three-dimensional euclidean space that we use to represent our physical world (see Sec. A.1.1). A frame of reference for position can then be thought of as a primary basis for the vector space V^3 (Slattery 1981, p. 606; Halmos 1958, p. 10). Let us view

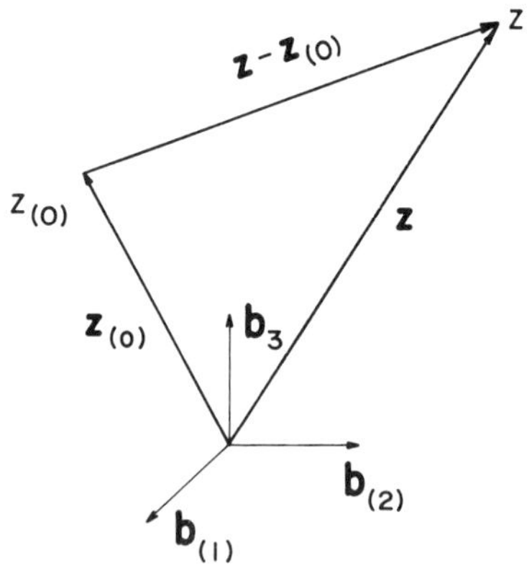

Figure 1.4.1-3: The points z and $z_{(0)}$ are located by the position vectors $\mathbf{z}$ and $\mathbf{z}_{(0)}$ with respect to the frame of reference for position ($\mathbf{b}_{(1)}$, $\mathbf{b}_{(2)}$, $\mathbf{b}_{(3)}$).

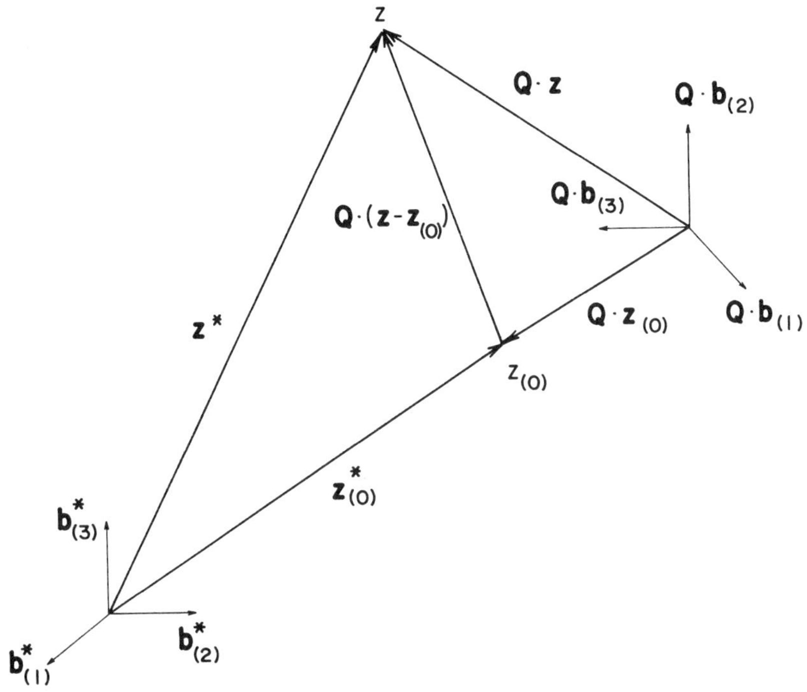

Figure 1.4.1-4: The points z and $z_{(0)}$ are located by the position vectors $\mathbf{z}^*$ and $z^*_{(0)}$ with respect to the frame of reference for position ($\mathbf{b}^*_{(1)}$, $\mathbf{b}^*_{(2)}$, $\mathbf{b}^*_{(3)}$). With respect to the starred frame of reference, the unstarred frame is seen as ($\mathbf{Q} \cdot \mathbf{b}_{(1)}$, $\mathbf{Q} \cdot \mathbf{b}_{(2)}$, $\mathbf{Q} \cdot \mathbf{b}_{(3)}$).

a typical point z in this space with respect to two such frames of reference: the $\mathbf{b}_{(i)}$ (i = 1, 2, 3) in Figure 1-3 and the $\mathbf{b}^*_{(j)}$ (j = 1, 2, 3) in Figure 1-4.

An orthogonal transformation preserves both lengths and angles (Slattery 1981, p. 633). Let **Q** be the orthogonal transformation that describes the rotation and (possibly) reflection that takes the $\mathbf{b}_{(i)}$ in Figure 1-3 into the vectors $\mathbf{Q} \cdot \mathbf{b}_{(i)}$ which are seen in Figure 1-4 with respect to the starred frame of reference for position. A reflection allows for the possibility that an observer in the new frame looks at the old frame through a mirror. Alternatively, a reflection allows for the possibility that two observers orient themselves oppositely, one choosing to work in terms of a right-handed frame of reference for position and the other in terms of a left-handed one. [For more on this point, I suggest that you read Truesdell (1966, p. 22), and Truesdell and Noll (1965, p. 24 and 47).]

The vector ($\mathbf{z} - \mathbf{z}_{(0)}$) in Figure 1-3 becomes $\mathbf{Q} \cdot (\mathbf{z} - \mathbf{z}_{(0)})$ when viewed in the starred frame shown in Figure 1-4. From Figure 1-4, it

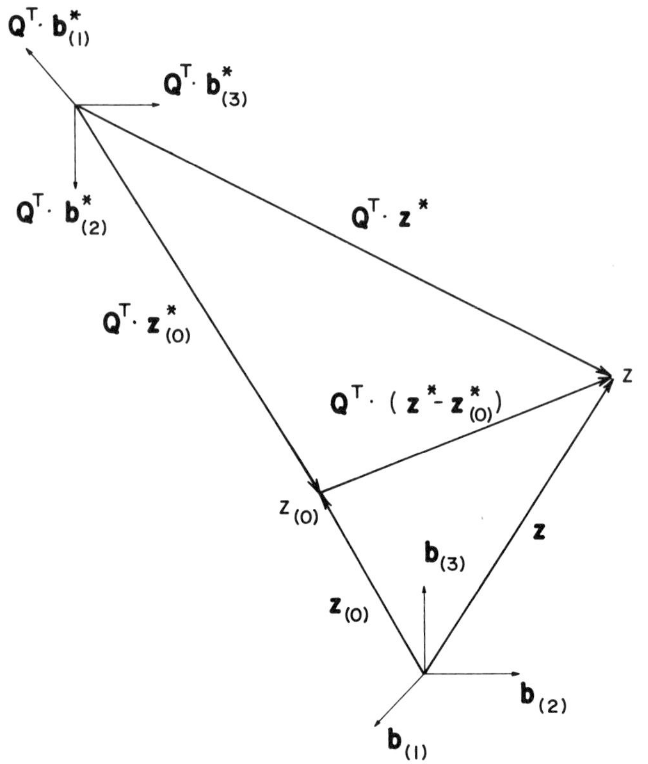

Figure 1.4.1-5: With respect to the unstarred frame of reference, the starred frame is seen as ($\mathbf{Q}^T \cdot \mathbf{b}^*_{(1)}$, $\mathbf{Q}^T \cdot \mathbf{b}^*_{(2)}$, $\mathbf{Q}^T \cdot \mathbf{b}^*_{(3)}$).

follows as well that

$$\mathbf{z}^* - \mathbf{z}^*_{(0)} = \mathbf{Q} \cdot (\mathbf{z} - \mathbf{z}_{(0)}) \tag{1-1}$$

Similarly, the vector ($\mathbf{z}^* - \mathbf{z}^*_{(0)}$) in Figure 1-4 is seen as $\mathbf{Q}^T \cdot (\mathbf{z}^* - \mathbf{z}^*_{(0)})$ when observed with respect to the unstarred frame in Figure 1-5. Figure 1-5 also makes it clear that

$$\mathbf{z} - \mathbf{z}_{(0)} = \mathbf{Q}^T \cdot (\mathbf{z}^* - \mathbf{z}^*_{(0)}) \tag{1-2}$$

Consider the collection of all pairs ($\mathbf{z}$, t) where $\mathbf{z}$ denotes position in euclidean space and t time. The collection of all such pairs may be referred to as space-time. A **change of frame** is a one-to-one mapping of space-time onto itself in such a manner that distances, time intervals, and temporal order remain unchanged.

Let $\mathbf{z}$ and t denote a position and time in the old frame, $\mathbf{z}^*$ and t^* the corresponding position and time in the new frame. We can extend the discussion above to conclude that the most general change of frame is of the form

$$\mathbf{z}^* = \mathbf{z}^*_{(0)}(t) + \mathbf{Q}(t) \cdot (\mathbf{z} - \mathbf{z}_{(0)}) \tag{1-3}$$

$$t^* = t - a \tag{1-4}$$

where we now allow the two frames discussed in Figures 1-3 and 1-4 to rotate and translate with respect to one another as functions of time. The quantity a is a real number. Equivalently, we could also write

$$\mathbf{z} = \mathbf{z}_{(0)} + \mathbf{Q}^T \cdot (\mathbf{z}^* - \mathbf{z}^*_{(0)}) \tag{1-5}$$

$$t = t^* + a \tag{1-6}$$

It is important to carefully distinguish between a frame of reference for position and a coordinate system. Any coordinate system whatsoever can be used to locate points in space with respect to three vectors defining a frame of reference for position and their intersection, although I recommend that admissible coordinate systems be restricted to those whose axes have a time-invariant orientation with respect to the frame. Let (x^1, x^2, x^3) be a curvilinear coordinate system associated with the frame of reference ($\mathbf{b}_{(1)}$, $\mathbf{b}_{(2)}$, $\mathbf{b}_{(3)}$); similarly, let (x^{*1}, x^{*2}, x^{*3}) be a curvilinear coordinate system associated with another frame of reference ($\mathbf{b}^*_{(1)}$, $\mathbf{b}^*_{(2)}$, $\mathbf{b}^*_{(3)}$). We will say that these two coordinate systems are the **same**, if the orientation of the natural basis fields $\mathbf{g}_i$ with respect to the vectors $\mathbf{b}_{(j)}$ is identical with the orientation of the natural basis fields $\mathbf{g}^*_i$ with respect to the vectors $\mathbf{b}^*_{(j)}$:

$$\mathbf{g}_i \cdot \mathbf{b}_{(j)} = \mathbf{g}^*_i \cdot \mathbf{b}^*_{(j)} \quad \text{for all i, j = 1, 2, 3} \tag{1-7}$$

We will generally find it convenient to use the same coordinate system in discussing two different frames of reference.

Let us now use the *same* rectangular cartesian coordinate system to discuss the change of frame illustrated in Figures 1-4 and 1-5. The orthogonal tensor **Q**

$$\mathbf{Q} = Q_{ij}\ \mathbf{e}^*_i\ \mathbf{e}_j \tag{1-8}$$

describes the rotation and (possibly) reflection that transforms the basis vectors $\mathbf{e}_j$ (j = 1, 2, 3) into the vectors

$$\mathbf{Q} \cdot \mathbf{e}_j = Q_{ij}\ \mathbf{e}^*_i \tag{1-9}$$

which are vectors expressed in terms of the starred frame of reference for position. The rectangular cartesian components of **Q** are defined by the angles between the $\mathbf{e}^*_i$ and the $\mathbf{Q} \cdot \mathbf{e}_j$:

$$Q_{ij} = \mathbf{e}_i^* \cdot (\mathbf{Q} \cdot \mathbf{e}_j) \tag{1-10}$$

The vector $(\mathbf{z} - \mathbf{z}_{(0)})$ in Figure 1-3 becomes

$$\mathbf{Q} \cdot (\mathbf{z} - \mathbf{z}_{(0)}) = Q_{ij} (z_j - z_{(0)j}) \mathbf{e}_i^* \tag{1-11}$$

when viewed in the starred frame shown in Figure 1-4. From Figure 1-4, it follows as well that

$$z_i^* \mathbf{e}_i^* = z_{(0)i}^* \mathbf{e}_i^* + Q_{ij} (z_j - z_{(0)j}) \mathbf{e}_i^* \tag{1-12}$$

Of particular interest to us is the form that a surface (see Sec. A.1.3)

$$\mathbf{z} = \mathbf{p}^{(\sigma)}(y^1, y^2) \tag{1-13}$$

takes under a change of frame. From (1-3), we have

$$\mathbf{z}^* = \mathbf{p}^{(\sigma)*}(y^1, y^2) = \mathbf{z}_{(0)}^* + \mathbf{Q} \cdot [\mathbf{p}^{(\sigma)}(y^1, y^2) - \mathbf{z}_{(0)}] \tag{1-14}$$

We see that it will usually be convenient to use the same set of surface coordinates in a new frame of reference.

Equation (1-14) indicates that the natural basis fields in the two frames of reference

$$\mathbf{a}_\alpha^* \equiv \frac{\partial \mathbf{p}^{(\sigma)*}}{\partial y^\alpha} \tag{1-15}$$

$$\mathbf{a}_\alpha \equiv \frac{\partial \mathbf{p}^{(\sigma)}}{\partial y^\alpha} \tag{1-16}$$

are related by

$$\mathbf{a}_\alpha^* = \mathbf{Q} \cdot \mathbf{a}_\alpha \tag{1-17}$$

Let

$$\mathbf{P}^* = \mathbf{a}_\alpha^* \mathbf{a}^{*\alpha} \tag{1-18}$$

and

$$\mathbf{P} = \mathbf{a}_\alpha \mathbf{a}^\alpha \tag{1-19}$$

be the projection tensors for the surface in the starred and unstarred frames of reference. For some purposes, (1-17) can be more conveniently written as

$$\mathbf{a}^*_\alpha = \mathbf{P}^* \cdot \mathbf{a}^*_\alpha$$
$$= \mathbf{P}^* \cdot \mathbf{Q} \cdot \mathbf{a}_\alpha$$
$$= \mathbf{P}^* \cdot \mathbf{Q} \cdot \mathbf{P} \cdot \mathbf{a}_\alpha \tag{1-20}$$

or

$$\mathbf{a}^*_\alpha = \boldsymbol{\mathcal{Q}} \cdot \mathbf{a}_\alpha \tag{1-21}$$

Here

$$\boldsymbol{\mathcal{Q}} \equiv \mathbf{P}^* \cdot \mathbf{Q} \cdot \mathbf{P} \tag{1-22}$$

is an orthogonal tangential transformation.

For more about changes of frame, I suggest that you refer to Truesdell (1966, p. 22), Truesdell and Noll (1965, p. 41), and Truesdell and Toupin (1960, p. 437).

References

Halmos, P. R., "Finite-Dimensional Vector Spaces," second edition, Van Nostrand, Princeton, NJ (1958).

Slattery, J. C., "Momentum, Energy, and Mass Transfer in Continua," McGraw-Hill, New York (1972); second edition, Robert E. Krieger, Malabar, FL 32950 (1981).

Truesdell, C., "The Elements of Continuum Mechanics," Springer-Verlag, New York (1966).

Truesdell, C., and R. A. Toupin, "Handbuch der Physik," vol. 3/1, edited by S. Flügge, Springer-Verlag, Berlin (1960).

Truesdell, C., and W. Noll, "Handbuch der Physik," vol. 3/3, edited by S. Flügge, Springer-Verlag, Berlin (1965).

1.4.2 Frame indifferent scalars, vectors, and tensors We speak of a quantity as being **frame indifferent**, if it remains unchanged or invariant under all changes of frame. From the last section, a superscript * denotes a quantity as observed in a new frame of reference.

A frame-indifferent scalar b does not change its value under a change of frame:

$$b^* = b \tag{2-1}$$

A frame-indifferent spatial vector remains the same directed line element under a change of frame in the sense that, if

$$\mathbf{v} = \mathbf{z}_1 - \mathbf{z}_2 \tag{2-2}$$

then

$$\mathbf{v}^* = \mathbf{z}_1^* - \mathbf{z}_2^* \tag{2-3}$$

From (1-3),

$$\mathbf{v}^* = \mathbf{Q} \cdot (\mathbf{z}_1 - \mathbf{z}_2) = \mathbf{Q} \cdot \mathbf{v} \tag{2-4}$$

A frame-indifferent second-order tensor is one that transforms frame-indifferent spatial vectors into frame-indifferent spatial vectors. If

$$\mathbf{v} = \mathbf{A} \cdot \mathbf{w} \tag{2-5}$$

$$\mathbf{v}^* = \mathbf{Q} \cdot \mathbf{v} \tag{2-6}$$

and

$$\mathbf{w}^* = \mathbf{Q} \cdot \mathbf{w} \tag{2-7}$$

then

$$\mathbf{v}^* = \mathbf{A}^* \cdot \mathbf{w}^* \tag{2-8}$$

This means that

$$\begin{aligned} \mathbf{Q} \cdot \mathbf{v} &= \mathbf{A}^* \cdot \mathbf{Q} \cdot \mathbf{w} \\ &= \mathbf{Q} \cdot \mathbf{A} \cdot \mathbf{w} \end{aligned} \tag{2-9}$$

or

$$\mathbf{A}^* = \mathbf{Q} \cdot \mathbf{A} \cdot \mathbf{Q}^T \tag{2-10}$$

Here $\mathbf{Q}^T$ is the transpose of the orthogonal tensor $\mathbf{Q}$ (Slattery 1981, p. 632).

Every tangential vector field **c** can be expressed as a linear combination of the natural basis vector fields for the surface coordinate system:

$$\mathbf{c} = c^{\alpha}\ \mathbf{a}_{\alpha} \tag{2-11}$$

In Sec. 1.4.1, we learned a natural basis vector field $\mathbf{a}_{\alpha}$ is transformed to

$$\mathbf{a}_{\alpha}^{*} = \mathbf{Q} \cdot \mathbf{a}_{\alpha} = \mathcal{Q} \cdot \mathbf{a}_{\alpha} \tag{2-12}$$

under a change of frame. A frame-indifferent tangential vector field must obey (2-4), like every spatial vector field. But in view of (2-12), it must also satisfy

$$\begin{aligned} \mathbf{c}^{*} = \mathbf{Q} \cdot \mathbf{c} &= c^{\alpha}\ \mathbf{Q} \cdot \mathbf{a}_{\alpha} \\ &= c^{\alpha}\ \mathcal{Q} \cdot \mathbf{a}_{\alpha} \\ &= c^{\alpha}\ \mathbf{a}_{\alpha}^{*} \end{aligned} \tag{2-13}$$

$$\mathbf{c}^{*} \equiv c^{\alpha}\ \mathbf{a}_{\alpha}^{*} = \mathcal{Q} \cdot \mathbf{c} \tag{2-14}$$

Using essentially the same argument that was employed in reaching (2-10), we find that a frame-indifferent tangential tensor field $\mathbf{T}$ becomes

$$\mathbf{T}^{*} = \mathbf{Q} \cdot \mathbf{T} \cdot \mathbf{Q}^{T} \tag{2-15}$$

or

$$\mathbf{T}^{*} = \mathcal{Q} \cdot \mathbf{T} \cdot \mathcal{Q}^{T} \tag{2-16}$$

under a change of frame.

Reference

Slattery, J. C., "Momentum, Energy, and Mass Transfer in Continua," McGraw-Hill, New York (1972); second edition, Robert E. Krieger, Malabar, FL 32950 (1981).

1.4.3 Equivalent motions We described the motion within a single phase in Sec. 1.1.1:

$$\mathbf{z} = \boldsymbol{\chi}_{\kappa}(\ \mathbf{z}_{\kappa},\ t\ ) \tag{3-1}$$

In Sec. 1.2.5 we represented the motion of a dividing surface by

$$\mathbf{z} = \boldsymbol{\chi}_{\kappa}^{(\sigma)}(y_{\kappa}^{1}, y_{\kappa}^{2}, t) \tag{3-2}$$

The forms of these relationships are understood to depend upon the reference configuration $\boldsymbol{\kappa}$ and upon the frame of reference. According to our discussion in Sec. 1.4.1, the same motions with respect to some new frame of reference (but the same reference configuration) take the forms

$$\mathbf{z}^{*} = \boldsymbol{\chi}_{\kappa}^{*}(\mathbf{z}_{\kappa}^{*}, t^{*}) = \mathbf{z}_{(0)}^{*}(t^{*}) + \mathbf{Q}(t) \cdot [\boldsymbol{\chi}_{\kappa} (\mathbf{z}_{\kappa}, t) - \mathbf{z}_{(0)}] \tag{3-3}$$

and

$$\begin{aligned} \mathbf{z}^{*} &= \boldsymbol{\chi}_{\kappa}^{*(\sigma)}(y_{\kappa}^{1}, y_{\kappa}^{2}, t^{*}) \\ &= \mathbf{z}_{(0)}^{*}(t^{*}) + \mathbf{Q}(t) \cdot [\boldsymbol{\chi}_{\kappa}^{(\sigma)} (y_{\kappa}^{1}, y_{\kappa}^{2}, t) - \mathbf{z}_{(0)}] \end{aligned} \tag{3-4}$$

We will say that any two motions $\boldsymbol{\chi}_{\kappa}$, $\boldsymbol{\chi}_{\kappa}^{(\sigma)}$ and $\boldsymbol{\chi}_{\kappa}^{*}$, $\boldsymbol{\chi}_{\kappa}^{*(\sigma)}$ that are related in this way are **equivalent motions** (Truesdell and Noll 1965, p. 42; Truesdell 1966, p. 22; Slattery 1981, p. 15).

Let us write (3-4) in an abbreviated form:

$$\mathbf{z}^{*} = \mathbf{z}_{(0)}^{*} + \mathbf{Q} \cdot (\mathbf{z} - \mathbf{z}_{(0)}) \tag{3-5}$$

The surface derivative of this equation gives

$$\mathbf{v}^{(\sigma)*} = \dot{\overline{\mathbf{z}_{(0)}^{*}}} + \dot{\mathbf{Q}} \cdot (\mathbf{z} - \mathbf{z}_{(0)}) + \mathbf{Q} \cdot \mathbf{v}^{(\sigma)} \tag{3-6}$$

where a dot over a letter indicates a time derivative. We have introduced here the surface velocity vectors in the new and old frames respectively as

$$\mathbf{v}^{(\sigma)*} \equiv \frac{d_{(s)}\mathbf{z}^{*}}{dt^{*}} = \frac{d_{(s)}\mathbf{z}^{*}}{dt} \tag{3-7}$$

and

$$\mathbf{v}^{(\sigma)} \equiv \frac{d_{(s)}\mathbf{z}}{dt} \tag{3-8}$$

From (3-5), we may write

$$\mathbf{z} - \mathbf{z}_{(0)} = \mathbf{Q}^{T} \cdot (\mathbf{z}^{*} - \mathbf{z}_{(0)}^{*}) \tag{3-9}$$

which allows us to express (3-6) as

$$\mathbf{v}^{(\sigma)*} - \dot{\overline{\mathbf{z}_{(0)}^{*}}} - \mathbf{Q} \cdot \mathbf{v}^{(\sigma)} = \dot{\mathbf{Q}} \cdot \mathbf{Q}^{T} \cdot (\mathbf{z}^{*} - \mathbf{z}_{(0)}^{*})$$

$$= \mathbf{A} \cdot (\mathbf{z}^* - \mathbf{z}^*_{(0)}) \tag{3-10}$$

Here

$$\mathbf{A} \equiv \dot{\mathbf{Q}} \cdot \mathbf{Q}^T \tag{3-11}$$

is the **angular velocity tensor of the starred frame with respect to the unstarred frame** (Truesdell 1966, p. 24).

Taking the surface derivative of (3-9), we find

$$\begin{aligned} \mathbf{v}^{(\sigma)} &= \overline{\dot{\mathbf{Q}^T}} \cdot (\mathbf{z}^* - \mathbf{z}^*_{(0)}) + \mathbf{Q}^T \cdot \left(\mathbf{v}^{(\sigma)*} - \overline{\dot{\mathbf{z}^*_{(0)}}} \right) \\ &= \dot{\mathbf{Q}}^T \cdot (\mathbf{z}^* - \mathbf{z}^*_{(0)}) + \mathbf{Q}^T \cdot \left(\mathbf{v}^{(\sigma)*} - \overline{\dot{\mathbf{z}^*_{(0)}}} \right) \end{aligned} \tag{3-12}$$

or

$$\begin{aligned} \mathbf{Q} \cdot \mathbf{v}^{(\sigma)} - \left(\mathbf{v}^{(\sigma)*} - \overline{\dot{\mathbf{z}^*_{(0)}}} \right) &= \mathbf{Q} \cdot \dot{\mathbf{Q}}^T \cdot (\mathbf{z}^* - \mathbf{z}^*_{(0)}) \\ &= \mathbf{A}^T \cdot (\mathbf{z}^* - \mathbf{z}^*_{(0)}) \end{aligned} \tag{3-13}$$

From this expression, we can identify $\mathbf{A}^T$ as the **angular velocity tensor of the unstarred frame with respect to the starred frame.**

Since $\mathbf{Q}$ is an orthogonal tensor (Slattery 1981, p. 632),

$$\mathbf{Q} \cdot \mathbf{Q}^T = \mathbf{I}^* \tag{3-14}$$

Differentiating this equation with respect to time, we have

$$\begin{aligned} \mathbf{A} &= \dot{\mathbf{Q}} \cdot \mathbf{Q}^T \\ &= - \mathbf{Q} \cdot \overline{\dot{\mathbf{Q}^T}} \\ &= - \mathbf{Q} \cdot \dot{\mathbf{Q}}^T \\ &= - \mathbf{A}^T \end{aligned} \tag{3-15}$$

In this way we see that the angular velocity tensor is skew symmetric.

The **angular velocity vector of the unstarred frame with respect to the starred frame** $\boldsymbol{\omega}$ is defined as (see Sec. A.1.2)

$$\boldsymbol{\omega} \equiv - \frac{1}{2} \boldsymbol{\varepsilon}^* : \mathbf{A}^T = \frac{1}{2} \boldsymbol{\varepsilon}^* : \mathbf{A} \tag{3-16}$$

Since

$$\boldsymbol{\omega} \wedge (\mathbf{z}^* - \mathbf{z}^*_{(0)}) = \boldsymbol{\varepsilon}^* : [(\mathbf{z}^* - \mathbf{z}^*_{(0)}) \boldsymbol{\omega}]$$

$$= \boldsymbol{\varepsilon}^* : \left[(\mathbf{z}^* - \mathbf{z}^*_{(0)}) \left(\frac{1}{2} \boldsymbol{\varepsilon}^* : \mathbf{A} \right) \right]$$

$$= e_{ijk} \, (z^*_k - z^*_{(0)k}) \left(\frac{1}{2} e_{jmn} \, A_{nm} \right) \mathbf{e}^*_i$$

$$= \frac{1}{2} (z^*_k - z^*_{(0)k})(A_{ik} - A_{ki}) \, \mathbf{e}^*_i$$

$$= (z^*_k - z^*_{(0)k}) \, A_{ik} \, \mathbf{e}^*_i$$

$$= \mathbf{A} \cdot (\mathbf{z}^* - \mathbf{z}^*_{(0)}) \tag{3-17}$$

we can write (3-10) as

$$\mathbf{v}^{(\sigma)} - \mathbf{Q} \cdot \mathbf{v}^{(\sigma)} = \overline{\dot{\mathbf{z}}^*_{(0)}} + \boldsymbol{\omega} \wedge (\mathbf{z}^* - \mathbf{z}^*_{(0)}) \tag{3-18}$$

The material derivative of (3-3) yields

$$\mathbf{v}^* = \overline{\dot{\mathbf{z}}^*_{(0)}} + \dot{\mathbf{Q}} \cdot (\mathbf{z} - \mathbf{z}_{(0)}) + \mathbf{Q} \cdot \mathbf{v} \tag{3-19}$$

The argument that we used to express (3-6) in the form of (3-18) may also be employed to write (3-19) in the more familiar form (Truesdell and Toupin 1960, p. 437)

$$\mathbf{v}^* - \mathbf{Q} \cdot \mathbf{v} = \overline{\dot{\mathbf{z}}^*_{(0)}} + \boldsymbol{\omega} \wedge (\mathbf{z}^* - \mathbf{z}^*_{(0)}) \tag{3-20}$$

Equations (3-18) and (3-20) clearly tell us that velocity is not a frame indifferent vector. In Exercise 1.4.3-1, we conclude that acceleration is another example of a vector that is not frame indifferent.

References

Slattery, J. C., "Momentum, Energy, and Mass Transfer in Continua," McGraw-Hill, New York (1972); second edition, Robert E. Krieger, Malabar, FL 32950 (1981).

Truesdell, C., "The Elements of Continuum Mechanics," Springer-Verlag, New York (1966).

Truesdell, C., and W. Noll, "Handbuch der Physik," vol. 3/3, edited by S. Flügge, Springer-Verlag, Berlin (1965).

Truesdell, C., and R. A. Toupin, "Handbuch der Physik," vol. 3/1, edited by S. Flügge, Springer-Verlag, Berlin (1960).

Exercises

1.4.3-1 *acceleration*

i) Determine that (Truesdell 1966, p. 24)

$$\frac{d_{(m)}\mathbf{v}^*}{dt} - \overset{\,\bullet\bullet}{\overline{\mathbf{z}^*_{(0)}}} - 2\mathbf{A} \cdot \left(\mathbf{v}^* - \overset{\bullet}{\overline{\mathbf{z}^*_{(0)}}} \right) - (\dot{\mathbf{A}} - \mathbf{A} \cdot \mathbf{A}) \cdot (\mathbf{z}^* - \mathbf{z}^*_{(0)})$$

$$= \mathbf{Q} \cdot \frac{d_{(m)}\mathbf{v}}{dt}$$

ii) Prove that (Truesdell and Toupin 1960, p. 440)

$$\frac{d_{(m)}\mathbf{v}^*}{dt} - \overset{\,\bullet\bullet}{\overline{\mathbf{z}^*_{(0)}}} - 2\mathbf{A} \cdot \mathbf{Q} \cdot \mathbf{v} - (\dot{\mathbf{A}} + \mathbf{A} \cdot \mathbf{A}) \cdot (\mathbf{z}^* - \mathbf{z}^*_{(0)})$$

$$= \mathbf{Q} \cdot \frac{d_{(m)}\mathbf{v}}{dt}$$

iii) Find that

$$\boldsymbol{\omega} \wedge [\boldsymbol{\omega} \wedge (\mathbf{z}^* - \mathbf{z}^*_{(0)})] = \mathbf{A} \cdot \mathbf{A} \cdot (\mathbf{z}^* - \mathbf{z}^*_{(0)})$$

$$\dot{\boldsymbol{\omega}} \wedge (\mathbf{z}^* - \mathbf{z}^*_{(0)}) = \dot{\mathbf{A}} \cdot (\mathbf{z}^* - \mathbf{z}^*_{(0)})$$

and

$$\boldsymbol{\omega} \wedge (\mathbf{Q} \cdot \mathbf{v}) = \mathbf{A} \cdot \mathbf{Q} \cdot \mathbf{v}$$

iv) Conclude that (Truesdell and Toupin 1960, p. 438)

$$\frac{d_{(m)}\mathbf{v}^*}{dt} - \overset{\,\bullet\bullet}{\overline{\mathbf{z}^*_{(0)}}} - \dot{\boldsymbol{\omega}} \wedge (\mathbf{z}^* - \mathbf{z}^*_{(0)}) - \boldsymbol{\omega} \wedge [\boldsymbol{\omega} \wedge (\mathbf{z}^* - \mathbf{z}^*_{(0)})]$$

$$- 2\, \boldsymbol{\omega} \wedge (\mathbf{Q} \cdot \mathbf{v}) = \mathbf{Q} \cdot \frac{d_{(m)}\mathbf{v}}{dt}$$

v) Supply a similar argument to show

$$\frac{d_{(s)}\mathbf{v}^{(\sigma)*}}{dt} - \overline{\ddot{\mathbf{z}}^{*}_{(0)}} - \dot{\boldsymbol{\omega}} \wedge (\mathbf{z}^{*} - \mathbf{z}^{*}_{(0)}) - \boldsymbol{\omega} \wedge [\boldsymbol{\omega} \wedge (\mathbf{z}^{*} - \mathbf{z}^{*}_{(0)})]$$

$$- 2\, \boldsymbol{\omega} \wedge (\mathbf{Q} \cdot \mathbf{v}^{(\sigma)}) = \mathbf{Q} \cdot \frac{d_{(s)}\mathbf{v}^{(\sigma)}}{dt}$$

1.4.3-2 Give an example of a scalar that is *not* frame indifferent. Hint: What vector is not frame indifferent?

1.4.3-3 *motion of a rigid body* Determine that the velocity distribution in a rigid body may be expressed as

$$\mathbf{v}^{*} = \mathbf{v}^{(\sigma)*} = \overline{\dot{\mathbf{z}}^{*}_{(0)}} + \boldsymbol{\omega} \wedge (\mathbf{z}^{*} - \mathbf{z}^{*}_{(0)})$$

What is the relation of the unstarred frame to the body in this case?

1.4.3-4 *deformation gradient, surface rate of deformation tensor, and surface vorticity tensor*

i) Let the motions $\chi_{\kappa}^{(\sigma)*}$ and $\chi_{\kappa}^{(\sigma)}$ be referred to the same intrinsic reference configuration. Determine that

$$\mathcal{F}^{*} = \mathbf{Q} \cdot \mathcal{F}$$

where $\mathcal{F}$ is the deformation gradient of the dividing surface (Sec. 1.2.5).

ii) Show that

$$\frac{d_{(s)}\mathcal{F}}{dt} = \nabla_{(\sigma)}\mathbf{v}^{(\sigma)} \cdot \mathcal{F}$$

iii) Take the surface material derivative of the result in (i) to find

$$\nabla_{(\sigma)}\mathbf{v}^{(\sigma)} \cdot \mathcal{F}^{*} = \mathbf{Q} \cdot \nabla_{(\sigma)}\mathbf{v}^{(\sigma)} \cdot \mathbf{Q}^{T} \cdot \mathcal{F}^{*} + \dot{\mathbf{Q}} \cdot \mathbf{Q}^{T} \cdot \mathcal{F}^{*}$$

iv) Note that $\mathcal{F}$ is not a singular tensor; it does not transform every vector into the zero vector. Conclude that

$$\mathbf{P}^* \cdot \nabla_{(\sigma)}\mathbf{v}^{(\sigma)*} = \mathcal{Q} \cdot \mathbf{P} \cdot \nabla_{(\sigma)}\mathbf{v}^{(\sigma)} \cdot \mathcal{Q}^T + \mathbf{P}^* \cdot \dot{\mathcal{Q}} \cdot \mathcal{Q}^T$$

v) Starting with

$$\mathcal{Q} \cdot \mathcal{Q}^T = \mathbf{P}^*$$

prove that

$$\mathbf{P}^* \cdot \dot{\mathcal{Q}} \cdot \mathcal{Q}^T = - \mathcal{Q} \cdot \dot{\mathcal{Q}}^T \cdot \mathbf{P}^* + \mathbf{P}^* \cdot \dot{\mathbf{P}}^* \cdot \mathbf{P}^*$$

vi) Use Exercise A.5.6-2 to show that

$$\mathbf{P}^* \cdot \dot{\mathbf{P}}^* \cdot \mathbf{P}^* = 0$$

and that therefore $\mathbf{P}^* \cdot \dot{\mathcal{Q}} \cdot \mathcal{Q}^T$ is skew-symmetric.

vii) The decomposition of a second-order tensor into symmetric and skew-symmetric portions is unique. Observe this implies that the surface rate of deformation tensor introduced in Sec. 1.2.8 is frame indifferent

$$\mathbf{D}^{(\sigma)*} = \mathcal{Q} \cdot \mathbf{D}^{(\sigma)} \cdot \mathcal{Q}^T$$

and that the **surface vorticity tensor**

$$\mathbf{W}^{(\sigma)} \equiv \frac{1}{2} [\mathbf{P} \cdot \nabla_{(\sigma)}\mathbf{v}^{(\sigma)} - (\nabla_{(\sigma)}\mathbf{v}^{(\sigma)})^T \cdot \mathbf{P}]$$

is not frame indifferent:

$$\mathbf{W}^{(\sigma)*} = \mathcal{Q} \cdot \mathbf{W}^{(\sigma)} \cdot \mathcal{Q}^T + \mathbf{P}^* \cdot \dot{\mathcal{Q}} \cdot \mathcal{Q}^T$$

1.4.3-5 *projection tensor* Prove that the projection tensor is frame indifferent:

$$\mathbf{P}^* = \mathcal{Q} \cdot \mathbf{P} \cdot \mathcal{Q}^T$$

1.4.3-6 *velocity difference* Show that at any position in space a difference in velocities measured with respect to the same frame is frame indifferent.

1.4.4 Principle of frame indifference A balloonist lays in his basket and observes the volume of his balloon as a function of temperature and pressure as he ascends through the atmosphere. His instruments transmit the local temperature and pressure to his friend on the ground, who is also able to measure the volume of the balloon as a function of time. Both of these men find the same relationship between volume, temperature, and pressure for the gas in the balloon.

A series of weights are successively added to one end of a spring, the other end of which is attached to the center of a horizontal turntable that rotates with a constant angular velocity. Two experimentalists observe the extension of the spring as the weights are added. One stands at the center of the turntable and rotates with it; his frame of reference is the axis of the turntable and a series of lines painted upon it. The second man stands next to the turntable; the walls of the laboratory form his frame of reference. Both of these observers come to the same conclusion regarding the behavior of the spring under stress.

We can summarize our feelings about these two experiments with the **principle of frame indifference** (Truesdell 1966, p. 97):

All physical laws, definitions, and descriptions of material behavior that hold in a dynamic process are the same for every observer, i.e. in every frame of reference.

It is easy to see that the postulate of conservation of mass, introduced in Section 1.3.1, satisfies this principle. Because we associated it with the material particles, both mass and its time derivative are automatically frame indifferent scalars.

I would like to emphasize that the principle of frame-indifference says nothing about two experiments conducted by one observer. For example, physical phenomena that are not observed in an experiment which rotates with an electric field may be detected when the same experiment rotates relative to the electric field (Lertes 1921; Grossetti 1958, 1959).

Noll (1958) gave a clear statement of the principle of frame-indifference as we now know it. But other writers before him had the same idea. Oldroyd (1950) in particular attracted considerable interest with his viewpoint. Truesdell and Noll (1965, p. 45) have gone back to the seventeenth century to trace the development of this idea through the literature of classical mechanics. Einstein's (1950, p. 61) principle of relativity makes the same statement for motions whose velocity is small compared to that of light.

References

Einstein, Albert, "The Meaning of Relativity," third edition, Princeton University Press, Princeton, NJ (1950).

Grossetti, E., *Nuovo Cimento* **10**, 193 (1958); *ibid.* **13**, 350 (1959).

Lertes, P., *Z. Phys.* **4**, 315 (1921); *ibid.* **6**, 56 (1921); *Phys. Z.* **22**, 621 (1921).

Noll, W., *Arch. Ration. Mech. Anal.* **2**, 197 (1958).

Oldroyd, J. G., *Proc. R. Soc. London A* **200**, 523 (1950).

Truesdell, C., and W. Noll, "Handbuch der Physik," vol. 3/3, edited by S. Flügge, Springer-Verlag, Berlin (1965).

Truesdell, C., "Six Lectures on Modern Natural Philosophy," Springer-Verlag, New York (1966).

Exercise

1.4.4-1 *conservation of mass* Although I did not raise this point in Sec. 1.3.1, mass is a primitive concept; it is not defined. The property of a body that we refer to as mass is described by a set of axioms. We began in Sec. 1.3.1 by saying

1) The mass of a body is independent of time.

There appears to be no motivation for dealing with negative masses, at least in classical mechanics.

2) Every body has a mass greater than (or equal to) zero.

The mass of a body should have nothing to do with the motion of the observer or experimentalist relative to the body.

3) The mass of a body should be frame indifferent:

$$\mathcal{M}^* = \mathcal{M}$$

From this last it follows immediately that axiom 1, conservation of mass, obeys the principle of frame indifference:

$$\frac{d\mathcal{M}^*}{dt^*} = \frac{d\mathcal{M}^*}{dt}$$

$$= \frac{d\mathcal{M}}{dt}$$

$$= 0$$

Although this implies that the equation of continuity and the jump mass balance also obey the principle of frame indifference, it is a worthwhile exercise to take the direct approach in the proof.

i) Starting with

$$\frac{d_{(m)}\rho}{dt} + \rho \operatorname{div} \mathbf{v} = 0$$

prove that

$$\frac{d_{(m)}\rho^*}{dt^*} + \rho^* \operatorname{div} \mathbf{v}^* = 0$$

ii) Begin with

$$\frac{d_{(s)}\rho^{(\sigma)}}{dt} + \rho^{(\sigma)} \operatorname{div}_{(\sigma)}\mathbf{v}^{(\sigma)} + [\rho(\mathbf{v} \cdot \boldsymbol{\xi} - v_{(\xi)})] = 0$$

to prove that

$$\frac{d_{(s)}\rho^{(\sigma)*}}{dt^*} + \rho^{(\sigma)*} \operatorname{div}_{(\sigma)}\mathbf{v}^{(\sigma*)} + [\rho^*(\mathbf{v}^* \cdot \boldsymbol{\xi}^* - v_{(\xi)}^*)] = 0$$

Notation for chapter 1

a	determinant, defined in Sec. A.2.1
$\mathbf{a}_\alpha$	natural basis vectors, introduced in Sec. A.2.1
$\mathbf{a}_{\kappa A}$	natural basis vectors used in reference configuration, introduced in Sec. A.2.1
$\mathbf{a}^\alpha$	dual basis vectors, introduced in Sec. A.2.3
A	angular velocity tensor of the starred frame with respect to the unstarred frame in Sec. 1.4.3
B	left Cauchy-Green tensor defined in Sec. 1.1.2. More commonly this denotes the second groundform tangential tensor field, introduced in Sec. A.5.3.
$\mathbf{B}_t$	left relative Cauchy-Green tensor, defined in Sec. 1.1.2
$\mathbf{B}^{(\sigma)}$	left surface Cauchy-Green tensor, defined in Sec. 1.2.6
$\mathbf{B}_t^{(\sigma)}$	left relative surface Cauchy-Green tensor, defined in Sec. 1.2.6
C	right Cauchy-Green tensor, defined in Sec. 1.1.2
$\mathbf{C}_t$	right relative Cauchy-Green tensor, defined in Sec. 1.1.2
$C^{(cl)}$	common line
$\mathbf{C}^{(\sigma)}$	right surface Cauchy-Green tensor, defined in Sec. 1.2.6
$\mathbf{C}_t^{(\sigma)}$	right relative surface Cauchy-Green tensor, defined in Sec. 1.2.6
dA	denotes that an area integration is to be performed
dm	denotes that an integration with respect to the measure of mass is to be performed
ds	denotes that a line integration with respect to arc length is to be performed
dV	denotes a volume integration is to be performed
D	rate of deformation tensor (Slattery 1981, p. 28)

$\mathbf{D}^{(\sigma)}$	surface rate of deformation tensor, introduced in Sec. 1.2.8
$\mathbf{e}_j$	basis vectors for rectangular cartesian coordinate system (Slattery 1981, pp. 607 and 614)
$\mathbf{e}_{\kappa j}$	basis vectors for rectangular cartesian coordinate system in terms of which the reference configuration of the body is described
$\mathbf{F}$	deformation gradient, defined in Sec. 1.1.1
$\mathbf{F}_t$	relative deformation gradient, defined in Sec. 1.1.1
$\boldsymbol{\mathcal{F}}$	surface deformation gradient, defined in Sec. 1.2.5
$\boldsymbol{\mathcal{F}}_t$	relative surface deformation gradient, defined in Sec. 1.2.5
H	mean curvature of dividing surface, introduced in Exercise A.5.3-3
$\mathbf{I}$	identity tensor that transforms every vector into itself
$J^{(\sigma)}$	defined in Sec. 1.3.3
K	total curvature of dividing surface, introduced in Exercise A.5.3-3
$\mathcal{M}$	mass of body
$\mathbf{n}$	unit normal outwardly directed with respect to a closed surface
$\mathbf{p}(\mathbf{z})$	position vector field, introduced in Sec. A.1.2
$\mathbf{p}^{(\sigma)}$	position vector field on Σ, introduced in Sec. 1.2.5
$\mathbf{P}$	projection tensor, defined in Sec. A.3.2
$\mathbf{P}_\kappa$	projection tensor for tangential vector fields on reference dividing surface
$\mathbf{Q}$	orthogonal transformation that describes the rotation and and (possibly) reflection that takes one frame into another in Sec. 1.4.1
$\boldsymbol{\mathcal{Q}}$	defined in Sec. 1.4.1
R	region occupied by body

$\mathbf{R}$	rotation tensor introduced in Sec. 1.1.2.
$\mathbf{R}_t$	relative rotation tensor, introduced in Sec. 1.1.2
$\mathfrak{R}$	surface rotation tensor, introduced in Sec. 1.2.6
$\mathfrak{R}_t$	relative surface rotation tensor, introduced in Sec. 1.2.6
$s_{(1)}, s_{(2)}$	arc lengths measured along curves 1 and 2 in the present configuration
$s_{\kappa(1)}, s_{\kappa(2)}$	arc lengths measured along curves 1 and 2 in the reference configuration
t	time
t_κ	time at which the reference configuration is defined
$\mathbf{u}$	time rate of change of spatial position following a surface point, introduced in Sec. 1.2.7
$\mathbf{u}^{(cl)}_{(v)}$	speed of displacement of common line, defined in Sec. 1.3.7
$\mathbf{U}$	right stretch tensor, introduced in Sec. 1.1.2
$\mathbf{U}_t$	right relative stretch tensor, introduced in Sec. 1.1.2
$\mathbf{U}^{(\sigma)}$	right surface stretch tensor, introduced in Sec. 1.2.6
$\mathbf{U}^{(\sigma)}_t$	right relative surface stretch tensor, introduced in Sec. 1.2.6
$\mathbf{v}$	velocity vector, defined in Sec. 1.1.1
$\mathbf{v}^{(\sigma)}$	surface velocity vector, defined in Sec. 1.2.5
$v^{(g)}_{(\xi)}$	speed of displacement, introduced in Sec. 1.2.7
$\mathbf{V}$	left stretch tensor, introduced in Sec. 1.1.2
$\mathbf{V}_t$	left relative stretch tensor, introduced in Sec. 1.1.2
$\mathbf{V}^{(\sigma)}$	left surface stretch tensor, introduced in Sec. 1.2.6
$\mathbf{V}^{(\sigma)}_t$	left relative surface stretch tensor, introduced in Sec. 1.2.6
x^i	curvilinear coordinates, introduced in Sec. A.1.2
$X^\alpha(\zeta^{(\sigma)})$	intrinsic configuration of Σ, introduced in Sec. 1.2.5

$X^{\alpha}(\zeta^{(\sigma)}, t)$	intrinsic motion of Σ, introduced in Sec. 1.2.5
$X^{\alpha}_{\kappa}(y^1_{\kappa}, y^2_{\kappa}, t)$	intrinsic deformation of Σ from reference configuration, introduced in Sec. 1.2.5
$X^{\alpha}_{t}(y^1, y^2, \overline{t})$	relative intrinsic deformation, introduced in Sec. 1.2.5
y^{α}	surface coordinates
y^{A}_{κ}	surface coordinates used in reference configuration
$\dot{\mathbf{y}}$	intrinsic surface velocity, defined in Sec. 1.2.7
$\mathbf{z}$	position vector
$\mathbf{z}_{\kappa}$	position vector denoting a place in the reference configuration
$\mathbf{z}_{(0)}$	reference position introduced in Sec. 1.4.1
z_j	rectangular cartesian coordinates
$z_{\kappa j}$	rectangular cartesian coordinates in terms of which the reference configuration of the body is defined

greek letters

δ_{mn}	kronecker delta
ζ	material particle, introduced in Sec. 1.1.1
$\zeta^{(\sigma)}$	surface particle, introduced in Sec. 1.2.5
Θ	contact angle
$\boldsymbol{\kappa}(\zeta)$	reference configuration of body, introduced in Sec. 1.1.1
$\boldsymbol{\kappa}^{(\sigma)}(y^1_{\kappa}, y^2_{\kappa})$	reference dividing surface, introduced in Sec. 1.2.5
$\boldsymbol{\kappa}^{(\sigma)}_{\kappa}(\zeta^{(\sigma)})$	reference configuration of Σ, introduced in Sec. 1.2.5
κ_1, κ_2	principal curvatures of dividing surface, introduced in Exercise A.5.3-3
λ	introduced in Sec. 1.3.2
λ^{+}, λ^{-}	introduced in Sec. 1.3.2

$\boldsymbol{\mu}$	unit vector tangent to dividing surface, normal to closed bounding curve, and outwardly directed with respect to this curve
$\mathbf{v}$	unit vector normal to common line and tangent to dividing surface
$\boldsymbol{\xi}$	unit normal to Σ, introduced in Sec. 1.2.7
ρ	mass density
$\rho^{(\sigma)}$	surface mass density, introduced in Sec. 1.3.1
Σ	dividing surface
$\boldsymbol{\chi}(\zeta)$	configuration of material body, introduced in Sec. 1.1.1
$\boldsymbol{\chi}(\zeta, t)$	motion of material body, introduced in Sec. 1.1.1
$\boldsymbol{\chi}_{\kappa}(z_{\kappa}, t)$	motion of material body, described in terms of the location of material particles in the reference configuration in Sec. 1.1.1
$\boldsymbol{\chi}_{\kappa}^{(\sigma)}(y_{\kappa}^{1}, y_{\kappa}^{2}, t)$	deformation of Σ, introduced in Sec. 1.2.5
$\boldsymbol{\omega}$	angular velocity vector of the unstarred frame with respect to the starred frame in Sec. 1.4.3

others

det	determinant operator
$\det_{(\sigma)}$	surface determinant operator, defined in Sec. A.3.7
div	divergence operator
$\text{div}_{(\sigma)}$	surface divergence operator, introduced in Secs. A.5.1 through A.5.6
grad	gradient with respect to the reference configuration
tr	trace operator
$\ldots_{,\alpha}$	surface covariant derivative, introduced in Sec. A.5.3
$\ldots^{(I)}$	denotes three-dimensional interfacial region, introduced in Sec. 1.3.2

$...^T$	transpose
$...^{-1}$	inverse
$...^*$	denotes an alternative frame of reference
$\dot{\overline{...}}$	an alternative expression for the material derivative
$\ddot{\overline{...}}$	an alternative expression for a second material derivative
(...)	jump for common line, defined in Sec. 1.3.7
[...]	jump for dividing surface, defined in Sec. 1.3.4
∇	gradient with respect to the current configuration
$\nabla_{(\sigma)}$	surface gradient operation, introduced in Secs. A.5.1 through A.5.6
$\frac{d_{(m)}}{dt}$	material derivative, defined in Sec. 1.1.1
$\frac{d_{(s)}}{dt}$	surface material derivative, defined in Sec. 1.2.5; see also Sec. 1.2.7

Reference

Slattery, J. C., "Momentum, Energy, and Mass Transfer in Continua," McGraw-Hill, NY (1972); second edition, Robert E. Krieger, Malabar, FL 32950 (1981).

2
Foundations for momentum transfer

We begin this chapter by looking at the implications of Euler's first and second laws at phase interfaces. Our emphasis is upon the forces that appear to act within a dividing surface. We have long been familiar with interfacial forces in static problems, where we speak of surface tension. Here we study the forms that these forces may take in dynamic problems.

We conclude our discussion with an outline of what must be said about real material behavior, if we are to account properly for momentum transfer at phase interfaces under conditions such that interfacial forces play an important role. Unfortunately, statistical mechanics has not as yet been used to relate the interfacial stress tensor either to the deformation or to the rate of deformation of the dividing surface. Within the bounds of continuum mechanics, we can indicate a number of rules that constitutive equations must satisfy (the principle of determinism, the principle of local action, the principle of frame indifference, ...), but from first principles we cannot derive an explicit relationship between stress and deformation. The approach here will be to make an assumption about the variables upon which the interfacial stress tensor depends and then to ask what can be said about the functional form of this relationship, if all of these rules are to be satisfied.

In this text, we are concerned with analyzing flows involving substantial interfacial transport phenomena. These are difficult problems to which it is helpful to bring a substantial knowledge of single-phase flows and of multiphase flows in which interfacial transport phenomena have been neglected. For additional reading, I recommend Slattery (1981), Bird, Steward, and Lightfoot (1960), Truesdell and Toupin (1960), Truesdell and Noll (1965), Truesdell (1966), and Batchelor (1967).

References

Batchelor, G. K., "An Introduction to Fluid Dynamics," Cambridge (1967).

Bird, R. B., W. E. Stewart, and E. N. Lightfoot, "Transport Phenomena," Wiley, New York (1960).

Slattery, J. C., "Momentum, Energy, and Mass Transfer in Continua," McGraw-Hill, New York (1972); second edition, Robert E. Krieger, Malabar, FL 32950 (1981).

Truesdell, C., and R. A. Toupin, "Handbuch der Physik," vol. 3/1, edited by S. Flügge, Springer-Verlag, Berlin (1960).

Truesdell, C., and W. Noll, "Handbuch der Physik," vol. 3/3, edited by S. Flügge, Springer-Verlag, Berlin (1960).

Truesdell, C. "The Elements of Continuum Mechanics," Springer-Verlag New York, New York (1966).

2.1 Force

2.1.1 What are forces? Forces are not defined. They are described by a set of properties or axioms.

Corresponding to each body B there is a distinct set of bodies B^e such that the union of B and B^e forms the universe. We refer to B^e as the exterior or the surroundings of the body B.

1. A system of forces is a function **f**(B, A) of pairs of bodies; the values of this function are vectors in V^3, the translation space of E^3.

The value of **f**(B, A) is called the force exerted on the body B by the body A.

2. For a specified body B, **f**(A, B^e) is an additive function defined over the subbodies A of B.

3. Conversely, for a specified body B, **f**(B, A) is an additive function defined over the subbodies A of B^e.

The forces acting upon a body should have nothing to do with the motion of the observer or experimentalist relative to the body.

4. Forces should be frame indifferent:

$$\mathbf{f}^* = \mathbf{Q} \cdot \mathbf{f} \tag{1-1}$$

The rate at which work is done on a body should be independent of the observer as well.

5. The rate at which work is done on a body by a system of forces acting upon it is frame indifferent[a].

---

a) We assume here that all work on the body is the result of forces acting on the body, including for example the electrostatic forces exerted by one portion of a curved, charged interface on another (Truesdell and Toupin 1960, pp. 538 and 546; Curtiss 1956; Livingston and Curtiss 1959; Dahler and Scriven 1961). It is possible to induce in a polar material a local source of moment of momentum with a rotating electric field (Lertes 1921; Grossetti 1958, 1959). In such a case it might also be necessary to account for the work done by the flux of moment of momentum at the bounding surface of the body. Effects of this type have not been investigated thoroughly, but they are thought to be negligibly small for all but unusual situations. Consequently, they are neglected here.

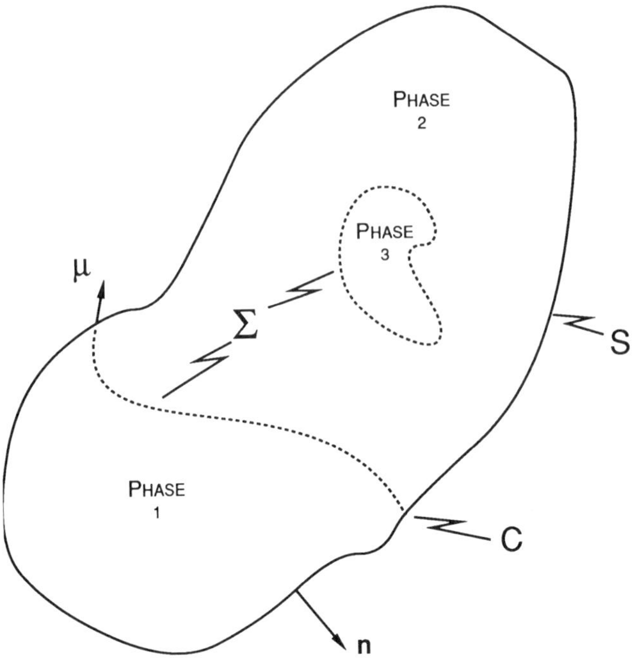

Figure 2.1.1-1: The region R occupied by a body B is bounded by a closed surface S. This body may be in the form of several phases, which adjoin one another at the dividing surfaces Σ. Those portions of Σ that are not closed are bounded by the closed curves C. The unit vector **n** is normal to S and outwardly directed with respect to R. The unit vector **μ** is normal to C, tangent to Σ, and outwardly directed with respect to R.

Let us examine the implications of this last statement. Thinking in terms of the body B shown in Figure 2.1.1-1, we can visualize that the force which B^e exerts upon B may act with different intensities within each phase and on S, Σ, and C. Let $\mathbf{f}_R$ be the force per unit volume which B^e imposes on B at each point within the several phases; $\mathbf{f}_S$ is the force per unit area acting on S; $\mathbf{f}_\Sigma$ the force per unit area on Σ; $\mathbf{f}_C$ the force per unit length on C. With this notation, the resultant force which B^e exerts upon B is

$$\mathbf{f}(\,B, B^e\,) \equiv \int_R \mathbf{f}_R \, dV + \int_S \mathbf{f}_S \, dA + \int_\Sigma \mathbf{f}_\Sigma \, dA + \int_C \mathbf{f}_C \, ds \tag{1-2}$$

and the rate at which work is done on B by these forces is

$$W \equiv \int_R \mathbf{v} \cdot \mathbf{f}_R \, dV + \int_S \mathbf{v} \cdot \mathbf{f}_S \, dA + \int_\Sigma \mathbf{v}^{(\sigma)} \cdot \mathbf{f}_\Sigma \, dA + \int_C \mathbf{v}^{(\sigma)} \cdot \mathbf{f}_C \, ds \tag{1-3}$$

Let us introduce as a shorthand notation

$$\mathbf{f}(\,B, B^e\,) \equiv \int_B d\mathbf{f} \tag{1-4}$$

where **f** is the vector measure[b] of this force. This allows us to write the power input to the body more conveniently as

$$W \equiv \int_B \mathbf{v} \cdot d\mathbf{f} \tag{1-5}$$

In these terms, axiom 5 becomes

$$W^* = W \tag{1-6}$$

or

$$\begin{aligned} W^* - W &= \int_B (\,\mathbf{v}^* \cdot d\mathbf{f}^* - \mathbf{v} \cdot d\mathbf{f}\,) \\ &= \int_B \{[\,\mathbf{Q} \cdot \mathbf{v} + \overline{\dot{\mathbf{z}}^*_{(0)}} + \dot{\mathbf{Q}} \cdot (\,\mathbf{z} - \mathbf{z}_{(0)}\,)] \cdot \mathbf{Q} \cdot d\mathbf{f} - \mathbf{v} \cdot d\mathbf{f}\,\} \\ &= \left(\mathbf{Q}^T \cdot \overline{\dot{\mathbf{z}}^*_{(0)}}\right) \cdot \int_B d\mathbf{f} + \operatorname{tr}\left(\mathbf{Q}^T \cdot \dot{\mathbf{Q}} \cdot \int_B (\,\mathbf{z} - \mathbf{z}_{(0)}\,)\, d\mathbf{f}\right) \\ &= \left(\mathbf{Q}^T \cdot \overline{\dot{\mathbf{z}}^*_{(0)}}\right) \cdot \int_B d\mathbf{f} \end{aligned}$$

---

b) The three components of the vector measure are scalar measures (Halmos 1950).

$$+ \operatorname{tr}\left(\mathbf{Q}^T \cdot \dot{\mathbf{Q}} \cdot \frac{1}{2} \int_B \{ (\mathbf{z} - \mathbf{z}_{(0)}) \, \mathbf{df} - [(\mathbf{z} - \mathbf{z}_{(0)}) \, \mathbf{df}]^T \} \right)$$

$$= 0 \tag{1-7}$$

In (1-7_1), we have used (1-5). In (1-7_2), we have employed an expression from Sec. 1.4.3 for $\mathbf{v}^*$, the velocity vector in the starred frame of reference. We have recognized that $\mathbf{Q}$ is a function only of time in (1-7_3). The skew-symmetry of (see Sec. 1.4.3)

$$\mathbf{Q}^T \cdot \dot{\mathbf{Q}} = - (\mathbf{Q}^T \cdot \dot{\mathbf{Q}})^T \tag{1-8}$$

led to the final rearrangement in (1-7_4) Equation (1-7) is true for every $\mathbf{Q}$. In particular, if $\mathbf{Q}$ is independent of time, (1-7) reduces to

$$\left(\mathbf{Q}^T \cdot \dot{\overline{\mathbf{z}^*_{(0)}}} \right) \cdot \int_B \mathbf{df} = 0 \tag{1-9}$$

But since $\mathbf{Q}^T \cdot \dot{\overline{\mathbf{z}^*_{(0)}}}$ is an arbitrary vector, this implies that the **sum of the forces acting on a body must be zero**:

$$\mathbf{f}(B, B^e) \equiv \int_B \mathbf{df} = 0 \tag{1-10}$$

In view of (1-10), equation (1-7) implies

$$\operatorname{tr}\left(\mathbf{Q}^T \cdot \dot{\mathbf{Q}} \cdot \frac{1}{2} \int_B \{ (\mathbf{z} - \mathbf{z}_{(0)}) \, \mathbf{df} - [(\mathbf{z} - \mathbf{z}_{(0)}) \, \mathbf{df}]^T \} \right) = 0 \tag{1-11}$$

or, since $\mathbf{Q}^T \cdot \dot{\mathbf{Q}}$ is an arbitrary skew-symmetric, second-order tensor,

$$\frac{1}{2} \int_B \{ (\mathbf{z} - \mathbf{z}_{(0)}) \, \mathbf{df} - [(\mathbf{z} - \mathbf{z}_{(0)}) \, \mathbf{df}]^T \} = 0 \tag{1-12}$$

From this we see that the **sum of the torques on a body must be zero** as well (see Sec. A.1.2),

$$\int_B (\mathbf{z} - \mathbf{z}_{(0)}) \wedge \mathbf{df} \equiv \boldsymbol{\varepsilon} : \int_B [(\mathbf{z} - \mathbf{z}_{(0)}) \, \mathbf{df}]^T$$

$$= - \boldsymbol{\varepsilon} : \frac{1}{2} \int_B \{ (\mathbf{z} - \mathbf{z}_{(0)}) \, \mathbf{df} - [(\mathbf{z} - \mathbf{z}_{(0)}) \, \mathbf{df}]^T \}$$

$$= 0 \tag{1-13}$$

Equations (1-10) and (1-13) lay the foundation for Euler's first and second laws, which are introduced in the next section.

Truesdell (1977) has laid out a set of axioms that characterize a system of forces. The discussion in this section as well as the next is an attempt to summarize the essential aspects of his ideas. I can also recommend a more popular treatment by Truesdell (1966, p. 94) of the concept of force in which he traces the historical development of the modern approach to this subject.

References

Curtiss, C. F., *J. Chem. Phys.* **24**, 225 (1956).

Dahler, J. S., and L. E. Scriven *Nature London* **192**, 36 (1961).

Grossetti, E., *Nuovo Cimeto* **10**, 193 (1958); *ibid.* **13**, 350 (1959).

Halmos, P. R., "Measure Theory," Van Nostrand Reinhold, New York (1950).

Lertes, Peter, *Z. Phys.* **4**, 315 (1921); *ibid.* **6**, 56 (1921); *Phys. Z.* **22**, 621 (1921).

Livingston, P. M., and C. F. Curtiss, *J. Chem. Phys.* **31**, 1643 (1959).

Truesdell, C., and R. A. Toupin, "Handbuch der Physik," vol. 3/1, edited by S. Flügge, Springer-Verlag, Berlin (1960).

Truesdell, C., "Six Lectures on Modern Natural Philosophy," Springer-Verlag, New York, New York (1966).

Truesdell, C., "A First Course in Rational Continuum Mechanics," vol. 1, Academic Press, New York (1977).

2.1.2 Euler's first and second laws In the last section, we did not distinguish between the force of inertia and the other forces that act upon a body.

We found that the force and torque which B^e exerts upon B are both zero. But to speak about B^e implies that we have all of the universe available to us for observation. In fact our experience is limited to a subcollection $\mathcal{S}$ of the universe which we can call **the great system**. The usual interpretation of this great system is that region of space out to the **fixed stars**.

Let the exterior $\mathcal{S}^e$ of the great system be a set of bodies distinct from $\mathcal{S}$ such that the union of $\mathcal{S}$ and $\mathcal{S}^e$ form the universe. We know nothing about the motions of the bodies included in $\mathcal{S}^e$ or about the forces among these bodies. We have no choice but to describe $\mathbf{f}(B, \mathcal{S}^e)$ only in terms of the body B and its motion. If we hypothesize that $\mathbf{f}(B, \mathcal{S}^e) = 0$ when the momentum of the body B is a constant, the simplest assumption we can make appears to be this.

6. The force exerted by the universe exterior to the great system upon any body B of the great system is the negative of the time rate of change of the momentum of the body with respect to a preferred frame of reference.

$$\mathbf{f}(B, \mathcal{S}^e) = -\frac{d}{dt}\int_B \mathbf{v}\, dm \tag{2-1}$$

We say that $\mathbf{f}(B, \mathcal{S}^e)$ is the **force of interia** acting on the body B; the preferred frame to which this axiom refers is said to be the **inertial frame of reference**. The fixed stars are usually interpreted as defining the inertial frame.

In the last section, we found that the sum of all of the forces acting upon a body must be zero. If we speak of the force imposed on a body by the great system as the **applied force**, this together with axiom 6 implies **Euler's first law:**

In an inertial frame of reference, the time rate of change of the momentum of a body is equal to the applied force.

Mathematically, this becomes[a]

$$\frac{d}{dt}\int_B \mathbf{v}\, dm = \mathbf{f}^{a} \tag{2-2}$$

where

---

a) It is understood here that a body does not exert a net force on itself:

$$\mathbf{f}(B, B) = 0.$$

$$\mathbf{f}^{a} = \mathbf{f}(B, B^{e}_{\mathcal{S}}) = \int_{B} d\mathbf{f}^{a} \tag{2-3}$$

Here $B^{e}_{\mathcal{S}}$ is that part of the exterior of B which is also part of the great system $\mathcal{S}$; $\mathbf{f}^{a}$ is the vector measure of the applied force. Remembering that mass and velocity are not continuous functions of volume and employing the notation introduced in Sec. 1.3.1, we can also express (2-2) as

$$\frac{d}{dt}\left(\int_{R} \rho\mathbf{v}\, dV + \int_{\Sigma} \rho^{(\sigma)}\mathbf{v}^{(\sigma)}\, dA \right) = \mathbf{f}^{a} \tag{2-4}$$

The transport theorem allows us to alternatively state this as (see Exercise 1.3.5-3)

$$\int_{R} \rho \frac{d_{(m)}\mathbf{v}}{dt}\, dV + \int_{\Sigma} \left\{ \rho^{(\sigma)} \frac{d_{(s)}\mathbf{v}^{(\sigma)}}{dt} + \left[\rho(\mathbf{v} - \mathbf{v}^{(\sigma)})(\mathbf{v} \cdot \boldsymbol{\xi} - v_{(\xi)}^{\{\xi\}}) \right] \right\} dA = \mathbf{f}^{a} \tag{2-5}$$

We also found in the last section that the sum of the torques acting on a body must be zero. Let us define the **applied torque** as that due to the applied force or the force on the body by the great system. Starting with (2-3) and (2-5), we can apply the transport theorem again to conclude

$$\int_{R} \rho\mathbf{z} \wedge \frac{d_{(m)}\mathbf{v}}{dt}\, dV + \int_{\Sigma} \left\{ \rho^{(\sigma)}\mathbf{z} \wedge \frac{d_{(s)}\mathbf{v}^{(\sigma)}}{dt} + \left[\rho\mathbf{z} \wedge (\mathbf{v} - \mathbf{v}^{(\sigma)})(\mathbf{v} \cdot \boldsymbol{\xi} - v_{(\xi)}^{\{\xi\}}) \right] \right\} dA$$

$$= \int_{R} \rho \frac{d_{(m)}}{dt}(\mathbf{z} \wedge \mathbf{v})\, dV + \int_{\Sigma} \left\{ \rho^{(\sigma)} \frac{d_{(s)}}{dt}(\mathbf{z} \wedge \mathbf{v}^{(\sigma)}) + \left[\rho\mathbf{z} \wedge (\mathbf{v} - \mathbf{v}^{(\sigma)})(\mathbf{v} \cdot \boldsymbol{\xi} - v_{(\xi)}^{\{\xi\}}) \right] \right\} dA$$

$$= \frac{d}{dt}\left(\int_R \rho \mathbf{z} \wedge \mathbf{v} \, dV + \int_\Sigma \rho^{(\sigma)} \mathbf{z} \wedge \mathbf{v}^{(\sigma)} \, dA \right)$$

$$= \int_B \mathbf{z} \wedge d\mathbf{f}^{\,a} \qquad (2\text{-}6)$$

or

$$\frac{d}{dt} \int_B \mathbf{z} \wedge \mathbf{v} \, dm = \int_B \mathbf{z} \wedge d\mathbf{f}^{\,a} \qquad (2\text{-}7)$$

This is a statement of **Euler's second law:**

In an inertial frame of reference, the time rate of change of the moment of momentum of a body is equal to the applied torque.

The ideas presented here have been abstracted from Truesdell (1977; 1966, p. 98). You might also look at Truesdell and Toupin (1960, p. 533) and Truesdell and Noll (1965, p. 43).

References

Truesdell, C., and R. A. Toupin, "Handbuch der Physik," vol. 3/1, edited by S. Flügge, Springer-Verlag, Berlin (1960).

Truesdell, C., and W. Noll, "Handbuch der Physik," vol. 3/3, edited by S. Flügge, Springer-Verlag, Berlin (1965).

Truesdell, C., "Six Lectures on Modern Natural Philosophy," Springer-Verlag, New York (1966).

Truesdell, C., "A First Course in Rational Continuum Mechanics," vol. 1, Academic Press, New York (1977).

2.1.3 Body forces and contact forces The applied forces on a body may be separated into two classes: body forces $\mathbf{f}_b^a$ and contact forces $\mathbf{f}_c^a$.

Body forces are presumed to be related to the masses of the bodies and are described as though they act directly on each material particle:

$$\mathbf{f}_b^a = \int_R \rho \mathbf{b} \, dV + \int_\Sigma \rho^{(\sigma)} \mathbf{b}^{(\sigma)} \, dA \tag{3-1}$$

Here R is the region occupied by the body, Σ the union of all the dividing surfaces, $\mathbf{b}$ the body force per unit mass acting on the material within a phase, and $\mathbf{b}^{(\sigma)}$ the body force per unit mass exerted on the dividing surfaces. The justification for associating masses with the dividing surfaces is discussed in detail in Sec. 1.3.2.

There are two types of body force densities.

External body force densities appear to be independent of the extent of the body. As another way of saying this, the same expression for $\mathbf{b}$ can be used to calculate the external body force on the body as a whole or any portion of the body. This implies that $\mathbf{b}$ can be expressed in terms of position, time, and a local description of the body and its motion. For example, the force density of a magnetic or electric field originating outside the body will not depend upon the extent of the body, although it may depend upon the local velocity and the local properties of the body. The force density of universal gravitation attributable to the rest of the universe does not depend upon size of the body, although it will be a function of position in the universe. In what follows, we will consider only those external body force densities that are known functions of position and time,

$$\mathbf{b} = \mathbf{b}(\mathbf{z}, t) \quad , \quad \mathbf{b}^{(\sigma)} = \mathbf{b}^{(\sigma)}(\mathbf{z}, t) \tag{3-2}$$

So long as we limit ourselves to the uniform field of gravity, which is of this form, we can say as well

$$\mathbf{b} = \mathbf{b}^{(\sigma)} \tag{3-3}$$

This point will be examined in more detail in the next section.

Mutual body force densities do depend upon the extent of the body (Truesdell and Toupin 1960, p. 533; Williams 1969 and 1970). The force density at one point within a body attributable by universal gravitation to the rest of the body clearly depends upon the extent of the body. The electrostatic force acting at a point in a body as the result of the charge distribution within the body is dependent upon the extent of the body. Within the immediate neighborhood of a phase interface or of a common line, we may wish to account for the effects of long range intermolecular forces (electrostatic, induction, and dispersion forces; see Exercises 2.1.3-1 and 2.1.3-2). A combination of mutual and external body forces is assumed in the development that follows.

Contact forces on the other hand are those applied forces that appear to be exerted on one body or another through their common surface of contact. They are presumed to be related to this surface, distributed over it, and independent of the masses of the bodies on either side. At first thought, we might say that the contact forces should be absolutely continuous functions of area. But this does not allow for the changing nature of the contact forces in the neighborhood of a phase interface. Referring to Figure 2.1.1-1, we must allow for concentrated contact forces in the curve C formed by the intersection of Σ with S. We will say

$$\mathbf{f}_c^a = \int_S \mathbf{t}(\,\mathbf{z}, S\,)\, dA + \int_C \mathbf{t}^{(\sigma)}(\,\mathbf{z}, C\,)\, ds \tag{3-4}$$

Here $\mathbf{t}(\,\mathbf{z}, S\,)$ is the **stress vector**, the contact force per unit area exerted upon a body at its bounding surface S; it is a function of position **z** on S. Similarly, $\mathbf{t}^{(\sigma)}(\,\mathbf{z}, C\,)$ is the **surface stress vector**, the contact force per unit length at the curve C; it is a function of position **z** on C.

It has always been assumed in classical continuum mechanics that within a single phase the stress is the same at **z** on all similarly-oriented surfaces with a common tangent plane at this point. This is known as **Cauchy's stress principle:**

There is a vector-valued function $\mathbf{t}(\,\mathbf{z}, \mathbf{n}\,)$ defined for all unit vectors **n** at any point **z** within a single phase such that

$$\mathbf{t}(\,\mathbf{z}, S\,) = \mathbf{t}(\,\mathbf{z}, \mathbf{n}\,) \tag{3-5}$$

Here **n** is the unit normal vector to S at **z** outwardly directed with respect to the body upon which the contact stress is applied.

We propose by analogy the **surface stress principle:**

There is a vector-valued function $\mathbf{t}^{(\sigma)}(\,\mathbf{z}, \boldsymbol{\mu}\,)$ defined for all unit tangent vectors $\boldsymbol{\mu}$ at any point **z** on a dividing surface Σ such that

$$\mathbf{t}^{(\sigma)}(\,\mathbf{z}, C\,) = \mathbf{t}^{(\sigma)}(\,\mathbf{z}, \boldsymbol{\mu}\,) \tag{3-6}$$

Here $\boldsymbol{\mu}$ is the unit tangent vector that is normal to C at **z** and outwardly directed with respect to the body upon which the contact stress is applied.

We are now in the position to return to Sec. 2.1.2 and express Euler's first and second laws in terms of these body and contact forces. **Euler's first law** becomes

$$\frac{d}{dt}\left(\int_R \rho \mathbf{v}\, dV + \int_\Sigma \rho^{(\sigma)} \mathbf{v}^{(\sigma)}\, dA \right)$$

$$= \int_S \mathbf{t}\, dA + \int_C \mathbf{t}^{(\sigma)}\, ds + \int_R \rho \mathbf{b}\, dV + \int_\Sigma \rho^{(\sigma)} \mathbf{b}^{(\sigma)}\, dA \tag{3-7}$$

and **Euler's second law** takes the form

$$\frac{d}{dt}\left(\int_R \rho \mathbf{z} \wedge \mathbf{v}\, dV + \int_\Sigma \rho^{(\sigma)} \mathbf{z} \wedge \mathbf{v}^{(\sigma)}\, dA \right)$$

$$= \int_S \mathbf{z} \wedge \mathbf{t}\, dA + \int_C \mathbf{z} \wedge \mathbf{t}^{(\sigma)}\, ds + \int_R \rho \mathbf{z} \wedge \mathbf{b}\, dV$$

$$+ \int_\Sigma \rho^{(\sigma)} \mathbf{z} \wedge \mathbf{b}^{(\sigma)}\, dA \qquad (3\text{-}8)$$

We know that Euler's first and second laws imply Cauchy's first and second laws at every point within each phase (Slattery 1981, p. 41). Our objective in the sections that follow will be to determine the restrictions on momentum transfer at the dividing surface required by (3-7) and (3-8). The approach will be very similar to that which we took in arriving at the jump mass balance Sec. 1.3.5.

References

Black, W., J. G. V. de Jongh, J. Th. G. Overbeck, and M. J. Sparnaay, *Trans. Faraday Soc.* **56**, 1597 (1960).

Churaev, N. V., *Colloid J. USSR* (Engl. Transl.) **36**(2), 283 (1974a).

Churaev, N. V., *Colloid J. USSR* (Engl. Transl.) **36**(2), 287 (1974b).

Derjaguin, B. V., Y. I. Rabinovich, and N. V. Churaev, *Nature (London)* **265**, 520 (1977).

Gregory, J., *Adv. Colloid Interface Sci.* **2**, 396 (1969).

Hirschfelder, J. O., C. F. Curtiss, and R. B. Bird, "Molecular Theory of Gases and Liquids," John Wiley, New York (1954).

Israelachvili, J. N., "Intermolecular and Surface Forces," Academic Press, New York (1985).

Kitchener, J. A., and A. P. Prosser, *Proc. R. Soc. Ser. A.* **242**, 403 (1957).

Miller, C. A., and E. Ruckenstein, *J. Colloid Interface Sci.* **48**, 368 (1974).

Ruckenstein, E., and R. K. Jain, *J. Chem. Soc. Faraday Trans. 2* **70**, 132 (1974).

Sheludko, A., D. Platikanov, and E. Manev, *Discuss. Faraday Soc.* **40**, 253 (1965).

Slattery, J. C., "Momentum, Energy, and Mass Transfer in Continua," McGraw-Hill, New York (1972); second edition, Robert E. Krieger, Malabar, FL 32950 (1981).

Sonntag, H., and K. Strenge, "Coagulation and Stability of Disperse Systems," Halsted Press, New York (1972).

Truesdell, C., and R. A. Toupin, "Handbuch der Physik," vol. 3/1, edited by S. Flügge, Springer-Verlag (1960).

Verwey, E. J. W., and J. Th. G. Overbeek, "Theory of the Stability of Lyophobic Colloids," Elsevier, Amsterdam (1948).

Williams, W. O., *Arch. Ration. Mech. Anal.* **34**, 245 (1969).

Williams, W. O., *Arch. Ration, Mech. Anal.* **36**, 270 (1970).

Exercises

2.1.3-1 *London-van der Waals forces* Consider for the moment a particulate model of a portion of a single phase that is outside the immediate neighborhood of the phase boundaries. The internal energy of this material is the sum of the kinetic energy of the molecules beyond the kinetic energy of the material as a whole and of the potential energy resulting from the positions of the molecules in the various intermolecular force fields (Sec. 5.6.2).

Within the immediate neighborhood of a phase boundary, a molecule of phase 1 is subjected to asymmetric intermolecular forces fields, since it sees the force fields of the molecules in phase 2 as well as the force fields of the molecules in phase 1. We could account for this asymmetry by recognizing that internal energy per unit mass is a function of distance from a phase interface, but it would be awkward, particularly in cases where there is more than one phase interface involved. Instead, it is normal practice to describe the internal energy as though the material were outside the immediate neighborhood of the phase boundary and to account for the asymmetry of the intermolecular force field by a mutual force. This mutual force $\mathbf{b}_m$ is normally represented in terms of a potential ϕ:

$$\mathbf{b}_m = - \nabla\phi$$

For the moment, let us neglect the electrostatic and induction contributions to the intermolecular forces, and let us consider only the effects of long-range dispersion (induced-dipole/induced-dipole) interactions

(Hirschfelder *et al.* 1954, p. 25). At separations that are relatively large compared with molecular dimensions, London (Hirschfelder *et al.* 1954, p. 30) proposed that the potential is proportional to the inverse sixth-power of the separation distance. More generally, the intermolecular potential energy function is assumed to be proportional to the inverse (m + 3)- power of the separation distance. The potential energy at any point within phase 2 is obtained as a sum of the integral of such a relationship over all of phase 2 and of the integrals of similar relationships over phases 1 and 3. We will refer to this as the **London-van der Waals force potential** resulting from the **London-van der Waals forces** acting within the immediate neighborhood of the phase interface.

Consider a thin film of phase 2 separated from the adjacent semi-infinite phases 1 and 3 by two parallel, planar interfaces. Confirm that at the interface separating phase 1 from 2 (Sheludko *et al.* 1965; Ruckenstein and Jain 1974)

$$\rho^{(2)}\phi^{(2)}_{LvW} = \phi_{\infty} + B/h^{m}$$

Here $\rho^{(2)}$ is the density of the film fluid and ϕ_{∞} is the interaction potential per unit volume of a semi-infinite film liquid in the limit as this interface is approached.

Typically we will be more interested in

$$\rho^{(2)}\phi^{(2)}_{LvW} - \rho^{(1)}\phi^{(1)}_{LvW} = \phi^{(21)}_{\infty} + B^{(21)}/h^{m} \tag{1}$$

and

$$\rho^{(2)}\phi^{(2)}_{LvW} - \rho^{(3)}\phi^{(3)}_{LvW} = \phi^{(23)}_{\infty} + B^{(23)}/h^{m} \tag{2}$$

For unretarded London-van der Waals forces (when the film thickness is less than 120 Å), $m \doteq 3$, $|B^{(2j)}| \sim 10^{-14}$ erg, and $A^{(2j)} \equiv 6\pi B^{(2j)}$ is referred to as the Hamaker constant; for retarded London-van der Waals forces (when the film thickness is larger than 400 Å), $m = 4$, and $|B^{(2j)}| \sim 10^{-19}$ erg cm (Kitchener and Prosser 1957; Black *et al.* 1960; Sheludko *et al.* 1965; Gregory 1969; Sonntag and Strenge 1972; Churaev 1974a,b; Derjaguin *et al.* 1977; Israelachvili 1985). We speak of

$$\pi^{(2j)}_{LvW} \equiv -B^{(2j)}/h^{m}$$

as the **disjoining pressure** of a flat film of thickness h at the interface separating the film from phase j.

We say that the disjoining pressure is **negative**, when $B^{(2j)}$ is positive. The interaction potential per unit volume of the film phase at the interface relative to that of the adjoining phase is larger than it would be, if the film phase were semi-infinite. A negative disjoining pressure acts to draw the two fluid-fluid interfaces together, preventing the formation of a stable film of uniform thickness. Note that $B^{(2j)}$ is always positive when phases 1 and 3 are identical.

We say that the disjoining pressure is **positive**, when B is negative. The interaction potential per unit volume of the film phase at the interface relative to that of the adjoining phase is smaller than it would be if the film phase were semi-infinite. A positive disjoining pressure acts to repel the two fluid-fluid interfaces, allowing the formation of a stable film of uniform thickness.

Miller and Ruchenstein (1974) have extended these results to a thin wedge-shaped film, but the error incurred in using (1) and (2) at interfaces having finite slopes appear to be small (see Sec. 3.2.7).

In order to determine whether it is reasonable to extend continuum mechanics to films sufficiently thin that the effects of the London-van der Waals forces become significant, consider Exercise 4.1.4-3. There we construct an estimate of interfacial tension based both upon an analysis of the extension of a thin film and upon a very simple discussion of molecular structure. Using prior estimates for the Hamaker constant, we find that the results are in good agreement with available experimental data. You should consider also the comparisons with experimental data given in Secs. 3.2.7 and 3.3.2, although the analyses upon which they are based involve more assumptions and the comparisons with experimental data are consequently more limited.

2.1.3-2 *electrostatic double-layer forces* Electrostatic double-layer forces are also mutual forces. Like the London-van der Waals forces discussed above, this mutual force per unit mass $\mathbf{b}_e$ is often represented in terms of a scalar potential ϕ_e:

$$\mathbf{b}_e = - \nabla \phi_e$$

Consider a thin film of phase 2 adjoining on each side a semi-infinite phase 1. The two interfaces are parallel planes. Phase 2 is a solution of a symmetric electrolyte (which dissociates into two ions, the valences of which are equal in magnitude). The film is sufficiently thick that we can neglect the effects attributable to interaction of the potential profiles associated with the two double layers. Confirm that for $\kappa h > 4$ (Verwey and Overbeek 1948)

$$\rho^{(2)} \phi_e^{(2)} = - D \exp(- \kappa h)$$

or

$$\rho^{(2)} \phi_e^{(2)} - \rho^{(1)} \phi_e^{(1)} = - D \exp(- \kappa h)$$

in which we have recognized that phase 1 is semi-infinite. Here

$$\kappa \equiv \left(\frac{8 \pi n z^2 e^2}{\varepsilon k T} \right)^{1/2}$$

is the inverse of the Debye length, h is the thickness of the film,

$$D \equiv 64nkT \tanh^2 \left(\frac{ze\psi}{4kT} \right)$$

e is the elementary charge, ε is the dielectric constant of the medium, k is the Boltzmann constant, T is the absolute temperature, n is the number density of either ionic species, z is the magnitude of the valence of either ionic species, and ψ is the electrostatic surface potential of the interfaces. We speak of

$$\pi_e \equiv D \exp(- \kappa h)$$

as the disjoining pressure attributable to the electrostatic double layer of a flat film of thickness h. When the electrostatic surface potentials have the same sign as in the case that we are discussing here, there is a positive disjoining pressure attributable to the electrostatic double layer that acts to repel the fluid-fluid interfaces from each other.

2.1.4 Euler's first law at dividing surfaces What does Euler's first law require at each point on a dividing surface?

Let

$$\mathbf{t} = \mathbf{t}(\mathbf{z}, \mathbf{n}) \tag{4-1}$$

be the stress that a material exerts upon a surface at a particular point $\mathbf{z}$ where the unit normal to the surface pointing into this material is $\mathbf{n}$. It can easily be established that $\mathbf{t}$ can be expressed as the result of the transformation of $\mathbf{n}$ by the stress tensor $\mathbf{T}$ (Slattery, 1981, p. 39)

$$\mathbf{t} = \mathbf{T} \cdot \mathbf{n} \tag{4-2}$$

To Euler's first law as stated in Sec. 2.1.3, let us apply the transport theorem for a material region containing a dividing surface (Exercise 1.3.5-3):

$$\int_R \rho \frac{d_{(m)}\mathbf{v}}{dt} \, dV + \int_\Sigma \left\{ \rho^{(\sigma)} \frac{d_{(s)}\mathbf{v}^{(\sigma)}}{dt} + [\rho(\mathbf{v} - \mathbf{v}^{(\sigma)})(\mathbf{v} \cdot \boldsymbol{\xi} - v_{(\xi)}^{\{g\}})] \right\} dA$$

$$= \int_S \mathbf{t} \, dA + \int_C \mathbf{t}^{(\sigma)} \, ds + \int_R \rho \mathbf{b} \, dV + \int_\Sigma \rho^{(\sigma)} \mathbf{b}^{(\sigma)} \, dA \tag{4-3}$$

Green's transformation (Slattery 1981, p. 661) allows us to say

$$\int_S \mathbf{t} \, dA = \int_S \mathbf{T} \cdot \mathbf{n} \, dA$$

$$= \int_R \text{div } \mathbf{T} \, dV + \int_\Sigma [\, \mathbf{T} \cdot \boldsymbol{\xi} \,] \, dA \tag{4-4}$$

and (4-3) can be rearranged as

$$\int_R \left(\rho \frac{d_{(m)}\mathbf{v}}{dt} - \text{div } \mathbf{T} - \rho \mathbf{b} \right) dV + \int_\Sigma \left\{ \rho^{(\sigma)} \frac{d_{(s)}\mathbf{v}^{(\sigma)}}{dt} - \rho^{(\sigma)}\mathbf{b}^{(\sigma)} \right.$$

$$\left. + \, [\, \rho(\, \mathbf{v} - \mathbf{v}^{(\sigma)} \,)(\, \mathbf{v} \cdot \boldsymbol{\xi} - v_{(\xi)} \,) - \mathbf{T} \cdot \boldsymbol{\xi} \,] \right\} dA$$

$$- \int_C \mathbf{t}^{(\sigma)} \, ds$$

$$= 0 \tag{4-5}$$

Equation (4-5) must be true for every body. In particular, for a body consisting of a single phase, the integral over the region R is zero, which implies

$$\rho \frac{d_{(m)}\mathbf{v}}{dt} = \text{div } \mathbf{T} + \rho \mathbf{b} \tag{4-6}$$

This is **Cauchy's first law** (Slattery 1981, p. 41). It expresses the constraint of Euler's first law at every point within a phase.

In view of (4-6), (4-5) further simplifies to

$$\int_\Sigma \left\{ \rho^{(\sigma)} \frac{d_{(s)}\mathbf{v}^{(\sigma)}}{dt} - \rho^{(\sigma)}\mathbf{b}^{(\sigma)} \right.$$

$$\left. + \, [\, \rho(\, \mathbf{v} - \mathbf{v}^{(\sigma)} \,)(\, \mathbf{v} \cdot \boldsymbol{\xi} - v_{(\xi)} \,) - \mathbf{T} \cdot \boldsymbol{\xi} \,] \right\} dA$$

$$- \int_C \mathbf{t}^{(\sigma)} \, ds$$

$$= 0 \tag{4-7}$$

We would like to write (4-7) as an integral over Σ that is equal to zero, because in this way we could determine the form of Euler's first law that must be true at every point on the dividing surface. But before we can do this, we must introduce the surface stress tensor.

Reference

Slattery, J. C., "Momentum, Energy, and Mass Transfer in Continua," McGraw-Hill, New York (1972); second edition, Robert E. Krieger, Malabar, FL 32950 (1981).

Exercise

2.1.4-1 *surface stress lemma* **Cauchy's stress lemma** states (Slattery 1981, p. 39)

$$\mathbf{t}(\mathbf{z}, \mathbf{n}) = -\mathbf{t}(\mathbf{z}, -\mathbf{n})$$

Consider two neighboring portions of a continuous dividing surface Σ and apply (4-7) to each portion and to their union. Deduce that on their common boundary

$$\mathbf{t}^{(\sigma)}(\mathbf{z}, \boldsymbol{\mu}) = -\mathbf{t}^{(\sigma)}(\mathbf{z}, -\boldsymbol{\mu})$$

In this way, we establish the **surface stress lemma:**

The surface stress vectors acting upon opposite sides of the same curve at a given point are equal in magnitude and opposite in direction.

2.1.5 Surface stress tensor We will prove in what follows that the surface stress vector may be expressed as

$$\mathbf{t}^{(\sigma)}(\mathbf{z}, \boldsymbol{\mu}) = \mathbf{T}^{(\sigma)} \cdot \boldsymbol{\mu} \tag{5-1}$$

Here $\mathbf{T}^{(\sigma)}$ is the **surface stress tensor;** $\boldsymbol{\mu}$ is the tangential vector that is normal to the curve on which $\mathbf{t}^{(\sigma)}$ acts and that is directed toward the portion of Σ exerting this force.

Previously the function $\mathbf{t}^{(\sigma)}(\mathbf{z}, \cdot)$ has been defined only for unit tangential vectors. Let us begin by extending its definition to all tangential vectors $\mathbf{a}$ at the point $\mathbf{z}$ on a dividing surface Σ:

$$\mathbf{t}^{(\sigma)}(\mathbf{z}, \mathbf{a}) \equiv \begin{cases} |\mathbf{a}|\, \mathbf{t}^{(\sigma)}\left(\mathbf{z}, \dfrac{\mathbf{a}}{|\mathbf{a}|}\right) & \text{if } \mathbf{a} \neq 0 \\ \mathbf{0} & \text{if } \mathbf{a} = 0 \end{cases} \tag{5-2}$$

If the scalar $A > 0$ and the vector $\mathbf{a} \neq 0$, then

$$\mathbf{t}^{(\sigma)}(\mathbf{z}, A\mathbf{a}) = |A\mathbf{a}|\, \mathbf{t}^{(\sigma)}\left(\mathbf{z}, \frac{A\mathbf{a}}{|A\mathbf{a}|}\right)$$

$$= A\, \mathbf{t}^{(\sigma)}(\mathbf{z}, \mathbf{a}) \tag{5-3}$$

If the scalar $A < 0$ and the vector $\mathbf{a} \neq 0$, we have

$$\mathbf{t}^{(\sigma)}(\mathbf{z}, A\mathbf{a}) = \mathbf{t}^{(\sigma)}(\mathbf{z}, -|A|\mathbf{a})$$

$$= |A|\, \mathbf{t}^{(\sigma)}(\mathbf{z}, -\mathbf{a})$$

$$= A\, \mathbf{t}^{(\sigma)}(\mathbf{z}, \mathbf{a}) \tag{5-4}$$

In the last line, we have used the surface stress lemma (Exercise 2.1.4-1). Our conclusion is that the function $\mathbf{t}^{(\sigma)}(\mathbf{z}, \cdot)$ is a homogeneous function of the tangential vectors at the point $\mathbf{z}$ on Σ.

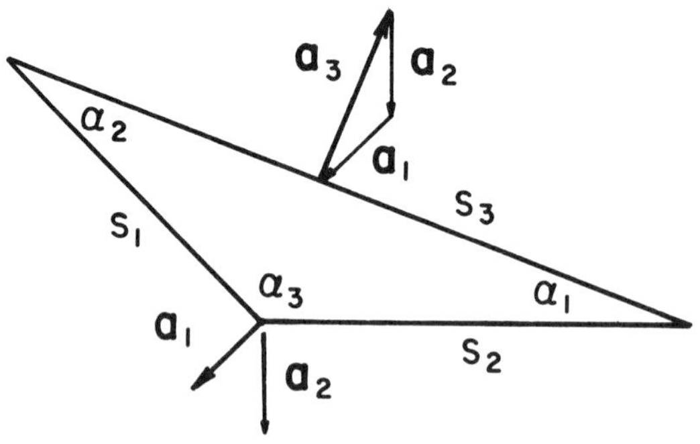

Figure 2.1.5-1: Intersection of Σ with the triangular prism bounded by P_1, P_2, and P_3.

Let us now prove that $\mathbf{t}^{(\sigma)}(\mathbf{z}, \cdot)$ is an additive function of the tangential vectors at the point $\mathbf{z}$ on Σ:

$$\mathbf{t}^{(\sigma)}(\mathbf{z}, \mathbf{a}_1 + \mathbf{a}_2) = \mathbf{t}^{(\sigma)}(\mathbf{z}, \mathbf{a}_1) + \mathbf{t}^{(\sigma)}(\mathbf{z}, \mathbf{a}_2) \tag{5-5}$$

If $\mathbf{a}_1$ and $\mathbf{a}_2$ are linearly dependent, this result follows immediately from the homogeneity of $\mathbf{t}^{(\sigma)}(\mathbf{z}, \cdot)$. We suppose that $\mathbf{a}_1$ and $\mathbf{a}_2$ are linearly independent.

Let us define the planes P_1 and P_2 to be normal to $\mathbf{a}_1$ and $\mathbf{a}_2$ at the point $\mathbf{z}$. Set

$$\mathbf{a}_3 \equiv - (\mathbf{a}_1 + \mathbf{a}_2) \tag{5-6}$$

Let P_3 be normal to $\mathbf{a}_3$ at the place $\mathbf{z} + \varepsilon\, \mathbf{a}_3$. The planes P_1, P_2, and P_3 bound a triangular prism. The intersection of this triangular prism with Σ is in turn bounded by three curves whose arc lengths are s_1, s_2, and s_3. Referring to Figure 2.1.5-1 and using the law of sines, we find that in the limit as $\varepsilon \to 0$:

$$\frac{|\mathbf{a}_1|}{\sin \alpha_1} = \frac{|\mathbf{a}_2|}{\sin \alpha_2} = \frac{|\mathbf{a}_3|}{\sin \alpha_3}$$

$$\frac{s_1}{\sin \alpha_1} = \frac{s_2}{\sin \alpha_2} = \frac{s_3}{\sin \alpha_3}$$

$$\frac{|\mathbf{a}_1|}{s_1} = \frac{|\mathbf{a}_2|}{s_2} = \frac{|\mathbf{a}_3|}{s_3} \tag{5-7}$$

Starting with (4-7) stated for the intersection of this triangular prism with Σ, we reason

$$0 = \text{limit } \varepsilon \to 0: \frac{|\mathbf{a}_3|}{s_3} \left(\int_C \mathbf{t}^{(\sigma)}(\mathbf{z}, \boldsymbol{\mu})\, ds - \int_\Sigma \left\{ \rho^{(\sigma)} \frac{d_{(s)} \mathbf{v}^{(\sigma)}}{dt} \right.\right.$$

$$\left.\left. - \rho^{(\sigma)}\mathbf{b}^{(\sigma)} + [\, \rho(\mathbf{v} - \mathbf{v}^{(\sigma)})(\mathbf{v} \cdot \boldsymbol{\xi} - v_{\{\xi\}}) - \mathbf{T} \cdot \boldsymbol{\xi}\,] \right\} dA \right)$$

$$= \text{limit } \varepsilon \to 0: \frac{|\mathbf{a}_3|}{s_3} \int_C \mathbf{t}^{(\sigma)}(\mathbf{z}, \boldsymbol{\mu})\, ds$$

$$= \text{limit } \varepsilon \to 0: \frac{|\mathbf{a}_3|}{s_3} \left[\mathbf{t}^{(\sigma)} \left(\mathbf{z}, \frac{\mathbf{a}_1}{|\mathbf{a}_1|} \right) s_1 + \mathbf{t}^{(\sigma)} \left(\mathbf{z}, \frac{\mathbf{a}_2}{|\mathbf{a}_2|} \right) s_2 \right.$$

$$+ \mathbf{t}^{(\sigma)}\left(\mathbf{z}, \frac{\mathbf{a}_3}{|\mathbf{a}_3|} \right) s_3 \Big]$$

$$= |\mathbf{a}_1|\, \mathbf{t}^{(\sigma)}\left(\mathbf{z}, \frac{\mathbf{a}_1}{|\mathbf{a}_1|} \right) + |\mathbf{a}_2|\, \mathbf{t}^{(\sigma)}\left(\mathbf{z}, \frac{\mathbf{a}_2}{|\mathbf{a}_2|} \right) + |\mathbf{a}_3|\, \mathbf{t}^{(\sigma)}\left(\mathbf{z}, \frac{\mathbf{a}_3}{|\mathbf{a}_3|} \right)$$

$$= \mathbf{t}^{(\sigma)}(\mathbf{z}, \mathbf{a}_1) + \mathbf{t}^{(\sigma)}(\mathbf{z}, \mathbf{a}_2) + \mathbf{t}^{(\sigma)}(\mathbf{z}, \mathbf{a}_3)$$

$$= \mathbf{t}^{(\sigma)}(\mathbf{z}, \mathbf{a}_1) + \mathbf{t}^{(\sigma)}(\mathbf{z}, \mathbf{a}_2) + \mathbf{t}^{(\sigma)}\Big(\mathbf{z}, - (\mathbf{a}_1 + \mathbf{a}_2)\Big) \tag{5-8}$$

and (5-5) follows.

Since every homogeneous, additive function of a vector can be represented by a linear transformation of the vector, (5-1) is proved. In particular, $\mathbf{T}^{(\sigma)}$ is a second-order surface tensor field (Sec. A.3.1), since $\mathbf{t}^{(\sigma)}(\mathbf{z}, \cdot)$ is defined only for tangential vectors. We will say more on this point in Sec. 2.1.7.

The proof I have given here is modeled after a similar one given by Truesdell (1977, Sec. III.4) for (4-2).

Reference

Truesdell, C., "A First Course in Rational Continuum Mechanics," vol. 1, Academic Press, New York (1977).

2.1.6 Jump momentum balance In the last section, we proved (5-1): the surface stress vector $\mathbf{t}^{(\sigma)}$ may be expressed as the second-order surface stress tensor $\mathbf{T}^{(\sigma)}$ operating on $\boldsymbol{\mu}$, the tangential vector that is normal to the curve on which $\mathbf{t}^{(\sigma)}$ acts and that is directed towards the portion of Σ exerting this force. Employing (5-1) as well as the surface divergence theorem (Sec. A.6.3), we have

$$\int_C \mathbf{t}^{(\sigma)}\, ds = \int_C \mathbf{T}^{(\sigma)} \cdot \boldsymbol{\mu}\, ds$$

$$= \int_\Sigma \text{div}_{(\sigma)} \mathbf{T}^{(\sigma)}\, ds \tag{6-1}$$

and (4-7) becomes

$$\int_{\Sigma} \left\{ \rho^{(\sigma)} \frac{d_{(s)} \mathbf{v}^{(\sigma)}}{dt} - \text{div}_{(\sigma)} \mathbf{T}^{(\sigma)} - \rho^{(\sigma)} \mathbf{b}^{(\sigma)} \right.$$

$$\left. + [\, \rho (\, \mathbf{v} - \mathbf{v}^{(\sigma)} \,)(\, \mathbf{v} \cdot \boldsymbol{\xi} - v_{(\xi)} \,) - \mathbf{T} \cdot \boldsymbol{\xi} \,] \right\} dA$$

$$= 0 \tag{6-2}$$

Since (6-2) must be true for every portion of a dividing surface no matter how large or small, we conclude (see Exercises 1.3.5-1 and -2)

$$\rho^{(\sigma)} \frac{d_{(s)} \mathbf{v}^{(\sigma)}}{dt} = \text{div}_{(\sigma)} \mathbf{T}^{(\sigma)} + \rho^{(\sigma)} \mathbf{b}^{(\sigma)}$$

$$- [\, \rho (\, \mathbf{v} - \mathbf{v}^{(\sigma)} \,)(\, \mathbf{v} \cdot \boldsymbol{\xi} - v_{(\xi)} \,) - \mathbf{T} \cdot \boldsymbol{\xi} \,] \tag{6-3}$$

We will refer to this as the **jump momentum balance**. It expresses the requirement of Euler's first law at every point on a dividing surface.

The usual assumption of continuum fluid mechanics is that the tangential components of velocity are continuous across a phase interface. We shall interpret this as meaning

$$\text{at } \Sigma: \mathbf{P} \cdot \mathbf{v} = \mathbf{P} \cdot \mathbf{v}^{(\sigma)} \tag{6-4}$$

Under these conditions, (6-3) simplifies to

$$\rho^{(\sigma)} \frac{d_{(s)} \mathbf{v}^{(\sigma)}}{dt} = \text{div}_{(\sigma)} \mathbf{T}^{(\sigma)} + \rho^{(\sigma)} \mathbf{b}^{(\sigma)}$$

$$- [\, \rho (\, \mathbf{v} \cdot \boldsymbol{\xi} - v_{(\xi)} \,)^2 \, \boldsymbol{\xi} - \mathbf{T} \cdot \boldsymbol{\xi} \,] \tag{6-5}$$

This is the form of the jump momentum balance that we will employ in the rest of this text. But before we leave this point, I would like to emphasize that (6-4) is a separate assumption that has nothing to do with Euler's first law.

If there is no mass transfer to or from the dividing surface,

$$\text{at } \Sigma: \mathbf{v} \cdot \boldsymbol{\xi} = v_{(\xi)} \tag{6-6}$$

and the jump momentum balance reduces to

$$\rho^{(\sigma)} \frac{d_{(s)} \mathbf{v}^{(\sigma)}}{dt} = \text{div}_{(\sigma)} \mathbf{T}^{(\sigma)} + \rho^{(\sigma)} \mathbf{b}^{(\sigma)} + [\, \mathbf{T} \cdot \boldsymbol{\xi} \,] \tag{6-7}$$

Exercises

2.1.6-1 *alternative derivation of the jump momentum balance* Write Euler's first law for a multiphase material body. Employ the transport theorem for a body containing a dividing surface (Sec. 1.3.4). Deduce the jump momentum balance (6-3) by allowing the material region to shrink around the dividing surface.

2.1.6-2 *alternative form of jump momentum balance* Use the jump mass balance in order to rearrange the jump momentum balance (6-3) in the form

$$\frac{\partial}{\partial t}\left(\rho^{(\sigma)} \mathbf{v}^{(\sigma)} \right) - \nabla_{(\sigma)}\left(\rho^{(\sigma)} \mathbf{v}^{(\sigma)}\right) \cdot \mathbf{u} + \text{div}_{(\sigma)}\left(\rho^{(\sigma)} \mathbf{v}^{(\sigma)} \mathbf{v}^{(\sigma)} \right)$$

$$= \text{div}_{(\sigma)} \mathbf{T}^{(\sigma)} + \rho^{(\sigma)} \mathbf{b}^{(\sigma)} - [\, \rho \mathbf{v}(\mathbf{v} \cdot \boldsymbol{\xi} - v_{\{\xi\}}) - \mathbf{T} \cdot \boldsymbol{\xi} \,]$$

2.1.6-3 *continuity of tangential components of velocity at a dividing surface under restricted circumstances* Let us ignore viscous effects within the adjoining bulk phases and within the dividing surface, so that we may write simply

$$\mathbf{T} = - P\mathbf{I}$$

and

$$\mathbf{T}^{(\sigma)} = \gamma \mathbf{P}$$

Do not ignore mass transfer across the phase interface, but do assume that there is no mass associated with the dividing surface.

i) Prove that under these restricted conditions the tangential components of the jump momentum balance require for $\alpha = 1, 2$

$$[\, \rho v_{\alpha} (\mathbf{v} \cdot \boldsymbol{\xi} - v_{\{\xi\}})] = 0$$

ii) From the jump mass balance

$$[\, \rho(\mathbf{v} \cdot \boldsymbol{\xi} - v_{\{\xi\}})] = 0$$

which allows us to conclude that for $\alpha = 1, 2$

$$[\, v_\alpha \,] = 0$$

or the tangential components of velocity are continuous across the dividing surface.

2.1.7 $\mathbf{T}^{(\sigma)}$ Is symmetric tangential tensor What does Euler's second law imply at each point on a dividing surface?

Let us begin by applying the transport theorem for a material region containing a dividing surface to Euler's second law as it was stated in Sec. 2.1.3:

$$\int_R \mathbf{z} \wedge \left(\rho \frac{d_{(m)}\mathbf{v}}{dt} \right) dV + \int_\Sigma \mathbf{z} \wedge \left\{ \rho^{(\sigma)} \frac{d_{(s)}\mathbf{v}^{(\sigma)}}{dt} + [\, \rho(\mathbf{v} - \mathbf{v}^{(\sigma)})(\mathbf{v} \cdot \boldsymbol{\xi} - v_{(\xi)})] \right\} dA$$

$$= \int_S \mathbf{z} \wedge \mathbf{t} \, dA + \int_C \mathbf{z} \wedge \mathbf{t}^{(\sigma)} \, ds$$

$$+ \int_R \mathbf{z} \wedge (\rho \mathbf{b}) \, dV + \int_\Sigma \mathbf{z} \wedge (\rho^{(\sigma)} \mathbf{b}^{(\sigma)}) \, dA \tag{7-1}$$

Green's transformation (Slattery 1981, p. 661) allows us to express the first terms on the right as

$$\int_S \mathbf{z} \wedge \mathbf{t} \, dA = \int_S \mathbf{z} \wedge (\mathbf{T} \cdot \mathbf{n}) \, dA$$

$$= \int_R \operatorname{div}(\mathbf{z} \wedge \mathbf{T}) \, dV + \int_\Sigma \mathbf{z} \wedge [\, \mathbf{T} \cdot \boldsymbol{\xi} \,] \, dA$$

$$= \int_R (\mathbf{z} \wedge \operatorname{div} \mathbf{T} + \boldsymbol{\varepsilon} : \mathbf{T}) \, dV + \int_\Sigma \mathbf{z} \wedge [\, \mathbf{T} \cdot \boldsymbol{\xi} \,] \, dA \tag{7-2}$$

The role of $\boldsymbol{\varepsilon}$ in forming vector products is summarized in Sec. A.1.2.

Recognizing Cauchy's first law (Sec. 2.1.4) as well as (7-2), we can write (7-1) as

$$\int_{\Sigma} \mathbf{z}\,\boldsymbol{\Lambda} \left\{ \rho^{(\sigma)} \frac{d_{(\sigma)}\mathbf{v}^{(\sigma)}}{dt} + [\, \rho(\, \mathbf{v} - \mathbf{v}^{(\sigma)}\,)(\, \mathbf{v} \cdot \boldsymbol{\xi} - v^{\{g\}}_{(\xi)}\,) - \mathbf{T} \cdot \boldsymbol{\xi}\,] - \rho^{(\sigma)}\mathbf{b}^{(\sigma)} \right\} dA$$

$$= \int_{C} \mathbf{z}\,\boldsymbol{\Lambda}\,\mathbf{t}^{(\sigma)}\, ds + \int_{R} \boldsymbol{\varepsilon} : \mathbf{T}\, dV \tag{7-3}$$

Equation (7-3) must be true for every body. In particular, for a body consisting of a single phase, the integral over the region R is zero, which implies (Slattery 1981, p. 21)

$$\boldsymbol{\varepsilon} : \mathbf{T} = 0 \tag{7-4}$$

Because $\boldsymbol{\varepsilon}$ is a skew-symmetric tensor (Sec. A.1.2 and Slattery 1981, p. 611), $\mathbf{T}$ must be symmetric:

$$\mathbf{T} = \mathbf{T}^{T} \tag{7-5}$$

The symmetry of the stress tensor is known as **Cauchy's second law**. It expresses the constraint of Euler's second law at each point within a phase.

After an application of the surface divergence theorem, the first term on the right of (7-3) becomes

$$\int_{C} \mathbf{z}\,\boldsymbol{\Lambda}\,\mathbf{t}^{(\sigma)}\, ds = \int_{C} \mathbf{z}\,\boldsymbol{\Lambda}\,(\, \mathbf{T}^{(\sigma)} \cdot \boldsymbol{\mu}\,)\, ds$$

$$= \int_{\Sigma} \mathrm{div}_{(\sigma)}(\, \mathbf{z}\,\boldsymbol{\Lambda}\,\mathbf{T}^{(\sigma)}\,)\, dA$$

$$= \int_{\Sigma} (\, \mathbf{z}\,\boldsymbol{\Lambda}\, \mathrm{div}_{(\sigma)}\mathbf{T}^{(\sigma)} + \boldsymbol{\varepsilon} : \mathbf{T}^{(\sigma)}\,)\, dA \tag{7-6}$$

Using (7-4) and (7-6), we may rearrange (7-3) in the form

$$\int_{\Sigma} \left(z\,\Lambda \left\{ \rho^{(\sigma)} \frac{d_{(s)}\,\mathbf{v}^{(\sigma)}}{dt} - \text{div}_{(\sigma)}\mathbf{T}^{(\sigma)} - \rho^{(\sigma)}\mathbf{b}^{(\sigma)} \right.\right.$$

$$\left. + \left[\, \rho(\,\mathbf{v} - \mathbf{v}^{(\sigma)}\,)(\,\mathbf{v} \cdot \boldsymbol{\xi} - v_{(\xi)}^{(\sigma)}\,) - \mathbf{T} \cdot \boldsymbol{\xi}\,\right] \right\}$$

$$\left. - \boldsymbol{\varepsilon} : \mathbf{T}^{(\sigma)} \right) dA$$

$$= 0 \tag{7-7}$$

In view of the jump momentum balance (6-3), this further simplifies to

$$\int_{\Sigma} \boldsymbol{\varepsilon} : \mathbf{T}^{(\sigma)}\, dA = 0 \tag{7-8}$$

Since (7-8) must be true for every portion of a dividing surface no matter how large or small, this means (see Exercises 1.3.5-1 and 1.3.5-2)

$$\boldsymbol{\varepsilon} : \mathbf{T}^{(\sigma)} = 0 \tag{7-9}$$

Because $\boldsymbol{\varepsilon}$ is a skew-symmetric tensor (Sec. A.1.2 and Slattery 1981, p. 611), $\mathbf{T}^{(\sigma)}$ must be symmetric:

$$\mathbf{T}^{(\sigma)} = \mathbf{T}^{(\sigma)T} \tag{7-10}$$

In Sec. 2.1.5, we observed that $\mathbf{T}^{(\sigma)}$ is a surface tensor field:

$$\begin{aligned} \mathbf{T}^{(\sigma)} &= \mathbf{T}^{(\sigma)} \cdot \mathbf{P} \\ &= \mathbf{T}^{(\sigma)T} \cdot \mathbf{P} \end{aligned} \tag{7-11}$$

In view of this, (7-10) tells us

$$\begin{aligned} \mathbf{T}^{(\sigma)} &= (\,\mathbf{T}^{(\sigma)T} \cdot \mathbf{P}\,)^{T} \\ &= \mathbf{P} \cdot \mathbf{T}^{(\sigma)} \\ &= \mathbf{P} \cdot \mathbf{T}^{(\sigma)} \cdot \mathbf{P} \end{aligned} \tag{7-12}$$

or $\mathbf{T}^{(\sigma)}$ is a tangential tensor field. Physically this means that the surface stress $\mathbf{t}^{(\sigma)}$ acts tangentially to the dividing surface and that $\mathbf{t}^{(\sigma)}$ is a tangential vector field.

To summarize, we have proved that $\mathbf{T}^{(\sigma)}$ is a symmetric tangential tensor field.

It is important to remember that the symmetry of the stress tensor and of the surface stress tensor is the result of our assumption that all work on the body is the result of forces, not torques, acting on the body (see Sec. 2.1.1, footnote a).

Reference

Slattery, J. C., "Momentum, Energy and Mass Transfer in Continua," McGraw-Hill, New York (1972); second edition, Robert E. Krieger, Malabar, FL 32950 (1981).

Exercises

2.1.7-1 *frame indifference of* $\mathbf{T}^{(\sigma)}$ In Sec. 2.1.1, we indicated that forces should be frame indifferent. Let $\boldsymbol{\mu}$ be the tangential vector that is normal to the curve on which the surface stress vector $\mathbf{t}^{(\sigma)}$ acts and that is directed towards the portion of the dividing surface exerting this force. Argue that $\boldsymbol{\mu}$ must be frame indifferent

$$\boldsymbol{\mu}^* = \mathcal{Q} \cdot \boldsymbol{\mu}$$

and conclude that the surface stress tensor is a frame indifferent tangential tensor field

$$\mathbf{T}^{(\sigma)*} = \mathcal{Q} \cdot \mathbf{T}^{(\sigma)} \cdot \mathcal{Q}^T$$

2.1.7-2 *frame indifference of* $\mathbf{T}$ Let $\mathbf{n}$ be the unit vector that is normal to the surface on which the stress vector $\mathbf{t}$ acts and that is directed towards the portion of the material exerting this force. Argue that $\mathbf{n}$ must be frame indifferent

$$\mathbf{n}^* = \mathbf{Q} \cdot \mathbf{n}$$

and conclude that the stress tensor is a frame indifferent tensor field

$$\mathbf{T}^* = \mathbf{Q} \cdot \mathbf{T} \cdot \mathbf{Q}^T$$

2.1.7-3 *frame indifferent forms of the jump momentum balance and the differential momentum balance* Use Exercise 1.4.3-1 to conclude that the frame indifferent form of the jump momentum balance is

$$\rho^{(\sigma)} \left\{ \frac{d_{(s)} \mathbf{v}^{(\sigma)}}{dt} - \ddot{\mathbf{z}}_{(0)} - \dot{\boldsymbol{\omega}} \wedge (\mathbf{z} - \mathbf{z}_{(0)}) - \boldsymbol{\omega} \wedge [\boldsymbol{\omega} \wedge (\mathbf{z} - \mathbf{z}_{(0)})] \right.$$

$$\left. - 2\, \boldsymbol{\omega} \wedge [\mathbf{v}^{(\sigma)} - \dot{\mathbf{z}}_{(0)} - \boldsymbol{\omega} \wedge (\mathbf{z} - \mathbf{z}_{(0)})] \right\}$$

$$= \text{div}_{(\sigma)} \mathbf{T}^{(\sigma)} + \rho^{(\sigma)} \mathbf{b}^{(\sigma)} - [\, \rho (\mathbf{v} \cdot \boldsymbol{\xi} - v_{(\xi)})^2 \boldsymbol{\xi} - \mathbf{T} \cdot \boldsymbol{\xi} \,]$$

and that the frame indifferent form of the differential momentum balance is

$$\rho \left\{ \frac{d_{(m)} \mathbf{v}}{dt} - \ddot{\mathbf{z}}_{(0)} - \dot{\boldsymbol{\omega}} \wedge (\mathbf{z} - \mathbf{z}_{(0)}) - \boldsymbol{\omega} \wedge [\boldsymbol{\omega} \wedge (\mathbf{z} - \mathbf{z}_{(0)})] \right.$$

$$\left. - 2\, \boldsymbol{\omega} \wedge [\mathbf{v} - \dot{\mathbf{z}}_{(0)} - \boldsymbol{\omega} \wedge (\mathbf{z} - \mathbf{z}_{(0)})] \right\}$$

$$= \text{div}\, \mathbf{T} + \rho \mathbf{b}$$

Here $\mathbf{z}_{(0)}$ is the position of a point that is fixed in the inertial frame of reference; $\boldsymbol{\omega}$ is the angular velocity of the inertial frame of reference with respect to the current frame of reference. The jump momentum balance and the differential momentum balance take these same forms in every frame of reference. In the inertial frame of reference for example,

$$\ddot{\mathbf{z}}_{(0)} = \dot{\boldsymbol{\omega}} = \boldsymbol{\omega} = 0$$

2.1.8 Surface velocity, surface stress, and surface body force In Sec. 1.3.2, I suggest how surface mass density may be interpreted when the dividing surface is thought of as a model for a three-dimensional interfacial region (see Sec. 1.2.4). What would be the corresponding interpretation for the surface stress tensor?

In Exercise 1.3.2-2, we examined the form that the general balance equation takes when the interface is represented by a three-dimensional region. Euler's first law is a special case for which

$$\psi \equiv \rho \mathbf{v} \qquad \psi^{(\sigma)} \equiv \rho^{(\sigma)} \mathbf{v}^{(\sigma)}$$

$$\boldsymbol{\phi} \equiv -\mathbf{T} \qquad \boldsymbol{\phi}^{(s)} \equiv -\mathbf{T}^{(\sigma)}$$

$$\boldsymbol{\zeta} \equiv \mathbf{b} \qquad \boldsymbol{\zeta}^{(\sigma)} \equiv \mathbf{b}^{(\sigma)} \tag{8-1}$$

Our conclusion is that

$$\rho^{(\sigma)}\,\mathbf{v}^{(\sigma)} = \int_{\lambda^-}^{\lambda^+} (\,\rho^{(I)}\mathbf{v}^{(I)} - \rho\mathbf{v}\,)(\,1 - \kappa_1\lambda\,)(\,1 - \kappa_2\lambda\,)\,d\lambda \tag{8-2}$$

$$\mathbf{T}^{(\sigma)} \equiv \left\{ \int_{\lambda^-}^{\lambda^+} (\,\mathbf{T}^{(I)} - \mathbf{T}\,)\,d\lambda \right\} \cdot \mathbf{P}$$

$$= \mathbf{P} \cdot \left\{ \int_{\lambda^-}^{\lambda^+} (\,\mathbf{T}^{(I)} - \mathbf{T}\,)\,d\lambda \right\} \cdot \mathbf{P} \tag{8-3}$$

$$\mathbf{b}^{(\sigma)} \equiv \frac{1}{\rho^{(\sigma)}} \int_{\lambda^-}^{\lambda^+} (\,\rho^{(I)}\mathbf{b}^{(I)} - \rho\mathbf{b}\,)(\,1 - \kappa_1\lambda\,)(\,1 - \kappa_2\lambda\,)\,d\lambda \tag{8-4}$$

The second line of (8-3) follows from the requirement that $\mathbf{T}^{(\sigma)}$ be a tangential tensor field (Sec. 2.1.7).

Note that $\mathbf{v}^{(\sigma)}$ is not defined by (8-2). Usually we will require the tangential components of $\mathbf{v}^{(\sigma)}$ to be continuous across the dividing surface,

$$\text{at } \Sigma : \mathbf{P} \cdot \mathbf{v}^{(\sigma)} = \mathbf{P} \cdot \mathbf{v} \tag{8-5}$$

The normal component of $\mathbf{v}^{(\sigma)}$ is identified as the speed of displacement of Σ, $v_{(\xi)}$. As a result, $\mathbf{v}^{(\sigma)}$ is specified by the location and motion of Σ. This is an important distinction, since we will sometimes wish to define the location of Σ in such a manner that $\rho^{(\sigma)} = 0$ (see Sec. 1.3.6).

Exercises

2.1.8-1 *Euler's second law for the interfacial region* Euler's first law implies (8-2), (8-3), and (8-4). Are these identifications consistent with Euler's second law?

i) Write Euler's second law for the interfacial region and rearrange in the form

$$\frac{d}{dt}\left\{ \int_{R} \mathbf{z} \wedge \rho\mathbf{v}\,dV + \int_{R^{(I)}} \mathbf{z} \wedge (\,\rho^{(I)}\mathbf{v}^{(I)} - \rho\mathbf{v}\,)\,dV \right\}$$

$$= \int_{S} \mathbf{z} \wedge \mathbf{t} \, dA + \int_{S^{(I)}} \mathbf{z} \wedge (\mathbf{t}^{(I)} - \mathbf{t}) \, dA$$

$$+ \int_{R} \mathbf{z} \wedge \rho \mathbf{b} \, dV + \int_{R^{(I)}} \mathbf{z} \wedge (\rho^{(I)} \mathbf{b}^{(I)} - \rho \mathbf{b}) \, dV \tag{1}$$

ii) Recognizing that the interfacial region is very thin, discuss the approximations involved in saying

$$\int_{R^{(I)}} \mathbf{z} \wedge (\rho^{(I)} \mathbf{v}^{(I)} - \rho \mathbf{v}) \, dV \doteq \int_{\Sigma} \mathbf{z} \wedge \rho^{(\sigma)} \mathbf{v}^{(\sigma)} \, dA \tag{2}$$

$$\int_{S^{(I)}} \mathbf{z} \wedge (\mathbf{t}^{(I)} - \mathbf{t}) \, dA \doteq \int_{C} \mathbf{z} \wedge \mathbf{t}^{(\sigma)} \, ds \tag{3}$$

$$\int_{R^{(I)}} \mathbf{z} \wedge (\rho^{(I)} \mathbf{b}^{(I)} - \rho \mathbf{b}) \, dV \doteq \int_{\Sigma} \mathbf{z} \wedge \rho^{(\sigma)} \mathbf{b}^{(\sigma)} \, dA \tag{4}$$

Equation (1) now takes the form of Euler's second law in Sec. 2.1.3, where the interface was represented as a dividing surface. Our conclusion is that (8-2), (8-3), and (8-4) are consistent with both Euler's first and second laws.

2.1.9 Euler's first law at common line What does Euler's first law require at each point on a common line?

Let us begin by restating Euler's first law (Sec. 2.1.3), expressing the stress vector $\mathbf{t}$ in terms of the stress tensor $\mathbf{T}$ (Sec. 2.1.4) and the surface stress vector $\mathbf{t}^{(\sigma)}$ in terms of the surface stress tensor $\mathbf{T}^{(\sigma)}$ (Sec. 2.1.5):

$$\frac{d}{dt} \left(\int_{R} \rho \mathbf{v} \, dV + \int_{\Sigma} \rho^{(\sigma)} \mathbf{v}^{(\sigma)} \, dA \right) = \int_{S} \mathbf{T} \cdot \mathbf{n} \, dA + \int_{C} \mathbf{T}^{(\sigma)} \cdot \boldsymbol{\mu} \, ds$$

$$+ \int_{R} \rho \mathbf{b} \, dV + \int_{\Sigma} \rho^{(\sigma)} \mathbf{b}^{(\sigma)} \, dA \tag{9-1}$$

The first integral on the right becomes after an application of Green's transformation (Slattery 1981, p. 661)

$$\int_S \mathbf{T} \cdot \mathbf{n} \, dA = \int_R \text{div} \, \mathbf{T} \, dV + \int_\Sigma [\, \mathbf{T} \cdot \boldsymbol{\xi} \,] \, dA \tag{9-2}$$

Using the surface divergence theorem (Exercise A.6.3-1), we can write the second integral on the right of (9-1) as

$$\int_C \mathbf{T}^{(\sigma)} \cdot \boldsymbol{\mu} \, ds = \int_\Sigma \text{div}_{(\sigma)} \mathbf{T}^{(\sigma)} \, dA + \int_{C^{(cl)}} (\, \mathbf{T}^{(\sigma)} \cdot \boldsymbol{\nu} \,) \, ds \tag{9-3}$$

After applying the transport theorem for a body containing intersecting dividing surfaces to the terms on the left of (9-1), we can use (9-2) and (9-3) to write (9-1) as

$$\begin{aligned} &\int_R \left(\rho \frac{d_{(m)} \mathbf{v}}{dt} - \text{div} \, \mathbf{T} - \rho \mathbf{b} \right) dV \\ &\quad + \int_\Sigma \Bigg\{ \rho^{(\sigma)} \frac{d_{(s)} \mathbf{v}^{(\sigma)}}{dt} - \text{div}_{(\sigma)} \mathbf{T}^{(\sigma)} - \rho^{(\sigma)} \mathbf{b}^{(\sigma)} \\ &\qquad + [\, \rho (\, \mathbf{v} - \mathbf{v}^{(\sigma)} \,)(\, \mathbf{v} \cdot \boldsymbol{\xi} - v_{(\xi)}^{\{g\}}) - \mathbf{T} \cdot \boldsymbol{\xi} \,] \Bigg\} dA \\ &\quad + \int_{C^{(cl)}} (\, \rho^{(\sigma)} \mathbf{v}^{(\sigma)} [\, \mathbf{v}^{(\sigma)} \cdot \boldsymbol{\nu} - u_{(\nu)}^{\{cl\}} \,] - \mathbf{T}^{(\sigma)} \cdot \boldsymbol{\nu} \,) \, ds \\ &= 0 \end{aligned} \tag{9-4}$$

In view of Cauchy's first law (Sec. 2.1.4) and the jump momentum balance (Sec. 2.1.6), (9-4) reduces to

$$\int_{C^{(cl)}} (\, \rho^{(\sigma)} \mathbf{v}^{(\sigma)} [\, \mathbf{v}^{(\sigma)} \cdot \boldsymbol{\nu} - u_{(\nu)}^{\{cl\}} \,] - \mathbf{T}^{(\sigma)} \cdot \boldsymbol{\nu} \,) \, ds = 0 \tag{9-5}$$

Since (9-1) is valid for every material body and every portion of a material body, (9-5) must be true for every portion of a common line. We conclude (see Exercises 1.3.8-1 and 1.3.8-2)

$$(\, \rho^{(\sigma)} \mathbf{v}^{(\sigma)} [\, \mathbf{v}^{(\sigma)} \cdot \boldsymbol{\nu} - u_{(\nu)}^{\{cl\}} \,] - \mathbf{T}^{(\sigma)} \cdot \boldsymbol{\nu} \,) = 0 \tag{9-6}$$

This can be called the **momentum balance at the common line.** It expresses the requirement of Euler's first law at every point on a common line.

If we neglect mass transfer at the common line

$$\text{at } C^{(cl)}: \quad \mathbf{v}^{(\sigma)} \cdot \boldsymbol{\nu} = u^{(cl)}_{(\nu)} \tag{9-7}$$

or if we neglect the effect of inertial forces at the common line, the momentum balance at the common line reduces to

$$(\mathbf{T}^{(\sigma)} \cdot \boldsymbol{\nu}) = 0 \tag{9-8}$$

Under static conditions, the surface stress acting at the common line may be expressed in terms of the surface tension and (9-8) becomes

$$(\gamma \boldsymbol{\nu}) = 0 \tag{9-9}$$

In the context of Figure 9-1, this says

$$\gamma^{(AB)} \boldsymbol{\nu}^{(AB)} + \gamma^{(AS)} \boldsymbol{\nu}^{(AS)} + \gamma^{(BS)} \boldsymbol{\nu}^{(BS)} = 0 \tag{9-10}$$

Here $\gamma^{(AB)}$, $\gamma^{(AS)}$, and $\gamma^{(BS)}$ are the surface tensions within the A-B, A-S, and B-S dividing surfaces in the limit as the common line is approached. Equations (9-9) and (9-10) have been referred to as the **Neumann triangle** (Buff 1960, p. 288).

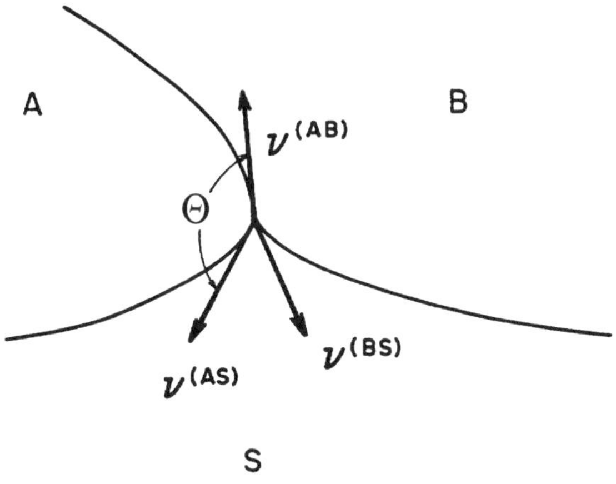

Figure 2.1.9-1: The common line formed by the intersection of the A-B, A-S, and B-S dividing surfaces. Here $\boldsymbol{\nu}^{(AB)}$, $\boldsymbol{\nu}^{(AS)}$, and $\boldsymbol{\nu}^{(BS)}$ are the unit vectors that are normal to the common line and that are both tangent to and directed into the A-B, A-S, and B-S dividing surfaces respectively; θ is the contact angle measured through phase A.

In deriving the transport theorem for a body containing intersecting dividing surfaces and therefore in deriving the momentum balance at the common line, we assume that **T** is piecewise continuous with piecewise continuous first derivatives (see Sec. 1.3.7). It will be important to keep this in mind in discussing the momentum balance at a common line on a rigid solid in the next section.

Note also that in stating (9-1) and in deriving the momentum balance at the common line, I have assumed that there is no momentum or force intrinsically associated with the common line. Gibbs (1948, p. 288 and 296) recognized the possibility of a force or line tension acting in the common line, but he gave it relatively little attention, suggesting that he considered it to be too small to be of importance. Buff and Saltsburg (1957; see also Buff 1960, p. 287) argued that, at least at equilibrium, the effect of the line tension would be too small to be of significance in macroscopically observable phenomena. Yet recent experimental studies suggest that the effects of line tension are observable (Platikanov *et al.* 1980; Kralchevsky *et al.* 1986. The observations of Gaydos and Neumann 1987 appear to have a better explanation, as indicated in Sec. 3.2.7). On balance, it seems premature to make a judgment on the matter at this writing.

References

Buff, F. P., "Handbuch der Physik," vol. 10, edited by S. Flügge, Springer-Verlag, Berlin (1960).

Buff, F. P., and H. Saltsburg, *J. Chem. Phys.* **26**, 23 (1957).

Gaydos, J., and A. W. Neumann, *J. Colloid Interface Sci.* **120**, 76 (1987).

Gibbs, J. W., "The Collected Works," vol. 1, Yale University Press, New Haven, Conn. (1948).

Kralchevsky, P. A., A. D. Nikolov, and I. B. Ivanov, *J. Colloid Interface Sci.* **112**, 132 (1986).

Platikanov, D., M. Nedyalkov, and V. Nasteva, *J. Colloid Interface Sci.* **75**, 620 (1980).

Slattery, J. C., "Momentum, Energy, and Mass Transfer in Continua," McGraw-Hill, New York (1972); second edition, Robert E. Krieger, Malabar, Fl 32950 (1981).

Exercise

2.1.9-1 *Euler's second law at common line* By an analysis analogous to (9-1) through (9-6) of the text, conclude that Euler's second law places no restrictions upon momentum transfer at the common line beyond that imposed by the momentum balance at the common line.

2.1.10 Momentum balance at common line on relatively rigid solid If the contact angle measured at the common line on a rigid solid through one of the fluid phases were either 0 or π, the solid would not be deformed by the stress exerted on the common line by the fluid-fluid interface, since this stress would act tangent to the solid surface.

If we assume that the contact angle is neither 0 nor π, usual practice is to ignore any deformation in the neighborhood of the common line, arguing that it would be less than the imperfections in a relatively high tensile strength solid surface that we are generally willing to ignore (Lester 1961; Wickham and Wilson 1975). We will ignore the deformation pictured in Figure 9-1 and idealize the neighborhood of the common line by Figure 10-1.

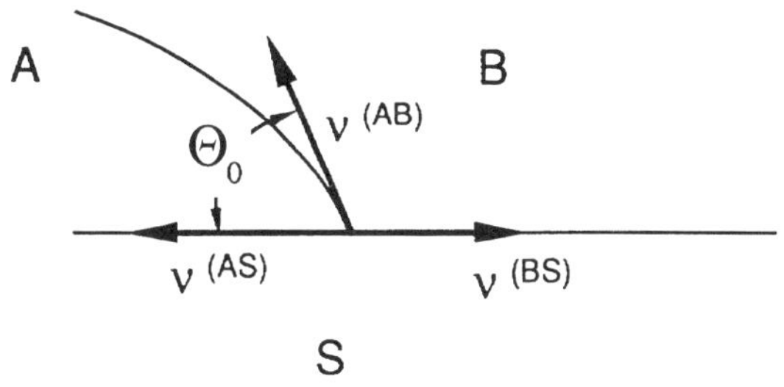

Figure 2.1.10-1: For a relatively rigid solid, the deformation shown in Figure 9-1 will normally not be apparent.

There is an important distinction to be made between choosing to ignore the deformation in a relatively high tensile strength solid and adopting the model of a perfectly rigid solid. It would appear that, with a perfectly rigid solid, **T** would no longer be piecewise continuous with piecewise continuous first derivatives as assumed in deriving the momentum balance at the common line. This assumes that **T** would be bounded in the adjoining fluid phases. [Every second-order tensor in three-dimensional space is bounded (Stakgold 1967, p. 140).]

In choosing to ignore the deformation of the solid in the neighborhood of the common line, we eliminate from consideration the effect of the mechanical properties of the solid and the effect of the normal component of the momentum balance at the common line. Only the component of (9-6) tangent to the solid surface is significant:

$$\mathbf{P}^{(AS)} \cdot \left(\rho^{(\sigma)} \mathbf{v}^{(\sigma)} [\, \mathbf{v}^{(\sigma)} \cdot \boldsymbol{\nu} - u^{(cl)}_{(\nu)} \,] - \mathbf{T}^{(\sigma)} \cdot \boldsymbol{\nu} \right) = 0 \tag{10-1}$$

Here $\mathbf{P}^{(AS)}$ is the projection tensor for the A-S dividing surface.

If, as is common, we neglect both mass transfer at the common line and any contribution to the surface stress tensor beyond the interfacial tension, (10-1) simplifies to

$$\gamma^{(AB)}\, \boldsymbol{\nu}^{(AS)} \cdot \boldsymbol{\nu}^{(AB)} + \gamma^{(AS)} - \gamma^{(BS)} = 0 \tag{10-2}$$

or

$$\gamma^{(AB)} \cos \Theta_0 + \gamma^{(AS)} - \gamma^{(BS)} = 0 \tag{10-3}$$

which is known as **Young's equation** (Young 1805). Here Θ_0 is the contact angle measured through phase A in Figure 10-1 and $\gamma^{(AB)}$ is the fluid-fluid interfacial tension. It is important to note that (10-3) and Θ_0 refer to the true common line (given a somewhat vague distinction in this continuum picture between a species adsorbed in a solid vapor interface on one side of the common line and a thin film of liquid on the other) rather than an apparent common line seen with 10× magnification. Observe also that, since $\gamma^{(AS)}$ and $\gamma^{(BS)}$ in general include corrections for the actual curvature of the fluid-solid interfaces, they are not thermodynamic interfacial tensions (see Sec. 5.8.6).[a] One should be cautious in interpreting $\gamma^{(AS)}$ and $\gamma^{(BS)}$ as thermodynamic quantities, unless $\Theta_0 = 0$ or π, in which case the solid would not be deformed.

Although the momentum balance at the common line (see Sec. 2.1.9) must always be satisfied, a value of Θ_0 can not be found such that Young's equation is satisfied for an arbitrary choice of three phases or therefore for an arbitrary choice of $\gamma^{(AB)}$, $\gamma^{(AS)}$, and $\gamma^{(BS)}$.[b] One possible answer to this

---

a) Gibbs (1948, p. 326; see also Dussan V. 1979) suggests that (10-3) can be derived by starting with the observation that energy is minimized at equilibrium. In Exercise 5.10.3-3, we observe that the Gibbs free energy is minimized as a function of time as equilibrium is approached for an isolated, isothermal, multiphase body consisting of isobaric fluids. In arriving at the final result of this exercise, both the Gibbs-Duhem equation and the surface Gibbs-Duhem equation are employed, which again means that all three phases are assumed to be fluids. As derived here, there is no recognition in these results that the internal energy of a solid is a function of its deformation from its reference configuration. On the basis of the discussion presented in this text, Young's equation is not implied by the observation that the Gibbs free energy is minimized at equilibrium for such a system.

dilemma is to remember that $\gamma^{(AS)}$ and $\gamma^{(BS)}$ are not thermodynamic quantities and they can not assume arbitrary values.

Since $\gamma^{(AS)}$ and $\gamma^{(BS)}$ are in general not material properties, it is not clear that they can be measured. For a gas-liquid-solid system for example, the common practice would be to measure the contact angles corresponding to two different liquids using perhaps 10× magnification. As pointed out above, (10-3) and Θ_0 refer to the true common line, not the apparent common line seen experimentally in this manner.

At the risk of needlessly emphasizing the point, in practice Θ_0 is not measured. Experimentalists report measurements of Θ determined at some distance from the common line, perhaps with 10× magnification. Here the fluid

---

b) It is common to express Young's equation in terms of the **spreading coefficient** (Adamson 1982; Miller and Neogi 1985)

$$S_{A/B} \equiv \gamma^{(BS)} - \gamma^{(AS)} - \gamma^{(AB)}$$

$$= \gamma^{(AB)}(\cos\Theta_0 - 1) \qquad (10\text{-}4)$$

If one interprets $\gamma^{(AS)}$ and $\gamma^{(BS)}$ as thermodynamic quantitities, it may not be possible to assign a contact angle Θ_0 such that (10-4) can be satisfied for an arbitrary selection of materials. The possibilities are these.

i) If $S_{A/B} > 0$, there is no value of Θ_0 that allows (10-4) to be satisfied. The usual interpretation is that fluid A will **spontaneously spread** over the solid, displacing fluid B. The solid prefers to be in contact with phase A, and the thin film of phase A can be expected to exhibit a positive disjoining pressure (see Exercise 2.1.3-1).

ii) If $S_{A/B} < -2\gamma^{(AB)}$, there again is no value of Θ_0 such that (10-4) can be satisfied. The interpretation is that a thin film of fluid A will **spontaneously dewet** the solid, being displaced by fluid B. In this case, the solid prefers to be in contact with phase B, and the thin film of fluid A can be expected to exhibit a negative disjoining pressure (see Exercise 2.1.3-1).

iii) If $-2\gamma^{(AB)} < S_{A/B} < 0$, there is a value of Θ_0 such that (10-4) can be satisfied, and the common line is stationary. In this case, fluid A is said to **partially wet** the solid.

In summary, if $S_{A/B} > 0$ or $S_{A/B} < -2\gamma^{(AB)}$, the usual interpretation is that the common line moves spontaneously over the solid; if $-2\gamma^{(AB)} < S_{A/B} < 0$, the common line is stationary.

For a different point of view, see Sec. 2.1.13.

films are sufficiently thick that the effects of mutual forces, such as London-van der Waals forces and electrostatic double-layer forces (see Exercises 2.1.3-1 and 2.1.3-2, Sec. 3.2.7, and Sec. 3.3.3) as well as structural forces (Israelachvili and McGuiggan 1988), can be neglected.

In summary, if $\gamma^{(BS)} - \gamma^{(AS)}$ can not be measured independently, Young's equation can not be used to determine Θ_0. At this writing, it appears that we will generally have little choice but to assume a value for Θ_0, as we have done in Secs. 3.2.7 and 3.3.3.

For another critique of Young's equation, see Dussan V. (1979).

References

Adamson, A. W., "Physical Chemistry of Surfaces," fourth edition, John Wiley, New York (1982).

Dussan V., E. B., *Ann Rev. Fluid Mech.* **11**, 371 (1979).

Gibbs, J. W., "The Collected Works," vol. 1, Yale University Press, New Haven, Conn. (1948).

Israelachvili, J. N., and P. M. McGuiggan, *Science* **241**, 795 (1988).

Lester, G. R., *J. Colloid Sci.* **16**, 315 (1961).

Miller, C. A., and P. Neogi, "Interfacial Phenomena," Marcel Dekker, New York (1985).

Stakgold, I., "Boundary Value Problems of Mathematical Physics," Vol. 1, Macmillan, New York (1967).

Wickham, G. R., and S. D. R. Wilson, *J. Colloid Interface Sci.* **51**, 189 (1975).

Young, T., *Philos. Trans. R. Soc. London* (4 to.) **95**, 65 (1805).

2.1.11 Factors influencing measured contact angles Remember that an experimentalist typically measures Θ, the apparent contact angle at some distance from the true common line. (See Secs. 2.1.10 and 3.2.7 for further discussion of this point.) The magnitude of Θ can be influenced by a number of factors.

Contact angle hysteresis The difference between advancing and receding contact angles is referred to as *contact angle hysteresis*. Either the roughness or heterogeneous composition of the surface is the usual cause (Dettre and Johnson 1964; Johnson and Dettre 1964a,b; Neumann and Good 1972; Eick *et al.* 1975; Morrow 1975; Huh and Mason 1977a; Joanny and de Gennes 1984; Schwartz and Garoff 1985; Morra *et al.* 1989). In some cases, the system requires a finite time to reach equilibrium (Elliott and Riddiford 1965), and the rate at which equilibrium is approached may itself depend upon the direction in which the common line recently has moved (Yarnold and Mason 1949). Depending upon the system, the approach to equilibrium may be characterized either by the adsorption (desorption) of a species in an interface or by the movement of an advancing (trailing) thin film. When roughness is absent and equilibrium is reached quickly, contact angle hysteresis appears to be eliminated (Morrow 1975).

Speed of displacement of common line The contact angle is known to depend upon the speed of displacement of the apparent common line as observed with 10× magnification (Ablett 1923; Rose and Heins 1962; Elliott and Riddiford 1965 and 1967[a]; Schonhorn *et al.* 1966; Ellison and Tejada 1968; Coney and Masica 1969; Schwartz and Tejada 1970, 1972; Hansen and Toong 1971; Radigan *et al.* 1974; Hoffman 1975; Burley and Kennedy 1976a,b; Kennedy and Burley 1977). Several distinct regimes are possible.

If the speed of displacement of the common line is sufficiently small, the advancing contact angle will be independent of it, since the solid within the immediate neighborhood of the common line nearly will have achieved equilibrium with the adjoining phases (Elliott and Riddiford 1967[a]).

If the speed of displacement of the common line is too large for equilibrium to be attained but too small for viscous effects within the adjoining fluid phases to play a significant role, the advancing contact angle will increase with increasing speed of displacement (Hansen and Miotto 1957; Elliott and Riddiford 1967; Ellison and Tejada 1968; Blake and Hayes 1969; Schwartz and Tejada 1970, 1972; Hoffman 1975; Chen 1988; Chen and Wada 1989). If we think of the **capillary number**

$$N_{ca} \equiv \frac{\mu u_{(v)}^{(cl)}}{\gamma} \tag{11-1}$$

---

a) Elliott and Riddiford's (1967) measurements of the receding contact angle are in error (Wilson 1975).

as being characteristic of the ratio of viscous forces to surface tension forces, we should be able to neglect the effect of viscous forces upon the contact angle when N_{ca} is sufficiently small (Hoffman 1975). Here μ is the viscosity of the more viscous fluid, $u_{\{v\}}^{(cl)}$ is the speed of displacement of the apparent common line or meniscus as observed experimentally, and γ is the fluid-fluid interfacial tension. Sometimes the advancing contact angle appears to approach an upper limit with increasing speed of displacement (Ellison and Tejada 1968; Schwartz and Tejada 1970, 1972; Hoffman 1975; Chen 1988; Fermigier and Jenffer 1988; Chen and Wada 1989). Under these conditions, the common line probably is moving too rapidly for the deviation from equilibrium to have a significant effect upon the contact angle but still too slowly for viscous effects within the adjoining phases to play a significant role.

If the speed of displacement of the common line is sufficiently large for the effects of the viscous forces to be important but too small for inertial effects within the adjoining fluid phases to be considered, the advancing contact angle will once again increase with increasing speed of displacement (Hoffman 1975; Cherry and Holmes 1969). If the **Reynolds number**, which characterizes the ratio of inertial forces to viscous forces,

$$N_{Re} \equiv \frac{\rho \, u_{\{v\}}^{(cl)} \, \ell}{\mu}$$

$$= N_{We}/N_{ca} \tag{11-2}$$

is sufficiently small or if the **Weber number**, which characterizes the ratio of inertial forces to surface tension forces,

$$N_{We} \equiv \frac{\rho (u_{\{v\}}^{(cl)})^2 \ell}{\gamma} \tag{11-3}$$

is sufficiently small for a fixed capillary number N_{ca}, we should be able to neglect the effects of the inertial forces (Hoffman 1975). Here ρ is the density of the more dense fluid and ℓ is a length characteristic of the particular geometry being considered.

With increasing speed of displacement, the contact angle approaches 180°. As this upper limit is reached, a continuous, visible film of the fluid originally in contact with the solid is entrained and the common line necessarily disappears (Burley and Kennedy 1976a,b; Kennedy and Burley 1977). This is referred to as **dynamic wetting**. Inertial forces have become more important than surface forces under these conditions.

Character of solid surface I mentioned the effect of roughness in connection with contact angle hysteresis. Roughness, scratches or grooves will affect both the advancing and receding contact angles (Morrow 1975; Huh and Mason 1977a; Oliver *et al.* 1977; Bayramli and Mason 1978; Mori *et al.* 1982; Cain *et al.* 1983). Chemical heterogeneities undoubtedly also play a significant role (Cassie 1948; Johnson and Dettre 1969; Neumann and Good 1972; Israelachvili and Gee 1989).

Geometry A number of techniques have been used to study the effect of the speed of displacement of the common line upon the contact angle.

1) Ablett (1923) used a partially immersed rotating cylinder. The depth of immersion was adjusted to give a flat phase interface, which in turn allowed the contact angle to be easily calculated.

2) Rose and Heins (1962), Blake *et al.* (1967), Hansen and Toong (1971), Hoffman (1975), and Legait and Sourieau (1985) have observed the advancing contact angle in the displacement of one fluid by another through a round cylindrical tube.

3) Coney and Masica (1969) studied displacement in a tube with a rectangular cross section.

4) Schonhorn *et al.* (1966), Radigan *et al.* (1974), Chen (1988), and Chen and Wada (1989) examined the spreading of sessile drops.

5) Ellison and Tejada (1968), Inverarity (1969), and Schwartz and Tejada (1970, 1972) photographed a wire entering a liquid-gas interface.

6) Burley and Kennedy (1976a,b) and Kennedy and Burley (1977) studied a plane tape entering a liquid-gas interface.

7) Johnson *et al.* (1977) and Cain *et al.* (1983) measured both advancing and receding contact angles using the Wilhelmy plate.

8) Elliott and Riddiford (1962, 1967) observed both advancing and receding contact angles in radial flow between two flat plates. Johnson et al. (1977) found considerably different results in examining similar systems with the Wilhelmy plate. Wilson (1975) has discussed the instabilities associated with radial flow between plates.

9) Ngan and Dussan V. (1982) measured the advancing dynamic contact angle during upward flow between parallel plates.

Unfortunately, entirely comparable experiments have rarely been carried out in two different geometries.

Gravity plays an important role in at least a portion of the experiments of Ablett (1923), Ellison and Tejada (1968), Schwartz and Tejada (1970, 1972), Johnson *et al.* (1977), Ngan and Dussan V. (1982), and Legait and Sourieau (1985). The principal motions are parallel with gravity.

Inverarity (1969) reports the equilibrium contact angle rather than the static advancing contact angle. Since these two contact angles are in general different, there is no basis for comparing his data with those of others.

Johnson *et al.* (1977) concluded that effects of adsorption on the solid in the immediate neighborhood of the moving common line might be present in the studies of Elliott and Riddiford (1962, 1967[a]).

Schonhorn *et al.* (1966) observed that the rate of spreading was unchanged when a drop was inverted. This demonstrated that the effect of gravity was negligible in their experiments.

The effects of both inertia and gravity are significant in the studies of Burley and Kennedy (1976a,b) and Kennedy and Burley (1977).

Coney and Masica (1969) considered displacement over a previously wet wall, which sets their work apart from others.

As one exception, Jiang *et al.* (1979) have successfully compared Hoffman's (1975) data with selected experiments of Schwartz and Tejada (1970), who measured the dynamic contact angle formed as a wire entered a liquid-gas interface (see Sec. 2.1.12).

The equilibrium contact angle observed with 10× magnification appears to be independent of measurement technique or geometry, so long as (see Sec. 3.2.7 for additional restrictions)

$$L/\ell \ll 1 \tag{11-4}$$

where ℓ is a length of the macroscopic system that characterizes the radius of curvature of the interface,

$$L \equiv \left(\frac{\gamma}{g \, \Delta\rho} \right)^{1/2} \tag{11-5}$$

is a characteristic length of the meniscus, g the acceleration of gravity, and $\Delta\rho$ the density difference between the two phases. Equation (11-4) appears to be valid for most measurements reported to date. Exceptions are the contact angles reported by Gaydos and Neumann (1987), which depend upon the diameter of the sessile drop used as we would predict. [We prefer the argument presented in Sec. 3.2.7 to the explanation of Gaydos and Neumann (1987) in terms of line tension (see Sec. 2.1.9).]

We find in Sec. 3.3.3 that the dynamic contact angle created as a thin film recedes under the influence of a negative disjoining pressure appears to be independent of geometry, so long as (11-4) is valid (see Sec. 3.3.3 for additional restrictions). We expect a similar argument to be developed for advancing contact angles. This expectation appears to be supported by the correlations of Jiang *et al.* (1979) and Chen (1988) for advancing contact angles (see Sec. 2.1.12). It is only in the data of Ngan and Dussan V. (1982) and of Legait and Sourieau (1985), for which (11-4) is not valid, that we see a dependence upon geometry.

Configuration of interface within immediate neighborhood of common line Chen and Wada (1989) and Heslot *et al.* (1989) have shown experimentally that the configuration of the interface within the immediate neighborhood of the common line is quite different from that which one would observe with a 10× microscope. For more on this point, see Sections 1.3.10, 2.1.13, 3.1.1, 3.2.7 and 3.3.3.

References

Ablett, R., *Philos. Mag.* (6) **46**, 244 (1923).

Bayramli, E., and S. G. Mason, *J. Colloid Interface Sci.* **66**, 200 (1978).

Blake, T. D., D. H. Everett, and J. M. Haynes, in "Wetting," S. C. I. Monograph No. 25, Society of Chemical Industry, London (1967).

Blake, T. D., and J. M. Haynes, *J. Colloid Interface Sci.* **30**, 421 (1969).

Burley, R.,, and B. S. Kennedy, *Chem. Eng. Sci.* **31**, 901 (1976).

Burley, R., and B. S. Kennedy, *Br. Polym. J.* **8**, 140 (1976).

Cain, J. B., D. W. Francis, R. D. Venter, and A. W. Neumann, *J. Colloid Interface Sci.* **94**, 123 (1983).

Cassie A. B. D., *Discuss. Faraday Soc.* **3**, 11 (1948).

Chen, J. D., *J. Colloid Interface Sci.* **122**, 60 (1988).

Chen, J. D., and N. Wada, *Phys. Rev. Letters* **62**, 3050 (1989).

Cherry, B. W., and C. M. Holmes, *J. Colloid Interface Sci.* **29**, 174 (1969).

Coney, T. A., and W. J. Masica, *NASA Tech. Note* **TN D-5115** (1969).

Dettre, R. H., and R. E. Johnson Jr., "Contact Angle, Wettability, and Adhesion," *Advances in Chemistry Series No. 43*, p. 136, American Chemical Society, Washington, D.C. (1964).

Dussan V., E. B., *J. Fluid Mech.* **77**, 665 (1976).

Eick, J. D., R. J. Good, and A. W. Neumann, *J. Colloid Interface Sci.* **53**, 235 (1975).

Elliott, G. E. P., and A. C. Riddiford, *Nature London* **195**, 795 (1962).

Elliott, G. E. P., and A. C. Riddiford, *Recent Progr. Surface Sci.* **2**, 111 (1965).

Elliott, G. E. P., and A. C. Riddiford, *J. Colloid Interface Sci.* **23**, 389

(1967).

Ellison, A. H., and S. B. Tejada, *NASA Contract Rep.* **CR 72441** (1968).

Fermigier, M. and P. Jenffer, *Annales de Physique (Colloque n° 2, suuplement au n° 3)* **13**, 37 (1988).

Gaydos, J., and A. W. Neumann, *J. Colloid Interface Sci.* **120**, 76 (1987).

Hansen, R. S., and M. Miotto, *J. Am. Chem. Soc.* **79**, 1765 (1957).

Hansen, R. J., and T. Y. Toong, *J. Colloid Interface Sci.* **36**, 410 (1971).

Heslot, F., N. Fraysse, and A. M. Cazabat, *Nature* **338**, 640 (1989).

Hoffman, R. L., *J. Colloid Interface Sci.* **50**, 228 (1975).

Huh, C., and S. G. Mason, *J. Colloid Interface Sci.* **60**, 11 (1977).

Huh, C., and S. G. Mason, *J. Fluid Mech.* **81**, 401 (1977).

Inverarity, G., Ph.D. dissertation, Victoria University of Manchester (1969).

Israelachvili, J. N., and M. L. Gee, *Langmuir* **5**, 288 (1989).

Jiang, T. S., S. G. Oh, and J. C. Slattery, *J. Colloid Interface Sci.* **69**, 74 (1979).

Joanny, J. F., and P. G. de Gennes, *J. Chem. Phys.* **81**, 552 (1984).

Johnson, R. E. Jr., and R. H. Dettre, "Contact Angle, Wettability, and Adhesion," *Advances in Chemistry Series No. 43*, p. 136, American Chemical Society, Washington, D.C. (1964).

Johnson, R. E. Jr., and R. H. Dettre, *J. Phys. Chem.* **68**, 1744 (1964).

Johnson, R. E. Jr., and R. H. Dettre, "Surface and Colloid Science," vol. 2, p. 85, edited by E. Matijevic, Wiley-Interscience, New York (1969).

Johnson, R. E. Jr., R. H. Dettre, and D. A. Brandreth, *J. Colloid Interface Sci.* **62**, 205 (1977).

Kennedy, B. S., and R. Burley, *J. Colloid Interface Sci.* **62**, 48 (1977).

Legait, B., and P. Sourieau, *J. Colloid Interface Sci.* **107**, 14 (1985).

Mori, Y. H., T. G. M. van de Ven, and S. G. Mason, *Colloids Surfaces* **4**, 1 (1982).

Morra, M., E. Occhiello, and F. Garbassi, *Langmuir* **5**, 872 (1989).

Morrow, N. R., *J. Can. Pet. Technol.* **14**, 42 (1975).

Neumann, A. W., and R. J. Good, *J. Colloid Interface Sci.* **38**, 341 (1972).

Ngan, C. G., and E. B. Dussan V., *J. Fluid Mech.* **118**, 27 (1982).

Oliver, J. F., C. Huh, and S. G. Mason, *J. Adhes.* **8**, 223 (1977).

Radigan, W., H. Ghiradella, H. L. Frisch, H. Schonhorn, and T. K. Kwei, *J. Colloid Interface Sci.* **49**, 241 (1974).

Rose, W., and R. W. Heins, *J. Colloid Sci.* **17**, 39 (1962).

Schonhorn, H., H. L. Frisch, and T. K. Kwei, *J. Appl. Phys.* **37**, 4967 (1966).

Schwartz, L. W., and S. Garoff, *Langmuir* **1**, 219 (1985).

Schwartz, A. M., and S. B. Tejada, *NASA Contract Rep.* **CR 72728** (1970).

Schwartz, A. M., and S. B. Tejada, *J. Colloid Interface Sci.* **38**, 359 (1972).

Wilson, S. D. R., *J. Colloid Interface Sci.* **51**, 532 (1975).

Yarnold, G. D., and B. J. Mason, *Proc. Phys. Soc. London* **B62**, 125 (1949).

2.1.12 Relationships for measured contact angles Although there has been some progress, it is not possible at the present time to rely upon theoretical models for a prediction of an equilibrium contact angle (Zisman 1964; Johnson and Dettre 1969; Davis 1975; Hough and White 1980; Wayner 1982; see also Sec. 3.2.7). If at all possible, experimental measurements should be made with the particular system with which you are concerned (Johnson and Dettre 1969; see also Exercises 3.2.4-1 and Section 3.2.6). [Contact angles are measured at some distance from the common line, perhaps with 10 $\times$ magnification. One should not expect to identify these measured contact angles with the true contact angle at the common line. For more on this point, see Secs. 2.1.10 and 3.2.7.]

The state of theoretical models that might be used for predicting a dynamic contact angle is no better (Hansen and Miotto 1957; Fritz 1965; Coney and Masica 1969; Cherry and Holmes 1969; Blake and Haynes 1969; see also Sec. 3.3.3). Experimental data are required.

In what follows, I suggest by way of an illustration how the utility of a set of experimental data can be considerably enhanced by using it to construct an empirical data correlation.

Hoffman (1975) has studied experimentally the displacement of one incompressible fluid B by another incompressible fluid A in a circular tube assuming a constant volumetric flow of A. In his experiments, he apparently saw no effects of adsorption or molecular rearrangement in the neighborhood of the common line (see previous section). The effect of inertial forces apparently was negligible. His tube was horizontal, effectively eliminating the influence of gravity. An interplay between viscous forces and surface forces probably resulted in the observed dependence of the advancing contact angle upon the speed of displacement of the common line. Equation (11-4) was valid, indicating that the effects of geometry or scale were negligible. This suggests

$$\cos \Theta = F(\cos \Theta^{(stat)}, \mu^{(A)}, \mu^{(B)}, \gamma^{(AB)}, u_{(v)}^{(cl)}) \tag{12-1}$$

where $\Theta^{(stat)}$ is the advancing contact angle Θ measured through phase A under static conditions, $\mu^{(A)}$ and $\mu^{(B)}$ are the viscosities of fluids A and B, $\gamma^{(AB)}$ is the interfacial tension between phases A and B, and $u_{(v)}^{(cl)}$ is the speed of displacement of the common line or meniscus measured relative to the solid surface. The Buckingham-Pi theorem (Brand 1957) further restricts the form of (12-1) to

Table 2.1.12-1: Comparison of (12-4) with portion of Hoffman's (1975) data upon which it is based.

Fluid	$\mu^{(A)}$ (dPa·s)	$\gamma^{(AB)}$ (mN/m)	$\Theta^{(stat)}$	tube diameter (cm)	no. points	% error[a]
GE silicone fluid SF-96	9.58	21.3	0°	0.195	17	15.0
Brookfield std. viscosity fluid	988	21.7	0°	0.195	23	3.1
Dow Corning 200 fluid	24,300	21	12°	0.195	3	3.0
Ashland Chem. Admex 760	1,093	43.8	69°	0.195	13	3.5
Santicizer 405	112	43.4	67°	0.195	8	5.7

a) Defined as $\dfrac{\Theta_{calc} - \Theta_{meas}}{\Theta_{calc}} \times 100$; ± unless noted otherwise.

$$\cos\Theta = F\left(\cos\Theta^{(\mathrm{stat})}, N_{ca}, \frac{\mu^{(B)}}{\mu^{(A)}}\right) \tag{12-2}$$

where we have introduced the capillary number

$$N_{ca} \equiv \frac{\mu^{(A)} u_{(v)}^{(cl)}}{\gamma^{(AB)}} \tag{12-3}$$

Hoffman (1975) actually restricted his experiments to liquids displacing air, for which $\mu^{(B)}/\mu^{(A)} \to 0$. As summarized in Table 2.1.12-1, a least-square-error fit of the data for the five fluids studied by Hoffman gave (Jiang *et al.* 1979)

$$\frac{\cos\Theta^{(\mathrm{stat})} - \cos\Theta}{\cos\Theta^{(\mathrm{stat})} + 1} = \tanh(4.96\, N_{ca}^{0.702}) \tag{12-4}$$

[The two parameters in (12-4) resulted from a nonlinear regression in which we minimized the sum of the squares of the percentage errors $(\Theta_{calc} - \Theta_{meas})/\Theta_{meas} \times 100$.] We also present in Table 2.1.12-2 a comparison be-

Table 2.1.12-2: Comparison of (12-4) with other data.

Ref.	Fluid	$\mu^{(A)}$ (dPa·s)	$\gamma^{(AB)}$ (mN/m)	$\Theta^{(\mathrm{stat})}$	tube diameter (cm)	no. points	% error[a]
Hansen and Toong (1971)	Nujol	1.77	30.4	22°[b]	0.238	8	+8.4
Hansen and Toong (1971)	Nujol	1.77	30.4	36°[b]	0.121	17	+8.8
Rose and Heins (1962)	Nujol	1.05	30.1	23°[b]	0.066 and 0.110	29	10.2
Rose and Heins (1962)	oleic acid	0.256	32.5	32°[b]	0.066 and 0.10	17	10.6

b) Given by Hoffman (1975).

tween (12-4) and the data of Hansen and Toong (1971) and of Rose and Heins (1962) for geometrically similar flows in which (11-4) was satisfied.

Table 2.1.12-3 shows a comparison of (12-4) with that portion of the Schwartz and Tejada (1970) data for which (11-4) is satisfied and

$$N_{We} \equiv \frac{\rho^{(A)} u_{(v)}^{(cl)2} \ell_0}{\gamma^{(AB)}} < 10^{-3}$$

Here N_{We} is the Weber number, which is the ratio of inertial forces to interfacial forces at the liquid-gas interface; $\rho^{(A)}$ is the density of the liquid; ℓ_0 a characteristic dimension of the system, chosen here to be the static meniscus height as estimated using the computations of Huh and Scriven (1971). The error is less than 15%, so long as

$$N_{Bo} \equiv \frac{\rho^{(A)} g \ell_0^2}{\gamma^{(AB)}} < 5 \times 10^{-2}$$

Here N_{Bo} is the Bond number, which is the ratio of gravity to interfacial forces, and g is the acceleration of gravity.

This suggests that (12-4) may be applicable to any macroscopic geometry so long as the effects of gravity, of inertia, and of adsorption all appear to be absent and (11-4) is satisfied. This is consistent with the conclu-

Table 2.1.12-3: Comparison of (12-4) with data for contact angle measured on cylinder entering liquid-gas interface (Schwartz and Tejada 1970). In all cases, $N_{We} < 10^{-3}$.

System[c]	N_{Bo}	$\Theta^{(stat)}$	no. points	% error
Water / Nylon	5.55×10^{-3}	70°	14	–13.9
DOS[d] / Teflon	2.07×10^{-2}	61°	19	–5.9
Methylene iodide / Nylon	6.04×10^{-2}	41°	11	–15.4
n-Octane / Teflon	9.59×10^{-2}	26°	1	–29.7
α - Bromonaphthalene / Nylon	1.20×10^{-1}	16°	7	–92.7

c) All solid surfaces are smooth.

d) Di(2-ethylhexyl) sebacate.

sion of Chen (1988), who reexamined this correlation and who presented additional data for the spontaneous spreading of silcone drops.

The data that are most obviously inconsistent with (12-4) are those of Ngan and Dussan V. (1982) and of Legait and Sourieau (1985), but these data do not satisfy (11-4). The dependence upon geometry seen in these data is expected.

References

Blake, T. D., and J. M. Haynes, *J. Colloid Interface Sci.* **30**, 421 (1969).

Brand, L., *Arch. Ration. Mech. Anal.* **1**, 35 (1957).

Chen, J. D., *J. Colloid Interface Sci.* **122**, 60 (1988).

Cherry, B. W., and C. M. Holmes, *J. Colloid Interface Sci.* **29**, 174 (1969).

Coney, T. A., and W. J. Masica, *NASA Tech. Note* **TN D-5115** (1969).

Davis, B. W., *J. Colloid Interface Sci.* **52**, 150 (1975).

Fritz, G., *Z. Angew. Phys.* **19**, 374 (1965).

Hansen, R. S., and M. Miotto, *J. Am. Chem. Soc.* **79**, 1765 (1957).

Hansen, R. J., and T. Y. Toong, *J. Colloid Interface Sci.* **36**, 410 (1971).

Hoffman, R. L., *J. Colloid Interface Sci.* **50**, 228 (1975).

Hough, D. B., and L. R. White, *Adv. Colloid Interface Sci.* **14**, 3 (1980).

Huh, C., and L. E. Scriven, *J. Colloid Interface Sci.* **35**, 85 (1971).

Jiang, T. S., S. G. Oh, and J. C. Slattery, *J. Colloid Interface Sci.* **69**, 74 (1979).

Johnson, R. E. Jr., and R. H. Dettre, "Surface and Colloid Science," vol. 2, p. 85, edited by E. Matijevic, Wiley-Interscience, New York (1969).

Legait, B., and P. Sourieau, *J. Colloid Interface Sci.* **107**, 14 (1985).

Ngan, C. G., and E. B. Dussan V., *J. Fluid Mech.* **118**, 27 (1982).

Rose, W., and R. W. Heins, *J. Colloid Sci.* **17**, 39 (1962).

Schwartz, A. M., and S. B. Tejada, *NASA Contract Rep.* **CR 72728** (1970).

Wayner, P. C., *J. Colloid Interface Sci.* **88**, 294 (1982).

Zisman, W. A., "Contact Angle, Wettability, and Adhesion," Advances in Chemistry Series No. 43, p. 1, American Chemical Society, Washington, D.C. (1964).

2.1.13 More comments concerning moving common lines and contact angles on rigid solids and their relation to the disjoining pressure In Sec. 1.3.10, I summarized the contradictions in our theoretical understanding of common lines moving across rigid solid surfaces, and I briefly discussed some possibilities open to us for modifying our approach to these problems. I would now like to offer a possible explanation for these contradictions as well as for some of the experimental phenomena that we have considered.

To keep the discussion focused, we will neglect the complications discussed in Sec. 2.1.11 attributable to the rate at which adsorption equilibrium is approached, to viscous forces, to inertial forces, to surface roughness, and to geometry. We will assume that the no-slip boundary condition is valid everywhere, in spite of the fact that the no-slip boundary condition at a moving common line leads to an unbounded force (Sec. 1.3.9). Our problem here is to resolve this apparent contradiction.

Following Li and Slattery (1990; see Sec. 3.3.3), our premise is that a common line does not move. As viewed on a macroscale, it appears to move as a succession of stationary common lines are formed on a microscale, driven by a negative disjoining pressure in the receding film. The amount of receding phase left stranded by the formation of this succession of common lines is too small to be easily detected [as in the experiments of Dussan V. and Davis (1974; see also Sec. 1.2.9)] and is likely to quickly disappear as the result of mass transfer to the adjacent phase. [The driving force for mass transfer would be $\Delta(\mu + \phi)$, where $\Delta\mu$ would be enhanced by the curvature of the interface and $\Delta\phi$ would be positive for a negative disjoining pressure.] This process can not occur at a common line, since this would imply a moving common line and unbounded forces.

On a macroscale, the apparent motion of the common line is driven by a combination of external forces (gravity), of convective forces, and of mutual forces. Consider the following examples.

Spontaneous spreading on a horizontal solid Consider the air-n-pentane-polytetrafloroethylene (PTFE) system. A drop of n-pentane will spread over a horizontal PTFE sheet, eventually creating a uniform film, the thickness of which is dictated by the strength of the positive (Hough and White 1980; see also Table 3.2.7-1) disjoining pressure of the film of pentane. (This assumes that the sheet is sufficiently large to accommodate the film.) As the air film collapses to form a succession of new common lines, the n-pentane advances, driven by the positive disjoining pressure in the organic phase pumping additional material into the neighborhood of the common line.

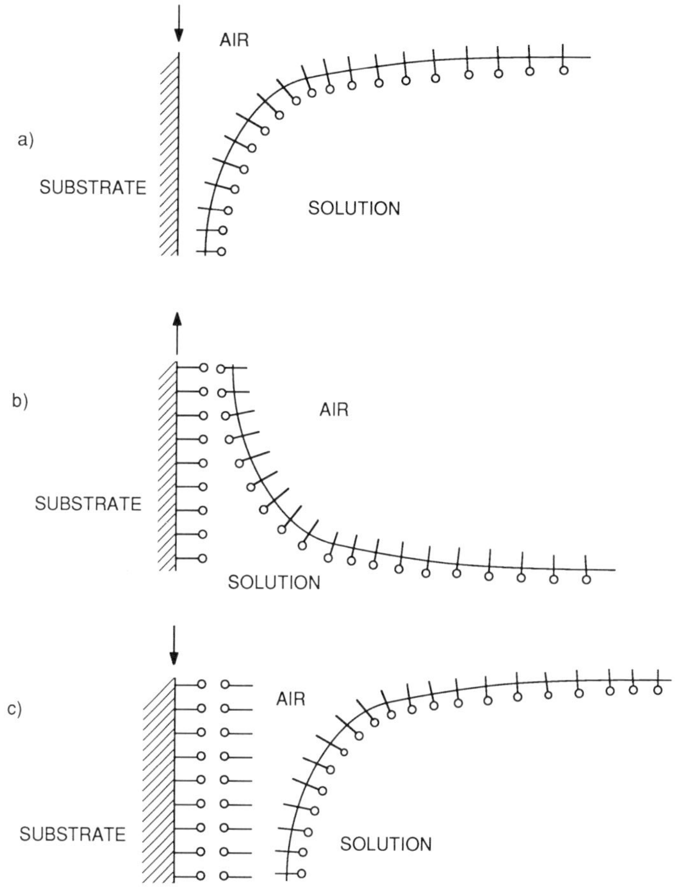

Figure 2.1.13-1: Idealized construction of Langmuir-Blodgett film

a) Substrate moves slowly down through monolayer deposited on solution-air interface. For deposition of initial monolayer, air film must exhibit negative disjoining pressure.

b) Initial monolayer having been deposited, substrate is slowly withdrawn through the monolayer on the solution-air interface. For deposition of second monolayer, solution film must exhibit negative disjoining pressure.

c) Substrate moves slowly down through the monolayer on the solution-air interface, and a third monolayer is deposited on the substrate.

The *precursor film* described in Sec. 1.2.9 (see also Sec. 1.3.10) is likely to be in part a thin film exhibiting a positive disjoining pressure. It should be expected to move at a different rate than the macroscopic film feeding it.

This view of spreading appears to be consistent with the observations of the edges of spreading drops by Chen and Wada (1989) and by Heslot *et al.* (1989).

In contrast, consider a system such as air-n-decane-PTFE. A thin film of n-decane exhibits a negative disjoining pressure (Hough and White 1980; see also Table 3.2.7-1). At equilibrium, n-decane forms a drop having a non-zero static contact angle $\Theta^{(stat)}$ measured through the liquid phase.

Spreading on a vertical solid A film of n-pentane will spread up a vertical PTFE sheet to a finite height at which the effect of the mutual forces creating the positive disjoining pressure in the neighborhood of the common line and pumping additional n-pentane into the neighborhood of the common line are balanced by the effect of gravity causing the film to drain (see Exercise 3.2.7-1).

It is uncertain whether this was the dominant effect observed by Bascom *et al.* (1964) (see Sec. 1.2.9).

Again in contrast, n-decane can be expected to form a static mensicus with a non-zero value of $\Theta^{(stat)}$.

Drainage of a film on a vertical wall Let us now consider an experiment in which a vertical sheet of PTFE is partially submerged in the liquid and partially withdrawn.

An n-pentane film, exhibiting a positive disjoining pressure, will at least initially drain, but the common line will not recede. It may actually advance as explained above (assuming that the equilibrium height of the film has not been exceeded).

The common line formed by a film of n-decane, exhibiting a negative disjoining pressure, will recede as the film drains.

Spreading by convection Now consider a variation on the above experiment in which the sheet of PTFE is withdrawn from the liquid pool and reversed, leaving a portion of the PTFE untouched by the liquid. In both cases, the liquid will run down the solid surface, displacing ahead of it a thin film of the air, which can be expected to exhibit a negative disjoining pressure between PTFE and either n-pentane or n-decane.

Langmuir-Blodgett film Finally, assume that for a given solid a thin film of phase B exhibits a negative disjoining pressure against phase A, and a thin film of phase A exhibits a negative disjoining pressure against phase B. Following the reasoning of Li and Slattery (1990; see Sec. 3.3.3), the common line can be forced to apparently move smoothly in both directions. In detail, any given common line is stationary and a sequence of common lines is formed as either displacement is performed.

These properties seem to be required for the successful construction of a Langmuir-Blodgett film by the successive deposition of monolayers

(Langmuir 1917; Blodgett 1935; Petty 1983). In its simplest form, a Langmuir-Blodgett film may be formed on a substrate by slowly oscillating a substrate through an insoluble monolayer deposited on a liquid-gas interface as sketched in Figure 13-1. Slow oscillation is required to allow the sequence of common lines to move across the substrate before the direction is changed.

The system must be such that a negative disjoining pressure is formed in the thin film of the retreating phase, forcing the collapse of the film as the common line moves apparently smoothly across the solid. Referring to Figure 13-1, we see that the orientation of the monolayers are reversed with each stroke. Let us assume either that the monolayers are formed from a nonionic species or that sufficient counter ions have been added to the solution to permit the effects of any repulsive electrostatic forces to be neglected. To the extent that the molecules in the thin film interact more strongly with the two bounding monolayers than either of thc adjoining bulk phases, the resulting London-van der Waals disjoining pressure may be similar to that in a thin film bounded by homophases: negative (see Sec. 2.1.3).

References

Bascom, W. D., R. D. Cottington, and C. R. Singleterry, in "Contact Angles, Wettability and Adhesion," Advances in Chemistry Series No. 43, p. 355, American Chemical Society, Washington, D.C. (1964).

Blodgett, K. B., *J. Am. Chem. Soc.* **57**, 1007 (1935).

Chen, J. D., and N. Wada, *Phys. Rev. Letters* **62**, 3050 (1989).

Dussan V., E. B., and S. H. Davis, *J. Fluid Mech.* **65**, 71 (1974).

Heslot, F., N. Fraysse, and A. M. Cazabat, *Nature* **338**, 640 (1989).

Hough, D. B., and L. R. White, *Adv. Colloid Interface Sci.* **14**, 3 (1980).

Langmuir, I., *J. Am. Chem. Soc.* **39**, 1848 (1917).

Li, D., and J. C. Slattery, *J. Colloid Interface Sci.* (1990).

Petty, M. C., *Endeavour (new series)* **7**, 65 (1983).

2.2 Behavior

2.2.1 Behavior of interfaces We have said nothing as yet about the behavior of materials. The postulate of mass conservation and the first and second laws of Euler are presumed to be true for all bodies. Yet we can all visualize that mercury-mercury vapor phase interfaces react differently to a disturbance than do water-water vapor interfaces. We must incorporate this information somewhere in our theoretical structure.

Let us consider the mathematical framework that we have constructed thus far. Assume that the description of the body force $\mathbf{b}$ is given: perhaps we have an essentially uniform gravitational field. To make things a little simpler, we can restrict ourselves to the motions of two phases in contact. As unknowns, we have in each phase the mass density ρ, the three components of velocity $\mathbf{v}$, and the six components of the symmetric stress tensor $\mathbf{T}$; in the dividing surface, we have the surface mass density $\rho^{(\sigma)}$, the three components of the surface velocity $\mathbf{v}^{(\sigma)}$ and the three components of the symmetric, tangential, surface stress tensor $\mathbf{T}^{(\sigma)}$. As equations, we have in each phase the equation of continuity and the three components of Cauchy's first law; in the dividing surface, we have the jump mass balance and the three components of the jump momentum balance. Altogether, we have 12 equations in 27 unknowns. This reinforces our intuitive feeling that further information is required.

This problem is well understood in the context of the motions of single-phase bodies. A constitutive equation relating $\mathbf{T}$ to the motion of the material is developed either empirically or in the context of statistical mechanics (Slattery 1981, p. 45). This gives us six additional equations for each phase, but we are still faced with 24 equations in 27 unknowns[a].

Perhaps the answer is obvious to you: we must relate $\mathbf{T}^{(\sigma)}$ to the motion and deformation of the dividing surface. One view of surface tension enters in this manner.

Before we discuss specific forms that constitutive equations for $\mathbf{T}^{(\sigma)}$ may take, let us see if our everyday experience in observing materials during deformations and motions will help us to lay down some rules governing such a relation. The principle of frame indifference was introduced in this way in Section 1.4.4; it will prove to be very useful as we continue. But there are other rules as well.

For example, it seems obvious that what happens to a body in the future is going to have no influence on its present state of stress. This suggests stating a **principle of determinism** (Truesdell 1966, p. 6; Truesdell and Noll 1965, p. 56):

---

a) Often these constitutive equations will be in terms of the thermodynamic pressure P, which can be related to ρ through an appropriate equation of state. We are then left with 26 equations in 29 unknowns.

The stress of a body is determined by the history of the motion that the body has undergone.

Our gross experience is that motion in one portion of a body does not *necessarily* have any effect on the state of stress in another portion of the body. With a slow-drying paint, I can run my brush or roller back and forth over one section of a surface without any evidence that the surface of the neighboring portions of wet paint are disturbed. From a somewhat different point of view, the physical idea of a contact force suggests that it should be determined by the motion in the immediate neighborhood of the point in question. We may state this as a **principle of local action** (Truesdell 1966, p. 6; Truesdell and Noll 1965, p. 56):

The motion of the material outside an arbitrarily small neighborhood of a material particle may be ignored in determining the stress acting at this particle.

It is possible to make further statements in much the same manner as above (Truesdell and Toupin 1960, p. 700; Truesdell and Noll 1965, p. 101).

The application of these principles in constructing constitutive equations for the stress tensor has already been discussed (Slattery 1981, p. 47; Coleman *et al.* 1966; Truesdell and Noll 1965, p. 56). In what follows, we show how they can be used in constructing constitutive equations for the surface stress tensor.

References

Coleman, B. D., H. Markovitz, and W. Noll, "Viscometric Flows of Non-Newtonian Fluids," Springer-Verlag, New York (1966).

Slattery, J. C., "Momentum, Energy, and Mass Transfer in Continua," McGraw-Hill, New York (1972); second edition, Robert E. Krieger, Malabar, FL 32950 (1981).

Truesdell, C., and R. A. Toupin, "Handbuch der Physik," vol. 3/1, edited by S. Flügge, Springer-Verlag, Berlin (1960).

Truesdell, C., and W. Noll, "Handbuch der Physik," vol. 3/3, edited by S. Flügge, Springer-Verlag, Berlin (1965).

Truesdell, C., "Six Lectures on Modern Natural Philosophy," Springer-Verlag, New York (1966).

2.2.2 Boussinesq surface fluid In the last section, we found that, in addition to the principle of frame indifference (Sec. 1.4.4), there are two other principles that every constitutive equation for the surface stress tensor should satisfy. Let us see how these three principles can help us in examining one relatively simple possibility.

We can satisfy the principle of determinism by requiring the surface stress to depend only upon a description of the present state of motion in the surface. Both the principle of determinism and the principle of local action are satisfied, if we assume that the surface stress at a point is a function of the surface velocity and the projection on the surface of the surface gradient of the surface velocity

$$\mathbf{T}^{(\sigma)} = \mathbf{H}(\ \mathbf{v}^{(\sigma)}, \mathbf{P} \cdot \nabla_{(\sigma)}\mathbf{v}^{(\sigma)}\ ) \tag{2-1}$$

Possible additional dependence of the surface stress upon the local thermodynamic state variables is not shown, since such dependence is not of primary concern as yet.

Every second-order tensor can be written as the sum of a symmetric tensor and a skew-symmetric tensor. In particular, the projection of the surface gradient of the surface velocity vector may be expressed as

$$\mathbf{P} \cdot \nabla_{(\sigma)}\mathbf{v}^{(\sigma)} = \mathbf{D}^{(\sigma)} + \mathbf{W}^{(\sigma)} \tag{2-2}$$

where $\mathbf{D}^{(\sigma)}$ is the ***surface rate of deformation*** tensor (see Sec. 1.2.8)

$$\mathbf{D}^{(\sigma)} \equiv \frac{1}{2}\left[\ \mathbf{P} \cdot \nabla_{(\sigma)}\mathbf{v}^{(\sigma)} + (\ \nabla_{(\sigma)}\mathbf{v}^{(\sigma)}\ )^T \cdot \mathbf{P}\ \right] \tag{2-3}$$

and $\mathbf{W}^{(\sigma)}$ will be referred to as the ***surface vorticity*** tensor

$$\mathbf{W}^{(\sigma)} \equiv \frac{1}{2}\left[\ \mathbf{P} \cdot \nabla_{(\sigma)}\mathbf{v}^{(\sigma)} - (\ \nabla_{(\sigma)}\mathbf{v}^{(\sigma)}\ )^T \cdot \mathbf{P}\ \right] \tag{2-4}$$

This allows us to rewrite (2-1) in the form

$$\mathbf{T}^{(\sigma)} \equiv \mathbf{H}(\ \mathbf{v}^{(\sigma)}, \mathbf{D}^{(\sigma)} + \mathbf{W}^{(\sigma)}\ ) \tag{2-5}$$

The principle of frame indifference (Sec. 1.4.4) and the frame indifference of the surface stress tensor (Exercise 2.1.7-1) require that in some new frame of reference

$$\begin{aligned} \mathbf{T}^{(\sigma)*} &= \mathcal{Q} \cdot \mathbf{T}^{(\sigma)} \cdot \mathcal{Q}^T \\ &= \mathbf{H}(\ \mathbf{v}^{(\sigma)*}, \mathbf{D}^{(\sigma)*} + \mathbf{W}^{(\sigma)*}\ ) \end{aligned} \tag{2-6}$$

This means that the function **H** must be such that

$$\mathcal{Q} \cdot \mathbf{H}(\, \mathbf{v}^{(\sigma)}, \mathbf{D}^{(\sigma)} + \mathbf{W}^{(\sigma)}\,) \cdot \mathcal{Q}^T = \mathbf{H}(\, \mathbf{v}^{(\sigma)*}, \mathbf{D}^{(\sigma)*} + \mathbf{W}^{(\sigma)*}\,) \tag{2-7}$$

From Sec. 1.4.3 and Exercise 1.4.3-4, we can also express this last as

$$\mathbf{H}(\, \mathbf{v}^{(\sigma)}, \mathbf{D}^{(\sigma)} + \mathbf{W}^{(\sigma)}\,) = \mathcal{Q}^T \cdot \mathbf{H}\Big[\, \dot{\overline{\mathbf{z}_0^*}} + \dot{\mathbf{Q}} \cdot (\, \mathbf{z} - \mathbf{z}_0\,) + \mathbf{Q} \cdot \mathbf{v}^{(\sigma)},$$
$$\mathcal{Q} \cdot \mathbf{D}^{(\sigma)} \cdot \mathcal{Q}^T + \mathcal{Q} \cdot \mathbf{W}^{(\sigma)} \cdot \mathcal{Q}^T + \mathbf{P}^* \cdot \dot{\mathcal{Q}} \cdot \mathcal{Q}^T \,\Big] \cdot \mathcal{Q} \tag{2-8}$$

Let us choose a particular change of frame such that

$$\mathbf{P}^* \cdot \dot{\mathcal{Q}} \cdot \mathcal{Q}^T = -\,\mathcal{Q} \cdot \mathbf{W}^{(\sigma)} \cdot \mathcal{Q}^T \tag{2-9}$$

and

$$\dot{\overline{\mathbf{z}_0^*}} = -\,\dot{\mathbf{Q}} \cdot (\mathbf{z} \; - \mathbf{z}_{(0)}) - \mathbf{Q} \cdot \mathbf{v}^{(\sigma)} \tag{2-10}$$

With this change of frame, we have from (2-8)

$$\mathbf{H}(\, \mathbf{v}^{(\sigma)}, \mathbf{D}^{(\sigma)} + \mathbf{W}^{(\sigma)}\,)$$
$$= \mathcal{Q}^T \cdot \mathbf{H}(\, \mathbf{0}, \mathcal{Q} \cdot \mathbf{D}^{(\sigma)} \cdot \mathcal{Q}^T + \mathbf{0}\,) \cdot \mathcal{Q}$$
$$= \mathcal{Q}^T \cdot \mathbf{H}(\, \mathbf{0}, \mathbf{D}^{(\sigma)*}\,) \cdot \mathcal{Q} \tag{2-11}$$

A second application of the principle of frame indifference gives

$$\mathcal{Q}^T \cdot \mathbf{H}(\, \mathbf{0}, \mathbf{D}^{(\sigma)*}\,) \cdot \mathcal{Q} = \mathbf{H}(\, \mathbf{0}, \mathbf{D}^{(\sigma)}\,) \tag{2-12}$$

which means

$$\mathbf{G}(\, \mathbf{D}^{(\sigma)}\,) \equiv \mathbf{H}(\, \mathbf{0}, \mathbf{D}^{(\sigma)}\,)$$
$$= \mathbf{H}(\, \mathbf{v}^{(\sigma)}, \mathbf{D}^{(\sigma)} + \mathbf{W}^{(\sigma)}\,) \tag{2-13}$$

We conclude that (2-5) must actually be of the form

$$\mathbf{T}^{(\sigma)} = \mathbf{G}(\, \mathbf{D}^{(\sigma)}\,) \tag{2-14}$$

where the function $\mathbf{G}$ must satisfy

$$\mathbf{G}(\, \mathbf{D}^{(\sigma)}\,) = \mathcal{Q}^T \cdot \mathbf{G}(\, \mathcal{Q} \cdot \mathbf{D}^{(\sigma)} \cdot \mathcal{Q}^T\,) \cdot \mathcal{Q} \tag{2-15}$$

In view of the physical interpretations developed in Sec. 1.2.8 for its components, the surface rate of deformation tensor $\mathbf{D}^{(\sigma)}$ is a reasonable independent variable in a constitutive equation for $\mathbf{T}^{(\sigma)}$.

Let me give you my physical interpretation of the use of the principle of frame indifference in the preceding paragraphs. Let us begin with an

of frame indifference in the preceding paragraphs. Let us begin with an observer in an arbitrary frame of reference who assumes that at the position $\mathbf{z}$ the surface stress tensor depends upon the surface velocity, the surface rate of deformation tensor, and the surface vorticity tensor: (2-5). The frame of reference for a second observer rotates and translates with the material in such a way that for him the surface velocity and the surface vorticity tensor are zero at this same point in space $\mathbf{z}^*$. This second experimentalist sees a dependence of the surface stress tensor upon only the surface rate of deformation tensor: (2-11). But the principle of material frame indifference requires that the description of the behavior of materials be the same for all observers. We consequently conclude that for all observers $\mathbf{T}^{(\sigma)}$ depends only upon the surface rate of deformation tensor.

The most general form that (2-14) can take when restricted by (2-15) is (Truesdell and Noll 1965, p. 32)

$$\mathbf{T}^{(\sigma)} = \kappa_0 \mathbf{P} + \kappa_1 \mathbf{D}^{(\sigma)} \tag{2-16}$$

where

$$\kappa_k = \kappa_k(\text{I}, \text{II}) \tag{2-17}$$

Here I and II are the two principal invariants of the surface rate of deformation tensor, i.e. the coefficients in the equation for the principal values of $\mathbf{D}^{(\sigma)}$:

$$\det{}_{(\sigma)}(\mathbf{D}^{(\sigma)} - m\,\mathbf{P}) = 0$$

$$m^2 - \text{I}\, m + \text{II} = 0 \tag{2-18}$$

We leave it as an exercise to show that

$$\text{I} \equiv \text{tr}\, \mathbf{D}^{(\sigma)} = \text{div}_{(\sigma)}\mathbf{v}^{(\sigma)} \tag{2-19}$$

$$\text{II} \equiv \det{}_{(\sigma)}\mathbf{D}^{(\sigma)} = \frac{1}{2}(\text{I}^2 - \overline{\text{II}}) \tag{2-20}$$

$$\overline{\text{II}} \equiv \text{tr}(\mathbf{D}^{(\sigma)} \cdot \mathbf{D}^{(\sigma)}) \tag{2-21}$$

Notice that this constitutive equation for the surface stress tensor automatically satisfies the implications for Euler's second law developed in Sec. 2.1.7, since the surface rate of deformation tensor is a symmetric tangential tensor.

It follows immediately from (2-16) that the most general *linear* relation between the surface stress tensor and the surface rate of deformation tensor which is consistent with the principle of frame indifference is

We shall later show that the jump Clausius-Duhem inequality for a dividing surface implies (Exercise 5.9.6-1)

$$\alpha = \gamma \tag{2-23}$$

$$\kappa \equiv \lambda + \varepsilon > 0 \tag{2-24}$$

and

$$\varepsilon > 0 \tag{2-25}$$

where γ is the thermodynamic surface tension. As a result, we normally write (2-22) in a form known as the linear **Boussinesq surface fluid model**:

$$\mathbf{T}^{(\sigma)} = [\, \gamma + (\, \kappa - \varepsilon\,)\mathrm{div}_{(\sigma)}\mathbf{v}^{(\sigma)}\,]\,\mathbf{P} + 2\varepsilon\,\mathbf{D}^{(\sigma)} \tag{2-26}$$

Here κ is referred to as the **surface dilatational viscosity** and ε as the **surface shear viscosity**.

A constitutive equation of the form (2-16) was first introduced by Hegde and Slattery (1971). Prompted by their recommendations, I prefer to write (2-16) in the form of (2-26) with

$$\kappa = \kappa(\, \mathrm{div}_{(\sigma)}\mathbf{v}^{(\sigma)}, |\nabla_{(\sigma)}\mathbf{v}^{(\sigma)}|\,) \tag{2-27}$$

$$\varepsilon = \varepsilon(\, \mathrm{div}_{(\sigma)}\mathbf{v}^{(\sigma)}, |\nabla_{(\sigma)}\mathbf{v}^{(\sigma)}|\,) \tag{2-28}$$

where

$$|\nabla_{(\sigma)}\mathbf{v}^{(\sigma)}| \equiv [\, 2\,\mathrm{tr}(\,\mathbf{D}^{(\sigma)} \cdot \mathbf{D}^{(\sigma)}\,)]^{1/2} \tag{2-29}$$

can be thought of as the **magnitude of the surface gradient of surface velocity.** In this context, we will refer to (2-26) as the **generalized Boussinesq surface fluid model**, to κ as the **apparent surface dilatational viscosity**, and to ε as the **apparent surface shear viscosity.**

The Boussinesq surface fluid model is the only special case of (2-16) that has received much attention in the literature. Of course, historically it was not developed in the manner that I have just described.

Hagen (1845) was apparently the first to suggest that the "viscosity" of the interfacial region differs from those of the adjoining phases. Plateau (1869) conducted some simple experiments using a floating needle rotating in a magnetic field. His interpretation of his results in terms of a "viscosity" of the liquid-gas interfacial region is almost certainly in error as the result of surface tension gradients developed in the interface.

Boussinesq's (1913) original statement of (2-26) is not particularly convenient, since it is expressed in terms of the principal axes of the surface rate of deformation tensor. Ericksen (1952) and Oldroyd (1955) wrote Boussinesq's equations in terms of an arbitrary surface coordinate system, but limited themselves to a stationary dividing surface. Scriven wrote Boussinesq's equations in terms of an arbitrary surface coordinate system, but limited themselves to a stationary dividing surface. Scriven (1960) gave them for a moving and deforming dividing surface in terms of an arbitrary surface coordinate system.

Careful measurements of the interfacial shear viscosity have been made by a number of workers (Brown *et al.* 1953; Burton and Mannheimer 1967; Mannheimer and Schechter 1970a; Wei and Slattery 1976; Deemer *et al.* 1980). The interfacial dilatational viscosity has been determined by Stoodt and Slattery (1984) for one system.

In some cases, nonlinear behavior has been observed (Brown *et al.* 1953; Mannheimer and Schechter 1970b; Pintar *et al.* 1971; Suzuki 1972; Wei and Slattery 1976; Jiang *et al.* 1983). This suggested the introduction of the generalized Boussinesq surface fluid, which describes the two interfacial viscosities by (2-27) and (2-28) as functions of the principal invariants of the surface rate of deformation tensor (Hegde and Slattery 1971). [The generalized Boussinesq surface fluid includes the special cases proposed by Mannheimer and Schechter (1970b) and by Pintar *et al.* (1971).]

References

Boussinesq, J., *Comptes Rendus des Seances de l' Académie des Sciences* **156**, 983, 1035, 1124 (1913); see Also *Ann. Chim. Phys.* (8) **29**, 349, 357, 364 (1913). The jump balance for surfaces of revolution are developed in *Comptes Rendus des Seances de l' Académie des Sciences* **157**, 7 (1913). The shape of a vertical jet striking a horizontal plate is investigated in *Comptes Rendus des Seances de l' Académie des Sciences* **157**, 89 (1913).

Brown, A. G., W. C. Thuman, and J. W. McBain, *J. Colloid Sci.* **8**, 491 (1953).

Burton, R. A., and R. J. Mannheimer, "Ordered Fluids and Liquid Crystals," Advances in Chemistry Series No. 63, p. 315, American Chemical Society (1967).

Deemer, A. R., J. D. Chen, M. G. Hegde, and J. C. Slattery, *J. Colloid Interface Sci.* **78**, 87 (1980).

Ericksen, J. L., *J. Ration. Mech. Anal.* **1**, 521 (1952).

Hagen, G. H. L., *Abhandlungen der Königlichen Akademie der Wissenschaften zu Berlin* (Math.), Berlin (1845), p. 41; *ibid.* (1846), p. 1; *Bericht über die zur Bekanntmachung geeigneten Verhandlungen der K. Preuss. Akademie der Wissenschaften zu Berlin*, Berlin (1845), p. 166; *ibid.* (1846), p. 154; *Ann. Phys. Chem.* **67**, 1, 152 (1846); *ibid.* **77**, 449 (1849).

Hegde, M. G., and J. C. Slattery, *J. Colloid Interface Sci.* **35**, 593 (1971).

(1983).

Mannheimer, R. J., and R. S. Schechter, *J. Colloid Interface Sci.* **32**, 195 (1970a).

Mannheimer, R. J., and R. S. Schechter, *J. Colloid Interface Sci.* **32**, 212 (1970b).

Oldroyd, J. G., *Proc. R. Soc. London A* **232**, 567 (1955).

Pintar, A. J., A. B. Israel, and D. T. Wasan, *J. Colloid Interface Sci.* **37**, 52 (1971).

Plateau, J., *Mémoires de l' Académie Royale de Sciences, des Lettres et des Beaux-Arts de Belgique* **37**, 1 (1869); *Ann. Chim. Phys.* (4) **17**, 260 (1869); *Phil. Mag.* (series 4) **38**, 445 (1869).

Scriven, L. E., *Chem. Eng. Sci.* **12**, 98 (1960). See also R. Aris, "Vectors, Tensors, and the Basic Equations of Fluid Mechanics," Prentice-Hall, Englewood Cliffs, NJ (1962), p. 232. For a list of typographical errors in these presentations, refer to J. C. Slattery, *Chem. Eng. Sci.* **19**, 379 (1964).

Stoodt, T. J., and J. C. Slattery, *AIChE J.* **30**, 564 (1984).

Suzuki, A., *Kolloid Z.Z. Polym.* **250**, 365 (1972).

Truesdell, C., and W. Noll, "Handbuch der Physik," Vol. 3/3, edited by S. Flügge, Springer-Verlag, Berlin (1965).

Wei, L. Y., and J. C. Slattery, "Colloid and Interface Science," Vol. IV, edited by M. Kerker, p. 399, Academic Press, New York (1976).

Exercise

2.2.2-1 *principal invariants of* $\mathbf{D}^{(\sigma)}$ Starting with (2-18_1), derive (2-18_2) with the coefficients given by (2-19) through (2-21).

2.2.3 Simple surface material The generalized Boussinesq surface fluid described in the previous section cannot account for the memory effects that have been observed in oscillatory flows (Mohan *et al.* 1976; Mohan and Wasan 1976; Addison and Schechter 1979). It says that only the current velocity gradient determines the surface stress at any point in an interface.

There have been specific models proposed that account for some memory effects (Mannheimer and Schechter 1970; Gardner *et al.* 1978). But the development of the literature describing the stress-deformation behavior of materials outside the immediate neighborhood of a phase interface indicates that progress might be enhanced, if, instead of starting with specific models in working towards an understanding of a broad class of materials, we define a general class of surface stress-deformation behavior in terms of which specific models might be introduced as required.

Our conception of material behavior outside the immediate neighborhood of phase interfaces was considerably clarified by the development of the Noll simple material (Noll 1958; Truesdell and Noll 1965, p. 60; Truesdell 1966, p. 35). This is a broad class of bulk stress-deformation behavior that is believed to include nearly all real materials, at least to the extent that they have been observed experimentally. Because it is a class of behaviors, it cannot be used to analyze specific flows and deformations without making further statements about the behavior of the material being studied. Several groups of flows, including the bases for most viscometer designs, have been analyzed for a subclass known as the incompressible Noll simple fluid (Coleman *et al.* 1966). The incompressible Noll simple fluid appears to describe the behaviors of most real liquids.

The success of these ideas suggests that we present in this and the next few sections an analogous development of a simple surface material, a simple surface solid, and a simple surface fluid. To make the comparison with the developments of Noll straightforward, I have made a special effort to preserve the same style of notation (see particularly Truesdell 1966, p. 35).

Let us begin with the basic principles given in Sec. 2.2.1. The principle of determinism is satisfied, if we say that the surface stress at the surface particle $\zeta^{(\sigma)}$ at time t is determined by the motions of the surface up to time t:

$$\mathbf{T}^{(\sigma)}(\ \zeta^{(\sigma)}, t\ ) = \underset{s=0}{\overset{\infty}{\boldsymbol{\mathfrak{A}}}}\ [\ \boldsymbol{\chi}^{(\sigma)}(\ \zeta^{(\sigma)}, t - s\ ), \zeta^{(\sigma)}, t\ ] \tag{3-1}$$

Here $\underset{s=0}{\overset{\infty}{\boldsymbol{\mathfrak{A}}}}$ is a tangential tensor-valued functional or rule of correspondence; to every history of a motion $\boldsymbol{\chi}^{(\sigma)}(\ \zeta^{(\sigma)}, t - s\ )$, surface particle $\zeta^{(\sigma)}$, and time t, it assigns a tangential tensor, the surface stress tensor $\mathbf{T}^{(\sigma)}$. The limits are meant to remind you that $\mathbf{T}^{(\sigma)}$ depends upon the motion at all past times corresponding to $0 \leq s < \infty$. Equation (3-1) says that the motion of the dividing surface up to and including the present time determines a unique symmetric surface stress tensor at each point on the surface; the way in which it does so may depend upon $\zeta^{(\sigma)}$ and t. The notation will be simplified somewhat, if for any function of time f(t − s) we introduce the **history** f^t

$$f^t \equiv f^t(s) \equiv f(t - s) \tag{3-2}$$

Using this notation, we may rewrite (3-1) in terms of the history of motion of the surface:

$$\mathbf{T}^{(\sigma)}(\zeta^{(\sigma)}, t) = \mathop{\mathcal{A}}_{s=0}^{\infty} [\boldsymbol{\chi}^{(\sigma)t}(\zeta^{(\sigma)}, s), \zeta^{(\sigma)}, t] \tag{3-3}$$

Applied to (3-3), the principle of local action says that the motion of the surface outside an arbitrarily small neighborhood of a surface particle may be ignored in determining the surface stress at this particle. Formally, if two motions $\bar{\boldsymbol{\chi}}^{(\sigma)t}$ and $\boldsymbol{\chi}^{(\sigma)t}$ coincide when $s \geq 0$ and $\zeta^{(\sigma)}$ belongs to an arbitrarily small neighborhood of $\zeta^{(\sigma)}$

$$\bar{\boldsymbol{\chi}}^{(\sigma)}(\zeta^{(\sigma)}, s) = \boldsymbol{\chi}^{(\sigma)t}(\zeta^{(\sigma)}, s) \tag{3-4}$$

then

$$\mathop{\mathcal{A}}_{s=0}^{\infty} [\bar{\boldsymbol{\chi}}^{(\sigma)t}(\zeta^{(\sigma)}, s), \zeta^{(\sigma)}, t] = \mathop{\mathcal{A}}_{s=0}^{\infty} [\boldsymbol{\chi}^{(\sigma)t}(\zeta^{(\sigma)}, s), \zeta^{(\sigma)}, t] \tag{3-5}$$

Intuitively, it seems reasonable that we should be able to learn whatever we would like about the behavior of a dividing surface by studying it in motions for which the surface deformation gradient (see Sec. 1.2.5) is a constant with respect to position in the reference configuration:

$$\begin{aligned} \mathbf{z} &= \boldsymbol{\chi}_{\kappa}^{(\sigma)}(y_{\kappa}^1, y_{\kappa}^2, t - s) \\ &= \boldsymbol{\mathcal{F}}(t - s) \cdot \mathbf{a}_{\kappa A}(y_{\kappa}^A - y_{\kappa 0}^A) + \mathbf{z}_0(t - s) \end{aligned} \tag{3-6}$$

We will refer to dividing surfaces for which this is indeed true as simple surface materials. Formally, a dividing surface described by (3-3) is called a **simple surface material**, if there exists a reference dividing surface $\kappa^{(\sigma)}$ (see Sec. 1.2.5) such that

$$\begin{aligned} \mathbf{T}^{(\sigma)}(\zeta^{(\sigma)}, t) &= \mathop{\mathcal{A}}_{s=0}^{\infty} [\boldsymbol{\chi}^{(\sigma)t}(\zeta^{(\sigma)}, s), \zeta^{(\sigma)}, t] \\ &= \mathop{\mathcal{B}_{\kappa}}_{s=0}^{\infty} [\boldsymbol{\mathcal{F}}^t(\zeta^{(\sigma)}, s), \zeta^{(\sigma)}] \end{aligned} \tag{3-7}$$

The surface stress at the place occupied by the surface particle $\zeta^{(\sigma)}$ at time t is determined by the history of the surface deformation gradients with respect to the reference dividing surface $\kappa^{(\sigma)}$ experienced by $\zeta^{(\sigma)}$ up to the

respect to the reference dividing surface $\kappa^{(\sigma)}$ experienced by $\zeta^{(\sigma)}$ up to the time t. We shall usually write (3-7) in one of several simpler forms

$$\mathbf{T}^{(\sigma)} = \underset{s=0}{\overset{\infty}{\mathcal{B}_\kappa}} [\, \mathcal{F}^t(s) \,] = \underset{s=0}{\overset{\infty}{\mathcal{B}_\kappa}} (\, \mathcal{F}^t \,) = \mathcal{B}(\, \mathcal{F}^t \,) \tag{3-8}$$

with the same meaning.

The form of (3-7) is further restricted by the principle of frame indifference (Sec. 1.4.4). This principle and the frame indifference of the surface stress tensor (Exercise 2.1.7-1) require that in some new frame of reference (3-7) becomes

$$\begin{aligned} \mathbf{T}^{(\sigma)*} &= \mathcal{Q} \cdot \mathbf{T}^{(\sigma)} \cdot \mathcal{Q}^T = \mathcal{B}(\, \mathcal{F}^{t\,*} \,) \\ &= \mathcal{B}(\, \mathcal{Q}^t \cdot \mathcal{F}^t \,) \end{aligned} \tag{3-9}$$

In writing this, we have noted from Exercise 1.4.3-4 the manner in which the surface deformation gradient transforms under a change in frame. This means that the functional $\mathcal{B}$ must be of such a form that

$$\mathcal{Q}(t) \cdot \mathcal{B}(\, \mathcal{F}^t \,) \cdot \mathcal{Q}(t)^T = \mathcal{B}(\, \mathcal{Q}^t \cdot \mathcal{F}^t \,) \tag{3-10}$$

Using the polar decomposition theorem, we can express (see Sec. 1.2.6).

$$\mathcal{F}^t = \mathcal{R}^t \cdot \mathbf{U}^{(\sigma)t} \tag{3-11}$$

Here $\mathcal{R}^t$ is the history of the surface rotation, a unique orthogonal tangential transformation from the space of tangential vector fields on the reference configuration of the dividing surface to the space of tangential vector fields on its current configuration. The history of the right surface stretch tensor $\mathbf{U}^{(\sigma)t}$ is a unique, positive-definite, symmetric, tangential tensor for the space of tangential vector fields on the reference configuration of the dividing surface. Eliminating $\mathcal{F}^t$ with (3-11), we can rearrange (3-10) as

$$\mathcal{B}(\, \mathcal{F}^t \,) = \mathcal{Q}(t)^T \cdot \mathcal{B}(\, \mathcal{Q}^t \cdot \mathcal{R}^t \cdot \mathbf{U}^{(\sigma)t} \,) \cdot \mathcal{Q}(t) \tag{3-12}$$

Since this relationship must hold for all $\mathcal{Q}^t$, $\mathcal{R}^t$, and $\mathbf{U}^{(\sigma)t}$, it must hold in particular for

$$\mathcal{Q}^t = \mathcal{R}^{t\,T} \tag{3-13}$$

in which case

$$\mathcal{B}(\, \mathcal{F}^t \,) = \mathcal{R}(t) \cdot \mathcal{B}(\, \mathbf{U}^{(\sigma)t} \,) \cdot \mathcal{R}(t)^T \tag{3-14}$$

Conversely, let us suppose that $\mathcal{B}$ does take this form. For an arbitrary history $\mathcal{Q}^t$, we have the polar decomposition (Sec. A.3.8)

$$\mathcal{Q}^t \cdot \mathcal{F}^t = (\mathcal{Q}^t \cdot \mathcal{R}^t) \cdot U^{(\sigma)t} \tag{3-15}$$

and

$$\mathcal{B}(\mathcal{Q}^t \cdot \mathcal{F}^t) = \mathcal{Q}(t) \cdot \mathcal{R}(t) \cdot \mathcal{B}(U^{(\sigma)t}) \cdot [\mathcal{Q}(t) \cdot \mathcal{R}(t)]^T$$

$$= \mathcal{Q}(t) \cdot \mathcal{B}(\mathcal{F}^t) \cdot \mathcal{Q}(t)^T \tag{3-16}$$

which means that (3-10) is satisfied. Our conclusion is that a constitutive equation for a simple surface material satisfies the principle of frame indifference, if and only if it can be written in the form

$$T^{(\sigma)} = \mathcal{R}(t) \cdot \mathcal{B}(U^{(\sigma)t}) \cdot \mathcal{R}(t)^T \tag{3-17}$$

Physically, this means that, although the history of the right surface stretch tensor $U^{(\sigma)t}$ may influence the present surface stress in any way at all, the history of the surface rotation $\mathcal{R}^t$ has no influence. It is only the presentation surface rotation $\mathcal{R}(t)$ that explicitly enters (3-17).

There are a number of other forms that (3-17) can take. One particularly useful expression is obtained by writing the history of the surface deformation gradient $\mathcal{F}^t$ in terms of the history of the relative surface deformation gradient $\mathcal{F}_t^t$ (see Sec. 1.2.5):

$$\mathcal{F}^t = \mathcal{F}_t^t \cdot \mathcal{F}(t) \tag{3-18}$$

Applying the polar decomposition theorem, we have

$$\mathcal{R}^t \cdot U^{(\sigma)t} = \mathcal{R}_t^t \cdot U_t^{(\sigma)t} \cdot \mathcal{R}(t) \cdot U^{(\sigma)}(t) \tag{3-19}$$

and its transpose

$$U^{(\sigma)t} \cdot \mathcal{R}^{tT} = U^{(\sigma)}(t) \cdot \mathcal{R}(t)^T \cdot U_t^{(\sigma)t} \cdot \mathcal{R}_t^{tT} \tag{3-20}$$

These relationships allow us to express the history of the right surface Cauchy-Green tensor as

$$C^{(\sigma)t} \equiv U^{(\sigma)t} \cdot U^{(\sigma)t}$$

$$= U^{(\sigma)t} \cdot \mathcal{R}^{tT} \cdot \mathcal{R}^t \cdot U^{(\sigma)t}$$

$$= U^{(\sigma)}(t) \cdot \mathcal{R}(t)^T \cdot C_t^{(\sigma)t} \cdot \mathcal{R}(t) \cdot U^{(\sigma)}(t) \tag{3-21}$$

where we have introduced the history of the right relative surface Cauchy-Green tensor

$$C_t^{(\sigma)t} \equiv U_t^{(\sigma)t} \cdot U_t^{(\sigma)t} \tag{3-22}$$

Equation (3-21) suggests that we introduce

Equation (3-21) suggests that we introduce

$$\mathfrak{K}(\mathbf{C}^{(\sigma)t}) \equiv \mathfrak{B}(\mathbf{U}^{(\sigma)t}) \tag{3-23}$$

and write as an alternative form for (3-17)

$$\begin{aligned} &\mathfrak{R}^{T}(t) \cdot \mathbf{T}^{(\sigma)} \cdot \mathfrak{R}(t) \\ &\quad = \mathfrak{L}[\, \mathfrak{R}^{T}(t) \cdot \mathbf{C}_{t}^{(\sigma)t} \cdot \mathfrak{R}(t) \,;\, \mathbf{C}^{(\sigma)}(t) \,] \\ &\quad \equiv \mathfrak{K}[\, \mathbf{U}^{(\sigma)}(t) \cdot \mathfrak{R}^{T}(t) \cdot \mathbf{C}_{t}^{(\sigma)t} \cdot \mathfrak{R}(t) \cdot \mathbf{U}^{(\sigma)}(t) \,] \end{aligned} \tag{3-24}$$

Equation (3-24) indicates that it is possible to express the effect of all the *past* deformation history by measuring the deformation with respect to the current configuration of the surface. However, in general a fixed reference configuration is required to account for the deformation of the surface at the present instant, as indicated by the appearance of $\mathbf{C}^{(\sigma)}(t)$.

References

Addison, J. V., and R. S. Schechter, *AIChE J.* **25**, 32 (1979).

Coleman, B. D., H. Markovitz, and W. Noll, "Viscometric Flows of Non-Newtonian Fluids," Springer-Verlag, New York (1966).

Gardner, J. W., J. Y. Addison, and R. S. Schechter, *AIChE J.* **24**, 400 (1978).

Mannheimer, R. J., and R. S. Schechter, *J. Colloid Interface Sci.* **32**, 225 (1970).

Mohan, V., B. K. Malviya, and D. T. Wasan, *Can. J. Chem. Eng.* **54**, 515 (1976).

Mohan, V., and D. T. Wasan, "Colloid and Interface Science," vol. IV, edited by M. Kerker, p. 439, Academic Press, New York (1976).

Noll, W., *Arch. Ration. Mech. Anal.* **2**, 197 (1958); see also "The Rational Mechanics of Materials," edited by C. Truesdell, Gordon and Breach, New York (1965).

Truesdell, C., and W. Noll, "Handbuch der Physik," vol. 3/3, edited by S. Flügge, Springer-Verlag, Berlin (1965).

Truesdell, C., "The Elements of Continuum Mechanics," Springer-Verlag, New York (1966).

2.2.4 Surface isotropy group An isotropic material is one that has no internal sense of direction. Another way of expressing this idea is to say that its behavior is unaffected by rotations or that rotations cannot be detected by experiments. Rotations performed before an experiment have no effect on its outcome. The behavior of the material relative to the reference configuration $\boldsymbol{\kappa}$ is the same as it would be relative to any other reference configuration obtained by a rotation.

But not all materials are isotropic. We are all familiar with the crystalline structure of some solids. These are substances for which there are only a finite number of symmetries with respect to any given reference configuration. There are only a finite number of transformations from any given reference configuration that cannot be detected by experiments.

Ideas such as these motivated Noll (1958; Truesdell and Noll 1965, p. 76; Truesdell 1966, p. 56) to investigate the class of transformations from any given reference configuration that leaves the response of a simple material to arbitrary deformation histories unchanged.

I wish to examine the class of transformations from any given reference configuration of a dividing surface that leaves the response of a simple surface material to arbitrary surface deformation histories unchanged.

How is the surface deformation gradient altered by a change of reference configuration? From Sec. 1.2.5, if

$$\mathbf{z} = \boldsymbol{\chi}_{\kappa_2}^{(\sigma)}(y^1_{\boldsymbol{\kappa}(2)}, y^2_{\boldsymbol{\kappa}(2)}, t) \tag{4-1}$$

describes the deformation from the reference configuration $\boldsymbol{\kappa}_2^{(\sigma)}$, the corresponding surface deformation gradient is (see Sec. 1.2.5)

$$\boldsymbol{\mathcal{F}} \equiv \operatorname{grad}_{(\boldsymbol{\kappa}2)} \boldsymbol{\chi}_{\kappa_2}^{(\sigma)}(y^1_{\boldsymbol{\kappa}(2)}, y^2_{\boldsymbol{\kappa}(2)}, t) \tag{4-2}$$

From the intrinsic configurations corresponding to $\boldsymbol{\kappa}_2^{(\sigma)}$ and $\boldsymbol{\kappa}_1^{(\sigma)}$ (see Sec. 1.2.5), we have

$$y^A_{\boldsymbol{\kappa}(2)} = K^A_{(2)}(\zeta^{(\sigma)}) \tag{4-3}$$

and

$$\zeta^{(\sigma)} = K^{-1}_{(1)}(y^1_{\boldsymbol{\kappa}(1)}, y^2_{\boldsymbol{\kappa}(1)}) \tag{4-4}$$

Equations (4-3) and (4-4) allow us to relate the intrinsic reference configurations corresponding to $\boldsymbol{\kappa}_1^{(\sigma)}$ and $\boldsymbol{\kappa}_2^{(\sigma)}$:

$$\begin{aligned} y^A_{\boldsymbol{\kappa}(2)} &= L^A(y^1_{\boldsymbol{\kappa}(1)}, y^2_{\boldsymbol{\kappa}(1)}) \\ &\equiv K^A_{(2)}[K^{-1}_{(1)}(y^1_{\boldsymbol{\kappa}(1)}, y^2_{\boldsymbol{\kappa}(1)})] \end{aligned} \tag{4-5}$$

We can write the mapping from $\boldsymbol{\kappa}_1^{(\sigma)}$ to $\boldsymbol{\kappa}_2^{(\sigma)}$ as (see Sec. 1.2.5)

$$\mathbf{z}_{\boldsymbol{\kappa}(2)} = \boldsymbol{\Lambda}(y^1_{\boldsymbol{\kappa}(1)}, y^2_{\boldsymbol{\kappa}(1)})$$

$$\equiv \kappa_2^{(\sigma)}[\, L^1(\, y^1_{\kappa(1)}, y^2_{\kappa(1)}\,), L^2(\, y^1_{\kappa(1)}, y^2_{\kappa(1)}\,)] \tag{4-6}$$

and the surface deformation gradient relative to $\kappa_1^{(\sigma)}$ becomes in terms of $\mathcal{F}$ defined by (4-2)

$$\text{grad}_{(\kappa 1)}\chi_{\kappa}{}^{(\sigma)}_{2}(\, y^1_{\kappa(2)}, y^2_{\kappa(2)}, t\,)$$

$$= \frac{\partial}{\partial y^A_{\kappa\,(2)}} \chi_{\kappa}{}^{(\sigma)}_{2}(\, y^1_{\kappa(2)}, y^2_{\kappa(2)}, t\,) \frac{\partial}{\partial y^B_{\kappa\,(1)}} L^A(\, y^1_{\kappa(1)}, y^2_{\kappa(1)}\,)\, \mathbf{a}^B_{\kappa\,(1)}$$

$$= \text{grad}_{(\kappa 2)}\chi_{\kappa}{}^{(\sigma)}_{2}(\, y^1_{\kappa(2)}, y^2_{\kappa(2)}, t\,) \cdot \text{grad}_{(\kappa 1)}\mathbf{\Lambda}(\, y^1_{\kappa(1)}, y^2_{\kappa(1)}\,)$$

$$= \mathcal{F} \cdot \mathcal{G} \tag{4-7}$$

Here

$$\mathcal{G} \equiv \text{grad}_{(\kappa 1)}\mathbf{\Lambda}(\, y^1_{\kappa(1)}, y^2_{\kappa(1)}\,)$$

$$= \frac{\partial}{\partial y^B_{\kappa\,(1)}} L^A(\, y^1_{\kappa(1)}, y^2_{\kappa(1)}\,)\, \mathbf{a}_{\kappa(2)A}\, \mathbf{a}^B_{\kappa\,(1)} \tag{4-8}$$

is the gradient of the mapping from $\kappa_1^{(\sigma)}$ to $\kappa_2^{(\sigma)}$.

Denote by $\mathcal{B}_{\kappa(1)}$ the response functional for a simple surface material with $\kappa_1^{(\sigma)}$ as the reference configuration; take $\mathcal{B}_{\kappa(2)}$ to be the response functional for the same simple surface material with $\kappa_2^{(\sigma)}$ as the reference. The relation between these two response functionals follows immediately from the definition of a simple surface material and (4-7):

$$T^{(\sigma)} = \mathcal{B}_{\kappa(2)}(\, \mathcal{F}^t\,)$$

$$= \mathcal{B}_{\kappa(1)}(\, \mathcal{F}^t \cdot \mathcal{G}) \tag{4-9}$$

Under what conditions does $\mathcal{G}$ represent a tangential transformation from one reference configuration to another that cannot be distinguished from the original by experimental studies of the surface stress tensor $T^{(\sigma)}$ as a function of $\mathcal{F}^t$, the history of the surface deformation gradient? It seems intuitively reasonable that the behavior of a dividing surface will change as material particles enter or leave it. We will restrict ourselves to changes of configuration involving no mass transfer to or from the adjoining phases. Since we can also expect that the response of a dividing surface will depend upon its surface mass density, let us limit ourselves to configurations with the same surface density. If these two conditions are to be satisfied, we see from Exercise 1.3.5-4 that we must consider only unimodular tangential transformations $\mathcal{H}$

$$\det_{(\kappa_1)}\mathcal{H} = \pm 1 \tag{4-10}$$

Let $\mathcal{H}$ be a time-independent unimodular transformation that satisfies

$$\mathcal{B}_{\kappa(1)}(\mathcal{F}^t \cdot \mathcal{H}) = \mathcal{B}_{\kappa(1)}(\mathcal{F}^t) \tag{4-11}$$

for all non-singular $\mathcal{F}^t$. If we interpret $\mathcal{H}$ as the gradient of the mapping from the reference configuration $\kappa_1^{(\sigma)}$ to the reference configuration $\kappa_2^{(\sigma)}$, then by (4-9)

$$\begin{aligned} T^{(\sigma)} &= \mathcal{B}_{\kappa(2)}(\mathcal{F}^t) \\ &= \mathcal{B}_{\kappa(1)}(\mathcal{F}^t) \end{aligned} \tag{4-12}$$

Equation (4-12) says that the response of the material to the deformation history $\mathcal{F}^t$ after transformation from the reference configuration $\kappa_1^{(\sigma)}$ to the reference configuration $\kappa_2^{(\sigma)}$ is the same as it would have been had the transformation not taken place. This means that $\mathcal{H}$ is a static unimodular tangential transformation which cannot be detected by an experiment.

If both $\mathcal{H}_1$ and $\mathcal{H}_2$ satisfy (4-11), then $\mathcal{H}_1 \cdot \mathcal{H}_2$ is also a solution:

$$\begin{aligned} \mathcal{B}_{\kappa}(\mathcal{F}^t \cdot \mathcal{H}_1 \cdot \mathcal{H}_2) &= \mathcal{B}_{\kappa}(\mathcal{F}^t \cdot \mathcal{H}_1) \\ &= \mathcal{B}_{\kappa}(\mathcal{F}^t) \end{aligned} \tag{4-13}$$

If $\mathcal{H}$ is a solution to (4-11), then

$$\begin{aligned} \mathcal{B}_{\kappa}(\mathcal{F}^t \cdot \mathcal{H}^{-1}) &= \mathcal{B}_{\kappa}(\mathcal{F}^t \cdot \mathcal{H}^{-1} \cdot \mathcal{H}) \\ &= \mathcal{B}_{\kappa}(\mathcal{F}^t) \end{aligned} \tag{4-14}$$

and $\mathcal{H}^{-1}$ is as well. The projection tensor $\mathbf{P}_\kappa$ also satisfies (4-11):

$$\mathcal{B}_{\kappa}(\mathcal{F}^t \cdot \mathbf{P}_\kappa) = \mathcal{B}_{\kappa}(\mathcal{F}^t) \tag{4-15}$$

Equations (4-13) through (4-15) imply that the collection of all time-independent unimodular tangential transformations which satisfy (4-11) forms a group. We will refer to this group as the **surface isotropy group** $g_\kappa^{(\sigma)}$ of the dividing surface at the surface particle whose place is $\mathbf{z}_\kappa$ in the reference configuration $\kappa^{(\sigma)}$. The surface isotropy group is the collection of all static unimodular tangential transformations from $\kappa^{(\sigma)}$ at $\mathbf{z}_\kappa$ that cannot be detected by experiment. We can also think of the surface isotropy group as the group of **surface material symmetries**. From its definition, $g_\kappa^{(\sigma)}$ is a subgroup of the unimodular group $u_\kappa^{(\sigma)}$ of tangential transformations:

$$g_\kappa^{(\sigma)} \subset u_\kappa^{(\sigma)} \tag{4-16}$$

How is the surface isotropy group altered under a change of reference configuration? Equation (4-8) defines $\mathcal{G}$, the gradient of the mapping Λ

from $\kappa_1^{(\sigma)}$ to $\kappa_2^{(\sigma)}$. Let $\mathcal{H}$ be a member of $g_{\kappa(1)}^{(\sigma)}$, the isotropy group for $\kappa^{(\sigma)}$. Equation (4-11) applied to $\mathcal{F}^t \cdot \mathcal{G}$ requires

$$\mathfrak{B}_{\kappa(1)}(\mathcal{F}^t \cdot \mathcal{G} \cdot \mathcal{H}) = \mathfrak{B}_{\kappa(1)}(\mathcal{F}^t \cdot \mathcal{G}) \tag{4-17}$$

By (4-9), this can be interpreted in terms of the response functional with respect to $\kappa_2^{(\sigma)}$,

$$\mathfrak{B}_{\kappa(1)}(\mathcal{F}^t \cdot \mathcal{G} \cdot \mathcal{H}) = \mathfrak{B}_{\kappa(2)}(\mathcal{F}^t) \tag{4-18}$$

or

$$\mathfrak{B}_{\kappa(1)}(\mathcal{F}^t \cdot \mathcal{G} \cdot \mathcal{H} \cdot \mathcal{G}^{-1} \cdot \mathcal{G}) = \mathfrak{B}_{\kappa(2)}(\mathcal{F}^t) \tag{4-19}$$

Again applying (4-9), we have

$$\mathfrak{B}_{\kappa(2)}(\mathcal{F}^t \cdot \mathcal{G} \cdot \mathcal{H} \cdot \mathcal{G}^{-1}) = \mathfrak{B}_{\kappa(2)}(\mathcal{F}^t) \tag{4-20}$$

This means that the isotropy group for $\kappa_2^{(\sigma)}$ is simply related to the isotropy group for $\kappa_1^{(\sigma)}$:

$$g_{\kappa(2)}^{(\sigma)} = \mathcal{G} \cdot g_{\kappa(1)}^{(\sigma)} \cdot \mathcal{G}^{-1} \tag{4-21}$$

Notice that, if

$$\mathcal{G} = \alpha\, \mathbf{P}_\kappa \tag{4-22}$$

the isotropy group $g_\kappa^{(\sigma)}$ is unchanged, so long as $\alpha \neq 0$. This should be interpreted as meaning that inversions and dilations leave surface material symmetries unchanged.

For all unimodular $\mathcal{H}$, $\mathcal{G} \cdot \mathcal{H} \cdot \mathcal{G}^{-1}$ is also unimodular and

$$\mathcal{G} \cdot u_{\kappa(1)}^{(\sigma)} \cdot \mathcal{G}^{-1} = u_{\kappa(2)}^{(\sigma)} \tag{4-23}$$

The largest isotropy group, the unimodular group $u_\kappa^{(\sigma)}$, corresponds to a simple surface material with maximum symmetry. No change of frame or reference can destroy this symmetry.

References

Noll, W., *Arch. Ration. Mech. Anal.* **2**, 197 (1958); see also "The Rational Mechanics of Materials," edited by C. Truesdell, Gordon and Breach, New York (1965).

Truesdell, C., "The Elements of Continuum Mechanics," Springer-Verlag, New York, (1966).

Truesdell, C., and W. Noll, “Handbuch der Physik,” vol. 3/3 edited by S. Flügge, Springer-Verlag, Berlin (1965).

2.2.5 Isotropic simple surface materials We are now in the position to extend Noll's (1958; Truesdell and Noll 1965, p. 78; Truesdell 1966, p. 60) discussion of isotropy to simple surface materials.

An isotropic simple surface material is one that has no internal sense of direction. Its behavior is unaffected by rotations. More formally, we will say that a simple surface material is **isotropic**, if there exists a reference configuration $\boldsymbol{\kappa}^{(\sigma)}$ for the dividing surface such that its isotropy group $g_{\boldsymbol{\kappa}}^{(\sigma)}$ with respect to $\boldsymbol{\kappa}^{(\sigma)}$ includes the full orthogonal group $\mathcal{O}_{\boldsymbol{\kappa}}$ of tangential transformations.

$$g_{\boldsymbol{\kappa}}^{(\sigma)} \supset \mathcal{O}_{\boldsymbol{\kappa}} \tag{5-1}$$

Such a reference configuration will be referred to as an **undistorted isotropic state** of the dividing surface.

No orthogonal transformation from an undistorted state can be detected experimentally. By (4-21), any orthogonal transformation carries one undistorted state into another.

Since the orthogonal group is the largest subgroup of the unimodular group (Brauer 1965; Noll 1965), (5-1) and (4-16) imply that

$$\text{either } g_{\boldsymbol{\kappa}}^{(\sigma)} = \mathcal{O}_{\boldsymbol{\kappa}} \text{ or } g_{\boldsymbol{\kappa}}^{(\sigma)} = u_{\boldsymbol{\kappa}}^{(\sigma)} \tag{5-2}$$

The surface isotropy group of an isotropic simple surface material is either the orthogonal group or the full unimodular group $u_{\boldsymbol{\kappa}}^{(\sigma)}$.

The constitutive equation for an isotropic simple surface material can be written in a particularly convenient form with respect to an undistorted isotropic state. In Sec. 2.2.3, we found as a consequence of the principle of frame indifference that for a simple surface material

$$\mathfrak{Q}^T(t) \cdot \mathbf{T}^{(\sigma)} \cdot \mathfrak{Q}(t) = \boldsymbol{\mathcal{L}}[\, \mathfrak{Q}(t)^T \cdot \mathbf{C}_t^{(\sigma)t} \cdot \mathfrak{Q}(t) \,;\, \mathbf{C}^{(\sigma)}(t) \,] \tag{5-3}$$

Equation (4-11) indicates that the surface stress remains unaltered, if at any given time t we replace $\boldsymbol{\mathcal{F}}^t$ by $\boldsymbol{\mathcal{F}}^t \cdot \mathfrak{Q}^T(t)$ in (5-3):

$$\begin{aligned}
&\boldsymbol{\mathcal{L}}[\, \mathfrak{Q}^T(t) \cdot \mathbf{C}_t^{(\sigma)t} \cdot \mathfrak{Q}(t) \,;\, \mathbf{C}^{(\sigma)}(t) \,] \\
&\quad = \boldsymbol{\mathcal{L}}[\, \mathfrak{Q}^T(t) \cdot \mathfrak{Q}(t) \cdot \mathbf{C}_t^{(\sigma)t} \cdot \mathfrak{Q}^T(t) \cdot \mathfrak{Q}(t) \,; \\
&\qquad\qquad \mathfrak{Q}(t) \cdot \mathbf{C}^{(\sigma)}(t) \cdot \mathfrak{Q}^T(t) \,] \\
&\quad = \boldsymbol{\mathcal{L}}[\, \mathbf{C}_t^{(\sigma)t} \,;\, \mathfrak{Q}(t) \cdot \mathbf{C}^{(\sigma)}(t) \cdot \mathfrak{Q}^T(t) \,]
\end{aligned}$$

$$= \mathcal{L}[\, \mathbf{C}_t^{(\sigma)t} \,;\, \mathbf{B}^{(\sigma)}(t) \,] \tag{5-4}$$

In arriving at this expression, we have observed from Sec. 1.2.6 that

$$\begin{aligned} \mathfrak{R}(t) \cdot \mathbf{C}^{(\sigma)}(t) \cdot \mathfrak{R}^T(t) &= \mathcal{F}(t) \cdot \mathcal{F}^T(t) \\ &= \mathbf{B}^{(\sigma)}(t) \end{aligned} \tag{5-5}$$

Here $\mathbf{B}^{(\sigma)}$ is the left surface Cauchy-Green tensor. As a consequence of (5-4), the constitutive equation for an isotropic simple surface material with respect to an undistorted isotropic state reduces to

$$\mathbf{T}^{(\sigma)} = \mathcal{L}[\, \mathbf{C}_t^{(\sigma)t} \,;\, \mathbf{B}^{(\sigma)}(t) \,] \tag{5-6}$$

and the dependence upon the surface rotation $\mathfrak{R}(t)$, seen in (5-3), drops out.

In some new frame of reference,

$$\mathcal{F}_t^{t\,*} = \mathcal{Q}^t \cdot \mathcal{F}_t^t \cdot \mathcal{Q}^T(t) \tag{5-7}$$

$$\mathcal{F}^*(t) = \mathcal{Q}(t) \cdot \mathcal{F}(t) \tag{5-8}$$

or

$$\mathbf{C}_t^{(\sigma)t*} = \mathcal{Q}(t) \cdot \mathbf{C}_t^{(\sigma)t} \cdot \mathcal{Q}^T(t) \tag{5-9}$$

$$\mathbf{B}^{(\sigma)*}(t) = \mathcal{Q}(t) \cdot \mathbf{B}^{(\sigma)}(t) \cdot \mathcal{Q}^T(t) \tag{5-10}$$

The restriction imposed by the principle of frame indifference in Sec. 2.2.3 consequently becomes for this case

$$\begin{aligned} &\mathcal{L}[\, \mathcal{Q}(t) \cdot \mathbf{C}_t^{(\sigma)t} \cdot \mathcal{Q}^T(t) \,;\, \mathcal{Q}(t) \cdot \mathbf{B}^{(\sigma)}(t) \cdot \mathcal{Q}^T(t) \,] \\ &\quad = \mathcal{Q}(t) \cdot \mathcal{L}[\, \mathbf{C}_t^{(\sigma)t} \,;\, \mathbf{B}^{(\sigma)}(t) \,] \cdot \mathcal{Q}^T(t) \end{aligned} \tag{5-11}$$

for all orthogonal tangential transformations $\mathcal{Q}(t)$.

References

Brauer, R., *Arch. Ration. Mech. Anal.* **18**, 97 (1965).

Noll, W., *Arch. Ration. Mech. Anal.* **2**, 197 (1958); see also "The Rational Mechanics of Materials," edited by C. Truesdell, Gordon and Breach, New York (1965).

Noll, W., *Arch. Ration. Mech. Anal.* **18**, 100 (1965).

Truesdell, C., "The Elements of Continuum Mechanics," Springer-Verlag,

New York (1966).

Truesdell, C., and W. Noll, "Handbuch der Physik," vol. 3/3, edited by S. Flügge, Springer-Verlag, Berlin (1965).

2.2.6 Simple surface solid A solid has some preferred configuration from which all changes of shape can be detected by experiment. A change of shape is a non-orthogonal transformation. Consequently, all non-orthogonal transformations from a preferred configuration for a solid can be detected by experiment (Noll 1958; Truesdell and Noll 1965, p. 81; Truesdell 1966, p. 61).

This conception of a solid can be extended to a simple surface material. A **simple surface solid** is a simple surface material that has a preferred reference configuration from which all changes of shape can be detected by experiment. All non-orthogonal tangential transformations from this preferred configuration can be detected by experiment. If $\boldsymbol{\kappa}^{(\sigma)}$ is the preferred reference configuration, the surface isotropy group $g_{\boldsymbol{\kappa}}^{(\sigma)}$ of a simple surface solid must be a subgroup of the full orthogonal group $\mathcal{O}_{\boldsymbol{\kappa}}$:

$$g_{\boldsymbol{\kappa}}^{(\sigma)} \subset \mathcal{O}_{\boldsymbol{\kappa}} \tag{6-1}$$

We can think of the preferred reference configuration $\boldsymbol{\kappa}^{(\sigma)}$ as an **undistorted solid state** of the dividing surface.

Let us examine the relationship between the isotropy groups corresponding to two undistorted solid states: $\overline{\boldsymbol{\kappa}}^{(\sigma)}$ and $\boldsymbol{\kappa}^{(\sigma)}$. We saw in Sec. 2.2.4 that

$$g_{\boldsymbol{\kappa}}^{(\sigma)} = \mathcal{G} \cdot g_{\overline{\boldsymbol{\kappa}}}^{(\sigma)} \cdot \mathcal{G}^{-1} \tag{6-2}$$

where $\mathcal{G}$ is the gradient of the mapping from $\overline{\boldsymbol{\kappa}}^{(\sigma)}$ to $\boldsymbol{\kappa}^{(\sigma)}$. In particular, if $\mathcal{H}$ is a member of $g_{\boldsymbol{\kappa}}^{(\sigma)}$ and $\overline{\mathcal{H}}$ a member of $g_{\overline{\boldsymbol{\kappa}}}^{(\sigma)}$, then

$$\mathcal{H} = \mathcal{G} \cdot \overline{\mathcal{H}} \cdot \mathcal{G}^{-1} \tag{6-3}$$

Let the polar decomposition of $\mathcal{G}$ take the form (see Sec. A.3.8)

$$\mathcal{G} = \mathcal{R} \cdot \mathrm{U} \tag{6-4}$$

Here $\mathcal{R}$ is an orthogonal tangential transformation from $\overline{\boldsymbol{\kappa}}^{(\sigma)}$ to $\boldsymbol{\kappa}^{(\sigma)}$; U is a positive-definite symmetric tangential tensor on $\overline{\boldsymbol{\kappa}}^{(\sigma)}$. Equation (6-3) may be rearranged to read

$$\mathcal{H} \cdot \mathcal{G} = \mathcal{G} \cdot \overline{\mathcal{H}} \tag{6-5}$$

or, in view of (6-4),

$$\mathcal{H}\cdot\mathcal{R}\cdot \mathrm{U} = \mathcal{R}\cdot \mathrm{U}\cdot\overline{\mathcal{H}}$$

$$= \mathcal{R}\cdot\overline{\mathcal{H}}\cdot(\overline{\mathcal{H}}\cdot \mathrm{U}\cdot\overline{\mathcal{H}}) \tag{6-6}$$

We can again invoke the polar decomposition theorem to find

$$\mathcal{H}\cdot\mathcal{R} = \mathcal{R}\cdot\overline{\mathcal{H}} \tag{6-7}$$

This means that

$$\mathcal{H} = \mathcal{R}\cdot\overline{\mathcal{H}}\cdot\mathcal{R}^{T} \tag{6-8}$$

or

$$g_{\kappa}^{(\sigma)} = \mathcal{R}\cdot g_{\overline{\kappa}}^{(\sigma)}\cdot\mathcal{R}^{T} \tag{6-9}$$

The isotropy groups appropriate to two undistorted solid states are orthogonal conjugates of one another.

If a simple surface solid is isotropic, there must also be a reference configuration $\kappa^{(\sigma)*}$ such that the corresponding isotropy group includes the full orthogonal group $\mathcal{O}_{\kappa*}$:

$$g_{\kappa*}^{(\sigma)} \supset \mathcal{O}_{\kappa*} \tag{6-10}$$

The reference configuration $\kappa^{(\sigma)*}$ is an undistorted isotropic state of the simple surface material; the reference configuration $\kappa^{(\sigma)}$ is an undistorted solid state. In Sec. 2.2.5, we saw that either $g_{\kappa*}^{(\sigma)} = u_{\kappa*}$ or $g_{\kappa*}^{(\sigma)} = \mathcal{O}_{\kappa*}$. If $g_{\kappa*}^{(\sigma)} = u_{\kappa*}$, then from (4-23) $g_{\kappa}^{(\sigma)} = u_{\kappa}$. Since this contradicts (6-1), we conclude

$$g_{\kappa*}^{(\sigma)} = \mathcal{O}_{\kappa*} \tag{6-11}$$

This means that $\kappa^{(\sigma)*}$ is not only an undistorted isotropic state, it is also an undistorted solid state. In view of (6-9), the isotropy group of an isotropic simple surface solid in any undistorted state $\kappa^{(\sigma)}$ is the orthogonal group $\mathcal{O}_{\kappa}$.

References

Noll, W., *Arch. Ration. Mech. Anal.* **2**, 197 (1958); see also "The Rational Mechanics of Materials," edited by C. Truesdell, Gordon and Breach, New York (1965).

Truesdell, C., "The Elements of Continuum Mechanics," Springer-Verlag, New York, (1966).

Truesdell, C., and W. Noll, "Handbuch der Physik," vol. 3/3, edited by S.

Flügge, Springer-Verlag, Berlin (1965).

2.2.7 Simple surface fluid We define a simple surface solid in terms of its surface isotropy group or surface material symmetries. We would like to develop a definition for a simple surface fluid on the same basis. We can be guided by Noll's conception of a simple fluid (Noll 1958; Truesdell and Noll 1965, p. 79; Truesdell 1966, p. 62).

Several physical ideas have been associated with the term "fluid" (Truesdell 1966, p. 62). Let us examine the two that seem most definite and see what they suggest for a simple surface fluid.

Lamb (1945, p. 1) states "The fundamental property of a fluid is that it cannot be in equilibrium in a state of stress such that the mutual action between two adjacent parts is oblique to the common surface." A fluid cannot support a shear stress when in equilibrium. We could take a simple surface material to be a fluid, if it cannot support a surface shear stress at equilibrium or

$$\mathbf{T}^{(\sigma)} = \gamma(\rho^{(\sigma)})\,\mathbf{P} + \mathcal{C}(\mathcal{F}) \tag{7-1}$$

where

$$\mathcal{C}(\mathbf{P}_\kappa) = \mathbf{0} \tag{7-2}$$

Since such a simple surface material could have any isotropy group at all, there is nothing to distinguish it from the simple surface solid.

Batchelor (1967, p. 1) feels "A portion of fluid . . . does not have a preferred shape . . . ." This might be interpreted that a fluid has no preferred reference configurations, so long as the density is unchanged (Truesdell 1966, p. 63). An alternative and equivalent interpretation is that a fluid does not alter its material response after an arbitrary deformation which does not change the density (Truesdell and Noll 1965, p. 79).

This latter interpretation suggests that we define a simple surface material to be a **simple surface fluid**, if for some reference configuration $\kappa^{(\sigma)}$ the surface isotropy group is the unimodular group

$$g_\kappa^{(\sigma)} = u_\kappa^{(\sigma)} \tag{7-3}$$

From (4-23), the surface isotropy group for every reference configuration is the corresponding unimodular group. Every simple surface fluid is isotropic and every configuration of a simple surface fluid is an undistorted isotropic state.

Since simple surface fluids are isotropic, (5-6) applies. For a surface fluid, $\mathbf{T}^{(\sigma)}$ is not influenced by a static deformation from one configuration to another, unless the deformation changes the surface density $\rho^{(\sigma)}$. In (5-6), the dependence upon $\mathbf{B}^{(\sigma)}(t)$ reduces to a dependence upon

$\det_{(\sigma)}\mathbf{B}^{(\sigma)}(t) = [\ \det_{(\sigma)}\boldsymbol{\mathcal{F}}(t)\ ]^2$ or $\rho^{(\sigma)}$ (see Exercise 1.3.5-4):

$$\mathbf{T}^{(\sigma)} = \boldsymbol{\mathcal{J}}(\ \mathbf{C}_t^{(\sigma)t}\ ;\ \rho^{(\sigma)}\ ) \tag{7-4}$$

This response functional must still satisfy (5-11),

$$\begin{aligned}&\boldsymbol{\mathcal{J}}(\ \boldsymbol{\mathcal{Q}}(t) \cdot \mathbf{C}_t^{(\sigma)t} \cdot \boldsymbol{\mathcal{Q}}^T(t)\ ;\ \rho^{(\sigma)}\ )\\ &\quad = \boldsymbol{\mathcal{Q}}(t) \cdot \boldsymbol{\mathcal{J}}(\ \mathbf{C}_t^{(\sigma)t}\ ;\ \rho^{(\sigma)}\ ) \cdot \boldsymbol{\mathcal{Q}}^T(t)\end{aligned} \tag{7-5}$$

If the surface fluid has been at rest for all past time,

$$\mathbf{C}_t^{(\sigma)t} = \mathbf{P}_\kappa \tag{7-6}$$

Equations (7-5) and (7-6) require

$$\boldsymbol{\mathcal{Q}}(t) \cdot \mathbf{T}^{(\sigma)} \cdot \boldsymbol{\mathcal{Q}}^T(t) = \mathbf{T}^{(\sigma)} \tag{7-7}$$

The only tangential tensor that satisfies this relationship for every orthogonal tangential transformation $\boldsymbol{\mathcal{Q}}(t)$ is (see Exercise A.3.6-1)

$$\mathbf{T}^{(\sigma)} = \gamma(\ \rho^{(\sigma)}\ )\mathbf{P} \tag{7-8}$$

The surface stress in a surface fluid at rest is a surface tension that depends only upon the surface density. This suggests that we write (7-5) more explicitly as

$$\mathbf{T}^{(\sigma)} = \gamma(\ \rho^{(\sigma)}\ )\mathbf{P} + \boldsymbol{\mathcal{S}}(\ \mathbf{K}_t^{(\sigma)t}\ ;\ \rho^{(\sigma)}\ ) \tag{7-9}$$

where

$$\mathbf{K}_t^{(\sigma)t} \equiv \mathbf{C}_t^{(\sigma)t} - \mathbf{P} \tag{7-10}$$

Remember that (7-6) still restricts the form of the response functional

$$\begin{aligned}&\boldsymbol{\mathcal{S}}(\ \boldsymbol{\mathcal{Q}}(t) \cdot \mathbf{K}_t^{(\sigma)t} \cdot \boldsymbol{\mathcal{Q}}^T(t)\ ;\ \rho^{(\sigma)}\ )\\ &\quad = \boldsymbol{\mathcal{Q}}(t) \cdot \boldsymbol{\mathcal{S}}(\ \mathbf{K}_t^{(\sigma)t}\ ;\ \rho^{(\sigma)}\ ) \cdot \boldsymbol{\mathcal{Q}}^T(t)\end{aligned} \tag{7-11}$$

and (7-2) requires

$$\boldsymbol{\mathcal{S}}(\ \mathbf{0}\ ;\ \rho^{(\sigma)}\ ) = \mathbf{0} \tag{7-12}$$

References

Noll, W., *Arch. Ration. Mech. Anal.* **2**, 197 (1958); see also "The Rational Mechanics of Materials," edited by C. Truesdell, Gordon and Breach, New York (1965).

Truesdell, C., and W. Noll, "Handbuch der Physik," vol. 3/3, edited by S. Flügge, Springer-Verlag, Berlin (1965).

Truesdell, C., "The Elements of Continuum Mechanics," Springer-Verlag, New York, (1966).

2.2.8 Fading memory and special cases of simple surface fluid For a simple surface material, the present surface stress is determined by the history of the surface deformation gradient. But our experience indicates that real materials have an imperfect memory and that not all past deformations are equally important in determining the present stress. This suggests the **principle of fading memory** (Truesdell and Noll 1965, p. 101):

Deformations that occurred in the distant past should have less influence in determining the present stress than those that occurred in the recent past.

With a requirement that $\mathcal{S}(\, \mathbf{K}_t^{(\sigma)t} \,;\, \rho^{(\sigma)} \,)$ in (7-10) be defined and sufficiently smooth in the neighborhood of the zero history of all possible deformations, we find [by analogy with the argument given by Truesdell and Noll (1965, p. 109) leading to the constitutive equation for finite linear viscoelasticity appropriate outside the immediate neighborhood of a phase interface]

$$\mathbf{S}^{(\sigma)} \equiv \mathbf{T}^{(\sigma)} - \gamma(\, P^{(\sigma)} \,)\, \mathbf{P}$$

$$= - \int_0^\infty \frac{1}{2}[\, \kappa(\, \rho^{(\sigma)}, s \,) - \varepsilon(\, \rho^{(\sigma)}, s \,)]\, \mathrm{tr}(\, \mathbf{K}_t^{(\sigma)t} \,)\, ds\, \mathbf{P}$$

$$- \int_0^\infty \varepsilon(\, \rho^{(\sigma)}, s \,)\, \mathbf{K}_t^{(\sigma)t}\, ds \tag{8-1}$$

for a sufficiently weak memory, for a sufficiently small deformation, or for a sufficiently slow deformation. Here $\kappa(\, \rho^{(\sigma)}, s \,)$ and $\varepsilon(\, \rho^{(\sigma)}, s \,)$ are scalar-valued functions of $\rho^{(\sigma)}$ and s. We will say that (8-1) describes the behavior of a **finite linear viscoelastic surface fluid.**

Since

$$\text{at } s = 0 : \ \mathbf{K}_t^{(\sigma)t} = \mathbf{0} \tag{8-2}$$

we have

$$\mathbf{K}_t^{(\sigma)t} = \int_0^s \frac{d_{(s)}}{ds} \mathbf{K}_t^{(\sigma)t} \, ds \tag{8-3}$$

Let us introduce

$$\overline{\mathbf{D}}^{(\sigma)} \equiv -\frac{1}{2}\frac{d_{(s)}}{ds}\mathbf{K}_t^{(\sigma)t}$$

$$= \frac{1}{2}[(\nabla_{(\sigma)}\overline{\mathbf{v}}^{(\sigma)})^T \cdot \boldsymbol{\mathcal{F}}_t^{(\sigma)t} + \boldsymbol{\mathcal{F}}_t^{(\sigma)tT} \cdot \nabla_{(\sigma)}\overline{\mathbf{v}}^{(\sigma)}] \tag{8-4}$$

in which $\overline{\mathbf{v}}^{(\sigma)}$ is the surface velocity vector at time $\overline{t}$. Note that

$$\text{at } s = 0 : \ \overline{\mathbf{D}}^{(\sigma)} = \mathbf{D}^{(\sigma)} \tag{8-5}$$

where $\mathbf{D}^{(\sigma)}$ is the surface rate of deformation tensor. In view of (8-3) and (8-4), (8-1) assumes the form

$$\mathbf{S}^{(\sigma)} = \left\{ \text{tr} \int_0^\infty [\,\kappa(\rho^{(\sigma)}, s) - \varepsilon(\rho^{(\sigma)}, s)] \int_0^s \overline{\mathbf{D}}^{(\sigma)} \, ds' \, ds \right\} \mathbf{P}$$

$$+ 2\int_0^\infty \varepsilon(\rho^{(\sigma)}, s) \int_0^s \overline{\mathbf{D}}^{(\sigma)} \, ds' \, ds \tag{8-6}$$

Let us introduce the following dimensionless variables:

$$s^* \equiv s/t_0 \qquad s^{**} \equiv s/t_p = s^* N_{De}$$

$$\kappa^* \equiv \frac{t_0^2}{\varepsilon_0}\kappa(\rho^{(\sigma)}, s) \qquad \varepsilon^* \equiv \frac{t_0^2}{\varepsilon_0}\varepsilon(\rho^{(\sigma)}, s)$$

$$\overline{\mathbf{D}}^{(\sigma)**} \equiv t_p \overline{\mathbf{D}}^{(\sigma)} \qquad \mathbf{S}^{(\sigma)**} \equiv \frac{t_p}{\varepsilon_0}\mathbf{S}^{(\sigma)} \tag{8-7}$$

Here t_0 is a characteristic relaxation time of the surface, t_p is a characteristic time for the process of deformation, ε_0 is a characteristic surface viscosity, and

$$N_{De} \equiv t_0/t_p \tag{8-8}$$

is the Deborah number. In terms of these dimensionless variables, (8-6) becomes

$$\mathbf{S}^{(\sigma)**} = \frac{1}{N_{De}} \left(\int_0^\infty (\kappa^* - \varepsilon^*) \operatorname{tr} \int_0^{s^* N_{De}} \overline{\mathbf{D}}^{(\sigma)**} \, ds^{**} \, ds^* \right) \mathbf{P}$$

$$+ \frac{2}{N_{De}} \int_0^\infty \varepsilon^* \int_0^{s^* N_{De}} \overline{\mathbf{D}}^{(\sigma)**} \, ds^{**} \, ds^* \tag{8-9}$$

Applying L'Hopital's rule, we find that in the limit $N_{De} \to 0$

$$\mathbf{S}^{(\sigma)**} = \int_0^\infty (\kappa^* - \varepsilon^*) \, s^* \, ds^* \operatorname{tr} \mathbf{D}^{(\sigma)**} \, \mathbf{P}$$

$$+ 2 \int_0^\infty \varepsilon^* \, s^* \, ds^{**} \, \mathbf{D}^{(\sigma)**} \tag{8-10}$$

or

$$\mathbf{S}^{(\sigma)} = (\kappa - \varepsilon) \operatorname{tr} \mathbf{D}^{(\sigma)} \, \mathbf{P} + 2 \, \varepsilon \, \mathbf{D}^{(\sigma)} \tag{8-11}$$

in which

$$\kappa \equiv \varepsilon_0 \int_0^\infty \kappa^* \, s^* \, ds^* \tag{8-12}$$

and

$$\varepsilon \equiv \varepsilon_0 \int_0^\infty \varepsilon^* \, s^* \, ds^* \tag{8-13}$$

Equation (8-11) is the **Boussinesq surface fluid model** introduced in Sec. 2.2.2. We can now view the Boussinesq surface fluid as the limiting case of the simple surface fluid with fading memory for which the relaxation time for the surface is so short or the flow so slow that memory effects disappear.

Reference

Truesdell C., and W. Noll, "Handbuch der Physik," vol. 3/3, edited by S. Flügge, Springer-Verlag, Berlin (1965).

2.2.9 Simple surface fluid crystals We adopt a suggestion of Truesdell's (1966, p. 64) in referring to all non-solid simple surface materials as **simple surface fluid crystals.**

For a simple surface fluid crystal, there is no reference configuration $\boldsymbol{\kappa}^{(\sigma)}$ for which

$$g_{\boldsymbol{\kappa}}^{(\sigma)} \subset \mathcal{O}_{\boldsymbol{\kappa}} \tag{9-1}$$

This is another way of saying that there is *some* undetectable change of shape from every reference configuration of a surface fluid crystal. There is also no reference configuration $\boldsymbol{\kappa}^{(\sigma)}$ for which

$$g_{\boldsymbol{\kappa}}^{(\sigma)} \supset \mathcal{O}_{\boldsymbol{\kappa}} \tag{9-2}$$

unless the surface fluid crystal is in fact a surface fluid. There are *some* rotations detectable from every reference configuration of a surface fluid crystal.

It is clear that a surface fluid crystal is a surface fluid, if and only if it is isotropic.

When sufficient experimental evidence becomes available to motivate pursuing this topic in more detail, I suggest that the reader consult Coleman (1965) and Wang (1965) (see also Truesdell and Noll 1965, p. 86).

References

Coleman, B. D., *Arch. Ration. Mech. Anal.* **20**, 41 (1965).

Truesdell, C., and W. Noll, "Handbuch der Physik," vol. 3/3, edited by S. Flügge, Springer-Verlag, Berlin (1965).

Truesdell, C., "The Elements of Continuum Mechanics," Springer-Verlag, New York (1966).

Wang, C. C., *Arch. Ration. Mech. Anal.* **20**, 1 (1965).

2.3 Structural models for interface

2.3.1 Concept (Deemer and Slattery 1978) The objective here is to study the effect of surfactant upon the stress-deformation behavior of the interface. Our approach is suggested by Einstein (1906, 1956), who proposed that a dilute suspension of neutrally buoyant solid spheres has nearly the same average stress-deformation behavior as a dilute solution of a solute, whose molecular weight is large in comparison with that of the solvent. More specifically, the equations of motion for the continuum represented by the suspension are obtained by taking the local volume average of the equations of motion for the individual phases in the suspension (Jiang *et al.* 1987).

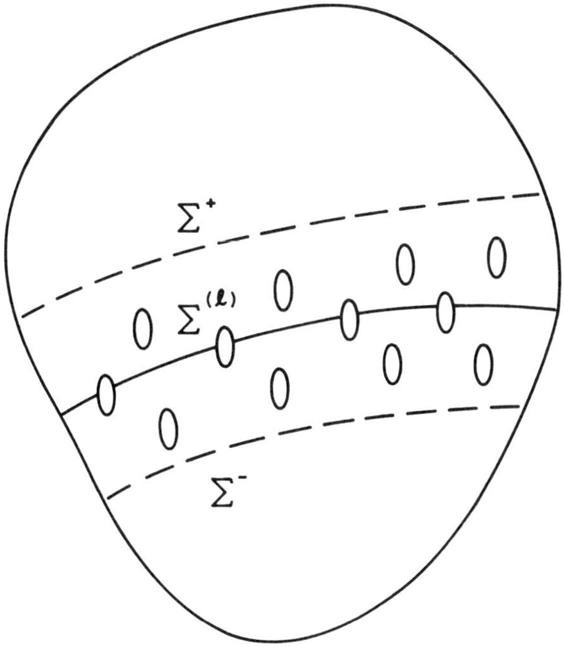

Figure 2.3.1-1: Region R occupied by a material body consisting of two phases. The interface is a three-dimensional region surrounding a singular surface $\Sigma^{(\ell)}$ that separates the two solvent phases. Rigid bodies, representing the surfactant molecules, are dispersed in some manner within this region. The lines formed by the intersection of $\Sigma^{(\ell)}$ with S are denoted by $C^{(\ell)}$; the dividing surface Σ is a simply connected extension of $\Sigma^{(\ell)}$.

We will make the following assumptions concerning the system.

i) There are only three components present: two mutually insoluble solvents and a surfactant that may be soluble in both solvents.

ii) The molecular weight of the surfactant is much larger than that of either of the solvents.

A structural or hydrodynamic model for the interface is sketched in Figure 1-1. The interface is a three-dimensional region surrounding a singular surface that separates the two solvent phases. Solid bodies, representing the surfactant molecules, are dispersed in some manner within this region. Some may intercept the singular surface. The interfacial region, modeled in this way, has the same average behavior as an interfacial region containing a solution of surfactant whose molecular weight is large in comparison with that of either solvent.

The general balance described in Exercise 1.3.2-2 may be extended to include this structural model for the interface.

$$\frac{d}{dt}\left[\int_R \psi \, dV + \int_{R^{(I)}} (\psi^{(I)} - \psi)\, dV + \int_{\Sigma^{(\ell)}} \psi_o^{(\sigma)}\, dA\right]$$

$$= -\int_S \boldsymbol{\phi} \cdot \mathbf{n}\, dA - \int_{S^{(I)}} (\boldsymbol{\phi}^{(I)} - \boldsymbol{\phi}) \cdot \mathbf{n}\, dA - \int_{C^{(\ell)}} \boldsymbol{\phi}_o^{(\sigma)} \cdot \boldsymbol{\mu}\, ds$$

$$+ \int_R \rho\zeta \, dV + \int_{R^{(I)}} (\rho^{(I)}\zeta^{(I)} - \rho\zeta)\, dV + \int_{\Sigma^{(\ell)}} \rho_o^{(\sigma)} \zeta_o^{(\sigma)}\, dA \quad (1\text{-}1)$$

Here $\Sigma^{(\ell)}$ is the liquid-liquid singular surface, $C^{(\ell)}$ are the lines formed by the intersection of $\Sigma^{(\ell)}$ with S, $\psi_o^{(\sigma)}$ is the density of the quantity per unit area on $\Sigma^{(\ell)}$, $\boldsymbol{\phi}_o^{(\sigma)}$ is the flux of the quantity (per unit length of line) through $C^{(\ell)}$, $\rho_o^{(\sigma)}$ the mass density (per unit area) on $\Sigma^{(\ell)}$, and $\zeta_o^{(\sigma)}$ the rate of production of the quantity per unit mass at each point on $\Sigma^{(\ell)}$. We can think of $\psi_o^{(\sigma)}$, $\boldsymbol{\phi}_o^{(\sigma)}$, $\rho_o^{(\sigma)}$, and $\zeta_o^{(\sigma)}$ as being the values of $\psi^{(\sigma)}$, $\boldsymbol{\phi}^{(\sigma)}$, $\rho^{(\sigma)}$, and $\zeta^{(\sigma)}$ appropriate to a clean interface. Let the dividing surface Σ be a simply connected extension of $\Sigma^{(\ell)}$. Using an argument essentially the same as that outlined in Exercise 1.3.2-2, we conclude that on $\Sigma^{(\ell)}$

$$\psi^{(\sigma,\ell)} \equiv \psi_o^{(\sigma)} + \int_{\lambda^-}^{\lambda^+} (\psi^{(I)} - \psi)(1 - \kappa_1\lambda)(1 - \kappa_2\lambda)\, d\lambda \quad (1\text{-}2)$$

$$\boldsymbol{\phi}^{(\sigma\ell)} \equiv \boldsymbol{\phi}_o^{(\sigma)} + \left[\int_{\lambda^-}^{\lambda^+} (\boldsymbol{\phi}^{(I)} - \boldsymbol{\phi})\, d\lambda\right] \cdot \mathbf{P} \quad (1\text{-}3)$$

and

$$\rho^{(\sigma,\ell)}\zeta^{(\sigma,\ell)} \equiv \rho_o^{(\sigma)}\, \zeta_o^{(\sigma)}$$

$$+\int_{\lambda^-}^{\lambda^+} (\rho^{(I)}\zeta^{(I)} - \rho\zeta)(1 - \kappa_1\lambda)(1 - \kappa_2\lambda)\, d\lambda \tag{1-4}$$

In contrast, on $\Sigma^{(s)} \equiv \Sigma - \Sigma^{(\ell)}$, that portion of Σ passing through the rigid bodies,

$$\psi^{(\sigma,s)} \equiv \int_{\lambda^-}^{\lambda^+} (\psi^{(\sigma)} - \psi)(1 - \kappa_1\lambda)(1 - \kappa_2\lambda)\, d\lambda \tag{1-5}$$

$$\phi^{(\sigma,s)} \equiv \left[\int_{\lambda^-}^{\lambda^+} (\phi^{(I)} - \phi)\, d\lambda\right] \cdot \mathbf{P} \tag{1-6}$$

and

$$\rho^{(\sigma,s)}\zeta^{(\sigma,s)} \equiv \int_{\lambda^-}^{\lambda^+} (\rho^{(I)}\zeta^{(I)} - \rho\zeta)(1 - \kappa_1\lambda)(1 - \kappa_2\lambda)\, d\lambda \tag{1-7}$$

In using (1-2) through (1-4) or (1-5) through (1-7), keep in mind that $\psi^{(I)}$, $\phi^{(I)}$, and $\rho^{(I)}\zeta^{(I)}$ are now discontinuous functions taking different forms in the fluid and solid phases.

References

Deemer, A. R., and J. C. Slattery, *Int. J. Multiphase Flow* **4**, 171 (1978).

Einstein, A., *Ann. Phys. Leipzig* **19**, 289 (1906).

Einstein, A., "Investigation on the Theory of the Brownian Movement," Dover, New York (1956).

Jiang, T. S., M. H. Kim, V. J. Kremesec Jr., and J. C. Slattery, *Chem. Eng. Commun.* **50,** 1 (1987).

2.3.2 Local area averages (Deemer and Slattery 1978) In introducing the structural model, we proposed that a suspension of rigid bodies distributed about a singular surface has the same *average* behavior as an interfacial region. All quantities such as mass, momentum, and stress are continuously distributed over a continuum dividing surface. Equations (1-2) through (1-4) and (1-5) through (1-7) do not have this feature. It must be an average of (1-2) and (1-5), of (1-3) and (1-6), of (1-4) and (1-7) with which we are to be concerned.

The problem is similar to that encountered with a structural or hydrodynamic model for a bulk solution, in which rigid bodies, representing large solute molecules, are dispersed in a solvent. The equations of motion for the continuum represented by the suspension are obtained by taking the local volume average of the equations of motion for the individual phases in the suspension (Batchelor 1970; Russel 1976; Jeffrey and Acrivos (1976)). This suggests that the general jump balance for the continuum dividing surface represented by an interfacial suspension is a local area average of the jump balance for the individual phases intercepted by Σ in Figure 1-1.

Let us make an additional assumption concerning the system.

iii) The characteristic length of a rigid body representing a surfactant molecule is everywhere very small compared with the radius of curvature of the dividing surface. At each point on the dividing surface, the surface may be considered flat over a region that is large compared with the region over which averages are to be defined.

Let us center upon each point on Σ in Figure 1-1 a circle $\mathcal{C}$. The diameter of the circle should not be so small that $\mathcal{C}$ encloses only a portion of $\Sigma^{(\ell)}$ or only a portion of $\Sigma^{(s)}$ at any point on Σ. On the other hand, the diameter of $\mathcal{C}$ must be sufficiently small that the region over which Σ is locally flat is large compared with the region enclosed by $\mathcal{C}$.

Let us define $\mathcal{S}$ to be the region enclosed by $\mathcal{C}$ at each point on Σ; $\mathcal{A}$ is the area of $\mathcal{S}$. We will denote by $\mathcal{C}^{(i)}$ the closed curve (or curves) bounding all of $\Sigma^{(i)}$ contained within $\mathcal{C}$; $\mathcal{S}^{(i)}$ is the region bounded by $\mathcal{C}^{(i)}$; $\mathcal{A}^{(i)}$ is the area of $\mathcal{S}^{(i)}$.

Assume that $B^{(i)}$ is some scalar, vector, or tensor associated with $\Sigma^{(i)}$. We can speak of the **local area average** for $\Sigma^{(i)}$ of $B^{(i)}$

$$\overline{B}^{(i)} \equiv \frac{1}{\mathcal{A}} \int_{\mathcal{S}^{(i)}} B^{(i)} \, dA \tag{2-1}$$

as well as the **intrinsic area average** for $\Sigma^{(i)}$ of $B^{(i)}$

$$\langle B \rangle^{(i)} \equiv \frac{1}{\mathcal{A}^{(i)}} \int_{\mathcal{S}^{(i)}} B^{(i)} \, dA \tag{2-2}$$

We will also require the **total local area average** of B,

$$\langle B \rangle \equiv \overline{B}^{(\ell)} + \overline{B}^{(s)} \tag{2-3}$$

In view of assumption (iii), we can prove (see Exercise 2.3.2-1)

$$\overline{\nabla_{(\sigma)}\mathrm{B}}^{(i)} = \nabla_{(\sigma)}\overline{\mathrm{B}}^{(i)} - \frac{1}{\mathcal{A}}\int_{\mathcal{C}^{(i)}-\mathcal{C}_e^{(i)}} \mathrm{B}^{(i)}\, \mathbf{v}^{(i)}\, ds \tag{2-4}$$

where

$$\mathcal{C}_e^{(i)} \equiv \mathcal{C} \cap \mathcal{C}^{(i)} \tag{2-5}$$

denotes that portion of $\mathcal{C}$ which coincides with $\mathcal{C}^{(i)}$. The unit vector $\mathbf{v}^{(i)}$ is normal to the curve $\mathcal{C}^{(i)} - \mathcal{C}_e^{(i)}$, tangent to and directed into $\mathcal{S}^{(i)}$. We can refer to (2-4) as the theorem for the **local area average of a surface gradient**. The theorem for the **local area average of a surface divergence**

$$\overline{\mathrm{div}_{(\sigma)}\mathbf{B}}^{(i)} = \mathrm{div}_{(\sigma)}\overline{\mathbf{B}}^{(i)} - \frac{1}{\mathcal{A}}\int_{\mathcal{C}^{(i)}-\mathcal{C}_e^{(ii)}} \mathbf{B}^{(i)} \cdot \mathbf{v}^{(i)}\, ds \tag{2-6}$$

follows immediately.

The next two sections illustrate how these ideas can be used.

References

Batchelor, G. K., *J. Fluid Mech.* **41**, 545 (1970).

Deemer, A. R., and J. C. Slattery, *Int. J. Multiphase Flow* **4**, 171 (1978).

Jeffrey, D. J., and A. Acrivos, *AIChE J.* **22**, 417 (1976).

Russel, W. B., *J. Colloid Interface Sci.* **55**, 590 (1976).

Slattery, J. C., "Momentum, Energy, and Mass Transfer in Continua," McGraw-Hill, New York (1972); second edition, Robert E. Krieger, Malibar, FL 32950 (1981).

Exercise

2.3.2-1 *theorem for the local area average of a surface gradient* Recognizing that we are dealing with a portion of the dividing surface that can be considered flat, construct a proof for (2-4) similar to that of the theorem for the local volume average of a gradient (Slattery 1981, p. 196).

2.3.3 Local area average of the jump mass balance from a structural model (Deemer and Slattery 1978) In developing the local area average of the jump mass balance, we will assume:

iv) The solid bodies are neutrally buoyant.

v) The intersection of Σ with a solid body does not change relative to the body as a function of time.

vi) The surface mass density $\rho_0^{(\sigma)}$ appropriate to a clean interface is a constant, independent of position and time on $\Sigma^{(\ell)}$.

vii) There is no mass transfer between the interface and either of the adjoining bulk phases.

In view of assumption (iv), the surface mass density can be computed on $\Sigma^{(\ell)}$ from (1-2) as

$$\rho^{(\sigma,\ell)} = \rho_0^{(\sigma)} + \int_{\lambda^-}^{\lambda^+} (\rho^{(I)} - \rho) \, d\lambda$$

$$= \rho_0^{(\sigma)} \tag{3-1}$$

and on $\Sigma^{(s)}$ from (1-5) as

$$\rho^{(\sigma,s)} = \int_{\lambda^-}^{\lambda^+} (\rho^{(I)} - \rho) \, d\lambda$$

$$= 0 \tag{3-2}$$

As a result, the total local area average of the surface mass density is

$$<\rho^{(\sigma)}> = \overline{\rho^{(\sigma)}}^{(\ell)}$$

$$= (1 - x)\rho_0^{(\sigma)} \tag{3-3}$$

in which x is introduced as the fraction of Σ occupied by the solid.

Using (3-1) and recognizing assumptions (vi) and (vii), we find that on $\Sigma^{(\ell)}$ the jump mass balance (Sec. 1.3.5) reduces to

$$\text{div}_{(\sigma)} \mathbf{v}^{(\sigma,\ell)} = 0 \tag{3-4}$$

Employing (3-2) and assumption (vii), we see that the jump mass balance is identically satisfied on $\Sigma^{(s)}$.

Because of assumption (v), area is preserved on $\Sigma^{(s)}$. Defining

$$\psi = \psi^{(\sigma,s)} \equiv 1$$

$$\phi = \phi^{(\sigma,s)} \equiv 0$$

$$\zeta = \zeta^{(\sigma,s)} \equiv 0 \tag{3-5}$$

and remembering assumption (vii), we observe that the general jump balance (Exercise 1.3.8-4) requires on $\Sigma^{(s)}$

$$\text{div}_{(\sigma)}\mathbf{v}^{(\sigma,s)} = 0 \tag{3-6}$$

The theorem for the local area average of a surface divergence (2-6) requires

$$\overline{\text{div}_{(\sigma)}\mathbf{v}^{(\sigma)}}^{(i)} = \text{div}_{(\sigma)}\ \overline{\mathbf{v}^{(\sigma)}}^{(i)} - \frac{1}{\mathcal{A}}\int_{\mathcal{C}^{(\ell s)}} \mathbf{v}^{(\sigma,i)} \cdot \boldsymbol{\nu}^{(i)}\, ds$$
$$= 0 \tag{3-7}$$

where $\mathcal{C}^{(\ell s)}$ are the curves formed by the intersections of the rigid bodies and Σ within $\mathcal{S}$. We will require $\mathbf{v}^{(\sigma)}$ to be a continuous function of position

$$\text{at } \mathcal{C}^{(\ell s)}: \ \mathbf{v}^{(\sigma,\ell)} = \mathbf{v}^{(\sigma,s)} \tag{3-8}$$

We conclude from (2-3), (3-7), and (3-8) that

$$\text{div}_{(\sigma)}\langle\mathbf{v}^{(\sigma)}\rangle = 0 \tag{3-9}$$

which we can refer to as the **total local area average of the jump mass balance** for the structural model.

Reference

Deemer, A. R., and J. C. Slattery, *Int. J. Multiphase Flow* **4**, 171 (1978).

2.3.4 Local area average of the jump momentum balance from a structural model (Deemer and Slattery 1978)

Let us make these further assumptions.

viii) The surface stress tensor $\mathbf{T}_0^{(\sigma)}$ appropriate to a clean interface is proportional to a constant surface tension γ_o, independent of position and time on $\Sigma^{(\ell)}$:

$$\mathbf{T}_o^{(\sigma)} = \gamma_o \, \mathbf{P} \tag{4-1}$$

The surface viscosity is taken to be zero for a clean interface.

ix) Inertial effects can be neglected in the interface.

x) The same external force (gravity) acts upon both the interface and the bulk phases. It can be expressed in terms of the potential energy per unit mass ϕ:

$$\mathbf{b}^{(\sigma)} = \mathbf{b} = -\nabla\phi \tag{4-2}$$

We will refer to a region in which the velocity distribution is disturbed by the presence of a body as the **disturbance neighborhood** of the body. If there are no bodies present, we will speak of the **undisturbed velocity** distribution.

xi) The suspension is sufficiently dilute that the disturbance neighborhoods associated with any two bodies do not overlap.

With assumption (viii), the surface stress tensor can be calculated on $\Sigma^{(\ell)}$ from (1-3) as

$$\mathbf{T}^{(\sigma\ell)} = \gamma_o \mathbf{P} + \mathbf{P} \cdot \left[\int_{\lambda^-}^{\lambda^+} (\mathbf{T}^{(i)} - \mathbf{T}) \, d\lambda \right] \cdot \mathbf{P} \tag{4-3}$$

and on $\Sigma^{(s)}$ from (1-6) as

$$\mathbf{T}^{(\sigma,s)} = \mathbf{P} \cdot \left[\int_{\lambda^-}^{\lambda^+} (\mathbf{T}^{(I)} - \mathbf{T}) \, d\lambda \right] \cdot \mathbf{P} \tag{4-4}$$

The total local area average of the surface stress tensor is consequently

$$\langle \mathbf{T}^{(\sigma)} \rangle = (1 - x) \, \gamma_o \mathbf{P} + \mathbf{P} \cdot \frac{1}{\mathcal{A}} \int_{\mathcal{S}} \left[\int_{\lambda^-}^{\lambda^+} (\mathbf{T}^{(I)} - \mathbf{T}) \, d\lambda \right] dA \cdot \mathbf{P} \tag{4-5}$$

where

$$\int_{\mathcal{S}} \left[\int_{\lambda^-}^{\lambda^+} (\mathbf{T}^{(I)} - \mathbf{T}) \, d\lambda \right] dA = \overline{\sum_{\text{bodies}}} \int_{R_{\text{disturb}}} (\mathbf{T}^{(I)} - \mathbf{T}) \, dV \tag{4-6}$$

Here $R_{disturb}$ is that region surrounding and including a single body in which $\mathbf{T}^{(I)}$ is not equal to the undisturbed stress distribution $\mathbf{T}$.

Russel (1976) has observed that

$$\begin{aligned}
&\text{div}[\, z(\, \mathbf{T} - \rho\phi\mathbf{I}\,)] \\
&\qquad = \mathbf{T} - \rho\phi\mathbf{I} + z\ \text{div}(\, \mathbf{T} - \rho\phi\mathbf{I}\,) \\
&\qquad = \mathbf{T} - \rho\phi\mathbf{I}
\end{aligned} \tag{4-7}$$

in which we have noted that Cauchy's first law (Sec. 2.1.4) takes the form

$$\text{div}\,(\, \mathbf{T} - \rho\phi\mathbf{I}\,) = 0 \tag{4-8}$$

for assumptions (ix) and (x). Equation (4-7) permits us to say that in the region R_{body} occupied by the body

$$\begin{aligned}
&\int_{R_{body}} (\, \mathbf{T}^{(I,s)} - \mathbf{T}\,)\, dV \\
&\quad = \int_{R_{body}} \{\, \text{div}[\, z(\, \mathbf{T}^{(I,s)} - \rho^{(I,s)}\phi\mathbf{I}\,)] + \rho^{(I,s)}\phi\,\mathbf{I} \\
&\qquad\qquad - \text{div}[\, z(\, \mathbf{T} - \rho\phi\mathbf{I}\,)] - \rho\phi\mathbf{I}\,\}\, dV \\
&\quad = \int_{R_{body}} \{\, \text{div}[\, z(\, \mathbf{T}^{(I,s)} - \mathbf{T}\,)] - z(\, \rho - \rho^{(I,s)}\,)\mathbf{b}\,\}\, dV \\
&\quad = \int_{R_{body}} \text{div}[\, z(\, \mathbf{T}^{(I,s)} - \mathbf{T}\,)]\, dV \\
&\quad = \int_{S_{body}} z(\, \mathbf{T}^{(I,s)} - \mathbf{T}\,) \cdot \mathbf{n}\, dA \\
&\quad = \int_{S_{body}} z(\, \mathbf{T}^{(I,\ell)} - \mathbf{T}\,) \cdot \mathbf{n}\, dA
\end{aligned} \tag{4-9}$$

in which we have introduced S_{body} as the closed bounding surface of the body. In the third line of this argument, we have reasoned

$$\int_{R_{body}} \{ \text{div}[\mathbf{z}(\rho - \rho^{(I,s)})\phi \mathbf{I}] + (\rho^{(I,s)} - \rho)\phi \mathbf{I} \} \, dV$$

$$= - \int_{R_{body}} \mathbf{z}(\rho - \rho^{(I,s)}) \, \mathbf{b} \, dV \tag{4-10}$$

This is zero for neutrally buoyant bodies either floating wholly within one phase or straddling Σ so long as the density distribution within the solid is such that locally $\rho^{(I,s)} = \rho$. For bodies of uniform density straddling Σ, we neglect its effect with respect to the first term on the right in the second line. In the fourth line, we have used Green's transformation (Slattery 1981, p. 661) and the jump momentum balance (Sec. 2.1.6) for the clean interface. In the fifth line, we have again employed the jump momentum balance, this time at the fluid-solid interface.

Equation (4-9) permits us to write

$$\int_{R_{disturb}} (\mathbf{T}^{(I)} - \mathbf{T}) \, dV = \int_{S_{body}} \mathbf{z}(\mathbf{T}^{(I,\ell)} - \mathbf{T}) \cdot \mathbf{n} \, dA$$

$$+ \int_{R_{disturb} - R_{body}} (\mathbf{T}^{(I,\ell)} - \mathbf{T}) \, dV \tag{4-11}$$

where $R_{disturb} - R_{body}$ is that portion of the disturbance region occupied by the liquid phases. This together with (4-5) and (4-6) gives

$$\langle \mathbf{T}^{(\sigma)} \rangle = (1 - x)\gamma_o \mathbf{P}$$

$$+ \frac{1}{\mathcal{A}} \sum_{\text{bodies}} \mathbf{P} \cdot \Bigg[\int_{S_{body}} \mathbf{z}(\mathbf{T}^{(I,\ell)} - \mathbf{T}) \cdot \mathbf{n} \, dA$$

$$+ \int_{R_{disturb} - R_{body}} (\mathbf{T}^{(I,\ell)} - \mathbf{T}) \, dV \Bigg] \cdot \mathbf{P} \tag{4-12}$$

With assumptions (ix) and (x), the jump momentum balance (Sec. 2.1.6) reduces to

$$\text{div}_{(\sigma)} \mathbf{T}^{(\sigma)} + \rho^{(\sigma)} \mathbf{b} + [\mathbf{T} \cdot \boldsymbol{\xi}] = 0 \tag{4-13}$$

The local area average of this for $\Sigma^{(i)}$ takes the form

$$\overline{\text{div}_{(\sigma)} \mathbf{T}^{(\sigma)}}^{(i)} + \overline{\rho^{(\sigma)} \mathbf{b}}^{(i)} + \overline{[\mathbf{T} \cdot \boldsymbol{\xi}]}^{(i)}$$

$$= \text{div}_{(\sigma)} \overline{\mathbf{T}^{(\sigma)}}^{(i)} - \frac{1}{\mathcal{A}} \int_{\mathcal{C}^{(\ell s)}} \mathbf{T}^{(\sigma,i)} \cdot \boldsymbol{\nu}^{(i)} \, ds$$

$$+ \overline{\rho^{(\sigma)}}^{(i)} \mathbf{b} + [\, \overline{\mathbf{T}}^{(i)} \cdot \boldsymbol{\xi} \,]$$

$$= 0 \tag{4-14}$$

in which (2-6) has been employed. We will require the surface stress vector to be a continuous function of position

$$\text{at } \mathcal{C}^{(\ell s)} : \ \mathbf{T}^{(\sigma,\ell)} \cdot \boldsymbol{\nu}^{(\ell)} + \mathbf{T}^{(\sigma,s)} \cdot \boldsymbol{\nu}^{(s)} = 0 \tag{4-15}$$

and we will assume $\mathcal{S}$ to be sufficiently small that

$$\text{at } \Sigma : \ \langle \mathbf{T} \rangle \doteq \mathbf{T} \tag{4-16}$$

With (3-3), this gives

$$\text{div}_{(\sigma)} \langle \mathbf{T}^{(\sigma)} \rangle + (1 - x) \rho_0^{(\sigma)} \mathbf{b} + [\, \mathbf{T} \cdot \boldsymbol{\xi} \,] = 0 \tag{4-17}$$

We will refer to this together with (4-12) as the **total local area average of the jump momentum balance** for the structural model.

References

Deemer, A. R., and J. C. Slattery, *Int. J. Multiphase Flow* **4**, 171 (1978).

Russel, W. B., *J. Colloid Interface Sci.* **55**, 590 (1976).

Slattery, J. C., "Momentum, Energy, and Mass Transfer in Continua," McGraw-Hill, New York (1972); second edition, Robert E. Krieger, Malabar, FL 32950 (1981).

2.3.5 A simple structural model (Deemer and Slattery 1978) In order to obtain more specific results that we might compare with experimental observations, we must describe in more detail the structure of the interface and the flow field to which it is subjected. Let us attribute to the interface a simple structure.

xii) The bodies used to represent surfactant molecules in the interfacial region are rigid spheres whose centers lie upon Σ as shown in Figure 5-1.

Let us assume that the flow field to which the interface is subjected is also simple.

xiii) In the flow field undisturbed by the presence of spheres, shear occurs only in planes parallel to Σ.

xiv) The undisturbed velocity distribution is a linear function of position (homogeneous) within the disturbance neighborhood of any sphere.

xv) Both bulk liquid phases can be described as incompressible Newtonian fluids.

Figure 2.3.5-1: Structural model for the interfacial region in which the surfactant molecules are represented as rigid spheres whose centers lie on the dividing surface.

In considering the stress distribution within an individual sphere, it will be convenient to work in terms of a frame of reference that is centered upon the sphere and that translates with the sphere. For the moment, we will not specify how this frame of reference rotates with respect to the sphere. *All vectors and tensors in this section refer to this translating and rotating frame of reference, unless explicitly noted otherwise.*

Let the subscript $_{(0)}$ indicate a quantity evaluated at the instantaneous position of the center of the sphere. At any instant of time, let us expand the undisturbed velocity distribution $\mathbf{v}$ on either side of Σ in a Taylor series with respect to the center of the sphere $\mathbf{z}_{(0)} = 0$ (Brenner 1958; Happel and Brenner 1973, p. 436)

$$\mathbf{v} = \mathbf{v}_{(0)} + \mathbf{D}_{(0)} \cdot \mathbf{z} + \mathbf{W}_{(0)} \cdot \mathbf{z} + \cdots \tag{5-1}$$

Here we have introduced the rate of deformation tensor

$$\mathbf{D} \equiv \frac{1}{2} [\, \nabla \mathbf{v} + (\nabla \mathbf{v})^T \,] \tag{5-2}$$

and the vorticity tensor

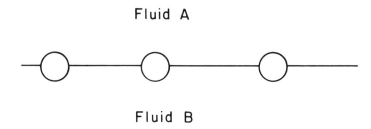

$$\mathbf{W} \equiv \frac{1}{2} [\nabla \mathbf{v} - (\nabla \mathbf{v})^T] \tag{5-3}$$

The higher order terms in (5-1) may be neglected in view of assumption (xiv).

Let us choose the frame of reference to rotate in such a way that

$$\mathbf{W}_{(0)} = 0 \tag{5-4}$$

In view of this, (5-1) requires that in the rotating and translating frame of reference on either side of Σ

$$\mathbf{v} = \mathbf{v}_{(0)} + \mathbf{D}_{(0)} \cdot \mathbf{z} + \cdots \tag{5-5}$$

With respect to a rectangular cartesian coordinate system (z_1, z_2, z_3) such that Σ lies in the plane $z_3 = 0$, assumption (xiii) requires that $\mathbf{D}_0$ have only one nonzero component:

$$D_{(0)12} \neq 0$$

$$D_{(0)11} = D_{(0)22} = D_{(0)33} = D_{(0)13} = D_{(0)23} = 0 \tag{5-6}$$

With this restriction, (5-5) satisfies the jump momentum balance (Sec. 2.1.6) and the requirement that velocity be continuous at Σ. Equation (5-5) describes the undisturbed velocity distribution everywhere in $R_{disturb}$ on both sides of Σ.

Assumption (xi) and (5-5) suggest

$$\text{as } r \to \infty : \ \mathbf{v}^{(I,\ell)} \to \mathbf{v}_{(0)} + \mathbf{D}_{(0)} \cdot \mathbf{z} \tag{5-7}$$

In the rotating and translating frame of reference, the sphere, whose radius is a, is seen to rotate as a solid body with an angular velocity $\mathbf{\Omega}$,

$$\text{at } r = a : \ \mathbf{v}^{(I,\ell)} = \mathbf{\Omega} \wedge \mathbf{z} \tag{5-8}$$

Here $\wedge$ indicates that a vector product (Slattery 1981, p. 653) is to be performed. The velocity distribution must be continuous

$$\text{at } \Sigma : \ \mathbf{v}^{(I,\ell)} \text{ is continuous} \tag{5-9}$$

and it must satisfy the jump momentum balance (Sec. 2.1.6)

$$\text{at } \Sigma : \ [\mathbf{T}^{(I,\ell)} \cdot \boldsymbol{\xi}] = 0 \tag{5-10}$$

With inertial effects neglected, a momentum balance states that the sum of the forces imposed upon the sphere by the fluid and by the external force (gravity) must be zero in the rotating and translating frame of reference,

$$\int_{S_{sphere}} \mathbf{T}^{(I,\ell)} \cdot \mathbf{n}\, dA + \int_{R_{sphere}} \rho^{(I,\ell)}\mathbf{b}\, dV = 0 \tag{5-11}$$

A moment of momentum balance similarly requires that the sum of the torques imposed upon the sphere by the fluid and by the external force must also be zero in this frame of reference:

$$\int_{S_{sphere}} \mathbf{z} \wedge (\mathbf{T}^{(I,\ell)} \cdot \mathbf{n})\, dA + \int_{R_{sphere}} \mathbf{z} \wedge \rho^{(I,S)}\mathbf{b}\, dV = 0 \tag{5-12}$$

Lamb's (1945, p. 596) solution of the equation of continuity and Cauchy's first law for the creeping flow of an incompressible Newtonian fluid was employed in the manner suggested by Happel and Brenner (1973, p. 65). The velocity and pressure distributions in the disturbance neighborhood of a single sphere consistent with boundary conditions (5-7) through (5-9), (5-11) and (5-12) take the form

$$\mathbf{v}^{(I,\ell)} = (1 - r^{\star\,-5})(\mathbf{D}_{(0)} \cdot \mathbf{z}) + \frac{5}{2}\frac{1}{a^2} (r^{\star\,-7} - r^{\star\,-5})(\mathbf{z} \cdot \mathbf{D}_{(0)} \cdot \mathbf{z})\, \mathbf{z} \tag{5-13}$$

$$p^{(I,\ell)} - p_0 + \rho^{(I,\ell)}\phi = -5\, \mu^{(I,\ell)} D_{(0)12}\, r^{\star\,-3} \sin^2\theta' \sin 2\phi' \tag{5-14}$$

where

$$r^\star \equiv \frac{r}{a} \tag{5-15}$$

Here θ' is the spherical coordinate measured from the z_3 axis; ϕ' is the spherical coordinate measured from the z_1 axis in the $z_1 - z_2$ plane. In arriving at this form of solution, we find that the center of the sphere moves with the local undisturbed velocity and that the angular velocity of the sphere relative to this frame of reference is zero.

The z_1 and z_2 (tangential) components of (5-10) are satisfied identically by (5-13). The z_3 (normal) component of (5-10) requires

$$\text{at } \Sigma : \; - p^{(I,\ell)} + \mu^{(I,\ell)} D_{33}^{(I,\ell)} \text{ is continuous} \tag{5-16}$$

From (5-13),

$$\text{at } \Sigma : \; D_{33}^{(I,\ell)} = \frac{5}{2} D_{(0)12} \left(r^{\star\,-5} - r^{\star\,-3} \right) \sin 2\phi' \tag{5-17}$$

Let us define potential energy such that

$$\text{at } \Sigma : \ \phi = 0 \tag{5-18}$$

So long as

$$\frac{\mu^{(I,\ell)} D_{(0)12}}{p_o} \ll 1 \tag{5-19}$$

(5-14), (5-17), and (5-18) imply that

$$\text{at } \Sigma : \ -p^{(I,\ell)} + \mu^{(I,\ell)} D_{33}^{(I,\ell)} = -p_o \tag{5-20}$$

which satisfies (5-16). For $\mu^{(I,\ell)} = 1$ mN s/m^2 (water) and $p_o = 10^5$ Pa (approximately atmospheric pressure), (5-19) requires $D_{(0)12} \ll 10^8\ s^{-1}$. This is well within the bounds of most experiments.

With the restriction (5-19), we can use (5-13) to compute

$$\int_{S_{sphere}} \mathbf{z}(\mathbf{T}^{(I,\ell)} - \mathbf{T}) \cdot \mathbf{n}\, dA + \int_{R_{disturb}-R_{sphere}} (\mathbf{T}^{(I,\ell)} - \mathbf{T})\, dV$$

$$= 2\pi\, a^3 (\mu^{(A)} + \mu^{(B)}) \tag{5-21}$$

with the understanding that S_{sphere} is the surface of the sphere and R_{sphere} is the region occupied by the sphere. The viscosities of the two adjoining phases are $\mu^{(A)}$ and $\mu^{(B)}$. This together with (4-12) yields the desired expression for the total local area average of the surface stress tensor

$$<\mathbf{T}^{(\sigma)}> = \gamma\, \mathbf{P} + 2\varepsilon <\mathbf{D}^{(\sigma)}> \tag{5-22}$$

where we have identified

$$\gamma \equiv (1 - x)\gamma_o \tag{5-23}$$

as the surface tension and

$$\varepsilon \equiv x\, a\, (\mu^{(A)} + \mu^{(B)}) \tag{5-24}$$

as the surface shear viscosity (Sec. 2.2.2). In reaching this result, we have said that locally the fraction of Σ occupied by the solid

$$x = \frac{1}{\mathcal{A}} \overline{\sum_{\text{bodies}}} (\pi a^2) \tag{5-25}$$

and that

$$<\mathbf{D}^{(\sigma)}> \doteq \mathbf{D}_{(0)} \tag{5-26}$$

This last is prompted by the common assumption that velocity is continuous across a dividing surface.

For the simple structural model of the interface shown in Figure 5-1 and restricted by assumptions (i) through (xv), the total local area average of the jump mass balance is given by (3-9). The total local area average of the jump momentum balance is described by (4-17), (5-22) and (5-23). The sphere radius a should be interpreted as the hydrodynamic radius of the surfactant molecule in the interface.

References

Brenner, H., *Phys. Fluids* **1**, 338 (1958).

Deemer, A. R., and J. C. Slattery, *Int. J. Multiphase Flow* **4**, 171 (1978).

Happel, J., and H. Brenner, "Low Reynolds Number Hydrodynamics," second edition, Noordhoff, Leyden (1973).

Lamb, H., "Hydrodynamics," Dover, New York (1945).

Slattery, J. C., "Momentum, Energy, and Mass Transfer in Continua," McGraw-Hill, New York (1972); second edition, Robert E. Krieger, Malabar, FL 32950 (1981).

2.3.6 Another simple structural model (Deemer and Slattery 1978)

The discussion given in the last section can be extended to another simple structural model for the interface in which assumptions (xii) and (xiv) are replaced by

xvi) The bodies used to represent surfactant molecules in the interfacial region are flexible chains of n rigid spheres. The chain begins with a sphere centered upon Σ and it extends into phase A as shown in Figure 6-1. The distance between the spheres is arbitrary but sufficiently large that the disturbance neighborhoods of the individual spheres do not overlap.

xvii) At some point in time, the chain assumes a configuration such that the centers of the spheres lie upon a straight line perpendicular to Σ.

xviii) The undisturbed velocity distribution is a linear function of position (homogeneous) within the union of the disturbance neighborhoods of the elements of the chain.

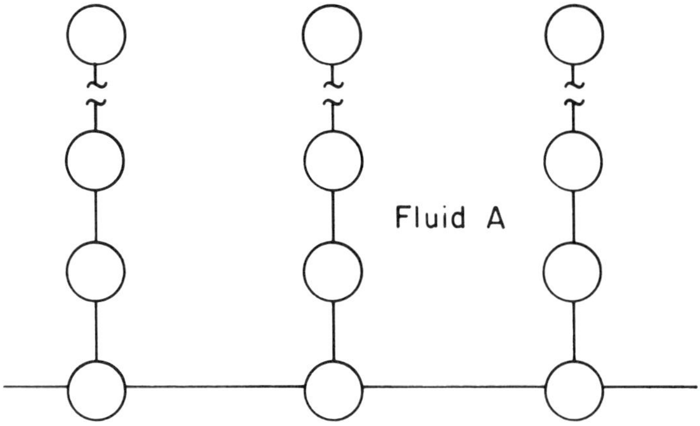

Figure 2.3.6-1: Structural model for the interfacial region in which the surfactant molecules are represented as flexible chains of n rigid spheres. The distance between the spheres is arbitrary but sufficiently large that the disturbance neighborhoods of the individual spheres do not overlap.

The velocity and pressure distributions within the disturbance neighborhood of the sphere centered upon Σ are again given by (5-13) and (5-14) with the restriction (5-19). The velocity and pressure distributions within the disturbance neighborhoods of a sphere lying wholly within phase A are described by equations analogous to (5-13) and (5-14) written with respect to a frame of reference in which the sphere is stationary.

Given assumption (xvii), how does the configuration of the chain change with time? Let us view the chain in the rotating and translating frame of reference appropriate to the first sphere, which is centered upon Σ. From (5-5), (5-6) and assumption (xviii), we see that the undisturbed velocity at the center of each sphere is the same. Since the center of a sphere moves with the local undisturbed velocity (Jiang *et al.* 1987), a chain whose spheres are aligned along a straight line perpendicular to Σ will maintain this configuration.

In the frame of reference chosen to rotate in such a manner that (5-4) is satisfied, assumption (xviii) requires the vorticity tensor to be zero everywhere within an individual disturbance neighborhood. This implies that the angular velocity of each sphere measured with respect to this frame of reference is zero. The spheres in the chain do not rotate relative to one another.

Reasoning as we did in the previous section, we conclude that the total local area average of the surface stress tensor is again given by (5-22), where we now identify as the surface shear viscosity

$$\varepsilon = x\, a[(2n - 1)\mu^{(A)} + \mu^{(B)}] \tag{6-1}$$

For the simple structural model of the interface shown in Figure 6-1 and restricted by assumptions (i) through (xi), (xiii), and (xv) through (xviii), the total local area average of the jump mass balance is (3-9). The total local area average of the jump momentum balance is given by (4-17), (5-22) and (6-1). The sphere radius a in this model should be interpreted as the hydrodynamic radius of one of the chain units in the surfactant molecule.

References

Deemer, A. R., and J. C. Slattery, *Int. J. Multiphase Flow* **4**, 171 (1978).

Jiang, T. S., M. H. Kim, V. J. Kremesec Jr., and J. C. Slattery, *Chem. Eng. Commun.* **50,** 1 (1987).

2.3.7 Comparison with previous results Equation (5-23) predicts that surface tension decreases with increasing concentration of surfactant, in agreement with common observations.

For very dilute solutions of surfactant in the interface, (5-24) and (6-1) predict surface shear viscosities that are less than the limit of sensitivity of any surface viscometer currently in use.

This is consistent with observation. Wasan *et al.* (1971) and Wei and Slattery (1976) studied a "clean" water-air interface with a deep channel surface viscometer and found that the surface shear viscosity was zero within the accuracy of their measurements. Gupta and Wasan (1974) and Poskanzer and Goodrich (1975) found that the surface shear viscosities were less than the sensitivities of their techniques for dilute solutions of non-interacting surfactants.

Cooper and Mann (1973) estimate on the basis of a kinetic theory calculation that, when the surface available per molecule is greater than 10^3 Å^2, the surface shear viscosity is of the order 10^{-12} mN s/m. Equation (5-24) gives on the order of 10^{-10} mN s/m when a = 10 Å and the surface available per molecule is 10^3 Å^2. Given some uncertainty regarding the basis for comparison, I regard these two estimates as being in reasonable agreement.

References

Cooper, E. R., and J. A. Mann, *J. Phys. Chem.* **77**, 3024 (1973).

Gupta, L., and D. T. Wasan, *Ind. Eng. Chem. Fundam.* **13**, 26 (1974).

Poskanzer, A. M., and F. C. Goodrich, *J. Phys. Chem.* **79**, 2122 (1975).

Wasan, D. T., L. Gupta, and M. K. Vora, *AIChE J.* **17**, 1287 (1971).

Wei, L. Y., and J. C. Slattery, "Colloid and Interface Science," Vol. IV, edited by M. Kerker, p. 399, Academic Press, New York (1976).

2.4 Summary

2.4.1 Summary of useful equations within bulk phases

Equation of continuity Conservation of mass requires that within the bulk phases (Sec. 1.3.5)

$$\frac{d_{(m)}\rho}{dt} + \rho \operatorname{div} \mathbf{v} = 0 \tag{1-1}$$

which is known as the **equation of continuity.**

Often we will describe liquids as being incompressible. For an **incompressible** fluid, ρ is a constant and the equation of continuity reduces to

$$\operatorname{div} \mathbf{v} = 0 \tag{1-2}$$

Cauchy's first law Euler's first law requires that within the bulk phases (Sec. 2.1.4)

$$\rho \frac{d_{(m)}\mathbf{v}}{dt} = - \nabla P + \operatorname{div} \mathbf{S} + \rho \mathbf{b} \tag{1-3}$$

which is referred to as **Cauchy's first law.** The **extra stress** tensor

$$\mathbf{S} \equiv \mathbf{T} + P\mathbf{I} \tag{1-4}$$

allows us to separate the stress attributable to the thermodynamic pressure P from that due to the deformation of the material.

In order to use Cauchy's first law, we must normally describe the stress-deformation behavior of the material. The **Newtonian fluid**

$$\mathbf{S} = \left(\lambda - \frac{2}{3} \mu \right) (\operatorname{div} \mathbf{v})\mathbf{I} + 2\mu\mathbf{D} \tag{1-5}$$

it is one of the simplest models for real viscous behavior. Here

$$\mathbf{D} = \frac{1}{2} [\nabla\mathbf{v} + (\nabla\mathbf{v})^T] \tag{1-6}$$

is the **rate of deformation** tensor, λ is the bulk dilatational viscosity, and μ is the bulk shear viscosity. For other models appropriate to non-linear or viscoelastic stress-deformation behavior, see Bird *et al.* (1977a,b).

For an **incompressible** Newtonian fluid, (1-5) simplifies to

$$\mathbf{S} = 2\mu\mathbf{D} \tag{1-7}$$

For an incompressible Newtonian fluid with constant bulk shear viscosity μ, Cauchy's first law becomes

$$\rho \frac{d_{(m)}\mathbf{v}}{dt} = -\nabla p + \mu \operatorname{div}(\nabla \mathbf{v}) + \rho \mathbf{b} \tag{1-8}$$

with the understanding that

$$p \equiv -\frac{1}{3} \operatorname{tr} \mathbf{T} \tag{1-9}$$

is the **mean pressure**. Often we will be willing to express **b** in terms of a **potential energy** per unit mass ϕ:

$$\mathbf{b} = -\nabla\phi \tag{1-10}$$

In this case, it generally will be convenient to rewrite (1-8) as

$$\rho \frac{d_{(m)}\mathbf{v}}{dt} = -\nabla \mathcal{P} + \mu \operatorname{div}(\nabla \mathbf{v}) \tag{1-11}$$

where

$$\mathcal{P} \equiv p + \rho\phi \tag{1-12}$$

is known as the **modified pressure**. (Note that the modified pressure can be introduced in this manner only for an incompressible fluid.)

Specific forms Specific forms of (1-1), (1-3), (1-6), and (1-8) are given elsewhere (Slattery 1981, p. 59; Bird *et al.* p. 83) for three common coordinate systems: rectangular cartesian, cylindrical, and spherical.

References

Bird, R. B., W. E. Stewart, and E. N. Lightfoot, "Transport Phenomena," John Wiley, New York (1960).

Bird, R. B., R. C. Armstrong, and O. Hassager, "Dynamics of Polymeric Liquids, Vol. 1 Fluid Mechanics," John Wiley, New York (1977a).

Bird, R. B., O. Hassager, R. C. Armstrong, C. F. Curtiss, "Dynamics of

Polymeric Liquids, Vol. 2 Kinetic Theory," John Wiley, New York (1977b).

Slattery, J. C., "Momentum, Energy, and Mass Transfer in Continua," McGraw-Hill, New York (1972); second edition, Robert E. Krieger, Malabar, FL 32950 (1981).

2.4.2 Summary of useful equations on dividing surfaces

Jump mass balance Conservation of mass requires that on the dividing surface (Sec. 1.3.5)

$$\frac{\partial \rho^{(\sigma)}}{\partial t} + \nabla_{(\sigma)}\rho^{(\sigma)} \cdot \dot{\mathbf{y}} + \rho^{(\sigma)}\text{div}_{(\sigma)}\mathbf{v}^{(\sigma)} + [\, \rho(\, \mathbf{v} \cdot \boldsymbol{\xi} - v_{\{\xi\}}\,)] = 0 \qquad (2\text{-}1)$$

which is known as the **jump mass balance.**

If there is **no mass transfer** to or from the dividing surface Σ,

$$\text{at } \Sigma : \;\; \mathbf{v} \cdot \boldsymbol{\xi} = v_{\{\xi\}} \qquad (2\text{-}2)$$

and (2-1) reduces to

$$\frac{\partial \rho^{(\sigma)}}{\partial t} + \nabla_{(\sigma)}\, \rho^{(\sigma)} \cdot \dot{\mathbf{y}} + \rho^{(\sigma)}\text{div}_{(\sigma)}\mathbf{v}^{(\sigma)} = 0 \qquad (2\text{-}3)$$

If

$$v_{\{\xi\}} = 0 \qquad (2\text{-}4)$$

we say that the dividing surface is **stationary**. In this case we can choose our coordinate system in such a way that

$$\mathbf{v}^{(\sigma)} = \dot{\mathbf{y}} \qquad (2\text{-}5)$$

Under these circumstances, (2-1) becomes

$$\frac{\partial \rho^{(\sigma)}}{\partial t} + \text{div}_{(\sigma)}(\, \rho^{(\sigma)}\mathbf{v}^{(\sigma)}\,) + [\, \rho\mathbf{v} \cdot \boldsymbol{\xi}\,] = 0 \qquad (2\text{-}6)$$

Occasionally, we are willing to say not only that there is no mass transfer to or from the dividing surface but also that either $\rho^{(\sigma)} = 0$ or $\rho^{(\sigma)}$

is independent of position and time on the dividing surface. As an example of this latter condition, consider a stationary interface over which any surfactant present is uniformly distributed. The jump mass balance (2-1) simplifies for these conditions to

$$\text{div}_{(\sigma)}\mathbf{v}^{(\sigma)} = 0 \tag{2-7}$$

Jump momentum balance Euler's first law requires that on the dividing surface (Sec. 2.1.6)

$$\rho^{(\sigma)}\left(\frac{\partial \mathbf{v}^{(\sigma)}}{\partial t} + \nabla_{(\sigma)}\mathbf{v}^{(\sigma)} \cdot \dot{\mathbf{y}}\right) = \nabla_{(\sigma)}\gamma + 2H\gamma\boldsymbol{\xi} + \text{div}_{(\sigma)}\mathbf{S}^{(\sigma)} + \rho^{(\sigma)}\mathbf{b}^{(\sigma)}$$

$$+ [- \rho(\mathbf{v} \cdot \boldsymbol{\xi} - v_{(\xi)}^{\{\xi\}})^2 \boldsymbol{\xi} + \mathbf{T} \cdot \boldsymbol{\xi}] \tag{2-8}$$

which is known as the **jump momentum balance**. In introducing the **surface extra stress** tensor

$$\mathbf{S}^{(\sigma)} \equiv \mathbf{T}^{(\sigma)} - \gamma\mathbf{P} \tag{2-9}$$

we are able to separate the surface stress attributable to the thermodynamic surface tension γ from the surface stress arising from deformation of the interface.

If there is **no mass transfer** to or from the dividing surface, (2-8) becomes in view of (2-2)

$$\rho^{(\sigma)}\left(\frac{\partial \mathbf{v}^{(\sigma)}}{dt} + \nabla_{(\sigma)}\mathbf{v}^{(\sigma)} \cdot \dot{\mathbf{y}}\right) = \nabla_{(\sigma)}\gamma + 2H\gamma\boldsymbol{\xi} + \text{div}_{(\sigma)}\mathbf{S}^{(\sigma)}$$

$$+ \rho^{(\sigma)}\mathbf{b}^{(\sigma)} + [\mathbf{T} \cdot \boldsymbol{\xi}] \tag{2-10}$$

If the dividing surface is **stationary**, (2-4) and (2-5) hold. The jump momentum balance (2-8) simplifies to

$$\rho^{(\sigma)}\left(\frac{\partial \mathbf{v}^{(\sigma)}}{\partial t} + \nabla_{(\sigma)}\mathbf{v}^{(\sigma)} \cdot \mathbf{v}^{(\sigma)}\right) = \nabla_{(\sigma)}\gamma + 2H\gamma\boldsymbol{\xi} + \text{div}_{(\sigma)}\mathbf{S}^{(\sigma)}$$

$$+ \rho^{(\sigma)}\mathbf{b}^{(\sigma)} + [- \rho(\mathbf{v} \cdot \boldsymbol{\xi})^2 \boldsymbol{\xi} + \mathbf{T} \cdot \boldsymbol{\xi}] \tag{2-11}$$

In order to use the jump momentum balance, we must normally say something about the stress-deformation behavior of the dividing surface. To date, the **Boussinesq surface fluid** model (Sec. 2.2.2)

$$\mathbf{S}^{(\sigma)} \equiv \mathbf{T}^{(\sigma)} - \gamma \mathbf{P} = (\kappa - \varepsilon)(\text{div}_{(\sigma)} \mathbf{v}^{(\sigma)}) \mathbf{P} + 2\varepsilon \mathbf{D}^{(\sigma)} \tag{2-12}$$

has been the constitutive equation for the surface stress tensor most commonly used. Here

$$\mathbf{D}^{(\sigma)} \equiv \frac{1}{2} [\mathbf{P} \cdot \nabla_{(\sigma)} \mathbf{v}^{(\sigma)} + (\nabla_{(\sigma)} \mathbf{v}^{(\sigma)})^T \cdot \mathbf{P}] \tag{2-13}$$

is the **surface rate of deformation** tensor (Sec. 1.2.8). For this model

$$\begin{aligned} \text{div}_{(\sigma)} \mathbf{S}^{(\sigma)} &= 2H(\kappa - \varepsilon)(\text{div}_{(\sigma)} \mathbf{v}^{(\sigma)}) \boldsymbol{\xi} \\ &\quad + \nabla_{(\sigma)}(\kappa - \varepsilon) \text{div}_{(\sigma)} \mathbf{v}^{(\sigma)} + (\kappa - \varepsilon) \nabla_{(\sigma)}(\text{div}_{(\sigma)} \mathbf{v}^{(\sigma)}) \\ &\quad + 2(\nabla_{(\sigma)} \varepsilon) \cdot \mathbf{D}^{(\sigma)} + 2\varepsilon \, \text{div}_{(\sigma)} \mathbf{D}^{(\sigma)} \end{aligned} \tag{2-14}$$

When we are willing to say not only that there is no mass transfer to or from the dividing surface but also that either $\rho^{(\sigma)} = 0$ or $\rho^{(\sigma)}$ is independent of position and time on the dividing surface, (2-7) implies

$$\text{div}_{(\sigma)} \mathbf{S}^{(\sigma)} = 2\varepsilon \, \text{div}_{(\sigma)} \mathbf{D}^{(\sigma)} \tag{2-15}$$

so long as we are also willing to say that ε is independent of position on the dividing surface.

Under **static** conditions, the jump momentum balance (2-8) reduces to

$$P^{(A)}(\boldsymbol{\xi}^{(A)} \cdot \boldsymbol{\xi}) + P^{(B)}(\boldsymbol{\xi}^{(B)} \cdot \boldsymbol{\xi}) = 2H\gamma \tag{2-16}$$

which is known as the **LaPlace equation**. Here $\boldsymbol{\xi}^{(A)}$ and $\boldsymbol{\xi}^{(B)}$ are the unit normals to the dividing surface pointing into phases A and B respectively; $\boldsymbol{\xi}$ is the unit normal such that $\boldsymbol{\xi} \cdot (\mathbf{a}_1 \wedge \mathbf{a}_2)$ is positive (Sec. A.2.6); the sign of the mean curvature H changes, if the direction of $\boldsymbol{\xi}$ is reversed (Exercise A.5.3-3 and symmetry of second groundform tangential tensor field **B**).

location of dividing surface The location of the dividing surface is defined for single-component systems in Sec. 1.3.6 and for multicomponent systems in Sec. 5.2.3.

specific forms For most problems, you will have to express the jump mass balance and the jump momentum balance in terms of the components of $\mathbf{v}^{(\sigma)}$ and $\mathbf{T}^{(\sigma)}$ starting with one of the above general forms. In doing so, I think you will find the results of Exercise 2.4.2-1 helpful.

Many problems involve interfaces that we are willing to describe as planes, cylinders, spheres, two-dimensional surfaces, or axially symmetric surfaces. For these configurations, you will find your work already finished in Tables 2.4.1-1 through -8. Note that these tables are presented in terms of the physical components of $\mathbf{v}^{(\sigma)}$ and $\mathbf{T}^{(\sigma)}$ (Secs. A.2.5 and

A.3.1; Slattery 1981, pp. 617, 629, and 647). It is common practice to use only the physical components of vectors and tensors when discussing applications in terms of orthogonal spatial coordinate systems and orthogonal surface coordinate systems. For this reason, we adopt a simpler notation for orthogonal surface coordinate systems that coincide with either two spatial cylindrical coordinates or two spatial spherical coordinates. For a plane surface described in terms of polar coordinates in Table 2.4.2-2, the physical components of $\mathbf{v}^{(\sigma)}$ are denoted as $v_r^{(\sigma)}$ and $v_\theta^{(\sigma)}$ rather than $v_{<1>}^{(\sigma)}$ and $v_{<2>}^{(\sigma)}$; the physical components of $\mathbf{T}^{(\sigma)}$ are indicated as $T_{rr}^{(\sigma)}$, and $T_{r\theta}^{(\sigma)}$.

Because of this change in notation, we do *not* employ the summation convention with the physical components of vector and tensor fields. The quantity $D_{rr}^{(\sigma)}$ is a single physical component for the spherical dividing surface and surface coordinate system described in Table 2.4.2-5.

References

Aris, R., "Vectors, Tensors, and the Basic Equations of Fluid Mechanics," Prentice-Hall, Englewood Cliffs, NJ (1962).

Giordano, R. M., personal communication, October 1979.

Scriven, L. E., *Chem. Eng. Sci.* **12**, 98 (1960).

Slattery, J. C., *Chem. Eng. Sci.* **19**, 379 (1964).

Slattery, J. C., "Momentum, Energy, and Mass Transfer in Continua," McGraw-Hill, New York (1972); second edition, Robert E. Krieger, Malabar, FL 32950 (1981).

Exercises

2.4.2-1 *components of (2-14)* In discussing specific applications, it is necessary to express (2-14) in terms of the components of $\mathbf{v}^{(\sigma)}$.

i) Let us express $\mathbf{v}^{(\sigma)}$ in terms of its tangential and normal components. Determine that

$$\text{div}_{(\sigma)}\mathbf{v}^{(\sigma)} = v^{(\sigma)\alpha}{}_{,\alpha} - 2H v_{(\xi)}^{(\sigma)} \tag{1}$$

$$\mathbf{D}^{(\sigma)} = \frac{1}{2}\left(v_{\alpha,\beta}^{(\sigma)} + v_{\beta,\alpha}^{(\sigma)} - 2 v_{(\xi)}^{(\sigma)} B_{\alpha\beta} \right)\mathbf{a}^\alpha \mathbf{a}^\beta \tag{2}$$

$$\mathrm{div}^{(\sigma)}\mathbf{D}^{(\sigma)} = \frac{1}{2}\left[\, v^{(\sigma)}_{\alpha,\,\beta\gamma} + v^{(\sigma)}_{\beta,\,\alpha\gamma} - 2(\, v^{(\xi)}B_{\alpha\beta}\,)_{,\gamma}\,\right]a^{\beta\gamma}\mathbf{a}^{\alpha}$$

$$+ (\, v^{(\sigma)}_{\beta,\,\gamma} - v^{(\xi)}B_{\beta\gamma}\,)B^{\beta\gamma}\boldsymbol{\xi} \tag{3}$$

and

$$\mathrm{div}_{(\sigma)}\mathbf{S}^{(\sigma)}$$

$$= \Big[(\, v^{(\sigma)\beta}_{,\beta} - 2Hv^{(\xi)}\,)\left(\frac{\partial \kappa}{\partial y^{\alpha}} - \frac{\partial \varepsilon}{\partial y^{\alpha}}\right)$$

$$+ (\, \kappa - \varepsilon\,)\frac{\partial}{\partial y^{\alpha}}(\, v^{(\sigma)\beta}_{,\beta} - 2Hv^{(\xi)}\,)$$

$$+ \frac{\partial \varepsilon}{\partial y^{\beta}}\, a^{\beta\gamma}(\, v^{(\sigma)}_{\gamma,\,\alpha} + v^{(\sigma)}_{\alpha,\,\gamma} - 2v^{(\xi)}B_{\gamma\alpha}\,)$$

$$+ \varepsilon(\, v^{(\sigma)}_{\alpha,\,\beta\gamma} + v^{(\sigma)}_{\beta,\,\alpha\gamma} - 2v^{(\xi)}_{,\gamma}B_{\alpha\beta} - 2v^{(\xi)}B_{\alpha\beta,\gamma}\,)a^{\beta\gamma}\Big]\mathbf{a}^{\alpha}$$

$$+ \Big[\, 2H(\, \kappa - \varepsilon\,)(\, v^{(\sigma)\beta}_{,\beta} - 2Hv^{(\xi)}\,)$$

$$+ \varepsilon(\, 2v^{(\sigma)}_{\gamma,\,\beta} - 2v^{(\xi)}B_{\beta\gamma}\,)B^{\beta\gamma}\Big]\boldsymbol{\xi} \tag{4}$$

ii) We may wish to write $\mathbf{v}^{(\sigma)}$ in terms of its spatial components. Show that

$$\mathrm{div}_{(\sigma)}\mathbf{v}^{(\sigma)} = v^{(\sigma)}_{i,\,\alpha}\frac{\partial x^{i}}{\partial y^{\beta}}\, a^{\alpha\beta} \tag{5}$$

$$\mathbf{D}^{(\sigma)} = \frac{1}{2}\left(\frac{\partial x^{i}}{\partial y^{\alpha}}\, v^{(\sigma)}_{i,\,\beta} + \frac{\partial x^{i}}{\partial y^{\beta}}\, v^{(\sigma)}_{i,\,\alpha}\right)\mathbf{a}^{\alpha}\mathbf{a}^{\beta} \tag{6}$$

$$\mathrm{div}_{(\sigma)}\mathbf{D}^{(\sigma)}$$

$$= \frac{1}{2}\left[\left(\frac{\partial x^{i}}{\partial y^{\alpha}}\, v^{(\sigma)}_{i,\,\beta}\right)_{,\gamma} + \left(\frac{\partial x^{i}}{\partial y^{\beta}}\, v^{(\sigma)}_{i,\,\alpha}\right)_{,\gamma}\right]a^{\beta\gamma}\mathbf{a}^{\alpha} + \frac{\partial x^{i}}{\partial y^{\beta}}\, v^{(\sigma)}_{i,\,\gamma}B^{\beta\gamma}\xi \tag{7}$$

and

$$\operatorname{div}_{(\sigma)} \mathbf{S}^{(\sigma)}$$

$$= \left\{ v^{(\sigma)}_{i,\beta} \frac{\partial x^i}{\partial y^\gamma} a^{\beta\gamma} \left(\frac{\partial \kappa}{\partial y^\alpha} - \frac{\partial \varepsilon}{\partial y^\alpha} \right) \right.$$

$$+ (\kappa - \varepsilon) \left(v^{(\sigma)}_{i,\beta} \frac{\partial x^i}{\partial y^\gamma} \right)_{,\alpha} a^{\beta\gamma}$$

$$+ \frac{\partial \varepsilon}{\partial y^\beta} a^{\beta\gamma} \left(\frac{\partial x^i}{\partial y^\gamma} v^{(\sigma)}_{i,\alpha} + \frac{\partial x^i}{\partial y^\alpha} v^{(\sigma)}_{i,\gamma} \right)$$

$$\left. + \varepsilon \left[\left(\frac{\partial x^i}{\partial y^\alpha} v^{(\sigma)}_{i,\beta} \right)_{,\gamma} + \left(\frac{\partial x^i}{\partial y^\beta} v^{(\sigma)}_{i,\alpha} \right)_{,\gamma} \right] a^{\beta\gamma} \right\}$$

$$\times \left\{ a^{\alpha\mu} \frac{\partial x^j}{\partial y^\mu} \mathbf{g}_j \right\}$$

$$+ \left[2H(\kappa - \varepsilon) v^{(\sigma)}_{i,\beta} \frac{\partial x^i}{\partial y^\gamma} a^{\beta\gamma} + 2\varepsilon \frac{\partial x^i}{\partial y^\beta} v^{(\sigma)}_{i,\gamma} B^{\beta\gamma} \right] \xi^j \mathbf{g}_j \qquad (8)$$

This last may also be written as (Slattery 1964, Eq. 4.7)

$$\operatorname{div}_{(\sigma)} \mathbf{S}^{(\sigma)}$$

$$= \left\{ v^{(\sigma)}_{i,\beta} \frac{\partial x^i}{\partial y^\gamma} \frac{\partial x^j}{\partial y^\mu} a^{\beta\gamma} a^{\alpha\mu} \left(\frac{\partial \kappa}{\partial y^\alpha} - \frac{\partial \varepsilon}{\partial y^\alpha} \right) \right.$$

$$+ \kappa \left(v^{(\sigma)}_{i,\beta\alpha} \frac{\partial x^i}{\partial y^\gamma} \frac{\partial x^j}{\partial y^\mu} a^{\beta\gamma} a^{\alpha\mu} - v^{(\sigma)}_{i,\beta} a^{\beta\gamma} \xi^i \xi^j_{,\gamma} \right)$$

$$+ \frac{\partial \varepsilon}{\partial y^\beta} a^{\beta\gamma} \left(\frac{\partial x^i}{\partial y^\gamma} v^{(\sigma)}_{i,\alpha} + \frac{\partial x^i}{\partial y^\alpha} v^{(\sigma)}_{i,\gamma} \right) a^{\alpha\mu} \frac{\partial x^j}{\partial y^\mu}$$

$$\left. + \varepsilon \left[v^{(\sigma)}_{i,\beta\gamma} \frac{\partial x^i}{\partial y^\alpha} \frac{\partial x^j}{\partial y^\mu} a^{\beta\gamma} a^{\alpha\mu} + 2H\, v^{(\sigma)}_{i,\alpha} \xi^i a^{\alpha\mu} \frac{\partial x^j}{\partial y^\mu} \right] \right\} \mathbf{g}_j$$

$$+\left\{ 2H(\kappa - \varepsilon) v^{(\sigma)}_{i,\beta} \frac{\partial x^i}{\partial y^\gamma} a^{\beta\gamma} - 2\varepsilon\, v^{(\sigma)}_{i,\gamma}\, \xi^i_{,\beta}\, a^{\beta\gamma} \right\} \xi^j \mathbf{g}_j \tag{9}$$

or (Aris 1962, Eq. 10.43.7 corrects the typographical errors in Scriven 1960, eq. 28)

$$\begin{aligned}
\text{div}_{(\sigma)} \mathbf{S}^{(\sigma)} \\
= \Bigg\{ & v^{(\sigma)}_{i,\beta} \frac{\partial x^i}{\partial y^\gamma} a^{\beta\gamma} a^{\mu\alpha} \left(\frac{\partial \kappa}{\partial y^\mu} - \frac{\partial \varepsilon}{\partial y^\mu} \right) \\
& + (\kappa + \varepsilon) \left(v^{(\sigma)}_{i,\beta} \frac{\partial x^i}{\partial y^\gamma} a^{\beta\gamma} \right)_{,\mu} a^{\mu\alpha} \\
& + \frac{\partial \varepsilon}{\partial y^\beta} a^{\beta\gamma} \left(\frac{\partial x^i}{\partial y^\gamma} v^{(\sigma)}_{i,\mu} + \frac{\partial x^i}{\partial y^\mu} v^{(\sigma)}_{i,\gamma} \right) a^{\mu\alpha} \\
& + \varepsilon \Big[2K\, v^{(\sigma)\alpha} + \varepsilon^{\alpha\beta} \varepsilon^{\mu\nu}\, v^{(\sigma)}_{\mu,\nu\beta} \\
& \qquad + 2\varepsilon^{\alpha\beta}\, \varepsilon^{\mu\nu}\, v^{(\sigma)}_{(\xi),\mu}\, B_{\beta\nu} \Big] \Bigg\} \frac{\partial x^j}{\partial y^\alpha} \mathbf{g}_j \\
& + \Bigg[2H(\kappa + \varepsilon) \left(v^{(\sigma)}_{i,\beta} \frac{\partial x^i}{\partial y^\gamma} a^{\beta\gamma} \right) \\
& \qquad + 2\varepsilon\, \varepsilon^{\beta\gamma} \varepsilon^{\mu\nu}\, v^{(\sigma)}_{i,\gamma} \frac{\partial x^i}{\partial y^\mu} B_{\beta\nu} \Bigg] \xi^j \mathbf{g}_j
\end{aligned} \tag{10}$$

In proving (10), note that (Exercise A.5.6-7)

$$v^{(\sigma)}_{\alpha,\beta\gamma} - v^{(\sigma)}_{\alpha,\gamma\beta} = K\, v^{(\sigma)\rho}\, \varepsilon_{\rho\alpha}\, \varepsilon_{\beta\gamma} \tag{11}$$

and (Exercise A.5.6-8)

$$B_{\alpha\beta,\gamma} = B_{\alpha\gamma,\beta} \tag{12}$$

Table 2.4.2-1: Stationary plane dividing surface viewed in a rectangular cartesian coordinate system (refer to Exercise A.2.1-4 and A.5.3-5)

dividing surface

$$z_3 = \text{a constant}$$

surface coordinates

$$y^1 \equiv z_1 \qquad y^2 \equiv z_2$$

jump mass balance

$$\frac{\partial \rho^{(\sigma)}}{\partial t} + \frac{\partial}{\partial z_1}\left(\rho^{(\sigma)} v_1^{(\sigma)} \right) + \frac{\partial}{\partial z_2}\left(\rho^{(\sigma)} v_2^{(\sigma)} \right) + [\, \rho v_3 \xi_3 \,] = 0$$

jump momentum balance

z_1 *component*

$$\rho^{(\sigma)} \left(\frac{\partial v_1^{(\sigma)}}{\partial t} + v_1^{(\sigma)} \frac{\partial v_1^{(\sigma)}}{\partial z_1} + v_2^{(\sigma)} \frac{\partial v_1^{(\sigma)}}{\partial z_2} \right)$$

$$= \frac{\partial \gamma}{\partial z_1} + \frac{\partial S_{11}^{(\sigma)}}{\partial z_1} + \frac{\partial S_{12}^{(\sigma)}}{\partial z_2} + \rho^{(\sigma)} b_1^{(\sigma)} + [\, T_{13} \xi_3 \,]$$

z_2 *component*

$$\rho^{(\sigma)} \left(\frac{\partial v_2^{(\sigma)}}{\partial t} + v_1^{(\sigma)} \frac{\partial v_2^{(\sigma)}}{\partial z_1} + v_2^{(\sigma)} \frac{\partial v_2^{(\sigma)}}{\partial z_2} \right)$$

$$= \frac{\partial \gamma}{\partial z_2} + \frac{\partial S_{21}^{(\sigma)}}{\partial z_1} + \frac{\partial S_{22}^{(\sigma)}}{\partial z_2} + \rho^{(\sigma)} b_2^{(\sigma)} + [\, T_{23} \xi_3 \,]$$

z_3 *component*

$$0 = \rho^{(\sigma)} b_3^{(\sigma)} + [\, -\rho (v_3)^2 \xi_3 + T_{33} \xi_3 \,]$$

Table 2.4.2-1 (cont.)

jump momentum balance for a Boussinesq surface fluid

z_1 *component*

$$\rho^{(\sigma)}\left(\frac{\partial v_1^{(\sigma)}}{\partial t} + v_1^{(\sigma)}\frac{\partial v_1^{(\sigma)}}{\partial z_1} + v_2^{(\sigma)}\frac{\partial v_1^{(\sigma)}}{\partial z_2}\right)$$

$$= \frac{\partial \gamma}{\partial z_1} + \left(\frac{\partial \kappa}{\partial z_1} - \frac{\partial \varepsilon}{\partial z_1}\right)\left(\frac{\partial v_1^{(\sigma)}}{\partial z_1} + \frac{\partial v_2^{(\sigma)}}{\partial z_2}\right)$$

$$+ (\kappa + \varepsilon)\left(\frac{\partial^2 v_1^{(\sigma)}}{\partial z_1^{\,2}} + \frac{\partial^2 v_2^{(\sigma)}}{\partial z_1 \partial z_2}\right) + 2\frac{\partial \varepsilon}{\partial z_1}\frac{\partial v_1^{(\sigma)}}{\partial z_1}$$

$$+ \frac{\partial \varepsilon}{\partial z_2}\left(\frac{\partial v_2^{(\sigma)}}{\partial z_1} + \frac{\partial v_1^{(\sigma)}}{\partial z_2}\right) + \varepsilon\left(\frac{\partial^2 v_1^{(\sigma)}}{\partial z_2^{\,2}} - \frac{\partial^2 v_2^{(\sigma)}}{\partial z_1 \partial z_2}\right)$$

$$+ \rho^{(\sigma)} b_1^{(\sigma)} + [\, T_{13}\xi_3 \,]$$

z_2 *component*

$$\rho^{(\sigma)}\left(\frac{\partial v_2^{(\sigma)}}{\partial t} + v_1^{(\sigma)}\frac{\partial v_2^{(\sigma)}}{\partial z_1} + v_2^{(\sigma)}\frac{\partial v_2^{(\sigma)}}{\partial z_2}\right)$$

$$= \frac{\partial \gamma}{\partial z_2} + \left(\frac{\partial \kappa}{\partial z_2} - \frac{\partial \varepsilon}{\partial z_2}\right)\left(\frac{\partial v_1^{(\sigma)}}{\partial z_1} + \frac{\partial v_2^{(\sigma)}}{\partial z_2}\right)$$

$$+ (\kappa + \varepsilon)\left(\frac{\partial^2 v_1^{(\sigma)}}{\partial z_1 \partial z_2} + \frac{\partial^2 v_2^{(\sigma)}}{\partial z_2^{\,2}}\right)$$

$$+ \frac{\partial \varepsilon}{\partial z_1}\left(\frac{\partial v_1^{(\sigma)}}{\partial z_2} + \frac{\partial v_2^{(\sigma)}}{\partial z_1}\right) + 2\frac{\partial \varepsilon}{\partial z_2}\frac{\partial v_2^{(\sigma)}}{\partial z_2}$$

$$+ \varepsilon\left(\frac{\partial^2 v_2^{(\sigma)}}{\partial z_1^{\,2}} - \frac{\partial^2 v_1^{(\sigma)}}{\partial z_1 \partial z_2}\right) + \rho^{(\sigma)} b_2^{(\sigma)} + [\, T_{23}\xi_3 \,]$$

Table 2.4.2-1 (cont.)

z_3 *component*

$$0 = \rho^{(\sigma)} b_3^{(\sigma)} + [\, -\rho(\, v_3 \,)^2 \,\xi_3 + T_{33}\xi_3 \,]$$

surface rate of deformation tensor

$$D_{11}^{(\sigma)} = \frac{\partial v_1^{(\sigma)}}{\partial z_1}$$

$$D_{22}^{(\sigma)} = \frac{\partial v_2^{(\sigma)}}{\partial z_2}$$

$$D_{12}^{(\sigma)} = D_{21}^{(\sigma)} = \frac{1}{2}\left(\frac{\partial v_1^{(\sigma)}}{\partial z_2} + \frac{\partial v_2^{(\sigma)}}{\partial z_1} \right)$$

Table 2.4.2-2: Stationary plane dividing surface viewed in a cylindrical coordinate system (refer to Exercises A.2.1-5 and A.5.3-6 and to Slattery 1981, pp. 618 and 644)

dividing surface

z = a constant

surface coordinates

$y^1 \equiv r \qquad y^2 \equiv \theta$

jump mass balance

$$\frac{\partial \rho^{(\sigma)}}{\partial t} + \frac{1}{r}\frac{\partial}{\partial r}\left(\rho^{(\sigma)} r v_r^{(\sigma)} \right) + \frac{1}{r}\frac{\partial}{\partial \theta}\left(\rho^{(\sigma)} v_\theta^{(\sigma)} \right) + [\, \rho v_z \xi_z \,] = 0$$

jump momentum balance

r *component*

$$\rho^{(\sigma)} \left(\frac{\partial v_r^{(\sigma)}}{\partial t} + v_r^{(\sigma)} \frac{\partial v_r^{(\sigma)}}{\partial r} + \frac{v_\theta^{(\sigma)}}{r} \frac{\partial v_r^{(\sigma)}}{\partial \theta} - \frac{\left(v_\theta^{(\sigma)} \right)^2}{r} \right)$$

$$= \frac{\partial \gamma}{\partial r} + \frac{1}{r}\frac{\partial (r S_{rr}^{(\sigma)})}{\partial r} + \frac{1}{r}\frac{\partial S_{r\theta}^{(\sigma)}}{\partial \theta} - \frac{1}{r} S_{\theta\theta}^{(\sigma)} + \rho^{(\sigma)} b_r^{(\sigma)} + [\, T_{rz} \xi_z \,]$$

θ *component*

$$\rho^{(\sigma)} \left(\frac{\partial v_\theta^{(\sigma)}}{\partial t} + v_r^{(\sigma)} \frac{\partial v_\theta^{(\sigma)}}{\partial r} + \frac{v_\theta^{(\sigma)}}{r} \frac{\partial v_\theta^{(\sigma)}}{\partial \theta} + \frac{v_r^{(\sigma)} v_\theta^{(\sigma)}}{r} \right)$$

$$= \frac{1}{r}\frac{\partial \gamma}{\partial \theta} + \frac{1}{r^2}\frac{\partial}{\partial r}\left(r^2 S_{\theta r}^{(\sigma)} \right) + \frac{1}{r}\frac{\partial S_{\theta\theta}^{(\sigma)}}{\partial \theta} + \rho^{(\sigma)} b_\theta^{(\sigma)} + [\, T_{\theta z} \xi_z \,]$$

z *component*

$$0 = \rho^{(\sigma)} b_z^{(\sigma)} + [\, - \rho v_z^2 \xi_z + T_{zz} \xi_z \,]$$

Table 2.4.2-2 (cont.)

jump momentum balance for a boussinesq surface fluid

r *component*

$$\rho^{(\sigma)}\left[\frac{\partial v_r^{(\sigma)}}{\partial t} + v_r^{(\sigma)}\frac{\partial v_r^{(\sigma)}}{\partial r} + \frac{v_\theta^{(\sigma)}}{r}\frac{\partial v_r^{(\sigma)}}{\partial \theta} - \frac{(v_\theta^{(\sigma)})^2}{r}\right]$$

$$= \frac{\partial \gamma}{\partial r} + \left[\frac{\partial \kappa}{\partial r} - \frac{\partial \varepsilon}{\partial r}\right]\left[\frac{1}{r}\frac{\partial (r v_r^{(\sigma)})}{\partial r} + \frac{1}{r}\frac{\partial v_\theta^{(\sigma)}}{\partial \theta}\right]$$

$$+ [\kappa + \varepsilon]\left[\frac{\partial}{\partial r}\left(\frac{1}{r}\frac{\partial (r v_r^{(\sigma)})}{\partial r}\right) + \frac{\partial}{\partial r}\left(\frac{1}{r}\frac{\partial v_\theta^{(\sigma)}}{\partial \theta}\right)\right]$$

$$+ 2\frac{\partial \varepsilon}{\partial r}\frac{\partial v_r^{(\sigma)}}{\partial r} + \frac{\partial \varepsilon}{\partial \theta}\left[\frac{\partial}{\partial r}\left(\frac{v_\theta^{(\sigma)}}{r}\right) + \frac{1}{r^2}\frac{\partial v_r^{(\sigma)}}{\partial \theta}\right]$$

$$+ \varepsilon\left[\frac{1}{r^2}\frac{\partial^2 v_r^{(\sigma)}}{\partial \theta^2} - \frac{2}{r^2}\frac{\partial v_\theta^{(\sigma)}}{\partial \theta} - \frac{\partial}{\partial r}\left(\frac{1}{r}\frac{\partial v_\theta^{(\sigma)}}{\partial \theta}\right)\right]$$

$$+ \rho^{(\sigma)} b_r^{(\sigma)} + [\, T_{rz}\xi_z \,]$$

Table 2.4.2-2 (cont.)

θ *component*

$$\rho^{(\sigma)}\left(\frac{\partial v_\theta^{(\sigma)}}{\partial t} + v_r^{(\sigma)}\frac{\partial v_\theta^{(\sigma)}}{\partial r} + \frac{v_\theta^{(\sigma)}}{r}\frac{\partial v_\theta^{(\sigma)}}{\partial \theta} + \frac{v_r^{(\sigma)} v_\theta^{(\sigma)}}{r}\right)$$

$$= \frac{1}{r}\frac{\partial \gamma}{\partial \theta} + \frac{1}{r}\left[\frac{\partial \kappa}{\partial \theta} - \frac{\partial \varepsilon}{\partial \theta}\right]\left[\frac{1}{r}\frac{\partial(r v_r^{(\sigma)})}{\partial r} + \frac{1}{r}\frac{\partial v_\theta^{(\sigma)}}{\partial \theta}\right]$$

$$+ [\,\kappa + \varepsilon\,]\left[\frac{1}{r^2}\frac{\partial^2(r v_r^{(\sigma)})}{\partial r \partial \theta} + \frac{1}{r^2}\frac{\partial^2 v_\theta^{(\sigma)}}{\partial \theta^2}\right]$$

$$+ \frac{\partial \varepsilon}{\partial r}\left[r\frac{\partial}{\partial r}\left(\frac{v_\theta^{(\sigma)}}{r}\right) + \frac{1}{r}\frac{\partial v_r^{(\sigma)}}{\partial \theta}\right] + \frac{2}{r^2}\frac{\partial \varepsilon}{\partial \theta}\left(\frac{\partial v_\theta^{(\sigma)}}{\partial \theta} + v_r^{(\sigma)}\right)$$

$$+ \varepsilon\left\{\frac{\partial}{\partial r}\left[\frac{1}{r}\frac{\partial(r v_\theta^{(\sigma)})}{\partial r}\right] + \frac{2}{r^2}\frac{\partial v_r^{(\sigma)}}{\partial \theta} - \frac{1}{r^2}\frac{\partial^2(r v_r^{(\sigma)})}{\partial r \partial \theta}\right\}$$

$$+ \rho^{(\sigma)} b_\theta^{(\sigma)} + [\, T_{\theta z}\xi_z \,]$$

z *component*

$$0 = \rho^{(\sigma)} b_z^{(\sigma)} + [\, -\rho v_z^{2}\xi_z + T_{zz}\xi_z \,]$$

surface rate of deformation tensor

$$D_{rr}^{(\sigma)} = \frac{\partial v_r^{(\sigma)}}{\partial r}$$

$$D_{\theta\theta}^{(\sigma)} = \frac{1}{r}\frac{\partial v_\theta^{(\sigma)}}{\partial \theta} + \frac{v_r^{(\sigma)}}{r}$$

$$D_{r\theta}^{(\sigma)} = D_{\theta r}^{(\sigma)} = \frac{1}{2}\left[\frac{1}{r}\frac{\partial v_r^{(\sigma)}}{\partial \theta} + r\frac{\partial}{\partial r}\left(\frac{v_\theta^{(\sigma)}}{r}\right)\right]$$

Table 2.4.2-3: Alternative form for stationary plane dividing surface viewed in a cylindrical coordinate system (refer to Exercises A.2.1-6 and A.5.3-7 and to Slattery 1981, pp. 618 and 644; Giordano 1979)

dividing surface

$$\theta = \text{a constant}$$

surface coordinates

$$y^1 \equiv r \qquad y^2 \equiv z$$

jump mass balance

$$\frac{\partial \rho^{(\sigma)}}{\partial t} + \frac{\partial(\rho^{(\sigma)} v_r^{(\sigma)})}{\partial r} + \frac{\partial(\rho^{(\sigma)} v_z^{(\sigma)})}{\partial z} + [\rho v_\theta^{(\sigma)} \xi_\theta] = 0$$

jump momentum balance

r component

$$\rho^{(\sigma)} \left(\frac{\partial v_r^{(\sigma)}}{\partial t} + v_r^{(\sigma)} \frac{\partial v_r^{(\sigma)}}{\partial r} + v_z^{(\sigma)} \frac{\partial v_r^{(\sigma)}}{\partial z} \right)$$

$$= \frac{\partial \gamma}{\partial r} + \frac{\partial S_{rr}^{(\sigma)}}{\partial r} + \frac{\partial S_{rz}^{(\sigma)}}{\partial z} + \rho^{(\sigma)} b_r^{(\sigma)} + [T_{r\theta} \xi_\theta]$$

θ component

$$0 = \rho^{(\sigma)} b_\theta^{(\sigma)} + [- \rho v_\theta^2 \xi_\theta + T_{\theta\theta} \xi_\theta]$$

z component

$$\rho^{(\sigma)} \left(\frac{\partial v_z^{(\sigma)}}{\partial t} + v_r^{(\sigma)} \frac{\partial v_z^{(\sigma)}}{\partial r} + v_z^{(\sigma)} \frac{\partial v_z^{(\sigma)}}{\partial z} \right)$$

$$= \frac{\partial \gamma}{\partial z} + \frac{\partial S_{zr}^{(\sigma)}}{\partial r} + \frac{\partial S_{zz}^{(\sigma)}}{\partial r} + \rho^{(\sigma)} b_z^{(\sigma)} + [T_{z\theta} \xi_\theta]$$

Table 2.4.2-3 (cont.)

jump momentum balance for a Boussinesq surface fluid

r *component*

$$\rho^{(\sigma)}\left(\frac{\partial v_r^{(\sigma)}}{\partial t}+v_r^{(\sigma)}\frac{\partial v_r^{(\sigma)}}{\partial r}+v_z^{(\sigma)}\frac{\partial v_r^{(\sigma)}}{\partial z}\right)$$

$$=\frac{\partial\gamma}{\partial r}+\left(\frac{\partial\kappa}{\partial r}-\frac{\partial\varepsilon}{\partial r}\right)\left(\frac{\partial v_r^{(\sigma)}}{\partial r}+\frac{\partial v_z^{(\sigma)}}{\partial z}\right)$$

$$+(\kappa+\varepsilon)\left(\frac{\partial^2 v_z^{(\sigma)}}{\partial r^2}+\frac{\partial^2 v_r^{(\sigma)}}{\partial r\partial z}\right)$$

$$+2\frac{\partial\varepsilon}{\partial r}\frac{\partial v_r^{(\sigma)}}{\partial r}+\frac{\partial\varepsilon}{\partial z}\left(\frac{\partial v_z^{(\sigma)}}{\partial r}+\frac{\partial v_r^{(\sigma)}}{\partial z}\right)$$

$$+\varepsilon\left(\frac{\partial^2 v_r^{(\sigma)}}{\partial z^2}-\frac{\partial^2 v_z^{(\sigma)}}{\partial r\partial z}\right)+\rho^{(\sigma)}b_r^{(\sigma)}+[\,T_{r\theta}\xi_\theta\,]$$

θ *component*

$$0=\rho^{(\sigma)}b_\theta^{(\sigma)}+[\,-\rho v_\theta^2\,\xi_\theta+T_{\theta\theta}\xi_\theta\,]$$

z *component*

$$\rho^{(\sigma)}\left(\frac{\partial v_z^{(\sigma)}}{\partial t}+v_r^{(\sigma)}\frac{\partial v_z^{(\sigma)}}{\partial r}+v_z^{(\sigma)}\frac{\partial v_z^{(\sigma)}}{\partial z}\right)$$

$$=\frac{\partial\gamma}{\partial z}+\left(\frac{\partial\kappa}{\partial z}-\frac{\partial\varepsilon}{\partial z}\right)\left(\frac{\partial v_r^{(\sigma)}}{\partial r}+\frac{\partial v_z^{(\sigma)}}{\partial z}\right)$$

$$+(\kappa+\varepsilon)\left(\frac{\partial^2 v_r^{(\sigma)}}{\partial z\partial r}+\frac{\partial^2 v_z^{(\sigma)}}{\partial z^2}\right)+\frac{\partial\varepsilon}{\partial r}\left(\frac{\partial v_r^{(\sigma)}}{\partial z}+\frac{\partial v_z^{(\sigma)}}{\partial r}\right)$$

$$+2\frac{\partial\varepsilon}{\partial z}\frac{\partial v_z^{(\sigma)}}{\partial z}+\varepsilon\left(\frac{\partial^2 v_z^{(\sigma)}}{\partial r^2}-\frac{\partial^2 v_r^{(\sigma)}}{\partial z\partial r}\right)+\rho^{(\sigma)}b_z^{(\sigma)}+[\,T_{z\theta}\xi_\theta\,]$$

Table 2.4.2-3 (cont.)

surface rate of deformation tensor

$$D_{rr}^{(\sigma)} = \frac{\partial v_r^{(\sigma)}}{\partial r}$$

$$D_{\theta\theta}^{(\sigma)} = \frac{\partial v_z^{(\sigma)}}{\partial z}$$

$$D_{rz}^{(\sigma)} = D_{zr}^{(\sigma)} = \frac{1}{2}\left(\frac{\partial v_r^{(\sigma)}}{\partial z} + \frac{\partial v_z^{(\sigma)}}{\partial r} \right)$$

Table 2.4.2-4: Cylindrical dividing surface viewed in a cylindrical coordinate system (refer to Exercises A.2.1-7 and A.5.3-8)

dividing surface

$$r = R(t)$$

surface coordinates

$$y^1 \equiv \theta \qquad y^2 \equiv z$$

$$a_{11} = R^2 \qquad a_{22} = 1$$

$$a_{12} = a_{21} = 0$$

unit normal

$$\xi_r = 1$$

$$\xi_\theta = \xi_z = 0$$

mean curvature

$$H = -\frac{1}{2R}$$

jump mass balance

$$\frac{\partial \rho^{(\sigma)}}{\partial t} + \frac{1}{R}\frac{\partial}{\partial \theta}\left(\rho^{(\sigma)} v_\theta^{(\sigma)} \right) + \frac{\partial}{\partial z}\left(\rho^{(\sigma)} v_z^{(\sigma)} \right) + \frac{1}{R}\rho v^{\{g\}}_{\{\xi\}} + [\, \rho v_r \xi_r \,] = 0$$

jump momentum balance

r *component*

$$-\frac{1}{R}\rho^{(\sigma)}\left(v_\theta^{(\sigma)} \right)^2 = -\frac{\gamma}{R} - \frac{1}{R} S_{\theta\theta}^{(\sigma)} + \rho^{(\sigma)} b_r^{(\sigma)} + [\, -\rho v_r^{\,2} \xi_r + T_{rr}\xi_r \,]$$

Table 2.4.2-4 (cont.)

θ *component*

$$\rho^{(\sigma)}\left(\frac{\partial v_\theta^{(\sigma)}}{\partial t}+\frac{1}{R}v_\theta^{(\sigma)}\frac{\partial v_\theta^{(\sigma)}}{\partial \theta}+v_z^{(\sigma)}\frac{\partial v_\theta^{(\sigma)}}{\partial z}-\frac{1}{R}v_\theta^{(\sigma)}v_{(\xi)}^{(\sigma)}\right)$$

$$=\frac{1}{R}\frac{\partial \gamma}{\partial \theta}+\frac{1}{R}\frac{\partial S_{\theta\theta}^{(\sigma)}}{\partial \theta}+\frac{\partial S_{\theta z}^{(\sigma)}}{\partial z}+\rho^{(\sigma)}b_\theta^{(\sigma)}+[\,T_{\theta r}\xi_r\,]$$

z *component*

$$\rho^{(\sigma)}\left(\frac{\partial v_z^{(\sigma)}}{\partial t}+\frac{1}{R}v_\theta^{(\sigma)}\frac{\partial v_z^{(\sigma)}}{\partial \theta}+v_z^{(\sigma)}\frac{\partial v_z^{(\sigma)}}{\partial z}\right)$$

$$=\frac{\partial \gamma}{\partial z}+\frac{1}{R}\frac{\partial S_{z\theta}^{(\sigma)}}{\partial \theta}+\frac{\partial S_{zz}^{(\sigma)}}{\partial z}+\rho^{(\sigma)}b_z^{(\sigma)}+[\,T_{zr}\xi_r\,]$$

jump momentum balance for a Boussinesq surface fluid

r *component*

$$-\frac{1}{R}\rho^{(\sigma)}(\,v_\theta^{(\sigma)}\,)^2=-\frac{\gamma}{R}-\frac{\kappa+\varepsilon}{R}\left(\frac{1}{R}\frac{\partial v_\theta^{(\sigma)}}{\partial \theta}+\frac{\partial v_z^{(\sigma)}}{\partial z}+\frac{1}{R}v_{(\xi)}^{(\sigma)}\right)$$

$$+\frac{2\varepsilon}{R}\frac{\partial v_z^{(\sigma)}}{\partial z}+\rho^{(\sigma)}b_r^{(\sigma)}+[\,-\rho v_r^{2}\,\xi_r+T_{rr}\xi_r\,]$$

Table 2.4.2-4 (cont.)

θ *component*

$$\rho^{(\sigma)}\left(\frac{\partial v_\theta^{(\sigma)}}{\partial t}+\frac{1}{R}v_\theta^{(\sigma)}\frac{\partial v_\theta^{(\sigma)}}{\partial\theta}+v_z^{(\sigma)}\frac{\partial v_\theta^{(\sigma)}}{\partial z}-\frac{1}{R}v_\theta^{(\sigma)}v_{(\xi)}^{(\sigma)}\right)$$

$$=\frac{1}{R}\frac{\partial\gamma}{\partial\theta}+\frac{1}{R}\left(\frac{\partial\kappa}{\partial\theta}-\frac{\partial\varepsilon}{\partial\theta}\right)\left(\frac{1}{R}\frac{\partial v_\theta^{(\sigma)}}{\partial\theta}+\frac{\partial v_z^{(\sigma)}}{\partial z}\right)$$

$$+(\kappa+\varepsilon)\frac{1}{R}\frac{\partial}{\partial\theta}\left(\frac{1}{R}\frac{\partial v_\theta^{(\sigma)}}{\partial\theta}+\frac{\partial v_z^{(\sigma)}}{\partial z}\right)+\frac{2}{R^2}\frac{\partial\varepsilon}{\partial\theta}\frac{\partial v_\theta^{(\sigma)}}{\partial\theta}$$

$$+\frac{\partial\varepsilon}{\partial z}\left(\frac{\partial v_\theta^{(\sigma)}}{\partial z}+\frac{1}{R}\frac{\partial v_z^{(\sigma)}}{\partial\theta}\right)+\varepsilon\left(\frac{\partial^2 v_\theta^{(\sigma)}}{\partial z^2}-\frac{1}{R}\frac{\partial^2 v_z^{(\sigma)}}{\partial\theta\partial z}\right)$$

$$+\rho^{(\sigma)}b_\theta^{(\sigma)}+[\,T_{\theta r}\xi_r\,]$$

z *component*

$$\rho^{(\sigma)}\left(\frac{\partial v_z^{(\sigma)}}{\partial t}+\frac{1}{R}v_\theta^{(\sigma)}\frac{\partial v_z^{(\sigma)}}{\partial\theta}+v_z^{(\sigma)}\frac{\partial v_z^{(\sigma)}}{\partial z}\right)$$

$$=\frac{\partial\gamma}{\partial z}+\left(\frac{\partial\kappa}{\partial z}-\frac{\partial\varepsilon}{\partial z}\right)\left(\frac{1}{R}\frac{\partial v_\theta^{(\sigma)}}{\partial\theta}+\frac{\partial v_z^{(\sigma)}}{\partial z}\right)$$

$$+(\kappa+\varepsilon)\frac{\partial}{\partial z}\left(\frac{1}{R}\frac{\partial v_\theta^{(\sigma)}}{\partial\theta}+\frac{\partial v_z^{(\sigma)}}{\partial z}\right)$$

$$+\frac{1}{R}\frac{\partial\varepsilon}{\partial\theta}\left(\frac{1}{R}\frac{\partial v_z^{(\sigma)}}{\partial\theta}+\frac{\partial v_\theta^{(\sigma)}}{\partial z}\right)+2\frac{\partial\varepsilon}{\partial z}\frac{\partial v_z^{(\sigma)}}{\partial z}$$

$$+\frac{\varepsilon}{R}\left(\frac{1}{R}\frac{\partial^2 v_z^{(\sigma)}}{\partial\theta^2}-\frac{\partial^2 v_\theta^{(\sigma)}}{\partial\theta\partial z}\right)+\rho^{(\sigma)}b_z^{(\sigma)}+[\,T_{zr}\xi_r\,]$$

Table 2.4.2-4 (cont.)

surface rate of deformation tensor

$$D^{(\sigma)}_{\theta\theta} = \frac{1}{R}\frac{\partial v^{(\sigma)}_{\theta}}{\partial \theta} + \frac{1}{R} v^{(\sigma)}_{(\xi)}$$

$$D^{(\sigma)}_{zz} = \frac{\partial v^{(\sigma)}_{z}}{\partial z}$$

$$D^{(\sigma)}_{\theta z} = D^{(\sigma)}_{z\theta} = \frac{1}{2}\left(\frac{\partial v^{(\sigma)}_{\theta}}{\partial z} + \frac{1}{R}\frac{\partial v^{(\sigma)}_{z}}{\partial \theta} \right)$$

Table 2.4.2-5: Spherical dividing surface viewed in a spherical coordinate system (refer to Exercises A.2.1-8 and A.5.3-9)

dividing surface

$$r = R(t)$$

surface coordinates

$$y^1 \equiv \theta \qquad y^2 \equiv \phi$$

$$a_{11} = R^2 \qquad a_{22} = R^2 \sin^2\theta$$

$$a_{12} = a_{21} = 0$$

unit normal

$$\xi_r = 1$$

$$\xi_\theta = \xi_\phi = 0$$

mean curvature

$$H = -\frac{1}{R}$$

jump mass balance

$$\frac{\partial \rho^{(\sigma)}}{\partial t} + \frac{1}{R \sin\theta}\frac{\partial}{\partial \theta}\left(\rho^{(\sigma)} v_\theta^{(\sigma)} \sin\theta \right) + \frac{1}{R \sin\theta}\frac{\partial}{\partial \phi}\left(\rho^{(\sigma)} v_\phi^{(\sigma)} \right)$$

$$+ \frac{2}{R}\rho^{(\sigma)} v_{(\xi)}^{(\sigma)} + [\, \rho v_r \xi_r \,] = 0$$

Table 2.4.2-5 (cont.)

jump momentum balance

r *component*

$$-\frac{1}{R}\rho^{(\sigma)}[(v_\theta^{(\sigma)})^2 + (v_\phi^{(\sigma)})^2] = -\frac{2\gamma}{R} - \frac{1}{R}(S_{\theta\theta}^{(\sigma)} + S_{\phi\phi}^{(\sigma)})$$

$$+ \rho^{(\sigma)} b_r^{(\sigma)} + [-\rho v_r^2 \xi_r + T_{rr}\xi_r]$$

θ *component*

$$\rho^{(\sigma)}\left[\frac{\partial v_\theta^{(\sigma)}}{\partial t} + \frac{1}{R} v_\theta^{(\sigma)} \frac{\partial v_\theta^{(\sigma)}}{\partial \theta} + \frac{v_\phi^{(\sigma)}}{R\ \sin\theta}\frac{\partial v_\theta^{(\sigma)}}{\partial \phi} - \frac{1}{R}(v_\phi^{(\sigma)})^2 \cot\theta\right.$$

$$\left. + \frac{1}{R} v_\theta^{(\sigma)} v_{(\xi)}^{(\sigma)}\right]$$

$$= \frac{1}{R}\frac{\partial\gamma}{\partial\theta} + \frac{1}{R\ \sin\theta}\frac{\partial}{\partial\theta}(S_{\theta\theta}^{(\sigma)}\sin\theta) + \frac{1}{R\ \sin\theta}\frac{\partial S_{\theta\phi}^{(\sigma)}}{\partial\phi}$$

$$-\frac{\cot\ \theta}{R} S_{\phi\phi}^{(\sigma)} + \rho^{(\sigma)} b_\theta^{(\sigma)} + [T_{\theta r}\xi_r]$$

ϕ *component*

$$\rho^{(\sigma)}\left(\frac{\partial v_\phi^{(\sigma)}}{\partial t} + \frac{1}{R} v_\theta^{(\sigma)} \frac{\partial v_\phi^{(\sigma)}}{\partial \theta} + \frac{v_\phi^{(\sigma)}}{R\ \sin\theta}\frac{\partial v_\phi^{(\sigma)}}{\partial \phi} + \frac{1}{R} v_\theta^{(\sigma)} v_\phi^{(\sigma)} \cot\theta\right.$$

$$\left. + \frac{1}{R} v_\phi^{(\sigma)} v_{(\xi)}^{(\sigma)}\right)$$

$$= \frac{1}{R\ \sin\theta}\frac{\partial\gamma}{\partial\phi} + \frac{1}{R\ \sin^2\theta}\frac{\partial}{\partial\theta}(S_{\phi\theta}^{(\sigma)}\sin^2\theta) + \frac{1}{R\ \sin\theta}\frac{\partial S_{\phi\phi}^{(\sigma)}}{\partial\phi}$$

$$+ \rho^{(\sigma)} b_\phi^{(\sigma)} + [T_{\phi r}\xi_r]$$

Table 2.4.2-5 (cont.)

jump momentum balance for a Boussinesq surface fluid

r *component*

$$-\frac{1}{R}\rho^{(\sigma)}[(v_\theta^{(\sigma)})^2 + (v_\phi^{(\sigma)})^2]$$

$$= -\frac{2\gamma}{R} - \frac{2\kappa}{R^2 \sin\theta}\left[\frac{\partial(v_\theta^{(\sigma)}\sin\theta)}{\partial\theta} + \frac{\partial v_\phi^{(\sigma)}}{\partial\phi}\right]$$

$$-\frac{4\kappa}{R^2}v_{(\xi)}^{(\sigma)} + \rho^{(\sigma)}b_r^{(\sigma)} + [-\rho v_r^2\xi_r + T_{rr}\xi_r]$$

Table 2.4.2-5 (cont.)

θ *component*

$$\rho^{(\sigma)}\left[\frac{\partial v_\theta^{(\sigma)}}{\partial t}+\frac{1}{R}v_\theta^{(\sigma)}\frac{\partial v_\theta^{(\sigma)}}{\partial \theta}+\frac{v_\phi^{(\sigma)}}{R\sin\theta}\frac{\partial v_\theta^{(\sigma)}}{\partial \phi}-\frac{1}{R}(v_\phi^{(\sigma)})^2\cot\theta\right.$$

$$\left.+\frac{1}{R}v_\theta^{(\sigma)}v_{(\xi)}^{\{g\}}\right]$$

$$=\frac{1}{R}\frac{\partial\gamma}{\partial\theta}+\frac{1}{R^2}\left(\frac{\partial\kappa}{\partial\theta}-\frac{\partial\varepsilon}{\partial\theta}\right)$$

$$\times\left\{\frac{1}{\sin\theta}\left[\frac{\partial}{\partial\theta}(v_\theta^{(\sigma)}\sin\theta)+\frac{\partial v_\phi^{(\sigma)}}{\partial\phi}\right]+2v_{(\xi)}^{\{g\}}\right\}$$

$$+(\kappa+\varepsilon)\frac{1}{R^2}\frac{\partial}{\partial\theta}\left[\frac{1}{\sin\theta}\frac{\partial(v_\theta^{(\sigma)}\sin\theta)}{\partial\theta}+\frac{1}{\sin\theta}\frac{\partial v_\phi^{(\sigma)}}{\partial\phi}\right]$$

$$+\frac{2}{R^2}\frac{\partial\varepsilon}{\partial\theta}\frac{\partial v_\theta^{(\sigma)}}{\partial\theta}+\frac{1}{R^2\sin^2\theta}\frac{\partial\varepsilon}{\partial\phi}\left[\frac{\partial v_\theta^{(\sigma)}}{\partial\phi}+\sin^2\theta\frac{\partial}{\partial\theta}\left(\frac{v_\phi^{(\sigma)}}{\sin\theta}\right)\right]$$

$$+\frac{\varepsilon}{R^2}\left[2v_\theta^{(\sigma)}+\frac{1}{\sin^2\theta}\frac{\partial^2 v_\theta^{(\sigma)}}{\partial\phi^2}-\frac{1}{\sin^2\theta}\frac{\partial^2(v_\phi^{(\sigma)}\sin\theta)}{\partial\phi\,\partial\phi}\right.$$

$$\left.+2v_{(\xi)}^{\{g\}}\cot\theta\right]+\rho^{(\sigma)}b_\theta^{(\sigma)}+[\,T_{\theta r}\xi_r\,]$$

Table 2.4.2-5 (cont.)

ϕ *component*

$$\rho^{(\sigma)}\left(\frac{\partial v_\phi^{(\sigma)}}{\partial t}+\frac{1}{R}v_\theta^{(\sigma)}\frac{\partial v_\phi^{(\sigma)}}{\partial \theta}+\frac{v_\phi^{(\sigma)}}{R\sin\theta}\frac{\partial v_\phi^{(\sigma)}}{\partial \phi}+\frac{1}{R}v_\theta^{(\sigma)}v_\phi^{(\sigma)}\cot\theta\right.$$

$$\left.+\frac{1}{R}v_\phi^{(\sigma)}v_{(\xi)}^{\{g\}}\right)$$

$$=\frac{1}{R\sin\theta}\frac{\partial\gamma}{\partial\phi}+\frac{1}{R^2\sin\theta}\left(\frac{\partial\kappa}{\partial\phi}-\frac{\partial\varepsilon}{\partial\phi}\right)$$

$$\times\left\{\frac{1}{\sin\theta}\left[\frac{\partial}{\partial\theta}\left(v_\theta^{(\sigma)}\sin\theta\right)+\frac{\partial v_\phi^{(\sigma)}}{\partial\phi}\right]+2v_{(\xi)}^{\{g\}}\right\}$$

$$+(\kappa+\varepsilon)\frac{1}{R^2\sin^2\theta}\frac{\partial}{\partial\phi}\left[\frac{\partial}{\partial\theta}\left(v_\theta^{(\sigma)}\sin\theta\right)+\frac{\partial v_\phi^{(\sigma)}}{\partial\phi}\right]$$

$$+\frac{1}{R^2\sin\theta}\frac{\partial\varepsilon}{\partial\theta}\left[\frac{\partial v_\theta^{(\sigma)}}{\partial\phi}+\sin^2\theta\frac{\partial}{\partial\theta}\left(\frac{v_\phi^{(\sigma)}}{\sin\theta}\right)\right]$$

$$+\frac{1}{R^2\sin^2\theta}\frac{\partial\varepsilon}{\partial\phi}\left(\frac{\partial v_\phi^{(\sigma)}}{\partial\phi}+v_\theta^{(\sigma)}\cos\theta\right)$$

$$+\frac{\varepsilon}{R^2}\left\{2v_\phi^{(\sigma)}+\frac{\partial}{\partial\theta}\left[\frac{1}{\sin\theta}\frac{\partial}{\partial\theta}\left(v_\phi^{(\sigma)}\sin\theta\right)\right]\right.$$

$$\left.-\frac{\partial}{\partial\theta}\left(\frac{1}{\sin\theta}\frac{\partial v_\theta^{(\sigma)}}{\partial\phi}\right)\right\}$$

$$+\rho^{(\sigma)}b_\phi^{(\sigma)}+[\,T_{\phi r}\xi_r\,]$$

Table 2.4.2-5 (cont.)

surface rate of deformation tensor

$$D^{(\sigma)}_{\theta\theta} = \frac{1}{R}\left(\frac{\partial v^{(\sigma)}_{\theta}}{\partial \theta} + v^{(\sigma)}_{\xi} \right)$$

$$D^{(\sigma)}_{\phi\phi} = \frac{1}{R \sin \theta} \frac{\partial v^{(\sigma)}_{\phi}}{\partial \phi} + \frac{\cot \theta}{R} v^{(\sigma)}_{\theta} + \frac{1}{R} v^{(\sigma)}_{\xi}$$

$$D^{(\sigma)}_{\theta\phi} = D^{(\sigma)}_{\phi\theta} = \frac{1}{2}\left[\frac{1}{R \sin \theta} \frac{\partial v^{(\sigma)}_{\theta}}{\partial \phi} + \frac{\sin \theta}{R} \frac{\partial}{\partial \theta}\left(\frac{v^{(\sigma)}_{\phi}}{\sin \theta} \right) \right]$$

Table 2.4.2-6: Deforming two-dimensional surface viewed in a rectangular cartesian coordinate system (refer to Exercise A.2.1-9, A.2.6-1, and A.5.3-10)

dividing surface

$$z_3 = h(z_1, t)$$

assumptions

$$\rho^{(\sigma)} = \rho^{(\sigma)}(z_1, t) \qquad \varepsilon = \varepsilon(z_1, t)$$

$$\gamma = \gamma(z_1, t) \qquad \kappa = \kappa(z_1, t)$$

rectangular cartesian spatial components of $\mathbf{v}^{(\sigma)}$:

$$v_1^{(\sigma)} = v_1^{(\sigma)}(z_1, t)$$

$$v_2^{(\sigma)} = 0$$

$$v_3^{(\sigma)} = v_3^{(\sigma)}(z_1, t)$$

surface coordinates

$$y^1 \equiv z_1 \qquad y^2 \equiv z_2$$

$$a_{11} = 1 + \left(\frac{\partial h}{\partial z_1}\right)^2$$

$$a_{22} = 1$$

$$a_{12} = a_{21} = 0$$

unit normal

$$\xi_1 = -\frac{\partial h}{\partial z_1}\left[1 + \left(\frac{\partial h}{\partial z_1}\right)^2\right]^{-1/2}$$

$$\xi_2 = 0$$

$$\xi_3 = \left[1 + \left(\frac{\partial h}{\partial z_1}\right)^2\right]^{-1/2}$$

Table 2.4.2-6 (cont.)

mean curvature

$$H = \frac{1}{2}\frac{\partial^2 h}{\partial {z_1}^2}\left[1 + \left(\frac{\partial h}{\partial z_1}\right)^2\right]^{-3/2}$$

components of u

Noting that

$$\mathbf{u} = u_i \mathbf{e}_i$$
$$= \overline{u}_{<\alpha>} \mathbf{a}_{<\alpha>} + u_{(\xi)} \boldsymbol{\xi}$$

we have

$$u_1 = u_2 = 0$$

$$u_3 = \frac{\partial h}{\partial t}$$

$$\overline{u}_{<1>} = \frac{\partial h}{\partial t}\frac{\partial h}{\partial z_1}\left[1 + \left(\frac{\partial h}{\partial z_1}\right)^2\right]^{-1/2}$$

$$\overline{u}_{<2>} = 0$$

$$u_{(\xi)} = \frac{\partial h}{\partial t}\left[1 + \left(\frac{\partial h}{\partial z_1}\right)^2\right]^{-1/2}$$

Table 2.4.2-6 (cont.)

components of surface velocity $\mathbf{v}^{(\sigma)}$

Observing that

$$\mathbf{v}^{(\sigma)} = v_i^{(\sigma)} \mathbf{e}_i$$

$$= \overline{v}^{(\sigma)}_{<\alpha>} \mathbf{a}_{<\alpha>} + v^{(\sigma)}_{(\xi)} \boldsymbol{\xi}$$

we find

$$\overline{v}^{(\sigma)}_{<1>} = \left[v_1^{(\sigma)} + \frac{\partial h}{\partial z_1} v_3^{(\sigma)} \right] \left[1 + \left(\frac{\partial h}{\partial z_1} \right)^2 \right]^{-1/2}$$

$$\overline{v}^{(\sigma)}_{<2>} = v_2^{(\sigma)} = 0$$

$$v^{(\sigma)}_{(\xi)} = \frac{\partial h}{\partial t} \left[1 + \left(\frac{\partial h}{\partial z_1} \right)^2 \right]^{-1/2}$$

$$= \left[- \frac{\partial h}{\partial z_1} v_1^{(\sigma)} + v_3^{(\sigma)} \right] \left[1 + \left(\frac{\partial h}{\partial z_1} \right)^2 \right]^{-1/2}$$

$$\frac{\partial h}{\partial t} = - \frac{\partial h}{\partial z_1} v_1^{(\sigma)} + v_3^{(\sigma)}$$

components of intrinsic surface velocity $\dot{\mathbf{y}}$

$$\dot{\mathbf{y}} = \mathbf{v}^{(\sigma)} - \mathbf{u}$$

$$= \dot{y}_{<\alpha>} \mathbf{a}_{<\alpha>}$$

$$\dot{y}_{<1>} = v_1^{(\sigma)} \left[1 + \left(\frac{\partial h}{\partial z_1} \right)^2 \right]^{1/2}$$

$$\dot{y}_{<2>} = 0$$

Table 2.4.2-6 (cont.)

jump mass balance

$$\frac{\partial \rho^{(\sigma)}}{\partial t} + \frac{\partial \rho^{(\sigma)}}{\partial z_1} v_1^{(\sigma)} + \rho^{(\sigma)} \left[\frac{1}{a_{11}} \frac{\partial}{\partial z_1} \left(v_1^{(\sigma)} + \frac{\partial h}{\partial z_1} v_3^{(\sigma)} \right) \right.$$

$$- \frac{1}{(a_{11})^2} \frac{\partial h}{\partial z_1} \frac{\partial^2 h}{\partial z_1{}^2} \left(v_1^{(\sigma)} + \frac{\partial h}{\partial z_1} v_3^{(\sigma)} \right)$$

$$\left. - \frac{1}{(a_{11})^2} \frac{\partial h}{\partial t} \frac{\partial^2 h}{\partial z_1{}^2} \right]$$

$$+ [\rho(\mathbf{v} \cdot \boldsymbol{\xi} - v_{(\xi)}^{(\sigma)})] = 0$$

jump momentum balance

Let

$$\mathbf{S}^{(\sigma)} = \overline{S}^{(\sigma)}_{<\alpha\beta>} \mathbf{a}_{<\alpha>} \mathbf{a}_{<\beta>}$$

z_1 *component*

$$\rho^{(\sigma)} \left(\frac{\partial v_1^{(\sigma)}}{\partial t} + \frac{\partial v_1^{(\sigma)}}{\partial z_1} v_1^{(\sigma)} \right)$$

$$= \frac{1}{a_{11}} \frac{\partial \gamma}{\partial z_1} + 2H\gamma\, \xi_1 + \frac{1}{(a_{11})^{1/2}} \frac{\partial}{\partial z_1} \left[\frac{\overline{S}^{(\sigma)}_{<11>}}{(a_{11})^{1/2}} \right]$$

$$+ \frac{1}{(a_{11})^{1/2}} \overline{S}^{(\sigma)}_{<12>} + \rho^{(\sigma)} b_1^{(\sigma)}$$

$$+ [-\rho(\mathbf{v} \cdot \boldsymbol{\xi} - v_{(\xi)}^{(\sigma)})^2\, \xi_1 + T_{11}\xi_1 + T_{13}\xi_3]$$

z_2 *component*

$$0 = \frac{1}{(a_{11})^{1/2}} \frac{\partial \overline{S}^{(\sigma)}_{<12>}}{\partial z_1} + \frac{\partial \overline{S}^{(\sigma)}_{<22>}}{\partial z_2} + \rho^{(\sigma)} b_2^{(\sigma)} + [T_{21}\xi_1 + T_{23}\xi_3]$$

Table 2.4.2-6 (cont.)

z_3 *component*

$$\rho^{(\sigma)}\left(\frac{\partial v_3^{(\sigma)}}{\partial t}+\frac{\partial v_3^{(\sigma)}}{\partial z_1}v_1^{(\sigma)}\right)=\frac{1}{a_{11}}\frac{\partial h}{\partial z_1}\frac{\partial \gamma}{\partial z_1}+2H\gamma\,\xi_3$$

$$+\frac{1}{(\,a_{11}\,)^{1/2}}\frac{\partial}{\partial z_1}\left[\frac{1}{(\,a_{11}\,)^{1/2}}\frac{\partial h}{\partial z_1}\overline{S}^{(\sigma)}_{<11>}\right]$$

$$+\frac{1}{(\,a_{11}\,)^{1/2}}\frac{\partial h}{\partial z_1}\frac{\partial \overline{S}^{(\sigma)}_{<12>}}{\partial z_2}+\rho^{(\sigma)}b_3^{(\sigma)}$$

$$+\left[-\rho(\,\mathbf{v}\cdot\boldsymbol{\xi}-v^{(\xi)}_{(\xi)}\,)^2\xi_3+T_{31}\xi_1+T_{33}\xi_3\,\right]$$

jump momentum balance for a Boussinesq surface fluid

z_1 *component*

$$\rho^{(\sigma)}\left(\frac{\partial v_1^{(\sigma)}}{\partial t}+\frac{\partial v_1^{(\sigma)}}{\partial z_1}v_1^{(\sigma)}\right)=\frac{1}{a_{11}}\frac{\partial \gamma}{\partial z_1}+2H\gamma\,\xi_1$$

$$+\frac{1}{(\,a_{11}\,)^2}\left(\frac{\partial v_1^{(\sigma)}}{\partial z_1}+\frac{\partial h}{\partial z_1}\frac{\partial v_3^{(\sigma)}}{\partial z_1}\right)\left[\frac{\partial(\kappa+\varepsilon)}{\partial z_1}\right.$$

$$\left.+\,2Ha_{11}\xi_1(\,\kappa+\varepsilon\,)\right]$$

$$+\frac{(\,\kappa+\varepsilon\,)}{a_{11}}\left\{\frac{\partial}{\partial z_1}\left[\frac{1}{a_{11}}\left(\frac{\partial v_1^{(\sigma)}}{\partial z_1}+\frac{\partial h}{\partial z_1}\frac{\partial v_3^{(\sigma)}}{\partial z_1}\right)\right]\right\}$$

$$+\rho^{(\sigma)}b_1^{(\sigma)}+\left[-\rho(\,\mathbf{v}\cdot\boldsymbol{\xi}-v^{(\xi)}_{(\xi)}\,)^2\xi_1+T_{11}\xi_1+T_{13}\xi_3\,\right]$$

z_2 *component*

$$0=\rho^{(\sigma)}b_2^{(\sigma)}+\left[\,T_{21}\xi_1+T_{23}\xi_3\,\right]$$

Table 2.4.2-6 (cont.)

z_3 *component*

$$\rho^{(\sigma)}\left(\frac{\partial v_{\langle 3\rangle}^{(\sigma)}}{\partial t}+\frac{\partial v_{\langle 3\rangle}^{(\sigma)}}{\partial z_1}v_{\langle 1\rangle}^{(\sigma)}\right)=\frac{1}{a_{11}}\frac{\partial h}{\partial z_1}\frac{\partial \gamma}{\partial z_1}+2H\gamma\,\xi_3$$

$$+\frac{\partial h}{\partial z_1}\left\{\frac{1}{(\,a_{11}\,)^2}\left(\frac{\partial v_{\langle 1\rangle}^{(\sigma)}}{\partial z_1}+\frac{\partial h}{\partial z_1}\frac{\partial v_{\langle 3\rangle}^{(\sigma)}}{\partial z_1}\right)\frac{\partial(\,\kappa+\varepsilon\,)}{\partial z_1}\right.$$

$$\left.+\frac{\kappa+\varepsilon}{a_{11}}\frac{\partial}{\partial z_1}\left[\frac{1}{a_{11}}\left(\frac{\partial v_{\langle 1\rangle}^{(\sigma)}}{\partial z_1}+\frac{\partial h}{\partial z_1}\frac{\partial v_{\langle 3\rangle}^{(\sigma)}}{\partial z_1}\right)\right]\right\}$$

$$+\left(\frac{2H\xi_3}{a_{11}}\frac{\partial v_{\langle 1\rangle}^{(\sigma)}}{\partial z_1}+\frac{\partial h}{\partial z_1}\frac{\partial v_{\langle 3\rangle}^{(\sigma)}}{\partial z_1}\right)(\,\kappa+\varepsilon\,)+\rho^{(\sigma)}b_{\langle 3\rangle}^{(\sigma)}$$

$$+\left[-\rho(\,\mathbf{v}\cdot\boldsymbol{\xi}-v_{(\xi)}^{(\sigma)}\,)^2\xi_3+T_{31}\xi_1+T_{33}\xi_3\,\right]$$

surface rate of deformation tensor

$$\mathbf{D}^{(\sigma)}=\bar{D}_{\langle\alpha\beta\rangle}^{(\sigma)}\mathbf{a}_{\langle\alpha\rangle}\mathbf{a}_{\langle\beta\rangle}$$

$$\bar{D}_{\langle 11\rangle}^{(\sigma)}=\frac{1}{a_{11}}\left(\frac{\partial v_{\langle 1\rangle}^{(\sigma)}}{\partial z_1}+\frac{\partial h}{\partial z_1}\frac{\partial v_{\langle 3\rangle}^{(\sigma)}}{\partial z_1}\right)$$

$$\bar{D}_{\langle 12\rangle}^{(\sigma)}=\bar{D}_{\langle 21\rangle}^{(\sigma)}=\bar{D}_{\langle 22\rangle}^{(\sigma)}=0$$

Table 2.4.2-7: Rotating and deforming axially symmetric surface viewed in a cylindrical coordinate system (refer to Exercises A.2.1-10, A.2.6-2, and A.5.3-11)

dividing surface

$$z = h(r, t)$$

assumptions

$$\rho^{(\sigma)} = \rho^{(\sigma)}(r, t) \qquad \varepsilon = \varepsilon(r, t)$$

$$\gamma = \gamma(r, t) \qquad \kappa = \kappa(r, t)$$

cylindrical components of $\mathbf{v}^{(\sigma)}$:

$$v_r^{(\sigma)} = v_r^{(\sigma)}(r, t)$$

$$v_\theta^{(\sigma)} = v_\theta^{(\sigma)}(r, t)$$

$$v_z^{(\sigma)} = v_z^{(\sigma)}(r, t)$$

surface coordinates

$$y^1 \equiv r \qquad y^2 \equiv \theta$$

$$a_{11} = 1 + \left(\frac{\partial h}{\partial r}\right)^2$$

$$a_{22} = r^2$$

$$a_{12} = a_{21} = 0$$

unit normal

$$\xi_r = -\frac{\partial h}{\partial r}\left[1 + \left(\frac{\partial h}{\partial r}\right)^2\right]^{-1/2}$$

$$\xi_\theta = 0$$

$$\xi_z = \left[1 + \left(\frac{\partial h}{\partial r}\right)^2\right]^{-1/2}$$

Table 2.4.2-7 (cont.)

second groundform tensor

$$B_{11} = \frac{\partial^2 h}{\partial r^2}\left[1 + \left(\frac{\partial h}{\partial r}\right)^2 \right]^{-1/2}$$

$$B_{22} = r\frac{\partial h}{\partial r}\left[1 + \left(\frac{\partial h}{\partial r}\right)^2 \right]^{-1/2}$$

$$B_{12} = B_{21} = 0$$

mean curvature

$$H = \frac{1}{2r}\left[r\frac{\partial^2 h}{\partial r^2} + \frac{\partial h}{\partial r} + \left(\frac{\partial h}{\partial r}\right)^3 \right]\left[1 + \left(\frac{\partial h}{\partial r}\right)^2 \right]^{-3/2}$$

$$= \frac{1}{2r}\frac{\partial}{\partial r}\left\{ r\frac{\partial h}{\partial r}\left[1 + \left(\frac{\partial h}{\partial r}\right)^2 \right]^{-1/2} \right\}$$

components of **u**

Noting that

$$\mathbf{u} = u_r \mathbf{g}_r + u_\theta \mathbf{g}_\theta + u_z \mathbf{g}_z$$

$$= \overline{u}_{<\alpha>} \mathbf{a}_{<\alpha>} + u_{(\xi)} \boldsymbol{\xi}$$

we have

Table 2.4.2-7 (cont.)

$$u_r = u_\theta = 0$$

$$u_z = \frac{\partial h}{\partial t}$$

$$\overline{u}_{<1>} = \frac{\partial h}{\partial t}\frac{\partial h}{\partial r}\left[1 + \left(\frac{\partial h}{\partial r}\right)^2\right]^{-1/2}$$

$$\overline{u}_{<2>} = 0$$

$$u_{(\xi)} = \frac{\partial h}{\partial t}\left[1 + \left(\frac{\partial h}{\partial r}\right)^2\right]^{-1/2}$$

components of surface velocity $\mathbf{v}^{(\sigma)}$

Observing that

$$\mathbf{v}^{(\sigma)} = v_r^{(\sigma)}\mathbf{g}_r + v_\theta^{(\sigma)}\mathbf{g}_\theta + v_z^{(\sigma)}\mathbf{g}_z$$

$$= \overline{v}_{<\alpha>}^{(\sigma)}\mathbf{a}_{<\alpha>} + v_{(\xi)}^{(\sigma)}\boldsymbol{\xi}$$

we find

$$\overline{v}_{<1>}^{(\sigma)} = \left[v_r^{(\sigma)} + \frac{\partial h}{\partial r}v_z^{(\sigma)}\right]\left[1 + \left(\frac{\partial h}{\partial r}\right)^2\right]^{-1/2}$$

$$\overline{v}_{<2>}^{(\sigma)} = v_\theta^{(\sigma)}$$

$$v_{(\xi)}^{(\sigma)} = \frac{\partial h}{\partial t}\left[1 + \left(\frac{\partial h}{\partial r}\right)^2\right]^{-1/2}$$

$$= \left[-\frac{\partial h}{\partial r}v_r^{(\sigma)} + v_z^{(\sigma)}\right]\left[1 + \left(\frac{\partial h}{\partial r}\right)^2\right]^{-1/2}$$

$$\frac{\partial h}{\partial t} = -\frac{\partial h}{\partial r}v_r^{(\sigma)} + v_z^{(\sigma)}$$

Table 2.4.2-7 (cont.)

components of intrinsic surface velocity $\dot{\mathbf{y}}$

$$\dot{\mathbf{y}} = \mathbf{v}^{(\sigma)} - \mathbf{u}$$

$$= \dot{y}_{<\alpha>} \mathbf{a}_{<\alpha>}$$

$$\dot{y}_{<1>} = v_r^{(\sigma)} \left[1 + \left(\frac{\partial h}{\partial r} \right)^2 \right]^{1/2}$$

$$\dot{y}_{<2>} = v_\theta^{(\sigma)}$$

jump mass balance

$$\frac{\partial \rho^{(\sigma)}}{\partial t} + \frac{\partial \rho^{(\sigma)}}{\partial r} v_r^{(\sigma)} + \rho^{(\sigma)} \left[\frac{1}{a_{11}} \left(\frac{\partial v_r^{(\sigma)}}{\partial r} + \frac{\partial h}{\partial r} \frac{\partial v_z^{(\sigma)}}{\partial r} \right) + \frac{1}{r} v_r^{(\sigma)} \right]$$

$$+ [\, \rho(\mathbf{v} \cdot \boldsymbol{\xi} - v_{(\xi)}^{(\sigma)})] = 0$$

jump momentum balance

Let

$$\mathbf{S}^{(\sigma)} = \overline{S}^{(\sigma)}_{<\alpha\beta>} \mathbf{a}_{<\alpha>} \mathbf{a}_{<\beta>}$$

r *component*

$$\rho^{(\sigma)} \left[\frac{\partial v_r^{(\sigma)}}{\partial t} + \frac{\partial v_r^{(\sigma)}}{\partial r} v_r^{(\sigma)} - \frac{1}{r} (v_\theta^{(\sigma)})^2 \right]$$

$$= \frac{1}{a_{11}} \frac{\partial \gamma}{\partial r} + 2H\gamma\, \xi_r + \frac{1}{r a_{11}} \frac{\partial}{\partial r} (r \overline{S}^{(\sigma)}_{<11>}) + \frac{1}{r(a_{11})^{1/2}} \frac{\partial \overline{S}^{(\sigma)}_{<12>}}{\partial \theta}$$

$$- \frac{1}{r} \overline{S}^{(\sigma)}_{<22>} + \frac{B_{11}}{a_{11}} \overline{S}^{(\sigma)}_{<11>} \xi_r + \rho^{(\sigma)} b_r^{(\sigma)}$$

$$+ [\, - \rho(\mathbf{v} \cdot \boldsymbol{\xi} - v_{(\xi)}^{(\sigma)})^2 \xi_r + T_{rr} \xi_r + T_{rz} \xi_z \,]$$

Table 2.4.2-7 (cont.)

θ *component*

$$\rho^{(\sigma)}\left(\frac{\partial v_{\theta}^{(\sigma)}}{\partial t} + \frac{\partial v_{\theta}^{(\sigma)}}{\partial r} v_{r}^{(\sigma)} + \frac{v_{r}^{(\sigma)} v_{\theta}^{(\sigma)}}{r}\right)$$

$$= \frac{1}{r^{2}(\ a_{11}\ )^{1/2}} \frac{\partial}{\partial r}\left(r^{2}\ \bar{S}^{(\sigma)}_{<12>}\ \right) + \frac{1}{r}\frac{\partial \bar{S}^{(\sigma)}_{<22>}}{\partial \theta} + \rho^{(\sigma)} b_{\theta}^{(\sigma)}$$

$$+ [\ T_{\theta r}\xi_{r} + T_{\theta z}\xi_{z}\ ]$$

z *component*

$$\rho^{(\sigma)}\left(\frac{\partial v_{z}}{\partial t} + \frac{\partial v_{z}^{(\sigma)}}{\partial r} v_{r}^{(\sigma)}\right)$$

$$= \frac{1}{a_{11}}\frac{\partial h}{\partial r}\frac{\partial \gamma}{\partial r} + 2H\gamma\,\xi_{z} + \frac{1}{r a_{11}}\frac{\partial h}{\partial r}\frac{\partial}{\partial r}\left(r\ \bar{S}^{(\sigma)}_{<11>}\ \right)$$

$$+ \frac{1}{r(\ a_{11}\ )^{1/2}}\frac{\partial h}{\partial r}\frac{\partial \bar{S}^{(\sigma)}_{<12>}}{\partial \theta} + \frac{B_{11}}{a_{11}}\bar{S}^{(\sigma)}_{<11>}\xi_{z} + \rho^{(\sigma)} b_{z}$$

$$+ [\ -\rho(\ \mathbf{v}\cdot\boldsymbol{\xi} - v_{(\xi)}\ )^{2}\xi_{z} + T_{zr}\xi_{r} + T_{zz}\xi_{z}\ ]$$

surface rate of deformation tensor

$$\mathbf{D}^{(\sigma)} = \bar{D}^{(\sigma)}_{<\alpha\beta>}\mathbf{a}_{<\alpha>}\mathbf{a}_{<\beta>}$$

$$\bar{D}^{(\sigma)}_{<11>} = \frac{1}{a_{11}}\left(\frac{\partial v_{r}^{(\sigma)}}{\partial r} + \frac{\partial h}{\partial r}\frac{\partial v_{z}^{(\sigma)}}{\partial r}\right)$$

$$\bar{D}^{(\sigma)}_{<22>} = \frac{1}{r} v_{r}^{(\sigma)}$$

$$\bar{D}^{(\sigma)}_{<12>} = \bar{D}^{(\sigma)}_{<21>} = \frac{r}{2(\ a_{11}\ )^{1/2}}\frac{\partial}{\partial r}\left(\frac{v_{\theta}^{(\sigma)}}{r}\right)$$

Table 2.4.2-8: Alternative form for rotating and deforming axially symmetric surface viewed in a cylindrical coordinate system (refer to Exercises A.2.1-11, A.2.6-3, and A.5.3-12)

dividing surface

$$r = c(z, t)$$

assumptions

$$\rho^{(\sigma)} = \rho^{(\sigma)}(z, t) \qquad \varepsilon = \varepsilon(z, t)$$

$$\gamma = \gamma(z, t) \qquad \kappa = \kappa(z, t)$$

cylindrical components of $\mathbf{v}^{(\sigma)}$:

$$v_r^{(\sigma)} = v_r^{(\sigma)}(z, t)$$

$$v_\theta^{(\sigma)} = v_\theta^{(\sigma)}(z, t)$$

$$v_z^{(\sigma)} = v_z^{(\sigma)}(z, t)$$

surface coordinates

$$y^1 \equiv z \qquad y^2 \equiv \theta$$

$$a_{11} = 1 + \left(\frac{\partial c}{\partial z}\right)^2$$

$$a_{22} = c^2$$

$$a_{12} = a_{21} = 0$$

unit normal

$$\xi_r = \left[1 + \left(\frac{\partial c}{\partial z}\right)^2\right]^{-1/2}$$

$$\xi_\theta = 0$$

$$\xi_z = -\frac{\partial c}{\partial z}\left[1 + \left(\frac{\partial c}{\partial z}\right)^2\right]^{-1/2}$$

Table 2.4.2-8 (cont.)

second groundform tensor

$$B_{11} = \frac{\partial^2 c}{\partial z^2}\left[1 + \left(\frac{\partial c}{\partial z}\right)^2\right]^{-1/2}$$

$$B_{22} = -c\left[1 + \left(\frac{\partial c}{\partial z}\right)^2\right]^{-1/2}$$

$$B_{12} = B_{21} = 0$$

mean curvature

$$H = \frac{1}{2c}\left[c\frac{\partial^2 c}{\partial z^2} - \left(\frac{\partial c}{\partial z}\right)^2 - 1\right]\left[1 + \left(\frac{\partial c}{\partial z}\right)^2\right]^{-3/2}$$

***components of* u**

Noting that

$$\mathbf{u} = u_r \mathbf{g}_r + u_\theta \mathbf{g}_\theta + u_z \mathbf{g}_z$$

$$= \overline{u}_{<\alpha>} \mathbf{a}_{<\alpha>} + u_{(\xi)} \boldsymbol{\xi}$$

we have

$$u_r = \frac{\partial c}{\partial t}$$

$$u_\theta = u_z = 0$$

$$\overline{u}_{<1>} = \frac{\partial c}{\partial t}\frac{\partial c}{\partial z}\left[1 + \left(\frac{\partial c}{\partial z}\right)^2\right]^{-1/2}$$

$$\overline{u}_{<2>} = 0$$

$$u_{(\xi)} = \frac{\partial c}{\partial t}\left[1 + \left(\frac{\partial c}{\partial z}\right)^2\right]^{-1/2}$$

Table 2.4.2-8 (cont.)

components of surface velocity $\mathbf{v}^{(\sigma)}$

Observing that

$$\mathbf{v}^{(\sigma)} = v_r^{(\sigma)} \mathbf{g}_r + v_\theta^{(\sigma)} \mathbf{g}_\theta + v_z^{(\sigma)} \mathbf{g}_z$$

$$= \overline{v}^{(\sigma)}_{<\alpha>} \mathbf{a}_{<\alpha>} + v^{(\sigma)}_{(\xi)} \boldsymbol{\xi}$$

we find

$$\overline{v}^{(\sigma)}_{<1>} = \left[\frac{\partial c}{\partial z} v_r^{(\sigma)} + v_z^{(\sigma)} \right] \left[1 + \left(\frac{\partial c}{\partial z} \right)^2 \right]^{-1/2}$$

$$\overline{v}^{(\sigma)}_{<2>} = v_\theta^{(\sigma)}$$

$$v^{(\sigma)}_{(\xi)} = \frac{\partial c}{\partial t} \left[1 + \left(\frac{\partial c}{\partial z} \right)^2 \right]^{-1/2}$$

$$= \left[v_r^{(\sigma)} - \frac{\partial c}{\partial z} v_z^{(\sigma)} \right] \left[1 + \left(\frac{\partial c}{\partial z} \right)^2 \right]^{-1/2}$$

$$\frac{\partial c}{\partial t} = v_r^{(\sigma)} - \frac{\partial c}{\partial z} v_z^{(\sigma)}$$

components of intrinsic surface velocity $\dot{\mathbf{y}}$

$$\dot{\mathbf{y}} = \mathbf{v}^{(\sigma)} - \mathbf{u}$$

$$= \dot{y}_{<\alpha>} \boldsymbol{a}_{<\alpha>}$$

$$\dot{y}_{<1>} = v_z^{(\sigma)} \left[1 + \left(\frac{\partial c}{\partial z} \right)^2 \right]^{1/2}$$

$$\dot{y}_{<2>} = v_\theta^{(\sigma)}$$

Table 2.4.2-8 (cont.)

jump mass balance

$$\frac{\partial \rho^{(\sigma)}}{\partial t} + \frac{\partial \rho^{(\sigma)}}{\partial z} v_z^{(\sigma)} + \rho^{(\sigma)} \left[\frac{1}{a_{11}} \left(\frac{\partial c}{\partial z} \frac{\partial v_r^{(\sigma)}}{\partial z} + \frac{\partial v_z^{(\sigma)}}{\partial z} \right) + \frac{1}{c} v_r^{(\sigma)} \right]$$

$$+ [\, \rho(\, \mathbf{v} \cdot \boldsymbol{\xi} - v_{(\xi)}^{(\sigma)}\,)] = 0$$

jump momentum balance

Let

$$\mathbf{S}^{(\sigma)} = \overline{S}^{(\sigma)}_{<\alpha\beta>} \mathbf{a}_{<\alpha>} \mathbf{a}_{<\beta>}$$

r *component*

$$\rho^{(\sigma)} \left[\frac{\partial v_r^{(\sigma)}}{\partial t} + \frac{\partial v_r^{(\sigma)}}{\partial z} v_z^{(\sigma)} - \frac{1}{c} (\, v_\theta^{(\sigma)}\,)^2 \right]$$

$$= \frac{1}{a_{11}} \frac{\partial c}{\partial z} \frac{\partial \gamma}{\partial z} + 2H\gamma\, \xi_r + \frac{1}{c a_{11}} \frac{\partial c}{\partial z} \frac{\partial (\, c \overline{S}^{(\sigma)}_{<11>}\,)}{\partial z}$$

$$+ \frac{1}{c(\, a_{11}\,)^{1/2}} \frac{\partial c}{\partial z} \frac{\partial \overline{S}^{(\sigma)}_{<12>}}{\partial \theta} - \frac{1}{c} \overline{S}^{(\sigma)}_{<22>} + \frac{B_{11}}{a_{11}} \overline{S}^{(\sigma)}_{<11>} \xi_r + \rho^{(\sigma)} b_r^{(\sigma)}$$

$$+ [\, - \rho(\, \mathbf{v} \cdot \boldsymbol{\xi} - v_{(\xi)}^{(\sigma)}\,)^2 \xi_r + T_{rr} \xi_r + T_{rz} \xi_z \,]$$

θ *component*

$$\rho^{(\sigma)} \left\{ \frac{\partial v_\theta^{(\sigma)}}{\partial t} + c \left[\frac{\partial}{\partial z} \left(\frac{v_\theta^{(\sigma)}}{c} \right) + \frac{v_\theta^{(\sigma)}}{c^2} \frac{\partial c}{\partial z} \right] v_z^{(\sigma)} + \frac{1}{c} v_r^{(\sigma)} v_z^{(\sigma)} \right\}$$

$$= \frac{1}{c^2 (\, a_{11}\,)^{1/2}} \frac{\partial (\, c^2 \overline{S}^{(\sigma)}_{<12>}\,)}{\partial z} + \frac{1}{c} \frac{\partial \overline{S}^{(\sigma)}_{<22>}}{\partial \theta} + \rho^{(\sigma)} b_\theta^{(\sigma)}$$

$$+ [\, T_{\theta r} \xi_r + T_{\theta z} \xi_z \,]$$

Table 2.4.2-8 (cont.)

z component

$$\rho^{(\sigma)}\left(\frac{\partial v_z^{(\sigma)}}{\partial t}+\frac{\partial v_z^{(\sigma)}}{\partial z}v_z^{(\sigma)}\right)=\frac{1}{a_{11}}\frac{\partial\gamma}{\partial z}+2H\gamma\,\xi_z+\frac{1}{ca_{11}}\frac{\partial}{\partial z}\left(c\overline{S}^{(\sigma)}_{<11>}\right)$$

$$+\frac{1}{c(\,a_{11}\,)^{1/2}}\frac{\partial\overline{S}^{(\sigma)}_{<12>}}{\partial\theta}+\frac{B_{11}}{a_{11}}\overline{S}^{(\sigma)}_{<11>}\xi_z+\rho^{(\sigma)}b_z^{(\sigma)}$$

$$+\left[\,-\rho(\,\mathbf{v}\cdot\boldsymbol{\xi}-v_{(\xi)}^{(\sigma)}\,)^2\xi_z+T_{zr}\xi_r+T_{zz}\xi_z\,\right]$$

surface rate of deformation tensor

$$\mathbf{D}^{(\sigma)}=\overline{D}^{(\sigma)}_{<\alpha\beta>}\mathbf{a}_{<\alpha>}\mathbf{a}_{<\beta>}$$

$$\overline{D}^{(\sigma)}_{<11>}=\frac{1}{a_{11}}\left(\frac{\partial c}{\partial z}\frac{\partial v_r^{(\sigma)}}{\partial z}+\frac{\partial v_z^{(\sigma)}}{\partial z}\right)$$

$$\overline{D}^{(\sigma)}_{<22>}=\frac{1}{c}v_r^{(\sigma)}$$

$$\overline{D}^{(\sigma)}_{<12>}=\overline{D}^{(\sigma)}_{<21>}=\frac{1}{2(\,a_{11}\,)^{1/2}}\left(\frac{\partial v_\theta^{(\sigma)}}{\partial z}-\frac{1}{c}\frac{\partial c}{\partial z}v_\theta^{(\sigma)}\right)$$

2.4.3 Summary of useful equations at common lines

Mass balance at common line Conservation of mass requires that at the common line (Sec. 1.3.8)

$$\left(\rho^{(\sigma)}[\, \mathbf{v}^{(\sigma)} \cdot \boldsymbol{\nu} - u_{\{\nu\}}^{(cl)} \,] \right) = 0 \tag{3-1}$$

which we will refer to as the **mass balance at the common line.**

If we are willing to say that there is **no mass transfer** at the common line $C^{(cl)}$ (see Secs. 1.3.8 through 1.3.10), then

$$\text{at } C^{(cl)}: \quad \mathbf{v}^{(\sigma)} \cdot \boldsymbol{\nu} = u_{\{\nu\}}^{(cl)} \tag{3-2}$$

Momentum balance at common line Euler's first law requires that at the common line (Sec. 2.1.9)

$$\left(\rho^{(\sigma)}\mathbf{v}^{(\sigma)}[\, \mathbf{v}^{(\sigma)} \cdot \boldsymbol{\nu} - u_{\{\nu\}}^{(cl)} \,] - \mathbf{T}^{(\sigma)} \cdot \boldsymbol{\nu} \right) = 0 \tag{3-3}$$

which we will refer to as the **momentum balance at the common line.**

If we are able to **neglect mass transfer** at the common line **or** if we **neglect** the effect of **inertial forces** at the common line, (3-3) reduces to

$$\left(\mathbf{T}^{(\sigma)} \cdot \boldsymbol{\nu} \right) = 0 \tag{3-4}$$

Under **static conditions,** the surface stress acting at the common line may be expressed in terms of the surface tension and (3-3) becomes

$$\left(\gamma \boldsymbol{\nu} \right) = 0 \tag{3-5}$$

In the context of Figure 2.1.9-1, this says

$$\gamma^{(AB)}\boldsymbol{\nu}^{(AB)} + \gamma^{(AS)}\boldsymbol{\nu}^{(AS)} + \gamma^{(BS)}\boldsymbol{\nu}^{(BS)} = 0 \tag{3-6}$$

Equations (3-5) and (3-6) are referred to as the **Neumann triangle** (Buff 1960, p. 288).

With reference to Figure 2.1.10-1, the momentum balance at a common line on a **rigid solid** may be expressed as

$$\mathbf{P}^{(AS)} \cdot \left(\rho^{(\sigma)}\mathbf{v}^{(\sigma)}[\, \mathbf{v}^{(\sigma)} \cdot \boldsymbol{\nu} - u_{\{\nu\}}^{(cl)} \,] - \mathbf{T}^{(\sigma)} \cdot \boldsymbol{\nu} \right) = 0 \tag{3-7}$$

Here $\mathbf{P}^{(AS)}$ is the projection tensor on the A-S dividing surface.

Often we will neglect both mass transfer at the common line and the effects of the viscous portion of the surface stress tensor. With these restrictions, (3-7) can be rearranged as (see Sec. 2.1.10)

$$\gamma^{(AB)} \cos \Theta_0 + \gamma^{(AS)} - \gamma^{(BS)} = 0 \tag{3-8}$$

which is known as **Young's equation** (Young 1805). Here Θ_0 is the contact angle measured through phase A at the common line in Figure 2.1.10-1 (rather than the static or dynamic contact angle measured by experimentalists at some distance from the common line, perhaps with 10 × magnification).

Either Young's equation should be used to determine Θ_0, having previously estimated or measured $\gamma^{(AS)}$ and $\gamma^{(BS)}$, or (10-3) should be employed to determine $\gamma^{(AS)}$ and $\gamma^{(BS)}$, having previously estimated or measured Θ_0. At this writing, it appears that we will generally have little choice but to assume a value for Θ_0, as we have done in Secs. 3.2.7 and 3.3.3. For more on this point, see Sec. 2.1.10.

References

Buff, F. P., "Handbuch der Physik," vol. 10, edited by S. Flügge, Springer-Verlag, Berlin (1960).

Young, T., *Philos. Trans. R. Soc. London* (4 to.) **95**, 65 (1805).

Notation for chapter 2

a	determinant, defined in Sec. A.2.1
$\mathbf{a}_{\alpha}$	natural basis vectors, introduced in Sec. A.2.1
$\mathbf{a}_{\kappa A}$	natural basis vectors for reference configuration
$\mathbf{a}_{<i>}$	physical basis vectors, introduced in Sec. A.2.5
$\mathbf{a}^{\alpha}$	dual basis vectors, introduced in Sec. A.2.3
$\mathbf{b}$	body force per unit mass exerted on the material within a phase, introduced in Sec. 2.1.3
$\mathbf{b}^{(\sigma)}$	body force per unit mass exerted on the dividing surface, introduced in Sec. 2.1.3
$\mathbf{B}$	second groundform tangential tensor field, introduced in Sec. A.5.3
$\mathbf{B}^{(\sigma)}$	left surface Cauchy-Green tensor, defined in Sec. 1.2.6
C	intersection of Σ and S as illustrated in Figure 2.2.2-1
$C^{(cl)}$	common line
$\mathbf{C}^{(\sigma)}$	right surface Cauchy-Green tensor, defined in Sec. 1.2.6
$\mathbf{C}_t^{(\sigma)}$	right relative surface Cauchy-Green tensor, defined in Sec. 1.2.6
dA	denotes that an area integration is to be performed
$d\mathbf{f}$	denotes that an integration with respect to the vector measure of force is to be performed
dm	denotes that an integration with respect to the measure of mass is to be performed
ds	denotes that a line integration with respect to arc length is to be performed. Also used to denote integration over past time as in Sec. 2.2.8.
dV	denotes that a volume integration is to be performed
$\mathbf{D}$	rate of deformation tensor (Slattery 1981, p. 28)

$\mathbf{D}^{(\sigma)}$ surface rate of deformation tensor, introduced in Sec. 1.2.8

$\mathbf{e}_j$ basis vectors for rectangular cartesian coordinate system (Slattery 1981, pp. 607 and 614)

$\mathbf{f}^a$ applied force, introduced in Sec. 2.1.2

$\mathbf{f}(B, A)$ system of forces, introduced in Sec. 2.1.1

$\mathcal{F}$ surface deformation gradient, defined in Sec. 1.2.5

$\mathcal{F}_t$ relative surface deformation gradient, defined in Sec. 1.2.5

$g_\kappa^{(\sigma)}$ surface isotropy group, introduced in Sec. 2.2.4

H mean curvature of dividing surface, introduced in Exercise A.5.3-3

$\mathbf{I}$ identity tensor that transforms every vector into itself

$\mathbf{K}_t^{(\sigma)t}$ defined in Sec. 2.2.7

$\mathbf{n}$ unit normal, outwardly directed with respect to a closed surface except where noted

N_{ca} capillary number, defined in Sec. 2.1.11

N_{We} Weber number, defined in Sec. 2.1.11

$\mathcal{O}_\kappa$ full orthogonal group, introduced in Sec. 2.2.5

p mean pressure, defined in Sec. 2.4.1

P thermodynamic pressure, defined in Sec. 5.8.3

$\mathbf{P}$ projection tensor, defined in Sec. A.3.2

$\mathbf{P}_\kappa$ projection tensor for tangential vector fields on reference dividing surface

$\mathcal{P}$ modified pressure, defined in Sec. 2.4.1

$\mathbf{Q}$ orthogonal transformation that describes the rotation and (possibly) reflection that takes one frame into another, introduced in Sec. 1.4.1

$\mathcal{Q}$ defined in Sec. 1.4.1

R region occupied by body

$\mathcal{R}$ surface rotation tensor, introduced in Sec. 1.2.6

s time in past measured relative to present, introduced in Sec. 2.2.3

S closed surface bounding body

$\mathbf{S}$ viscous portion of stress tensor, defined in Sec. 2.4.1

$\mathbf{S}^{(\sigma)}$ viscous portion of surface stress tensor, defined in Sec. 2.4.2

t time

$\mathbf{t}$ stress vector, introduced in Sec. 2.1.3

$\mathbf{t}^{(\sigma)}$ surface stress vector, introduced in Sec. 2.1.3

$\mathbf{T}$ stress tensor, introduced in Sec. 2.1.4

$\mathbf{T}^{(\sigma)}$ surface stress tensor, introduced in Sec. 2.1.5

$u_\kappa^{(\sigma)}$ unimodular group, introduced in Sec. 2.2.4

$\mathbf{u}$ time rate of change of spatial position following a surface point, introduced in Sec. 1.2.7

$\mathbf{u}_{(v)}^{(cl)}$ speed of displacement of common line, defined in Sec. 1.3.7

$\mathbf{U}^{(\sigma)}$ right surface stretch tensor, introduced in Sec. 1.2.6

$\mathbf{U}_t^{(\sigma)}$ right relative surface stretch tensor, introduced in Sec. 1.2.6

$\mathbf{v}$ velocity vector, defined in Sec. 1.1.1

$\mathbf{v}^{(\sigma)}$ surface velocity vector, defined in Sec. 1.2.5

$v_{(\xi)}^{(\xi)}$ speed of displacement, introduced in Sec. 1.2.7

W rate at which work is done on body, introduced in Sec. 2.1.1

$\mathbf{W}$ vorticity tensor, defined in Sec. 2.3.5

$\mathbf{W}^{(\sigma)}$ surface vorticity tensor, defined in Sec. 2.2.2

x^i curvilinear coordinates, introduced in Sec. A.1.2

y^α surface coordinates

y_κ^A surface coordinates used in reference configuration

$\dot{\mathbf{y}}$ intrinsic surface velocity, defined in Sec. 1.2.7

$\mathbf{z}$ position vector

$\mathbf{z}_\kappa$ position vector denoting a place in the reference configuration

$\mathbf{z}_{(0)}$ reference position, introduced in Sec. 1.4.1

z_j rectangular cartesian coordinates

greek letters

γ thermodynamic interfacial tension, defined in Sec. 5.8.6

ε surface shear viscosity, introduced in Sec. 2.2.2

$\boldsymbol{\varepsilon}$ defined in Sec. A.1.2

$\zeta^{(\sigma)}$ surface particle, introduced in Sec. 1.2.5

Θ dynamic contact angle (see Secs. 2.1.10 and 3.3.3)

$\Theta^{(stat)}$ limit of Θ measured under static conditions (see Secs. 2.1.12 and 3.2.7)

Θ_0 contact angle measured at the common line (see Sec. 2.1.10)

κ surface dilatational viscosity, introduced in Sec. 2.2.2

$\boldsymbol{\kappa}^{(\sigma)}(y_\kappa^1, y_\kappa^2)$ reference dividing surface, introduced in Sec. 1.2.5

κ_1, κ_2 principal curvatures of dividing surface, introduced in Exercise A.5.3-3

λ introduced in Sec. 1.3.2. Also used as the bulk viscosity of a Newtonian fluid in Sec. 2.4.1.

λ^+, λ^- introduced in Sec. 1.3.2

$\mathbf{\Lambda}$	indicates that a vector product is to be performed (Slattery 1981, p. 653), except where noted
μ	shear viscosity of Newtonian fluid, introduced in Sec. 2.4.1
$\boldsymbol{\mu}$	unit vector normal to closed bounding curve that is both tangent and outwardly directed with respect to this curve, except where noted
$\boldsymbol{\nu}$	unit vector normal to common line and tangent to dividing surface
$\boldsymbol{\xi}$	unit normal to Σ introduced in Sec. 1.2.7
ρ	mass density
$\rho^{(\sigma)}$	surface mass density, introduced in Sec. 1.3.1
Σ	dividing surface
$\boldsymbol{\chi}_{\kappa}^{(\sigma)}(y_{\kappa}^1, y_{\kappa}^2, t)$	deformation of Σ, introduced in Sec. 1.2.5
ϕ	potential energy per unit mass, defined in Secs. 2.1.3 and 2.3.4
$\boldsymbol{\omega}$	angular velocity vector of the unstarred frame with respect to the starred frame in Sec. 1.4.3

others

$\mathbf{0}$	transforms every vector into the zero vector
$\det_{(\sigma)}$	surface determinant operator, defined in Sec. A.3.7
div	divergence operator
$\text{div}_{(\sigma)}$	surface divergence operator, introduced in Secs. A.5.1 through A.5.6
grad	gradient with respect to the reference configuration
tr	trace operator
$\cdots_{,\alpha}$	surface covariant derivative, introduced in Sec. A.5.3
$\cdots_{\boldsymbol{\kappa}}$	denotes reference configuration as in Sec. 1.2.5

$\ldots^{(I)}$	denotes three-dimensional interfacial region, introduced in Sec. 1.3.2
$\ldots^{t}$	denotes history of variable, introduced in Sec. 2.2.3
$\ldots^{T}$	transpose
$\ldots^{-1}$	inverse
$\ldots^{*}$	denotes an alternative frame of reference
$\ldots^{\star}$	denotes dimensionless variable
$\dot{\overline{\ldots}}$	an alternative expression for the material derivative
$\ddot{\overline{\ldots}}$	an alternative expression for a second material derivative
$(\ldots)$	jump for common line, defined in Sec. 1.3.7
$[\ldots]$	jump for dividing surface, defined in Sec. 1.3.4
∇	gradient with respect to the current configuration
$\nabla_{(\sigma)}$	surface gradient operation, introduced in Secs. A.5.1 through A.5.6
$\frac{d_{(m)}}{dt}$	material derivative, defined in Sec. 1.1.1
$\frac{d_{(s)}}{dt}$	surface material derivative, defined in Sec. 1.2.5; see also Sec. 1.2.7

Reference

Slattery, J. C., "Momentum, Energy, and Mass Transfer in Continua," McGraw-Hill, New York (1972); second edition, Robert E. Krieger, Malabar, FL 32950 (1981).

3

Applications of the differential balances to momentum transfer

Let us now see what must be done in order to arrive at detailed descriptions for the configurations and motions of adjoining multiple phases within specified boundaries consistent with the principles we have developed in the preceding two chapters.

In writing this chapter, I have assumed that you have already been introduced to the more standard problems of fluid mechanics involving the flow of a single phase (Slattery 1981, chapter 3; Whitaker 1968, chapter 5; Bird *et al.* 1960, chapter 3) and that you wish to learn more about analyzing multiphase flows when interfacial effects are important.

References

Bird, R. B., W. E. Stewart, and E. N. Lightfoot, "Transport Phenomena," John Wiley, New York (1960).

Slattery, J. C., "Momentum, Energy, and Mass Transfer in Continua," McGraw-Hill, New York (1972); second edition, Robert E. Krieger, Malabar, FL 32950 (1981).

Whitaker, S., "Introduction to Fluid Mechanics," Prentice-Hall, Englewood Cliffs, NJ (1968).

3.1 Philosophy

3.1.1 Structure of problem It is relatively easy to outline the problem that must be solved in order to describe the configurations and motions of several adjoining phases.

i) *Within each phase*, we must satisfy the equation of continuity, Cauchy's first law, and an appropriate model for bulk stress-deformation behavior such as the Newtonian fluid (Sec. 2.4.1; Slattery 1981, p. 59).

ii) *On each dividing surface*, we must satisfy the jump mass balance, the jump momentum balance, and an appropriate model for surface stress-deformation behavior such as the Boussinesq surface fluid (Sec. 2.4.2).

iii) *At each common line*, there are mass and momentum balances to be obeyed (Sec. 2.4.3).

iv) *Everywhere*, velocity must be finite.

v) *Within each phase*, the stress tensor is bounded, since every second order tensor in three space is bounded (Stakgold 1967, p. 140).

vi) *On each dividing surface*, the surface stress tensor is bounded, by the same argument.

vii) *Within each phase in the neighborhood of a common line*, it is important to describe the mutual forces (London-van der Waals forces, electrostatic forces; see Exercises 2.1.3-1 and 2.1.3-2) accurately. I will have more to say about this below.

viii) *At each dividing surface*, the tangential components of velocity are continuous. This is often referred to as the no-slip boundary condition. It is a consequence of assuming nearly local equilibrium at the phase interface (see Sec. 5.10.3), analogous with our assumptions that temperature (see Sec. 6.1.1) or chemical potential (see Sec. 6.1.2) are continuous across an interface. Lacking contrary experimental evidence, this condition is applied even when there is a phase change, the normal component of velocity is not continuous at the interface, and the adjacent phases are clearly not in local equilibrium. A careful examination of the no-slip condition has been given by Goldstein (1938, p. 676). It is known to fail, when one of the phases is a rarefied gas.

common lines on* relatively *rigid solids Problems involving common lines on relatively high tensile strength rigid solids require special consideration, because we will normally decide to ignore the deformation of the solid within the immediate neighborhood of the common line. We can summarize our discussions in Secs. 2.1.10 and 2.1.13 with these additional comments.

ix) *A common line can not move.* As viewed on a macroscale, it appears to move relative to the solid as a succession of stationary common lines are formed on a microscale, driven by a negative disjoining pressure in the receding film (Li and Slattery 1990; see also Sec. 3.3.3). The solid prefers to be in contact with the advancing phase. The amount of receding phase left stranded by the formation of this succession of common lines is too small to be easily detected, as in the experiments of Dussan V. and Davis (1974; see also Sec. 1.2.9).

This colaescence process can not occur at a common line, since this would imply a moving common line and unbounded forces. The argument that unbounded forces appear at a moving common line is unaffected by the existence of mutual forces (London-van der Waals forces and electrostatic double-layer forces; see Sec. 1.3.9).

On a macroscale, the apparent motion of the common line is driven by a combination of external forces (gravity), of convective forces, and of mutual forces (London-van der Waals forces and electrostatic double-layer forces). Various examples are given in Sec. 2.1.13.

x) *The mass balance at a common line on a rigid solid is satisfied identically*, since all velocities measured relative to the rigid solid go to zero at the common line. This assumes that any mass transfer across the common line resulting from differences in chemical potential can be neglected.

xi) *The momentum balance at a common line on a rigid solid reduces to Young's equation* (Sec. 2.1.10)

$$\gamma^{(AB)} \cos \Theta_0 + \gamma^{(AS)} - \gamma^{(BS)} = 0 \tag{1-1}$$

where $\gamma^{(ij)}$ denotes the interfacial tension acting in the i-j interface of Figure 2.1.10-1 and Θ_0 is the contact angle measured through phase A at the common line (rather than the static or dynamic contact angle measured by experimentalists at some distance from the common line, perhaps with 10 × magnification). Either Young's equation should be used to determine Θ_0, having previously estimated or measured $\gamma^{(AS)} - \gamma^{(BS)}$, or it should be employed to determine $\gamma^{(AS)} - \gamma^{(BS)}$, having previously estimated or measured Θ_0. As explained in Sec. 2.1.10, it appears at this writing that we will generally have little choice but to assume a value for Θ_0 as we have done in Secs. 3.2.7 and 3.3.3.

The preceding discussion assumes that one is willing to solve a complete problem, generally involving an *inner solution* for the immediate neighborhood of the common line, in which the effects of mutual forces such as London-van der Waals forces and electrostatic forces are included, and an *outer solution* that does not include the common line, in which the effects of mutual forces are excluded. The inner solution is commonly avoided by using prior experimental measurements or a data correlation such as that given in Sec. 2.1.12 to specify the dynamic contact angle at the apparent common line, the line at which the experimentalist measures the static or dynamic contact angle with perhaps 10 × magnification. This is practical, only when the effects of gravity and of inertia can be ignored and

$$L/\ell << 1 \tag{1-2}$$

where ℓ is a dimension of the macroscopic system that characterizes the radius of curvature of the interface,

$$L \equiv \left(\frac{\gamma}{g\ \Delta\rho} \right)^{1/2} \tag{1-3}$$

is a characteristic length of the meniscus, g the acceleration of gravity, and $\Delta\rho$ the density difference between the phases. It is only under these conditions that I expect the static and dynamic contact angles to be independent of geometry or scale (see Secs. 2.1.11, 2.1.12, 3.2.7, and 3.3.3).

For such an analysis of a dynamic problem, it is important to keep in mind that, since the outer solution does not include the common line, any prediction of unbounded forces at the apparent common line has no physical significance.

slip at apparent common lines on* relatively *rigid solids As an alternative to accounting for the effects of mutual forces, one can presume instead that there is slip within the immediate neighborhood of the common line. We saw in Sec. 1.3.10 that allowing slip within the neighborhood of a moving common line eliminates the inconsistencies posed by its motion.

Several descriptions of slip have been suggested.

Perhaps the simplest is that we allow complete slip for a distance ℓ following an advancing common line, but no slip thereafter (Huh and Mason 1977):

$$\text{for } 0 \leq z \leq \ell\text{:} \quad \mathbf{S} \cdot \mathbf{n} = 0 \tag{1-4}$$

$$\text{for } z > \ell\text{:} \quad \Delta\mathbf{v} = 0 \tag{1-5}$$

Here z denotes position in the solid-fluid interface measured normal to the moving common line; $\mathbf{n}$ is the unit normal to the solid-fluid interface pointing into the fluid; $\Delta\mathbf{v}$ is the difference between the velocity of the fluid in the limit as the solid is approached and the velocity of the solid.

The velocity difference also can be thought of as being proportional to the shear stress that the fluid exerts upon the boundary:

$$\Delta\mathbf{v} = \frac{1}{\beta} \mathbf{S} \cdot \mathbf{n} \tag{1-6}$$

We will refer to β as the slip coefficient, observing that as $\beta \to \infty$ we recover the conventional no-slip condition. The slip coefficient β may be a constant or it may be a function of the magnitude of the velocity difference $\Delta \mathbf{v}$ (Dussan V. 1976). Hocking (1976) has developed estimates for β in the context of displacement over a wavy surface. For a Newtonian fluid, the slip length μ/β, a measure of the distance over which slip is postulated to be important, varies with the slip model, but it is assumed to be of the order of molecular dimensions. The slip ratio, the ratio of the slip length to a macroscopic length scale of the system, is invariably small.

In order to analyze a flow field, one can seek a perturbation solution valid only for small values of the slip ratio. This expansion is singular, since the lowest order approximation, which corresponds to the no-slip problem, predicts unbounded forces. By using matched asymptotic expansions, the flow field separates into two regions: the inner region, scaled by the slip length and valid in the immediate neighborhood of the common line, and the outer region, scaled by a macroscopic dimension of the system and valid throughout the remainder of the flow field (Hocking 1977; Kafka and Dussan V. 1979; Dussan V. 1979).

While this approach will not be developed further here, it does have both the advantage and the disadvantage that the mechanics of the common line region are simplified. Any discussion of the form of the long-range intermolecular forces within the neighborhood of the common line or of the description of the interfacial separation (see Secs. 3.2.7 and 3.3.3 as well as Exercise 4.1.4-3) is avoided. Unfortunately, these links with the chemistry of the system are lost as well. For interesting examples of its use, see Davis (1983), Dussan V. and Davis (1986), and Young and Davis (1987).

References

Davis, S. H., *J. Appl. Mech.* **105,** 977 (1983).

Dussan V., E. B., *J. Fluid Mech.* **77**, 665 (1976).

Dussan V., E. B., *Ann. Rev. Fluid Mech.* **11**, 371 (1979).

Dussan V., E. B., and S. H. Davis, *J. Fluid Mech.* **65**, 71 (1974).

Dussan V., E. B., and S. H. Davis, *J. Fluid Mech.* **173**, 115 (1986).

Goldstein, S., "Modern Developments in Fluid Mechanics," Oxford University Press, London (1938).

Hocking, L. M., *J. Fluid Mech.* **76**, 801 (1976).

Hocking, L. M., *J. Fluid Mech.* **79**, 209 (1977).

Huh, C., and S. G. Mason, *J. Fluid Mech.* **81**, 401 (1977).

Kafka, F. Y., and E. B. Dussan V., *J. Fluid Mech.* **95**, 539 (1979).

Li, D., and J. C. Slattery, *J. Colloid Interface Sci.* (1990).

Slattery, J. C., "Momentum, Energy, and Mass Transfer in Continua," McGraw-Hill, New York (1972); second edition, Robert E. Krieger, Malabar, FL 32950 (1981).

Stakgold, I., "Boundary Value Problems of Mathematical Physics," Vol. 1, Macmillan, New York (1967).

Young, G. W., and S. H. Davis, *J. Fluid Mech.* **174**, 327 (1987).

3.1.2 Approximations Although it is relatively easy to outline as I have done in the preceding section the relationships and conditions that must be satisfied in describing any multiphase flow, this is not sufficient, if we are to describe real physical phenomena. One must decide precisely what problem is to be solved and what approach is to be taken in seeking a solution. I think in terms of the approximations that must be introduced. Approximations are always made, when we describe real physical problems in mathematical terms.

We can distinguish at least four classes of approximations.

a) *idealization of physical problem.* Usually the physical problem with which we are concerned is too difficult to describe in detail. One answer is to replace it with a problem that has most of the important features of our original problem but that is sufficiently simple for us to analyze. Physical boundaries are replaced by geometric surfaces (Sec. 3.2.2). Viscous effects may be neglected in a gas phase with respect to those in a liquid phase (Sec. 3.4.1). The motion of boundaries is simplified (Sec. 3.5.1). The disturbance caused by a tracer particle floating in an interface may be ignored (Sec. 3.5.1). Bulk stress-deformation behavior may be described by the incompressible Newtonian fluid (Sec. 3.4.1). Interfacial stress-deformation behavior may be represented by the Boussinesq surface fluid model (Sec. 3.4.1).

b) *limiting case.* Even after such an idealization of our original physical problem, it may still be too difficult. We may wish to consider a limiting case in which one or more of the parameters appearing in our boundary-value problem approach zero. This is equivalent to asking for a solution correct to the zeroth perturbation in these parameters. This can result in simplifications both in Cauchy's first law (Slattery 1981, chapter 3) and in the jump momentum balance (Sec. 3.7.1).

c) *integral average.* In many cases, our requirements do not demand detailed solutions. Perhaps some type of integral average is of primary interest. I explore this approach in Chapter 4.

d) *mathematical approximation.* Often we will introduce a mathematical approximation in seeking a solution to a boundary-value problem. This is inherent with numerical solutions.

While we should attempt to hold approximations to a minimum, even greater effort should be devoted to recognizing those that we are forced to introduce. For it is in this way that we shall better understand the experimental limitations of our theories.

Reference

Slattery, J. C., "Momentum, Energy, and Mass Transfer in Continua," McGraw-Hill, New York (1972); second edition, Robert E. Krieger, Malabar, FL 32950 (1981).

3.2 In the absence of deformation

3.2.1 Classes of problems In Sec. 3.2, we will talk about phase interfaces in the absence of deformation. In all cases, there is a frame of reference with respect to which there is no motion anywhere within the system being considered. Usually this will be the laboratory frame of reference, but we will see that this is not necessary.

Because of this lack of motion with respect to some preferred frame of reference, the interface and the adjoining phases are static, but they should not necessarily be taken to be at equilibrium. Long after motion can be said to be absent from a system, its chemical composition may be slowly shifting as whatever surfactants present distribute themselves between the interface and the adjoining phases. You might expect this point to be made in Chapter 6, where we explicitly account for mass transfer. But in fact it is the rare experiment in which surfactants, perhaps present only in the form of unwanted contaminants, have been eliminated.

We can distinguish at least two classes of problems.

In the simplest class, we neglect the effect of all external forces. Pressure is independent of position within each phase and the jump momentum balance requires

$$[\, P\boldsymbol{\xi} \,] = 2H\gamma\boldsymbol{\xi} \tag{1-1}$$

which implies that the mean curvature H is independent of position on the interface. These problems are discussed further in Exercise 3.2.1-1 and illustrated in the next section.

A second and often more difficult class of problems includes those in which the effect of one or more external forces can not be neglected. In Sec. 3.2.3, we consider the spinning drop measurement of interfacial tension in which the drop shape is controlled by the Coriolis force and the effect of gravity is neglected. In Exercise 3.2.6-1, I discuss a rotating interface in which the effect both of gravity and of the Coriolis force play important roles in determining its shape.

Reference

Concus, P. R., and R. Finn, *Acta Mathematica* **132**, 177 (1974).

Exercise 3.2.1-1 *axially symmetric interface in absence of external forces* (Concus and Finn 1974) In Exercise A.5.3-11, we find that for an axially symmetric surface in cylindrical coordinates

$$z = h(r) \tag{1}$$

the mean curvature H may be described by

$$H = \frac{1}{2r}[\, rh'' + h' + (h')^3 \,][\, 1 + (h')^2 \,]^{-3/2}$$

$$= \frac{1}{2r}\frac{d}{dr}\left(\frac{rh'}{[\, 1 + (h')^2 \,]^{1/2}}\right) \tag{2}$$

where

$$h' \equiv \frac{dh}{dr} \tag{3}$$

We confine our attention here to a static interface in the absence of gravity, in which case H is a constant. We assume that the surface coordinate system has been chosen such that $H > 0$.

Integrating (2), we find

$$\frac{h'}{[\, 1 + (h')^2 \,]^{1/2}} = Hr + Br^{-1} \tag{4}$$

Notice that a solution of (4) can exist only in an interval of r for which

$$|\, Hr + Br^{-1} \,| < 1 \tag{5}$$

There are three possibilities.

i) If $B > 0$, prove that solutions exist only if

$$B < \frac{1}{4H} \tag{6}$$

Prove that there is a solution in the interval (r_1, r_2)

$$0 < r_1 < \frac{1}{2H} < r_2 < \frac{1}{H} \tag{7}$$

where r_1 and r_2 assume all values in the indicated ranges as B varies from zero to its upper bound and

$$\text{as } r \to r_1 :\; h' \to \infty \tag{8}$$

$$\text{as } r \to r_2 :\; h' \to \infty \tag{9}$$

An example is illustrated in Figure 1-1.

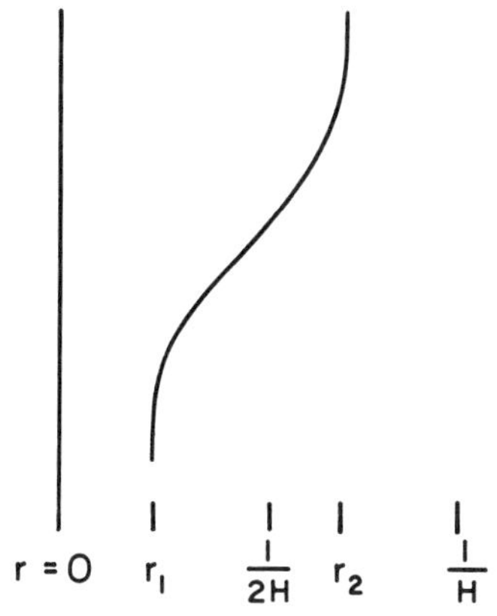

Figure 3.2.1-1: Section of rotationally symmetric interface for $B > 0$.

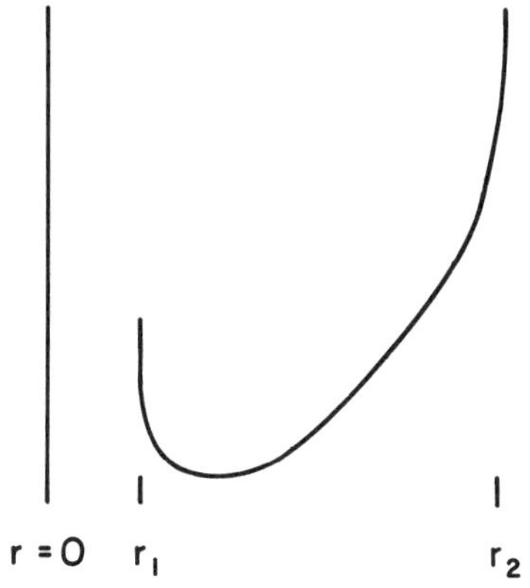

Figure 3.2.1-2: Section of rotationally symmetric interface for $B < 0$.

ii) If $B = 0$, prove that the unique solution of (4) is a sphere of radius (in spherical coordinates)

$$r = \frac{1}{H} \tag{10}$$

iii) If $B < 0$, prove that there is a solution in the interval (r_1, r_2)

$$0 < r_1 < \infty \tag{11}$$

$$\frac{1}{H} < r_2 < \infty \tag{12}$$

where r_1 and r_2 increase through all values of their ranges as B decreases from 0 to $-\infty$ and

$$\text{as } r \to r_1 \;:\; h' \to -\infty \tag{13}$$

$$\text{as } r \to r_2 \;:\; h' \to \infty \tag{14}$$

One case is shown in Figure 1-2.

3.2.2 Displacement of residual oil: a static analysis (Oh and Slattery 1979) An oil reservoir is a permeable layer of rock partially filled with oil. The remainder of the pore space is filled with water or water and gas. The oil, water, and gas are not uniformly dispersed through the reservoir rock. There will be some stratification resulting from the different densities of the phases.

After an oil field is discovered by sinking one or more wells into the oil-bearing rock formation, there are at least three different stages that can be distinguished in the production process.

In the primary stage of production, oil and brine are driven into a well from the surrounding rock by the relatively large difference between the initial field pressure and the pressure in the well. Perhaps 10-20% of the oil originally in place is recovered in this manner (Geffen 1973).

In the secondary stage of production, water or steam is pumped into a selected pattern of wells in a field forcing a portion of the remaining oil into other production wells and trapping the rest as residual oil. The void volume in a permeable rock may be thought of as many intersecting pores of varying diameters. Consider a particular pore that bifurcates to form two pores of unequal diameters, which in turn rejoin at some point downstream. A segment of oil (residual oil) will be trapped in that pore through which the oil is displaced by the water more slowly. During secondary recovery, 10-40% of the oil initially in place will be produced, but 40-80% of the oil originally in place is left behind in the field (Geffen 1973).

A number of tertiary recovery techniques have been proposed (Geffen 1973). We will focus our attention here on the recovery of residual oil by a low interfacial tension waterflood[a]. It has been estimated that in carefully selected, well designed, good performing operations an additional 30% of the oil originally in place might be recovered by a low interfacial tension waterflood (Geffen 1973).

The crude oil-water interfacial tension can be reduced by orders of magnitude when either a mixture of surfactants or an alkaline solution is added to the waterflood.

Significant recovery of residual oil has been reported in laboratory tests using waterfloods to which small amounts of a petroleum sulfonate had been added (Foster 1973; Gale and Sanvik 1973; Hill *et al.* 1973). Depending upon the particular petroleum sulfonate used and upon the salt concentration, crude oil-water interfacial tensions as low as 10^{-2} to 10^{-4} dyne/cm have been obtained.

Some crude oils contain organic acids that can react with an alkaline waterflood to produce a mixture of surfactants at the crude oil-water interface that reduce the interfacial tension. Significant recovery of residual oil has been reported in laboratory tests using low interfacial tension and alkaline waterfloods (Jennings *et al.* 1974; Cooke *et al.* 1974). Earlier studies of alkaline waterflooding emphasized the effects of wettability upon the recovery of residual oil rather than the effects of interfacial tension (Leach *et al.* 1962; Mungan 1966; Emery *et al.* 1970).

---

a) There are at least two types of surfactant waterfloods (Gogarty 1976).

One type employs a small volume (3 to 20% of the pore volume) of a more concentrated surfactant solution. The concentration of the surfactant solution is sufficiently high to ensure that it is miscible with the crude oil in all proportions. As a small slug of this surfactant solution moves through the porous structure mixing with and displacing the crude oil, surfactant is lost by adsorption on the rock and the solution is diluted with the connate water present in the structure. As the concentration of surfactant falls, the mixture can move from a single phase region to a multiphase region on its phase diagram. We can expect that such a miscible displacement carried out with a small slug of surfactant solution will revert to an immiscible displacement at some point in the reservoir (Healy and Reed 1974; Healy et al. 1975; Gilliland and Conley 1975).

In the type with which we are concerned here, a large volume (15 to 60% of the pore volume or more) of a dilute surfactant solution is used. The crude oil is nearly insoluble in the surfactant solution and we speak of an *immiscible* displacement.

Both types of surfactant waterfloods are normally followed by a more viscous dilute polymer waterflood, in order to ensure stability of the displacement on a macroscopic scale.

This effect of interfacial tension in tertiary displacement was foreseen by Dombrowski and Brownell (1954) and by Moore and Slobod (1956), who worked in the context of a secondary oil recovery.

Our objective in this and later sections (Secs. 4.1.6 through 4.1.9 and 7.1.5) is a more detailed understanding of the variables affecting the displacement of residual oil.

Analysis for single pore Let us determine for a single irregular pore partially filled with residual oil the critical value for the pressure drop across the pore, below which the residual oil cannot be displaced but instead will assume a static configuration. We will begin by calculating under static conditions the pressure drop as a function of the pore geometry, the volume of the oil segment, the position of the oil segment in the pore, the crude oil-water interfacial tension, the advancing contact angle, and the receding contact angle. The critical pressure drop is its maximum value as a function the position of the oil segment in the pore.

The real physical problem is idealized by the following assumptions.

i) The pore is axially symmetric; its radius is a sinusoidal function of axial position as shown in Figure 2-1.

ii) There is only one segment of phase 2 (oil) in the pore.

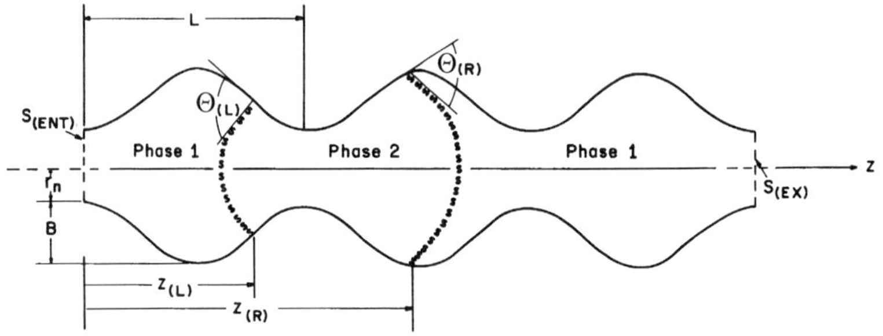

Figure 3.2.2-1: The model containing a segment of phase 2 (oil) surrounded by phase 1 (water). There is incipient displacement from left to right in the sense that $\Theta_{(L)}$ is the critical advancing contact angle and $\Theta_{(R)}$ is the critical receding contact angle.

iii) Both phase 1 (water) and phase 2 are incompressible fluids. The volumes of the various phases do not change as the segment of phase 2 assumes a sequence of static configurations in the pore.

iv) We consider here only the outer problem, and all contact angles are static contact angles as would be measured by an experimentalist (see Sec. 3.1.1). There is incipient motion from left to right in Figure 2-1, in the sense that $\Theta_{(L)} = \Theta_{(a)}$ and $\Theta_{(R)} = \Theta_{(r)}$. When the contact angle measured through phase 1 exceeds a critical value $\Theta_{(a)}$ known as the critical advancing contact angle, the common line moves and phase 1 advances. When the contact angle measured through phase 1 is less than a critical value $\Theta_{(r)}$ known as the critical receding contact angle, phase 1 recedes. For intermediate contact angles, the common line is stationary. This phenomenon is known as contact angle hysteresis.

v) The interfacial tension, the advancing contact angle, and the receding contact angle are independent of the position of the segment of phase 2.

vi) The force of gravity can be neglected with respect to the pressure drop over the pore.

With reference to Figure 2-1, the jump momentum balance applied to the advancing (left) interface requires

$$p^{(1)}_{(ent)} - p^{(2)} = - 2H^{(1,2)}\gamma \tag{2-1}$$

where $p^{(1)}_{(ent)}$ is the pressure in phase 1 at the entrance and $p^{(2)}$ is the pressure within the segment of phase 2.

Since we are neglecting the effects of gravity, $p^{(1)}_{(ent)}$ and $p^{(2)}$ are constants in the limit as this interface is approached and the mean curvature $H^{(1,2)}$ is independent of position on the interface. The rotationally symmetric interface is a segment of a sphere (Exercise 3.2.1-1). Referring to Figure 2-2, we see that

$$H^{(1,2)} = \left. \frac{\cos\beta}{r_w} \right|_{z = z_{(L)}} \tag{2-2}$$

where r_w is the local radius of the pore.

When the slope dr_w/dz of the pore wall is positive as shown in Figure 2-2A,

$$\begin{aligned} \cos\beta &= \cos(\Theta + \alpha) \\ &= (\cos\Theta - \sin\Theta\tan\alpha)\cos\alpha \end{aligned} \tag{2-3}$$

Since $dr_w/dz \geq 0$ and $\alpha \leq \pi/2$, we have

$$\cos\alpha = (1 + \tan^2\alpha)^{-1/2} \tag{2-4}$$

and

$$\cos\beta = \frac{(\cos\Theta - \sin\Theta\tan\alpha)}{(1 + \tan^2\alpha)^{1/2}} \tag{2-5}$$

When the slope of the pore wall is negative as shown in Figure 2-2B,

$$\cos \beta = \cos (\Theta + \alpha - \pi)$$

$$= - \cos (\Theta + \alpha)$$

$$= - (\cos \Theta - \sin \alpha \tan \alpha) \cos \alpha \tag{2-6}$$

Since $dr_w/dz < 0$ and $\pi/2 < \alpha < \pi$, we can say

$$\cos \alpha = - (1 + \tan^2 \alpha)^{-1/2} \tag{2-7}$$

and (2-5) again applies.

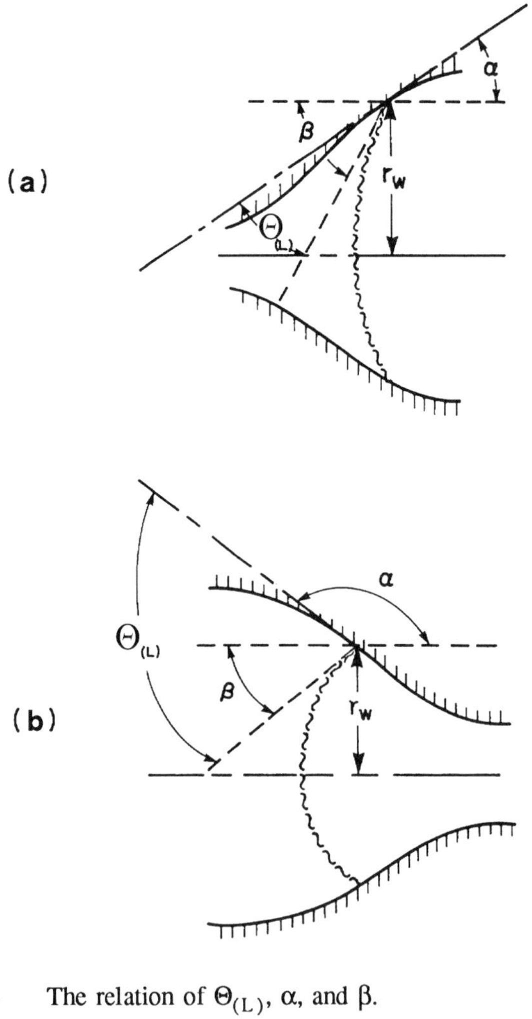

Figure 3.2.2-2 The relation of $\Theta_{(L)}$, α, and β.

From Figure 2-1, we see that the local radius r_w of the pore can be expressed as a function of axial position z:

$$r_w = r_n + \frac{B}{2}\left[1 - \cos\left(\frac{2\pi z}{L}\right)\right] \tag{2-8}$$

Therefore,

$$\tan\alpha = \frac{dr_w}{dz}$$

$$= \frac{\pi B}{L}\sin\left(\frac{2\pi z}{L}\right) \tag{2-9}$$

From (2-2), (2-5), and (2-9), it follows that

$$H^{(1,2)} = \left\{\cos\Theta_{(L)} - \left(\frac{B\pi}{L}\right)\sin\Theta_{(L)}\sin\left(\frac{2\pi z_{(L)}}{L}\right)\right\}$$

$$\left\{r_n + \frac{B}{2}\left[1 - \cos\left(\frac{2\pi z_{(L)}}{L}\right)\right]\right\}^{-1} \tag{2-10}$$

$$\left\{1 + \left(\frac{B\pi}{L}\right)^2\sin^2\left(\frac{2\pi z_{(L)}}{L}\right)\right\}^{-1/2} \tag{2-10}$$

In the same manner, the jump momentum balance applied to the receding (right) interface in Figure 2-1 requires

$$p^{(2)} - p^{(1)}_{(ex)} = -2H^{(2,1)}\gamma \tag{2-11}$$

with the understanding that $p^{(1)}_{(ex)}$ is the pressure in the water downstream of the segment of phase 2. For this interface, it is necessary only to substitute in (2-10) $(\pi - \Theta_{(R)})$ for $\Theta_{(L)}$ and $z_{(R)}$ for $z_{(L)}$ in order to obtain

$$H^{(2,1)} = -\left\{\cos\Theta_{(R)} - \left(\frac{B\pi}{L}\right)\sin\Theta_{(R)}\sin\left(\frac{2\pi z_{(R)}}{L}\right)\right\}$$

$$\left\{r_n + \frac{B}{2}\left[1 - \cos\left(\frac{2\pi z_{(R)}}{L}\right)\right]\right\}^{-1}$$

$$\left\{ 1 + \left(\frac{B\pi}{L} \right)^2 \sin^2 \left(\frac{2\pi z_{(R)}}{L} \right) \right\}^{-1/2} \tag{2-12}$$

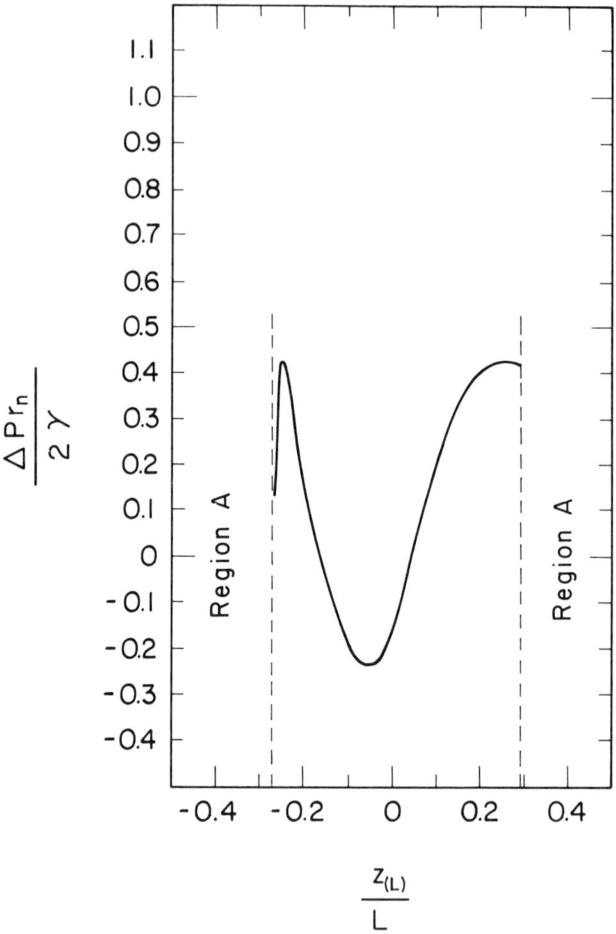

Figure 3.2.2-3: The pressure drop required to hold a segment of residual oil in a static configuration as a function of the position of its advancing (left) interface in Figure 2-1 for typical water-wet behavior. In particular, $\Theta_{(L)} = 40°$, $\Theta_{(R)} = 4°$, $B/r_n = 1.5$, $L/r_n = 6$, and $V^{(2)} = 80\ r_n^3$. In region A, the receding interface can not touch the pore wall, because of the discontinuous movement illustrated in Figure 2-6C.

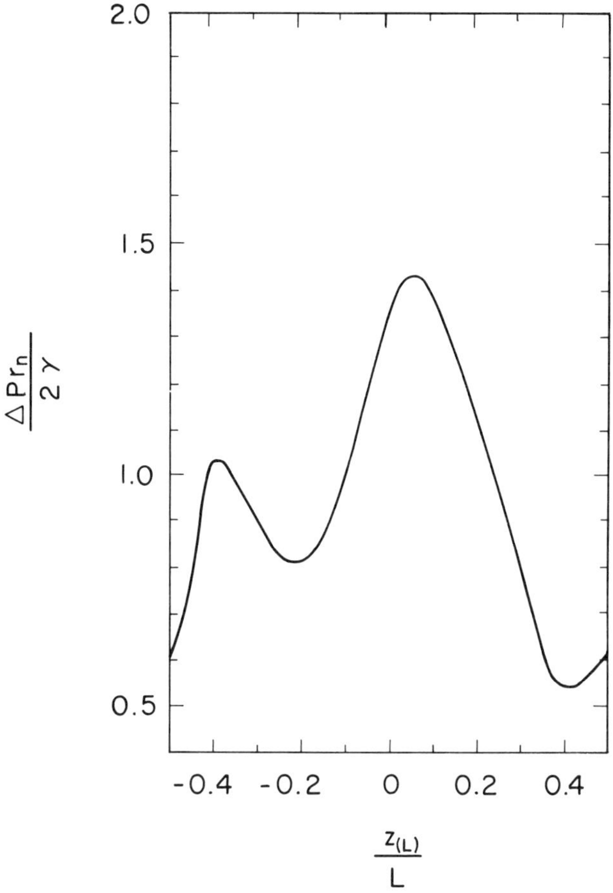

Figure 3.2.2-4: The pressure drop required to hold a segment of residual oil in a static configuration as a function of the position of its advancing (left) interface in Figure 2-1 for typical intermediate wettability behavior. In particular, $\Theta_{(L)} = 140°$, $\Theta_{(R)} = 38°$, $B/r_n = 1.5$, $L/r_n = 6$, and $V^{(2)} = 80\ r_n^3$.

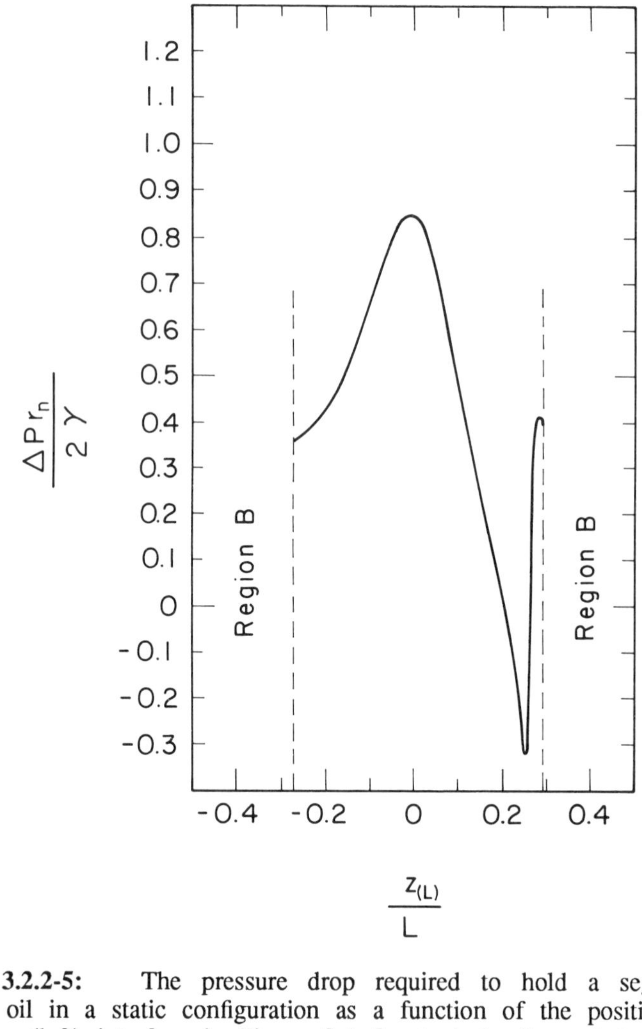

Figure 3.2.2-5: The pressure drop required to hold a segment of residual oil in a static configuration as a function of the position of its advancing (left) interface in Figure 2-1 for typical oil-wet behavior. In particular, $\Theta_{(L)} = 175°$, $\Theta_{(R)} = 130°$, $B/r_n = 1.5$, $L/r_n = 6$, and $V^{(2)} = 80\ r_n^3$. In region B, the receding interface can not touch the pore wall because of the discontinuous movement shown in Figure 2-8C.

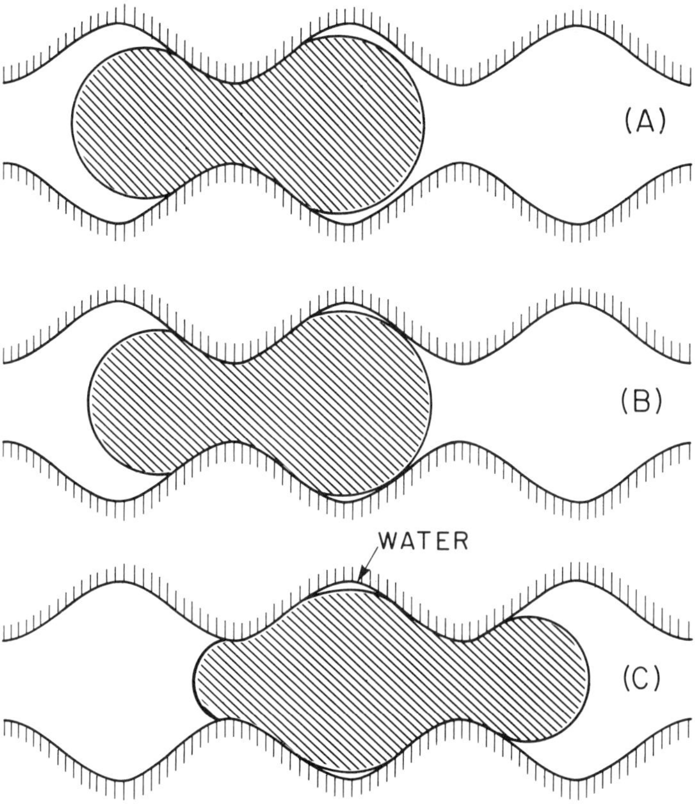

Figure 3.2.2-6: Sequence of static configurations as an oil segment moves through a water-wet pore over a ring of water trapped on the wall of the pore.

Adding (2-1) and (2-11), we find

$$\frac{\Delta p r_n}{2\gamma} = - r_n (H^{(1,2)} + H^{(2,1)}) \tag{2-13}$$

after having introduced

$$\Delta p \equiv p^{(1)}_{(ent)} - p^{(1)}_{(ex)} \tag{2-14}$$

as the pressure drop over the oil segment.

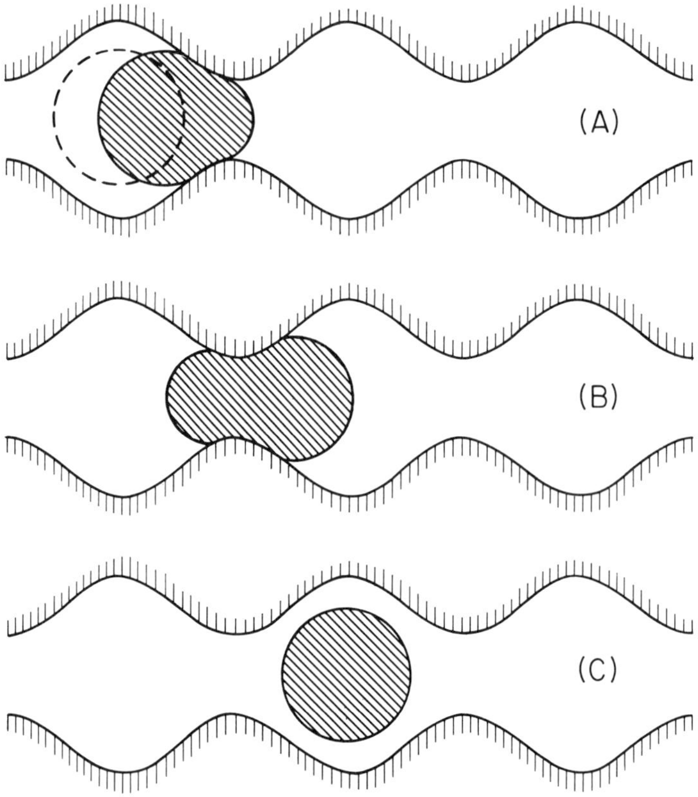

Figure 3.2.2-7: A small oil segment may be trapped in the form of a spherical drop as it is displaced through a water-wet pore.

For given values of $\Theta_{(L)}$, $\Theta_{(R)}$, the volume $V^{(2)}$ of the segment of phase 2, and the pore dimensions r_n and B, we can calculate from (2-13) $\Delta p r_n / 2\gamma$ as a function of $z_{(L)}/L$. Examples of these computations are shown in Figures 2-3 through 2-5. In each case, the maximum value of $\Delta p r_n / 2\gamma$ is the critical value below which the segment of phase 2 can assume a static configuration and will not be displaced from the pore.

Discontinuous movements in single pore In carrying out these computations, we observed that some oil globules, either because of their size or because of the magnitudes of their advancing and receding contact angles, cannot assume a static configuration at every point in the pore. If we think of an oil segment slowly being displaced through a sequence of

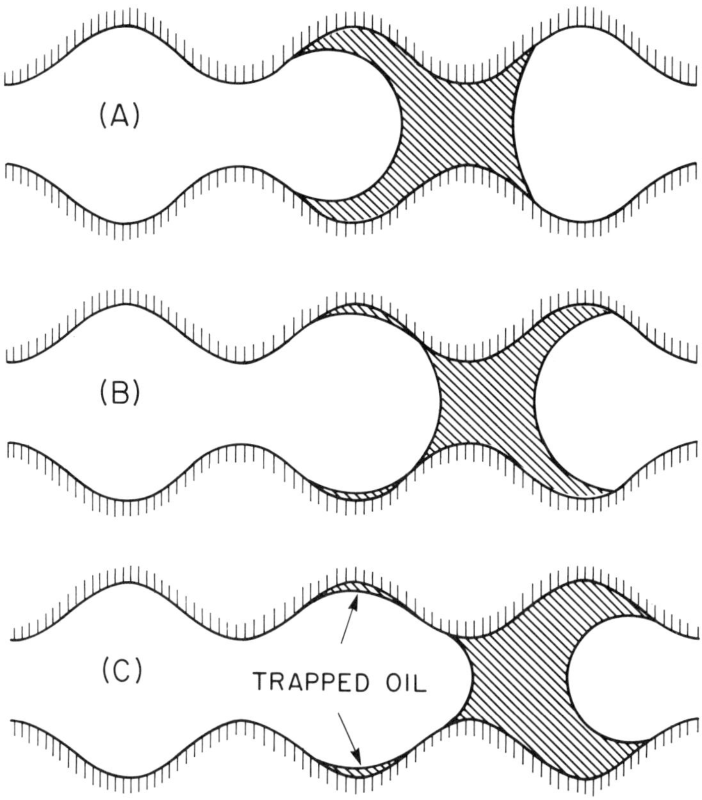

Figure 3.2.2-8: Sequence of static configurations as an oil segment moves through an oil-wet pore, leaving a portion of itself behind trapped on the wall.

static configurations, at some point it might appear to move discontinuously from one configuration to another, because intermediate static configurations fail to exist. Examples of the manner in which discontinuous movement can occur are shown in Figures 2-6 through 2-9.

Figure 2-6 pictures an oil segment moving from left to right under water-wet conditions. Notice that there will be a discontinuity in its movement as it passes over the trapped ring of water. This form of discontinuous movement was observed in constructing Figure 2-3.

In Figure 2-7, a smaller oil segment is isolated as a spherical drop at one point in a displacement under water-wet conditions. Very small drops that have been isolated in this manner are immediately displaced through the neck of the pore. Larger isolated drops remain trapped.

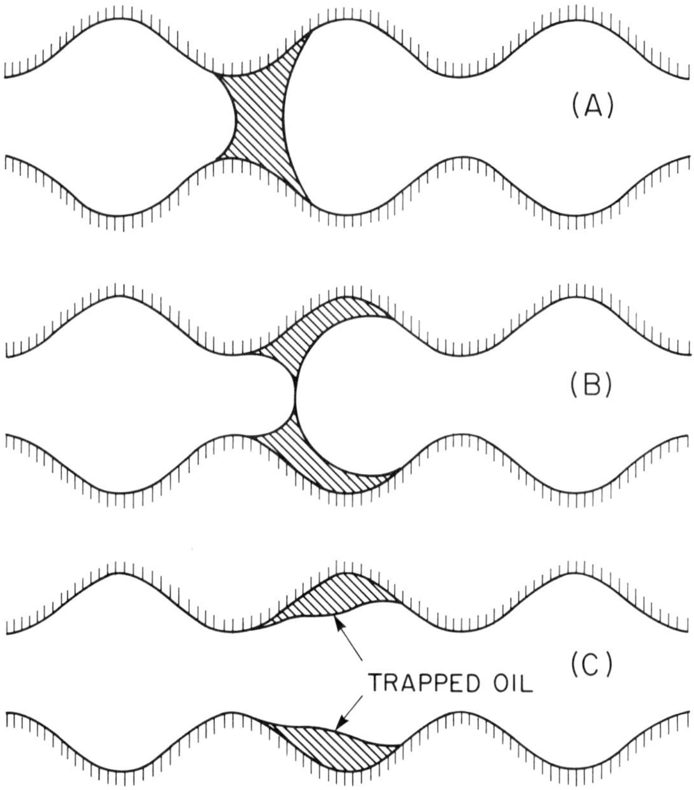

Figure 3.2.2-9: A smaller oil segment can not assume a static configuration at all positions in an oil-wet pore. It will be trapped on the wall of the pore.

Large and very small oil segments will be efficiently displaced under water-wet conditions. The remobilization of those small oil segments left behind, trapped as isolated spherical drops, depends upon other considerations. If the pressure drop over the pore is sufficiently large, the thin aqueous film separating the leading surface of the drop from the pore wall may rupture, allowing the drop to return to the configuration shown in Figure 2-7A. Alternatively, if the aqueous film is stable, the drop may deform under a sufficiently large pressure drop and squeeze through the neck of the pore, never touching the pore wall. Fortunately, even if the pressure drop over the pore is not sufficiently large to bring either of these mechanisms into play, larger oil segments being swept downstream can

Table 3.2.2-1: Maximum value of $\Delta p r_n/2\gamma$ as a function of $z_{(L)}/L$ for several values of $V^{(2)}/r_n^3$ assuming a typical water-wet behavior and Morrow's (1975) class III contact angle hysteresis. In particular, $\Theta_E = \Theta_{(L)} = 40°$, $\Theta_{(R)} = 4°$, $B/r_n = 1.5$, and $L/r_n = 6$.

$V^{(2)}/r_n^3$	$(\Delta p r_n/2\gamma)_{max}$
46.00*	0.323
50	0.676
56.48	0.884**
60	0.755
70	0.523
80	0.431
90	0.376
100	0.339
105.42***	0.323

* V_D/r_n^3
** maximum value of $(\Delta p r_n/2\gamma)_{max}$
*** $(V_D + V_{UP} - V_{CW})/r_n^3$

coalesce with the smaller isolated spherical drops for an efficient displacement. This suggests that under water-wet conditions everything possible should be done to promote coalescence.

Figure 2-8 shows a residual oil globule in a sequence of static configurations as it advances from left to right through an oil-wet pore. It appears that some residual oil will be left behind and that the volume of the initial oil segment will be reduced as it moves through the pore. A discontinuous movement of this kind was found in carrying out the computations for Figure 2-5.

Figure 2-9 indicates what can happen to a smaller oil segment under oil-wet conditions. It is so small that it cannot maintain a static configuration at all positions in the pore. It must rupture, again trapping oil on the boundary of the pore.

We can think of oil being displaced under oil-wet conditions by alternately being spread over the pore walls and then remobilized. Oil-wet conditions appear to be unfavorable for efficient displacement of residual oil.

Trapping of residual oil by these discontinuous movements can be eliminated by carrying out the displacement under intermediate wettability conditions (see for example Figure 2-4).

Table 3.2.2-2: Maximum value of $\Delta p r_n / 2\gamma$ as a function of $z_{(L)}/L$ for several values of $V^{(2)}/r_n^3$ assuming a typical intermediate wettability behavior and Morrow's (1975) class III contact angle hysteresis. In particular, $\Theta_E = 89°$, $\Theta_{(L)} = 140°$, $\Theta_{(R)} = 38°$, $B/r_n = 1.5$, and $L/r_n = 6$.

$V^{(2)}/r_n^3$	$(\Delta p r_n / 2\gamma)_{max}$
10.16*	1.52
20	1.41
30	1.34
40	1.38
50	1.48
60	1.69
62.91	1.75**
70	1.59
73.19***	1.52

* V_R / r_n^3
** maximum value of $(\Delta p r_n / 2\gamma)_{max}$
*** $(V_R + V_{UP})/r_n^3$

Table 3.2.2-3: Maximum value of $\Delta p r_n / 2\gamma$ as a function of $z_{(L)}/L$ for several values of $V^{(2)}/r_n^3$ assuming a typical oil-wet behavior and Morrow's (1975) class III contact angle hysteresis. In particular, $\Theta_E = 137°$, $\Theta_{(L)} = 175°$, $\Theta_{(R)} = 130°$, $B/r_n = 1.5$, and $L/r_n = 6$.

$V^{(2)}/r_n^3$	$(\Delta p r_n / 2\gamma)_{max}$
32.21*	0.557
40	0.513
50	0.594
60	0.776
68.88	0.979**
70	0.978
80	0.853
90	0.604
95.24***	0.557

* V_R / r_n^3
** maximum value of $(\Delta p r_n / 2\gamma)_{max}$
*** $(V_R + V_{UP})/r_n^3$

Effect of oil segment volume The critical value of $\Delta p r_n/2\gamma$ required to recover a residual oil globule trapped in a single pore is $(\Delta p r_n/2\gamma)_{max}$, the maximum value of $\Delta p r_n/2\gamma$ as a function of $z_{(L)}/L$ necessary to hold the oil in a static configuration with incipient motion to the right in Figure 2-1. For given contact angles and for a given pore geometry, $(\Delta p r_n/2\gamma)_{max}$ is a function of the volume of the oil segment. The functional dependence of $(\Delta p r_n/2\gamma)_{max}$ upon oil volume will not be smooth because of discontinuous movements.

Table 2-1 presents $(\Delta p r_n/2\gamma)_{max}$ as a function of dimensionless oil volume $V^{(2)}/r_n^3$ for a typical water-wet case, assuming that an oil segment whose volume is less than V_D will not be displaced, at least not in the manner described (see ***Discontinuous movements...***). Here V_D is the maximum volume of an oil segment that becomes a drop at some point in the displacement (see Figure 2-7). For oil volumes larger than $(V_D + V_{UP} - V_{CW})$, $(\Delta p r_n/2\gamma)_{max}$ is a periodic function of oil volume with a period $(V_{UP} - V_{CW})$. By V_{UP} we mean the volume of a unit pore, the volume between $z_{(L)}/L = 0$ and 1 in Figure 2-1; V_{CW} is the volume of the connate water in a unit pore (see Figure 2-6C).

For intermediate wettabilities, an oil segment whose volume is less than either V_D or V_R will not be displaced in the manner described (see ***Discontinuous movements...***). By V_R, we mean the maximum volume of an oil globule that ruptures and is left trapped on the pore wall at some point in the displacement (see Figure 2-9 as well as Tables 2-2 and 2-3). Table 2-2 gives $(\Delta p r_n/2\gamma)_{max}$ as a function of dimensionless oil volume for a typical case of intermediate wettability in which V_R happens to be larger than V_D. For the advancing and receding contact angles specified, there is no connate water trapped during the displacment (see Figure 2-6C). For oil volumes larger than $(V_R + V_{UP})$, $(\Delta p r_n/2\gamma)_{max}$ is a periodic function of oil volume with a period V_{UP}.

Table 2-3 gives $(\Delta p r_n/2\gamma)_{max}$ as a function of dimensionless oil volume for typical oil-wet case, assuming that an oil segment whose volume is less than V_R will not be displaced in the manner described (see ***Discontinuous movements...***). For oil volumes larger than $(V_R + V_{UP})$, $(\Delta p r_n/2\gamma)_{max}$ is a periodic function of oil volume with a period V_{UP}.

Effect of wettability Morrow (1975) studied contact angle hysteresis on polytetrafluorethylene (PTFE) surfaces. He notes very little hysteresis in the equilibrium contact angle Θ_E observed on specially prepared smooth PTFE surfaces. He was able to establish an experimental relationship (his class III contact angle hysteresis) between Θ_E and the advancing and receding contact angles observed on roughened PTFE surfaces. Advancing and receding contact angles at these rough surfaces agreed well with the apparent advancing and receding contact angles obtained from capillary pressure data in six different types of consolidated

Table 3.2.2-4: Maximum values of $\Delta p r_n/2\gamma$ as a function of $z_{(L)}/L$ for values of Θ_E assuming Morrow's (1975) class III contact angle hysteresis. In particular, $B/r_n = 1.5$, $L/r_n = 6$, and $V^{(2)}/r_n^3 = 80$.

Θ_E^* (degrees)	$\Theta_{(L)}^*$ (degrees)	$\Theta_{(R)}^*$ (degrees)	$(\Delta p r_n/2\gamma)_{max}$
24	10	0	0.381
30	20	2	0.387
40	40	4	0.431
49	60	7	0.637
59	80	11	0.870
69	100	14	1.11
79	120	21	1.31
89	140	38	1.44
100	160	64	1.42
117	170	90	1.25
137	175	130	0.853
150	178	160	0.544
156	180	170	0.534

* Θ_E : Equilibrium contact angle measured through the aqueous phase.

$\Theta_{(L)}$: Advancing contact angle measured through the aqueous phase in Figure 2-1.

$\Theta_{(R)}$: Receding contact angle measured through the aqueous phase in Figure 2-1.

PTFE porous media (Morrow 1976). In the absence of better information, we assume here that Morrow's (1975) class III contact angle hysteresis is applicable.

Table 2-4 presents the maximum value of $(\Delta p r_n/2\gamma)_{max}$ as a function of $z_{(L)}/L$ for several values of the equilibrium contact angle Θ_E. Table 2-5 gives the maximum value of $(\Delta p r_n/2\gamma)_{max}$ as a function of both $z_{(L)}/L$ and $V^{(2)}/r_n^3$ for these same values of Θ_E.

For efficient recovery of residual oil, $(\Delta p r_n/2\gamma)_{max}$ should be as small as possible. For a given pressure drop across a globule of residual oil, there is a critical interfacial tension above which the globule cannot be displaced but will instead assume a static configuration in a pore. This critical interfacial tension should be as large as possible for efficient operation of a low interfacial tension waterflood.

Table 3.2.2-5: Maximum values of $\Delta p r_n/2\gamma$ as a function of both $z_{(L)}/L$ and $V^{(2)}/r_n^3$ for several values of Θ_E assuming Morrow's (1975) class III contact angle hysteresis. In particular, $B/r_n = 1.5$ and $L/r_n = 6$.

Θ_E^* (degrees)	$\Theta_{(L)}^*$ (degrees)	$\Theta_{(R)}^*$ (degrees)	$(\Delta p r_n/2\gamma)_{max}$
24	10	0	0.659
30	20	2	0.725
40	40	4	0.884
49	60	7	1.08
59	80	11	1.28
69	100	14	1.50
79	120	21	1.68
89	140	38	1.75
100	160	64	1.64
117	170	90	1.40
137	175	130	0.979
150	178	160	0.721
156	180	170	0.664

* Θ_E : Equilibrium contact angle measured through the aqueous phase.

$\Theta_{(L)}$: Advancing contact angle measured through the aqueous phase in Figure 2-1.

$\Theta_{(R)}$: Receding contact angle measured through the aqueous phase in Figure 2-1.

Tables 2-4 and 2-5 suggest that, when one considers only the effect of wetting, water-wet behavior may be preferable for tertiary oil recovery. But the advantage of water-wet behavior seen from this point of view is not large. It is difficult to compare with the advantage that trapping by discontinuous movements can be eliminated with an intermediate wettability condition.

The definition of the state of wetting behavior is arbitrary, especially when contact angle hysteresis exists (Treiber *et al.* 1972; Melrose and Brandner 1974; Morrow 1976). In what follows, an equilibrium contact angle of 40° measured through the aqueous phase [with a critical advancing contact angle of 40° and a critical receding contact angle of 4° (Morrow 1975)] is selected as typical of water-wet behavior; an equilibrium contact angle of 89° [with a critical advancing contact angle of 140° and a critical

Table 3.2.2-6: Maximum value of $\Delta p r_n / 2\gamma$ as a function of both $z_{(L)}/L$ and $V^{(2)}/r_n^3$ for several values of B/r_n and for three typical wettability conditions assuming Morrow's (1975) class III contact angle hysteresis. In particular, $L/r_n = 6$.

	$(\Delta p r_n / 2\gamma)_{max}$		
B/r_n	$\Theta_E = 40°$ $\Theta_{(L)} = 40°$ $\Theta_{(R)} = 4°$ (water-wet)	$\Theta_E = 89°$ $\Theta_{(L)} = 140°$ $\Theta_{(R)} = 38°$ (intermediate)	$\Theta_E = 137°$ $\Theta_{(L)} = 175°$ $\Theta_{(R)} = 130°$ (oil-wet)
0.0*	0.232	1.55	0.353
0.5	0.558	1.65	0.667
1.0	0.756	1.71	0.859
1.5	0.884	1.75	0.979
2.0	0.957	1.78	1.05
2.5	1.00	1.80	1.10
3.0	1.03	1.82	1.14

* Straight cylinder

receding contact angle of 38° (Morrow 1975)] as typical of intermediate wettability; an equilibrium contact angle of 137° [with a critical advancing contact angle of 175° and a critical receding contact angle of 130° (Morrow 1975)] as typical of oil-wet behavior.

Effect of pore geometry Table 2-6 shows $(\Delta p r_n / 2\gamma)_{max}$ as a function of both $z_{(L)}/L$ and $V^{(2)}/r_n^3$ for several values of B/r_n and for three typical wettability conditions. With reference to Figure 2-1, these results show that, as the ratio of the bulge radius $(B + r_n)$ to the neck radius r_n increases, $(\Delta p r_n / 2\gamma)_{max}$ also increases.

The effect of changing the pore wettability is more significant for small values of B/r_n. This suggests that the alteration of wettability may be effective for a low interfacial tension waterflow in a sandstone, which would typically be characterized by a small value of B/r_n (Dullien and Batra 1970; Dullien *et al.* 1972; Batra and Dullien 1973).

Table 3.2.2-7: Maximum value of $\Delta p r_n / 2\gamma$ as a function of both $z_{(L)}/L$ and $V^{(2)}/r_n^3$ for several values of B/r_n and for three typical wettability conditions, assuming no contact angle hysteresis. In particular, $L/r_n = 6$.

	$(\Delta p r_n / 2\gamma)_{max}$		
B/r_n	$\Theta_E = \Theta_{(L)} = \Theta_{(R)} = 40°$ (water-wet)	$\Theta_E = \Theta_{(L)} = \Theta_{(R)} = 89°$ (intermediate)	$\Theta_E = \Theta_{(L)} = \Theta_{(R)} = 137°$ (oil-wet)
0.0*	0.0	0.0	0.0
0.5	0.376	0.414	0.380
1.0	0.606	0.666	0.613
1.5	0.756	0.820	0.764
2.0	0.853	0.923	0.861
2.5	0.917	0.997	0.925
3.0	0.961	1.06	0.970

* Straight cylinder

Effect of contact angle hysteresis Table 2-7 is prepared in the same manner as Table 2-6, with the exception that contact angle hysteresis is eliminated.

Notice that, when contact angle hysteresis is eliminated, the advantage of water-wet behavior is lost for a given permeable structure. This supports Morrow's (1976) observation that, for a given permeable structure, it is more important to account for contact angle hysteresis than for the details of the pore geometry when considering the effects of wettability alteration.

However, pore geometry is important. For a given wettability condition, $(\Delta p r_n / 2\gamma)_{max}$ increases as the ratio of the bulge radius $(B + r_n)$ to the neck radius r_n increases, whether or not there is contact angle hysteresis.

References

Batra, V. K., and F. A. L. Dullien, *Soc. Pet. Eng. J.* **13**, 256 (1973).

Cooke, C. E., Jr., R. E. Williams, and P. A. Kolodzie, *J. Pet. Technol.* **26**, 1365 (1974).

Dombrowski, H. S., and L. E. Brownell, *Ind. Eng. Chem.* **46**, 1207 (1954).

Dullien, F. A. L., and V. K. Batra, *Ind. Eng. Chem.* **62** (10), 25 (1970).

Dullien, F. A. L., G. K. Dhawan, N. Gurak, and L. Babjak, *Soc. Pet. Eng. J.* **12**, 289 (1972).

Emery, L. W., N. Mungan, and R. W. Nicholson, *J. Pet. Technol.* **22**, 1569 (1970).

Foster, W. R., *J. Pet. Technol.* **25**, 205 (1973).

Gale, W. W., and E. I. Sanvik, *Soc. Pet. Eng. J.* **13**, 191 (1973).

Geffen, T. M., *Oil Gas J.* **66** (May 7, 1973).

Gilliland, H. E., and F. R. Conley, *Proc. 9th World Pet. Congr.* **4**, 259 (1975).

Gogarty, W. B., *J. Pet. Technol.* **28**, 93 (1976).

Healy, R. N., and R. L. Reed, *Soc. Pet. Eng. J.* **14**, 491 (1974).

Healy, R. N., R. L. Reed, and C. W. Carpenter Jr., *Soc. Pet. Eng. J.* **15**, 87 (1975).

Hill, H. J., J. Reisberg, and G. L. Stegemeier, *J. Pet. Technol.* **25**, 186 (1973).

Jennings, H. Y., Jr., C. E. Johnson Jr., and C. D. McAuliffe, *J. Pet. Technol.* **26**, 1344 (1974).

Leach R. O., O. R. Wagner, and H. W. Wood, *J. Pet. Technol.* **14**, 206 (1962).

Melrose, J. C., and C. F. Brandner, *J. Can. Pet. Technol.* **54** (Oct.-Dec. 1974).

Moore, T. F., and R. L. Slobod, *Prod. Monthly*, 20 (August 1956).

Morrow, N. R., *J. Can. Pet. Technol.* **14**, 42 (1975).

Morrow, N. R., *J. Can. Pet. Technol.* **15**, 49 (1976).

Mungan, N., *J. Pet. Technol.* **18**, 247 (1966).

Oh, S. G., and J. C. Slattery, *Soc. Pet. Eng. J.* **19**, 83 (1979).

Treiber, L. E., D. L. Archer, and W. W. Owens, *Soc. Pet. Eng. J.* **12**, 531 (1972).

3.2.3 Spinning drop interfacial tensiometer (Slattery and Chen 1978) Although not restricted to this limit, only three experiments have been demonstrated as being suitable for measuring ultra low interfacial tensions (less than 10^{-2} dyne/cm): the spinning drop (Vonnegut 1942; Princen *et al.* 1967; Ryden and Albertsson 1971; Cayias *et al.* 1975; Manning and Scriven 1977), the micropendant drop (Hill *et al.* 1973), and the sessile drop (Wilson *et al.* 1976).

The spinning drop experiment sketched in Figure 3-1 is perhaps the simplest to construct and to operate. At steady state, the lighter drop phase (A), the heavier continuous phase (B), and the tube all rotate as rigid bodies with the same angular velocity Ω.

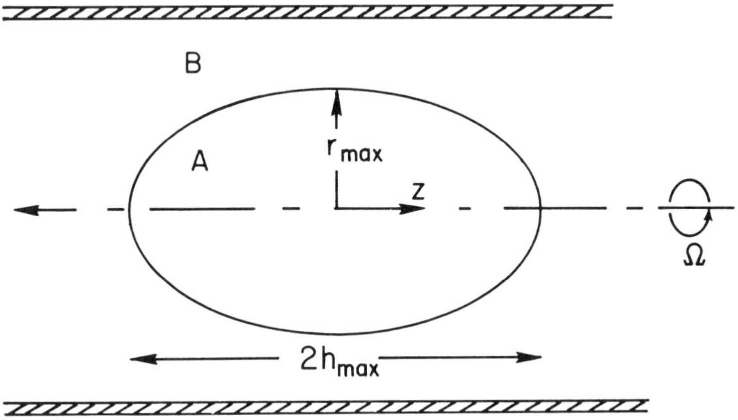

Figure 3.2.3-1: At steady state, the drop phase A, the continuous phase B, and the tube all rotate as rigid bodies with the same angular velocity Ω.

Although the analysis for the spinning drop presented by Princen *et al.* (1967) and by Cayias *et al.* (1975) is correct, it is inconvenient to use because their results are given in terms of the radius of a sphere having the same volume as the drop. An experimentalist normally measures the maximum diameter ($2r_{max}$ after correction for refraction) and the length ($2h_{max}$) of the drop.

The revised solution that follows avoids this difficulty. It is more than a simple rearrangement of their result, since the radius of curvature of the drop at its axis of rotation (*a* in their notation) is treated in their development as though it were known.

From Cauchy's first law for each phase (i = A, B)

$$\mathcal{P}^{(i)} - \mathcal{P}_o^{(i)} = \frac{1}{2}\rho^{(i)}\Omega^2 r^2 \tag{3-1}$$

where $\mathcal{P}_o^{(i)}$ is the modified pressure of phase i on the axis of the cylinder. The jump momentum balance requires

$$\mathcal{P}^{(A)} - \mathcal{P}^{(B)} = -2H\gamma \tag{3-2}$$

where we have assumed that the Bond number

$$N_{Bo} \equiv \frac{(\rho^{(B)} - \rho^{(A)})g r_o^2}{\gamma}$$

$$\ll 1 \tag{3-3}$$

Here H is the mean curvature of the interface, γ is the interfacial tension, and r_o is a characteristic length of the system that will be defined shortly. If we describe the configuration of the interface by

$$z = h(r) \tag{3-4}$$

the mean curvature takes the form (Table 2.4.2-7)

$$H = \frac{1}{2r}\frac{d}{dr}\left\{ r\frac{dh}{dr}\left[1 + \left(\frac{dh}{dr}\right)^2\right]^{-1/2}\right\} \tag{3-5}$$

It is convenient to introduce the change of variable (Princen *et al.* 1967)

$$\tan\theta = -\frac{dh}{dr} \tag{3-6}$$

in terms of which (3-5) becomes

$$-2H = \frac{1}{r}\frac{d}{dr}(r\sin\theta) \tag{3-7}$$

Given (3-1) and (3-7), we can rearrange (3-2) as

$$\frac{1}{r^*}\frac{d}{dr^*}(\, r^* \sin\theta \,) = A - r^{*2} \tag{3-8}$$

where we have found it convenient to define

$$r^* \equiv \frac{r}{r_o} \tag{3-9}$$

$$A \equiv \frac{1}{\gamma}(\, \mathcal{P}_o^{(A)} - \mathcal{P}_o^{(B)} \,)\, r_o \tag{3-10}$$

and

$$r_o \equiv \left(\frac{(\rho^{(B)} - \rho^{(A)})\Omega^2}{2\gamma} \right)^{-1/3} \tag{3-11}$$

Equation (3-8) can be integrated once consistent with the boundary conditions

$$\text{at } r^* = 0 : \ \theta = 0 \tag{3-12}$$

and

$$\text{at } r^* = r^*_{max} : \ \theta = \frac{\pi}{2} \tag{3-13}$$

to find

$$\sin\theta = A\,\frac{r^*}{2} - \frac{r^{*3}}{4} \tag{3-14}$$

with the requirement

$$A = \frac{2}{r^*_{max}} + \frac{r^{*\,2}_{max}}{2} \tag{3-15}$$

We can now visualize integrating (3-6) to learn

$$h^*_{max} = \int_0^{r^*_{max}} \frac{\sin\theta \, dr^*}{(1 - \sin^2\theta)^{1/2}} \tag{3-16}$$

in which $\sin\theta$ is given by (3-14) and h^* is defined by analogy with (3-9). With the change of variable

$$q \equiv \frac{A}{2} - \frac{r^{*2}}{4} \tag{3-17}$$

(3-16) can be written as

$$h^*_{max} = \int_{q_1}^{A/2} \frac{q \, dq}{[(q - q_1)(q - q_2)(q - q_3)]^{1/2}} \tag{3-18}$$

with

$$q_1 = \frac{1}{r^*_{max}} \tag{3-19}$$

$$q_2 = \frac{r^{*2}_{max}}{8} + \frac{1}{2}\left(\frac{r^{*4}_{max}}{16} + r^*_{max}\right)^{1/2} \tag{3-20}$$

$$q_3 = \frac{r^{*2}_{max}}{8} - \frac{1}{2}\left(\frac{r^{*4}_{max}}{16} + r^*_{max}\right)^{1/2} \tag{3-21}$$

Let us consider two limiting cases. For a cylindrical drop

$$\sin\theta = 1 \tag{3-22}$$

and it follows immediately from (3-8) and (3-15) that

$$r^*_{max} = 2^{1/3} \tag{3-23}$$

For a spherical drop whose dimensionless radius is R^*

$$\sin\theta = \frac{r^*}{R^*} \tag{3-24}$$

and we conclude from (3-8) and (3-15) that

Table 3.2.3-1: r^*_{max} and V^* as functions of r_{max}/h_{max} from (3-27) and (3-32)[a]

r_{max}/h_{max}	r^*_{max}	V^*
1.0	0.	0.
0.9997	0.1	0.0042
0.9980	0.2	0.0336
0.9932	0.3	0.1139
0.9840	0.4	0.2725
0.9687	0.5	0.5406
0.9459	0.6	0.9571
0.9140	0.7	1.5745
0.8710	0.8	2.4714
0.8148	0.9	3.7782
0.7415	1.0	5.7457
0.6432	1.1	8.9812
0.4928	1.2	15.9687
0.3332	1.25	29.0379
0.3268	1.251	29.8101
0.3198	1.252	30.6741
0.3122	1.253	31.6546
0.3038	1.254	32.7883
0.2945	1.255	34.1321
0.2837	1.256	35.7818
0.2708	1.257	37.9195
0.2543	1.258	40.9617
0.2297	1.259	46.2969
0.2262	1.2591	47.1309
0.2225	1.2592	48.0733
0.2183	1.2593	49.1567
0.2136	1.2594	50.4307
0.2081	1.2595	51.9769
0.2016	1.2596	53.9444
0.1932	1.2597	56.6522

a) This table, a minor revision of that presented by Slattery and Chen (1978), is due to Chen (1979), who gives the computer program as well as a discussion of the truncation errors.

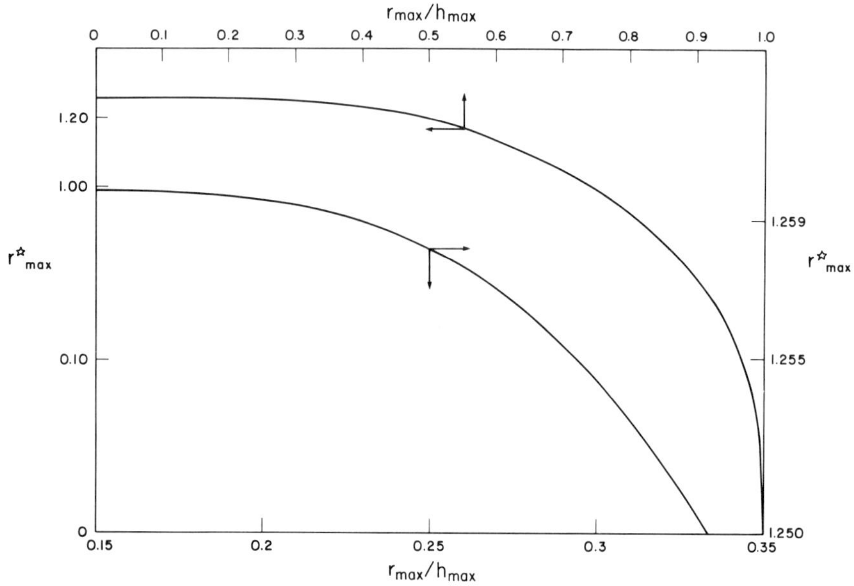

Figure 3.2.3-2: $r^{\star}_{max}$ as a function of r_{max}/h_{max} from (3-27).

$$r^{\star}_{max} = 0 \tag{3-25}$$

This suggests that we confine our attention to the range $0 \leq r^{\star}_{max} \leq 2^{1/3}$ for which

$$q_1 > q_2 > q_3 \tag{3-26}$$

With this restriction, (3-18) can be integrated to find (Gröbner and Hofreiter 1965, p. 78)

$$h^{\star}_{max} = \frac{2}{(q_1 - q_3)^{1/2}} \Big[q_1 F(\phi_o, k) - (q_1 - q_3) E(\phi_o, k) + (q_1 - q_3)(\tan \phi_o) \sqrt{a - k^2 \sin^2 \phi_o} \Big] \tag{3-27}$$

in which

$$E(\phi_o, k) \equiv \int_0^{\phi_o} \sqrt{1 - k^2 \sin^2 \psi} \, d\psi \tag{3-28}$$

$$F(\phi_o, k) \equiv \int_0^{\phi_o} \frac{d\psi}{\sqrt{1 - k^2 \sin^2\psi}} \tag{3-29}$$

$$\phi_o \equiv \arcsin\left(\frac{A - 2q_1}{A - 2q_2}\right)^{1/2} \tag{3-30}$$

$$k^2 \equiv \frac{q_2 - q_3}{q_1 - q_3} \tag{3-31}$$

Following a suggestion of Princen *et al.* (1967), the volume of the drop becomes

$$V^* \equiv \frac{V}{r_o^3} = \frac{8}{3}\pi\left[\left(\frac{1}{r^*_{max}} + \frac{r^{*2}_{max}}{4}\right)h^*_{max} - 1\right] \tag{3-32}$$

We computed $r_{max}/h_{max} = r^*_{max}/h^*_{max}$ from (3-27) and V^* from (3-32) as functions of r^*_{max}. In these computations, the incomplete elliptic integrals (3-28) and (3-29) were expanded in series (Byrd and Friedman 1971, p. 300). The results are presented in Table 3-1 and Figure 3-2 in a form convenient for experimentalists: r^*_{max} and V^* as functions of r_{max}/h_{max}.

In order to use the spinning drop technique to determine interfacial tension, one must measure both r_{max} and h_{max} for given values of $(\rho^{(B)} - \rho^{(A)})$ and Ω. Having r_{max}/h_{max}, one determines r^*_{max} from Table 3-1 or Figure 3-2. When $r_{max}/h_{max} < 0.25$, (3-23) can be used with less than 0.4% error in the interfacial tension. The interfacial tension can then be found by rearranging the definition for r^*_{max}:

$$\gamma = \frac{1}{2}\left(\frac{r_{max}}{r^*_{max}}\right)^3 (\rho^{(B)} - \rho^{(A)})\,\Omega^2 \tag{3-33}$$

Princen *et al.* (1967) and Cayias *et al.* (1975) expressed their results as functions of a dimensionless shape parameter α, which here becomes

$$\alpha = \left(\frac{2}{A}\right)^3 \tag{3-34}$$

While their results can not be readily rearranged in the form given here, the relation between α and r_{max}/h_{max} calculated from (3-34) and Table 3-1 agrees with that given by Princen *et al.* (1967).

There are several cautions to be observed in using this technique.

1) From (3-3) and (3-11),

$$N_{Bo} = \frac{2^{2/3}(\rho^{(B)} - \rho^{(A)})^{1/3} g}{\gamma^{1/3}\Omega^{4/3}}$$

$$<< 1 \tag{3-35}$$

This implies that the angular velocity Ω should be as large as possible. Unfortunately, so long as $(\rho^{(B)} - \rho^{(A)}) \neq 0$, we can approach the rigid body motion assumed here only asymptotically as $N_{Bo} \to 0$ or as $\Omega \to \infty$. For any finite value of Ω, the axis of the drop or bubble will be slightly displaced from the axis of the tube, resulting in convective mixing within the two phases (Manning and Scriven 1977). In addition to negating rigid body motion, these convection patterns have the potential of creating interfacial tension gradients in the interface leading to a loss of stability, if there is a surfactant present.

If Ω is too large, the drop or bubble phase will contact the ends of the tube, introducing end effects not taken into account in this analysis.

It may be possible to find a compromise by studying γ as a function of Ω (Currie and Van Nieuwkoop 1982).

2) Let us assume that the phases are equilibrated before they are introduced into the apparatus. As Ω begins to increase towards its final value, the drop lengthens and the area of the drop surface increases. If there is no surfactant present, the system reaches a steady state rapidly. If there is a surfactant present, the system approaches a steady state more slowly: as surfactant is adsorbed in the interface, the interfacial tension falls, the interfacial area of the drop increases allowing more surfactant to come into the interface, ...

The problem is obvious, when the phases are not pre-equilibrated (Rubin and Radke 1980; Brown and Radke 1980). But it could also be a serious problem, if the primary source of surfactant is a dilute solution forming the drop phase. If sufficient surfactant were removed from the drop phase by adsorption in the expanded interface, the equilibrium compositions of the two phases could be changed.

These may have been the conditions studied by Kovitz (1977-79). The drop phases were dilute solutions of surfactant in octane and the continuous phases were aqueous solutions of NaCl. After pre-equilibration, the drop phase remained the primary source of surfactant. He observed that days or weeks might be required to achieve an equilibrium interfacial tension with the spinning drop experiment compared with minutes or hours in static experiments such as the sessile drop and pendant drop. Frequently an equilibrium interfacial tension could never be measured with the spinning drop interfacial tensiometer, either because the drop disappeared entirely or because the drop phase evolved into a shape that was clearly not amenable to data reduction. [This may have been the result of buoyancy-induced convection becoming more important as the single-phase region was approached on the phase diagram (Manning and Scriven 1977). Lee and Tadros (1982) observed interfacial instabilities leading to disintegration of the interface and the formation of droplets that they attributed to interfacial turbulence.] When an equilibrium interfacial tension was measured, it often differed by an order of magnitude from the values found with static experiments [meniscal breakoff interfacial tensiometer (Sec. 3.2.4), du Noüy ring (Exercise 3.2.4-2), sessile drop (Sec. 3.2.6)]. In contrast, for single component systems he found excellent agreement between measurements made with the spinning drop and those made with static experiments.

3) Vibrations should be carefully reduced, since they promote the development of instabilities (Manning and Scriven 1977).

4) The spinning tube should be horizontal, in order to prevent migration of the bubble or drop to one of the ends.

In spite of all of these potential problems, the spinning drop technique has been used successfully in the measurement of interfacial (and surface) tensions having a wide range of values.

Wüstneck and Wärnheim (1988) report excellent agreement between measurements made with a du Noüy ring (Exercise 3.2.4-2) and those made with a spinning drop for an aqueous sodium dodecylsulphate solution against air (their figure 2). Their surfactant was confined to the continuous phase, eliminating any significant change in the equilibrium phase compositions.

In the experiments of Ryden and Albertsson (1971) and Torza (1975), significant amounts of surfactant were not present.

References

Brown, J. B., and C. J. Radke, *Chem. Eng. Sci.* **35**, 1458 (1980)

Byrd, P. F., and M. D. Friedman, "Handbook of Elliptic Integrals for Engineers and Scientists," second edition, Springer-Verlag, Berlin (1971).

Cayias, J. L., R. S. Schechter, and W. H. Wade, in "Adsorption at Interfaces," ACS Symposium Series no. 8, edited by K. L. Mittal, p. 234, American Chemical Society, Washington, D.C. (1975).

Chen, J. D., M. S. Thesis, Department of Chemical Engineering, Northwestern University, Evanston, IL 60208-3120 (1979).

Currie, P. K., and J. Van Nieuwkoop, *J. Colloid Interface Sci.* **87**, 301 (1982).

Gröbner, W., and N. Hofreiter, "Integraltafel", fourth edition, vol. 1,

Springer-Verlag, Wien (1965).

Hill, H. J., J. Reisberg, and G. L. Stegemeier, *J. Pet. Technol.* 186 (Feb. 1973).

Kovitz, A. A., unpublished observations, Department of Mechanical and Nuclear Engineering, Northwestern University, Evanston, IL 60208-3120 (1977-79).

Lee, G. W. J., and Th. F. Tadros, *Colloids Surfaces* **5**, 105 (1982).

Manning, C. D., and L. E. Scriven, *Rev. Sci. Instrum.* **48**, 1699 (1977).

Princen, H. M., I. Y. Z. Zia, and S. G. Mason, *J. Colloid Interface Sci.* **23**, 99 (1967).

Rubin, E., and C. J. Radke, *Chem. Eng. Sci.* **35**, 1129 (1980).

Ryden, J., and P. A. Albertsson, *J. Colloid Interface Sci.* **37**, 219 (1971).

Slattery, J. C., and J. D. Chen, *J. Colloid Interface Sci.* **64**, 371 (1978).

Torza, S., *Rev. Sci. Instrum.* **46**, 778 (1975).

Vonnegut, B., *Rev. Sci. Instrum.* **13**, 6 (1942).

Wilson, P. M., C. L. Murphy, and W. R. Foster, "The Effects of Sulfonate Molecular Weight and Salt Concentration on the Interfacial Tension of Oil-Brine-Surfactant Systems," Society of Petroleum Engineers Symposium on Improved Recovery, March 22-24, 1976.

Wüstneck, R., and T. Wärnheim, *Colloid Polym. Sci.* **266**, 926 (1988).

3.2.4 Meniscal breakoff interfacial tensiometer Figure 4-1 shows two geometries that may prove to be the basis for a useful technique for measuring interfacial tension. A circular knife-edge is positioned so as to just touch the A-B phase interface. In Figure 4-1(a), the general level of the A-B interface is slowly lowered until it breaks away from the knife-edge; in Figure 4-1(b), the A-B interface is slowly raised until breakoff occurs. From the measured value of $H = H_{max}$ at breakoff, the interfacial tension can be calculated.

Let us describe the configuration of the interface in cylindrical coordinates by

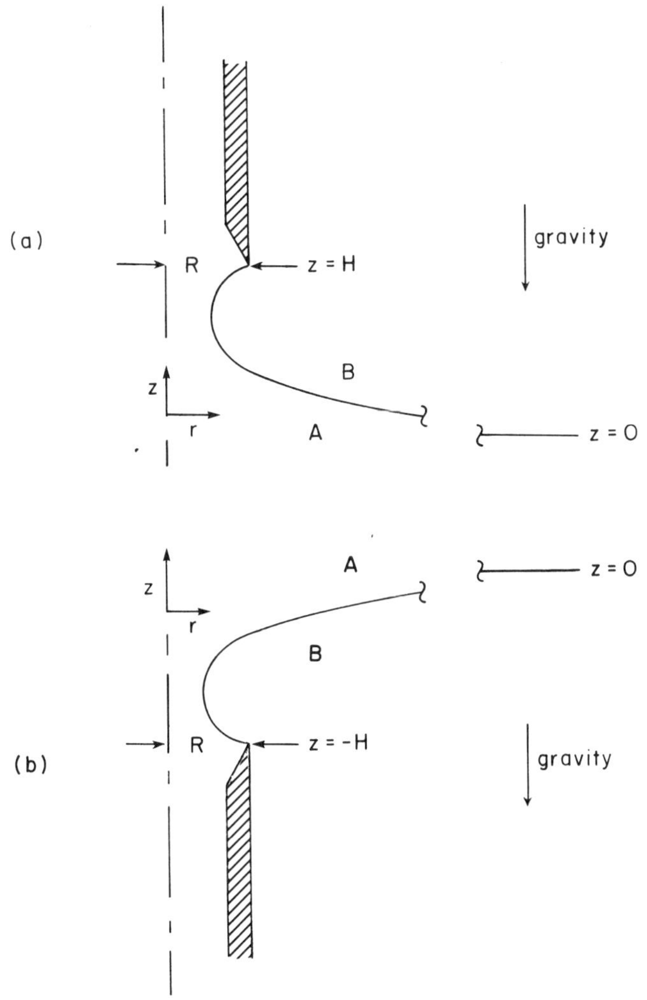

Figure 3.2.4-1: Meniscal breakoff interfacial tensiometer

(a) When lower phase preferentially wets the knife-edge, the A-B interface is slowly lowered until breakoff occurs.

(b) When upper phase preferentially wets the knife-edge, the A-B interface is slowly raised until breakoff occurs.

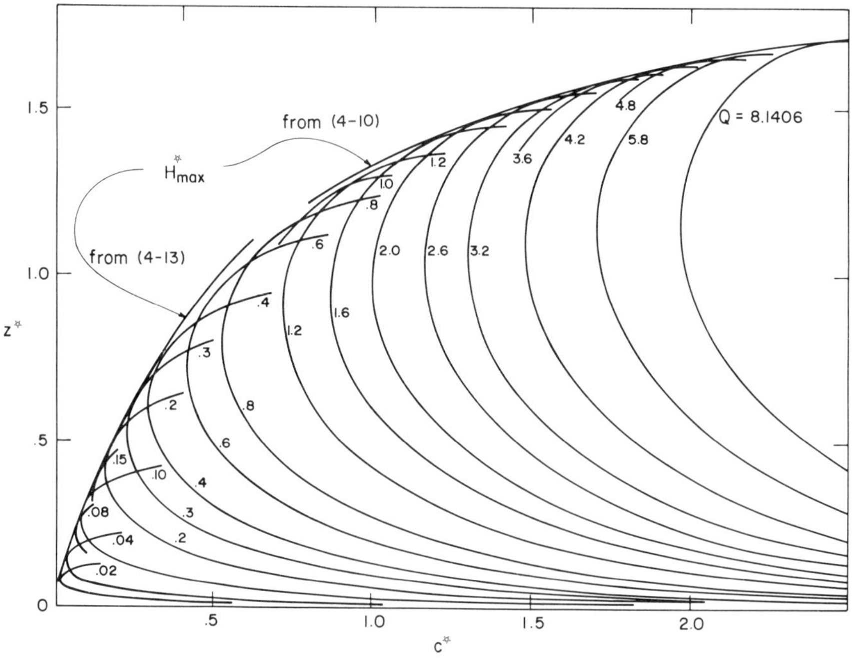

Figure 3.2.4-2: One parameter family of solutions to (4-6) consistent with boundary conditions (4-7) and (4-8) presented by Kovitz (1975). The curves are labeled by the value of an arbitrary parameter Q introduced in developing the numerical solution. The knife-edge is located at $c^* = R^*$, $z^* = H^*$. As the knife-edge is slowly raised to H^*_{max} (for fixed R^*, H^* increases), the interface assumes a sequence of static configurations. For $H^* > H^*_{max}$, no solutions exist and the interface ruptures.

$$r = c(z) \tag{4-1}$$

From either the r- or z-component of the jump momentum balance at the A-B phase interface, we have (Table 2.4.2-8)

$$p_0^{(B)} - p_0^{(A)} + (\rho^{(A)} - \rho^{(B)}) gz$$
$$= \frac{\gamma}{c} \left[c \frac{d^2 c}{dz^2} - \left(\frac{dc}{dz} \right)^2 - 1 \right] \left[1 + \left(\frac{dc}{dz} \right)^2 \right]^{-3/2} \tag{4-2}$$

Here $p_0^{(A)}$ is the pressure within phase A as the interface approached in a region sufficiently far away from the knife-edge that the curvature of the interface is zero; $p_0^{(B)}$ is the pressure within phase B as the interface is approached in the same region. Equation (4-2) requires that sufficiently far away from the knife-edge

$$\text{at } z = 0 : \; p_0^{(A)} = p_0^{(B)} \tag{4-3}$$

Equation (4-2) can be more conveniently written in terms of the dimensionless variables

$$c^* \equiv c \left(\frac{\gamma}{(\rho^{(A)} - \rho^{(B)})g} \right)^{-1/2} \tag{4-4}$$

$$z^* \equiv z \left(\frac{\gamma}{(\rho^{(A)} - \rho^{(B)})g} \right)^{-1/2} \tag{4-5}$$

as

$$z^* = \frac{1}{c^*} \left[c^* \frac{d^2 c^*}{dz^{*2}} - \left(\frac{dc^*}{dz^*} \right)^2 - 1 \right] \left[1 + \left(\frac{dc^*}{dz^*} \right)^2 \right]^{-3/2} \tag{4-6}$$

If our objective is to describe the elevated meniscus shown in Figure 4-1(a), we must solve (4-6) consistent with the boundary conditions

$$\text{as } z^* \to 0 \; : \; c^* \to \infty \tag{4-7}$$

$$\text{at } z^* = H^* : \; c^* = R^* \tag{4-8}$$

In order to describe the depressed meniscus pictured in Figure 4-1(b), we must find a solution for (4-6) consistent with (4-7) and

$$\text{at } z^* = -H^* : \; c^* = R^* \tag{4-9}$$

Notice that the elevated meniscus problem takes the same form as that for the depressed meniscus, when z is replaced by $-z$ and $(\rho^{(A)} - \rho^{(B)})$ by $(\rho^{(B)} - \rho^{(A)})$. For this reason, we will confine our attention for the moment to the elevated meniscus.

Both Padday and Pitt (1973) and Kovitz (1975) have considered this problem. Selected members of the family of meniscus profiles as calculated by Kovitz (1975) are shown in Figure 4-2; they are identified by the value of a parameter Q introduced in his numerical solution. For a selected value of R^* (a given knife-edge diameter and chemical system), the interface can assume a continuous sequence of configurations corresponding to increasing the elevation H^* from 0 to some maximum value H^*_{max}, beyond which static solutions do not exist and the interface ruptures.

Figure 3.2.4-3: Photographs by Kovitz (1975) of an air-water interface for $R^\star = 1.12$.

(a) $H^\star = -\ 0.24$. The knife-edge is below the air-water interface, which intersects the rod with a finite contact angle.

(b) $H^\star = 0.24$. The air-water interface intersects the knife-edge. The contact angle boundary condition is removed.

(c) $H^\star = 1.20$

(d) $H^\star = 1.25$

(e) $H^\star = 1.30$

(f) $H^\star = 1.36$. See the comparison with theory in Figure 4-4.

Breakoff occurred at $H^\star = 1.38$.

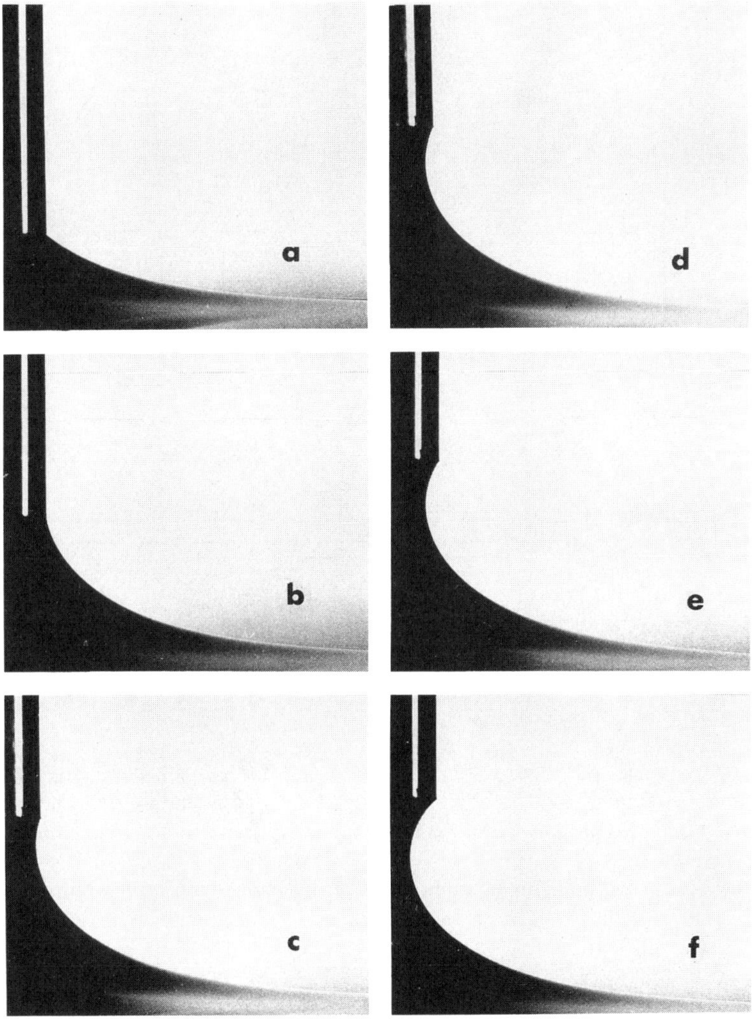
a
d
b
e
c
f

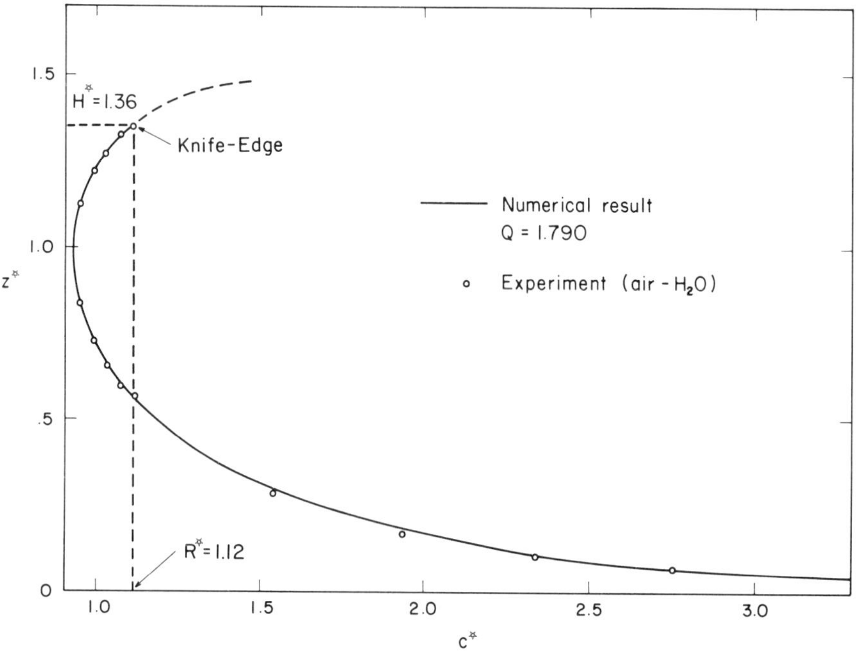

Figure 3.2.4-4: Comparison between air-water interface shown in Figure 4-3(f) and the numerical computations illustrated in Figure 4-2 (Kovitz 1975).

A sequence of such static configurations for an air-water interface is presented in Figure 4-3. As an example, Figure 4-4 shows that the profile pictured in Figure 4-3(f) compares well with Kovitz's numerical computations.

Figure 4-2 indicates the possibility of two different configurations passing through a given knife-edge denoted by a point ($R^\star$, $H^\star$). One of these two configurations always touches the envelope $H^\star_{max}$ before contacting the knife-edge. Kovitz (1975) has never observed such a configuration in his experiments and believes it to be unstable.

Kovitz (1975) provides analytical expressions for various portions of the envelope curve $H^\star_{max} = H^\star_{max}\,(R^\star)$ in Figure 4-2. For large $R^\star$,

$$H^\star_{max} = 2 - \frac{2}{3} R^{\star -1} - \frac{1}{9} R^{\star -2} + O(\,R^{\star -3}\,) \tag{4-10}$$

which may be used with less than 1% error for $R^\star \geq 1.9$. For $0.2 \leq R^\star \leq 1.9$, a polynomial fit to the numerical results yields

$$R^* = \sum_{n=1}^{6} C_n \left[\tan\left(\frac{\pi}{4} H^*_{max} \right) \right]^n \tag{4-11}$$

where

$$C_1 = 0.18080 \qquad C_2 = 0.84317$$

$$C_3 = -0.69494 \qquad C_4 = 0.29211$$

$$C_5 = -0.06233 \qquad C_6 = 0.00532 \tag{4-12}$$

For sufficiently small R^*,

$$R^* = 4\, e^{-(1+\sigma)}\, e^{-\sec\phi}\, (1 - \cos\phi)^{-1}$$

$$H^*_{max} = 4\, e^{-(1+\sigma)}\, e^{-\sec\phi} \left(\frac{1 + \cos\phi}{1 - \cos\phi} \right) \tan\phi \qquad (4\text{-}13)^a$$

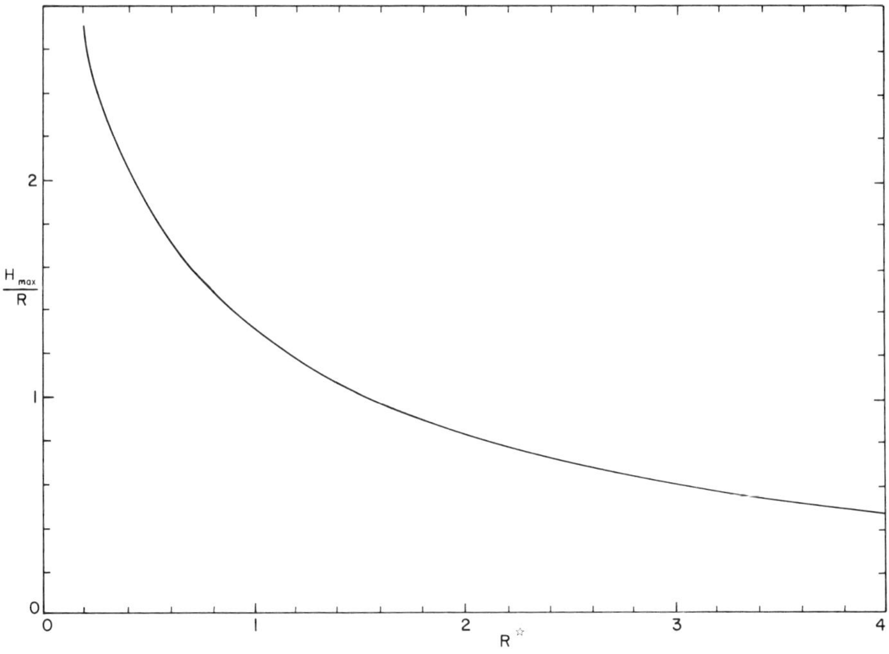

Figure 3.2.4-5: H_{max}/R as a function of R^* from (4-10) and (4-11).

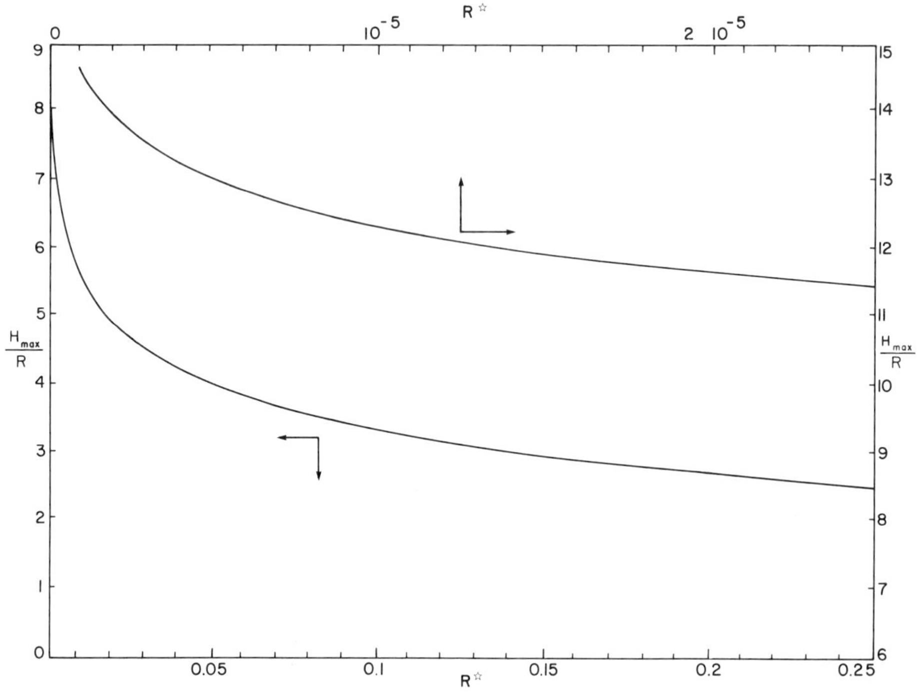

Figure 3.2.4-6: H_{max}/R as a function of R^* from (4-11), (4-13), and (4-16).

in which

$$\sigma = 0.5772157... \tag{4-14}$$

is the Euler constant (and $\cot\phi = dc^*/dz^*$). Equations (4-13) must be solved simultaneously for various values of

$$\phi < \frac{\pi}{2} \tag{4-15}$$

to obtain H^*_{max} as a function of R^*. In the limit as $R^* \to 0$, $\phi \to \pi/2$, and we can approximate with less than 0.4% error for $R^* < 10^{-5}$:

---

a) This corrects Kovitz's (1975) equation [46].

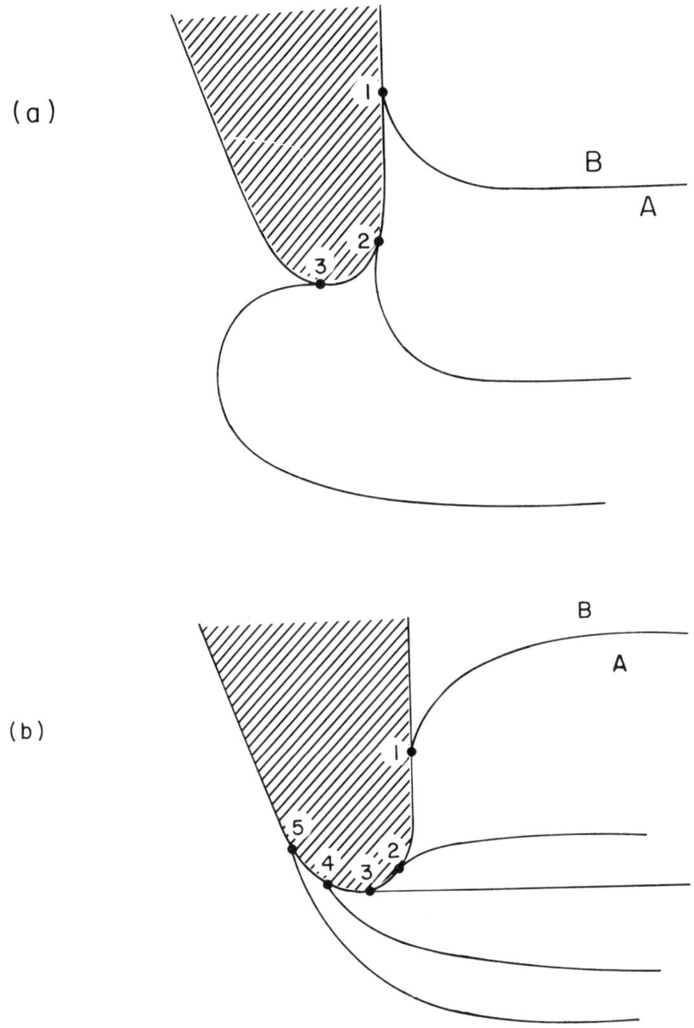

Figure 3.2.4-7: Configurations in the immediate neighborhood of knife-edge as it is raised to form an elevated meniscus.

(a) Lower phase A wets the knife-edge.

(b) Upper phase B wets the knife-edge.

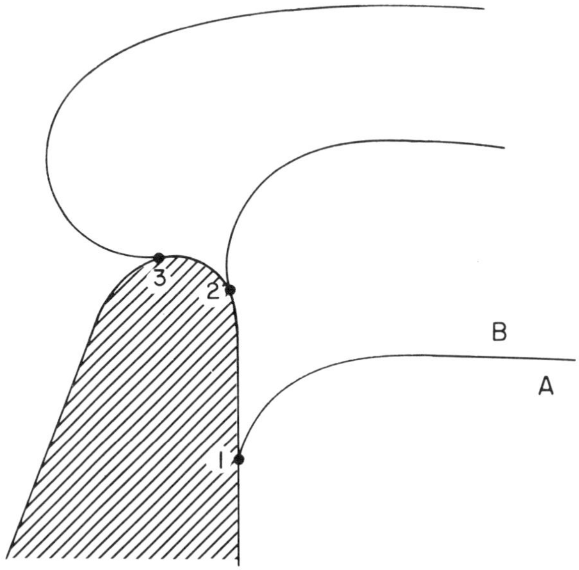

Figure 3.2.4-8: Configurations in the immediate neighborhood of knife-edge wet by upper phase B as the knife-edge is lowered to form a depressed meniscus.

$$H^*_{max} = R^* \ln\left(\frac{4}{R^*} e^{-(1+\sigma)} \right) \tag{4-16}$$

Figures 4-5 and 4-6 provide convenient summaries of these results.

In order to use the meniscal breakoff technique to determine interfacial tension, one must measure H_{max} for known values of R and $(\rho^{(A)} - \rho^{(B)})$. This gives H_{max}/R which can be used to find the corresponding value of R^* either from (4-10) through (4-16) or from Figures 4-5 and 4-6. The definition for R^* can then be used to solve immediately for γ :

$$\gamma = \left(\frac{R}{R^*} \right)^2 (\rho^{(A)} - \rho^{(B)}) g \tag{4-17}$$

As Figure 4-7 suggests, a knife-edge is not simply a geometric circle. When it is examined closely, we see that a contact angle boundary condition must be satisfied at the solid (Kovitz and Yannimaras 1976). The preceding discussion presumes the radius of curvature of the knife-edge to be sufficiently small that the outer solution (outside the immediate neighborhood of the common line; see Sec. 3.1.1) is independent of the static contact angle. Figure 4-7 suggests a sequence of configurations assumed by the interface as the knife-edge is raised to form an elevated meniscus. In Figure 4-7(a), the lower phase wets the knife-edge and all of the configurations shown in Figure 4-2, including the limiting configuration corresponding to $H^* = H^*_{max}$, can be assumed by the interface. Comparing Figure 4-7(b) with Figure 4-2, we see that when the upper phase wets the knife-edge the limiting configuration for which $H^* = H^*_{max}$ can not be assumed by the interface. Breakoff will occur prematurely at a value of H^* less than H^*_{max}. When the upper phase wets the knife-edge, a depressed meniscus must be used as suggested in Figure 4-8.

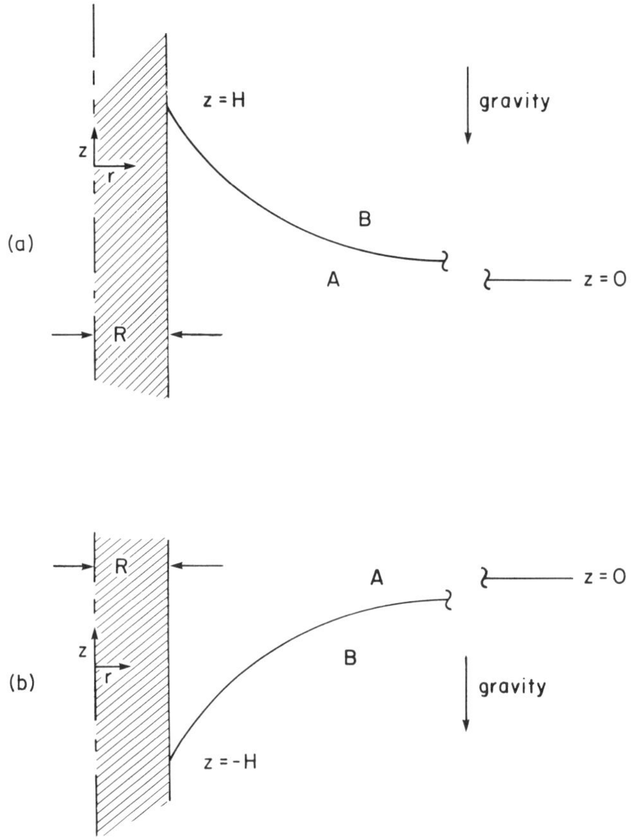

Figure 3.2.4-9: Contact angle limited menisci on exterior of rod

(a) when lower phase preferentially wets the rod

(b) when upper phase preferentially wets the rod.

The menisci formed on the exterior of rods as in Figures 4-9 and 4-3(a) are special cases of those formed at knife-edges. Equation (4-6) must again be solved consistent with (4-7), but boundary condition (4-8) is replaced by a specification of the contact angle between the rod and the meniscus. Figure 4-2 again applies. Given this contact angle, computations such as those illustrated in Figure 4-2 can be used to determine the elevation or depression H of the common line. Note that for a given rod there is only one solution. There is nothing comparable to the breakoff of menisci from knife-edges. The rod meniscus problem has received considerable attention in the literature (LaPlace 1839; Freud and Freud 1930; White and Tallmadge 1965; Huh and Scriven 1969; Hildebrand and Tallmadge 1970; Padday and Pitt 1972 and 1973). It has been used as a technique for measuring the contact angle (Bartell *et al.* 1963; Johnson and Dettre 1969).

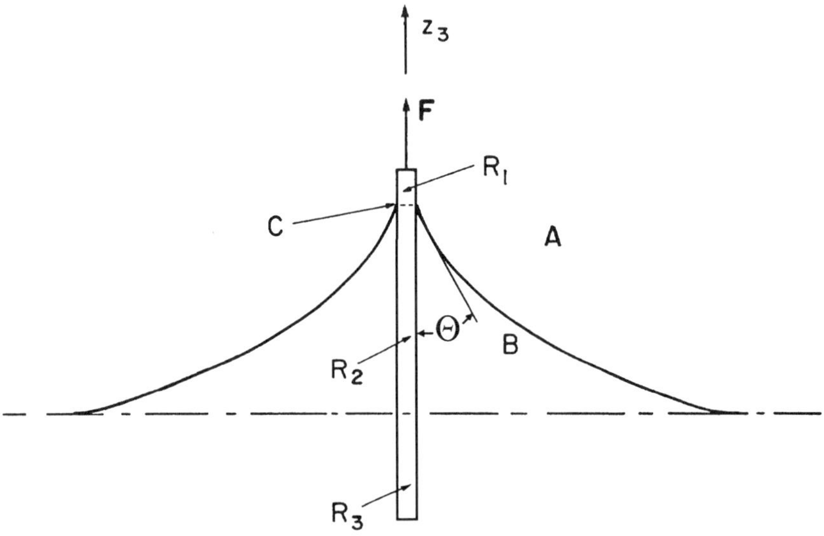

Figure 3.2.4-10: Wilhelmy plate being used to measure the interfacial tension between phases A and B.

References

Bartell, F. E., J. L. Culbertson, and M. A. Miller, *J. Phys. Chem.* **40**, 881 (1936).

du Noüy, P. Lecomte, *J. Gen. Physiol.* **1**, 521 (1919).

Fox, H. W., and C. H. Chrisman Jr., *J. Phys. Chem.* **56**, 284 (1952).

Freud, B. B., and H. Z. Freud, *J. Am. Chem. Soc.* **52**, 1772 (1930).

Harkins, W. D., and H. F. Jordan, *J. Am. Chem. Soc.* **52**, 1751 (1930).

Hildebrand, M. A., and J. A. Tallmadge, *J. Fluid Mech.* **44**, 811 (1970).

Huh, C., and L. E. Scriven, *J. Colloid Interface Sci.* **30**, 323 (1969).

Johnson, R. E., Jr. and R. H. Dettre, "Surface and Colloid Science", vol. 2, edited by E. Matijevic, Wiley-Interscience, New York (1969).

Kovitz, A. A., *J. Colloid Interface Sci.* **50**, 125 (1975); *ibid.* **52**, 412 (1975).

Kovitz, A. A., and D. Yannimaras, "Measurement of Interfacial Tension by the Rod-in-Free-Surface Meniscal Breakoff Method," ERDA Symposium on Enhanced Oil and Gas Recovery, Tulsa, OK (Sept. 9-10, 1976).

LaPlace, P. S. de, "Mecanique celeste," translated by N. Bowditch, vol. 4, page 948, Little and Brown, Boston (1839).

Padday, J. F., and A. Pitt, *J. Colloid Interface Sci.* **38**, 323 (1972).

Padday, J. F., and A R. Pitt, *Philos. Trans. R. Soc. London* **275A**, 489 (1973).

White, D. A., and J. A. Tallmadge, *J. Fluid Mech.* **23**, 325 (1965).

Wilhelmy, L., *Ann. Phys.* **119**, 177 (1863).

Zuidema, H. H., and G. W. Waters, *Ind. Eng. Chem.* **13**, 312 (1941).

Exercises

3.2.4-1 *Wilhelmy plate* Figure 4-10 shows a Wilhelmy (1863) plate being used to measure the interfacial tension between phases A and B. Here **F** is the maximum force that can be exerted on the plate in the opposite direction from gravity while maintaining the static partially submerged configuration shown, R_1 is the region occupied by the plate above the common line, R_2 the region below the common line C but above the level of the interface very far away from the plate, R_3 the remaining region occupied by the plate, and Θ the contact angle measured through phase B.

Determine that the difference between the z_3-component of the maximum force that can be exerted on the partially submerged plate and the z_3-component of the force required to suspend the plate in phase A is

$$F_3 - (\rho^{(S)} - \rho^{(A)})(V_1 + V_2 + V_3) g$$
$$= \gamma L \cos \Theta - (\rho^{(B)} - \rho^{(A)}) V_3 g$$

where V_1, V_2, and V_3 are the volumes of regions R_1, R_2, and R_3 and L is the length of the common line.

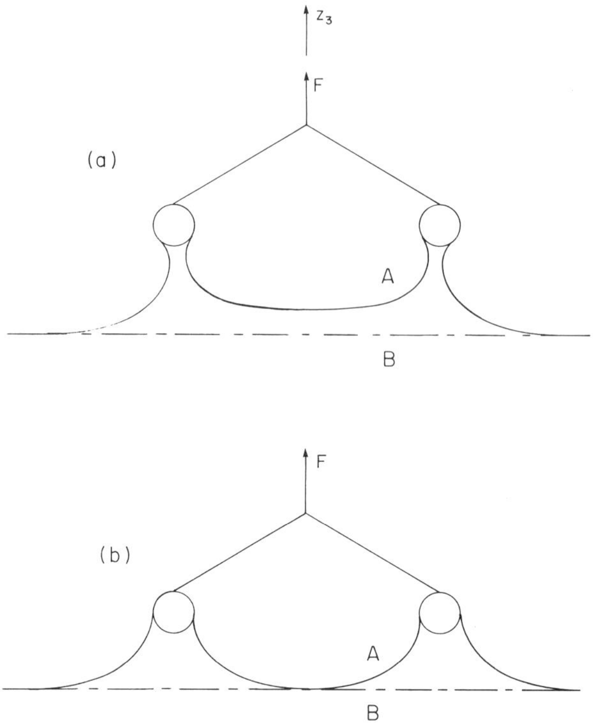

Figure 3.2.4-11:

(a) du Noüy ring being used to measure the interfacial tension between phases A and B.

(b) The limiting case of a du Noüy ring so large that the configuration of the interface inside the ring is very nearly the same as that outside the ring.

Usually the Wilhelmy plate experiment is arranged in such a manner that $\Theta = 0$ and $V_3 = 0$.

3.2.4-2 *du Noüy ring* Figure 4-11a shows a du Noüy (1919) ring being used to measure the interfacial tension between phases A and B. Here **F** is the maximum force that can be exerted on the ring in the opposite direction from gravity while maintaining the static partially submerged configuration show.

Consider the limiting case shown in Figure 4-11b. The radius R_r of the ring is so large that the configuration of the interface inside the ring is very nearly the same as that outside the ring. Assume that the contact angle measured through phase B is zero. Determine that the difference between the z_3-component of the maximum force that can be exerted on the partially submerged ring and the z_3-component of the force required to suspend the ring in phase A is

$$F_3 - (\rho^{(S)} - \rho^{(A)})(2\pi^2 R_r R_w^2)g$$
$$= 4\pi R_r \gamma + (\rho^{(B)} - \rho^{(A)})(4\pi R_r R_w h - \pi^2 R_r R_w^2)g$$

where R_w is the radius of the wire and h is the elevation of the center of the wire above the level of the interface very far away from the ring.

Normally one will use a smaller ring as shown in Figure 4-11 and the meniscus inside the ring will be substantially different from that outside. This more general case has been analyzed by Freud and Freud (1930), but the empirical data correlations available (Harkins and Jordan 1930; Zuidema and Waters 1941; Fox and Chrisman 1952) are more convenient to use.

3.2.5 Pendant drop The pendant drop and emerging bubble are shown in Figure 5-1. Interfacial tension can be deduced by comparing their measured profiles with those expected theoretically.

If we describe the configuration of the interface in cylindrical coordinates by

$$r = c(z) \tag{5-1}$$

we have from either the r- or z- component of the jump momentum balance at the A-B phase interface (Table 2.4.2-8)

$$p_H^{(B)} - p_H^{(A)} - (\rho^{(A)} - \rho^{(B)}) g (H - z)$$
$$= \frac{\gamma}{c} \left[c \frac{d^2 c}{dz^2} - \left(\frac{dc}{dz} \right)^2 - 1 \right] \left[1 + \left(\frac{dc}{dz} \right)^2 \right]^{-3/2} \tag{5-2}$$

Here $p_H^{(A)}$ is the pressure within phase A as the interface is approached within the plane of the knife-edge; $p_H^{(B)}$ is the pressure within phase B as the interface is approached within the same plane. Equation (5-2) may be more conveniently expressed in terms of the dimensionless variables introduced in Sec. 3.2.4 as

$$- G + z^* = \frac{1}{c^*} \left[c^* \frac{d^2 c^*}{dz^{*2}} - \left(\frac{dc^*}{dz^*} \right)^2 - 1 \right] \left[1 + \left(\frac{dc^*}{dz^*} \right)^2 \right]^{-3/2} \tag{5-3}$$

in which we have introduced as a definition

$$G \equiv [\, p_H^{(A)} - p_H^{(B)} + (\rho^{(A)} - \rho^{(B)})\, gH \,] \times [\, \gamma\, (\rho^{(A)} - \rho^{(B)})\, g \,]^{-1/2} \qquad (5\text{-}4)$$

If our objective is to describe the pendant drop shown in Figure 5-1(a), we must solve (5-3) consistent with the boundary conditions

$$\text{at } z^* = 0 : \; c^* = 0 \qquad (5\text{-}5)$$

$$\text{at } z^* \to 0 : \; \frac{dc^*}{dz^*} \to \infty \qquad (5\text{-}6)$$

$$\text{at } z^* = H^* : \; c^* = R^* \qquad (5\text{-}7)$$

In order to describe the emerging bubble pictured in Figure 5-1(b), we must find a solution for (5-3) consistent with (5-5), (5-6), and

$$\text{at } z^* = -H^* : \; c^* = R^* \qquad (5\text{-}8)$$

Notice that the emerging bubble problem takes the same form as that for the pendant drop when z is replaced by $-z$ and $(\rho^{(A)} - \rho^{(B)})$ by $(\rho^{(B)} - \rho^{(A)})$. This means that we can confine our attention to the pendant drop.

Padday and Pitt (1973), Pitts (1974), and Kovitz (1974) have discussed the static stability of a pendant drop or the range of variables for which a static configuration of the pendant drop exists. Selected members of the family of meniscus profiles as calculated by Kovitz (1974) are shown in Figure 5-2. They are identified by their corresponding value of G, found by noting that the intersection of a curve with the ordinate corresponds to the apex of the drop at $c^* = z^* = 0$. The knife-edge tip of the capillary is located at some point $(R^*, H^* - G)$.

For a given value of R^*, there is a limited range of $H^* - G$ for which static pendant drops exist. If $(R^*, H^* - G)$ falls below the outer envelope $A_1\ B_1\ C_1\ E_2$, no stable pendant drop configuration is possible. Experimentally, we see a drop form and continuously elongate until breakoff occurs. If $(R^*, H^* - G) > 0$ falls above the envelope $A_2\ B_2\ C_2\ D_2\ E_2$, again no stable static drop configuration exists and the fluid will continuously drip from the capillary tip. Within these bounds, one or more configurations are possible for each value of $(R^*, H^* - G)$.

When $(R^*, H^* - G)$ falls within the region bounded by $H^* - G < 0$ but greater than the *inner* envelope $A_2\ B_2\ C_2$, there is only one solution. This solution corresponds to a stable configuration which will occur naturally.

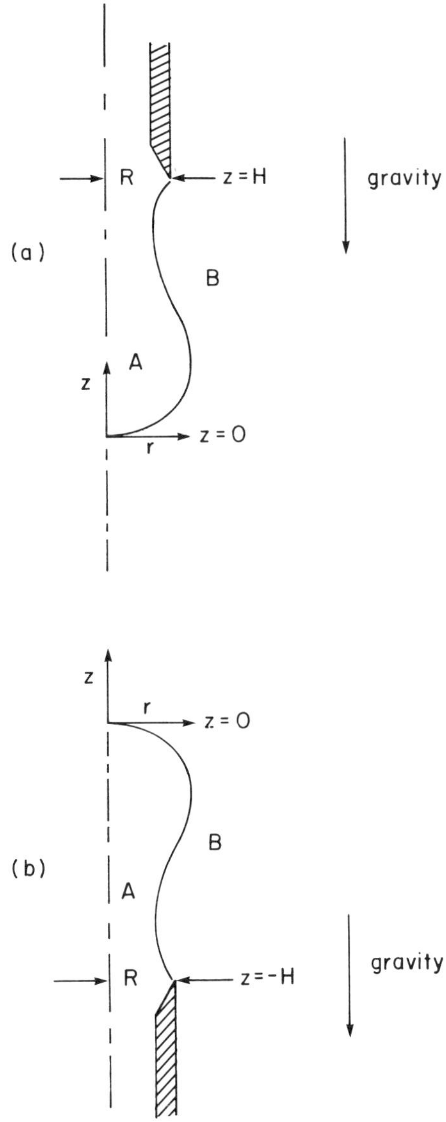

Figure 3.2.5-1:

(a) Pendant drop in which the density of the drop is greater than that of the continuous phase.

(b) Emerging bubble in which the density of the bubble is less than that of the continuous phase.

Figure 3.2.5-2: One parameter family of solutions to (5-3) consistent with boundary conditions (5-5) through (5-7) presented by Kovitz (1974). The curves are identified by the value of the parameter G, found by noting that the intersection of a curve with the ordinate corresponds to the apex of the drop $c^\star = z^\star = 0$. The knife-edge tip of the capillary is located at $c^\star = R^\star$, $z^\star = H^\star$. The dashed curve A_2 B_2 C_2 E_2 represents the envelope $H^\star - G_{min}$ or the minimum value of G, below which there are no solutions for a given $R^\star$ and dripping occurs. The dashed curve A_1 B_1 C_1 E_1 represents the envelope $H^\star - G_{max}$ or the maximum value of G, above which there are no solutions for a given $R^\star$ and breakoff of an elongating drop occurs.

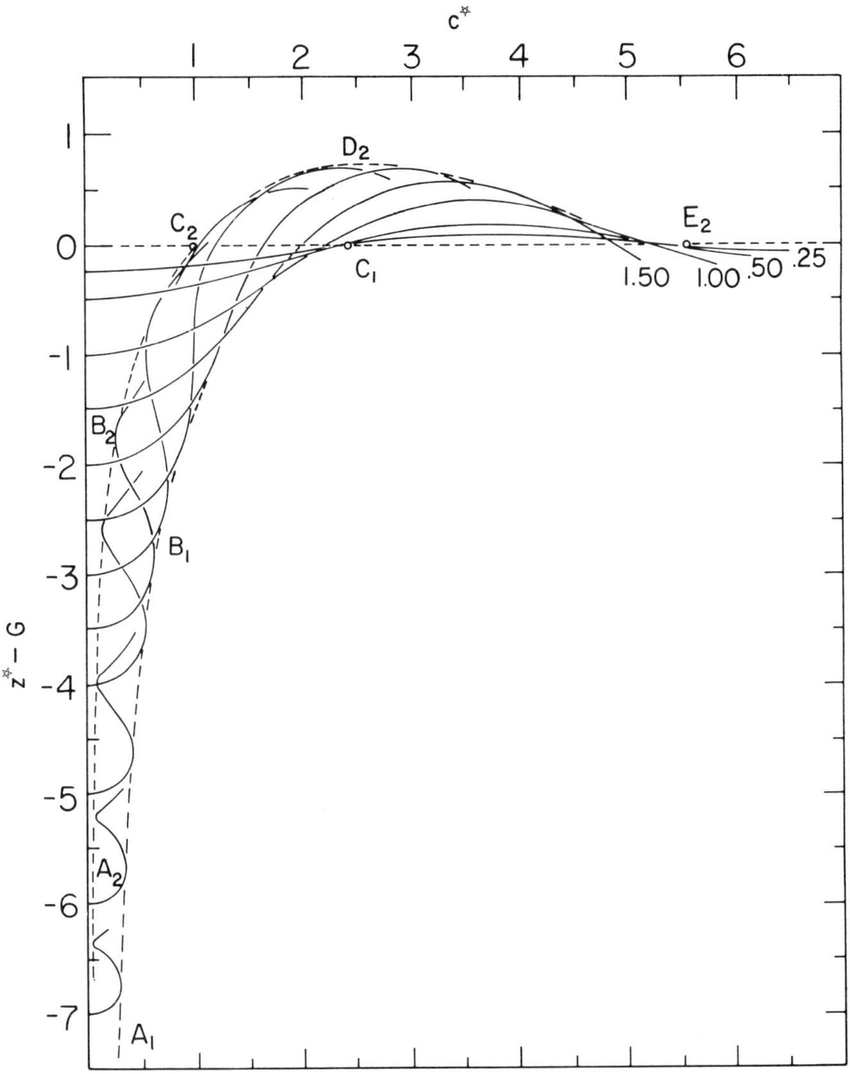
c*
1
2
3
4
5
6
1
0
-1
-2
-3
-4
-5
-6
-7
z* - G
D2
C2
E2
C1
1.50
1.00
.50
.25
B2
B1
A2
A1

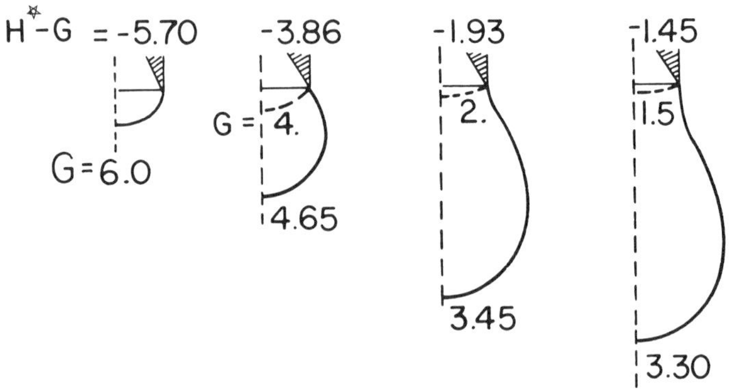

Figure 3.2.5-3: Sequence of pendant drops for $R^* = 0.34$ as H^* increases (Kovitz 1974).

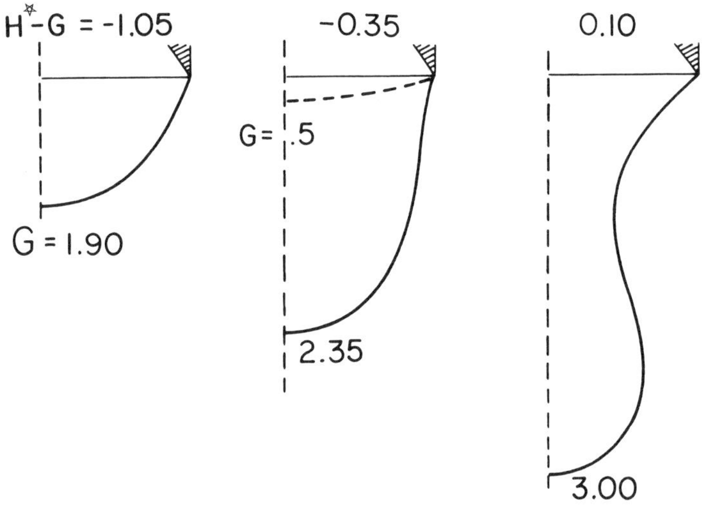

Figure 3.2.5-4: Sequence of pendant drops for $R^* = 1.18$ as H^* increases (Kovitz 1974).

When (R^*, $H^* - G$) falls within the region between the inner and outer envelopes and $H^* - G < 0$, there is a finite number (greater than 2) of drop configurations (corresponding to different values of G). The drop configuration corresponding to the smallest value of G (smallest pressure difference at the drop apex) touches neither the inner nor the outer envelope. Experiment (Kovitz 1974) confirms that it is stable and that it will occur naturally. The configuration corresponding to the next larger value of G is tangent to the outer envelope. It is metastable in the sense that it can be observed experimentally but, under any perturbation, it either contracts to the stable configuration or elongates continuously until it ruptures (Kovitz 1974). All drop configurations for G greater than these two smallest values are tangent to both envelopes before reaching the capillary tip. Kovitz (1974) has never observed these profiles experimentally.

When (R^*, $H^* - G$) falls within the region for which $H^* - G > 0$ but less than the continuation of the inner envelope C_2 D_2 E_2, there is no stable configuration. There is only the single metastable configuration. All other drop profiles touch both the inner and outer envelopes before reaching the knife-edge capillary tip.

Figure 5-3 shows the range of $H^* - G$ for which at least two pendant drop configurations exist when $R^* = 0.34$. At $H^* - G = 5.70$, the outer envelope is intersected and there is only the one stable drop configuration corresponding to $G = 6.0$. For $H^* - G < -5.70$, no solutions exist. At the other extreme, the inner envelope is intersected at $H^* - G = -1.45$ for both $G = 1.50$ and 3.30. For $0 > H^* - G > -1.45$, only one solution exists. For $R^* = 0.34$, $H^* - G = -3.86$, the configuration for which $G = 4$ is stable; the configuration for which $G = 4.65$ is metastable.

Figure 5-4 illustrates another case corresponding to $R^* = 1.18$. In this case, the intersection of the inner envelope occurs for $H^* - G = 0.10 > 0$ for which only one metastable drop configuration exists ($G = 3$).

Limiting cases of the inner and outer envelopes in Figure 5-2 are derived by Hida and Miura (1970), by Hida and Nakanishi (1970), and by Kovitz (1974).

There are two ways in which the pendant drop or emerging bubble can be used to measure interfacial tension.

The preferred way of measuring interfacial tension with the pendant drop is to compare the experimentally observed drop configuration with those predicted theoretically. Tabular numerical solutions of (5-3) consistent with (5-5) and (5-6) have been presented by a variety of workers. The most recent tables are those prepared by Hartland and Hartley (1976) and by Padday (1971; these tables are available on microfiches from Kodak Limited, Research Laboratories, Wealdstone, Harrow, Middlesex, England; see also Padday and Pitt 1972). Padday (1969, pp. 111 and 152) summarizes with comments the earlier tables prepared by Andreas *et al.* (1938), by Niederhauser and Bartell (1950, p. 114), by Fordham (1948), by Mills (1953), and by Stauffer (1965). [In the tables of Hartland and Hartley (1976),

$$B = \frac{2}{G} \tag{5-9}$$

The tables of Padday and Pitt (1972, 1973) are based upon a rearrangement of (5-3)

$$2 + \beta\, z^{**} = \frac{1}{c^{**}} \left[-\, c^{**} \frac{d^2 c^{**}}{dz^{**2}} + \left(\frac{dc^{**}}{dz^{**}} \right)^2 + 1 \right]$$

$$\times \left[1 + \left(\frac{dc^{**}}{dz^{**}} \right)^2 \right]^{-3/2} \tag{5-10}$$

where b is the radius of curvature at the apex of the drop and

$$c^{**} \equiv \frac{c}{b} \tag{5-11}$$

$$z^{**} \equiv \frac{z}{b} \tag{5-12}$$

$$\beta \equiv \frac{b^2\ (\ \rho^{(B)} - \rho^{(A)}\ )\ g}{\gamma} \tag{5-13}$$

In arriving at (5-10), we have noted that at $z^{**} = 0$

$$-2Hb = \frac{1}{c^{**}} \left[-\, c^{**} \frac{d^2 c^{**}}{dz^{**2}} + \left(\frac{dc^{**}}{dz^{**}} \right)^2 + 1 \right] \left[1 + \left(\frac{dc^{**}}{dz^{**}} \right)^2 \right]^{-3/2}$$

$$= 2$$

$$= G \left[\frac{\gamma}{(\ \rho^{(A)} - \rho^{(B)}\ )\ g} \right]^{-1/2} b \tag{5-14}$$

From (5-13) and (5-14), we see that the shape parameter β is negative for a pendant drop and simply related to G:

$$\beta = -\frac{4}{G^2} \qquad (5\text{-}15)]$$

Perhaps the simplest technique is to slowly increase the volume of the drop until rupture occurs at the intersection of $c^* = R^*$ with the outer envelope. Unfortunately, the volume of this drop is not identically equal to the volume of the drop which breaks away and which would be measured experimentally. A correction must be employed (Padday 1969, p. 131). This is referred to as the **drop weight technique**, since experimentally one would measure the weight of the drop which breaks away.

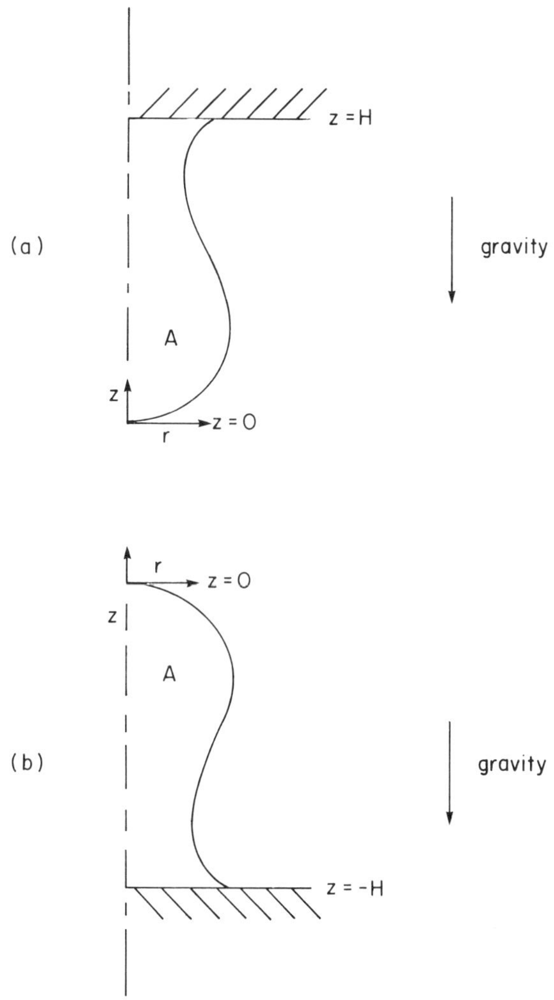

Figure 3.2.5-5:

(a) Contact-angle-limited pendant drop in which the density of the drop is greater than that of the continuous phase.

(b) Contact-angle-limited emerging bubble in which the density of the bubble is less than that of the continuous phase.

The pendant drop and emerging bubble shown in Figure 5-1 may be thought of as radius-limited, in the sense that the drop configurations must pass through the knife-edge tip of the capillary. Special cases of this problem are the contact-angle-limited pendant drop and emerging bubble shown in Figure 5-5. Equation (5-3) must again be solved consistent with boundary conditions (5-5) and (5-6), but boundary condition (5-7) is now replaced by a specification of the contact angle at $z^* = H^*$. Figure 5-2 again applies. Given the contact angle between the meniscus and the solid wall from which the drop is suspended, Figure 5-2 can be used to determine the configuration of the drop for a specified value $H^* - G$. Notice that for a given value of $H^* - G$, only one solution is possible. Furthermore, the drop phase must wet the wall. Multiple solutions are possible as $H^* - G$ varies (as the volume of the drop varies).

As in the previous section, the radius of the knife-edge tip of the capillary is presumed here to be sufficiently small that the outer solution (outside the immediate neighborhood of the common line where the effects of mutual forces such as London-van der Waals forces can be ignored; see Sec. 3.1.1) is independent of the static contact angle. It is only necessary that the bubble or drop wet the capillary.

References

Andreas, J. M., E. A. Hauser, and W. B. Tucker, *J. Phys. Chem.* **42**, 1001 (1938).

Fordham, S., *Proc. R. Soc. London* **194A**, 1 (1948).

Hartland, S., and R. W. Hartley, "Axisymmetric Fluid-Liquid Interfaces", Elsevier, Amsterdam (1976).

Hida, K., and H. Miura, *Nippon Koku Uchu Gakkai-Shi* **18**, 433 (1970), as quoted by Kovitz (1974).

Hida, K., and T. Nakanishi, *J. Phys. Soc. Japan* **28**, 1336 (1970), as quoted by Kovitz (1974).

Kovitz, A. A., Proc. International Colloquium on Drops and Bubbles, Ed. D. J. Collins, M. S. Plesset, and M. M. Saffren, Vol. 2, p.304, California Institute of Technology, Pasadena, CA (1974).

Mills, O. S., *Br. J. Appl. Phys.* **4**, 247 (1953).

Niederhauser, D. O., and F. E. Bartell, "Report of Progress-Fundamental Research on Occurrence and Recovery of Petroleum, 1948-49", American Petroleum Institute, Baltimore (1950), as quoted by Padday (1969, p. 112).

Padday, J. F., "Surface and Colloid Science", vol. 1, edited by E. Matijevic, Wiley-Interscience, New York (1969).

Padday, J. F., *Philos. Trans. R. Soc. London* **269A**, 265 (1971).

Padday, J. F., and A. Pitt, *J. Colloid Interface Sci.* **38**, 323 (1972).

Padday, J. F., and A. R. Pitt, *Philos. Trans. R. Soc. London* **275A**, 489 (1973).

Pitts, E., *J. Fluid Mech.* **63**, 487 (1974).

Stauffer, C. E., *J. Phys. Chem.* **69**, 1933 (1965).

3.2.6 Sessile drop The captive bubble and sessile drop are shown in Figure 6-1. They are directly analogous with the contact angle limited pendant drop and emerging bubble sketched in Figure 5-5. The only difference is that the densities of the two phases are interchanged. As with the pendant drop and emerging bubble, interfacial tension can be inferred by comparing their measured profiles with those expected experimentally.

Let us proceed as we did in Sec. 3.2.5 to describe the configuration of the interface in cylindrical coordinates. Rather than (5-3), we have from either the r- or z- component of the jump momentum balance at the A-B phase interface

$$-G - z^{*}$$

$$= \frac{1}{c^{*}} \left[c^{*} \frac{d^{2}c^{*}}{dz^{*2}} - \left(\frac{dc^{*}}{dz^{*}} \right)^{2} - 1 \right] \left[1 + \left(\frac{dc^{*}}{dz^{*}} \right)^{2} \right]^{-3/2} \tag{6-1}$$

where

$$G \equiv [\, p_{H}^{(A)} - p_{H}^{(B)} - (\, \rho^{(B)} - \rho^{(A)}\,)\, gH \,] \times$$
$$[\, \gamma\, (\, \rho^{(B)} - \rho^{(A)}\,)\, g \,]^{-1/2} \tag{6-2}$$

$$c^{*} \equiv c \left[\frac{\gamma}{(\, \rho^{(B)} - \rho^{(A)}\,)\, g} \right]^{-1/2} \tag{6-3}$$

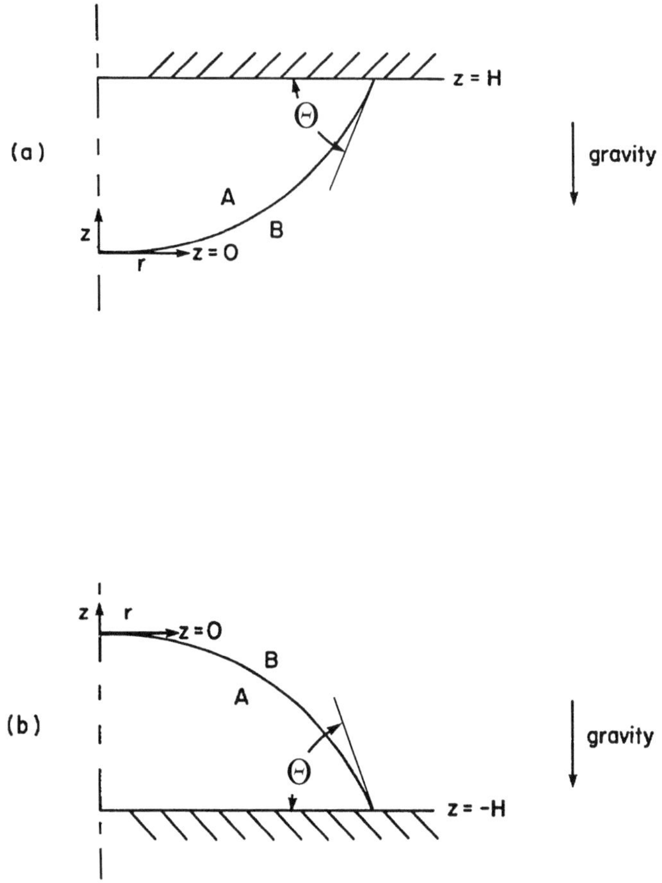

Figure 3.2.6-1

(a) Captive bubble in which density of the drop is less than that of the continuous phase.

(b) Sessile drop in which density of the drop is greater than that of the continuous phase.

$$\text{at } z^\star = H^\star : \frac{dc^\star}{dz^\star} = \cot \Theta^{(stat)} \tag{6-7}$$

in which $\Theta^{(stat)}$ is the static contact angle measured through phase A.

In order to describe the sessile drop sketched in Figure 6-1(b), we must determine a solution for (6-1) consistent with (6-5), (6-6) and

$$\text{at } z^* = -H^* : \frac{dc^*}{dz^*} = -\cot \Theta^{(stat)} \tag{6-8}$$

The sessile drop problem takes the same form as that for the captive bubble when z is replaced by – z and $(\rho^{(B)} - \rho^{(A)})$ by $(\rho^{(A)} - \rho^{(B)})$, which means that it is not necessary to give separate discussions for these two problems.

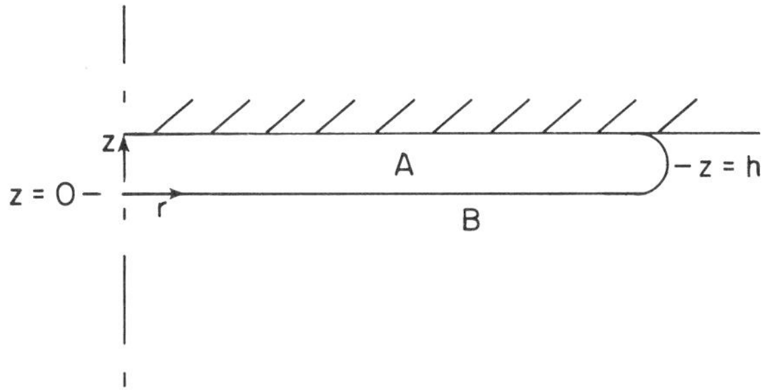

Figure 3.2.6-2: Large flat captive bubble that does not wet the solid.

In the limit of the large flat captive bubble shown in Figure 6-2, (6-1) simplifies to

$$-G - z^* = \frac{d^2c^*}{dz^{*2}} \left[1 + \left(\frac{dc^*}{dz^*} \right)^2 \right]^{-3/2} \tag{6-9}$$

If the bubble does not wet the solid surface upon which it rests, then there is an axial position h such that

$$\text{at } z^* = h^* : \frac{dc^*}{dz^*} = 0 \tag{6-10}$$

Since the bubble is relatively flat at its apex, we will require

$$\text{as } z^* \to 0 : \frac{dc^*}{dz^*} \to \infty$$

$$\frac{d^2c^*}{dz^{*2}} \text{ remains finite} \tag{6-11}$$

This permits us to conclude from (6-9) that

$$G = 0 \tag{6-12}$$

Finally, we can integrate (6-9) consistent with (6-10) and (6-11) to find

$$-\int_0^{h^*} z^* \, dz^* = \int_\infty^0 \left[1 + \left(\frac{dc^*}{dz^*} \right)^2 \right]^{-3/2} d\left(\frac{dc^*}{dz^*} \right)$$

$$\frac{h^{*2}}{2} = 1 \tag{6-13}$$

or

$$\gamma = \frac{1}{2} h^2 \left(\rho^{(B)} - \rho^{(A)} \right) g \tag{6-14}$$

While this is a simple expression, it is relatively difficult to locate the position h of the maximum diameter.

For greater accuracy, it is preferable to consider a smaller bubble and to compare the experimental and theoretical profiles at several axial positions. In these cases, we must rely upon the tabular numerical solutions of (6-1) consistent with (6-5) through (6-7) which are available. The most recent tables are those prepared by Hartland and Hartley (1976; see footnote a of Sec. 3.2.5) and by Padday[a] (1971; these tables are available on microfiches from Kodak Limited, Research Laboratories,

---

a) These tables were prepared with (6-1) in the form of (5-10). Note now however that

at $z^{**} = 0$:

$$-2Hb = \frac{1}{c^{**}} \left[-c^{**} \frac{d^2 c^{**}}{dz^{**2}} + \left(\frac{dc^{**}}{dz^{**}} \right)^2 + 1 \right] \left[1 + \left(\frac{dc^{**}}{dz^{**}} \right)^2 \right]^{-3/2}$$

$$= 2$$

$$= G \left[\frac{\gamma}{\left(\rho^{(B)} - \rho^{(A)} \right) g} \right]^{-1/2} b \tag{6-15}$$

(cont.)

Wealdstone, Harrow, Middlesex, England; see also Padday and Pitt, 1972). Padday (1969, pp. 106 and 151) summarizes with comments the earlier tables prepared by Bashforth and Adams (1892) and by Blaisdell (1940). The contact angle limited captive bubble and sessile drop are stable in all configurations as are the analogous contact angle limited pendant drop and emerging bubble discussed in Sec. 3.2.5.

Besides their use in determining interfacial tension, the contact angle-limited captive bubble and sessile drop can also be used to measure the contact angle (Johnson and Dettre 1969).

The contact angle limited captive bubble and sessile drop in Figure 6-1 may be thought of as special cases of the radius-limited captive bubble and sessile drop shown in Figure 6-3. Equation (6-1) is again to be solved consistent with boundary conditions (6-5) and (6-6), but boundary condition (6-7) is now to be replaced by a specification of the bubble radius at $z = H$. The radius limited captive bubble is analogous to the radius limited pendant drop in the sense that not all configurations satisfying these boundary conditions are stable (Padday and Pitt 1973; Oliver *et al.* 1977).

Finally, note that the analysis presented here is for the outer solution (outside the immediate neighborhood of the common line, where the effects of mutual forces such as London-van der Waals forces and electrostatic double-layer forces can be ignored; see Sec. 3.1.1). The static contact angle is one that an experimentalist might measure with perhaps $10 \times$ magnification.

References

Bashforth, F., and J. C. Adams, "An Attempt to Test the Theory of Capillary Action," Cambridge University Press and Deighton, Bell, and Co. (1892).

Blaisdell, B. E., *J. Math. Phys.* **19**, 186, 217 and 228 (1940).

---

a) (cont.) which implies

$$\beta = \frac{4}{G^2} \tag{6-16}$$

The shape factor β is positive for the captive bubble-sessile drop but negative for the pendant drop-emerging bubble.

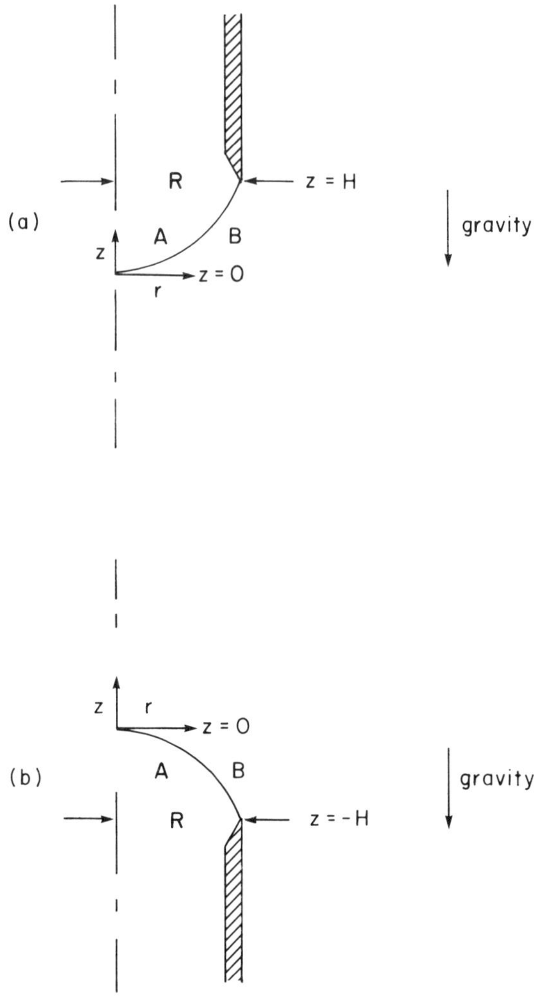

Figure 3.2.6-3

(a) Radius limited captive bubble in which density of bubble is less than that of continuous phase.

(b) Radius limited sessile drop in which density of drop is greater than that of continuous phase.

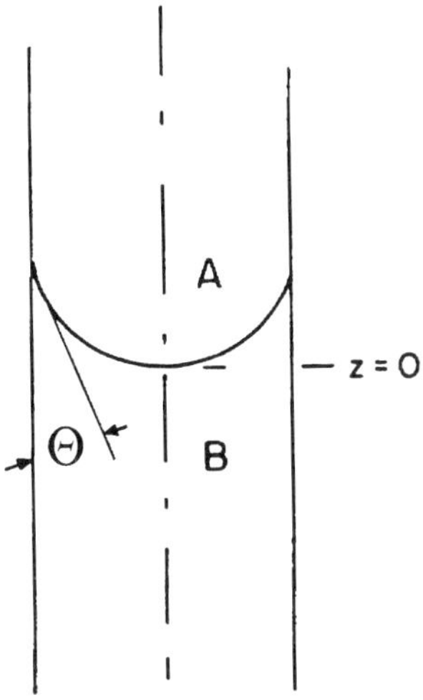

Figure 3.2.6-4: Interface formed between phases A and B in a vertical tube that rotates with a constant angular velocity Ω.

Hartland, S., and R. W. Hartley, "Axisymmetric Fluid-Liquid Interfaces," Elsevier, Amsterdam (1976).

Johnson, R. E. Jr., and R. H. Dettre, "Surface and Colloid Science," vol. 2, edited by E. Matijevic, Wiley-Interscience, New York (1969).

Oliver, J. F., C. Huh, and S. G. Mason, *J. Colloid Interface Sci.* **59**, 568 (1977).

Padday, J. F., "Surface and Colloid Science," vol. 1, edited by E. Matijevic, Wiley-Interscience, New York (1969).

Padday, J. F., *Philos. Trans. R. Soc. London* **269A**, 265 (1971).

Padday, J. F., and A. Pitt, *J. Colloid Interface Sci.* **38**, 323 (1972).

Padday, J. F., and A. R. Pitt, *Philos. Trans. R. Soc. London* **275A**, 489 (1973).

Princen, H. M., and M. P. Aronson, in "Colloid and Interface Science vol. III: Adsorption, Catalysis, Solid Surfaces, Wetting, Surface Tension, and Water," p. 359, Academic Press, New York (1976).

Wasserman, M. L., and J. C. Slattery, *Proc. Phys. Soc. London* **84**, 795 (1964).

Exercise

3.2.6-1 *rotating menisci* Figure 6-4 shows an interface formed in a vertical tube of radius R that rotates with a constant angular velocity Ω. The contact angle measured through phase B is Θ.

Assume that the configuration of the interface in cylindrical coordinates is described by $z = h(r)$. Determine that h can be found by satisfying

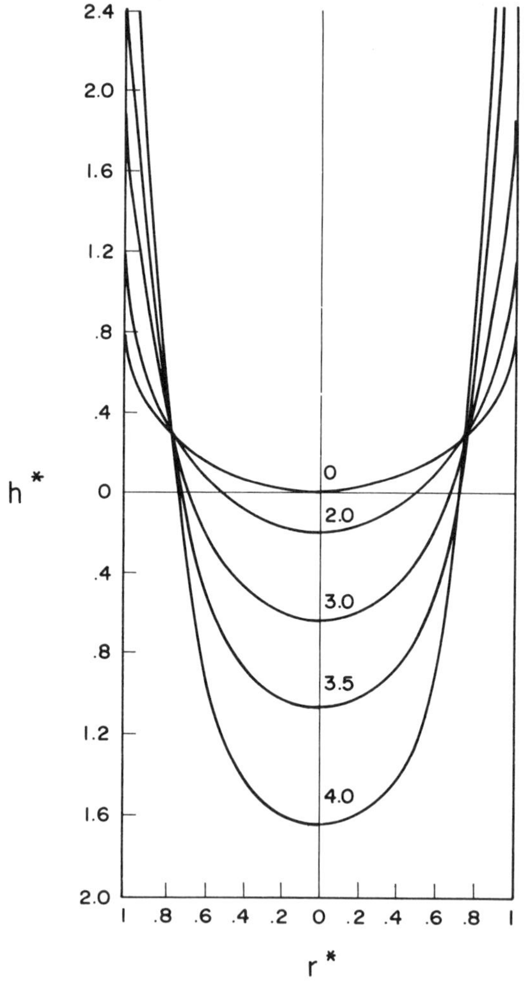

Figure 3.2.6-5: Computed profiles of rotating menisci from Princen and Aronson (1976) for $\Theta = 0$, $(\rho^{(B)} - \rho^{(A)})gR^2/\gamma = 1.3389$, and various values of $\Omega(R/g)^{1/2}$.

$$\frac{1}{r^*}\left[r^*\frac{d^2h^*}{dr^{*2}}+\frac{dh^*}{dr^*}+\left(\frac{dh^*}{dr^*}\right)^3\right]\left[1+\left(\frac{dh^*}{dr^*}\right)^2\right]^{-3/2}$$

$$=\frac{(\rho^{(B)}-\rho^{(A)})\,gR^2}{\gamma}\left(h^*-\frac{R\Omega^2}{2g}r^{*2}\right)+C \tag{1}$$

subject to the boundary conditions

$$\text{at } r^*=0:\ h^*=0\,,\ \frac{dh^*}{dr^*}=0 \tag{2}$$

$$\text{at } r^*=1:\ \frac{dh^*}{dr^*}=(\sin\Theta)^{-1} \tag{3}$$

Here

$$r^*\equiv\frac{r}{R}\,,\quad z^*\equiv\frac{z}{R}$$

$$C\equiv(p_0^{(A)}-p_0^{(B)})\frac{R}{\gamma} \tag{4}$$

Here $p_0^{(A)}$ is the pressure as ($z = 0$, $r = 0$) is approached within phase A, and $p_0^{(B)}$ is the pressure as this same point is approached within phase B.

This problem was first solved by Wasserman and Slattery (1964). Further numerical results have been presented by Princen and Aronson (1976), a portion of which are shown in Figure 6-5.

3.2.7 Static common line, contact angle, and film configuration (with D. Li) It has been recognized for a long time that the molecular interactions play an important role within the immediate neighborhood of a three-phase line of contact or common line where the meniscal film is very thin. The configuration of the fluid-fluid interface in the common line region and static contact angle depend on the sign and the strength of the disjoining pressure.

The influence of molecular interactions upon the configuration of the fluid-fluid interface in the neighborhood of the common line was first analyzed by Rayleigh (1890). He has recognized that, in two dimensions sufficiently far away from the common line, the fluid-fluid interface approaches a plane inclined at a angle $\Theta^{(stat)}$ to the solid surface. As the common line is approached, the curvature of the interface is determined by a force balance to form a different angle Θ_0 with the solid surface. More recently, the configuration of the interface near the common line has been investigated in the context of statistical mechanics (Berry 1974; Benner et al. 1982) and by molecular dynamics simulations (Saville 1977). See Dussan V. (1979) for a further review.

Dzyaloshinskii *et al.* (1960) [see also Martynov *et al.* (1977)] argued that

$$\cos \Theta^{(stat)} = 1 + \frac{1}{\gamma}\int_{h_0}^{\infty} \Pi(h)\, dh \tag{7-1}$$

where γ is surface tension, $\Pi(h)$ is the disjoining pressure (see Sec. 2.1.3), h is the film thickness, and h_0 is the thickness of a film covering the solid. Their arguments [as well as those of Padday and Uffindell (1968), of Israelachvili (1973), and of Hough and White (1980)] depend upon assigning a thermodynamic significance to the solid-fluid interfacial tensions appearing in Young's equation, which, as I explain in Sec. 2.1.10, is appropriate only in the limit $\Theta_0 = 0$. It is clear from (7-1) that, for the formation of a static film configuration, the disjoining pressure must be negative. When the disjoining pressure is positive, a static configuration can not exist, unless gravity is taken into account (Martynov *et al.* 1977). [Martynov *et al.* (1977) attributes (7-1) to Derjaguin (1940). We have not been able to identify this result in his work.]

Miller and Ruckenstein (1974; Jameson and del Cerro 1976) derived a different relation by minimizing free energy. They avoided using a description of London-van der Waals forces derived for films of uniform thickness (see Sec. 2.1.3 and assumption vi below), but they did assume that the interface was a plane.

Martynov *et al.* (1977) have explicitly recognized $\Theta_0 = 0$ in showing that the Laplace equation written in terms of the disjoining pressure leads to (7-1). Wayner (1980, 1982) obtained both the configuration of the interface and $\Theta^{(stat)}$ for a negative disjoining pressure attributable to unretarded London-van der Waals forces.

For a positive disjoining pressure, Joanny and de Gennes (1984; 1986) determined the configuration of the interface assuming that the slope of the film profile and the static contact angle are very small. However, their result shows that the slope becomes unbounded as the common line is approached, contradicting their original assumption.

Figure 7-1 shows in two dimensions the three-phase line of contact or common line formed at equilibrium between phases A, B, and a solid S. The static contact angle $\Theta^{(stat)}$ might be measured by an experimentalist at some distance from the common line, using perhaps 10 $\times$ magnification. Our objective is to determine $\Theta^{(stat)}$ as a function of Θ_0, the contact angle at the common line, for a particular description of the London-van der Waals forces acting in the neighborhood of the common line. We will be particularly concerned to determine the influence of the geometry of the macroscopic system upon the result.

We will make a number of assumptions.

i) The solid is rigid, and its surface is smooth and planar. Its orientation with respect to gravity is arbitrary.

ii) The system is static, and equilibrium has been established. There are no interfacial tension gradients developed in the system.

iii) All external and mutual body forces (see Sec. 2.1.3) can be represented as the gradient of a potential energy per unit mass ϕ

$$\mathbf{b} = -\nabla\phi \tag{7-2}$$

iv) We will account for both London-van der Waals forces and gravity within the immediate neighborhood of the common line, neglecting the effects of any electrostatic double-layer.

v) Israelachvili (1973) suggests that the distance between the two interfaces can not go to zero due to the finite size of the constituent atoms and that a finite interfacial separation d must be recognized. He recommends that d be viewed as the mean distance between the centers of individual atoms and estimated as

$$d = 0.91649[\ M/(\ \rho^{(A)} n_a N\ )]^{1/3} \tag{7-3}$$

Here M is the molecular weight, n_a the number of atoms per molecule, N Avogadro's constant (6.023×10^{23} mol^{-1}), and $\rho^{(A)}$ the mass density of liquid A. In arriving at (7-3), the atoms have been assumed to be in a close packing arrangement. For a molecule consisting of a repeating unit, such as a n-alkane with $-CH_2-$ being the repeating unit, we suggest a simple picture in which the molecules are arranged in such a manner that the *repeating units* are in a close packing arrangement. Retracing the argument of Israelachvili (1973), we have instead

$$d = 0.91649[\ M/(\ \rho^{(A)} n_u N\ )]^{1/3} \tag{7-4}$$

where n_u is the number of repeating units in the molecule. This last is similar to the suggestion of Padday and Uffindell (1968), who replaced M/n_u by the $-CH_2-$ group weight and took the coefficient to be unity.

For an interface between phases A and S, we suggest

$$d = (\ d^{(A)} + d^{(S)}\ )/2 \tag{7-5}$$

where, if appropriate, $d^{(A)}$ and $d^{(S)}$ are computed using (7-4).

More recently, Israelachvili (1985, p. 157) has recommended using d = 1.65 Å in estimating the surface tension (see Exercise 4.1.4-3).

vi) By a common description of the difference in the potential energies attributable to London-van der Waals forces and to gravity at a phase interface (see Exercise 2.1.3-1)

$$\rho^{(A)}\phi^{(A)} - \rho^{(B)}\phi^{(B)}$$

$$= \phi_{\infty}^{(AB)} + \frac{B^{(AB)}}{h'^{m}} - (\rho^{(A)} - \rho^{(B)})g(z_1 \sin\alpha - h' \cos\alpha) \qquad (7\text{-}6)$$

in which h' is the thickness of the film and the last term on the right accounts for the effect of gravity. We will follow Wayner (1980, 1982) in recognizing that the minimum film is a monolayer. Somewhat simplistically, we will estimate the thickness of the monolayer as d, which now might be thought of as the distance between the centers of the last layer of solid atoms and the centers of the first layer of liquid atoms or repeating units. This suggests that, in the continuum model of mechanics used here, we measure the film thickness h from the centers of the first layer of liquid atoms or repeating units and that we replace (7-6) by

$$\rho^{(A)}\phi^{(A)} - \rho^{(B)}\phi^{(B)}$$

$$= \phi_{\infty}^{(AB)} + \frac{B^{(AB)}}{(h + d)^{m}}$$

$$- (\rho^{(A)} - \rho^{(B)})g(z_1 \sin\alpha - h \cos\alpha) \qquad (7\text{-}7)$$

In this simplistic model, the common line can be visualized as running through the centers of the liquid atoms or repeating units on the leading edge of the monolayer. As the common line is approached, the London-van der Waals forces remain bounded as the thickness of the liquid film approaches zero.

Since (7-7) is derived for a thin film of uniform thickness (see Sec. 2.1.3), we might expect our results to be limited to $dh/dz_1 \to 0$ and $\tan \Theta^{(stat)} \to 0$. As we indicate below, the error introduced by this approximation appears to be surprisingly small.

vii) Since we restrict ourselves to the case of unretarded London-van der Waals forces [for films less than 120 Å thick; see (7-21)], we can expect our results to be limited to $dh/dz_1 \to 0$ and $\tan \Theta^{(stat)} \to 0$.

viii) As the ratio $d^{(A)}/d^{(S)} \to 1$, we anticipate that a continuous monolayer of phase A will be deposited and $\Theta_0 \to 0$. As $d^{(A)}/d^{(S)}$ deviates from unity, we might expect the monolayer of phase A to be less regular and Θ_0 to be greater than 0.

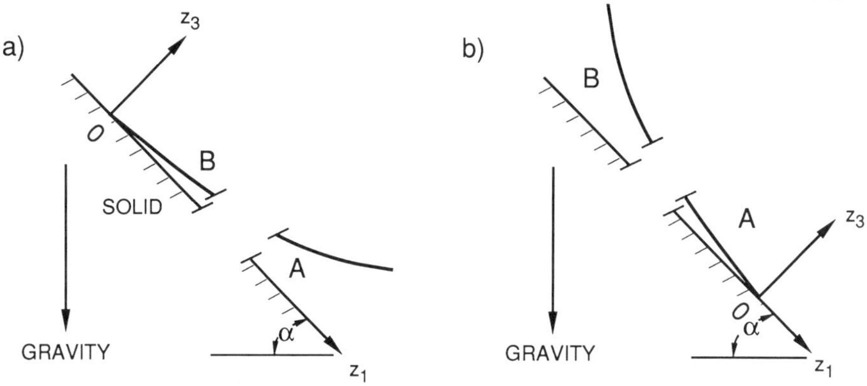

Figure 3.2.7-1: A static contact angle.

a) A disjoining pressure acts in a thin film of phase A. An experimentalist measures an acute contact angle $\Theta^{(stat)}$ at some distance from the common line.

b) A disjoining pressure acts in a thin film of phase B. An experimentalist measures an obtuse contact angle $\Theta^{(stat)}$ at some distance from the common line.

Referring to Figure 7-1 and Table 2.4.2-6, our initial objective will be to determine the configuration of the A-B interface:

$$z_3 = h(z_1) \tag{7-8}$$

Both the z_1 and z_3 components of the jump momentum balance reduce to

$$p^{(B)} - p^{(A)} = \pm \gamma \frac{d^2 h}{dz_1^2} \left[1 + \left(\frac{dh}{dz_1} \right)^2 \right]^{-3/2} \tag{7-9}$$

where the + sign is required for the case of the acute angle $\Theta^{(stat)}$ shown in Figure 7-1a and the – sign for the case of the obtuse angle $\Theta^{(stat)}$ shown in Figure 7-1b. Here $p^{(A)}$ and $p^{(B)}$ are the pressures in phases A and B evaluated at the interface separating these phases. In view of assumptions iii and iv, we conclude from Cauchy's first law (see Sec. 2.4.1) that

$$\begin{aligned} \mathcal{P}^{(A)} &\equiv p^{(A)} + \rho^{(A)}\phi^{(A)} \\ &= C_1 \end{aligned} \tag{7-10}$$

$$\begin{aligned} \mathcal{P}^{(B)} &\equiv p^{(B)} + \rho^{(B)}\phi^{(B)} \\ &= C_2 \end{aligned} \tag{7-11}$$

in which C_1 and C_2 are constants. This permits us to write (7-9) as

$$C_2 - C_1 + \rho^{(A)}\phi^{(A)} - \rho^{(B)}\phi^{(B)} = \pm \gamma \frac{d^2h}{dz_1^2}\left[1 + \left(\frac{dh}{dz_1}\right)^2\right]^{-3/2} \tag{7-12}$$

or, in view of (7-7),

$$\begin{aligned} C_2 - C_1 + \phi_\infty^{(AB)} \pm \frac{B^{(AB)}}{(h + d)^m} \\ - (\rho^{(A)} - \rho^{(B)})g(z_1 \sin\alpha - h\cos\alpha) \\ = \pm \gamma \frac{d^2h}{dz_1^2}\left[1 + \left(\frac{dh}{dz_1}\right)^2\right]^{-3/2} \end{aligned} \tag{7-13}$$

For the fourth term on the left of (7-13), the + sign is again required for the case of the acute angle $\Theta^{(stat)}$ in Figure 7-1a resulting in a thin film of phase A, and the – sign for the case of the obtuse angle $\Theta^{(stat)}$ in Figure 7-1b resulting in a thin film of phase B.

Let us introduce as dimensionless variables

$$h^* \equiv \frac{h}{L} \qquad z_1^* \equiv \frac{z_1}{L}$$

$$d^* \equiv \frac{d}{L} \qquad \phi_\infty^{(AB)*} \equiv \frac{\phi_\infty^{(AB)}L}{\gamma}$$

$$B^* \equiv \frac{B^{(AB)}}{\gamma L^{m-1}} \qquad H^* \equiv HL \tag{7-14}$$

in which

$$2H^* \equiv \frac{d^2h^*}{dz_1^{*2}}\left[1 + \left(\frac{dh^*}{dz_1^*}\right)^2\right]^{-3/2} \tag{7-15}$$

and L is defined such that the Bond number

$$\begin{aligned} N_{Bo} &\equiv (\rho^{(A)} - \rho^{(B)})\frac{gL^2}{\gamma} \\ &= 1 \end{aligned} \tag{7-16}$$

This permits us to express (7-13) as

$$C_2^* - C_1^* + \phi_\infty^{(AB)*} \pm \frac{B^*}{(h^* + d^*)^m} - z_1^* \sin\alpha + h^* \cos\alpha$$

$$= \pm 2H^* \tag{7-17}$$

Since (see Exercise 2.1.3-1)

$$|B^*| << 1 \tag{7-18}$$

our objective is to develop a solution for (7-17) that is correct in the limit $|B^*| \to 0$ or a perturbation solution that is correct to the zeroth order in $|B^*|$.

Outside the immediate neighborhood of the common line where the meniscal film is no longer very thin, (7-17) reduces in this limit to

$$C_2^* - C_1^* + \phi_\infty^{(AB)*} - z_1^* \sin\alpha + h^* \cos\alpha = \pm 2H^* \tag{7-19}$$

This must be solved consistent with

$$\text{as } z_1^* \to 0: \quad \frac{dh^*}{dz_1^*} \to \tan \Theta^{(stat)} \tag{7-20}$$

and two other boundary conditions for the particular macroscopic system with which one is concerned, similar to the three boundary conditions imposed on the rotationally symmetric sessile drop in Sec. 3.2.6. We will refer to this as the *outer problem*, in the sense that it is outside the immediate neighborhood of the common line where the meniscal film is no longer very thin and the effects of the London-van der Waals forces can be neglected with respect to those of gravity. Note that (7-20) is not imposed at the common line, but as the common line is approached in the outer solution.

Within the immediate neighborhood of the common line where the meniscal film is relatively thin, the effects of the London-van der Waals forces must be preserved. We will restrict our consideration here to the case of unretarded London-van der Waals forces (for films less than 120 Å thick; see Exercise 2.1.3-1) for which

$$B^{(AB)} = \frac{1}{6\pi}\left(A^{(FF)} - A^{(FS)} \right)$$

$$m = 3 \tag{7-21}$$

Here $A^{(FF)}$ is the Hamaker constant for the interaction of the film fluid with itself; $A^{(FS)}$ is the Hamaker constant for the interaction of the film fluid with the solid. This suggests that we introduce expanded variables within the immediate neighborhood of the common line:

$$h^{**} \equiv \frac{h}{L} | B^* |^{-1/2} \qquad z_1^{**} \equiv \frac{z_1}{L} | B^* |^{-1/2}$$

$$d^{**} \equiv \frac{d}{L} | B^* |^{-1/2} \qquad \phi_\infty^{(AB)**} \equiv \frac{\phi_\infty^{(AB)} L}{\gamma} | B^* |^{1/2} \tag{7-22}$$

In terms of these expanded variables, (7-17) becomes

$$C_2^{**} - C_1^{**} + \phi_\infty^{(AB)**} \pm \frac{B^*}{|B^*|} \frac{1}{(h^{**} + d^{**})^3}$$

$$- | B^* | (z_1^{**} \sin \alpha - h^{**} \cos \alpha) = \pm 2H^{**} \tag{7-23}$$

in which

$$2H^{**} \equiv \frac{d^2 h^{**}}{dz_1^{**2}} \left[1 + \left(\frac{dh^{**}}{dz_1^{**}} \right)^2 \right]^{-3/2} \tag{7-24}$$

Alternatively, we can write (7-23) as

$$\pm \frac{B^*}{|B^*|} \frac{1}{(h^{**} + d^{**})^3} - | B^* |^{1/2} [(z_1^* - z_{1\infty}^*) \sin \alpha - (h^* - h_\infty^*) \cos \alpha]$$

$$= \pm 2H^{**} \mp 2 | B^* |^{1/2} H_\infty^* \tag{7-25}$$

Here the subscript $\ldots_\infty$ indicates a quantity seen as the common line is approached in the outer solution. Equation (7-23) or (7-25) must be solved consistent with

$$\text{at } z_1^{**} = 0: \quad h^{**} = 0 \tag{7-26}$$

$$\text{at } z_1^{**} = 0: \quad \frac{dh^{**}}{dz_1^{**}} = \tan \Theta_0 \tag{7-27}$$

where Θ_0 is the contact angle at the common line. This is known as the *inner problem*, in the sense that it describes the configuration of the fluids within the immediate neighborhood of the common line where the meniscal film is relatively thin and the effects of the London-van der Waals forces must be included.

Finally, the inner and outer solutions must be consistent in some intermediate or *matching region*

$$\text{as } |B^*| \to 0 \text{ and } z_1^{**} \to \pm\infty: \quad H^{**} \to |B^*|^{1/2} H_\infty^* \tag{7-28}$$

Equation (7-28) provides the third boundary condition required to evaluate the constant ($C_2^{**} - C_1^{**} + \phi_\infty^{(AB)**}$) in (7-23) or $|B^*|^{1/2}(z_{1\infty}^* \sin\alpha - h_\infty^* \cos\alpha \pm 2H_\infty^*)$ in (7-25).

In this intermediate or matching region

$$\text{as } |B^*| \to 0 \text{ and } z_1^{**} \to \pm\infty: \quad \frac{dh^{**}}{dz_1^{**}} \to \left(\frac{dh^*}{dz_1^*} \right)_\infty = \tan \Theta^{(stat)} \tag{7-29}$$

Equation (7-29) allows us to compute finally the specific form of the relationship

$$\tan \Theta^{(stat)} = g(d^{**}, \Theta_0, \alpha, |B^*|^{1/2} H_\infty^*) \tag{7-30}$$

Observe that in general $\Theta^{(stat)}$ is not a material parameter as a result of its dependence upon the outer solution through H_∞^*. This means that $\Theta^{(stat)}$ can be expected to be dependent upon the measurement technique or, equivalently, geometry. This is supported by the observations of Gaydos and Neumann (1987), who found that $\Theta^{(stat)}$ was a function of the diameter of the sessile drop used. (In our opinion, their explanation in terms of line tension is incorrect. See also Sec. 2.1.9.)

Physically, H_∞ can be expected to be proportional to $1/\ell$, where ℓ is a characteristic dimension of the macroscopic system. The dimension of the meniscus as seen in the outer solution is characterized by L. In the limits $|B^*| \to 0$ and $H_\infty^* \sim L/\ell \ll 1$ as is true for most experiments, (7-28) reduces to

$$\text{as } |B^*| \to 0 \text{ and } z_1^{**} \to \pm\infty: \quad H^{**} \to 0 \tag{7-31}$$

and (7-30) takes the form

$$\tan \Theta^{(stat)} = g(d^{**}, \Theta_0, \alpha) \tag{7-32}$$

In these limits, $\Theta^{(stat)}$ depends upon the technique used for its measurement or the geometry of the macroscopic problem only through α or only to the extent that the effects of gravity are important in the inner problem.

***Static contact angle for a negative disjoining pressure* (A > 0) *in the limit* $H_\infty^* \ll 1$** Our expectation is that, in the limits $H_\infty^* = LH_\infty \ll 1$ and $|B^*| \to 0$, the effects of gravity can be neglected with respect to those of the London-van der Waals forces in the inner problem. In view of this assumption and (7-31), equation (7-25) reduces to

$$2H^{**} = \frac{1}{(h^{**} + d^{**})^3} \tag{7-33}$$

or [see (7-24)]

$$\frac{1}{2}\frac{d}{dh^{**}}\left(\frac{dh^{**}}{dz_1^{**}}\right)^2 \left[1 + \left(\frac{dh^{**}}{dz_1^{**}}\right)^2\right]^{-3/2} = \frac{1}{(h^{**} + d^{**})^3} \tag{7-34}$$

Integrating (7-34) consistent with (7-26) and (7-27), we find

$$2(1 + \tan^2\Theta_0)^{-1/2} - 2\left[1 + \left(\frac{dh^{**}}{dz_1^{**}}\right)^2\right]^{-1/2}$$

$$= \left[\frac{1}{d^{**2}} - \frac{1}{(h^{**} + d^{**})^2}\right] \tag{7-35}$$

Finally, (7-29) requires

$$(1 + \tan^2\Theta_0)^{-1/2} - (1 + \tan^2\Theta^{(stat)})^{-1/2} = \frac{1}{2d^{**2}} \tag{7-36}$$

or

$$|\cos\Theta_0| - |\cos\Theta^{(stat)}| = \frac{1}{12\pi\gamma d^2}(A^{(FF)} - A^{(FS)}) \tag{7-37}$$

where we have recognized that

$$\text{as } |B^*| \to 0 \text{ and } z_1^{**} \to \pm\infty: \ h^{**} \to \infty \tag{7-38}$$

The interpretation of (7-37) is that, for a negative disjoining pressure ($A^{(FF)} - A^{(FS)} > 0$) in phase A (Figure 7-1a), $\Theta_0 < \Theta^{(stat)}$; for a negative disjoining pressure in phase B (Figure 7-1b), $\Theta_0 > \Theta^{(stat)}$ (recognizing that contact angles are measured through phase A in this discussion). Equation (7-37) is a special case of (7-1) given by Dzyaloshinskii *et al.* (1960). This specific form appears to have been first derived by Padday and Uffindell (1968), who limited themselves to $\Theta_0 = 0$ by their use of Young's equation (see Sec. 2.1.10), in the same way as did Dzyaloshinskii *et al.* (1960). [Israelachvili (1973) and Hough and White (1980) give similar derivations with the same restriction to $\Theta_0 = 0$. Wayner (1982), who explicitly restricted himself to $\Theta_0 = 0$, gave a derivation similar to that

Table 3.2.7-1: Comparisons of (7-37) with the experimental measurements of $\Theta^{(stat)}$ for n-alkanes reported by Fox and Zisman (1950) and with the estimates of Hough and White (1980). Fox and Zisman (1950) studied sessile drops of the various n-alkanes on polytetrafloroethylene (PTFE) against air at 20.0 ± 0.1 °C and 50% relative humidity. In using (7-37), we have assumed that $\Theta_0 = 0$, we have assumed Hamaker constants computed by Hough and White (1980) together with the combining rules discussed by Israelachvili (1985, p. 154), we have employed the values of γ measured by Jasper and Kring (1955), and we have estimated d by 1.65 Å as recommended by Israelachvili (1985, p. 157) as well as by (7-3) through (7-5).

n	$d^{(A)}$ (Å)[a]	$d^{(A)}$ (Å)[b]	A (10^{-13} erg)[c]	$\Theta_{meas.}$[d]	$\Theta_{calc.}$[e]	$\Theta_{calc.}$[f]	$\Theta_{calc.}$[g]	$\Theta_{calc.}$[h]
5	2.054	3.088	–0.01	wetting				
6	2.028	3.029	0.15	12	12.3	18.1	23.0	21.6
7	2.012	2.991	0.28	21	16.2	23.8	30.2	29.1
8	1.999	2.962	0.38	26	18.3	26.9	34.0	33.5
9	1.990	2.939	0.48	32	20.1	29.5	37.3	36.9
10	1.982	2.921	0.56	35	21.3	31.3	39.6	39.8
11	1.976	2.906	0.59	39	21.6	31.7	39.9	40.7
12	1.971	2.894	0.68	42	22.9	33.6	42.4	43.1
14	1.962	2.874	0.71	44	23.0	33.6	42.3	44.1
16	1.956	2.860	0.79	46	23.9	35.0	44.0	45.8

a) Calculated using (7-3).

b) Calculated using (7-4).

c) $A \equiv A^{(FF)} - A^{(FS)}$

d) Measured by Fox and Zisman (1950).

e) Calculated from (7-37) with (7-5) and with (7-4) used to estimate $d^{(A)}$ and $d^{(S)}$. For PTFE, we estimated $d^{(S)} = 3.107$ Å.

f) Calculated from (7-37) with (7-5) and with (7-3) used to estimate $d^{(A)}$ and $d^{(S)}$. For PTFE, we estimated $d^{(S)} = 2.154$ Å.

g) Calculated using d = 1.65 Å, as recommended by Israelachvili (1985); see assumption v.

h) Calculated by Hough and White (1980).

presented here.]

In Table 3.2.7-1, we show comparisons of (7-37) with the experimental measurements of $\Theta^{(stat)}$ for n-alkanes reported by Fox and Zisman (1950) for sessile drops. We have employed the values of A computed by Hough and White (1980) for a thin film of phase A, and we have arbitrarily chosen $\Theta_0 = 0$ (see assumption viii).

When either 1.65 Å (Israelachvili 1985, p. 157) or (7-3) and (7-5) is used to estimate d, (7-37) gives the wrong trend compared with the experimental data, in the sense that we would expect better agreement with experimental data in the limit $\Theta^{(stat)} \to 0$ (see assumption vii). The estimates of Hough and White (1980) have the same problem. With (7-4) and (7-5), (7-37) gives better comparisons with the data as $\Theta^{(stat)} \to 0$. For larger values of $\Theta^{(stat)}$, it may be that we are not justified in neglecting the retardation of the London-van der Waals forces (see assumption vii) or that Θ_0 is greater than 0 (see assumption viii).

Equation (7-6), and therefore (7-7), was derived with the assumption that the liquid film has a uniform thickness. In order to test the effect of this restriction, we corrected (7-37) using the result of Miller and Ruckenstein (1974) for the London-van der Waals potential in a wedge-shaped film and conservatively assuming the film had a uniform slope equal to $\tan \Theta^{(stat)}$. For the range of observations in Table 3.2.7-1, the correction was less than 2%.

Static film configuration for a negative disjoining pressure **(A > 0)** ***in the limit*** $\mathbf{H_\infty^\star << 1}$ To obtain the configuration of the liquid-gas interface within the immediate neighborhood of the common line, (7-34) must be solved consistent with (7-26) and (7-27). Considering assumption vii, we will restrict our attention to

$$\tan^2 \Theta^{(stat)} << 1 \tag{7-39}$$

We see from (7-36) that $\tan^2 \Theta_0 < \tan^2 \Theta^{(stat)}$. In view of (7-39), this suggests that we assume

$$\left(\frac{dh^{**}}{dz_1^{**}} \right)^2 << 1 \tag{7-40}$$

and write (7-34) as

$$\frac{d^2 h^{**}}{dz_1^{**2}} = (h^{**} + d^{**})^{-3} \tag{7-41}$$

Integrating once with (7-26) and (7-27), we have

$$\left(\frac{dh^{**}}{dz_1^{**}} \right)^2 = d^{**-2} + \tan^2 \Theta_0 - (h^{**} + d^{**})^{-2} \tag{7-42}$$

Recognizing (7-29), we realize

$$\tan^2\Theta^{(stat)} = d^{**-2} + \tan^2\Theta_0 \tag{7-43}$$

The solution of (7-42) consistent with (7-25) and (7-43) is

$$\left[\tan^2\Theta^{(stat)}(h^{**} + d^{**})^2 - 1 \right]^{1/2}$$

$$= \pm z_1^{**} \tan^2\Theta^{(stat)} + (d^{**2} \tan^2\Theta^{(stat)} - 1)^{1/2} \tag{7-44}$$

This gives the thickness h^{**} as a function of z_1^{**}. The + sign is for the case of acute angle $\Theta^{(stat)}$ shown in Figure 7-1a and the – sign for the obtuse angle $\Theta^{(stat)}$ shown in Figure 7-1b.

If we assume that $\Theta_0 = 0$ or π, (7-44) becomes in view of (7-43)

$$h^{**} = (z_1^{**2}/d^{**2} + d^{**2})^{1/2} - d^{**} \tag{7-45}$$

or

$$h = \left(\frac{1}{6\pi\gamma d^2} (A^{(FF)} - A^{(FS)}) z_1^2 + d^2 \right)^{1/2} - d \tag{7-46}$$

or

$$h = (z_1^2 \tan^2\Theta^{(stat)} + d^2)^{1/2} - d \tag{7-47}$$

which is consistent with the numerical solution of Wayner (1980).

***Static film configuration for a positive disjoining pressure* (A < 0) *in the limit* H_∞^* << 1** For a positive disjoining pressure ($A^{(FF)} - A^{(FS)} < 0$), the effects of gravity can not be neglected with respect to those of the London-van der Waals forces. A very thin meniscal film will spread over an inclined solid plane until the effects of the London-van der Waals forces are balanced by those of gravity (Kayser *et al.* 1985).

However, there is one portion of a static meniscus on a vertical wall that can be described in a relatively simple manner. From the experimental observations of Bascom *et al.* (1964) and of Kayser *et al.* (1985), we estimate that the effects of curvature are important in the inner solution for the meniscus only very close to the common line. In most of the meniscal region described by the inner solution, the effects of curvature can be neglected with respect to those of gravity. Under these conditions, (7-25) reduces to

$$-\frac{1}{(h^{**} + d^{**})^3} + |B^*| (z_1^{**} - z_{1\infty}^{**}) = 0 \tag{7-48}$$

or

$$-\frac{|B^*|}{h^{*3}} + z_1^* - z_{1\infty}^* = 0 \tag{7-49}$$

or

$$h = \left[\frac{6\pi(\rho^{(A)} - \rho^{(B)})g}{|A^{(FF)} - A^{(FS)}|}(z_1 - z_{1\infty}) \right]^{-1/3} \tag{7-50}$$

outside the neighborhood of the common line and outside the neighborhood of the pool of liquid from which the meniscus has risen. In arriving at this result, we have assumed that in this region $h^* >> d^*$.

Equation (7-50) appears to be consistent with the experimental observations of Kayser *et al.* (1985) for SF_6 on fused silica.

References

Bascom, W. D., R. D. Cottington, and C. R. Singleterry, in "Contact Angles, Wettability and Adhesion," *Advances in Chemistry Series No. 43*, p. 355, American Chemical Society, Washington, D. C. (1964).

Berry, M. V., *J. Phys. A: Math., Nucl. Gen.* **7**, 231 (1974).

Benner, R. E. Jr., L. E. Scriven, and H. Ted Davis, *Faraday Symposium, The Royal Soc. Chem.* **16**, 169 (1982).

Derjaguin, B. V., *Zh. Fiz. Khim.* **14**, 137 (1940).

Dzyaloshinskii, I. E., E. M. Lifshitz, and L. P. Pitaevskii, *Soviet Physics JETP* **37**, 161 (1960).

Dussan V., E. B., *Ann. Rev. Fluid Mech.* **11**, 371 (1979).

Fox, H. W., and W. A. Zisman, *J. Colloid Sci.* **5**, 514 (1950).

Gaydos, J., and A. W. Neumann, *J. Colloid Interface Sci.* **120**, 76 (1987)

Hough, D. B., and L. R. White, *Adv. Colloid Interface Sci.* **14**, 3 (1980).

Israelachvili, J. N., *J. Chem. Soc. Faraday Trans. II* **69**, 1729 (1973).

Israelachvili, J. N., "Intermolecular and Surface Forces," Academic Press (1985).

Jameson, G. J., and M. C. G. del Cerro, *J. Chem. Faraday Trans. I* **72**, 883 (1976).

Jasper, J. J., and E. V. Kring, *J. Phys. Chem.* **59**, 1019 (1955).

Joanny, J. F., and P. G. de Gennes, *C. R. Acad. Sc. Paris*, **299**, 279 (1984).

Joanny, J. F., and P. G. de Gennes, *J. Colloid Interface Sci.* **111**, 94 (1986).

Kayser, R. F., J. W. Schmidt, and M. R. Moldover, *Phys. Rev. Lett.* **54**, 707 (1985).

Martynov, G. A., V. M. Starov, and N. V. Churaev, *Colloid J. USSR* **39**, 472 (1977).

Miller, C. A., and E. Ruckenstein, *J. Colloid Interface Sci.* **48**, 368 (1974).

Padday, J. F., and N. D. Uffindell, *J. Phys. Chem.* **72**, 1407 (1968).

Rayleigh, Lord, *Philos. Mag.* **30**, 285 (1890).

Saville, G., *J. Chem. Soc. Faraday Trans.* II **73**, 1122 (1978).

Wayner, P. C., *J. Colloid Interface Sci.* **77**, 495 (1980).

Wayner, P. C., *J. Colloid Interface Sci.* **88**, 294 (1982).

3.3 In the absence of viscous surface forces

3.3.1 Coalescence: drainage and stability of thin films (with J. D. Chen) The rate at which drops or bubbles, suspended in a liquid, coalesence is important to the preparation and stability of emulsions, of foams, and of dispersions; to liquid-liquid extractions; to the formation of an oil bank during the displacement of oil from a reservoir rock; to mineral flotation.

On a smaller scale, when two drops (or bubbles) are forced to approach one another in a liquid phase or when a drop is driven through a liquid phase to a solid surface, a thin liquid film forms between the two interfaces and begins to drain. As the thickness of the draining film becomes sufficiently small (about 1000 Å), the effects of the London-van der Waals forces and of electrostatic double-layer forces become significant. Depending upon the sign and the magnitude of the disjoining pressure attributable to the London-van der Waals forces and the repulsive force of electrostatic double-layer forces, there may be a critical thickness at which the film becomes unstable, ruptures and coalescence occurs. (See Exercise 2.1.3-1 for the definition of the disjoining pressure.)

In examining prior studies, it is helpful to consider separately those pertinent to the early stage of thinning in which the effects of any disjoining pressure are negligible and those describing the latter stage of thinning in which the disjoining pressure may be controlling. Since the literature is extensive, I will also separate studies of coalescence on solid planes from those at fluid-fluid interfaces. [Comprehensive reviews of thin films are given by Kitchener (1964), Sheludko (1967), Clunie *et al.* (1971), Buscall and Ottewill (1975), and Ivanov (1988).] Reviews concerned with the drainage and stability of thin liquid films are given by Liem and Woods (1974b), Ivanov and Jain (1979), Jain *et al.* (1979), Ivanov (1980), Ivanov and Dimitrov (1988), and Maldarelli and Jain (1988).]

Initial stage of thinning: at a solid plane For the moment, let us confine our attention to the initial stage of thinning in which the effects of London-van der Waals forces and of electrostatic forces can be neglected.

When a drop or bubble approaches a solid plane, the thin liquid film formed is dimpled as the result of the radial pressure distribution developed (Derjaguin and Kussakov 1939; Elton 1948; Evans 1954; Platikanov 1964; Hartland 1969a; Aronson and Princen 1975; Nakamura and Uchida 1980). It is only in the limit as the approach velocity approaches zero that the dimple disappears.

Frenkel and Mysels (1962; see also Buevich and Lipkina 1975) have developed approximate expressions for the thickness of the dimpled film at the center and at the rim or barrier ring as functions of time. Their expression for the thinning rate at the rim is nearly equal to that predicted by the simple analysis of Reynolds (1886) for two plane parallel discs and it is in reasonable agreement with some of the measurements of Platikanov (1964; Lin and Slattery 1982a; Chen and Slattery 1982). Their predicted thinning rate at the center of the film is lower than that seen experimentally (Platikanov 1964; Lin and Slattery 1982a; Chen and Slattery 1982).

However, it should be kept in mind that their result contains an adjustable parameter (an initial time) that has no physical significance.

Hartland (1969a) developed a more detailed model for the evolution of the thinning film assuming that the shape of the drop beyond the rim did not change with time and that the configuration of the fluid-fluid interface at the center was a spherical cap. He proposed that the initial film profile be taken directly from experimental data.

Hartland and Robinson (1977) assumed that the fluid-fluid interface consists of two parabolas, the radius of curvature at the apex varying with time in the central parabola and a constant in the peripheral parabola. A priori knowledge was required of the radial position outside the dimple rim at which the film pressure equaled the hydrostatic pressure.

For the case of a small spherical drop or bubble approaching a solid plane, the development of Lin and Slattery (1982a) is an improvement over those of Hartland (1969a) and of Hartland and Robinson (1977). Since the initial and boundary conditions are more complete, less a priori experimental information is required. The results are in reasonable agreement with Platikanov's (1964) data for the early stage of thinning in which any effects of London-van der Waals forces and of electrostatic forces can generally be neglected. [Since the theory is for small spherical drops, it cannot be compared with Hartland's (1969a) data for large drops.]

Initial stage of thinning: at a fluid-fluid interface Let us continue to confine our attention to the initial stage of thinning in which the effects of London-van der Waals forces and of electrostatic forces can be neglected.

As a drop (or bubble) is forced to approach a fluid-fluid interface, the minimum film thickness is initially at the center. As thinning proceeds, the minimum film thickness moves to the rim or barrier ring (Allan *et al.* 1961; MacKay and Mason 1963; Hodgson and Woods 1969) and a dimpled film is formed. Allan *et al.* (1961) found that the dimpling developed when the film was 0.3 to 1.2 μm thick. The rim radius is a function of time (Allan *et al.* 1961; MacKay and Mason 1963; Hodgson and Woods 1969).

Princen (1963) extended the Frankel and Mysels (1962) theory to estimate the thinning rate both at the center and at the rim as a small drop (or bubble) approaches a fluid-fluid interface. His prediction for the thinning rate at the rim is again nearly equal to that given by the simple analysis of Reynolds (1886). But, just as in the Frankel and Mysels (1962) theory, there is an adjustable parameter (an initial time) that has no physical significance.

Hartland (1970) developed a more detailed analysis to predict film thickness as a function both of time and of radial position. He assumed that both fluid-fluid interfaces were equidistant from a spherical *equilibrium* surface at all times and that the film shape immediately outside the rim was independent of time. The initial film profile had to be given experimentally.

Lin and Slattery (1982b) developed a more complete hydrodynamic theory for the thinning of a liquid film between a small, nearly spherical drop and a fluid-fluid interface that included an a priori estimate of the initial profile. Their theory is in reasonable agreement with data of Woods and Burrill (1972), Burrill and Woods (1973b), and Liem and Woods (1974a) for the early stage of thinning in which any effects of London-van der Waals forces and of electrostatic forces generally can be neglected. [Because their theory assumes that the drop is small and nearly spherical, it could not be compared with Hartland's (1967, 1968, 1969b) data for larger drops.]

Barber and Hartland (1976) included the effects of the interfacial viscosities in describing the thinning of a liquid film bounded by partially mobile parallel planes. The impact of their argument is diminished by the manner in which they combine the effects of the interfacial viscosities with that of an interfacial tension gradient.

Latter stage of thinning: at a solid plane Experimentally we find that, when a small bubble is pressed against a solid plane, the dimpled liquid film formed may either rupture at the rim (Evans 1954; Aronson and Princen 1975) or become a flat, (apparently) stable film (Derjaguin and Kussakov 1939; Evans 1953, 1954; Read and Kitchener 1967, 1969; Blake and Kitchener 1972; Schulze 1972; Schulze and Cichos 1972b; Blake 1975).

When the thickness of the draining film becomes sufficiently small (about 1000 Å), the effects of the London-van der Waals forces and of electrostatic double-layer forces become significant. Depending upon the properties of the system and the thickness of the film on the solid, the London-van der Waals forces can contribute either a positive or negative component to the disjoining pressure (Dzyaloshinskii *et al.* 1961; Sonntag and Strenge 1972; Blake 1975). An electrostatic double layer also may contribute either a positive or negative component to the disjoining pressure. A positive disjoining pressure will slow the rate of thinning at the rim and stabilize the thinning film. A negative disjoining pressure will enhance the rate of thinning at the rim and destabilize the film.

When London-van der Waals forces are dominant, with time the film either ruptures or becomes flat. Platikanov (1964) and Blake (1975) showed experimentally that a positive London-van der Waals disjoining pressure can be responsible for the formation of a flat, equilibrium wetting film on a solid surface. Schulze (1975) found that, for an aqueous film of dodecylamine between a bubble and a quartz surface at a high KCl concentration (0.01 to 0.1 N), the effects of the electrostatic double-layer forces are negligible compared with those of the London-van der Waals forces. If the electrostatic double-layer disjoining pressure is positive, the wetting film thickness decreases with increasing electrolyte concentration. Read and Kitchener (1967) showed that for a dilute KCl aqueous film ($< 10^{-3}$ N) on a silica surface thicker than 300 Å, the electrostatic double-layer forces dominate. As the KCl concentration increases from 2×10^{-5} to 10^{-4} N, they found that the wetting film thickness decreases from 1200 to 800 Å due to the decrease in the electrostatic double-layer repulsion. Similarly, Schulze and Cichos (1972b) found that the thickness of the aqueous wetting film between a bubble and a quartz plate decreases with increasing KCl concentration in the aqueous film. The wetting film thickness varied from almost 500 Å in 10^{-5} N KCl solutiuon to about 100

Å in 0.1 N KCl solution. They concluded that, for film thicknesses greater than 200 Å, the contribution of London-van der Waals forces to the disjoining pressure could be neglected in aqueous electrolyte solutions of less than 0.01 N concentration.

A positive contribution to the disjoining pressure by electrostatic double-layer forces can be changed to a negative one by the adsorption in one of the interfaces of an ion having the opposite charge. For example, a stable aqueous wetting film may be formed in the absence of an ionic surfactant, but the film may rupture in its presence (Schulze and Cichos 1972a; Schulze 1975; Aronson *et al.* 1978).

When the contributions to the disjoining pressure of both the London-van der Waals forces and the electrostatic double-layer forces are negative, the dimpled film always ruptures, and the rupture thickness decreases with increasing electrolyte concentration (Blake and Kitchener 1972; Aronson and Princen 1975; Schulze 1975).

When these two components to the disjoining pressure have different signs, the dimpled film will either rupture or become flat, depending upon which contribution is dominant. Schulze and Cichos (1972a) explained that the existence of a 0.1 N KCl aqueous film on a quartz surface is due to the existence of a positive London-van der Waals disjoining pressure. They also explained the rupture caused by the addition of $AlCl_3$ is due to the adsorption of Al^{+3} ions in the quartz surface, resulting in a negative electrostatic double-layer contribution to the disjoining pressure.

Because of mathematical difficulties, especially when using the dimpled film model, only London-van der Waals forces have been taken into account in most analyses of the stability of thin liquid films on solid planes. Only Jain and Ruckenstein (1976) and Chen (1984) have included the effect of electrostatic double-layer forces in their disjoining pressure. Chen (1984) has discussed the ineraction of positive and negative components to the disjoining pressure on the dynamics of a dimpled film. The numerical results are consistent with the experimental observations mentioned above.

Ruckenstein and Jain (1974) and Jain and Ruckenstein (1976) studied the stability of an unbounded, stagnant, plane, liquid film. Jain and Ivanov (1980) considered a ring-shaped film. Their results suggest that the critical thickness decreases as the rim radius increases and as the width of the ring decreases. Blake and Kitchener (1972) found experimentally that films of smaller diameter were more stable to ambient vibrations than larger films.

Williams and Davis (1982) included the effects of the London-van der Waals forces in studying the evolution of an unbounded static film subjected to sinusoidal initial disturbances.

Buevich and Lipkina (1978) have extended their results for a dimpled thinning film (Buevich and Lipkina 1975) to include the effects of London-van der Waals forces. They studied the thinning rate only at the rim. They found that the rim thicknesses can reach zero in a finite time. Although they did not mention it explicitly, this is true only for a negative disjoining pressure.

Chen and Slattery (1982) have extended the development of Lin and Slattery (1982a) for dimpled films to include the effects of the London-van der Waals forces as a small spherical drop or bubble approaches a solid plane. When the disjoining pressure is negative, there is a critical film thickness at the rim at which the film begins to thin rapidly, leading to the rupture of the film and coalescence. Unfortunately, there are no experimental data with which to compare their predicted coalescence time, the time during which a small drop or bubble appears to rest at a phase interface before it coalesces under the influence of London-van der Waals forces. The inclusion of a positive disjoining pressure results in better descriptions of the film profiles measured by Platikanov (1964) for air bubbles pressed against glass plates. Chen (1984) takes into account electrostatic forces as well.

Although not a solid plane, Li and Slattery (1990) have extended these developments to the problem of attachment (coalescence) of a solid sphere to a bubble, including the effects of electrostatic forces as well as London-van der Waals forces. This is the central problem in mineral flotation (Evans 1954; Laskowski and Kitchener 1969; Derjaguin *et al.* 1984; Fuerstenau and Urbina 1987).

Latter stage of thinning: at a fluid-fluid interface For a drop or bubble forced to approach its homophase, the contribution of the London-van der Waals forces to the disjoining pressure is always negative. The effect is to enhance the rate of thinning at the rim and destabilize the film. This effect becomes significant, when the film thickness is of the order of 1000 Å and increases with decreasing film thickness. When the film thickness at the rim is reduced to a few hundred angstroms, the London-van der Waals forces become sufficiently strong that the film ruptures at the rim.

Burrill and Woods (1973a,b) studied experimentally the coalescence of small oil drops at an interface between oil and an aqueous solution of sodium lauryl sulfate and KC1. They observed that nearly all of the films ruptured at the rim and that the rim thickness at which rupture occurred was between 300 and 500 Å (Burrill and Woods 1973a).

The addition of more KC1 to the aqueous solution resulted in more rapid drainage to rupture (Burrill and Woods 1973b), which was probably attributable to diminished repulsive forces of the electrostatic double-layer.

The effect of surfactant on the coalescence time has been studied by Cockbain and McRoberts (1953), Nielson *et al.* (1958), Biswas and Haydon (1962), Hodgson and Lee (1969), Hodgson and Woods (1969), Komasawa and Otake (1970), Lang and Wilke (1971), Burrill and Woods (1973a,b), Davis and Smith (1976), Lee and Tadros (1982), and Kitamura *et al.* (1988). For a single-component system (clean or in the absence of surfactant), the coalescence times were very short. With the addition of a small amount of surfactant, the coalescence time normally increased dramatically. Under these conditions, the interface could be expected to be less mobile as the result of the effect of either interfacial tension gradients or the interfacial viscosities. Alternatively, an ionic surfactant system could add a positive component to the disjoining pressure. Variations on this theme might be attributable to adsorption competition between species in a mixed surfactant system (Kitamura *et al.* 1988), to a complex adsorbed film of mixed surfactants (Becher 1966; Davis and Smith 1976), to interfacial turbulence (Lee and Tadros 1982), or to an alteration of the disjoining pressure by the reagent system which for example resulted in a pH change.

The effect of drop size on the coalescence time has been studied by Gillespie and Rideal (1956), Elton and Picknett (1957), Nielsen *et al.* (1958), Charles and Mason (1960), MacKay and Mason (1963), Jeffreys and Hawksley (1965), Hodgson and Lee (1969), Hodgson and Woods (1969), Komasawa and Otake (1970), Lang and Wilke (1971), Woods and Burrill (1972), Burrill and Woods (1973b), and Davis and Smith (1976). All workers, with the exceptions of Nielsen *et al.* (1958), Hodgson and Lee (1969), Lange and Wilke (1971), and Davis and Smith (1976), found that the rest time increased with drop size.

MacKay and Mason (1963) extended the Reynolds (1886) equation for plane parallel disks to include the effect of London-van der Waals forces. They found that the film thickness can become zero in a finite time, when the disjoining pressure is negative. A similar conclusion was reached by Hodgson and Woods (1969), who employed a cylindrical drop model.

Flumerfelt *et al.* (1982) have extended the analysis of Barber and Hartland (1976) to include the effects of the London-van der Waals forces in considering the thinning of a liquid film bounded by partially mobile parallel planes. As do Barber and Hartland (1976), they combine the effects of the interfacial viscosities with that of the interfacial tension gradient.

Because of their simple geometry, the stability of plane parallel thinning films, both radially unbounded and radially bounded, have received the most attention (Vrij 1966; Vrij and Overbeek 1968; Sheludko and Manev 1968; Ivanov *et al.* 1970; Ivanov and Dimitrov 1974; Manev *et al.* 1974; Gumerman and Homsy 1975; Jain *et al.* 1979; Ivanov *et al.* 1979; Ivanov 1980; Maldarelli and Jain 1988).

The critical thickness of a free, circular, plane parallel, thinning film decreases with increasing surfactant concentration (Sheludko and Manev 1968; Ivanov *et al.* 1970; Manev *et al.* 1974; Ivanov 1980) and decreasing radius (Sheludko and Manev 1968; Ivanov *et al.* 1970; Manev *et al.* 1974; Gumerman and Homsy 1975; Ivanov 1980). This means that smaller films containing more concentrated surfactant are more stable than larger films containing less concentrated surfactant.

We can expect that the stability of a thinning film formed between a drop (or bubble) and a fluid-fluid interface is somewhat different from that predicted for a plane parallel film. For example, Gumerman and Homsy (1975) predict that a free, circular, plane parallel, thinning film will rupture at its center, where the minimum thickness occurs during a fluctuation according to their analysis. But as a drop or bubble approaches a fluid-fluid interface, the minimum thickness is at the rim and rupture occurs off center (Charles and Mason 1960; Burrill and Woods 1973a).

Chen *et al.* (1984) have followed Buevich and Lipkina (1975, 1978) to obtain an expression for the rate of thinning at the rim as a bubble or drop approaches a fluid-fluid interface. For a bubble or drop approaching its homophase, the London-van der Waals disjoining pressure will be negative, leading to the development of an instability, rupture, and coalescence.

Motivated by the experimental observation that rupture occurs off-center (Charles and Mason 1960; Burrill and Woods 1973a), Chen *et al.* (1984) have constructed a linear stability analysis of this thinning equation to predict the rest time or coalescence time for a bubble at a fluid-fluid interface.

Hahn *et al.* (1985; see Sec. 3.3.2) have extended the development of Lin and Slattery (1982b) in providing a more complete description for the effects of the London-van der Waals forces as a small spherical drop or bubble approaches a fluid-fluid interface. The general trends predicted agree with those derived from the more approximate theory of Chen *et al.* (1984): the coalescence time increases as the bubble or drop diameter increases, as the viscosity of the draining film increases, as the interfacial tension decreases, as the strength of the London-van der Waals forces decreases, and as the density difference between the two phases increases. Their predicted coalescence times are upper bounds in the sense that they do not allow for the development of asymmetric drainage and of instabilities leading to premature rupture as observed by some experimentalists. (They would not necessarily give upper bounds for systems in which electrostatic double-layer forces played an important role.) Their predictions appear to be more accurate than those of Chen *et al.* (1984), when compared with the experimental observations of Allan *et al.* (1961), MacKay and Mason (1963), and Woods and Burrill (1972; Burrill and Woods 1973b).

Chen *et al.* (1988) have extended the study of Hahn *et al.* (1985) to include both the effects of London-van der Waals forces and of electrostatic double-layer forces. Unfortunately, there are no experimental data with which to compare the results.

In including the effects of the surface viscosities (see Sec. 3.4.1) as well as those of the London-van der Waals forces for a liquid film bounded by partially mobile parallel planes, Hahn and Slattery (1985) have modified the development of Barber and Hartland (1976; see also Flumerfelt *et al.* 1982). They find that the dependence of the coalescence time upon bubble radius, the viscosity of the draining film, the surface tension, the strength of the London-van der Waals forces, and the density difference between the two phases as described above is moderated or even reversed by the inclusion of the effects of the surface viscosities.

Hahn and Slattery (1986) have followed Lin and Slattery (1982b) and Hahn *et al.* (1985) in constructing a more complete analysis for the effects of the surface viscosities upon the coalescence time. Their qualitative conclusions are the same as those of Hahn and Slattery (1985) described above, but they observe that the neglect of film dimpling by Hahn and Slattery (1985) leads to serious errors. This is consistent with the better agreement found between the computations of Hahn and Slattery (1986) and the experimental observations of Li and Slattery (1988).

The effect of interfacial viscoelasticity upon coalescence can be seen in part through the experimental studies of Cockbain and McRoberts (1953), Nielsen *et al.* (1958), Biswas and Haydon (1962), and Glass *et al.* (1970).

References

Allan, R. S., G. E. Charles, and S. G. Mason, *J. Colloid Sci.* **16**, 150 (1961).

Aronson, M. P., and H. M. Princen, *J. Colloid Interface Sci.* **52**, 345 (1975).

Aronson, M. P., M. F. Petko, and H. M. Princen, *J. Colloid Interface Sci.* **65**, 296 (1978).

Barber, A. D., and S. Hartland, *Can. J. Chem. Eng.* **54**, 279 (1976).

Becher, P., "Emulsions: Theory and Practice," second edition, pp. 106-110 Reinhold, New York (1966).

Biswas, B., and D. A. Haydon, *Kolloid Z. Z. Polym.* **185**, 31 (1962).

Blake, T. D., *J. Chem. Soc. Faraday Trans. 1* **71**, 192 (1975).

Blake, T. D., and J. A. Kitchener, *J. Chem. Soc. Faraday Trans. 1* **68**, 1435 (1972).

Buevich, Yu. A., and E. Kh. Lipkina, *Z. Prikl. Mekh. Tekh. Fiziki* **2**, 80 (1975).

Buevich, Yu. A., and E. Kh. Lipkina, *Kolloid. Z.* **40**, 201 (1978).

Burrill, K. A., and D. R. Woods, *J. Colloid Interface Sci.* **42**, 15 (1973a).

Burrill, K. A., and D. R. Woods, *J. Colloid Interface Sci.* **42**, 35 (1973b).

Buscall, R., and R. H. Ottewill, in "Colloid Science," vol. 2, ed. by E. H. Everett, p. 191, The Chemical Society, London (1975).

Charles, G. E., and S. G. Mason, *J. Colloid Sci.* **15**, 236 (1960).

Chen, J. D., *J. Colloid Interface Sci.* **98**, 329 (1984).

Chen, J. D., P. S. Hahn, and J. C. Slattery, *AIChE J.* **30**, 622 (1984).

Chen, J. D., P. S. Hahn, and J. C. Slattery, *AIChE J.* **34**, 140 (1988).

Chen, J. D., and J. C. Slattery, *AIChE J.* **28**, 955 (1982); **29**, 174, 526 (1983).

Clunie, J. S., J. F. Goodman, and B. T. Ingram, *Surface Colloid Sci.* **3**, 167 (1971).

Cockbain, E. G., and T. S. McRoberts, *J. Colloid Sci.* **8**, 440 (1953).

Davis, S. S., and A. Smith, *Colloid Polym. Sci.* **254**, 82 (1976).

Derjaguin, B. V., S. S. Dukhin, and N. N. Rulyov, *Surface Colloid Sci.* **13**, edited by E. Matijevic and R. J. Good, 71 (1984).

Derjaguin, B., and M. Kussakov, *Acta Physicochim. USSR* **10**, 25 (1939).

Dzyaloshinskii, I. E., E. M. Lifshitz, and L. P. Pitaevskii, *Adv. Phys.* **10**, 165 (1961).

Elton, G. A. H., *Proc. R. Soc. London A* **194**, 275 (1948).

Elton, G. A. H., and R. G. Picknett, *Proceedings of Second International Congress of Surface Acitivity* **1**, 287, Academic Press, New York (1957).

Evans, L. F., *Nature London* **172**, 776 (1953).

Evans, L. F., *Ind. Eng. Chem.* **46**, 2420 (1954).

Flumerfelt, R. W., J. P. Oppenheim, and J. R. Son, *AIChE Symp. Ser.* **78** (212), 113 (1982).

Frankel, S. P., and K. J. Mysels, *J. Phys. Chem.* **66**, 190 (1962).

Fuerstenau, D. W., and R. H. Urbina, in "Reagents in Mineral Technology," edited by P. Somasundaran and B. M. Moudgil, p. 1, Marcel Dekker, New York (1987).

Gillespie, T., and E. K. Rideal, *Trans. Faraday Soc.* **52**, 173 (1956).

Glass, J. E., R. D. Lundberg, and F. F. Bailey Jr., *J. Colloid Interface Sci.* **33**, 491 (1970).

Gumerman, R. J., and G. M. Homsy, *Chem. Eng. Commun.* **2**, 27 (1975).

Hahn, P. S., J. D. Chen, and J. C. Slattery, *AIChE J.* **31**, 2026 (1985).

Hahn, P. S., and J. C. Slattery, *AIChE J.* **31**, 950 (1985).

Hahn, P. S., and J. C. Slattery, *AIChE Symp. Ser.* **82** (252), 100 (1986).

Hartland, S., *Trans. Inst. Chem. Eng.* **45**, T102 (1967).

Hartland, S., *Trans. Inst. Chem. Eng.* **46**, T275 (1968).

Hartland, S., *Chem. Eng. Prog. Symp. Ser.* **65** (91), 82 (1969a).

Hartland, S., *Chem. Eng. Sci.* **24**, 611 (1969b).

Hartland, S., *Chem. Eng. J. Lausanne* **1**, 67 (1970).

Hartland, S., and J. D. Robinson, *J. Colloid Interface Sci.* **60**, 72 (1977).

Hodgson, T. D., and J. C. Lee, *J. Colloid Interface Sci.* **30**, 94 (1969).

Hodgson, T. D., and D. R. Woods, *J. Colloid Interface Sci.* **30**, 429 (1969).

Ivanov, I. B., *Pure Appl. Chem.* **52**, 1241 (1980).

Ivanov, I. B., "Thin Liquid Films," Marcel Dekker, New York (1988).

Ivanov, I. B., and D. S. Dimitrov, *Colloid Polym. Sci.* **252**, 982 (1974).

Ivanov, I. B., and D. S. Dimitrov, in "Thin Liquid Films," edited by I. B. Ivanov, p. 379, Marcel Dekker, New York (1988).

Ivanov, I. B., and R. K. Jain, in "Dynamics and Instability of Fluid Interfaces," ed. by S. Sorensen, p. 120, Springer-Verlag, New York (1979).

Ivanov, I. B., R. K. Jain, P. Somasundaran, and T. T. Traykov, in "Solution Chemistry of Surfactants," ed. by K. L. Mittal, vol. 2, p. 817, Plenum Press, New York (1979).

Ivanov, I. B., B. Radoev, E. Manev, and A. Sheludko, *Trans. Faraday Soc.* **66**, 1262 (1970).

Jain, R. K., and I. B. Ivanov, *J. Chem. Soc. Faraday Trans. 2* **76**, 250 (1980).

Jain, R. K., I. B. Ivanov, C. Maldarelli, and E. Ruckenstein, in "Dynamics and Instability of Fluid Interfaces," ed. by T. S. Sorensen, pp. 140-167, Springer-Verlag, New York (1979).

Jain, R. K., and E. Ruckenstein, *J. Colloid Interface Sci.* **54**, 108 (1976).

Jeffreys, G. V., and J. L. Hawksley, *AIChE J.* **11**, 413 (1965).

Kitamura, Y., T. Ohta, and T. Takahashi, *Sekiyu Gakkaishi* **31**, 244 (1988).

Kitchener, J. A., in "Recent Progress in Surface Science", ed. by J. F. Danielli, K. G. A. Pankhurst, and A. C. Riddiford, vol. 1, p. 51, Academic Press, New York (1964).

Komasawa, I., and T. Otake, *J. Chem. Eng. Japan* **3**, 243 (1970).

Lang, S. B., and C. R. Wilke, *Ind. Eng. Chem. Fundam.* **10**, 329, 341

(1971).

Laskowski, J., and J. A. Kitchener, *J. Colloid Interface Sci.* **29**, 670 (1969).

Lee, G. W. J., and Th. F. Tadros, *Colloids Surf.* **5**, 129 (1982).

Li, D., and J. C. Slattery, *AIChE J.* **34**, 862 (1988).

Li, D., and J. C. Slattery, *I & E C Research* (1990).

Liem, A. J. S., and D. R. Woods, *Can. J. Chem. Eng.* **52**, 222 (1974a).

Liem, A. J. S., and D. R. Woods, *AIChE Symposium Ser. No. 144*, **70**, 8 (1974b).

Lin, C. Y., and J. C. Slattery, *AIChE J.* **28**, 147 (1982a).

Lin, C. Y., and J. C. Slattery, *AIChE J.* **28**, 786 (1982b).

MacKay, G. D. M., and S. G. Mason, *Can. J. Chem. Eng.* **41**, 203 (1963).

Maldarelli, C., and R. K. Jain, in "Thin Liquid Films," edited by I. B. Ivanov, p. 497, Marcel Dekker, New York (1988).

Manev, E., A. Sheludko, and D. Exerowa, *Colloid Polym. Sci.* **252**, 586 (1974).

Nakamura, M., and K. Uchida, *J. Colloid Interface Sci.* **78**, 479 (1980).

Nielsen, L. E., R. Wall, and G. Adams, *J. Colloid Sci.* **13**, 441 (1958).

Platikanov, D., *J. Phys. Chem.* **68**, 3619 (1964).

Princen, H. M., *J. Colloid Sci.* **18**, 178 (1963).

Read, A. D., and J. A. Kitchener, in "Wetting," Soc. Chem. Ind. (London) Monograph **25**, 300 (1967).

Read, A. D., and J. A. Kitchener, *J. Colloid Interface Sci.* **30**, 391 (1969).

Reynolds, O., *Philos. Trans. R. Soc. London A* **177**, 157 (1886).

Ruckenstein, E., and R. K. Jain, *J. Chem. Soc. Faraday Trans. II* **70**, 132 (1974).

Schulze, H. J., *Naturwissenschaften* **59**, 119 (1972).

Schulze, H. J., *Colloid Polym. Sci.* **253**, 730 (1975).

Schulze, H. J., and C. Cichos, *Z. Phys. Chem.* (Leipzig) **251**, 145 (1972a).

Schulze, H. J., and C. Cichos, *Z. Phys. Chem.* (Leipzig) **251**, 252 (1972b).

Sheludko, A., *Adv. Colloid Interface Sci.* **1**, 391 (1967).

Sheludko, A., and E. Manev, *Trans. Faraday Soc.* **64**, 1123 (1968).

Sonntag, H., and K. Strenge, "Coagulation and Stability of Disperse Systems," translated by R. Kondor, Halsted Press, New York (1972).

Vrij, A., *Discuss. Faraday Soc.* **42**, 23 (1966).

Vrij, A., and J. Th. G. Overbeek, *J. Am. Chem. Soc.* **90**, 3074 (1968).

Williams, M. B., and S. H. Davis, *J. Colloid Interface Sci.* **90**, 220 (1982).

Woods, D. R., and K. A. Burrill, *J. Electroanal. Chem.* **37**, 191 (1972).

3.3.2 Effects of London-van der Waals forces on the thinning and rupture of a dimpled liquid film as a small drop or bubble approaches a fluid-fluid interface (Lin and Slattery 1982b; Hahn *et al.* 1985)
Figure 2-1 shows the liquid film formed as a small drop or bubble approaches a fluid-fluid interface. Our objective is to determine the shape of this film as it drains with time.

In carrying out this computation, we will make a number of assumptions.

i) Viewed in the cylindrical coordinate system of Figure 2-1, the two interfaces bounding the draining liquid film are axisymmetric ($i = 1, 2$):

$$z = h_i(r, t) \tag{2-1}$$

ii) The dependence of h_i ($i = 1, 2$) upon r is sufficiently weak that

$$\left(\frac{\partial h_i}{\partial r}\right)^2 << 1 \tag{2-2}$$

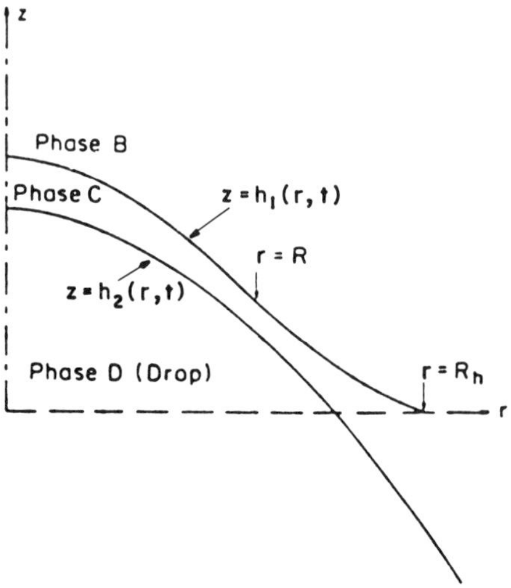

Figure 3.3.2-1: A symmetric drop or bubble (phase D) moves through a liquid (phase C) as it approaches a fluid-fluid interface (between phases C and B). The configuration of the drop-fluid interface is given by $z = h_2(r, t)$; that of the fluid-fluid interface by $z = h_1(r, t)$.

iii) Introducing

$$h = h(r, t) \equiv h_1 - h_2 \tag{2-3}$$

let R be the rim radius of the drop such that

$$\text{at } r = R = R(t) : \frac{\partial h}{\partial r} = 0 \tag{2-4}$$

The Reynolds lubrication theory approximation applies in the sense that, if

$$h_o \equiv h(0, 0) \tag{2-5}$$

and

$$R_o \equiv R(0) \tag{2-6}$$

we will require

$$\left(\frac{h_o}{R_o} \right)^2 << 1 \tag{2-7}$$

iv) There is surfactant present in both interfaces. The resulting interfacial tension gradients are sufficiently large that the tangential components of velocity **v** are zero (i = 1, 2)

$$\text{at } z = h_i : \ \mathbf{P} \cdot \mathbf{v} = 0 \tag{2-8}$$

Here **P** is the projection tensor that transforms every vector on an interface into its tangential components. The interfacial tension gradient required to create such an **immobile** interface is very small (Sheludko 1967; Lin and Slattery 1982a,b; Hahn *et al.* 1985; see also Exercise 3.3.2-1). We will consequently assume that at the same radial positions the interfacial tensions in the two interfaces are equal. In this limit, the results developed will apply both to a liquid drop approaching a liquid-liquid interface and to a gas bubble approaching a gas-liquid interface, since all circulation within phases B and D in Figure 2-1 is suppressed.

v) The effect of mass transfer is neglected.

vi) The pressure p_o within the drop is independent of time and position. The pressure within phase B is equal to the local hydrostatic pressure p_h, which is also assumed to be a constant.

vii) The liquid is an incompressible, Newtonian fluid, the viscosity of which is a constant.

viii) All inertial effects are neglected.

ix) The effects of gravity and of electrostatic double-layer forces are neglected within the draining liquid film.

x) The pressure within the draining film approaches its local hydrostatic value beyond the rim where the Reynolds lubrication theory approximation (assumption iii) is still valid. At this point ($r = R_h$), the two principal curvatures of the drop are constants independent of time,

$$\text{at } r = R_h : \ \frac{\partial h}{\partial r} = \left(\frac{\partial h}{\partial r} \right)_{t=0} \tag{2-9}$$

$$\text{at } r = R_h : \ \frac{\partial^2 h}{\partial r^2} = \left(\frac{\partial^2 h}{\partial r^2} \right)_{t=0} \tag{2-10}$$

xi) Experimental observations (Allan *et al.* 1961; Hodgson and Woods 1969) suggest that there is a time at which the thinning rate at the rim is equal to the thinning rate at the center. At time t = 0 in our computations, the thinning rate is independent of radial position. We will also assume that for t > 0 the thinning rate at the center is always greater than the thinning rate at the rim, so long as the effects of any disjoining pressure are negligible.

xii) The drop is sufficiently small that it may be assumed to be spherical. This is equivalent to assuming that the Bond number

$$N_{Bo} \equiv \frac{\Delta\rho g R_d^2}{\gamma_o} << 1 \qquad (2\text{-}11)$$

Here

$$\Delta\rho \equiv \rho^{(D)} - \rho^{(C)} \qquad (2\text{-}12)$$

is the magnitude of the density difference between the drop phase and the continuous phase, g the magnitude of the acceleration of gravity, R_d the radius of the drop, and γ_o the equilibrium interfacial tension.

xiii) Within each phase, the mutual force per unit mass $\mathbf{b}_m$ known as the London-van der Waals force is representable in terms of a scalar potential ϕ:

$$\mathbf{b}_m = - \nabla\phi \qquad (2\text{-}13)$$

If the fluid-fluid interfaces were parallel planes, we could say (see Exercise 2.1.3-1)

$$\text{at } z = h_1: \; \rho^{(C)}\phi^{(C)} - \rho^{(B)}\phi^{(B)} = \phi_\infty^{(CB)} + \frac{B^{(CB)}}{h^m}$$

$$\text{at } z = h_2: \; \rho^{(C)}\phi^{(C)} - \rho^{(D)}\phi^{(D)} = \phi_\infty^{(CD)} + \frac{B^{(CD)}}{h^m} \qquad (2\text{-}14)$$

Here $\rho^{(i)}$ is the density and $\phi^{(i)}$ the interaction potential per unit volume of phase i; $\phi_\infty^{(ij)}$ and $B^{(ij)}$ are constants. Note that $B^{(ij)}$ is always positive when two homophases approach each other. In this case, the interaction potential per unit volume of the continuous phase at the interface relative to that of the adjoining phase is larger than it would be, if the continuous phase were semi-infinite. This corresponds to a negative disjoining pressure that acts to draw the two fluid-fluid interfaces together. A stable film of uniform thickness can never be formed in this case.

xiv) Because the dependence of h upon r is weak (assumption ii), we will assume that the local value of the interaction energy per unit volume of the film liquid at the fluid-fluid interface is equal to that of a flat film of the same thickness. For more on this point, see Exercise 2.1.3-1.

xv) Since the critical film thicknesses measured or predicted by Allan *et al.* (1961), MacKay and Mason (1963), Vrij (1966), Ivanov *et al.* (1970), Burrill and Woods (1973a), and Chen *et al.* (1984) are normally larger than about 120 Å, we expect that m = 4 in (2-13), corresponding to a representation of retarded London-van der Waals forces, is more appropriate (see Exercise 2.1.3-1). Since we lack more detailed information, we will assume $|B| = 10^{-19}$ erg cm in our analysis as suggested in Exercise 2.1.3-1.

In constructing this development, we will find it convenient to work in terms of the dimensionless variables.

$$r^* \equiv \frac{r}{R_o} \qquad z^* \equiv \frac{z}{h_o}$$

$$h_i^* \equiv \frac{h_i}{h_o} \qquad H_i^* \equiv H_i R_o \quad (i = 1, 2)$$

$$h^* \equiv h_1^* - h_2^* \qquad p^* \equiv \frac{p^{(C)}}{\rho^{(C)} v_o^2}$$

$$p_o^* \equiv \frac{p_o}{\rho^{(C)} v_o^2} \qquad p_h^* \equiv \frac{p_h}{\rho^{(C)} v_o^2}$$

$$v_r^* \equiv \frac{v_r}{v_o} \qquad v_z^* \equiv \frac{R_o v_z}{h_o v_o}$$

$$t^* \equiv \frac{t v_o}{R_o} \qquad \gamma^* \equiv \frac{\gamma}{\gamma_o}$$

$$\phi^{(i)*} \equiv \frac{\phi^{(i)}}{v_o^2} \qquad \mathcal{P}^* \equiv \frac{p^{(C)} + \rho^{(C)}\phi^{(C)}}{\rho^{(C)} v_o^2}$$

$$\mathcal{P}_o^* \equiv \frac{p_o + \rho^{(D)}\phi^{(D)}}{\rho^{(C)} v_o^2} \qquad \mathcal{P}_h^* \equiv \frac{p_h + \rho^{(B)}\phi^{(B)}}{\rho^{(C)} v_o^2} \tag{2-15}$$

and dimensionless Reynolds, Weber, and capillary numbers

$$N_{Re} \equiv \frac{\rho^{(C)} v_o R_o}{\mu} \qquad N_{We} \equiv \frac{\rho^{(C)} v_o^2 R_o}{\gamma_o}$$

$$N_{ca} \equiv \frac{\mu v_o}{\gamma_o} \tag{2-16}$$

Here H_i (i = 1, 2) are the mean curvatures of the interfaces. The characteristic speed v_o will be defined later.

Equation (2-1) suggests that we seek a solution in which the velocity distribution takes the form

$$v_r^* = v_r^*(r^*, z^*, t^*)$$

$$v_z^* = v_z^*(r^*, z^*, t^*)$$

$$v_\theta^* = 0 \tag{2-17}$$

Under these circumstances, the equation of continuity for an incompressible fluid requires (Sec. 1.3.5; Slattery 1981, p. 60)

$$\frac{1}{r^*}\frac{\partial}{\partial r^*}(r^* v_r^*) + \frac{\partial v_z^*}{\partial z^*} = 0 \tag{2-18}$$

In the limit of assumption iii, the r^*-, θ-, and z^*- components of Cauchy's first law for an incompressible, Newtonian fluid with a constant viscosity reduces for creeping flow to (Sec. 2.4.1; Slattery 1981, p. 62)

$$\frac{\partial \mathcal{P}^*}{\partial r^*} = \frac{1}{N_{Re}}\left(\frac{R_o}{h_o}\right)^2 \frac{\partial^2 v_r^*}{\partial z^{*2}} \tag{2-19}$$

$$\frac{\partial \mathcal{P}^*}{\partial \theta^*} = 0 \tag{2-20}$$

$$\frac{\partial \mathcal{P}^*}{\partial z^*} = \frac{1}{N_{Re}}\frac{\partial^2 v_z^*}{\partial z^{*2}} \tag{2-21}$$

Here we have neglected the effects of gravity within the draining liquid film (assumption ix), and we have represented the London-van der Waals mutual force in terms of the scalar potential ϕ (assumption xiii). Equations (2-19) and (2-21) imply

$$\frac{\partial \mathcal{P}^*}{\partial z^*} << \frac{\partial \mathcal{P}^*}{\partial r^*} \tag{2-22}$$

and the dependence of $\mathcal{P}^*$ upon z^* can be neglected. Note that the scaling argument used to neglect inertial effects (assumption viii) in arriving at (2-19) and (2-21) is presumed not to be the one ultimately used here. For this reason, we will regard the magnitude of N_{Re} and the definition of v_o to be as yet unspecified.

The jump mass balance (Sec. 1.3.5; the overall jump mass balance of Sec. 5.4.1) is satisfied identically, since we define the position of the dividing surface by choosing $\rho^{(\sigma)} = 0$ (Sec. 1.3.6) and since the effect of mass transfer is neglected (assumption v). The jump mass balance for surfactant (Sec. 5.2.1) is not required here, since we assume that the interfacial tension gradient developed in the interface is so small that its effect and the effect of a concentration gradient developed in the interface can be neglected (assumption iv).

With assumptions v and vi, the jump momentum balance (Sec. 2.4.2) for the interface between phases B and C reduces to

$$\nabla_{(\sigma)}\gamma + 2H_1\, \gamma\, \boldsymbol{\xi} - (\mathbf{T} + p_h \mathbf{I}) \cdot \boldsymbol{\xi} = 0 \tag{2-23}$$

Here $\boldsymbol{\xi}$ is the unit normal to the interface pointing out of the liquid film. Under the conditions of assumptions ii and iii, the r- and z-components of (2-23) assume the forms at $z^* = h_1^*$ (see Table 2.4.2-7)

$$\frac{\partial \gamma^*}{\partial r^*} - 2\frac{h_o}{R_o} H_1^* \gamma^* \frac{\partial h_1^*}{\partial r^*} - N_{We} \frac{h_o}{R_o} (p^* - p_h^*) \frac{\partial h_1^*}{\partial r^*}$$

$$- N_{ca} \frac{R_o}{h_o} \frac{\partial v_r^*}{\partial z^*} = 0 \tag{2-24}$$

and

$$\frac{h_o}{R_o} \frac{\partial h_1^*}{\partial r^*} \frac{\partial \gamma^*}{\partial r^*} + 2H_1^* \gamma^* + N_{We} (p^* - p_h^*)$$

$$- 2N_{ca} \frac{\partial v_z^*}{\partial z^*} + N_{ca} \frac{\partial v_r^*}{\partial z^*} \frac{\partial h_1^*}{\partial r^*} = 0 \tag{2-25}$$

The θ-component is satisfied identically. Adding $(h_o/R_o)(\partial h_1^*/\partial r^*)$ times (2-25) to (2-24) and recognizing assumptions ii and iii, we have

$$\text{at } z^* = h_1^* : \quad \frac{\partial \gamma^*}{\partial r^*} - N_{ca} \frac{R_o}{h_o} \frac{\partial v_r^*}{\partial z^*} = 0 \tag{2-26}$$

and (2-24) implies[a]

$$\text{at } z^* = h_1^* : \quad 2H_1^* \gamma^* + N_{We}(p^* - p_h^*) = 0 \tag{2-27}$$

In a similar fashion, we can also see that the jump momentum balance for the interface between phases C and D reduces to

$$\text{at } z^* = h_2^* : \quad \frac{\partial \gamma^*}{\partial r^*} + N_{ca} \frac{R_o}{h_o} \frac{\partial v_r^*}{\partial z^*} = 0 \tag{2-28}$$

$$\text{at } z^* = h_2^* : \quad 2H_2^* \gamma^* - N_{We}(p^* - p_o^*) = 0 \tag{2-29}$$

We will recognize assumptions (iii) and (iv) to say

$$\text{at } z^* = h_i^* : \quad v_r^* = 0 \quad (i = 1, 2) \tag{2-30}$$

and we will employ (2-26) and (2-28) to calculate the interfacial tension gradient required to create the immobile interfaces assumed here.

---

a) In arriving at (2-26) and (2-27), it was not necessary to make any statement about the relative magnitudes of N_{ca} and N_{We} or the definition of v_o. But some statement is necessary, in order to establish consistency with (2-25).

Substituting (2-26) into (2-25), we have

$$2H_1^* \gamma^* + N_{We}(p^* - p_h^*) + 2N_{ca} \left(\frac{\partial v_r^*}{\partial z^*} \frac{\partial h_1^*}{\partial r^*} - \frac{\partial v_z^*}{\partial z^*} \right) = 0$$

It follows from (2-49), (2-50), and (2-53) that

$$| \, 2H_1^* \gamma^* \, | >> \left| \, 2N_{ca} \left(\frac{\partial v_r^*}{\partial z^*} \frac{\partial h_1^*}{\partial r^*} - \frac{\partial v_z^*}{\partial z^*} \right) \right|$$

in agreement with (2-27).

Since we neglect the effect of mass transfer on the velocity distribution (assumption v; see Table 2.4.2-7),

$$\text{at } z^* = h_i^* : \; v_z^* = \frac{\partial h_i^*}{\partial t^*} + \frac{\partial h_i^*}{\partial r^*} v_r^* \quad (i = 1, 2) \tag{2-31}$$

Note that

$$\text{at } r^* = R^* : \; \frac{\partial h^*}{\partial r^*} = 0 \tag{2-32}$$

$$\text{at } r^* = 0 : \; \frac{\partial h^*}{\partial r^*} = \frac{\partial h_1^*}{\partial r^*} = 0 \tag{2-33}$$

and

$$\text{at } r^* = 0 : \; \frac{\partial p^*}{\partial r^*} = \frac{\partial \mathcal{P}^*}{\partial r^*} = 0 \tag{2-34}$$

Equations (2-27), (2-29), and (2-34) together with assumptions ii and iv imply

$$\text{at } r^* = 0 : \; \frac{\partial(H_1^* - H_2^*)}{\partial r^*} = \frac{1}{2}\frac{h_o}{R_o}\left(- \frac{1}{r^{*2}}\frac{\partial h^*}{\partial r^*} + \frac{1}{r^*}\frac{\partial^2 h^*}{\partial r^{*2}} + \frac{\partial^3 h^*}{\partial r^{*3}} \right)$$

$$= 0 \tag{2-35}$$

An application of L'Hospital's rule shows us that

$$\text{at } r^* = 0 : \; \frac{\partial^3 h^*}{\partial r^{*3}} = 0 \tag{2-36}$$

or alternatively

$$\text{at } r^* = 0 : \; \frac{\partial^2 h^*}{\partial r^{*2}} = \frac{1}{r^*}\frac{\partial h^*}{\partial r^*} \tag{2-37}$$

According to assumption x, there is a point $r^* = R_h^* > R^*$ where the pressure p^* within the draining film approaches the local hydrostatic pressure in the neighborhood of the drop and the effects of the London-van der Waals force disappear,

$$\text{at } r^* \to R_h^* : \; p^* \to p_h^*, \; h_l^* \to 0 \tag{2-38}$$

Assumption x also requires that

$$\text{at } r^* = R_h^* : \; \frac{\partial h^*}{\partial r^*} = \left(\frac{\partial h^*}{\partial r^*} \right)_{t^*=0} \tag{2-39}$$

$$\text{at } r^* = R_h^* : \; \frac{\partial^2 h^*}{\partial r^{*2}} = \left(\frac{\partial^2 h^*}{\partial r^{*2}} \right)_{t^*=0} \tag{2-40}$$

The initial time is to be chosen by requiring (assumption xi)

$$\text{at } t^* = 0 : \; \frac{\partial h^*}{\partial t^*} = \text{constant} \tag{2-41}$$

Since the drop is sufficiently small to be assumed spherical (assumption xii), the jump momentum balance requires [see (2-29)]

$$p_h^* - p_o^* = - \frac{2}{N_{We} R_d^*} \tag{2-42}$$

where R_d is the radius of the drop. Because surface tension is assumed to be nearly independent of position by assumption iv, (2-28) implies that the effects of viscous forces can be neglected in the jump momentum balance[a].

An integral momentum balance for the drop requires [for more details, see Lin and Slattery (1982a)]

$$N_{ca} \int_0^{R_h^*} (p^* - p_h^*) \, r^* \, dr^* = \frac{(R_o)^2 \Delta \rho g}{\gamma_o} \frac{2}{3} (R_d^*)^3 \tag{2-43}$$

If R_f denotes the value of the dimple radius as time becomes large, we would expect from (2-27) and (2-29) that

$$\text{as } t^* \to \infty : \; p^* - p_o^* \to \frac{1}{2} (p_h^* - p_o^*) \quad \text{for } 0 \le r^* \le R_f^* \tag{2-44}$$

and from (2-38) that

$$\text{as } t^* \to \infty : \ p^* \to p_h^* \qquad \text{for } r^* > R_f^* \tag{2-45}$$

Recognizing (2-42), (2-44), and (2-45), we find that (2-43) gives (Chappelear 1961)

$$\text{as } t^* \to \infty : \ R^* \to R_f^* = \left(\frac{4}{3} \frac{\Delta\rho g R_o^2}{\gamma_o} \right)^{1/2} R_d^{*2} \tag{2-46}$$

Given R_d, we determine R_f by requiring (2-46) to be satisfied (since R_o drops out of this equation). We identify $R_o = R_f/R_f^*$.

For the sake of simplicity, let us define our characteristic speed

$$v_o \equiv \frac{\mu}{\rho^{(C)} R_o} \tag{2-47}$$

which means

$$N_{Re} = 1, \quad N_{We} = N_{ca} = \frac{\mu^2}{\rho^{(C)} R_o \gamma_o} \tag{2-48}$$

Note that we have not used this definition for v_o or this definition for N_{Re} in scaling the Navier-Stokes equation to neglect inertial effects (assumption viii). The scaling argument required to suggest a priori under what circumstances inertial effects can be ignored would be different, based perhaps on the initial speed of displacement of one of the fluid-fluid interfaces calculated at the center of the film.

Our objective in what follows is to obtain a solution to (2-18) and (2-19) consistent with (2-27), (2-29) through (2-33), (2-36) through (2-41), and the second portion of assumption xi. Given R_d, we determine R_f by requiring that, as $t^* \to \infty$ or just prior to the development of an instability and coalescence, (2-46) be satisfied; we identify $R_o = R_f/R_f^*$. Note that, in addition to physical properties, only one parameter is required: R_d.

Solution Integrating (2-19) twice consistent with (2-30), we find in view of (2-48)

$$v_r^* = \frac{1}{2} \left(\frac{h_o}{R_o} \right)^2 \frac{\partial \mathcal{P}^*}{\partial r^*} [\, z^{*2} - (\, h_1^* + h_2^* \,)\, z^* + h_1^* h_2^* \,] \tag{2-49}$$

Substituting (2-49) into (2-18) and integrating once, we have

$$v_z^* = -\frac{1}{2} \left(\frac{h_o}{R_o} \right)^2 \left\{ \left[\frac{\partial^2 \mathcal{P}^*}{\partial r^{*2}} + \frac{1}{r^*} \frac{\partial \mathcal{P}^*}{\partial r^*} \right] \right.$$

$$\times \left[\frac{z^{*3}}{3} - \frac{1}{2} (h_1^* + h_2^*) z^{*2} + h_1^* h_2^* z^* \right]$$

$$+ \frac{\partial \mathcal{P}^*}{\partial r^*} \left[- \left(\frac{\partial h_1^*}{\partial r^*} + \frac{\partial h_2^*}{\partial r^*} \right) \frac{z^{*2}}{2} \right.$$

$$\left. \left. + \left(\frac{\partial h_1^*}{\partial r^*} h_2^* + h_1^* \frac{\partial h_2^*}{\partial r^*} \right) z^* \right] \right\} - C(r^*) \tag{2-50}$$

in which $C(r^*)$ is an as yet undetermined function of r^*.

With (2-49) and (2-50), equation (2-31) tells us ($i = 1, 2$)

$$- \frac{\partial h_i^*}{\partial t^*} = \frac{1}{2} \left(\frac{h_o}{R_o} \right)^2 \left\{ \left[\frac{\partial^2 \mathcal{P}^*}{\partial r^{*2}} + \frac{1}{r^*} \frac{\partial \mathcal{P}^*}{\partial r^*} \right] \right.$$

$$\times \left[\frac{1}{3} h_i^{*3} - \frac{1}{2} (h_1^* + h_2^*) h_i^{*2} + h_1^* h_2^* h_i^* \right]$$

$$+ \frac{\partial \mathcal{P}^*}{\partial r^*} \left[- \left(\frac{\partial h_1^*}{\partial r^*} + \frac{\partial h_2^*}{\partial r^*} \right) \frac{h_i^{*2}}{2} \right.$$

$$\left. \left. + \left(\frac{\partial h_1^*}{\partial r^*} h_2^* + h_1^* \frac{\partial h_2^*}{\partial r^*} \right) h_i^* \right] \right\} - C(r^*) \tag{2-51}$$

and the difference of these two expressions gives

$$\frac{\partial h^*}{\partial t^*} = \left(\frac{h_o}{R_o} \right)^2 \left\{ \frac{1}{12} \left[\frac{\partial^2 \mathcal{P}^*}{\partial r^{*2}} + \frac{1}{r^*} \frac{\partial \mathcal{P}^*}{\partial r^*} \right] h^{*3} \right.$$

$$\left. + \frac{1}{4} h^{*2} \frac{\partial h^*}{\partial r^*} \frac{\partial \mathcal{P}^*}{\partial r^*} \right\} \tag{2-52}$$

Taking the difference between (2-27) and (2-29), recognizing (2-14), (2-42), and (2-48), and applying the appropriate expressions for the dimensionless mean curvatures H_i^* ($i = 1, 2$), we see

$$N_{ca} (p^* - p_o^*) + \frac{1}{R_d^*}$$

$$= N_{ca}(\mathcal{P}^* - \mathcal{P}_o^*) - N_{ca}\left[\phi^{(C)*}(h_2^*) - \frac{\rho^{(D)}}{\rho^{(C)}}\phi^{(B)*}(h_2^*)\right] + \frac{1}{R_d^*}$$

$$= N_{ca}(\mathcal{P}^* - \mathcal{P}_o^*) - \left(\Phi_\infty^* + \frac{h_o^*}{R_o^*}\frac{B^*}{h^{*m}}\right) + \frac{1}{R_d^*}$$

$$= -\frac{1}{2}\frac{h_o}{R_o}\frac{1}{r^*}\frac{\partial}{\partial r^*}\left(r^*\frac{\partial h^*}{\partial r^*}\right) \tag{2-53}$$

In writing this, we have taken $\gamma^* = 1$ by assumption iv and we have defined

$$\Phi_\infty^* \equiv \frac{R_o}{\gamma_o}\Phi_\infty \qquad B^* \equiv \frac{R_o^2 B^{(CD)}}{\gamma_o h_o^{m+1}} \tag{2-54}$$

Inserting (2-53) into (2-52), we discover

$$-\frac{\partial h^*}{\partial t^{*\prime}} = \frac{1}{3}h^{*3}\left(\frac{1}{r^{*3}}\frac{\partial h^*}{\partial r^*} - \frac{1}{r^{*2}}\frac{\partial^2 h^*}{\partial r^{*2}} + \frac{2}{r^*}\frac{\partial^3 h^*}{\partial r^{*3}} + \frac{\partial^4 h^*}{\partial r^{*4}}\right)$$

$$+ h^{*2}\frac{\partial h^*}{\partial r^*}\left(-\frac{1}{r^{*2}}\frac{\partial h^*}{\partial r^*} + \frac{1}{r^*}\frac{\partial^2 h^*}{\partial r^{*2}} + \frac{\partial^3 h^*}{\partial r^{*3}}\right)$$

$$+ \frac{2}{3}m^*B^*\left[\frac{1}{r^*}\frac{1}{h^{*m-2}}\frac{\partial h^*}{\partial r^*} + \frac{1}{h^{*m-2}}\frac{\partial^2 h^*}{\partial r^{*2}}\right.$$

$$\left. - \frac{m-2}{h^{*m-1}}\left(\frac{\partial h^*}{\partial r^*}\right)^2\right] \tag{2-55}$$

where

$$t^{*\prime} \equiv \frac{t\mu}{8\rho^{(C)}R_o^2 N_{ca}}\left(\frac{h_o}{R_o}\right)^3 \tag{2-56}$$

Note that, after an application of L'Hospital's rule with full recognition of (2-33), (2-36), and (2-37), we obtain

$$\text{limit } r^* \to 0: \quad -\frac{\partial h^*}{\partial t^{*\prime}} = \frac{8}{9} h^{*3} \frac{\partial^4 h^*}{\partial r^{*4}} + \frac{4}{3} \frac{mB^*}{h^{*m-2}} \frac{\partial^2 h^*}{\partial r^{*2}} \tag{2-57}$$

Our first objective is to calculate the initial dependence of h^* upon radial position consistent with assumption xi. Recognizing that the rate of thinning is independent of radial position at the initial time, we can use (2-54) and (2-56) to say at $t^{*\prime} = 0$

$$\frac{8}{3}\left(\frac{\partial^2 h^*}{\partial r^{*4}}\right)_{r^*=0} + 4mB^*\left(\frac{\partial^2 h^*}{\partial r^{*2}}\right)_{r^*=0}$$

$$= h^{*3}\left(\frac{1}{r^{*3}}\frac{\partial h^*}{\partial r^*} - \frac{1}{r^{*2}}\frac{\partial^2 h^*}{\partial r^{*2}} + \frac{2}{r^*}\frac{\partial^3 h^*}{\partial r^{*3}} + \frac{\partial^4 h^*}{\partial r^{*4}}\right)$$

$$+ 3h^{*2}\frac{\partial h^*}{\partial r^*}\left(-\frac{1}{r^{*2}}\frac{\partial h^*}{\partial r^*} + \frac{1}{r^*}\frac{\partial^2 h^*}{\partial r^{*2}} + \frac{\partial^3 h^*}{\partial r^{*3}}\right)$$

$$+ 2m^*B^*\left[\frac{1}{r^*}\frac{1}{h^{*m-2}}\frac{\partial h^*}{\partial r^*} + \frac{1}{h^{*m-2}}\frac{\partial^2 h^*}{\partial r^{*2}}\right.$$

$$\left. - \frac{m-2}{h^{*m-1}}\left(\frac{\partial h^*}{\partial r^*}\right)^2\right] \tag{2-58}$$

We require the result be consistent with (2-33), (2-36), and (2-32) in the form

$$\text{at } r^* = 1: \quad \frac{\partial h^*}{\partial r^*} = 0 \tag{2-59}$$

and with (2-38) expressed as

$$\text{as } r^* \to R_h^*: \quad \frac{1}{r^*}\frac{\partial}{\partial r^*}\left(r^*\frac{\partial h^*}{\partial r^*}\right) \to \frac{2}{R_d}\frac{R_o^2}{h_o} \tag{2-60}$$

where p_h^* has been determined by (2-42) and p^* by (2-53).

In order to integrate a finite-difference form of (2-58), we replace (2-60) by

$$\text{at } r^* = 1 : \frac{\partial^2 h^*}{\partial r^{*2}} = C \tag{2-61}$$

where C, which is the difference between the sum of the principal curvatures for interface 1 and the sum of the principal curvatures for interface 2, is a free parameter, the value of which will be determined shortly.

For each value of C, we can determine for a given value of B a tentative initial configuration of the film by integrating (2-58) consistent with (2-33), (2-37), (2-59), and (2-61). The dimensionless radial position at which the pressure gradient becomes negligible is tentatively identified as R_h, subject to later verification that assumption ii is still satisfied at this point.

Equation (2-55) can be integrated consistent with each of these tentative initial configurations, (2-33), (2-37), (2-9) and (2-10), the latter two boundary conditions first having been made dimensionless. Equation (2-32) permits us to identify R^* as a function of time; R_f^* is the value of R^* as $t^{*\prime} \to \infty$, which can be obtained from our numerical computation. We employed the Crank-Nicolson technique (Myers 1971); accuracy was checked by decreasing the time and space intervals. We used $\Delta r^* = 0.02$ and $\Delta t^{*\prime} = 0.02$ to 0.05.

For a drop freely approaching a liquid-liquid interface under the influence of gravity and of a disjoining pressure that is greater than or equal to zero, R_d is measured and R_f is determined by (Chappelear 1961; Princen 1963; Lin and Slattery 1982b)

$$\text{as } t^{*\prime} \to \infty : \; R \to R_f = \left(\frac{4}{3} \frac{\Delta\rho g}{\gamma_o} \right)^{1/2} (R_d)^2 \tag{2-62}$$

From Figures 2-2 through 2-7, as well as Figures 2 through 5 of Chen and Slattery (1982), we see that R_f^* is independent of the magnitude and sign of the disjoining pressure. There is no reason to believe that the initial film radius R_o should be dependent upon the magnitude or sign of the disjoining pressure. It follows that R_f must also be independent of the magnitude and sign of the disjoining pressure. This implies that (2-62) is valid for the case of a negative disjoining pressure as well, in the limit as the time at which the film ruptures is approached.

Having determined R_f from (2-62) and $R_f^* = 1.1$ from our numerical computation, we can identify $R_o = R_f/R_f^*$. From our numerical computation, we obtain $1/r^*[\partial/\partial r^*(r^* \partial h^*/\partial r^*)] = 12.58$ which allows us to determine h_o from (2-60).

In addition to requiring that at time $t^{*\prime} = 0$ the thinning rate is independent of radial position, assumption xi demands that for $t^{*\prime} > 0$ the thinning rate at the center is always greater than the thinning rate at the rim, so long as the effects of any disjoining pressure are negligible. Our numerical computations indicate that, for sufficiently small $B^* > 0$, there is a minimum value of C such that the thinning rate at the center is always greater than the thinning rate at the rim in the early state of the thinning process where the effects of the disjoining pressure can be neglected. For each minimum value of C, there is a corresponding maximum value of h_o for which the thinning rate at the center is always greater than the thinning rate at the rim for $t^{*\prime} > 0$. We will choose this maximum value of h_o as our initial film thickness at the center.

We must now check whether assumption ii is satisfied at R_h^*; if it is satisfied here, it will be satisfied everywhere. It is desirable to choose R_h^* as large as possible, in order to make the pressure gradient at this point clearly negligible. But if R_h^* is assigned too large a value, assumption ii will be violated. If assumption ii cannot be satisfied by reducing R_h^*, C must be increased. For this reason, $h^*(r^*, t^{*\prime})$ is weakly dependent upon the bubble radius and the other physical parameters that enter (2-62).

We have from (2-29), (2-48), and (2-53)

$$\frac{1}{r^*}\frac{\partial}{\partial r^*}\left(r^*\frac{\partial h_2^*}{\partial r^*}\right) = -\frac{1}{2}\frac{1}{r^*}\frac{\partial}{\partial r^*}\left(r^*\frac{\partial h^*}{\partial r^*}\right) - \frac{R_o}{h_o}\frac{1}{R_d^*} \tag{2-63}$$

Given h^*, we can integrate this consistent with (2-33) and (2-38) to determine h_2^*, the interface between phases C and D in Figure 2-1. Having found h_2^* and h^*, we can compute h_1^* by difference.

Finally, we can examine assumption iv that the interfacial tension gradient required to achieve an immobile interface is very small. Given (2-49) and (2-53), and either (2-26) or (2-28), we can reason that

$$\frac{\partial\gamma^*}{\partial r^*} = -\frac{1}{2}\left(\frac{R_o}{h_o}\right)^2 h^*\left\{\frac{1}{2}\frac{\partial}{\partial r^*}\left[\frac{1}{r^*}\frac{\partial}{\partial r^*}\left(r^*\frac{\partial h^*}{\partial r^*}\right)\right] + \frac{mB^*}{h^{*\,m+1}}\right\} \tag{2-64}$$

This can be integrated consistent with

$$\text{at } r^* = R_h^*: \ \gamma = 1 \tag{2-65}$$

which is in effect a definition for γ_o. Lin and Slattery (1982b) have found that the results of their integrations are consistent with assumption iv for cases where the effects of the London-van der Waals forces are negligible. They did not consider examples in which the effects of the London-van der Waals forces were important. See Exercise 3.3.2-1 for a different point of view.

Results: film profiles Figures 2-2 through 2-7 show the dimensionless film thickness h^* as a function r^* and $t^{*\prime}$ for m = 4 and varying values of $B^* \geq 0$. In each case the initial profiles are identical, suggesting that, for sufficiently small values of B^*, the effects of the disjoining pressure can be neglected during the early stage of the thinning process. Figure 2-8 also shows that the London-van der Waals forces are ineffective in the early stages of drainage, effective in the intermediate stages, and dominant in the latter stages. For relatively small values of

B^*(= 10^{-7} to 10^{-8}) the increase in the thinning rate that leads to film rupture is much smaller than that for larger values of B^*(= 10^{-4} to 10^{-5}). For very small values of B^*, the film becomes very thin at the rim before the effect of the London-van der Waals forces is noticed.

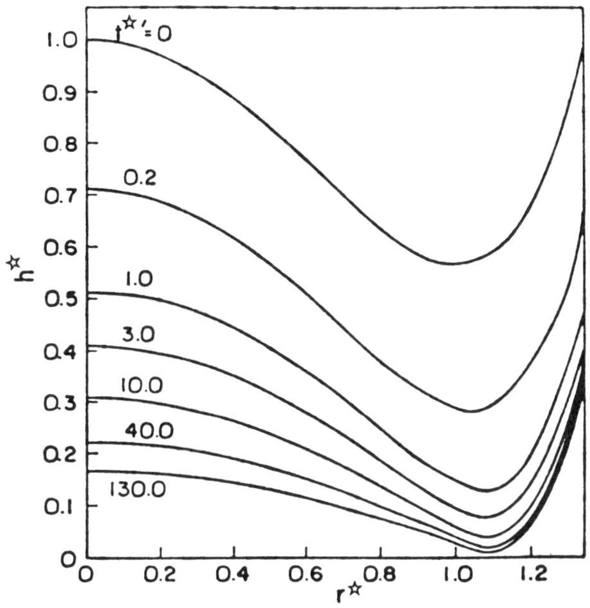

Figure 3.3.2-2: Dimensionless film thickness, h^*, as function of dimensionless radial position and dimensionless time for $R_h^* = 1.69$, $B^* = 0$, $m = 4$, $C = 5.05$.

There have been very few experimental studies of the film configuration as a function of time. Hartland (1967, 1968) observed large drops for which the analysis developed here is not applicable (assumption xii). Hodgson and Woods (1969), Woods and Burrill (1972), Burrill and Woods (1973a,b) and Liem and Woods (1974a) found five distinct drainage patterns with small drops. Most of their results show asymmetric patterns, indicating instabilities. We could not expect good agreement between this portion of their data and the theory developed here, since we have assumed symmetric profiles (assumption i). We have carried out our computations for only two of their systems.

Since insufficient data were given to identify our $t^{*\prime} = 0$ [see assumption xi and (2-57)], we related our time scale to the experimental time scales by matching in each case the initial measurement of the film thickness at the rim. As we shall demonstrate, we found it satisfactory to identify $R_h^* = 1.69$ for both systems. [Remember that R_h^* should be sufficiently large for the pressure gradient to be negligible and yet sufficiently small for assumption ii to remain valid.] Since the film thickness at the rim is generally larger than 120Å, we used m = 4 to describe the fully retarded London-van der Waals interaction potential in the film (Black *et al.* 1960; Churaev 1974a,b). Because no values of B are available for these systems, we chose in both cases as an order of magnitude approximately $B = 10^{-19}$ erg cm (assumption xv).

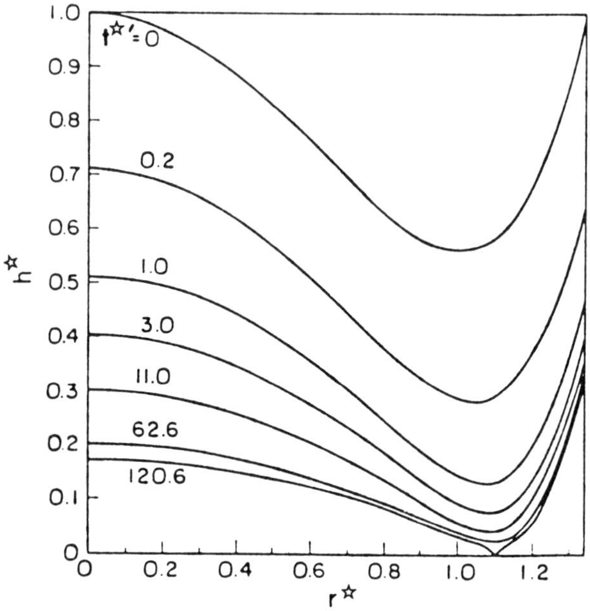

Figure 3.3.2-3: Dimensionless film thickness, h^*, as a function of dimensionless radial position and dimensionless time for $R_h^* = 1.69$, $B^* = 10^{-8}$, m = 4, C = 5.05.

Cyclohexanol-water Burrill and Woods (1973b) report observations for a cyclohexanol drop approaching a water-cyclohexanol interface: $R_d = 0.62$ mm, $\gamma_o = 3.39$ mN/m, $\mu = 1.0$ mPa s, $\Delta\rho = 5.1 \times 10^{-2}$ g/cm³. The water contained 10^{-4} g/L sodium lauryl sulfate and 0.05 N KC1. During the first 60 s of elapsed time, the film alternately drained symmetrically and asymmetrically. Thereafter, it drained symmetrically, until it ruptured.

We find that at our $t^{*\prime} = 0$ [Burrill and Woods (1973b) t = 38 s] when the thinning rate is independent of position

$$R_o = \frac{R_f}{1.1} = 1.44 \times 10^{-2} \text{ cm}$$

$$h_o = h(0, 0) = 5.30 \times 10^{-4} \text{ cm}$$

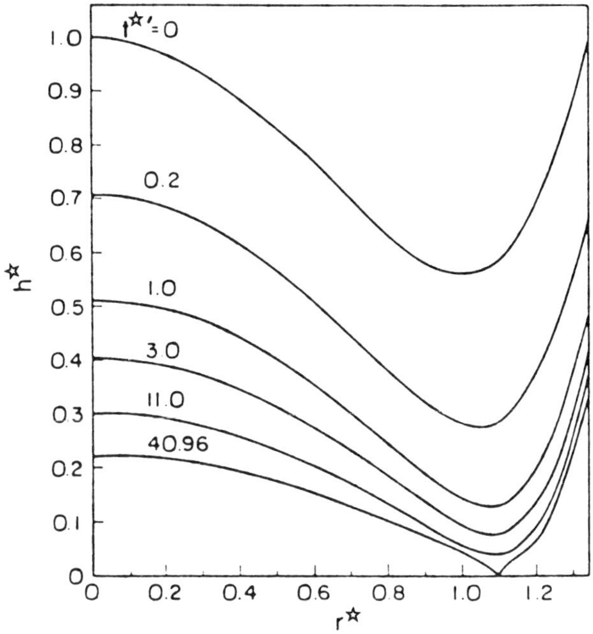

Figure 3.3.2-4: Dimensionless film thickness, $h^\star$, as a function of dimensionless radial position and dimensionless time for $R_h^\star = 1.69$, $B^\star = 10^{-7}$, $m = 4$, $C = 5.05$.

and

$$\left(\frac{h_o}{R_o}\right)^2 = 1.36 \times 10^{-3}$$

which means that the Reynolds lubrication theory approximation (assumption iii) is applicable. At $r^\star = R_h^\star = 1.69$,

$$\left(\frac{\partial h}{\partial r}\right)^2 = 5.03 \times 10^{-2}$$

and assumption ii is still satisfied at this point.

Burrill and Woods (1973b) measured film thickness as a function of time only at the center and at the rim. Figure 2-9 compares our computations ($R_h^\star = 1.69$, $B^\star = 1.26 \times 10^{-7}$, $m = 4$, $C = 5.05$), Lin and Slattery's (1982b) computations ($R_h^\star = 1.69$, $B^\star = 0$, $C = 5.05$), Burrill and Woods' (1973b) data, and Princen's (1963) estimation. [For a discussion of

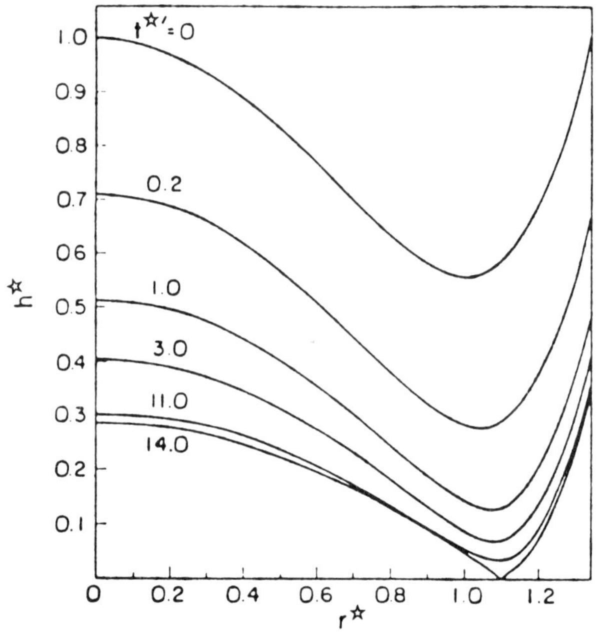

Figure 3.3.2-5: Dimensionless film thickness, h^*, as a function of dimensionless radial position and dimensionless time for $R_h^* = 1.69$, $B^* = 10^{-6}$, $m = 4$, $C = 5.05$.

the limitations of Princen's (1963) approximation, see Lin and Slattery (1982b).] Both our time scale and Princen's (1963) time scale were adjusted by matching the theoretical predictions to the first experimental measurement at the rim. At shorter times ($t < 100$ s) when the film thickness is larger than the distance over which the London-van der Waals forces are effective, our predictions are identical with those of Lin and Slattery (1982b). But at longer times ($t > 120$ s), our theory predicts smaller rim thicknesses, larger center thicknesses, and slower draining than that of Lin and Slattery (1982b). This is attributable to the London-van der Waals component of the negative disjoining pressure, the effect of which is more pronounced at longer times when the film thickness at the rim is smaller.

Toluene-water Liem and Woods (1974a) studied a toluene drop approaching a water-toluene interface: $R_d = 0.842$ mm, $\gamma_o = 33.5$ mN/m, $\mu = 1.0$ mPa s, $\Delta\rho = 1.33 \times 10^{-1}$ g/cm^3. The water contained 10^{-4} M palmitic acid.

At our $t = 0$ [Liem and Woods (1974a) $t = -0.99$ s],

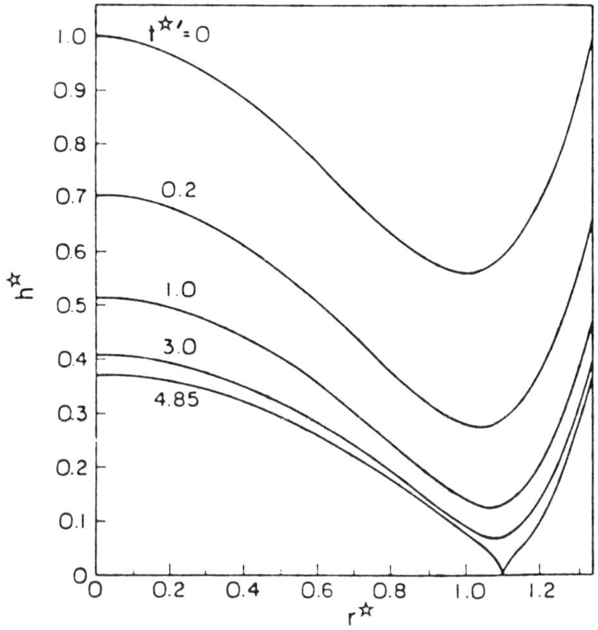

Figure 3.3.2-6: Dimensionless film thickness, h^*, as a function of dimensionless radial position and dimensionless time for $R_h^* = 1.69$, $B^* = 10^{-5}$, $m = 4$, $C = 5.05$.

$$R_o = \frac{R_f}{1.1} = 1.47 \times 10^{-2} \text{ cm}$$

$$h_o = 4.06 \times 10^{-4} \text{ cm}$$

and

$$\left(\frac{h_o}{R_o}\right)^2 = 7.64 \times 10^{-4}$$

consistent with the Reynolds lubrication theory approximation (assumption iii). At $r^* = R_h^* = 1.69$,

$$\left(\frac{\partial h}{\partial r}\right)^2 = 2.83 \times 10^{-2}$$

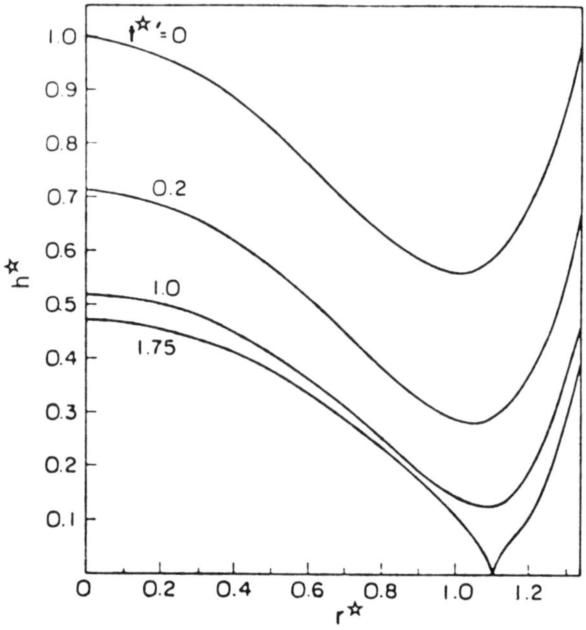

Figure 3.3.2-7: Dimensionless film thickness, h^*, as a function of dimensionless radial position and dimensionless time for $R_h^* = 1.69$, $B^* = 10^{-4}$, $m = 4$, $C = 5.05$.

which indicates that assumption ii was not violated.

Figure 2-10 compares our computations ($R_h^* = 1.69$, $B^* = 5.80 \times 10^{-8}$, $m = 4$, $C = 5.05$), Lin and Slattery's (1982b) computations ($R_h^* = 1.69$, $B^* = 0$, $C = 5.505$), Liem and Woods' (1974a) data, and Princen's (1963) estimation.

Discussion There are several possible explanations for the difference between our computations and the experimental data shown in Figures 2-9 and 2-10.

In most of the experimental data of Burrill and Woods (1973b) and Liem and Woods (1974a), there is evidence of asymmetric drainage patterns, indicating instabilities. To the extent that this was true also for the data discussed here, we could not expect good agreement between our theory and the experimental data, since our theory assumes symmetric profiles (assumption i). For more about the development of asymmetric instabilities, see below.

Since the experimental film thickness at the rim is substantially less than 400 Å at long times, it may not be justified to assume $m = 4$ (assumption xv).

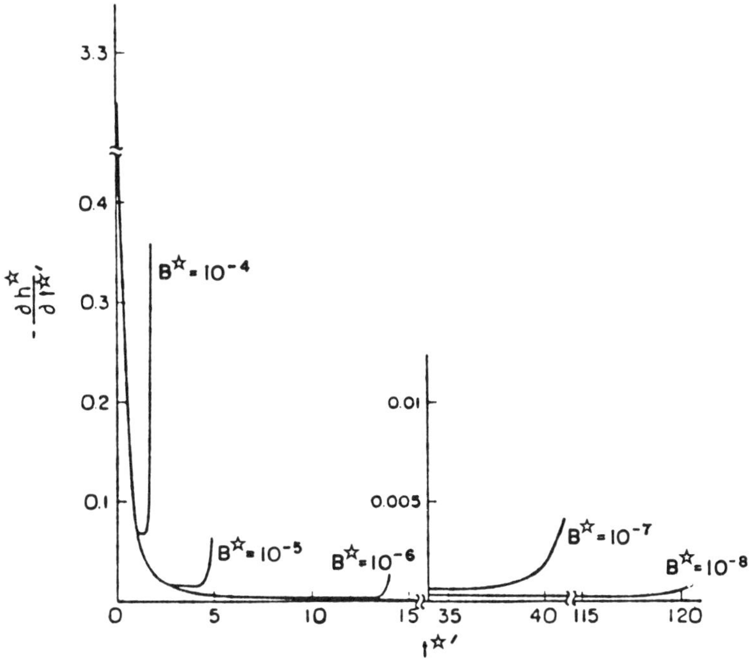

Figure 3.3.2-8: Dimensionless thinning rate $\partial h^*/\partial t^{*\prime}$ at the rim as a function of dimensionless time for different values of B^*.

We know that the value for B we have assumed is only an approximation (assumption xv).

The interfaces may not have been entirely immobile as presumed both in our theory (assumption iv) and in the estimation of Princen (1963).

Results: coalescence times Let us define $t_c^{*\prime}$ as the dimensionless time at which the film thickness at the rim goes to zero or the dimensionless time at which the film ruptures. We will refer to this as our dimensionless coalescence time.

Our numerical computations give the graphical relationship between $t_c^{*\prime}$ and B^* shown in Figure 2-11. Alternatively, we can express this relationship as

$$t_c^{*\prime} = 2.5 \times 10^{-2}\, B^{*\,-0.46} \tag{2-66}$$

In using this relationship, we recommend that (2-62) be employed to identify

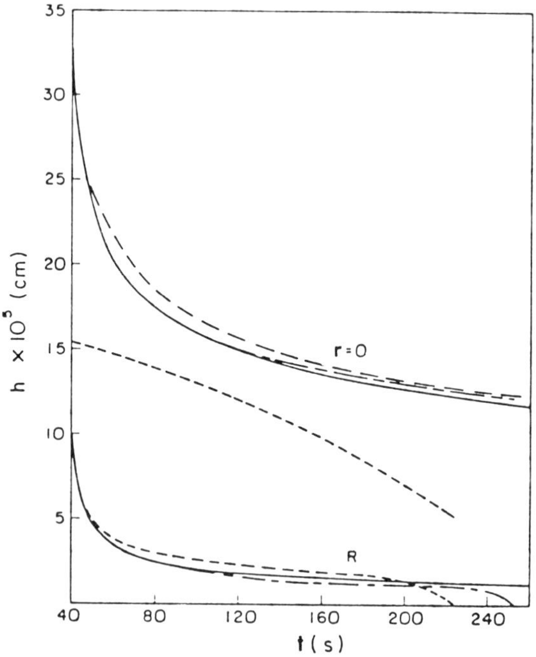

Figure 3.3.2-9: Comparisons of the film thickness at the center and at the rim calculated using the present theory ($R_h^\star = 1.69$, $B^\star = 1.26 \times 10^{-7}$, $m = 4$, $C = 5.05$, ——— – ———). Lin and Slattery's theory (1982b, $R_h^\star = 1.69$, $B^\star = 0$, $C = 5.05$, ————), and Princen's estimation (1963, $n = 2$, —— —— ——), with those measured by Burrill and Woods (1973b, ——) for a cyclohexanol-water system. Here t represents the experimental time scale of Burrill and Woods (1973b).

$$R_o = \frac{R_f}{1.1} = 1.05 \left(\frac{\Delta \rho g}{\gamma_o} \right)^{1/2} R_d^2 \tag{2-67}$$

The dimensionless mean curvature at $r^\star = R_h^\star$ was generated in our computation to be 12.58. This together with (2-60) and (2-67) fix the initial film thickness at the center:

$$h_o = 0.175 \frac{\Delta \rho g (R_d)^3}{\gamma_o} \tag{2-68}$$

In view of (2-56), (2-67), and (2-68), equation (2-66) may be rearranged as

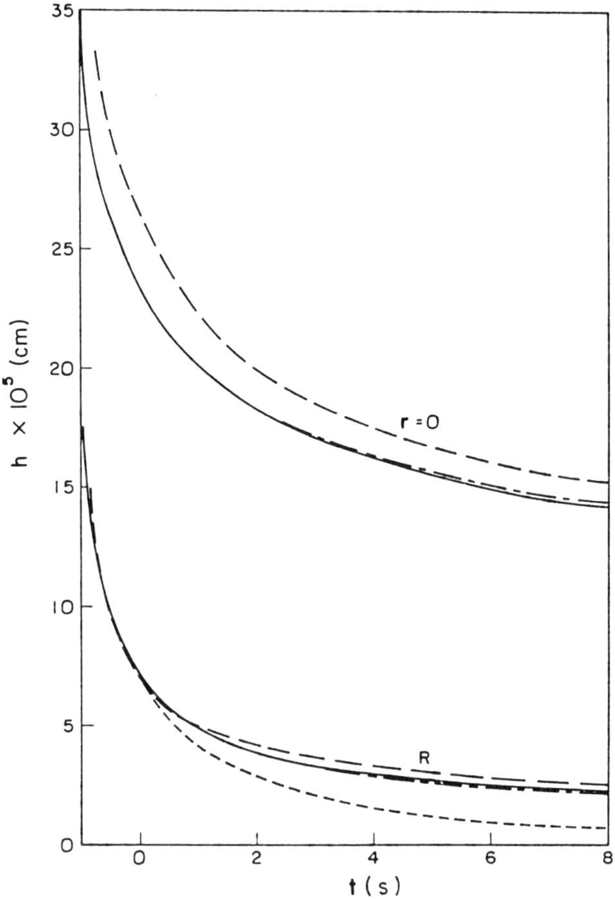

Figure 3.3.2-10: Comparisons of the film thickness at the center and at the rim calculated using the present theory ($R_h^* = 1.69$, $B^* = 5.80 \times 10^{-8}$, $m = 4$, $C = 5.05$, ——— – ———). Lin and Slattery's theory (1982b, $R_h^* = 1.69$, $B^* = 0$, $C = 5.05$, ————), and Princen's estimation (1963, $n = 2$, ——— ——— ———), with those measured by Burrill and Woods (1974, ——) for a toluene-water system. Here t represents the experimental time scale of Liem and Woods (1974a).

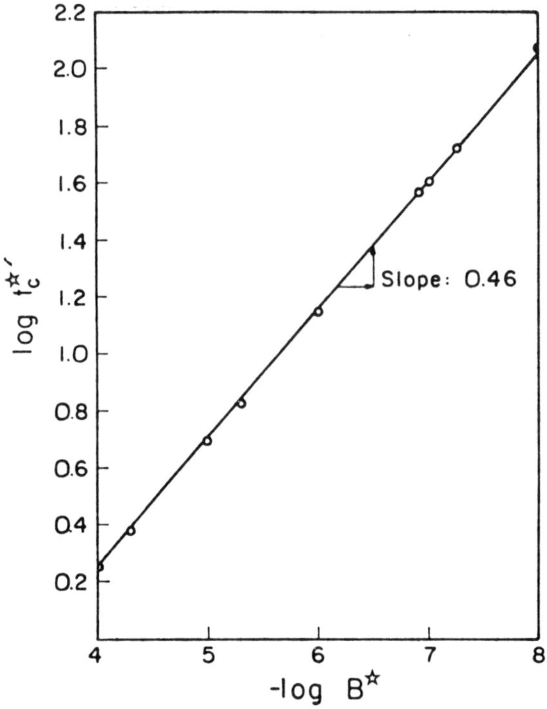

Figure 3.3.2-11: Dependence of $t_c^{*\prime}$ on B^*.

$$t_c = 0.79 \frac{\mu(R_d)^{4.06}(\Delta\rho g)^{0.84}}{\gamma_o^{1.38} B^{0.46}} \tag{2-69}$$

Allan *et al.* (1961) report that, instead of (2-62), their experimental measurements of R_f^* were better described as though the drop were approaching an immobile, plane interface. This implies as an alternative to (2-67) (Derjaguin and Kussakov 1939; Lin and Slattery 1982a)

$$R_o = \frac{R_f}{1.1} = 0.742 \left(\frac{\Delta\rho g}{\gamma_o}\right)^{1/2} (R_d)^2 \tag{2-70}$$

The corresponding initial film thickness at the center is

$$h_o = 8.75 \times 10^{-2} \frac{\Delta\rho g R_d^3}{\gamma_o} \tag{2-71}$$

Using (2-56), (2-70), and (2-71), we find that (2-66) gives

$$\bar{t}_c = 0.44 \frac{\mu(R_d)^{4.06} (\Delta\rho g)^{0.84}}{\gamma_o^{1.38} B^{0.46}} \tag{2-72}$$

[We use the overbar here to remind ourselves that (2-70) is being employed rather than (2-67).]

Table 2-1 lists the theoretical expressions for the coalescence time developed by Chen *et al.* (1984) for their model, for the plane parallel disc model, and for the cylindrical drop model.

The experimentally observed rest time or coalescence time is the elapsed time during which an experimentalist perceives a drop to be resting on an interface before coalescence occurs. It does not begin at what we have defined to be t = 0 in our computations, but the difference must be very small. Marrucci (1969) describes coalescence as occurring in two stages. During the first stage, the dimpled film is formed very rapidly; during the second, the film drains until it ruptures and coalescence occurs. Our computations, as well as those described by Chen *et al.* (1984), are intended to approximate this second stage, neglecting the very small additional time required to execute the first stage.

For lack of better information, we again choose $B = 10^{-19}$ erg cm and m = 4 (assumption xv) in comparing these various theoretical results with the available experimental data.

Comparing our results in the forms of (2-69) and (2-72) with those developed by Chen *et al.* (1984) in Table 2-1, we see that the general trends are similar. The functional dependence of coalescence time upon system properties such as drop diameter, viscosity of the draining film, interfacial tension, and the London-van der Waals constant B was discussed by Chen *et al.* (1984). The cylindrical drop model predicts that the coalescence time is independent of the density difference in contrast with the other models, but the available experimental data do not permit discrimination on this basis.

Equations (2-69) and (2-72) are in good agreement with the experimental data of MacKay and Mason (1963) in Table 2-2. Since they used all of their materials in the condition in which they were received and since it is very difficult to produce and maintain uncontaminated aqueous solutions, we follow the suggestion of Platikanov (1964) in assuming that some surface-active material was present and that both interfaces are immobile.

The comparison with the experimental data of Woods and Burrill (1972; Burrill and Woods 1973b) in Table 2-3 is mixed. In going from left to right in Table 2-3, they experienced increasing difficulties with asymmetric instabilities, which are not taken into account in our analysis. Our computations also neglect the effect of any electrostatic double layer (assumption ix). They added KCl in an attempt to minimize this effect, but clearly there was some difference between the two cyclohexanol-water

Table 3.3.2-1: Other expressions for coalescence time from a linear stability analysis of Chen *et al.* (1984)[b]

Coalescence Time	Remarks
$t_{c(4)} = 1.070 \dfrac{\mu R_d^{3.4}}{\gamma^{1.2} B^{0.4}} (\Delta\rho g)^{0.6}$ $\bar{t}_{c(4)} = 0.705 \dfrac{\mu R_d^{3.4}}{\gamma^{1.2} B^{0.4}} (\Delta\rho g)^{0.6}$	rim of dimpled film
$t_{c(M)} = 1.046 \dfrac{\mu R_d^{4.5}}{\gamma^{1.5} B^{0.5}} (\Delta\rho g)$ $\bar{t}_{c(M)} = 0.370 \dfrac{\mu R_d^{4.5}}{\gamma^{1.5} B^{0.5}} (\Delta\rho g)$	plane parallel disc model (Mackay and Mason 1963)
$t_{c(H)} = 5.202 \dfrac{\mu R_d^{1.75}}{\gamma^{0.75} B^{0.25}}$	cylindrical drop model (Hodgson and Woods 1969)

systems shown that was not taken into account here. Burrill and Woods (1973a) pointed out that the cause of film rupture was often seen to be dust, dirt, or local depressions in film thickness. Since our computation assumes symmetric drainage and fully immobile interfaces free of dust, our estimate of the coalescence time could be expected at best to be only an upper bound, given the proper magnitude of B and assuming that electrostatic forces could be neglected.

b) Overbar indicates that (2-70) was used rather than (2-67).

Table 3.3.2-2: Comparisons with MacKay and Mason's (1963) experimental data

System	1	2	3	4	5
γ (mN/m)	14.0	34.1	19.1	34.1	19.1
$\Delta\rho$(gm/cm^3)	0.0538	0.1145	0.0483	0.1145	0.0483
μ(mN s/m^2) $\times 10^{-2}$	0.0522	0.0463	0.192	0.01	0.01
R_d(cm)	0.0325	0.0425	0.0425	0.0425	0.0425
N_{Bo}	0.0040	0.0059	0.0045	0.0059	0.0045
t_{exp}(S)	8.1	17.6	>90	< 4	< 4
$t_{c(4)}$(s) [$\overline{t}_{c(4)}$(s)]	8.8 [5.8]	10.5 [6.9]	51.6 [34]	2.3 [1.5]	2.7 [1.8]
$t_{c(M)}$(s) [$\overline{t}_{c(M)}$(s)]	34.8 [12.3]	58.0 [20.5]	242 [85.7]	12.4 [4.4]	12.7 [4.5]
$t_{c(H)}$(s)	5.2	3.8	24.5	0.8	1.3
t_c(s) [$\overline{t}_c$(s)]	15.0 [8.4]	21.8 [12.8]	97.7 [54.4]	4.7 [2.6]	5.1 [2.8]

drop - bulk phase

System 1: H_2O - cinnamaldehyde
System 2: H_2O - diphenyl sulphide
System 3: H_2O - dibutyl phthalate
System 4: diphenyl sulphide - H_2O
System 5: dibutyl phthalate - H_2O

All theories other than the cylindrical drop model overestimate the coalescence times observed by Allan *et al.* (1961) in Table 2-4. We do not attach any significance to the fact that the cylindrical drop model gives a better representation of the experimental data. The analysis of Hodgson and Woods (1969) was unrealistic (Chen *et al.* 1984). Perhaps our estimate of B was too small for this system (assumption xv) or the interfaces are partially mobile.

Table 3.3.2-3: Comparisons with data of Woods and Burrill (1972; Burrill and Woods 1973b)

System	1		2	3	4
Conc. SLS(gm/l)[c]	10^{-4}	10^{-6}	10^{-6}	10^{-6}	10^{-6}
Normality KC1	0.05	0.01	0.01	0.01	0.01
γ (mN/m)	3.93		20.5(25)[e]	28.9	35(33.5)[e]
$\Delta\rho$(gm/cm^3)	0.051		0.0097	0.053	0.133
μ(mN s/m^2) $\times 10^{-2}$	0.01		0.01	0.1	0.01
R_d(cm)	0.062		0.1061	0.1061	0.1061
N_{Bo}	0.049		0.0052	0.020	0.042
t_{exp}(s)	~240	<110	< 15	< 16	< 12
$t_{c(4)}$(s) [$\bar{t}_{c(4)}$(s)]	67.5 [44.5]		21.3 [14.1]	39.2 [25.8]	54.1 [35.6]
$t_{c(M)}$(s) [$\bar{t}_{c(M)}$(s)]	780 [276]		141.3 [50.0]	455.2 [161.5]	859.4 [304.0]
$t_{c(H)}$(s)	8.1		6.0	4.6	4.0
t_c(s) [$\bar{t}_c$(s)]	219 [122]		49.2 [27.4]	127 [71.0]	212 [118]
Remarks	Composite of even and uneven drainage		Uneven drainage at t=1(s)	Uneven drainage from t=4(s)	Uneven drainage at t=8(s)

drop - bulk phase

System 1: cyclohexanol - water
System 2: anisol - water
System 3: CA[d] - water
System 4: toluene - water

---

c) Sodium lauryl sulfate

Notice in Tables 2-2 through 2-4 that our estimate t_c for the coalescence time is always two to three times larger than the comparable estimate $t_{c(4)}$ by Chen *et al.* (1984). This might suggest that Chen *et al.* (1984) have more accurately accounted for the development of instabilities in their analysis. We do not believe that this is the case. The thinning at the rim driven by the London-van der Waals forces observed both in these computations and in those of Chen *et al.* (1984) is not the result of an instability, because the film is assumed to be axisymmetric. In contrast, the asymmetric thinning experimentally observed in three dimensions (Liem and Woods 1974a,b; Hartland 1969; Hodgson and Woods 1969; Burrill and Woods 1973a,b; Salajan *et al.* 1987) is likely the result of an instability.

More generally, several conclusions can be drawn.

1. Coalescence time increases with increasing bubble or drop diameter.

This is consistent with (2-69) and (2-72) as well as all of the results of Chen *et al.* (1984) shown in Table 2-1.

There have been several studies of the effect of drop size on the coalescence time (Gillespie and Rideal 1956; Elton and Picknett 1957; Nielsen *et al.* (1958); Charles and Mason 1960; MacKay and Mason 1963; Jeffreys and Hawksley 1965; Hodgson and Lee 1969; Hodgson and Woods 1969; Komasawa and Otake 1970; Lang and Wilke 1971; Woods and Burrill 1972; Burrill and Woods 1973b; Davis and Smith 1976). All workers, except Hodgson and Lee (1969), Lang and Wilke (1971), and Davis and Smith (1976), found that the coalescence time increased with drop size.

Previous workers disagree on the functional relation between coalescence time and drop size. For a water drop approaching a water-enzene interface, Charles and Mason (1960) report $t_c \sim R_d^{3.15}$ ($0.124 \leq N_{Bo} \leq 0.306$); Jeffreys and Hawksley (1965) $t_c \sim R_d$ ($0.140 \leq N_{Bo} \leq 0.430$); Lang and Wilke (1971) say t_c is independent of R_d ($0.153 \leq N_{Bo} \leq 0.604$). For a benzene drop approaching a water-benzene interface, Komasawa and Otake (1970) report $t_c \sim R_d^{2.1}$ ($0.158 \leq N_{Bo} \leq 0.455$). But note that none of these experimental studies satisfies the requirement of assumption xii of this analysis that $N_{Bo} << 1$. For comparison, we predict from (2-69) and (2-72) that $t_c \sim R_d^{4.06}$ for $N_{Bo} << 1$. The predictions of Chen *et al.* (1984) in Table 2-1 vary.

---

d) Mixture of 0.84 mole fraction anisole with 0.16 mole fraction cyclohexane

e) Liem and Woods (1974a) report different values from those given by Woods and Burrill (1972; Burrill and Woods 1973a). Here we use only the first values.

Table 3.3.2-4: Comparisons with data of Allan *et al.* (1961) for rising nitrogen bubbles at 20°C

System	1	2	3	4	5	6	7	8
γ(mN/m)	63.6	64.1	64.5	61.5	52.0	35.3	34.5	32.0
ρ(gm/cm^3)	1.2535	1.2367	1.2221	1.2371	1.2362	1.0447	1.0378	1.0295
μ(mN s/m^2) $\times 10^{-2}$	8.05	2.65	1.15	2.75	2.61	8.02	2.63	1.13
R_d(cm)	0.0285	0.0285	0.0285	0.0285	0.0285	0.0275	0.0275	0.0275
N_{Bo}	0.016	0.015	0.015	0.016	0.019	0.022	0.022	0.024
t_{exp}(s)	340	18.1	14.9	28.0	> 360	> 360	69.1	40.0
$t_{c(4)}$(s) [$\bar{t}_{c(4)}$(s)]	938 [618]	302 [199]	131 [86]	329 [217]	382 [252]	1507 [993]	442 [291]	247 [163]
$t_{c(M)}$(s) [$\bar{t}_{c(M)}$(s)]	7186 [2542]	2307 [816]	981 [347]	2547 [901]	3109 [1100]	12289 [4347]	4144 [1466]	1976 [699]
$t_{c(H)}$(s)	207	68	29	73	78	301	101	46
t_c(s) [$\bar{t}_c$(s)]	2370 [1320]	763 [425]	325 [181]	839 [467]	1003 [558]	3949 [2199]	1329 [740]	629 [350]

System 1: 97% aqueous glycerol (no emulsifier)
System 2: 91% aqueous glycerol (no emulsifier)
System 3: 85% aqueous glycerol (no emulsifier)
System 4: 91% aqueous glycerol (with 0.0025% Tween 20)
System 5: 91% aqueous glycerol (with 0.0250% Tween 20)
System 6: UCON oil mixture (1)
System 7: UCON oil mixture (2)
System 8: UCON oil mixture (3)

2. The coalescence time increases as the viscosity of the draining film increases, as the interfacial tension decreases, and as the strength of the London-van der Waals forces (denoted by B) decreases.

This is also consistent with (2-69) and (2-72) as well as the results of Chen *et al.* (1984) shown in Table 2-1.

All of the data in Table 2-4 support the coalescence time being proportional to the viscosity of the draining film. The data for aqueous glycerol solutions with Tween 20 is consistent with the coalescence time increasing as the interfacial tension decreases, although none of the models can explain the large difference without permitting B to change. There are no experimental data that would allow us to directly check the dependence of coalescence time upon B.

3. The coalescence time increases as the density difference between the two phases increases.

This is consistent with (2-69) and (2-72) as well as all of the results of Chen *et al.* (1984) shown in Table 2-1, with the exception of the cylindrical drop model.

Although this result appears to be supported by the data of Davis and Smith (1976) and of Salajan *et al.* (1987), no firm conclusion can be drawn. For example, in the experiments of Davis and Smith (1976; see their Table 7), for oil drops of n-alkane (C_6 to C_{16}) the density difference increased, the Hamaker constant decreased (Hough and White 1980), the interfacial tension decreased, and the electrophoretic mobility increased. All of these changes tended to favor increasing the coalescence time.

Cautions In applying these results, it is important that all of the assumptions made in this analysis are observed: $N_{Bo} << 1$, the drop or bubble is sufficiently small to be considered spherical outside the dimpled region, mass transfer can be neglected, ...

When comparing these results with experimental data, remember that, when the composition of the system is changed, many properties can be affected: interfacial tension, London-van der Waals forces, electrostatic double-layer force, interfacial viscoelasticity, density differences, ... This makes the analyses of experiments difficult and somethimes confusing. For example, Kitamura *et al.* (1988) found that coalescence time decreases with decreasing interfacial tension. They lowered the interfacial tension by adding alcohol to the aqueous phase in order to lower the adsorption of surfactant Span 80 in the interface. Some reports of experiments show a good correlation between drop stability (coalescence time) and interfacial viscoelasticity, while others indicate the opposite (Cockbain and McRoberts 1953; Nielsen *et al.* 1958; Biswas and Haydon 1962; Glass *et al.* 1970). The conclusion is that the effect of a single parameter is very difficult to establish by changing the composition of a system.

There may be similar difficulties in comparing emulsion stability with drop stability. Davis and Smith (1976) found no relationship between the two, but the surfactant concentrations that they used in the two studies differed by two orders of magnitude. In contrast, Kitamura *et al.* (1988; see also Takahashi *et al.* 1986 and Kitamura *et al.* 1987) found a good correlation for similar reagent concentrations. We would expect a good correlation, since the results for coalescence of two equal-sized drops (Chen 1985) are very similar to those for coalescence of a drop with a plane interface.

Finally, keep in mind that in the above discussion we have examined only the effects of London-van der Waals forces. We have considered neither electrostatic double-layer forces nor other types of mutual forces which have been described (Churaev and Derjaguin 1985; Israelachvili and McGuiggan 1988).

References

Allan, R. S., G. E. Charles, and S. G. Mason, *J. Colloid Sci.* **16**, 150 (1961).

Barber, A. D., and S. Hartland, *Can. J. Chem. Eng.* **54**, 279 (1976).

Biswas, B., and D. A. Haydon, *Kolloid Z. Z. Polym.* **185**, 31 (1962).

Black, W., J. G. V. de Jongh, J. Th. G. Overbeck, and M. J. Sparnaay, *Trans. Faraday Soc.* **56**, 1597 (1960).

Blake, T. D., *J. Chem. Soc. Faraday Trans. I* **71**, 192 (1975).

Buevich, Yu. A., and E. Kh. Lipkina, *Z. Prikl. Mekh. Tekh. Fiziki* **2**, 80 (1975).

Buevich, Yu. A., and E. Kh. Lipkina, *Kolloid Z.* **40**, 201 (1978).

Burrill, K. A., and D. R. Woods, *J. Colloid Interface Sci.* **42**, 15 (1973a).

Burrill, K. A., and D. R. Woods, *J. Colloid Interface Sci.* **42**, 35 (1973b).

Buscall, R., and R. H. Ottewill, in "Colloid Science," vol. 2, ed. by E. H. Everett, p. 191, The Chemical Society, London (1975).

Chappelear, D. C., *J. Colloid Sci.* **16**, 186 (1961).

Charles, G. E., and S. G. Mason, *J. Colloid Sci.* **15**, 236 (1960).

Chen, J. D., *J. Colloid Interface Sci.* **98**, 329 (1984).

Chen, J. D., *J. Colloid Interface Sci.* **107**, 209 (1985).

Chen, J. D., P. S. Hahn, and J. C. Slattery, *AIChE J.* **30**, 622 (1984).

Chen, J. D., P. S. Hahn, and J. C. Slattery, *AIChE J.* **34**, 140 (1988).

Chen, J. D., and J. C. Slattery, *AIChE J.* **28**, 955 (1982); **29**, 174, 526 (1983).

Churaev, N. V., *Colloid J. USSR* (Engl. Transl.) **36**(2), 283 (1974a).

Churaev, N. V., *Colloid J. USSR* (Engl. Transl.) **36**(2), 287 (1974b).

Churaev, N. V., and B. V. Derjaguin, *J. Colloid Interface Sci.* **103**, 542 (1985).

Cockbain, E. G., and T. S. McRoberts, *J. Colloid Sci.* **8**, 440 (1953).

Davis, S. S., and A. Smith, *Colloid Polymer Sci.* **254**, 82 (1976).

Derjaguin, B. V., and M. Kussakov, *Acta Physicochim. USSR* **10**, 25 (1939).

Elton, G. A. H., and R. G. Picknett, *Proceedings of Second International Congress of Surface Activity* **1**, 287, Academic Press, New York (1957).

Flumerfelt, R. W., J. P. Oppenheim, and J. R. Son, *AIChE Symp. Ser.* **78** (212), 113 (1982).

Gillespie, T., and E. K. Rideal, *Trans. Faraday Soc.* **52**, 173 (1956).

Glass, J. E., R. D. Lundberg, and F. F. Bailey Jr., *J. Colloid Interface Sci.* **33**, 491 (1970).

Hahn, P. S., J. D. Chen, and J. C. Slattery, *AIChE J.* **31**, 2026 (1985).

Hahn, P. S., and J. C. Slattery, *AIChE J.* **31**, 950 (1985).

Hahn, P. S., and J. C. Slattery, *AIChE Symp. Ser.* **82** (252), 100 (1986).

Hartland, S., *Trans. Inst. Chem. Eng.* **45**, T102 (1967).

Hartland, S., *Trans. Inst. Chem. Eng.* **46**, T275 (1968).

Hartland, S., *Chem. Eng. Sci.* **24**, 611 (1969).

Hodgson, T. D., and J. C. Lee, *J. Colloid Interface Sci.* **30**, 94 (1969).

Hodgson, T. D., and D. R. Woods, *J. Colloid Interface Sci.* **30**, 429 (1969).

Hough, D. B., and L. R. White, *Adv. Colloid Interface Sci.* **14**, 3 (1980).

Israelachvili, J. N., and P. M. McGuiggan, *Science* **241**, 795 (1988).

Ivanov, I. B., B. Radoev, E. Manev, and A. Sheludko, *Trans. Faraday Soc.* **66**, 1262 (1970).

Jeffreys, G. V., and J. L. Hawksley, *AIChE J.* **11**, 413 (1965).

Kitamura, Y., M. Asano, and T. Takahashi, *Sekiyu Gakkaishi* **30**, 230 (1987).

Kitamura, Y., T. Ohta, and T. Takahashi, *Sekiyu Gakkaishi* **31**, 244 (1988).

Komasawa, I., and T. Otake, *J. Chem. Eng. Japan* **3**, 243 (1970).

Lang, S. B., and C. R. Wilke, *Ind. Eng. Chem. Fundam.* **10**, 341 (1971).

Li, D., and J. C. Slattery, *AIChE J.* **34**, 862 (1988).

Li, D., and J. C. Slattery, *I & E C Research* (1990).

Liem, A. J. S., and D. R. Woods, *Can. J. Chem. Eng.* **52**, 222 (1974a).

Liem, A. J. S., and D. R. Woods, *AIChE Symposium Ser. No. 144,* **70**, 8 (1974b).

Lin, C. Y., and J. C. Slattery, *AIChE J.* **28**, 147 (1982a).

Lin, C. Y., and J. C. Slattery, *AIChE J.* **28**, 786 (1982b).

MacKay, G. D. M., and S. G. Mason, *Can. J. Chem. Eng.* **41**, 203 (1963).

Marrucci, G., *Chem. Eng. Sci.* **24**, 975 (1969).

Myers, G. E., "Analytical Methods in Conduction Heat Transfer," p. 274, McGraw-Hill, New York (1971).

Nielsen, L. E., R. Wall, and G. Adams, *J. Colloid Sci.* **13**, 441 (1958).

Platikanov, D., *J. Phys. Chem.* **68**, 3619 (1964).

Princen, H. M., *J. Colloid Sci.* **18**, 178 (1963).

Salajan, M., I. Demeter-Vodnar, E. Gavrila, and E. Chifu, *Revista de Chimie* **38**, 882 (1987).

Sheludko, A., *Adv. Colloid Interface Sci.* **1**, 391 (1967).

Slattery, J. C., "Momentum, Energy, and Mass Transfer in Continua," McGraw-Hill, New York (1972); second edition, Robert E. Krieger, Malabar, FL 32950 (1981).

Takahashi, T., Y. Kitamura, and S. Okagaki, *Kagaku Kogaku Ronbunshu* **12**, 327 (1986).

Vrij, A., *Discuss. Faraday Soc.* **42**, 23 (1966).

Woods, D. R., and K. A. Burrill, *J. Electroanal. Chem.* **37**, 191 (1972).

Exercises

3.3.2-1 *magnitude of the interfacial tension gradient* We have simplified our analysis by adopting the lubrication theory approximation (assumption iii). Alternatively, we could say that we have carried out a first-order perturbation analysis with

$$k^* \equiv \frac{h_o}{R_o} \tag{1}$$

as the perturbation parameter. One advantage of this latter point of view is that we can more easily make a statement about the magnitude of the interfacial tension gradient required to achieve an immobile interface, when the effects of surface viscosities are neglected.

Let us assume

$$\frac{\partial}{\partial r^*}, \ \frac{\partial}{\partial z^*} = O(\ 1\ ) \quad \text{as } k^* \to 0 \tag{2}$$

$$v_r^*, \ v_z^* = O(\ k^{*\alpha}\ ) \quad \text{as } k^* \to 0 \tag{3}$$

$$N_{ca} = O(\ k^{*\beta}\ ) \qquad \text{as } k^* \to 0 \tag{4}$$

Equation (2-19) implies that

$$p^* = O(\ k^{*\alpha - 2}\ ) \qquad \text{as } k^* \to 0 \tag{5}$$

From the z^*-component of the jump momentum balance, we observe that

$$\alpha + \beta = 3 \tag{6}$$

We conclude from the r^*-component of the jump momentum balance (Lin and Slattery 1982b, Eqs. 15 and 17)

$$\frac{\partial \gamma^*}{\partial r^*} = O(k^{*2}) \tag{7}$$

which can be safely neglected in our lubrication theory or first-order perturbation analysis.

This is consistent with the analysis of Lin and Slattery (1982b) and assumption iv.

3.3.2-2 *more on coalescence at fluid-fluid interfaces* As an alternative to the more complete discussion of coalescence given by Lin and Slattery (1982b) and Hahn *et al.* (1985), consider the approximate analysis of Chen *et al.* (1984) for thinning at the rim of a dimpled film that follows a suggestion of Buevich and Lipkina (1975, 1978). The results are shown in Table 2-1.

3.3.2-3 *effects of the interfacial viscosities at fluid-fluid interfaces* Follow Hahn and Slattery (1985) in modifying the development of Barber and Hartland (1976; see also Flumerfelt *et al.* 1982) to include the effects of the surface viscosities as well as those of the London-van der Waals forces for a liquid film bounded by partially mobile parallel planes.

Hahn and Slattery (1985) find that the dependence of the coalescence time upon bubble radius, the viscosity of the draining film, the surface tension, the strength of the London-van der Waals forces, and the density difference between the two phases as described by Chen *et al.* (1984) and Hahn *et al.* (1985) is moderated or even reversed by the inclusion of the effects of the surface viscosities.

3.3.2-4 *more on the effects of the interfacial viscosities at fluid-fluid interfaces* Note that Hahn and Slattery (1986) have followed Lin and Slattery (1982) and Hahn *et al.* (1985) in constructing a more complete analysis for the effects of the surface viscosities upon the coalescence time. Their qualitative conclusions are the same as those of Hahn and Slattery (1985), but they observe that the neglect of film dimpling by Hahn and Slattery (1985) led to serious errors. This is consistent with the better agreement found between the computations of Hahn and Slattery (1986) and the experimental observations of Li and Slattery (1988).

3.3.2-5 *effects of electrostatic forces at fluid-fluid interfaces* Consider the analysis of Chen *et al.* (1988) that extends the discussion in the text by including the effects of electrostatic double-layer forces as well as London-van der Waals forces.

3.3.2-6 *coalescence at fluid-solid interfaces* Follow Lin and Slattery

(1982a) and Chen and Slattery (1982) in constructing an analysis for coalescence at fluid-solid interfaces that is similar to the one given in the text for fluid-fluid interfaces.

3.3.2-7 *effects of electrostatic forces at fluid-solid interfaces* Follow Chen (1984) in extending the discussion in Exercise 3.3.2-6 to include the effects of electrostatic forces as well as London-van der Waals forces.

3.3.2-8 *mineral flotation* Follow Li and Slattery (1990) in extending the discussion in Exercise 3.3.2-5 to the attachment (coalescence) of a solid sphere to a bubble, including the effects of electrostatic forces as well as London-van der Waals forces. This is the central problem in mineral flotation.

3.3.2-9 *balance between capillary pressure and disjoining pressure in stable equilibrium film* In the systems studied by Platikanov (1964), a bubble was pressed against a solid surface through a continuous liquid phase and the draining liquid film that was formed evolved into a uniform film as equilibrium was approached. Prove that the existence of this uniform film was the result of a balance between the positive disjoining pressure $-B/h_\infty^4$ and the capillary pressure $2\gamma_o/R_b$ (Sheludko 1967; Buscall and Ottewill 1975).

Blake (1975) demonstrated that a positive disjoining pressure attributable only to London-van der Waals forces was sufficient to form such an equilibrium film. Using only values of B and the final film thickness h_∞ obtained from the computations described by Chen and Slattery (1982), we find in Table 2-5 good comparisons between the capillary pressure and the London-van der Waals disjoining pressure for the three systems studied by Platikanov (1964).

Table 3.3.2-5: Comparison of capillary pressure and disjoining pressure (Chen and Slattery 1982).

System	$\frac{2\gamma_o}{R_o}$ (Pa)	$\frac{-B}{h_\infty^4}$ (Pa)
0.1 N KC1 solution-air	60.6	61.7
aniline-air	30.8	31.1
ethanol-air	18.6	18.6

3.3.2-10 *coalescence between two equal-sized spherical drops or bubbles* Follow Chen (1985) in developing an analysis for the coalescence of two equal-sized spherical drops or bubbles.

3.3.3 Moving common line, contact angle, and film configuration (Li and Slattery 1990) A three-phase line of contact (or common line) is the curve formed by the intersection of two dividing interfaces. The movement of a common line over a rigid solid plays an important role in the spreading phenomena which occur during a wide variety of coating and displacement processes: the manufacture of coated paper, the manufacture of photographic film, the application of soil repellents to carpeting, the deposition of photo resists during the manufacture of microelectronic circuits, and the displacement of oil from an oil reservoir.

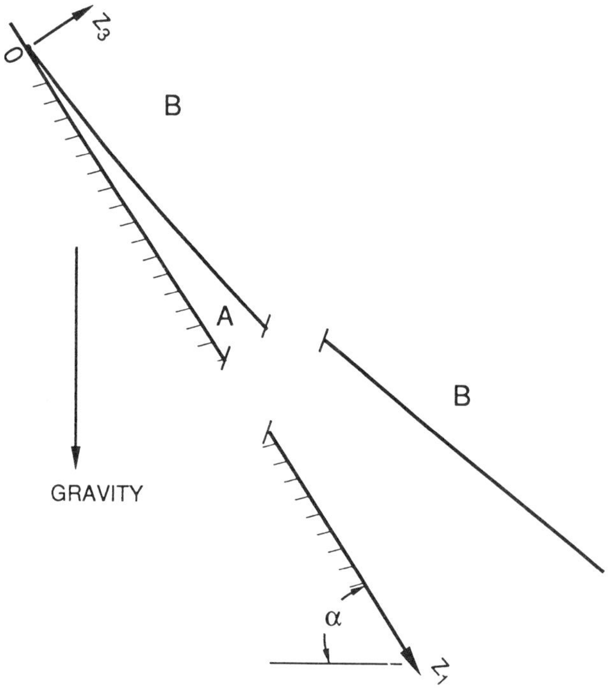

Figure 3.3.3-1: A draining film of liquid A formed as a plate is withdrawn from a pool of A into gas B at a constant velocity V.

In Sec. 1.3.10, I summarized the contradictions in our theoretical understanding of common lines moving across rigid solid surfaces, and I briefly discussed some possibilities open to us for analyzing these problems. In Sec. 2.1.13, I proposed a possible explanation for these contradictions as well as for some of the experimental phenomena that we have considered.

Our premise is that a common line does not move. As viewed on a macroscale, it appears to move as a succession of stationary common lines are formed on a microscale, driven by a negative disjoining pressure in the receding film. The amount of receding phase left stranded by the formation of this succession of common lines is too small to be easily detected, as in the experiments of Dussan V. and Davis (1974; see also Sec.

1.2.9).

The objective of the analysis that follows is to illustrate this thesis by computing the dynamic contact angle Θ, which might be measured by an experimentalist at some distance from the common line using perhaps 10 × magnification. We saw in Sec. 2.1.12 that experimentally observed values of Θ measured through an *advancing* liquid phase against air can be correlated as a function of the static contact angle $\Theta^{(stat)}$ and a capillary number N'_{ca} based upon the speed of displacement of the apparent common line (as seen by an experimentalist using perhaps 10 × magnification). We would expect that a similar correlation could be constructed for Θ measured through a *receding* liquid phase against air, although few measurements of the receding contact angle are currently available [see for example Johnson *et al.* (1977), who used the Wilhelmy plate].

Figure 3-1 shows a draining film of liquid A formed as a plate is withdrawn from a pool of A into gas B at a constant speed V. We will assume that the thin film of liquid A on the solid exhibits a negative disjoining pressure against gas B, consistent with our discussion in Sec. 2.1.13 of receding films.

We will make a number of assumptions.

i) The solid is rigid, and its surface is smooth and planar. Its orientation with respect to gravity is arbitrary.

ii) Viewed in the rectangular cartesian coordinate system of Figure 3-1 in which the origin is fixed on the original common line or one of the common lines formed during the immediate past, the interface bounding the draining film of liquid A takes the form

$$z_3 = h(z_1, t) \tag{3-1}$$

and the plate is stationary.

iii) The dependence of h upon z_1 is sufficiently weak that

$$\left(\frac{\partial h}{\partial z_1}\right)^2 << 1 \tag{3-2}$$

iv) The Reynolds lubrication theory approximation applies in the sense that

$$\left(\frac{h_o}{L_o}\right)^2 << 1 \tag{3-3}$$

where L_o and h_o are respectively characteristic lengths in the z_1 and z_3 directions that will be defined later.

v) If there is no surfactant present, the interfacial tension is a constant independent of position on the interface, and the interfacial viscosities are zero. We will refer to such an interface as being *mobile.*

vi) If there is surfactant present, the tangential components of velocity at the liquid-gas interface are zero

$$\text{at } z_3 = h: \quad \mathbf{P} \cdot \mathbf{v} = 0 \tag{3-4}$$

Here **P** is the projection tensor that transforms every vector on an interface into its tangential components. The interfacial tension gradient required to create such an *immobile* interface is very small (Sheludko 1967; Lin and Slattery 1982a,b; Hahn *et al.* 1985; see also Exercise 3.3.2-1).

vii) The effect of mass transfer is neglected.

viii) The pressure p_o within the surrounding gas B is independent of time and position. Viscous effects within the gas phase are neglected.

ix) The liquid A is an incompressible Newtonian fluid, the viscosity of which is a constant.

x) All inertial effects are neglected.

xi) At the solid surface

$$z_3 = 0: \quad \mathbf{v} = 0 \tag{3-5}$$

xii) All external and mutual body forces (see Sec. 2.1.3) can be represented as the gradient of a potential energy per unit mass ϕ

$$\mathbf{b} = -\nabla\phi \tag{3-6}$$

xiii) We will account for both London-van der Waals forces and gravity within the immediate neighborhood of the common line, neglecting electrostatic double-layer effects.

xiv) Israelachvili (1973) suggests that the distance between the two interfaces can not go to zero due to the finite size of the constituent atoms and that a finite interfacial separation d must be recognized. He recommends that d be viewed as the mean distance between the centers of individual atoms and estimated as

$$d = 0.91649[\, M/(\, \rho^{(A)} n_a N\,)]^{1/3} \tag{3-7}$$

Here M is the molecular weight, n_a the number of atoms per molecule, N Avogadro's constant (6.023×10^{23} mol^{-1}), and $\rho^{(A)}$ the mass density of liquid A. In arriving at (3-7), the atoms have been assumed to be in a close packing arrangement. For a molecule consisting of a repeating unit, such as a n-alkane with $-CH_2-$ being the repeating unit, the comparison with experimental data in Exercise 4.1.4-3 suggests a simplistic picture in which the molecules are arranged in such a manner that the *repeating units* are in a close packing arrangement. Retracing the argument of Israelachvili (1973), we have instead

$$d = 0.91649[\ M/(\ \rho^{(A)} n_u N\ )]^{1/3} \tag{3-8}$$

where n_u is the number of repeating units in the molecule. This last is similar to the suggestion of Padday and Uffindell (1968), who replaced M/n_u by the $-CH_2-$ group weight and took the coefficient to be unity.

For the interface between phases A and S, we suggest

$$d = (\ d^{(A)} + d^{(S)}\ )/2 \tag{3-9}$$

where, if appropriate, $d^{(A)}$ and $d^{(S)}$ are computed using (3-8).

More recently, Israelachvili (1985, p. 157) has recommended using d = 1.65 Å in estimating the surface tension for n-alkanes (see Exercise 4.1.4-3).

xv) If the fluid-fluid and fluid-solid interfaces were parallel planes, the difference in the potential energies attributable to London-van der Waals forces and to gravity at the fluid-fluid interface could be described as (see Exercise 2.1.3-1)

$$\rho^{(A)}\phi^{(A)} - \rho^{(B)}\phi^{(B)}$$

$$= \phi_\infty^{(AB)} + \frac{B^{(AB)}}{h'^m} - (\ \rho^{(A)} - \rho^{(B)}\ )g(\ z_1 \sin\alpha - h' \cos\alpha\ ) \tag{3-10}$$

in which h' is the thickness of the film and the last term on the right accounts for the effect of gravity. We will follow Wayner (1980, 1982) in recognizing that the minimum film is a monolayer. Somewhat simplistically, we will estimate the thickness of the monolayer as d, which now might be thought of as the distance between the centers of the last layer of solid atoms and the centers of the first layer of liquid atoms or repeating units. This suggests that, in the continuum model of mechanics used here, we measure the film thickness h from the centers of the first layer of liquid atoms or repeating units and that we replace (3-10) by

$$\rho^{(A)}\phi^{(A)} - \rho^{(B)}\phi^{(B)}$$

$$= \phi_\infty^{(AB)} + \frac{B^{(AB)}}{(\ h + d\ )^m}$$

$$- (\ \rho^{(A)} - \rho^{(B)}\ )g(\ z_1 \sin\alpha - h \cos\alpha\ ) \tag{3-11}$$

In this simplistic model, the common line can be visualized as running through the centers of the liquid atoms or repeating units on the leading edge of the monolayer. As the common line is approached, the London-van der Waals forces remain bounded as the thickness of the liquid film approaches zero. Because the disjoining pressure is assumed to be negative here, $B^{(AB)} > 0$.

Because the dependence of h upon z_1 is weak (assumption iii), we will assume that (3-11) applies here.

xvi) Again because the dependence of h upon z_1 is weak (assumption iii), we expect our results to be limited to tan $\Theta \to 0$, which suggests that

$$\Theta_0 = 0 \tag{3-12}$$

In constructing this development, we will find it convenient to work in terms of the dimensionless variables

$$z_1^* \equiv \frac{z_1}{L_o} \qquad z_3^* \equiv \frac{z_3}{h_o}$$

$$h^* \equiv \frac{h}{h_o} \qquad d^* \equiv \frac{d}{h_o}$$

$$H^* \equiv HL_o \qquad t^* \equiv \frac{t v_o}{L_o}$$

$$v_1^* \equiv \frac{v_1}{v_o} \qquad v_3^* \equiv \frac{L_o}{h_o}\frac{v_3}{v_o}$$

$$\gamma^* \equiv \frac{\gamma}{\gamma_o} \qquad \phi^{(i)*} \equiv \frac{\phi^{(i)}}{v_o^2}$$

$$p^* \equiv \frac{p}{\rho^{(A)} v_o^2} \qquad p_o^* \equiv \frac{p_o}{\rho^{(A)} v_o^2}$$

$$\mathcal{P}^* \equiv \frac{p + \rho^{(A)}\phi^{(A)}}{\rho^{(A)} v_o^2} \qquad \mathcal{P}_o^* \equiv \frac{p_o + \rho^{(A)}\phi^{(A)}}{\rho^{(A)} v_o^2} \tag{3-13}$$

and the dimensionless Reynolds, Weber, capillary and Bond numbers

$$N_{Re} \equiv \frac{\rho^{(A)} v_o L_o}{\mu} \qquad N_{We} \equiv \frac{\rho^{(A)} v_o^2 L_o}{\gamma_o}$$

$$N_{ca} \equiv \frac{\mu v_o}{\gamma_o} \qquad N_{Bo} \equiv \frac{(\rho^{(A)} - \rho^{(B)}) g L_o^2}{\gamma_o} \tag{3-14}$$

Here H is the mean curvature of the liquid-gas interface, γ the surface tension, γ_o the equilibrium surface tension, μ the viscosity of liquid A, g the magnitude of the acceleration of gravity, p_o the pressure in the gas phase, and p the pressure in the liquid phase. We will identify

$$L_o \equiv \left[\frac{\gamma_o}{(\rho^{(A)} - \rho^{(B)}) g} \right]^{1/2} \tag{3-15}$$

and

$$h_o \equiv L_o \tan \Theta^{(stat)} \tag{3-16}$$

which means

$$N_{Bo} = 1 \tag{3-17}$$

The characteristic speed v_o will be defined later.

Equation (3-1) suggests that we seek a solution in which the velocity distribution takes the form

$$v_1^\star = v_1^\star(z_1^\star, z_3^\star, t^\star)$$

$$v_3^\star = v_3^\star(z_1^\star, z_3^\star, t^\star) \tag{3-18}$$

Under these circumstances, the equation of continuity for an incompressible fluid says

$$\frac{\partial v_1^\star}{\partial z_1^\star} + \frac{\partial v_3^\star}{\partial z_3^\star} = 0 \tag{3-19}$$

In the limit of assumption iv, Cauchy's first law for an incompressible, Newtonian fluid with a constant viscosity reduces for creeping flow to

$$\frac{\partial \mathcal{P}^\star}{\partial z_1^\star} = \frac{1}{N_{Re}} \left(\frac{L_o}{h_o} \right)^2 \frac{\partial^2 v_1^\star}{\partial z_3^{\star 2}} \tag{3-20}$$

$$\frac{\partial \mathcal{P}^*}{\partial z_3^*} = \frac{1}{N_{Re}} \frac{\partial^2 v_3^*}{\partial z_3^{*2}} \tag{3-21}$$

The z_2-component requires $\mathcal{P}^*$ to be independent of z_2^*. Equations (3-20) and (3-21) imply that

$$\frac{\partial \mathcal{P}^*}{\partial z_3^*} << \frac{\partial \mathcal{P}^*}{\partial z_1^*} \tag{3-22}$$

and the dependence of $\mathcal{P}^*$ upon z_3^* can be neglected. Note that the scaling argument used to neglect inertial effects (assumption x) in arriving at (3-20) and (3-21) is presumed not be the one shown here. For this reason, we will regard the magnitude of N_{Re} and the definition of v_o to be as yet unspecified.

The jump mass balance (Sec. 1.3.5; the overall jump mass balance of Sec. 5.4.1) is satisfied identically, since we define the position of the dividing surface by choosing $\rho^{(\sigma)} = 0$ (Sec. 1.3.6) and since the effect of mass transfer is neglected (assumption vii). The jump mass balance for surfactant (Sec. 5.2.1) is not required here, since we assume either that the interfacial tension is constant (assumption v) or that the interfacial tension gradient is so small that its effect and the effect of a concentration gradient developed in the interface can be neglected (assumption vi).

With assumptions vii, viii, and x, the jump momentum balance (Sec. 2.4.2) for the interface between phases A and B reduces to

$$\nabla_{(\sigma)}\gamma + 2H\,\gamma\,\boldsymbol{\xi} - (\,\mathbf{T} + p_o\mathbf{I}\,)\cdot\boldsymbol{\xi} = 0 \tag{3-23}$$

Here $\boldsymbol{\xi}$ is the unit normal to the interface pointing out of the liquid film. Under the conditions of assumptions iii and iv, the z_1- and z_3- components of (3-23) assume the forms at $z_3^* = h^*$ (see Table 2.4.2-6)

$$\frac{\partial \gamma^*}{\partial z_1^*} - 2H^*\gamma^* \frac{h_o}{L_o}\frac{\partial h^*}{\partial z_1^*} - N_{We}(\,p^* - p_o^*\,)\frac{h_o}{L_o}\frac{\partial h^*}{\partial z_1^*} - N_{ca}\frac{L_o}{h_o}\frac{\partial v_1^*}{\partial z_3^*} = 0 \tag{3-24}$$

and

$$\frac{h_o}{L_o}\frac{\partial \gamma^*}{\partial z_1^*}\frac{\partial h^*}{\partial z_1^*} + 2H^*\gamma^* + N_{We}(\,p^* - p_o^*\,) - 2N_{ca}\frac{\partial v_3^*}{\partial z_3^*}$$

$$+ N_{ca}\frac{\partial v_1^*}{\partial z_3^*}\frac{\partial h^*}{\partial z_1^*} = 0 \tag{3-25}$$

The z_2-component is satisfied identically. Adding $(\,h_o/L_o\,)(\,\partial h^*/\partial z_1^*\,)$

times (3-25) to (3-24) and recognizing assumptions iii and iv, we have

$$\frac{\partial \gamma^*}{\partial z_1^*} - N_{ca} \frac{L_o}{h_o} \frac{\partial v_1^*}{\partial z_3^*} = 0 \tag{3-26}$$

and (3-24) implies[a]

$$2H^* \gamma^* + N_{We}(p^* - p_o^*) = 0 \tag{3-27}$$

For a mobile interface, (3-26) requires (see assumptions v and viii)

$$\text{at } z_3^* = h^*: \quad \frac{\partial v_1^*}{\partial z_3^*} = 0 \tag{3-28}$$

For an immobile interface (see assumptions iv and vi)

$$\text{at } z_3^* = h^*: \quad v_1^* = 0 \tag{3-29}$$

and we can employ (3-26) to calculate the interfacial tension gradient required to create an immobile interface.

Since we neglect the effect of mass transfer on the velocity distribution (assumption vii; see Table 2.4.2-6),

---

a) In arriving at (3-26) and (3-27), it was not necessary to make any statement about the relative magnitudes of N_{ca} and N_{We} or the definition of v_o. But some statement is necessary, in order to establish consistency with (3-25).

Substituting (3-26) into (3-25), we have

$$2H^* \gamma^* + N_{We}(p^* - p_o^*) + 2N_{ca} \left(\frac{\partial v_1^*}{\partial z_3^*} \frac{\partial h^*}{\partial z_1^*} - \frac{\partial v_3^*}{\partial z_3^*} \right) = 0$$

It follows from (3-32) through (3-34) and (3-36) that

$$| 2H^* \gamma^* | >> \left| 2N_{ca} \left(\frac{\partial v_1^*}{\partial z_3^*} \frac{\partial h^*}{\partial z_1^*} - \frac{\partial v_3^*}{\partial z_3^*} \right) \right|$$

in agreement with (3-27).

$$\text{at } z_3^\star = h^\star: \; v_3^\star = \frac{\partial h^\star}{\partial t^\star} + \frac{\partial h^\star}{\partial z_1^\star} v_1^\star \tag{3-30}$$

Finally, (3-5) requires at the solid surface

$$\text{at } z_3^\star = 0: \; v_1^\star = v_3^\star = 0 \tag{3-31}$$

Our objective in what follows is to obtain a solution to (3-19) and (3-20) consistent with (3-27), (3-30), (3-31) and either (3-28) or (3-29).

Solution Integrating (3-20) twice consistent with (3-31) and either (3-28) or (3-29), we find that

$$v_1^\star = N_{Re} \left(\frac{h_o}{L_o} \right)^2 \frac{\partial \mathcal{P}^\star}{\partial z_1^\star} \left(\frac{1}{2} z_3^{\star 2} - \frac{1}{n} h^\star z_3^\star \right) \tag{3-32}$$

where

n = 1 for a mobile interface
n = 2 for an immobile interface (3-33)

Substituting (3-32) into (3-19) and integrating once consistent with (3-31), we have

$$v_3^\star = -\frac{1}{2} N_{Re} \left(\frac{h_o}{L_o} \right)^2 \left[\frac{\partial^2 \mathcal{P}^\star}{\partial z_1^{\star 2}} \left(\frac{1}{3} z_3^{\star 3} - \frac{1}{n} h^\star z_3^{\star 2} \right) - \frac{1}{n} \frac{\partial \mathcal{P}^\star}{\partial z_1^\star} \frac{\partial h^\star}{\partial z_1^\star} z_3^{\star 2} \right] \tag{3-34}$$

Equations (3-30), (3-32) and (3-34) tell us

$$-\frac{\partial h^\star}{\partial t^\star} = -N_{Re} \frac{3-n}{6n} \left(\frac{h_o}{L_o} \right)^2 \frac{\partial}{\partial z_1^\star} \left(h^{\star 3} \frac{\partial \mathcal{P}^\star}{\partial z_1^\star} \right) \tag{3-35}$$

Equation (3-27), with the appropriate expression for the dimensionless mean curvature H, says

$$N_{We} (\mathcal{P}^\star - \mathcal{P}_o^\star) - N_{We} \left(\phi^{(A)\star} - \frac{\rho^{(B)}}{\rho^{(A)}} \phi^{(B)\star} \right) = -\frac{h_o}{L_o} \frac{\partial^2 h^\star}{\partial z_1^{\star 2}} \tag{3-36}$$

In terms of dimensionless variables, (3-11) becomes

$$\phi^{(A)*} - \frac{\rho^{(B)}}{\rho^{(A)}}\phi^{(B)*} = \frac{1}{N_{We}}\left[\phi_\infty^{(AB)*} + \frac{h_o}{L_o}\frac{B^*}{(h^* + d^*)^m} - z_1^* \sin\alpha + \frac{h_o}{L_o} h^* \cos\alpha\right] \tag{3-37}$$

where we have recognized (3-17), we have defined

$$\phi_\infty^{(AB)*} \equiv \frac{L_o}{\gamma_o}\phi_\infty^{(AB)} \qquad B^* \equiv \frac{L_o^2 B^{(AB)}}{\gamma_o h_o^{m+1}} \tag{3-38}$$

and we have set $\gamma^* = 1$ either by assumption v or by assumption vi. In view of (3-37), we can differentiate (3-36) to find

$$\frac{\partial \mathcal{P}^*}{\partial z_1^*} = -\frac{1}{N_{We}}\frac{h_o}{L_o}\left[\frac{\partial^3 h^*}{\partial z_1^{*3}} + \frac{mB^*}{(h^* + d^*)^{m+1}}\frac{\partial h^*}{\partial z_1^*} + \frac{L_o}{h_o}\sin\alpha - \frac{\partial h^*}{\partial z_1^*}\cos\alpha\right] \tag{3-39}$$

Equation (3-35) together with (3-39) defines the configuration of the film as a function of time and position.

Since (see Exercise 2.1.3-1)

$$B^* << 1 \tag{3-40}$$

our objective is to develop a solution that is correct in the limit $B^* \to 0$ or a perturbation solution that is correct to the zeroth order in B^*. Outside the immediate neighborhood of the common line, (3-39) reduces in this limit to

$$\frac{\partial \mathcal{P}^*}{\partial z_1^*} = -\frac{1}{N_{We}}\frac{h_o}{L_o}\left(\frac{\partial^3 h^*}{\partial z_1^{*3}} + \frac{L_o}{h_o}\sin\alpha - \frac{\partial h^*}{\partial z_1^*}\cos\alpha\right) \tag{3-41}$$

Equation (3-35) together with (3-41) must be solved consistent with

$$\text{as } z_1^* \to 0: \quad \frac{dh^*}{dz_1^*} \to k \tag{3-42}$$

as well as three other boundary conditions and an initial condition for the particular macroscopic flow with which one is concerned. Here

$$k \equiv \frac{\tan \Theta}{\tan \Theta^{(stat)}} \tag{3-43}$$

where Θ is the receding dynamic contact angle corresponding to the speed of the common line. Note that $k = 1$ corresponds to the static problem. We will refer to this as the *outer problem*, in the sense that it is outside the immediate neighborhood of the common line. Equation (3-42) is imposed not at the common line, but as the common line is approached in the outer solution.

Within the immediate neighborhood of the common line, the effects of the London-van der Waals forces must be preserved. We will restrict our consideration here to the case of unretarded London-van der Waals forces (for films less than about 120 Å thick; see Exercise 2.1.3-1) for which

$$B^{(AB)} = \frac{1}{6\pi} \left(A^{(AA)} - A^{(AS)} \right)$$

$$m = 3 \tag{3-44}$$

Here $A^{(AA)}$ is the Hamaker constant for the interaction of the film fluid with itself; $A^{(AS)}$ is the Hamaker constant for the interaction of the film fluid with the solid. This suggests that we introduce expanded variables within the immediate neighborhood of the common line:

$$h^{**} \equiv \frac{h^*}{B^{*1/2}} \qquad d^{**} \equiv \frac{d^*}{B^{*1/2}}$$

$$z_1^{**} \equiv \frac{z_1^*}{B^{*1/2}} \qquad t^{**} \equiv \frac{t^*}{B^{*1/2}} \tag{3-45}$$

In terms of these variables, (3-35) and (3-39) require

$$-\frac{\partial h^{**}}{\partial t^{**}} = N_{ca} \frac{3-n}{6n} \left(\frac{h_o}{L_o} \right)^3 \left\{ h^{**3} \frac{\partial^4 h^{**}}{\partial z_1^{**4}} + 3h^{**2} \frac{\partial h^{**}}{\partial z_1^{**}} \frac{\partial^3 h^{**}}{\partial z_1^{**3}} \right.$$

$$+ 3h^{**3} \left[\frac{1}{(h^{**} + d^{**})^4} \frac{\partial^2 h^{**}}{\partial z_1^{**2}} \right.$$

$$\left. - \frac{1 - 3d^{**}/h^{**}}{(h^{**} + d^{**})^5} \left(\frac{\partial h^{**}}{\partial z_1^{**}} \right)^2 \right]$$

$$+ 3B^* \frac{L_o}{h_o} \sin\alpha\, h^{**2} \frac{\partial h^{**}}{\partial z_1^{**}} - 3B^* \cos\alpha\, h^{**2} \left(\frac{\partial h^{**}}{\partial z_1^{**}} \right)^2$$

$$- B^* \cos\alpha\, h^{**3} \frac{\partial^2 h^{**}}{\partial z_1^{**2}} \Bigg\} \tag{3-46}$$

Since $B^* \to 0$, the effect of gravity can be neglected in the inner region.

For the sake of simplicity, let us define the characteristic speed as

$$v_o \equiv \frac{3-n}{6n} \frac{\gamma_o}{\mu} \left(\frac{h_o}{L_o} \right)^3 \tag{3-47}$$

permitting us to write (3-46) as

$$- \frac{\partial h^{**}}{\partial t^{**}} = h^{**3} \frac{\partial^4 h^{**}}{\partial z_1^{**4}} + 3h^{**2} \frac{\partial h^{**}}{\partial z_1^{**}} \frac{\partial^3 h^{**}}{\partial z_1^{**3}}$$

$$+ 3h^{**3} \left[\frac{1}{(h^{**} + d^{**})^4} \frac{\partial^2 h^{**}}{\partial z_1^{**2}} \right.$$

$$\left. - \frac{1 - 3d^{**}/h^{**}}{(h^{**} + d^{**})^5} \left(\frac{\partial h^{**}}{\partial z_1^{**}} \right)^2 \right] \tag{3-48}$$

Here we have neglected the effects of gravity as indicated following (3-46).

Note that (see Exercise 3.3.3-1)

$$d^{**} = 1 \tag{3-49}$$

or

$$\tan \Theta^{(stat)} = \left(\frac{A}{6\pi \gamma_o d^2} \right)^{1/2} \tag{3-50}$$

This last also follows from the result of Sec. 3.2.7 in the limit $\Theta^{(stat)} \to 0$. In view of (3-49), equation (3-48) simplifies to

$$- \frac{\partial h^{**}}{\partial t^{**}} = h^{**3} \frac{\partial^4 h^{**}}{\partial z_1^{**4}} + 3h^{**2} \frac{\partial h^{**}}{\partial z_1^{**}} \frac{\partial^3 h^{**}}{\partial z_1^{**3}}$$

$$+ 3h^{**3} \left[\frac{1}{(h^{**} + 1)^4} \frac{\partial^2 h^{**}}{\partial z_1^{**2}} \right.$$

$$- \frac{1 - 3/h^{**}}{(h^{**} + 1)^5} \left(\frac{\partial h^{**}}{\partial z_1^{**}} \right)^2 \Bigg] \tag{3-51}$$

Equation (3-51) must be solved consistent with

$$\text{at } z_1^{**} = 0: \ \ h^{**} = 0 \tag{3-52}$$

and, in view of (3-12),

$$\text{at } z_1^{**} = 0: \ \ \frac{dh^{**}}{dz_1^{**}} = 0 \tag{3-53}$$

In addition, the inner and outer solutions must be consistent in some intermediate region

$$\text{as } B^* \to 0 \text{ and } z_1^{**} \to \infty: \ \ \frac{\partial^2 h^{**}}{\partial z_1^{**2}} \to B^{*1/2} \left(\frac{\partial^2 h^*}{\partial z_1^{*2}} \right)_\infty \tag{3-54}$$

$$\text{as } B^* \to 0 \text{ and } z_1^{**} \to \infty: \ \ \frac{\partial h^{**}}{\partial z_1^{**}} \to \left(\frac{\partial h^*}{\partial z_1^*} \right)_\infty = k \tag{3-55}$$

Here the subscript $..._\infty$ indicates a quantity in the outer solution seen as the common line is approached.

Finally, we require an initial configuration of the interface. At first thought, it would appear most natural to take as our initial configuration the static configuration. This would correspond to considering the problem of start-up from rest with a step-change in the speed of the wall from 0 to V and a corresponding step-change in $\partial h^{**}/\partial z_1^{**}$ from 1 to k as $B^* \to 0$ and $z_1^{**} \to \infty$. The difficulty is that this would be inconsistent with our neglect of inertial effects in assumption x. Instead, we will say that

$$\text{at } t^{**} = 0: \ \ h^{**} = (k^2 z_1^{**2} + 1)^{1/2} - 1 \tag{3-56}$$

which is consistent with (3-52) through (3-55). We will not define the experimental technique required to achieve this initial condition. Rather, we shall assume that for $t^{**} >> 1$ the effect of the initial condition is negligible.

From the simultaneous solutions of the inner and outer problems, one can compute the dimensionless **speed of displacement of the apparent common line** (determined by the tangent to the interface as $B^* \to 0$ and $z_1^{**} \to \infty$ in the inner solution or $z_1^* \to z_{1\infty}^*$ in the outer solution)

$$u^{(cl)*} \equiv \frac{dz_1^{(cl)*}}{dt^*} = \frac{dz_1^{(cl)**}}{dt^{**}} \tag{3-57}$$

and determine the specific form of the relationship

$$u^{(cl)*} = g\left[k, B^{*1/2} \left(\frac{\partial^2 h^*}{\partial z_1^{*2}} \right)_\infty , t^{**} \right] \tag{3-58}$$

In general the dynamic contact angle Θ can be expected to be dependent upon the measurement technique or, equivalently, the geometry of the macroscopic system as a result of its dependence upon the curvature in the outer solution as the common line is approached. [This is analogous with the observations of Ngan and Dussan V. (1982) and of Legait and Sourieau (1985) for advancing common lines.]

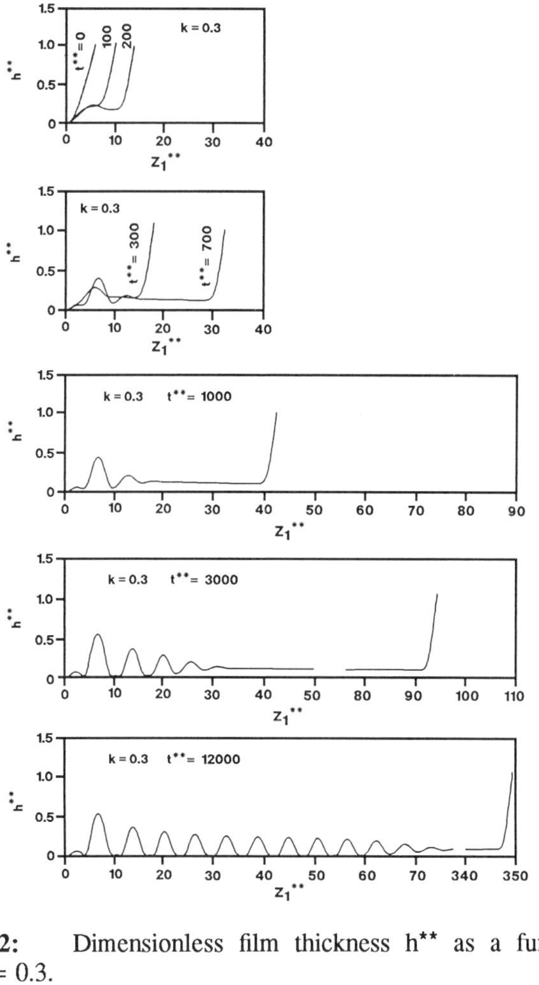

Figure 3.3.3-2: Dimensionless film thickness h^{**} as a function of z_1^{**} and t^{**} for $k = 0.3$.

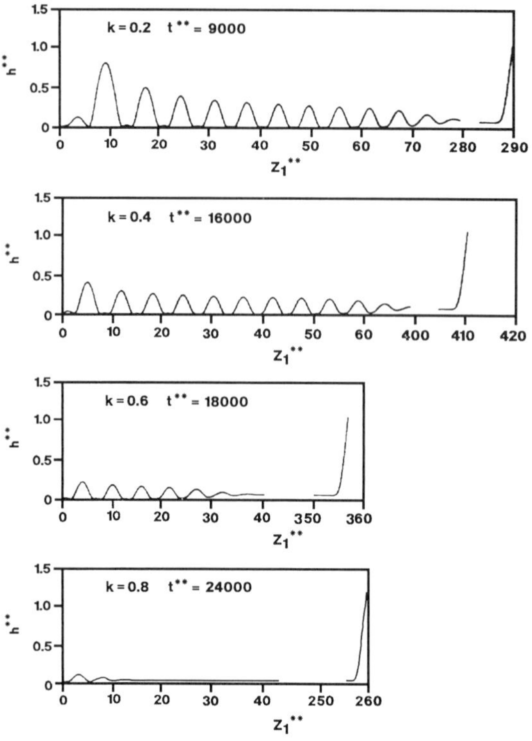

Figure 3.3.3-3: Dimensionless film thickness h^{**} as a function of z_1^{**} for different values of k.

Physically, the maximum value of the dimensional curvature in the outer solution as the common line is approached can be expected to be proportional to $1/\ell$, where ℓ is a characteristic dimension of the macroscopic system. In the limiting case such that $B^* \to 0$,

$$\left(\frac{\partial^2 h^*}{\partial z_1^{*2}}\right)_\infty = \frac{L_o^2}{h_o}\left(\frac{\partial^2 h}{\partial z_1^2}\right)_\infty \sim \frac{h_o}{\ell}\left(\frac{L_o}{h_o}\right)^2 << 1 \tag{3-59}$$

and (3-3) are all satisfied, as is true for most experiments used to measure Θ, (3-54) reduces to

$$\text{as } B^* \to 0 \text{ and } z_1^{**} \to \infty : \frac{\partial^2 h^{**}}{\partial z_1^{**2}} \to 0 \tag{3-60}$$

and (3-58) takes the form

$$u^{(cl)*} = g(k, t^{**}) \tag{3-61}$$

In this limit, the relationship between $u^{(cl)*}$ and k no longer depends upon the technique used for its measurement or the geometry of the macroscopic problem (see Sec. 2.1.12).

In what follows, we consider only the solution of this limiting case. Equation (3-51) is solved consistent with (3-52), (3-53), (3-55), (3-56), and (3-60) using the Crank-Nicolson method (Myers 1971).

Results and discussion Figure 3-2 shows the dimensionless film thickness h^{**} as a function of z_1^{**} and t^{**} for $k = 0.3$. As t^{**} increases, the bulk of phase A recedes, and a very thin liquid film is left behind. The slope of the surface of the bulk of phase A is the receding contact angle Θ, which remains almost unchanged with t^{**}. We will refer to the junction of the thin film and bulk of phase A (identified by the intersection of the tangent to this portion of the surface with the z_1^{**} axis) as the **apparent common line** and to its position as $z_1^{(cl)**}$. As t^{**} increases, waves form in the thin film driven by the negative disjoining pressure with coalescence occurring at the troughs of these waves, and a small residue of phase A is left behind. This coalescence process can not occur at a common line, since this would imply a moving common line and unbounded forces.

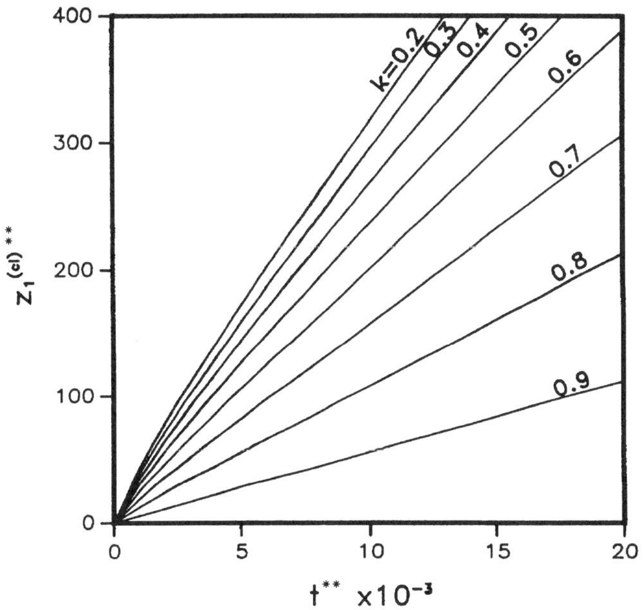

Figure 3.3.3-4: Position $z_1^{(cl)**}$ of apparent common line as a function of time t^{**}.

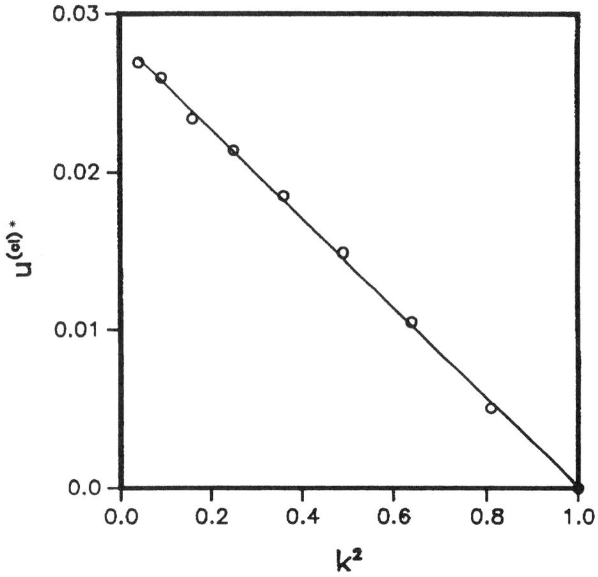

Figure 3.3.3-5: Dependence of $u^{(cl)*}$ on k^2.

Figure 3-3 shows h^{**} as a function of z_1^{**} and t^{**} for different values of k. The average thickness of the thin film and therefore the residue of phase A left behind increases as k decreases or as the receding contact angle Θ decreases.

Notice that the thickness of the residual film in Figures 3-2 and 3-3 is less than molecular dimensions. We believe that this should be interpreted as a continuum description of what in reality is an incomplete monomolecular film.

Figure 3-4 presents the position $z_1^{(cl)**}$ of the apparent common line as a function of t^{**} for different values of k. After a short initial period attributable to the initial configuration (3-56), $u^{(cl)*}$ becomes a constant. The relationship between $u^{(cl)*}$ and k is shown in Figure 3-5. Alternatively, we can express this relationship as

$$u^{(cl)*} = 0.028(\ 1 - k^2\ ) \tag{3-62}$$

or

$$u^{(cl)*} = 0.028 \tan^{-2} \Theta^{(stat)}(\ \tan^2 \Theta^{(stat)} - \tan^2 \Theta\ ) \tag{3-63}$$

We will define a new capillary number based on the speed of the apparent common line as

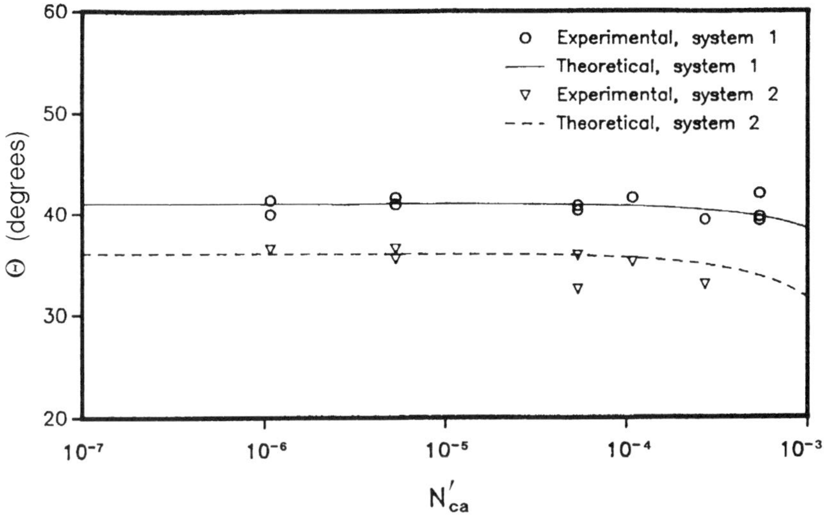

Figure 3.3.3-6: Comparisons of (3-67) with the experimental data of Johnson *et al.* (1977): system 1, hexadecane on teflon; system 2, hexadecane on siliconed glass.

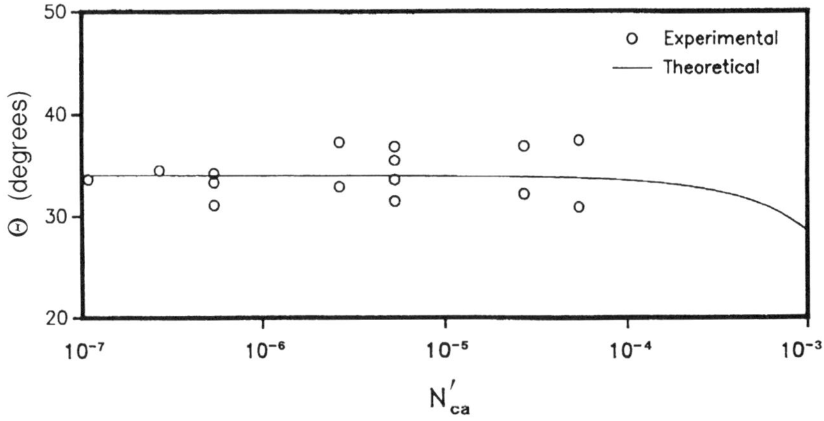

Figure 3.3.3-7: Comparisons of (3-67) with the experimental data of Johnson *et al.* (1977): water on monolayer of trimethyloctadecylammonium chloride on glass.

$$N'_{ca} \equiv \frac{\mu u^{(cl)}}{\gamma_0} \tag{3-64}$$

According to (3-13), the dimensionless speed of the apparent common line can be also written as

$$u^{(cl)*} = \frac{u^{(cl)}}{v_0} \tag{3-65}$$

With (3-15), (3-16), (3-47), (3-63) and (3-65), equation (3-64) becomes

$$N'_{ca} = \frac{3 - n}{6n} \tan^3 \Theta^{(stat)} u^{(cl)*} \tag{3-66}$$

or

$$N'_{ca} = 0.028 \frac{3 - n}{6n} \tan \Theta^{(stat)} \left(\tan^2 \Theta^{(stat)} - \tan^2 \Theta \right) \tag{3-67}$$

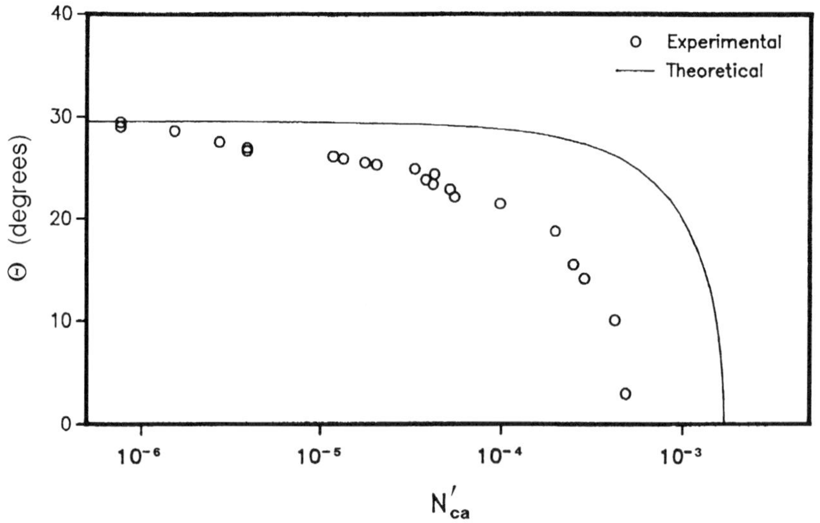

Figure 3.3.3-8: Comparisons of (3-67) with the experimental data of Hopf and Stechemesser (1988): distilled water on monolayer of octadecylamine deposited on quartz by the Langmuir-Blodgett technique.

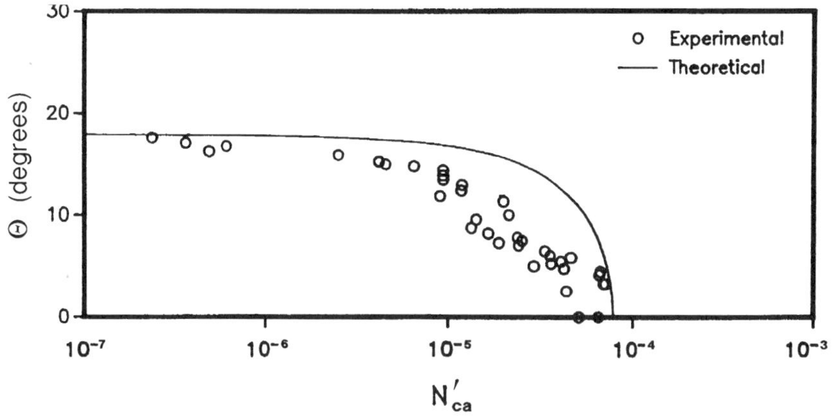

Figure 3.3.3-9: Comparisons of (3-67) with the experimental data of Hopf and Stechemesser (1988): surfactant solution (10^{-5} M dodecylamine + 10^{-2} M KCl at pH = 4.0 adjusted by adding KCl solution) on quartz.

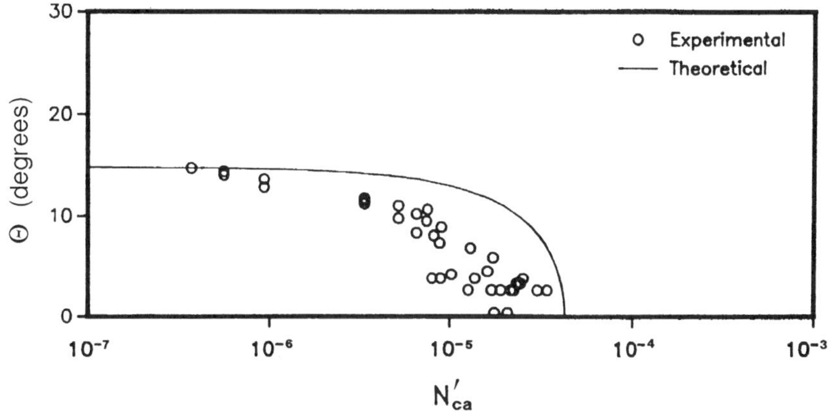

Figure 3.3.3-10: Comparisons of (3-67) with the experimental data of Hopf and Stechemesser (1988): surfactant solution (10^{-5} M dodecylamine + 10^{-2} M KCl at pH = 10.0 adjusted by adding KOH solution) on quartz.

Equation (3-67) gives the relationship between the static contact angle $\Theta^{(stat)}$, the receding contact angle Θ, and the capillary number N'_{ca}. When N'_{ca} increases, Θ decreases. In the limit as $\Theta^{(stat)} \to 0$, the apparent common line will not recede.

There have been very few experimental studies of the receding contact angle Θ as a function of the speed of displacement of the apparent common line. Figures 3-6 and 3-7 compare our prediction (3-67) with the experimental data of Johnson *et al.* (1977), who used the Wilhelmy plate. Since surfactant was not used in the their experiments, we assumed that the surface is mobile (assumption v) and n = 1. The static contact angles $\Theta^{(stat)}$, which were not measured, were estimated by extrapolating $u^{(cl)} \to 0$ in each case. Because of assumptions iii, iv, and xvi, we compare (3-67) only with their observations for smaller Θ. Unfortunately, their $u^{(cl)}$ were sufficiently small that Θ was nearly a constant.

Figures 3-8 through 3-10 compare (3-67) with the experimental data of Hopf and Stechemesser (1988), who measured the receding contact angle as a function of the speed of movement of the common line after rupture of the thin liquid film between a quartz plate and a gas bubble generated at the tip of a capillary. The difference between our prediction and the experimental data is likely attributable to the role of inertial effects in their experiment.

Elliott and Riddiford (1967), Black (1972), Johnson *et al.* (1977), and Petrov and Radoev (1981) reported measurements of the receding contact angle as a function of the speed of displacement of the common line for large contact angles. Because of assumptions iii, iv, and xvi, (3-67) is not applicable to these cases.

References

Black, T. D., *VDI-Berichte Nr.* **182**, 117 (1972).

Dussan V., E. B., and S. H. Davis, *J. Fluid Mech.* 65, **71** (1974).

Elliott, G. E. P., and A. C. Riddiford, *J. Colloid Interface Sci.* **23**, 389 (1967).

Hahn, P. S., J. D. Chen, and J. C. Slattery, *AIChE J.* **31**, 2126 (1985).

Hopf, W., and H. Stechemesser, *Colloids Surf.* **33**, 25 (1988).

Israelachvili, J. N., *J. Chem. Soc. Faraday Trans. II* **69**, 1729 (1973).

Israelachvili, J. N., "Intermolecular and Surface Forces," p. 157, Academic Press (1985).

Johnson, R. E. Jr., R. H. Dettre, and D. A. Brandreth, *J. Colloid Interface Sci.* **62**, 205 (1977).

Legait, B., and P. Sourieau, *J. Colloid Interface Sci.* **107**, 14 (1985).

Li, D., and J. C. Slattery, *J. Colloid Interface Sci.* (1990).

Lin, C. Y., and J. C. Slattery, *AIChE J.* **28**, 147 (1982a).

Lin, C. Y., and J. C. Slattery, *AIChE J.* **28**, 786 (1982b).

Myers, G. E., "Analytical Methods in Conduction Heat Transfer," p. 274, McGraw-Hill, New York (1971).

Ngan, C. G., and E. B. Dussan V., *J. Fluid Mech.* **118**, 27 (1982).

Padday, J. F., and N. D. Uffindell, *J. Phys. Chem.* **72**, 1407 (1968).

Petrov, J. G., and B. P. Radoev, *Colloid Polym. Sci.* **259**, 753 (1981).

Sheludko, A., *Adv. Colloid Interface Sci.* **1**, 391 (1967).

Wayner, P. C., *J. Colloid Interface Sci.* **77**, 495 (1980).

Wayner, P. C., *J. Colloid Interface Sci.* **88**, 294 (1982).

Exercise

3.3.3-1 *derivation of (3-49)*

i) Starting with (3-36) and (3-37), show that under static conditions

$$\frac{d^2h^{**}}{dz_1^{**2}} = \frac{1}{(h^{**} + d^{**})^3} \tag{1}$$

ii) Integrating (1) consistent with (3-55) and recognizing that for the static problem considered here k = 1, we find

$$\left(\frac{dh^{**}}{dz_1^{**}} \right)^2 = - \frac{1}{(h^{**} + d^{**})^2} + 1 \tag{2}$$

iii) Use (3-52) and (3-53) to arrive at (3-49).

3.4 Boussinesq surface fluid

3.4.1 Knife-edge surface viscometer (Lifshutz *et al.* 1971) The knife-edge surface viscometer introduced by Brown *et al.* (1953) for gas-liquid phase interfaces evolved from several earlier designs (Foust and Harkins 1938; Chaminade *et al.* 1950; Ellis *et al.* 1955). Its principal elements are sketched in Figure 1-1. The circular knife-edge is positioned at the gas-liquid phase interface in such a manner that to the eye the interface is optically flat. As the pan rotates with a constant angular velocity Ω, one measures $\mathcal{T}_z$, the z-component of the torque required to hold the circular knife-edge stationary.

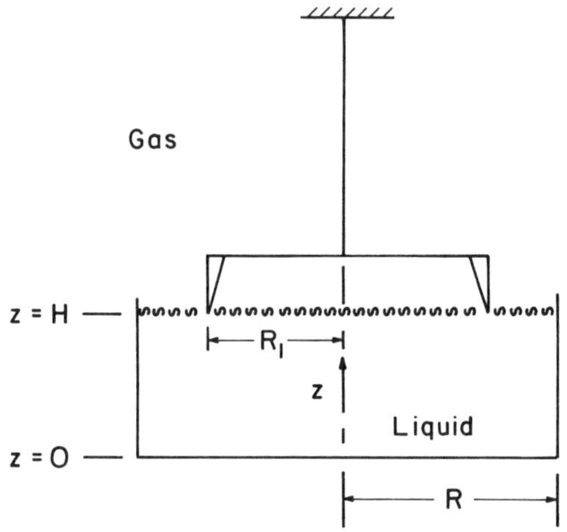

Figure 3.4.1-1: Knife-edge surface viscometer. As the pan rotates with a constant angular velocity Ω, one measures $\mathcal{T}_z$, the z-component of the torque required to hold the circular knife-edge stationary.

Our objective in what follows is to calculate $\mathcal{T}_z$ as a function of the gas-liquid surface shear viscosity ε, Ω, and all of the parameters required to describe the geometry. The result provides us with the basic analysis of the knife-edge surface viscometer (Lifshutz *et al.* 1971; Mannheimer and Burton 1970).

For reference, let me list the assumptions that we shall make in carrying out this analysis.

i) Viscous effects are neglected in the gas phase.

ii) All inertial effects are neglected.

iii) There is no mass transfer across the liquid-gas interface.

iv) The viscometer operates under steady-state conditions in which all variables are independent of time.

v) The liquid phase is an incompressible Newtonian fluid.

vi) The stress-deformation behavior of the gas-liquid phase interface may be represented by the Boussinesq surface fluid model (Sec. 2.2.2).

vii) All physical properties are constants.

viii) The circular knife-edge is positioned in such a manner that the phase interface is a plane.

ix) The effect of the knife edge is to say that, at a circle of radius R_1 in the phase interface, velocity is zero.

If the pan rotates with a constant angular velocity Ω, then

$$\text{at } z = 0 : \; v_\theta = r\Omega \,, \; v_r = v_z = 0 \tag{1-1}$$

and

$$\text{at } r = R : \; v_\theta = R\Omega \,, \; v_r = v_z = 0 \tag{1-2}$$

where R is the radius of the pan. The contact between the knife edge and the phase interface is idealized as a perfect circle. Since the knife-edge is stationary, we require

$$\text{at } r = R_1 \,, \; z = H : \; \mathbf{v} = 0 \tag{1-3}$$

These boundary conditions suggest that we seek a solution of the form

$$v_r = v_z = 0 \tag{1-4}$$

$$v_\theta^\star \equiv \frac{v_\theta}{R\Omega} - r^\star$$

$$= v_\theta^\star(r^\star, z^\star) \tag{1-5}$$

where for convenience I have introduced

$$r^\star \equiv \frac{r}{R} \qquad z^\star \equiv \frac{z}{R} \tag{1-6}$$

In these terms, (1-1) through (1-3) become

$$\text{at } z^\star = 0 : \; v_\theta^\star = 0 \tag{1-7}$$

$$\text{at } r^* = 1 : \quad v_\theta^* = 0 \tag{1-8}$$

and

$$\text{at } r^* = R_1^* , z^* = H^* : \quad v_\theta^* = - R_1^* \tag{1-9}$$

The equation of continuity is satisfied identically by (1-4) and (1-5) (see Slattery 1981, p. 60).

The r- and z- components of Cauchy's first law for an incompressible Newtonian fluid require (Slattery 1981, p. 62)

$$p^* \equiv \frac{p}{p_o}$$

$$= - \frac{\rho g z}{p_o} + C \tag{1-10}$$

Here p_o is the ambient pressure in the gas phase, g is the acceleration of gravity, and C is a constant of integration. The θ-component simplifies to

$$\frac{\partial}{\partial r^*} \left[\frac{1}{r^*} \frac{\partial}{\partial r^*} (r^* v_\theta^*) \right] + \frac{\partial^2 v_\theta^*}{\partial z^{*2}} = 0 \tag{1-11}$$

From Table 2.4.2-2, the jump mass balance for the liquid-gas phase interface (Sec. 1.3.5; the overall jump mass balance of Sec. 5.4.1) is satisfied identically, since we define the position of the dividing surface by choosing $\rho^{(\sigma)} = 0$ (Sec. 1.3.6) and since the effect of mass transfer is neglected (assumption iii).

Returning to Table 2.4.2-2, we see that the r- and z- components of the jump momentum balance for a Boussinesq surface fluid say only that

$$\text{at } z^* = H^* , \text{ for } 0 \le r^* < R_1^* , R_1^* < r^* < 1 : \quad p^* = 1 \tag{1-12}$$

The θ-component reduces to

$$\text{at } z^* = H^* , \text{ for } 0 \le r^* \le R_1^* , R_1^* < r^* < 1 :$$

$$\frac{\partial v_\theta^*}{\partial z^*} = N \frac{\partial}{\partial r^*} \left[\frac{1}{r^*} \frac{\partial}{\partial r^*} (r^* v_\theta^*) \right] \tag{1-13}$$

in which I have introduced

$$N \equiv \frac{\varepsilon}{\mu R} \tag{1-14}$$

We can use (1-12) to evaluate in (1-10)

$$C = 1 + \frac{\rho g H}{p_o} \tag{1-15}$$

We must now solve (1-11) consistent with the boundary conditions (1-7) through (1-9) and (1-13). Our objective will be to use this solution in order to determine $\mathcal{T}_z$, the torque that the fluid exerts upon the knife edge:

$$\mathcal{T}_z^* \equiv \frac{\mathcal{T}_z}{2\pi R^3 \mu \Omega}$$

$$= N\, R_1^{*2} \left(\left.\frac{\partial v_\theta^*}{\partial r^*}\right|_{r^* = R_1^* +} - \left.\frac{\partial v_\theta^*}{\partial r^*}\right|_{r^* = R_1^* -} \right) \tag{1-16}$$

where R_1^* + indicates the limit as R_1^* is approached from above and R_1^* – the limit as R_1^* is approached from below. In arriving at (1-16), I have observed that, since the knife edge is idealized as a line in the dividing surface, the bulk fluid can not directly contribute to the torque. The torque is solely the result of the interfacial stresses exerted upon either side of the knife edge.

Let us denote by

$$\overline{v}(\xi_i, z^*) \equiv \int_0^1 r^*\, v_\theta^*(r^*, z^*)\, J_1(\xi_i r^*)\, dr^* \tag{1-17}$$

the finite Hankel transform of v_θ^* (Irving and Mullineux 1960, p. 157). By ξ_i, I mean a root of

$$J_1(\xi_i) = 0 \qquad \text{for } i = 1, 2, \ldots \tag{1-18}$$

The finite Hankel transform of (1-11) is

$$\frac{\partial^2 \overline{v}}{\partial z^{*2}} - (\xi_i)^2\, \overline{v} = 0 \tag{1-19}$$

In arriving at the second term on the left, two integrations by parts were required as well as (1-8) and (1-18). The transform of (1-7) is

$$\text{at } z^* = 0 : \ \overline{v} = 0 \tag{1-20}$$

The transform of (1-13) reduces after considerable rearrangement to

$$\text{at } z^* = H^* : \ \frac{\partial \overline{v}}{\partial z^*} = -\frac{\mathcal{T}_z^*}{R_1^*} J_1(\xi_i R_1^*) - (\xi_i)^2 N \overline{v} \tag{1-21}$$

In order to obtain this, two integrations by parts were used as well as (1-8), (1-16) and (1-18).

The solution to (1-19) that satisfies boundary conditions (1-20) and (1-21) can be readily shown to be

$$\overline{v} = A_i \sinh(\xi_i z^*) \tag{1-22}$$

in which I have introduced

$$A_i \equiv -\frac{\mathcal{T}_z^*}{R_1^*} \frac{J_1(\xi_i R_1^*)}{(\xi_i)^2 N \sinh(\xi_i H^*) + \xi_i \cosh(\xi_i H^*)} \tag{1-23}$$

Taking the inverse transformation (Irving and Mullineux 1959, p. 157), we have

$$v_\theta^* = 2 \sum_{i=1}^{\infty} \frac{A_i \sinh(\xi_i z^*)}{[J_0(\xi_i)]^2} J_1(\xi_i r^*) \tag{1-24}$$

The as yet unknown $\mathcal{T}_z^*$ can now be determined by requiring (1-9) to be satisfied (Lifshutz *et al.* 1971; Mannheimer and Burton 1970):

$$\mathcal{T}_z^* \equiv \frac{\mathcal{T}_z}{2\pi R^3 \mu \Omega}$$

$$= R_1^{*2} \left\{ 2 \sum_{i=1}^{\infty} \frac{[J_1(\xi_i R_1^*)]^2}{[J_0(\xi_i)]^2 \, [(\xi_i)^2 N + \xi_i \coth(\xi_i H^*)]} \right\}^{-1} \tag{1-25}$$

This series is not well represented by its first few terms, since we are attempting to represent a discontinuity in the derivative of the surface velocity by a series of infinitely differentiable functions. As many as 5000 terms may be required for adequate convergence. Specific results are shown in Figures 1-2 and 1-3.

As I suggested at the beginning, this geometry is of interest as the basis for a surface viscometer. One measures $\mathcal{T}_z^*$ and uses (1-25) to compute the corresponding value of N. Figures 1-2 and 1-3 indicate that there are at least five aspects of (1-25) of experimental interest.

a) The torque $\mathcal{T}_z$ is proportional to Ω.

b) For large values of N, $\mathcal{T}_z^\star$ is a linear function of N, and the contribution of the bulk viscous forces to $\mathcal{T}_z$ can be neglected with respect to the contribution of the surface viscous forces.

c) For large values of $H^\star$, $\mathcal{T}_z^\star$ is independent of $H^\star$.

d) For small values of N, $\mathcal{T}_z^\star$ is nearly independent of N.

e) As $R_1^\star \to 1$, $\mathcal{T}_z^\star$ increases, but the limiting value of N below which the instrument should not be used does not change appreciably.

The results are relatively insensitive to the width of a finely ground knife-edge (Briley *et al.* 1976). See Sec. 4.5.7 for further details.

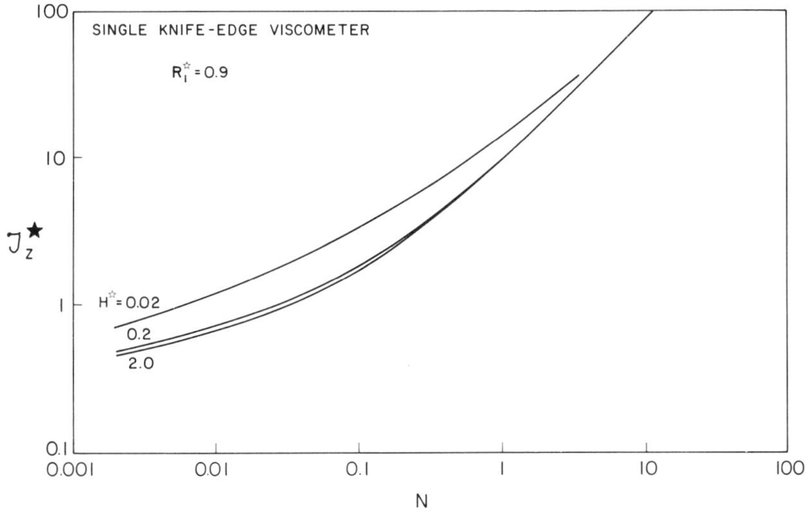

Figure 3.4.1-2: Effect of $H^\star$ upon torque for knife-edge surface viscometer from (1-25).

References

Briley, P. B., A. R. Deemer, and J. C. Slattery, *J. Colloid Interface Sci.* **56**, 1 (1976).

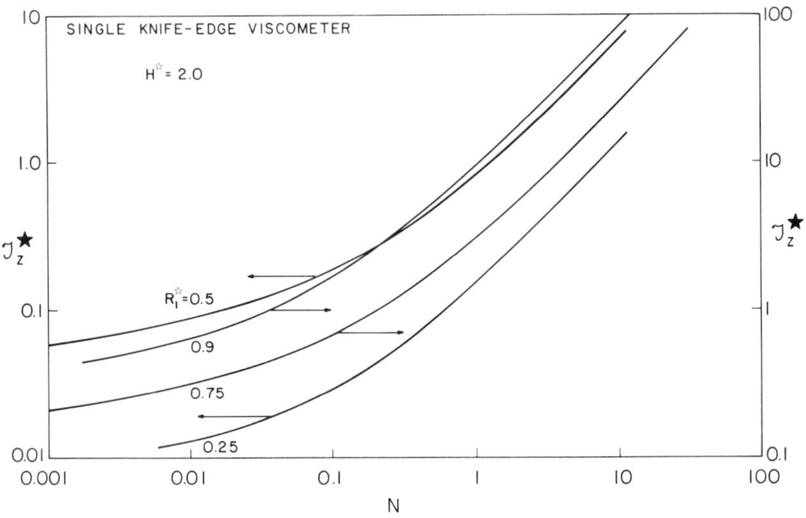

Figure 3.4.1-3: Effect of $R_1^\star$ upon torque for knife-edge surface viscometer from (1-25).

Brown, A. G., W. C. Thuman, and J. W. McBain, *J. Colloid Sci.* **8**, 491 (1953).

Chaminade, R., D. G. Dervichian, and M. Joly, *J. Chim. Phys.* **47**, 883 (1950).

Chen, J. D., personal communication (1978).

Davies, J. T., *Proc. Intern. Congr. Surface Activity, 2nd London* **1**, 220 (1957).

Deemer, A. R., J. D. Chen, M. G. Hegde, and J. C. Slattery, *J. Colloid Interface Sci.* **78**, 87 (1980).

Ellis, S. C., A. F. Lanham, and K. G. Parkhurst, *J. Sci. Inst.* **32**, 70 (1955).

Foust, L., and W. D. Harkins, *J. Phys. Chem.* **42**, 897 (1938).

Gladden, G. P., and E. L. Neustadter, *J. Inst. Petrol. London* **58**, 351 (1972).

Gladden, G. P., and E. L. Neustadter, *Chem. phys. chem. Anwedungstech. Grenzflaechenaktiven Stoffe. Ber. Int. Kongr., 6th* **2**, 535 (1973).

Irving, J., and N. Mullineux, "Mathematics in Physics and Engineering," Academic Press, New York (1960).

Lifshutz, N., M. G. Hegde, and J. C. Slattery, *J. Colloid Interface Sci.* **37**, 73 (1971).

Mannheimer, R. J., and R. A. Burton, *J. Colloid Interface Sci.* **32**, 73 (1970).

Mannheimer, R. J., and R. S. Schechter, *J. Colloid Interface Sci.* **32**, 195 (1970).

Oh, S. G., and J. C. Slattery, *J. Colloid Interface Sci.* **67**, 516 (1978).

Shotton, E., K. Wibberley, B. Warburton, S. S. David, and P. L. Finlay, *Rheol. Acta* **10**, 142 (1971).

Slattery, J. C., "Momentum, Energy, and Mass Transfer in Continua," McGraw-Hill, New York (1972); second edition, Robert E. Krieger, Malabar, FL 32950 (1981).

Wei, L. Y., and J. C. Slattery, in "Colloid and Interface Science vol. IV: Hydrosols and Rheology," edited by M. Kerker, p. 399, Academic Press, New York (1976).

Wibberley, K., *J. Pharm. Pharmacol.* **14**, 87T (1962).

Exercises

3.4.1-1 *double knife-edge surface viscometer* (Lifschutz *et al.* 1971) Using the approach described in the text, analyze the double knife-edge surface viscometer pictured in Figure 1-4. Determine that

$$v_\theta^\star = -\frac{\mathcal{T}_{z1}^\star}{R_1^\star} S_1(r^\star, z^\star) - \frac{\mathcal{T}_{z2}^\star}{R_2^\star} S_2(r^\star, z^\star) \tag{1}$$

where

$$S_j(r^\star, z^\star)$$

$$\equiv 2 \sum_{i=1}^{\infty} \frac{J_1(\xi_i R_j^\star)\ \sinh(\xi_i z^\star)\ J_1(\xi_i r^\star)}{[J_0(\xi_i)]^2 [(\xi_i)^2 N \sinh(\xi_i H^\star) + \xi_i \cosh(\xi_i H^\star)]} \tag{2}$$

$$\mathcal{T}^{*}_{z1} \equiv \frac{\mathcal{T}_{z1}}{2\pi R^3 \mu\Omega}$$

$$= \frac{R_1^{*2}\ S_2(R_2^*, H^*) - R_1^* R_2^*\ S_2(R_1^*, H^*)}{S_1(R_1^*, H^*) S_2(R_2^*, H^*) - S_1(R_2^*, H^*) S_2(R_1^*, H^*)} \tag{3}$$

$$\mathcal{T}^{*}_{z2} \equiv \frac{\mathcal{T}_{z2}}{2\pi R^3 \mu\Omega}$$

$$= \frac{R_2^{*2}\ S_1(R_1^*, H^*) - R_1^* R_2^*\ S_1(R_2^*, H^*)}{S_1(R_1^*, H^*) S_2(R_2^*, H^*) - S_1(R_2^*, H^*) S_2(R_1^*, H^*)} \tag{4}$$

Here $\mathcal{T}_{z1}$ is the torque exerted upon the inner knife edge (at $r = R_1$) and $\mathcal{T}_{z2}$ is the torque exerted upon the outer knife edge (at $r = R_2$). It follows immediately that

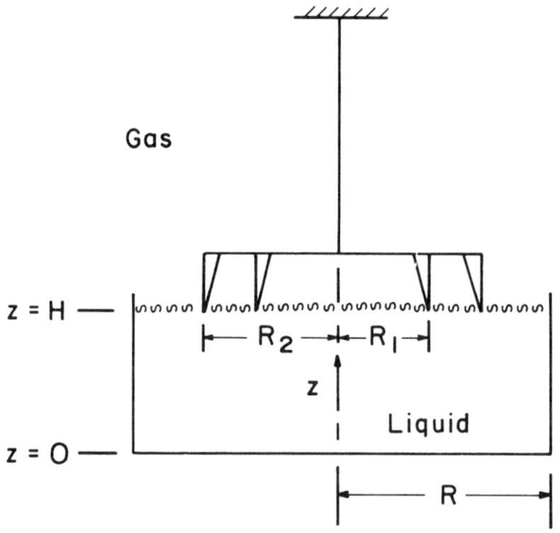

Figure 3.4.1-4: Double knife-edge surface viscometer. The pan rotates with a constant angular velocity Ω while the knife-edges are held stationary. One measures either $\mathcal{T}_z$, the z-component of the torque required to hold the knife-edges stationary, or the surface velocity at the centerline between the knife-edges.

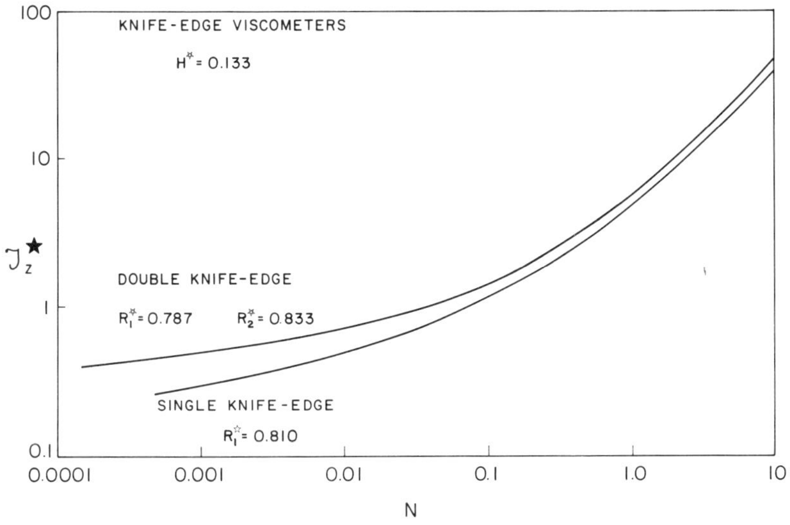

Figure 3.4.1-5: Torque predicted by (5) of Exercise 3.4.1-1 for a double knife-edge surface viscometer and the torque predicted by (1-25) for a comparable single knife-edge instrument.

$$\mathcal{T}_z^\star \equiv \frac{\mathcal{T}_z}{2\pi R^3 \mu \Omega}$$

$$= \mathcal{T}_{z1}^\star + \mathcal{T}_{z2}^\star \tag{5}$$

The observations made in the text concerning (1-25) apply equally well to (5). Figure 1-5 shows the torque predicted by (5) for a double knife-edge viscometer and the torque predicted by (1-25) for a comparable single knife-edge instrument. The double knife-edge viscometer is somewhat less sensitive and for that reason has not been used in this form.

Davies (1957) suggested that, rather than measuring the torque required to hold the knife-edges stationary, one might measure the surface velocity at the centerline between the knife edges, $v_{\theta c}$. From (1),

$$\frac{\Omega_c}{\Omega} = - \frac{2\mathcal{T}_{z1}^\star}{R_1^\star[\, R_1^\star + R_2^\star \,]} S_1\left(\frac{R_1^\star + R_2^\star}{2}, H^\star \right)$$

$$- \frac{2\mathcal{T}_{z2}^\star}{R_2^\star[\, R_1^\star + R_2^\star \,]} S_2\left(\frac{R_1^\star + R_2^\star}{2}, H^\star \right) \tag{6}$$

which is shown for a typical case in Figure 1-6. Given a measured value of $v_{\theta c}$, one could calculate the corresponding value of N from (6). For the case presented in Figure 1-6, this would be practical only if N > 0.01.

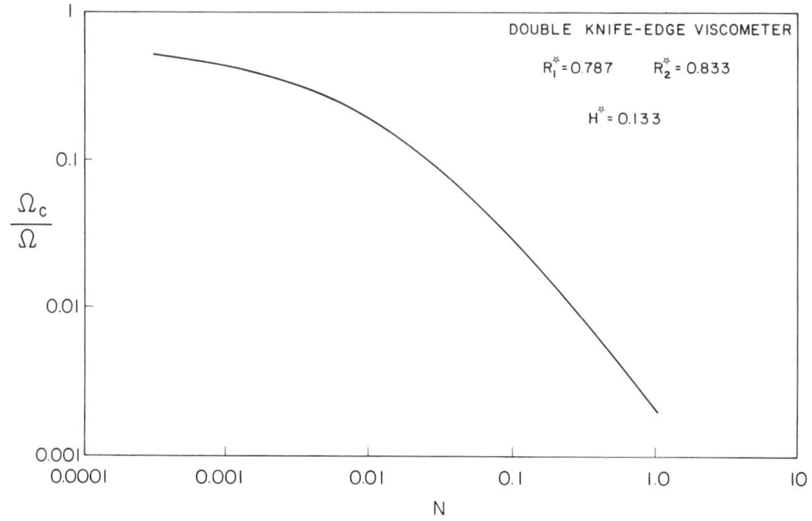

Figure 3.4.1-6: Centerline surface velocity predicted by (1) of Exercise 3.4.1-1 for a double knife-edge surface viscometer.

3.4.1-2 *multiple knife-edge surface viscometer* (Briley *et al.* 1976) The multiple knife-edge surface viscometer is an extension of the double knife-edge surface viscometer shown in Figure 1-4, in which we now have M concentric circular knife edges, the ith edge being located at $r = R_i$. Using the approach described in the text, determine that for this instrument

$$v_\theta^* = -\sum_{j=1}^{M} \frac{1}{R_j^*} \mathcal{T}_{zj}^* \, S_j(r^*, z^*) \tag{1}$$

where the $\mathcal{T}_{zj}^*$ (j = 1, 2, ..., M) are determined by solving simultaneously the set of linear equations (k = 1, 2, ..., M)

$$R_k^* = \sum_{j=1}^{M} \frac{1}{R_j^*} \mathcal{T}_{zj}^* \, S_j(R_k^*, H^*) \tag{2}$$

The total torque exerted upon the set of knife edges by the interface is given by

$$\mathcal{T}_z^* \equiv \frac{\mathcal{T}_z}{2\pi R^3 \mu \Omega}$$

$$= \sum_{j=1}^{M} \mathcal{T}_{zj}^* \tag{3}$$

3.4.1-3 *deep channel surface viscometer* Consider the deep channel surface viscometer described in Sec. 3.5.1 and Figure 3.5.1-1. Assuming that the interface can be described by the linear Boussinesq surface fluid, determine that the surface velocity distribution takes the form (Mannheimer and Schechter 1970)

$$v_\theta^{(\sigma)*} = \sum_{n=1}^{\infty} \left[F_n \phi_1 (\lambda_n r^*) \right] \left[\cosh(\lambda_n a^*) + \sinh(\lambda_n a^*) \left(\varepsilon^* \lambda_n + \frac{\mu^{(2)}}{\mu^{(1)}} \coth(\lambda_n b^*) \right) \right]^{-1} \tag{1}$$

The notation of Sec. 3.5.1 is adopted here with the addition of

$$\varepsilon^* \equiv \frac{\varepsilon}{\mu^{(1)} R(1 - k)} \tag{2}$$

Given the measured center line velocity, we can use (1) to compute the corresponding dimensionless surface shear viscosity ε^*. With this value of ε^*, (1) can be employed again to calculate the corresponding surface velocity distribution. If the surface velocity distribution calculated in this manner agrees with the measured velocity distribution, the surface stress-deformation behavior may be described by the linear Boussinesq surface fluid with a constant surface shear viscosity.

The deep channel takes its name from the observation that, if

$$a^* \equiv \frac{a}{R(1 - k)} \tag{3}$$

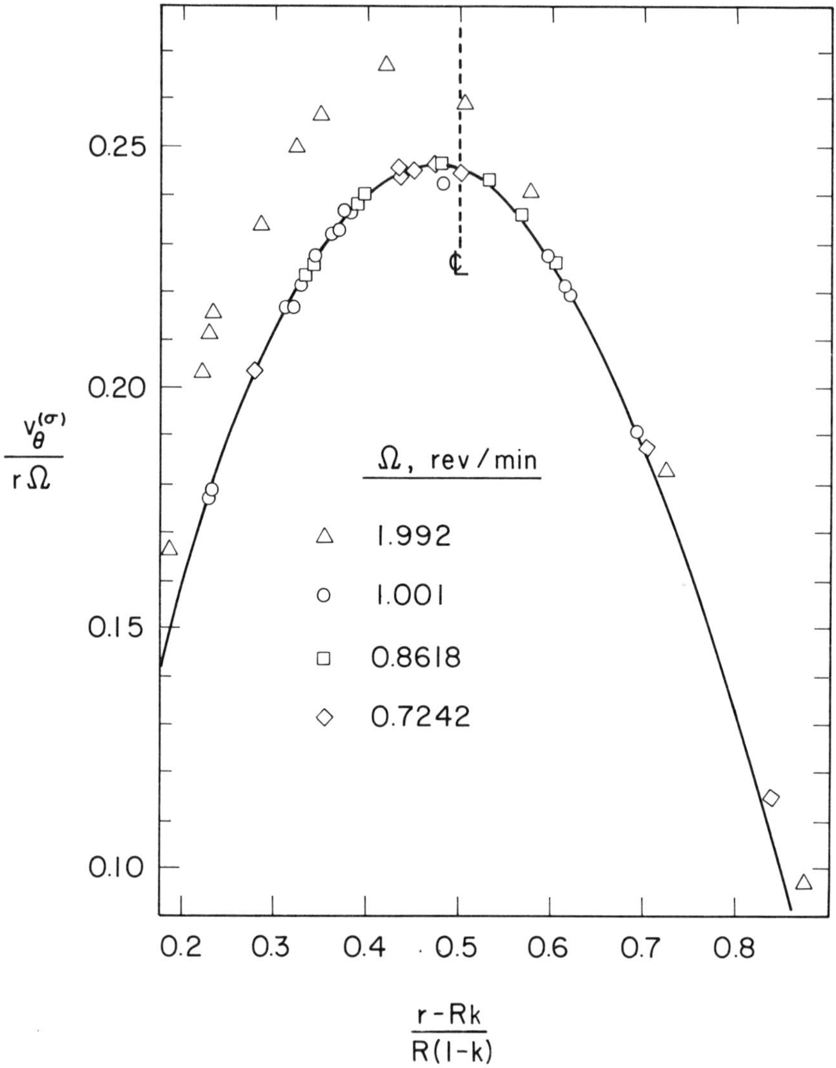

Figure 3.4.1-7: Velocity distributions at a distilled water-air phase interface corresponding to several values of Ω (Wei and Slattery 1976). In all experiments, a = 1.07 cm, b = 3.07 cm, R = 6.450 cm, and k = 0.7736. The solid curve is predicted from the result of Exercise 3.4.1-3 for $\varepsilon = 0$.

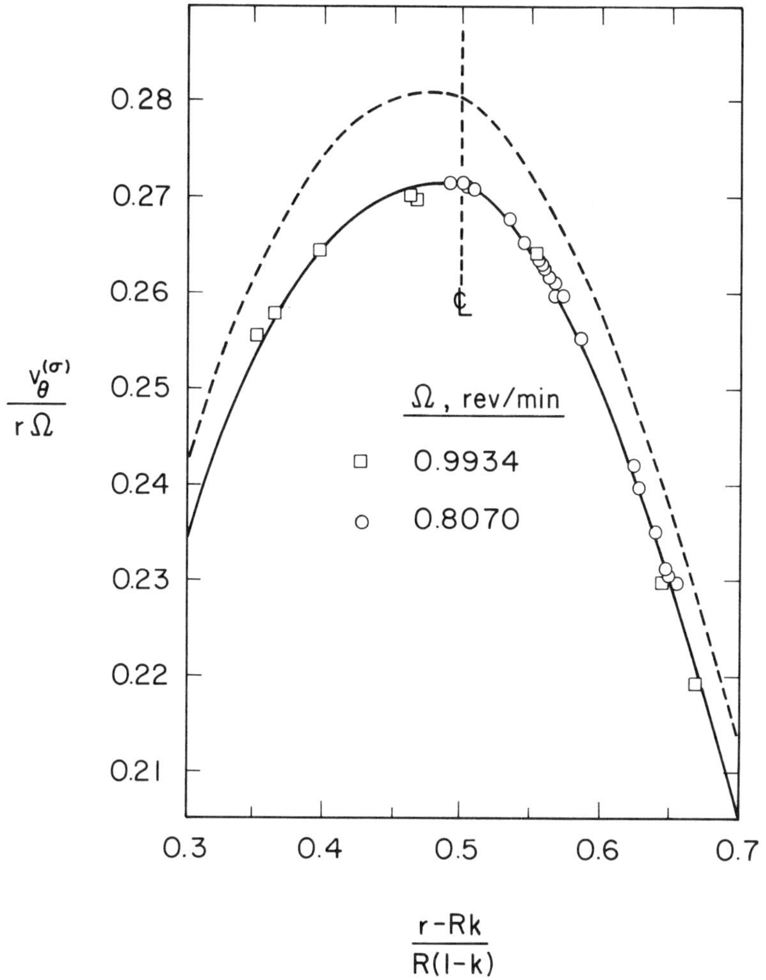

Figure 3.4.1-8: Velocity distributions at an interface between a 6% aqueous solution of potassium oleate and air corresponding to two values of Ω (Wei and Slattery 1976). In both experiments, a = 1.003 cm, b = 3.14 cm, R = 6.450 cm, and k = 0.7736. The solid curve is predicted from the result of Exercise 3.4.1-3 for $\varepsilon = 1.60 \times 10^{-4}$ mN s/m [Chen (1978) gave the revised data analysis shown following the suggestion of Deemer *et al.* (1980)]; the dashed curve is predicted for $\varepsilon = 0$.

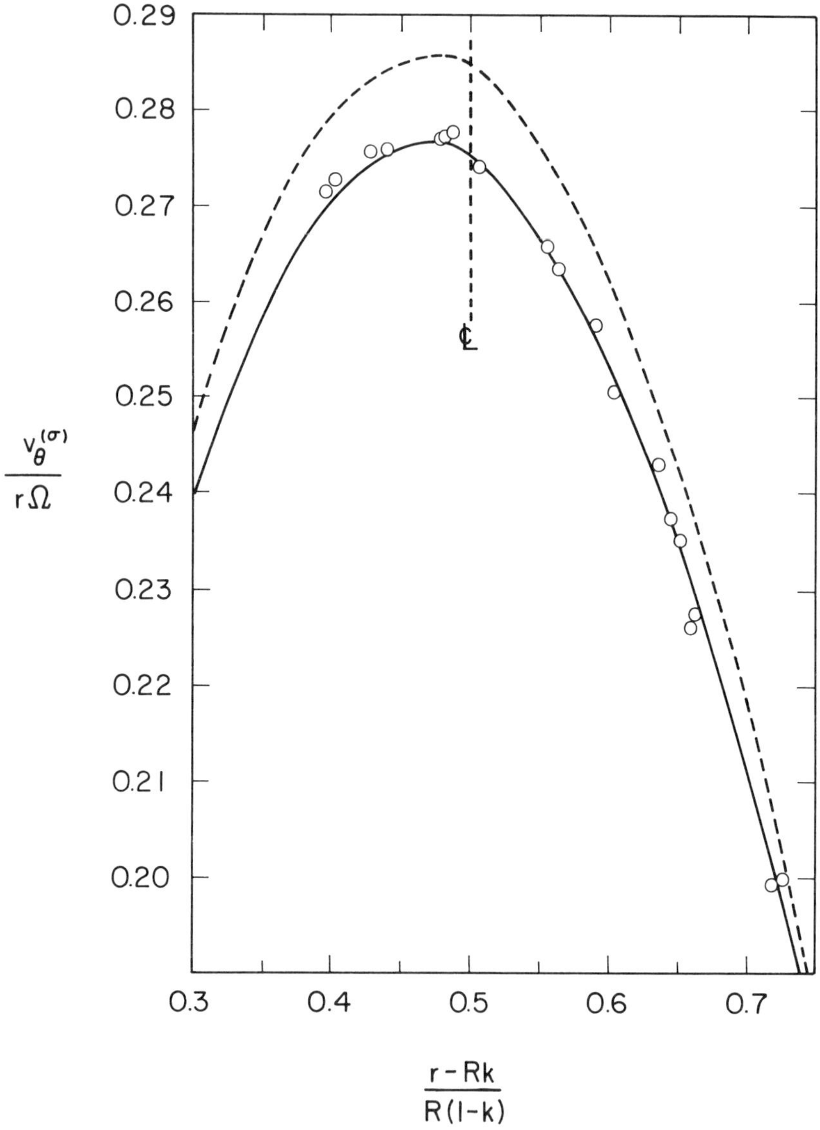

Figure 3.4.1-9: Velocity distribution at an interface between a 6% aqueous solution of potassium oleate and air (Wei and Slattery 1976). In this experiment, Ω = 1.004 rev/min, a = 0.996 cm, b = 3.14 cm, R = 6.450 cm, and k = 0.7736. The solid curve is predicted from the result of Exercise 3.4.1-3 for ε = 1.60 $\times$ 10^{-4} mN s/m [Chen (1978) gave the revised data analysis shown following the suggestion of Deemer *et al.* (1980)]; the dashed curve is predicted for ε = 0.

is sufficiently large, it is necessary to retain only the first term in the series on the right side of (1). This leads to a relatively simple expression for the surface shear viscosity in terms on the dimensionless center line surface velocity $v_{\theta c}^{(\sigma)*}$:

$$\varepsilon^* = \frac{1}{\lambda_1}\left[\frac{F_1}{v_{\theta c}^{(\sigma)}\ \sinh(\lambda_1 a^*)}\phi\left(\lambda_1 \frac{1+k}{2(1-k)}\right) - \coth(\lambda_1 a^*) - \frac{\mu^{(2)}}{\mu^{(1)}}\coth(\lambda_1 b^*)\right] \tag{4}$$

The velocity distributions measured by Wei and Slattery (1976) at a distilled water-air interface are shown in Figure 1-7 compared with (1), with the assumption that $\varepsilon = 0$. There is good agreement, with the exception of the data for $\Omega = 1.992$ rev/min. or for

$$N_{Re} \equiv \frac{\rho^{(1)}\Omega R^2}{2\mu^{(1)}}(1-k^2) = 186 \tag{5}$$

We believe that these latter data are beginning to show the effects of inertia. Mannheimer and Schechter (1970) observed a similar effect at $N_{Re} = 130$ with a narrower channel.

Figures 1-8 and 1-9 show that an interface between air and a 6% aqueous solution of potassium oleate is well described by the linear Boussinesq surface fluid model and that its surface shear viscosity is significantly different from zero [Wei and Slattery 1976; Chen (1978) gave the revised data analysis shown following the suggestion of Deemer *et al.* (1980)].

3.4.1-4 *more on deep channel surface viscometer* When phase 2 in Figure 3.5.1-1 is opaque, the velocity distribution in the fluid-fluid interface must be viewed from below. Alternatively, one can use the configuration shown in Figure 3.5.1-4, where phase 3 is a gas.

Assuming that both fluid-fluid interfaces can be described by the linear Boussinesq surface fluid, determine that the surface velocity distribution at the interface between phases 2 and 3 takes the form (Deemer *et al.* 1980)

$$v_\theta^{(\sigma,23)*} = \sum_{n=1}^{\infty} B_n\left[\cosh(\lambda_n c^*) - P_n \sinh(\lambda_n c^*)\right]\phi_1(\lambda_n r^*) \tag{1}$$

in which

$$B_n \equiv \left[2\,(1-k) \left\{ k^2\, \phi_0\left(\frac{\lambda_n k}{1-k}\right) - \phi_0\left(\frac{\lambda_n}{1-k}\right) \right\} \right]$$

$$\times \left[\lambda_n \sinh(\lambda_n a^*) \left\{ \left[\phi_0\left(\frac{\lambda_n k}{1-k}\right) \right]^2 - k^2 \left[\phi_0\left(\frac{\lambda_n k}{1-k}\right) \right]^2 \right\} \right.$$

$$\left. \times \left\{ \lambda_n \varepsilon^{(12)*} + \frac{\mu^{(2)}}{\mu^{(1)}} P_n + \coth(\lambda_n a^*) \right\} \right]^{-1} \tag{2}$$

$$P_n \equiv \left\{ \varepsilon^{(23)*} \lambda_n \cosh(\lambda_n c^*) + \sinh(\lambda_n c^*) \right.$$

$$\left. + \frac{\mu^{(3)}}{\mu^{(2)}} \cosh(\lambda_n c^*) \coth(\lambda_n [\, b^* - c^* \,]) \right\}$$

$$\times \left\{ \varepsilon^{(23)*} \lambda_n \sinh(\lambda_n c^*) + \cosh(\lambda_n c^*) \right.$$

$$\left. + \frac{\mu^{(3)}}{\mu^{(2)}} \sinh(\lambda_n c^*) \coth(\lambda_n [\, b^* - c^* \,]) \right\}^{-1} \tag{3}$$

$$\varepsilon^{(12)*} \equiv \frac{\varepsilon^{(12)}}{\mu^{(1)} R(1-k)} \tag{4}$$

and

$$\varepsilon^{(23)*} \equiv \frac{\varepsilon^{(23)}}{\mu^{(2)} R(1-k)} \tag{5}$$

The notation is otherwise consistent with Exercise 3.4.1-3 and Sec. 3.5.1.

Given the center line velocity measured in the interface between phases 2 and 3 and an independent measurement of $\varepsilon^{(23)}$, we can use (1) to compute the corresponding dimensionless interfacial shear viscosity $\varepsilon^{(12)*}$ for the liquid-liquid interface. With this value of $\varepsilon^{(12)*}$, (1) can be employed again to calculate the corresponding surface velocity distribution $v^{(g,23)}$. If $v^{(g,23)}$ calculated in this manner agrees with the measured velocity distribution in the interface between phases 2 and 3, we can conclude that the interfacial stress-deformation behavior for the interface

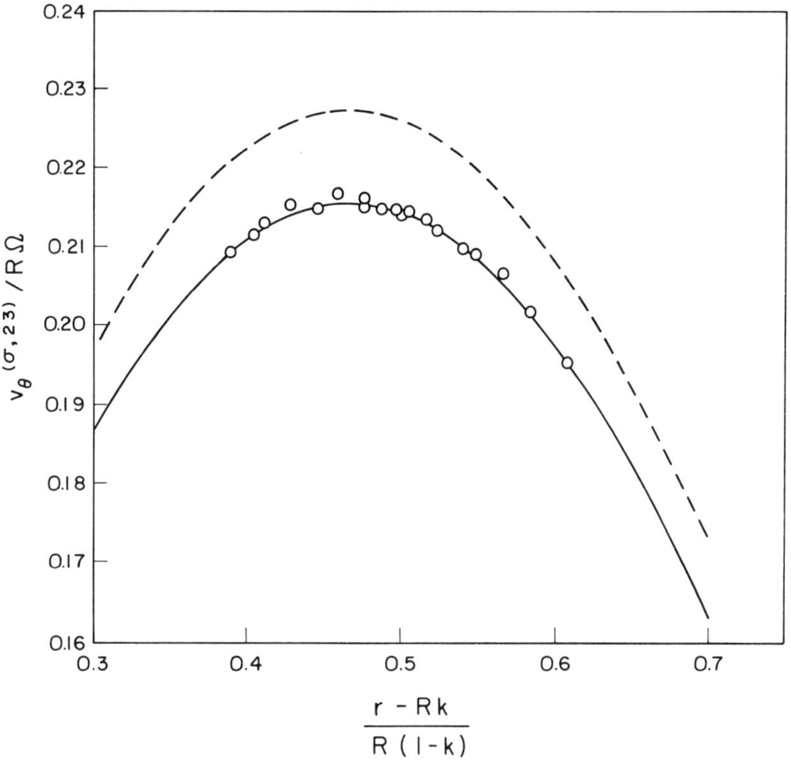

Figure 3.4.1-10: Velocity distribution at an interface between air and decane (Deemer *et al.* 1980). With reference to Figure 3.5.1-4, phase 1 is an aqueous solution containing 1% by weight sodium chloride and 0.267% by weight Petrostep 450, phase 2 is decane, and phase 3 is air. In this experiment, $\Omega = 0.895$ rev/min, a = 1.028 cm, b = 5.6 cm, c = 0.1074 cm, R = 6.683 cm, and k = 0.7787. The solid curve is predicted from the result of Exercise 3.4.1-4 for $\varepsilon^{(12)} = 3.13 \times 10^{-4}$ mN s/m and $\varepsilon^{(23)} = 0$; the dashed curve is predicted for $\varepsilon^{(12)} = \varepsilon^{(23)} = 0$.

between phases 1 and 2 may be described by the linear Boussinesq surface fluid with a constant interfacial shear viscosity.

If a^* is sufficiently large, it is necessary to retain only the first term in the series on the right side of (1), in which case

$$\varepsilon^{(12)*} = \frac{1}{\lambda_1} \left\{ 2\,(1-k) \left[k^2\, \phi_0\!\left(\frac{\lambda_1 k}{1-k} \right) - \phi_0\!\left(\frac{\lambda_1}{1-k} \right) \right] \right.$$
$$\left. \times \left[\cosh(\lambda_1 c^*) - P_1 \sinh(\lambda_1 c^*) \right] \phi_1\!\left(\frac{\lambda_1}{2} \left[\frac{1+k}{1-k} \right] \right) \right\}$$

$$\times \left\{ v_{\theta c}^{(\sigma,23)*} \lambda_1 \sinh(\lambda_1 a^*) \left\{ \left[\phi_0 \left(\frac{\lambda_1}{1 - k} \right) \right]^2 \right. \right.$$

$$\left. \left. - k^2 \left[\phi_0 \left(\frac{\lambda_1 k}{1 - k} \right) \right]^2 \right\} - \frac{\mu^{(2)}}{\mu^{(1)}} P_1 - \coth(\lambda_1 a^*) \right\}^{-1} \tag{6}$$

Here $v_{\theta c}^{(\sigma,23)*}$ is the dimensionless center line velocity measured in the interface between phases 2 and 3.

Deemer *et al.* (1980) report three similar sets of data, one of which is shown in Figure 1-10. Their conclusion is that the liquid-liquid interface studied is well represented by the linear Boussinesq surface fluid and that its interfacial shear viscosity is significantly different from zero.

3.4.1-5 *zero-thickness disk interfacial viscometer* (Oh and Slattery 1978) For the zero-thickness disk interfacial viscometer shown in Figure 1-11, determine that the dimensionless interfacial velocity distribution $v_\theta^{(\sigma)*}$ and the dimensionless torque $\mathcal{T}_z^*$ exerted by the fluids on the stationary disk take the respective forms

$$v_\theta^{(\sigma)} \equiv \frac{v_\theta^{(\sigma)}}{R\Omega} - r^*$$

$$= -\sum_{i=1}^{\infty} \frac{1}{\xi_i^2} \left(\frac{A_i}{N} + \frac{B_i}{NY} \right) \left[J_1(\xi_i r^*) \right.$$

$$\left. - \frac{R_2^*}{r^*} \left(\frac{1 - r^{*2}}{1 - R_2^{*2}} \right) J_1(\xi_i R_2^*) \right] - \frac{R_2^{*2}}{r^*} \left(\frac{1 - r^{*2}}{1 - R_2^{*2}} \right) \tag{1}$$

and

$$\mathcal{T}_z^* \equiv \frac{\mathcal{T}_z}{2\pi R^3 \mu^{(1)} \Omega}$$

$$= -\sum_{i=1}^{\infty} \left[\frac{1}{\xi_i^2} \left(A_i + \frac{B_i}{Y} \right) \right] \left[2R_2^* \frac{J_1(\xi_i R_2^*)}{(1 - R_2^{*2})} \right] + \frac{2R_2^{*2} N}{(1 - R_2^{*2})} \tag{2}$$

in which

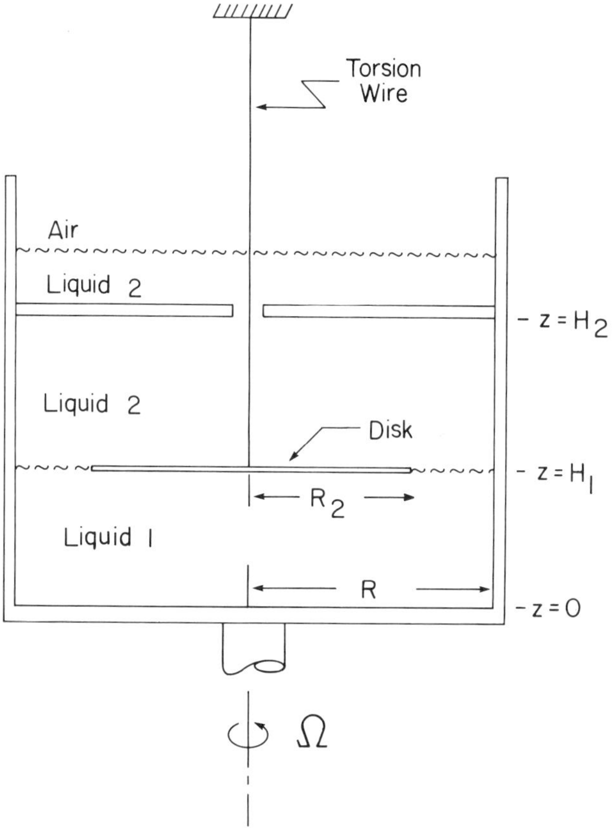

Figure 3.4.1-11: Zero-thickness disk interfacial viscometer. The torque $-\mathcal{T}_z$ required to hold the disk stationary is measured as a function of the angular velocity Ω of the dish.

$$A_i \equiv \left(\frac{2\xi_i}{\tanh(\ \xi_i H_1^\star\ )\ [\ J_0(\ \xi_i\ )]^2} \right)$$

$$\times \left(\int_{R_2^\star}^{1} \alpha\ J_1(\ \xi_i \alpha\ )\ v_\theta^{(\sigma)\star}(\ \alpha\ )\ d\alpha\ - \frac{1}{\xi_i} R_2^{\star 2}\ J_2(\ \xi_i R_2^\star\ ) \right) \tag{3}$$

$$B_i \equiv \left(\frac{2\,\xi_i}{\tanh(\ \xi_i [\ H_2^\star\ -\ H_1^\star\ ])\ [\ J_0(\ \xi_i\ )]^2} \right)$$

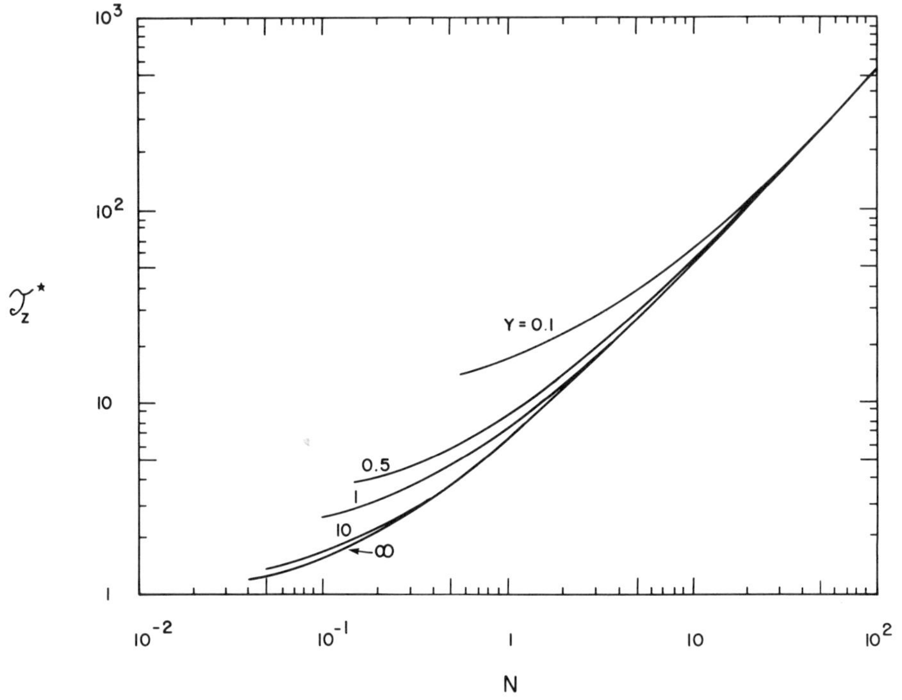

Figure 3.4.1-12: Dimensionless torque $\mathcal{T}_z^\star$ as a function of N for zero-thickness disk interfacial viscometer with $R_2^\star = 0.8536$ and $H_1^\star = H_2^\star - H_1^\star = 0.2252$.

$$\times \left(\int_{R_2^\star}^{1} \alpha \, J_1(\,\xi_i \alpha\,) \, v_\theta^{(\sigma)\star}(\,\alpha\,) \, d\alpha - \frac{1}{\xi_i} R_2^{\star 2} \, J_2(\,\xi_i R_2^\star\,) \right) \tag{4}$$

$$r^\star \equiv \frac{r}{R} \qquad z^\star \equiv \frac{z}{R} \tag{5}$$

$$N \equiv \frac{\varepsilon}{\mu^{(1)} R} \tag{6}$$

$$Y \equiv \frac{\mu^{(1)}}{\mu^{(2)}} \tag{7}$$

$$J_1(\,\xi_i\,) = 0 \qquad \text{for } i = 1, 2, \ldots \tag{8}$$

Table 3.4.1-1: Comparison for liquid-gas interface ($Y \to \infty$) of $\mathcal{T}_z^*$ from (2) of Exercise 3.4.1-5 with upper and lower bounds for $\mathcal{T}_z^*$ from Briley et al. (1976). It is assumed that $R_2^* = 0.8536$ and $H_1^* = 0.2252$.

	$\mathcal{T}_z^*$		
N	upper bound	lower bound	(2)
10	54.74	54.72	54.74
1	6.410	6.384	6.407
0.1	1.560	1.529	1.559

Unfortunately, (1) does not provide us with an explicit expression for the surface velocity distribution $v_\theta^{(\sigma)*}$. From (3) and (4), we see that prior knowledge of $v_\theta^{(\sigma)*}$ has been assumed. Oh and Slattery (1978) suggest that (1) be used as a basis for an iterative solution for $v_\theta^{(\sigma)*}$. They begin with a relatively simple first estimate for $v_\theta^{(\sigma)*}$, calculate A_i and B_i from (3) and (4), compute an improved estimate from (1), and then repeat the process until the deviation between successive estimates for $v_\theta^{(\sigma)*}$ is less than 0.05% at each of 100 equally spaced radial positions.

When the depths of the two fluids are equal

$$H_1^* = H_2^* - H_1^* \tag{9}$$

$$A_i = B_i \tag{10}$$

and (1) simplifies to

$$v_\theta^{(\sigma)*} = \frac{1}{N}\left(1 - \frac{1}{Y}\right)\sum_{i=1}^{\infty}\frac{A_i}{\xi_i^2}\left[J_1(\xi_i r^*) - \frac{R_2^*}{r^*}\left(\frac{1 - r^{*2}}{1 - R_2^{*2}}\right)J_1(\xi_i R_2^*)\right] - \frac{R_2^{*2}}{r^*}\left(\frac{1 - r^{*2}}{1 - R_2^{*2}}\right) \tag{11}$$

and (2) becomes

$$\mathcal{T}_z^* = -\left(1 + \frac{1}{Y}\right)\sum_{i=1}^{\infty}\frac{A_i}{\xi_i^2}\left[2R_2^*\frac{J_1(\xi_i R_2^*)}{(1 - R_2^{*2})}\right] + \frac{2R_2^{*2}N}{(1 - R_2^{*2})} \tag{12}$$

The iterative computation for $\mathcal{T}_z^*$ as a function of N is now required only for one value of Y, say $Y \to \infty$. For any other value of Y, it is necessary

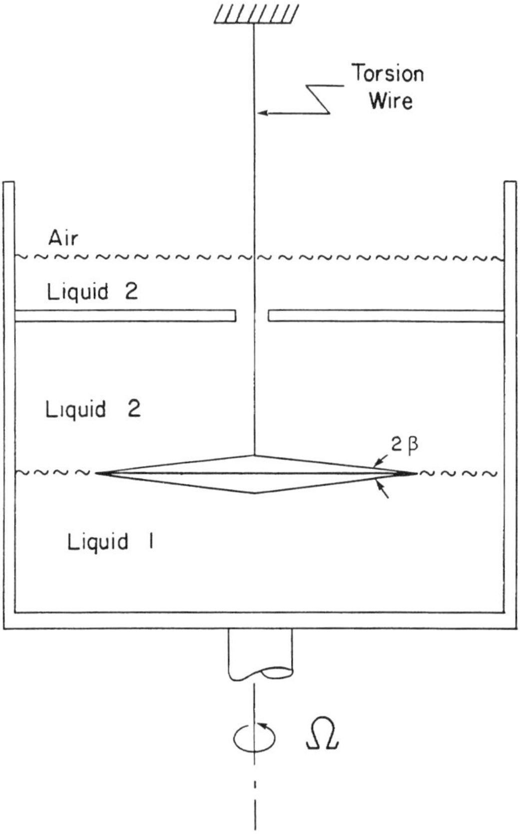

Figure 3.4.1-13: Thin biconical bob interfacial viscometer. The torque $-\mathcal{T}_z$ required to hold the bob stationary is measured as a function of the angular velocity Ω of the dish.

only to identify $N = (1 + Y^{-1})N^{\infty}$ and $\mathcal{T}_z^{*} = (1 + Y^{-1})\mathcal{T}_z^{*\infty}$. It was in this manner that Figure 1-12 was constructed.

In the limit as the viscosity ratio $Y \to \infty$, we recover the results for a liquid-gas interface. Table 3.4.1-1 shows that the torque computed from (2) in this limit is consistent with the upper and lower bounds computed by Briley *et al.* (1976; see also Sec. 4.4.7). In preparing this table, Oh and Slattery (1978) found that the estimate obtained for $\mathcal{T}_z^{*}$ at any particular iteration was accurate to five significant figures, so long as at least 500 terms were retained in (1) and 1000 terms in (2). These truncations were adopted in preparing Figure 1-12.

From the viewpoint of the experimentalist, there are several interesting aspects of this interfacial viscometer.

a) The torque $\mathcal{T}_z$ is proportional to Ω.

b) For large values of N, $\mathcal{T}_z^\star$ is a linear function of N, and the contribution of the bulk viscous forces to $\mathcal{T}_z^\star$ can be neglected with respect to the contribution of the surface viscous forces.

c) For large values of $H_1^\star$ and $H_2^\star - H_1^\star$, $\mathcal{T}_z^\star$ is independent of their values.

d) For small values of N, $\mathcal{T}_z^\star$ is nearly independent of N.

e) As $R_2^\star \to 1$, $\mathcal{T}_z^\star$ increases, but the limiting value of N below which the instrument should not be used does not change appreciably.

3.4.1-6 *Thin biconical bob interfacial viscometer* (Oh and Slattery 1978) The thin biconical bob interfacial viscometer shown in Figure 1-13 had been proposed and used (Wibberley 1962, Shotton *et al.* 1971, Gladden and Neustadter 1972, 1973) prior to the analysis by Oh and Slattery (1978).

For a sufficiently thin biconical bob (in the limit as $\beta \to 0$ in Figure 1-13), the ratio of the torque exerted by the bulk phases upon the bob to that exerted upon a disk having the same radius approaches the ratio of their surface areas: $\cos^{-1}\beta$. Show that, with this approximation, the torque on the biconical bob can be estimated as

$$\mathcal{T}_z^\star = -\sum_{i=1}^{\infty}\left[\frac{1}{\xi_i^2}\left(A_i + \frac{1}{Y}B_i\right)\right]\left[2R_2^\star\frac{J_1(\xi_i R_2^\star)}{(1 - R_2^{\star 2})} + \frac{2R_2^{\star 2}N}{(1 - R_2^{\star 2})}\right.$$
$$\left. + \left(\frac{1}{\cos\beta} - 1\right)\xi_i R_2^{\star 2}\, J_2(\xi_i R_2^\star)\right] + \frac{2R_2^{\star 2}N}{(1 - R_2^{\star 2})} \qquad (1)$$

where all of the notation is identical to that used in Exercise 3.4.1-5. This represents a very small correction to (2) of Exercise 3.4.1-5: less than 1% for $\beta \leq 20°$ when N can be determined by measuring $\mathcal{T}_z^\star$.

3.5 Generalized Boussinesq surface fluid

3.5.1 Deep channel surface viscometer The deep channel surface viscometer introduced by Burton and Mannheimer (1967; Osborne 1968; Mannheimer and Schechter 1968, 1970a; Pintar *et al.* 1971) was originally intended as a device for measuring the surface shear viscosity of an interface described by the linear Boussinesq surface fluid (see Exercise 3.4.1-3). In Figure 1-1, the walls are stationary concentric cylinders; the floor of the viscometer moves with a constant angular velocity. The experimentalist is required to measure the velocity of one or more particles in the interface in order to estimate the center-line surface velocity.

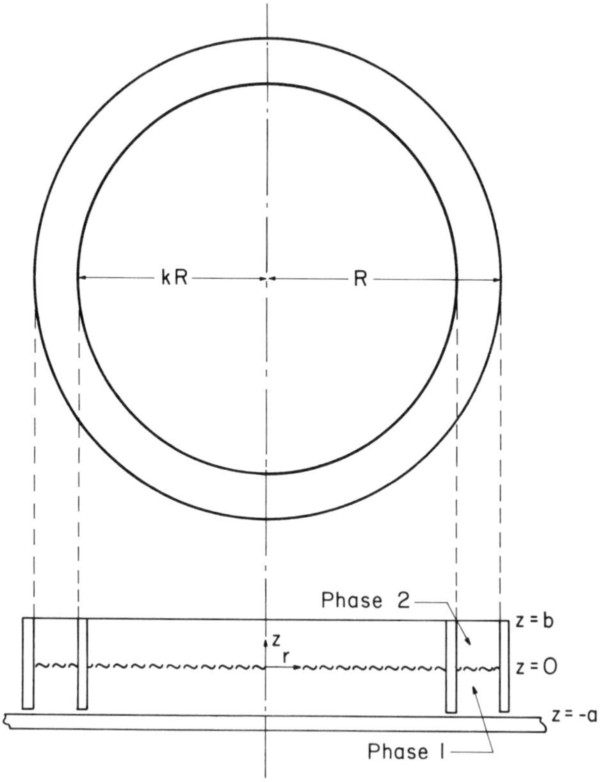

Figure 3.5.1-1: Deep Channel surface viscometer. The walls of the channel are stationary; the floor rotates with a constant angular velocity Ω.

It has been demonstrated to be a practical device for measuring the surface shear viscosity at liquid-gas interfaces (Mannheimer an Schechter 1970a; Pintar *et al.* 1971). Some measurements have also been made with liquid-liquid systems (Wasan *et al.* 1971; Deemer *et al.* 1980).

In the presence of large surfactant molecules, the surface shear viscosity may not be a constant (Brown *et al.* 1953; Pintar *et al.* 1971; Suzuki 1972; Wei and Slattery 1976). Three suggestions have been made for using the deep channel surface viscometer to measure the apparent surface shear viscosity of liquid-gas interfaces described by the generalized Boussinesq surface fluid (Sec. 2.2.2). Mannheimer and Schechter (1970b) assume that the interfacial stress-deformation behavior can be described by a Bingham plastic model in which no flow is possible until a critical interfacial stress has been exceeded. Pintar *et al.* (1971) give an approximate analysis in which a Powell-Eyring model is used for the interface. Hegde and Slattery (1971; Wei and Slattery 1976) show how the apparent surface shear viscosity can be determined from measurements of the velocity distribution in the interface without assuming a specific form for the interfacial stress-deformation behavior. In what follows, we summarize this suggestion of Hegde and Slattery (1971; Wei and Slattery 1976).

We make the following assumptions.

i) All inertial effects are neglected.

ii) There is no mass transfer across the liquid-gas interface.

iii) The viscometer operates under steady-state conditions in which all variables are independent of time.

iv) Both the liquid and the gas phases may be described as incompressible, Newtonian fluids.

v) The stress-deformation behavior of the liquid-gas interface may be represented by the generalized Boussinesq surface fluid.

vi) All physical properties are constants.

vii) The interface is a horizontal plane.

We have the following boundary conditions imposed by the geometry. The two walls and the ceiling of the annular channel are stationary,

$$\text{at } r = kR : \quad \mathbf{v}^{(1)} = \mathbf{v}^{(2)} = 0 \tag{1-1}$$

$$\text{at } r = R : \quad \mathbf{v}^{(1)} = \mathbf{v}^{(2)} = 0 \tag{1-2}$$

$$\text{at } z = b : \quad \mathbf{v}^{(2)} = 0 \tag{1-3}$$

Here $\mathbf{v}^{(1)}$ denotes the velocity distribution in the liquid phase and $\mathbf{v}^{(2)}$ the velocity distribution in the gas phase. The floor of the channel moves with a constant angular velocity Ω,

$$\text{at } z = -a : \quad v_\theta^{(1)} = r\,\Omega, \quad v_r^{(1)} = v_z^{(1)} = 0 \tag{1-4}$$

The tangential components of velocity are continuous across the gas-liquid phase interface

$$\text{at } z = 0: \quad v_\theta^{(1)} = v_\theta^{(2)} = v_\theta^{(\sigma)}$$

$$v_r^{(1)} = v_r^{(2)} = v_r^{(\sigma)} \tag{1-5}$$

Since there is no mass transfer across the phase interface,

$$\text{at } z = 0: \quad v_z^{(1)} = v_z^{(2)} = 0 \tag{1-6}$$

The jump mass balance for the liquid-gas interface (Sec. 1.3.5; the overall jump mass balance of Sec. 5.4.1) is satisfied identically, since we define the position of the dividing surface by choosing $\rho^{(\sigma)} = 0$ (Sec. 1.3.6) and since the effect of mass transfer is neglected (assumption ii).

In a standard analysis of this geometry, we would impose the jump momentum balance as a boundary condition to be satisfied and we would solve for the velocity distribution at the phase interface. Here we assume instead that we have measured the θ-component of the velocity distribution in the phase interface,

$$\text{at } z = 0: \quad v_\theta^{(\sigma)} = h(r) \tag{1-7}$$

At the conclusion of our discussion, we will use the jump momentum balance to determine the surface stress distribution in the phase interface.

The boundary conditions (1-1) through (1-6) suggest that we seek a velocity distribution which in each phase takes the form

$$v_\theta = v_\theta(r, z) \qquad v_r = v_z = 0 \tag{1-8}$$

in the cylindrical coordinate system suggested in Figure 1-1. This form of velocity distribution satisfies the equation of continuity (Slattery 1981, p. 60) identically. The equation of motion for an incompressible Newtonian fluid implies (Slattery 1981, p. 62)

$$\frac{\partial}{\partial r^\star}\left[\frac{1}{r^\star}\frac{\partial}{\partial r^\star}(r^\star v_\theta^\star)\right] + \frac{\partial^2 v_\theta^\star}{\partial z^{\star 2}} = 0 \tag{1-9}$$

where we have introduced as dimensionless variables

$$v_\theta^\star \equiv \frac{v_\theta}{R\Omega} \tag{1-10}$$

$$r^\star \equiv \frac{r}{R(1-k)} \tag{1-11}$$

$$z^* \equiv \frac{z}{R(1-k)} \tag{1-12}$$

The solution to (1-9) that is consistent with boundary conditions (1-1) through (1-7) is

$$v_\theta^{(1)*} = \sum_{n=1}^{\infty} [A_n \sinh(\lambda_n z^*) + B_n \cosh(\lambda_n z^*)]\, \phi_1(\lambda_n r^*) \tag{1-13}$$

$$v_\theta^{(2)*} = \sum_{n=1}^{\infty} B_n [-\coth(\lambda_n b^*) \sinh(\lambda_n z^*) + \cosh(\lambda_n z^*)]\, \phi_1(\lambda_n r^*) \tag{1-14}$$

where

$$A_n = B_n \coth(\lambda_n a^*) - F_n \operatorname{csch}(\lambda_n a^*) \tag{1-15}$$

$$B_n = 2(1-k)^2 h_n \left\{ \left[\phi_0\left(\frac{\lambda_n}{1-k}\right)\right]^2 - k^2 \left[\phi_0\left(\frac{\lambda_n k}{1-k}\right)\right]^2 \right\}^{-1} \tag{1-16}$$

$$F_n \equiv (1-k)\frac{2}{\lambda_n} \left[k^2 \phi_0\left(\frac{\lambda_n k}{1-k}\right) - \phi_0\left(\frac{\lambda_n}{1-k}\right)\right]$$

$$\times \left\{ \left[\phi_0\left(\frac{\lambda_n}{1-k}\right)\right]^2 - k^2 \left[\phi_0\left(\frac{\lambda_n k}{1-k}\right)\right]^2 \right\}^{-1} \tag{1-17}$$

$$h_n \equiv \int_{\frac{k}{1-k}}^{\frac{1}{1-k}} r^* \phi_1(\lambda_n r^*)\, h^*(r^*)\, dr^* \tag{1-18}$$

$$\phi_p(\lambda_n r^*) \equiv Y_1\left(\frac{\lambda_n k}{1-k}\right) J_p(\lambda_n r^*) - J_1\left(\frac{\lambda_n k}{1-k}\right) Y_p(\lambda_n r^*) \tag{1-19}$$

The λ_n ($n = 1, 2, \ldots$) are the roots of (Jahnke and Emde 1945, p. 205)

$$\phi_1\left(\frac{\lambda_n}{1-k}\right) = 0 \tag{1-20}$$

If we take our surface coordinates as r and θ, it follows that there is only one nonzero component of the dimensionless surface rate of deformation tensor

$$D_{r\theta}^{(\sigma)*} \equiv \frac{(1-k)}{\Omega} D_{r\theta}^{(\sigma)} = \frac{r^*}{2} \frac{d}{dr^*} \left(\frac{h^*}{r^*} \right) \tag{1-21}$$

and only one nonzero component of the dimensionless surface shear stress

$$T_{r\theta}^{(\sigma)*} \equiv \frac{T_{r\theta}^{(\sigma)}}{\mu^{(1)} R\Omega} \tag{1-22}$$

The r- and z- components of the jump momentum balance are satisfied identically; the θ-component requires (Table 2.4.2-2)

$$\text{at } z^* = 0 : \frac{1}{r^{*2}} \frac{d}{dr^*} \left(r^{*2} T_{r\theta}^{(\sigma)*} \right) = \frac{\partial v_\theta^{(\sigma)*}}{\partial z^*} - \frac{\mu^{(2)}}{\mu^{(1)}} \frac{\partial v_\theta^{(2)*}}{\partial z^*} \tag{1-23}$$

Let us define c^* to be that value of r^* such that

$$\text{at } r^* = c^* : \frac{d}{dr^*} \left(\frac{h^*}{r^*} \right) = 0 \tag{1-24}$$

It follows from the constitutive equation for the generalized Boussinesq surface fluid (Sec. 2.2.2) and (1-21) that

$$\text{at } r^* = c^* : T_{r\theta}^{(\sigma)*} = 0 \tag{1-26}$$

In view of (1-13) and (1-14), we can integrate (1-23) consistent with boundary condition (1-25) to find

$$T_{r\theta}^{(\sigma)*} = \sum_{n=1}^{\infty} \left[A_n + B_n \frac{\mu^{(2)}}{\mu^{(1)}} \coth(\lambda_n b^*) \right] \times \left[\phi_2(\lambda_n r^*) - \left(\frac{c^*}{r^*} \right)^2 \phi_2(\lambda_n c^*) \right] \tag{1-26}$$

We are now prepared to use the experimental surface velocity distribution $h^*(r^*)$ in (1-21) and (1-26) in order to determine the surface shear viscosity (Sec. 2.2.2)

$$\varepsilon = \mu^{(1)} (1 - k) R \frac{T_{r\theta}^{(g)*}}{2D_{r\theta}^{(g)*}} \tag{1-27}$$

as a function of the magnitude of the surface velocity gradient

$$\left| \nabla_{(\sigma)} \mathbf{v}^{(\sigma)} \right| = \frac{\Omega r^*}{1 - k} \left| \frac{d}{dr^*} \left(\frac{h^*}{r^*} \right) \right| \tag{1-28}$$

An experiment gives us the surface velocity distribution in the form of discrete measurements rather than a continuous function g(r) (see for example Figure 1-2). From (1-5), (1-13), and (1-14),

$$v_\theta^{(\sigma)*} = \sum_{n=1}^{\infty} B_n \phi_1 (\lambda_n r^*) \tag{1-29}$$

which suggests that we assume

$$h^*(r^*) = \sum_{n=1}^{N} \widetilde{B}_n \phi_1 (\lambda_n r^*) \tag{1-30}$$

The coefficients $\widetilde{B}_n$ can be determined by a least-square-error fit of (1-30) to the experimental data. For the optimum value of N, we can immediately identify $B_n = \widetilde{B}_n$. This permits us to calculate ε as a function of $|\nabla_{(\sigma)} \mathbf{v}^{(\sigma)}|$ using (1-15), (1-21), and (1-26) through (1-28).

Figure 1-2 presents the velocity distribution measured by Wei and Slattery (1976) at an interface between air and an n-octadecanol monolayer (20.5 $Å^2$/molecule) over distilled water. Given the center-line velocity and assuming that the surface stress-deformation behavior is described by the linear Boussinesq surface fluid, the result of Exercise 3.4.1-3 requires $\varepsilon^* = 0.324$ and $\varepsilon = 4.42 \times 10^{-3}$ dyne s/cm. The corresponding surface velocity distribution predicted by the result of Exercise 3.4.1-3 is shown in Figure 1-2 as the dashed curve. This interface is not well described by the linear Boussinesq surface fluid.

Its behavior is better described by the generalized Boussinesq surface fluid, which allows the apparent surface shear viscosity to be a function of the surface rate of deformation. The apparent surface shear viscosity must be determined from calculated values of the surface shear stress as a

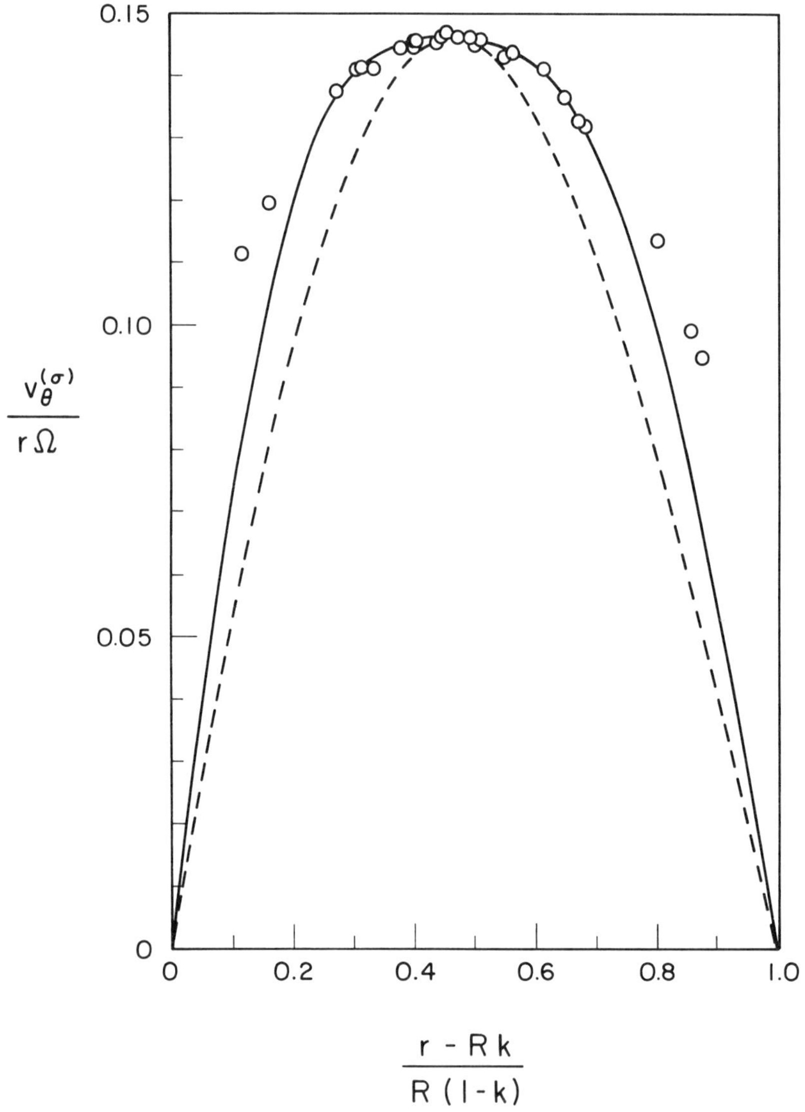

Figure 3.5.1-2: Velocity distribution at an interface between air and a n-octadecanol monolayer (20.5 Å^2/molecule) over distilled water (Wei and Slattery 1976). In this experiment, Ω = 1.004 rev/min, a = 0.998 cm ($a^\star$ = 0.683), b = 3.14 cm ($b^\star$ = 2.15), and $\mu^{(2)}/\mu^{(1)}$ = 0.0182. The solid curve represents a least-square-error fit of (1-30) with N = 4 to the experimental data. The dashed line is predicted from the result of Exercise 3.4.1-3 for ε = 4.42×10^{-3} dyne s/cm ($\varepsilon^\star$ = 0.324).

function of $|\nabla_{(\sigma)} \mathbf{v}^{(\sigma)}|$. Success depends upon the accuracy with which the velocity distribution is determined, since the computation of $|\nabla_{(\sigma)} \mathbf{v}^{(\sigma)}|$

involves a derivative of these data. Some uncertainties were unavoidable in our experimental measurements. We recommend choosing the largest value of N for which the same relationship between ε and $|\nabla_{(\sigma)} \mathbf{v}^{(\sigma)}|$ is obtained for both positive and negative values of the surface velocity gradient, in this case N = 4. The resulting representation for the velocity distribution is shown in Figure 1-2. Equations (1-26) through (1-28) were then used to compute $|T^{(\sigma)}_{r\theta}|$ and ε as functions of $|\nabla_{(\sigma)} \mathbf{v}^{(\sigma)}|$ These functions are shown in Figure 1-3.

Note that in Figure 1-3 the apparent surface shear viscosity decreases as $|\nabla_{(\sigma)} \mathbf{v}^{(\sigma)}|$ increases. There is a direct analogy in bulk stress-deformation behavior: the apparent viscosity is also commonly observed to decrease with increasing rate of deformation.

For the deep channel surface viscometer shown in Figure 3.5.1-1, it will normally be more convenient to view from the top the velocity distribution in the fluid-fluid interface. But with an opaque upper phase such as a crude oil, this will be impossible. Either the apparatus must be arranged for viewing from the bottom or one must use the configuration shown in Figure 1-4, where phase 3 is a gas. Deemer *et al.* (1980) have extended to this geometry the analysis described above for a generalized Boussinesq surface fluid. Their results for a linear Boussinesq surface fluid are summarized in Exercise 3.4.1-4.

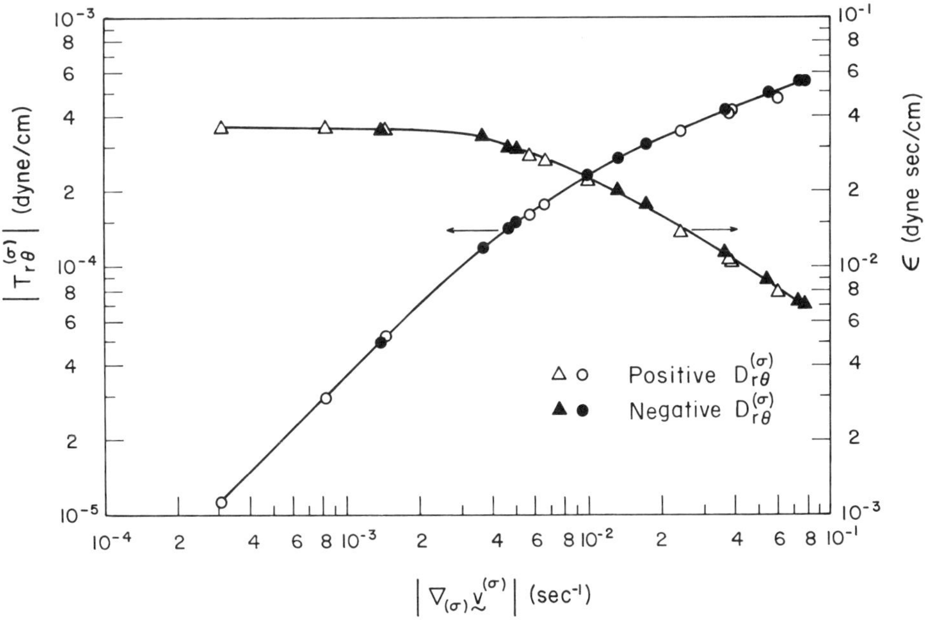

Figure 3.5.1-3: The absolute value of the surface shear stress $|T^{(\sigma)}_{r\theta}|$ and the apparent shear viscosity ε as functions of $|\nabla_{(\sigma)} \mathbf{v}^{(\sigma)}|$ for $\text{div}_{(\sigma)} \mathbf{v}^{(\sigma)} = 0$ at an interface between air and a n-octadecanol monolayer (20.5 Å^2/molecule) over distilled water.

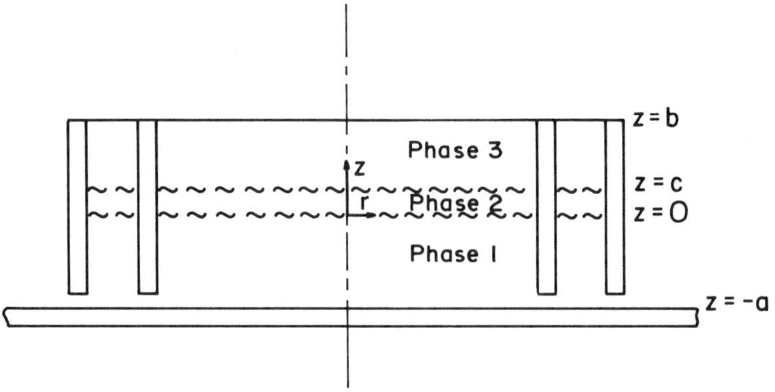

Figure 3.5.1-4: Deep channel surface viscometer for liquid-liquid-gas system.

References

Brown, A. G., W. C. Thuman, and J. W. McBain, *J. Colloid Sci.* **8**, 491 (1953).

Burton, R. A., and R. J. Mannheimer, "Ordered Fluids and Liquid Crystals," Advances in Chemistry Series No. 63, p. 315, American Chemical Society, Washington DC (1967).

Deemer, A. R., J. D. Chen, M. G. Hegde, and J. C. Slattery, *J. Colloid Interface Sci.* **78**, 87 (1980).

Hegde, M. G., and J. C. Slattery, *J. Colloid Interface Sci.* **35**, 593 (1971).

Jahnke, E., and F. Emde, "Tables of Functions," 4th edition, Dover, New York (1945).

Mannheimer, R. J., and R. S. Schechter, *J. Colloid Interface Sci.* **27**, 324 (1968).

Mannheimer, R. J., and R. S. Schechter, *J. Colloid Interface Sci.* **32**, 195 (1970a).

Mannheimer, R. J., and R. S. Schechter, *J. Colloid Interface Sci.* **32**, 212 (1970b).

Osborne, M. F. M., *Kolloid. Z. Z. Polym.* **224**, 150 (1968).

Pintar, A. J., A. B. Israel, and D. T. Wasan, *J. Colloid Interface Sci.* **37**, 52

(1971).

Slattery, J. C., "Momentum, Energy, and Mass Transfer in Continua," McGraw-Hill, New York (1972); second edition, Robert E. Krieger, Malabar, FL 32950 (1981).

Suzuki, A., *Kolloid. Z. Z. Polym.* **250**, 365 (1972).

Wasan, D. T., L. Gupta, and M. K. Vora, *AIChE J.* **17**, 1287 (1971).

Wei, L. Y., and J. C. Slattery, in "Colloid and Interface Science vol. IV: Hydrosols and Rheology," edited by M. Kerker, p. 399, Academic Press, New York (1976).

3.6 Simple surface fluid

3.6.1 Curvilineal surface flows (Ramamohan and Slattery 1984) We have shown in Sec. 2.2.7 that under isothermal conditions the constitutive equation for the simple surface fluid has the form

$$\mathbf{S}^{(\sigma)} \equiv \mathbf{T}^{(\sigma)} - \gamma\,(\,\rho^{(\sigma)}\,)\,\mathbf{P} = \boldsymbol{\mathcal{S}}(\,\mathbf{K}_t^{(\sigma)t}\,;\,\rho^{(\sigma)}\,) \tag{1-1}$$

Where $\mathbf{T}^{(\sigma)}$ is the surface stress tensor, $\gamma(\rho^{(\sigma)})$ is the surface tension that depends only upon the surface mass density $\rho^{(\sigma)}$, $\mathbf{P}$ is the projection tensor that transforms every vector on the dividing surface into its tangential component, and

$$\mathbf{K}_t^{(\sigma)t} \equiv \mathbf{C}_t^{(\sigma)t} - \mathbf{P} \tag{1-2}$$

in which

$$\mathbf{C}_t^{(\sigma)t} \equiv \boldsymbol{\mathcal{F}}_t^{tT} \cdot \boldsymbol{\mathcal{F}}_t^{t} \tag{1-3}$$

is the history of the relative right surface Cauchy-Green tensor and $\boldsymbol{\mathcal{F}}_t^{t}$ is the history of the relative surface deformation gradient. The form of the response functional is restricted by

$$\boldsymbol{\mathcal{S}}(\,\boldsymbol{\mathcal{Q}}(t)\cdot\mathbf{K}_t^{(\sigma)t}\cdot\boldsymbol{\mathcal{Q}}^T(t)\,;\,\rho^{(\sigma)}\,)$$

$$= \boldsymbol{\mathcal{Q}}(t)\cdot\boldsymbol{\mathcal{S}}(\,\mathbf{K}_t^{(\sigma)t}\,;\,\rho^{(\sigma)}\,)\cdot\boldsymbol{\mathcal{Q}}^T(t) \tag{1-4}$$

and

$$\boldsymbol{\mathcal{S}}(\,\mathbf{0}\,;\,\rho^{(\sigma)}\,) = \mathbf{0} \tag{1-5}$$

Here

$$\boldsymbol{\mathcal{Q}} \equiv \mathbf{P}^*\cdot\mathbf{Q}\cdot\mathbf{P} \tag{1-6}$$

with the understanding that $\mathbf{Q}$ is the orthogonal tensor that describes the rotation and (possibly) reflection which transforms vectors in the old frame of reference into vectors in the new frame of reference and $\mathbf{P}^*$ is the projection tensor in the new frame of reference. We denote by t the current time.

We can develop for this form of material behavior a class of solutions, known as the curvilineal surface flows, that can be analyzed without assuming specific forms for the memory functions. In Secs. 3.6.2 and 3.6.3, I show that the deep channel surface viscometer and the oscillating deep channel surface viscometer belong to this class of flows.

Any flow in which the time rate of change of position measured relative to an orthogonal surface coordinate system $(y^1,\ y^2)$ following a surface material particle (see Sec. 1.2.5) takes the form

$$\dot{y}^1 = 0$$

$$\dot{y}^2 = v(y^1 , t) \tag{1-7}$$

will be referred to as a curvilineal surface flow, so long as the components of the projection tensor along the surface pathlines of the flow are constants. If (y^1, y^2) denote the surface coordinates of a surface particle at the current time t and if ($\bar{y}^1$, $\bar{y}^2$) the surface coordinates of a surface particle at some earlier time τ, the surface pathlines of the flow are given by

$$\bar{y}^1 = y^1$$

$$\bar{y}^2 = y^2 + \int_t^{\tau} v(y^1 , \tau') \, d\tau' \tag{1-8}$$

Note that it is possible to have a curvilineal surface flow in which the configuration of the surface changes with respect to time.

For a curvilineal surface flow, the relative surface deformation gradient takes the form

$$\begin{aligned}
\mathcal{F}_t &= \frac{\partial \boldsymbol{\chi}_t^{(\sigma)}}{\partial y^\alpha} \mathbf{a}^\alpha \\
&= \frac{\partial \bar{\mathbf{z}}}{\partial \bar{y}^\beta} \frac{\partial \bar{y}^\beta}{\partial y^\alpha} \mathbf{a}^\alpha \\
&= \frac{\partial \bar{y}^\beta}{\partial y^\alpha} \bar{\mathbf{a}}_\beta \, \mathbf{a}^\alpha \\
&= \sum_{\alpha=1}^{2} \sum_{\beta=1}^{2} \left(\frac{\bar{a}_{\beta\beta}}{a_{\alpha\alpha}} \right)^{1/2} \frac{\partial \bar{y}^\beta}{\partial y^\alpha} \bar{\mathbf{a}}_{<\beta>} \, \mathbf{a}_{<\alpha>}
\end{aligned} \tag{1-9}$$

Here (α = 1, 2; no sum on α)

$$\begin{aligned}
\mathbf{a}_{<\alpha>} &= \frac{1}{a_{\alpha\alpha}{}^{1/2}} \mathbf{a}_\alpha \\
&= \frac{1}{a_{\alpha\alpha}{}^{1/2}} \frac{\partial \mathbf{z}}{\partial y^\alpha}
\end{aligned} \tag{1-10}$$

are the physical basis vectors for our surface coordinate system in the current configuration of the surface and ($\alpha = 1, 2$; no sum on α)

$$a_{\alpha\alpha} \equiv \mathbf{a}_\alpha \cdot \mathbf{a}_\alpha \tag{1-11}$$

are the diagonal components of the projection tensor $\mathbf{P}$ at the current time t. The overbar denotes that $\overline{\mathbf{a}}_{<\beta>}$ refers to the configuration assumed by the surface at some prior time τ. In writing (1-9), we have recognized that by definition the components of the projection tensor are constants along the surface pathlines in a curvilineal surface flow. The matrix of the components of the relative surface deformation gradient $[\boldsymbol{\mathcal{F}}_t]$ with respect to the basis $\overline{\mathbf{a}}_{<\alpha>}\ \mathbf{a}_{<\beta>}$ ($\alpha, \beta = 1, 2$) is

$$[\boldsymbol{\mathcal{F}}_t] = \begin{bmatrix} 1 & 0 \\ \left(\frac{a_{22}}{a_{11}}\right)^{1/2} \int_t^\tau \frac{\partial v}{\partial y^1}\, d\tau' & 1 \end{bmatrix}$$

$$= \begin{bmatrix} 1 & 0 \\ 0 & 1 \end{bmatrix}$$

$$+ \begin{bmatrix} 0 & 0 \\ \left(\frac{a_{22}}{a_{11}}\right)^{1/2} \int_t^\tau \frac{\partial v}{\partial y^1}\, d\tau' & 0 \end{bmatrix} \tag{1-12}$$

Since both $\mathbf{a}_{<\alpha>}$ ($\alpha = 1, 2$) and $\overline{\mathbf{a}}_{<\beta>}$ ($\beta = 1, 2$) are orthonormal bases, they are related by an orthogonal tangential transformation $\mathbf{Q}^{(\sigma)}(\tau)$:

$$\overline{\mathbf{a}}_{<\alpha>} = \mathbf{Q}^{(\sigma)}(\tau) \cdot \mathbf{a}_{<\alpha>} \tag{1-13}$$

An orthogonal tangential transformation is one that preserves lengths and angles in sending every tangential vector field on a surface Σ into another tangential vector field on a surface $\overline{\Sigma}$; it sends every spatial vector field normal to the surface Σ into the zero vector. From (1-12) and (1-13), we have

$$\mathbf{Q}^{(\sigma)T}(\tau) \cdot \boldsymbol{\mathcal{F}}_t = \mathbf{P} + \{ k(\tau) - k(t) \}\, \mathbf{N}^{(\sigma)} \tag{1-14}$$

where

$$k(\tau) \equiv \left(\frac{a_{22}}{a_{11}}\right)^{1/2} \int_0^\tau \frac{\partial v}{\partial y^1}\, d\tau' \tag{1-15}$$

and $\mathbf{N}^{(\sigma)}$ is the tangential tensor whose matrix with respect to the basis

$\mathbf{a}_{<\alpha>} \mathbf{a}_{<\beta>}$ ($\alpha, \beta = 1, 2$) is

$$[\mathbf{N}^{(\sigma)}] = \begin{bmatrix} 0 & 0 \\ 1 & 0 \end{bmatrix} \tag{1-16}$$

Note that

$$\mathbf{N}^{(\sigma)2} \equiv \mathbf{N}^{(\sigma)} \cdot \mathbf{N}^{(\sigma)} = \mathbf{0} \tag{1-17}$$

We define the exponential of a tangential tensor $\mathbf{A}^{(\sigma)}$ by the convergent power series expansion

$$\exp(\mathbf{A}^{(\sigma)}) \equiv \mathbf{P} + \sum_{n=1}^{\infty} \frac{1}{n!} \mathbf{A}^{(\sigma)n} \tag{1-18}$$

Recognizing (1-17) and (1-18), we may express (1-14) as

$$\mathbf{Q}^{(\sigma)T} \cdot \mathcal{F}_t^t = \exp(g(s)\, \mathbf{N}^{(\sigma)}) \tag{1-19}$$

in which

$$g(s) \equiv k(t - s) - k(t)$$

$$= k(\tau) - k(t) \tag{1-20}$$

In view of (1-3) and (1-19), equation (1-2) takes the form

$$\mathbf{K}_t^{(\sigma)t} = \exp(g(s)\, \mathbf{N}^{(\sigma)T}) \cdot \exp(g(s)\, \mathbf{N}^{(\sigma)}) - \mathbf{P} \tag{1-21}$$

allowing us to write (1-1) as

$$\mathbf{S}^{(\sigma)} = \mathcal{N}(g(s) \,;\, \mathbf{N}^{(\sigma)}) \tag{1-22}$$

Here we have introduced

$$\mathcal{N}(g(s) \,;\, \mathbf{N}^{(\sigma)})$$

$$= \mathcal{S}\Big(\exp(g(s)\, \mathbf{N}^{(\sigma)T}) \cdot \exp(g(s)\, \mathbf{N}^{(\sigma)}) - \mathbf{P} \,;\, \rho^{(\sigma)} \Big) \tag{1-23}$$

The form of this response functional is restricted by (1-4)

$$\mathcal{Q}(t) \cdot \mathcal{N}(g(s) \,;\, \mathbf{N}^{(\sigma)})$$

$$= \mathcal{N}\left(\alpha\, g(s) \,;\, \frac{1}{\alpha}\, \boldsymbol{\mathcal{Q}}(t) \cdot \mathcal{N}^{(\sigma)} \cdot \boldsymbol{\mathcal{Q}}^{T}(t) \right) \tag{1-24}$$

for all real α.

Since $\mathbf{N}^{(\sigma)}$ has a specific form, it follows from (1-18) that the components of $\mathbf{S}^{(\sigma)}$ are functionals of g(s) only:

$$S^{(\sigma)}_{<12>} = \underset{s=0}{\overset{\infty}{t_0}}\left(g(s) \right) \tag{1-25}$$

$$S^{(\sigma)}_{<11>} = \underset{s=0}{\overset{\infty}{t_1}}\left(g(s) \right) \tag{1-26}$$

$$S^{(\sigma)}_{<22>} = \underset{s=0}{\overset{\infty}{t_2}}\left(g(s) \right) \tag{1-27}$$

The three scalar functionals

$$\underset{s=0}{\overset{\infty}{t_0}} \qquad \underset{s=0}{\overset{\infty}{t_1}} \qquad \underset{s=0}{\overset{\infty}{t_2}}$$

are independent of the basis and depend only on the interface under consideration. For this reason, we will refer to them as surface material functionals. They completely characterize the mechanical behavior of the interface in curvilineal surface flows.

In (1-25), let

$$\alpha = -1 \tag{1-28}$$

and

$$[\,\boldsymbol{\mathcal{Q}}\,] = \begin{bmatrix} 1 & 0 \\ 0 & -1 \end{bmatrix} \tag{1-29}$$

with respect to the appropriate basis. We conclude that

$$\boldsymbol{\mathcal{Q}} \cdot \mathbf{S}^{(\sigma)} \cdot \boldsymbol{\mathcal{Q}}^{T} = \mathcal{N}\left(-\,g(s) \,;\, \mathbf{N}^{(\sigma)} \right) \tag{1-30}$$

and

$$[\,\boldsymbol{\mathcal{Q}} \cdot \mathbf{S}^{(\sigma)} \cdot \boldsymbol{\mathcal{Q}}^{T}\,] = \begin{bmatrix} S^{(\sigma)}_{<11>} & -S^{(\sigma)}_{<12>} \\ -S^{(\sigma)}_{<12>} & S^{(\sigma)}_{<22>} \end{bmatrix} \tag{1-31}$$

This implies that the three surface material functionals must satisfy

$$\overset{\infty}{\underset{s=0}{t_0}}(-g(s)) = -\overset{\infty}{\underset{s=0}{t_0}}(g(s)) \tag{1-32}$$

$$\overset{\infty}{\underset{s=0}{t_1}}(-g(s)) = \overset{\infty}{\underset{s=0}{t_1}}(g(s)) \tag{1-33}$$

and

$$\overset{\infty}{\underset{s=0}{t_2}}(-g(s)) = \overset{\infty}{\underset{s=0}{t_2}}(g(s)) \tag{1-34}$$

In view of (1-5), they are further restricted by

$$\overset{\infty}{\underset{s=0}{t_0}}(0) = \overset{\infty}{\underset{s=0}{t_1}}(0) = \overset{\infty}{\underset{s=0}{t_2}}(0) = 0 \tag{1-35}$$

The general form of this discussion was suggested by the analogous treatment of curvilineal flows of incompressible simple fluids outside the immediate neighborhood of a phase interface given by Truesdell and Noll (1965).

References

Ramamohan, T. R., and J. C. Slattery, *Chem. Eng. Commun.* **26**, 219 (1984).

Truesdell, C., and W. Noll, in "Handbuch der Physik," vol. 3/3, edited by S. Flügge, p. 432, Springer-Verlag, Berlin (1965).

3.6.2 More about deep channel surface viscometer (Ramamohan and Slattery 1984) Assuming a generalized Boussinesq surface fluid, I showed in Sec. 3.5.1 how the apparent surface shear viscosity can be determined from measurements of the velocity distribution in the interface of the deep channel surface viscometer shown in Figure 3.5.1-1 without assuming a specific form for the interfacial stress-deformation behavior. In what follows, we show how the approach described there can be generalized to measure the surface material function

$$\overset{\infty}{t}_{\substack{0\\ s=0}} .$$

The assumptions made in this analysis are identical with those introduced in Sec. 3.5.1, with the exception of

v') The stress-deformation behavior of the gas-liquid interface may be represented by the simple surface fluid model.

The initial steps in this analysis also follow those made in Sec. 3.5.1 through equation (3.5.1-21).

From (1-15), (1-20), and (3.5.1-9),

$$g(s) = -\frac{\Omega}{1 - k} s\, r^* \frac{\partial}{\partial r^*}\left(\frac{h^*}{r^*}\right) \tag{2-1}$$

Let us define c^* to be that value of r^* such that

$$\text{at } r^* = c^* : \frac{d}{dr^*}\left(\frac{h^*}{r^*}\right) = 0 \tag{2-2}$$

It follows from (1-35), (2-1), and (2-2) that

$$\text{at } r^* = c^* : S^{(\sigma)*}_{r\theta} = S^{(\sigma)*}_{rr} = S^{(\sigma)*}_{\theta\theta} = 0 \tag{2-3}$$

in which we have introduced the dimensionless surface stress

$$S^{(\sigma)*} \equiv \frac{S^{(\sigma)}}{\mu^{(1)} R\Omega} \tag{2-4}$$

The θ-component of the jump momentum balance requires (Table 2.4.2-2)

$$\text{at } z^* = 0 : \frac{1}{r^{*2}} \frac{d}{dr^*}\left(r^{*2}\, S^{(\sigma)*}_{r\theta} \right) = \frac{\partial v_\theta{}^{(1)*}}{\partial z^*} - \frac{\mu^{(2)}}{\mu^{(1)}} \frac{\partial v_\theta{}^{(2)*}}{\partial z^*} \tag{2-5}$$

In view of (3.5.1-14) and (3.5.1-15), we can integrate (2-5) consistent with (2-3) to find

$$S_{r\theta}^{(g)*} = \sum_{n=1}^{\infty} \left[A_n + B_n \frac{\mu^{(2)}}{\mu^{(1)}} \coth(\lambda_n b^*) \right]$$

$$\times \left[\phi_2(\lambda_n r^*) - \left(\frac{c^*}{r^*} \right)^2 \phi_2(\lambda_n c^*) \right] \tag{2-6}$$

Employing (1-25), (1-32), and (2-1), we can write

$$S_{r\theta}^{(g)} = \overset{\infty}{\underset{s=0}{t_0}} (g(s))$$

$$= \overset{\infty}{\underset{s=0}{t_0}} \left(- \frac{\Omega}{1 - k} s \, r^* \frac{d}{dr^*} \left[\frac{h^*}{r^*} \right] \right)$$

$$= \frac{\Omega}{1 - k} r^* \frac{d}{dr^*} \left(\frac{h^*}{r^*} \right) \varepsilon_{app} \tag{2-7}$$

Here we have found it convenient to define

$$\varepsilon_{app} \equiv \varepsilon_{app}\left(| \nabla_{(\sigma)} \mathbf{v}^{(\sigma)} | \right)$$

$$\equiv \left[\frac{\Omega}{1 - k} r^* \frac{d}{dr^*} \left(\frac{h^*}{r^*} \right) \right]^{-1} \overset{\infty}{\underset{s=0}{t_0}} \left(- \frac{\Omega}{1 - k} s \, r^* \frac{d}{dr^*} \left[\frac{h^*}{r^*} \right] \right) \tag{2-8}$$

and to observe that

$$| \nabla_{(\sigma)} \mathbf{v}^{(\sigma)} | = \frac{\Omega r^*}{1 - k} \left| \frac{d}{dr^*} \left(\frac{h^*}{r^*} \right) \right| \tag{2-9}$$

Note that

$$\varepsilon_{app} \equiv \frac{S_{r\theta}^{(g)}}{r \frac{d}{dr} \left(\frac{h}{r} \right)}$$

$$= \frac{\mu^{(1)} R (1 - k) S_{r\theta}^{(g)*}}{r^* \frac{d}{dr^*} \left(\frac{h^*}{r^*} \right)} \tag{2-10}$$

has the physical significance of an apparent surface shear viscosity.

The r-component of the jump momentum balance requires (Table 2.4.2-2)

$$\text{at } z^{*} = 0 : \frac{d}{dr^{*}}\left(r^{*} S_{rr}^{(\sigma)*} \right) = S_{\theta\theta}^{(\sigma)*} \tag{2-11}$$

Since $S_{rr}^{(\sigma)*}$ and $S_{\theta\theta}^{(\sigma)*}$ cannot be directly related to the velocity distribution, we are unable to say anything about the surface material functionals

$$\overset{\infty}{\underset{s=0}{t_1}} \quad \text{and} \quad \overset{\infty}{\underset{s=0}{t_2}}$$

from such an experiment. The z-component of the jump momentum balance simply demands that pressure be continuous across the interface.

An experiment gives us the surface velocity distribution in the form of discrete measurements rather than a continuous function h(r). From (3.5.1-5), (3.5.1-14), and (3.5.1-15),

$$v_{\theta}^{(\sigma)*} = \sum_{n=1}^{\infty} B_n \, \phi_1 (\lambda_n r^{*}) \tag{2-12}$$

which suggests that we assume

$$h^{*}(r^{*}) = \sum_{n=1}^{N} \tilde{B}_n \, \phi_1 (\lambda_n r^{*}) \tag{2-13}$$

The coefficients $\tilde{B}_n$ can be determined by a least-square-error fit of (2-12) to the experimental data. For the optimum value of N, we can immediately identify $B_n = \tilde{B}_n$. This permits us to calculate ε_{app} as a function of $|\nabla_{(\sigma)} v^{(\sigma)}|$ using (3.5.1-16), (3.5.1-17), (2-9), and (2-10). Calculating ε_{app} is equivalent to determining the surface material function

$$\overset{\infty}{\underset{s=0}{t_0}}$$

The practicality of this analysis is demonstrated experimentally in Sec. 3.5.1.

Reference

Ramamohan, T. R., and J. C. Slattery, *Chem. Eng. Commun.* **26**, 219 (1984).

3.6.3 Oscillating deep channel surface viscometer (Ramamohan and Slattery 1984) In order to investigate some effects of memory, Mohan et al. (1976), Mohan and Wasan (1976), and Addison and Schechter (1979) suggested minor modifications of the deep channel surface viscometer described in Sec. 3.5.1. Mohan *et al.* (1976) and Mohan and Wasan (1976) observed the effects of surface viscoelasticity with a deep channel surface viscometer, the floor of which periodically started and stopped. Following a proposal by Mohan and Wasan (1976), Addison and Schechter (1979) demonstrated the effects of surface viscoelasticity with a deep channel surface viscometer, the floor of which was subjected to a forced, small-amplitude, sinusoidal oscillation.

Our objective in what follows will be to show how the oscillating deep channel surface viscometer can be used to measure one of the memory functions in the finite linear viscoelastic surface fluid (see Sec. 2.2.8).

We make the following assumptions.

i) There is no mass transfer across the liquid-gas interface.

ii) The viscometer operates in a periodic manner such that all variables are periodic functions of time.

iii) The liquid phase may be described as an incompressible, Newtonian fluid; viscous effects in the gas phase are neglected.

iv) The stress-deformation behavior of the liquid-gas interface may be represented by the simple surface fluid model.

v) The amplitude of oscillation is sufficiently small that all convective inertial terms in the Navier-Stokes equation can be neglected and that the behavior of the simple surface fluid reduces to that of a finite linear viscoelastic surface fluid (see Sec. 2.2.8).

vi) All physical properties are constants.

vii) The interface is a horizontal plane.

Referring to Figure 3-1, we have the following boundary conditions imposed upon the liquid phase by the geometry. The two walls of the annular channel are stationary

$$\text{at } r = kR : \quad \mathbf{v} = 0 \tag{3-1}$$

$$\text{at } r = R : \quad \mathbf{v} = 0 \tag{3-2}$$

and the floor of the channel oscillates sinusoidally

$$\text{at } z = 0 : \quad v_\theta = r\,\Omega_a \cos \omega t$$

$$v_r = v_z = 0 \tag{3-3}$$

The tangential components of velocity are continuous at the gas-liquid phase interface

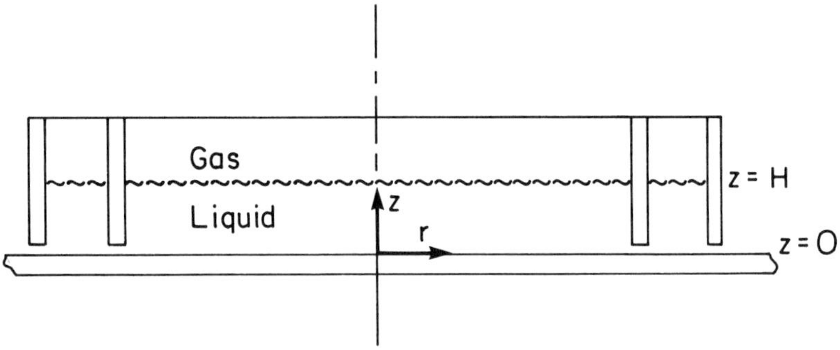

Figure 3.6.3-1: Oscillating deep channel surface viscometer. The walls of the channel are stationary; the angular velocity of the floor oscillates according to (3-3).

$$\text{at } z = H : \quad v_r = v_r^{(\sigma)}$$

$$v_\theta = v_\theta^{(\sigma)} \tag{3-4}$$

Since there is no mass transfer to or from the phase interface,

$$\text{at } z = H : \quad v_z = 0 \tag{3-5}$$

These boundary conditions suggest that we seek a velocity distribution within the liquid phase of the form

$$v_\theta = v_\theta(\, r, z, t\,) \qquad v_r = v_z = 0 \tag{3-6}$$

This form of velocity distribution satisfies the equation of continuity (Slattery 1981, p. 60) identically. The θ-component of the Navier-Stokes equation for an incompressible Newtonian fluid implies (Slattery 1981, p. 62)

$$N \frac{\partial v_\theta^*}{\partial t^*} = \frac{\partial}{\partial r^*}\left[\frac{1}{r^*} \frac{\partial}{\partial r^*} (r^* v_\theta^*) \right] + \frac{\partial^2 v_\theta^*}{\partial z^{*2}} \tag{3-7}$$

where we have introduced as dimensionless variables

$$v_\theta^* \equiv \frac{v_\theta}{v_o} \qquad t^* \equiv \omega t$$

$$r^* \equiv \frac{r}{R} \qquad z^* \equiv \frac{z}{R} \tag{3-8}$$

and

$$N \equiv \frac{\rho R^2 \omega}{\mu} \tag{3-9}$$

We will leave the characteristic speed v_o undefined, since its choice will not directly influence our results. The r- and z- components of the Navier-Stokes equation are used to determine the pressure distribution within each phase, with which we will not be concerned here.

The jump mass balance (Sec. 1.3.5; the overall jump mass balance of Sec. 5.4.1) is satisfied identically, since we define the position of the dividing surface by choosing $\rho^{(\sigma)} = 0$ (Sec. 1.3.6) and since the effect of mass transfer is neglected (assumption i).

In terms of these variables, the r- and θ- components of the jump momentum balance require (see Table 2.4.2-2)

$$\text{at } z^* = H^* : \frac{\partial}{\partial r^*} (r^* S_{rr}^{(\sigma)*}) = S_{\theta\theta}^{(\sigma)*} \tag{3-10}$$

$$\text{at } z^* = H^* : \frac{1}{r^{*2}} \frac{\partial}{\partial r^*} (r^{*2} S_{r\theta}^{(\sigma)*}) = \frac{\partial v_\theta^*}{\partial z^*} \tag{3-11}$$

in which

$$S^{(\sigma)*} \equiv \frac{S^{(\sigma)}}{\mu\, v_o} \tag{3-12}$$

From (1-16) through (1-18) and (1-21), the matrix of $\mathbf{K}_t^{(\sigma)t}$ written with respect to the physical basis in cylindrical coordinates is

$$[\ \mathbf{K}_t^{(\sigma)t}\ ] = g(s)\begin{bmatrix} 0 & 1 \\ 1 & 0 \end{bmatrix} + \{\ g(s)\ \}^2 \begin{bmatrix} 1 & 0 \\ 0 & 0 \end{bmatrix} \tag{3-13}$$

Equations (1-15), (1-20), and (3-6) tell us that

$$g(\ s\ ) = -\ N_\omega r^* \int_0^{s^*} \frac{\partial}{\partial r^*}\left(\frac{v_\theta^*}{r^*}\right) ds^* \tag{3-14}$$

where we have introduced

$$N_\omega \equiv \frac{v_o}{R\omega} \tag{3-15}$$

By assumption v, the amplitude of oscillation is sufficiently small that the behavior of the simple surface fluid reduces to that of a finite linear viscoelastic surface fluid (see Sec. 2.2.8)

$$\mathbf{S}^{(\sigma)} = -\int_0^\infty \frac{1}{2}[\ \kappa(\ \rho^{(\sigma)},\ s\ ) - \varepsilon(\ \rho^{(\sigma)},\ s\ )\ ]\ \text{tr}\ \mathbf{K}_t^{(\sigma)t}\ ds\ \mathbf{P}$$

$$-\int_0^\infty \varepsilon(\ \rho^{(\sigma)},\ s\ )\ \mathbf{K}_t^{(\sigma)t}\ ds \tag{3-16}$$

Here $\kappa(\ \rho^{(\sigma)},\ s\ )$ and $\varepsilon(\ \rho^{(\sigma)},\ s\ )$ are scalar-valued functions of $\rho^{(\sigma)}$ and s. It follows from (3-13), (3-14), and (3-16) that in particular

$$S_{r\theta}^{(\sigma)} = -\int_0^\infty \varepsilon(\ \rho^{(\sigma)},\ s\ )\ g(s)\ ds$$

$$= N_\omega\ r^* \int_0^\infty \varepsilon(\ \rho^{(\sigma)},\ s\ ) \left[\int_0^{s^*} \frac{\partial}{\partial r^*}\left(\frac{v_\theta^*}{r^*}\right) ds^*\ \right] ds \tag{3-17}$$

We seek a first-order perturbation solution to this problem, taking as our perturbation parameter

$$N_\Omega \equiv \frac{R\Omega_a}{v_o} \tag{3-18}$$

This has the effect of eliminating the remaining convective inertial term in the r-component of the Navier-Stokes equation, allowing the resulting pressure distribution to be consistent with the existence of a flat phase interface. It also eliminates the second term in (3-13), permitting us to conclude from (3-16) that

$$S_{rr}^{(\sigma)} = S_{\theta\theta}^{(\sigma)} = 0 \tag{3-19}$$

and that (3-10) is satisfied identically.

We will assume that at the first perturbation the solution takes the form

$$v_\theta^* = N_\Omega \, w(r^*, z^*) \exp(it^*) \tag{3-20}$$

From (3-1) through (3-3), (3-7), (3-11), and (3-20), we see that

$$iNw = \frac{\partial}{\partial r^*}\left[\frac{1}{r^*} \frac{\partial}{\partial r^*}(r^* w) \right] + \frac{\partial^2 w}{\partial z^{*2}} \tag{3-21}$$

must be solved consistent with the conditions

$$\text{at } z^* = 0 : \quad w = r^* \tag{3-22}$$

$$\text{at } r^* = k : \quad w = 0 \tag{3-23}$$

$$\text{at } r^* = 1 : \quad w = 0 \tag{3-24}$$

$$\text{at } z^* = H^* : \quad - \frac{\alpha}{r^{*2}} \frac{\partial}{\partial r^*} \left[r^{*3} \frac{\partial}{\partial r^*} \left(\frac{w}{r^*} \right) \right] = \frac{\partial w}{\partial z^*} \tag{3-25}$$

in which we have introduced for convenience

$$\alpha \equiv \frac{1}{\mu R \omega} \int_0^\infty \varepsilon(\rho^{(\sigma)}, s) \, [\, 1 - \exp(- i \, s^*)] \, ds \tag{3-26}$$

This problem can be solved by separation of variables. Its solution is

$$w = \sum_{j=1}^{\infty} \{ A_j \sinh[(iN + \lambda_j^2)^{1/2} z^*]$$

$$+ B_j \cosh[(iN + \lambda_j^2)^{1/2} z^*] \} \, c_1(\lambda_j r^*) \tag{3-27}$$

where

$$c_m(\lambda_j r^*) \equiv J_m(\lambda_j r^*) - \frac{J_m(\lambda_j k)}{Y_m(\lambda_j k)} Y_m(\lambda_j r^*) \tag{3-28}$$

and λ_j (j = 1, 2, ...) are solutions of

$$J_1(\lambda_j) - \frac{J_1(\lambda_j k)}{Y_1(\lambda_j k)} Y_1(\lambda_j) = 0 \tag{3-29}$$

The coefficients in (3-27) are given by

$$\begin{aligned} A_j = B_j \{ & -\alpha \lambda_j^2 \cosh[(iN + \lambda_j^2)^{1/2} H^*] \\ & + (iN + \lambda_j^2)^{1/2} \sinh[(iN + \lambda_j^2)^{1/2} H^*]\} \\ \times \{ & \alpha\lambda_j^2 \sinh[(iN + \lambda_j^2)^{1/2} H^*] \\ & - (iN + \lambda_j^2)^{1/2} \cosh[(iN + \lambda_j^2)^{1/2} H^*]\}^{-1} \end{aligned} \tag{3-30}$$

$$B_j = -\frac{\beta_j}{\lambda_j}[c_0(\lambda_j) - k^2 c_0(\lambda_j k)] \tag{3-31}$$

with

$$\beta_j \equiv 2\{[c_0(\lambda_j)]^2 - k^2 [c_0(\lambda_j k)]^2\}^{-1} \tag{3-32}$$

For comparison with experimental data, we require the real portion of (3-20). This suggests that we write

$$(iN + \lambda_j^2)^{1/2} = a_{1j} + i\, b_{1j} \tag{3-33}$$

with

$$a_{1j} \equiv (N^2 + \lambda_j^4)^{1/4} \cos \theta_{1j} \tag{3-34}$$

$$b_{1j} \equiv (N^2 + \lambda_j^4)^{1/4} \cos \theta_{1j} \tag{3-35}$$

$$\theta_{1j} \equiv \frac{1}{2} \tan^{-1}\left(\frac{N}{\lambda_j^2}\right) \tag{3-36}$$

that

$$\cosh[(iN + \lambda_j^2)^{1/2} z^*] = a_{2j}(z^*) + i\, b_{2j}(z^*) \tag{3-37}$$

with

$$a_{2j}(z^\star) \equiv \cosh(a_{1j}z^\star) \cos(b_{1j}z^\star) \tag{3-38}$$

$$b_{2j}(z^\star) \equiv \sinh(a_{1j}z^\star) \sin(b_{1j}z^\star) \tag{3-39}$$

that

$$\sinh[(iN + \lambda_j^2)^{1/2} z^\star] = a_{3j}(z^\star) + i\, b_{3j}(z^\star) \tag{3-40}$$

with

$$a_{3j}(z^\star) \equiv \sinh(a_{1j}z^\star) \cos(b_{1j}z^\star) \tag{3-41}$$

$$b_{3j}(z^\star) \equiv \cosh(a_{1j}z^\star) \sin(b_{1j}z^\star) \tag{3-42}$$

that

$$\alpha = a_4(\omega) + i\, b_4(\omega) \tag{3-43}$$

with

$$a_4(\omega) \equiv \frac{1}{\mu R\omega} \int_0^\infty \varepsilon(\rho^{(\sigma)}, s) \,[\, 1 - \cos(s^\star)]\, ds \tag{3-44}$$

$$b_4(\omega) \equiv \frac{1}{\mu R\omega} \int_0^\infty \varepsilon(\rho^{(\sigma)}, s) \sin(s^\star)\, ds \tag{3-45}$$

that

$$\begin{aligned} A_j &= a_{7j} + b_{7j} \\ &= \frac{a_{5j} + i\, b_{5j}}{a_{6j} + i\, b_{6j}} \end{aligned} \tag{3-46}$$

with

$$\begin{aligned} a_{5j} \equiv &- B_j\lambda_j^2 [\, a_4\ a_{2j}(H^\star) - b_4 b_{2j}(H^\star) \,] \\ &+ B_j \,[\, a_{1j}\ a_{3j}(H^\star) - b_{1j}\ b_{3j}(H^\star) \,] \end{aligned} \tag{3-47}$$

$$b_{5j} \equiv - B_j\lambda_j^2 [\, a_4\ b_{2j}(H^\star) + b_4 a_{2j}(H^\star) \,]$$

$$+ B_j [a_{1j} b_{3j}(H^*) + b_{2j} a_{3j}(H^*)] \tag{3-48}$$

$$a_{6j} \equiv \lambda_j^2 [a_4 a_{3j}(H^*) - b_4 b_{3j}(H^*)] - a_{1j} a_{2j}(H^*) + b_{1j} b_{2j}(H^*) \tag{3-49}$$

$$b_{6j} \equiv \lambda_j^2 [a_4 b_{3j}(H^*) + b_4 a_{3j}(H^*)] - a_{1j} b_{2j}(H^*) - b_{1j} a_{2j}(H^*) \tag{3-50}$$

$$a_{7j} \equiv \frac{a_{5j} a_{6j} + b_{5j} b_{6j}}{a_{6j}^2 + b_{6j}^2} \tag{3-51}$$

$$b_{7j} \equiv \frac{a_{6j} b_{5j} - a_{5j} b_{6j}}{a_{6j}^2 + b_{6j}^2} \tag{3-52}$$

and that

$$w = \sum_{j=1}^{\infty} [a_{8j}(z^*) + i\, b_{8j}(z^*)]\, c_1(\lambda_j r^*) \tag{3-53}$$

with

$$a_{8j}(z^*) \equiv a_{7j} a_{3j}(z^*) - b_{7j} b_{3j}(z^*) + B_j a_{2j}(z^*) \tag{3-54}$$

$$b_{8j}(z^*) \equiv b_{7j} a_{3j}(z^*) + a_{7j} b_{3j}(z^*) + B_j b_{2j}(z^*) \tag{3-55}$$

The conclusion is that the real portion of (3-20) takes the form

$$\mathrm{Re}(v_\theta^*) = N_\Omega \sum_{j=1}^{\infty} [a_{8j}(z^*) \cos t^* - b_{8j}(z^*) \sin t^*]\, c_1(\lambda_j r^*) \tag{3-56}$$

The velocity of a small particle floating on the center-line of the interface may be expressed as

$$\mathrm{Re}(v_\theta^*) = V \cos(t^* + \phi) \tag{3-57}$$

in which

$$V \equiv N_\Omega \left\{ \left[\sum_{j=1}^{\infty} a_{8j}(H^*)\, c_1 \left(\lambda_j \frac{k+1}{2} \right) \right]^2 + \left[\sum_{j=1}^{\infty} b_{8j}(H^*)\, c_1 \left(\lambda_j \frac{k+1}{2} \right) \right]^2 \right\}^{1/2} \tag{3-58}$$

is the amplitude of the oscillation of its velocity and

$$\phi \equiv \tan^{-1} \left\{ \left[\sum_{j=1}^{\infty} b_{8j}(H^*)\, c_1 \left(\lambda_j \frac{k+1}{2} \right) \right] \times \left[\sum_{j=1}^{\infty} a_{8j}(H^*)\, c_1 \left(\lambda_j \frac{k+1}{2} \right) \right]^{-1} \right\} \tag{3-59}$$

is the phase lag.

Let us visualize an experiment in which we measure this amplitude V and phase lag ϕ as functions of the frequency ω. Given V and ϕ for a particular frequency, we can solve for $a_4(\omega)$ and $b_4(\omega)$ using (3-34) through (3-36), (3-38), (3-39), (3-41), (3-42), (3-44), (3-45), (3-47) through (3-52), (3-54), (3-55), (3-58), and (3-59). If we could do this for a sufficiently wide range of frequencies, we could return to (3-45) and take an inverse Fourier sine transform to determine the memory function $\varepsilon(\rho^{(\sigma)}, s)$.

In carrying out these experiments as a function of frequency ω, it is helpful to note that

$$\text{limit } \omega \to 0: \quad \mu R \omega\, a_4(\omega) \to 0 \tag{3-60}$$

$$\text{limit } \omega \to 0: \quad \mu R \omega\, b_4(\omega) \to 0 \tag{3-61}$$

$$\text{limit } \omega \to \infty: \quad \mu R \omega\, a_4(\omega) \to \text{a constant} \tag{3-62}$$

$$\text{limit } \omega \to \infty: \quad \mu R \omega\, b_4(\omega) \to 0 \tag{3-63}$$

To arrive at (3-62) and (3-63), we have used the Riemann-Lebesgue lemma (Bender and Orszag 1978).

Addison and Schechter (1979) measured V as a function of frequency, but they did not measure the phase lag ϕ. In order to analyze their data, they followed Gardner *et al.* (1978). The analogous treatment here would be to assume a form for the memory function $\varepsilon(\rho^{(\sigma)}, s)$ involving several parameters. These parameters could be determined by fitting (3-58) to the experimental data.

References

Addison, J. V., and R. S. Schechter, *AIChE J.* **25**, 32 (1979).

Bender, C. M., and S. A. Orszag, "Advanced Mathematical Methods for Scientists and Engineers," McGraw-Hill, New York (1978).

Gardner, J. W., J. V. Addison, and R. S. Schechter, *AIChE J.* **24**, 400 (1978).

Mohan, V., B. K. Malviya, and D. T. Wasan, *Can. J. Chem. Eng.* **54**, 515 (1976).

Mohan, V., and D. T. Wasan, in "Colloid and Interface Science," vol. IV, edited by M. Kerker, p. 439, Academic Press, New York (1976).

Ramamohan, T. R., and J. C. Slattery, *Chem. Eng. Commun.* **26**, 219 (1984).

Slattery, J. C., "Momentum, Energy, and Mass Transfer in Continua," McGraw-Hill, New York (1972); second edition, Robert E. Krieger, Malabar, FL 32950 (1981).

3.7 Limiting case

3.7.1 When effects of interfacial viscosities dominate In the preceding sections, we see examples of problems that can be solved exactly. Most problems are not of this nature. After we have made what seems to be reasonable initial assumptions, the boundary-value problem with which we are faced may be very difficult. Before beginning an extensive numerical solution, you are advised to consider what may be learned from limiting cases, a number of which have been described elsewhere (Slattery 1981).

Let us restrict our attention to isothermal, two-phase flows of incompressible Newtonian fluids in which no interfacial concentration gradients are developed and all interfacial properties are constants. We will continue to define the location of the dividing surface by requiring $\rho^{(\sigma)} = 0$ (Sec. 1.3.6).

We will find it helpful to introduce as dimensionless variables

$$y^{\alpha *} \equiv \frac{y^{\alpha}}{\ell_o} \qquad \alpha = 1, 2$$

$$z_i^* \equiv \frac{z_i}{\ell_o} \qquad i = 1, 2, 3$$

$$\mathcal{P}^{(j)*} \equiv \frac{\ell_o \mathcal{P}^{(j)}}{\gamma} \qquad \phi^* \equiv \frac{\phi}{g \ell_o}$$

$$\mathbf{v}^{(j)*} \equiv \frac{\mathbf{v}^{(j)}}{v_o} \qquad \mathbf{S}^{(j)*} \equiv \frac{\ell_o \mathbf{S}^{(j)}}{\mu^{(j)} v_o}$$

$$\mathbf{v}^{(\sigma)*} \equiv \frac{\mathbf{v}^{(\sigma)}}{v_o} \qquad \mathbf{S}^{(\sigma)*} \equiv \frac{\ell_o \mathbf{S}^{(\sigma)}}{(\kappa_o + \varepsilon_o) v_o}$$

$$\dot{\mathbf{y}}^* \equiv \frac{\dot{\mathbf{y}}}{v_o} \qquad H^* \equiv \ell_o H$$

$$\mathbf{b}^{(\sigma)*} \equiv \frac{\mathbf{b}^{(\sigma)}}{g} \tag{1-1}$$

Here ℓ_o is a characteristic length, $\mathcal{P}^{(j)}$ the modified pressure for phase j

$$\mathcal{P}^{(j)} \equiv p^{(j)} + \rho^{(j)} \phi \tag{1-2}$$

ϕ the potential energy per unit mass in terms of which we assume that the external force per unit mass **b** may be expressed as

$$\mathbf{b} \equiv - \nabla \phi \tag{1-3}$$

g the magnitude of the acceleration of gravity, v_o a characteristic speed, κ_o a characteristic interfacial dilatational viscosity, and ε_o a characteristic interfacial shear viscosity.

In terms of these dimensionless variables, we can express the equation of continuity for phase j (Sec. 2.4.1) as

$$\text{div } \mathbf{v}^* = 0 \tag{1-4}$$

Cauchy's first law for the incompressible Newtonian phase j (Sec. 2.4.1) as

$$N_{Re}^{(j)} N_{ca} \frac{\mu^{(j)}}{\mu^{(2)}} (\nabla \mathbf{v}^{(j)*} \cdot \mathbf{v}^{(j)*})$$

$$= - \nabla \mathcal{P}^{(j)*} + N_{ca} \frac{\mu^{(j)}}{\mu^{(2)}} \text{div}(\nabla \mathbf{v}^{(j)*}) \tag{1-5}$$

the jump mass balance (Sec. 2.4.2) as

$$\left[\frac{\rho}{\rho^{(\sigma)}} (\mathbf{v}^* \cdot \boldsymbol{\xi} - v_{(\xi)}^{\{\mathbf{g}\}*}) \right] = 0 \tag{1-6}$$

and the jump momentum balance (Sec. 2.4.2) as

$$\frac{2}{N_{\kappa+\varepsilon} N_{ca}} H^* \boldsymbol{\xi} + \text{div}_{(\sigma)} \mathbf{S}^{(\sigma)*}$$

$$+ \left[- \frac{N_{Re}}{N_{\kappa+\varepsilon}} \frac{\mu}{\mu^{(2)}} (\mathbf{v}^* \cdot \boldsymbol{\xi} - v_{(\xi)}^{\{\mathbf{g}\}*})^2 \boldsymbol{\xi} \right.$$

$$\left. - \frac{1}{N_{\kappa+\varepsilon} N_{ca}} \mathcal{P}^* \boldsymbol{\xi} + \frac{1}{N_{\kappa+\varepsilon}} \frac{\mu}{\mu^{(2)}} \mathbf{S}^* \cdot \boldsymbol{\xi} \right]$$

$$+ \frac{N_{Bo}}{N_{\kappa+\varepsilon} N_{ca}} \phi^* \boldsymbol{\xi}^{(1)} = 0 \tag{1-7}$$

Here $\boldsymbol{\xi}^{(1)}$ is the unit normal pointing into phase 1. By N_{ca}, $N_{\kappa+\varepsilon}$, $N_{Re}^{(j)}$, and $N_{Bo}^{(j)}$ I mean the capillary number, the dimensionless sum of the characteristic interfacial viscosities, the Reynolds number for phase j, and the Bond number for phase j:

$$N_{ca} \equiv \frac{\mu^{(2)} v_o}{\gamma} \qquad N_{\kappa+\varepsilon} \equiv \frac{\kappa_o + \varepsilon_o}{\mu^{(2)} \ell_o}$$

$$N_{Re}^{(j)} \equiv \frac{\rho^{(j)} v_o \ell_o}{\mu^{(j)}} \qquad N_{Bo} \equiv \frac{(\rho^{(1)} - \rho^{(2)}) g \ell_o^2}{\gamma} \tag{1-8}$$

Remember that the definition of a characteristic quantity is arbitrary, although a particular choice may have the advantage of simplifying the form of a boundary-value problem. Our understanding here is that phase 1 is the more viscous, more dense phase; phase 2 the less viscous, less dense phase.

If we restrict our attention to problems in which

$$N_{Re}^{(j)} N_{ca} \frac{\mu^{(j)}}{\mu^{(2)}} << 1$$

$$N_{ca} \frac{\mu^{(j)}}{\mu^{(2)}} << 1$$

$$\frac{1}{N_{\kappa+\varepsilon}} \frac{\mu^{(j)}}{\mu^{(2)}} << 1$$

$$N_{\kappa+\varepsilon} N_{ca} << 1$$

$$\frac{|N_{Bo}|}{N_{\kappa+\varepsilon} N_{ca}} << 1$$

$$\frac{N_{Re}^{(2)}}{N_{\kappa+\varepsilon}} << 1 \tag{1-9}$$

Cauchy's first law for the incompressible Newtonian phase j in the form of (1-5) reduces to

$$\nabla \mathcal{P}^{(j)*} = 0 \tag{1-10}$$

the normal component of the jump momentum balance (1-7) to

$$0 = 2H^* \boldsymbol{\xi} - [\, \mathcal{P}^* \boldsymbol{\xi} \,] \tag{1-11}$$

and the tangential component of (1-7) to

$$\mathbf{P} \cdot \text{div}_{(\sigma)} \mathbf{S}^{(\sigma)\star} = 0 \tag{1-12}$$

The physical interpretation of $N_{Re}^{(j)} N_{ca} << 1$ is that bulk inertial forces can be neglected with respect to interfacial tension forces, $N_{ca} << 1$ that bulk viscous forces can be neglected with respect to interfacial tension forces, $N_{\kappa+\varepsilon}^{-1} << 1$ that bulk viscous forces can be neglected with respect to interfacial viscous forces, $N_{\kappa+\varepsilon} N_{ca} << 1$ that interfacial viscous forces can be neglected with respect to interfacial tension forces, $N_{Bo} << 1$ that the effect of gravity can be neglected with respect to interfacial tension in the normal component of the jump momentum balance, and finally $N_{Re}^{(j)} N_{\kappa+\varepsilon}^{-1}$ << 1 that bulk inertial forces can be neglected with respect to surface viscous forces.

Notice also that the effect of inequalities (1-9) is to seek a solution correct to the zeroth order in these parameters.

The sections that follow illustrate this class of problems.

Reference

Slattery, J. C., "Momentum, Energy, and Mass Transfer in Continua," McGraw-Hill, New York (1972); second edition, Robert E. Krieger, Malabar, FL 32950 (1981).

Exercise

3.7.1-1 *for horizontal, plane interface* For the case of a horizontal, plane interface, H = 0 and the interfacial tension will not directly enter the analysis. I suggest that you retain the same dimensionless variables defined in the text with the exception of

$$\mathcal{P}^{(j)\star} \equiv \frac{\mathcal{P}^{(j)}}{\rho^{(j)} g \ell_o} \tag{1}$$

to express the equation of continuity for phase j as (1-4), Cauchy's first law for the incompressible Newtonian phase j as

$$N_{Fr}(\nabla \mathbf{v}^{(j)\star} \cdot \mathbf{v}^{(j)\star}) = - \nabla \mathcal{P}^{(j)\star} + \frac{N_{Fr}}{N_{Re}^{(j)}} \text{div}(\nabla \mathbf{v}^{(j)\star}) \tag{2}$$

and the jump mass balance in the form of (1-6). The jump momentum balance takes the form

$$\text{div}_{(\sigma)}\mathbf{S}^{(\sigma)*} + \left[-\frac{N_{Re}}{N_{\kappa+\varepsilon}} \frac{\mu}{\mu^{(2)}} (\mathbf{v}^* \cdot \boldsymbol{\xi} - v_{(\xi)}^{\{\xi\}*})^2 \boldsymbol{\xi} \right.$$

$$\left. - \frac{N_{Re}}{N_{\kappa+\varepsilon} N_{Fr}} \frac{\mu}{\mu^{(2)}} \mathcal{P}^* \boldsymbol{\xi} + \frac{1}{N_{\kappa+\varepsilon}} \frac{\mu}{\mu^{(2)}} \mathbf{S}^* \cdot \boldsymbol{\xi} \right]$$

$$= 0 \tag{3}$$

Here the Froude number

$$N_{Fr} \equiv \frac{v_o^2}{g\ell_o} \tag{4}$$

and the potential energy of the interface is defined to be zero.

Let us further restrict our attention to problems in which

$$N_{Fr} << 1$$

$$\frac{N_{Fr}}{N_{Re}^{(j)}} << 1$$

$$\frac{N_{Re}^{(j)}}{N_{\kappa+\varepsilon}} \frac{\mu^{(j)}}{\mu^{(2)}} << 1$$

$$\frac{N_{\kappa+\varepsilon} N_{Fr}}{N_{Re}} \frac{\mu^{(2)}}{\mu^{(j)}} << 1$$

$$\frac{1}{N_{\kappa+\varepsilon}} \frac{\mu^{(j)}}{\mu^{(2)}} << 1 \tag{5}$$

Cauchy's first law for the incompressible Newtonian phase j in the form of (2) reduces to (1-10), the normal component of the jump momentum balance (3) to

$$[\, \mathcal{P}^* \boldsymbol{\xi} \,] = 0 \tag{6}$$

and the tangential component of (3) to (1-12). The physical interpretation of $N_{Fr} << 1$ is that bulk inertial forces can be neglected with respect to gravity, $N_{Fr}/N_{Re}^{(j)} << 1$ that bulk viscous forces can be neglected with respect to gravity, $N_{Re}^{(j)} N_{\kappa+\varepsilon}^{-1} << 1$ that bulk inertial forces can be neglected with respect to interfacial viscous forces, and $N_{\kappa+\varepsilon} N_{Fr} N_{\kappa+\varepsilon}^{-1} <<$ 1 that interfacial viscous forces can be neglected with respect to gravity.

As in the text, the effect of inequalities (5) is to seek a solution correct to the zeroth order in these parameters.

3.7.2 Displacement in a capillary (Giordano and Slattery 1983)

Figure 2-1 shows phase 1 displacing phase 2 in a cylindrical capillary of radius R. The common line C, formed by the intersection of the fluid-fluid interface Σ with the capillary wall, moves with a speed of displacement V in a frame of reference fixed with respect to the capillary. The contact angle Θ is measured through phase 1. The imaginary surfaces $S^{(1)}$ and $S^{(2)}$ are fixed with respect to Σ.

Our objective in Sec. 4.1.9 will be to estimate the speed of displacement as a function of the pressure drop between $S^{(1)}$ and $S^{(2)}$. In this section, we will restrict our attention to determining the velocity distribution in the fluid-fluid interface.

Rather than working in terms of a frame of reference fixed with respect to the capillary, we will find it more convenient to construct our solution in a frame of reference fixed with respect to the common line C. In this frame of reference, the interface is fixed in space and the capillary moves to the left in Figure 2-1 with a speed V.

The following physical assumptions will be made in our development.

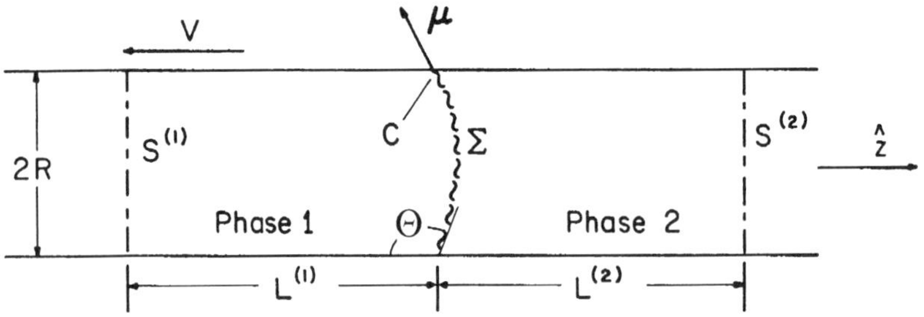

Figure 3.7.2-1: Phase 1 displacing phase 2 in a cylindrical capillary viewed with respect to a frame of reference in which the common line C is stationary.

i) We will confine our attention to the limiting case described in the previous section. In describing this limiting case, we will identify our characteristic length ℓ_o and characteristic speed v_o as

$$\ell_o \equiv R$$

$$v_o \equiv V \tag{2-1}$$

ii) The displacement is stable in the sense that, in the frame of reference fixed with respect to the common line, the flow is independent of time.

iii) The fluid-fluid interfacial stress-deformation behavior can be represented by the linear Boussinesq surface fluid model (Sec. 2.2.2).

iv) The speed of displacement V is sufficiently small that the surfactant concentration in the dividing surface may be considered nearly independent of position and that interfacial tension and the two interfacial viscosities may be treated as constants. This is reasonable in the context of oil displacement, the application developed in Sec. 4.1.9 (Giordano and Slattery 1983, Appendix A). In the analysis that follows, we are interested only in the flow outside the immediate neighborhood of the common line. The slip ratio (Sec. 3.1.1) is assumed to be so small, that we seek a perturbation solution correct to the zeroth order in this parameter. For these reasons, we need not specify the details of the slip model (Sec. 3.1.1).

From the limiting form of Cauchy's first law (1-10), we see that the modified pressure is independent of position within the adjacent bulk phases. Recognizing that a contact angle boundary condition must be satisfied in the limit as the common line is approached in the outer solution, we see that the limiting form of the normal component of the jump momentum balance (1-11) implies that the interface is a spherical segment.

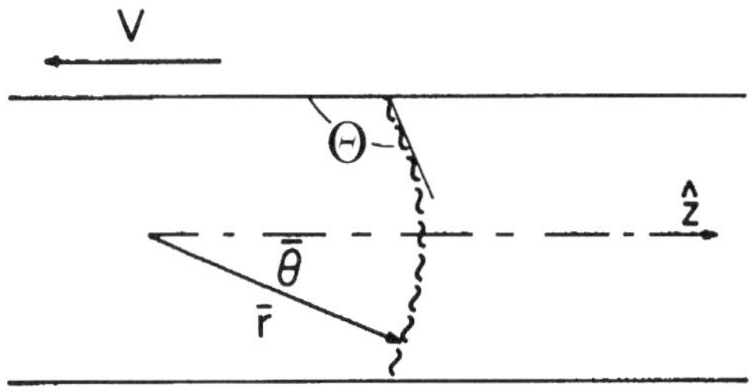

Figure 3.7.2-2: Spherical coordinate system for frame of reference in which the common line at $\bar{\theta} = |\Theta - \pi/2|$ is stationary. When $N_{ca} = 0$, the spherical phase interface coincides with a coordinate surface.

In terms of the spherical coordinate system described in Figure 2-2, we will assume that for this limiting case the surface velocity distribution takes the form

$$v_{\bar{\theta}}^{(\sigma)} = v_{\bar{\theta}}^{(\sigma)}(\bar{\theta})$$

$$v_{\bar{r}}^{(\sigma)} = v_{\bar{\phi}}^{(\sigma)} = 0 \tag{2-2}$$

The tangential component of the jump momentum balance (1-12) consequently reduces to (Table 2.4.2-5)

$$\frac{d}{d\bar{\theta}}\left[\frac{1}{\sin\bar{\theta}}\frac{d}{d\bar{\theta}}\left(v_{\bar{\theta}}^{(\sigma)*}\sin\bar{\theta}\right)\right]+\frac{2\varepsilon}{\kappa+\varepsilon}v_{\bar{\theta}}^{(\sigma)*}=0 \tag{2-3}$$

We will require

$$\text{at } \bar{\theta}=\left|\Theta-\frac{\pi}{2}\right| : \quad v_{\bar{\theta}}^{(\sigma)*}=A \tag{2-4}$$

leaving A for the moment unspecified. The solution to (2-3) consistent with (2-4) is

$$v_{\bar{\theta}}^{(\sigma)*}=A\,\frac{P_{\nu}^{-1}(\cos\bar{\theta})}{P_{\nu}^{-1}(\sin\Theta)} \tag{2-5}$$

in which P_{ν}^{-1} is the associated Legendre function of the first kind (Abramowitz and Stegun 1965) and ν satisfies

$$\nu(\nu+1)=\frac{2\varepsilon}{\kappa+\varepsilon} \tag{2-6}$$

solution for A In order to determine A in (2-4), we must examine the implications of conservation of mass and of momentum within the adjacent bulk phases.

Let us begin by thinking in terms of a cylindrical coordinate system ($\hat{r}$, $\hat{\theta}$, $\hat{z}$) in which the axis of the tube coincides with the $\hat{z}$ coordinate axis and the $\hat{z}$ axis points in the direction of displacement. Let

$$\hat{r}^{*}\equiv\frac{\hat{r}}{R} \qquad \hat{z}^{*}\equiv\frac{\hat{z}}{R} \tag{2-7}$$

be the corresponding dimensionless coordinates. We seek a two-dimensional solution for the velocity distribution in the two bulk phases (j = 1, 2):

$$v_{\hat{r}}^{(j)*}=v_{\hat{r}}^{(j)*}(\hat{r},\hat{z})$$

$$v_{\hat{z}}^{(j)*}=v_{\hat{z}}^{(j)*}(\hat{r},\hat{z})$$

$$v_{\hat{\theta}}^{(j)*}=0 \tag{2-8}$$

Since our particular concern is to identify A, we will focus our attention upon the outer solution for the velocity distribution in the limit as the common line is approached. Let δ be a length much greater than the slip length that characterizes a small neighborhood surrounding the common line, in the majority of which the outer solution must be used to describe the velocity distribution. Define

$$x^* \equiv -\frac{\hat{z}^*}{N_\delta} \qquad y^* \equiv \frac{1 - \hat{r}^*}{N_\delta} \tag{2-9}$$

where

$$N_\delta \equiv \frac{\delta}{R} \tag{2-10}$$

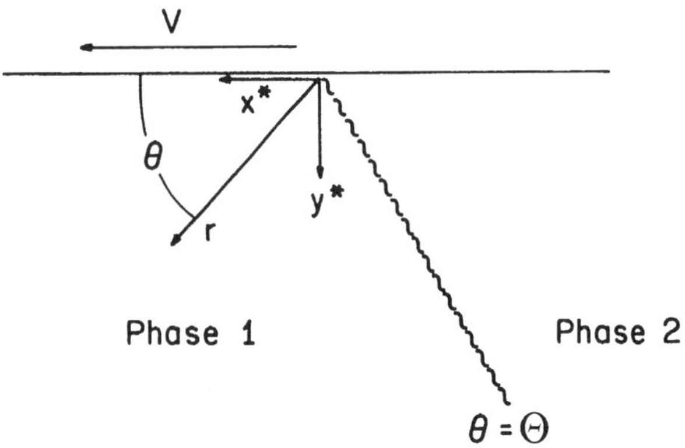

Figure 3.7.2-3: Polar coordinate system centered on common line viewed with respect to a frame of reference in which the common line is stationary.

To the lowest order in N_δ, the displacement problem simplifies to the flow shown in Figure 2-3: flow in the immediate neighborhood of the common line formed as a flat plate passes through a fluid-fluid interface at an angle Θ such that the interface remains a plane.

This problem is more conveniently analyzed in terms of the dimensionless polar coordinates (r^*, θ) centered on the common line in Figure 2-3 and defined such that

$$x^* = r^* \cos\theta$$

$$y^* = r^* \sin\theta \tag{2-11}$$

The equation of continuity for each of the incompressible bulk phases may be identically satisfied by expressing the velocity distributions in terms of $\psi^{(j)}$ (j = 1, 2), the lowest order terms in expansions for the stream functions in terms of N_δ:

$$v_r^{(j)*} = -\frac{1}{r^*}\frac{\partial \psi^{(j)}}{\partial \theta}$$

$$v_\theta^{(j)*} = \frac{\partial \psi^{(j)}}{\partial r^*} \tag{2-12}$$

The only nonzero component of the curl of the Navier-Stokes equation may consequently be expressed as (Slattery 1981, p. 66)

$$\left[\frac{1}{r^*}\frac{\partial}{\partial r^*}\left(r^*\frac{\partial}{\partial r^*}\right) + \frac{1}{r^{*2}}\frac{\partial^2}{\partial \theta^2}\right]^2 \psi^{(j)} = 0 \tag{2-13}$$

The frame of reference is defined such that the common line is stationary and the tube wall is in motion, which implies

$$\text{at } \theta = 0: \quad \frac{1}{r^*}\frac{\partial \psi^{(1)}}{\partial \theta} = -1 \tag{2-14}$$

$$\text{at } \theta = \pi: \quad \frac{1}{r^*}\frac{\partial \psi^{(2)}}{\partial \theta} = 1 \tag{2-15}$$

The tangential component of velocity must be continuous at the fluid-fluid interface,

$$\text{at } \theta = \Theta: \quad \frac{\partial \psi^{(1)}}{\partial \theta} = \frac{\partial \psi^{(2)}}{\partial \theta} \tag{2-16}$$

Very far upstream of the fluid-fluid interface, we have Poiseuille flow. Since we are free to specify $\psi^{(1)} = 0$ at one point on the tube wall, this implies

$$\text{as } \hat{z}^* \to -\infty: \text{ the stream function} \to -\frac{\hat{r}^*}{2}\left(1 - \hat{r}^{*2}\right) \tag{2-17}$$

Because the normal component of velocity is zero on the tube wall, it follows that

$$\text{at } \theta = 0: \quad \psi^{(1)} = 0 \tag{2-18}$$

Equation (2-17) tells us that along the center streamline the stream function equals zero. We require that this center streamline be continuous with all of the streamlines in the fluid-fluid phase interface consistent with the two-dimensional flow (2-8) and that the component of velocity normal to the interface be zero:

$$\text{at } \theta = \Theta: \quad \psi^{(1)} = \psi^{(2)} = 0 \tag{2-19}$$

Reversing this same argument in phase 2, we conclude

$$\text{at } \theta = \pi: \quad \psi^{(2)} = 0 \tag{2-20}$$

To the lowest order in N_δ, the normal component of the jump momentum balance at the fluid-fluid interface is satisfied identically. The tangential component reduces to (Table 2.4.2-3)

$$\text{at } \theta = \Theta: \quad \frac{N_{\kappa+\varepsilon}}{N_\delta} \frac{\partial^2}{\partial r^{*2}} \left(\frac{1}{r^*} \frac{\partial \psi^{(1)}}{\partial \theta} \right) + \frac{1}{r^{*2}} \frac{\partial^2 \psi^{(2)}}{\partial \theta^2} - \frac{N_\mu}{r^{*2}} \frac{\partial^2 \psi^{(1)}}{\partial \theta^2} = 0 \tag{2-21}$$

The work of Huh and Scriven (1971) suggests that

$$\psi_p^{(1)} = r^* \left(C_1 \sin\theta + C_2 \theta \sin\theta + C_3 \theta \cos\theta \right) \tag{2-22}$$

$$\psi_p^{(2)} = r^* \left(C_4 \sin\theta + C_5 \cos\theta + C_6 \theta \sin\theta + C_7 \theta \cos\theta \right) \tag{2-23}$$

$$C_1 \equiv -1 - C_3$$

$$C_2 \equiv \frac{\sin^2\Theta}{D} \left\{ \sin^2\Theta - (\Theta - \pi)^2 + \frac{1}{N_\mu} \left[\Theta(\Theta - \pi) - \sin^2\Theta \right] \right\}$$

$$C_3 \equiv \frac{\sin 2\Theta}{2D} \left\{ \sin^2\Theta - (\Theta - \pi)^2 \right.$$

$$+\frac{1}{N_\mu}\left[\Theta(\Theta-\pi)-\sin^2\Theta-\pi\tan\Theta\right]\Big\}$$

$$C_4 \equiv -1-\pi C_6 - C_7$$

$$C_5 \equiv \pi C_7$$

$$C_6 \equiv \frac{\sin^2\Theta}{D}\left[\sin^2\Theta-\Theta(\Theta-\pi)+\frac{1}{N_\mu}(\Theta^2-\sin^2\Theta)\right]$$

$$C_7 \equiv \frac{\sin 2\Theta}{2D}\left[\sin^2\Theta-\Theta(\Theta-\pi)-\pi\tan\Theta+\frac{1}{N_\mu}(\Theta^2-\sin^2\Theta)\right]$$

$$D \equiv \left(\frac{1}{2}\sin 2\Theta-\Theta\right)[(\Theta-\pi)^2-\sin^2\Theta]$$

$$+\frac{1}{N_\mu}\left(\Theta-\pi-\frac{1}{2}\sin 2\Theta\right)(\Theta^2-\sin^2\Theta) \tag{2-24}$$

is a particular solution to (2-13) through (2-16) and (2-18) through (2-21).
The general solution takes the form

$$\psi^{(j)} = \psi_p^{(j)} + \sum_{m=1}^{\infty} B_m^{(j)}\phi_m^{(j)} \tag{2-25}$$

where the eigenfunctions $\phi_m^{(j)}$ are solutions of (2-13) consistent with the homogeneous boundary conditions

$$\text{at } \theta = 0: \quad \frac{\partial\psi^{(1)}}{\partial\theta} = 0 \tag{2-26}$$

$$\text{at } \theta = \pi: \quad \frac{\partial\psi^{(2)}}{\partial\theta} = 0 \tag{2-27}$$

as well as (2-16) and (2-18) through (2-21). The constants $B_m^{(j)}$ are arbitrary, since boundary conditions have not been specified as $r^* \to \infty$. The eigensolutions may be found by separation of variables. In the limit as $r^* \to 0$, the eigenvalues are roots of (Giordano 1980)

$$\{\sin(\lambda_m\Theta)\sin([\lambda_m-2]\Theta)-\lambda_m(\lambda_m-2)\sin^2\Theta\}$$

$$\times\{\sin(\lambda_m[\pi-\Theta])\sin([\lambda_m-2][\pi-\Theta])$$

$$- \lambda_m (\lambda_m - 2) \sin^2\Theta \} = 0 \qquad (2\text{-}28)$$

In order that velocities be bounded, we discard all roots whose real part is less than 1. There are no roots of (2-28) whose real part equals 1. This means that the particular solution $\psi_p^{(j)}$ (j = 1, 2), which is $O(r^*)$, dominates the eigensolutions as the common line is approached.

From (2-12) and (2-22) through (2-25), we conclude that

$$A \equiv - \text{limit } r^* \to 0: \; v_r^{(\sigma)*}$$

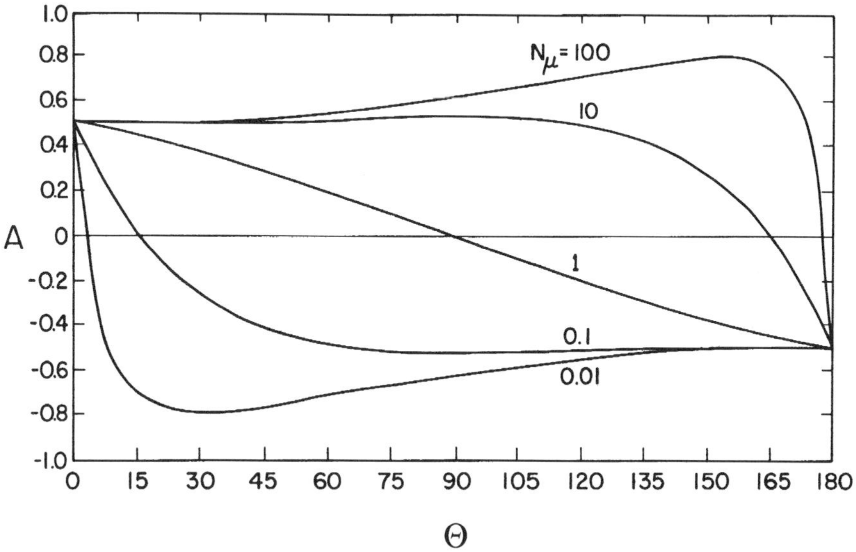

Figure 3.7.2-4: Function A defined by (2-29).

$$= - \{ N_\mu (\Theta \cos \Theta - \sin \Theta)[\sin^2 \Theta - (\Theta - \pi)^2]$$

$$+ [(\Theta - \pi) \cos \Theta - \sin \Theta](\Theta^2 - \sin^2\Theta)\}$$

$$\times \{ N_\mu (\sin \Theta \cos \Theta - \Theta)[(\Theta - \pi)^2 - \sin^2 \Theta]$$

$$+ (\Theta - \pi - \sin \Theta \cos \Theta)(\Theta^2 - \sin^2 \Theta)\}^{-1} \qquad (2\text{-}29)$$

Figure 2-4 shows A as a function of N_μ and Θ.

The values for A in the limits as $\Theta \to 0°$ and $\Theta \to 180°$ initially appear to be incorrect in Figure 2-4, since we would expect the interfacial velocity to approach the velocity of the wall. The answer is that this is a result of an analysis for the outer solution of the velocity distribution. Compatibility with the velocity boundary condition at the wall would be imposed in the inner solution.

References

Abramowitz, M., and I. A. Stegun, "Handbook of Mathematical Functions," Dover, New York (1965).

Giordano, R. M., Ph.D. dissertation, Northwestern University, Evanston, Illinois (1980).

Giordano, R. M., and J. C. Slattery, *AIChE J.* **29**, 483 (1983).

Huh, C., and L. E. Scriven, *J. Colloid Interface Sci.* **35**, 85 (1971).

Slattery, J. C., "Momentum, Energy, and Mass Transfer in Continua," McGraw-Hill, New York (1972); second edition, Robert E. Krieger, Malabar, FL 32950 (1981).

3.7.3 Several interfacial viscometers suitable for measuring generalized Boussinesq surface fluid behavior (Jiang *et al.* 1983) Up to the present time, the deep-channel is the only interfacial viscometer for which complete analyses of non-linear surface stress-deformation behavior have appeared. (For a review, see Secs. 3.5.1 and 3.6.2).

The knife-edge (Mannheimer and Burton 1970; Lifshutz *et al.* 1971), disk (Oh and Slattery 1978), and thin biconical bob (Oh and Slattery 1978) interfacial viscometers are sketched in Figures 3.4.1-1, 3.4.1-11, and 3.4.1-13. As described in Sec. 3.4.1, they were originally intended for measuring the surface shear viscosity of an interface described by the linear Boussinesq surface fluid. In each case, one measures the torque required to hold the bob stationary as the dish of radius R rotates with a constant angular velocity Ω. The knife-edge and dish interfacial viscometers are appropriate only for gas-liquid interfaces. The thin biconical bob interfacial viscometer can be used for both gas-liquid and liquid-liquid interfaces.

In what follows, we develop for these three rotational viscometers two new techniques for measuring nonlinear interfacial stress-deformation behavior described by the generalized Boussinesq surface fluid model. With one technique, a single bob is used. With the second technique, two bobs having somewhat different radii are employed.

We will make the following physical assumptions in this development.

i) This analysis is constructed in the context of the limiting case described in Exercise 3.7.1-1. In describing this limiting case, we will identify our characteristic length ℓ_0 and characteristic speed v_0 as

$$\ell_o \equiv R$$

$$v_o \equiv R\Omega \tag{3-1}$$

ii) The fluid-fluid interfacial stress-deformation behavior can be represented by the generalized Boussinesq surface fluid model (Sec. 2.2.2).

iii) Any surfactant present in the phase interface is uniformly distributed over the entire surface. The interfacial tension is independent of position in the dividing surface. The interfacial viscosities are dependent upon position only to the extent of their dependence upon the local rate of surface deformation.

Because we are concerned with the limiting case of Exercise 3.7.1-1, the velocity and stress distributions within the bulk phases will not be necessary for our present purposes. For that reason, we will say no more about the equation of continuity and Cauchy's first law.

In solving (1-12) consistent with the generalized Boussinesq surface fluid model, we must observe that the bob is stationary

$$\text{at } r = \frac{R}{s} : \ \mathbf{v}^{(\sigma)} = 0 \tag{3-2}$$

and that the dish rotates with a constant angular velocity Ω

$$\text{at } r = R : \ v_\theta^{(\sigma)} = R\Omega, \ v_r^{(\sigma)} = 0 \tag{3-3}$$

where

$$s \equiv \frac{R}{R_b} \tag{3-4}$$

and R_b is the radius of the bob measured in the plane of the interface.

Our objective is to compute the relationship between Ω and $\mathcal{T}_z$, the z component of the torque exerted on the bob by the two phases and the interface:

$$\mathcal{T}_z^* \equiv \frac{\mathcal{T}_z}{2\pi\mu^{(1)}\Omega R^3}$$

$$= N_\varepsilon s^{-2} \ S_{r\theta}^{(\sigma)*}\Big|_{r^* = s^{-1}} + \frac{1}{2\pi R^2}\int_{S_b} (\ a_z^{(1)*} + a_z^{(2)*}\ )\ dA \tag{3-5}$$

Here S_b is the surface of the bob and

$$\mathbf{a}^{(j)*} \equiv \mathbf{p}^* \wedge (\mathbf{T}^{(j)*} \cdot \mathbf{n}^{(j)}) \tag{3-6}$$

in which $\mathbf{p}$ is the position vector measured with respect to some point on the axis of the bob and $\mathbf{n}^{(j)}$ is the unit normal to the surface of the bob pointing into phase j. For the limiting case of Exercise 3.7.1-1, (3-5) reduces to

$$\mathcal{T}_z = 2\pi R_b^2 S_{r\theta}^{(\sigma)} |_{r=R_b} \tag{3-7}$$

Equations (3-2) and (3-3) suggest that we seek a solution to (1-12) of the form

$$v_\theta^{(\sigma)} = v_\theta^{(\sigma)}(r) \qquad v_r^{(\sigma)} = 0 \tag{3-8}$$

This implies that there is only one nonzero component of the surface rate of deformation tensor (see Table 2.4.2-2)

$$D_{r\theta}^{(\sigma)} = \frac{r}{2} \frac{d\omega^{(\sigma)}}{dr} \tag{3-9}$$

and

$$\text{div}_{(\sigma)} \mathbf{v}^{(\sigma)} = 0 \tag{3-10}$$

where we have introduced as the local angular velocity

$$\omega^{(\sigma)} \equiv \frac{v_\theta^{(\sigma)}}{r} \tag{3-11}$$

This in turn means that there is only one nonzero component $S_{r\theta}^{(\sigma)}$ of the viscous portion of the surface stress tensor and (1-12) reduces to

$$\frac{d}{dr} (r^2 S_{r\theta}^{(\sigma)}) = 0 \tag{3-12}$$

This can be integrated once consistent with (3-7) to find

$$S_{r\theta}^{(\sigma)} = \frac{\mathcal{T}_z}{2\pi r^2} \tag{3-13}$$

Because of (3-10), the generalized Boussinesq surface fluid model takes the form (Sec. 2.2.2)

$$\tau = \lambda\, \varepsilon(\lambda) \tag{3-14}$$

with

$$\tau \equiv \left[\frac{1}{2} \operatorname{tr}(\mathbf{S}^{(\sigma)} \cdot \mathbf{S}^{(\sigma)}) \right]^{1/2}$$

$$= \frac{\mathcal{T}_z}{2\pi r^2} \tag{3-15}$$

and

$$\lambda \equiv | \nabla_{(\sigma)} \mathbf{v}^{(\sigma)} | \equiv [\, 2 \operatorname{tr}(\mathbf{D}^{(\sigma)} \cdot \mathbf{D}^{(\sigma)})]^{1/2}$$

$$= r \frac{d\omega^{(\sigma)}}{dr} \tag{3-16}$$

Here $\varepsilon(\lambda)$ is the apparent interfacial shear viscosity. Recognizing that there should be a one-to-one relationship between $S_{r\theta}^{(\sigma)}$ and $D_{r\theta}^{(\sigma)}$, we assume that (3-14) can be inverted to be written as

$$\beta(\tau) \equiv \frac{1}{\varepsilon} = \frac{\lambda}{\tau} = \frac{r}{\tau} \frac{d\omega^{(\sigma)}}{dr} \tag{3-17}$$

Here we have introduced the apparent interfacial fluidity $\beta(\tau)$, which is the inverse of the apparent interfacial shear viscosity ε. Equation (3-17) can be integrated once consistent with (3-2) and (3-3) to find

$$\Omega = \int_{R/s}^{R} \frac{\beta(\tau)\tau}{r} \, dr \tag{3-18}$$

or in view of (3-15)

$$\Omega = \frac{1}{2} \int_{\frac{\mathcal{T}_z}{2\pi R^2}}^{\frac{s^2 \mathcal{T}_z}{2\pi R^2}} \beta(\, \tau \,)\, d\tau \tag{3-19}$$

In what follows, we develop two techniques based on (3-19) that allow us to interpret experimental data in the form of the apparent interfacial shear viscosity $\varepsilon(\lambda)$. These techniques are analogous to those used with the concentric cylinder viscometer for measuring the apparent shear viscosities of non-Newtonian fluids (Krieger and Maron 1952; Krieger and Elrod 1953; Krieger and Maron 1954; Krieger 1968; Krieger 1969; Yang and Krieger 1978).

single-bob technique Differentiating (3-19), we find

$$\frac{d\Omega}{d\mathcal{T}_z} = \frac{1}{4\pi R^2}\left[s^2\, \beta\left(\frac{s^2\,\mathcal{T}_z}{2\pi R^2} \right) - \beta\left(\frac{\mathcal{T}_z}{2\pi R^2} \right)\right] \tag{3-20}$$

or

$$F(\,\mathcal{T}_z\,) = \beta\left(\frac{s^2\,\mathcal{T}_z}{2\pi R^2} \right) - s^{-2}\,\beta\left(\frac{\mathcal{T}_z}{2\pi R^2} \right) \tag{3-21}$$

where

$$F(\,\mathcal{T}_z\,) \equiv 4\pi s^{-2} R^2 \frac{d\Omega}{d\mathcal{T}_z} \tag{3-22}$$

Equation (3-21) allows us to observe

$$s^{-2n} F(\,s^{-2n}\mathcal{T}_z\,)$$

$$= s^{-2n}\,\beta\left(\frac{s^{-2(n-1)}\mathcal{T}_z}{2\pi R^2} \right) - s^{-2(n+1)}\,\beta\left(\frac{s^{-2n}\,\mathcal{T}_z}{2\pi R^2} \right) \tag{3-23}$$

and consequently

$$\beta\left(\frac{s^2\,\mathcal{T}_z}{2\pi R^2} \right) = \sum_{n=0}^{\infty} s^{-2n} F(\,s^{-2n}\mathcal{T}_z\,) \tag{3-24}$$

This is a slowly convergent series, which may be evaluated asymptotically using the Euler-Maclaurin sum formula (Jeffreys and Jeffreys 1956; Abramowitz and Stegun 1964) as

$$\sum_{n=0}^{\infty} s^{-2n} F(\,s^{-2n}\mathcal{T}_z\,)$$

$$= \int_0^{\infty} s^{-2x}\, F(\,s^{-2x}\mathcal{T}_z\,)\,dx + \frac{1}{2}F(\,\mathcal{T}_z\,)$$

$$+\sum_{k=1}^{m} \frac{B_{2k}}{(2k)!}\left\{ \left(\frac{d}{dx}\right)^{(2k-1)} [s^{-2x}\ F(s^{-2x}\mathcal{T}_z)]\right\}_{x=0}^{x=\infty}$$

$$+\ \delta_m \tag{3-25}$$

where

$$\delta_m \equiv \sum_{i=1}^{\infty} \int_i^{i+1} B_{2m+1}(x-i) \left(\frac{d}{dx}\right)^{(2m+1)} \left[s^{-2x}\ F(s^{-2x}\mathcal{T}_z)\right] dx \tag{3-26}$$

The Bernouilli polynomials $B_n(x)$ are defined by a generating function

$$\frac{t\,e^{xt}}{e^t - 1} = \sum_{n=0}^{\infty} B_n(x) \frac{t^n}{n!} \tag{3-27}$$

and the Bernouilli numbers $B_n \equiv B_n(0)$ by

$$\frac{t}{e^t - 1} = \sum_{n=0}^{\infty} B_n \frac{t^n}{n!} \tag{3-28}$$

Making the transformation of variables

$$y \equiv s^{-2x}\mathcal{T}_z \tag{3-29}$$

and recognizing (3-22), we can readily evaluate the first integral on the right of (3-25) as

$$\int_0^{\infty} s^{-2x}\ F(s^{-2x}\mathcal{T}_z)\ dx$$

$$= \frac{2\pi R^2}{\mathcal{T}_z s^2 \ln s} \int_0^{\mathcal{T}_z} \frac{d\Omega}{dy} dy$$

$$= \frac{2\pi R^2 \Omega}{\mathcal{T}_z s^2 \ln s} \tag{3-30}$$

The derivatives appearing in (3-25) are

$$\left\{\left(\frac{d}{dx}\right)^{(2k-1)} [\ s^{-2x}\ F(s^{-2x}\mathcal{T}_z)]\right\}_{x=0}^{x=\infty}$$

$$= 2^{2k} \left\{ \frac{2\pi R^2}{s^2 \mathcal{T}_z} \left[\frac{d}{d(\ln \tau_z)} \right]^{2k} \Omega \right\} (\ln s)^{2k-1} \tag{3-31}$$

In view of (3-22), (3-25), (3-30) and (3-31), equation (3-24) becomes

$$\beta(\tau_b) = \frac{\Omega}{\tau_b \ln s} \left[1 + N \ln s + \sum_{k=1}^{\infty} \frac{2^{2k} B_{2k}}{(2k)!} h_{2k} (\ln s)^{2k} \right] \tag{3-32}$$

with the assumption that, in the limit as $m \to \infty$, $\delta_m \to 0$. In (3-32), we have defined

$$\tau_b \equiv \frac{s^2 \mathcal{T}_z}{2\pi R^2} \tag{3-33}$$

$$h_n \equiv \frac{1}{\Omega} \left(\frac{d}{d(\ln \tau_z)} \right)^n \Omega \tag{3-34}$$

and

$$N \equiv h_1 \equiv \frac{d(\ln \Omega)}{d(\ln \tau_z)} \tag{3-35}$$

By making use of (3-28) and recognizing that

$$B_0 = 1, \ B_1 = -\frac{1}{2}, \ B_{2k+1} = 0 \ \text{ for } k \geq 1 \tag{3-36}$$

we can write (3-32) as

$$\beta(\tau_b) = \frac{2N\Omega}{\tau_b(1 - s^{-2N})} \left[1 + \frac{1 - s^{-2N}}{2N \ln s} \times \sum_{k=1}^{\infty} \frac{2^{2k} B_{2k}}{(2k)!} (h_{2k} - N^{2k})(\ln s)^{2k} \right] \tag{3-37}$$

To evaluate the correction, we note from (3-34) and (3-35) that

$$h_n = \left(N + \frac{d}{d(\ln \tau_z)} \right) h_{n-1}$$

$$= \left(N + \frac{d}{d(\ln \tau_z)} \right)^{n-1} N \tag{3-38}$$

Recognizing the noncommutability of the operators N and $d/d(\ell n \mathcal{T}_z)$, we can expand (3-38) to obtain

$$h_n = N^n \left\{ 1 + n! \left(\frac{N^{-2} N^{(1)}}{2!(n-2)!} + \frac{N^{-3} N^{(2)}}{3!(n-3)!} \right.\right.$$

$$\left.\left. + \frac{N^{-4} [N^{(3)} + 3(N^{(1)})^2]}{4!(n-4)!} + \dots \right) \right\} \tag{3-39}$$

Here

$$N^{(n)} \equiv \left(\frac{d}{d(\ln \tau_z)} \right)^n N \tag{3-40}$$

We can decompose the correction terms in (3-37) into subseries:

$$\sum_{k=1}^{\infty} \frac{2^{2k} B_{2k}}{(2k)!} (h_{2k} - N^{2k})(\ln s)^{2k}$$

$$= \sum_{k=1}^{\infty} B_{2k}(2N \ln s)^{2k} \left(\frac{N^{-2} N^{(1)}}{2!(2k-2)!} + \frac{N^{-3} N^{(2)}}{3!(2k-3)!} \right.$$

$$\left. + \frac{N^{-4}[N^{(3)} + 3 N^{(1)2}]}{4!(2k-4)!} + \dots \right) \tag{3-41}$$

In view of (3-28) and (3-36), we can rewrite (3-37) as

$$\frac{1}{\varepsilon} = \beta(\tau_b)$$

$$= \frac{2N\Omega}{\tau_b(1 - s^{-2N})} \left\{ 1 + N^{-2} N^{(1)} f_1(2N \ln s) \right.$$

$$+ N^{-3} N^{(2)} f_2(2N \ln s)$$

$$\left. + N^{-4}[N^{(3)} + 3 N^{(1)2}] f_3(2N \ln s) + \dots \right\} \tag{3-42}$$

where

$$f_n(x) \equiv \frac{x^n(1 - e^{-x})}{(n+1)!} \left(\frac{d}{dx}\right)^{n+1} \left(\frac{x}{e^x - 1} + \frac{x}{2} - 1\right) \tag{3-43}$$

Figure 3-1 shows the functions $f_n(x)$ for $n = 1, 2, 3$. All $f_n(n = 1, ...)$ vanish for small and large x.

In order to determine the fluidity $1/\varepsilon = \beta$, we must measure $\mathcal{T}_z$ as a function of Ω. Equation (3-42) is a rapidly converging series, which reduces to power-law behavior in the limit where N is a constant. Even for large deviations from power-law behavior and large radius ratios, we can expect the correction to be small (Krieger 1969).

For a linear Boussinesq surface fluid,

$$N = 1 \tag{3-44}$$

and (3-42) reduces to

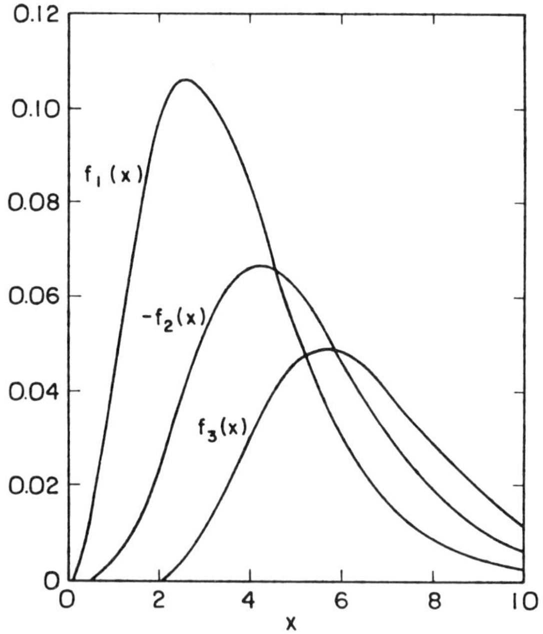

Figure 3.7.3-1: Functions $f_n(x)$ $(n = 1, 2, 3)$ defined by (3-43).

$$\frac{1}{\varepsilon} = \frac{2\Omega}{\tau_b(\ 1\ -\ s^{-2}\ )} \tag{3-45}$$

which is consistent with the result of Oh and Slattery (1978) in the limit of Exercise 3.7.1-1.

double-bob technique In the single-bob technique, only one interfacial viscometer is used. Let us now assume that several geometrically similar interfacial viscometers are available, so that s may be considered as an independent variable.

Differentiating (3-19), we discover

$$\left(\frac{\partial \Omega}{\partial s}\right)_{\tau_z} = \frac{s\mathcal{T}_z}{2\pi R^2}\ \beta\left(\frac{s^2\,\mathcal{T}_z}{2\pi R^2}\right) \tag{3-46}$$

or

$$\beta(\ \tau_b\ ) = \frac{s}{\tau_b}\left(\frac{\partial \Omega}{\partial s}\right)_{\tau_z} \tag{3-47}$$

A reasonable approximation can be obtained using only two bobs corresponding to radius ratios s_α and s_β. For the same value of $\mathcal{T}_z$, one would measure an angular velocity Ω_α with $s = s_\alpha$ and Ω_β with $s = s_\beta$. Expanding Ω_α and Ω_β in Taylor series about $s = s_m$

$$s_m \equiv \frac{1}{2}(\ s_\alpha + s_\beta\ ) \tag{3-48}$$

and taking the difference, we find

$$\left(\frac{\partial \Omega}{\partial s}\right)_{\tau_z}\Bigg|_{s_m} = \frac{\Omega_\alpha - \Omega_\beta}{s_\alpha - s_\beta} - \frac{1}{24}\left(\frac{\partial^3 \Omega}{\partial s^3}\right)_{\tau_z}\Bigg|_{s_m}(\ s_\alpha - s_\beta\ )^2 + \dots \tag{3-49}$$

or

$$\frac{1}{\varepsilon} = \beta(\ \tau_m\ )$$

$$= \frac{s_m}{\tau_m}\left[\frac{\Omega_\alpha - \Omega_\beta}{s_\alpha - s_\beta}\right] - \frac{s_m}{24\,\tau_m}\left(\frac{\partial^3 \Omega}{\partial s^3}\right)_{\tau_z}\Bigg|_{s_m}(\ s_\alpha - s_\beta\ )^2 + \dots \tag{3-50}$$

in which

$$\tau_m \equiv \frac{s_m{}^2 \, \mathcal{T}_z}{2 \pi R^2} \tag{3-51}$$

By making $(s_\alpha - s_\beta)$ sufficiently small, the fluidity $1/\varepsilon = \beta$ can be approximated by the first term of (3-50).

Table 3.7.3-1: Dimensions of the dish and bobs used

	radius (cm)
Dish	10.00
Disk No. 1	8.724
Disk No. 2	7.828
Knife-Edge	8.566
Biconical Bob	8.987

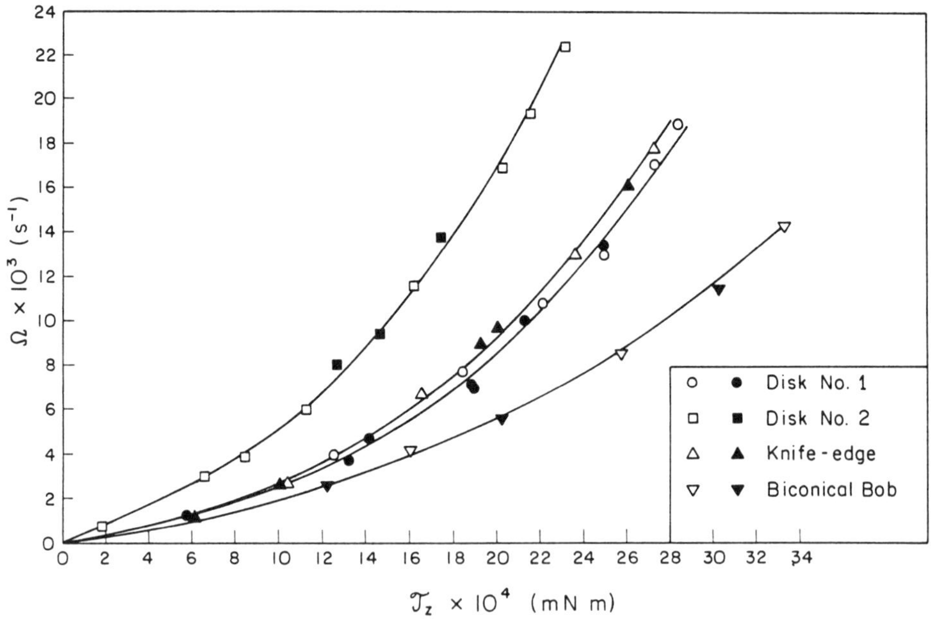

Figure 3.7.3-2: Angular velocity as a function of torque for the interface between air and an aqueous solution of 0.1 wt.% dodecyl sodium sulfate (Eastman Kodak cat. no. 5967). The points indicated by open symbols were obtained by increasing the angular velocity; the points denoted by closed symbols by decreasing the angular velocity.

experimental study using rotational interfacial viscometers In order to test these two techniques, a series of experiments were carried out using the disk, knife-edge, and thin biconical bob interfacial viscometers. The dimensions of the bobs and the dish are given in Table 3-1. The bob was suspended by a torsion wire. Small deflections of the torsion wire from rest (when the angular velocity of the dish is zero) were determined by reflecting a low-power laser beam off a small mirror mounted on the bob. In this way, the torque required to hold the bob stationary as the dish rotated with a constant angular velocity was measured.

The interface between air and an aqueous solution of 0.1 wt.% dodecyl sodium sulfate was studied with these rotational interfacial viscometers at room temperature (23 ± 1°C). The solution was prepared by gently mixing dodecyl sodium sulfate (Catalog no. 5967, Eastman Kodak Co., Rochester, NY) and doubly distilled water for about two hours. The temperature was slightly elevated to facilitate solubilization.

No attempt was made to further purify the dodecyl sodium sulfate, which was almost certainly contaminated from the manufacturing process with a small amount of dodecyl alcohol. The interface between air and an aqueous solution of pure dodecyl sodium sulfate exhibits a very small interfacial viscosity. We wished to study an interface exhibiting a relatively large interfacial viscosity, which was conveniently provided by the

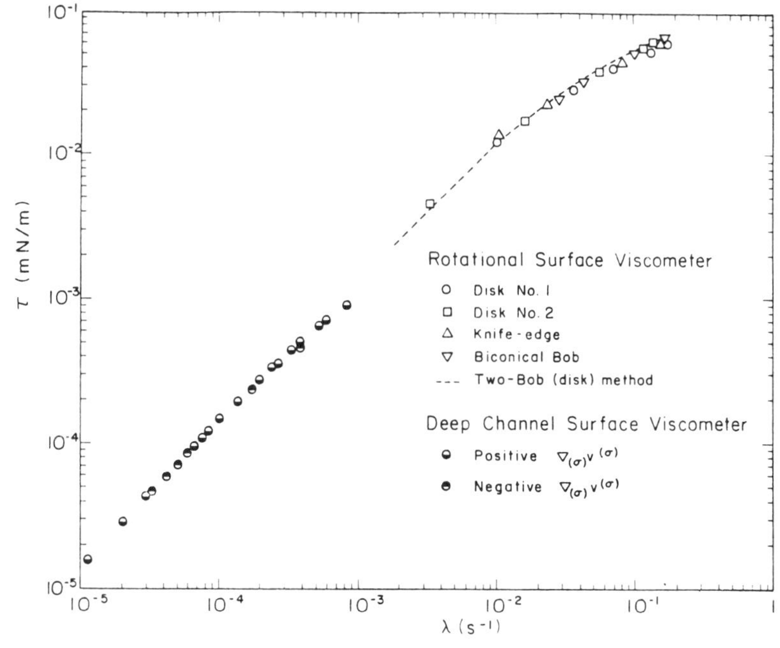

Figure 3.7.3-3: $\mathcal{T}$ as a function of λ for $\text{div}_{(\sigma)} \mathbf{v}^{(\sigma)} = 0$ at an interface between air and 0.1 wt.% aqueous solution of dodecyl sodium sulfate.

interaction between the dodecyl sodium sulfate and the dodecyl alcohol. Our objective here was not to provide archival data for a well-defined interface, but rather to demonstrate the consistency of a variety of experimental techniques.

The interfacial tension was measured with a Wilhelmy plate interfacial tensiometer. The equilibrium interfacial tension (γ = 21.1 mN/m) was reached one hour after the interface had been formed; it remained constant for at least one week.

All interfacial stress-deformation measurements were made after the interfacial tension had reached this equilibrium value. The equilibrium interfacial stress-deformation behavior was reached 10-24 hours after the interface had been formed; the time required to reach equilibrium decreased with the dish in motion. All experimental measurements were repeatable and independent of whether the angular velocity of the dish was being increased or decreased in sequence as shown in Figure 3-2.

The measured equilibrium interfacial stress-deformation behavior is shown in Figures 3-3 and 3-4. Four sets of data are analyzed using the single-bob technique (3-42). The data for the two disks are also analyzed by the double-bob technique (3-50). For the single-bob technique, only the leading term of (3-42) is significant. The first correction terms is always below 1%.

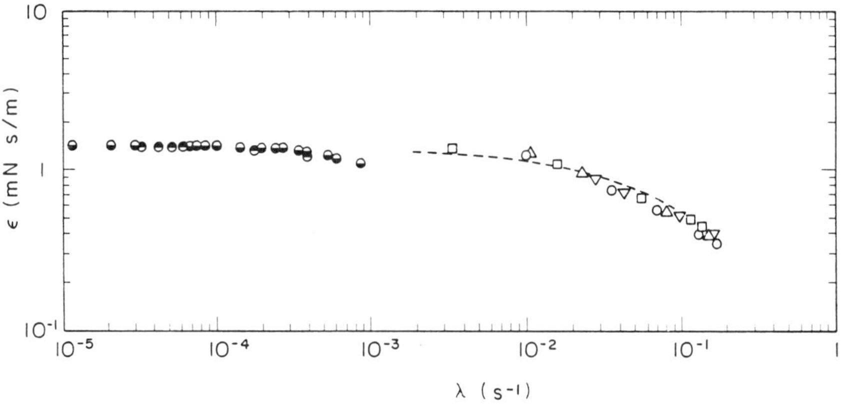

Figure 3.7.3-4: Apparent interfacial shear viscosity ε as a function of λ for $\text{div}_{(\sigma)}\mathbf{v}^{(\sigma)} = 0$ at an interface between air and 0.1 wt.% aqueous solution of dodecyl sodium sulfate.

Note that the requirements of the limiting case of Exercise 3.7.1-1 are satisfied, since $N_\varepsilon \doteq 14$ and the angular velocities shown in Figure 3-2 are very small.

experimental study using deep-channel interfacial viscometer In order to further test the consistency of these measurements, we studied this same system in the deep-channel interfacial viscometer sketched in Figure 3.5.1-1. With the disk rotating at a constant angular velocity, one measures the velocity distribution in the interface. This is a fundamentally different viscometer geometry, the analysis of which is given in Sec. 3.5.1 and does not require the limit described in Exercise 3.7.1-1.

In this study, the inside radius of the outer channel was 6.681 cm; the outside radius of the inner channel wall was 5.203 cm; the depth of the liquid in the channel was 1.14 cm. Several 10μm particles were placed on the interface. With the dish rotating at a constant angular velocity, the velocity distribution at the interface was determined by measuring at several radial positions the times required for a particle to travel a predetermined distance. Measurements were made two to three days after the interface had been formed.

The results shown in Figures 3-3 and 3-4 are consistent with those obtained using the rotational viscometers. This is the first time that consistent measurements of interfacial stress-deformation behavior have been reported for different interfacial viscometers.

References

Abramowitz, M., and I. A. Stegun, "Handbook of Mathematical Functions," sec. 23, National Bureau of Standards Applied Mathematics Series No. 55, U.S. Government Printing Office, Washington, DC., 1964.

Jeffreys, H., and B. S. Jeffreys, "Methods of Mathematical Physics," third edition, p. 279, Cambridge University Press, Cambridge, 1956.

Jiang, T. S., J. D. Chen, and J. C. Slattery, *J. Colloid Interface Sci.* **96**, 7 (1983).

Krieger, I. M., and S. H. Maron, *J. Appl. Phys.* **23**, 147 (1952).

Krieger, I. M., and H. Elrod, *J. Appl. Phys.* **24**, 134 (1953).

Krieger, I. M., and S. H. Maron, *J. Appl. Phys.* **25**, 72 (1954).

Krieger, I. M., *Trans. Soc. Rheol.* **12**, 5 (1968).

Krieger, I. M., *Proc. Fifth Int. Congr. Rheol.* **1**, 511 (1969).

Lifshutz, N., M. G. Hegde and J. C. Slattery, *J. Colloid Interface Sci.* **37**, 73 (1971).

Mannheimer, R. J., and R. A. Burton, *J. Colloid Interface Sci.* **32**, 73 (1970).

Oh, S. G., and J. C. Slattery, *J. Colloid Interface Sci.* **67**, 516 (1978).

Yang, T. M. T., and I. M. Krieger, *J. Rheol.* **22**, 413 (1978).

Notation for chapter 3

$\mathbf{a}^{\alpha}$	dual basis vectors, introduced in Sec. A.2.3
$\mathbf{a}_{\beta}$	natural basis vectors, introduced in Sec. A.2.1
$\mathbf{a}_{<\alpha>}$	physical basis vectors, introduced in Sec. A.2.5
$a_{\alpha\alpha}$	diagonal components of the projection tensor **P**, introduced in Sec. A.3.2; see also Sec. A.2.1
$\mathbf{b}$	external force per unit mass, introduced in Sec. 2.1.3
$\mathbf{b}_m$	mutual force per unit mass, introduced in Sec. 2.1.3
$C_t^{(\sigma)t}$	history of the relative right surface Cauchy-Green tensor, introduced in Sec. 1.2.6; see also Sec. 2.2.3 where history of a function of time is defined
$\mathbf{D}^{(\sigma)}$	surface rate of deformation tensor, introduced in Sec. 1.2.8
$\mathcal{F}_t^t$	history of the relative surface deformation gradient, introduced in Sec. 1.2.5; see also Sec. 2.2.3 where history of a function of time is defined
g	magnitude of the acceleration of gravity
H	mean curvature defined in Sec. A.5.3
$\mathbf{I}$	identity tensor that transforms every vector into itself
$K_t^{(\sigma)t}$	defined in Sec. 2.2.7
$\mathbf{n}$	unit normal to surface
N_{Bo}	dimensionless Bond number, defined in Sec. 3.2.3
N_{ca}	dimensionless capillary number, defined in Sec. 3.7.1
N_{Fr}	dimensionless Froude number, defined in Sec. 3.7.1
N_{Re}	dimensionless Reynolds number, defined in Sec. 3.7.1
$N_{\kappa+\varepsilon}$	dimensionless sum of the characteristic interfacial viscosities, defined in Sec. 3.7.1
$\mathbf{p}(z)$	position vector field, introduced in Sec. A.1.2

p mean pressure, defined in Sec. 2.4.1

P thermodynamic pressure, defined in Sec. 5.8.3

$\mathbf{P}$ projection tensor, defined in Sec. A.3.2

$\mathcal{P}$ modified pressure, defined in Sec. 2.4.1

$\mathbf{Q}$ orthogonal tensor that describes the rotation and (possibly) reflection that takes one frame into another, introduced in Sec. 1.4.1

$\mathcal{Q}$ defined in Sec. 1.4.1

r cylindrical or spherical coordinate

$\mathbf{S}$ viscous portion of stress tensor, defined in Sec. 2.4.1

$\mathbf{S}^{(\sigma)}$ viscous portion of surface stress tensor, defined in Sec. 2.4.2

t time

$\mathbf{T}$ stress tensor

$\mathbf{T}^{(\sigma)}$ surface stress tensor

$\mathbf{v}$ velocity vector

$\mathbf{v}^{(\sigma)}$ surface velocity vector

$v_{\xi}^{(\sigma)}$ speed of displacement, introduced in Sec. 1.2.7

y^{α} surface coordinates ($\alpha = 1, 2$), introduced in Sec. A.2.1

z cylindrical coordinate

$\mathbf{z}$ position vector

greek letters

γ interfacial tension, defined in Sec. 5.8.6

ε surface shear viscosity, introduced in Sec. 2.2.2

θ cylindrical coordinate

κ	surface dilatational viscosity, introduced in Sec. 2.2.2
μ	viscosity
ξ	unit normal to interface
π	3.14158......; disjoining pressure, introduced in Sec. 2.1.3
ρ	mass density
$\rho^{(\sigma)}$	surface mass density
ϕ	potential energy per unit mass, introduced in Sec. 2.4.1; see also Sec. 2.1.3

others

div	divergence operator
$\text{div}_{(\sigma)}$	surface divergence operator, introduced in Secs. A.5.1 through A.5.6
tr	trace operator
$...^t$	denotes history of a function of time, introduced in Sec. 2.2.3
$...^T$	transpose
$...^{-1}$	inverse
$...^*$	denotes an alternative frame of reference
$...^*$	indicates dimensionless quantity
$...^{(i)}$	variable associated with phase i
[...]	jump for dividing surface, defined in Sec. 1.3.4
∇	gradient operator
$\nabla_{(\sigma)}$	surface gradient operation, introduced in Secs. A.5.1 through A.5.6

4
Applications of integral averaging to momentum transfer

In the context of discussing approximations in Sec. 3.1.2, I mentioned that not every interesting problem should be attacked by attempting to find a simultaneous solution to the equation of continuity, Cauchy's first law, the jump mass balance, and the jump momentum balance. Some problems are really too difficult to be solved in this manner. In other cases, the amount of effort required for such a solution is not justified, when the end purpose for which the solution is being developed is taken into account.

In the majority of fluid-mechanics problems, the quantity of ultimate interest is an integral. Perhaps it is a volume flow rate or a force exerted on a surface. This suggests that I set aside an entire chapter in order to exploit approaches to problems in which the independent variables are integrals or integral averages.

I begin by developing the integral balances for arbitrary systems. I conclude by illustrating how variational and extremum principles can be effectively used in solving problems that may be stated in terms of these integral balances. I have chosen not to consider local volume averaging here, since interfacial phenomena are not the dominant features of most of the applications developed to date. I hope to make local volume averaging the subject of a separate discussion.

I encourage those of you who are primarily interested in energy and mass transfer to pay close attention to this chapter. The ideas developed here are taken over, almost without change, and applied to energy and mass transfer in Chapter 7.

References

Slattery, J. C., "Momentum, Energy, and Mass Transfer in Continua," McGraw-Hill, New York (1972); second edition, Robert E. Krieger, Malabar, FL 32950 (1981).

4.1 Integral balances

4.1.1 Introduction By an integral balance, I mean an equation that describes accumulation of mass, momentum, or some other quantity within a system in terms of influx and outflow. The implication is that the system over which the balance is made need not be a collection of material particles. The system might be the mixture of fuel and air within a carburetor or the oil and water within a single pore of a permeable rock. In these two examples, the system will normally not be a material one. In operation, fuel and air will be both entering and leaving the carburetor. During production, the water may displace the oil from the pore.

Integral balances are useful in situations where for various reasons we may wish to make a statement about the system as a whole without worrying about a detailed description of the motions of the fluids within the system. In Secs. 4.1.6 and 4.1.7 we employ integral momentum balances in a qualitative discussion of the relative effects of interfacial tension and the interfacial viscosities during the production of oil. To the extent that we are satisfied with the answers obtained, we are able to avoid a detailed solution of displacement in an irregular capillary.

Since the integral balances are statements concerning arbitrary systems rather than material bodies, our fundamental postulates are not immediately applicable. The mass of an arbitrary system is not conserved. The mass of fuel in a carburetor may vary as a function of time.

However our fundamental postulates imply statements that must be obeyed at every point within the region occupied by the system. At every point within a phase, conservation of mass implies the equation of continuity; at every point on a dividing surface, the jump mass balance; at every point on a common line, the mass balance at a common line. An integral balance is formed by integrating these point statements over the system's region. The result is rearranged into an expression for accumulation by means of the generalized transport theorem (Exercise 1.3.4-1 or Exercise 1.3.7-1).

I will not repeat here my previous development of the integral balances (Slattery 1981). That discussion includes the effects of turbulence, the construction of relevant empiricisms, and a number of simple examples. But it does not include interfacial effects. It is these interfacial effects that are of primary interest to us here and upon which I will focus my attention in the following sections.

Reference

Slattery, J. C., "Momentum, Energy, and Mass Transfer in Continua," McGraw-Hill, New York (1972); second edition, Robert E. Krieger, Malabar, FL 32950 (1981).

4.1.2 Integral mass balance The generalized transport theorem (Exercise 1.3.7-1) gives us an expression for the time rate of change of mass within an arbitrary system:

$$\frac{d}{dt}\left(\int_{R_{(sys)}} \rho \, dV + \int_{\Sigma_{(sys)}} \rho^{(\sigma)} \, dA \right)$$

$$= \int_{R_{(sys)}} \frac{\partial \rho}{\partial t} \, dV + \int_{S_{(sys)}} \rho \mathbf{v}_{(sys)} \cdot \mathbf{n} \, dA + \int_{C_{(sys)}} \rho^{(\sigma)} \mathbf{v}_{(sys)} \cdot \boldsymbol{\mu} \, ds$$

$$+ \int_{\Sigma_{(sys)}} \left(\frac{\partial \rho^{(\sigma)}}{\partial t} - \nabla_{(\sigma)} \rho^{(\sigma)} \cdot \mathbf{u} - 2H\rho^{(\sigma)} v_{(\xi)} - [\, \rho v_{(\xi)} \,] \right) dA$$

$$- \int_{C^{(cl)}_{(sys)}} (\, \rho^{(\sigma)} u^{(cl)}_{(v)} \,) \, ds \qquad (2\text{-}1)$$

Here $R_{(sys)}$ is the region occupied by the system, $S_{(sys)}$ is the closed surface bounding the system, $\Sigma_{(sys)}$ are the dividing surfaces contained within the arbitrary system, $C_{(sys)}$ are the lines formed by the intersection of these dividing surfaces with $S_{(sys)}$, $C^{(cl)}_{(sys)}$ are the common lines contained within the system, $\mathbf{n}$ is the unit vector normal to $S_{(sys)}$ and outwardly directed with respect to the system, and $\boldsymbol{\mu}$ is the unit vector normal to $C_{(sys)}$ that is both tangent and outwardly directed with respect to the dividing surface. On $S_{(sys)}$, $\mathbf{v}_{(sys)} \cdot \mathbf{n}$ is the normal component of the velocity of this boundary; on $C_{(sys)}$, $\mathbf{v}_{(sys)} \cdot \boldsymbol{\mu}$ is the component of the velocity of this curve in the direction $\boldsymbol{\mu}$.

Integrating the equation of continuity (Sec. 1.3.5) over $R_{(sys)}$, we have

$$\int_{R_{(sys)}} \left[\frac{\partial \rho}{\partial t} + \operatorname{div}(\, \rho \mathbf{v} \,) \right] dV$$

$$= \int_{R_{(sys)}} \frac{\partial \rho}{\partial t} \, dV + \int_{S_{(sys)}} \rho \mathbf{v} \cdot \mathbf{n} \, dA - \int_{\Sigma_{(sys)}} [\, \rho \mathbf{v} \cdot \boldsymbol{\xi} \,] \, dA$$

$$= 0 \qquad (2\text{-}2)$$

The integral of the jump mass balance (Sec. 1.3.5) over $\Sigma_{(sys)}$ gives

$$\int_{\Sigma_{(sys)}} \left\{ \frac{\partial \rho^{(\sigma)}}{\partial t} - \nabla_{(\sigma)} \rho^{(\sigma)} \cdot \mathbf{u} + \operatorname{div}_{(\sigma)}(\, \rho^{(\sigma)} \mathbf{v}^{(\sigma)} \,) \right.$$

$$+ [\, \rho(\mathbf{v} \cdot \boldsymbol{\xi} - v_{\{\xi\}})] \Big\} \, dA$$

$$= \int_{\Sigma_{(sys)}} \Big\{ \frac{\partial \rho^{(\sigma)}}{\partial t} - \nabla_{(\sigma)} \rho^{(\sigma)} \cdot \mathbf{u} - 2H \rho^{(\sigma)} v_{\{\xi\}}$$

$$+ [\, \rho(\mathbf{v} \cdot \boldsymbol{\xi} - v_{\{\xi\}})] \Big\} \, dA$$

$$+ \int_{C_{(sys)}} \rho^{(\sigma)} \mathbf{v}^{(\sigma)} \cdot \boldsymbol{\mu} \, ds - \int_{C^{(cl)}_{(sys)}} (\rho^{(\sigma)} \mathbf{v}^{(\sigma)} \cdot \boldsymbol{\nu}) \, ds$$

$$= 0 \tag{2-3}$$

The integral of the mass balance at the common line (Sec. 1.3.8) over $C^{(cl)}_{(sys)}$ says

$$\int_{C^{(cl)}_{(sys)}} (\rho^{(\sigma)} [\, \mathbf{v}^{(\sigma)} \cdot \boldsymbol{\nu} - u^{(cl)}_{(\nu)}]) \, ds = 0 \tag{2-4}$$

Subtracting (2-2) through (2-4) from (2-1), we conclude that

$$\frac{d}{dt} \Big(\int_{R_{(sys)}} \rho \, dV + \int_{\Sigma_{(sys)}} \rho^{(\sigma)} \, dA \Big)$$

$$= \int_{S_{(ent\ ex)}} \rho(\mathbf{v} - \mathbf{v}_{(sys)}) \cdot (- \mathbf{n}) \, dA$$

$$+ \int_{C_{(ent\ ex)}} \rho^{(\sigma)} (\mathbf{v}^{(\sigma)} - \mathbf{v}_{(sys)}) \cdot (- \boldsymbol{\mu}) \, ds \tag{2-5}$$

where $S_{(ent\ ex)}$ and $C_{(ent\ ex)}$ denote the entrance and exit portions of $S_{(sys)}$ and $C_{(sys)}$ respectively. This is known as the **integral mass balance** for an arbitrary system. It says that the time rate of change of mass in the system is equal to the net rate at which mass is brought into the system through the entrances and exits.

For more on the integral mass balance excluding interfacial effects, see

Slattery (1981, p. 219) and Bird *et al.* (1960, p. 209).

References

Bird, R. B., W. E. Stewart, and E. N. Lightfoot, "Transport Phenomena," John Wiley, New York (1960).

Slattery, J. C., "Momentum, Energy, and Mass Transfer in Continua," McGraw-Hill, New York (1972); second edition, Robert E. Krieger, Malabar, FL 32950 (1981).

4.1.3 Integral momentum balance The integral momentum balance should discuss the time rate of change of momentum in an arbitrary system. From the generalized transport theorem (Exercise 1.3.7-1),

$$
\begin{aligned}
\frac{d}{dt}\left(\int_{R_{(sys)}} \rho \mathbf{v} \, dV + \int_{\Sigma_{(sys)}} \rho^{(\sigma)} \mathbf{v}^{(\sigma)} \, dA \right)
&= \int_{R_{(sys)}} \frac{\partial(\rho \mathbf{v})}{\partial t} dV + \int_{\Sigma_{(sys)}} \rho \mathbf{v} (\mathbf{v}_{(sys)} \cdot \mathbf{n}) \, dA \\
&\quad + \int_{C_{(sys)}} \rho^{(\sigma)} \mathbf{v}^{(\sigma)} \mathbf{v}_{(sys)} \cdot \boldsymbol{\mu} \, ds \\
&\quad + \int_{\Sigma_{(sys)}} \left\{ \frac{\partial(\rho^{(\sigma)} \mathbf{v}^{(\sigma)})}{\partial t} - \nabla_{(\sigma)}(\rho^{(\sigma)} \mathbf{v}^{(\sigma)}) \cdot \mathbf{u} \right. \\
&\qquad \left. - 2H\rho^{(\sigma)} \mathbf{v}^{(\sigma)} v_{\{\xi\}} - [\rho \mathbf{v} \, v_{\{\xi\}}] \right\} dA \\
&\quad - \int_{C_{(sys)}^{(cl)}} (\rho^{(\sigma)} \mathbf{v}^{(\sigma)} u_{\{v\}}^{(cl)}) \, ds
\end{aligned}
\tag{3-1}
$$

Integrating Cauchy's first law (Sec. 2.1.4) over $R_{(sys)}$, we find

$$\int_{R_{(sys)}} \left[\frac{\partial(\rho \mathbf{v})}{\partial t} + \text{div}(\rho \mathbf{v}\mathbf{v}) - \text{div}\, \mathbf{T} - \rho \mathbf{b} \right] dV$$

$$= \int_{R_{(sys)}} \left[\frac{\partial(\rho \mathbf{v})}{\partial t} - \rho \mathbf{b} \right] dV + \int_{S_{(sys)}} [\rho \mathbf{v}(\mathbf{v} \cdot \mathbf{n}) - \mathbf{T} \cdot \mathbf{n}]\, dA$$

$$- \int_{\Sigma_{(sys)}} [\rho \mathbf{v}(\mathbf{v} \cdot \boldsymbol{\xi}) - \mathbf{T} \cdot \boldsymbol{\xi}]\, dA$$

$$= 0 \tag{3-2}$$

From the integral of the jump momentum balance (Sec. 2.1.6) over $\Sigma_{(sys)}$, we see

$$\int_{\Sigma_{(sys)}} \left\{ \frac{\partial(\rho^{(\sigma)} \mathbf{v}^{(\sigma)})}{\partial t} - \nabla_{(\sigma)}(\rho^{(\sigma)} \mathbf{v}^{(\sigma)}) \cdot \mathbf{u} + \text{div}_{(\sigma)}(\rho^{(\sigma)} \mathbf{v}^{(\sigma)} \mathbf{v}^{(\sigma)}) \right.$$

$$- \text{div}_{(\sigma)} \mathbf{T}^{(\sigma)} - \rho^{(\sigma)} \mathbf{b}^{(\sigma)}$$

$$\left. + [\rho \mathbf{v}(\mathbf{v} \cdot \boldsymbol{\xi} - v_{\{\xi\}}) - \mathbf{T} \cdot \boldsymbol{\xi}] \right\} dA$$

$$= \int_{\Sigma_{(sys)}} \left\{ \frac{\partial(\rho^{(\sigma)} \mathbf{v}^{(\sigma)})}{\partial t} - \nabla_{(\sigma)}(\rho^{(\sigma)} \mathbf{v}^{(\sigma)}) \cdot \mathbf{u} - 2H \rho^{(\sigma)} \mathbf{v}^{(\sigma)} v_{\{\xi\}} \right.$$

$$\left. - \rho^{(\sigma)} \mathbf{b}^{(\sigma)} + [\rho \mathbf{v}(\mathbf{v} \cdot \boldsymbol{\xi} - v_{\{\xi\}}) - \mathbf{T} \cdot \boldsymbol{\xi}] \right\} dA$$

$$+ \int_{C_{(sys)}} (\rho^{(\sigma)} \mathbf{v}^{(\sigma)} \mathbf{v}^{(\sigma)} - \mathbf{T}^{(\sigma)}) \cdot \boldsymbol{\mu}\, ds$$

$$- \int_{C_{(sys)}^{(cl)}} ([\rho^{(\sigma)} \mathbf{v}^{(\sigma)} \mathbf{v}^{(\sigma)} - \mathbf{T}^{(\sigma)}] \cdot \boldsymbol{\nu})\, ds$$

$$= 0 \tag{3-3}$$

The integral of the momentum balance at the common line (Sec. 2.1.9) requires

$$\int_{C^{(cl)}_{(sys)}} \left(\rho^{(\sigma)} \mathbf{v}^{(\sigma)} [\, \mathbf{v}^{(\sigma)} \cdot \boldsymbol{\nu} - u^{(cl)}_{(\nu)} \,] - \mathbf{T}^{(\sigma)} \cdot \boldsymbol{\nu} \right) ds = 0 \tag{3-4}$$

Subtracting (3-2) through (3-4) from (3-1), we conclude that

$$\frac{d}{dt}\left(\int_{R_{(sys)}} \rho \mathbf{v} \, dV + \int_{\Sigma_{(sys)}} \rho^{(\sigma)} \mathbf{v}^{(\sigma)} \, dA \right)$$

$$= \int_{R_{(sys)}} \rho \mathbf{b} \, dV + \int_{\Sigma_{(sys)}} \rho^{(\sigma)} \mathbf{b}^{(\sigma)} \, dA$$

$$+ \int_{S_{(sys)}} [\, \rho \mathbf{v}(\, \mathbf{v} - \mathbf{v}_{(sys)} \,) - \mathbf{T} \,] \cdot (\, -\mathbf{n} \,) \, dA$$

$$+ \int_{C_{(sys)}} [\, \rho^{(\sigma)} \mathbf{v}^{(\sigma)} (\, \mathbf{v}^{(\sigma)} - \mathbf{v}_{(sys)} \,) - \mathbf{T}^{(\sigma)} \,] \cdot (\, -\boldsymbol{\mu} \,) \, ds \tag{3-5}$$

which is one form of the **integral momentum balance**. It says that the time rate of change of momentum in the system is equal to the net rate at which momentum is brought into the system with material crossing the boundaries plus the sum of the forces acting on the system. Its physical meaning may be clarified by writing (3-5) as

$$\frac{d}{dt}\left(\int_{R_{(sys)}} \rho \mathbf{v} \, dV + \int_{\Sigma_{(sys)}} \rho^{(\sigma)} \mathbf{v}^{(\sigma)} \, dA \right)$$

$$= \int_{S_{(ent\ ex)}} \rho \mathbf{v}(\, \mathbf{v} - \mathbf{v}_{(sys)} \,) \cdot (\, -\mathbf{n} \,) \, dA$$

$$+ \int_{C_{(ent\ ex)}} \rho^{(\sigma)} \mathbf{v}^{(\sigma)} (\, \mathbf{v}^{(\sigma)} - \mathbf{v}_{(sys)} \,) \cdot (\, -\boldsymbol{\mu} \,) \, ds$$

$$- \mathcal{F} + \int_{R_{(sys)}} \rho \mathbf{b} \, dV + \int_{\Sigma_{(sys)}} \rho^{(\sigma)} \mathbf{b}^{(\sigma)} \, dA$$

$$+ \int_{S_{(ent\ ex)}} (\mathbf{T} + p_0 \mathbf{I}) \cdot \mathbf{n} \, dA + \int_{C_{(ent\ ex)}} \mathbf{T}^{(\sigma)} \cdot \boldsymbol{\mu} \, ds \qquad (3\text{-}6)$$

where

$$\mathcal{F} \equiv \int_{S_{(sys)} - S_{(ent\ ex)}} (\mathbf{T} + p_0 \mathbf{I}) \cdot (-\mathbf{n}) \, dA$$

$$+ \int_{C_{(sys)} - C_{(ent\ ex)}} \mathbf{T}^{(\sigma)} \cdot (-\boldsymbol{\mu}) \, ds \qquad (3\text{-}7)$$

denotes the **force that the system exerts upon the impermeable portion of its bounding surface** beyond the force attributable to the ambient or reference pressure p_0. The first two integrals on the right describe the net rate at which momentum is brought into the system at the entrance and exit surfaces $S_{(ent\ ex)}$; the fourth and fifth terms, the external force acting on the system; the last two terms, viscous and interfacial forces acting at $S_{(ent\ ex)}$.

For more on the integral momentum balance excluding interfacial effects, see Slattery (1981, p. 225) and Bird *et al.* (1960, p. 210).

References

Bird, R. B., W. E. Stewart, and E. N. Lightfoot, "Transport Phenomena," John Wiley, New York (1960).

Slattery, J. C., "Momentum, Energy, and Mass Transfer in Continua," McGraw-Hill, New York (1972); second edition, Robert E. Krieger, Malabar, FL 32950 (1981).

Washburn, E. W., *Phys. Rev.* **17**, 273 (1921).

West, G. D., *Proc. R. Soc. London A* **86**, 20 (1912).

Exercise

4.1.3-1 *Washburn equation* (West 1912; Washburn 1921) Let us consider the displacement of phase 2 by phase 1 in the straight capillary shown in Figure 3-1. Let us assume:

a) The bulk phases can be described as incompressible Newtonian fluids.

b) The entrance and exit surfaces denoted by the dashed lines in Figure 3-1, are moving with the speed of displacement of the interface. They are located sufficiently far from the interface that we have Poiseuille flow (Slattery 1981, p. 73) at these cross sections.

c) The displacement is stable in the sense that, in a frame of reference fixed with respect to the common line, the flow is independent of time.

d) The interfacial stress can be represented by an interfacial tension γ that is independent of position,

$$\mathbf{T}^{(\sigma)} = \gamma \mathbf{P}$$

e) The effect of inertia can be neglected.

f) The effects of gravity are ignored.

For a frame of reference in which the tube is stationary, consider the system enclosed by the dashed lines in Figure 3-1 that moves with the interface. Include within this system the moving common line and the solid-fluid interfaces.

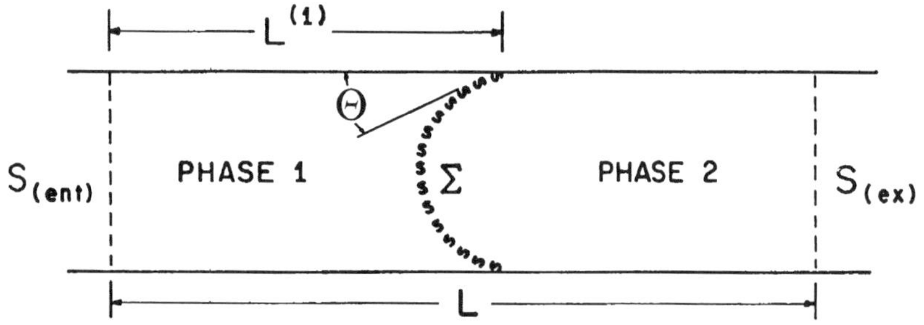

Figure 4.1.3-1 Displacement of phase 2 by phase 1. In a frame of reference that is fixed with respect to the tube, the entrance $S_{(ent)}$ and the exit $S_{(ex)}$ for the system are stationary; the interface Σ moves from left to right. The contact angle Θ is measured through phase 1.

i) Deduce from the axial component of the integral momentum balance that

$$(\mathcal{P}_{(ent)} - \mathcal{P}_{(ex)})\pi R^2 = \mathcal{F}_z$$

where

$$\mathcal{F}_z = - \int_{S_{(wall)}} S_{rz} \, dA - 2\pi R \gamma \cos \Theta$$

R is the radius of the tube, Θ is the contact angle measured through phase 1, and $S_{(wall)}$ is that portion of the tube wall bounding the system.

ii) Use the Poiseuille velocity distribution to approximate the axial component of the force that phases 1 and 2 exert on the wall of the capillary. Conclude that the speed of displacement of the interface

$$V = \frac{1}{\mu^{(2)}} \left[\Delta p + \frac{2\gamma}{R} \cos \Theta \right] \left[\frac{8}{R} \left(N_{\mu} \frac{L^{(1)}}{R} + \frac{L - L^{(1)}}{R} \right) \right]^{-1}$$

where

$$N_{\mu} \equiv \frac{\mu^{(1)}}{\mu^{(2)}}$$

$$\Delta p \equiv p_{(ent)} - p_{(ex)}$$

This is usually referred to as the Washburn equation (West 1912; Washburn 1921).

For a different derivation of the Washburn equation including the effects of gravity and for a discussion leading to its experimental verification, see Exercise 4.1.9-1.

4.1.4 Integral mechanical energy balance The integral mechanical energy balance should give us the time rate of change of kinetic and potential energy in an arbitrary system. From the generalized transport theorem (Exercise 1.3.7-1)

$$\frac{d}{dt} \left[\int_{R_{(sys)}} \rho \left(\frac{1}{2} v^2 + \phi \right) dV + \int_{\Sigma_{(sys)}} \rho^{(\sigma)} \left(\frac{1}{2} v^{(\sigma)2} + \phi \right) dA \right]$$

$$= \int_{R_{(sys)}} \frac{\partial}{\partial t} \left[\rho \left(\frac{1}{2} v^2 + \phi \right) \right] dV$$

$$+ \int_{S_{(sys)}} \rho \left(\frac{1}{2} v^2 + \phi \right) (\mathbf{v}_{(sys)} \cdot \mathbf{n}) \, dA$$

$$+ \int_{C_{(sys)}} \rho^{(\sigma)} \left(\frac{1}{2} v^{(\sigma)2} + \phi \right) (\mathbf{v}_{(sys)} \cdot \boldsymbol{\mu}) \, ds$$

$$+ \int_{\Sigma_{(sys)}} \left\{ \frac{\partial}{\partial t} \left[\rho^{(\sigma)} \left(\frac{1}{2} v^{(\sigma)2} + \phi \right) \right] \right.$$

$$- \nabla_{(\sigma)} \left[\rho^{(\sigma)} \left(\frac{1}{2} v^{(\sigma)2} + \phi \right) \right] \cdot \mathbf{u}$$

$$- 2H\rho^{(\sigma)} \left(\frac{1}{2} v^{(\sigma)2} + \phi \right) v_{\{\xi\}}^{\{\xi\}}$$

$$\left. - \left[\rho^{(\sigma)} \left(\frac{1}{2} v^{(\sigma)2} + \phi \right) v_{\{\xi\}}^{\{\xi\}} \right] \right\} dA$$

$$- \int_{C^{(cl)}_{(sys)}} \left[\rho^{(\sigma)} \left(\frac{1}{2} v^{(\sigma)2} + \phi \right) u^{(cl)}_{(v)} \right] ds \tag{4-1}$$

where ϕ is the potential energy per unit mass in terms of which

$$\mathbf{b} = - \nabla\phi \tag{4-2}$$

The scalar product of Cauchy's first law (Sec 2.1.5) with velocity $\mathbf{v}$

$$\mathbf{v} \cdot \left(\rho \frac{d_{(m)}\mathbf{v}}{dt} - \text{div } \mathbf{T} - \rho \mathbf{b} \right) = 0 \tag{4-3}$$

can be integrated over $R_{(sys)}$ to find (Exercise 4.1.4-1):

$$\int_{R_{(sys)}} \mathbf{v} \cdot \left(\rho \frac{d_{(m)}\mathbf{v}}{dt} - \text{div } \mathbf{T} - \rho \mathbf{b} \right) dV$$

$$= \int_{R_{(sys)}} \left\{ \frac{\partial}{\partial t}\left[\rho\left(\frac{1}{2} v^2 + \phi \right)\right] - P \operatorname{div} \mathbf{v} + \operatorname{tr}(\mathbf{S} \cdot \nabla \mathbf{v}) \right\} dV$$

$$+ \int_{S_{(sys)}} \left[\rho\left(\frac{1}{2} v^2 + \phi \right)(\mathbf{v} \cdot \mathbf{n}) - \mathbf{v} \cdot (\mathbf{T} \cdot \mathbf{n})\right] dA$$

$$- \int_{\Sigma_{(sys)}} \left[\rho\left(\frac{1}{2} v^2 + \phi \right)(\mathbf{v} \cdot \boldsymbol{\xi}) - \mathbf{v} \cdot (\mathbf{T} \cdot \boldsymbol{\xi})\right] dA$$

$$= 0 \tag{4-4}$$

Here

$$\mathbf{S} \equiv \mathbf{T} + P\mathbf{I} \tag{4-5}$$

is the extra stress or viscous portion of the stress tensor. The scalar product of the jump momentum balance (Sec 2.1.7) with surface velocity $\mathbf{v}^{(\sigma)}$

$$\mathbf{v}^{(\sigma)} \cdot \left\{ \rho^{(\sigma)} \frac{d_{(s)} \mathbf{v}^{(\sigma)}}{dt} - \operatorname{div}_{(\sigma)} \mathbf{T}^{(\sigma)} - \rho^{(\sigma)} \mathbf{b}^{(\sigma)} \right.$$

$$\left. + \left[\rho(\mathbf{v} - \mathbf{v}^{(\sigma)})(\mathbf{v} \cdot \boldsymbol{\xi} - v_{\{\xi\}}^{\{g\}}) - \mathbf{T} \cdot \boldsymbol{\xi} \right] \right\}$$

$$= 0 \tag{4-6}$$

can be integrated over $\Sigma_{(sys)}$ to learn (Exercise 4.1.4-2):

$$\int_{\Sigma_{(sys)}} \left\{ \frac{\partial}{\partial t}\left[\rho^{(\sigma)}\left(\frac{1}{2} v^{(\sigma)2} + \phi \right)\right] - \nabla_{(\sigma)}\left[\rho^{(\sigma)}\left(\frac{1}{2} v^{(\sigma)2} + \phi \right)\right] \cdot \mathbf{u} \right.$$

$$- 2H\rho^{(\sigma)}\left(\frac{1}{2} v^{(\sigma)2} + \phi \right) v_{\{\xi\}}^{\{g\}}$$

$$+ \gamma \operatorname{div}_{(\sigma)} \mathbf{v}^{(\sigma)} + \operatorname{tr}(\mathbf{S}^{(\sigma)} \cdot \nabla_{(\sigma)} \mathbf{v}^{(\sigma)})$$

$$+ \left[\rho\left(\frac{1}{2} v^{(\sigma)2} + \phi \right)(\mathbf{v} \cdot \boldsymbol{\xi} - v_{\{\xi\}}^{\{g\}})\right]$$

$$+ [\, \rho \mathbf{v}^{(\sigma)} \cdot (\mathbf{v} - \mathbf{v}^{(\sigma)})(\mathbf{v} \cdot \boldsymbol{\xi} - v_{\{\xi\}}^{\{g\}})$$

$$- \mathbf{v}^{(\sigma)} \cdot \mathbf{T} \cdot \boldsymbol{\xi} \,] \Big\} \, dA$$

$$+ \int_{C_{(sys)}} \Big[\rho^{(\sigma)} \Big(\frac{1}{2} v^{(\sigma)2} + \phi \Big) (\mathbf{v}_{(sys)} \cdot \boldsymbol{\mu})$$

$$- \mathbf{v}_{(sys)} \cdot \mathbf{T}^{(\sigma)} \cdot \boldsymbol{\mu} \Big] \, ds$$

$$- \int_{C^{(cl)}_{(sys)}} \Big(\rho^{(\sigma)} \Big[\frac{1}{2} v^{(\sigma)2} + \phi \Big] [\mathbf{v}^{(\sigma)} \cdot \boldsymbol{\nu}]$$

$$- \mathbf{v}_{(sys)} \cdot \mathbf{T}^{(\sigma)} \cdot \boldsymbol{\nu} \Big) \, ds$$

$$= 0 \tag{4-7}$$

In arriving at this result, I have said that the same external force acts on both the bulk phases and the dividing surface

$$\mathbf{b}^{(\sigma)} = \mathbf{b} = -\nabla\phi \tag{4-8}$$

and I have introduced

$$\mathbf{S}^{(\sigma)} \equiv \mathbf{T}^{(\sigma)} - \gamma \mathbf{P} \tag{4-9}$$

as the extra surface stress tensor or the viscous portion of the surface stress tensor.

Subtracting (4-4) and (4-7) from (4-1), we conclude that

$$\frac{d}{dt} \Big[\int_{R_{(sys)}} \rho \Big(\frac{1}{2} v^2 + \phi \Big) \, dV + \int_{\Sigma_{(sys)}} \rho^{(\sigma)} \Big(\frac{1}{2} v^{(\sigma)2} + \phi \Big) \, dA \Big]$$

$$= \int_{R_{(sys)}} \Big[P \operatorname{div} \mathbf{v} - \operatorname{tr}(\mathbf{S} \cdot \nabla \mathbf{v}) \Big] \, dV$$

$$+ \int_{S_{(sys)}} \Big[- \rho \Big(\frac{1}{2} v^2 + \phi \Big) (\mathbf{v} - \mathbf{v}_{(sys)}) \cdot \mathbf{n} + \mathbf{v} \cdot (\mathbf{T} \cdot \mathbf{n}) \Big] \, dA$$

$$+ \int_{C_{(sys)}} \Big[- \rho^{(\sigma)} \Big(\frac{1}{2} v^{(\sigma)2} + \phi \Big) (\mathbf{v}^{(\sigma)} - \mathbf{v}_{(sys)}) \cdot \boldsymbol{\mu}$$

$$+ \mathbf{v}^{(\sigma)} \cdot \mathbf{T}^{(\sigma)} \cdot \boldsymbol{\mu} \Big] \, ds$$

$$+ \int_{\Sigma_{(sys)}} \Big\{ \Big[\frac{1}{2} \rho \, | \, \mathbf{v} - \mathbf{v}^{(\sigma)} |^2 \, (\mathbf{v} \cdot \boldsymbol{\xi} - v_{\{\xi\}}^{\{g\}})$$

$$- (\mathbf{v} - \mathbf{v}^{(\sigma)}) \cdot \mathbf{T} \cdot \boldsymbol{\xi} \Big] - \gamma \, \text{div}_{(\sigma)} \mathbf{v}^{(\sigma)}$$

$$- \text{tr}(\mathbf{S}^{(\sigma)} \cdot \nabla_{(\sigma)} \mathbf{v}^{(\sigma)}) \Big\} \, dA$$

$$+ \int_{C_{(sys)}^{(cl)}} \Big(\frac{1}{2} \rho^{(\sigma)} v^{(\sigma)2} [\, \mathbf{v}^{(\sigma)} \cdot \boldsymbol{\nu} - u_{(\nu)}^{(cl)} \,]$$

$$- \mathbf{v}^{(\sigma)} \cdot \mathbf{T}^{(\sigma)} \cdot \boldsymbol{\nu} \Big) \, ds \tag{4-10}$$

which is one form of the **integral mechanical energy balance.** Its physical meaning becomes clearer when it is rearranged as

$$\frac{d}{dt} \Big[\int_{R_{(sys)}} \rho \Big(\frac{1}{2} v^2 + \phi \Big) \, dV + \int_{\Sigma_{(sys)}} \rho^{(\sigma)} \Big(\frac{1}{2} v^{(\sigma)2} + \phi \Big) \, dA \Big]$$

$$= \int_{S_{(ent\ ex)}} \rho \Big(\frac{1}{2} v^2 + \phi + \frac{P - p_0}{\rho} \Big) (\mathbf{v} - \mathbf{v}_{(sys)}) \cdot (- \mathbf{n}) \, dA$$

$$+ \int_{C_{(ent\ ex)}} \rho^{(\sigma)} \Big(\frac{1}{2} v^{(\sigma)2} + \phi - \frac{\gamma}{\rho^{(\sigma)}} \Big) (\mathbf{v}^{(\sigma)} - \mathbf{v}_{(sys)})$$

$$\cdot (- \boldsymbol{\mu}) \, ds$$

$$- \mathcal{W} - \mathcal{E} + \int_{R_{(sys)}} (P - p_0) \, \text{div} \, \mathbf{v} \, dV$$

$$+\int_{S_{(\text{ent ex})}} [-(P-p_0)(\mathbf{v}_{(\text{sys})}\cdot\mathbf{n}) + \mathbf{v}\cdot\mathbf{S}\cdot\mathbf{n}]\, dA$$

$$+\int_{C_{(\text{ent ex})}} [\gamma(\mathbf{v}_{(\text{sys})}\cdot\boldsymbol{\mu}) + \mathbf{v}^{(\sigma)}\cdot\mathbf{S}^{(\sigma)}\cdot\boldsymbol{\mu}]\, ds$$

$$+\int_{\Sigma_{(\text{sys})}} \Big\{\Big[\frac{1}{2}\rho\,|\mathbf{v}-\mathbf{v}^{(\sigma)}|^2 (\mathbf{v}\cdot\boldsymbol{\xi} - v_{(\xi)}^{\{g\}})$$

$$-(\mathbf{v}-\mathbf{v}^{(\sigma)})\cdot(\mathbf{T}+p_0\mathbf{I})\cdot\boldsymbol{\xi}\Big]$$

$$-\gamma\,\text{div}_{(\sigma)}\mathbf{v}^{(\sigma)}\Big\}\, dA$$

$$+\int_{C_{(\text{sys})}^{(\text{cl})}} \Big(\frac{1}{2}\rho^{(\sigma)}v^{(\sigma)2}\,[\mathbf{v}^{(\sigma)}\cdot\boldsymbol{\nu} - u_{(\nu)}^{\{\text{cl}\}}]$$

$$-\mathbf{v}^{(\sigma)}\cdot\mathbf{T}^{(\sigma)}\cdot\boldsymbol{\nu}\Big)\, ds \qquad (4\text{-}11)$$

with the understanding that

$$\mathcal{W} \equiv \int_{S_{(\text{sys})}-S_{(\text{ent ex})}} \mathbf{v}\cdot(\mathbf{T}+p_0\mathbf{I})\cdot(-\mathbf{n})\, dA$$

$$+\int_{C_{(\text{sys})}-C_{(\text{ent ex})}} \mathbf{v}^{(\sigma)}\cdot\mathbf{T}^{(\sigma)}\cdot(-\boldsymbol{\mu})\, ds \qquad (4\text{-}12)$$

denotes the **rate at which work is done by the system on the surroundings** at the moving impermeable surfaces of the system (beyond any work done on these surfaces by the ambient pressure p_0) and

$$\mathcal{E} \equiv \int_{R_{(\text{sys})}} \text{tr}(\mathbf{S}\cdot\nabla\mathbf{v})\, dV + \int_{\Sigma_{(\text{sys})}} \text{tr}(\mathbf{S}^{(\sigma)}\cdot\nabla_{(\sigma)}\mathbf{v}^{(\sigma)})\, dA \qquad (4\text{-}13)$$

is the **rate at which mechanical energy is dissipated** by the action of viscous forces both within the bulk phases and in the dividing surfaces. On the right side of (4-11)

$$\int_{S_{(ent\ ex)}} \rho\left(\frac{1}{2}v^2 + \phi\right)(\mathbf{v} - \mathbf{v}_{(sys)}) \cdot (-\mathbf{n})\, dA$$

$$+ \int_{C_{(ent\ ex)}} \rho^{(\sigma)}\left(\frac{1}{2}v^{(\sigma)2} + \phi\right)(\mathbf{v}^{(\sigma)} - \mathbf{v}_{(sys)}) \cdot (-\boldsymbol{\mu})\, ds$$

is the net rate at which kinetic and potential energy is brought into the system with any material that moves across the boundary. By

$$\int_{S_{(ent\ ex)}} \rho\left(\frac{P - p_0}{\rho}\right)(\mathbf{v} - \mathbf{v}_{(sys)}) \cdot (-\mathbf{n})\, dA$$

$$+ \int_{C_{(ent\ ex)}} \rho^{(\sigma)}\left(-\frac{\gamma}{\rho^{(\sigma)}}\right)(\mathbf{v}^{(\sigma)} - \mathbf{v}_{(sys)}) \cdot (-\boldsymbol{\mu})\, ds$$

$$+ \int_{S_{(ent\ ex)}} -(P - p_0)(\mathbf{v}_{(sys)} \cdot \mathbf{n})\, dA$$

$$+ \int_{C_{(ent\ ex)}} \gamma(\mathbf{v}_{(sys)} \cdot \boldsymbol{\mu})\, ds$$

$$= -\int_{S_{(ent\ ex)}} (P - p_0)(\mathbf{v} \cdot \mathbf{n})\, dA + \int_{C_{(ent\ ex)}} \gamma(\mathbf{v}^{(\sigma)} \cdot \boldsymbol{\mu})\, ds$$

I mean the net rate at which pressure (beyond the reference or ambient pressure p_0) and surface tension does work on the system at the entrances and exits.

Often we will be willing to neglect

$$\int_{R_{(sys)}} (P - p_0)\, \text{div}\, \mathbf{v}\, dV$$

the work done by pressure (beyond p_0) in dilating the bulk phases,

$$\int_{\Sigma_{(sys)}} \gamma\, \text{div}_{(\sigma)} \mathbf{v}^{(\sigma)}\, dA$$

the work done by surface tension in dilating the dividing surfaces,

$$\int_{S_{(ent\ ex)}} \mathbf{v} \cdot \mathbf{S} \cdot \mathbf{n}\ dA + \int_{C_{(ent\ ex)}} \mathbf{v}^{(\sigma)} \cdot \mathbf{S}^{(\sigma)} \cdot \boldsymbol{\mu}\ ds$$

the work done on the system at the entrances and exits by bulk and surface viscous forces,

$$\int_{\Sigma_{(sys)}} \Big[\frac{1}{2} \rho \,|\, \mathbf{v} - \mathbf{v}^{(\sigma)} \,|^2 \,(\mathbf{v} \cdot \boldsymbol{\xi} - v_{\{\xi\}}^{\{\xi\}}) - (\mathbf{v} - \mathbf{v}^{(\sigma)}) \cdot (\mathbf{T} + p_0 \mathbf{I}) \cdot \boldsymbol{\xi} \Big]\, dA$$

the net rate at which kinetic energy is created and work is done as the result of mass transfer at the dividing surfaces $\Sigma_{(sys)}$, and

$$\int_{C^{(cl)}_{(sys)}} \Big(\frac{1}{2} \rho^{(\sigma)} v^{(\sigma)2} (\mathbf{v}^{(\sigma)} \cdot \boldsymbol{\nu} - u^{\{cl\}}_{\{\nu\}}) - \mathbf{v}^{(\sigma)} \cdot \mathbf{T}^{(\sigma)} \cdot \boldsymbol{\nu} \Big)\, ds$$

the rate at which kinetic energy is created and work is done at the common lines $C^{(cl)}_{(sys)}$. Under these circumstances, (4-11) reduces to

$$\frac{d}{dt} \Big[\int_{R_{(sys)}} \rho \Big(\frac{1}{2} v^2 + \phi \Big)\, dV + \int_{\Sigma_{(sys)}} \rho^{(\sigma)} \Big(\frac{1}{2} v^{(\sigma)2} + \phi \Big)\, dA \Big]$$

$$= \int_{S_{(ent\ ex)}} \rho \Big(\frac{1}{2} v^2 + \phi + \frac{P - p_0}{\rho} \Big) (\mathbf{v} - \mathbf{v}_{(sys)}) \cdot (- \mathbf{n})\, dA$$

$$+ \int_{C_{(ent\ ex)}} \rho^{(\sigma)} \Big(\frac{1}{2} v^{(\sigma)2} + \phi - \frac{\gamma}{\rho^{(\sigma)}} \Big) (\mathbf{v}^{(\sigma)} - \mathbf{v}_{(sys)})$$

$$\cdot (- \boldsymbol{\mu})\, ds$$

$$- \mathcal{W} - \mathcal{E} + \int_{S_{(ent\ ex)}} [- (P - p_0) \mathbf{v}_{(sys)} \cdot \mathbf{n}]\, dV$$

$$+ \int_{C_{(ent\ ex)}} \gamma\, \mathbf{v}_{(sys)} \cdot \boldsymbol{\mu}\, ds \tag{4-14}$$

For more on the integral mechanical energy balance excluding interfacial effects, see Slattery (1981, pp. 238 and 436) and Bird *et al.* (1960, p. 211).

References

Bird, R. B., W. E. Stewart, and E. N. Lightfoot, "Transport Phenomena," John Wiley, New York (1960).

Israelachvili, J. N., *J. Chem. Soc. Faraday Trans. II* **69**, 1729 (1973).

Israelachvili, J. N., "Intermolecular and Surface Forces," p. 157, Academic Press (1985).

Jasper, J. J., and E. V. Kring, *J. Phys. Chem.* **59**, 1019 (1955).

Hough, D. B., and L. R. White, *Adv. Colloid Interface Sci.* **14**, 3 (1980).

Padday, J. F., and N. D. Uffindell, *J. Phys. Chem.* **72**, 1407 (1968).

Richards, T. W., and E. K. Carver, *J. Am. Chem. Soc.* **43**, 827 (1921).

Slattery, J. C., "Momentum, Energy, and Mass Transfer in Continua," McGraw-Hill, New York (1972); second edition, Robert E. Krieger, Malabar, FL 32950 (1981).

Exercises

4.1.4-1 *derivation of (4-4)* As steps in deriving (4-4), prove that

i) $$\int_{R_{(sys)}} \rho \mathbf{v} \cdot \frac{d_{(m)}\mathbf{v}}{dt}\, dV$$

$$= \int_{R_{(sys)}} \frac{\partial}{\partial t}\left(\frac{1}{2} \rho v^2 \right) dV + \int_{S_{(sys)}} \frac{1}{2} \rho v^2 (\, \mathbf{v} \cdot \mathbf{n}\,)\, dA$$

$$- \int_{\Sigma_{(sys)}} \left[\frac{1}{2} \rho v^2 (\mathbf{v} \cdot \boldsymbol{\xi}) \right] dA \qquad (1)$$

ii) $$- \int_{R_{(sys)}} \mathbf{v} \cdot \operatorname{div} \mathbf{T} \, dV$$

$$= \int_{R_{(sys)}} \left[- P \operatorname{div} \mathbf{v} + \operatorname{tr}(\mathbf{S} \cdot \nabla \mathbf{v}) \right] dV$$

$$- \int_{S_{(sys)}} \mathbf{v} \cdot (\mathbf{T} \cdot \mathbf{n}) \, dA + \int_{S_{(sys)}} [\mathbf{v} \cdot (\mathbf{T} \cdot \boldsymbol{\xi})] \, dA \qquad (2)$$

iii) $$- \int_{R_{(sys)}} \rho \mathbf{v} \cdot \mathbf{b} \, dV$$

$$= \int_{R_{(sys)}} \frac{\partial (\rho \phi)}{\partial t} dV + \int_{S_{(sys)}} \rho \phi (\mathbf{v} \cdot \mathbf{n}) \, dA$$

$$- \int_{\Sigma_{(sys)}} [\rho \phi (\mathbf{v} \cdot \boldsymbol{\xi})] \, dA \qquad (3)$$

4.1.4-2 *derivation of (4-7)* In deriving (4-7), first prove that

i) $$\int_{\Sigma_{(sys)}} \rho^{(\sigma)} \mathbf{v}^{(\sigma)} \cdot \frac{d_{(s)} \mathbf{v}^{(\sigma)}}{dt} dA$$

$$= \int_{\Sigma_{(sys)}} \left\{ \frac{\partial}{\partial t} \left(\frac{1}{2} \rho^{(\sigma)} \mathbf{v}^{(\sigma)2} \right) - \nabla_{(\sigma)} \left(\frac{1}{2} \rho^{(\sigma)} \mathbf{v}^{(\sigma)2} \right) \cdot \mathbf{u} \right.$$

$$+ \frac{1}{2} \mathbf{v}^{(\sigma)2} \, [\rho (\mathbf{v} \cdot \boldsymbol{\xi} - v_{(\xi)})]$$

$$\left. - H \rho^{(\sigma)} \mathbf{v}^{(\sigma)2} v_{(\xi)} \right\} dA$$

$$+ \int_{C_{(sys)}} \frac{1}{2}\rho^{(\sigma)}\mathbf{v}^{(\sigma)2}\ \mathbf{v}^{(\sigma)} \cdot \boldsymbol{\mu}\ ds$$

$$- \int_{C^{(cl)}_{(sys)}} \left(\frac{1}{2}\rho^{(\sigma)}\mathbf{v}^{(\sigma)2}\ \mathbf{v}^{(\sigma)} \cdot \boldsymbol{\nu}\right) ds \tag{1}$$

ii) $$\int_{\Sigma_{(sys)}} \mathbf{v}^{(\sigma)} \cdot \operatorname{div}_{(\sigma)}\mathbf{T}^{(\sigma)}\ dA$$

$$= \int_{\Sigma_{(sys)}} [\ \gamma \operatorname{div}_{(\sigma)}\mathbf{v}^{(\sigma)} + \operatorname{tr}(\ \mathbf{S}^{(\sigma)} \cdot \nabla_{(\sigma)}\mathbf{v}^{(\sigma)}\ )]\ dA$$

$$- \int_{C_{(sys)}} \mathbf{v}^{(\sigma)} \cdot (\ \mathbf{T}^{(\sigma)} \cdot \boldsymbol{\mu}\ )\ ds$$

$$+ \int_{C^{(cl)}_{(sys)}} (\ \mathbf{v}^{(\sigma)} \cdot [\ \mathbf{T}^{(\sigma)} \cdot \boldsymbol{\nu}\ ])\ ds \tag{2}$$

iii) $$- \int_{\Sigma_{(sys)}} \rho^{(\sigma)}\mathbf{v}^{(\sigma)} \cdot \mathbf{b}^{(\sigma)}\ dA$$

$$= \int_{\Sigma_{(sys)}} \left\{ \frac{\partial(\ \rho^{(\sigma)}\phi\ )}{\partial t} - \phi\nabla_{(\sigma)}\rho^{(\sigma)} \cdot \mathbf{u} - \rho^{(\sigma)}\frac{\partial\phi}{\partial t} \right.$$

$$+ \rho^{(\sigma)}(\ \nabla\phi \cdot \boldsymbol{\xi}\ )v_{(\xi)} - 2H\rho^{(\sigma)}\phi v_{(\xi)}$$

$$\left. + [\ \rho\phi(\ \mathbf{v} \cdot \boldsymbol{\xi} - v_{(\xi)}\ )] \right\} dA$$

$$+ \int_{C_{(sys)}} \rho^{(\sigma)}\phi(\ \mathbf{v}^{(\sigma)} \cdot \boldsymbol{\mu}\ )\ ds$$

$$- \int_{C^{(cl)}_{(sys)}} (\ \rho^{(\sigma)}\phi[\ \mathbf{v}^{(\sigma)} \cdot \boldsymbol{\nu}\ ])\ ds \tag{3}$$

iv) $$\left(\frac{\partial \phi}{\partial t}\right)_{y^1, y^2} = \left(\frac{\partial \phi}{\partial t}\right)_{x^1, x^2, x^3} + \nabla \phi \cdot \mathbf{u}$$

$$= \nabla_{(\sigma)} \phi \cdot \mathbf{u} + (\nabla \phi \cdot \boldsymbol{\xi})(\mathbf{u} \cdot \boldsymbol{\xi})$$

4.1.4-3 *estimating interfacial tension* (with D. Li and M. J. Kim) The vertical thin liquid film shown in Figure 4-1 is extended slowly as a function of time by moving the upper solid support. The lower solid support is stationary. The adjoining phase is a gas.

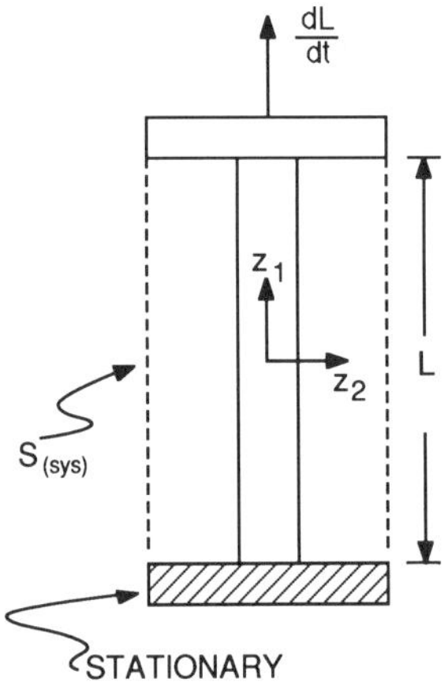

Figure 4.1.4-1: A vertical thin film is extended slowly as a funtion of time by moving the upper solid support. The lower solid support is stationary.

Consider the constant volume system indicated by the dashed lines that includes the liquid film as well as a portion of the adjoining gas phases. We will find it convenient to make the following assumptions.

a) The expansion of the film is sufficiently slow that the effects of inertia, and therefore kinetic energy, may be neglected.

b) There are no external or mutual forces associated with the liquid-gas interfaces.

c) Since the system is chosen to have a constant volume, it has no entrances or exits.

d) The liquid is assumed to be incompressible.

e) There is no mass transfer across the liquid-gas interface.

f) The liquid-gas interfaces are planes. There are at least two ways in which this may be achieved. 1) The solid supports could be such that the dynamic contact angle is 90°. 2) The system could be chosen to include a constant volume of the liquid film but to exclude the three-phase lines of contact on the upper (moving) solid support as well as the adjacent liquid bounded by curved surfaces.

g) The rate at which mechanical energy is dissipated by the action of viscous forces both within the bulk phases and in the dividing surfaces can be neglected.

h) Israelachvili (1973) suggests that the distance between the two interfaces can not go to zero due to the finite size of the constituent atoms and that a finite interfacial separation d must be recognized. He recommends that d be viewed as the mean distance between the centers of individual atoms and estimated as

$$d = 0.91649[\, M/(\, \rho^{(L)} n_a N\,)]^{1/3} \tag{1}$$

Here M is the molecular weight, n_a the number of atoms per molecule, N Avogadro's constant (6.023×10^{23} mol^{-1}), and $\rho^{(L)}$ the mass density of the liquid. In arriving at (1), the atoms have been assumed to be in a close packing arrangement. For a molecule consisting of a repeating unit, such as an n-alkane with $-CH_2-$ being the repeating unit, we suggest a simple picture in which the molecules are arranged such that the *repeating units* are in a close packing arrangement. Retracing the argument of Israelachvili (1973), we have instead

$$d = 0.91649[\, M/(\, \rho^{(L)} n_u N\,)]^{1/3} \tag{2}$$

where n_u is the number of repeating units in the molecule. This is similar to the suggestion of Padday and Uffindell (1968), who replaced M/n_u by the $-CH_2-$ group weight and took the coefficient to be unity.

j) The potential energy of the gas phase attributable to London-van der Waals forces can be neglected with respect to that of the liquid phase.

k) Following the discussion in Exericise 2.1.3-1, we will neglect the London-van der Waals interactions between the gas and liquid phases and compute the potential energy per unit volume attributable to both London-van der Waals forces and gravity at any point within the liquid phase as

$$\rho^{(L)}\phi^{(L)} = -4 \int_{-(h+d^{(LG)})/2}^{(h+d^{(LG)})/2} \int_0^\infty \int_0^\infty \frac{A^{(LL)}}{\pi^2 r^6} \, dx\, dy\, dz + \rho^{(L)} g z_1$$

$$= \frac{4A^{(LL)}}{3\pi} \Big[(h + d^{(LG)} + 2z_2)^{-3}$$

$$+ (h + d^{(LG)} - 2z_2)^{-3} \Big] + \rho^{(L)} g z_1 \quad (3)$$

Here

$$r \equiv [x^2 + y^2 + (z - z_2)^2]^{1/2} \quad (4)$$

is the separation distance between a specified point and any other point within the liquid phase, h is the film thickness measured from the center of the first layer of repeating units to the center of the last layer of repeating units, and $d^{(LG)}$ is the liquid-gas interfacial separation, which must be distinguished from the liquid-liquid interfacial separation d. [Equations (1) and (2) are intended to be used for interfaces separating condensed phases and can not be employed in estimating $d^{(LG)}$.] This means that the volume of the liquid film at any point in time is $L(h + d^{(LG)})$, where L is the instantaneous length of the film.

We assume that there are no electrostatic effects.

i) For this system, show that

$$\frac{d}{dt} \int_{R_{(sys)}} \rho\phi \, dV = - \mathcal{W} - \int_{\Sigma_{(sys)}} \gamma \, \text{div}_{(\sigma)} \mathbf{v}^{(\sigma)} \, dA \quad (5)$$

ii) Neglect the stresses that the bulk fluid exerts on the moving upper solid support and compute

$$\mathcal{W} = 2\gamma \frac{dL}{dt} \quad (6)$$

iii) Assume that all new surface is created at the moving upper solid support and that elsewhere the film's surface is immobile or that the tangential components of the surface velocity are zero. Conclude that

$$\int_{\Sigma_{(sys)}} \gamma \, \text{div}_{(\sigma)} \mathbf{v}^{(\sigma)} \, dA = 0 \quad (7)$$

iv) Considering a unit width of film having a length L, determine that in view of (6) and (7)

$$\frac{d}{dt}\int_{-h/2}^{h/2}\int_{0}^{L}\frac{4A^{(LL)}}{3\pi}\Big[(h + d^{(LG)} + 2z_2)^{-3}$$

$$+ (h + d^{(LG)} - 2z_2)^{-3}\Big]\,dz_1\,dz_2$$

$$+\frac{d}{dt}\Big[\frac{1}{2}(\rho^{(L)} - \rho^{(G)})ghL^2\Big]$$

$$= \frac{d}{dt}\left[\frac{8A^{(LL)}h(h + d^{(LG)})L}{3\pi d^{(LG)2}(2h + d^{(LG)})^2} + \frac{1}{2}(\rho^{(L)} - \rho^{(G)})ghL^2\right]$$

$$= 2\gamma\frac{dL}{dt} \tag{8}$$

v) Because the liquid film has a constant volume $L(h + d^{(LG)})$, argue that

$$\frac{dh}{dt} = -(h + d^{(LG)})\frac{1}{L}\frac{dL}{dt} \tag{9}$$

vi) There is a critical value of $h = h_c$ at which the film ruptures. Assume that

$$h_c = d \tag{10}$$

Determine that as $h \to h_c$, we have

$$\gamma = \frac{4A^{(LL)}}{3\pi}\left[\frac{d(d + d^{(LG)})}{d^{(LG)2}(2d + d^{(LG)})^2} - \frac{d + d^{(LG)}}{(2d + d^{(LG)})^3}\right]$$

$$+\frac{1}{4}(d - d^{(LG)})(\rho^{(L)} - \rho^{(G)})gL \tag{11}$$

Table 4.1.4-1: Comparisons of (12) from Exercise 4.1.4-3 with the experimental measurements of γ for n-alkanes at 20.0 ± 0.08 °C reported by Jasper and Kring (1955) and with (13) recommended by Israelachvili (1985, p. 157). In these comparisons, we have employed the values of $A^{(LL)}$ computed by Hough and White (1980) and we have assumed $d^{(LG)} = \alpha d$, where α has been found by minimizing the sum of the squared errors.

n	d(Å)[a]	d(Å)[b]	A[c] (× 10^{-13} erg)	γ_{meas}[d] (mN/m)	γ_{calc}[e] (mN/m)	γ_{calc}[f] (mN/m)	γ_{calc}[g] (mN/m)
5	2.054	3.088	3.74	16.05	16.68	18.22	17.36
6	2.028	3.029	4.06	18.40	18.82	19.78	19.33
7	2.012	2.991	4.31	20.14	20.49	21.00	20.85
8	1.999	2.962	4.49	21.62	21.76	21.87	22.01
9	1.990	2.939	4.66	22.85	22.94	22.70	23.05
10	1.982	2.921	4.81	23.83	23.97	23.43	23.98
11	1.976	2.906	4.87	24.66	24.52	23.72	22.43
12	1.971	2.894	5.03	25.35	25.54	24.50	25.36
13	1.966	2.883	5.04	25.99	25.78	24.55	25.54
14	1.962	2.874	5.09	26.56	26.20	24.80	25.90
15	1.959	2.867	5.15	27.07	26.64	25.09	26.28
16	1.956	2.860	5.22	27.47	27.14	25.43	26.72

a) Calculated using (1).

b) Calculated using (2).

c) $A \equiv A^{(LL)}$

d) Measured by Jasper and Kring (1955).

e) Calculated using (12) and (2). For this case, we found $\alpha = 1.132$

f) Calculated using (13) and $D_0 = 1.65$Å (Israelachvili 1985, p. 157).

g) Calculated using (12) and (1) (replacing n_u by n_a). For this case, we found $\alpha = 1.391$.

vii) Argue that normally gravity could be neglected to obtain

$$\gamma = \frac{4A^{(LL)}}{3\pi}\left[\frac{d(d + d^{(LG)})}{d^{(LG)2}(2d + d^{(LG)})^2} - \frac{d + d^{(LG)}}{(2d + d^{(LG)})^3}\right] \tag{12}$$

As shown in Table 4.1.4-1, (12) used together with (2) is in excellent agreement with the experimental measurements of Jasper and Kring (1955), when we choose $d^{(LG)} = \alpha d$ and $\alpha = 1.132$. [The experimental study of Richards and Carver (1921) indicates that the difference between surface tensions of the n-alkanes against their own vapor and the surface tensions against air is less than 0.5%.] When (12) is used together with (1) (replacing n_u by n_a), the slope for γ as a function of n is less satisfactory.

Table 4.1.4-1 also compares (12) with

$$\gamma = \frac{A^{(LL)}}{24\pi D_0^2} \tag{13}$$

using $D_0 = 1.65$Å recommended by Israelachvili (1985, p. 157). It represents the data well, although the slope of γ as a function of n is somewhat better predicted by (12).

4.1.5 Integral moment of momentum balance The integral moment of momentum balance tells us about the time rate of change of the moment of momentum in an arbitrary system. From the generalized transport theorem (Exercise 1.3.7-1)

$$\frac{d}{dt}\left(\int_{R_{(sys)}} \mathbf{z} \wedge \rho \mathbf{v} \, dV + \int_{\Sigma_{(sys)}} \mathbf{z} \wedge \rho^{(\sigma)} \mathbf{v}^{(\sigma)} \, dA\right)$$

$$= \int_{R_{(sys)}} \frac{\partial}{\partial t}(\mathbf{z} \wedge \rho \mathbf{v}) \, dV + \int_{S_{(sys)}} (\mathbf{z} \wedge \rho \mathbf{v})(\mathbf{v}_{(sys)} \cdot \mathbf{n}) \, dA$$

$$+ \int_{C_{(sys)}} (\mathbf{z} \wedge \rho^{(\sigma)} \mathbf{v}^{(\sigma)})(\mathbf{v}_{(sys)} \cdot \boldsymbol{\mu}) \, ds$$

$$+ \int_{\Sigma_{(sys)}} \left\{ \frac{\partial}{\partial t}(\mathbf{z} \wedge \rho^{(\sigma)} \mathbf{v}^{(\sigma)}) - \nabla_{(\sigma)}(\mathbf{z} \wedge \rho^{(\sigma)} \mathbf{v}^{(\sigma)}) \cdot \mathbf{u} \right.$$

$$- 2H(\mathbf{z} \wedge \rho^{(\sigma)}\mathbf{v}^{(\sigma)})v_{(\xi)} - [(\mathbf{z} \wedge \rho\mathbf{v})v_{(\xi)}]\Big\} \, dA$$

$$- \int_{C^{(cl)}_{(sys)}} ([\mathbf{z} \wedge \rho^{(\sigma)}\mathbf{v}^{(\sigma)}]u^{(cl)}_{(v)}) \, ds \tag{5-1}$$

The vector cross product of Cauchy's first law (Sec. 2.1.4) with the position vector $\mathbf{z}$ can be integrated over $R_{(sys)}$ to find

$$\int_{R_{(sys)}} \mathbf{z} \wedge \left(\rho\frac{d_{(m)}\mathbf{v}}{dt} - \text{div } \mathbf{T} - \rho\mathbf{b} \right) dV$$

$$= \int_{R_{(sys)}} \left[\frac{d}{dt}(\mathbf{z} \wedge \rho\mathbf{v}) - \mathbf{z} \wedge \rho\mathbf{b} \right] dV$$

$$+ \int_{S_{(sys)}} [(\mathbf{z} \wedge \rho\mathbf{v})(\mathbf{v} \cdot \mathbf{n}) - \mathbf{z} \wedge (\mathbf{T} \cdot \mathbf{n})] \, dA$$

$$- \int_{\Sigma_{(sys)}} [(\mathbf{z} \wedge \rho\mathbf{v})(\mathbf{v} \cdot \boldsymbol{\xi}) - \mathbf{z} \wedge (\mathbf{T} \cdot \boldsymbol{\xi})] \, dA$$

$$= 0 \tag{5-2}$$

The vector cross product of the jump momentum balance (Sec. 2.1.6) with $\mathbf{z}$ gives after integration over $\Sigma_{(sys)}$

$$\int_{\Sigma_{(sys)}} \mathbf{z} \wedge \Big\{ \rho^{(\sigma)} \frac{d_{(s)}\mathbf{v}^{(\sigma)}}{dt} - \text{div}_{(\sigma)}\mathbf{T}^{(\sigma)} - \rho^{(\sigma)}\mathbf{b}^{(\sigma)}$$

$$+ [\rho(\mathbf{v} - \mathbf{v}^{(\sigma)})(\mathbf{v} \cdot \boldsymbol{\xi} - v_{(\xi)}) - \mathbf{T} \cdot \boldsymbol{\xi}] \Big\} \, dA$$

$$= \int_{\Sigma_{(sys)}} \Big\{ \frac{\partial}{\partial t}(\mathbf{z} \wedge \rho^{(\sigma)}\mathbf{v}^{(\sigma)}) - \nabla_{(\sigma)}(\mathbf{z} \wedge \rho^{(\sigma)}\mathbf{v}^{(\sigma)}) \cdot \mathbf{u}$$

$$- 2H(\mathbf{z} \wedge \rho^{(\sigma)}\mathbf{v}^{(\sigma)})v_{(\xi)} - \mathbf{z} \wedge \rho^{(\sigma)}\mathbf{b}$$

$$+ [(\mathbf{z} \wedge \rho\mathbf{v})(\mathbf{v} \cdot \boldsymbol{\xi} - v_{(\xi)}) - \mathbf{z} \wedge (\mathbf{T} \cdot \boldsymbol{\xi})]\Big\} \, dA$$

$$+ \int_{C_{(sys)}} [(\mathbf{z} \wedge \rho^{(\sigma)}\mathbf{v}^{(\sigma)})(\mathbf{v}^{(\sigma)} \cdot \boldsymbol{\mu}) - \mathbf{z} \wedge (\mathbf{T}^{(\sigma)} \cdot \boldsymbol{\mu})] \, ds$$

$$- \int_{C^{(cl)}_{(sys)}} ([\mathbf{z} \wedge \rho^{(\sigma)}\mathbf{v}^{(\sigma)}] u^{(\sigma)}_{(\nu)} - \mathbf{z} \wedge [\mathbf{T}^{(\sigma)} \cdot \boldsymbol{\nu}]) \, ds$$

$$= 0 \tag{5-3}$$

The integral over $C^{(cl)}_{(sys)}$ of the vector cross product of the momentum balance at the common line with $\mathbf{z}$ says

$$\int_{C^{(cl)}_{(sys)}} \mathbf{z} \wedge (\rho^{(\sigma)}\mathbf{v}^{(\sigma)}[\mathbf{v}^{(\sigma)} \cdot \boldsymbol{\nu} - u^{(\sigma)}_{(\nu)}] - \mathbf{T}^{(\sigma)} \cdot \boldsymbol{\nu}) \, ds$$

$$= 0 \tag{5-4}$$

Subtracting (5-2) through (5-4) from (5-1), we have

$$\frac{d}{dt}\left(\int_{R_{(sys)}} \mathbf{z} \wedge \rho\mathbf{v} \, dV + \int_{\Sigma_{(sys)}} \mathbf{z} \wedge \rho^{(\sigma)}\mathbf{v}^{(\sigma)} \, dA \right)$$

$$= \int_{R_{(sys)}} \mathbf{z} \wedge \rho\mathbf{b} \, dV + \int_{\Sigma_{(sys)}} \mathbf{z} \wedge \rho^{(\sigma)}\mathbf{b}^{(\sigma)} \, dA$$

$$+ \int_{S_{(sys)}} \mathbf{z} \wedge [\rho\mathbf{v}(\mathbf{v} - \mathbf{v}_{(sys)}) - \mathbf{T}] \cdot (- \mathbf{n}) \, dA$$

$$+ \int_{C_{(sys)}} \mathbf{z} \wedge [\rho^{(\sigma)}\mathbf{v}^{(\sigma)}(\mathbf{v}^{(\sigma)} - \mathbf{v}_{(sys)}) - \mathbf{T}^{(\sigma)}]$$

$$\cdot (- \boldsymbol{\mu}) \, ds \tag{5-5}$$

which is a form of the **integral moment of momentum balance**. Its physical meaning is clearer after rearrangement in the form

$$\frac{d}{dt}\left(\int_{R_{(sys)}} \mathbf{z} \wedge \rho\mathbf{v} \, dV + \int_{\Sigma_{(sys)}} \mathbf{z} \wedge \rho^{(\sigma)}\mathbf{v}^{(\sigma)} \, dA \right)$$

$$= \int_{S_{(\text{ent ex})}} \mathbf{z} \wedge \rho \mathbf{v}(\mathbf{v} - \mathbf{v}_{(sys)}) \cdot (-\mathbf{n})\, dA$$

$$+ \int_{C_{(\text{ent ex})}} (\mathbf{z} \wedge \rho^{(\sigma)} \mathbf{v}^{(\sigma)})(\mathbf{v}^{(\sigma)} - \mathbf{v}_{(sys)}) \cdot (-\boldsymbol{\mu})\, ds$$

$$- \mathcal{T} + \int_{R_{(sys)}} \mathbf{z} \wedge \rho \mathbf{b}\, dV + \int_{\Sigma_{(sys)}} \mathbf{z} \wedge \rho^{(\sigma)} \mathbf{b}^{(\sigma)}\, dA$$

$$+ \int_{S_{(\text{ent ex})}} [-(P - p_0)(\mathbf{z} \wedge \mathbf{n}) + \mathbf{z} \wedge (\mathbf{S} \cdot \mathbf{n})]\, dA$$

$$+ \int_{C_{(\text{ent ex})}} [\gamma(\mathbf{z} \wedge \boldsymbol{\mu}) + \mathbf{z} \wedge (\mathbf{S}^{(\sigma)} \cdot \boldsymbol{\mu})]\, ds \tag{5-6}$$

where

$$\mathcal{T} \equiv \int_{S_{(sys)} - S_{(\text{ent ex})}} \mathbf{z} \wedge [(\mathbf{T} + p_0 \mathbf{I}) \cdot (-\mathbf{n})]\, dA$$

$$+ \int_{C_{(sys)} - C_{(\text{ent ex})}} \mathbf{z} \wedge [\mathbf{T}^{(\sigma)} \cdot (-\boldsymbol{\mu})]\, ds \tag{5-7}$$

is the **torque** that the system exerts upon the impermeable portion of its bounding surface beyond the torque attributable to the ambient pressure p_0. On the right side of (5-6),

$$\int_{S_{(\text{ent ex})}} (\mathbf{z} \wedge \rho \mathbf{v})(\mathbf{v} - \mathbf{v}_{(sys)}) \cdot (-\mathbf{n})\, dA$$

$$+ \int_{C_{(\text{ent ex})}} (\mathbf{z} \wedge \rho^{(\sigma)} \mathbf{v}^{(\sigma)})(\mathbf{v}^{(\sigma)} - \mathbf{v}_{(sys)}) \cdot (-\boldsymbol{\mu})\, ds$$

is the net rate at which moment of momentum is carried into the system with material moving across its boundary;

$$\int_{R_{(sys)}} \mathbf{z} \wedge \rho \mathbf{b} \, dV + \int_{\Sigma_{(sys)}} \mathbf{z} \wedge \rho^{(\sigma)} \mathbf{b}^{(\sigma)} \, dA$$

is the torque exerted on the system by the external forces;

$$\int_{S_{(ent\ ex)}} [- (P - p_0)(\mathbf{z} \wedge \mathbf{n}) \, dA + \int_{C_{(ent\ ex)}} \gamma (\mathbf{z} \wedge \boldsymbol{\mu}) \, ds$$

is the torque applied to the system at the entrances and exits by pressure (beyond P_0) and by surface tension.

Usually it will be acceptable to neglect

$$\int_{S_{(ent\ ex)}} \mathbf{z} \wedge (\mathbf{S} \cdot \mathbf{n}) \, dA + \int_{C_{(ent\ ex)}} \mathbf{z} \wedge (\mathbf{S}^{(\sigma)} \cdot \boldsymbol{\mu}) \, ds$$

the torque applied to the system at its entrances and exits by the viscous forces. Under these conditions, (5-6) reduces to

$$\frac{d}{dt}\left(\int_{R_{(sys)}} \mathbf{z} \wedge \rho \mathbf{v} \, dV + \int_{\Sigma_{(sys)}} \mathbf{z} \wedge \rho^{(\sigma)} \mathbf{v}^{(\sigma)} \, dA \right)$$

$$= \int_{S_{(ent\ ex)}} (\mathbf{z} \wedge \rho \mathbf{v})(\mathbf{v} - \mathbf{v}_{(sys)}) \cdot (- \mathbf{n}) \, dA$$

$$+ \int_{C_{(ent\ ex)}} (\mathbf{z} \wedge \rho^{(\sigma)} \mathbf{v}^{(\sigma)})(\mathbf{v}^{(\sigma)} - \mathbf{v}_{(sys)}) \cdot (- \boldsymbol{\mu}) \, ds - \mathcal{T}$$

$$+ \int_{R_{(sys)}} \mathbf{z} \wedge \rho \mathbf{b} \, dV + \int_{\Sigma_{(sys)}} \mathbf{z} \wedge \rho^{(\sigma)} \mathbf{b}^{(\sigma)} \, dA$$

$$- \int_{S_{(ent\ ex)}} (P - p_0)(\mathbf{z} \wedge \mathbf{n}) \, dA$$

$$+ \int_{C_{(ent\ ex)}} \gamma (\mathbf{z} \wedge \boldsymbol{\mu}) \, ds \qquad (5\text{-}8)$$

For more about the integral moment of momentum balance excluding interfacial effects, see Slattery (1981, p. 254).

Reference

Slattery, J. C., "Momentum, Energy, and Mass Transfer in Continua," McGraw-Hill, New York (1972); second edition, Robert E. Krieger, Malabar, FL 32950 (1981).

4.1.6 Entrapment of residual oil (Slattery 1974) Sec. 3.2.2, I outline the various stages in which oil is produced from a petroleum reservoir. In the secondary stage of petroleum recovery, brine or steam is pumped into a selected pattern of wells in a field forcing a portion of the oil into other wells intended for production. In this manner, 10-40% of the oil initially in place will be recovered, but 40-80% of the oil originally found in the reservoir will be left behind (Geffen 1973). Our objective in what follows is to better understand the manner in which this residual oil is trapped during the secondary stage of production.

The void volume in a permeable rock, such as that in which oil is found, may be thought of as many intersecting pores of varying diameters. Visualize two neighboring pore networks having different mean radii that offer parallel paths for displacement. If the oil is displaced by the water more rapidly in the network having the larger mean pore radius, a portion of the oil in the network having the smaller mean pore radius will be bypassed and trapped. In this manner, large blobs or ganglia of residual oil occupying many neighboring pores can be trapped in the permeable rock structure accounting for the large percentage of residual oil that can remain in a field at the conclusion of waterflooding.

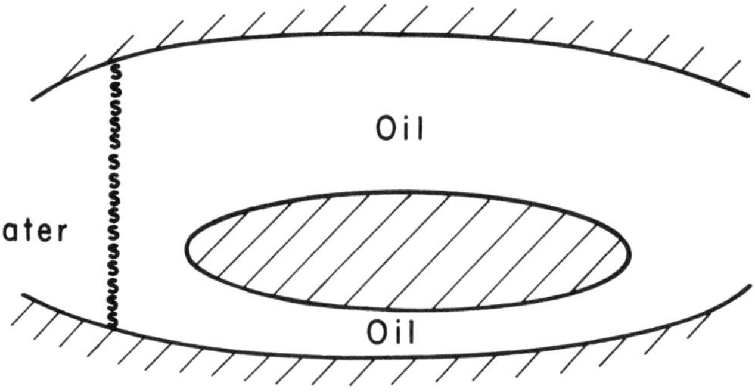

Figure 4.1.6-1 Water-oil interface approaching bifurcation in pore.

This mechanism for entrapment may be easier to understand on a smaller scale. Figure 6-1 shows a particular pore that bifurcates to form two pores of unequal diameters, which in turn rejoin at some point downstream. In what follows, we will confine our attention to the displacement of oil by water in this bifurcated pore as the flow proceeds from left to right.

Several observations are helpful before making an analysis.

If the water-oil interface advances through the small pore more rapidly, residual oil will tend to be trapped in the large pore. If the interface moves through the large pore more rapidly, residual oil will be trapped in the small pore.

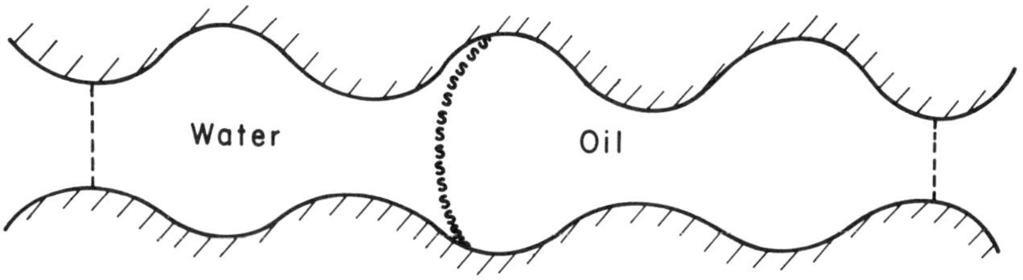

Figure 4.1.6-2 In a water-wet pore, a water-oil interface will creep forward from a neck toward a bulge, until an instability develops and the next Haines jump occurs.

The character of the trapping process will probably depend upon whether we are considering an oil-wet or a water-wet permeable structure.

In reality, the pores are not as regular as I have indicated in Figure 6-1. They might be better described by the irregular channels shown in Figures 6-2 and 6-3.

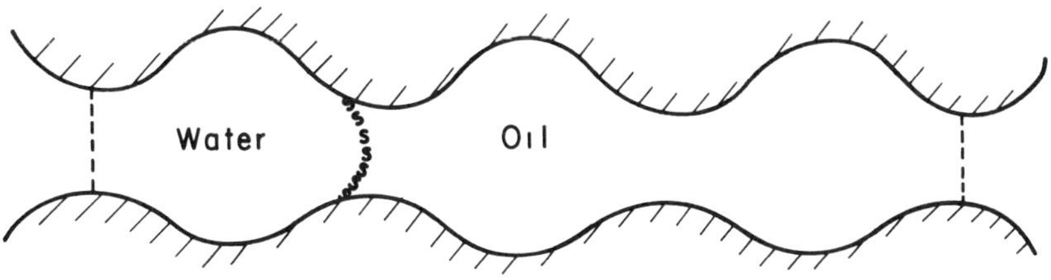

Figure 4.1.6-3 In an oil-wet pore, a water-oil interface will creep forward from a buldge toward a neck.

The displacement of a water-oil interface will not occur smoothly as a function of time under these conditions. Haines (1930; Miller and Miller 1956; Melrose 1965; Heller 1968) observed that an interface advances through an irregular pore in a series of jumps. Heller (1968) used high-speed photography to follow the Haines jumps executed by water-air interfaces advancing in the individual pores of a permeable structure. He found that an individual Haines jump takes place within less than 10^{-5}s. (For more on similar phenomena within the context of the displacement of residual oil, see Secs. 3.2.2 and 4.1.7.)

Following a jump, an oil-water interface creeps forward a short distance until an instability develops and the next jump occurs. The limits of the creeping motion between Haines jumps may be approximately identified with the limits of static stability discussed in Sec. 7.2.5. In a water-wet pore, a water-oil interface will creep forward from a neck toward a bulge (see Figure 6-2). In an oil-wet pore, a water-oil interface will creep forward from a bulge toward a neck (Figure 6-3).

Experimental evidence suggests that the water-oil interface will not advance smoothly as a function of time even during this creeping motion. Poynting and Thomson (1902, p. 142) forced mercury up a capillary tube and then gradually reduced the pressure. Instead of falling, the mercury at first adjusted itself to the reduced pressure by altering the curvature of the air-mercury interface. When the pressure gradient finally grew too large, the mercury fell a short distance in the tube before stopping, repeating the deformation of the interface, and falling again. Consequently, they observed the interface to move through the tube in a series of irregular movements. Similarly, Jacobs (1966) reported erratic movements in the displacement of dilute solutions of blood serum with air in a capillary.

These irregular, sticking movements can be understood in terms of the contact angle hysteresis to be expected on contaminated or roughened surfaces. Let us consider the air-mercury interface moving over glass. When the contact angle measured through the air is greater than the advancing contact angle, the air advances over the glass. When the contact angle measured through the air is less than the receding contact angle, the air retreats. When the contact angle is less than the advancing contact angle but greater than the receding contact angle, the common line is stationary. In the Poynting and Thomson (1902, p.142) experiment, the air-mercury interface slowly deformed until the contact angle measured through the air equaled the advancing contact angle and the interface moved. Inertia carried the interface to a position where the contact angle was less than the advancing contact angle (but greater than the receding contact angle). Time was required for the interface to slowly deform until the contact angle again equaled the advancing contact angle and the common line could again advance.

This suggests that we reexamine the creeping motion between Haines jumps as a water-oil interface advances through a pore. It may consist of one or more **pauses** during which the contact angle measured through the aqueous phase increases, followed by an irregular motion when this contact angle finally equals the critical value required for advance.

If the water-oil interface is eventually trapped as it moves through a pore, the point of trapping will occur as the interface creeps forward between Haines jumps. For that reason, an advancing water-oil interface will be trapped as it approaches a bulge in a water-wet pore. In an oil-wet pore, it will be trapped as it approaches a neck.

approximate analysis The rate limiting steps in the episodic advancement of a water-oil interface through an irregular pore are the periods in which the interface creeps forward between Haines jumps.

In this analysis, we will go even further and assume that the rate limiting steps are the pauses that occur during this creeping motion as the contact angle measured through the aqueous phases increases towards its critical advancing value. Let us estimate the effects of interfacial behavior upon the slow deformation that occurs during one of these pauses. In doing so, we shall not concern ourselves either with the Haines jump or with the development of the instability that leads to the jump.

Consider as a system all of the oil and water in the pore between the fixed entrance and exit surfaces shown as dashed lines in Figures 6–2 and 6–3. Let us write the z component of the momentum balance for this system making the following assumptions.

1. Neglect all inertial effects. The water-oil interface deforms very slowly during a pause in the creeping motion. Inertial effects are important only during a Haines jump.

2. The pore is horizontal.

3. The interfacial stress is described by the linear Boussinesq surface fluid model (Sec. 2.2.2).

4. Both the pore wall and the water-oil phase interface are rotationally symmetric.

5. Neglect mass transfer to and from the water-oil interface.

6. Neglect the viscous forces acting at the entrance and exit cross-sections. This would be completely justified, if we were dealing with Poiseuille flow in a cylindrical tube of uniform diameter.

In view of the first two restrictions, the z component of the integral momentum balance (Sec. 4.1.3) becomes

$$0 = -\mathbf{g}_z \cdot \int_{S_{(wall)}} (\mathbf{T} + p_0 \mathbf{I}) \cdot (-\mathbf{n}) \, dA$$

$$- \mathbf{g}_z \cdot \int_{C} \mathbf{T}^{(\sigma)} \cdot (-\boldsymbol{\mu}) \, ds$$

$$+ \mathbf{g}_z \cdot \int_{S_{(ent)} + S_{(ex)}} (\mathbf{T} + p_0 \mathbf{I}) \cdot \mathbf{n} \, dA \tag{6-1}$$

Here $S_{(wall)}$ is that portion of the boundary of the system coinciding with the pore wall; C is the intersection of the water-oil interface with the pore wall; $S_{(ent)}$ is the entrance to the system, the surface shown as a dashed line on the left sides of Figures 6-2 and 6-3; $S_{(ex)}$ is the exit surface, shown as a dashed line on the right sides of Figures 6-2 and 6-3.

We will find it convenient to distinguish between the pressure and the viscous forces acting on the pore wall,

$$\mathbf{g}_z \cdot \int_{S_{(wall)}} (\mathbf{T} + p_0 \mathbf{I}) \cdot (- \mathbf{n}) \, dA$$

$$= F^{(p)} + F^{(v)}_{(o)} + F^{(v)}_{(w)} \tag{6-2}$$

Here

$$F^{(p)} \equiv \mathbf{g}_z \cdot \int_{S_{(wall)}} (p - p_0) \mathbf{n} \, dA \tag{6-3}$$

is the z component of the pressure force exerted by the oil and water phases on the pore wall $S_{(wall)}$,

$$F^{(v)}_{(o)} \equiv \mathbf{g}_z \cdot \int_{S_{(wall,oil)}} \mathbf{S} \cdot (- \mathbf{n}) \, dA \tag{6-4}$$

is the z component of the viscous force exerted by the oil on that portion of the pore wall $S_{(wall,oil)}$ in contact with oil, and

$$F^{(v)}_{(w)} \equiv \mathbf{g}_z \cdot \int_{S_{(wall,water)}} \mathbf{S} \cdot (- \mathbf{n}) \, dA \tag{6-5}$$

is the z component of the viscous force exerted by the water on that portion of the pore wall $S_{(wall,water)}$ in contact with water.

For the linear Boussinesq surface fluid model (Sec. 2.2.2)

$$\mathbf{T}^{(\sigma)} = [\gamma + (\kappa - \varepsilon) \, \text{div}_{(\sigma)} \mathbf{v}^{(\sigma)}] \mathbf{P} + 2\varepsilon \, \mathbf{D}^{(\sigma)} \tag{6-6}$$

which means

$$\text{on C:} \quad \mathbf{g}_z \cdot \mathbf{T}^{(\sigma)} \cdot \boldsymbol{\mu} = [\gamma + (\kappa - \varepsilon) \, \text{div}_{(\sigma)} \mathbf{v}^{(\sigma)}] \mu_z$$

$$+ 2\varepsilon \, \mathbf{g}_z \cdot \mathbf{D}^{(\sigma)} \cdot \boldsymbol{\mu} \tag{6-7}$$

Viewing our rotationally symmetric water-oil interface (assumption 4) with respect to a cylindrical coordinate system

$$z = f(y^1 , t)$$

$$y^1 = r$$

$$y^2 = \theta \tag{6-8}$$

we can reason that

$$\text{on } C: \; \mathbf{g}_z \cdot \mathbf{D}^{(\sigma)} \cdot \boldsymbol{\mu} = (\mathbf{g}_z \cdot \mathbf{a}_\alpha) D^{(\sigma)\alpha\beta} (\mathbf{a}_\beta \cdot \boldsymbol{\mu})$$

$$= \frac{\partial f}{\partial y^1} D^{(\sigma)11} \mu_1 \tag{6-9}$$

and

$$D^{(\sigma)}_{22} = r \left(v_r^{(\sigma)} + \frac{\partial v_\theta^{(\sigma)}}{\partial \theta} \right) \tag{6-10}$$

I suggest that we say

$$\text{on C:} \; D^{(\sigma)}_{22} = r \left(v_r^{(\sigma)} + \frac{\partial v_\theta^{(\sigma)}}{\partial \theta} \right)$$

$$= 0 \tag{6-11}$$

arguing that, since the common line is stationary during a pause in the creeping motion, there is little reason to expect appreciable mass transfer through the common line to the wall. This in turn permits us to observe

$$\text{on C:} \quad D^{(\sigma)11} = a^{11} \, \text{div}_{(\sigma)} \mathbf{v}^{(\sigma)} \tag{6-12}$$

where

$$\text{div}_{(\sigma)} \mathbf{v}^{(\sigma)} = \text{tr} \, \mathbf{D}^{(\sigma)}$$

$$= a^{11} D^{(\sigma)}_{11} + a^{22} D^{(\sigma)}_{22} \tag{6-13}$$

Noting

$$\text{on } C: \quad \frac{\partial f}{\partial y^1} a^{11} \mu_1 = \mu_z \tag{6-14}$$

we see from (6-9) and (6-12) that

$$\text{on } C: \quad \mathbf{g}_z \cdot \mathbf{T}^{(\sigma)} \cdot \boldsymbol{\mu} = [\, \gamma + (\kappa - \varepsilon)\, \text{div}_{(\sigma)} \mathbf{v}^{(\sigma)} \,] \mu_z \tag{6-15}$$

Since we are neglecting mass transfer to and from the water-oil interface (assumption 5), the jump mass balance (Sec. 1.3.5) requires

$$\begin{aligned} \text{div}_{(\sigma)} \mathbf{v}^{(\sigma)} &= - \frac{1}{\rho^{(\sigma)}} \frac{d_{(s)} \rho^{(\sigma)}}{dt} \\ &= - \frac{d_{(s)} \ln \rho^{(\sigma)}}{dt} \end{aligned} \tag{6-16}$$

We conclude from (6-15) and (6-16) that

$$\begin{aligned} &\mathbf{g}_z \cdot \int_C \mathbf{T}^{(\sigma)} \cdot (- \boldsymbol{\mu})\, ds \\ &= -2\pi R_{(wo)} \left[\gamma + (\kappa + \varepsilon) \frac{d_{(s)} \ln \rho^{(\sigma)}}{dt} \right] \mu_z \end{aligned} \tag{6-17}$$

in which $R_{(wo)}$ is the radius of the pore at the line of contact with the water-oil interface.

Since we are neglecting the effect of viscous forces at the entrance and exit of the system (assumption 6),

$$\begin{aligned} &\mathbf{g}_z \cdot \int_{S_{(ent)} + S_{(ex)}} (\mathbf{T} + p_0 \mathbf{I}) \cdot \mathbf{n}\, dA \\ &= \int_{S_{(ent)}} (p - p_0)\, dA - \int_{S_{(ex)}} (p - p_0)\, dA \\ &= \Delta p\, \pi R^2 \end{aligned} \tag{6-18}$$

in which we have introduced the pressure drop over the system

$$\Delta p \equiv p_{(ent)} - p_{(ex)} \tag{6-19}$$

Literally, R is the radius of the pore at the entrance and exit surfaces in

Figures 6-2 and 6-3. Since the locations of these entrance and exit surfaces are arbitrary beyond requiring their radii to be equal, R may be interpreted as a radius characteristic of the pore, such as the mean pore radius.

With the aid of (6-2), (6-17), and (6-18), we can write (6-1) in the more convenient form

$$- F^{(p)} - F^{\{v\}}_{\{o\}} - F^{\{v\}}_{\{w\}} + 2\pi R_{(wo)} \left[\gamma - (\kappa + \varepsilon) \frac{d_{(s)} \ln \rho^{(\sigma)}}{dt} \right] \mu_z$$

$$+ \Delta p \, \pi R^2 = 0 \tag{6-20}$$

Let us introduce as dimensionless variables

$$t^* \equiv \frac{t v_{(o)}}{\ell} \tag{6-21}$$

$$F^{(p)*} \equiv \frac{F^{(p)}}{|\nabla \langle p \rangle| \pi R^2 \ell} \tag{6-22}$$

$$F^{\{v\}*}_{\{o\}} \equiv \frac{F^{\{v\}}_{\{o\}}}{2\pi L_{(o)} \mu_{(o)} v_{(o)}} \tag{6-23}$$

$$F^{\{v\}*}_{\{w\}} \equiv \frac{F^{\{v\}}_{\{w\}}}{2\pi (\ell - L_{(o)}) \mu_{(w)} v_{(o)}} \tag{6-24}$$

where ℓ is the length of the pores shown in Figure 6-1, $L_{(o)}$ is the length of the oil segment remaining in a pore at any given time, $v_{(o)}$ is the average z component of velocity for the oil in the pore, $\mu_{(o)}$ is the viscosity of the oil, and $\mu_{(w)}$ is the viscosity of the water. The pore length ℓ is viewed as a constant here, since in Figure 6-1 we wish to compare two pores that have the same length but possibly different diameters. In (6-21), we have chosen as the characteristic time a rough measure of the time required for the water-oil interface to travel the length of the pore. The pressure drag is characterized by $|\nabla \langle p \rangle| \pi R^2 \ell$ in (6-22), where $|\nabla \langle p \rangle|$ is the magnitude of the gradient of the total local volume average of pressure (see Slattery 1981, p. 196). The viscous drag of the oil in (6-23) is crudely proportional to the product of $2\pi R L_{(o)}$, a measure of the area of the pore wall covered by oil, and $\mu_{(o)} v_{(o)}/R$, a measure of the viscous shear force exerted by the oil on the pore wall. The viscous drag of the water in (6-24) is in a similar manner taken to be proportional to the product of $2\pi R (\ell - L_{(o)})$, a measure of the area of the pore wall covered by water, and $\mu_{(w)} v_{(o)}/R$, a measure of the viscous shear force exerted by the water on the pore wall. Equation (6-20) can be rearranged in terms of these variables to read

$$v^*_{(o)} \equiv \frac{v_{(o)}\mu_{(o)}}{|\nabla\langle p\rangle|\ell^2}$$

$$= \frac{R^{*2}A - R^*N_\gamma B}{L^*_{(o)}D + (1 - L^*_{(o)})\frac{\mu_{(w)}}{\mu_{(o)}}E + R^*N_{\kappa+\varepsilon}F}$$

$$= \frac{R^{*2}A - R^*N_\gamma B}{\tilde{D} + R^*N_{\kappa+\varepsilon}F} \tag{6-25}$$

In writing this, I have introduced as definitions

$$R^* \equiv \frac{R}{\ell} \tag{6-26}$$

$$L^*_{(o)} \equiv \frac{L_{(o)}}{\ell} \tag{6-27}$$

$$N_\gamma \equiv \frac{\gamma}{|\nabla\langle p\rangle|\ell^2} \tag{6-28}$$

$$N_{\kappa+\varepsilon} \equiv \frac{\kappa+\varepsilon}{\ell\mu_{(o)}} \tag{6-29}$$

$$A \equiv \frac{\Delta p}{|\nabla\langle p\rangle|\ell} - F^{(p)*} \tag{6-30}$$

$$B \equiv -2\frac{R_{(wo)}}{R}\mu_z \tag{6-31}$$

$$D \equiv 2F^{(v)*}_{(o)} \tag{6-32}$$

$$E \equiv 2F^{(v)*}_{(w)} \tag{6-33}$$

$$F \equiv 2 \frac{R_{(wo)}}{R} \frac{d_{(s)} \ln \rho^{(\sigma)}}{dt^*} \mu_z \tag{6-34}$$

$$\tilde{D} \equiv L^*_{(o)} D + (1 - L^*_{(o)}) \frac{\mu_{(w)}}{\mu_{(o)}} E \tag{6-35}$$

We will assume that we have chosen our dimensionless variables such that the magnitudes of A, B, D, E, and F are all of order unity (they differ by less than a factor of ten from unity).

Let us examine the signs of some of the terms that appear in (6-25). We will assume that the flow is always in a positive z direction and therefore

$$R^*A \geq N_\gamma B \tag{6-36}$$

For oil-wet rock, B is positive; for water-wet, negative. The dimensionless scalars D and E are positive. In the water-wet case shown in Figure 6-2, the area of the water-oil interface decreases as a function of time. With the assumption that there is no mass transfer from the adjoining phases, the surface mass density must increase as a function of time. Since μ_z is positive, we have

$$\frac{d_{(s)} \ln \rho^{(\sigma)}}{dt^*} \mu_z \geq 0$$

$$F \geq 0 \tag{6-37}$$

A similar argument for the oil-wet case shown in Figure 6-3 indicates that inequality (6-37) is valid there as well.

Equation (6-25) describes the average velocity in the oil phase during a pause in the creeping motion between Haines jumps. Since we visualize that these pauses are the rate limiting steps, we can use (6-25) to draw several conclusions concerning the relative rates of displacement in the two portions of the bifurcated pore pictured in Figure 6-1 and concerning the subsequent entrapment of residual oil by water.

distribution From (6-25),

$$\frac{\partial v^*_{(o)}}{\partial R^*} = \frac{R^*A(\tilde{D} + R^*N_{\kappa+\varepsilon}F) + \tilde{D} (R^*A - N_\gamma B)}{(\tilde{D} + R^*N_{\kappa+\varepsilon}F)^2} \tag{6-38}$$

If this derivative is positive, the oil will be displaced more rapidly in the large pore of Figure 6-1 and residual oil will be trapped in the smaller pore.

If this derivative is negative, the water-oil interface will advance more rapidly through the small pore, trapping residual oil in the larger pore. In view of inequalities (6-36) and (6-37), the sign of this derivative will depend on the sign and magnitude of A.

For an oil-wet rock, inequality (6-36) requires that A be positive.

1. In the displacement of oil by water in an oil-wet permeable structure, residual oil will be trapped in the smaller of those pores initially containing oil.

In a water-wet rock, if A is sufficiently large (it may be negative) that $\partial v^{*}_{(o)}/\partial R^{*}$ is positive, we will speak of a **free or forced displacement** of the oil by water. These terms are motivated by our estimate that, if we assume pressure to be roughly independent of position within each phase,

$$A \doteq \frac{\Delta p}{|\nabla \langle p \rangle| \ell} - \frac{\Delta p(\pi R^2 - \pi R^2_{(wo)})}{|\nabla \langle p \rangle| \pi R^2 \ell}$$

$$= \frac{\Delta p}{|\nabla \langle p \rangle| \ell} \left(\frac{R_{(wo)}}{R} \right)^2 \tag{6-39}$$

If Δp is negative, we say that it is a **free displacement;** if it is positive, a **forced displacement**.

2. In the free or forced displacement of oil by water in a water-wet structure, residual oil will be trapped in the smaller of those pores initially containing oil.

In a water-wet rock, let A and $\partial v^{*}_{(o)}/\partial R^{*}$ be negative, although inequality (6-36) is still observed. The flow of water into the rock is slowed by the pressure gradient imposed locally. We will refer to this condition as **restricted displacement** of the oil by water.

3. In a restricted displacement of oil by water in a water-wet structure, residual oil will be trapped in the larger pores.

Statements 2 and 3 concerning water-wet structures may be supported by the experiments of Handy and Datta (1966).

effect of pressure gradient The effect of increasing the magnitude of the pressure while holding the oil-water interfacial tension constant can be seen from

$$\frac{\partial(v^{*}_{(o)} N_{\gamma}^{-1})}{\partial N_{\gamma}^{-1}} = \frac{R^{*2} A - R^{*} N_{\gamma} B}{\tilde{D} + R^{*} N_{\kappa+\varepsilon} F} \tag{6-40}$$

in which we have recognized that ($R^{*2} - R^* N_\gamma B$) should be nearly independent of the magnitude of the pressure gradient. The result is predictable. The rate of displacement is always increased by increasing the magnitude of the pressure gradient. Of greater concern to us here is the efficiency with which the displacement will proceed. In particular, for which of the pores shown in Figure 6-1 is the effect more significant?

In order to answer this question, consider

$$\frac{\partial^2 (v^*_{(o)} N_\gamma^{-1})}{\partial N_\gamma^{-1} \, \partial R^*} = \frac{\partial v^*_{(o)}}{\partial R^*} \tag{6-41}$$

Since this derivative has the same sign as $\partial v^*_{(o)}/\partial R^*$, we would expect more oil to be trapped under all conditions. This lead us to conclude

4. The fraction of oil trapped in either oil-wet or water-wet rock will be enhanced by increasing the magnitude of the pressure gradient.

effect of viscosity ratio The effect of changing the water-oil viscosity ratio can be computed from (6-25):

$$\frac{\partial v^*_{(o)}}{\partial (\mu_{(w)}/\mu_{(o)})} = - \frac{E (1 - L^*_{(o)})(R^{*2} A - R^* N_\gamma B)}{(\tilde{D} + R^* N_{\kappa+\varepsilon} F)^2} \tag{6-42}$$

It follows from (6-36) that this derivative is always negative. The rate of displacement of oil by water would be decreased by adding a small amount of polymer to the waterflood and in that way increasing $\mu_{(w)}/\mu_{(o)}$. This is true, but is the effect more significant in the larger or smaller pores?

Let us examine instead

$$\frac{\partial^2 v^*_{(o)}}{\partial (\mu_{(w)}/\mu_{(o)}) \, \partial R^*} = - \frac{E(1 - L^*_{(o)})}{\tilde{D} + R^* N_{\kappa+\varepsilon} F} \frac{\partial v^*_{(o)}}{\partial R^*}$$

$$+ \frac{E(1 - L^*_{(o)})(R^* A - N_\gamma B) R^* N_{\kappa+\varepsilon} F}{(\tilde{D} + R^* N_{\kappa+\varepsilon} F)^3} \tag{6-43}$$

For the moment, let us ignore the effects of the surface viscosities and let us consider the effect of only the first term on the right of (6-43). The sign of this term must be opposite of $\partial v^*_{(o)}/\partial R^*$, which leads us to conclude that

5. The fraction of oil trapped in either oil-wet or water-wet rock will be decreased by increasing the water-oil viscosity ratio so long as the effects of the interfacial viscosities can be ignored.

This suggests that recovery efficiency might be enhanced by flooding with a polymer solution, but it ignores possible side effects such as adsorption of the polymer on the rock structure and chemical decomposition of the polymer with time under reservoir conditions. The effect can be a large one, since

$$\frac{\partial^2 v^*_{(o)}}{\partial(\mu_{(w)}/\mu_{(o)})\partial R^*}$$

is of the same order as $\partial v^*_{(o)}/\partial R^*$ (estimating that the magnitudes of D, E, F, and $L_{(o)}$ are of the order unity and that $\mu_{(w)}/\mu_{(o)}$ and $N_{\kappa+\varepsilon}$ are small).

The second term on the right of (6-43) is proportional to the surface viscosities, and it is always positive. For large surface viscosities, the second term on the right of (6-43) is on the order of (N_γ/R^*) times the first and the influence of the surface viscosities may be significant (R^* is neglected with respect to N_γ in making this estimate). This indicates

6. In an oil-wet permeable structure and in the free or forced displacement of oil by water in a water-wet structure, the water-oil surface viscosities should be as small as possible for maximum advantage from increasing the water-oil viscosity ratio.

7. In a restricted displacement of oil by water in a water-wet structure, the water-oil surface viscosities should be as large as possible for maximum advantage from increasing the water-oil viscosity ratio.

These effects of the surface viscosities have not been investigated experimentally.

effect of surface tension Returning to (6-25), we see that

$$\frac{\partial v^*_{(o)}}{\partial N_\gamma} = \frac{-R^* B}{\tilde{D} + R^* N_{\kappa+\varepsilon} F} \tag{6-44}$$

In an oil-wet structure, B is positive; a reduction in the interfacial tension leads to an increase in the displacement velocities in both of the pores in Figure 6-1. In a water-wet structure, B is negative; the displacement velocities in both pores are decreased when the surface tension is reduced. But in which pore is the effect greater?

Let us consider

$$\frac{\partial^2 v^*_{(o)}}{\partial N_\gamma \partial R^*} = \frac{-B\tilde{D}}{(\tilde{D} + R^* N_{\kappa+\varepsilon} F)^2} \tag{6-45}$$

Note that the magnitude of $\partial v^*_{(o)}/\partial R^*$ is on the order of N_γ times that of $\partial^2 v^*_{(o)}/\partial N_\gamma \partial R^*$ (estimating that the magnitudes of A, B, D, and E are of the order unity and neglecting $R^{*2} N_{\kappa+\varepsilon}$ with respect to N_γ)

8. The fraction of oil trapped in either oil-wet or water-wet rock structures will be relatively independent of the water-oil interfacial tension unless N_γ is small (unless the surface tension is small).

Equation (6-45) implies that the effect of reducing the water-oil surface tension in an oil-wet rock is to increase the displacement velocity in the large pore of Figure 6-1 relative to that in the smaller pore.

9. The entrapment of residual oil in the smaller pores of an oil-wet rock will be reinforced by reducing the surface tension. The fraction of the oil trapped will not decrease, even in N_γ is small.

From (6-45), a reduction of the water-oil surface tension in a water-wet structure tends to decrease the displacement velocity in the large pore of Figure 6-1 relative to that in the smaller pore.

10. In the free or forced displacement of oil by water in a water-wet structure, the residual saturation will be decreased by lowering the surface tension, if N_γ is sufficiently small.

11. In a restricted displacement of oil by water in a water-wet structure, the entrapment of residual oil in the large pores will be enhanced by lowering the surface tension. The fraction of oil trapped will not decrease, even if N_γ is small.

Conclusion 8 is supported by the observations of Wagner and Leach (1966)[a] who studied displacement using the equilibrium vapor and liquid phases of the methane-n-pentane system. They found that, independent of whether it was the gas or liquid phase being displaced, the residual saturation was unchanged as the surface tension was lowered to 0.1 mN/m.

For $\gamma < 0.1$ mN/m, Wagner and Leach (1966) found a sharp change in the dependence of the residual saturation upon γ, which suggests a change in the nature of the physical process. For the forced displacement of the non-wetting phase by the wetting phase, the residual saturation generally decreased as the surface tension was lowered, in agreement with conclusion 10. But for the displacement of the wetting phase by the non-wetting phase, the residual saturation also decreased as the surface tension was lowered, in contrast with conclusion 9. I discuss in more detail in the next section the possibility that there may have been another physical process that was important in their experiments for $\gamma < 0.1$ mN/m : a portion of the oil initially trapped by the mechanism described here may have been subsequently displaced under the existing pressure gradient.

effect of surface viscosities In order to investigate the effect of the surface viscosities, let us consider

$$\frac{\partial^2 v^*_{(o)}}{\partial N_{\kappa+\varepsilon}\partial R^*}$$

$$= [- R^{*2} AF(\tilde{D} + R^* N_{\kappa+\varepsilon} F) - 2R^*\tilde{D}F(R^*A - N_\gamma B)]$$

$$\times (\tilde{D} + R^* N_{\kappa+\varepsilon} F)^{-3}$$

$$= \Big\{ R^{*2} AF(\tilde{D} + R^* N_{\kappa+\varepsilon} F)$$

$$- 2 R^*F[R^*A(\tilde{D} + R^* N_{\kappa+\varepsilon} F) + \tilde{D}(R^*A - N_\gamma B)]\Big\}$$

$$\times \{ \tilde{D} + R^* N_{\kappa+\varepsilon} F \}^{-3} \tag{6-46}$$

We can begin by observing that the magnitude of $\partial v^*_{(o)}/\partial R^*$ is about $N_{\kappa+\varepsilon}$ times $\partial^2 v^*_{(o)}/\partial N_{\kappa+\varepsilon}\partial R^*$ if $N_{\kappa+\varepsilon}$ is large and the same order of magnitude as $\partial^2 v^*_{(o)}/\partial N_{\kappa+\varepsilon}\partial R^*$ if $N_{\kappa+\varepsilon}$ is small.

12. The fraction of the oil trapped in either oil-wet or water-wet structures will be relatively independent of the water-oil surface viscosities, unless $N_{\kappa+\varepsilon}$ is small (unless the surface viscosities are small).

Since inequality (6-36) requires A to be positive for an oil-wet rock, $\partial^2 v^*_{(o)}/\partial N_{\kappa+\varepsilon}\partial R^*$, will be always negative. When the water-oil surface viscosities are reduced, the displacement velocity in the larger pore of Figure 6-1 is increased in relation to that in the smaller pore.

13. The entrapment of residual oil in the smaller pores of an oil-wet rock will be enhanced by reducing the surface viscosities. The fraction of the oil trapped will not decrease, even in $N_{\kappa+\varepsilon}$ is small.

Let us now consider a free or forced displacement of oil by water in a water-wet rock. The derivative $\partial v^*_{(o)}/\partial R^*$ must always be positive, although

---

a) In the paper of Wagner and Leach (1966), the drawings shown as Figures 11, 12 and 13 appear to be respectively Figures 13, 11, and 12.

A may be either positive or negative. If A is positive, $\partial^2 v^*_{(o)}/\partial N_{\kappa+\varepsilon}\partial R^*$ will be negative. If A is negative, (6-38) and (6-46) require that $\partial^2 v^*_{(o)}/\partial N_{\kappa+\varepsilon}\partial R^*$ again be negative.

14. In a free or forced displacement of oil by water in a water-wet structure, the entrapment of residual oil in the smaller pores will be reinforced by reducing the surface viscosities. The fraction of the oil trapped will not decrease, even if $N_{\kappa+\varepsilon}$ is small.

For a restricted displacement of oil by water in a water-wet structure, $\partial^2 v^*_{(o)}/\partial N_{\kappa+\varepsilon}\partial R^*$ may be either positive or negative and no general conclusions can be drawn.

The available experimental evidence is not definitive. Kimbler *et al.* (1966) reported for a water-wet porous bed three different waterfloods with qualitatively different surface viscosities. With a rigid oil-water interface, they found appreciably more residual oil than with an expanded one. Unfortunately, they do not report the corresponding surface tension. Their observations might be explained by a reduction in the water-oil surface tension (conclusion 10).

References

Geffen, T. M., *Oil Gas J.* 66 (May 7, 1973).

Haines, W. B., *J. Agric. Sci.* **20,** 97 (1930).

Handy, L. L., and P. Datta, *Soc. Pet. Eng. J.* **6,** 261 (1966).

Heller, J. P., *Soil Sci. Soc. Am. Proc.* **32,** 778 (1968).

Jacobs, H. R., *Biorheology* **3,** 117 (1966).

Kimbler, O. K., R. L. Reed, and I. H. Silberberg, *Soc. Pet. Eng. J.* **6,** 153 (1966).

Melrose, J. C., *Soc. Pet. Eng. J.* **5,** 259 (1965).

Miller, E. E., and R. D. Miller, *J. Appl. Phys.* **27,** 324 (1956).

Poynting, J. H., and J. J. Thomson, "A Text-Book of Physics - Properties of Matter," Charles Griffin, London (1902).

Slattery, J. C., *AIChE J.* **20,** 1145 (1974).

Slattery, J. C., "Momentum, Energy, and Mass Transfer in Continua," McGraw-Hill (1972); second edition, Robert E. Krieger, Malabar, FL 32950 (1981).

Wagner, O. R., and R. O. Leach, *Soc. Pet. Eng. J.* **6**, 335 (1966).

4.1.7 Displacement of residual oil (Slattery 1974) Under what conditions will a segment of oil, that has been isolated during secondary recovery, continue to be displaced?

Several observations are in order before beginning our analysis.

The majority of residual oil is trapped in the form of blobs or ganglia occupying a number of neighboring pores. It is with the hope of understanding the displacement of these larger blobs of residual oil that we confine our attention here to the simpler problem of the displacement of small blobs or **segments** of oil occupying only a portion of a single pore.

In the last section, we were concerned to understand a mechanism by which residual oil may be trapped during a waterflood. For that reason, we focused our attention on the relative motions of oil-water interfaces in two capillaries having the same length but different mean radii. In what follows, we consider only the particular capillary in which an oil segment finds itself trapped.

We can anticipate that the displacement process will depend to some extent upon whether we are considering an oil-wet or a water-wet porous structure.

The pores in a permeable rock are irregular channels whose diameters vary with axial position.

We suggest in our static analysis of the displacement of residual oil in Sec. 3.2.2 that an oil segment is displaced through an irregular pore in a series of jumps. We might expect that these jumps would have the same character as those observed by Haines (1930; Miller and Miller 1956; Melrose 1965; Heller 1968) within the context of secondary recovery. This appears to be confirmed by the observed motions of blobs occupying a number of neighboring pores (Ng *et al.* 1978).

Following a jump, an oil segment will creep forward a short distance until an instability develops and the next jump occurs. As in Sec. 4.1.6, we will approximately identify the limits of the creeping motion between Haines jumps with the limits of static stability discussed in Sec. 7.1.5. Figure 7-1 shows a small oil segment undergoing a creeping motion through a water-wet pore. In Figure 7-2, a small oil segment is shown in creeping motion through an oil-wet pore. Keep in mind that the portion of the pore in which an oil segment undergoes this creeping motion will depend upon the volume of the oil segment, the wettability condition, and the contact angle hysteresis.

Experiments suggest that the oil segment will not advance smoothly as a function of time even during this creeping motion. Yarnold (1938) observed that, when a liquid segment is slowly displaced through a straight glass capillary, it moves in an irregular, episodic fashion, appearing to periodically stick to the tube wall. It may be that Ng *et al.* (1978) observed this same phenomenon as the various menisci of a large blob episodically adjusted their positions prior to a spontaneous advance or jump.

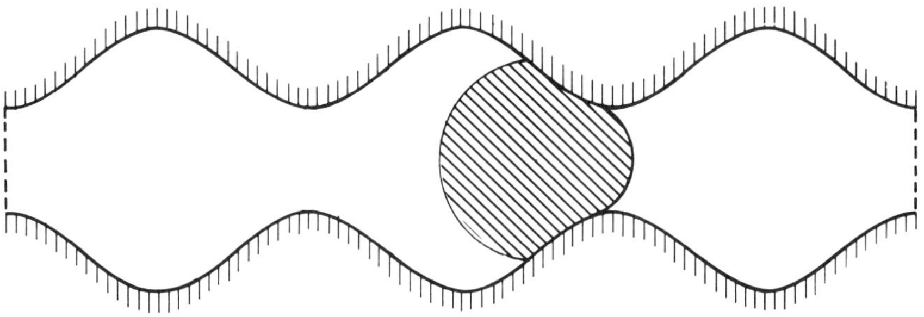

Figure 4.1.7-1 Oil segment in creeping motion through a water-wet pore following a jump (see Figure 3.2.2-3). Morrow's (1975) class III contact angle hysteresis is assumed with advancing and receding contact angles measured through the aqueous phase taken to be 40° and 4° respectively. In the context of Sec. 4.1.8, this depicts a water segment creeping through a gas-wet pore.

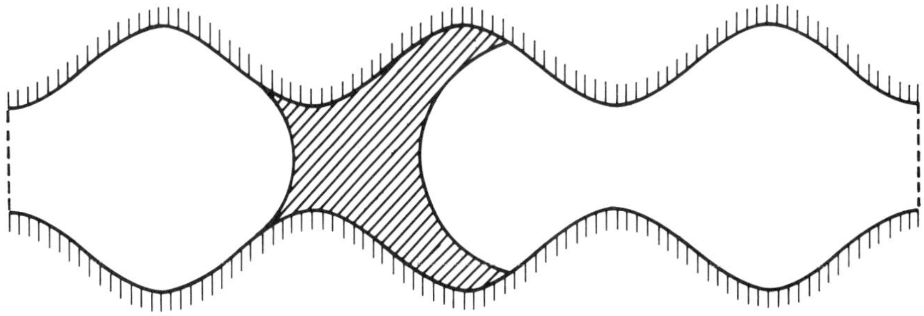

Figure 4.1.7-2 Oil segment in creeping motion through a oil-wet pore following a jump (see Figure 3.2.2-5). Morrow's (1975) class III contact angle hysteresis is assumed with advancing and receding contact angles measured through the aqueous phase taken to be 175° and 130° respectively. In the context of Sec. 4.1.8, this shows a water segment creeping through a water-wet pore.

These irregular, sticking motions can be understood in terms of the contact angle hysteresis to be expected on contaminated or roughened surfaces. Let us consider the oil segment shown in Figures 7-1 and 7-2. The oil advances when either of two conditions is satisfied.

i) When the contact angle measured through the water at the leading interface is less than the receding contact angle, the water retreats and the common line formed by the intersection of the leading interface with the pore wall advances.

ii) When the contact angle measured through the water at the trailing interface is greater than the advancing contact angle, the water advances and the common line formed by the intersection of the trailing interface with the pore wall advances.

An oil segment that appears to be stuck to the pore wall will slowly deform until one of these conditions is satisfied and it advances. Inertia will carry the segment to a position where neither of these conditions is satisfied, where both of the common lines are stationary, and where the segment again appears to stick to the wall. Time is required for the segment to slowly deform until one of the conditions is again satisfied and the common lines can again advance.

This suggests that we reexamine the creeping motion between jumps as an oil segment advances through a pore. It may consist of one or more pauses during which the contact angle measured through the aqueous phase at the leading interface decreases and the contact angle measured through the aqueous phase at the trailing interface increases. These pauses are followed by an irregular motion when either of these contact angles finally equals their critical values required for advance.

The implication here is that the displacing fluid is immiscible with the oil. If it is miscible with the oil in some portions of the reservoir, it is likely to be immiscible with it in other portions, either because it has become mixed with existing reservoir fluids (brine, gas) or because one of its components (surfactant, polymer) has been lost by adsorption on the pore walls of the rock.

Our discussion here is within the context of the displacement of residual oil by an aqueous (surfactant, polymer) solution. But all of our conclusions are equally valid for oil displacement by a gas. Some of the special problems associated with a gas drive are examined in the next section.

approximate analysis The rate limiting steps in the episodic advance of an oil segment through an irregular pore are the periods in which the segment creeps forward between jumps.

In this analysis, we will go even further and assume that the rate limiting steps are the pauses which occur during this creeping motion as the contact angle measured through the aqueous phase at the leading interface decreases and the contact angle measured through the aqueous phase at the trailing interface increases. Let us estimate the effects of interfacial behavior upon the slow deformation that occurs during one of these pauses.

In carrying out this analysis, we shall not be concerned either with the jump or with the development of any instability that leads to the jump.

Consider as a system all of the oil and water in the pore between the fixed entrance and exit surfaces shown as dashed lines in Figures 7-1 and 7-2. Reasoning as we did in arriving at (6-25), we can write the z component of the momentum balance for this system as

$$v_{(o)}^{**} \equiv \frac{v_{(o)}\mu_{(o)}}{|\nabla\langle p\rangle|\ell^{**}R^2}$$

$$= \frac{A - N_{\gamma}^{**}G}{\ell^{**}\tilde{D} + N_{(\kappa+\varepsilon)}^{**}H/\ell^{**}} \tag{7-1}$$

I have introduced as definitions here

$$\ell^{**} \equiv \frac{\ell}{R} \tag{7-2}$$

$$N_{\gamma}^{**} \equiv \frac{\gamma}{|\Delta\langle p\rangle\ell^{**}R^2} \tag{7-3}$$

$$N_{\kappa+\varepsilon}^{**} \equiv \frac{\kappa + \varepsilon}{R\mu_{(o)}} \tag{7-4}$$

$$G \equiv B_{(\ell)} + B_{(t)} \tag{7-5}$$

$$H \equiv F_{(\ell)} + F_{(t)} \tag{7-6}$$

Here B and F are defined by (6-31) and (6-34); the subscripts $_{(\ell)}$ and $_{(t)}$ refer to the leading and trailing interfaces respectively. Note that (6-25) was used in comparing pores of different mean radii but the same length ℓ. For that reason, there I chose ℓ as the characteristic length in defining dimensionless variables. Here we are talking about a single pore, and it seems more natural (though not necessary) to choose as our characteristic pore dimension the radius R.

We will again assume that we have chosen our dimensionless variables such that the magnitudes of A, D, E, G, and H are all of order unity (they differ by less than a factor of ten from unity).

Let us begin by examining the signs of some of the terms that appear in (7-1). We will assume that the flow is always in the positive z direction and consequently

$$A - N_{\gamma}^{**}\, G \geq 0 \tag{7-7}$$

For a water-wet rock, we assume that the leading interface is controlling ($-\mu_{(\ell)z} >> \mu_{(t)z}$ in Figure 7-1) and[a]

$$G > 0 \tag{7-8}$$

For an oil-wet rock, we estimate that the trailing interface is controlling ($- \mu_{(t)z} >> \mu_{(t)z}$ in Figure 7-2) and inequality (7-8) still applies. Inequalities (7-7) and (7-8) imply

$$A > 0 \tag{7-9}$$

Considerations similar to those that lead to inequalities (6-35) indicate that for both oil-wet and water-wet rock

$$H \geq 0 \tag{7-10}$$

Equation (7-1) describes the average velocity within an oil segment during a pause in the creeping motion between jumps. Since we visualize that these pauses are the rate limiting steps, we can say that everything done to enhance the average velocity within the oil segment will also increase the rate of displacement of residual oil.

Several conclusions can be drawn.

critical value for interfacial tension From (7-1) and inequality (7-7), we see that

1. The dimensionless interfacial tension $N^{**}_{(\gamma)}$ must be less than some critical value before any residual oil can be recovered by displacement.

---

a) For a capillary of uniform diameter

$$\mu_{(\ell)z} = - \cos(\pi - \Theta_A)$$

$$\mu_{(t)z} = \cos(\pi - \Theta_R)$$

where Θ_A and Θ_R are respectively the advancing and receding contact angles measured through the oil phase. Therefore

$$G = 2 \cos(\pi - \Theta_A) - 2 \cos(\pi - \Theta_R)$$

which is in agreement with inequality (7-8), since the receding contact angle will always be smaller than the advancing contact angle.

Inequality (7-7) also tells us that the critical value for N_γ^{**} has a magnitude of order unity. This is something we can check. As an approximation we can say

$$N_\gamma^{**} \equiv \frac{\gamma}{|\nabla\langle p\rangle| \ell^{**} R^2}$$

$$= \left(\frac{\gamma\psi}{|\nabla\langle p\rangle| k} \right)_{crit} \left(\frac{k}{\psi \ell^{**} R^2} \right) \tag{7-11}$$

where k is the permeability of the structure to a single phase and ψ is the porosity of the structure. Oh and Slattery (1979) estimate that for a typical water-wet oil-bearing sandstone, $(|\nabla\langle p\rangle|\, k/\gamma\psi)_{crit} = 3.07 \times 10^{-3}$. Let us choose R to be the neck radius for the irregular pore. For k = 250 millidarcies (2.47×10^{-13} m^2), a reasonable value for R appears to be 10 microns (Batra and Dullien 1973). Let us take $\ell^{**} = 6$, the length of a repeating unit of irregular pore (Payatakes *et al.* 1973). For an oil-bearing sandstone, typically $\psi = 0.2$. Our conclusion is that the critical value of $N_\gamma^{**} = 0.67$ in agreement with our initial estimate.

This critical interfacial tension is predicted in Sec. 3.2.2. The existence of such a critical interfacial tension has been confirmed in several experimental studies (Moore and Slobod 1956; Taber 1969; Taber *et al.* 1973; Foster 1973; Ehrlich *et al.* 1974; Abrams 1975).

effect of pressure gradient Let us assume that N_γ^{**} is less than the critical value. We see from

$$\frac{\partial(\, v_{(o)}\, N_\gamma^{**-1}\,)}{\partial N_\gamma^{-1}} = \frac{A - N_\gamma^{**} G}{\ell^{**}\tilde{D} + N_{\kappa+\varepsilon}^{**} H / \ell^{**}} \tag{7-12}$$

the effect of increasing the magnitude of the pressure gradient while holding the oil-water interfacial tension constant. We have recognized here that $(A - N_\gamma^{**} G)$ should be nearly independent of the magnitude of the pressure gradient. Since $v_{(o)}^{**}$ is assumed to be positive, this derivative must be positive and we have

2. For values of N_γ^{**} less than the critical value, the rate of displacement of residual oil will increase as the applied pressure gradient increases.

This is confirmed experimentally (Wagner and Leach 1966; Taber 1969; Taber *et al.* 1973).

effect of interfacial tension We can also ask about the effect of lowering the interfacial tension while maintaining the pressure gradient constant:

$$\frac{\partial v^{**}_{(o)}}{\partial N^{**}_{\gamma}} = \frac{-\ G}{\ell^{**}\tilde{D} + N^{**}_{\kappa+\varepsilon}\ H/\ell^{**}} \tag{7-13}$$

Because of inequality (7-8), we conclude

3. For values of N^{**}_{γ} less than the critical value, the rate of displacement of residual oil will increase as the oil-water interfacial tension is decreased.

We discussed in conjunction with conclusions 9 and 10 of Sec. 4.1.6 the anomalous behavior observed by Wagner and Leach (1966). When the interfacial tension was decreased below a critical value, the residual saturation was found to decrease both in their simulated oil-wet tests and in their simulated water-wet tests. This is in contrast with conclusion 8 of Sec. 4.1.6. But their findings are in agreement with conclusion 3 above, if we assume that some of the simulated oil phase, after initially being trapped, is subsequently displaced. This would imply that some of the simulated oil was recovered after breakthrough. Their experiment was not designed to sense recovery after breakthrough in the simulated water-wet tests.

effect of surface viscosites The effect of lowering the surface viscosities is indicated by

$$\frac{\partial v^{**}_{(o)}}{\partial N^{**}_{\kappa+\varepsilon}} = -\frac{(\ A - N^{**}_{\gamma}G\ )(\ H/\ell^{**}\ )}{(\ \ell^{**}\tilde{D} + N_{\kappa+\varepsilon}\ H/\ell^{**}\ )^2} \tag{7-14}$$

This means

4. For values of N^{**}_{γ} less than the critical value, the rate of displacement of residual oil will increase as the sum of the oil-water surface viscosities is decreased.

Once we are below the critical interfacial tension, how does the effect of decreasing the interfacial viscosities compare with that obtained when the interfacial tension is decreased? For purposes of comparison, let us talk about relative or percentage changes. Starting with (7-13) and (7-14), consider

$$N^{**}_{\gamma}\ \frac{\partial v^{**}_{(o)}}{\partial N^{**}_{\gamma}} = \frac{\partial v^{**}_{(o)}}{\partial \ln N^{**}_{\gamma}}$$

$$= -\frac{G N^{**}_{\gamma}}{\ell^{**}\tilde{D} + N^{**}_{\kappa+\varepsilon}H/\ell^{**}} \tag{7-15}$$

and

$$N^{**}_{\kappa+\varepsilon} \frac{\partial v^{**}_{(o)}}{\partial N^{**}_{\kappa+\varepsilon}} = \frac{\partial v^{**}_{(o)}}{\partial \ln N^{**}_{\kappa+\varepsilon}}$$

$$= - \frac{N^{**}_{\kappa+\varepsilon}(A - N^{**}_{\gamma}G)(H/\ell^{**})}{(\ell^{**}\tilde{D} + N^{**}_{\kappa+\varepsilon}H/\ell^{**})^2} \tag{7-16}$$

This tells us that

5. When N^{**}_{γ} is less than the critical value and $N^{**}_{\kappa+\varepsilon}$ is large, equal percentage reductions of the interfacial tension and the interfacial viscosities are equally important. When $N^{**}_{\kappa+\varepsilon}$ is small, it is more important to reduce the interfacial tension.

At the present time, there are no experimental studies of the effect of the interfacial viscosities upon the rate of displacement of residual oil by surfactant solutions in porous media. Experimental studies of displacement in cylindrical capillaries are consistent with the conclusions drawn here (Stoodt and Slattery 1984; see also Sec. 4.1.9). We are also encouraged by the agreement between experimental observations and the predictions of a similar theory developed in the next section regarding the effects of the surface viscosities upon foam displacement.

References

Abrams, A., *Soc. Pet. Eng. J.* **15,** 437 (1975).

Batra, V. K., and F. A. L. Dullien, *Soc. Pet. Eng. J.* **13,** 256 (1973).

Ehrlich, R., H. H. Hasiba, and P. Raimondi, *J. Pet. Technol.* **26,** 1335 (1974).

Foster, W. R., *J. Pet. Technol.* **25**, 205 (1973).

Haines, W. B., *J. Agric. Sci.* **20,** 97 (1930).

Heller, J. P., *Soil Sci. Soc. Am. Proc.* **32,** 778 (1968).

Melrose, J. C., *Soc. Pet. Eng. J.* **5,** 259 (1965).

Miller, E. E., and R. D. Miller, *J. Appl. Phys.* **27,** 324 (1956).

Moore, T. F., and R. L. Slobod, *Prod. Mon.* **20,** 20 (Aug. 1956).

Morrow, N. R., *J. Can. Pet. Technol.* **14**, 42 (1975).

Ng, K. M., H. T. Davis, and L. E. Scriven, *Chem. Eng. Sci.* **33,** 1009 (1978).

Oh, S. G., and J. C. Slattery, *Soc. Pet. Eng. J.* **19,** 83 (1979).

Payatakes, A. C., C. Tien, and R. M. Turian, *AIChE J.* **19,** 58 (1973).

Slattery, J. C., *AIChE J.* **20,** 1145 (1974).

Stoodt, T. J., and J. C. Slattery, *AIChE J.* **30,** 564 (1984).

Taber, J. J., *Soc. Pet. Eng. J.* **9,** 3 (1969).

Taber, J. J., J. C. Kirby, and F. U. Schroeder, *AIChE Symp. Ser.* **69** (127), 53 (1973).

Wagner, O. R., and R. O. Leach, *Soc. Pet. Eng. J.* **6,** 335 (1966).

Yarnold, G. D., *Proc. Phys. Soc. London* **50,** 540 (1938).

4.1.8 Displacement of residual oil by a stable foam (Slattery 1979)

In the last section, we examined the effects of several parameters upon the displacement of residual oil. The discussion was set within the context of displacement by an aqueous (surfactant, polymer) solution. But all of the conclusions there are equally valid for a gas drive.

A major disadvantage of a gas drive is that the mobility of the gas (the ratio of the relative permeability of the gas to its viscosity) is much larger than the mobility of the oil. This results in an unstable displacement, with the gas fingering ahead through the oil and water and bypassing the majority of it.

It has been suggested that this unfavorable mobility ratio could be reversed by incorporating the gas in a foam which would be used to displace the crude oil and water (Bernard 1963, Holm 1970; Raza 1970; Minssieux 1974). In practice, either an aqueous surfactant solution and gas could be injected simultaneously or injection of surfactant solution could alternate with injection of gas, allowing the foam to form in place within the porous rock as the gas fingered through the surfactant solution.

The void volume in a permeable rock, such as that in which oil is found, is composed of many intersecting irregular pores, the mean diameter of which may be on the order of 20 μm (Batra and Dullien 1973). It is unlikely that, within such a small scale pore network, a foam has the same ordered structure of thin films intersecting in Plateau borders that we observe when the same liquid and gas are shaken together in a graduate cylinder. Within the pore network, the aqueous surfactant solution from which the foam is formed is likely to be dispersed in the form of thin films or segments separating bubbles of gas. This conception of the foam structure is supported by Mast's (1972) observations.

We must further distinguish between stable and unstable foams. Mast (1972) observed that, in an unstable foam, liquid and gas are transported through the progressive rupture and regeneration of the foam structure. In a stable foam, the liquid and gas are displaced as a body.

In what follows, we consider the displacement of only stable foams. The case of an unstable foam is considered in Exercise 4.1.8-1.

Our objective here is to study the factors controlling the efficiency of a foam displacement, both its local efficiency and its sweep efficiency or conformance. The local efficiency describes the degree to which the foam is able to displace residual oil from all of the pores within the immediate neighborhood of any point within the rock. The sweep efficiency or conformance describes the degree to which the foam is able to move through all of the pores with nearly the same speed, minimizing any tendency to finger ahead in one macroscopic region and bypass the residual oil in another.

Visualize two neighboring pore networks having different mean radii that offer parallel paths for displacement. If the oil is displaced by the foam more rapidly in the network having the larger mean pore radius, a portion of the oil in the network having the smaller mean pore radius will be bypassed and lost.

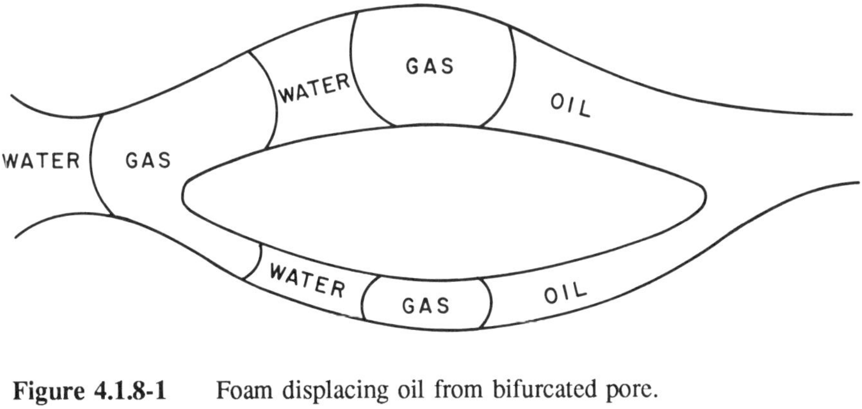

Figure 4.1.8-1 Foam displacing oil from bifurcated pore.

It is easier to discuss this mechanism for foam displacement on a smaller scale. Figure 8-1 shows a pore that bifurcates to form two pores of unequal diameters, which rejoin downstream. Assume that displacement takes place from left to right.

In constructing our detailed analysis, several observations will be useful.

We must be able to displace the water segments in both the large and small pores, if the local efficiency of the foam displacement is to be improved.

If the water segment advances through the small pore more rapidly, residual oil will tend to be bypassed in the large pore. If the water segment moves through the large pore more rapidly, residual oil will tend to be bypassed in the small pore. We can improve both the local and the sweep efficiencies of a foam displacement to the extent that we can cause the water segments to move through the large and small pores at nearly the same speeds.

Everything said in Sec. 4.1.7 about the manner in which an oil segment moves through an irregular pore applies equally well here to the displacement of water segment in a stable foam. A water segment moves through a pore in a series of jumps. Between each jump, it undergoes a slow creeping motion that may be interrupted by one or more pauses, during which the leading and trailing phase interfaces appear to stick to the pore wall. During such a pause, the contact angle measured through the gas phase at the leading interface decreases and the contact angle measured through the gas phase at the trailing interface increases, until the critical values required for motion of the leading and trailing common lines are achieved and the segment again begins to creep forward in the pore.

approximate analysis As in Sec. 4.1.7, I will assume here that the rate limiting steps in the episodic advance of a water segment through an irregular pore are these pauses in the creeping motion between jumps.

Consider as a system all of the water and gas in the pore between the fixed entrance and exit surfaces shown as dashed lines in Figures 7-1 and 7-2. The z component of the momentum balance for this system is essentially the same as (7-1).

$$v^{*}_{(w)} \equiv \frac{v_{(w)}\,\mu_{(w)}}{|\nabla\langle p\rangle|\ \ell^{2}}$$

$$= \frac{R^{*2}\ A - R^{*}N_{\gamma}\ G}{L^{*}_{(w)}D + (\ 1\ -\ L^{*}_{(w)}\ )\ \frac{\mu_{(g)}}{\mu_{(w)}}\ E\ +\ R^{*}N_{\kappa+\varepsilon}H}$$

$$= \frac{R^{*2}A\ -\ R^{*}N_{\gamma}\,G}{\tilde{D}\ +\ R^{*}N_{\kappa+\varepsilon}\ H} \tag{8-1}$$

with the understanding that

$$L^{*}_{(w)} \equiv \frac{L_{(w)}}{\ell} \tag{8-2}$$

$$t^{*} \equiv \frac{t v_{(w)}}{\ell} \tag{8-3}$$

$$N_{\kappa+\varepsilon} \equiv \frac{\kappa + \varepsilon}{\ell \mu_{(w)}} \tag{8-4}$$

$$D \equiv 2\, F_{\{w\}}^{\{v\}*} \tag{8-5}$$

$$E \equiv 2\, F_{\{g\}}^{\{v\}*} \tag{8-6}$$

$$\tilde{D} \equiv L^{*}_{(w)} D + (\, 1 - L^{*}_{(w)}\,)\, \frac{\mu_{(g)}}{\mu_{(w)}}\, E \tag{8-7}$$

$$F_{\{g\}}^{\{v\}*} \equiv \frac{F_{\{w\}}^{\{v\}}}{2\pi L_{(w)}\, \mu_{(w)} v_{(w)}} \tag{8-8}$$

$$F_{\{g\}}^{\{v\}*} \equiv \frac{F_{\{g\}}^{\{v\}}}{2\pi(\, \ell - L_{(w)}\,)\mu_{(g)} v_{(w)}} \tag{8-9}$$

Here $v_{(w)}$ is the average z component of the velocity in the aqueous phase; $\mu_{(g)}$ and $\mu_{(w)}$ are the viscosities of the gas and aqueous phases; $L_{(w)}$ is the length of the aqueous surfactant solution segment; the surface tension and surface viscosities are those appropriate to the gas-aqueous surfactant solution interface. As in Sec. 4.1.6, I wish to compare pores of different mean radii but the same length ℓ. For that reason, I have chosen ℓ as the characteristic length in defining dimensionless variables.

We argue as in Sec. 4.1.7 that we have chosen our dimensionless variables such that the magnitudes of A, D, E, G, H, and D are all of order unity and that

$$R^{*}A - N_{\gamma}G \geq 0 \tag{8-10}$$

$$A \geq 0 \tag{8-11}$$

$$G \geq 0 \tag{8-12}$$

$$H \geq 0 \tag{8-13}$$

Equation (8-1) describes the average velocity within the aqueous phase during a pause in the creeping motion between jumps. Since we visualize that these pauses are the rate limiting steps, we can use (8-1) to draw several conclusions concerning the relative rates of displacement in the two portions of the bifurcated pore pictured in Figure 8-1 and the efficiency with which a stable foam will displace a gas.

critical value for interfacial tension From (8-1) and (8-10), we observe that

1. The dimensionless interfacial tension N_γ must be less than some critical value before any residual oil can be recovered by displacement. The critical value of N_γ has a magnitude of order unity.

Minssieux (1974) reported a critical pressure gradient below which a continuous foam flow could not be achieved. The corresponding value of N_γ is estimated from his data to be a little less than unity.

We can write

$$N_\gamma \equiv \frac{\gamma}{|\nabla\langle p\rangle|\ \ell^2}$$

$$= \frac{\gamma R^{*2}}{|\nabla\langle p\rangle|\ R^2} \tag{8-14}$$

For a typical displacement $|\nabla\langle p\rangle| \sim 1$ psi/ft (2.26×10^4 N/m^3). Let us take $R^{*-1} = 6$, the length of a repeating unit of irregular pore (Payatakes et al. 1973). Let us choose R to be the neck radius for the irregular pore. For a rock whose permeability is 250 millidarcies (2.46×10^{-13} m^2), a reasonable value for R appears to be 10μm (Batra and Dullien 1973). Taking $N_\gamma = 1$, we estimate the critical value for the surface tension γ above which a stable foam could not be displaced through the porous rock structure as 8×10^{-2} mN/m.

For most foams, the gas-aqueous surfactant solution surface tension is considerably larger. This appears to explain why stable foams are frequently described as blocking agents to seal leaks in gas storage reservoirs (Bernard and Holm 1970; Albrecht and Marsden 1970), to prevent gas coning in the neighborhood of a production well (Raza 1970), or to reduce the flow through high permeability zones in heterogeneous reservoirs (Holm 1970).

For Holm's (1968) experiments, N_γ (based upon the mean pore diameter) is estimated to be greater than unity. The flows recorded in these experiments may have proceeded by a different mechanism than the one assumed here. He observed that the gas flowed as the result of bubbles or short liquid segments breaking and reforming. We can visualize the liquid segments thinned by draining, became unstable, and finally broke. There is nothing in Holm's (1968) description or Mast's (1972) characterization of flow in smaller pores to suggest that these liquid segments broke and reformed as the result of jumps.

distribution of bypassed oil From (1),

$$\frac{\partial v^*_{(w)}}{\partial R^*} = \frac{R^* A(\tilde{D} + R^* N_{\kappa+\varepsilon} H) + \tilde{D}(R^* A - N_\gamma G)}{(\tilde{D} + R^* N_{\kappa+\varepsilon} H)^2} \tag{8-15}$$

which in view of (8-10) and (8-11) must be positive. This means that the water segment shown in the larger pore of Figure 8-1 will be displaced more rapidly. The implication is that

2. For values of N_γ less than the critical value, a stable foam will preferentially flow through the larger pores, failing to displace portions of the residual oil in smaller pores.

In order to improve the displacement efficiency of a stable foam, we must reduce $\partial v^*_{(w)}/\partial R^*$, forcing the water segments to move through the large and small pores at more nearly the same speed.

effect of pressure gradient The effect of increasing the magnitude of the pressure gradient while holding the gas-water surface tension constant can be seen from

$$\frac{\partial(v^*_{(w)} N^{-1})}{\partial N_\gamma^{-1}} = \frac{R^{*2} A - R^* N_\gamma G}{\tilde{D} + R^* N_{\kappa+\varepsilon} H} \tag{8-16}$$

in which we have observed that $(R^{*2}A - R^*N_\gamma G)$ should be nearly independent of the magnitude of the pressure gradient. Because $v^*_{(w)}$ is assumed to be positive, the rate at which a water segment is displaced increases as the magnitude of the pressure gradient is increased. Of greater concern here is the efficiency with which the displacement proceeds. In particular, for which of the pores shown in Figure 8-1 is the effect more significant?

Consider

$$\frac{\partial^2(v^*_{(w)} N_\gamma^{-1})}{\partial N_\gamma^{-1}} = \frac{\partial v^*_{(w)}}{\partial R^*} \tag{8-17}$$

this derivative is always positive, suggesting

3. For values of N_γ less than the critical value, the displacement efficiency of a stable foam can be increased by decreasing the magnitude of the pressure gradient. The optimum value for N_γ is only slightly less than the critical value.

effect of viscosity ratio The effect of changing the gas-aqueous solution viscosity ratio is seen from (8-1) to be

$$\frac{\partial(v^*_{(w)} \mu_{(g)} / \mu_{(w)})}{\partial(\mu_{(g)} / \mu_{(w)})}$$

$$= \frac{(R^{*2} A - R^* N_\gamma G)(L^*_{(w)} D + R^* N_{\kappa+\varepsilon} H)}{(\tilde{D} + R^* N_{\kappa+\varepsilon} H)^2} \tag{8-18}$$

Inequality (8-10) requires this derivative to be positive. The speed at which an aqueous segment is displaced decreases as the viscosity of the aqueous phase is increased (as $\mu_{(g)}/\mu_{(w)}$ is decreased).

In order to examine the effect upon efficiency, consider

$$\frac{\partial^2 (v^*_{(w)} \mu_{(g)} / \mu_{(w)})}{\partial (\mu_{(g)} / \mu_{(w)}) \partial R^*}$$

$$= \Big[R^*(R^*A - N_\gamma G)(1 - L^*_{(w)}) \frac{\mu_{(g)}}{\mu_{(w)}} N_{\kappa+\varepsilon} EH$$

$$+ (R^*A - N_\gamma G)(L^*_{(w)} D + R^* N_{\kappa+\varepsilon} H) \tilde{D}$$

$$+ R^*A(L^*_{(w)} D + R^* N_{\kappa+\varepsilon} H)(\tilde{D} + R^* N_{\kappa+\varepsilon} H) \Big]$$

$$\times [\tilde{D} + R^* N_{\kappa+\varepsilon} H]^{-3} \tag{8-19}$$

This derivative is positive, which allows us to conclude

4. For values of N_γ less than the critical value, the displacement efficiency of a stable foam can be raised by increasing the viscosity of the aqueous solution (decreasing $\mu_{(g)}/\mu_{(w)}$).

This suggests that the displacement efficiency of a stable foam might be enhanced, if it were formed from an aqueous solution of an appropriate polymer. The effect can be a large one, since

$$\frac{\partial^2 (v^*_{(w)} \mu_{(g)} / \mu_{(w)})}{\partial (\mu_{(g)} / \mu_{(w)}) \partial R^*}$$

is on the order of

$$\frac{\mu_{(w)}}{\mu_{(g)}} \frac{\partial (v^*_{(w)} \mu_{(g)} / \mu_{(w)})}{\partial R^*}$$

(estimating that the magnitudes of A, D, E, G, and H are of the order unity).

effect of surface tension Returning to (8-1), we see that

$$\frac{\partial v^*_{(w)}}{\partial N_\gamma} = \frac{R^* G}{\tilde{D} + R^* N_{\kappa+\varepsilon} H} \tag{8-20}$$

Since this derivative must always be negative, we observe

5. For values of N_γ less than the critical value, the rate of displacement of a stable foam can be enhanced by decreasing the gas-water surface tension.

This agrees with Kanda and Schechter's (1976) observation that breakthrough time is increased by increasing the gas-water surface tension.

In order to investigate the effect upon efficiency, consider

$$\frac{\partial^2 v^*_{(w)}}{\partial N_\gamma \partial R^*} = \frac{-\ G\tilde{D}}{(\ \tilde{D} + R^* N_{\kappa+\varepsilon} H\ )^2} \tag{8-21}$$

This must also be negative, which means

6. For values of N_γ less than the critical value, the displacement efficiency of a stable foam can be raised by increasing the gas-aqueous solution surface tension. The optimum value for N_γ is only slightly less than the critical value.

Since the magnitude of $\partial v^*_{(w)}/\partial R^*$ is on the order of $R^{*2} N_{\kappa+\varepsilon}$ times that of $\partial^2 v^*_{(w)}/\partial N_\gamma \partial R^*$ (estimating N_γ to be of order unity), the effect will be largest when the surface viscosities are small.

In contrast, Kanda and Schechter (1976) state that in their experimental studies the efficiency of a foam displacement was raised by decreasing the surface tension. From their data, it appears that N_γ based upon the mean pore diameter was considerably larger than unity. Consequently, they may have seen displacement only in the larger pores. As the interfacial tension was decreased, displacement took place in a larger fraction of the pores, resulting in an enhanced displacement efficiency.

effect of surface viscosities From (8-1),

$$\frac{\partial v^*_{(w)}}{\partial N_{\kappa+\varepsilon}} = -\ R^{*2} H(\ R^* A - N_\gamma G\ )(\ \tilde{D} + R^* N_{\kappa+\varepsilon} H\ )^{-2} \tag{8-22}$$

Since this derivative must always be negative, we can say

7. For values of N_γ less than the critical value, the rate of displacement of a stable foam can be enhanced by decreasing the gas-water surface viscosities.

This agrees with Kanda and Schechter's (1976) observation that breakthrough time increased as the surface shear viscosity increased. As suggested above, the foam may not have been displaced within the smaller pores in their experiments. The interpretation is that breakthrough time increased for the larger pores in which the foam was displaced.

Since

$$\frac{\partial^2 v^*_{(w)}}{\partial N_{\kappa+\varepsilon}\partial R^*} = - [\, R^{*2} AH(\tilde{D} + R^* N_{\kappa+\varepsilon} H)$$

$$+ 2R^*\tilde{D}H(R^*A - N_\gamma G)][\, \tilde{D} + R^* N_{\kappa+\varepsilon} H \,]^{-3} \qquad (8\text{-}23)$$

must also be negative, we conclude

8. For values of N_γ less than the critical value, the displacement efficiency of a stable foam can be raised by increasing the gas-aqueous solution surface viscosities.

This also agrees with the observations of Kanda and Schechter (1976). Since the magnitude of $\partial v^*_{(w)}/\partial R^*$ is on the order of $N_{\kappa+\varepsilon}$ times that of $\partial^2 v^*_{(w)}/\partial N_{\kappa+\varepsilon}\partial R^*$, the effect will be larger when the surface viscosities are small.

References

Albrecht, R. A., and S. S. Marsden, *Soc. Pet. Eng. J.* **10,** 51 (1970).

Batra, V. K., and F. A. L. Dullien, *Soc. Pet. Eng. J.* **13,** 256 (1973).

Bernard, G. G., *Prod. Mon.* **27** (1), 18 (1963).

Bernard, G. G., and L. W. Holm, *Soc. Pet. Eng. J.* **10,** 9 (1970).

Hahn, P. S., T. R. Ramamohan, and J. C. Slattery, *AIChE J.* **31,** 1029 (1985).

Holm, L. W., *Soc. Pet. Eng. J.* **10,** 359 (1968).

Holm, L. W., *J. Pet. Technol.*, 1499 (Dec., 1970).

Kanda, M., and R. S. Schechter, "On the Mechanism of Foam Formation in Porous Media," SPE 6200, Society of Petroleum Engineers, P.O. Box 833836, Richardson, TX 75083-3836 (1976).

Mast, R. F., "Microscopic Behavior of Foam in Porous Media," SPE 3997, Society of Petroleum Engineers, P.O. Box 833836, Richardson, TX 75083-3836 (1972).

Minssieux, L., *J. Pet. Technol.* 100 (Jan., 1974).

Payatakes, A. C., C. Tien, and R. M. Turian, *AIChE J.* **19,** 58 (1973).

Raza, S. H., *Soc. Pet. Eng. J.* **10,** 328 (1970).

Slattery, J. C., *AIChE J.* **25,** 283 (1979).

Exercise

4.1.8-1 *displacement of residual oil by an unstable foam* Holm (1968) suggested that foams capable of displacement in porous media are unstable in the sense that liquid and gas are transported through the progressive rupture and regeneration of the foam structure. This is supported by the analysis in the text, which shows that there is a critical interfacial tension above which the foam can not move. For most foams, the interfacial tension is considerably larger than this critical value.

Follow Hahn *et al.* (1985) in developing a *qualitative* analysis for the displacement of an unstable foam in a porous structure. This will form the basis for a discussion of two aspects of mobility control using unstable foams: speed of displacement and displacement efficiency. Although the assumptions are quite different, this qualitative analysis takes the same form as that developed in the text.

Hahn *et al.* (1985) have drawn a number of conclusions.

An unstable foam will preferentially displace through an intermediate range of pore sizes, bypassing any residual oil present both in the larger pores and in those pores whose diameter is less than the critical diameter required for foam formation.

The speed of displacement of an unstable foam will be enhanced by making it less stable (increasing the magnitude of the negative disjoining pressure attributable to London-van der Waals forces), increasing the surface tension, decreasing the surface viscosities, increasing its quality (volume fraction of gas), increasing the bubble size, and decreasing the viscosity of the foaming agent solution. Naturally, there is a maximum quality consistent with the existence of a foam.

The displacement efficiency of an unstable foam will be enhanced by making it more stable (reducing the magnitude of the negative disjoining pressure attributable to the London-van der Waals forces), decreasing the surface tension, decreasing the surface viscosities, decreasing its quality, decreasing the bubble size, and increasing the viscosity of the foaming agent solution. But note that there are both a minimum quality and a minimum bubble size below which unstable foams are not formed.

The ideal objective would be to increase both the speed of displacement of an unstable foam as well as its displacement efficiency. This is possible only by decreasing the surface viscosities. With all of the other variables, whenever either the speed of displacement or the displacement efficiency is increased, the other is decreased.

Whether one alters the concentration of the existing surfactant or introduces a new surfactant, it is nearly impossible to reduce the surface viscosities without also changing other physical properties such as the surface tension. For example, if the surface tension is increased, the displacement efficiency will tend to decrease; if the surface tension is decreased, the speed of displacement will tend to decrease. If both the speed of displacement and the displacement efficiency are reduced, the effects of the surface viscosities must be dominant. The question as to whether they will be dominant in any given situation is beyond the scope of this discussion. It can be answered only by a quantitative analysis.

4.1.9 Capillary rise (Giordano and Slattery 1983; Stoodt and Slattery 1984) The experiment with which we are concerned here is shown schematically in Figure 9-1. A precision-bore, redrawn, Pyrex, capillary tube of radius R is mounted vertically. A length of flexible Teflon tubing is attached to the bottom by means of a Teflon fitting. For time $t < 0$, the equilibrium position of the liquid-gas interface in the capillary is $L_0^{(1)}$ above the bottom of the capillary and the equilibrium position of the liquid-gas interface in the Teflon tubing is h_0 below the bottom of the capillary tube. At time $t = 0$, h_0 is reduced to h and the liquid-gas interface in the capillary rises to a new equilibrium position

$$L_{eq}^{(1)} = \frac{2\gamma \cos \Theta}{R(\rho^{(1)} - \rho^{(2)})g} - h \tag{9-1}$$

in which Θ is the contact angle measured through the displacing liquid phase, $\rho^{(1)}$ the density of the liquid phase, $\rho^{(2)}$ the density of the displaced gas phase, and g the acceleration of gravity.

Let us take as our system all of the liquid and gas in the capillary and Teflon tubing. We will make the following physical assumptions in analyzing the flow in this system.

i) Both the displacing and displaced fluids are incompressible and Newtonian.

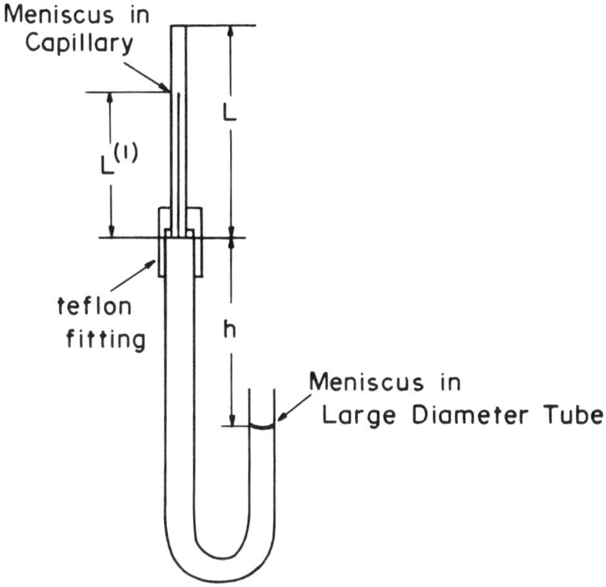

Figure 4.1.9-1 Vertical capillary used to study capillary rise. The height $L^{(1)}$ of liquid in the capillary is measured as a funciton of time. The time rate of change of $L^{(1)}$ is determined by the value of h chosen.

ii) The fluid-fluid interfacial stress-deformation behavior can be represented by the linear Boussinesq surface fluid model (see Sec. 2.2.2).

iii) The location of the dividing surface can be arbitrarily specified for any choice of reference state (see Secs. 1.3.6 and 5.2.3). Take the reference state to be the static state corresponding to the limit in which the speed of displacement of the fluid-fluid interface is zero. Let $\rho_0^{(\sigma)}$ be the total surface mass density in this reference state. The dividing surface in this reference state is located such that it is sensibly coincident with the interface and

$$\frac{\rho_0^{(\sigma)}}{\rho^{(2)}R} << 1 \tag{9-2}$$

iv) Inertial forces are neglected with respect to viscous forces within the bulk phases, which implies

$$N_{Re}^{(j)} \equiv \frac{\rho^{(j)} V R}{\mu^{(j)}} << 1 \tag{9-3}$$

where $N_{Re}^{(j)}$ ($j = 1,2$) is the Reynolds number for phase j, V the speed of displacement of the phase interface in the capillary and $\mu^{(j)}$ the viscosity of phase j.

v) We require that the Bond number

$$N_{Bo} \equiv \frac{| \rho^{(1)} - \rho^{(2)} | g R^2}{\gamma} << 1 \tag{9-4}$$

which means that the effect of gravity upon the configuration of the interface can be neglected.

vi) In the dividing surface, interfacial viscous forces dominate bulk viscous forces,

$$N_{\kappa+\varepsilon} \equiv \frac{\kappa + \varepsilon}{\mu^{(2)} R} >> 1 \tag{9-5}$$

Here κ is the interfacial dilatational viscosity and ε the interfacial shear viscosity.

vii) The capillary number N_{ca} is small,

$$N_{ca} \equiv \frac{\mu^{(2)} V}{\gamma} << 1 \tag{9-6}$$

viii) The speed of displacement V is sufficiently small that the surfactant concentration in the dividing surface may be considered nearly independent of position and that interfacial tension and the two interfacial viscosities may be treated as constants [see Giordano and Slattery 1983, Appendix A].

ix) The displacement is stable in the sense that, in a frame of reference fixed with respect to the common line, the flow is independent of time.

x) The rate at which work is done by viscous forces at the entrance and exit for the system is neglected. This would be exact, if the gas were in Poiseuille flow at these surfaces.

xi) The rate of viscous dissipation of mechanical energy within the bulk phases in the capillary is estimated as though these phases are in Poiseuille flow. Outside the capillary, the rate of viscous dissipation of mechanical energy is neglected.

xii) In forming the integral mechanical energy balance, the configuration of the fluid-fluid interface in the capillary is represented as a spherical segment (correct to the lowest order in N_{ca}). The interface in the Teflon tubing is taken to be a plane.
xiii) The area integral of the interfacial viscous dissipation in the inner region near the common line is neglected in comparison with the dissipation in the rest of the interface. This appears reasonable, since in any well posed problem the viscous dissipation must be everywhere bounded and the area of the interface in the inner region is neglected compared to the remainder of the interface. (See Sec. 3.3.3 for explanation of inner and outer regions).

xiv) The effect of the viscous surface stress acting on the common line is neglected with the assumption that the contact angle measured at the common line is unaffected by small perturbations of the capillary number N_{ca} from zero. This assumption is supported by experimental observations of the contact angle at a small distance from the common line (see Sec. 2.1.11 for a review).

Under these conditions, the integral mechanical balance of Sec. 4.1.4 readily reduces for this system to

$$
\begin{aligned}
\frac{d}{dt}\int_{R_{(sys)}} \rho\phi \, dV
&= \int_{S_{(ent\ ex)}} [-\mathcal{P}\mathbf{v}\cdot\mathbf{n} + \mathbf{v}\cdot\mathbf{S}\cdot\mathbf{n}] \, dA \\
&\quad - \int_{R_{(sys)}} \operatorname{tr}(\mathbf{S}\cdot\nabla\mathbf{v}) \, dV \\
&\quad + \int_{\Sigma} [-\gamma \operatorname{div}_{(\sigma)}\mathbf{v}^{(\sigma)} + \operatorname{tr}(\mathbf{S}^{(\sigma)}\cdot\nabla\mathbf{v}^{(\sigma)})] \, dA \\
&\quad + \int_{C} [\gamma\,\mathbf{v}^{(\sigma)}\cdot\boldsymbol{\mu} + \mathbf{v}^{(\sigma)}\cdot\mathbf{S}^{(\sigma)}\cdot\boldsymbol{\mu}] \, ds
\end{aligned}
\tag{9-7}
$$

The term on the left of (9-7) may be evaluated as

$$
\frac{d}{dt}\int_{R_{(sys)}} \rho\phi \, dV = \pi R^2 (\rho^{(1)} - \rho^{(2)}) g (L^{(1)} + h) V \tag{9-8}
$$

Here we have recognized that

$$V = \frac{dL^{(1)}}{dt} = \frac{A_t}{\pi R^2} \frac{dh}{dt} \tag{9-9}$$

in which A_t is the cross-sectional area of the Teflon tubing.

In view of assumption x, the first term on the right of (9-7) may be written as

$$\int_{S_{(ent\ ex)}} [- \mathcal{P} \mathbf{v} \cdot \mathbf{n} + \mathbf{v} \cdot \mathbf{S} \cdot \mathbf{n}] \, dA$$

$$= \Delta \mathcal{P} \, V \pi R^2 \tag{9-10}$$

where

$$\Delta \mathcal{P} \equiv \mathcal{P}_{(ent)} - \mathcal{P}_{(ex)} \tag{9-11}$$

is the difference in modified pressure between the entrance and exit surfaces in Figure 9-1.

With assumption xi, the second term on the right of (9-7) simplifies to

$$\int_{R_{(sys)}} \text{tr}(\mathbf{S} \cdot \nabla \mathbf{v}) \, dV = 8\pi\mu^{(1)} L^{(1)} V^2 + 8\pi\mu^{(2)} (L - L^{(1)}) V^2 \tag{9-12}$$

with the understanding that $\mu^{(1)}$ is the viscosity of the displacing liquid phase and $\mu^{(2)}$ is the viscosity of the displaced gas phase.

The surface divergence theorem (Sec. A.6.3) requires

$$\int_{\Sigma} \text{div}_{(\sigma)} \mathbf{v}^{(\sigma)} \, dA = \int_{C} \mathbf{v}^{(\sigma)} \cdot \boldsymbol{\mu} \, ds - \int_{\Sigma} 2H \, \mathbf{v}^{(\sigma)} \cdot \boldsymbol{\xi} \, dA \tag{9-13}$$

With the approximation that the configuration of the interface can be represented as a spherical segment (assumption xii), we note

$$\int_{\Sigma} 2H\gamma \mathbf{v}^{(\sigma)} \cdot \boldsymbol{\xi} \, dA = 2\pi R \gamma V \cos \Theta \tag{9-14}$$

By assumptions i through ix, the velocity distribution in the phase interface is described by the limiting case analyzed in Secs. 3.5.1 and 3.5.2. Using assumption xiii, we find that the rate of viscous dissipation of mechanical energy in the interface becomes

$$\int_{\Sigma} \text{tr}(\, \mathbf{S}^{(\sigma)} \cdot \mathbf{D}^{(\sigma)}\,)\, dA = \frac{2\pi}{R^2}(\,\kappa + \varepsilon\,) V^2 A^2 G \tag{9-15}$$

Here A is determined in Sec. 3.5.2. and

$$G \equiv (1 - \nu)\,\frac{P_{\nu-1}^{-1}(\,\sin\Theta\,)}{P_{\nu}^{-1}(\,\sin\Theta\,)} + (\,1 - \nu^2\,)\sin\Theta \tag{9-16}$$

in which $P_{\nu-1}^{-1}$ is the associated Legendre function of the first kind (Abramowitz and Stegun 1965) and ν satisfies

$$\nu(\,\nu + 1\,) = \frac{2\varepsilon}{\kappa + \varepsilon} \tag{9-17}$$

Figure 9-2 shows G as a function of κ/ε and Θ. It is noteworthy that

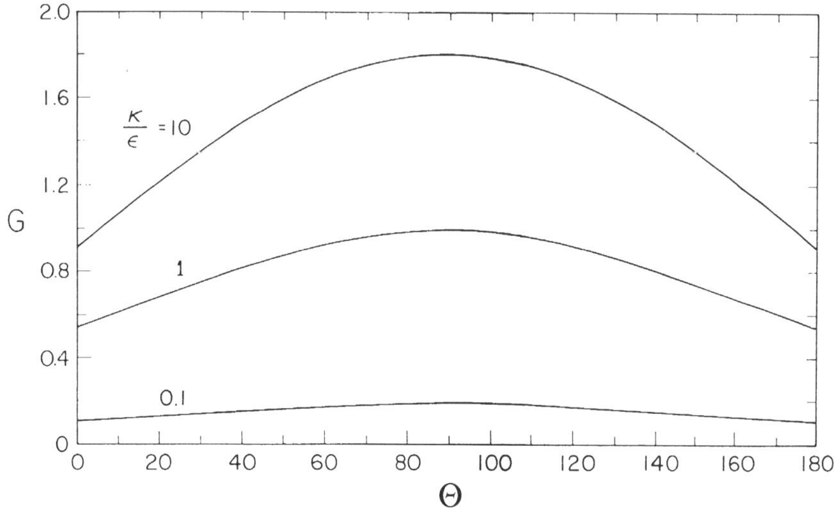

Figure 4.1.9-2 Function G defined by (9-16).

$$\text{limit}\ \frac{\kappa}{\varepsilon} \to \infty : \ G \to 1 + \sin \Theta \tag{9-18}$$

and

$$\text{limit}\ \frac{\kappa}{\varepsilon} \to 0 : \ G \to 0 \tag{9-19}$$

Assumption xiv leads us to say

$$\int_C \mathbf{v}^{(\sigma)} \cdot \mathbf{S}^{(\sigma)} \cdot \boldsymbol{\mu}\ ds \doteq 0 \tag{9-20}$$

With (9-8), (9-10), (9-12) through (9-15), and (9-20), equation (9-7) may be written as

$$V = \frac{dL^{(1)}}{dt}$$

$$= \frac{1}{\mu^{(2)}} \left[\nabla \mathcal{P} - (\rho^{(1)} - \rho^{(2)})(L^{(1)} + h)g + \frac{2\gamma}{R} \cos \Theta \right]$$

$$\times \left[\frac{8}{R} \left(N_\mu \frac{L^{(1)}}{R} + \frac{L - L^{(1)}}{R} \right) + \frac{2}{R} N_{\kappa+\varepsilon} A^2 G \right]^{-1} \tag{9-21}$$

where we have introduced

$$N_\mu \equiv \frac{\mu^{(1)}}{\mu^{(2)}} \tag{9-22}$$

Integrating (9-21), we conclude

$$\frac{tR}{\mu^{(2)}}(\rho^{(1)} - \rho^{(2)})g + \frac{8}{R} (N_\mu - 1)(L^{(1)} - L_0^{(1)})$$

$$= \left[\frac{8L}{R} + 2N_{\kappa+\varepsilon} A^2 G + \frac{8}{R} (N_\mu - 1) L_{eq}^{(1)} \right]$$

$$\times \ln \left(\frac{L_{eq}^{(1)} - L_0^{(1)}}{L_{eq}^{(1)} - L^{(1)}} \right) \tag{9-23}$$

For the purpose of more clearly identifying surface viscous effects, it will be convenient to rearrange (9-23) as

$$Z = S \ln X \tag{9-24}$$

where

$$Z \equiv \frac{tR^2}{\mu^{(2)}} (\rho^{(1)} - \rho^{(2)})g + 8(N_\mu - 1)(L^{(1)} - L_0^{(1)})$$

$$- [8L + 8(N_\mu - 1)L_{eq}^{(1)}] \ln X \tag{9-25}$$

$$S \equiv 2R\, N_{\kappa+\varepsilon} A^2 G \tag{9-26}$$

$$X \equiv \frac{L_{eq}^{(1)} - L_0^{(1)}}{L_{eq}^{(1)} - L^{(1)}} \tag{9-27}$$

In the absence of interfacial viscous effects, we will find it helpful to express (9-23) as

$$Y = R^2 t \tag{9-28}$$

in which we have introduced

$$Y \equiv \frac{\mu^{(2)}}{(\rho^{(1)} - \rho^{(2)})g} \{[8L + 8(N_\mu - 1)L_{eq}^{(1)}] \ln X$$

$$- 8(N_\mu - 1)(L^{(1)} - L_0^{(1)})\} \tag{9-29}$$

experimental results (Stoodt and Slattery 1984) The data for three runs, in which a doubly distilled water-air interface is displaced through tube 1, are plotted in Figure 9-3 as suggested by (9-28). Figure 9-4 is a similar plot of data for two runs, in which an octane-air interface is displaced through tube 2. These data plots confirm both that our optically measured values for the tube radii are correct and that the surface viscosities for these two interfaces are either zero or below the sensitivity of this experiment, as we would expect for relatively clean interfaces.

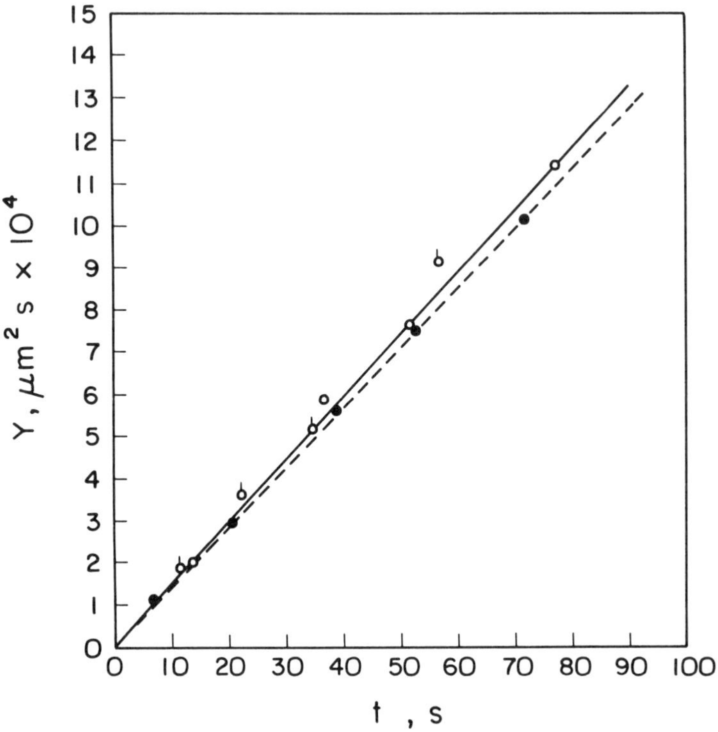

Figure 4.1.9-3 Data for three runs in which a doubly distilled water-air interface is displaced through tube 1, plotted as suggest by (9-26). The dashed line corresponds to the optically measured R = 37.5 μm; the solid line is the least squares fit of (9-26) to the data, suggesting R = 38.5 μm.

As another check on our experiment technique, we compared the equilibrium heights measured in these experiments with those predicted by (9-1), using the optically measured tube radii and assuming $\Theta = 0$. For the doubly distilled water-air interface in tube 1, the measured value of $L_{eq}^{(1)}$ was 0.387 compared with 0.395 from (9-1). For the octane-air interface in tube 2, we observed 0.123 compared with 0.129 from (9-1).

Surface viscous effects were expected and observed for the interface between air and an aqueous solution of 0.1 wt% dodecyl sodium sulfate. Figure 9-5 displays for a selected run the deviation from (9-23) with $N_{\kappa+\varepsilon} = 0$.

A least squares fit of (9-24) to the data for two runs in tube 1 shown in Figure 9-6. The conclusion is that with 95% confidence limits

$$A^2 G(\kappa + \varepsilon) = 25.3 \pm 0.8 \text{ mN s/m}$$

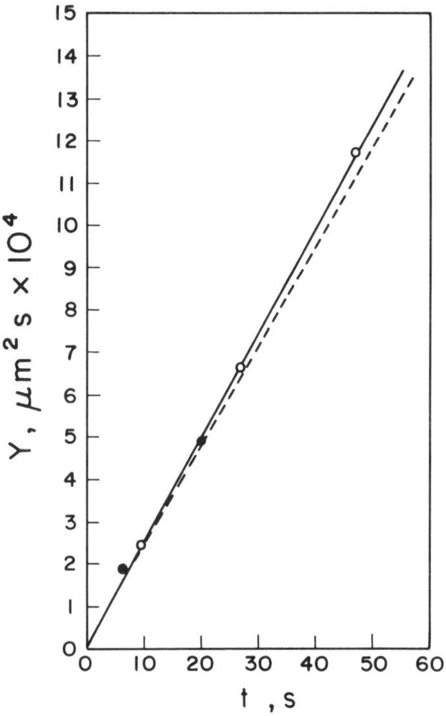

Figure 4.1.9-4 Data for two runs in which an octane-air interface is displaced through tube 2, plottted as suggested by (9-26). The dashed line corresponds to the optically measured R = 48.6 μm; the solid line is the least square fit of (9-26) to the data, suggesting R = 49.7 μm.

A similar least squares fit of (9-24) to the data for two runs in tube 2 is given in Figure 9-7 with the conclusion that

$$A^2 G(\kappa + \varepsilon) = 27.0 \pm 0.4 \text{ mN s/m}$$

For this system, N_{μ} = 50.8. With 0° ≤ Θ ≤ 30°, A = 0.5 (see Sec. 3.5.2). For the same system, Jiang *et al.* (1983; see also Sec. 3.5.3) have measured $\varepsilon \sim 1$ mN s/m. If we assume Θ = 0, (9-18) implies that G = 1. We conclude that for the two runs in tube 1

$$\kappa + \varepsilon = 101 \pm 3 \text{ mN s/m}$$

and for the two runs in tube 2

$$\kappa + \varepsilon = 108 \pm 2 \text{ mN s/m}$$

There is no apparent effect of tube diameter and the surface dilatational viscosity appears to be two orders of magnitude larger than the surface shear viscosity.

In the displacement of a solution-air interface, the equilibrium height $L_{eq}^{(1)}$ was approached slowly. Comparison with (9-1) indicated that the equilibrium contact angle was substantially larger than the apparent dynamic contact angle.

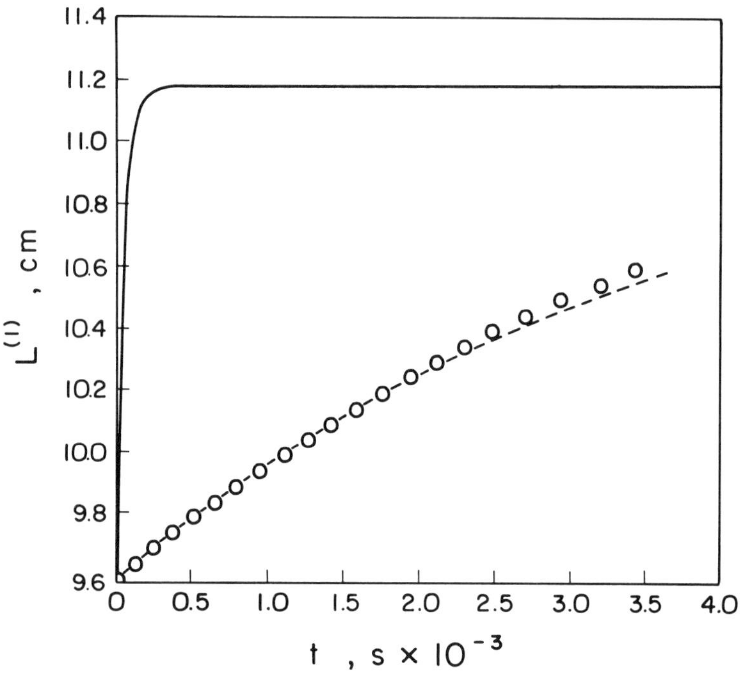

Figure 4.1.9-5 Displacement of solution-air interface through tube 1, as observed in run II. The solid line corresponds to the Washburn (West 1912; Washburn 1921) equation (9-12) with $N_{\kappa+\varepsilon} = 0$; the dashed line is (9-19) with $\kappa + \varepsilon = 101$ mN s/m. In both cases, the contact angle Θ is assumed to be zero.

References

Abramowitz, M., and I. A. Stegun, "Handbook of Mathematical Functions," Dover, New York (1965).

Fisher, L. R., and P. D. Lark, *J. Colloid Interface Sci.* **69,** 486 (1979).

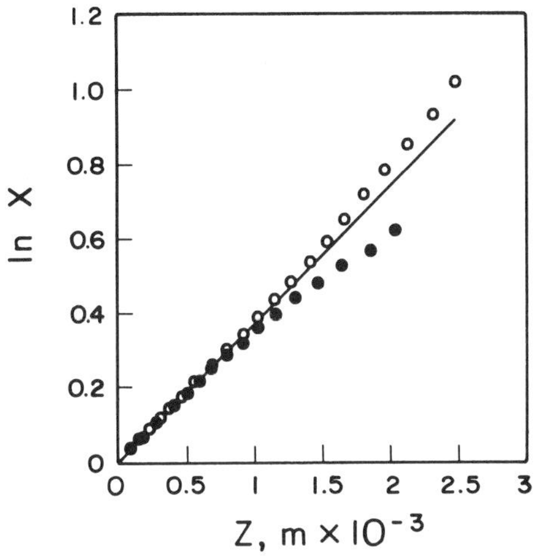

Figure 4.1.9-6 Data for two runs in which a solution-air interface is displaced through tube 1, plotted as suggested by (9-22). The solid line is the least squares fit of (9-22) to the data, suggesting $\kappa + \varepsilon = 101 \pm 3$ mN s/m.

Giordano, R. M., and J. C. Slattery, *AIChE J.* **29,** 483 (1983).

Jiang, T. S., J. D. Chen, and J. C. Slattery, *J. Colloid Interface Sci.* **96,** 7 (1983).

Sen, B. L., and J. C. Slattery, Appendix B of Giordano and Slattery (1983).

Slattery, J. C., *AIChE J.* **20,** 1145 (1974).

Slattery, J. C., *AIChE J.* **25,** 283 (1979).

Stoodt, T. J., and J. C. Slattery, *AIChE J.* **30,** 564 (1984).

Washburn, E. W., *Phys. Rev.* **17,** 273 (1921).

West, G. D., *Proc. R. Soc. London A* **86,** 20 (1912).

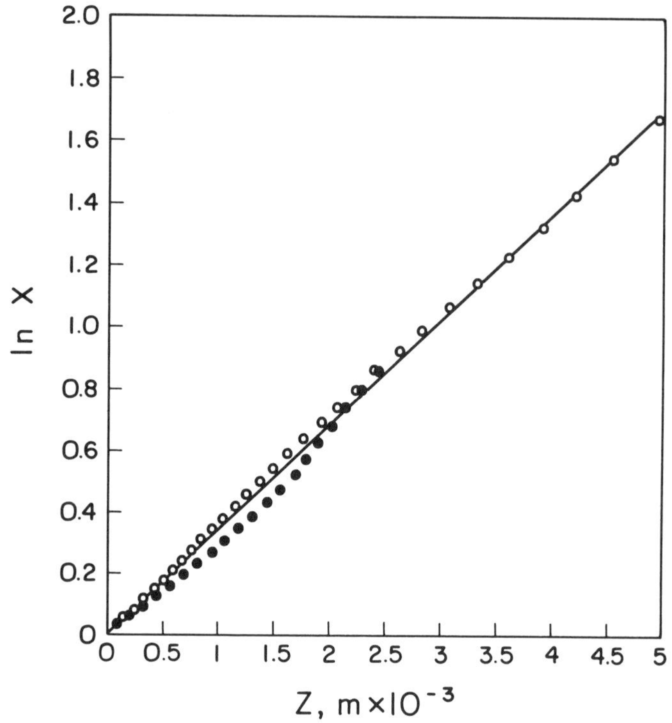

Figure 4.1.9-7 Data for two runs in which a solution-air interface is displaced through tube 2, plotted as suggested by (9-22). The solid line is the least squares fit of (9-22) to the data, suggesting $\kappa + \varepsilon = 108 \pm 2$ mN s/m.

Exercises

4.1.9-1 *more on Washburn equation* (West 1912; Washburn 1921) In Exercise 4.1.3-1, we derived the Washburn equation starting from the axial component of the integral momentum balance. Equation (9-21) is a generalization of the Washburn equation, in which we have accounted for the effect of the interfacial viscosities.

Conclude that, in the absence of the effects of the interfacial viscosities, the derivation of the Washburn equation requires only assumptions i, iii, iv, and viii through xii of the text as well as assumption d of Exercise 4.1.3-1.

Figures 9-3 and 9-4 constitute experimental verifications of the Washburn equation for systems in which the effects of the interfacial viscosities can be neglected. For further experimental studies, see Fisher and Lark (1979).

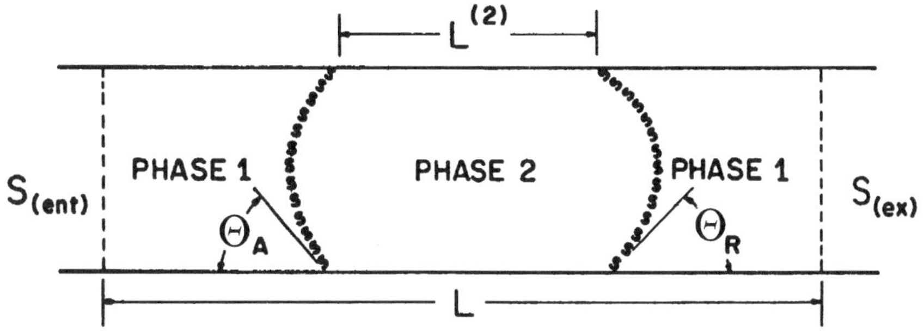

Figure 4.1.9-8 *Segment of phase 2 displaced by phase 1.* In a frame of reference that is fixed with respect to the tube, the entrance $S_{(ent)}$ and the exit $S_{(ex)}$ for the system are stationary; the interfaces move from left to right. The contact angles are measured through phase 1.

4.1.9-2 *displacement of segment* (Giordano and Slattery 1983) Follow the derivation of (9-21) but neglect the effects of gravity in considering the displacement of the segment of phase 2 in Figure 9-8. Conclude that the speed of displacement of the segment

$$V = \frac{1}{\mu^{(2)}} \left[\Delta p + \frac{2\gamma}{R} \left(\cos \Theta_A - \cos \Theta_R \right) \right]$$

$$\times \left[\frac{8}{R} \left(N_\mu \frac{L - L^{(2)}}{R} + \frac{L^{(2)}}{R} \right) \right]^{-1}$$

where

$$A_A = A(\Theta_A , N_\mu)$$

$$A_R = A(\pi - \Theta_R , N_\mu^{-1})$$

and

$$G_A = G(\Theta_A , \kappa/\varepsilon)$$

$$G_R = G(\pi - \Theta_R , \kappa/\varepsilon)$$

The subscript A refers to the advancing (left) interface in Figure 9-8 and the subscript R refers to the receding (right) interface.

4.1.9-3 *relative effects of interfacial viscosities upon displacement of residual oil* (Sen and Slattery 1983) As explained in Sec. 4.1.7, Slattery (1974, 1979) constructed qualitative analyses of the displacement process assuming that the rate limiting steps are the pauses during the creeping motion between jumps. His arguments were based upon the axial component of an integral momentum balance for that period of time in which the common line is stationary.

We are now in the position to repeat those arguments, assuming that the rate limiting steps are the creeping motions between jumps and that for simplicity the integral mechanical energy balance of Exercise 4.1.9-2 applies. The results are substantially the same with one exception. Whereas Slattery (1974, 1979) was not able to distinguish between the effects of the interfacial dilatational and shear viscosities, we can with the result of Exercise 4.1.9-2.

As an example, let us consider the effect upon the rate of displacement of a liquid segment when the concentration $\rho_{(s)}^{(1)}$ of surfactant in phase 1 is changed. From Exercise 4.1.9-2,

$$\frac{dV^*}{d\rho_{(s)}^{(1)}} = \frac{\partial V^*}{\partial N_\gamma} \frac{dN_\gamma}{d\rho_{(s)}^{(1)}} + \frac{\partial V^*}{\partial N_\kappa} \frac{dN_\kappa}{d\rho_{(s)}^{(1)}} + \frac{\partial V^*}{\partial N_\varepsilon} \frac{dN_\varepsilon}{d\rho_{(s)}^{(1)}} + \dots \tag{1}$$

where

$$V^* \equiv \frac{V \mu^{(2)}}{|\Delta p| R} \tag{2}$$

$$\frac{\partial V^*}{\partial N_\kappa} = -\frac{V^*}{d^*} \left\{ 2(A_A^2 G_A + A_R^2 G_R) + 2\left(\frac{\kappa}{\varepsilon} + 1\right) \left[A_A^2 \frac{\partial G_A}{\partial(\kappa/\varepsilon)} + A_R^2 \frac{\partial G_R}{\partial(\kappa/\varepsilon)} \right] \right\} \tag{3}$$

$$\frac{\partial V^*}{\partial N_\varepsilon} = -\frac{V^*}{d^*} \left\{ 2(A_A^2 G_A + A_R^2 G_R) - 2\frac{\kappa}{\varepsilon}\left(\frac{\kappa}{\varepsilon} + 1\right) \left[A_A^2 \frac{\partial G_A}{\partial(\kappa/\varepsilon)} + A_A^2 \frac{\partial G_R}{\partial(\kappa/\varepsilon)} \right] \right\} \tag{4}$$

$$N_\gamma \equiv \frac{\gamma}{|\Delta p| R} \qquad N_\kappa \equiv \frac{\kappa}{\mu^{(2)} R} \qquad N_\varepsilon \equiv \frac{\varepsilon}{\mu^{(2)} R} \tag{5}$$

and

$$d \equiv 8\left(N_\mu \frac{L^{(1)}}{R} + \frac{L^{(2)}}{R}\right) + 2N_{\kappa+\varepsilon}\left(A_A^2\, G_A + A_R^2\, G_R \right) \tag{6}$$

The effects of the two interfacial viscosities are characterized by the relative magnitudes of $\partial V^\star/\partial N_\kappa$ and $\partial V^\star/\partial N_\varepsilon$. There are several cases to consider.

If κ/ε remains a constant as κ and ε change, we see from (3) and (4) that

$$\frac{\partial V^\star/\partial N_\kappa}{\partial V^\star/\partial N_\varepsilon} = 1 \tag{7}$$

This may be considered to be the case treated by Slattery (1974, 1979).

If κ/ε changes, say as the result of a change in $\rho^{(\sigma)}_{(S)}$, equations (3) and (4) together with Figure 9-2 demand

$$\frac{\partial V^\star/\partial N_\kappa}{\partial V^\star/\partial N_\varepsilon} > 1 \tag{8}$$

If $\kappa/\varepsilon \leq 1$ numerical computations show

$$\frac{\partial V^\star/\partial N_\kappa}{\partial V^\star/\partial N_\varepsilon} >> 1 \tag{9}$$

If $\kappa/\varepsilon = 0$, it follows from (9-19), (3) and (4) that

$$\frac{\partial V^\star}{\partial N_\varepsilon} = 0 \tag{10}$$

and

$$\text{limit } \frac{\kappa}{\varepsilon} \to 0 : \quad \frac{\partial V^\star/\partial N_\kappa}{\partial V^\star/\partial N_\varepsilon} \to \infty \tag{11}$$

Our conclusion is that, while both interfacial viscosities play important roles, the interfacial dilatational viscosity may be the more important of the two. The experiments described in the text also suggest that it may be the larger of the two.

4.2 Approximate solutions

4.2.1 A few special techniques In the preceding sections, the emphasis has been on alternative descriptions of the physical aspects of flows, in order to simplify the corresponding mathematical solutions.

Now I would like to direct your attention to the approximate solution of physically motivated mathematical problems. Although this is not the place to survey the varied techniques that are available for obtaining approximate solutions to differential equations, the variational principle and the two bounding principles developed in the sections that follow are not readily accessible from the more general discussions in the literature. They are applicable to only a few classes of the problems with which we are concerned, and they take specialized forms when applied to these problems. I have placed their discussion here, since they further illustrate the use of integral averages.

4.3 Variational principle

4.3.1 Introduction Before I intoduce the application that I have in mind, let me explain the term *variational principle*.

Let us assume that we seek the solution of a particular differential equation consistent with a set of boundary conditions. With some arbitrary definition of *approximate*, we can propose an n parameter family of approximate or ***trial*** solutions for this problem. Perhaps they all satisfy the boundary conditions, but not necessarily the differential equation. A ***variational principle*** corresponding to this boundary value problem and this family of trial solutions would say that the best of all these trial solutions is the one corresponding to which a particular integral, the **variational integral**, achieves a stationary value (often either a minimum or maximum) as a function of these n parameters.

Perhaps the concept of a variational principle can be clarified with a simple example. Let us assume that we are given a function

$$I(y) \equiv \int_{x_1}^{x_2} F(\, x, y, y' \,)\, dx \tag{1-1}$$

of

$$y = y(x) \tag{1-2}$$

The prime is used here as shorthand for a derivative with respect to x:

$$y' \equiv \frac{dy}{dx} \tag{1-3}$$

We wish to find the boundary value problem that y(x) must satisfy, if I(y) is to achieve a stationary value.

To be more specific, let Y(x) be the exact solution to this boundary value problem and let $\eta(x)$ be some arbitrary function. The function

$$y(\, x, \varepsilon \,) = Y(x) + \varepsilon\, \eta(x) \tag{1-4}$$

may be viewed as a one parameter family of trial solutions with the property that

$$\text{at } \varepsilon = 0 : \frac{dI}{d\varepsilon} = 0 \tag{1-5}$$

In order to take advantage of this result, we can compute

$$\frac{dI}{d\varepsilon} = \int_{x_1}^{x_2} \left[\left(\frac{\partial F}{\partial y}\right)_{y'} \frac{\partial y}{\partial \varepsilon} + \left(\frac{\partial F}{\partial y'}\right)_{y} \frac{\partial y'}{\partial \varepsilon} \right] dx$$

$$= \int_{x_1}^{x_2} \left[\left(\frac{\partial F}{\partial y}\right)_{y'} \frac{\partial y}{\partial \varepsilon} + \left(\frac{\partial F}{\partial y'}\right)_{y} \frac{d}{dx}\left(\frac{\partial y}{\partial \varepsilon}\right) \right] dx \qquad (1\text{-}6)$$

In order to keep the notation as simple as possible, I denote here by d/dx a derivation with respect to x holding ε constant. It is customary to introduce the notation

$$\delta I \equiv \frac{dI}{d\varepsilon}$$

$$\delta y \equiv \frac{\partial y}{\partial \varepsilon}$$

$$\delta y' \equiv \frac{\partial y'}{\partial \varepsilon} \qquad (1\text{-}7)$$

which are referred to respectively as the **variations** of I, y and y'. We can then rewrite (1-6) as

$$\delta I = \int_{x_1}^{x_2} \left[\left(\frac{\partial F}{\partial y}\right)_{y'} \delta y + \left(\frac{\partial F}{\partial y'}\right)_{y} \delta y' \right] dx$$

$$= \int_{x_1}^{x_2} \left[\left(\frac{\partial F}{\partial y}\right)_{y'} \delta y + \left(\frac{\partial F}{\partial y'}\right)_{y} \frac{d}{dx}(\delta y) \right] dx \qquad (1\text{-}8)$$

Integrating the last term on the right by parts, we conclude that at $\varepsilon = 0$:

$$0 = \delta I$$

$$= \int_{x_1}^{x_2} \left[\left(\frac{\partial F}{\partial y}\right)_{y'} \frac{d}{dx}\left(\frac{\partial F}{\partial y'}\right)_{y} \right] \delta y \, dx + \left[\left(\frac{\partial F}{\partial y'}\right)_{y} \delta y \right]_{x=x_1}^{x=x_2} \qquad (1\text{-}9)$$

There are two cases to be considered.

If y(x) is required to satify

$$\text{at } x = x_1 : \; y = a$$

$$\text{at } x = x_2 : \; y = b \tag{1-10}$$

then the variation of y at x_1 and x_2 is zero and (1-9) simplifies to at $\varepsilon = 0$:

$$\begin{aligned} 0 &= \delta \mathbf{I} \\ &= \int_{x_1}^{x_2} \left[\left(\frac{\partial F}{\partial y} \right)_{y'} - \frac{d}{dx} \left(\frac{\partial F}{\partial y'} \right)_y \right] \delta y \, dx \end{aligned} \tag{1-11}$$

Since the variation of y is otherwise arbitrary, we can conlude that

$$\left(\frac{\partial F}{\partial y} \right)_{y'} - \frac{d}{dx} \left(\frac{\partial F}{\partial y'} \right)_y = 0 \tag{1-12}$$

This is referred to as the **Euler equation** corresponding to the variational integral **I**. The solution to (1-12) with boundary conditions (1-10) is the same function for which **I**(y) achieves a stationary value.

Now let us assume that we are *not* given a priori the boundary conditions (1-10). Since the variation of y is arbitray, (1-12) again follows from (1-9) as the Euler equation corresponding of **I**. In addition we find

$$\text{at } x = x_1, x_2 : \; \left(\frac{\partial F}{\partial y'} \right)_{y'} = 0 \tag{1-13}$$

This can be referred to as the **natural boundary conditions**. In this case, the solution to the differential equation (1-12) consistent with the natural boundary conditions (1-13) is the same function for which I(y) attains a stationary value.

In summary, we can state as a variational principle that the same function for which I(y) attains a stationary value also satisfies the differential equation (1-12) consistent with any combination of the boundary conditions (1-10) and (1-13).

Now let us see how we would go about using this variational principle.

Assume first that we wish to obtain an approximate solution to a differential equation that can be expressed in the form of (1-12) consistent with the boundary conditions (1-10). We must first propose a trial function for y(x) that satisfies the boundary conditions (1-10) and that involves one or more free parameters. Substituting this trial function into (1-1), we solve for the values of these free parameters for which **I**(y) achieves a stationary value. In this way we determine that member of the family of trial functions that most nearly approximates the exact solution to the differential equation (1-12).

If the boundary value problem with which we are concerned takes the form of (1-12) and (1-13), there are no restrictions upon our choice of trial function. It is necessary only that the trial function involve one or more free parameters. The best values for these free parameters are those for which I(y) attains a stationary value. The corresponding member of the family of trial functions is the best approximation for the exact solution to the differential equation (1-12) consistent with the boundary conditions (1-13).

In order to say whether this stationary value is a minimum, maximum, or inflection point, we would have to examine the value of the second variation of I: $\delta^2 I$. This may not be necessary, so long as it is practical to calculate the first derivatives of I with respect to each of the free parameters in trial function. When these first derivatives are set equal to zero, we have a set of equations that in principle can be solved simultaneously for the values of the parameters corresponding to the stationary value of I. This is an effective approach, when the trial function can be chosen in such a manner that these equations are all linear. On the other hand, a numerical search for the optimum values of these parameters will be practical, only if we can say beforehand that the stationary value of I is either a minimum or a maximum.

For a more complete introductory treatment of the calculus of variations, I recommend Hildebrand (1952).

Reference

Hildebrand, F. B., "Methods of Applied Mathematics," Prentice-Hall, Englewood Cliffs, N. J. (1952).

4.3.2 Variational principle for fluid statics [suggested by Ehrlich (1974)] A variational principle appropriate to static multiphase bodies may be useful in studying the static positions and configurations of phase interfaces.

For an isolated, isothermal, multiphase body totally enclosed by fixed, impermeable, adiabatic walls, the Helmholtz free energy reaches a local minimum as a function of time at equilibrium (Exercise 5.10.3-1). In reaching this conclusion, we must neglect the effects of inertial forces and of external forces. If in addition we are willing to limit out attention to disturbances in which

i) mass transfer across phase interfaces can be neglected,

ii) the concentration distributions in the adjacent phases and in the interface are not altered as equilibrium is approached, and

iii) as equilibrium is approached, P is independent of time within each phase and γ is independent of time on each dividing surface

we conclude that

$$J \equiv - \int_R P \, dV + \int_\Sigma \gamma \, dA \tag{2-1}$$

also reaches a local minimum as a function of time at equilibrium (Exercise 5.10.3-1). Here Σ denotes only the internal phase interfaces of the body; it does not include the system boundaries.

This suggests that J may be an appropriate variational integral for static multiphase bodies. Keep in mind that the context in which we will discuss J as a variational integral differs from that described in Exercise 5.10.3-1. We are not thinking of the body approaching equilibrium as a function of time. Rather we are attempting to describe a body that is already at equilibrium, and we wish to ask how J changes as we search through a family of trial interface configurations for the one that most nearly approximates the exact configuration.

I think that it is helpful to begin by visualizing that we wish to write trial functions for the configuration of each phase

$$\mathbf{z} = \boldsymbol{\chi}(\zeta) \tag{2-2}$$

and for the configuration of the interfaces

$$\mathbf{z} = \boldsymbol{\chi}^{(\sigma)}(\zeta^{(\sigma)}) \tag{2-3}$$

Considering pressure and surface tension to be explicit functions of position, we can employ the theorem of Exercise 4.3.2-1 to say

$$\delta J = - \int_R [\nabla P \cdot \delta\boldsymbol{\chi} + P \, \text{div}(\delta\boldsymbol{\chi})] \, dV$$

$$+ \int_\Sigma [\nabla_{(\sigma)}\gamma \cdot \delta\boldsymbol{\chi}^{(\sigma)} + \gamma \, \text{div}_{(\sigma)}(\delta\boldsymbol{\chi}^{(\sigma)})] \, dA \tag{2-4}$$

The first integral on the right can be rearranged by Green's transformation (Slattery 1981, p. 661):

$$- \int_R [\ \nabla P \cdot \delta\boldsymbol{\chi} + P\ \text{div}(\ \delta\boldsymbol{\chi}\ )]\ dV$$

$$= - \int_R \text{div}(\ P\ \delta\boldsymbol{\chi}\ )\ dV$$

$$= \int_\Sigma \boldsymbol{[}\ P\boldsymbol{\xi}\ \boldsymbol{]} \cdot \delta\boldsymbol{\chi}^{(\sigma)}\ dA \tag{2-5}$$

The surface divergence theorem (Sec. A.6.3) can be used to more conveniently express the second integral on the right of (2-4) as

$$\int_\Sigma [\ \nabla_{(\sigma)}\gamma \cdot \delta\boldsymbol{\chi}^{(\sigma)} + \gamma\ \text{div}_{(\sigma)}(\ \delta\boldsymbol{\chi}^{(\sigma)}\ )]\ dA$$

$$= \int_\Sigma \text{div}_{(\sigma)}(\ \gamma\ \delta\boldsymbol{\chi}^{(\sigma)}\ )\ dA$$

$$= \int_\Sigma [\ \text{div}_{(\sigma)}(\ \gamma\mathbf{P} \cdot \delta\boldsymbol{\chi}^{(\sigma)}\ ) - 2H\gamma\boldsymbol{\xi} \cdot \delta\boldsymbol{\chi}^{(\sigma)}\ ]\ dA$$

$$= - \int_{C^{(cl)}} \boldsymbol{(}\ \gamma\boldsymbol{\nu}\ \boldsymbol{)} \cdot \delta\boldsymbol{\chi}^{(\sigma)}\ ds - \int_\Sigma 2H\gamma\boldsymbol{\xi} \cdot \delta\boldsymbol{\chi}^{(\sigma)}\ dA \tag{2-6}$$

where $C^{(cl)}$ denotes the common line and the boldface parentheses are defined in Sec. 1.3.7.

Referring to Sec. 4.3.1, we see that (2-4), (2-5), and (2-6) imply

at $\varepsilon = 0$:

$$0 = \delta J$$

$$= \int_\Sigma (\ \boldsymbol{[}\ P\boldsymbol{\xi}\ \boldsymbol{]} - 2H\gamma\boldsymbol{\xi}\ ) \cdot \delta\boldsymbol{\chi}^{(\sigma)}\ dA - \int_{C^{(cl)}} \boldsymbol{(}\ \gamma\boldsymbol{\nu}\ \boldsymbol{)} \cdot \delta\boldsymbol{\chi}^{(\sigma)}\ ds \tag{2-7}$$

and the corresponding Euler equations are

$$\text{at } \Sigma\text{:}\quad \boldsymbol{[}\ P\boldsymbol{\xi}\ \boldsymbol{]} - 2H\gamma\boldsymbol{\xi} = 0 \tag{2-8}$$

and

$$\text{at } C^{(cl)}: \ (\gamma \nu) = 0 \tag{2-9}$$

Equation (2-8) is the jump momentum balance at the interior phase interfaces of the body (Sec. 2.4.2). Equation (2-9) is the Neumann triangle, the force balance that must be satisfied at the common line (Sec. 2.4.3).

Now let us examine the form of the boundary value problem that arises in fluid statics, when the effects of the external forces are neglected. The interfacial tension will be a specified constant within each dividing surface. Cauchy's first law

$$-\nabla P = 0 \tag{2-10}$$

requires the pressure distribution to be a constant within each phase consistent with (2-8), (2-9), and any other boundary condition (such as a specification of pressure at one point on the boundary where a pressure gauge might be located).

We are now ready to state our **variational principle for fluid statics:**

The arrangement of dividing surfaces for which J attains a stationary value is a static solution of Cauchy's first law and the required boundary conditions.

References

Ehrlich, R., Gulf Research and Development Co., P.O. Drawer 2038, Pittsburgh, PA 15238, personal communication (1974).

Slattery, J.C., "Momentum, Energy, and Mass Transfer in Continua," McGraw-Hill, New York (1972); second edition, Robert E. Krieger, Malabar, FL 32950 (1981).

Exercise

Exercise 4.3.2-1 *variation following a body* We wish to calculate the variation of

$$K \equiv \int_R \Psi \, dV + \int_\Sigma \Psi^{(\sigma)} \, dA$$

a quantity associated with a body.

Let us define

$$\delta\Psi \equiv \left(\frac{\partial\Psi}{\partial\varepsilon}\right)_{\zeta}$$

$$\delta\boldsymbol{\chi} \equiv \left(\frac{\partial\boldsymbol{\chi}}{\partial\varepsilon}\right)_{\zeta}$$

as variations of Ψ and of position following a material particle. Similarly

$$\delta\Psi^{(\sigma)} \equiv \left(\frac{\partial\Psi^{(\sigma)}}{\partial\varepsilon}\right)_{\zeta^{(\sigma)}}$$

$$\delta\boldsymbol{\chi}^{(\sigma)} \equiv \left(\frac{\partial\boldsymbol{\chi}^{(\sigma)}}{\partial\varepsilon}\right)_{\zeta^{(\sigma)}}$$

are variations of $\Psi^{(\sigma)}$ and of position following a surface particle. Guided by our development of the transport theorem for a body containing intersecting dividing surfaces in Sec. 1.3.7, prove that

$$\delta K = \int_R [\,\delta\Psi + \Psi\,\text{div}(\,\delta\boldsymbol{\chi}\,)]\,dv$$

$$+ \int_\Sigma [\,\delta\Psi^{(\sigma)} + \Psi^{(\sigma)}\,\text{div}_{(\sigma)}(\,\delta\boldsymbol{\chi}^{(\sigma)}\,)]\,dA$$

Hint: In deriving this expression, we must require

$$\text{at } \Sigma : \ \delta\boldsymbol{\chi} \cdot \boldsymbol{\xi} = \left(\frac{\partial \mathbf{p}^{(\sigma)}}{\partial\varepsilon}\right)_{y^1, y^2} \cdot \boldsymbol{\xi}$$

and

$$\text{at } C^{(cl)} : \ \delta\boldsymbol{\chi}^{(\sigma)} \cdot \mathbf{v} = \left(\frac{\partial \mathbf{p}^{(cl)}}{\partial\varepsilon}\right)_{y} \cdot \mathbf{v}$$

where more generally the common line is the locus of a point whose position is a function of one parameter y,

$$\mathbf{z} = \mathbf{p}^{(cl)}(\,y\,)$$

4.3.3 An example: a spherical cap The utility of this variational principle for fluid statics can be better appreciated after considering a specific example.

Phases A and B are separated by a solid plate. At one point in the plate, there is a circular hole, the edge of which is ground to a knife-edge. Figure 3-1 shows an interface between phases A and B located at the circle formed by this knife-edge. Given the pressure $P^{(A)}$ in phase A and the pressure $P^{(B)}$ in phase B and neglecting the effects of gravity, let us determine the configuration of the A-B interface.

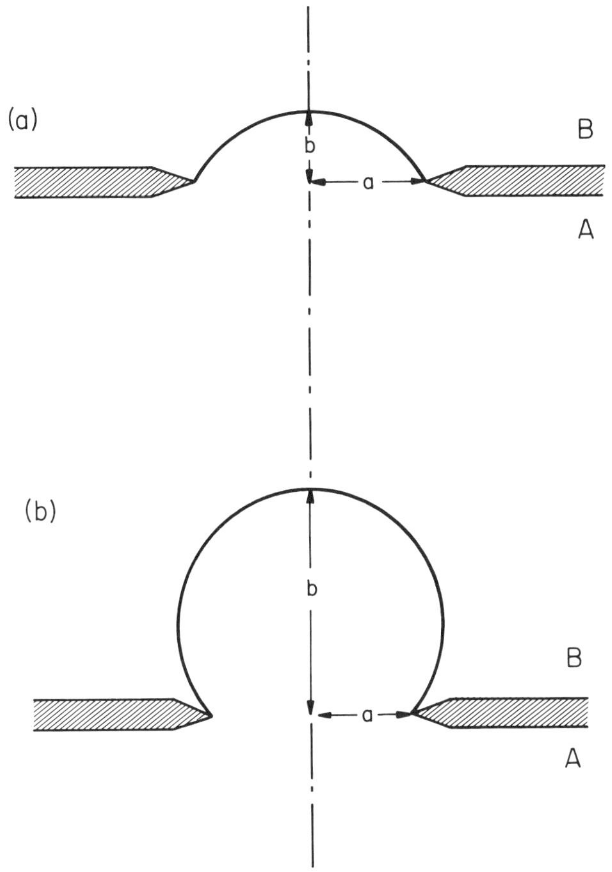

Figure 4.3.3-1: Spherical cap
(a) Interface in the form of a "dish" shaped spherical cap separates phases A and B.
(b) Interface is a "bubble" shaped cap.

Because the knife-edge is axially symmetric and because the configuration of the interface is assumed to be unaffected by gravity, the interface is a segment of a sphere (Concus and Finn 1974). Under these conditions,

$$J \equiv - \int_R P\,dV + \int_\Sigma \gamma\,dA$$

$$= - \left(P^{(A)} - P^{(B)} \right) \left(\frac{\pi b}{6} \right) \left(3a^2 + b^2 \right) + \gamma\pi \left(a^2 + b^2 \right) \tag{3-1}$$

Here we have noted that for a spherical cap the volume V_{cap}, the area A_{cap}, and the radius R are

$$V_{cap} = \frac{\pi b}{6} \left(3a^2 + b^2 \right) \tag{3-2}$$

$$A_{cap} = \pi \left(a^2 + b^2 \right) \tag{3-3}$$

$$R = \frac{a^2 + b^2}{2b} \tag{3-4}$$

By the variational principle of the preceding section, any static configuration of the spherical cap must be consistent with

$$\frac{dJ}{db} = 0 \tag{3-5}$$

The result is

$$\frac{R}{a} = \frac{2}{N} \tag{3-6}$$

or

$$\frac{b}{a} = \frac{2}{N} \pm \left(\frac{4}{N^2} - 1 \right)^{1/2} \tag{3-7}$$

where

$$N \equiv \frac{\left(P^{(A)} - P^{(B)} \right) a}{\gamma} \tag{3-8}$$

Note that there are no solutions for $N > 2$.

Observe that there are two solutions: the dish shaped interface shown in Figure 3-1a and the bubble shaped interface shown in Figure 3-1b.

In Sec. 4.3.2 and in Exercise 5.10.3-1, I point out that, under a set of assumptions we might be willing to make here, J is minimized as a function of time at equilibrium for an isolated body. You can readily establish for yourself that J is a maximum as a function of b for the bubble solution, but a minimum as a function of b for the dish solution. We would expect only the dish solution to be experimentally observable.

The most serious assumption made in reaching this conclusion is to ignore the relation between pressure and volume in the two adjoining bulk phases. This is equivalent to assuming that these two phases are nearly unbounded. This assumption has been eliminated by Dyson (1980), who directly minimized the Helmholtz free energy (Exercise 5.10.3-1). It is interesting that, when the adjoining bulk phases are bounded, both the dish and the bubble solutions can be observed experimentally (Kovitz 1974).

References

Boys, C.V., "Soap-Bubbles," Dover, New York (1959).

Concus, P., and R. Finn, *Acta math.* **132,** 177 (1974).

Dyson, D.C., *J. Colloid Interface Sci.* **76,** 277 (1980).

Kovitz. A.A., Proceedings of the International Colloquium on Drops and Bubbles, California Institute of Technology and Jet Propulsion Laboratory, Pasadena, CA., August 28-30 (1974).

Exercise

4.3.3-1 *two spherical caps* Referring to Figure 3-2, phase A is isolated from phase B within an axially symmetric tube, the ends of which are ground to knife-edges. The ends of the tube are sealed by A-B interfaces located at the circles formed by the knife-edges. Given the pressure $P^{(A)}$ in phase A and the pressure $P^{(B)}$ in phase B and neglecting the effects of gravity, let us determine the configuration of the A-B interfaces.

i) Reason that

$$J_0 = J = -\left(P^{(A)} - P^{(B)} \right)\left[\frac{\pi b_1}{6} \left(3a_1^2 + b_1^2 \right) + \frac{\pi b_2}{6} \left(3a_2^2 + b_2^2 \right) \right]$$

$$+ \gamma\left[\pi\left(a_1^2 + b_1^2 \right) + \pi\left(a_2^2 + b_2^2 \right)\right] + \text{constant}$$

ii) Set

$$\frac{dJ_0}{db_1} = 0$$

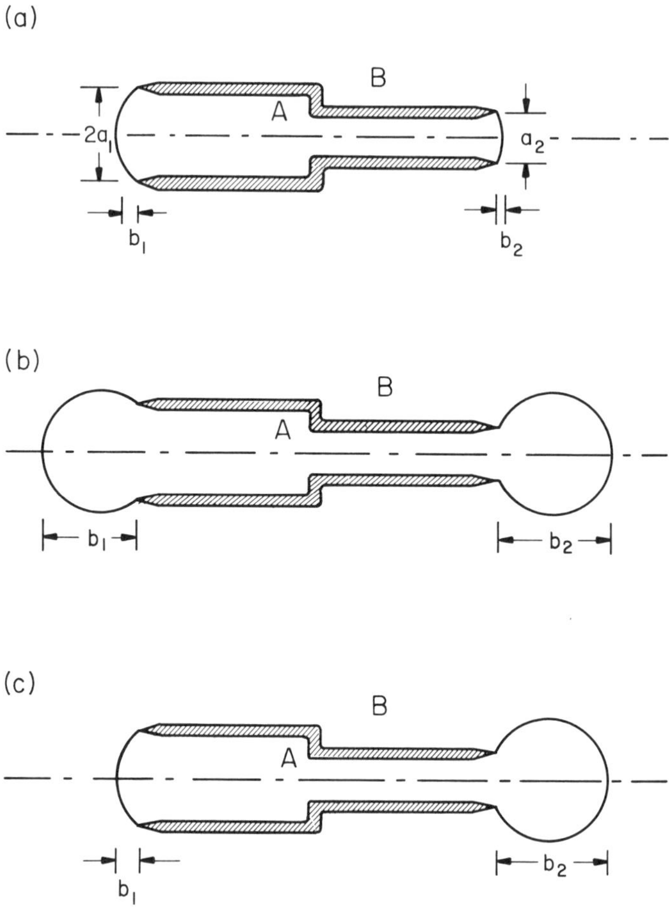

Figure 4.3.3-2: Axially symmetric tube with A-B interfaces at both ends.

(a) Both interfaces are dish shaped.

(b) Both interfaces are bubble shaped. This can not exist at equilibrium and it will not be observed experimentally.

(c) One interface is dish shaped and one is bubble shaped.

with the constraint

$$\frac{\pi b_1}{6}\left(3a_1^2 + b_1^2 \right) + \frac{\pi b_2}{6}\left(3a_2^2 + b_2^2 \right) = \text{constant}$$

to conclude that

$$b_1 = b_2 \frac{(a_1^2 + b_1^2)}{(a_2^2 + b_2^2)}$$

or

$$R_1 = R_2$$

iii) Determine that

$$\frac{d^2 J_0}{db_1^2} = \frac{d^2 J}{db_1^2} = 2\pi\gamma\left[1 - \frac{b_1}{R_1} + \left(\frac{b_1}{b_2}\right)^2\left(1 - \frac{b_2}{R_2}\right)\right]$$

At equilibrium, under the set of assumption given in Sec. 4.3.2 and in Exercise 5.10.3-1, both b_1/R_1 and b_2/R_2 can not be greater than 1. Experimentally, we can not observe that the configuration shown in Figure 3-2b in which both interfaces are bubble shaped (Boys 1959, p. 50).

In reaching this conclusion, we have ignored for the sake of simplicity the relations between pressure and volume in the three phases.

4.4 Extremum principles

4.4.1 Extremum principles for multiphase flows We have seen in the preceding sections that static problems involving multiple fluid phases can be solved approximately by using a variational principle. No variational principles have been described in the literature for multiphase flows.

For many situations, the velocity and pressure distributions are of interest primarily in calculating the rate of viscous dissipation of mechanical energy or the rate at which work is expended. The sections that follow develop upper and lower bounds for the rate of dissipation of mechanical energy resulting from the action of viscous forces both in the bulk phases and in the phase interfaces, at least for some simple classes of material behavior (Hopke and Slattery 1975).

The class of problems for which these bounding principles are applicable is defined by a number of assumptions.

i) The velocity distribution is independent of time. The speed of displacement of the phase interfaces is consequently zero.

ii) There is no mass transfer to or from the phase interfaces. The normal component of velocity at the interfaces is zero.

iii) The tangential components of velocity are continuous across the interfaces. The velocity vector is continuous across all the interfaces.

iv) The bulk phases are incompressible. The equation of continuity within each bulk phase reduces to

$$\text{div } \mathbf{v} = 0 \tag{1-1}$$

v) The external force per unit mass **b** may be represented by the gradient of a scalar potential ϕ (potential energy per unit mass)

$$\mathbf{b} = -\nabla\phi \tag{1-2}$$

vi) All inertial effects are neglected. Within the bulk phases, Cauchy's first law takes the form

$$\text{div}(\mathbf{T} - \rho\phi\,\mathbf{I}) = 0 \tag{1-3}$$

vii) The stress-deformation behavior of the bulk phases can be described by one of the generalized newtonian models for incompressible fluids (Slattery 1981, p. 51):

$$\mathbf{S} \equiv \mathbf{T} + p\mathbf{I}$$

$$= 2\eta \mathbf{D} \tag{1-4}$$

where

$$\eta = \eta(D) \tag{1-5}$$

and

$$D \equiv [\ \mathrm{tr}(\ \mathbf{D} \cdot \mathbf{D}\ )]^{1/2} \tag{1-6}$$

An alternative form of this same class of models requires

$$2\mathbf{D} = \varphi \mathbf{S} \tag{1-7}$$

with

$$\varphi = \varphi(S) \tag{1-8}$$

and

$$S \equiv [\ \mathrm{tr}(\ \mathbf{S} \cdot \mathbf{S}\ )]^{1/2} \tag{1-9}$$

Some additional assumptions are noted in Secs. 4.4.5 and 4.4.6.

The material developed in Secs. 4.4.2 through 4.4.6 represents an extension to multiphase systems (Hopke and Slattery 1975) of bounding principles originally proposed for single phase systems by Hill (1956) and by Hill and Power (1956) (see also Slattery 1981, p. 262).

References

Hill, R., *J. Mech. Phys. Solids* **5,** 66 (1956).

Hill, R., and G. Power, *Q. J. Mech. Appl. Math.* **9,** 313 (1956).

Hopke, S. W., and J. C. Slattery, *Int. J. Multiphase Flow* **1,** 727 (1975).

Slattery, J. C., "Momentum, Energy, and Mass Transfer in Continua," McGraw-Hill, New York (1972); second edition, Robert E. Krieger, Malabar, FL 32950 (1981).

4.4.2 The primary velocity extremum principle As I mentioned in the last section, we are restricting our attention here to bulk phases whose stress-deformation behavior can be described by one of the generalized newtonian models for incompressible fluids. In particular, let us assume that the model takes the form of (1-4). For this model of material behavior, we can introduce a scalar potential function

$$E = E(D) \equiv \int_0^{D^2} \eta \, dD^2 \tag{2-1}$$

such that[a]

---

a) The space $\mathcal{L}$ of all second-order tensors is nine dimensional (Slattery 1981, p. 626). The symmetric second-order tensors form a six-dimensional subspace $\mathcal{S}$. Let $\varepsilon = \varepsilon(\mathbf{A})$ be a scalar-valued function of one second-order tensor $\mathbf{A}$ in either $\mathcal{L}$ or $\mathcal{S}$. Let A_{ij} be the rectangular cartesian components of $\mathbf{A}$. Then $\varepsilon(\mathbf{A})$ may also be viewed as a real-valued function of the nine real variables A_{ij}, if $\mathbf{A}$ belongs to $\mathcal{L}$, or of the six real variables A_{ij} $(i \leq j)$, if $\mathbf{A}$ belongs to $\mathcal{S}$.

If $\mathbf{A}$ belongs to $\mathcal{L}$, we define the gradient of ε as (Truesdell and Noll 1965, p. 24)

$$\frac{\partial \varepsilon}{\partial \mathbf{A}} \equiv \frac{\partial \varepsilon}{\partial A_{ij}} \mathbf{e}_i \mathbf{e}_j$$

$$= \frac{\partial \varepsilon}{\partial \overline{A}_{ij}} \overline{\mathbf{g}}_i \overline{\mathbf{g}}_j$$

$$= \frac{\partial \varepsilon}{\partial \overline{A}_{<ij>}} \overline{\mathbf{g}}_{<i>} \overline{\mathbf{g}}_{<j>}$$

Here $\overline{A}_{ij}$ and $\overline{A}_{<ij>}$ are respectively the covariant and physical components of $\mathbf{A}$ with respect to some curvilinear coordinate system.

If $\mathbf{A}$ belongs to $\mathcal{S}$, then the domain of definition for ε must be extended to $\mathcal{L}$ by setting

$$\tilde{\varepsilon}(\mathbf{A}) \equiv \varepsilon\left(\frac{1}{2} [\, \mathbf{A} + \mathbf{A}^T \,]\right)$$

The derivatives of $\tilde{\varepsilon}$ are taken with respect to $\mathbf{A}$ as though $\mathbf{A}$ belongs to $\mathcal{L}$, after which we define (cont.)

$$\frac{\partial E}{\partial \mathbf{D}} = 2\eta \mathbf{D} \tag{2-2}$$

Upon comparison with (1-4), we conclude that

$$\mathbf{S} = \frac{\partial E}{\partial \mathbf{D}} \tag{2-3}$$

Let us now visualize a particular flow for which **v** is the actual velocity distribution. By this I mean that **v** satisfies the equation of motion, the equation of continuity, the jump mass balance, the jump momentum balance, and all of the required boundary conditions. Let **v*** be some trial or approximate velocity distribution, the properties of which we

---

a) (continued from previous page)

$$\frac{\partial \varepsilon}{\partial \mathbf{A}} \equiv \frac{\partial \tilde{\varepsilon}}{\partial \mathbf{A}}$$

and again restrict **A** to $\mathcal{S}$.

As an example, if **A** belongs to $\mathcal{S}$ and

$$\varepsilon(\mathbf{A}) \equiv A_{12}$$

then

$$\tilde{\varepsilon} \equiv \frac{1}{2} (A_{12} + A_{21})$$

$$\frac{\partial \tilde{\varepsilon}}{\partial A_{12}} = \frac{\partial \tilde{\varepsilon}}{\partial A_{21}} = \frac{1}{2}$$

and

$$\frac{\partial \varepsilon}{\partial \mathbf{A}} \equiv \frac{\partial \tilde{\varepsilon}}{\partial \mathbf{A}} = \frac{1}{2} (\mathbf{e}_1 \mathbf{e}_2 + \mathbf{e}_2 \mathbf{e}_1)$$

shall specify later. Let us expand E(D) in a truncated Taylor series with a remainder term:

$$E(D^*) - E(D) = \frac{\partial E}{\partial D_{ij}}\bigg|_{D} (D^*_{ij} - D_{ij})$$

$$+ \frac{1}{2} \frac{\partial^2 E}{\partial D_{ij}\, \partial D_{mn}}\bigg|_{\bar{D}} (D^*_{ij} - D_{ij})(D^*_{mn} - D_{mn}) \tag{2-4}$$

For simplicity, we speak in terms of a rectangular cartesian coordinate system; the D_{ij} are the rectangular cartesian components of **D**. Here $\bar{D}$ is a suitable value between D and D*. Recognizing that E is a function of D, we can calculate

$$\frac{\partial E}{\partial D_{ij}} = \frac{1}{D} \frac{dE}{dD} D_{ij} \tag{2-5}$$

and

$$\frac{\partial^2 E}{\partial D_{ij}\, \partial D_{mn}} = \left(\frac{1}{D^2} \frac{d^2 E}{dD^2} - \frac{1}{D^3} \frac{dE}{dD} \right) D_{mn} D_{ij} + \frac{1}{D} \frac{dE}{dD} \delta_{im} \delta_{jn} \tag{2-6}$$

These relationships allow us to express (2-4) as

$$E(D^*) - E(D) = \frac{1}{D} \frac{dE}{dD}\bigg|_{D} D_{ij}(D^*_{ij} - D_{ij})$$

$$+ \frac{1}{2} \left(\frac{1}{\bar{D}^2} \frac{d^2 E}{dD^2}\bigg|_{\bar{D}} - \frac{1}{\bar{D}^3} \frac{dE}{dD}\bigg|_{\bar{D}} \right) [\bar{D}_{ij} (D^*_{ij} - D_{ij})]^2$$

$$+ \frac{1}{2} \frac{1}{\bar{D}} \frac{dE}{dD}\bigg|_{\bar{D}} (D^*_{ij} - D_{ij})(D^*_{ij} - D_{ij}) \tag{2-7}$$

The Schwarz inequality (Halmos 1958, p. 125) requires

$$[\bar{D}_{ij}(D^*_{ij} - D_{ij})]^2 \leq \bar{D}_{mn} \bar{D}_{mn} (D^*_{ij} - D_{ij})(D^*_{ij} - D_{ij})$$

$$= \bar{D}^2 (D^*_{ij} - D_{ij})(D^*_{ij} - D_{ij}) \tag{2-8}$$

From (2-1), we find

$$2\eta = \frac{1}{D}\frac{dE}{dD} \tag{2-9}$$

All known experimental data indicate that the apparent viscosity η should be positive:

$$2\eta = \frac{1}{D}\frac{dE}{dD} \geq 0 \tag{2-10}$$

Inequalities (2-8) and (2-10) permit us to learn from (2-7) that

$$E(D^*) - E(D) \geq \frac{1}{D}\frac{dE}{dD}\bigg|_D D_{ij}(D^*_{ij} - D_{ij})$$

$$+ \frac{1}{2}\frac{1}{\bar{D}^2}\frac{d^2 E}{dD^2}\bigg|_{\bar{D}} [\bar{D}_{ij} (D^*_{ij} - D_{ij})]^2 \tag{2-11}$$

Equations (1-4) and (2-9) tell us that

$$S \equiv [\operatorname{tr}(\mathbf{S} \cdot \mathbf{S})]^{1/2}$$

$$= \frac{dE}{dD} \tag{2-12}$$

or

$$\frac{dS}{dD} = \frac{d^2 E}{dD^2} \tag{2-13}$$

Experimentally, we know that without exception

$$\frac{dS}{dD} = \frac{d^2 E}{dD^2} \geq 0 \tag{2-14}$$

In view of this, inequality (2-11) implies

$$E(D^*) - E(D) \geq \frac{1}{D}\frac{dE}{dD}\bigg|_D D_{ij}(D^*_{ij} - D_{ij})$$

$$= \frac{\partial E}{\partial D_{ij}}(D^*_{ij} - D_{ij})$$

$$= \text{tr}\left[\frac{\partial E}{\partial \mathbf{D}} \cdot (\mathbf{D}^* - \mathbf{D}) \right]$$

$$= \text{tr}[\mathbf{S} \cdot (\mathbf{D}^* - \mathbf{D})] \tag{2-15}$$

Let us define $\mathbf{v}^*$ to be an approximate or trial velocity distribution that satisfies the equation of continuity for an incompressible fluid

$$\text{div } \mathbf{v}^* = 0 \tag{2-16}$$

as well as any explicit boundary conditions on velocity,

$$\text{on } S_v : \ \mathbf{v}^* = \mathbf{v} \tag{2-17}$$

In particular, since there is assumed to be no mass transfer to or from the dividing surfaces and since the tangential components of velocity are continuous across the interface (see Sec. 4.4.1), $\mathbf{v}^*$ must be continuous everywhere.

Our primary interest is in the integral of inequality (2-15) over an arbitrary region R occupied by multiple phases:

$$\int_R \{ E(\mathbf{D}^*) - E(\mathbf{D}) - \text{tr}[\mathbf{S} \cdot (\mathbf{D}^* - \mathbf{D})]\} \, dV \geq 0 \tag{2-18}$$

Looking at the third term on the left, we can reason that

$$\int_R \text{tr}(\mathbf{S} \cdot \mathbf{D}) \, dV = \int_R \text{tr}(\mathbf{S} \cdot \nabla \mathbf{v}) \, dV$$

$$= \int_R \text{tr}[(\mathbf{T} - \rho\phi \mathbf{I}) \cdot \nabla \mathbf{v}] \, dV$$

$$= \int_R \text{div}[(\mathbf{T} - \rho\phi \mathbf{I}) \cdot \mathbf{v}] \, dV$$

$$- \int_R \mathbf{v} \cdot \text{div}(\mathbf{T} - \rho\phi \mathbf{I}) \, dV$$

$$= \int_R \text{div}[(\mathbf{T} - \rho\phi \mathbf{I}) \cdot \mathbf{v}] \, dV$$

$$= \int_S \mathbf{v} \cdot (\mathbf{T} - \rho\phi\mathbf{I}) \cdot \mathbf{n} \, dA$$

$$- \int_\Sigma \mathbf{v} \cdot [\mathbf{T} \cdot \boldsymbol{\xi} - \rho\phi\boldsymbol{\xi}] \, dA \tag{2-19}$$

In the first line, I have used the symmetry of the extra-stress tensor; in the second, the equation of continuity for an incompressible fluid; in the third, an integration by parts; in the fourth, Cauchy's first law for creeping motion (1-3); in the fifth Green's transformation (Slattery 1981, p. 661). Here I have introduced S as the closed surface bounding the region R; Σ is the collection of dividing surfaces (representing phase interfaces) contained within R. In arriving at this result, I have also observed that velocity $\mathbf{v}$ is continuous across Σ (see Sec. 4.4.1). In precisely the same way, we can also calculate that

$$\int_R \mathrm{tr}(\mathbf{S} \cdot \mathbf{D}^*) \, dV = \int_S \mathbf{v}^* \cdot (\mathbf{T} \cdot \rho\phi\mathbf{I}) \cdot \mathbf{n} \, dA$$

$$- \int_\Sigma \mathbf{v}^* \cdot [\mathbf{T} \cdot \boldsymbol{\xi} - \rho\phi\boldsymbol{\xi}] \, dA \tag{2-20}$$

Using (2-17), (2-19), and (2-20), we can rearrange (2-18) to conclude (Hopke and Slattery 1975)

$$\int_R E(\mathbf{D}) \, dV \le \int_R E(\mathbf{D}^*) \, dV$$

$$+ \int_{S - S_v} (\mathbf{v} - \mathbf{v}^*) \cdot (\mathbf{T} - \rho\phi\mathbf{I}) \cdot \mathbf{n} \, dA$$

$$- \int_\Sigma (\mathbf{v} - \mathbf{v}^*) \cdot [\mathbf{T} \cdot \boldsymbol{\xi} - \rho\phi\boldsymbol{\xi}] \, dA \tag{2-21}$$

Here $S - S_v$ is that portion of the closed surface S bounding R upon which velocity is not explicitly specified. Inequality (2-21) is the **primary velocity extremum principle**. For more explicit forms, see Secs. 4.4.5 and 4.4.6.

The physical significance of the velocity extremum principle will be developed in Secs. 4.4.4 through 4.4.6.

References

Halmos, P. R., "Finite-dimensional Vector Spaces," second edition, Van Nostrand, Princeton, N.J. (1958).

Hopke, S. W., and J. C. Slattery, *Int. J. Multiphase Flow* **1,** 727 (1975).

Slattery, J. C., "Momentum, Energy, and Mass Transfer in Continua," McGraw-Hill, New York (1972); second edition, Robert E. Krieger, Malabar, FL 32950 (1981).

Truesdell, C., and W. Noll, "Handbuch der Physik," vol. 3/3, edited by S. Flügge, Springer-Verlag, Berlin (1965).

4.4.3 The primary stress extremum principle Let us continue to confine ourselves to bulk phases whose stress-deformation behavior can be described by one of the generalized newtonian models for the incompressible fluids, but let us now assume that the model takes the alternative form (1-7). For this model of material behavior, we can introduce a scalar potential function

$$E_c = E_c(S) \equiv \int_0^{S^2} \frac{1}{4}\,\varphi\, dS^2 \tag{3-1}$$

such that (see Sec. 4.4.2, footnote a)

$$\frac{\partial E_c}{\partial S} = \frac{1}{2}\,\varphi S \tag{3-2}$$

Comparison with (1-7) shows that

$$\mathbf{D} = \frac{\partial E_c}{\partial S} \tag{3-3}$$

As mentioned in introducing inequalities (2-10) and (2-14), all known experimental data are consistent with saying that

$$\varphi \geq 0 \tag{3-4}$$

and

$$\frac{dD}{dS} \geq 0 \tag{3-5}$$

By an argument directly analogous to that used in deriving (2-15), these assumptions lead us to say

$$E_c(S^*) - E_c(S) \geq \text{tr}[\, \mathbf{D} \cdot (\, \mathbf{S}^* - \mathbf{S}\,)] \tag{3-6}$$

Let us define $\mathbf{T}^*$ to be a trial or approximate stress distribution that satisfies Cauchy's first law for creeping motion (see Sec. 4.4.1)

$$\text{div}(\, \mathbf{T}^* - \rho^* \phi \mathbf{I}\,) = 0 \tag{3-7}$$

in which ρ^* denotes the density distribution consistent with the trial or approximate configuration Σ^* of the dividing surfaces. The trial stress distribution $\mathbf{T}^*$ does not necessarily satisfy all of the required boundary conditions on the stress tensor, but we will require it to satisfy the jump momentum balance at Σ^*. I will have more to say on this point in Secs. 4.4.5 and 4.4.6. The trial extra-stress tensor $\mathbf{S}^*$ is defined by

$$\mathbf{S}^* \equiv \mathbf{T}^* - \frac{1}{3}(\, \text{tr}\, \mathbf{T}^*\,)\mathbf{I} \tag{3-8}$$

This amounts to identifying p^* as the mean pressure (Slattery 1981, p. 49).

Our primary interest is in the integral of inequality (3-6) over an arbitrary region R occupied by multiple phases

$$\int_R \{\, E_c(\, S^*\,) - E_c(S) - \text{tr}[\, \mathbf{D} \cdot (\, \mathbf{S}^* - \mathbf{S}\,)]\} \, dV \geq 0 \tag{3-9}$$

From (3-1) and (1-7), we know that

$$\frac{dE_c}{dS} = \frac{1}{2}\varphi S = D \tag{3-10}$$

From (2-9) and (1-4),

$$\frac{dE}{dD} = 2\eta\, D = S \tag{3-11}$$

These expression can be used to reason that

$$\int_0^S \frac{dE_c}{dS}\, dS + \int_0^D \frac{dE}{dD}\, dD = \int_0^S D\, dS + \int_0^D S\, dD$$

$$= \int_0^{SD} d(\, SD\,) \tag{3-12}$$

or

$$E_c + E = SD = 2\eta D^2 = \text{tr}(\, \mathbf{S} \cdot \mathbf{D}\,) \tag{3-13}$$

We can then use this relationship to eliminate $E_c(S)$ from (3-9) and write

$$\int_R E(D)\, dV \geq - \int_R E_c(\, S^*\,)\, dV + \int_R \text{tr}(\, \mathbf{D} \cdot \mathbf{S}^*\,)\, dV \tag{3-14}$$

We can employ an argument similar to that used in deriving (2-19) to reason that

$$\int_R \text{tr}(\, \mathbf{D} \cdot \mathbf{S}^*\,)\, dV = \int_S \mathbf{v} \cdot (\, \mathbf{T}^* - \rho^* \phi \mathbf{I}\,) \cdot \mathbf{n}\, dA$$

$$- \int_{\Sigma^*} \mathbf{v} \cdot [\, \mathbf{T}^* \cdot \boldsymbol{\xi} - \rho^* \phi \boldsymbol{\xi}\,]\, dA \tag{3-15}$$

Here $\boldsymbol{\xi}^*$ is the unit normal to Σ^*. This allows us to conclude from (3-14) that (Hopke and Slattery 1975)

$$\int_R E(D)\, dV \geq - \int_R E_c(\, S^*\,)\, dV + \int_S \mathbf{v} \cdot (\, \mathbf{T}^* - \rho^* \phi \mathbf{I}\,) \cdot \mathbf{n}\, dA$$

$$- \int_{\Sigma^*} \mathbf{v} \cdot [\, \mathbf{T}^* \cdot \boldsymbol{\xi}^* - \rho^* \phi \boldsymbol{\xi}^*\,]\, dA \tag{3-16}$$

We will refer to inequality (3-16) as the **primary stress extremum principle**. For more explicit forms, see Secs. 4.4.5 and 4.4.6.

The physical significance of the stress extremum principle will become clear in Secs. 4.4.4 through 4.4.6.

References

Hopke, S. W., and J. C. Slattery, *Int. J. Multiphase Flow* **1,** 727 (1975).

Slattery, J. C., "Momentum, Energy, and Mass Transfer in Continua," McGraw-Hill, New York (1972); second edition, Robert E. Krieger, Malabar, FL 32950 (1981).

4.4.4 Physical interpretation of E If for a fixed value of p

$$t^p \, E(D) = E(tD) \tag{4-1}$$

no matter what value t assumes, then E is called a homogeneous function of degree p. Euler's theorem on homogeneous functions says that (Kaplan 1973, p.139)

$$pE = D \frac{dE}{dD} \tag{4-2}$$

Using (2-3) and (2-5), we discover

$$\begin{aligned} E &= \frac{1}{p} D_{ij} \frac{\partial E}{\partial D_{ij}} \\ &= \frac{1}{p} D_{ij} \, S_{ij} \\ &= \frac{1}{p} \operatorname{tr}(\mathbf{S} \cdot \mathbf{D}) \end{aligned} \tag{4-3}$$

The function E is proportional to the rate of dissipation of energy in the fluid per unit volume.

The practicality of this expression is that E is homogeneous for some simple models of fluid behavior. In particular, p = 2 for a newtonian fluid. For more on this point, see Slattery (1981, pp. 269 and 270).

When E is a homogeneous function of degree p, the primary velocity and stress extremum principles discussed in Secs. 4.4.2 and 4.4.3 require respectively (Hopke and Slattery 1975)

$$\frac{1}{p} \int_R \operatorname{tr}(\mathbf{S} \cdot \mathbf{D}) \, dV \leq \int_R E(\mathbf{D}^*) \, dV$$

$$+ \int_{S - S_v} (\mathbf{v} - \mathbf{v}^*) \cdot (\mathbf{T} \cdot \rho\phi\mathbf{I}) \cdot \mathbf{n} \, dA$$

$$- \int_{\Sigma} (\mathbf{v} - \mathbf{v}^*) \cdot [\mathbf{T} \cdot \boldsymbol{\xi} - \rho\phi\boldsymbol{\xi}] \, dA \tag{4-4}$$

and

$$\frac{1}{p} \int_R \operatorname{tr}(\mathbf{S} \cdot \mathbf{D}) \, dV \geq - \int_R E_c(\mathbf{S}^*) \, dV$$

$$+ \int_S \mathbf{v} \cdot (\mathbf{T}^* - \rho^*\phi\mathbf{I}) \cdot \mathbf{n} \, dA$$

$$- \int_{\Sigma^*} \mathbf{v} \cdot [\mathbf{T}^* \cdot \boldsymbol{\xi}^* - \rho^*\phi\boldsymbol{\xi}^*] \, dA \tag{4-5}$$

For this restricted class of fluids, the velocity and stress extremum principles give bounds for the rate of viscous dissipation of mechanical energy within the bulk phases.

In order to apply inequalities (4-4) and (4-5) to multiphase systems, something more must be said about the integrals over the dividing surfaces appearing in these relationships. The next two sections discuss two classes of problems for which usable results can be obtained.

References

Hopke, S. W., and J. C. Slattery, *Int. J. Multiphase Flow* **1,** 727 (1975).

Kaplan, W., "Advanced Calculus, second edition, Addison-Wesley, Reading, MA (1973).

Slattery, J. C., "Momentum, Energy, and Mass Transfer in Continua," McGraw-Hill, New York (1972); second edition, Robert E. Krieger, Malabar, FL 32950 (1981).

4.4.5 Extremum principles for uniform surface tension The primary velocity and stress extremum principles developed in the preceding sections take a more useful form when we make these additional assumptions.

viii') The effects of the external force are neglected within the dividing surface.

ix') Interfacial tension is independent of position within the dividing surface. Since inertial effects are neglected as well, the jump momentum balance requires (Sec. 2.3.1)

$$\text{at } \Sigma : \ [\, \mathbf{T}^* \cdot \boldsymbol{\xi} \,] = -\, 2H\gamma\, \boldsymbol{\xi} \tag{5-1}$$

We will insist as well that the trial stress distribution $\mathbf{T}^*$ introduced in Sec. 4.4.3 obey a similar relationship at the trial configuration for the dividing surface,

$$\text{at } \Sigma^* : \ [\, \mathbf{T}^* \cdot \boldsymbol{\xi}^* \,] = -\, 2H^*\gamma\, \boldsymbol{\xi}^* \tag{5-2}$$

x') The actual dividing surface Σ belongs to a known family of parallel surfaces, from which the trial surface Σ^* is to be chosen. For example, we may have experimental evidence to suggest that the interface is a plane parallel to the plane $z = 0$ in rectangular cartesian coordinates.

xi') The actual velocity distribution $\mathbf{v}$ is everywhere tangent to this family of parallel surfaces,

$$\text{at } \Sigma^* : \ \mathbf{v} \cdot \boldsymbol{\xi}^* = 0 \tag{5-3}$$

Under these circumstances, we will also be able to require that

$$\text{at } \Sigma : \ \mathbf{v}^* \cdot \boldsymbol{\xi} = 0 \tag{5-4}$$

In view of (5-1), (5-4), and assumption (ii) (see Sec. 4.4.1), the velocity extremum principle in the form of (4-4) becomes (Hopke and Slattery 1975)

$$\frac{1}{p} \int_R \operatorname{tr}(\, \mathbf{S} \cdot \mathbf{D} \,)\, dV \leq \int_R E(\, \mathbf{D}^* \,)\, dV$$

$$+ \int_{S - S_v} (\, \mathbf{v} - \mathbf{v}^* \,) \cdot (\, \mathbf{T} - \rho\phi\mathbf{I} \,) \cdot \mathbf{n}\, dA \tag{5-5}$$

Similarly, (5-2) and (5-3) require that the stress extremum principle in the form of (4-5) simplifies to (Hopke and Slattery 1975)

$$\frac{1}{p} \int_R \operatorname{tr}(\mathbf{S} \cdot \mathbf{D})\, dV \geq - \int_R E_c(\mathbf{S}^*)\, dV$$

$$+ \int_S \mathbf{v} \cdot (\mathbf{T}^* - \rho^* \phi \mathbf{I}) \cdot \mathbf{n}\, dA \tag{5-6}$$

References

Hopke, S. W. and J. C. Slattery *Int. J. Multiphase Flow* **1,** 727 (1975).

Slattery, J. C., "Momentum, Energy, and Mass Transfer in Continua," McGraw-Hill, New York (1972); second edition, Robert E. Krieger, Malabar, FL 32950 (1981).

4.4.6 Extremum principles for more general interfacial stresses

There is another class of problems for which the primary extremum principles developed in Secs. 4.4.2 through 4.4.4 take a more useful form. This class is delineated by the following assumptions (made in addition to those given in Sec. 4.4.1).

viii") The configuration and location of the phase interface must be known a priori. For example, the interface may be a horizontal plane, the elevation of which is not a function of the flow.

This means that the normal component of the trial velocity distribution $\mathbf{v}^*$ will be required to be zero at Σ.

ix") The surface mass density $\rho^{(\sigma)}$ is a constant. Since there is no mass transfer to or from the phase interfaces, the jump mass balance reduces to (Sec. 1.3.5)

$$\operatorname{div}_{(\sigma)} \mathbf{v}^{(\sigma)} = 0 \tag{6-1}$$

Physically this means that there is no local dilation of the phase interface. We will require that the trial surface velocity distribution $\mathbf{v}^{(\sigma)*}$ obey a similar relationship at Σ:

$$\operatorname{div}_{(\sigma)} \mathbf{v}^{(\sigma)*} = 0 \tag{6-2}$$

x") The effects of the external force are neglected within the dividing surface. Since inertial effects are neglected as well, the jump momentum balance requires (Sec. 2.1.7)

$$[\,\mathbf{T}\cdot\boldsymbol{\xi}\,] = -\,\mathrm{div}_{(\sigma)}\mathbf{T}^{(\sigma)} \tag{6-3}$$

We will insist as well that the symmetric trial stress distribution $\mathbf{T}^*$ and the symmetric trial surface stress distribution $\mathbf{T}^{(\sigma)*}$ satisfy a similar relationship at Σ:

$$[\,\mathbf{T}^*\cdot\boldsymbol{\xi}\,] = -\,\mathrm{div}_{(\sigma)}\mathbf{T}^{(\sigma)*} \tag{6-4}$$

xi") The surface stress-deformation behavior can be described by one of the generalized Boussinesq surface fluid models (Sec. 2.2.2), which in view of (6-1) take the form

$$\begin{aligned}\mathbf{S}^{(\sigma)} &\equiv \mathbf{T}^{(\sigma)} - \gamma\,\mathbf{P}\\ &= \varepsilon\,\mathbf{D}^{(\sigma)}\end{aligned} \tag{6-5}$$

where

$$\varepsilon = \varepsilon(\,D^{(\sigma)}\,) \tag{6-6}$$

and

$$D^{(\sigma)} \equiv [\,\mathrm{tr}(\,\mathbf{D}^{(\sigma)}\cdot\mathbf{D}^{(\sigma)}\,)]^{1/2} \tag{6-7}$$

Using arguments analogous to those employed in Secs. 4.4.2 and 4.4.3, we can prove that, for this description of material behavior, there are two scalar potential functions

$$E^{(\sigma)} = E^{(\sigma)}(\,D^{(\sigma)}\,) \equiv \int_0^{D^{(\sigma)2}} \varepsilon\,dD^{(\sigma)2} \tag{6-8}$$

and

$$E_c^{(\sigma)} = E_c^{(\sigma)}(S^{(\sigma)}) \equiv \int_0^{S^{(\sigma)2}} \frac{1}{4\varepsilon}\,dS^{(\sigma)2} \tag{6-9}$$

such that (see Sec. 4.4.2, footnote a)

$$\mathbf{S}^{(\sigma)} = \frac{\partial E^{(\sigma)}}{\partial \mathbf{D}^{(\sigma)}} \tag{6-10}$$

and

$$\mathbf{D}^{(\sigma)} = \frac{\partial E_c^{(\sigma)}}{\partial \mathbf{S}^{(\sigma)}} \tag{6-11}$$

In (6-9), I have defined

$$S^{(\sigma)} \equiv [\, \text{tr}(\, \mathbf{S}^{(\sigma)} \cdot \mathbf{S}^{(\sigma)} \,)]^{1/2} \tag{6-12}$$

Both of these potential functions can be required to be convex:

$$E^{(\sigma)}(\, \mathbf{D}^{(\sigma)*} \,) - E^{(\sigma)}(\, \mathbf{D}^{(\sigma)} \,) \geq \text{tr}\left[\frac{\partial E^{(\sigma)}}{\partial \mathbf{D}^{(\sigma)}} \cdot (\, \mathbf{D}^{(\sigma)*} - \mathbf{D}^{(\sigma)} \,)\right] \tag{6-13}$$

$$E_c^{(\sigma)}(\, \mathbf{S}^{(\sigma)*} \,) - E_c^{(\sigma)}(\, \mathbf{S}^{(\sigma)} \,) \geq \text{tr}\left[\frac{\partial E_c^{(\sigma)}}{\partial \mathbf{S}^{(\sigma)}} \cdot (\, \mathbf{S}^{(\sigma)*} - \mathbf{S}^{(\sigma)} \,)\right] \tag{6-14}$$

Sufficient conditions for their convexity are

$$2\varepsilon = \frac{1}{D^{(\sigma)}} \frac{dE^{(\sigma)}}{dD^{(\sigma)}} \geq 0 \tag{6-15}$$

and

$$\frac{dS^{(\sigma)}}{dD^{(\sigma)}} = \frac{d^2 E^{(\sigma)}}{dD^{(\sigma)\,2}} \geq 0 \tag{6-16}$$

It follows from (6-8) and (6-9) that (see Sec. 4.4.3)

$$E^{(\sigma)} + E_c^{(\sigma)} = S^{(\sigma)} D^{(\sigma)} = \text{tr}(\, \mathbf{S}^{(\sigma)} \cdot \mathbf{D}^{(\sigma)} \,) \tag{6-17}$$

Recognizing (6-10), let us integrate (6-13) over the phase interface,

$$\int_{\Sigma} \{\, E^{(\sigma)}(\, \mathbf{D}^{(\sigma)*} \,) - E^{(\sigma)}(\, \mathbf{D}^{(\sigma)} \,)$$
$$- \text{tr}[\, \mathbf{S}^{(\sigma)} \cdot (\, \mathbf{D}^{(\sigma)*} - \mathbf{D}^{(\sigma)} \,)]\} \, dA \geq 0 \tag{6-18}$$

We will now regard $\mathbf{D}^{(\sigma)*}$ as being defined in terms of the trial surface velocity distribution $\mathbf{v}^{(\sigma)*}$. From (6-1) and the symmetry of $\mathbf{T}^{(\sigma)}$, it follows that

$$\text{tr}(\, \mathbf{S}^{(\sigma)} \cdot \mathbf{D}^{(\sigma)} \,) = \text{tr}(\, \mathbf{T}^{(\sigma)} \cdot \mathbf{D}^{(\sigma)} \,)$$

$$= \text{tr}(\, \mathbf{T}^{(\sigma)} \cdot \nabla_{(\sigma)} \mathbf{v}^{(\sigma)} \,) \tag{6-19}$$

By an application of the surface divergence theorem (Sec. A.6.3), we are able to say

$$\int_{\Sigma} \text{tr}(\, \mathbf{S}^{(\sigma)} \cdot \mathbf{D}^{(\sigma)} \,)\, dA = \int_{C} \mathbf{v}^{(\sigma)} \cdot \mathbf{T}^{(\sigma)} \cdot \boldsymbol{\mu}\, ds$$

$$- \int_{\Sigma} \mathbf{v}^{(\sigma)} \cdot \text{div}_{(\sigma)} \mathbf{T}^{(\sigma)}\, dA \tag{6-20}$$

Here C denotes the closed curve bounding the phase interface Σ; $\boldsymbol{\mu}$ is the outwardly directed unit normal to C; ds indicates that a line integration is to be performed. In a similar manner, we can show

$$\int_{\Sigma} \text{tr}(\, \mathbf{S}^{(\sigma)} \cdot \mathbf{D}^{(\sigma)*} \,)\, dA = \int_{C} \mathbf{v}^{(\sigma)*} \cdot \mathbf{T}^{(\sigma)} \cdot \boldsymbol{\mu}\, ds$$

$$- \int_{\Sigma} \mathbf{v}^{(\sigma)*} \cdot \text{div}_{(\sigma)} \mathbf{T}^{(\sigma)}\, dA \tag{6-21}$$

By means of (6-20) and (6-21), we may express (6-18) as

$$\int_{\Sigma} E^{(\sigma)}(\, \mathbf{D}^{(\sigma)} \,)\, dA \leq \int_{\Sigma} E^{(\sigma)}(\, \mathbf{D}^{(\sigma)*} \,)\, dA$$

$$+ \int_{C} (\, \mathbf{v} - \mathbf{v}^* \,) \cdot \mathbf{T}^{(\sigma)} \cdot \boldsymbol{\mu}\, ds$$

$$- \int_{\Sigma} (\, \mathbf{v} - \mathbf{v}^* \,) \cdot (\, \text{div}_{(\sigma)} \mathbf{T}^{(\sigma)} \,)\, dA \tag{6-22}$$

in which we have recognized that

$$\text{at } \Sigma : \quad \mathbf{v}^{(\sigma)} = \mathbf{v} \tag{6-23}$$

and

$$\text{at } \Sigma : \quad \mathbf{v}^{(\sigma)*} = \mathbf{v}^* \tag{6-24}$$

Equation (6-14), simplified by means of (6-11), may now be integrated over Σ to obtain

$$\int_{\Sigma} \{ E_c^{(\sigma)}(\mathbf{S}^{(\sigma)*}) - E_c^{(\sigma)}(\mathbf{S}^{(\sigma)}) - \text{tr}[\mathbf{D}^{(\sigma)} \cdot (\mathbf{S}^{(\sigma)*} - \mathbf{S}^{(\sigma)})]\} \, dA \geq 0 \tag{6-25}$$

in which I have introduced the trial surface extra-stress tensor

$$\mathbf{S}^{(\sigma)*} \equiv \mathbf{T}^{(\sigma)*} - \frac{1}{2} (\text{tr } \mathbf{T}^{(\sigma)*})\mathbf{P} \tag{6-26}$$

Because of (6-17), we may also express (6-25) as

$$\int_{\Sigma} [E_c^{(\sigma)}(\mathbf{S}^{(\sigma)*}) + E^{(\sigma)}(\mathbf{D}^{(\sigma)}) - \text{tr}(\mathbf{D}^{(\sigma)} \cdot \mathbf{S}^{(\sigma)*})] \, dA \geq 0 \tag{6-27}$$

Using an argument analogous to that employed in proving (6-20), we find

$$\int_{\Sigma} \text{tr}(\mathbf{D}^{(\sigma)} \cdot \mathbf{S}^{(\sigma)*}) \, dA = \int_{C} \mathbf{v} \cdot \mathbf{T}^{(\sigma)*} \cdot \boldsymbol{\mu} \, ds - \int_{\Sigma} \mathbf{v} \cdot \text{div}_{(\sigma)} \mathbf{T}^{(\sigma)*} \, dA \tag{6-28}$$

This allows us to conclude from (6-27) that

$$\int_{\Sigma} E^{(\sigma)}(\mathbf{D}^{(\sigma)}) \, dA \geq - \int_{\Sigma} E_c^{(\sigma)}(\mathbf{S}^{(\sigma)*}) \, dA + \int_{C} \mathbf{v} \cdot \mathbf{T}^{(\sigma)*} \cdot \boldsymbol{\mu} \, ds - \int_{\Sigma} \mathbf{v} \cdot (\text{div}_{(\sigma)} \mathbf{T}^{(\sigma)*}) \, dA \tag{6-29}$$

If for a fixed value of p

$$t^p \, E^{(\sigma)}(\mathbf{D}^{(\sigma)}) = E^{(\sigma)}(t\mathbf{D}^{(\sigma)}) \tag{6-30}$$

independent of the value t assumes, then $E^{(\sigma)}$ is said to be a homogeneous function of degree p. From Euler's theorem for homogeneous functions

(Kaplan 1973), p. 139)

$$p\, E^{(\sigma)} = D^{(\sigma)} \frac{dE^{(\sigma)}}{dD^{(\sigma)}} \tag{6-31}$$

The potential function $E^{(\sigma)}$ is an explicit function of $D^{(\sigma)}$, which means

$$\frac{\partial E}{\partial D^{(\sigma)}_{ij}} = \frac{1}{D^{(\sigma)}} \frac{dE^{(\sigma)}}{dD^{(\sigma)}} D^{(\sigma)}_{ij} \tag{6-32}$$

Using (6-10) and (6-32), we can rearrange (6-31) to read

$$\begin{aligned} E^{(\sigma)} &= \frac{1}{p} D^{(\sigma)}_{ij} \frac{\partial E^{(\sigma)}}{\partial D^{(\sigma)}_{ij}} \\ &= \frac{1}{p} D^{(\sigma)}_{ij} S^{(\sigma)}_{ij} \\ &= \frac{1}{p} \operatorname{tr}(\mathbf{S}^{(\sigma)} \cdot \mathbf{D}^{(\sigma)}) \end{aligned} \tag{6-33}$$

For the linear Boussinesq surface fluid model (Sec. 2.2.2), ε is a constant and $E^{(\sigma)}$ is a homogeneous function of degree 2:

$$E^{(\sigma)} = \varepsilon\, D^{(\sigma)2} \tag{6-34}$$

Another example is provided by the surface power model fluid for which

$$\begin{aligned} \varepsilon &= \varepsilon_0 (|\nabla_{(\sigma)} \mathbf{v}^{(\sigma)}|)^{n-1} \\ &= \varepsilon_0 (2D^{(\sigma)2})^{(n-1)/2} \\ &= \varepsilon_0 \left(\frac{S^{(\sigma)2}}{2\varepsilon_0^2} \right)^{(n-1)/2n} \end{aligned} \tag{6-35}$$

In writing this expression, I have recognized that

$$S^{(\sigma)2} = 2\varepsilon_0^2 (2D^{(\sigma)2})^n \tag{6-36}$$

It follows immediately that for the surface power model fluid

$$E^{(\sigma)} = \frac{\varepsilon_0}{n+1} (2\, D^{(\sigma)2})^{(n+1)/2} \tag{6-37}$$

$$E_c^{(\sigma)} = \frac{\varepsilon_0^n}{n+1}\left(\frac{S^{(\sigma)\,2}}{2\,\varepsilon_0^2}\right)^{(n+1)/2n} \tag{6-38}$$

and $E^{(\sigma)}$ is a homogeneous function of degree n+1.

With the assumption that $E^{(\sigma)}$ is a homogeneous function of degree p, (6-33) allows us to express (6-22) as

$$\frac{1}{p}\int_{\Sigma} \text{tr}(\,\mathbf{S}^{(\sigma)} \cdot \mathbf{D}^{(\sigma)}\,)\,dA \le \int_{\Sigma} E^{(\sigma)}(\,D^{(\sigma)*}\,)\,dA$$

$$+ \int_{C} (\,\mathbf{v} - \mathbf{v}^*\,) \cdot \mathbf{T}^{(\sigma)} \cdot \boldsymbol{\mu}\,ds$$

$$- \int_{\Sigma} (\,\mathbf{v} - \mathbf{v}^*\,) \cdot (\,\text{div}_{(\sigma)}\mathbf{T}^{(\sigma)}\,)\,dA \tag{6-39}$$

and (6-29) as

$$\frac{1}{p}\int_{\Sigma} \text{tr}(\,\mathbf{S}^{(\sigma)} \cdot \mathbf{D}^{(\sigma)}\,)\,dA \ge -\int_{\Sigma} E_c^{(\sigma)}(\,S^{(\sigma)*}\,)\,dA$$

$$+ \int_{C} \mathbf{v} \cdot \mathbf{T}^{(\sigma)*} \cdot \boldsymbol{\mu}\,ds - \int_{\Sigma} \mathbf{v} \cdot (\,\text{div}_{(\sigma)}\,\mathbf{T}^{(\sigma)*}\,)\,dA \tag{6-40}$$

When we restrict ourselves to homogeneous potential functions $E^{(\sigma)}$ that are of the same degree p as E, the velocity extremum principle in the form of (4-4) can be added to (6-39) to learn (Hopke and Slattery 1975)

$$\frac{1}{p}\left[\int_{R} \text{tr}(\,\mathbf{S} \cdot \mathbf{D}\,)\,dV + \int_{\Sigma} \text{tr}(\,\mathbf{S}^{(\sigma)} \cdot \mathbf{D}^{(\sigma)}\,)\,dA\right]$$

$$\le \int_{R} E(\,D^*\,)\,dV + \int_{\Sigma} E^{(\sigma)}(\,D^{(\sigma)*}\,)\,dA$$

$$+ \int_{S-S_v} (\,\mathbf{v} - \mathbf{v}^*\,) \cdot (\,\mathbf{T} - \rho\phi\mathbf{I}\,) \cdot \mathbf{n}\,dA$$

$$+ \int_{C} (\,\mathbf{v} - \mathbf{v}^*\,) \cdot \mathbf{T}^{(\sigma)} \cdot \boldsymbol{\mu}\,ds \tag{6-41}$$

In arriving at this result, I have used the jump momentum balance (6-3) and I have observed that the normal components of both $\mathbf{v}$ and $\mathbf{v}^*$ are zero at Σ. Similarly, when the stress extremum principle in the from of (4-5) is added to (6-40) under these conditions, we have (Hopke and Slattery 1975)

$$\frac{1}{p}\left[\int_R \mathrm{tr}(\,\mathbf{S}\cdot\mathbf{D}\,)\,dV + \int_\Sigma \mathrm{tr}(\,\mathbf{S}^{(\sigma)}\cdot\mathbf{D}^{(\sigma)}\,)\,dA\right]$$

$$\geq -\int_R E_c(\,\mathbf{S}^*\,)\,dV - \int_\Sigma E_c^{(\sigma)}(\,\mathbf{S}^{(\sigma)*}\,)\,dA$$

$$+\int_S \mathbf{v}\cdot(\,\mathbf{T}^* - \rho\phi\mathbf{I}\,)\cdot\mathbf{n}\,dA$$

$$+\int_C \mathbf{v}\cdot\mathbf{T}^{(\sigma)*}\cdot\boldsymbol{\mu}\,ds \tag{6-42}$$

Here I have noted that the trial stress distributions $\mathbf{T}^*$ and $\mathbf{T}^{(\sigma)*}$ must be consistent with the jump momentum balance (6-4) at Σ. For these restricted classes of bulk and surface stress-deformation behavior, the velocity and stress extremum principles give bounds for the rate of viscous dissipation of mechanical energy within the bulk phases and within the dividing surfaces.

Inequalities (6-41) and (6-42) are most useful, when the bulk phases are incompressible newtonian fluids and the surface stress-deformation behavior is described by the linear Boussinesq surface fluid model. In this case $p = 2$ and the upper bound (6-41) for the rate of dissipation of energy becomes

$$\frac{1}{2}\left[\int_R \mathrm{tr}(\,\mathbf{S}\cdot\mathbf{D}\,)\,dV + \int_\Sigma \mathrm{tr}(\,\mathbf{S}^{(\sigma)}\cdot\mathbf{D}^{(\sigma)}\,)\,dA\right]$$

$$\leq \mu\int_R \mathrm{tr}(\,\mathbf{D}^*\cdot\mathbf{D}^*\,)\,dV + \varepsilon\int_\Sigma \mathrm{tr}(\,\mathbf{D}^{(\sigma)*}\cdot\mathbf{D}^{(\sigma)*}\,)\,dA$$

$$+\int_{S-S_v} (\,\mathbf{v}-\mathbf{v}^*\,)\cdot(\,\mathbf{T}-\rho\phi\mathbf{I}\,)\cdot\mathbf{n}\,dA$$

$$+\int_C (\,\mathbf{v}-\mathbf{v}^*\,)\cdot\mathbf{T}^{(\sigma)}\cdot\boldsymbol{\mu}\,ds \tag{6-43}$$

Similarly, the lower bound (6-42) for the rate of dissipation of energy takes the form

$$\frac{1}{2}\left[\int_R \mathrm{tr}(\mathbf{S}\cdot\mathbf{D})\,dV + \int_\Sigma \mathrm{tr}(\mathbf{S}^{(\sigma)}\cdot\mathbf{D}^{(\sigma)})\,dA\right]$$

$$\geq -\frac{1}{4\mu}\int_R \mathrm{tr}(\mathbf{S}^{*}\cdot\mathbf{S}^{*})\,dV - \frac{1}{4\varepsilon}\int_\Sigma \mathrm{tr}(\mathbf{S}^{(\sigma)*}\cdot\mathbf{S}^{(\sigma)*})\,dA$$

$$+\int_S \mathbf{v}\cdot(\mathbf{T}^{*} - \rho^{*}\phi\mathbf{I})\cdot\mathbf{n}\,dA + \int_C \mathbf{v}\cdot\mathbf{T}^{(\sigma)*}\cdot\boldsymbol{\mu}\,ds \qquad (6\text{-}44)$$

In the next section, I illustrate their use with a particular example.

References

Hopke, S. W., and J. C. Slattery, *Int. J. Multiphase Flow* **1,** 727 (1975).

Kaplan, W., "Advanced Calculus," second edition, Addison-Wesley, Reading, MA (1973).

Exercises

4.4.6-1 *basic extremum principles* (Hopke and Slattery 1975) Determine that

$$\int_R E(D)\,dV + \int_\Sigma E^{(\sigma)}(D^{(\sigma)})\,dA \leq \int_R E(D^{*})\,dV$$

$$+\int_\Sigma E^{(\sigma)}(D^{(\sigma)*})\,dA$$

$$+\int_{S - S_v} (\mathbf{v} - \mathbf{v}^{*})\cdot(\mathbf{T} - \rho\phi\mathbf{I})\cdot\mathbf{n}\,dA$$

$$+\int_C (\mathbf{v} - \mathbf{v}^{*})\cdot\mathbf{T}^{(\sigma)}\cdot\boldsymbol{\mu}\,ds \qquad (1)$$

and

$$\int_R E(D)\, dV + \int_\Sigma E^{(\sigma)}(\, D^{(\sigma)}\,)\, dA \geq - \int_R E_c(\, S^*\,)\, dV$$

$$- \int_\Sigma E_c^{(\sigma)}(\, S^{(\sigma)*}\,)\, dA + \int_S \mathbf{v} \cdot (\, \mathbf{T}^* - \rho\phi\mathbf{I}\,) \cdot \mathbf{n}\, dA$$

$$+ \int_C \mathbf{v} \cdot \mathbf{T}^{(\sigma)*} \cdot \boldsymbol{\mu}\, ds \qquad (2)$$

These inequalities can be used directly, if one is principally interested in approximating either the velocity distribution or the stress distribution. Of a given family of trial velocity distributions, the "best" approximation of the true velocity distribution is that member which minimizes the right side of inequality (1). Of a given family of trial stress distributions, the "best" approximation of the true stress distribution is that member which maximizes the right side of inequality (2).

4.4.6-2 *E and $E^{(\sigma)}$ are homogeneous functions of different degress* (Hopke and Slattery 1975) Let us assume that we are concerned with newtonian bulk fluid behavior, for which E is a homogeneous function of degree 2, and with nonlinear interfacial behavior. In particular, let us assume that the interface may be described by a surface power model fluid, for which $E^{(\sigma)}$ is a homogeneous function of degree n + 1 and for which n ≤ 1. Reason that

$$\int_R E(D)\, dV + \int_\Sigma E^{(\sigma)}(\, D^{(\sigma)}\,)\, dA \geq \frac{1}{2} \Big[\int_R \mathrm{tr}(\, \mathbf{S} \cdot \mathbf{D}\,)\, dV$$

$$+ \int_\Sigma \mathrm{tr}(\, \mathbf{S}^{(\sigma)} \cdot \mathbf{D}^{(\sigma)}\,)\, dA \Big] \qquad (1)$$

and

$$\int_R E(D)\, dV + \int_\Sigma E^{(\sigma)}(\, D^{(\sigma)}\,)\, dA \leq \frac{1}{n+1} \Big[\int_R \mathrm{tr}(\, \mathbf{S} \cdot \mathbf{D}\,)\, dV$$

$$+ \int_\Sigma \mathrm{tr}(\, \mathbf{S}^{(\sigma)} \cdot \mathbf{D}^{(\sigma)}\,)\, dA \Big] \qquad (2)$$

These two inequalities can in turn be combined with those developed in Exercise 4.4.6-1 to provide upper and lower bounds for the rate of viscous dissipation of mechanical energy.

4.4.6-3 *bounding principles with no restriction on the surface mass density distribution* Often we will not be willing to make assumption ix", which says that the surface mass density $\rho^{(\sigma)}$ is a constant. Another pair of bounding principles can be developed, if we are willing to say instead

ix''') The interfacial tension γ is a constant.

At the same time, we will eliminate the restriction (6-2) on the trial surface velocity distribution $\mathbf{v}^{(\sigma)*}$ as well.

Although it would be possible to allow for nonlinear stress-deformation behavior, for simplicity let us replace vii by

vii''') The bulk phase can be described as incompressible Newtonian fluids.

and xi'' by

xi''') The surface stress-deformation behavior can be described by the linear Boussinesq surface fluid model

$$\mathbf{T}^{(\sigma)} = [\, \gamma + (\, \kappa - \varepsilon\,)\, \text{div}_{(\sigma)}\mathbf{v}^{(\sigma)}\,]\, \mathbf{P} + 2\varepsilon\, \mathbf{D}^{(\sigma)}$$

$$= \gamma\, \mathbf{P} + \mathbf{S}^{\{1\}} + \mathbf{S}^{\{2\}}$$

where

$$\mathbf{S}^{\{1\}} \equiv (\, \kappa - \varepsilon\,)\text{div}_{(\sigma)}\mathbf{v}^{(\sigma)}\, \mathbf{P}$$

$$\mathbf{S}^{\{2\}} \equiv 2\varepsilon\, \mathbf{D}^{(\sigma)}$$

We will additionally require

xii''') The surface dilatational viscosity κ must be greater than or equal to the surface shear viscosity ε :

$$\kappa - \varepsilon \geq 0$$

This goes beyond the requirements of the jump Clausius-Duhem inequality explained in Exercise 5.9.6-1, which says only that $\kappa \geq 0$ and $\varepsilon \geq 0$.

Note that the results of Sec. 4.4.4 are immediately applicable. We can write by analogy with (6-39) and (6-40):

$$\frac{1}{2}\int_{\Sigma} \text{tr}(\ \mathbf{S}\{\substack{\varsigma\\2}\} \cdot \mathbf{D}^{(\sigma)}\ )\ dA \le \int_{\Sigma} E\{\substack{\varsigma\\2}\}(\ \mathbf{D}^{(\sigma)*}\ )\ dA$$

$$+ \int_{C} (\ \mathbf{v} - \mathbf{v}^*\ ) \cdot \mathbf{S}\{\substack{\varsigma\\2}\} \cdot \boldsymbol{\mu}\ ds - \int_{\Sigma} (\ \mathbf{v} - \mathbf{v}^*\ ) \cdot \text{div}_{(\sigma)}\mathbf{S}\{\substack{\varsigma\\2}\}\ \ dA \quad (1)$$

$$\frac{1}{2}\int_{\Sigma} \text{tr}(\ \mathbf{S}\{\substack{\varsigma\\2}\} \cdot \mathbf{D}^{(\sigma)}\ )\ dA \ge - \int_{\Sigma} E\{\substack{\varsigma\\2}\}_c(\ \mathbf{S}\{\substack{\varsigma\\2}\}^*\ )\ dA$$

$$+ \int_{C} \mathbf{v} \cdot \mathbf{S}\{\substack{\varsigma\\2}\}^* \cdot \boldsymbol{\mu}\ ds - \int_{\Sigma} \mathbf{v} \cdot \text{div}_{(\sigma)}\mathbf{S}\{\substack{\varsigma\\2}\}^*\ \ dA \quad (2)$$

Observe that

$$\mathbf{S}\{\substack{\varsigma\\1}\} = M\ \mathbf{P}$$

where

$$M \equiv (\ \kappa - \varepsilon\ )L$$

$$L \equiv \text{div}_{(\sigma)}\mathbf{v}^{(\sigma)}$$

Let us introduce two scalar potential functions

$$E\{\substack{\varsigma\\1}\} \equiv \frac{1}{2}(\ \kappa - \varepsilon\ )L^2$$

and

$$E\{\substack{\varsigma\\1}\}_c \equiv \frac{M^2}{2(\ \kappa\ - \varepsilon\ )}$$

in terms of which we can write

$$M = \frac{\partial E\{\substack{\varsigma\\1}\}}{\partial L}$$

and

$$L = \frac{\partial E\{\sigma\}_c}{\partial M}$$

Using assumption xii''', prove that

$$\frac{1}{2}\int_{\Sigma} \mathrm{tr}(\, \mathbf{S}\{\sigma\} \cdot \mathbf{D}^{(\sigma)}\,)\, dA \le \int_{\Sigma} E\{\sigma\}(\, L^* \,)\, dA$$

$$+ \int_{C} (\, \mathbf{v} - \mathbf{v}^* \,) \cdot \mathbf{S}\{\sigma\} \cdot \boldsymbol{\mu}\, ds$$

$$- \int_{\Sigma} (\, \mathbf{v} - \mathbf{v}^* \,) \cdot \mathrm{div}_{(\sigma)} \mathbf{S}\{\sigma\}\, dA \tag{3}$$

and

$$\frac{1}{2}\int_{\Sigma} \mathrm{tr}(\, \mathbf{S}\{\sigma\} \cdot \mathbf{D}^{(\sigma)}\,)\, dA \ge - \int_{\Sigma} E\{\sigma\}_c(M^*)\, dA + \int_{C} \mathbf{v} \cdot \mathbf{S}\{\sigma\}^* \cdot \boldsymbol{\mu}\, ds$$

$$- \int_{\Sigma} \mathbf{v} \cdot \mathrm{div}_{(\sigma)} \mathbf{S}\{\sigma\}^*\, dA \tag{4}$$

From (4-4), (1), and (3), we find

$$\frac{1}{2}\int_{R} \mathrm{tr}(\, \mathbf{S} \cdot \mathbf{D}\,)\, dV + \frac{1}{2}\int_{\Sigma} \mathrm{tr}(\, \mathbf{S}^{(\sigma)} \cdot \mathbf{D}^{(\sigma)}\,)\, dA$$

$$\le \mu \int_{R} \mathrm{tr}(\, \mathbf{D}^* \cdot \mathbf{D}^* \,)\, dV + \varepsilon \int_{\Sigma} \mathrm{tr}(\, \mathbf{D}^{(\sigma)*} \cdot \mathbf{D}^{(\sigma)*} \,)\, dA$$

$$+ \frac{(\, \kappa - \varepsilon \,)}{2} \int_{\Sigma} (\, \mathrm{div}_{(\sigma)} \mathbf{v}^{(\sigma)*} \,)^2\, dA$$

$$+ \int_{S - S_v} (\, \mathbf{v} - \mathbf{v}^* \,) \cdot (\, \mathbf{T} - \rho\phi \mathbf{I} \,) \cdot \mathbf{n}\, dA$$

$$+ \int_{C} (\, \mathbf{v} - \mathbf{v}^* \,) \cdot \mathbf{S}^{(\sigma)} \cdot \boldsymbol{\mu}\, ds \tag{5}$$

having noted that

$$\text{at } \Sigma: \ [\, \mathbf{T} \cdot \boldsymbol{\xi} \,] + \mathrm{div}_{(\sigma)} \mathbf{T}^{(\sigma)} = 0$$

and

$$\text{at } \Sigma: \ \mathbf{v} \cdot \boldsymbol{\xi} = \mathbf{v}^* \cdot \boldsymbol{\xi} = 0$$

From (4-5), (2), and (4), we deduce in a similar fashion that

$$\begin{aligned}
&\frac{1}{2} \int_R \mathrm{tr}(\, \mathbf{S} \cdot \mathbf{D} \,)\, dV + \frac{1}{2} \int_\Sigma \mathrm{tr}(\, \mathbf{S}^{(\sigma)} \cdot \mathbf{D}^{(\sigma)} \,)\, dA \\
&\quad \geq - \frac{1}{4\mu} \int_\Sigma \mathrm{tr}(\, \mathbf{S}^* \cdot \mathbf{S}^* \,)\, dV - \frac{1}{4\varepsilon} \int_\Sigma \mathrm{tr}\left(\mathbf{S}\genfrac{\{}{\}}{0pt}{}{\sigma}{2}^* \cdot \mathbf{S}\genfrac{\{}{\}}{0pt}{}{\sigma}{2}^* \right) dA \\
&\qquad - \frac{1}{2(\, \kappa - \varepsilon \,)} \int_\Sigma M^{*2}\, dA + \int_S \mathbf{v} \cdot (\, \mathbf{T}^* - \rho\phi \mathbf{I} \,) \cdot \mathbf{n}\, dA \\
&\qquad + \int_C \left[\, \mathbf{v} \cdot \left(M^* \mathbf{P} + \mathbf{S}\genfrac{\{}{\}}{0pt}{}{\sigma}{2}^* \right) \cdot \boldsymbol{\mu} \,\right] ds
\end{aligned} \tag{6}$$

with the additional requirements

$$\text{at } \Sigma: \ [\, \mathbf{T}^* \cdot \boldsymbol{\xi} \,] + \mathrm{div}_{(\sigma)} \mathbf{T}^{(\sigma)*} = 0$$

where

$$\mathbf{T}^{(\sigma)*} \equiv (\, \gamma + M^* \,)\mathbf{P} + \mathbf{S}\genfrac{\{}{\}}{0pt}{}{\sigma}{2}^*$$

4.4.7 An example: blunt knife-edge surface viscometer (Briley *et al.* 1976) In Sec. 3.4.1, I introduced the knife-edge surface viscometer assuming that the contact between the knife-edge and the liquid-gas interface is a geometric circle. In practice, this is only an approximation that appears even more questionable when a knife-edge is examined under a microscope.

The *perfect* knife-edge of Sec. 3.4.1 may be thought of as a limiting case of the blunt knife-edge (or flat ring) shown in Figure 7-1. Goodrich and Allen (1972) neglected the presence of the pan in Figure 7-1 and examined the velocity distribution generated in a semi-infinite body of liquid by a rotating, blunt knife-edge in the liquid-gas interface. Their analysis did not allow them to calculate the torque acting upon the knife-edge.

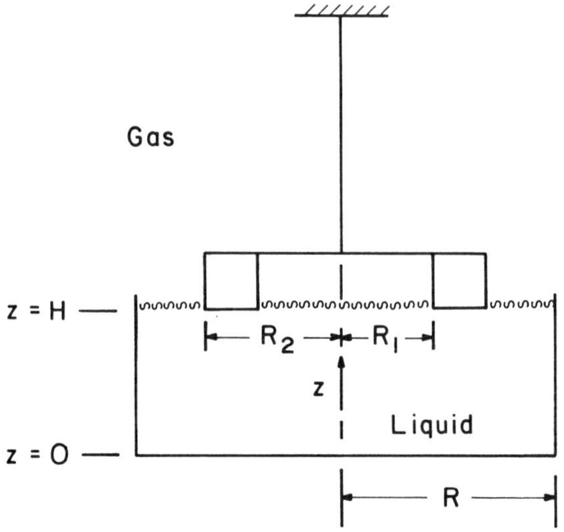

Figure 4.4.7-1 Blunt knife-edge surface viscometer. As the pan rotates with a constant velocity Ω, one measures $\mathcal{T}_z$, the z component of the torque required to hold the circular blunt knife-edge stationary.

The objective of this analysis is to calculate the effect of the thickness of the knife-edge upon $\mathcal{T}_z$, the z component of the torque required to hold the circular knife-edge stationary.

We will require the following assumptions in this analysis.

i) Viscous effects are neglected in the gas phase.

ii) All inertial effects are neglected.

iii) There is no mass transfer across the liquid-gas interface.

iv) The viscometer operates under steady-state conditions in which all variables are independent of time.

v) The liquid phase is an incompressible Newtonian fluid.

vi) The stress-deformation behavior of the gas-liquid phase interface may be represented by the Boussinesq surface fluid model (Sec. 2.2.2).

vii) All physical properties are constants.

viii) The circular knife-edge is positioned in such a manner that the phase interface is a plane.

ix) A solution of the same form found in Sec. 3.4.1 exists:

$$v_r = v_z = 0 \tag{7-1}$$

$$v_\theta^* \equiv \frac{v_\theta}{R\Omega} - r^*$$

$$= v_\theta^* \,(r^*, z^*) \tag{7-2}$$

where

$$r^* \equiv \frac{r}{R}$$

$$z^* \equiv \frac{z}{R} \tag{7-3}$$

and Ω is the magnitude of the angular velocity of the pan.

As boundary conditions, we must require that the angular velocity is a constant independent of position on the surface of the pan

$$\text{at } z^* = 0 : \; v_\theta^* = 0 \tag{7-4}$$

$$\text{at } r^* = 1 : \; v_\theta^* = 0 \tag{7-5}$$

and that the knife-edge is stationary

$$\text{at } z^* = \frac{H}{R} \text{ for } R_1^* \leq r^* \leq R_2^*: \; v_\theta^* = - r^* \tag{7-6}$$

From Table 2.4.2-2, the r and z components of the jump momentum balance for a Boussinesq surface say only that

$$\text{at } z^* = H^* \text{ for } 0 \leq r^* < R_1^* \text{ and } R_2^* < r^* < 1: \; p^* \equiv \frac{p}{p_0} = 1 \tag{7-7}$$

The θ component reduces to

$$\text{at } z^* = H^* \;\; \text{for } 0 \leq r^* < R_1^* \text{ and } R_2^* < r^* < 1:$$

$$\frac{\partial v_\theta^*}{\partial z^*} = N \frac{\partial}{\partial r^*} \left[\frac{1}{r^*} \frac{\partial}{\partial r^*} (r^* v_\theta^*) \right] \tag{7-8}$$

in which I have introduced

$$N \equiv \frac{\varepsilon}{\mu R} \tag{7-9}$$

We have not been able to find a solution to Cauchy's first law for a Newtonian fluid, (3.4.1-10) and (3.4.1-11), that is of the form (7-1) and (7-2) and that is consistent with boundary conditions (7-4) through (7-8).

Fortunately, it is not the detailed velocity distribution within the liquid in which we are interested. Our primary concern is with $\mathcal{T}_z$, the z component of the torque required to hold the knife-edge stationary. Let us see how we can use the bounding principles described in the previous section to obtain upper and lower bounds on $\mathcal{T}_z$ as a function of the gas-liquid surface shear viscosity ε, Ω, the thickness of the knife-edge, and all of the other parameters required to describe the geometry.

From the integral mechanical energy balance (Sec. 4.3.4), the rate at which work is done by the liquid on the rotating pan is the negative of the rate at which energy is dissipated by viscous forces acting both within the liquid and within the interface:

$$\begin{aligned} \mathcal{W} &= - \mathcal{E} \\ &\equiv - \int_R \mathrm{tr}(\mathbf{S} \cdot \mathbf{D}) \, dV - \int_\Sigma \mathrm{tr}(\mathbf{S}^{(\sigma)} \cdot \mathbf{D}^{(\sigma)}) \, dA \end{aligned} \tag{7-10}$$

Since the pan rotates as a solid body,

$$\text{on the pan:} \quad \mathbf{v} = \mathbf{\Omega} \wedge \mathbf{z} \tag{7-11}$$

where only the z component of the angular velocity $\mathbf{\Omega}$ is nonzero. This permits us to rearrange the expression for the rate at which work is done by the liquid on the rotating pan as

$$\begin{aligned} \mathcal{W} &= \int_{S_{pan}} \mathbf{v} \cdot (\mathbf{T} + p_0 \mathbf{I}) \cdot (- \mathbf{n}) \, dA \\ &= \int_{S_{pan}} (\mathbf{\Omega} \wedge \mathbf{z}) \cdot (\mathbf{T} + p_0 \mathbf{I}) \cdot (- \mathbf{n}) \, dA \end{aligned}$$

$$= - \boldsymbol{\Omega} \cdot \boldsymbol{\mathcal{T}}$$

$$= - \Omega \, \mathcal{T}_z \tag{7-12}$$

Here **n** is the unit normal that is outwardly directed with respect to the closed surface bounding the liquid and

$$\boldsymbol{\mathcal{T}} \equiv \int_{S_{pan}} \mathbf{z} \wedge (\mathbf{T} + p_0 \mathbf{I}) \cdot \mathbf{n} \, dA \tag{7-13}$$

is the torque required to hold the circular knife-edge stationary (it is equal in magnitude and opposite in sign to the torque exerted on the knife-edge by the fluid).

From (7-10) and (7-12), we see that $\mathcal{T}_z$ is proportional to the rate at which energy is dissipated by viscous forces acting both within the liquid and within the interface

$$\Omega \, \mathcal{T}_z = \int_R \text{tr}(\mathbf{S} \cdot \mathbf{D}) \, dV + \int_\Sigma \text{tr}(\mathbf{S}^{(\sigma)} \cdot \mathbf{D}^{(\sigma)}) \, dA \tag{7-14}$$

for which upper and lower bounds can be obtained using (6-43) and (6-44).

Upper bound Let us choose the liquid as the system to which inequality (6-43) is to be applied. In view of (7-14), (6-43) reduces to

$$\frac{1}{2} \Omega \, \mathcal{T}_z \leq \mu \int_R \text{tr}(\mathbf{D}^* \cdot \mathbf{D}^*) \, dV + \varepsilon \int_\Sigma \text{tr}(\mathbf{D}^{(\sigma)*} \cdot \mathbf{D}^{(\sigma)*}) \, dA \tag{7-15}$$

in which we have noted that, since velocity is specified a priori on all of S (the pan and the knife-edge), $\mathbf{v}^* = \mathbf{v}$ on all of these surfaces.

If we choose a trial velocity distribution $\mathbf{v}^*$ having the form of (7-1) and (7-2), the equation of continuity for the liquid phase (Slattery 1981, p. 59) is satisfied identically. The jump mass balance for the liquid-gas interface (Table 2.4.2-2) is also identically satisfied by this form of trial velocity distribution under the stated assumptions that the flow is steady-state and that there is no mass transfer across the liquid-gas interface. It is necessary to insure only that all of the velocity boundary conditions are obeyed.

This is easily accomplished, if we form a separate trial velocity distribution for each of three regions in Figure 7-1. In region I, we propose

$$\text{for } 0 \leq r^* < R_1^*: \quad v_\theta^{**} \equiv A(r^*) \, B(z^*) + C(r^*) \, D(z^*) \tag{7-16}$$

where $A(r^*)$, $B(z^*)$, $C(r^*)$, and $D(z^*)$ are polynomial functions chosen such that

$$A(0) = C(0) = 0 \tag{7-17}$$

$$A(R_1^*) = -R_1^* \qquad B(H^*) = 1 \qquad D(H^*) = 0 \tag{7-18}$$

$$B(0) = D(0) = 0 \tag{7-19}$$

Equation (7-17) ensure that the dimensionless trial distribution goes to zero on the axis of the pan; (7-18) and (7-19) are required by the velocity boundary conditions at the knife-edge (7-6) and at the floor of the pan (7-4). In region II, we say

$$\text{for } R_2^* < r^* \le 1: \; v_\theta^{**} \equiv H(r^*)\, I(z^*) + J(r^*)\, K(z^*) \tag{7-20}$$

The polynomial functions $E(z^*)$, $F(r^*)$, and $G(z^*)$ must satisfy

$$E(0) = G(0) = 0 \tag{7-21}$$

$$E(H^*) = 1 \qquad G(H^*) = 0 \tag{7-22}$$

which say that the trial velocity distribution must satisfy the velocity boundary conditions at the floor of the pan (7-4) and at the knife-edge (7-6). In region III,

$$\text{for } R_2^* < r^* \le 1: \; v_\theta^{**} \equiv H(r^*)\, I(z^*) + J(r^*)\, K(z^*) \tag{7-23}$$

where the polynomial functions $H(r^*)$, $I(z^*)$, $J(r^*)$, and $K(z^*)$ conform to the requirements of (7-4) through (7-6):

$$H(R_2^*) = -R_2^* \qquad I(H^*) = 1 \qquad K(H^*) = 0 \tag{7-24}$$

$$H(1) = J(1) = 0 \tag{7-25}$$

$$I(0) = K(0) = 0 \tag{7-26}$$

In addition, the trial velocity distribution must be continuous at $r^* = R_1^*$ and at $r^* = R_2^*$, which means

$$E(z^*) = B(z^*) - \frac{1}{R_1^*} C(R_1^*)\, D(z^*) + \frac{1}{R_1^*} F(R_1^*)\, G(z^*) \tag{7-27}$$

and

$$E(z^*) = I(z^*) - \frac{R}{R_2} J(R_2^*)\, K(z^*) + \frac{R}{R_2} F(R_2^*)\, G(z^*) \tag{7-28}$$

The polynomial coefficients in (7-16), (7-20), and (7-23) were optimized by requiring that the upper bound for $\mathcal{T}_z$, given by (7-15), be a minimum (Briley 1972; Deemer 1976).

Lower bound Let us now consider inequality (6-44), regarding the liquid as the system to which it is to be applied. Because of (7-14), (6-44) becomes

$$\frac{1}{2}\,\Omega\,\mathcal{T}_z \geq -\frac{1}{4\mu}\int_R \operatorname{tr}(\,\mathbf{S}^* \cdot \mathbf{S}^*\,)\,dV - \frac{1}{4\varepsilon}\int_\Sigma \operatorname{tr}(\,\mathbf{S}^{(\sigma)*} \cdot \mathbf{S}^{(\sigma)*}\,)\,dA$$

$$+ \int_{S_{dish}} \mathbf{v} \cdot \mathbf{S}^* \cdot \mathbf{n}\,dA + \int_{C_{dish}} \mathbf{v} \cdot \mathbf{S}^{(\sigma)*} \cdot \boldsymbol{\mu}\,ds \tag{7-29}$$

In arriving at this result, we have noted that the knife-edge is stationary and that the normal component of velocity is zero on the dish.

A lower bound for $\mathcal{T}_z$ could be computed using inequality (7-29) and a trial stress distribution chosen in much the same manner as the trial velocity distribution described above. However, we had difficulty devising a trial stress distribution that gave a lower bound for $\mathcal{T}_z$ which was acceptably close to our upper bound.

The multiple knife-edge surface viscometer is described in Exercise 3.4.1-2 and pictured in Figure 3.4.1-4 for the case of two zero-thickness knife-edges. The exact solution for the stress distribution in this instrument is an appropriate trial stress distribution for use in inequality (7-29): it satisfies Cauchy's first law for creeping motion (3-7) everywhere within the liquid and the jump momentum balance (6-3) everywhere on the liquid-gas phase interface. Let us now prove that the torque required to hold a multiple knife-edge stationary is a lower bound for the torque required to hold a blunt knife-edge stationary, so long as the multiple knife-edge instrument and the blunt knife-edge instrument are otherwise identical.

Let $\mathcal{T}_z^{(L)}$ be the torque required to hold the multiple knife-edge stationary. From (7-14), we see

$$\frac{1}{2}\,\Omega\,\mathcal{T}_z^{(L)} = \frac{1}{4\mu}\int_R \operatorname{tr}(\,\mathbf{S}^* \cdot \mathbf{S}^*\,)\,dV$$

$$+ \frac{1}{4\varepsilon}\int_{\Sigma^m} \operatorname{tr}(\,\mathbf{S}^{(\sigma)*} \cdot \mathbf{S}^{(\sigma)*}\,)\,dA \tag{7-30}$$

When (7-29) is applied to the multiple knife-edge instrument, it becomes an equality and, in view of (7-30), requires

$$\Omega\, \mathcal{T}_z^{(L)} = \int_{S_{dish}} \mathbf{v} \cdot \mathbf{S}^* \cdot \mathbf{n}\, dA + \int_{C_{dish}} \mathbf{v} \cdot \mathbf{S}^{(\sigma)*} \cdot \boldsymbol{\mu}\, ds \tag{7-31}$$

Adding (7-30) to (7-29) and subtracting (7-31), we find

$$\frac{1}{2}\Omega(\mathcal{T}_z - \mathcal{T}_z^{(L)}) \geq \frac{1}{4\varepsilon} \int_{\Sigma^m - \Sigma} \mathrm{tr}(\mathbf{S}^{(\sigma)*} \cdot \mathbf{S}^{(\sigma)*})\, dA \tag{7-32}$$

Here Σ refers to the liquid-gas interface in the blunt knife-edge geometry. Since the integral on the right of (7-32) must be positive, we conclude that

$$\mathcal{T}_z > \mathcal{T}_z^{(L)} \tag{7-33}$$

In words, the exact result for the torque exerted on a multiple knife-edge forms a lower bound for the torque on a blunt one whose inner and outer radii are the same.

Results The influence of the knife-edge radius and the importance of the pan depth have already been investigated in the context of the single, zero-thickness instrument (Sec. 3.4.1). Let us concentrate out attention here upon the effect of the thickness of the knife-edge.

We report in Tables 7-1 and 7-2 upper and lower bounds for the dimensionless torque

$$\mathcal{T}_z^\star \equiv \frac{\mathcal{T}_z}{2\pi R^3 \mu \Omega} \tag{7-34}$$

exerted on three blunt knife-edges of different dimensionless thicknesses. The arithmetic averages of the bounds are compared in Figure 7-2. It is feasible to fabricate knife-edges that correspond to these dimensionless thicknesses, so long as the pan radius is 10 cm or greater. In every case, the average dimensionless radius $R_{av}^\star \equiv (R_1^\star + R_2^\star)/2 = 0.8536$ and dimensionless depth $H^\star = 0.2252$ were chosen to be the same as those used by Brown *et al.* (1953). The programs for these computations are given by Deemer (1976).

In Exercise 7-1, we report an analytic expression for the torque exerted on a blunt knife-edge in the limit as $N \longrightarrow \infty$, $H^\star \longrightarrow \infty$ and $R_2^{\star -1} \longrightarrow \infty$:

$$\mathcal{T}_z^{(\infty)*} = 2N\, R_2^{\star 2} + \frac{4}{3\pi} R_2^{\star 3} - \frac{R_1^{\star 4}}{4R_2^\star}\; {}_2F_1\left(\frac{3}{2}, \frac{1}{2}; 3 ; [\, R_1/R_2\,]^2 \right) \tag{7-35}$$

Here

$${}_2F_1\left(\frac{3}{2}, \frac{1}{2}; 3 ; [\, R_1/R_2\,]^2 \right)$$

is a Gauss hypergeometric series (Abramowitz and Stegun 1965, p. 556). Our numerical results are seen to converge to this expression in Table 7-4.

Table 4.4.7-1: Torque required to hold the blunt knife-edge stationary for $\Delta R^* \equiv R_2^* - R_1^* = 0.005$[a]

N	Number of constants optimized for upper bound	$\mathcal{T}_z^{(U)*}$	$\mathcal{T}_z^{(L)*}$	$\mathcal{T}_z^{(av)*}$	maximum percentage error
10	31	55.92	55.92	55.92	0.000
1	38	6.464	6.461	6.463	0.0309
0.1	45	1.302	1.296	1.299	0.231
0.01	59	0.5980	0.5762	0.5871	1.86
0.001	59	0.4801	0.4182	0.4492	6.89
0.0001	59	0.4773	0.3800	0.4287	11.3

a) $H^* = 0.2252$, $R_1^* = 0.8511$, $R_2^* = 0.8561$, $(R_1^* + R_2^*)/2 = 0.8536$

For very large values of N, the interfacial forces are more important than the bulk viscous forces in computing the torque. Consequently, the torque is not very sensitive to the thickness of the knife-edge. Some dependence upon thickness remains in Tables 7-1 through 7-3 and in Figure 7-2, because we have chosen to compare knife-edges whose average radii are the same. As the thickness increases while the average radius remains constant, the outer radius moves closer to the pan, increasing the velocity gradients in the interface and the interfacial forces acting on the outer edge of the knife-edge.

For very small values of N, the torque on the knife-edge attributable to interfacial forces is negligibly small in comparison with that due to the bulk viscous forces. As a result, for small values of N the torque is nearly independent of N and highly dependent upon the thickness of the knife-edge. For these reason, the useful range of this instrument is N > 0.01.

Table 4.4.7-2: Torque required to hold the blunt knife-edge stationary for $\Delta R^* \equiv R_2^* - R_1^* = 0.01$[b]

N	Number of constants optimized for upper bound	$\mathcal{T}_z^{(U)*}$	$\mathcal{T}_z^{(L)*}$	$\mathcal{T}_z^{(av)*}$	maximum percentage error
10	31	57.16	57.15	57.16	0.0175
1	38	6.599	6.597	6.598	0.0152
0.1	45	1.334	1.329	1.332	0.225
0.01	59	0.6323	0.6119	0.6221	1.64
0.001	59	0.5198	0.4670	0.4934	5.35
0.0001	59	0.5064	0.4288	0.4676	8.30

Table 4.4.7-3: Torque required to hold the blunt knife-edge stationary for $\Delta R^* \equiv R_2^* - R_1^* = 0.02$[c]

N	Number of constants optimized for upper bound	$\mathcal{T}_z^{(U)*}$	$\mathcal{T}_z^{(L)*}$	$\mathcal{T}_z^{(av)*}$	maximum percentage error
10	31	59.76	59.76	59.76	0.000
1	38	6.884	6.882	6.883	0.0145
0.1	45	1.398	1.394	1.396	0.143
0.01	59	0.6880	0.6729	0.6805	1.11
0.001	59	0.5789	0.5372	0.5581	3.74
0.0001	59	0.5585	0.4933	0.5259	6.20

b) $H^* = 0.2252$, $R_1^* = 0.8486$, $R_2^* = 0.8586$, $(R_1^* + R_2^*)/2 = 0.8536$
c) $H^* = 0.2252$, $R_1^* = 0.8436$, $R_2^* = 0.8636$, $(R_1^* + R_2^*)/2 = 0.8536$

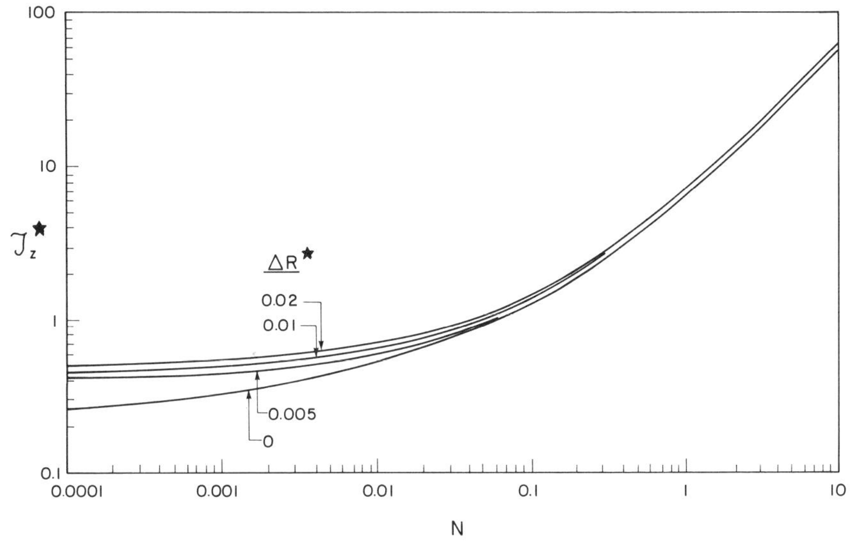

Figure 4.4.7-2 The arithmetic average of the upper and lower bounds for the dimensionless torque $\mathcal{T}_z^*$ as a function of N and $\Delta R^* \equiv R_2^* - R_1^*$ for the blunt knife-edge surface viscometer. The average dimensionless radius $R_{av}^* \equiv (R_1^* + R_2^*)/2 = 0.8536$ and dimensionless depth $H^* = 0.2252$ were chosen to be the same as those used by Brown *et al.* (1953). The curve for the zero-thickness knife-edge was calculated using the exact solution determined in Sec. 3.4.1.

Although we do not have an exact solution, the maximum percentage error involved in using $\mathcal{T}_z^{(av)*}$ is seen in Tables 7-1 through 7-3 to always be less than 2% for $N > 0.01$ and to be negligibly small for $N > 0.1$. For $N > 1$, the error made in interpreting data for a finely ground knife-edge by the zero-thickness solution (Sec. 3.4.1) should be less than 2%.

Table 4.4.7-4: Comparison of upper[d] and lower bounds for torque with (7-35)[e]

$R_2^{*-1} = H^*$	N	$\mathcal{T}_z^{(U)*}$	$\mathcal{T}_z^{(L)*}$	$\mathcal{T}_z^{(av)*}$	$\mathcal{T}_z^{(\infty)*}$	Percentage Difference
2.000	10	6.810	6.799	6.805	5.009	35.86
2.000	100	66.81	66.72	66.77	50.01	33.51
2.000	1000	666.8	665.9	666.4	500.0	33.28
2.500	10	3.886	3.879	3.883	3.204	21.19
2.500	100	38.17	38.11	38.14	32.00	19.19
2.500	1000	381.0	380.4	380.7	320.0	18.97
3.333	10	2.013	2.008	2.011	1.805	11.41
3.333	100	19.82	19.78	19.80	18.01	9.94
3.333	1000	197.8	197.5	197.7	180.0	9.83
5.000	10	0.8452	0.8419	0.8436	0.8006	5.37
5.000	100	8.345	8.321	8.333	8.001	4.15
5.000	1000	83.35	83.11	83.23	80.00	4.04
10.00	10	0.2051	0.2024	0.2038	0.2001	1.85
10.00	100	2.024	2.011	2.018	2.000	0.90
10.00	1000	20.21	20.09	20.15	20.00	0.75

d) 59 constants optimized
e) $\Delta R^* = 0.001$

References

Abramowitz, M., and I. A. Stegun, "Handbook of Mathematical Functions," 7th corrected printing, Dover, New York (1965).

Briley, P. B., MS Thesis, Northwestern University (1972). For revised programs, see Deemer (1976).

Briley, P. B., A. R. Deemer, and J. C. Slattery, *J. Colloid Interface Sci.* **56,** 1 (1976).

Brown, A. G., W. C. Thuman, and J. W. McBain, *J. Colloid Sci.* **8,** 491 (1953).

Deemer, A. R., PhD dissertation, Northwestern University (1976).

Goodrich, F. C., and L. H. Allen, *J. Colloid Interface Sci.* **40,** 329 (1972).

Slattery, J. C., "Momentum, Energy, and Mass Transfer in Continua," McGraw-Hill, New York (1972); second edition, Robert E. Krieger, Malabar, FL 32950 (1981).

Exercise

4.4.7-1 *blunt knife-edge viscometer in the limit where bulk viscous forces can be neglected with respect to interfacial viscous forces* (Briley *et al.* 1976) Let us confine our attention to a blunt knife-edge rotating in a liquid-gas interface under conditions such that bulk viscous forces can be neglected with respect to interfacial viscous forces. An incompressible, Newtonian liquid occupies the semi-infinite half-space $z < 0$. The physical restrictions (1) through (9) stated in the text again apply. Determine that, as the knife-edge is rotated at a constant angular velocity, the torque exerted on the knife-edge is

$$\frac{\mathcal{T}_z}{2\pi\Omega\mu(R_1)^3} = 2\left(\frac{R_2}{R_1}\right)^2\left(\frac{\varepsilon}{\mu R_1}\right) + \frac{4}{3\pi}\left(\frac{R_2}{R_1}\right)^3$$

$$- \frac{R_1}{4R_2}\,{}_2F_1\left(\frac{3}{2}, \frac{1}{2}; 3; [R_1/R_2]^2\right)$$

where

$${}_2F_1\left(\frac{3}{2}, \frac{1}{2}; 3; [R_1/R_2]^2\right)$$

is a Gauss hypergeometric series (Abramowitz and Stegun 1965, p. 556).

Notation for chapter 4

$\mathbf{a}_\alpha$	natural basis vectors, introduced in Sec. A.2.1
$\mathbf{b}$	body force per unit mass exerted on the material within a phase, introduced in Sec. 2.1.3
$\mathbf{b}^{(\sigma)}$	body force per unit mass exerted on the dividing surface, introduced in Sec. 2.1.3
$C_{(ent\ ex)}$	entrance and exit portions of $C_{(sys)}$
$C_{(sys)}$	lines formed by the intersection of the dividing surfaces with $S_{(sys)}$
$C_{(sys)}^{(cl)}$	common lines contained within the system
dA	denotes that an area integration is to be performed
dV	denotes that a volume integration is to be performed
ds	denotes that a line integration with respect to arc length is to be performed
$\mathbf{D}$	rate of deformation tensor (Slattery 1981, p. 28)
$\mathbf{D}^{(\sigma)}$	surface rate of deformation tensor, introduced in Sec. 1.2.8
$\mathbf{e}_j$	basis vectors for rectangular cartesian coordinate system (Slattery 1981, pp. 607 and 614)
E	scalar potential function, defined in Sec. 4.4.2
E_c	scalar potential function, defined in Sec. 4.4.3
$\mathcal{E}$	rate at which mechanical energy is dissipated by the action of viscous forces both within the bulk phases and in the dividing surfaces, defined in Sec. 4.1.4
$\boldsymbol{\mathcal{F}}$	force that the system exerts upon the impermeable portion of its bounding surface beyond the force attributable to the ambient or reference pressure p_0, defined in Sec. 4.1.3
g	magnitude of the acceleration of gravity
$\mathbf{g}_z$	physical basis vector for curvilinear coordinate system (Slattery 1981, p. 617)

H	mean curvature of dividing surface, introduced in Exercise A.5.3-3
I	identity tensor that transforms every vector into itself
k	permeability of the structure to a single phase
n	outwardly directed unit vector normal to $S_{(sys)}$
N_{Bo}	dimensionless Bond number, introduced in Sec. 4.1.9
N_{ca}	dimensionless capillary number, introduced in Sec. 4.1.9
$N_{Re}^{(j)}$	dimensionless Reynolds number for phase j, introduced in Sec. 4.1.9
N_{γ}	dimensionless interfacial tension, introduced in Sec. 4.1.6
$N_{\kappa+\varepsilon}$	dimensionless sum of the characteristic interfacial viscosities, defined in Sec. 4.1.6
N_{μ}	dimensionless viscosity ratio, defined in Sec. 4.1.9
p	mean pressure, defined in Sec. 2.4.1
p_0	ambient pressure
$\langle p \rangle$	total local volume average of pressure (Slattery 1981, p. 196)
P	thermodynamic pressure, defined in Sec. 5.8.3
P	projection tensor, defined in Sec. A.3.2
$\mathcal{P}$	modified pressure, defined in Sec. 2.4.1
r	cylindrical coordinate
$R_{(sys)}$	region occupied by the system
S	viscous portion of stress tensor, defined in Sec. 2.4.1
$\mathbf{S}^{(\sigma)}$	viscous portion of surface stress tensor, defined in Sec. 2.4.2
$S_{(ent\ ex)}$	entrance and exit portions of $S_{(sys)}$

$S_{(sys)}$	closed surface bounding system
t	time
$\mathbf{T}$	stress tensor, introduced in Sec. 2.1.4
$\mathbf{T}^{(\sigma)}$	surface stress tensor, introduced in Sec. 2.1.5
$\mathcal{T}$	torque that the system exerts upon the impermeable portion of its bounding surface beyond the torque attributable to the ambient pressure p_0, defined in Sec. 4.1.5
$\mathbf{u}$	time rate of change of spatial position following a surface point, introduced in Sec. 1.2.7
$u_{(v)}^{(cl)}$	speed of displacment of common line, defined in Sec. 1.3.7
$\mathbf{v}$	velocity vector, defined in Sec. 1.1.1
$\mathbf{v}_{(sys)}$	local velocity of the boundary of the system
$\mathbf{v}^{(\sigma)}$	surface velocity vector, defined in Sec. 1.2.5
$v_{(\xi)}^{(g)}$	speed of displacment, introduced in Sec. 1.2.7
$\mathcal{W}$	rate at which work is done by the system on the surroundings at the moving impermeable surfaces of the system (beyond any work done on these surfaces by the ambient pressure p_0), defined in Sec. 4.1.4
z	cylindrical coordinate
$\mathbf{z}$	position vector

greek letters

γ	thermodynamic interfacial tension, defined in Sec. 5.8.6
ε	surface shear viscosity, introduced in Sec. 2.2.2
η	apparent viscosity, defined in Sec. 4.4.1
θ	cylindrical coordinate
Θ	dynamic contact angle (see Secs. 2.1.10 and 3.3.3)
κ	surface dilatational viscosity, introduced in Sec. 2.2.2

$\boldsymbol{\Lambda}$	indicates that a vector product is to be performed (Slattery 1981, p. 653)
$\boldsymbol{\mu}$	unit vector normal to $C_{(sys)}$ that is both tangent and outwardly directed with respect to the dividing surface
$\mathbf{v}$	unit vector normal to common line and tangent to dividing surface
$\boldsymbol{\xi}$	unit normal to Σ, introduced in Sec. 1.2.7
π	3.14159...
ρ	mass density
$\rho^{(\sigma)}$	surface mass density, introduced in Sec. 1.3.1
$\Sigma_{(sys)}$	dividing surface within system
ϕ	potential energy per unit mass, defined in Secs. 2.1.3 and 2.3.4
φ	defined in Sec. 4.4.1
ψ	porosity
$\boldsymbol{\Omega}$	angular velocity, introduced in Sec. 4.4.7

others

div	divergence operator
$\text{div}_{(\sigma)}$	surface divergence operator, introduced in Secs. A.5.1 through A.5.6
tr	trace operator
$...^{(i)}$	variable associated with phase i
$...^{T}$	transpose
$...^{-1}$	inverse
$...^{*}$	denote a trial or approximation solution
$...^{*}$	indicates dimensionless quantity

(...)	jump for common line, defined in Sec. 1.3.7
[...]	jump for dividing surface, defined in Sec. 1.3.4
∇	gradient operator with respect to the current configuration
$\nabla_{(\sigma)}$	surface gradient operation, introduced in Secs. A.5.1 through A.5.6
$\frac{d_{(m)}}{dt}$	material derivative, defined in Sec. 1.1.1
$\frac{d_{(s)}}{dt}$	surface material derivative, defined in Sec. 1.2.5; see also Sec. 1.2.7

Reference

Slattery, J. C., "Momentum, Energy, and Mass Transfer in Continua," McGraw-Hill, NY (1972); second edition, Robert E. Krieger, Malabar, FL 32950 (1981).

5
Foundations for simultaneous momentum, energy, and mass transfer

Up to this point, I have been concerned with bodies composed of a single species. Such a body is a special type of multicomponent material. We are fortunate that, with a few relatively minor modifications, all we have said in chapters 1 and 2 about single-component systems can be applied to multicomponent ones.

I have chosen to delay discussion of the foundations for energy transfer until we were ready to talk about multicomponent bodies. This is thermodynamics, but not precisely the subject presented to us by Gibbs, since he was concerned with materials at equilibrium. We deal here with non-equilibrium situations in which momentum, energy, and mass are being transferred.

I conclude by examining the behavior of interfaces undergoing simultaneous momentum, energy, and mass transfer. Note in particular the restrictions placed upon interfacial behavior by the jump Clausius-Duhem inequality and by the requirement that an isolated body should be capable of achieving an intrinsically stable equilibrium.

5.1 Viewpoint

5.1.1 Viewpoint in considering multicomponent materials Up to this point we have been concerned with single-component materials or materials of uniform composition. Hereafter, we shall be treating a body consisting of N species, which is undergoing an arbitrary number of homogeneous and heterogeneous chemical reactions.

Our first task is to choose a model for an N-component material that allows the various species to move relative to one another at varying rates. The simplest approach is to view each species as a continuum in just the same way as we regarded a single-component body as a continuum in Chapter 1. This superposition of N constituent continua is consistent with our usual practice of identifying compositions with each point in a multicomponent mixture.

One feature of this model initially may seem confusing. At any point in space occupied by the mixture, N material particles (one from each of the continua representing individual species) coexist. The confusion seems to arise from a dangerous and incorrect identification of a material particle in a continua of species A with a molecule of species A.

Our particular interest here is multiphase, multicomponent bodies. For these bodies, the interfacial region is viewed as a superposition of N interfacial regions, one for each of the N constituent continua. But there is only one dividing surface, the multicomponent dividing surface, that represents all of the N interfacial regions and therefore the interfacial region for the multicomponent body. Possible definitions for the location of a reference dividing surface are discussed in Sec. 5.2.3.

5.1.2 Body, motion, and material coordinates of species A The ideas that we introduced in Sec. 1.1.1 for a single-component material may be extended easily to a particular species A in an N-constituent mixture. For the moment, we shall confine our attention to a body consisting of a single phase.

As for a single-component body, we have four equivalent descriptions of the motion of a body of species A. I will discuss only the material and the referential descriptions, allowing the reader to construct for himself the spatial and the relative.

material description A **body** of species A is a set; any element $\zeta_{(A)}$ of the set is called a particle of species A or a **material particle** of species A. A one-to-one continuous mapping of this set onto a region of the three-dimensional euclidean space (E^3, V^3) exists and is called a **configuration** of the body of constituent A:

$$\mathbf{z} = \boldsymbol{\chi}_{(A)}(\zeta_{(A)}) \tag{2-1}$$

$$\zeta_{(A)} = \chi_{(A)}^{-1}(\mathbf{z}) \tag{2-2}$$

The point $\mathbf{z} = \boldsymbol{\chi}_{(A)}(\zeta_{(A)})$ of (E^3, V^3) is called the place occupied by the particle $\zeta_{(A)}$; $\zeta_{(A)}$ is the particle at the place $\mathbf{z}$ in (E^3, V^3).

For the moment, we will not concern ourselves directly with a body of an N-component mixture, but rather with the N constituent bodies. It is the superposition of these N constituent bodies that forms the model for the N-component mixture.

A **motion** of a body of species A is a one-parameter family of configurations; the real parameter t is time. We write

$$\mathbf{z} = \boldsymbol{\chi}_{(A)}(\zeta_{(A)}, t) \tag{2-3}$$

and

$$\zeta_{(A)} = \chi_{(A)}^{-1}(\mathbf{z}, t) \tag{2-4}$$

Let B be any quantity: scalar, vector, or tensor. We shall have occasion to talk about the time derivative of B following the motion of a material particle of A:

$$\frac{d_{(A)}B}{dt} \equiv \left(\frac{\partial B}{\partial t}\right)_{\zeta_{(A)}} \tag{2-5}$$

We shall refer to $d_{(A)}B/dt$ as the **material derivative** of B **following species A**. For example, the **velocity of species A**, $\mathbf{v}_{(A)}$, represents the time rate of change of position of a material particle of A,

$$\mathbf{v}_{(A)} \equiv \frac{d_{(A)}\mathbf{z}}{dt} = \left(\frac{\partial \boldsymbol{\chi}_{(A)}}{dt}\right)_{\zeta_{(A)}} \tag{2-6}$$

In the material description, we deal directly with the abstract particles in terms of which the body of species A is defined.

referential description The concept of a material particle of A is abstract. The material particles have not been defined and we have no way of directly identifying them. We are able to observe only spatial descriptions of a body. This suggests that we identify the material particles $\zeta_{(A)}$ by their positions in some particular configuration, for example the configuration assumed at time t = 0. The place of a particle of species A in the **reference configuration** $\boldsymbol{\kappa}_{(A)}$ will be denoted by

$$\mathbf{z}_{\kappa(A)} = \boldsymbol{\kappa}_{(A)}(\zeta_{(A)}) \tag{2-7}$$

The particle at the place $\mathbf{z}_{\kappa(A)}$ in the configuration $\boldsymbol{\kappa}_{(A)}$ may be expressed as

$$\zeta_{(A)} = \kappa_{(A)}^{-1}(\mathbf{z}_{\kappa(A)}) \tag{2-8}$$

If $\boldsymbol{\chi}_{(A)}$ is a motion of the body of species A, then

$$\mathbf{z} = \boldsymbol{\chi}_{\kappa(A)}(\mathbf{z}_{\kappa(A)}, t) \equiv \boldsymbol{\chi}_{(A)}[\kappa_{(A)}^{-1}(\mathbf{z}_{\kappa(A)}), t] \tag{2-9}$$

and

$$\mathbf{z}_{\kappa(A)} = \chi_{\kappa(A)}^{-1}(\mathbf{z}, t) \equiv \boldsymbol{\kappa}_{(A)}[\chi_{(A)}^{-1}(\mathbf{z}, t)] \tag{2-10}$$

These expressions describe the motion of component A in terms of the reference configuration $\boldsymbol{\kappa}_{(A)}$. We say that they define a family of **deformations** from $\boldsymbol{\kappa}_{(A)}$. The subscript $\ldots_{\boldsymbol{\kappa}}$ is to remind you that the form of $\boldsymbol{\chi}_{\kappa(A)}$ depends upon the choice of reference configuration.

The physical meaning of the material derivative introduced in (2-5) may now be clarified, if we think of it as a derivative with respect to time holding position in the reference configuration $\boldsymbol{\kappa}_{(A)}$ fixed:

$$\frac{d_{(A)}B}{dt} = \left(\frac{\partial B}{\partial t}\right)_{\mathbf{z}_{\kappa(A)}} = \left(\frac{\partial B}{\partial t}\right)_{z_{\kappa(A)1},\, z_{\kappa(A)2},\, z_{\kappa(A)3}} \tag{2-11}$$

In particular, the velocity of species A becomes

$$\mathbf{v}_{(A)} = \left(\frac{\partial \boldsymbol{\chi}_{\kappa(A)}}{\partial t}\right)_{\mathbf{z}_{\kappa(A)}} = \left(\frac{\partial \boldsymbol{\chi}_{\kappa(A)}}{\partial t}\right)_{z_{\kappa(A)1},\, z_{\kappa(A)2},\, z_{\kappa(A)3}} \tag{2-12}$$

5.1.3 Motion of multicomponent dividing surface The discussion given in the last section applies without change to any portion of a body of species A whose current configuration lies within a region occupied by a single phase of the multicomponent material. This means that we must give special attention only to the dividing surface.

In Sec. 1.2.5, we used four methods to describe the motion of a single component dividing surface. This can be our guide in discussing the motion of a multicomponent dividing surface. I again will develop only the material and referential descriptions, leaving the construction of the spatial and the relative descriptions for the reader.

I will consider in detail only one of the species A in a multicomponent body consisting of N constituents.

material description A multicomponent dividing surface Σ in (E^3, V^3) is the locus of a point whose position is a function of two parameters y^1 and y^2 (see Sec. A.1.3):

$$\mathbf{z} = \mathbf{p}^{(\sigma)}(y^1 , y^2) \tag{3-1}$$

The two surface coordinates y^1 and y^2 uniquely determine a point on the surface.

There exists a many-to-one mapping of a portion of the body of species A onto Σ,

$$y^\alpha = X^\alpha_{(A)}(\zeta^{\{\sigma\}}_{\{A\}}) \qquad \alpha = 1,2 \tag{3-2}$$

This mapping is the **intrinsic configuration** for the surface particles of species A on Σ. We visualize that there may be many material particles of A occupying any given point on Σ. The set of all material particles of A at any point on Σ is denoted by $\zeta^{\{\sigma\}}_{\{A\}}$ and will be referred to as a **surface particle** of species A. Consequently, (3-2) and

$$\zeta^{\{\sigma\}}_{\{A\}} = X^{-1}_{(A)}(y^1 , y^2) \tag{3-3}$$

may also be thought of as a one-to-one mapping of the set of surface particles of A onto Σ. From (3-2), the point (y^1, y^2) on Σ is the place occupied by $\zeta^{\{\sigma\}}_{\{A\}}$; (3-3) tells us which surface particle of A is at the place (y^1, y^2).

Equations (3-1) and (3-2) together give us an expression for the **configuration** of the surface particles of A in (E^3, V^3):

$$\begin{aligned} \mathbf{z} &= \boldsymbol{\chi}^{\{\sigma\}}_{\{A\}}(\zeta^{\{\sigma\}}_{\{A\}}) \\ &\equiv \mathbf{p}^{(\sigma)}[X^1_{(A)}(\zeta^{\{\sigma\}}_{\{A\}}), X^2_{(A)}(\zeta^{\{\sigma\}}_{\{A\}})] \end{aligned} \tag{3-4}$$

The mapping $\chi^{\{\sigma\}}_{\{A\}}$ does not have an inverse; although there is a position in space corresponding to every surface particle of species A, the converse is not true.

A moving and deforming multicomponent dividing surface Σ in (E^3, V^3) is the locus of a point whose position is a function of two surface coordinates and time t,

$$\mathbf{z} = \mathbf{p}^{(\sigma)}(y^1 , y^2, t) \tag{3-5}$$

An **intrinsic motion** of the surface particles of species A on Σ is a one-parameter family of intrinsic configurations:

$$y^\alpha = X^\alpha_{(A)}(\zeta^{\{\sigma\}}_{\{A\}}, t) \qquad \alpha = 1, 2 \tag{3-6}$$

$$\zeta^{\{\sigma\}}_{\{A\}} = X^{-1}_{(A)}(y^1 , y^2, t) \tag{3-7}$$

Equations (3-6) and (3-7) tell us how the surface particles of A move from point to point on the dividing surface independently of how the surface itself is moving. Equations (3-5) and (3-6) together give us an expression for the **motion** of the surface particles of A in (E^3, V^3),

$$
\begin{aligned}
\mathbf{z} &= \boldsymbol{\chi}_{\{{\sigma \atop A}\}}(\zeta_{\{{\sigma \atop A}\}}, t) \\
&\equiv \mathbf{p}^{(\sigma)}[\, X^1_{(A)}(\zeta_{\{{\sigma \atop A}\}}, t), X^2_{(A)}(\zeta_{\{{\sigma \atop A}\}}, t), t \,]
\end{aligned}
\tag{3-8}
$$

The motion of the surface particles of A in (E^3, V^3) is a one-parameter family of configurations. It is the result both of the movement of the surface Σ itself (3-5) and of the intrinsic motion of the surface particles of A within Σ (3-6 and 3-7).

Let B be any quantity: scalar, vector or tensor. We can talk about the time derivative of B following the motion of a surface particle of species A,

$$
\frac{d_{(A,s)}B}{dt} \equiv \left(\frac{\partial B}{\partial t} \right)_{\zeta_{\{{\sigma \atop A}\}}}
\tag{3-9}
$$

We shall refer to $d_{(A,s)}B/dt$ as the **surface material derivative** of B **following species** A. For example, the **surface velocity of species A,** $\mathbf{v}_{\{{\sigma \atop A}\}}$, represents the time rate of change of position of a surface particle of species A,

$$
\mathbf{v}_{\{{\sigma \atop A}\}} \equiv \frac{d_{(A,s)}\mathbf{z}}{dt} \equiv \left(\frac{\partial \boldsymbol{\chi}_{\{{\sigma \atop A}\}}}{dt} \right)_{\zeta_{\{{\sigma \atop A}\}}}
\tag{3-10}
$$

In the **material description,** we deal directly with the abstract particles in terms of which the body of species A is defined.

referential description The surface particles of species A are primitive concepts, in the sense that they have not been defined. We have no way of directly identifying them. We are able to observe only spatial descriptions of a dividing surface. For example, at some **reference time** t_κ the dividing surface takes the form

$$
\begin{aligned}
\mathbf{z}_\kappa &= \boldsymbol{\kappa}^{(\sigma)}(y^1_\kappa, y^2_\kappa) \\
&\equiv \mathbf{p}^{(\sigma)}(y^1_\kappa, y^2_\kappa, t_\kappa)
\end{aligned}
\tag{3-11}
$$

which we can call the **reference dividing surface**. This suggests that we identify surface particles of A by their **reference intrinsic configuration** with respect to this reference dividing surface:

$$
y^M_\kappa = K^M_{(A)}(\zeta_{\{{\sigma \atop A}\}})
$$

$$\equiv X^{M}_{(A)}(\zeta^{(\sigma)}_{(A)}, t_{\kappa}) \tag{3-12}$$

$$\zeta^{(\sigma)} = K^{-1}_{(A)}(y^1_{\kappa}, y^2_{\kappa})$$

$$\equiv X^{-1}_{(A)}(y^1_{\kappa}, y^2_{\kappa}, t_{\kappa}) \tag{3-13}$$

Equations (3-11) and (3-12) together describe the **reference configuration** of the surface particles of A in (E^3, V^3),

$$\mathbf{z}_{\kappa} = \boldsymbol{\kappa}^{(\sigma)}_{K(A)}(\zeta^{(\sigma)}_{(A)})$$

$$\equiv \boldsymbol{\kappa}^{(\sigma)}[K^1_{(A)}(\zeta^{(\sigma)}_{(A)}), K^2_{(A)}(\zeta^{(\sigma)}_{(A)})] \tag{3-14}$$

One could introduce a different reference dividing surface for each species in the N component mixture. I have tried to suggest by the way in which I have introduced notation that this will generally be an unnecessary complication. The same reference dividing surface can be used for all species in the mixture.

We can now use (3-12) and (3-13) to describe the **intrinsic deformation** of species A from the reference intrinsic configuration:

$$y^{\alpha} = X^{\alpha}_{K(A)}(y^1_{\kappa}, y^2_{\kappa}, t)$$

$$\equiv X^{\alpha}_{(A)}[K^{-1}_{(A)}(y^1_{\kappa}, y^2_{\kappa}), t] \tag{3-15}$$

$$y^M_{\kappa} = K^M_{X(A)}(y^1, y^2, t)$$

$$\equiv K^M_{(A)}[X^{-1}_{(A)}(y^1, y^2, t)] \tag{3-16}$$

These are really alternative expressions for the intrinsic motion of species A. Equation (3-15) gives the position on the dividing surface at time t of the surface particle of species A that was at (y^1_{κ}, y^2_{κ}) at the reference time t_{κ}. The surface particle of A that currently is at (y^1, y^2) was at (y^1_{κ}, y^2_{κ}) according to (3-16). The **deformation** of species A from the reference configuration in (E^3, V^3) may be expressed as

$$\mathbf{z} = \boldsymbol{\chi}^{(\sigma)}_{K(A)}(y^1_{\kappa}, y^2_{\kappa}, t)$$

$$\equiv \boldsymbol{\chi}^{(\sigma)}_{(A)}[K^{-1}_{(A)}(y^1_{\kappa}, y^2_{\kappa}), t] \tag{3-17}$$

$$\mathbf{z}_{\kappa} = \boldsymbol{\chi}^{(\sigma)}_{X(A)}(y^1, y^2, t)$$

$$\equiv \boldsymbol{\chi}^{(\sigma)}_{(A)}[X^{-1}_{(A)}(y^1, y^2, t), t_{\kappa}] \tag{3-18}$$

The physical meaning of the surface material derivative introduced in (3-9) may now be clarified, if we think of it as a derivative with respect to time of a quantity B associated with species A holding position in the reference intrinsic configuration fixed:

$$\frac{d_{(A,s)}B}{dt} = \left(\frac{\partial B}{\partial t}\right)_{y^1_\kappa,\, y^2_\kappa} \tag{3-19}$$

In particular, the surface velocity of species A becomes

$$\mathbf{v}^{(\sigma)}_{(A)} \equiv \frac{d_{(A,s)}\mathbf{z}}{dt} = \left(\frac{\partial \boldsymbol{\chi}^{(\sigma)}_{\kappa(A)}}{\partial t}\right)_{y^1_\kappa,\, y^2_\kappa} \tag{3-20}$$

5.1.4 More about surface velocity of species A Let B be any quantity: scalar, vector, or tensor. Let us assume that B is an explicit function of position (y^1, y^2) on the dividing surface and time:

$$B = B(y^1, y^2, t) \tag{4-1}$$

The surface material derivative of B following species A, introduced in Sec. 5.1.3, becomes

$$\begin{aligned}
\frac{d_{(A,s)}B}{dt} &\equiv \left(\frac{\partial B}{\partial t}\right)_{\zeta^{(\sigma)}_{(A)}} \\
&\equiv \left(\frac{\partial B}{\partial t}\right)_{y^1_\kappa,\, y^2_\kappa} \\
&= \frac{\partial B(y^1, y^2, t)}{\partial t} + \frac{\partial B(y^1, y^2, t)}{\partial y^\alpha}\frac{\partial X^\alpha_{\kappa(A)}(y^1_\kappa, y^2_\kappa, t)}{\partial t} \\
&= \left(\frac{\partial B}{\partial t}\right)_{y^1_\kappa, y^2_\kappa} + \nabla_{(\sigma)}B \cdot \dot{\mathbf{y}}_{(A)}
\end{aligned} \tag{4-2}$$

where we have used (3-15) in defining the **intrinsic surface velocity of species A**

$$\begin{aligned}
\dot{\mathbf{y}}_{(A)} &= \dot{y}^\alpha_{(A)}\mathbf{a}_\alpha \\
&\equiv \frac{\partial X^\alpha_{\kappa(A)}(y^1_\kappa, y^2_\kappa, t)}{\partial t}\mathbf{a}_\alpha
\end{aligned}$$

$$\equiv \frac{d_{(A,s)}y^{\alpha}}{dt}\,\mathbf{a}_{\alpha} \tag{4-3}$$

We will have particular interest in the surface velocity of species A, the time rate of change of spatial position following a surface particle of A:

$$\mathbf{v}^{(\sigma)}_{(A)} \equiv \left(\frac{\partial \boldsymbol{\chi}^{(\sigma)}_{\kappa\,(A)}}{\partial t}\right)_{y^1_{\boldsymbol{\kappa}},\, y^2_{\boldsymbol{\kappa}}}$$

$$\equiv \frac{d_{(A,s)}\mathbf{z}}{dt} \tag{4-4}$$

From (3-5), any multicomponent dividing surface in (E^3, V^3) is the locus of a point whose position may be a function of the two surface coordinates and time:

$$\mathbf{z} = \mathbf{p}^{(\sigma)}(\, y^1, y^2, t\,) \tag{4-5}$$

Applying (4-2) to (4-5), we find

$$\mathbf{v}^{(\sigma)}_{(A)} = \frac{\partial \mathbf{p}^{(\sigma)}(\, y^1, y^2, t\,)}{\partial t} + \nabla_{(\sigma)}\, \mathbf{p}^{(\sigma)}(\, y^1, y^2, t\,) \cdot \dot{\mathbf{y}}_{(A)}$$

$$= \mathbf{u} + \dot{\mathbf{y}}_{(A)} \tag{4-6}$$

Here we have noted that in Sec. 1.2.7 $\mathbf{u}$ is defined to be the time rate of change of spatial position following a surface point (y^1, y^2)

$$\mathbf{u} \equiv \frac{\partial \mathbf{p}^{(\sigma)}(\, y^1, y^2, t\,)}{\partial t} \tag{4-7}$$

and that the tangential gradient of the position vector is the projection tensor $\mathbf{P}$ (see Sec. A.5.1)

$$\nabla_{(\sigma)}\mathbf{p}^{(\sigma)}(y^1, y^2, t) = \mathbf{P} \tag{4-8}$$

Note that $\dot{\mathbf{y}}_{(A)}$ does not represent the tangential component of $\mathbf{v}^{(\sigma)}_{(A)}$. In general, $\mathbf{u}$ has both normal and tangential components; there is only one choice of surface coordinates for which

$$\mathbf{u} = \mathbf{v}^{(\sigma)}_{(\xi)}\boldsymbol{\xi} \tag{4-9}$$

where $v_{(\xi)}$ is the **speed of displacement** of the surface. For more on this point, see Sec. 1.2.7.

5.2 Mass balance

5.2.1 Species mass balance I said in Sec. 1.3.1 that mass is conserved. This is certainly an idea that we shall wish to preserve in talking about multicomponent bodies. I will come back to this point in Sec. 5.4.1.

But what about the masses of the individual species of which the multicomponent mixture is composed? In general they are not conserved. Individual species may be continually forming and decomposing as the result of chemical reactions. This concept is formalized by the **species mass balance**:

The time rate of change of the mass of each species A (A = 1, 2,..., N) in a multicomponent mixture is equal to the rate at which the mass of species A is produced by chemical reactions.

Let $\rho_{(A)}$ denote the mass density of species A (mass per unit volume); $\rho^{(\sigma)}_{(A)}$ the surface mass density of species (mass per unit area); $r_{(A)}$ the rate at which mass of species A is produced per unit volume as the result of homogeneous chemical reactions; $r^{(\sigma)}_{(A)}$ the rate at which mass of species A is produced per unit area by heterogeneous chemical reactions (chemical reactions on the dividing surface). Let $R_{(A)}$ be the region occupied by body of species A (a set of material particles of species A). The species mass balance for A states that

$$\frac{d}{dt}\left(\int_{R_{(A)}} \rho_{(A)}\, dV + \int_{\Sigma} \rho^{(\sigma)}_{(A)}\, dA\right) = \int_{R_{(A)}} r_{(A)}\, dV + \int_{\Sigma} r^{(\sigma)}_{(A)}\, dA \qquad (1\text{-}1)$$

Remember that the limits on these integrations will usually be functions of time.

It is for this reason that in Exercise 5.2.1-2 we have modified the transport theorem of Sec. 1.3.4 to apply to a collection of particles of species A. Applying it to the left side of (1-1), we find

$$\int_{R_{(A)}} \left(\frac{d_{(A)}\rho_{(A)}}{dt} + \rho_{(A)} \operatorname{div} \mathbf{v}_{(A)}\right) dV$$

$$+ \int_{\Sigma} \left(\frac{d_{(A,s)}\rho^{(\sigma)}_{(A)}}{dt} + \rho^{(\sigma)}_{(A)} \operatorname{div}_{(\sigma)}\mathbf{v}^{(\sigma)}_{(A)}\right.$$

$$\left. + \left[\rho_{(A)}\left(\mathbf{v}_{(A)} \cdot \boldsymbol{\xi} - v^{(\sigma)}_{(\xi)}\right)\right]\right) dA$$

$$= \int_{R_{(A)}} r_{(A)}\, dV + \int_{\Sigma} r^{(\sigma)}_{(A)}\, dA \qquad (1\text{-}2)$$

or

$$\int_{R_{(A)}} \left(\frac{d_{(A)} \rho_{(A)}}{dt} + \rho_{(A)} \operatorname{div} \mathbf{v}_{(A)} - r_{(A)} \right) dV$$

$$+ \int_{\Sigma} \left(\frac{d_{(A,s)} \rho^{(\sigma)}_{(A)}}{dt} + \rho^{(\sigma)}_{(A)} \operatorname{div}_{(\sigma)} \mathbf{v}^{(\sigma)}_{(A)} - r^{(\sigma)}_{(A)} \right.$$

$$\left. + [\, \rho_{(A)} (\mathbf{v}_{(A)} \cdot \boldsymbol{\xi} - v^{(\sigma)}_{(\xi)}) \,] \right) dA = 0 \tag{1-3}$$

Equation (1-3) must be true for every body. In particular, for a body consisting of a single phase, the integral over the region $R_{(A)}$ is zero, which implies (Slattery 1981, p. 456 for alternative forms)

$$\frac{d_{(A)} \rho_{(A)}}{dt} + \rho_{(A)} \operatorname{div} \mathbf{v}_{(A)} = r_{(A)} \tag{1-4}$$

This is the **equation of continuity for species A**. It expresses the constraint of the species mass balance at every point within a phase.

In view of (1-4), (1-3) reduces to

$$\int_{\Sigma} \left(\frac{d_{(A,s)} \rho^{(\sigma)}_{(A)}}{dt} + \rho^{(\sigma)}_{(A)} \operatorname{div}_{(\sigma)} \mathbf{v}^{(\sigma)}_{(A)} - r^{(\sigma)}_{(A)} \right.$$

$$\left. + [\, \rho_{(A)} (\mathbf{v}_{(A)} \cdot \boldsymbol{\xi} - v^{(\sigma)}_{(\xi)}) \,] \right) dA = 0 \tag{1-5}$$

But (1-1) is valid for every body of species A and every portion of such a body, no matter how large or small. Since (1-5) must be true for every portion of a multicomponent dividing surface, we conclude (see Exercises 1.3.5-1 and 1.3.5-2)

$$\frac{d_{(A,s)} \rho^{(\sigma)}_{(A)}}{dt} + \rho^{(\sigma)}_{(A)} \operatorname{div}_{(\sigma)} \mathbf{v}^{(\sigma)}_{(A)} - r^{(\sigma)}_{(A)} + [\, \rho_{(A)} (\mathbf{v}_{(A)} \cdot \boldsymbol{\xi} - v^{(\sigma)}_{(\xi)}) \,]$$

$$= 0 \tag{1-6}$$

This is the **jump mass balance for species A**. It expresses the species mass balance at every point on a multicomponent dividing surface.

A useful alternative form of the jump mass balance for A is

$$\frac{\partial \rho^{(\sigma)}_{(A)}}{\partial t} - \nabla_{(\sigma)}\rho^{(\sigma)}_{(A)} \cdot \mathbf{u} + \text{div}_{(\sigma)}\left(\rho^{(\sigma)}_{(A)}\mathbf{v}^{(\sigma)}_{(A)} \right) - r^{(\sigma)}_{(A)}$$

$$+ \left[\rho_{(A)}\left(\mathbf{v}_{(A)} \cdot \boldsymbol{\xi} - v^{(\sigma)}_{(\xi)} \right)\right] = 0 \tag{1-7}$$

Often we will find it more convenient to work in terms of the surface molar density of species A

$$c^{(\sigma)}_{(A)} \equiv \frac{\rho^{(\sigma)}_{(A)}}{M_{(A)}} \tag{1-8}$$

where $M_{(A)}$ is the molecular weight of A. In this case, (1-6) and (1-7) become respectively

$$\frac{d_{(A,s)}c^{(\sigma)}_{(A)}}{dt} + c^{(\sigma)}_{(A)} \, \text{div}_{(\sigma)}\mathbf{v}^{(\sigma)}_{(A)} - \frac{r^{(\sigma)}_{(A)}}{M_{(A)}} + \left[c_{(A)}\left(\mathbf{v}_{(A)} \cdot \boldsymbol{\xi} - v^{(\sigma)}_{(\xi)} \right)\right]$$

$$= 0 \tag{1-9}$$

and

$$\frac{\partial c^{(\sigma)}_{(A)}}{dt} - \nabla_{(\sigma)}c^{(\sigma)}_{(A)} \cdot \mathbf{u} + \text{div}_{(\sigma)}\left(c^{(\sigma)}_{(A)}\mathbf{v}^{(\sigma)}_{(A)}\right) - \frac{r^{(\sigma)}_{(A)}}{M_{(A)}}$$

$$+ \left[c_{(A)}\left(\mathbf{v}_{(A)} \cdot \boldsymbol{\xi} - v^{(\sigma)}_{(A)} \right)\right] = 0 \tag{1-10}$$

The major problem of heterogeneous catalysis is to describe $r^{(\sigma)}_{(A)}$, especially at fluid-solid phase interfaces. The usual empirical approach for a fluid-solid phase interface is to say that $r^{(\sigma)}_{(A)}$ is proportional to some power of the concentration of A in the dividing surface

$$\frac{r^{(\sigma)}_{(A)}}{M_{(A)}} = k\left[c^{(\sigma)}_{(A)} \right]^n \tag{1-11}$$

or, if we assume local equilibrium between the dividing surface and an adjacent bulk phase,

$$\text{at the dividing surface:} \quad \frac{r^{(\sigma)}_{(A)}}{M_{(A)}} = k'\left[c_{(A)} \right]^n \tag{1-12}$$

In what follows, we shall assume that we already have a constitutive equation or empirical data correlation for $r^{\sigma}_{(A)}$ as a function of composition, temperature, ...

Reference

Slattery, J. C., "Momentum, Energy, and Mass Transfer in Continua," McGraw-Hill, New York (1972); second edition, Robert E. Krieger, Malabar, FL 32950 (1981).

Exercises

5.2.1-1 *surface transport theorem for species A* Let $\Psi^{\sigma}_{(A)}$ be any scalar-, vector-, or tensor-valued function of time and position associated with species A on the multicomponent dividing surface. Show that

$$\frac{d}{dt}\int_{\Sigma} \Psi^{\sigma}_{(A)}\, dA = \int_{\Sigma} \left(\frac{d_{(A,s)}\Psi^{\sigma}_{(A)}}{dt} + \Psi^{\sigma}_{(A)}\, \text{div}_{(\sigma)} \mathbf{v}^{\sigma}_{(A)} \right) dA$$

This is the **surface transport theorem for species A.**

Hint: Review the derivation of the surface transport theorem in Sec. 1.3.3.

5.2.1-2 *transport theorem for body of species A containing dividing surface* Let $\Psi_{(A)}$ be any scalar-, vector-, or tensor-valued function of time and position in the region $R_{(A)}$ occupied by the body of species A; $\Psi^{\sigma}_{(A)}$ is any scalar-, vector-, or tensor-valued function of time and position associated with species A on the multicomponent dividing surface. Derive the following **transport theorem for a body of species A containing a dividing surface**:

$$\frac{d}{dt}\left(\int_{R_{(A)}} \Psi_{(A)}\, dV + \int_{\Sigma} \Psi^{\sigma}_{(A)}\, dA \right)$$

$$= \int_{R_{(A)}} \left(\frac{d_{(A)}\Psi_{(A)}}{dt} + \Psi_{(A)}\, \text{div}\, \mathbf{v}_{(A)} \right) dV$$

$$+ \int_{\Sigma} \left(\frac{d_{(A,s)}\Psi^{\sigma}_{(A)}}{dt} + \Psi^{\sigma}_{(A)}\, \text{div}_{(\sigma)} \mathbf{v}^{\sigma}_{(A)} \right.$$

$$+ \left[\Psi_{(A)} (\mathbf{v}_{(A)} \cdot \boldsymbol{\xi} - \mathbf{v}^{\{\xi\}}) \right] \Big) \, dA$$

The notation has the same meaning as in Sec. 1.3.4 with the understanding that we are now considering a multicomponent dividing surface.

5.2.1-3 *alternative derivation of the jump mass balance for species A* Write the species mass balance for a body of A, employing the transport theorem of Exercise 5.2.1-2. Deduce the jump mass balance for species A (1-6) by allowing the region occupied by A to shrink around the multicomponent dividing surface

5.2.1-4 *preliminary view of overall jump mass balance* For every multicomponent, multiphase body, we can define the **total mass density**

$$\rho \equiv \sum_{A=1}^{N} \rho_{(A)}$$

the **total surface mass density**

$$\rho^{(\sigma)} \equiv \sum_{A=1}^{N} \rho^{\{\sigma\}}_{(A)}$$

the **mass-averaged velocity**

$$\mathbf{v} \equiv \frac{1}{\rho} \sum_{A=1}^{N} \rho_{(A)} \mathbf{v}_{(A)}$$

and the **mass-averaged surface velocity**

$$\mathbf{v}^{(\sigma)} \equiv \frac{1}{\rho^{(\sigma)}} \sum_{A=1}^{N} \rho^{\{\sigma\}}_{(A)} \mathbf{v}^{\{\sigma\}}_{(A)}$$

i) Show that the sum of (1-4) over all N species may be written as (Slattery 1981, p. 452)

$$\frac{\partial \rho}{\partial t} + \operatorname{div}(\rho \mathbf{v}) = \sum_{A=1}^{N} r_{(A)}$$

Our conception of conservation of mass tells us that no mass can be created by homogeneous chemical reactions within a phase:

$$\sum_{A=1}^{N} r_{(A)} = 0$$

Consequently, we have that

$$\frac{\partial \rho}{\partial t} + \text{div}(\rho \mathbf{v}) = 0$$

This **overall equation of continuity** is formally identical to the equation of continuity for a single-component material (Slattery 1981, p. 458).

ii) Show that the overall equation of continuity may also be expressed as

$$\frac{d_{(\mathbf{v})}\rho}{dt} + \rho \, \text{div} \, \mathbf{v} = 0$$

where

$$\frac{d_{(\mathbf{v})}\rho}{dt} \equiv \frac{\partial \rho}{\partial t} + \nabla \rho \cdot \mathbf{v}$$

iii) Reason that the sum of (1-7) over all N species is

$$\frac{\partial \rho^{(\sigma)}}{\partial t} - \nabla_{(\sigma)}\rho^{(\sigma)} \cdot \mathbf{u} + \text{div}_{(\sigma)}(\rho^{(\sigma)}\mathbf{v}^{(\sigma)}) + [\, \rho(\mathbf{v} \cdot \boldsymbol{\xi} - v_{(\xi)}^{(\sigma)})]$$

$$= \sum_{A=1}^{N} r_{(A)}^{(\sigma)}$$

Our intuitive view of mass conservation, formalized in Sec. 5.4.1, says that no mass can be created by chemical reactions at a multicomponent dividing surface:

$$\sum_{A=1}^{N} r_{(A)}^{(\sigma)} = 0$$

As a result,

$$\frac{\partial \rho^{(\sigma)}}{\partial t} - \nabla_{(\sigma)}\rho^{(\sigma)} \cdot \mathbf{u} + \text{div}_{(\sigma)}(\, \rho^{(\sigma)}\mathbf{v}^{(\sigma)}\,) + [\, \rho(\, \mathbf{v} \cdot \boldsymbol{\xi} - v_{\{\xi\}}^{\{g\}}\,)] = 0$$

This **overall jump mass balance** is identical in form with the jump mass balance derived in Sec. 1.3.5 for a single-component material.

iv) Conclude that the overall jump mass balance may also be written as

$$\frac{d_{(\mathbf{v}^{(\sigma)})}\rho^{(\sigma)}}{dt} + \rho^{(\sigma)}\text{div}_{(\sigma)}\mathbf{v}^{(\sigma)} + [\, \rho(\, \mathbf{v} \cdot \boldsymbol{\xi} - v_{\{\xi\}}^{\{g\}}\,)] = 0$$

where

$$\frac{d_{(\mathbf{v}^{(\sigma)})}\rho^{(\sigma)}}{dt} = \frac{\partial \rho^{(\sigma)}}{\partial t} + \nabla_{(\sigma)}\rho^{(\sigma)} \cdot (\, \mathbf{v}^{(\sigma)} - \mathbf{u}\,)$$

5.2.1-5 *interpretations for* $\rho_{\{A\}}^{\{\sigma\}}$ *and* $r_{\{A\}}^{\{\sigma\}}$ Let us now think of the multicomponent dividing surface as a model for a three-dimensional interfacial region (see Sec. 1.2.4). Using the notation and ideas developed in Sec. 1.3.2, conclude that

$$\rho_{\{A\}}^{\{\sigma\}} \equiv \int_{\lambda^-}^{\lambda^+} (\, \rho_{\{A\}}^{\{I\}} - \rho_{(A)}\,)(\, 1 - \kappa_1 \lambda\,)(\, 1 - \kappa_2 \lambda\,)\, d\lambda$$

and

$$r_{\{A\}}^{\{\sigma\}} \equiv \int_{\lambda^-}^{\lambda^+} (\, r_{\{A\}}^{\{I\}} - r_{(A)}\,)(\, 1 - \kappa_1 \lambda\,)(\, 1 - \kappa_2 \lambda\,)\, d\lambda$$

This interpretation for the surface mass density of A makes it clear that $\rho_{\{A\}}^{\{\sigma\}}$ may be either positive or negative, depending upon the precise location of the multicomponent dividing surface Σ. The location of Σ is discussed in Sec. 5.2.3.

5.2.1-6 *transport theorem for body of species A containing intersecting dividing surfaces* Let $\Psi_{(A)}$ be any scalar-, vector-, or tensor-valued function of time and position in the region $R_{(A)}$ occupied by the body of species A; $\Psi_{\{A\}}^{\{\sigma\}}$ is any scalar-, vector-, or tensor-valued function of time and position associated with species A on the multicomponent dividing surfaces. Derive the following **transport theorem for a body of species A containing intersecting dividing surfaces** (see Sec. 1.3.7):

$$\frac{d}{dt}\left(\int_{R_{(A)}} \Psi_{(A)} \, dV + \int_{\Sigma} \Psi^{(\sigma)}_{(A)} \, dA \right)$$

$$= \int_{R_{(A)}} \left(\frac{d_{(A)} \Psi_{(A)}}{dt} + \Psi_{(A)} \operatorname{div} \mathbf{v}_{(A)} \right) dV$$

$$+ \int_{\Sigma} \left(\frac{d_{(A,s)} \Psi^{(\sigma)}_{(A)}}{dt} + \Psi^{(\sigma)}_{(A)} \operatorname{div}_{(\sigma)} \mathbf{v}^{(\sigma)}_{(A)} \right.$$

$$\left. + \left[\Psi_{(A)} \left(\mathbf{v}_{(A)} \cdot \boldsymbol{\xi} - v^{(\xi)} \right) \right] \right) dA$$

$$+ \int_{C^{(cl)}} \left(\Psi^{(\sigma)}_{(A)} \left[\mathbf{v}^{(\sigma)}_{(A)} \cdot \boldsymbol{\nu} - u^{(c\ell)}_{(v)} \right] \right) ds$$

5.2.1-7 *species mass balance at multicomponent common line* Use the transport theorem of Exercise 5.2.1-6 to determine that at the common line

$$\left(\rho^{(\sigma)}_{(A)} \left[\mathbf{v}^{(\sigma)}_{(A)} \cdot \boldsymbol{\nu} - u^{(c\ell)}_{(v)} \right] \right) = 0$$

This will be known as the **mass balance for species A at a multicomponent common line.**

Note that in deriving this balance equation we have not accounted for the mass of species A that might be intrinsically associated with the common line. This point does not appear to have received any attention in the literature. For related references, see the comments concluding Sec. 2.1.9.

5.2.2 Concentrations, velocities, and mass fluxes One of the most confusing aspects of mass-transfer problems is that there are several sets of terminology in common use. The various possibilities have been explored in detail for solutions outside the immediate neighborhood of the interface (Slattery 1981, p. 452; Bird *et al.* 1960, p. 498; see also Tables 5.2.2-1 through 5.2.2-4), but some repetition may be justified in the context of the dividing surface.

In discussing the concentration of species A on a multicomponent dividing surface, one is free at least in principle to refer to the surface mass density $\rho^{(\sigma)}_{(A)}$, the surface molar density $c^{(\sigma)}_{(A)}$, the surface mass fraction

$\omega_{(A)}^{(\sigma)}$, or the surface mole fraction $x_{(A)}^{(\sigma)}$. The relations between these quantities are examined in Tables 5.2.2-1 and 5.2.2-2. Potential difficulties with using $\omega_{(A)}^{(\sigma)}$ or $x_{(A)}^{(\sigma)}$ are pointed out in the next section.

It is natural to develop some conception for the mean motion of the material in the dividing surface. We can talk about either the surface mass-averaged velocity

$$\mathbf{v}^{(\sigma)} \equiv \sum_{A=1}^{N} \omega_{(A)}^{(\sigma)} \mathbf{v}_{(A)}^{(\sigma)} \tag{2-1}$$

or the surface molar-averaged velocity

$$\mathbf{v}^{\star(\sigma)} \equiv \sum_{A=1}^{N} x_{(A)}^{(\sigma)} \mathbf{v}_{(A)}^{(\sigma)} \tag{2-2}$$

Relations between these two average velocities are given in Table 5.2.2-3.

In Sec. 5.2.1, we showed that the jump mass balance for species A takes the form

$$\frac{\partial \rho_{(A)}^{(\sigma)}}{\partial t} - \nabla_{(\sigma)} \rho_{(A)}^{(\sigma)} \cdot \mathbf{u} + \text{div}_{(\sigma)}\left(\rho_{(A)}^{(\sigma)} \mathbf{v}_{(A)}^{(\sigma)} \right) - r_{(A)}^{(\sigma)}$$

$$+ \left[\rho_{(A)}\left(\mathbf{v}_{(A)} \cdot \boldsymbol{\xi} - v_{(\xi)}^{(\sigma)} \right)\right] = 0 \tag{2-3}$$

The quantity

$$\mathbf{n}_{(A)}^{(\sigma)} \equiv \rho_{(A)}^{(\sigma)} \mathbf{v}_{(A)}^{(\sigma)} \tag{2-4}$$

can be thought of as the surface mass flux of species A with respect to the frame of reference. The alternative ways that we have of looking at concentrations and velocities suggest the introduction of the surface molar flux with respect to the frame of reference, as well as mass and molar fluxes with respect to the mass- and molar-averaged velocities. A tabulation of the relations between these quantities is given in Table 5.2.2-4.

Just as we may write the jump mass balance for A in terms of $\mathbf{n}_{(A)}^{(\sigma)}$

$$\frac{\partial \rho_{(A)}^{(\sigma)}}{\partial t} - \nabla_{(\sigma)} \rho_{(A)}^{(\sigma)} \cdot \mathbf{u} + \text{div}_{(\sigma)} \mathbf{n}_{(A)}^{(\sigma)} - r_{(A)}^{(\sigma)}$$

$$+ \left[\mathbf{n}_{(A)} \cdot \boldsymbol{\xi} - \rho_{(A)} v_{(\xi)}^{(\sigma)} \right] = 0 \tag{2-5}$$

we may rearrange it in terms of all the other mass and molar flux vectors. For example, the surface mass flux of species A with respect to the surface mass-averaged velocity is defined as

$$\mathbf{j}^{(\sigma)}_{(A)} \equiv \rho^{(\sigma)}_{(A)} \left(\mathbf{v}^{(\sigma)}_{(A)} - \mathbf{v}^{(\sigma)} \right) \tag{2-6}$$

Equation (2-3) may be rewritten as

$$\frac{\partial \rho^{(\sigma)}_{(A)}}{\partial t} - \nabla_{(\sigma)}\rho^{(\sigma)}_{(A)} \cdot \mathbf{u} + \text{div}_{(\sigma)}\left(\rho^{(\sigma)}_{(A)}\, \mathbf{v}^{(\sigma)} \right) + \text{div}_{(\sigma)}\mathbf{j}^{(\sigma)}_{(A)} - r^{(\sigma)}_{(A)}$$

$$+ \left[\mathbf{j}_{(A)} \cdot \boldsymbol{\xi} + \rho_{(A)}\left(\mathbf{v} \cdot \boldsymbol{\xi} - v_{(\xi)} \right)\right] = 0 \tag{2-7}$$

From the overall jump mass balance (Exercise 5.2.1-4 and Sec. 5.4.1), we find

$$\frac{\partial \rho^{(\sigma)}_{(A)}}{\partial t} - \nabla_{(\sigma)}\rho^{(\sigma)}_{(A)} \cdot \mathbf{u} + \text{div}_{(\sigma)}\left(\rho^{(\sigma)}_{(A)}\, \mathbf{v}^{(\sigma)} \right)$$

$$= \frac{\partial \rho^{(\sigma)}_{(A)}}{\partial t} + \nabla_{(\sigma)}\rho^{(\sigma)}_{(A)} \cdot \left(\mathbf{v}^{(\sigma)} - \mathbf{u} \right) + \rho^{(\sigma)}_{(A)}\, \text{div}_{(\sigma)}\mathbf{v}^{(\sigma)}$$

$$= \frac{d_{(\mathbf{v}^{(\sigma)})}\rho^{(\sigma)}_{(A)}}{dt} - \omega^{(\sigma)}_{(A)} \frac{d_{(\mathbf{v}^{(\sigma)})}\rho^{(\sigma)}}{dt} - \omega^{(\sigma)}_{(A)} \left[\rho\left(\mathbf{v} \cdot \boldsymbol{\xi} - v_{(\xi)} \right)\right]$$

$$= \rho^{(\sigma)} \frac{d_{(\mathbf{v}^{(\sigma)})}\omega^{(\sigma)}_{(A)}}{dt} - \omega^{(\sigma)}_{(A)} \left[\rho\left(\mathbf{v} \cdot \boldsymbol{\xi} - v_{(\xi)} \right)\right] \tag{2-8}$$

If B is any scalar, vector, or tensor, I use the notation

$$\frac{d_{(\mathbf{v}^{(\sigma)})}B}{dt} = \frac{\partial B}{\partial t} + \nabla_{(\sigma)}B \cdot \left(\mathbf{v}^{(\sigma)} - \mathbf{u} \right) \tag{2-9}$$

to denote the derivative with respect to time following a fictitious particle that moves with the local surface mass-averaged velocity $\mathbf{v}^{(\sigma)}$. Equations (2-7) and (2-8) give

$$\rho^{(\sigma)} \frac{d_{(\mathbf{v}^{(\sigma)})}\omega^{(\sigma)}_{(A)}}{dt} + \text{div}_{(\sigma)}\mathbf{j}^{(\sigma)}_{(A)} - r^{(\sigma)}_{(A)} + \left[\mathbf{j}_{(A)} \cdot \boldsymbol{\xi} \right.$$

$$\left. + \rho\left(\omega_{(A)} - \omega^{(\sigma)}_{(A)} \right)\left(\mathbf{v} \cdot \boldsymbol{\xi} - v_{(\xi)} \right)\right] = 0 \tag{2-10}$$

Several other alternative forms of the jump mass balance for species A are presented in Table 5.2.2-5.

To assist the reader in formulating problems, (2-5) is illustrated for four specific dividing surfaces in Sec. 5.11.1.

References

Bird, R. B., W. E. Stewart, and E. N. Lightfoot, "Transport Phenomena," seventh corrected printing, Wiley, New York (1960).

Slattery, J. C., "Momentum, Energy, and Mass Transfer in Continua," McGraw-Hill, New York (1972); second edition, Robert E. Krieger, Malabar, FL 32950 (1981).

Exercise

5.2.2-1 *frame indifference of mass flux vectors* A velocity difference is a frame indifferent vector (Slattery 1981, p. 16). Prove that $\mathbf{j}_{(A)}$, $\mathbf{j}^{(\sigma)}_{(A)}$, $\mathbf{j}^{\star}_{(A)}$, $\mathbf{j}^{\star(\sigma)}_{(A)}$, $\mathbf{J}_{(A)}$, $\mathbf{J}^{(\sigma)}_{(A)}$, $\mathbf{J}^{\star}_{(A)}$, and $\mathbf{J}^{\star(\sigma)}_{(A)}$ are frame indifferent vectors. Conclude that $\mathbf{n}_{(A)}$, $\mathbf{n}^{(\sigma)}_{(A)}$, $\mathbf{N}_{(A)}$, and $\mathbf{N}^{(\sigma)}_{(A)}$ are not frame indifferent vectors.

Table 5.2.2-1: *notation for concentrations*

$\rho_{(A)}$	mass density of A
$\rho^{(\sigma)}_{(A)}$	surface mass density of A
$\rho \equiv \sum_{A=1}^{N} \rho_{(A)}$	total mass density
$\rho^{(\sigma)} \equiv \sum_{A=1}^{N} \rho^{(\sigma)}_{(A)}$	total surface mass density
$\omega_{(\sigma)} \equiv \dfrac{\rho_{(A)}}{\rho}$	mass fraction of A
$\omega^{(\sigma)}_{(A)} \equiv \dfrac{\rho^{(\sigma)}_{(A)}}{\rho^{(\sigma)}}$	surface mass fraction of A
$c_{(A)} \equiv \dfrac{\rho_{(A)}}{M_{(A)}}$	molar density of A
$c^{(\sigma)}_{(A)} \equiv \dfrac{\rho^{(\sigma)}_{(A)}}{M_{(A)}}$	surface molar density of A
$c \equiv \sum_{A=1}^{N} c_{(A)}$	total molar density
$c^{(\sigma)} \equiv \sum_{A=1}^{N} c^{(\sigma)}_{(A)}$	total surface molar density
$x_{(A)} \equiv \dfrac{c_{(A)}}{c}$	mole fraction of A
$x^{(\sigma)}_{(A)} \equiv \dfrac{c^{(\sigma)}_{(A)}}{c^{(\sigma)}}$	surface mole fraction of A
$M \equiv \dfrac{\rho}{c} = \sum_{A=1}^{N} x_{(A)} M_{(A)}$	molar-averaged molecular weight

Table 5.2.2-1: *notation for concentrations (cont.)*

$$\frac{1}{M} = \frac{c}{\rho} = \sum_{A=1}^{N} \frac{\omega_{(A)}}{M_{(A)}}$$

$$M^{(\sigma)} \equiv \frac{\rho^{(\sigma)}}{c^{(\sigma)}} = \sum_{A=1}^{N} x_{(A)}^{(\sigma)} M_{(A)}$$ molar-averaged molecular weight of surface

$$\frac{1}{M^{(\sigma)}} = \frac{c^{(\sigma)}}{\rho^{(\sigma)}} = \sum_{A=1}^{N} \frac{\omega_{(A)}^{(\sigma)}}{M_{(A)}}$$

Table 5.2.2-2: *relations between mass and mole fractions*

$$x_{(A)} = \frac{\omega_{(A)}/M_{(A)}}{\sum_{B=1}^{N} \omega_{(B)}/M_{(B)}} \qquad \omega_{(B)} = \frac{x_{(A)}/M_{(A)}}{\sum_{B=1}^{N} x_{(B)}/M_{(B)}}$$

$$x_{(A)}^{(\sigma)} = \frac{\omega_{(A)}^{(\sigma)}/M_{(A)}}{\sum_{B=1}^{N} \omega_{(B)}^{(\sigma)}/M_{(B)}} \qquad \omega_{(B)}^{(\sigma)} = \frac{x_{(A)}^{(\sigma)}/M_{(A)}}{\sum_{B=1}^{N} x_{(B)}^{(\sigma)}/M_{(B)}}$$

for a binary system: for a binary system:

$$dx_{(A)} = \left(\frac{\rho}{c}\right)^2 \frac{d\omega_{(A)}}{M_{(A)}M_{(B)}} \qquad d\omega_{(A)} = \left(\frac{c}{\rho}\right)^2 M_{(A)}M_{(B)}\, dx_{(A)}$$

$$dx_{(A)}^{(\sigma)} = \left(\frac{\rho^{(\sigma)}}{c^{(\sigma)}}\right)^2 \frac{d\omega_{(A)}^{(\sigma)}}{M_{(A)}M_{(B)}} \qquad d\omega_{(A)}^{(\sigma)} = \left(\frac{c^{(\sigma)}}{\rho^{(\sigma)}}\right)^2 M_{(A)}M_{(B)}\, dx_{(A)}^{(\sigma)}$$

Table 5.2.2-3: *various velocities and relations between them*

$\mathbf{v}_{(A)}$	velocity of species A
$\mathbf{v}_{(A)}^{(\sigma)}$	surface velocity of species A
$\mathbf{v} \equiv \sum_{A=1}^{N} \omega_{(A)} \mathbf{v}_{(A)}$	mass-averaged velocity
$\mathbf{v}^{(\sigma)} \equiv \sum_{A=1}^{N} \omega_{(A)}^{(\sigma)} \mathbf{v}_{(A)}^{(\sigma)}$	mass-averaged surface velocity
$\mathbf{v}^{\star} \equiv \sum_{A=1}^{N} x_{(A)} \mathbf{v}_{(A)}$	molar-averaged velocity
$\mathbf{v}^{(\sigma)\star} \equiv \sum_{A=1}^{N} x_{(A)}^{(\sigma)} \mathbf{v}_{(A)}^{(\sigma)}$	molar-averaged surface velocity
$\mathbf{v} - \mathbf{v}^{\star} \equiv \sum_{A=1}^{N} \omega_{(A)} (\mathbf{v}_{(A)} - \mathbf{v}^{\star})$	
$\mathbf{v}^{(\sigma)} - \mathbf{v}^{(\sigma)\star} \equiv \sum_{A=1}^{N} \omega_{(A)}^{(\sigma)} (\mathbf{v}_{(A)}^{(\sigma)} - \mathbf{v}^{(\sigma)\star})$	
$\mathbf{v}^{\star} - \mathbf{v} \equiv \sum_{A=1}^{N} x_{(A)} (\mathbf{v}_{(A)} - \mathbf{v})$	
$\mathbf{v}^{(\sigma)\star} - \mathbf{v}^{(\sigma)} \equiv \sum_{A=1}^{N} x_{(A)}^{(\sigma)} (\mathbf{v}_{(A)}^{(\sigma)} - \mathbf{v}^{(\sigma)})$	

Table 5.2.2-4: *mass and molar fluxes*

quantity	with respect to frame of reference	with respect to $\mathbf{v}$ or $\mathbf{v}^{(\sigma)}$	with respect to $\mathbf{v}^{\star}$ or $\mathbf{v}^{\star(\sigma)}$
mass flux of A	$\mathbf{n}_{(A)} \equiv \rho_{(A)}\,\mathbf{v}_{(A)}$	$\mathbf{j}_{(A)} \equiv \rho_{(A)}(\mathbf{v}_{(A)} - \mathbf{v})$	$\mathbf{j}^{\star}_{(A)} \equiv \rho_{(A)}(\mathbf{v}_{(A)} - \mathbf{v}^{\star})$
surface mass flux of A	$\mathbf{n}\left\{\begin{smallmatrix}\sigma\\A\end{smallmatrix}\right\} \equiv \rho\left\{\begin{smallmatrix}\sigma\\A\end{smallmatrix}\right\}\mathbf{v}\left\{\begin{smallmatrix}\sigma\\A\end{smallmatrix}\right\}$	$\mathbf{j}\left\{\begin{smallmatrix}\sigma\\A\end{smallmatrix}\right\} \equiv \rho\left\{\begin{smallmatrix}\sigma\\A\end{smallmatrix}\right\}(\mathbf{v}\left\{\begin{smallmatrix}\sigma\\A\end{smallmatrix}\right\} - \mathbf{v}^{(\sigma)})$	$\mathbf{j}\left\{\begin{smallmatrix}\sigma\\A\end{smallmatrix}\right\}^{\star} \equiv \rho\left\{\begin{smallmatrix}\sigma\\A\end{smallmatrix}\right\}(\mathbf{v}\left\{\begin{smallmatrix}\sigma\\A\end{smallmatrix}\right\} - \mathbf{v}^{(\sigma)\star})$
molar flux of A	$\mathbf{N}_{(A)} \equiv c_{(A)}\,\mathbf{v}_{(A)}$	$\mathbf{J}_{(A)} \equiv c_{(A)}(\mathbf{v}_{(A)} - \mathbf{v})$	$\mathbf{J}^{\star}_{(A)} \equiv c_{(A)}(\mathbf{v}_{(A)} - \mathbf{v}^{\star})$
surface molar flux of A	$\mathbf{N}\left\{\begin{smallmatrix}\sigma\\A\end{smallmatrix}\right\} \equiv c\left\{\begin{smallmatrix}\sigma\\A\end{smallmatrix}\right\}\mathbf{v}\left\{\begin{smallmatrix}\sigma\\A\end{smallmatrix}\right\}$	$\mathbf{J}\left\{\begin{smallmatrix}\sigma\\A\end{smallmatrix}\right\} \equiv c\left\{\begin{smallmatrix}\sigma\\A\end{smallmatrix}\right\}(\mathbf{v}\left\{\begin{smallmatrix}\sigma\\A\end{smallmatrix}\right\} - \mathbf{v}^{(\sigma)})$	$\mathbf{J}\left\{\begin{smallmatrix}\sigma\\A\end{smallmatrix}\right\}^{\star} \equiv c\left\{\begin{smallmatrix}\sigma\\A\end{smallmatrix}\right\}(\mathbf{v}\left\{\begin{smallmatrix}\sigma\\A\end{smallmatrix}\right\} - \mathbf{v}^{(\sigma)\star})$
sum of mass fluxes	$\sum_{A=1}^{N} \mathbf{n}_{(A)} = \rho\mathbf{v}$	$\sum_{A=1}^{N} \mathbf{j}_{(A)} = 0$	$\sum_{A=1}^{N} \mathbf{j}^{\star}_{(A)} = \rho(\mathbf{v} - \mathbf{v}^{\star})$
sum of surface mass fluxes	$\sum_{A=1}^{N} \mathbf{n}\left\{\begin{smallmatrix}\sigma\\A\end{smallmatrix}\right\} = \rho^{(\sigma)}\,\mathbf{v}^{(\sigma)}$	$\sum_{A=1}^{N} \mathbf{j}\left\{\begin{smallmatrix}\sigma\\A\end{smallmatrix}\right\} = 0$	$\sum_{A=1}^{N} \mathbf{j}\left\{\begin{smallmatrix}\sigma\\A\end{smallmatrix}\right\}^{\star} = \rho^{(\sigma)}(\mathbf{v}^{(\sigma)} - \mathbf{v}^{(\sigma)\star})$
sum of molar fluxes	$\sum_{A=1}^{N} \mathbf{N}_{(A)} = c\mathbf{v}^{\star}$	$\sum_{A=1}^{N} \mathbf{J}_{(A)} = c(\mathbf{v}^{\star} - \mathbf{v})$	$\sum_{A=1}^{N} \mathbf{J}^{\star}_{(A)} = 0$

Table 5.2.2-4: *mass and molar fluxes (cont.)*

quantity	with respect to frame of reference	with respect to $\mathbf{v}$ or $\mathbf{v}^{(\sigma)}$	with respect to $\mathbf{v}^{\star}$ or $\mathbf{v}^{\star(\sigma)}$
sum of surface molar fluxes	$\sum_{A=1}^{N} \mathbf{N}_{(A)}^{(\sigma)} = c^{(\sigma)} \mathbf{v}^{\star(\sigma)}$	$\sum_{A=1}^{N} \mathbf{J}_{(A)}^{(\sigma)} = c^{(\sigma)} (\mathbf{v}^{(\sigma)\star} - \mathbf{v}^{(\sigma)})$	$\sum_{A=1}^{N} \mathbf{J}_{(A)}^{(\sigma)\star} = 0$
fluxes in terms of $\mathbf{n}_{(A)}$	$\mathbf{N}_{(A)} = \frac{\mathbf{n}_{(A)}}{M_{(A)}}$	$\mathbf{j}_{(A)} = \mathbf{n}_{(A)} - \omega_{(A)} \sum_{B=1}^{N} \mathbf{n}_{(B)}$	$\mathbf{j}_{(A)}^{\star} = \mathbf{n}_{(A)} - M_{(A)} x_{(A)} \sum_{B=1}^{N} \frac{\mathbf{n}_{(B)}}{M_{(B)}}$
surface fluxes in terms of $\mathbf{n}_{(A)}^{(\sigma)}$	$\mathbf{N}_{(A)}^{(\sigma)} = \frac{\mathbf{n}_{(A)}^{(\sigma)}}{M_{(A)}}$	$\mathbf{j}_{(A)}^{(\sigma)} = \mathbf{n}_{(A)}^{(\sigma)} - \omega_{(A)}^{(\sigma)} \sum_{B=1}^{N} \mathbf{n}_{(B)}^{(\sigma)}$	$\mathbf{j}_{(A)}^{(\sigma)\star} = \mathbf{n}_{(A)}^{(\sigma)} - M_{(A)} x_{(A)}^{(\sigma)} \sum_{B=1}^{N} \frac{\mathbf{n}_{(B)}^{(\sigma)}}{M_{(B)}}$
fluxes in terms of $\mathbf{N}_{(A)}$	$\mathbf{n}_{(A)} = M_{(A)} \mathbf{N}_{(A)}$	$\mathbf{J}_{(A)} = \mathbf{N}_{(A)} - \frac{\omega_{(A)}}{M_{(A)}} \sum_{B=1}^{N} M_{(B)} \mathbf{N}_{(B)}$	$\mathbf{J}_{(A)}^{\star} = \mathbf{N}_{(A)} - x_{(A)} \sum_{B=1}^{N} \mathbf{N}_{(B)}$
surface fluxes in terms of $\mathbf{N}_{(A)}^{(\sigma)}$	$\mathbf{n}_{(A)}^{(\sigma)} = M_{(A)} \mathbf{N}_{(A)}^{(\sigma)}$	$\mathbf{J}_{(A)}^{(\sigma)} = \mathbf{N}_{(A)}^{(\sigma)} - \frac{\omega_{(A)}^{(\sigma)}}{M_{(A)}} \sum_{B=1}^{N} M_{(B)} \mathbf{N}_{(B)}^{(\sigma)}$	$\mathbf{J}_{(A)}^{(\sigma)\star} = \mathbf{N}_{(A)}^{(\sigma)} - x_{(A)}^{(\sigma)} \sum_{B=1}^{N} \mathbf{N}_{(B)}^{(\sigma)}$

Table 5.2.2-4: *mass and molar fluxes (cont.)*

quantity	with respect to frame of reference	with respect to $\mathbf{v}$ or $\mathbf{v}^{(\sigma)}$	with respect to $\mathbf{v}^{\star}$ or $\mathbf{v}^{\star(\sigma)}$
fluxes in terms of $\mathbf{j}_{(A)}$ and $\mathbf{v}$	$\mathbf{n}_{(A)} = \mathbf{j}_{(A)} + \rho_{(A)}\mathbf{v}$	$\mathbf{J}_{(A)} = \frac{\mathbf{j}_{(A)}}{M_{(A)}}$	$\mathbf{j}^{\star}_{(A)} = \mathbf{j}_{(A)} - \omega_{(A)} M \sum_{B=1}^{N} \frac{\mathbf{j}_{(B)}}{M_{(B)}}$ for a binary system: $\mathbf{j}^{\star}_{(A)} = \frac{M}{M_{(B)}} \mathbf{j}_{(A)}$
surface fluxes in terms of $\mathbf{j}^{(\sigma)}_{(A)}$ and $\mathbf{v}^{(\sigma)}$	$\mathbf{n}^{(\sigma)}_{(A)} = \mathbf{j}^{(\sigma)}_{(A)} + \rho^{(\sigma)}_{(A)}\mathbf{v}^{(\sigma)}$	$\mathbf{J}^{(\sigma)}_{(A)} = \frac{\mathbf{j}^{(\sigma)}_{(A)}}{M_{(A)}}$	$\mathbf{j}^{(\sigma)\star}_{(A)} = \mathbf{j}^{(\sigma)}_{(A)} - \omega^{(\sigma)}_{(A)} M^{(\sigma)} \sum_{B=1}^{N} \frac{\mathbf{j}^{(\sigma)}_{(B)}}{M_{(B)}}$ for a binary system: $\mathbf{j}^{(\sigma)\star}_{(A)} = \frac{M^{(\sigma)}}{M_{(B)}} \mathbf{j}^{(\sigma)}_{(A)}$
surface fluxes in terms of $\mathbf{J}^{\star}_{(A)}$ and $\mathbf{v}^{\star}$	$\mathbf{N}_{(A)} = \mathbf{J}^{\star}_{(A)} + c_{(A)}\mathbf{v}^{\star}$	$\mathbf{J}_{(A)} = \mathbf{J}^{\star}_{(A)} - \frac{x_{(A)}}{M} \sum_{B=1}^{N} M_{(B)} \mathbf{J}^{\star}_{(B)}$ for a binary system: $\mathbf{J}_{(A)} = \frac{M_{(B)}}{M} \mathbf{J}^{\star}_{(A)}$	

Table 5.2.2-4: *mass and molar fluxes (cont.)*

quantity	with respect to frame of reference	with respect to $\mathbf{v}$ or $\mathbf{v}^{(\sigma)}$	with respect to $\mathbf{v}^{\star}$ or $\mathbf{v}^{\star(\sigma)}$
surface fluxes in terms of $\mathbf{J}\{^{\sigma}_{A}\}^{\star}$ and $\mathbf{v}^{\star(\sigma)}$	$\mathbf{N}\{^{\sigma}_{A}\} = \mathbf{J}\{^{\sigma}_{A}\}^{\star} + c\{^{\sigma}_{A}\}\mathbf{v}^{\star(\sigma)}$	$\mathbf{J}\{^{\sigma}_{A}\} = \mathbf{J}\{^{\sigma}_{A}\}^{\star} - \dfrac{x\{^{\sigma}_{A}\}}{M^{(\sigma)}} \sum_{B=1}^{N} M_{(B)}\mathbf{J}\{^{\sigma}_{B}\}^{\star}$ for a binary system: $\mathbf{J}\{^{\sigma}_{A}\} = \dfrac{M_{(B)}}{M^{(\sigma)}}\mathbf{J}\{^{\sigma}_{A}\}^{\star}$	$\mathbf{j}\{^{\sigma}_{A}\}^{\star} = M_{(A)}\mathbf{J}\{^{\sigma}_{A}\}^{\star}$

Table 5.2.2-5: *forms of the jump mass balance for species A*

$$\frac{\partial \rho^{(\sigma)}_{(A)}}{\partial t} - \nabla_{(\sigma)}\rho^{(\sigma)}_{(A)} \cdot \mathbf{u} + \operatorname{div}_{(\sigma)}\mathbf{n}^{(\sigma)}_{(A)} - r^{(\sigma)}_{(A)} + [\, \mathbf{n}_{(A)} \cdot \boldsymbol{\xi} - \rho_{(A)} v^{(\sigma)}_{(\xi)} \,] = 0 \quad \text{(A)}$$

$$\frac{\partial c^{(\sigma)}_{(A)}}{\partial t} - \nabla_{(\sigma)}c^{(\sigma)}_{(A)} \cdot \mathbf{u} + \operatorname{div}_{(\sigma)}\mathbf{N}^{(\sigma)}_{(A)} - \frac{r^{(\sigma)}_{(A)}}{M_{(A)}} + [\, \mathbf{N}_{(A)} \cdot \boldsymbol{\xi} - c_{(A)} v^{(\sigma)}_{(\xi)} \,] = 0 \quad \text{(B)}$$

$$\rho^{(\sigma)} \frac{d_{(\mathbf{v}^{(\sigma)})}\omega^{(\sigma)}_{(A)}}{dt} + \operatorname{div}_{(\sigma)}\mathbf{j}^{(\sigma)}_{(A)} - r^{(\sigma)}_{(A)}$$
$$+ [\, \mathbf{j}_{(A)} \cdot \boldsymbol{\xi} + \rho(\, \omega_{(A)} - \omega^{(\sigma)}_{(A)} \,)(\, \mathbf{v} \cdot \boldsymbol{\xi} - v^{(\sigma)}_{(\xi)} \,)] = 0 \quad \text{(C)}\dagger$$

$$c^{(\sigma)} \frac{d_{(\mathbf{v}^{\star(\sigma)})}x^{(\sigma)}_{(A)}}{dt} + \operatorname{div}_{(\sigma)}\mathbf{J}^{\star(\sigma)}_{(A)} - \frac{r^{(\sigma)}_{(A)}}{M_{(A)}} + x^{(\sigma)}_{(A)} \sum_{A=1}^{N} \frac{r^{(\sigma)}_{(A)}}{M_{(A)}}$$
$$+ [\, \mathbf{J}^{\star}_{(A)} \cdot \boldsymbol{\xi} + c(\, x_{(A)} - x^{(\sigma)}_{(A)} \,)(\, \mathbf{v}^{\star} \cdot \boldsymbol{\xi} - v^{(\sigma)}_{(\xi)} \,)] = 0 \quad \text{(D)}\ddagger$$

$$\frac{d_{(\mathbf{v}^{(\sigma)})}\rho^{(\sigma)}_{(A)}}{dt} + \rho^{(\sigma)}_{(A)} \operatorname{div}_{(\sigma)}\mathbf{v}^{(\sigma)} + \operatorname{div}_{(\sigma)}\mathbf{j}^{(\sigma)}_{(A)} - r^{(\sigma)}_{(A)}$$
$$+ [\, \mathbf{j}_{(A)} \cdot \boldsymbol{\xi} + \rho_{(A)}(\, \mathbf{v} \cdot \boldsymbol{\xi} - v^{(\sigma)}_{(\xi)} \,)] = 0 \quad \text{(E)}\dagger$$

$$\frac{d_{(\mathbf{v}^{\star(\sigma)})}c^{(\sigma)}_{(A)}}{dt} + c^{(\sigma)}_{(A)} \operatorname{div}_{(\sigma)}\mathbf{v}^{(\sigma)\star} + \operatorname{div}_{(\sigma)}\mathbf{j}^{(\sigma)\star}_{(A)} - \frac{r^{(\sigma)}_{(A)}}{M_{(A)}}$$
$$+ [\, \mathbf{J}^{\star}_{(A)} \cdot \boldsymbol{\xi} + c_{(A)}(\, \mathbf{v}^{\star} \cdot \boldsymbol{\xi} - v^{(\sigma)}_{(\xi)} \,)] = 0 \quad \text{(F)}\ddagger$$

---

† $\dfrac{d_{(\mathbf{v}^{(\sigma)})}\psi}{dt} \equiv \dfrac{\partial \psi}{\partial t} + \nabla_{(\sigma)}\psi \cdot (\, \mathbf{v}^{(\sigma)} - \mathbf{u} \,)$

‡ $\dfrac{d_{(\mathbf{v}^{\star(\sigma)})}\psi}{dt} \equiv \dfrac{\partial \psi}{\partial t} + \nabla_{(\sigma)}\psi \cdot (\, \mathbf{v}^{\star(\sigma)} - \mathbf{u} \,)$

5.2.3 Location of multicomponent dividing surface The remarks that I made in Sec. 1.3.6 with respect to locating dividing surfaces within single component bodies are directly applicable here to multicomponent bodies. The problem is that it is not sufficient to say the dividing surface is sensibly coincident with the phase interface. There are an infinite number of choices that could be said to satisfy this requirement.

Gibbs (1948, p. 234) suggested that the dividing surface be located such that the surface molar density of one of the components B is zero:

$$c_{(B)}^{(\sigma,B)} = 0 \tag{3-1}$$

Using this convention, $c_{(A)}^{(\sigma,B)}$ designates the surface molar density of species A. Often species B is chosen to be the solvent or the species whose concentration is largest. A major disadvantage with this convention is that the surface molar density of one or more of the species may be negative. A major advantage is that the surface mass or molar densities can be measured with relative ease, when one of the adjoining phases is an ideal solution (see Sec. 5.8.7).

The discussion in Sec. 1.3.6 suggests that the dividing surface be located such that the total surface mass density is zero:

$$\begin{aligned} \rho^{(\sigma,\rho)} &\equiv \sum_{A=1}^{N} \rho_{(A)}^{(\sigma,\rho)} \\ &= 0 \end{aligned} \tag{3-2}$$

With this convention, $\rho_{(A)}^{(\sigma,\rho)}$ is the surface mass density of species A. Not only does this convention allow surface mass densities of individual species to be negative, it also prevents the introduction of mass fractions.

We might choose the dividing surface such that the total surface molar density is zero:

$$\begin{aligned} c^{(\sigma,c)} &\equiv \sum_{A=1}^{N} c_{(A)}^{(\sigma,c)} \\ &= 0 \end{aligned} \tag{3-3}$$

In addition to permitting the surface molar densities of individual species to be negative, it rules out the use of mole fractions.

Lucassen-Reynders and van den Tempel (1967; Lucassen-Reynders 1976) have suggested that the dividing surface be located such that the total surface molar density is a constant $c^{(\sigma)\infty}$ independent of composition

$$c^{(\sigma,L)} \equiv \sum_{A=1}^{N} c_{(A)}^{(\sigma,L)}$$

$$= c^{(\sigma)\infty} \tag{3-4}$$

and that $c^{(\sigma)\infty}$ be sufficiently large to ensure positive values for all individual surface molar densities $c_{(A)}^{(\sigma, L)}$ and surface mole fractions $x_{(A)}^{(\sigma, L)}$. It is this convention that we will find most appropriate in discussing specific models for adsorption behavior in Sec. 5.8.7.

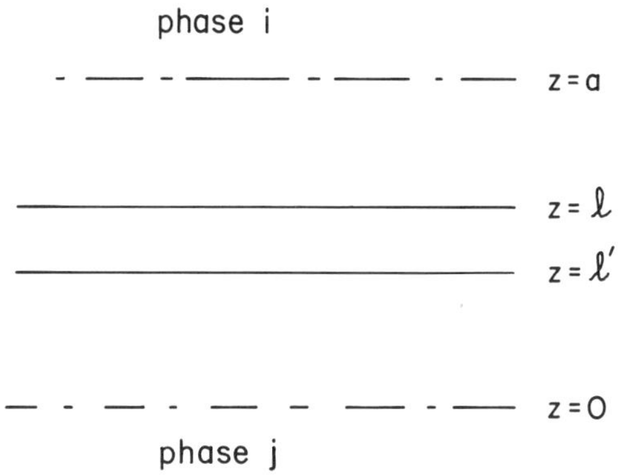

Figure 5.2.3-1: Alternative dividing surfaces $z = \ell$ and $z = \ell'$ within the interfacial region which is bounded by the planes $z = 0$ and $z = a$ and which separates phases i and j.

The surface mass or molar densities associated with two conventions are easily related, so long as the dividing surfaces are locally parallel (separated by a uniform distance). Referring to Figure 5.2.3-1, we have

$$\ell c_{(A)}^{(j)} + (a - \ell) c_{(A)}^{(i)} + c_{(A)}^{(\sigma)} = \ell' c_{(A)}^{(j)} + (a - \ell') c_{(A)}^{(i)} + c_{(A)}^{(\sigma)\prime} \tag{3-5}$$

or

$$c_{(A)}^{(\sigma)\prime} = c_{(A)}^{(\sigma)} + (\ell' - \ell)(c_{(A)}^{(i)} - c_{(A)}^{(j)}) \tag{3-6}$$

In particular, the surface molar density $c_{(A)}^{(\sigma, L)}$ in the Lucassen-Reynders and van den Tempel convention can be expressed in terms of $c_{(A)}^{(\sigma, B)}$ in the Gibbs convention as

$$c_{(A)}^{(\sigma, L)} = c_{(A)}^{(\sigma, B)} + (\ell^{(L)} - \ell^{(B)})(c_{(A)}^{(i)} - c_{(A)}^{(j)}) \tag{3-7}$$

Consider a two component system in which phase i is water vapor and phase j is a dilute aqueous solution of a single surfactant species A:

$$(c_{(A)}^{(j)} - c_{(A)}^{(i)}) << (c_{(B)}^{(j)} - c_{(B)}^{(i)})$$

In such a case, (3-6) suggests that the surface molar density of surfactant may be nearly independent of the convention chosen and nearly equal to value found using the Gibbs convention (see Sec. 5.8.7). In contrast, the surface molar density of water may be quite different from the value zero determined using the Gibbs Convention.

Other conventions have been proposed for locating the dividing surface, for example the **surface of tension** in Sec. 5.8.6, footnote (a). Guggenheim and Adam (1933) give an interesting review of many of these conventions.

References

Gibbs, J. W., "The Collected Works," vol. 1, Yale University Press, New Haven, CT (1948).

Guggenheim, E. A., and N. K. Adam, *Proc. R. Soc. London A* **139**, 218 (1933).

Lucassen-Reynders, E. H., and M. van den Tempel, "Proceedings of the Fourth International Congress on Surface Active Substances," vol. 2, p. 779, Gordon and Breach, New York (1967).

Lucassen-Reynders, E. H., *Progress in Surface and Membrane Science* **10**, 253 (1976), edited by D. A. Cadenhead and J. F. Danielli.

5.3 Further comments on viewpoint

5.3.1 Further comments on viewpoint of multicomponent materials In the preceding sections, we have visualized a multicomponent mixture as a superposition of material bodies, one corresponding to each of the N species present. This a convenient outlook in discussing a mass balance for each of these species.

In the sections that follow, it will be more convenient to think of the multicomponent mixture as a whole in considering mass conservation, forces, energy transmission, the energy balance, and the Clausius-Duhem inequality.[a] Exercise 5.2.1-4 suggests how we may proceed. We define a **mass-averaged material particle** of a multicomponent mixture to be a artificial) particle which moves with the mass-averaged velocity of the mixture (see Table 5.2.2-3). A **multicomponent body** is defined to be a set, the elements of which are mass-averaged material particles. A

---

a) It is not necessary that forces, energy transmission, the energy balance, and the Clausius-Duhem inequality be introduced in the context of the multicomponent mixture as a whole. Axioms can be stated for a species as a single body (Truesdell 1957, 1969; Truesdell and Toupin 1960; Fick 1855; Maxwell 1867; Stefan 1871; Nachbar *et al.* 1959; Kelly 1964). Complete theories for various types of materials have been presented in this context (Truesdell 1962; Eringen and Ingram 1965; Green and Naghdi 1965, 1967a, 1967b, 1968, 1969; Crochet and Naghdi 1966; Green and Steel 1966; Mills 1966; Bowen 1967, 1968, 1969; Ingram and Eringen 1967; Dunwoody and Müller 1968; Müller 1968; Bowen and Wiese 1969; Doria 1969; Bowen and Garcia 1970; Craine *et al.* 1970; Dunwoody 1970; Gurtin and de La Penha 1970; Gurtin 1971).

If one takes the position that the best theory is the simplest one consistent with experimental observation, then this approach is open to question. For example, in talking about the momentum balance for a particular species, a stress tensor is introduced. The stress-deformation behavior of the species is described by a constitutive equation that relates this component stress tensor to the motion of the body. Yet to date no one has suggested how this component stress might be measured or how the stress-deformation behavior of a component can be distinguished from that of the body as a whole. The most apparent practical use of the component stress tensor is in the definition of a stress tensor for the multicomponent body as a whole.

On the other hand, Truesdell (1962) has given a very interesting discussion of diffusion in multicomponent mixtures using the component momentum balances.

It would be premature to recommend this approach for general use. But, since this is an active area of research, it also would be unwise to speculate on how it will develop.

mass-averaged surface particle is a (artificial) particle that moves with the mass-averaged surface velocity of the mixture. The notation $d_{(v)}/dt$ indicates a derivative with respect to time following a mass-averaged material particle; $d_{(v(\sigma))}/dt$ is a derivative with respect to time following a mass-averaged surface particle. By $R_{(v)}$ we mean the region of space occupied by a set of mass-averaged material particles; $S_{(v)}$ denotes the closed bounding surface of this region; $C_{(v)}$ is the curve of intersection between the multicomponent dividing surface Σ and $S_{(v)}$.

References

Bowen, R. M., *Arch. Ration. Mech. Anal.* **24**, 370 (1967).

Bowen, R. M., *J. Chem. Phys.* **49**, 1625 (1968).

Bowen, R. M., *Arch. Ration. Mech. Anal.* **34**, 97 (1969).

Bowen, R. M., and J. C. Wiese, *Int. J. Eng. Sci* **7**, 689 (1969).

Bowen, R. M., and D. J. Garcia, *Int. J. Eng. Sci.* **8**, 63 (1970).

Craine, R. E., A. E. Green, and P. M. Naghdi, *Q. J. Mech. Appl. Math.* **23**, 171 (1970).

Crochet, M. J., and P. M. Naghdi, *Int. J. Eng. Sci.* **4**, 383 (1966).

Doria, M. L., *Arch. Ration. Mech. Anal.* **32**, 343 (1969).

Dunwoody, N. T., and I. Müller, *Arch. Ration. Mech. Anal.* **29**, 344 (1968).

Dunwoody, N. T., *Arch. Ration. Mech. Anal.* **38**, 348 (1970).

Eringen, A. C., and J. D. Ingram, *Int. J. Eng. Sci.* **3**, 197 (1965).

Fick, A., *Ann. Phys. Chem.* **94**, 59 (1855).

Green, A. E., and P. M. Naghdi, *Int. J. Eng. Sci.* **3**, 231 (1965).

Green, A. E., and P. M. Naghdi, *Arch. Ration. Mech. Anal.* **24**, 243 (1967).

Green, A. E., and P. M. Naghdi, *Arch. Ration. Mech. Anal.* **27**, 175 (1967).

Green, A. E., and P. M. Naghdi, *Int. J. Eng. Sci.* **6**, 631 (1968).

Green, A. E., and P. M. Naghdi, *Q. J. Mech. Appl. Math.* **22**, 427 (1969).

Green, A. E., and T. R. Steel, *Int. J. Eng. Sci.* 4, 483 (1966).

Gurtin, M. E., and G. M. de La Penha, *Arch. Ration. Mech. Anal.* **36**, 390 (1970).

Gurtin, M. E., *Arch. Ration. Mech. Anal.* **43**, 198 (1971).

Ingram, J. D., and A. C. Eringen, *Int. J. Eng. Sci.* **5**, 289 (1967).

Kelly, P. D., *Int. J. Eng. Sci.* **2**, 129 (1964).

Maxwell, J. C., *Philos. Trans. R. Soc. London* **157**, 49 (1867); *Philos. Mag.* **32**, 390 (1866); *ibid.* **35**, 129, 185 (1868).

Mills, N., *Int. J. Eng. Sci.* **4**, 97 (1966).

Müller, I., *Arch. Ration. Mech. Anal.* **28**, 1 (1968).

Nachbar, W., F. Williams, and S. S. Penner, *Q. Appl. Math.* **17**, 43 (1959).

Stefan, J., *Sitzungsberichte der Mathematisch-Naturwissenschaftlichen Classe der kaiserlichen Akademie der Wissenschaften, Wien* **63** (Abth. 2), 63 (1871).

Truesdell, C., and R. A. Toupin, "Handbuch der Physik," vol. 3/1, edited by S. Flügge, Springer-Verlag, Berlin (1960).

Truesdell, C., *J. Chem. Phys.* **37**, 2336 (1962).

Truesdell, C., *Rend. Accad. Lincei* **22**, 33, 158 (1957); see also "Rational Mechanics of Materials," Intl. Sci. Rev. Ser. 292-305, Gordon and Breach, New York (1965).

Truesdell, C., "Rational Thermodynamics," McGraw-Hill, New York (1969).

Exercises

5.3.1-1 *surface transport theorem for multicomponent dividing surface* Let $\psi^{(\sigma)}$ be any scalar-, vector-, or tensor-valued function of time and position on a multicomponent dividing surface. Show that

$$\frac{d}{dt}\int_{\Sigma} \psi^{(\sigma)}\, dA = \int_{\Sigma} \left(\frac{d_{(\mathbf{v}^{(\sigma)})}\psi^{(\sigma)}}{dt} + \psi^{(\sigma)}\, \text{div}_{(\sigma)}\mathbf{v}^{(\sigma)} \right) dA$$

This is the **surface transport theorem for a multicomponent dividing surface.**

5.3.1-2 *transport theorem for multicomponent body containing dividing surface* Let ψ be any scalar-, vector-, or tensor-valued function of time and position in the region $R_{(v)}$ occupied by a multicomponent body; $\psi^{(\sigma)}$ is any scalar-, vector-, or tensor-valued function of time and position on a multicomponent dividing surface. Derive the following transport theorem for a multicomponent body containing a dividing surface:

$$\frac{d}{dt}\left(\int_{R_{(v)}} \psi \, dV + \int_{\Sigma} \psi^{(\sigma)} \, dA\right) = \int_{R_{(v)}} \left(\frac{d_{(v)}\psi}{dt} + \psi \operatorname{div} \mathbf{v}\right) dV$$

$$+ \int_{\Sigma} \left(\frac{d_{(v^{(\sigma)})}\psi^{(\sigma)}}{dt} + \psi^{(\sigma)} \operatorname{div}_{(\sigma)} \mathbf{v}^{(\sigma)} + [\psi(\mathbf{v} \cdot \boldsymbol{\xi} - v_{\{\xi\}})]\right) dA$$

5.3.1-3 *transport theorem for multicomponent body containing intersecting dividing surface* Let ψ be any scalar-, vector-, or tensor-valued function of time and position in the region $R_{(v)}$ occupied by a multicomponent body; $\psi^{(\sigma)}$ is any scalar-, vector-, or tensor-valued function of time and position on a multicomponent dividing surface. Derive the following **transport theorem for a multicomponent body containing intersecting dividing surfaces** (see Sec. 1.3.7):

$$\frac{d}{dt}\left(\int_{R_{(v)}} \psi \, dV + \int_{\Sigma} \psi^{(\sigma)} \, dA\right) = \int_{R_{(v)}} \left(\frac{d_{(v)}\psi}{dt} + \psi \operatorname{div} \mathbf{v}\right) dV$$

$$+ \int_{\Sigma} \left(\frac{d_{(v^{(\sigma)})}\psi^{(\sigma)}}{dt} + \psi^{(\sigma)} \operatorname{div}_{(\sigma)} \mathbf{v}^{(\sigma)} + [\psi(\mathbf{v} \cdot \boldsymbol{\xi} - v_{\{\xi\}})]\right) dA$$

$$+ \int_{C^{(cl)}} \left(\psi^{(\sigma)}[\mathbf{v}^{(\sigma)} \cdot \boldsymbol{\nu} - u_{\{\nu\}}^{(cl)}]\right) ds$$

5.4 Mass

5.4.1 Conservation of mass We can now restate our first postulate:

conservation of mass: The mass of a multicomponent body is independent of time.

We can express the mass $\mathcal{M}$ of a multicomponent body as

$$\mathcal{M} = \int_{R_{(v)}} \rho \, dV + \int_{\Sigma} \rho^{(\sigma)} \, dA \tag{1-1}$$

where ρ is the total mass density

$$\rho \equiv \sum_{A=1}^{N} \rho_{(A)} \tag{1-2}$$

and $\rho^{(\sigma)}$ is the total surface mass density

$$\rho^{(\sigma)} \equiv \sum_{A=1}^{N} \rho^{(\sigma)}_{(A)} \tag{1-3}$$

Our statement that mass is conserved says that

$$\frac{d\mathcal{M}}{dt} = \frac{d}{dt}\left(\int_{R_{(v)}} \rho \, dV + \int_{\Sigma} \rho^{(\sigma)} \, dA \right) = 0 \tag{1-4}$$

The transport theorem of Exercise 5.3.1-2 allows us to rewrite (1-4) as

$$\int_{R_{(v)}} \left(\frac{d_{(v)}\rho}{dt} + \rho \operatorname{div} \mathbf{v} \right) dV + \int_{\Sigma} \left(\frac{d_{(v^{(\sigma)})}\rho^{(\sigma)}}{dt} + \rho^{(\sigma)} \operatorname{div}_{(\sigma)} \mathbf{v}^{(\sigma)} + [\rho (\mathbf{v} \cdot \boldsymbol{\xi} - v^{(\sigma)}_{(\xi)})] \right) dA = 0 \tag{1-5}$$

Equation (1-5) must be true for every body. In particular, for a body consisting of a single phase, the integral over the region $R_{(v)}$ must be zero, which implies (Slattery 1981, p. 22)

$$\frac{d_{(\mathbf{v})}\rho}{dt} + \rho \operatorname{div} \mathbf{v} = 0 \tag{1-6}$$

This is known as the **overall equation of continuity**. It expresses the constraint of conservation of mass at every point within a phase.

In view of the overall equation of continuity,(1-5) reduces to

$$\int_{\Sigma} \left(\frac{d_{(\mathbf{v}^{(\sigma)})}\rho^{(\sigma)}}{dt} + \rho^{(\sigma)} \operatorname{div}_{(\sigma)}\mathbf{v}^{(\sigma)} + [\, \rho(\, \mathbf{v} \cdot \boldsymbol{\xi} - \mathrm{v}\{\xi\}\,)] \right) dA = 0 \tag{1-7}$$

Since this must be true for every multicomponent dividing surface and for every portion of a multicomponent dividing surface, we conclude (see Exercises 1.3.5-1 and 1.3.5-2)

$$\frac{d_{(\mathbf{v}^{(\sigma)})}\rho^{(\sigma)}}{dt} + \rho^{(\sigma)} \operatorname{div}_{(\sigma)}\mathbf{v}^{(\sigma)} + [\, \rho(\, \mathbf{v} \cdot \boldsymbol{\xi} - \mathrm{v}\{\xi\}\,)] = 0 \tag{1-8}$$

This is the **overall jump mass balance** for a multicomponent dividing surface. Not surprisingly, it has the same form as the jump mass balance for a single-component material developed in Sec. 1.3.5.

Equation (1-8) confirms the intuitive feeling expressed in Exercise 5.2.1-4 that no mass can be created by chemical reactions:

$$\sum_{A=1}^{N} r_{(A)} = 0 \tag{1-9}$$

$$\sum_{A=1}^{N} r^{(\sigma)}_{(A)} = 0 \tag{1-10}$$

Two commonly used forms of the overall jump mass balance are given in Table 5.4.1-1.

Table 5.4.1-1: Forms of overall jump mass balance

$$\frac{d_{(\mathbf{v}^{(\sigma)})}\rho^{(\sigma)}}{dt} + \rho^{(\sigma)} \operatorname{div}_{(\sigma)}\mathbf{v}^{(\sigma)} + [\, \rho(\, \mathbf{v} \cdot \boldsymbol{\xi} - \mathrm{v}\{\xi\}\,)] = 0 \tag{A}$$

$$\frac{d_{(\mathbf{v}^{\star(\sigma)})}c^{(\sigma)}}{dt} + c^{(\sigma)} \operatorname{div}_{(\sigma)}\mathbf{v}^{\star(\sigma)} + [\, c(\, \mathbf{v}^{\star} \cdot \boldsymbol{\xi} - \mathrm{v}\{\xi\}\,)] = \sum_{A=1}^{N} \frac{r^{(\sigma)}_{(A)}}{M_{(A)}} \tag{B}$$

Reference

Slattery, J. C., "Momentum, Energy, and Mass Transfer in Continua," McGraw-Hill, New York (1972); second edition, Robert E. Krieger, Malabar, FL 32950 (1981).

Exercises

5.4.1-1 *alternative form of transport theorem for multicomponent body containing dividing surface* Assuming that mass is conserved, prove that

$$\frac{d}{dt}\left(\int_{R_{(v)}} \rho \hat{\psi}\, dV + \int_{\Sigma} \rho^{(\sigma)} \hat{\psi}^{(\sigma)}\, dA \right) = \int_{R_{(v)}} \rho \frac{d_{(v)}\hat{\psi}}{dt}\, dV$$

$$+ \int_{\Sigma} \left\{ \rho^{(\sigma)} \frac{d_{(v^{(\sigma)})}\hat{\psi}^{(\sigma)}}{dt} + [\, \rho(\hat{\psi} - \hat{\psi}^{(\sigma)})(\mathbf{v} \cdot \boldsymbol{\xi} - v_{(\xi)}^{\{g\}})] \right\} dA$$

5.4.1-2 Derive (B) of Table 5.4.1-1 starting from the jump mass balance for an individual species.

5.4.1-3 *overall mass balance at multicomponent common line* Use the transport theorem of Exercise 5.3.1-3 to determine that at the common line

$$(\rho^{(\sigma)}[\, \mathbf{v}^{(\sigma)} \cdot \mathbf{v} - u_{(v)}^{\{cl\}} \,]) = 0$$

This will be known as the **overall mass balance at a multicomponent common line.**

As in Sec. 1.3.8, I have not accounted for mass that might be intrinsically associated with the common line. For related references, see the comments concluding Sec. 2.1.9.

5.4.1-4 *alternative form for overall mass balance at multicomponent common line* Starting with the mass balance for an individual species at a multicomponent common line (Exercise 5.2.1-7), determine that an alternative form for the overall mass balance at the common line (Exercise 5.4.1-3) is

$$\left(c^{(\sigma)} \left[\mathbf{v}^{\star(\sigma)} \cdot \mathbf{v} - u_{(\mathbf{v})}^{(cl)} \right] \right) = 0$$

5.4.1-5 *alternative form of transport theorem for multicomponent body containing intersecting dividing surfaces* Assuming that mass is conserved, prove that

$$\frac{d}{dt} \left(\int_{R_{(v)}} \rho \hat{\psi} \, dV + \int_{\Sigma} \rho^{(\sigma)} \hat{\psi}^{(\sigma)} \, dA \right) = \int_{R_{(v)}} \rho \frac{d_{(v)} \hat{\psi}}{dt} \, dV$$

$$+ \int_{\Sigma} \left\{ \rho^{(\sigma)} \frac{d_{(v(\sigma))} \hat{\psi}^{(\sigma)}}{dt} + \left[\rho \left(\hat{\psi} - \hat{\psi}^{(\sigma)} \right) \left(\mathbf{v} \cdot \boldsymbol{\xi} - v_{(\xi)}^{(\xi)} \right) \right] \right\} dA$$

$$+ \int_{C^{(cl)}} \left(\rho^{(\sigma)} \hat{\psi}^{(\sigma)} \left[\mathbf{v}^{(\sigma)} \cdot \mathbf{v} - u_{(\mathbf{v})}^{(cl)} \right] \right) ds$$

5.5 Force

5.5.1 Euler's first and second laws In Secs. 2.1.1 and 2.1.2, we described the properties of forces by a set of six axioms. We will now recognize that these six axioms apply to multicomponent bodies as defined in Sec. 5.3.1. Note that, since a single-component body is a special case of a multicomponent body, there is no contradiction with either the spirit or the letter of our previous discussions.

With this change of viewpoint, we can repeat the reasoning of Sec. 2.1.2 to arrive at **Euler's first law**:

In an inertial frame of reference, the time rate of change of the momentum of a multicomponent body is equal to the applied force.

and **Euler's second law**:

In an inertial frame of reference, the time rate of change of the moment of momentum of a multicomponent body is equal to the applied torque.

5.5.2 Jump momentum balance Our discussion of stress and body forces in Secs. 2.1.2 and 2.1.3 continues to apply, but we must now recognize that each species may be subject to different external forces. Consider for example an aqueous solution of acetic acid in an electric field. The acetic acid will be partially dissociated, which means that we must consider four separate species: the hydrogen ion, the acetate ion, acetic acid, and water. The force of the electric field upon the hydrogen ions will be equal in magnitude and opposite in direction to the force that it exerts upon the acetate ions. The electric field exerts no force directly upon either the water or the acetic acid. Yet all four species are under the influence of gravity.

With this thought in mind, our statement of Euler's first law may be written as

$$\frac{d}{dt}\left(\int_{R_{(v)}} \rho \mathbf{v} \, dV + \int_{\Sigma} \rho^{(\sigma)} \mathbf{v}^{(\sigma)} dA \right) = \int_{S_{(v)}} \mathbf{t} \, dA + \int_{C_{(v)}} \mathbf{t}^{(\sigma)} \, ds$$
$$+ \int_{R_{(v)}} \sum_{A=1}^{N} \rho_{(A)} \mathbf{b}_{(A)} \, dV + \int_{\Sigma} \sum_{A=1}^{N} \rho^{(\sigma)}_{(A)} \mathbf{b}^{(\sigma)}_{(A)} \, dA \qquad (2\text{-}1)$$

Here $\mathbf{b}_{(A)}$ denotes the body force per unit mass acting upon species A within a phase; $\mathbf{b}^{(\sigma)}_{(A)}$ is the body force per unit mass acting upon species A in a dividing surface.

If we introduce the stress tensor **T** as we did in Sec. 2.1.4 and the surface stress tensor $\mathbf{T}^{(\sigma)}$ as in Sec. 2.1.5, we may use the transport theorem of Exercise 5.4.1-1 to find

$$\int_{R_{(v)}} \left(\rho \frac{d_{(v)}\mathbf{v}}{dt} - \operatorname{div} \mathbf{T} - \sum_{A=1}^{N} \rho_{(A)} \mathbf{b}_{(A)} \right) dV$$

$$+ \int_{\Sigma} \left(\rho^{(\sigma)} \frac{d_{(v(\sigma))}\mathbf{v}^{(\sigma)}}{dt} - \operatorname{div}_{(\sigma)} \mathbf{T}^{(\sigma)} - \sum_{A=1}^{N} \rho^{(\sigma)}_{(A)} \mathbf{b}^{(\sigma)}_{(A)} \right.$$

$$\left. + \left[\rho(\mathbf{v} - \mathbf{v}^{(\sigma)})(\mathbf{v} \cdot \boldsymbol{\xi} - v^{(\xi)}) - \mathbf{T} \cdot \boldsymbol{\xi} \right] \right) dA = 0 \qquad (2\text{-}2)$$

Equation (2-2) must be true for every body. In particular, for a body consisting of a single phase, the integral over the region $R_{(v)}$ is zero, which implies (Slattery 1981, p. 22)

$$\rho \frac{d_{(v)}\mathbf{v}}{dt} = \operatorname{div} \mathbf{T} + \sum_{A=1}^{N} \rho_{(A)} \mathbf{b}_{(A)} \qquad (2\text{-}3)$$

This is **Cauchy's first law** for a multicomponent mixture (Slattery 1981, p. 460). It expresses the constraint of Euler's first law at every point within a phase.

In view of Cauchy's first law, (2-2) reduces to

$$\int_{\Sigma} \left(\rho^{(\sigma)} \frac{d_{(v(\sigma))}\mathbf{v}^{(\sigma)}}{dt} - \operatorname{div}_{(\sigma)} \mathbf{T}^{(\sigma)} - \sum_{A=1}^{N} \rho^{(\sigma)}_{(A)} \mathbf{b}^{(\sigma)}_{(A)} \right.$$

$$\left. + \left[\rho(\mathbf{v} - \mathbf{v}^{(\sigma)})(\mathbf{v} \cdot \boldsymbol{\xi} - v^{(\xi)}) - \mathbf{T} \cdot \boldsymbol{\xi} \right] \right) dA = 0 \qquad (2\text{-}4)$$

This must be true for every multicomponent dividing surface and for every portion of a multicomponent dividing surface. We conclude (see Exercises 1.3.5-1 and 1.3.5-2)

$$\rho^{(\sigma)} \frac{d_{(v(\sigma))}\mathbf{v}^{(\sigma)}}{dt} - \operatorname{div}_{(\sigma)} \mathbf{T}^{(\sigma)} - \sum_{A=1}^{N} \rho^{(\sigma)}_{(A)} \mathbf{b}^{(\sigma)}_{(A)}$$

$$+ \left[\rho(\mathbf{v} - \mathbf{v}^{(\sigma)})(\mathbf{v} \cdot \boldsymbol{\xi} - v^{(\xi)}) - \mathbf{T} \cdot \boldsymbol{\xi} \right] dA = 0 \qquad (2\text{-}5)$$

which we will refer to as the **jump momentum balance** for a multicomponent dividing surface. It expresses the requirement of Euler's first law at every point on a multicomponent dividing surface.

It is usual in continuum fluid mechanics to say that the tangential components of velocity are continuous across a phase interface. With this assumption, (2-5) simplifies to

$$\rho^{(\sigma)} \frac{d_{(\mathbf{v}^{(\sigma)})}\mathbf{v}^{(\sigma)}}{dt} - \text{div}_{(\sigma)}\mathbf{T}^{(\sigma)} - \sum_{A=1}^{N} \rho_{(A)}^{(\sigma)}\, \mathbf{b}_{(A)}^{(\sigma)}$$

$$+ [\, \rho(\, \mathbf{v} \cdot \boldsymbol{\xi} - v_{(\xi)}^{(\sigma)}\,)^2\, \boldsymbol{\xi} - \mathbf{T} \cdot \boldsymbol{\xi}\,]\, dA = 0 \tag{2-6}$$

We will employ this form of the jump momentum balance in the remainder of the text.

If we care to introduce the mass-averaged surface body force per unit mass

$$\mathbf{b}^{(\sigma)} \equiv \sum_{A=1}^{N} \omega_{(A)}^{(\sigma)}\, \mathbf{b}_{(A)}^{(\sigma)} \tag{2-7}$$

we can express (2-6) in the same for as the jump momentum balance for a single-component material:

$$\rho^{(\sigma)} \frac{d_{(\mathbf{v}^{(\sigma)})}\mathbf{v}^{(\sigma)}}{dt} - \text{div}_{(\sigma)}\mathbf{T}^{(\sigma)} - \rho^{(\sigma)}\mathbf{b}^{(\sigma)}$$

$$+ [\, \rho(\, \mathbf{v} \cdot \boldsymbol{\xi} - v_{(\xi)}^{(\sigma)}\,)^2\, \boldsymbol{\xi} - \mathbf{T} \cdot \boldsymbol{\xi}\,]\, dA = 0 \tag{2-8}$$

Reference

Slattery, J. C., "Momentum, Energy, and Mass Transfer in Continua," McGraw-Hill, New York (1972); second edition, Robert E. Krieger, Malabar, FL 32950 (1981).

Exercise

5.5.2-1 *momentum balance at multicomponent common line* Determine that the momentum balance at a multicomponent common line takes the same form as for a single-component material (see Sec. 2.1.9 and Exercise 5.3.1-3).

5.5.3 $\mathbf{T}^{(\sigma)}$ is symmetric, tangential tensor As noted in the last section, the treatment of stress and body forces in Secs. 2.1.2 and 2.1.3 is readily extended to multicomponent bodies. The fact that each of the species may be subject to different external forces is only a minor complication. Keeping this in mind, we may express Euler's second law more formally as

$$\frac{d}{dt}\left(\int_{R_{(v)}} \rho \mathbf{z} \wedge \mathbf{v}\, dV + \int_{\Sigma} \rho^{(\sigma)} \mathbf{z} \wedge \mathbf{v}^{(\sigma)}\, dA\right) = \int_{S_{(v)}} \mathbf{z} \wedge \mathbf{t}\, dA$$

$$+ \int_{C_{(v)}} \mathbf{z} \wedge \mathbf{t}^{(\sigma)}\, ds + \int_{R_{(v)}} \sum_{A=1}^{N} \rho_{(A)}\, \mathbf{z} \wedge \mathbf{b}_{(A)}\, dV$$

$$+ \int_{\Sigma} \sum_{A=1}^{N} \rho_{(A)}^{(\sigma)}\, \mathbf{z} \wedge \mathbf{b}_{(A)}^{(\sigma)}\, dA \tag{3-1}$$

Starting with this statement and taking precisely the same approach outlined in Sec. 2.1.7 for a single-component body, we reach two conclusions. Euler's second law implies **Cauchy's second law**, symmetry of the stress tensor, at each point within a phase of a multicomponent mixture (Slattery 1981, p. 461):

$$\mathbf{T} = \mathbf{T}^T \tag{3-2}$$

More important for our work here, we find that the surface stress tensor $\mathbf{T}^{(\sigma)}$ is a **symmetric tangential tensor field**:

$$\mathbf{T}^{(\sigma)} = \mathbf{T}^{(\sigma)T} \tag{3-3}$$

Reference

Slattery, J. C., "Momentum, Energy, and Mass Transfer in Continua," McGraw-Hill, New York (1972); second edition, Robert E. Krieger, Malabar, FL 32950 (1981).

5.6 Energy

5.6.1 Rate of energy transmission Like forces, energy transmission rates are described by a set of properties or axioms. They are not defined.

Corresponding to each body B there is a distinct set of bodies B^e such that the union of B and B^e forms the universe. We refer to B^e as the exterior or the surroundings of the body B.

1. A system of energy transmission rates is a scalar-valued function $\mathcal{Q}(B, C)$ of pairs of bodies.

The value of $\mathcal{Q}(B, C)$ is called the **rate of energy transmission** from body C to body B.

2. For a specified body B, $\mathcal{Q}(C, B^e)$ is an additive function defined over the sub-bodies C of B.

3. Conversely, for a specified body B, $\mathcal{Q}(B, C)$ is an additive function defined over the sub-bodies C of B^e.

The rate of energy transmission to a body should have nothing to do with the motion of the observer or experimentalist relative to the body.

4. Energy transmission rates should be frame indifferent

$$\mathcal{Q}^* = \mathcal{Q} \tag{1-1}$$

5.6.2 Energy balance I think that you will be better able to visualize our next step, if you consider for a moment a particulate model of a real material. The molecules are all in relative motion with respect to the material which they compose. They have kinetic energy associated with them beyond the kinetic energy of the material as a whole. They also possess potential energy as the result of their positions in the various intermolecular force fields. It is these forms of kinetic and potential energy that we associate with the internal energy of the material.

But this does not define internal energy in our continuum model for a real material. Like mass, force, and the rate of energy transmission, internal energy is not defined in the context of continuum mechanics. Instead, we describe its properties. We state as *axioms* those properties that we feel are true for all materials. Those statements about material behavior that we feel are applicable only to a restricted class of materials are termed *constitutive equations*.

We might be tempted to postulate that the time rate of change of the internal energy of a multicomponent body is equal to the rate at which work is done on the body by the system of forces acting upon it plus the rate of energy transmission to it. This appears to be simple to put in quantitative terms for single-phase bodies. But it is an awkward statement for multiphase bodies, when mass transfer is permitted (see Exercise 5.6.2-1).

Instead, we take as an axiom applicable to all materials the **energy balance**:

In an inertial frame of reference, the time rate of change of the internal and kinetic energy of a multicomponent body is equal to the rate at which work is done on the body by the system of contact, external and mutual forces acting upon it plus the rate of energy transmission to the body. (We have assumed that all work on the body is the result of forces acting on the body. See Sec. 2.1.1 footnote a.)

In an inertial frame of reference, the energy balance says

$$\frac{d}{dt}\left[\int_{R_{(\mathbf{v})}} \rho\left(\hat{U} + \frac{1}{2}v^2\right) dV + \int_{\Sigma} \rho^{(\sigma)}\left(\hat{U}^{(\sigma)} + \frac{1}{2}v^{(\sigma)2}\right) dA\right]$$

$$= \int_{S_{(\mathbf{v})}} \mathbf{v} \cdot (\mathbf{T} \cdot \mathbf{n})\, dA + \int_{C_{(\mathbf{v})}} \mathbf{v}^{(\sigma)} \cdot (\mathbf{T}^{(\sigma)} \cdot \boldsymbol{\mu})\, ds$$

$$+ \int_{R_{(\mathbf{v})}} \sum_{A=1}^{N} \rho_{(A)} \mathbf{v}_{(A)} \cdot \mathbf{b}_{(A)}\, dV$$

$$+ \int_{\Sigma} \sum_{A=1}^{N} \rho_{(A)}^{(\sigma)}\, \mathbf{v}_{(A)}^{(\sigma)} \cdot \mathbf{b}_{(A)}^{(\sigma)}\, dA + Q(B, B^e) \tag{2-1}$$

where $\hat{U}$ denotes the internal energy per unit mass within a phase, $\hat{U}^{(\sigma)}$ the internal energy per unit mass associated with a dividing surface, and $Q(B, B^e)$ the rate of energy transmission to the body B from the surroundings.

Exercise

5.6.2-1 *energy balance for single phase*

a) As a lemma of the energy balance, prove the **energy balance for a single-phase body**:

The time rate of change of the internal energy of a body is equal to the rate at which work is done on the body by the system of forces acting upon it plus the rate of energy transmission to the body from the surroundings.

Include in this system of forces the force of inertia.

b) What happens when we attempt to extend this lemma to multiphase bodies, allowing arbitrary mass transfer at all phase interfaces?

5.6.3 Radiant and contact energy transmission

The rate of energy transmission $\mathcal{Q}(\, B, B^e\,)$ to a body may be separated into the rate of radiant energy transmission $\mathcal{Q}_r(\, B, B^e\,)$ and the rate of contact energy transmission $\mathcal{Q}_c(\, B, B^e\,)$:

$$\mathcal{Q}(\, B, B^e\,) = \mathcal{Q}_r(\, B, B^e\,) + \mathcal{Q}_c(\, B, B^e\,) \tag{3-1}$$

The analogy with body forces and contact forces in Sec. 2.1.3 is strong and obvious.

The **rate of radiant energy transmission** is presumed to be related to the masses of the bodies and is described as though it acts directly on each material particle:

$$\mathcal{Q}_r(\, B, B^e\,) = \int_{R_{(v)}} \rho\, Q\, dV + \int_{\Sigma} \rho^{(\sigma)}\, Q^{(\sigma)}\, dA \tag{3-2}$$

Here Q is the rate of radiant energy transmission per unit mass to the material within a phase and $Q^{(\sigma)}$ is the rate of radiant energy transmission per unit mass to the material in the dividing surface.

The **rate of contact energy transmission** on the other hand describes the rate at which energy is transmitted from one body to another through their common surface of contact. It is presumed to be related to this surface, distributed over it, and independent of the masses of the bodies on either side. It is not an absolutely continuous function of area, since this does not allow for the changing nature of the rate of contact energy transmission in the neighborhood of a phase interface. Referring to Figure 2.1.1-1, we must allow for the rate of contact energy transmission through the curve C formed by the intersection of Σ with S. We write

$$\mathcal{Q}_c(\, B, B^e\,) = \int_{S_{(v)}} Q_c(\, \mathbf{z}, S_{(v)}\,)\, dA + \int_{C_{(v)}} Q_c^{(\sigma)}(\, \mathbf{z}, C_{(v)}\,)\, ds \tag{3-3}$$

Here $Q_c(\, \mathbf{z}, S_{(v)}\,)$ is the energy flux, the rate of contact energy transmission per unit area to a body at its bounding surface $S_{(v)}$; it is a function of position $\mathbf{z}$ on $S_{(v)}$. Similarly, $Q_c^{(\sigma)}(\, \mathbf{z}, C_{(v)}\,)$ is the surface energy flux, the rate of contact energy transmission per unit length at the curve $C_{(v)}$; it is a function of position $\mathbf{z}$ on $C_{(v)}$.

It is always assumed in classical continuum mechanics that within a single phase the energy flux is the same at $\mathbf{z}$ on all similarly-oriented surfaces with a common tangent plane at this point. This axiom can be referred to as the **energy flux principle**:

There is a scalar-valued function $Q_c(\,\mathbf{z}, \mathbf{m}\,)$ defined for all unit vectors $\mathbf{m}$ at any point $\mathbf{z}$ within a single phase such that

$$Q_c(\,\mathbf{z}, S_{(v)}\,) = Q_c(\,\mathbf{z}, \mathbf{n}\,) \tag{3-4}$$

Here $\mathbf{n}$ is the unit normal vector to $S_{(v)}$ at $\mathbf{z}$ outwardly directed with respect to the body receiving the energy. We propose an analogous axiom, the **surface energy flux principle**:

There is a scalar-valued function $Q_c^{(\sigma)}(\,\mathbf{z}, \boldsymbol{\mu}\,)$ defined for all unit tangent vectors $\mathbf{v}$ at any point $\mathbf{z}$ on a dividing surface Σ such that

$$Q_c^{(\sigma)}(\,\mathbf{z}, C_{(v)}\,) = Q_c^{(\sigma)}(\,\mathbf{z}, \boldsymbol{\mu}\,) \tag{3-5}$$

Here $\boldsymbol{\mu}$ is the unit tangent vector that is normal to $C_{(v)}$ at $\mathbf{z}$ and outwardly directed with respect to the body receiving the energy.

Returning to Sec. 5.6.2, we can now express the energy balance in terms of the rates of radiant and contact energy transmission.

$$\frac{d}{dt}\left[\int_{R_{(v)}} \rho\left(\hat{U} + \frac{1}{2}v^2\right) dV + \int_{\Sigma} \rho^{(\sigma)}\left(\hat{U}^{(\sigma)} + \frac{1}{2}v^{(\sigma)2}\right) dA\right]$$

$$= \int_{S_{(v)}} \mathbf{v}\cdot(\,\mathbf{T}\cdot\mathbf{n}\,)\, dA + \int_{C_{(v)}} \mathbf{v}^{(\sigma)}\cdot(\,\mathbf{T}^{(\sigma)}\cdot\boldsymbol{\mu}\,)\, ds$$

$$+ \int_{R_{(v)}} \sum_{A=1}^{N} \rho_{(A)}\, \mathbf{v}_{(A)}\cdot\mathbf{b}_{(A)}\, dV$$

$$+ \int_{\Sigma} \sum_{A=1}^{N} \rho_{(A)}^{(\sigma)}\, \mathbf{v}_{(A)}^{(\sigma)}\cdot\mathbf{b}_{(A)}^{(\sigma)}\, dA$$

$$+ \int_{S_{(v)}} Q_c(\mathbf{z}, \mathbf{n})\, dA + \int_{C_{(v)}} Q_c^{(\sigma)}(\mathbf{z}, \boldsymbol{\mu})\, ds$$

$$+ \int_{R_{(v)}} \rho Q \, dV + \int_{\Sigma} \rho^{(\sigma)} Q^{(\sigma)} \, dA \tag{3-6}$$

5.6.4 Jump energy balance It is easily established that the energy flux may be represented in the form of a scalar product (Slattery 1981, p. 293)

$$Q_c(\mathbf{z}, \mathbf{n}) = - \mathbf{q} \cdot \mathbf{n} \tag{4-1}$$

where **q** is known as the **energy flux** vector.

Let us now rearrange the energy balance given in (3-6), applying the transport theorem (Exercise 5.4.1-1) to the terms on the left and Green's transformation (Slattery 1981, p. 661) to the two integrals over $S_{(v)}$. After ordering the terms, we can express the result as

$$\int_{R_{(v)}} \left[\rho \frac{d_{(v)}}{dt} \left(\hat{U} + \frac{1}{2} v^2 \right) - \operatorname{div} (\mathbf{T} \cdot \mathbf{v}) - \sum_{A=1}^{N} \rho_{(A)} \mathbf{v}_{(A)} \cdot \mathbf{b}_{(A)} \right.$$

$$\left. + \operatorname{div} \mathbf{q} - \rho Q \right] dV$$

$$+ \int_{\Sigma} \left\{ \rho^{(\sigma)} \frac{d_{(\mathbf{v}^{(\sigma)})}}{dt} \left(\hat{U}^{(\sigma)} + \frac{1}{2} v^{(\sigma)2} \right) \right.$$

$$- \operatorname{div}_{(\sigma)}(\mathbf{T}^{(\sigma)} \cdot \mathbf{v}^{(\sigma)})$$

$$- \sum_{A=1}^{N} \rho_{(A)}^{(\sigma)} \mathbf{v}_{(A)}^{(\sigma)} \cdot \mathbf{b}_{(A)}^{(\sigma)} - \rho^{(\sigma)} Q^{(\sigma)}$$

$$+ \left[\rho \left(\hat{U} + \frac{1}{2} v^2 - \hat{U}^{(\sigma)} - \frac{1}{2} v^{(\sigma)2} \right) (\mathbf{v} \cdot \boldsymbol{\xi} - v_{(\xi)}^{(\sigma)}) \right.$$

$$\left. \left. - \mathbf{v} \cdot \mathbf{T} \cdot \boldsymbol{\xi} + \mathbf{q} \cdot \boldsymbol{\xi} \right] \right\} dA$$

$$- \int_{C_{(v)}} Q_c^{(\sigma)}(\mathbf{z}, \boldsymbol{\mu}) \, ds$$

$$= 0 \tag{4-2}$$

Equation (4-2) must be true for every body that does not contain intersecting dividing surfaces. In particular, for a body consisting of a single phase, the integral over the region $R_{(v)}$ is zero, which implies (Slattery 1981, p. 22)

$$\rho \frac{d_{(v)}}{dt} \left(\hat{U} + \frac{1}{2} v^2 \right) = \text{div}(\mathbf{T} \cdot \mathbf{v}) + \sum_{A=1}^{N} \rho_{(A)} \mathbf{v}_{(A)} \cdot \mathbf{b}_{(A)}$$

$$- \text{div} \, \mathbf{q} + \rho Q \tag{4-3}$$

This is the **differential energy balance.**[a] It expresses the restrictions of the energy balance at every point within a phase.

In view of (4-3), (4-2) reduces to

$$\int_{\Sigma} \left\{ \rho^{(\sigma)} \frac{d_{(v^{(\sigma)})}}{dt} \left(\hat{U}^{(\sigma)} + \frac{1}{2} v^{(\sigma)2} \right) - \text{div}_{(\sigma)}(\mathbf{T}^{(\sigma)} \cdot \mathbf{v}^{(\sigma)}) \right.$$

$$- \sum_{A=1}^{N} \rho_{(A)}^{(\sigma)} \mathbf{v}_{(A)}^{(\sigma)} \cdot \mathbf{b}_{(A)}^{(\sigma)} - \rho^{(\sigma)} Q^{(\sigma)}$$

$$+ \left[\rho \left(\hat{U} + \frac{1}{2} v^2 - \hat{U}^{(\sigma)} - \frac{1}{2} v^{(\sigma)2} \right) (\mathbf{v} \cdot \boldsymbol{\xi} - v_{(\xi)}^{(\sigma)}) \right.$$

$$\left. \left. - \mathbf{v} \cdot \mathbf{T} \cdot \boldsymbol{\xi} + \mathbf{q} \cdot \boldsymbol{\xi} \right] \right\} dA$$

---

a) In a prior treatment, I began with a similar energy balance axiom appropriate for all bodies, but the kinetic energy per unit mass at each point in the body was described as the sum of the kinetic energies of the individual species (Slattery 1981, Eq. 8.3.5-1). This meant that the resulting differential energy balance (Slattery 1981, Eq. 8.3.5-2) included a small additional term: the sum of the relative kinetic energies of the individual species. A similar additional term arose in the jump energy balance (Slattery 1981, Exercise 8.3.5-1), which also neglected all interfacial effects.

$$- \int_{C_{(v)}} Q_c^{(\sigma)}(\mathbf{z}, \boldsymbol{\mu}) \, ds$$

$$= 0 \tag{4-4}$$

This is the requirement of the energy balance for dividing surfaces. The surface energy flux may also be interpreted as a scalar product (see Exercise 5.6.4-2)

$$Q_c^{(\sigma)}(\mathbf{z}, \boldsymbol{\mu}) = - \mathbf{q}^{(\sigma)} \cdot \boldsymbol{\mu} \tag{4-5}$$

The vector $\mathbf{q}^{(\sigma)}$ is called the **surface energy flux** vector. After applying the surface divergence theorem to the line integral over $C_{(v)}$, we find that (4-4) becomes

$$\int_{\Sigma} \Bigg\{ \rho^{(\sigma)} \frac{d_{(v^{(\sigma)})}}{dt} \left(\hat{U}^{(\sigma)} + \frac{1}{2} v^{(\sigma)2} \right) - \text{div}_{(\sigma)}(\mathbf{T}^{(\sigma)} \cdot \mathbf{v}^{(\sigma)})$$

$$- \sum_{A=1}^{N} \rho_{(A)}^{(\sigma)} \mathbf{v}_{(A)}^{(\sigma)} \cdot \mathbf{b}_{(A)}^{(\sigma)} + \text{div}_{(\sigma)} \mathbf{q}^{(\sigma)} - \rho^{(\sigma)} Q^{(\sigma)}$$

$$+ \Bigg[\rho \left(\hat{U} + \frac{1}{2} v^2 - \hat{U}^{(\sigma)} - \frac{1}{2} v^{(\sigma)2} \right) (\mathbf{v} \cdot \boldsymbol{\xi} - v_{(\xi)}^{(\sigma)})$$

$$- \mathbf{v} \cdot \mathbf{T} \cdot \boldsymbol{\xi} + \mathbf{q} \cdot \boldsymbol{\xi} \Bigg] \Bigg\} dA$$

$$= 0 \tag{4-6}$$

This means that at every point on Σ (see Exercises 1.3.5-1 and 1.3.5-2)

$$\rho^{(\sigma)} \frac{d_{(v^{(\sigma)})}}{dt} \left(\hat{U}^{(\sigma)} + \frac{1}{2} v^{(\sigma)2} \right) = \text{div}_{(\sigma)}(\mathbf{T}^{(\sigma)} \cdot \mathbf{v}^{(\sigma)})$$

$$+ \sum_{A=1}^{N} \rho_{(A)}^{(\sigma)} \mathbf{v}_{(A)}^{(\sigma)} \cdot \mathbf{b}_{(A)}^{(\sigma)} - \text{div}_{(\sigma)} \mathbf{q}^{(\sigma)} + \rho^{(\sigma)} Q^{(\sigma)}$$

$$+ \Bigg[- \rho \left(\hat{U} + \frac{1}{2} v^2 - \hat{U}^{(\sigma)} - \frac{1}{2} v^{(\sigma)2} \right) (\mathbf{v} \cdot \boldsymbol{\xi} - v_{(\xi)}^{(\sigma)})$$

$$+ \mathbf{v} \cdot \mathbf{T} \cdot \boldsymbol{\xi} - \mathbf{q} \cdot \boldsymbol{\xi} \Big] \tag{4-7}$$

which will be known as the **jump energy balance.**[a]

This result can be simplified. Let us take the scalar product of the jump momentum balance of Sec. 5.5.2 with $\mathbf{v}^{(\sigma)}$

$$\rho^{(\sigma)} \frac{d_{(\mathbf{v}^{(\sigma)})}}{dt} \left(\frac{1}{2} v^{(\sigma)2} \right) = \text{div}_{(\sigma)}(\mathbf{T}^{(\sigma)} \cdot \mathbf{v}^{(\sigma)}) - \text{tr}(\mathbf{T}^{(\sigma)} \cdot \nabla_{(\sigma)} \mathbf{v}^{(\sigma)})$$

$$+ \sum_{A=1}^{N} \rho_{(A)}^{(\sigma)} \mathbf{v}_{(A)}^{(\sigma)} \cdot \mathbf{b}_{(A)}^{(\sigma)} + \Big[- \rho \mathbf{v}^{(\sigma)} \cdot (\mathbf{v} - \mathbf{v}^{(\sigma)})(\mathbf{v} \cdot \boldsymbol{\xi} - v_{(\xi)}^{(\sigma)})$$

$$+ \mathbf{v}^{(\sigma)} \cdot \mathbf{T} \cdot \boldsymbol{\xi} \Big] \tag{4-8}$$

and subtract it from (4-7):

$$\rho^{(\sigma)} \frac{d_{(\mathbf{v}^{(\sigma)})} \hat{U}^{(\sigma)}}{dt} = \text{tr}(\mathbf{T}^{(\sigma)} \cdot \nabla_{(\sigma)} \mathbf{v}^{(\sigma)}) + \sum_{A=1}^{N} \mathbf{j}_{(A)}^{(\sigma)} \cdot \mathbf{b}_{(A)}^{(\sigma)}$$

$$- \text{div}_{(\sigma)} \mathbf{q}^{(\sigma)} + \rho^{(\sigma)} Q^{(\sigma)} + \Big[- \rho \Big(\hat{U} - \hat{U}^{(\sigma)} + \frac{1}{2} \left| \mathbf{v} - \mathbf{v}^{(\sigma)} \right|^2 \Big)$$

$$(\mathbf{v} \cdot \boldsymbol{\xi} - v_{(\xi)}^{(\sigma)}) + (\mathbf{v} - \mathbf{v}^{(\sigma)}) \cdot \mathbf{T} \cdot \boldsymbol{\xi} - \mathbf{q} \cdot \boldsymbol{\xi} \Big] \tag{4-9}$$

This is the **reduced form** of the **jump energy balance.**

We usually assume in continuum fluid mechanics that the tangential components of velocity are continuous across a phase interface

$$\text{at } \Sigma : \quad \mathbf{P} \cdot \mathbf{v} = \mathbf{P} \cdot \mathbf{v}^{(\sigma)} \tag{4-10}$$

in which case (4-9) becomes

$$\rho^{(\sigma)} \frac{d_{(\mathbf{v}^{(\sigma)})} \hat{U}^{(\sigma)}}{dt} = \text{tr}(\mathbf{T}^{(\sigma)} \cdot \nabla_{(\sigma)} \mathbf{v}^{(\sigma)}) + \sum_{A=1}^{N} \mathbf{j}_{(A)}^{(\sigma)} \cdot \mathbf{b}_{(A)}^{(\sigma)}$$

$$- \text{div}_{(\sigma)} \mathbf{q}^{(\sigma)} + \rho^{(\sigma)} Q^{(\sigma)} + \Big[- \rho (\hat{U} - \hat{U}^{(\sigma)})(\mathbf{v} \cdot \boldsymbol{\xi} - v_{(\xi)}^{(\sigma)})$$

$$-\frac{1}{2}\rho(\mathbf{v}\cdot\boldsymbol{\xi} - v_{(\xi)}^{\{\mathrm{g}\}})^3 + (\mathbf{v}\cdot\boldsymbol{\xi} - v_{(\xi)}^{\{\mathrm{g}\}})\boldsymbol{\xi}\cdot\mathbf{T}\cdot\boldsymbol{\xi} - \mathbf{q}\cdot\boldsymbol{\xi}\Big] \tag{4-11}$$

This is the expression of the reduced form of the jump energy balance that we will employ in the remainder of the text.

If there is no mass transfer to or from the dividing surface,

$$\text{at } \Sigma : \ \mathbf{v}\cdot\boldsymbol{\xi} = v_{(\xi)}^{\{\mathrm{g}\}} \tag{4-12}$$

and (4-11) further simplifies to

$$\rho^{(\sigma)} \frac{d_{(\mathbf{v}^{(\sigma)})}\hat{U}^{(\sigma)}}{dt} = \mathrm{tr}(\,\mathbf{T}^{(\sigma)}\cdot\nabla_{(\sigma)}\mathbf{v}^{(\sigma)}\,) + \sum_{A=1}^{N} \mathbf{j}_{(A)}^{\{\sigma\}}\cdot\mathbf{b}_{(A)}^{\{\sigma\}}$$

$$-\,\mathrm{div}_{(\sigma)}\mathbf{q}^{(\sigma)} + \rho^{(\sigma)}Q^{(\sigma)} - [\,\mathbf{q}\cdot\boldsymbol{\xi}\,] \tag{4-13}$$

If our concern is primarily with interphase energy transfer rather than with the energy associated with the dividing surface, we will normally neglect all interfacial effects and write the jump energy balance as

$$\Big[-\rho\hat{U}(\,\mathbf{v}\cdot\boldsymbol{\xi} - v_{(\xi)}^{\{\mathrm{g}\}}\,) - \frac{1}{2}\rho(\,\mathbf{v}\cdot\boldsymbol{\xi} - v_{(\xi)}^{\{\mathrm{g}\}}\,)^3$$

$$+\,(\,\mathbf{v}\cdot\boldsymbol{\xi} - v_{(\xi)}^{\{\mathrm{g}\}}\,)\boldsymbol{\xi}\cdot\mathbf{T}\cdot\boldsymbol{\xi} - \mathbf{q}\cdot\boldsymbol{\xi}\Big] = 0 \tag{4-14}$$

There is no experimental evidence to suggest that interfacial effects in the jump energy balance significantly affect interphase energy transfer. On the other hand, it is clear that interfacial effects in the jump momentum balance may affect the velocity distribution within the immediate neighborhood of the interface and in this way indirectly influence the interphase energy transfer. It will also be obvious shortly that, even if the interfacial energy has no practical effect upon interphase energy transfer, it is not zero, since interfacial tension is a derivative of $\hat{U}^{(\sigma)}$ (see Sec. 5.8.6). Normally we will be willing to neglect the effect of mass transfer upon the interchange of kinetic energy and upon the work done at the phase interface with respect to its effect upon the interchange of internal energy at the interface. In the case, (4-14) simplifies to

$$[\,\rho\hat{U}(\,\mathbf{v}\cdot\boldsymbol{\xi} - v_{(\xi)}^{\{\mathrm{g}\}}\,) + \mathbf{q}\cdot\boldsymbol{\xi}\,] = 0 \tag{4-15}$$

If there is no mass transfer, (4-14) further reduces to

$$[\, \mathbf{q} \cdot \boldsymbol{\xi} \,] = 0 \tag{4-16}$$

Reference

Slattery, J. C., "Momentum, Energy, and Mass Transfer in Continua," McGraw-Hill, New York (1972); second edition, Robert E. Krieger, Malabar, FL 32950 (1981).

Exercises

5.6.4-1 *surface energy flux lemma* Consider two neighboring portions of a continuous dividing surface and apply (4-4) to each portion and to their union. Deduce that on their common boundary

$$Q_c^{(\sigma)}(\mathbf{z}, \boldsymbol{\mu}) = -\, Q_c^{(\sigma)}(\mathbf{z}, -\boldsymbol{\mu})$$

In this way, we establish the **surface energy flux lemma**:

The surface energy fluxes to opposite sides of the same curve at a given point are equal in magnitude and opposite in sign.

5.6.4-2 *surface energy flux vector* By a development which parallels that given in Sec. 2.1.6, show that the surface contact energy flux may be expressed as

$$Q_c^{(\sigma)}(\mathbf{z}, \boldsymbol{\mu}) = -\, \mathbf{q}^{(\sigma)} \cdot \boldsymbol{\mu}$$

where $\mathbf{q}^{(\sigma)}$ is the *surface energy flux* vector.

5.6.4-3 *frame indifference of the energy flux vectors* Determine that the energy flux and the surface energy flux are frame indifferent vectors:

$$\mathbf{q}^* = \mathbf{q}$$

$$\mathbf{q}^{(\sigma)*} = \mathbf{q}^{(\sigma)}$$

5.6.4-4 *energy balance at multicomponent common line* Starting with the energy balance in the form of (3-6), use the transport theorem of Exercise 5.4.1-5 to conclude that at the common line

$$\left(\rho^{(\sigma)} \left[\hat{U}^{(\sigma)} + \frac{1}{2} v^{(\sigma)2} \right] [\mathbf{v}^{(\sigma)} \cdot \boldsymbol{\nu} - u^{(cl)}_{(\nu)}] + \mathbf{q}^{(\sigma)} \cdot \boldsymbol{\nu} - \mathbf{v}^{(\sigma)} \cdot \mathbf{T}^{(\sigma)} \cdot \boldsymbol{\nu} \right) = 0$$

This will be known as the **energy balance at a multicomponent common line**.

If you are concerned with the energy balance at a common line on a rigid solid surface, you may wish to consider rewriting this balance equation in a form analogous to that outlined in Sec. 2.1.10 for the momentum balance at a common line.

Note that I have not accounted for energy or any energy flux that might be associated with the common line. The need for these quantities does not appear to have been established in the literature. For related references, see the comments concluding Sec. 2.1.9.

5.7 Entropy

5.7.1 Clausius-Duhem inequality Let us resketch the physical picture for internal energy in the context of a particulate model for a real material. As the result of their relative motion, the molecules possess kinetic energy with respect to the material as a whole. They also have potential energy as the result of their relative positions in the various force fields acting among the molecules. We think of this kinetic and potential energy as the internal energy of the material.

Still working in the context of a particulate model, we can see that the internal energy of a material is not sufficient to specify its state. Consider two samples of the same material, both having the same internal energy. One has been compressed and cooled; its molecules are in close proximity to one another, but they move slowly. The other has been expanded and heated; the molecules are not very close to one another, but they move rapidly. As I have pictured them, these two samples can be distinguished by their division of internal energy between the kinetic energy and potential energy of the molecules.

But there is another way of describing the difference between these two samples. We can imagine that the molecules in the compressed and cooled material appear in more orderly arrays than do those in the expanded and heated material. There is a difference in the degree of disorder.

Entropy is the term that we shall adopt for a measure of the disorder in a material. Like internal energy, it is not defined. We will instead describe its properties with an axiom applicable to all materials and with a series of constitutive equations appropriate to a restricted class of materials.

Some familiar observations suggest what we should say about entropy. Let us begin by thinking about some situations in which the surroundings do relatively little work on a body.

Even on a cold day, the air in my office becomes noticeably warmer because of the sunshine through my window. During the winter, my office is heated by the energy transmission from the radiator. In the summer, it is cooled by the energy transmission to the coils in my air conditioner. Since the volume of air in the room is a constant, we can say that any energy transferred to the room increases the kinetic energy of the air molecules and entropy of the air.

On the basis of these observations, we might be inclined to propose as a fundamental postulate that "the time rate of change of the entropy of a multicomponent body is locally proportional to the rate of energy transmission to the body." Unfortunately, this is not entirely consistent with other experiments.

We have all observed how a child's balloon filled with helium will markedly decrease in diameter overnight. The helium diffuses out through the elastomer membrane faster than oxygen and nitrogen diffuse in. As the result of diffusion, an ordered system in which the helium was carefully segregated from the surrounding air becomes disordered. Diffusion leads to an increase in the entropy of the system consisting of the air in the room and the helium in the balloon.

Perhaps the fundamental postulate proposed above should be corrected to say that "the time rate of change of the entropy of a multicomponent body is locally proportional to the rates of energy transmission and of mass transfer for all species to the body." The relationship between the rate of contact entropy transmission and the rate of contact energy transmission to the body would have to be such as to account for any effects attributable to mass transfer between the multicomponent body and the surroundings.

But even this improved postulate does not explain other observations. Let us consider some experiments in which the energy transfer and mass transfer between the multicomponent body and the surroundings seem less important. Open a paper clip and repeatedly twist the ends with respect to one another until the metal breaks. The metal is warm to the touch. It is easy to confirm that the grease-packed front-wheel bearings on an automobile (with power transmitted to the rear axle) become hot during a highway trip. Since the paper clip and the grease-packed bearings have roughly constant volumes, we can estimate the kinetic energy portions of their entropies have increased as the result of the systems of forces acting upon them. More important, in every situation that I can recall where there is negligible energy and mass transfer with the surroundings, the entropy of a body always increases as the result of work done. It does not matter whether the work is done by the body on the surroundings or by the surroundings on the body.

It appears that we have two choices open to us in trying to summarize our observations.

We might say that "the time rate of change of the entropy of a multicomponent body is equal to the rate of entropy transmission to the body plus a multiple of the absolute value of the rate at which work is done on the body by the surroundings." I can not say whether this would lead to a self-consistent theory, but it is clear that it would be awkward to work in terms of the absolute value of the rate at which work is done.

As the literature has developed, it appears preferable to instead state as an axiom the **Clausius-Duhem inequality**:

The minimum time rate of change of the entropy of a multicomponent body is equal to the rate of entropy transmission to the body.

I realize that this is not a directly useful statement as it stands. In order to make it one, we must be able to describe the rate of entropy transmission to the body in terms of the rates of energy transmission and of mass transfer to the body.

For a lively and rewarding discussion of the Clausius-Duhem inequality (the **second law of thermodynamics** if you prefer), I encourage you to read Truesdell (1969). As will become plain shortly, I have also been influenced here by Gurtin and Vargas (1971).

References

Gurtin, M. E., and A. S. Vargas, *Arch. Ration. Mech. Anal.* **43**, 179 (1971).

Truesdell, C., "Rational Thermodynamics," McGraw-Hill, New York (1969).

5.7.2 Radiant and contact entropy transmission Entropy transmission rates can be described by a set of properties or axioms that are very similar to those describing energy transmission rates (see Sec. 5.6.1).

Corresponding to each body B there is a distinct set of bodies B^e such that the union of B and B^e forms the universe. We refer to B^e as the exterior or the surroundings of the body B.

1. A system of entropy transmission rates is a scalar-valued function $\mathcal{E}(B, C)$ of pairs of bodies.

The value of $\mathcal{E}(B, C)$ is called the rate of entropy transmission from body C to body B.

2. For a specified body B, $\mathcal{E}(C, B^e)$ is an additive function defined over the sub-bodies C of B.

3. Conversely, for a specified body B, $\mathcal{E}(B, C)$ is an additive function defined over the sub-bodies C of B^e.

The rate of entropy transmission to a body should have nothing to do with the motion of the observer or experimentalist relative to the body.

4. Entropy transmission rates should be frame indifferent:

$$\mathcal{E}^* = \mathcal{E} \tag{2-1}$$

By analogy with our discussion of energy transmission, we will separate the rate of entropy transmission $\mathcal{E}(B, B^e)$ to a body into the rate of radiant entropy transmission $\mathcal{E}_r(B, B^e)$ and the rate of contact entropy transmission $\mathcal{E}_c(B, B^e)$:

$$\mathcal{E}(B, B^e) = \mathcal{E}_r(B, B^e) + \mathcal{E}_c(B, B^e) \tag{2-2}$$

The **rate of radiant entropy transmission** is presumed to be related to the masses of the bodies and is described as though it acts directly on each material particle:

$$\mathcal{E}_r(B, B^e) = \int_{R_{(v)}} \rho E \, dV + \int_{\Sigma} \rho^{(\sigma)} E^{(\sigma)} \, dA \tag{2-3}$$

Here E is the rate of radiant entropy transmission per unit mass to the material within a phase and $E^{(\sigma)}$ is the rate of radiant entropy transmission per unit mass to the material in the dividing surface. The experimental observations noted in Sec. 5.7.1 suggest that the rates of radiant energy and entropy transmission have the same sign.

5. The rates of radiant energy and entropy transmission are proportional:

$$E = \frac{Q}{T} \tag{2-4}$$

$$E^{(\sigma)} = \frac{Q^{(\sigma)}}{T^{(\sigma)}} \tag{2-5}$$

The proportionality factors T and $T^{(\sigma)}$ are positive scalar fields known respectively as the **temperature** and the **surface temperature**.

The **rate of contact entropy transmission** on the other hand describes the rate at which entropy is transmitted from one body to another through their common surface of contact. It is presumed to be related to this surface, distributed over it, and independent of the masses of the bodies on either side. It is not an absolutely continuous function of area, since this does not allow for the changing nature of the rate of contact entropy transmission in the neighborhood of a phase interface. Referring to Figure 2.1.1-1, we must allow for the rate of contact entropy transmission through the curve $C_{(v)}$ formed by the intersection of Σ with $S_{(v)}$. We write

$$\mathcal{E}_c (B, B^e) = \int_{S_{(v)}} E_c(\mathbf{z}, S_{(v)}) \, dA + \int_{C_{(v)}} E_c^{(\sigma)}(\mathbf{z}, C_{(v)}) \, ds \tag{2-6}$$

Here $E_c(\mathbf{z}, S_{(v)})$ is the entropy flux, the rate of contact entropy transmission per unit area to a body at its bounding surface $S_{(v)}$; it is a function of position $\mathbf{z}$ on $S_{(v)}$. Similarly, $E_c^{(\sigma)}(\mathbf{z}, C_{(v)})$ is the surface entropy flux, the rate of contact entropy transmission per unit length at the curve $C_{(v)}$; it is a function of position $\mathbf{z}$ on $C_{(v)}$.

Our treatment of energy transmission suggests that we state two further axioms.

6. **entropy flux principle:** There is a scalar-valued function $E_c(\mathbf{z}, \mathbf{m})$ defined for all unit vectors $\mathbf{m}$ at any point $\mathbf{z}$ within a single phase such that

$$E_c(\mathbf{z}, S_{(v)}) = E_c(\mathbf{z}, \mathbf{n}) \tag{2-7}$$

Here $\mathbf{n}$ is the unit normal vector to $S_{(v)}$ at $\mathbf{z}$ outwardly directed with respect to the body receiving the entropy.

7. **surface entropy flux principle:** There is a scalar-valued function $E_c^{(\sigma)}(\mathbf{z}, \mathbf{v})$ defined for all unit tangent vectors $\mathbf{v}$ at any point $\mathbf{z}$ on a dividing surface Σ such that

$$E_c^{(\sigma)}(\mathbf{z}, C_{(v)}) = E_c^{(\sigma)}(\mathbf{z}, \boldsymbol{\mu}) \tag{2-8}$$

Here $\boldsymbol{\mu}$ is the unit tangent vector that is normal to $C_{(v)}$ at $\mathbf{z}$ and outwardly directed with respect to the body receiving the entropy.

5.7.3 Jump Clausius-Duhem inequality We can now state the Clausius-Duhem inequality of Sec. 5.7.1 more formally as

$$\text{minimum} \frac{d}{dt}\left(\int_{R_{(v)}} \rho \hat{S} \, dV + \int_{\Sigma} \rho^{(\sigma)} \hat{S}^{(\sigma)} \, dA \right) = \int_{R_{(v)}} \rho \frac{Q}{T} \, dV$$
$$+ \int_{\Sigma} \rho^{(\sigma)} \frac{Q^{(\sigma)}}{T^{(\sigma)}} \, dA + \int_{S_{(v)}} E_c(\mathbf{z}, \mathbf{n}) \, dA$$
$$+ \int_{C_{(v)}} E_c^{(\sigma)}(\mathbf{z}, \boldsymbol{\mu}) \, ds \tag{3-1}$$

or

$$\frac{d}{dt}\left(\int_{R_{(v)}} \rho \hat{S} \, dV + \int_{\Sigma} \rho^{(\sigma)} \hat{S}^{(\sigma)} \, dA \right) \geq \int_{R_{(v)}} \rho \frac{Q}{T} \, dV + \int_{\Sigma} \rho^{(\sigma)} \frac{Q^{(\sigma)}}{T^{(\sigma)}} \, dA$$
$$+ \int_{S_{(v)}} E_c(\mathbf{z}, \mathbf{n}) \, dA + \int_{C_{(v)}} E_c^{(\sigma)}(\mathbf{z}, \boldsymbol{\mu}) \, ds \tag{3-2}$$

Here $\hat{S}$ is the entropy per unit mass within a phase and $\hat{S}^{(\sigma)}$ is the entropy per unit mass associated with a dividing surface.

It can be established that the entropy flux may be represented in the form of a scalar product (Exercise 5.7.3-2)

$$E_c(\mathbf{z}, \mathbf{n}) = -\frac{1}{T} \mathbf{e} \cdot \mathbf{n} \tag{3-3}$$

where $\mathbf{e}$ will be referred to as the **thermal energy flux vector**. The vector $\mathbf{e}/T$ is the entropy flux; the explicit dependence upon temperature is suggested by the relation between the rates of radiant energy and entropy transmission in Sec. 5.7.2. Let us rearrange (3-2), applying the transport theorem (Exercise 5.4.1-1) to the terms on the left and Green's transformation to the integral over $S_{(v)}$. After ordering the terms, we have for a body that does not include intersecting dividing surfaces (see Exercise 5.7.3-9)

$$\text{minimum}\left(\int_{R_{(v)}} \rho \frac{d_{(v)}\hat{S}}{dt}\, dV + \int_{\Sigma} \left\{ \rho^{(\sigma)} \frac{d_{(v(\sigma))}\hat{S}^{(\sigma)}}{dt} + [\rho(\hat{S} - \hat{S}^{(\sigma)})(\mathbf{v}\cdot\boldsymbol{\xi} - v_{(\xi)})] \right\} dA \right)$$

$$= \int_{R_{(v)}} \left\{ -\operatorname{div}\left(\frac{\mathbf{e}}{T}\right) + \rho\frac{Q}{T}\right\} dV + \int_{\Sigma} \left\{ \rho^{(\sigma)} \frac{Q^{(\sigma)}}{T^{(\sigma)}} - \left[\frac{1}{T}\mathbf{e}\cdot\boldsymbol{\xi}\right]\right\} dA$$
$$+ \int_{C_{(v)}} E_c^{(\sigma)}(\mathbf{z}, \boldsymbol{\mu})\, ds \tag{3-4}$$

or

$$\int_{R_{(v)}} \left[\rho \frac{d_{(v)}\hat{S}}{dt} + \operatorname{div}\left(\frac{\mathbf{e}}{T}\right) - \rho\frac{Q}{T}\right] dV$$
$$+ \int_{\Sigma} \left\{ \rho^{(\sigma)} \frac{d_{(v(\sigma))}\hat{S}^{(\sigma)}}{dt} - \rho^{(\sigma)} \frac{Q^{(\sigma)}}{T^{(\sigma)}} + \left[\rho(\hat{S} - \hat{S}^{(\sigma)})(\mathbf{v}\cdot\boldsymbol{\xi} - v_{(\xi)}) + \frac{1}{T}\mathbf{e}\cdot\boldsymbol{\xi}\right]\right\} dA$$
$$- \int_{C_{(v)}} E_c^{(\sigma)}(\mathbf{z}, \boldsymbol{\mu})\, ds \geq 0 \tag{3-5}$$

Inequality (3-5) applies to every body that does not include intersecting dividing surfaces. For a body consisting of a single phase, the integral over the region $R_{(v)}$ must be greater than or equal to zero, which implies (see Exercise 5.7.3-3)

$$\rho \frac{d_{(v)}\hat{S}}{dt} + \operatorname{div}\left(\frac{\mathbf{e}}{T}\right) - \rho\frac{Q}{T} \geq 0 \tag{3-6}$$

This is the **differential Clausius-Duhem inequality**, which expresses the restrictions of the Clausius-Duhem inequality at every point within a phase.

In Exercise 5.7.3-4, we see that, since they apply to every multiphase body, (3-4) and (3-5) imply respectively

$$\text{minimum} \int_{\Sigma} \left\{ \rho^{(\sigma)} \frac{d_{(v(\sigma))}\hat{S}^{(\sigma)}}{dt} + [\, \rho(\, \hat{S} - \hat{S}^{(\sigma)}\,)(\, \mathbf{v} \cdot \boldsymbol{\xi} - v_{\{\xi\}}\,)] \right\} dA$$

$$= \int_{\Sigma} \left\{ \rho^{(\sigma)} \frac{Q^{(\sigma)}}{T^{(\sigma)}} - \left[\frac{1}{T} \mathbf{e} \cdot \boldsymbol{\xi} \right] \right\} dA + \int_{C_{(v)}} E_c^{(\sigma)}(\, \mathbf{z}, \boldsymbol{\mu}\,)\, ds \tag{3-7}$$

and

$$\int_{\Sigma} \left\{ \rho^{(\sigma)} \frac{d_{(v(\sigma))}\hat{S}^{(\sigma)}}{dt} - \rho^{(\sigma)} \frac{Q^{(\sigma)}}{T^{(\sigma)}} + \left[\rho(\, \hat{S} - \hat{S}^{(\sigma)}\,)(\, \mathbf{v} \cdot \boldsymbol{\xi} - v_{\{\xi\}}\,) \right.\right.$$

$$\left.\left. + \frac{1}{T} \mathbf{e} \cdot \boldsymbol{\xi} \right] \right\} dA - \int_{C_{(v)}} E_c^{(\sigma)}(\, \mathbf{z}, \boldsymbol{\mu}\,)\, ds \geq 0 \tag{3-8}$$

Equation (3-7) and inequality (3-8) state the requirement of the Clausius-Duhem inequality for the dividing surface as a whole.

The surface entropy flux may also be interpreted as a scalar product (see Exercise 5.7.3-6):

$$E_c^{(\sigma)}(\, \mathbf{z}, \boldsymbol{\mu}\,) = -\frac{1}{T^{(\sigma)}} \mathbf{e}^{(\sigma)} \cdot \boldsymbol{\mu} \tag{3-9}$$

We denote by $\mathbf{e}^{(\sigma)}$ the **surface thermal energy flux** vector. If we assume Σ does not include any intersecting dividing surfaces, we can apply the surface divergence theorem to the line integral over $C_{(v)}$ in (3-8) and write

$$\int_{\Sigma} \left\{ \rho^{(\sigma)} \frac{d_{(v(\sigma))}\hat{S}^{(\sigma)}}{dt} + \text{div}_{(\sigma)}\left(\frac{\mathbf{e}^{(\sigma)}}{T^{(\sigma)}} \right) - \rho^{(\sigma)} \frac{Q^{(\sigma)}}{T^{(\sigma)}} \right.$$

$$\left. + \left[\rho(\, \hat{S} - \hat{S}^{(\sigma)}\,)(\, \mathbf{v} \cdot \boldsymbol{\xi} - v_{\{\xi\}}\,) + \frac{1}{T} \mathbf{e} \cdot \boldsymbol{\xi} \right] \right\} dA \geq 0 \tag{3-10}$$

Using the argument developed in Exercise 5.7.3-7, we discover that (3-10) also implies that at each point on the dividing surface

$$\rho^{(\sigma)} \frac{d_{(v^{(\sigma)})}\hat{S}^{(\sigma)}}{dt} + \text{div}_{(\sigma)}\left(\frac{\mathbf{e}^{(\sigma)}}{T^{(\sigma)}}\right) - \rho^{(\sigma)}\frac{Q^{(\sigma)}}{T^{(\sigma)}}$$

$$+ \left[\rho(\hat{S} - \hat{S}^{(\sigma)})(\mathbf{v}\cdot\boldsymbol{\xi} - v_{(\xi)}^{\{g\}}) + \frac{1}{T}\mathbf{e}\cdot\boldsymbol{\xi}\right] \geq 0 \qquad (3\text{-}11)$$

We will refer to this as the **jump Clausius-Duhem inequality**.

References

Slattery, J. C., "Momentum, Energy, and Mass Transfer in Continua," McGraw-Hill, New York (1972); second edition, Robert E. Krieger, Malabar, FL 32950 (1981).

Truesdell, C., "A First Course in Rational Continuum Mechanics," Department of Mechanics, The Johns Hopkins University, Baltimore, Maryland 21218 (1972).

Exercises

5.7.3-1 *entropy flux lemma* Consider two neighboring portions of a continuous body. Apply the Clausius-Duhem inequality in the form of (3-1) to each portion and to their union. Deduce that on their common boundary

$$E_c(\mathbf{z}, \mathbf{n}) = -E_c(\mathbf{z}, -\mathbf{n})$$

In this way, we establish the **entropy flux lemma**:

The entropy fluxes to opposite sides of the same surface at a given point within a phase are equal in magnitude and opposite in sign.

5.7.3-2 *thermal energy flux vector* Starting with the Clausius-Duhem inequality in the form of (3-1), construct a proof for (3-3). Use as a model for your development the proof that the stress vector can be expressed in terms of a stress tensor (Slattery 1981, p. 39; Truesdell 1972, Sec. III.3).

5.7.3-3 *proof of (3-6)* Inequality (3-5) implies that for a body consisting of a single phase

$$\int_{R_{(v)}} \left[\rho \frac{d_{(v)}\hat{S}}{dt} + \operatorname{div}\left(\frac{\mathbf{e}}{T}\right) - \rho \frac{Q}{T} \right] dV \geq 0$$

i) Construct a sequence of portions of this body with monotonically decreasing volumes, all the members of which include the same arbitrary point.

ii) Normalize the sequence by dividing each member by its volume.

iii) Take the limit of this normalized sequence to conclude that (3-6) holds at any arbitrary point within a phase.

5.7.3-4 *proof of (3-7) and (3-8)* Equation (3-4) and inequality (3-5) apply to every multiphase body that does not include intersecting dividing surfaces.

i) Construct a sequence of portions of a multiphase body with monotonically decreasing volumes, all the members of which include the same portion Σ of dividing surface.

ii) Take the limit of this sequence to conclude that for any arbitrary portion Σ of dividing surface (3-7) and (3-8) hold.

5.7.3-5 *surface entropy flux lemma* Consider two neighboring portions of a continuous dividing surface and apply (3-7) to each portion and to their union. Deduce that on their common boundary

$$E_c^{(\sigma)}(\mathbf{z}, \boldsymbol{\mu}) = -E_c^{(\sigma)}(\mathbf{z}, -\boldsymbol{\mu})$$

In this way, we establish the **surface entropy flux lemma**:

The surface entropy fluxes to opposite sides of the same curve at a given point are equal in magnitude and opposite in sign.

5.7.3-6 *surface thermal energy flux vector* Starting with the Clausius Duhem inequality in the form of (3-7), construct a proof for (3-9). Use as a model for your development the proof that the surface stress vector can be expressed in terms of a surface stress tensor (Sec. 2.1.5).

5.7.3-7 *proof of (3-11)* Inequality (3-10) is valid for every portion of a dividing surface.

i) Construct a sequence of portions of this dividing surface with monotonically decreasing areas, all the members of which include the same arbitrary point.

ii) Normalize the sequence by dividing each member by its surface area.

iii) Take the limit of this normalized sequence to conclude that (3-11) holds at any arbitrary point on the surface.

5.7.3-8 *frame indifference of the thermal energy flux vectors*
Determine that the thermal energy flux and the surface thermal energy flux are frame indifferent vectors:

$$\mathbf{e}^* = \mathbf{e}$$

$$\mathbf{e}^{(\sigma)*} = \mathbf{e}^{(\sigma)}$$

5.7.3-9 *Clausius-Duhem inequality at multicomponent common line*

i) Use the transport theorem of Exercise 5.4.1-5 as well as (3-3) and (3-9) to write the Clausius-Duhem inequality (3-1) as

$$\text{minimum} \left(\int_{R_{(v)}} \rho \frac{d_{(v)}\hat{S}}{dt} dV + \int_{\Sigma} \left\{ \rho^{(\sigma)} \frac{d_{(v^{(\sigma)})}\hat{S}^{(\sigma)}}{dt} \right.\right.$$

$$\left. + \left[\rho (\hat{S} - \hat{S}^{(\sigma)})(\mathbf{v} \cdot \boldsymbol{\xi} - v_{\{\xi\}}) \right] \right\} dA$$

$$\left. + \int_{C^{(cl)}} \left(\rho^{(\sigma)}\hat{S}^{(\sigma)}[\mathbf{v}^{(\sigma)} \cdot \mathbf{v} - u_{\{\mathbf{v}\}}^{(cl)}] \right) ds \right)$$

$$= \int_{R_{(v)}} \left[- \text{div}\left(\frac{\mathbf{e}}{T} \right) + \rho \frac{Q}{T} \right] dV$$

$$+ \int_{\Sigma} \left\{ - \text{div}_{(\sigma)}\left(\frac{\mathbf{e}^{(\sigma)}}{T^{(\sigma)}} \right) + \rho^{(\sigma)} \frac{Q^{(\sigma)}}{T^{(\sigma)}} - \left[\frac{\mathbf{e}}{T} \cdot \boldsymbol{\xi} \right] \right\} dA$$

$$- \int_{C^{(cl)}} \left(\frac{1}{T^{(\sigma)}} \mathbf{e}^{(\sigma)} \cdot \mathbf{v} \right) ds$$

In a similar fashion, rearrange (3-2) in the form

$$\int_{R_{(v)}} \left[\rho \frac{d_{(v)}\hat{S}}{dt} + \operatorname{div}\left(\frac{\mathbf{e}}{T}\right) - \rho \frac{Q}{T} \right] dV$$

$$+ \int_{\Sigma} \left\{ \rho^{(\sigma)} \frac{d_{(\mathbf{v}^{(\sigma)})}\hat{S}^{(\sigma)}}{dt} + \operatorname{div}_{(\sigma)}\left(\frac{\mathbf{e}^{(\sigma)}}{T^{(\sigma)}}\right) - \rho^{(\sigma)} \frac{Q^{(\sigma)}}{T^{(\sigma)}} \right.$$

$$\left. + \left[\rho(\hat{S} - \hat{S}^{(\sigma)})(\mathbf{v} \cdot \boldsymbol{\xi} - v_{(\xi)}^{\{g\}}) + \frac{1}{T} \mathbf{e} \cdot \boldsymbol{\xi} \right] \right\} dA$$

$$+ \int_{C^{(cl)}} \left(\rho^{(\sigma)}\hat{S}^{(\sigma)}[\mathbf{v}^{(\sigma)} \cdot \boldsymbol{\nu} - u_{(\nu)}^{\{cl\}}] \right.$$

$$\left. + \frac{1}{T^{(\sigma)}} \mathbf{e}^{(\sigma)} \cdot \boldsymbol{\nu} \right) ds \geq 0$$

ii) Following the suggestion of Exercise 5.7.3-4, construct a sequence of portions of a multiphase body with monotonically decreasing volumes, all the members of which include the same portion Σ of the dividing surface. Take the limit of this sequence to conclude that for any arbitrary portion Σ of dividing surface

$$\text{minimum} \left(\int_{\Sigma} \left\{ \rho^{(\sigma)} \frac{d_{(\mathbf{v}^{(\sigma)})}\hat{S}^{(\sigma)}}{dt} + [\rho(\hat{S} - \hat{S}^{(\sigma)})(\mathbf{v} \cdot \boldsymbol{\xi} - v_{(\xi)}^{\{g\}})] \right\} dA \right.$$

$$\left. + \int_{C^{(cl)}} (\rho^{(\sigma)}\hat{S}^{(\sigma)}[\mathbf{v}^{(\sigma)} \cdot \boldsymbol{\nu} - u_{(\nu)}^{\{cl\}}]) \, ds \right)$$

$$= \int_{\Sigma} \left\{ - \operatorname{div}_{(\sigma)}\left(\frac{\mathbf{e}^{(\sigma)}}{T^{(\sigma)}}\right) + \rho^{(\sigma)} \frac{Q^{(\sigma)}}{T^{(\sigma)}} - \left[\frac{\mathbf{e}}{T} \cdot \boldsymbol{\xi} \right] \right\} dA$$

$$- \int_{C^{(cl)}} \left(\frac{1}{T^{(\sigma)}} \mathbf{e}^{(\sigma)} \cdot \boldsymbol{\nu} \right) ds$$

and

$$\int_{\Sigma} \left\{ \rho^{(\sigma)} \frac{d_{(\mathbf{v}^{(\sigma)})}\hat{S}^{(\sigma)}}{dt} + \operatorname{div}_{(\sigma)}\left(\frac{\mathbf{e}^{(\sigma)}}{T^{(\sigma)}}\right) - \rho^{(\sigma)} \frac{Q^{(\sigma)}}{T^{(\sigma)}} \right.$$

$$+ \left[\rho(\hat{S} - \hat{S}^{(\sigma)})(\mathbf{v} \cdot \boldsymbol{\xi} - v_{(\xi)}^{\{\xi\}}) + \frac{1}{T} \mathbf{e} \cdot \boldsymbol{\xi} \right] \Bigg\} dA$$

$$+ \int_{C^{(cl)}} \left(\rho^{(\sigma)} \hat{S}^{(\sigma)} [\mathbf{v}^{(\sigma)} \cdot \boldsymbol{\nu} - u_{(\nu)}^{\{cl\}}] + \frac{1}{T^{(\sigma)}} \mathbf{e}^{(\sigma)} \cdot \boldsymbol{\nu} \right) ds \geq 0$$

iii) Construct a sequence of portions of the multicomponent dividing surface with monotonically decreasing areas, all the members of which include the same portion $C^{(cl)}$ of the common line. Take the limit of this sequence to conclude that for any arbitrary portion $C^{(cl)}$ of the common line

$$\text{minimum} \left\{ \int_{C^{(cl)}} (\rho^{(\sigma)} \hat{S}^{(\sigma)} [\mathbf{v}^{(\sigma)} \cdot \boldsymbol{\nu} - u_{(\nu)}^{\{cl\}}]) \, ds \right\}$$

$$= - \int_{C^{(cl)}} \left(\frac{1}{T^{(\sigma)}} \mathbf{e}^{(\sigma)} \cdot \boldsymbol{\nu} \right) ds$$

and

$$\int_{C^{(cl)}} \left(\rho^{(\sigma)} \hat{S}^{(\sigma)} [\mathbf{v}^{(\sigma)} \cdot \boldsymbol{\nu} - u_{(\nu)}^{\{cl\}}] + \frac{1}{T^{(\sigma)}} \mathbf{e}^{(\sigma)} \cdot \boldsymbol{\nu} \right) ds \geq 0$$

iv) Following the suggestion of Exercise 5.7.3-7, construct a sequence of portions of this common line with monotonically decreasing lengths, all the members of which include the same arbitrary point. Normalize the sequence by dividing each member by its length. Take the limit of this normalized sequence to conclude that at any arbitrary point on the common line

$$\left(\rho^{(\sigma)} \hat{S}^{(\sigma)} [\mathbf{v}^{(\sigma)} \cdot \boldsymbol{\nu} - u_{(\nu)}^{\{cl\}}] + \frac{1}{T^{(\sigma)}} \mathbf{e}^{(\sigma)} \cdot \boldsymbol{\nu} \right) \geq 0$$

This can be referred to as the **Clausius-Duhem inequality at a multicomponent common line**.

If you are concerned with the Clausius-Duhem inequality at a common line on a rigid solid surface, you may wish to consider rewriting this inequality in a manner analogous to that outlined in Secs. 2.1.10 for the momentum balance at a common line.

Note that I have not accounted for entropy or for an entropy flux that could be associated with the common line. The need for these quantities does not appear to have been established in the literature. For related references, see the comments concluding Sec. 2.1.9.

5.8 Behavior as restricted by Clausius-Duhem inequalities

5.8.1 Behavior of multicomponent materials Up to this point in this chapter, we have been concerned with the form and implications of axioms stated for all materials. But materials and phase interfaces do not all behave in the same manner. The response of phase interfaces may depend upon their temperature and composition. We have all had the experience of adding soap to water to find that foam is formed upon agitation. The amount of foam formed depends both upon the amount of agitation and upon the amount of soap added. It also depends upon the temperature. A standard technique for breaking a foam is to elevate the temperature.

The primary idea to be exploited in this section is that any material or any phase interface should be capable of undergoing all processes that are consistent with our fundamental axioms. In particular, we use the differential Clausius-Duhem inequality and the jump Clausius-Duhem inequality to restrict the form of descriptions for material behavior. (Contrast this philosophy with one in which these inequalities are used to define the class of processes consistent with a given set of statements about material behavior.)

In the sections that follow, I begin by discussing bulk behavior, the behavior of the material not in the phase interface. I have two reasons for doing this. First, the context in which bulk behavior is introduced here is different from that which I have previously recommended (Slattery 1981, p. 283). Second, its more familiar ideas can serve as a model for the development of interfacial behavior.

Reference

Slattery, J. C., "Momentum, Energy, and Mass Transfer in Continua," McGraw-Hill, New York (1972); second edition, Robert E. Krieger, Malabar, FL 32950 (1981).

5.8.2 Bulk behavior: implications of Clausius-Duhem inequality Let us begin by investigating the restrictions that the differential Clausius-Duhem inequality (Sec. 5.7.3)

$$\rho T \frac{d_{(v)}\hat{S}}{dt} + T \operatorname{div}\left(\frac{\mathbf{e}}{T}\right) - \rho Q \geq 0 \tag{2-1}$$

places upon the form of descriptions for bulk material behavior. The approach is suggested by Gurtin and Vargas (1971).

The differential energy balance (Sec. 5.6.4)

$$\rho \frac{d_{(v)}}{dt}\left(\hat{U} + \frac{1}{2} v^2 \right) = \operatorname{div}(\mathbf{T} \cdot \mathbf{v}) + \sum_{B=1}^{N} \rho_{(B)} \mathbf{v}_{(B)} \cdot \mathbf{b}_{(B)}$$

$$- \operatorname{div} \mathbf{q} + \rho Q \qquad (2\text{-}2)$$

can be used to eliminate Q in (2-1). But before we do so, let us simplify (2-2). The scalar product of the velocity vector **v** with Cauchy's first law for a multicomponent mixture (Sec. 5.5.2)

$$\rho \mathbf{v} \cdot \frac{d_{(v)}\mathbf{v}}{dt} = \mathbf{v} \cdot \operatorname{div} \mathbf{T} + \sum_{B=1}^{N} \rho_{(B)}\, \mathbf{v} \cdot \mathbf{b}_{(B)}$$

$$\rho \frac{d_{(v)}}{dt}\left(\frac{1}{2} v^2 \right) = \operatorname{div}(\mathbf{T} \cdot \mathbf{v}) - \operatorname{tr}(\mathbf{T} \cdot \mathbf{D}) + \sum_{B=1}^{N} \rho_{(B)} \mathbf{v} \cdot \mathbf{b}_{(B)} \qquad (2\text{-}3)$$

can be subtracted from (2-2) to yield

$$\rho \frac{d_{(v)}\hat{U}}{dt} = \operatorname{tr}(\mathbf{T} \cdot \mathbf{D}) + \sum_{B=1}^{N} \mathbf{j}_{(B)} \cdot \mathbf{b}_{(B)} - \operatorname{div} \mathbf{q} + \rho Q \qquad (2\text{-}4)$$

We can now use this form of the differential energy balance to eliminate Q in the differential Clausius-Duhem inequality (2-1) and write

$$\rho \frac{d_{(v)}\hat{U}}{dt} - \rho T \frac{d_{(v)}\hat{S}}{dt} - \operatorname{tr}(\mathbf{T} \cdot \mathbf{D}) + \operatorname{div}(\mathbf{q} - \mathbf{e}) + \frac{1}{T} \mathbf{e} \cdot \nabla T$$

$$- \sum_{B=1}^{N} \mathbf{j}_{(B)} \cdot \mathbf{b}_{(B)} \leq 0 \qquad (2\text{-}5)$$

We will find it more convenient to work in terms of the Helmholtz free energy per unit mass

$$\hat{A} \equiv \hat{U} - T\hat{S} \qquad (2\text{-}6)$$

in terms of which (2-5) becomes

$$\rho \frac{d_{(v)}\hat{A}}{dt} + \rho \hat{S} \frac{d_{(v)}T}{dt} - \operatorname{tr}(\mathbf{T} \cdot \mathbf{D}) + \operatorname{div}(\mathbf{q} - \mathbf{e}) + \frac{1}{T} \mathbf{e} \cdot \nabla T$$

$$-\sum_{B=1}^{N} \mathbf{j}_{(B)} \cdot \mathbf{b}_{(B)} \leq 0 \tag{2-7}$$

In order to make further progress, we must restrict ourselves to a class of constitutive equations. Let us assume that

$$\hat{A} = \hat{A}(\boldsymbol{\Lambda}), \quad \mathbf{e} = \mathbf{e}(\boldsymbol{\Lambda}), \quad \mathbf{q} = \mathbf{q}(\boldsymbol{\Lambda}), \quad \mathbf{j}_{(A)} = \mathbf{j}_{(A)}(\boldsymbol{\Lambda}),$$

$$r_{(A)} = r_{(A)}(\boldsymbol{\Lambda}), \quad \mathbf{T} = \mathbf{T}(\boldsymbol{\Lambda}, \boldsymbol{\chi}^t) \tag{2-8}$$

where

$$\boldsymbol{\Lambda} = (\hat{V}, T, \omega_{(1)}, \omega_{(2)}, \ldots, \omega_{(N-1)}, \nabla\hat{V}, \nabla T, \nabla\omega_{(1)}, \nabla\omega_{(2)}, \ldots,$$

$$\nabla\omega_{(N-1)}, \mathbf{b}_{(1)}, \mathbf{b}_{(2)}, \ldots, \mathbf{b}_{(N)}) \tag{2-9}$$

is a set of independent variables common to all of these constitutive equations and $\boldsymbol{\chi}^t$ is the history of the motion of the multicomponent material (see Sec. 5.3.1 for viewpoint). The reason for introducing the dependence upon the $\mathbf{b}_{(A)}$ will become clear in Secs. 5.9.2 through 5.9.8; in particular, Sec. 5.9.5 develops the well known dependence of the bulk mass flux vector upon the body forces (Slattery 1981, pages 477 and 525).

Using the chain rule, we can say from $(2\text{-}8_1)$ that[a]

---

a) Let f(**v**) be a scalar function of a vector. The derivative of f with respect to **v** is a vector denoted by $\partial f/\partial \mathbf{v}$ and defined by its scalar product with any arbitrary vector **a**:

$$\frac{\partial f}{\partial \mathbf{v}} \cdot \mathbf{a} \equiv \text{limit } s \to 0: \ \frac{1}{s}[\, f(\mathbf{v} + s\mathbf{a}) - f(\mathbf{v})]$$

In a rectangular cartesian coordinate system, this last takes the form (Slattery 1981, p. 614)

$$\frac{\partial f}{\partial \mathbf{v}} \cdot \mathbf{a} = \frac{\partial f}{\partial v_i} a_i$$

For the particular case $\mathbf{a} = \mathbf{e}_j$ (continued on next page)

$$\frac{d_{(v)}\hat{A}}{dt} = \frac{\partial \hat{A}}{\partial \hat{V}} \frac{d_{(v)}\hat{V}}{dt} + \frac{\partial \hat{A}}{\partial T} \frac{d_{(v)}T}{dt} + \sum_{B=1}^{N-1} \frac{\partial \hat{A}}{\partial \omega_{(B)}} \frac{d_{(v)}\omega_{(B)}}{dt}$$

$$+ \frac{\partial \hat{A}}{\partial \nabla \hat{V}} \cdot \frac{d_{(v)}\nabla \hat{V}}{dt} + \frac{\partial \hat{A}}{\partial \nabla T} \cdot \frac{d_{(v)}\nabla T}{dt} + \sum_{B=1}^{N-1} \frac{\partial \hat{A}}{\partial \nabla \omega_{(B)}} \cdot \frac{d_{(v)}\nabla \omega_{(B)}}{dt}$$

$$+ \sum_{B=1}^{N} \frac{\partial \hat{A}}{\partial \mathbf{b}_{(B)}} \cdot \frac{d_{(v)}\mathbf{b}_{(B)}}{dt} \tag{2-10}$$

This together with the overall equation of continuity (Sec. 5.4.1)

$$\rho \frac{d_{(v)}\hat{V}}{dt} = \text{div } \mathbf{v} \tag{2-11}$$

and the equation of continuity for species B (Sec. 5.2.1)

---

a) (continued from previous page)

$$\frac{\partial f}{\partial \mathbf{v}} \cdot \mathbf{e}_j = \frac{\partial f}{\partial v_j}$$

and we conclude

$$\frac{\partial f}{\partial \mathbf{v}} = \frac{\partial f}{\partial v_i} \mathbf{e}_i$$

More generally,

$$\frac{\partial f}{\partial \mathbf{v}} = \frac{\partial f}{\partial \overline{v}_i} \overline{\mathbf{g}}_i = \frac{\partial f}{\partial \overline{v}_{<i>}} \overline{\mathbf{g}}_{<i>}$$

where $\overline{v}_i$ and $\overline{v}_{<i>}$ are, respectively, the covariant and physical components of $\mathbf{v}$ with respect to some curvilinear coordinate system.

$$\rho \frac{d_{(v)} \omega_{(B)}}{dt} = -\operatorname{div} \mathbf{j}_{(B)} + r_{(B)} \tag{2-12}$$

allow us to rearrange (2-7) in the form

$$\begin{aligned}
&\rho\left[\left(\frac{\partial \hat{A}}{\partial T} + \hat{S}\right) \frac{d_{(v)}T}{dt} + \frac{\partial \hat{A}}{\partial \nabla \hat{V}} \cdot \frac{d_{(v)}\nabla \hat{V}}{dt} + \frac{\partial \hat{A}}{\partial \nabla T} \cdot \frac{d_{(v)}\nabla T}{dt}\right. \\
&\qquad \left. + \sum_{B=1}^{N-1} \frac{\partial \hat{A}}{\partial \nabla \omega_{(B)}} \cdot \frac{d_{(v)} \nabla \omega_{(B)}}{dt} + \sum_{B=1}^{N} \frac{\partial \hat{A}}{\partial \mathbf{b}_{(B)}} \cdot \frac{d_{(v)} \mathbf{b}_{(B)}}{dt}\right] \\
&\quad - \operatorname{tr}\left[\left(\mathbf{T} - \frac{\partial \hat{A}}{\partial \hat{V}} \mathbf{I}\right) \cdot \mathbf{D}\right] + \operatorname{div}\left(\mathbf{q} - \mathbf{e} - \sum_{B=1}^{N-1} \frac{\partial \hat{A}}{\partial \omega_{(B)}} \mathbf{j}_{(B)}\right) \\
&\quad + \sum_{B=1}^{N-1} \mathbf{j}_{(B)} \cdot \nabla\left(\frac{\partial \hat{A}}{\partial \omega_{(B)}}\right) + \sum_{B=1}^{N-1} \frac{\partial \hat{A}}{\partial \omega_{(B)}} r_{(B)} \\
&\quad + \frac{1}{T} \mathbf{e} \cdot \nabla T - \sum_{B=1}^{N} \mathbf{j}_{(B)} \cdot \mathbf{b}_{(B)} \leq 0
\end{aligned} \tag{2-13}$$

It is a simple matter to construct T, $\hat{V}$, $\omega_{(B)}$, ∇T, $\nabla \hat{V}$, $\nabla \omega_{(B)}$, and $\mathbf{b}_{(B)}$ fields such that at any given point within a phase at any specified time

$$\frac{d_{(v)}T}{dt}, \quad \frac{d_{(v)}\nabla T}{dt}, \quad \frac{d_{(v)}\nabla \hat{V}}{dt}, \quad \frac{d_{(v)}\nabla \omega_{(B)}}{dt}, \quad \frac{d_{(v)} \mathbf{b}_{(B)}}{dt}$$

take arbitrary values (Gurtin and Vargas 1971). We conclude that

$$\hat{S} = -\frac{\partial \hat{A}}{\partial T} \tag{2-14}$$

$$\hat{A} = \hat{A}(\hat{V}, T, \omega_{(1)}, \omega_{(2)}, \ldots, \omega_{(N-1)}) \tag{2-15}$$

and

$$- \operatorname{tr}\left[\left(\mathbf{T} - \frac{\partial \hat{A}}{\partial \hat{V}} \mathbf{I}\right) \cdot \mathbf{D}\right] + \operatorname{div}\left(\mathbf{q} - \mathbf{e} - \sum_{B=1}^{N-1} \frac{\partial \hat{A}}{\partial \omega_{(B)}} \mathbf{j}_{(B)}\right)$$

$$+\sum_{B=1}^{N-1} \mathbf{j}_{(B)} \cdot \nabla\left(\frac{\partial \hat{A}}{\partial \omega_{(B)}}\right) + \sum_{B=1}^{N-1} \frac{\partial \hat{A}}{\partial \omega_{(B)}} r_{(B)}$$

$$+\frac{1}{T}\mathbf{e} \cdot \nabla T - \sum_{B=1}^{N} \mathbf{j}_{(B)} \cdot \mathbf{b}_{(B)} \leq 0 \tag{2-16}$$

For simplicity, let us introduce

$$\mathbf{k} \equiv \mathbf{q} - \mathbf{e} - \sum_{B=1}^{N-1} \frac{\partial \hat{A}}{\partial \omega_{(B)}} \mathbf{j}_{(B)} \tag{2-17}$$

and write inequality (2-16) as

$$\operatorname{div} \mathbf{k} + f(\boldsymbol{\Lambda}, \boldsymbol{\chi}^{t}) \leq 0 \tag{2-18}$$

The vector $\mathbf{k}$ is frame indifferent (Exercises 5.2.2-1, 5.6.4-3, and 5.7.3-8)

$$\mathbf{k}^{*} = \mathbf{Q} \cdot \mathbf{k} \tag{2-19}$$

From (2-8), we see that it is a function only of $\boldsymbol{\Lambda}$

$$\mathbf{k} = \mathbf{h}(\boldsymbol{\Lambda}) \tag{2-20}$$

By the principle of frame indifference, this same form of relationship must hold in every frame of reference

$$\mathbf{k}^{*} = \mathbf{h}(\mathbf{Q} \cdot \boldsymbol{\Lambda}) \tag{2-21}$$

where (body forces are postulated to be frame indifferent in Sec. 2.1.1)

$$\mathbf{Q} \cdot \boldsymbol{\Lambda} \equiv (\hat{V}, T, \omega_{(1)}, \omega_{(2)}, \ldots, \omega_{(N-1)}, \mathbf{Q} \cdot \nabla\hat{V}, \mathbf{Q} \cdot \nabla T,$$
$$\mathbf{Q} \cdot \nabla\omega_{(1)}, \mathbf{Q} \cdot \nabla\omega_{(2)}, \ldots, \mathbf{Q} \cdot \nabla\omega_{(N-1)}, \mathbf{Q} \cdot \mathbf{b}_{(1)},$$
$$\mathbf{Q} \cdot \mathbf{b}_{(2)}, \ldots, \mathbf{Q} \cdot \mathbf{b}_{(N)}) \tag{2-22}$$

Equations (2-19) through (2-21) imply $\mathbf{h}(\boldsymbol{\Lambda})$ is an isotropic function:

$$\mathbf{h}(\mathbf{Q} \cdot \boldsymbol{\Lambda}) = \mathbf{Q} \cdot \mathbf{h}(\boldsymbol{\Lambda}) \tag{2-23}$$

Applying the chain rule to (2-20), we have

$$\text{div } \mathbf{k} = \text{tr}\left(\frac{\partial \mathbf{h}}{\partial \nabla \hat{\mathrm{V}}} \cdot \nabla^2 \hat{\mathrm{V}} \right) + \text{tr}\left(\frac{\partial \mathbf{h}}{\partial \nabla \mathrm{T}} \cdot \nabla^2 \mathrm{T} \right)$$

$$+ \sum_{B=1}^{N-1} \text{tr}\left(\frac{\partial \mathbf{h}}{\partial \nabla \omega_{(B)}} \cdot \nabla^2 \omega_{(B)} \right) + \sum_{B=1}^{N} \text{tr}\left(\frac{\partial \mathbf{h}}{\partial \mathbf{b}_{(B)}} \cdot \nabla \mathbf{b}_{(B)} \right)$$

$$+ g(\mathbf{\Lambda}) \tag{2-24}$$

We can construct compatible T, $\hat{V}$, $\omega_{(B)}$, ∇T, $\nabla \hat{V}$, $\nabla \omega_{(B)}$, and $\mathbf{b}_{(B)}$ fields such that at any given point within a phase at any specified time

$$\nabla^2 \hat{V}, \quad \nabla^2 T, \quad \nabla^2 \omega_{(B)}, \quad \nabla \mathbf{b}_{(B)}$$

take arbitrary values (Gurtin and Vargas 1971). In view of (2-18),

$$\begin{aligned} \text{tr}\left(\frac{\partial \mathbf{h}}{\partial \nabla \hat{\mathrm{V}}} \cdot \nabla^2 \hat{\mathrm{V}} \right) &= \text{tr}\left(\frac{\partial \mathbf{h}}{\partial \nabla \mathrm{T}} \cdot \nabla^2 \mathrm{T} \right) \\ &= \text{tr}\left(\frac{\partial \mathbf{h}}{\partial \nabla \omega_{(B)}} \cdot \nabla^2 \omega_{(B)} \right) \\ &= \text{tr}\left(\frac{\partial \mathbf{h}}{\partial \mathbf{b}_{(B)}} \cdot \nabla \mathbf{b}_{(B)} \right) \\ &= 0 \end{aligned} \tag{2-25}$$

which implies that $\partial \mathbf{h}/\partial \mathbf{b}_{(B)}$ as well as the symmetric parts of

$$\frac{\partial \mathbf{h}}{\partial \nabla \hat{V}}, \quad \frac{\partial \mathbf{h}}{\partial \nabla T}, \quad \frac{\partial \mathbf{h}}{\partial \nabla \omega_{(B)}}$$

are all zero. Gurtin (1971, lemma 6.2; see also Gurtin and Vargas 1971, lemma 10.2) has proved that, when the symmetric portions of the derivatives of an isotropic vector function $\mathbf{h}(\mathbf{\Lambda})$ with respect to each of the independent vectors are all zero, the function itself is zero. In this case, we conclude

$$\mathbf{k} = 0 \tag{2-26}$$

or

$$\mathbf{e} = \mathbf{q} - \sum_{B=1}^{N-1} \frac{\partial \hat{A}}{\partial \omega_{(B)}} \mathbf{j}_{(B)} \tag{2-27}$$

This last in turn implies that (2-16) reduces to

$$- \operatorname{tr}\left[\left(\mathbf{T} - \frac{\partial \hat{A}}{\partial \hat{V}}\mathbf{I}\right) \cdot \mathbf{D}\right] + \sum_{B=1}^{N-1} \mathbf{j}_{(B)} \cdot \nabla\left(\frac{\partial \hat{A}}{\partial \omega_{(B)}}\right) + \sum_{B=1}^{N-1} \frac{\partial \hat{A}}{\partial \omega_{(B)}} r_{(B)}$$

$$+ \frac{1}{T}\mathbf{e} \cdot \nabla T - \sum_{B=1}^{N} \mathbf{j}_{(B)} \cdot \mathbf{b}_{(B)} \leq 0 \qquad (2\text{-}28)$$

Equation (2-27) and inequality (2-28) take more familiar forms when expressed in terms of the chemical potential. We will return to this point in Sec. 5.8.4. But before we do so, let us examine the implications of (2-15).

References

Gurtin, M. E., *Arch. Ration. Mech. Anal.* **43**, 198 (1971).

Gurtin, M. E., and A. S. Vargas, *Arch. Ration. Mech. Anal.* **43**, 179 (1971).

Slattery, J. C., "Momentum, Energy, and Mass Transfer in Continua," McGraw-Hill, New York (1972); second edition, Robert E. Krieger, Malabar, FL 32950 (1981).

5.8.3 Bulk behavior: implications of caloric equation of state In the last section, we began with some broad statements about material behavior and concluded that, if the Clausius-Duhem inequality was to be obeyed,

$$\hat{A} = \hat{A}(\hat{V}, T, \omega_{(1)}, \omega_{(2)}, \ldots, \omega_{(N-1)}) \qquad (3\text{-}1)$$

or

$$\check{A} = \check{A}(T, \rho_{(1)}, \rho_{(2)}, \ldots, \rho_{(N)}) \qquad (3\text{-}2)$$

where hereafter $\vee$ over a symbol should be read *per unit volume*. We will refer to these statements as alternative forms of the caloric equation of state. At the same time, we proved

$$\hat{S} = -\left(\frac{\partial \hat{A}}{\partial T}\right)_{\hat{V},\omega_{(B)}} = -\frac{1}{\rho}\left(\frac{\partial \check{A}}{\partial T}\right)_{\rho_{(B)}} \tag{3-3}$$

Let us introduce the thermodynamic pressure

$$P \equiv -\left(\frac{\partial \hat{A}}{\partial \hat{V}}\right)_{T,\omega_{(B)}} \tag{3-4}$$

and the chemical potential

$$\mu_{(B)} \equiv \left(\frac{\partial \check{A}}{\partial \rho_{(B)}}\right)_{T,\rho_{(C)}(C \neq B)} \tag{3-5}$$

The differentials of (3-1) and (3-2) may consequently be expressed as

$$d\hat{A} = -P\, d\hat{V} - \hat{S}\, dT + \sum_{B=1}^{N-1} \left(\frac{\partial \hat{A}}{\partial \omega_{(B)}}\right)_{\hat{V},T,\omega_{(C)}(C \neq B,N)} d\omega_{(B)} \tag{3-6}$$

and

$$d\check{A} = -\check{S}\, dT + \sum_{B=1}^{N} \mu_{(B)}\, d\rho_{(B)} \tag{3-7}$$

Equation (3-7) may be rearranged to read

$$d\left(\frac{\hat{A}}{\hat{V}}\right) = -\frac{\hat{S}}{\hat{V}}\, dT + \sum_{B=1}^{N} \mu_{(B)}\, d\left(\frac{\omega_{(B)}}{\hat{V}}\right) \tag{3-8}$$

or

$$d\hat{A} = -\left(-\frac{\hat{A}}{\hat{V}} + \sum_{B=1}^{N} \mu_{(B)}\rho_{(B)}\right) d\hat{V} - \hat{S}\, dT + \sum_{B=1}^{N-1} \left(\mu_{(B)} - \mu_{(N)}\right) d\omega_{(B)} \tag{3-9}$$

Comparison of coefficients in (3-6) and (3-9) gives

$$\left(\frac{\partial \hat{A}}{\partial \omega_{(B)}} \right)_{\hat{V}, T, \omega_{(C)} (C \neq B, N)} = \mu_{(B)} - \mu_{(N)} \tag{3-10}$$

as well as **Euler's equation**

$$\hat{A} = - P\hat{V} + \sum_{B=1}^{N} \mu_{(B)} \omega_{(B)} \tag{3-11}$$

Equations (3-6) and (3-10) yield the **Gibbs equation**:

$$d\hat{A} = - P \, d\hat{V} - \hat{S} \, dT + \sum_{B=1}^{N-1} (\mu_{(B)} - \mu_{(N)}) \, d\omega_{(B)} \tag{3-12}$$

The **Gibbs-Duhem equation** follows immediately by subtracting (3-12) from the differential of (3-11):

$$\hat{S} \, dT - \hat{V} \, dP + \sum_{B=1}^{N} \omega_{(B)} \, d\mu_{(B)} = 0 \tag{3-13}$$

I would like to emphasize that Euler's equation, the Gibbs equation, and the Gibbs-Duhem equation all apply to dynamic processes, so long as the statements about behavior made in Sec. 5.8.2 are applicable to the materials being considered.

References

Adamson, A. W., "Physical Chemistry of Surfaces," third edition, Wiley, New York (1976).

Defay, R., I. Prigogine, A. Bellemans, and D. H. Everett, "Surface Tension and Adsorption," Wiley, New York (1966).

Scheidegger, A. E., "The Physics of Flow through Porous Media," third edition, U. of Toronto Press (1974).

Slattery, J. C., "Momentum, Energy, and Mass Transfer in Continua," McGraw-Hill, New York (1972); second edition, Robert E. Krieger, Malabar, FL 32950 (1981).

Thomson, W. (Lord Kelvin), *Philos. Mag.* [4] **42**, 448 (1871).

Exercises

5.8.3-1 *specific variables per unit mole* Let $c_{(A)}$ denote moles of species A per unit volume and $x_{(A)}$ the mole fraction of species A. Denote by ~ over a symbol that we are dealing with a quantity per unit mole.

i) Show that entropy per unit mole may be expressed as

$$\tilde{S} = -\left(\frac{\partial \tilde{A}}{\partial T}\right)_{\hat{U}, x_{(B)}} = -\frac{1}{c}\left(\frac{\partial \check{A}}{\partial T}\right)_{c_{(B)}}$$

where

$$c \equiv \frac{1}{\tilde{V}}$$

is the total mole density.

ii) Determine that the thermodynamic pressure is also given by

$$P = -\left(\frac{\partial \tilde{A}}{\partial \tilde{V}}\right)_{T, x_{(B)}}$$

iii) It is common to use another chemical potential defined on a mole basis:

$$\mu_{(A)}^{(m)} \equiv \left(\frac{\partial \check{A}}{\partial c_{(A)}}\right)_{T, c_{(B)}(B \neq A)}$$

Derive the analogs to (3-11) through (3-13) in terms of this molar chemical potential.

5.8.3-2 *partial mass variables* Let $\hat{\Phi}$ be any intensive (per unit mass) variable

$$\hat{\Phi} = \hat{\Phi}(T, P, \omega_{(1)}, \omega_{(2)}, \ldots, \omega_{(N-1)})$$

We will define the partial mass variable $\overline{\overline{\Phi}}_{(A)}$ by requiring (Slattery 1981, p. 289)

$$\overline{\overline{\Phi}}_{(A)} - \overline{\overline{\Phi}}_{(N)} \equiv \left(\frac{\partial \hat{\Phi}}{\partial \omega_{(A)}}\right)_{T, P, \omega_{(B)}(B \neq A, N)}$$

and

$$\hat{\Phi} = \sum_{A=1}^{N} \overline{\overline{\Phi}}_{(A)} \omega_{(A)}$$

Determine that

$$\sum_{A=1}^{N} \left(\frac{\partial \overline{\overline{\Phi}}_{(B)}}{\partial \omega_{(B)}} \right)_{T,P,\omega_{(C)}(C \neq B,N)} \omega_{(A)} = 0$$

5.8.3-3 *partial molal variables* Let $\tilde{\Phi}$ be any intensive (per unit mole) variable

$$\tilde{\Phi} = \tilde{\Phi}(T, P, x_{(1)}, x_{(2)}, \ldots, x_{(N-1)})$$

We will define the partial molal variable $\overline{\overline{\Phi}}_{(A)}^{(m)}$ by requiring (Slattery 1981, p. 289)

$$\overline{\overline{\Phi}}_{(A)}^{(m)} - \overline{\overline{\Phi}}_{(N)}^{(m)} \equiv \left(\frac{\partial \tilde{\Phi}}{\partial x_{(A)}} \right)_{T,P,x_{(B)}(B \neq A,N)}$$

and

$$\tilde{\Phi} = \sum_{A=1}^{N} \overline{\overline{\Phi}}_{(A)}^{(m)} x_{(A)}$$

Determine that

$$\sum_{A=1}^{N} \left(\frac{\partial \overline{\overline{\Phi}}_{(A)}^{(m)}}{\partial x_{(B)}} \right)_{T,P,x_{(C)}(C \neq B,N)} x_{(A)} = 0$$

5.8.3-4 *relation between partial mass and partial molal variables* Prove that

$$\overline{\overline{\Phi}}_{(A)}^{(m)} = M_{(A)} \overline{\overline{\Phi}}_{(A)}$$

5.8.3-5 Determine that

i) $$\overline{\overline{S}}_{(A)} = -\left(\frac{\partial \mu_{(A)}}{\partial T}\right)_{P,\omega_{(B)}}$$

ii) $$\overline{\overline{V}}_{(A)} = \left(\frac{\partial \mu_{(A)}}{\partial P}\right)_{T,\omega_{(B)}}$$

(Hint: See Slattery (1981, p. 287) and the Gibbs-Duhem equation.)

5.8.3-7 *Kelvin's equation* Consider a single-component liquid at equilibrium with its vapor. The pressure $P^{(v)}$ within the vapor phase is known as the vapor pressure.

For many materials, we have readily available to us in the literature the vapor pressure $P_0^{(v)}$ measured under conditions such that the mean curvature H of the vapor-liquid interface is zero. We wish to determine the dependence of vapor pressure upon H.

i) For a spherical liquid drop surrounded by vapor, observe that

$$P^{(\ell)} - P^{(v)} = 2H\gamma$$

ii) Considering a sequence of liquid drops at equilibrium, establish (see Sec. 5.10.3)

$$\frac{d\mu^{(v)}}{dH} = \frac{d\mu^{(\ell)}}{dH}$$

iii) Conclude from the Gibbs-Duhem equation that

$$\hat{V}^{(v)} \frac{dP^{(v)}}{dH} = \hat{V}^{(\ell)} \frac{dP^{(\ell)}}{dH}$$

or

$$\frac{\hat{V}^{(v)} - \hat{V}^{(\ell)}}{\hat{V}^{(\ell)}} dP^{(v)} = 2\gamma\, dH$$

iv) Use the ideal gas law to conclude that

$$\frac{RT}{M\hat{V}^{(\ell)}}\ln\left(\frac{P^{(v)}}{P_0^{(v)}}\right) - \left(P^{(v)} - P_0^{(v)}\right) = \frac{2\gamma}{R_d}$$

where M is molecular weight, R the gas law constant, and R_d the radius of the drop.

v) Conclude that, if $\hat{V}^{(v)} >> \hat{V}^{(\ell)}$, this reduces to Kelvin's (1871) equation

$$\ln\left(\frac{P^{(v)}}{P_0^{(v)}}\right) = \frac{2\gamma M\hat{V}^{(\ell)}}{R_d\ RT}$$

The vapor pressure in equilibrium with small liquid drops is larger than that observed over flat phase interfaces. This explains why vapors supersaturate and why large drops will grow faster during a condensation than small drops (Adamson 1976; Defay *et al.* 1966).

vi) Repeat the same analysis for a spherical vapor bubble whose radius is R_b to conclude

$$\ln\left(\frac{P^{(v)}}{P_0^{(v)}}\right) = -\frac{2\gamma M\hat{V}^{(\ell)}}{R_b\ RT}$$

The vapor pressure within small bubbles is smaller than that reported over flat phase interfaces. This explains why liquids supersaturate and why there is a tendency for small bubbles to collapse during a vaporization.

vii) Now consider a vapor-liquid interface in a capillary tube of radius R_c. Let Θ be the contact angle measured through the liquid phase. Again assuming that $\hat{V}^{(v)} >> \hat{V}^{(\ell)}$, reason that

$$\ln\left(\frac{P^{(v)}}{P_0^{(v)}}\right) = -\frac{2\gamma M\hat{V}^{(\ell)}}{R_c\ RT}\cos\Theta$$

The vapor pressure in equilibrium with a wetting liquid in small capillaries is smaller than that observed over flat phase interfaces, which explains the condensation of a vapor in the pores of a solid (Adamson 1976; Defay *et al.* 1966; Scheidegger 1974).

5.8.4 Bulk behavior: more on implications of Clausius-Duhem inequality In Sec. 5.8.2, we found that, as the result of the Clausius-Duhem inequality, the thermal energy flux vector **e** is related to the energy flux vector $\mathbf{q}$ by

$$\mathbf{e} = \mathbf{q} - \sum_{B=1}^{N-1} \left(\frac{\partial \hat{A}}{\partial \omega_{(B)}} \right)_{\hat{V}, T, \omega_{(C)} (C \neq B, N)} \mathbf{j}_{(B)} \tag{4-1}$$

In the last section, we introduced the chemical potential and learned

$$\left(\frac{\partial \hat{A}}{\partial \omega_{(B)}} \right)_{\hat{V}, T, \omega_{(C)} (C \neq B, N)} = \mu_{(B)} - \mu_{(N)} \tag{4-2}$$

Since (Table 5.2.2-4)

$$\sum_{B=1}^{N} \mathbf{j}_{(B)} = 0 \tag{4-3}$$

we have (see also Exercise 5.8.4-1)

$$\mathbf{e} = \mathbf{q} - \sum_{B=1}^{N} \mu_{(B)} \mathbf{j}_{(B)} \tag{4-4}$$

The Clausius-Duhem inequality reduced in Sec. 5.8.2 to

$$\begin{aligned} &- \operatorname{tr}\left[\left(\mathbf{T} - \left(\frac{\partial \hat{A}}{\partial \hat{V}} \right)_{T, \omega_{(B)}} \mathbf{I} \right) \cdot \mathbf{D} \right] \\ &\quad + \sum_{B=1}^{N-1} \mathbf{j}_{(B)} \cdot \nabla \left(\frac{\partial \hat{A}}{\partial \omega_{(B)}} \right)_{\hat{V}, T, \omega_{(C)} (C \neq B, N)} \\ &\quad + \sum_{B=1}^{N-1} \left(\frac{\partial \hat{A}}{\partial \omega_{(B)}} \right)_{\hat{V}, T, \omega_{(C)} (C \neq B, N)} r_{(B)} \\ &\quad + \frac{1}{T} \mathbf{e} \cdot \nabla T - \sum_{B=1}^{N} \mathbf{j}_{(B)} \cdot \mathbf{b}_{(B)} \leq 0 \end{aligned} \tag{4-5}$$

Noting that (Sec. 5.4.1)

$$\sum_{B=1}^{N} r_{(B)} = 0 \tag{4-6}$$

we may express this in terms of the thermodynamic pressure (Sec. 5.8.3)

$$P \equiv -\left(\frac{\partial \hat{A}}{\partial \hat{V}}\right)_{T,\omega_{(B)}} \tag{4-7}$$

and the chemical potentials (4-2) (see also Exercise 5.8.4-2):

$$\frac{1}{T}\mathbf{e}\cdot\nabla T - \text{tr}[(\mathbf{T}+P\mathbf{I})\cdot\mathbf{D}] + \sum_{B=1}^{N}\mathbf{j}_{(B)}\cdot(\nabla\mu_{(B)} - \mathbf{b}_{(B)})$$

$$+\sum_{B=1}^{N}\mu_{(B)}\, r_{(B)} \leq 0 \tag{4-8}$$

Reference

Slattery, J. C., "Momentum, Energy, and Mass Transfer in Continua," McGraw-Hill, New York (1972); second edition, Robert E. Krieger, Malabar, FL 32950 (1981).

Exercises

5.8.4-1 *alternative expression for* **e** The **Gibbs free energy** per unit mass $\hat{G}$ is defined in terms of the **enthalpy** per unit mass

$$\hat{H} \equiv \hat{U} + P\hat{V}$$

by

$$\begin{aligned}\hat{G} &\equiv \hat{H} - T\hat{S}\\ &= \hat{U} + P\hat{V} - T\hat{S}\\ &= \hat{A} + P\hat{V}\end{aligned}$$

i) Use the Gibbs equation to conclude

$$\left(\frac{\partial \hat{G}}{\partial \omega_{(B)}}\right)_{T,P,\omega_{(C)}(C\neq B,N)} = \mu_{(B)} - \mu_{(N)}$$

ii) The **partial mass Gibbs free energies** are defined by (Exercise 5.8.3-2)

$$\overline{\overline{G}}_{(B)} - \overline{\overline{G}}_{(N)} \equiv \left(\frac{\partial \hat{G}}{\partial \omega_{(B)}} \right)_{T,P,\omega_{(C)}(C \neq B,N)}$$

and the requirement that

$$\hat{G} = \sum_{B=1}^{N} \overline{\overline{G}}_{(B)} \omega_{(B)}$$

Prove that[a]

$$\mathbf{e} = -\sum_{B=1}^{N} \overline{\overline{G}}_{(B)} \mathbf{j}_{(B)}$$

as well as (4-4).

5.8.4-2 *alternative form for (4-8)* The partial mass entropies are defined by (Exercise 5.8.3-2)

---

a) I have previously expressed the thermal energy flux as (Slattery 1981, 469)

$$\mathbf{e} = \mathbf{q} - \sum_{B=1}^{N} \overline{\overline{H}}_{(B)} \mathbf{j}_{(B)} + T \sum_{B=1}^{N} \overline{\overline{S}}_{(B)} \mathbf{j}_{(B)}$$

and denoted by a special symbol (The term involving the relative kinetic energies of the individual species does not appear here. See footnote a of Sec. 5.6.4)

$$\boldsymbol{\varepsilon} \equiv \mathbf{q} - \sum_{B=1}^{N} \overline{\overline{H}}_{(B)} \mathbf{j}_{(B)}$$

Here $\overline{\overline{H}}_{(B)}$ and $\overline{\overline{S}}_{(B)}$ are respectively the partial mass enthalpy and the partial mass entropy (Exercise 5.8.3-2).

$$\overline{\overline{S}}_{(B)} - \overline{\overline{S}}_{(N)} \equiv \left(\frac{\partial \hat{S}}{\partial \omega_{(B)}} \right)_{T,P,\omega_{(C)}(C \neq B,N)}$$

and the requirement that

$$\hat{S} = \sum_{B=1}^{N} \overline{\overline{S}}_{(B)} \omega_{(B)}$$

Let us define

$$\mathbf{d}_{(A)} \equiv \frac{\rho_{(A)}}{cRT} \left(\nabla\mu_{(A)} - \mathbf{b}_{(A)} + \overline{\overline{S}}_{(A)} \nabla T - \frac{1}{\rho} \nabla P + \sum_{B=1}^{N} \omega_{(B)} \mathbf{b}_{(B)} \right)$$

i) Prove that (4-8) may be expressed as

$$\frac{1}{T}\left(\mathbf{e} - \sum_{B=1}^{N} T\overline{\overline{S}}_{(B)}\mathbf{j}_{(B)} \right) \cdot \nabla T - \text{tr}[(\mathbf{T} + P\mathbf{I}) \cdot \mathbf{D}]$$

$$+ \sum_{B=1}^{N} \frac{cRT}{\rho_{(B)}} \mathbf{j}_{(B)} \cdot \mathbf{d}_{(B)} + \sum_{B=1}^{N} \mu_{(B)} r_{(B)} \leq 0$$

ii) Noting (Exercise 5.8.3-5)

$$\overline{\overline{S}}_{(A)} = - \left(\frac{\partial \mu_{(A)}}{\partial T} \right)_{P,\omega_{(B)}}$$

$$\overline{\overline{V}}_{(A)} = \left(\frac{\partial \mu_{(A)}}{\partial P} \right)_{T,\omega_{(B)}}$$

determine that

$$\mathbf{d}_{(A)} = \frac{\rho_{(A)}}{cRT} \left[\sum_{\substack{B=1 \\ B \neq A}}^{N} \left(\frac{\partial \mu_{(A)}}{\partial \omega_{(B)}} \right)_{T,P,\omega_{(C)}(C \neq A,B)} \nabla \omega_{(B)} \right.$$

$$\left. + \left(\overline{\overline{V}}_{(A)} - \frac{1}{\rho} \right) \nabla P - \left(\mathbf{b}_{(A)} - \sum_{B=1}^{N} \omega_{(B)} \mathbf{b}_{(B)} \right) \right]$$

iii) From Exercise 5.8.3-1, we see that

$$\mu_{(A)}^{(m)} = M_{(A)}\mu_{(A)}$$

where $M_{(A)}$ is the molecular weight of species A. We also know that (Exercise 5.8.3-4)

$$\overline{\overline{V}}_{(A)}^{(m)} = M_{(A)}\overline{\overline{V}}_{(A)}$$

Here $\overline{\overline{V}}_{(A)}^{(m)}$ is the partial molal volume defined by (Exercise 5.8.3-3)

$$\overline{\overline{V}}_{(A)}^{(m)} - \overline{\overline{V}}_{(N)}^{(m)} \equiv \left(\frac{\partial \tilde{V}}{\partial x_{(A)}}\right)_{T,P,x_{(B)}(B \neq A,N)}$$

and

$$\tilde{V} = \sum_{A=1}^{N} \overline{\overline{V}}_{(A)}^{(m)} x_{(A)}$$

Reason that we can also write

$$\mathbf{d}_{(A)} = \frac{x_{(A)}}{RT}\left[\sum_{\substack{B=1\\B\neq A}}^{N} \left(\frac{\partial \mu_{(A)}^{(m)}}{\partial x_{(B)}}\right)_{T,P,x_{(C)}(C\neq A,B)} \nabla x_{(B)} \right.$$
$$+ M_{(A)}\left(\frac{\overline{\overline{V}}_{(A)}^{(m)}}{M_{(A)}} - \frac{1}{\rho}\right)\nabla P$$
$$\left. - M_{(A)}\left(\mathbf{b}_{(A)} - \sum_{B=1}^{N} \omega_{(B)}\mathbf{b}_{(B)}\right)\right]$$

iv) It will be useful to observe that, as the result of the Gibbs-Duhem equation,

$$\sum_{B=1}^{N} \mathbf{d}_{(B)} = 0$$

5.8.5 Surface behavior: implications of jump Clausius-Duhem inequality Let us begin by examining the restrictions that the jump Clausius-Duhem inequality (Sec. 5.7.3)

$$\rho^{(\sigma)} \frac{d_{(\mathbf{v}^{(\sigma)})}\hat{S}^{(\sigma)}}{dt} + \text{div}_{(\sigma)}\left(\frac{\mathbf{e}^{(\sigma)}}{T^{(\sigma)}}\right) - \rho^{(\sigma)} \frac{Q^{(\sigma)}}{T^{(\sigma)}}$$

$$+ \left[\rho(\hat{S} - \hat{S}^{(\sigma)})(\mathbf{v}\cdot\boldsymbol{\xi} - v_{(\xi)}) + \frac{1}{T}\mathbf{e}\cdot\boldsymbol{\xi}\right] \geq 0 \tag{5-1}$$

places upon the form of descriptions for surface material behavior.

The jump energy balance (Sec. 5.6.4)

$$\rho^{(\sigma)} \frac{d_{(\mathbf{v}^{(\sigma)})}\hat{U}^{(\sigma)}}{dt} = \text{tr}(\mathbf{T}^{(\sigma)}\cdot\mathbf{D}^{(\sigma)}) + \sum_{B=1}^{N} \mathbf{j}_{(B)}\cdot\mathbf{b}_{(B)}$$

$$- \text{div}_{(\sigma)}\mathbf{q}^{(\sigma)} + \rho^{(\sigma)} Q^{(\sigma)}$$

$$+ \left[-\rho\left(\hat{U} - \hat{U}^{(\sigma)} + \frac{1}{2}|\mathbf{v} - \mathbf{v}^{(\sigma)}|^2\right)(\mathbf{v}\cdot\boldsymbol{\xi} - v_{(\xi)})\right.$$

$$\left. + (\mathbf{v} - \mathbf{v}^{(\sigma)})\cdot\mathbf{T}\cdot\boldsymbol{\xi} - \mathbf{q}\cdot\boldsymbol{\xi}\right] \tag{5-2}$$

can be used to eliminate $Q^{(\sigma)}$ in (5-1):

$$\rho^{(\sigma)} \frac{d_{(\mathbf{v}^{(\sigma)})}\hat{A}^{(\sigma)}}{dt} + \rho^{(\sigma)}\hat{S}^{(\sigma)} \frac{d_{(\mathbf{v}^{(\sigma)})}\hat{T}^{(\sigma)}}{dt} - \text{tr}(\mathbf{T}^{(\sigma)}\cdot\mathbf{D}^{(\sigma)})$$

$$+ \text{div}_{(\sigma)}(\mathbf{q}^{(\sigma)} - \mathbf{e}^{(\sigma)}) + \frac{1}{T^{(\sigma)}}\mathbf{e}^{(\sigma)}\cdot\nabla_{(\sigma)}T^{(\sigma)}$$

$$- \sum_{B=1}^{N} \mathbf{j}_{(B)}\cdot\mathbf{b}_{(B)}$$

$$+ \left[\rho\left(\hat{U} - \hat{U}^{(\sigma)} + \frac{1}{2}|\mathbf{v} - \mathbf{v}^{(\sigma)}|^2\right)(\mathbf{v}\cdot\boldsymbol{\xi} - v_{(\xi)})\right.$$

$$- (\mathbf{v} - \mathbf{v}^{(\sigma)})\cdot\mathbf{T}\cdot\boldsymbol{\xi} + \mathbf{q}\cdot\boldsymbol{\xi}$$

$$- \rho \, T^{(\sigma)} (\hat{S} - \hat{S}^{(\sigma)})(\mathbf{v} \cdot \boldsymbol{\xi} - \mathbf{v}_{(\xi)}) + \frac{T^{(\sigma)}}{T} \mathbf{e} \cdot \boldsymbol{\xi} \Big]$$

$$\leq 0 \tag{5-3}$$

In arriving at this result, I have introduced the **surface Helmholtz free energy** per unit mass:

$$\hat{A}^{(\sigma)} \equiv \hat{U}^{(\sigma)} - T^{(\sigma)} \hat{S}^{(\sigma)} \tag{5-4}$$

To proceed further, we must restrict ourselves to a class of constitutive equations. Let us assume that

$$\hat{A}^{(\sigma)} = \hat{A}^{(\sigma)}(\boldsymbol{\Lambda}^{(\sigma)}) \qquad \mathbf{e}^{(\sigma)} = \mathbf{e}^{(\sigma)}(\boldsymbol{\Lambda}^{(\sigma)})$$

$$\hat{\mathbf{q}}^{(\sigma)} = \mathbf{q}^{(\sigma)}(\boldsymbol{\Lambda}^{(\sigma)}) \qquad \mathbf{j}_{(A)}^{(\sigma)} = \mathbf{j}_{(A)}^{(\sigma)}(\boldsymbol{\Lambda}^{(\sigma)})$$

$$r_{(A)}^{(\sigma)} = r_{(A)}^{(\sigma)}(\boldsymbol{\Lambda}^{(\sigma)}) \qquad \mathbf{T}^{(\sigma)} = \mathbf{T}^{(\sigma)}(\boldsymbol{\Lambda}^{(\sigma)}, \boldsymbol{\chi}^{(\sigma)t}) \tag{5-5}$$

where

$$\begin{aligned}\boldsymbol{\Lambda}^{(\sigma)} \equiv \Lambda^{(\sigma)}(& \hat{\mathcal{A}}, T^{(\sigma)}, \omega_{(1)}^{(\sigma)}, \omega_{(2)}^{(\sigma)}, \ldots, \omega_{(N-1)}^{(\sigma)}, \nabla_{(\sigma)}\hat{\mathcal{A}}, \nabla_{(\sigma)}T^{(\sigma)}, \\ & \nabla_{(\sigma)}\omega_{(1)}^{(\sigma)}, \nabla_{(\sigma)}\omega_{(2)}^{(\sigma)}, \ldots, \nabla_{(\sigma)}\omega_{(N-1)}^{(\sigma)}, \\ & \mathbf{b}_{(1)}^{(\sigma)}, \mathbf{b}_{(2)}^{(\sigma)}, \ldots, \mathbf{b}_{(N)}^{(\sigma)})\end{aligned} \tag{5-6}$$

is a set of independent variables common to all of these constitutive equations,

$$\hat{\mathcal{A}} \equiv \frac{1}{\rho^{(\sigma)}} \tag{5-7}$$

and $\boldsymbol{\chi}^{(\sigma)t}$ is the history of the motion of the multicomponent dividing surface (see Sec. 5.3.1 for viewpoint). The reason for introducing the dependence upon the $\mathbf{b}_{(A)}^{(\sigma)}$ will be clear in Sec. 5.8.8.

Using the chain rule, we can say from (5-5$_1$) that (see footnote a of Sec. 5.8.2)

$$\frac{d_{(\mathbf{v}^{(\sigma)})}\hat{A}^{(\sigma)}}{dt} = \frac{\partial \hat{A}^{(\sigma)}}{\partial \hat{\mathcal{A}}} \frac{d_{(\mathbf{v}^{(\sigma)})}\hat{\mathcal{A}}}{dt} + \frac{\partial \hat{A}^{(\sigma)}}{\partial T^{(\sigma)}} \frac{d_{(\mathbf{v}^{(\sigma)})}T^{(\sigma)}}{dt}$$

$$+ \sum_{B=1}^{N-1} \frac{\partial \hat{A}^{(\sigma)}}{\partial \omega_{(B)}^{(\sigma)}} \frac{d_{(\mathbf{v}^{(\sigma)})}\omega_{(B)}^{(\sigma)}}{dt} + \frac{\partial \hat{A}^{(\sigma)}}{\partial \nabla_{(\sigma)}\hat{\mathcal{A}}} \cdot \frac{d_{(\mathbf{v}^{(\sigma)})}\nabla_{(\sigma)}\hat{\mathcal{A}}}{dt}$$

$$+ \frac{\partial \hat{A}^{(\sigma)}}{\partial \nabla_{(\sigma)} T^{(\sigma)}} \cdot \frac{d_{(v^{(\sigma)})} \nabla_{(\sigma)} T^{(\sigma)}}{dt}$$

$$+ \sum_{B=1}^{N-1} \frac{\partial \hat{A}^{(\sigma)}}{\partial \nabla_{(\sigma)} \omega_{(B)}^{(\sigma)}} \cdot \frac{d_{(v^{(\sigma)})} \nabla_{(\sigma)} \omega_{(B)}^{(\sigma)}}{dt}$$

$$+ \sum_{B=1}^{N} \frac{\partial \hat{A}^{(\sigma)}}{\partial \mathbf{b}_{(B)}^{(\sigma)}} \cdot \frac{d_{(v^{(\sigma)})} \mathbf{b}_{(B)}^{(\sigma)}}{dt} \tag{5-8}$$

This together with the overall jump mass balance (Sec. 5.4.1)

$$\rho^{(\sigma)} \frac{d_{(v^{(\sigma)})} \hat{\mathcal{A}}}{dt} = \text{div}_{(\sigma)} \mathbf{v}^{(\sigma)} + \frac{1}{\rho^{(\sigma)}} [\, \rho (\mathbf{v} \cdot \boldsymbol{\xi} - v_{(\xi)}^{(\sigma)})] \tag{5-9}$$

and the jump mass balance for species B (Sec. 5.2.1)

$$\rho^{(\sigma)} \frac{d_{(v^{(\sigma)})} \omega_{(B)}^{(\sigma)}}{dt} = - \text{div}_{(\sigma)} \mathbf{j}_{(B)}^{(\sigma)} + r_{(B)}^{(\sigma)}$$

$$- [\, \mathbf{j}_{(B)} \cdot \boldsymbol{\xi} + \rho (\omega_{(B)} - \omega_{(B)}^{(\sigma)})(\mathbf{v} \cdot \boldsymbol{\xi} - v_{(\xi)}^{(\sigma)})] \tag{5-10}$$

allow us to rearrange (5-3) in the form

$$\rho^{(\sigma)} \left[\left(\frac{\partial \hat{A}^{(\sigma)}}{\partial T^{(\sigma)}} + \hat{S}^{(\sigma)} \right) \frac{d_{(v^{(\sigma)})} T^{(\sigma)}}{dt} + \frac{\partial \hat{A}^{(\sigma)}}{\partial \nabla_{(\sigma)} \hat{\mathcal{A}}} \cdot \frac{d_{(v^{(\sigma)})} \nabla_{(\sigma)} \hat{\mathcal{A}}}{dt} \right.$$

$$+ \frac{\partial \hat{A}^{(\sigma)}}{\partial \nabla_{(\sigma)} T^{(\sigma)}} \cdot \frac{d_{(v^{(\sigma)})} \nabla_{(\sigma)} T^{(\sigma)}}{dt}$$

$$+ \sum_{B=1}^{N-1} \frac{\partial \hat{A}^{(\sigma)}}{\partial \nabla_{(\sigma)} \omega_{(B)}^{(\sigma)}} \cdot \frac{d_{(v^{(\sigma)})} \nabla_{(\sigma)} \omega_{(B)}^{(\sigma)}}{dt}$$

$$+ \sum_{B=1}^{N} \frac{\partial \hat{A}^{(\sigma)}}{\partial \mathbf{b}_{(B)}^{(\sigma)}} \cdot \frac{d_{(v^{(\sigma)})} \mathbf{b}_{(B)}^{(\sigma)}}{dt}$$

$$- \text{tr} \left[\left(\mathbf{T}^{(\sigma)} - \frac{\partial \hat{A}^{(\sigma)}}{\partial \hat{\mathcal{A}}} \mathbf{P} \right) \cdot \mathbf{D}^{(\sigma)} \right]$$

$$+ \operatorname{div}_{(\sigma)} \left[\mathbf{q}^{(\sigma)} - \mathbf{e}^{\sigma)} - \sum_{B=1}^{N-1} \frac{\partial \hat{A}^{(\sigma)}}{\partial \omega_{(B)}^{(\sigma)}} \mathbf{j}_{(B)}^{(\sigma)} \right]$$

$$+ \sum_{B=1}^{N-1} \mathbf{j}_{(B)}^{(\sigma)} \cdot \nabla_{(\sigma)} \left(\frac{\partial \hat{A}^{(\sigma)}}{\partial \omega_{(B)}^{(\sigma)}} \right)$$

$$+ \sum_{B=1}^{N-1} \frac{\partial \hat{A}^{(\sigma)}}{\partial \omega_{(B)}^{(\sigma)}} r_{(B)}^{(\sigma)} + \frac{1}{T^{(\sigma)}} \mathbf{e}^{(\sigma)} \cdot \nabla_{(\sigma)} T^{(\sigma)}$$

$$- \sum_{B=1}^{N} \mathbf{j}_{(B)}^{(\sigma)} \cdot \mathbf{b}_{(B)}^{(\sigma)} + \frac{1}{\rho^{(\sigma)}} \frac{\partial \hat{A}^{(\sigma)}}{\partial \hat{\mathcal{A}}} [\, \rho(\mathbf{v} \cdot \boldsymbol{\xi} - v_{(\xi)}^{(\sigma)})]$$

$$- \sum_{B=1}^{N-1} \frac{\partial \hat{A}^{(\sigma)}}{\partial \omega_{(B)}^{(\sigma)}} [\, j_{(B)} \cdot \boldsymbol{\xi} + \rho(\omega_{(B)} - \omega_{(B)}^{(\sigma)})(\mathbf{v} \cdot \boldsymbol{\xi} - v_{(\xi)}^{(\sigma)})]$$

$$+ \Big[\rho \left(\hat{U} - \hat{U}^{(\sigma)} + \frac{1}{2} \mid \mathbf{v} - \mathbf{v}^{(\sigma)} \mid^2 \right) (\mathbf{v} \cdot \boldsymbol{\xi} - v_{(\xi)}^{(\sigma)})$$

$$- (\mathbf{v} - \mathbf{v}^{(\sigma)}) \cdot \mathbf{T} \cdot \boldsymbol{\xi} + \mathbf{q} \cdot \boldsymbol{\xi}$$

$$- \rho T^{(\sigma)} (\hat{S} - \hat{S}^{(\sigma)})(\mathbf{v} \cdot \boldsymbol{\xi} - v_{(\xi)}^{(\sigma)}) - \frac{T^{(\sigma)}}{T} \mathbf{e} \cdot \boldsymbol{\xi} \Big]$$

$$\leq 0 \qquad (5\text{-}11)$$

It is a simple matter to construct $T^{(\sigma)}$, $\hat{\mathcal{A}}$, $\omega_{(B)}^{(\sigma)}$, $\nabla_{(\sigma)} T^{(\sigma)}$, $\nabla_{(\sigma)} \hat{\mathcal{A}}$, $\nabla_{(\sigma)} \omega_{(B)}^{(\sigma)}$, and $\mathbf{b}_{(B)}^{(\sigma)}$ fields such that at any given point on the surface at any specified time

$$\frac{d_{(\mathbf{v}(\sigma))} T^{(\sigma)}}{dt}, \quad \frac{d_{(\mathbf{v}(\sigma))} \nabla_{(\sigma)} \hat{\mathcal{A}}}{dt}, \quad \frac{d_{(\mathbf{v}(\sigma))} \nabla_{(\sigma)} T^{(\sigma)}}{dt}, \quad \frac{d_{(\mathbf{v}(\sigma))} \nabla_{(\sigma)} \omega_{(B)}^{(\sigma)}}{dt},$$

$$\frac{d_{(\mathbf{v}(\sigma))} \mathbf{b}^{(\sigma)}}{dt}$$

take arbitrary values. We conclude that

$$\hat{S}^{(\sigma)} = -\frac{\partial \hat{A}^{(\sigma)}}{\partial T^{(\sigma)}} \tag{5-12}$$

$$\hat{A}^{(\sigma)} = \hat{A}^{(\sigma)}(\hat{\mathcal{A}}, T^{(\sigma)}, \omega_{(1)}^{(\sigma)}, \omega_{(2)}^{(\sigma)}, \ldots, \omega_{(N-1)}^{(\sigma)}) \tag{5-13}$$

and

$$\begin{aligned}
&-\text{tr}\left[\left(\mathbf{T}^{(\sigma)} - \frac{\partial \hat{A}^{(\sigma)}}{\partial \hat{\mathcal{A}}}\mathbf{P}\right) \cdot \mathbf{D}^{(\sigma)}\right] \\
&\quad + \text{div}_{(\sigma)}\left(\mathbf{q}^{(\sigma)} - \mathbf{e}^{(\sigma)} - \sum_{B=1}^{N-1} \frac{\partial \hat{A}^{(\sigma)}}{\partial \omega_{(B)}^{(\sigma)}} \mathbf{j}_{(B)}^{(\sigma)}\right) \\
&\quad + \sum_{B=1}^{N-1} \mathbf{j}_{(B)}^{(\sigma)} \cdot \nabla_{(\sigma)}\left(\frac{\partial \hat{A}^{(\sigma)}}{\partial \omega_{(B)}^{(\sigma)}}\right) \\
&\quad + \sum_{B=1}^{N-1} \frac{\partial \hat{A}^{(\sigma)}}{\partial \omega_{(B)}^{(\sigma)}} r_{(B)}^{(\sigma)} + \frac{1}{T^{(\sigma)}} \mathbf{e}^{(\sigma)} \cdot \nabla_{(\sigma)} T^{(\sigma)} \\
&\quad - \sum_{B=1}^{N} \mathbf{j}_{(B)}^{(\sigma)} \cdot \mathbf{b}_{(B)}^{(\sigma)} + \frac{1}{\rho^{(\sigma)}} \frac{\partial \hat{A}^{(\sigma)}}{\partial \mathcal{A}} [\, \rho(\mathbf{v} \cdot \boldsymbol{\xi} - v_{(\xi)}^{(\sigma)})] \\
&\quad - \sum_{B=1}^{N-1} \frac{\partial \hat{A}^{(\sigma)}}{\partial \omega_{(B)}^{(\sigma)}} [\, \mathbf{j}_{(B)} \cdot \boldsymbol{\xi} + \rho(\omega_{(B)} - \omega_{(B)}^{(\sigma)})(\mathbf{v} \cdot \boldsymbol{\xi} - v_{(\xi)}^{(\sigma)})] \\
&\quad + \Big[\rho\Big(\hat{U} - \hat{U}^{(\sigma)} + \frac{1}{2} \,|\, \mathbf{v} - \mathbf{v}^{(\sigma)} \,|^2\Big)(\mathbf{v} \cdot \boldsymbol{\xi} - v_{(\xi)}^{(\sigma)}) \\
&\qquad - (\mathbf{v} - \mathbf{v}^{(\sigma)}) \cdot \mathbf{T} \cdot \boldsymbol{\xi} + \mathbf{q} \cdot \boldsymbol{\xi} \\
&\qquad - \rho T^{(\sigma)}(\hat{S} - \hat{S}^{(\sigma)})(\mathbf{v} \cdot \boldsymbol{\xi} - v_{(\xi)}^{(\sigma)}) - \frac{T^{(\sigma)}}{T}\mathbf{e} \cdot \boldsymbol{\xi}\Big] \\
&\leq 0
\end{aligned} \tag{5-14}$$

For simplicity, let us introduce

$$\mathbf{k}^{(\sigma)} \equiv \mathbf{q}^{(\sigma)} - \mathbf{e}^{(\sigma)} - \sum_{B=1}^{N-1} \frac{\partial \hat{A}^{(\sigma)}}{\partial \omega_{(B)}^{(\sigma)}} \mathbf{j}_{(B)}^{(\sigma)} \tag{5-15}$$

and write inequality (5-14) as

$$\mathrm{div}_{(\sigma)}\mathbf{k}^{(\sigma)} + f(\,\mathbf{\Lambda}^{(\sigma)}, \boldsymbol{\chi}^{(\sigma)t}, \mathbf{\Lambda}, \boldsymbol{\chi}^{t}\,) \leq 0 \tag{5-16}$$

The vector $\mathbf{k}^{(\sigma)}$ is frame indifferent (Exercises 5.2.2-1, 5.6.4-3, and 5.7.3-8)

$$\mathbf{k}^{(\sigma)*} = \mathbf{Q} \cdot \mathbf{k}^{(\sigma)} \tag{5-17}$$

From (5-5), we see that it is a function only of $\mathbf{\Lambda}^{(\sigma)}$

$$\mathbf{k}^{(\sigma)} = \mathbf{h}(\,\mathbf{\Lambda}^{(\sigma)}\,) \tag{5-18}$$

By the principle of frame indifference, this same form of relationship must hold in every frame of reference

$$\mathbf{k}^{(\sigma)*} = \mathbf{h}(\,\mathbf{Q} \cdot \mathbf{\Lambda}^{(\sigma)}\,) \tag{5-19}$$

where (body forces are postulated to be frame indifferent in Sec. 2.1.1)

$$\begin{aligned}\mathbf{Q} \cdot \mathbf{\Lambda}^{(\sigma)} \equiv (\,& \hat{\mathcal{A}}, T^{(\sigma)}, \omega^{(\sigma)}_{(1)}, \omega^{(\sigma)}_{(2)}, \dots, \omega^{(\sigma)}_{(N-1)}, \mathbf{Q} \cdot \nabla_{(\sigma)}\hat{\mathcal{A}}, \\ & \mathbf{Q} \cdot \nabla_{(\sigma)}T^{(\sigma)}, \mathbf{Q} \cdot \nabla_{(\sigma)}\omega^{(\sigma)}_{(1)}, \mathbf{Q} \cdot \nabla_{(\sigma)}\omega^{(\sigma)}_{(2)}, \dots, \\ & \mathbf{Q} \cdot \nabla_{(\sigma)}\omega^{(\sigma)}_{(N-1)}, \mathbf{Q} \cdot \mathbf{b}^{(\sigma)}_{(1)}, \mathbf{Q} \cdot \mathbf{b}^{(\sigma)}_{(2)}, \dots, \mathbf{Q} \cdot \mathbf{b}^{(\sigma)}_{(N)}\,)\end{aligned} \tag{5-20}$$

Equations (5-17) through (5-19) imply $\mathbf{h}(\,\mathbf{\Lambda}^{(\sigma)}\,)$ is an isotropic function:

$$\mathbf{h}(\,\mathbf{Q} \cdot \mathbf{\Lambda}^{(\sigma)}\,) = \mathbf{Q} \cdot \mathbf{h}(\,\mathbf{\Lambda}^{(\sigma)}\,) \tag{5-21}$$

Applying the chain rule to (5-18), we have

$$\begin{aligned}\mathrm{div}_{(\sigma)}\mathbf{k}^{(\sigma)} = {} & \mathrm{tr}\left(\frac{\partial \mathbf{h}}{\partial \nabla_{(\sigma)}\hat{\mathcal{A}}} \cdot \nabla^{2}_{(\sigma)}\hat{\mathcal{A}} \right) + \mathrm{tr}\left(\frac{\partial \mathbf{h}}{\partial \nabla_{(\sigma)} T^{(\sigma)}} \cdot \nabla^{2}_{(\sigma)}T^{(\sigma)} \right) \\ & + \sum_{B=1}^{N-1} \mathrm{tr}\left(\frac{\partial \mathbf{h}}{\partial \nabla_{(\sigma)}\,\omega^{(\sigma)}_{(B)}} \cdot \nabla^{2}_{(\sigma)}\omega^{(\sigma)}_{(B)} \right) \\ & + \sum_{B=1}^{N} \mathrm{tr}\left(\frac{\partial \mathbf{h}}{\partial \mathbf{b}^{(\sigma)}_{(B)}} \cdot \nabla_{(\sigma)}\mathbf{b}^{(\sigma)}_{(B)} \right) + g(\,\mathbf{\Lambda}^{(\sigma)}\,)\end{aligned} \tag{5-22}$$

We can construct compatible $T^{(\sigma)}$, $\hat{\mathcal{A}}$, $\omega^{(\sigma)}_{(B)}$, $\nabla_{(\sigma)}T^{(\sigma)}$, $\nabla_{(\sigma)}\hat{\mathcal{A}}$, $\nabla_{(\sigma)}\omega^{(\sigma)}_{(B)}$ and $\mathbf{b}^{(\sigma)}_{(B)}$ fields such that at any given point within a phase at any specified time

$$\nabla^2_{(\sigma)}\hat{\mathcal{A}}, \quad \nabla^2_{(\sigma)}\mathrm{T}^{(\sigma)}, \quad \nabla^2_{(\sigma)}\omega_{(\mathrm{B})}^{\{\mathrm{g}\}}, \quad \nabla_{(\sigma)}\mathbf{b}_{(\mathrm{B})}^{\{\mathrm{g}\}}$$

take arbitrary values. In view of (5-16),

$$\begin{aligned}
\mathrm{tr}\left(\frac{\partial \mathbf{h}}{\partial \nabla_{(\sigma)}\hat{\mathcal{A}}} \cdot \nabla^2_{(\sigma)}\hat{\mathcal{A}}\right) &= \mathrm{tr}\left(\frac{\partial \mathbf{h}}{\partial \nabla_{(\sigma)}\mathrm{T}^{(\sigma)}} \cdot \nabla^2_{(\sigma)}\mathrm{T}^{(\sigma)}\right) \\
&= \mathrm{tr}\left(\frac{\partial \mathbf{h}}{\partial \nabla_{(\sigma)}\,\omega_{(\mathrm{B})}^{\{\mathrm{g}\}}} \cdot \nabla^2_{(\sigma)}\mathrm{T}_{(\mathrm{B})}^{\{\mathrm{g}\}}\right) \\
&= \mathrm{tr}\left(\frac{\partial \mathbf{h}}{\partial \mathbf{b}_{(\mathrm{B})}^{\{\mathrm{g}\}}} \cdot \nabla_{(\sigma)}\mathbf{b}_{(\mathrm{B})}^{\{\mathrm{g}\}}\right) \\
&= 0
\end{aligned} \tag{5-23}$$

which implies that

$$\frac{\partial \mathbf{h}}{\partial \mathbf{b}_{(\mathrm{B})}^{\{\mathrm{g}\}}}$$

as well as the symmetric parts of

$$\frac{\partial \mathbf{h}}{\partial \nabla_{(\sigma)}\hat{\mathcal{A}}}, \quad \frac{\partial \mathbf{h}}{\partial \nabla_{(\sigma)}\mathrm{T}^{(\sigma)}}, \quad \frac{\partial \mathbf{h}}{\partial \nabla_{(\sigma)}\,\omega_{(\mathrm{B})}^{\{\mathrm{g}\}}}$$

are all zero. Gurtin (1971, lemma 6.2; see also Gurtin and Vargas 1971, lemma 10.2) has proved that, when the symmetric portions of the derivatives of an isotropic vector function $\mathbf{h}(\mathbf{\Lambda}^{(\sigma)})$ with respect to each of the independent vectors are all zero, the function itself is zero. In this case, we conclude

$$\mathbf{k}^{(\sigma)} = 0 \tag{5-24}$$

or

$$\mathbf{e}^{(\sigma)} = \mathbf{q}^{(\sigma)} - \sum_{\mathrm{B}=1}^{\mathrm{N}-1} \frac{\partial \hat{\mathrm{A}}^{(\sigma)}}{\partial \omega_{(\mathrm{B})}^{\{\mathrm{g}\}}}\,\mathbf{j}_{(\mathrm{B})}^{\{\mathrm{g}\}} \tag{5-25}$$

This last in turn implies that (5-14) reduces to

$$-\,\mathrm{tr}\left[\left(\mathbf{T}^{(\sigma)} - \frac{\partial \hat{\mathrm{A}}^{(\sigma)}}{\partial \hat{\mathcal{A}}}\mathbf{P}\right) \cdot \mathbf{D}^{(\sigma)}\right] + \sum_{\mathrm{B}=1}^{\mathrm{N}-1} \mathbf{j}_{(\mathrm{B})}^{\{\mathrm{g}\}} \cdot \nabla_{(\sigma)}\left(\frac{\partial \hat{\mathrm{A}}^{(\sigma)}}{\partial \omega_{(\mathrm{B})}^{\{\mathrm{g}\}}}\right)$$

$$+ \sum_{B=1}^{N-1} \frac{\partial \hat{A}^{(\sigma)}}{\partial \omega_{(B)}^{(\sigma)}} r_{(B)}^{(\sigma)} + \frac{1}{T^{(\sigma)}} \mathbf{e}^{(\sigma)} \cdot \nabla_{(\sigma)} T^{(\sigma)}$$

$$- \sum_{B=1}^{N} \mathbf{j}_{(B)}^{(\sigma)} \cdot \mathbf{b}_{(B)}^{(\sigma)} + \frac{1}{\rho^{(\sigma)}} \frac{\partial \hat{A}^{(\sigma)}}{\partial \mathcal{A}} [(\mathbf{v} \cdot \boldsymbol{\xi} - v_{(\xi)}^{(\sigma)})]$$

$$- \sum_{B=1}^{N-1} \frac{\partial \hat{A}^{(\sigma)}}{\partial \omega_{(B)}^{(\sigma)}} [\mathbf{j}_{(B)} \cdot \boldsymbol{\xi} + \rho(\omega_{(B)} - \omega_{(B)}^{(\sigma)})(\mathbf{v} \cdot \boldsymbol{\xi} - v_{(\xi)}^{(\sigma)})]$$

$$+ \Big[\rho \Big(\hat{U} - \hat{U}^{(\sigma)} + \frac{1}{2} | \mathbf{v} - \mathbf{v}^{(\sigma)} |^2 \Big)(\mathbf{v} \cdot \boldsymbol{\xi} - v_{(\xi)}^{(\sigma)})$$

$$- (\mathbf{v} - \mathbf{v}^{(\sigma)}) \cdot \mathbf{T} \cdot \boldsymbol{\xi} + \mathbf{q} \cdot \boldsymbol{\xi}$$

$$- \rho T^{(\sigma)} (\hat{S} - \hat{S}^{(\sigma)})(\mathbf{v} \cdot \boldsymbol{\xi} - v_{(\xi)}^{(\sigma)}) - \frac{T^{(\sigma)}}{T} \mathbf{e} \cdot \boldsymbol{\xi} \Big]$$

$$\leq 0 \qquad (5\text{-}26)$$

In Sec. 5.8.8, I express (5-25) and (5-26) in terms of the surface chemical potential. But before I do so, let us examine the implications of (5-13).

References

Gurtin, M. E., *Arch. Ration. Mech. Anal.* **43**, 198 (1971).

Gurtin, M. E., and A. S. Vargas, *Arch. Ration. Mech. Anal.* **43**, 179 (1971).

5.8.6 Surface behavior: implications of surface caloric equation of state In the last section, we began with some broad statements about material behavior and concluded that, if the jump Clausius-Duhem inequality was to be obeyed,

$$\hat{A}^{(\sigma)} = \hat{A}^{(\sigma)}(\mathcal{A}, T^{(\sigma)}, \omega_{(1)}^{(\sigma)}, \omega_{(2)}^{(\sigma)}, \ldots, \omega_{(N-1)}^{(\sigma)}) \qquad (6\text{-}1)$$

or

$$\overline{A}^{(\sigma)} = \overline{A}^{(\sigma)}(T^{(\sigma)}, \rho_{(1)}, \rho_{(2)},..., \rho_{(N)}) \tag{6-2}$$

where hereafter $^{-}$ over a symbol should be read "per unit area". We will refer to these statements as alternative forms of the **surface caloric equation of state**. At the same time, we proved

$$\hat{S}^{(\sigma)} = - \left(\frac{\partial \hat{A}^{(\sigma)}}{\partial \hat{\mathcal{A}}} \right)_{\hat{\mathcal{A}},\omega_{(B)}} = - \frac{1}{\rho^{(\sigma)}} \left(\frac{\partial \overline{A}^{(\sigma)}}{\partial T^{(\sigma)}} \right)_{\rho_{(B)}} \tag{6-3}$$

Let us introduce the **thermodynamic surface tension**

$$\gamma \equiv \left(\frac{\partial \hat{A}^{(\sigma)}}{\partial \hat{\mathcal{A}}} \right)_{T^{(\sigma)},\omega_{(B)}} \tag{6-4}$$

and the **surface chemical potential**

$$\mu_{(B)} \equiv \left(\frac{\partial \overline{A}^{(\sigma)}}{\partial \rho_{(B)}} \right)_{T^{(\sigma)},\rho_{(C)}(C \neq B)} \tag{6-5}$$

The differentials of (6-1) and (6-2) may consequently be expressed as

$$d\hat{A}^{(\sigma)} = \gamma \, d\hat{\mathcal{A}} - \hat{S}^{(\sigma)} \, dT^{(\sigma)} + \sum_{B=1}^{N-1} \left(\frac{\partial \hat{A}^{(\sigma)}}{\partial \omega_{(B)}} \right)_{\hat{\mathcal{A}},T^{(\sigma)},\omega_{(C)}(C \neq B,N)} d\omega_{(B)} \tag{6-6}$$

and

$$d\overline{A}^{(\sigma)} = - \overline{S}^{(\sigma)} \, dT^{(\sigma)} + \sum_{B=1}^{N} \mu_{(B)} \, d\rho_{(B)} \tag{6-7}$$

Equation (6-7) may be rearranged to read

$$d\left(\frac{\hat{A}^{(\sigma)}}{\hat{\mathcal{A}}} \right) = - \frac{\hat{S}^{(\sigma)}}{\hat{\mathcal{A}}} dT^{(\sigma)} + \sum_{B=1}^{N} \mu_{(B)} \, d\left(\frac{\omega_{(B)}}{\hat{\mathcal{A}}} \right) \tag{6-8}$$

or

$$d\hat{A}^{(\sigma)} = -\left(-\frac{\hat{A}^{(\sigma)}}{\hat{\mathcal{A}}} + \sum_{B=1}^{N} \mu_{(B)}^{(\sigma)} \rho_{(B)}^{(\sigma)} \right) d\hat{\mathcal{A}} - \hat{S}^{(\sigma)}\, dT^{(\sigma)} + \sum_{B=1}^{N-1} \left(\mu_{(B)}^{(\sigma)} - \mu_{(N)}^{(\sigma)} \right) d\omega_{(B)}^{(\sigma)} \tag{6-9}$$

Comparison of coefficients in (6-6) and (6-9) gives

$$\left(\frac{\partial \hat{A}^{(\sigma)}}{\partial \omega_{(B)}^{(\sigma)}} \right)_{\hat{\mathcal{A}},T^{(\sigma)},\omega_{(C)}^{(\sigma)}(C \neq B,N)} = \mu_{(B)}^{(\sigma)} - \mu_{(N)}^{(\sigma)} \tag{6-10}$$

as well as the **surface Euler equation**

$$\hat{A}^{(\sigma)} = \gamma \hat{\mathcal{A}} + \sum_{B=1}^{N} \mu_{(B)}^{(\sigma)} \omega_{(B)}^{(\sigma)} \tag{6-11}$$

Equations (6-6) and (6-10) yield the **surface Gibbs equation**[a]:

$$d\hat{A}^{(\sigma)} = \gamma\, d\hat{\mathcal{A}} - \hat{S}^{(\sigma)}\, dT^{(\sigma)} + \sum_{B=1}^{N-1} \left(\mu_{(B)}^{(\sigma)} - \mu_{(N)}^{(\sigma)} \right) d\omega_{(B)}^{(\sigma)} \tag{6-12}$$

---

a) The complete parallel of Gibbs (1948, p. 225) discussion would include a dependence upon the two principal radii of curvature in (5-6), (6-1), and (6-2), resulting for example in a surface tension that depended upon curvature from (6-4). Gibbs (1948, p. 228) arrived at a result analogous to (6-12) only for a particular choice of dividing surface at which the partial derivatives of energy with respect to the two principal radii of curvature are identically zero. This dividing surface is normally referred to as the **surface of tension** (Buff 1951, 1956; Hill 1952; Kondo 1956; Ono and Kondo 1960).

I have chosen to neglect this possible dependence upon the principal radii of curvature of the surface, since any effect appears to be negligible under normal circumstances. Gibbs (1948, p. 227) is of the opinion that "... a surface may be regarded as nearly plane, when the radii of curvature are very large in proportion to the thickness of the non-homogeneous film." Buff (1956) estimates that any correction to surface tension for curvature is eight orders of magnitude smaller in usual applications.

The **surface Gibbs-Duhem equation** follows immediately by subtracting (6-12) from the differential of (6-11):

$$\hat{S}^{(\sigma)}\, dT^{(\sigma)} + \hat{\mathcal{A}}\, d\gamma + \sum_{B=1}^{N} \omega_{(B)}^{(\sigma)}\, d\mu_{(B)}^{(\sigma)} = 0 \tag{6-13}$$

References

Buff, F. P., *J. Chem. Phys.* **19**, 1591 (1951).

Buff, F. P., *J. Chem. Phys.* **25**, 146 (1956).

Gibbs, J. W., "The Collected Works," vol. 1, Yale University Press, New Haven, Conn. (1948).

Hill, T. L., *J. Chem. Phys.* **20**, 141 (1952).

Kondo, S., *J. Chem. Phys.* **25**, 662 (1956).

Ono, S., and S. Kondo, "Handbuch der Physik," vol. 10, edited by S. Flügge, Springer-Verlag, Berlin (1960).

Slattery, J. C., "Momentum, Energy, and Mass Transfer in Continua," McGraw-Hill, New York (1972); second edition, Robert E. Krieger, Malabar, FL 32950 (1981).

Exercises

5.8.6-1 *specific variables per unit mole* Let $c_{(A)}^{(\sigma)}$ denote moles of species A per unit area in the interface and $x_{(A)}^{(\sigma)}$ the mole fraction of species A. Denote by ~ over a symbol that we are dealing with a quantity per unit mole.

i) Show that the surface entropy per unit mole may be expressed as

$$\tilde{S}^{(\sigma)} = -\left(\frac{\partial \tilde{A}^{(\sigma)}}{\partial T^{(\sigma)}}\right)_{\tilde{\mathcal{A}},\,\omega_{(B)}^{(\sigma)}} = -\frac{1}{c^{(\sigma)}}\left(\frac{\partial \overline{A}^{(\sigma)}}{\partial T^{(\sigma)}}\right)_{c_{(B)}^{(\sigma)}}$$

where

$$c^{(\sigma)} \equiv \frac{1}{\tilde{\mathcal{A}}}$$

is the total mole density in the interface

ii) Determine that the thermodynamic surface tension is also given by

$$\gamma = \left(\frac{\partial \tilde{A}^{(\sigma)}}{\partial \tilde{\mathcal{A}}} \right)_{T^{(\sigma)}, x_{(B)}}$$

iii) It is common to use another chemical potential defined on a molar basis:

$$\mu_{(A)}^{(m)} \equiv \left(\frac{\partial \overline{A}^{(\sigma)}}{\partial c_{(A)}} \right)_{T^{(\sigma)}, c_{(B)} (B \neq A)}$$

Derive the analogs to (6-11) through (6-13) in terms of this molar chemical potential.

5.8.6-2 *Maxwell relations for interface* Let us define

$$\hat{A}^{(\sigma)} \equiv \hat{U}^{(\sigma)} - T^{(\sigma)} \hat{S}^{(\sigma)}$$

$$\hat{H}^{(\sigma)} \equiv \hat{U}^{(\sigma)} - \gamma \hat{\mathcal{A}}$$

$$\hat{G}^{(\sigma)} \equiv \hat{H}^{(\sigma)} - T^{(\sigma)} \hat{S}^{(\sigma)}$$

We refer to $\hat{A}^{(\sigma)}$ as **surface Helmholtz free energy** per unit mass, $\hat{H}^{(\sigma)}$ as **surface enthalpy** per unit mass, and $\hat{G}^{(\sigma)}$ as **surface Gibbs free energy** per unit mass. Determine that

i) $$\left(\frac{\partial T^{(\sigma)}}{\partial \hat{\mathcal{A}}} \right)_{\hat{S}^{(\sigma)}, \omega_{(B)}} = \left(\frac{\partial \gamma}{\partial \hat{S}^{(\sigma)}} \right)_{\hat{\mathcal{A}}, \omega_{(B)}}$$

$$\left(\frac{\partial T^{(\sigma)}}{\partial \omega_{(A)}} \right)_{\hat{\mathcal{A}}, \hat{S}^{(\sigma)}, \omega_{(B)} (B \neq A, N)} = \left(\frac{\partial (\mu_{(A)} - \mu_{(N)})}{\partial \hat{S}^{(\sigma)}} \right)_{\hat{\mathcal{A}}, \omega_{(B)}}$$

$$\left(\frac{\partial \gamma}{\partial \omega_{(A)}} \right)_{\hat{\mathcal{A}}, \hat{S}^{(\sigma)}, \omega_{(B)} (B \neq A, N)} = \left(\frac{\partial (\mu_{(A)} - \mu_{(N)})}{\partial \hat{\mathcal{A}}} \right)_{\hat{S}^{(\sigma)}, \omega_{(B)}}$$

$$\left(\frac{\partial (\mu_{(A)} - \mu_{(N)})}{\partial \omega_{(B)}} \right)_{\hat{\mathcal{A}}, \hat{S}^{(\sigma)}, \omega_{(C)} (C \neq B, N)}$$

$$= \left(\frac{\partial(\, \mu\{{g \atop B}\} - \mu\{{g \atop N}\}\,)}{\partial \omega\{{g \atop A}\}} \right)_{\hat{\mathcal{A}},\hat{S}^{(\sigma)},\omega\{{g \atop C}\}(C \neq A,N)}$$

ii) $$- \left(\frac{\partial \hat{S}^{(\sigma)}}{\partial \hat{\mathcal{A}}} \right)_{T^{(\sigma)},\omega\{{g \atop B}\}} = \left(\frac{\partial \gamma}{\partial T^{(\sigma)}} \right)_{\hat{\mathcal{A}},\omega\{{g \atop B}\}}$$

$$- \left(\frac{\partial \hat{S}^{(\sigma)}}{\partial \omega\{{g \atop A}\}} \right)_{\hat{\mathcal{A}},T^{(\sigma)},\omega\{{g \atop B}\}(B \neq A,N)} = \left(\frac{\partial(\, \mu\{{g \atop A}\} - \mu\{{g \atop N}\}\,)}{\partial T^{(\sigma)}} \right)_{\hat{\mathcal{A}},\omega\{{g \atop B}\}}$$

$$\left(\frac{\partial \gamma}{\partial \omega\{{g \atop A}\}} \right)_{\hat{\mathcal{A}},T^{(\sigma)}\omega\{{g \atop B}\}(B \neq A,N)} = \left(\frac{\partial(\, \mu\{{g \atop A}\} - \mu\{{g \atop N}\}\,)}{\partial \hat{\mathcal{A}}} \right)_{T^{(\sigma)},\omega\{{g \atop B}\}}$$

$$\left(\frac{\partial(\, \mu\{{g \atop A}\} - \mu\{{g \atop N}\}\,)}{\partial \omega\{{g \atop B}\}} \right)_{\hat{\mathcal{A}},T^{(\sigma)},\omega\{{g \atop C}\}(C \neq B,N)}$$

$$= \left(\frac{\partial(\, \mu\{{g \atop B}\} - \mu\{{g \atop N}\}\,)}{\partial \omega\{{g \atop A}\}} \right)_{\hat{\mathcal{A}},T^{(\sigma)},\omega\{{g \atop C}\}(C \neq A,N)}$$

iii) $$\left(\frac{\partial T^{(\sigma)}}{\partial \gamma} \right)_{\hat{S}^{(\sigma)},\omega\{{g \atop B}\}} = - \left(\frac{\partial \hat{\mathcal{A}}}{\partial \hat{S}^{(\sigma)}} \right)_{\gamma,\omega\{{g \atop B}\}}$$

$$\left(\frac{\partial T^{(\sigma)}}{\partial \omega\{{g \atop A}\}} \right)_{\gamma,\hat{S}^{(\sigma)},\omega\{{g \atop B}\}(B \neq A,N)} = \left(\frac{\partial(\, \mu\{{g \atop A}\} - \mu\{{g \atop N}\}\,)}{\partial \hat{S}^{(\sigma)}} \right)_{\gamma,\omega\{{g \atop B}\}}$$

$$- \left(\frac{\partial \hat{\mathcal{A}}}{\partial \omega\{{g \atop A}\}} \right)_{\gamma,\hat{S}^{(\sigma)},\omega\{{g \atop B}\}(B \neq A,N)} = \left(\frac{\partial(\, \mu\{{g \atop A}\} - \mu\{{g \atop N}\}\,)}{\partial \gamma} \right)_{\hat{S}^{(\sigma)},\omega\{{g \atop B}\}}$$

$$\left(\frac{\partial(\, \mu\{{g \atop A}\} - \mu\{{g \atop N}\}\,)}{\partial \omega\{{g \atop B}\}} \right)_{\gamma,\hat{S}^{(\sigma)},\omega\{{g \atop C}\}(C \neq B,N)}$$

$$= \left(\frac{\partial(\, \mu\{{g \atop B}\} - \mu\{{g \atop N}\}\,)}{\partial \omega\{{g \atop A}\}} \right)_{\gamma,\hat{S}^{(\sigma)},\omega\{{g \atop C}\}(C \neq A,N)}$$

iv) $$\left(\frac{\partial \hat{S}^{(\sigma)}}{\partial \gamma}\right)_{T^{(\sigma)},\omega\{{}^{\mathfrak{g}}_{B}\}} = \left(\frac{\partial \hat{\mathcal{A}}}{\partial T^{(\sigma)}}\right)_{\gamma,\omega\{{}^{\mathfrak{g}}_{B}\}}$$

$$-\left(\frac{\partial \hat{S}^{(\sigma)}}{\partial \omega\{{}^{\mathfrak{g}}_{A}\}}\right)_{T^{(\sigma)},\gamma,\omega\{{}^{\mathfrak{g}}_{B}\}(B \neq A,N)} = \left(\frac{\partial\left(\mu\{{}^{\mathfrak{g}}_{A}\} - \mu\{{}^{\mathfrak{g}}_{N}\}\right)}{\partial T^{(\sigma)}}\right)_{\gamma,\omega\{{}^{\mathfrak{g}}_{B}\}}$$

$$-\left(\frac{\partial \hat{\mathcal{A}}}{\partial \omega\{{}^{\mathfrak{g}}_{A}\}}\right)_{T^{(\sigma)},\gamma,\omega\{{}^{\mathfrak{g}}_{B}\}(B \neq A,N)} = \left(\frac{\partial\left(\mu\{{}^{\mathfrak{g}}_{A}\} - \mu\{{}^{\mathfrak{g}}_{N}\}\right)}{\partial \gamma}\right)_{T^{(\sigma)},\omega\{{}^{\mathfrak{g}}_{B}\}}$$

$$\left(\frac{\partial\left(\mu\{{}^{\mathfrak{g}}_{A}\} - \mu\{{}^{\mathfrak{g}}_{N}\}\right)}{\partial \omega\{{}^{\mathfrak{g}}_{B}\}}\right)_{T^{(\sigma)},\gamma,\omega\{{}^{\mathfrak{g}}_{C}\}(C \neq B,N)}$$

$$= \left(\frac{\partial\left(\mu\{{}^{\mathfrak{g}}_{B}\} - \mu\{{}^{\mathfrak{g}}_{N}\}\right)}{\partial \omega\{{}^{\mathfrak{g}}_{A}\}}\right)_{T^{(\sigma)},\gamma,\omega\{{}^{\mathfrak{g}}_{C}\}(C \neq A,N)}$$

5.8.6-3 *more Maxwell relations for interface* Following the definitions introduced in Exercise 5.8.6-2, we have that

$$\overline{A}^{(\sigma)} = \overline{U}^{(\sigma)} - T^{(\sigma)}\overline{S}^{(\sigma)}$$

$$\overline{H}^{(\sigma)} = \overline{U}^{(\sigma)} - \gamma$$

$$\overline{G}^{(\sigma)} = \overline{H}^{(\sigma)} - T^{(\sigma)}\overline{S}^{(\sigma)}$$

Determine that

i) $$\left(\frac{\partial T^{(\sigma)}}{\partial \rho\{{}^{\mathfrak{g}}_{A}\}}\right)_{\overline{S}^{(\sigma)},\rho\{{}^{\mathfrak{g}}_{B}\}(B \neq A)} = \left(\frac{\partial \mu\{{}^{\mathfrak{g}}_{A}\}}{\partial \overline{S}^{(\sigma)}}\right)_{\rho\{{}^{\mathfrak{g}}_{B}\}}$$

$$\left(\frac{\partial \mu\{{}^{\mathfrak{g}}_{A}\}}{\partial \rho\{{}^{\mathfrak{g}}_{B}\}}\right)_{\overline{S}^{(\sigma)},\rho\{{}^{\mathfrak{g}}_{C}\}(C \neq B)} = \left(\frac{\partial \mu\{{}^{\mathfrak{g}}_{B}\}}{\partial \rho\{{}^{\mathfrak{g}}_{A}\}}\right)_{\overline{S}^{(\sigma)},\rho\{{}^{\mathfrak{g}}_{C}\}(C \neq A)}$$

ii) $$-\left(\frac{\partial \overline{S}^{(\sigma)}}{\partial \rho\{{}^{\mathfrak{g}}_{A}\}}\right)_{T^{(\sigma)},\rho\{{}^{\mathfrak{g}}_{B}\}(B \neq A)} = \left(\frac{\partial \mu\{{}^{\mathfrak{g}}_{A}\}}{\partial T^{(\sigma)}}\right)_{\rho\{{}^{\mathfrak{g}}_{B}\}}$$

$$\left(\frac{\partial \mu\{{}^{\mathfrak{g}}_{A}\}}{\partial \rho\{{}^{\mathfrak{g}}_{B}\}}\right)_{T^{(\sigma)},\rho\{{}^{\mathfrak{g}}_{C}\}(C \neq B)} = \left(\frac{\partial \mu\{{}^{\mathfrak{g}}_{B}\}}{\partial \rho\{{}^{\mathfrak{g}}_{A}\}}\right)_{T^{(\sigma)},\rho\{{}^{\mathfrak{g}}_{C}\}(C \neq A)}$$

iii) $$\left(\frac{\partial T^{(\sigma)}}{\partial \gamma}\right)_{\overline{S}^{(\sigma)}, \rho_{(B)}^{(\sigma)}} = 0$$

$$\left(\frac{\partial \mu_{(A)}^{(\sigma)}}{\partial \gamma}\right)_{\overline{S}^{(\sigma)}, \rho_{(B)}^{(\sigma)}} = 0$$

iv) $$\left(\frac{\partial \overline{S}^{(\sigma)}}{\partial \gamma}\right)_{T^{(\sigma)}, \rho_{(B)}^{(\sigma)}} = 0$$

$$\left(\frac{\partial \mu_{(A)}^{(\sigma)}}{\partial \gamma}\right)_{T^{(\sigma)}, \rho_{(B)}^{(\sigma)}} = 0$$

5.8.6-4 *heat capacities for interface* We define the **surface heat capacity** per unit mass **at constant surface tension** as

$$\hat{c}_{\gamma}^{(\sigma)} \equiv T^{(\sigma)} \left(\frac{\partial \hat{S}^{(\sigma)}}{\partial T^{(\sigma)}}\right)_{\gamma, \omega_{(B)}^{(\sigma)}}$$

and the **surface heat capacity** per unit mass **at constant specific area** as

$$\hat{c}_{\mathcal{A}}^{(\sigma)} \equiv T^{(\sigma)} \left(\frac{\partial \hat{S}^{(\sigma)}}{\partial T^{(\sigma)}}\right)_{\hat{\mathcal{A}}, \omega_{(B)}^{(\sigma)}}$$

i) Determine that

$$\hat{c}_{\gamma}^{(\sigma)} = \left(\frac{\partial \hat{H}^{(\sigma)}}{\partial T^{(\sigma)}}\right)_{\gamma, \omega_{(B)}^{(\sigma)}}$$

and

$$\hat{c}_{\mathcal{A}}^{(\sigma)} = \left(\frac{\partial \hat{U}^{(\sigma)}}{\partial T^{(\sigma)}}\right)_{\hat{\mathcal{A}}, \omega_{(B)}^{(\sigma)}}$$

ii) Prove that

$$\rho^{(\sigma)}\hat{c}_{\gamma}^{(\sigma)} = \rho^{(\sigma)}\hat{c}_{\mathcal{A}}^{(\sigma)} - \left(\frac{\partial \gamma}{\partial T^{(\sigma)}}\right)_{\hat{\mathcal{A}},\omega_{(B)}^{(\sigma)}} \left(\frac{\partial \ln \hat{\mathcal{A}}}{\partial \ln T^{(\sigma)}}\right)_{\gamma,\omega_{(B)}^{(\sigma)}}$$

5.8.6-5 *partial mass surface variables* Let $\hat{\Phi}^{(\sigma)}$ be any intensive (per unit mass) surface variable

$$\hat{\Phi}^{(\sigma)} = \hat{\Phi}^{(\sigma)}\left(T^{(\sigma)}, \gamma, \omega_{(1)}^{(\sigma)}, \omega_{(2)}^{(\sigma)}, \ldots, \omega_{(N-1)}^{(\sigma)} \right)$$

We will define the partial mass surface variable $\overline{\overline{\Phi}}^{(\sigma)}$ by requiring (see Exercise 5.8.3-2 and Slattery 1981, p. 289)

$$\overline{\overline{\Phi}}_{(A)}^{(\sigma)} - \overline{\overline{\Phi}}_{(N)}^{(\sigma)} \equiv \left(\frac{\partial \hat{\Phi}^{(\sigma)}}{\partial \omega_{(A)}^{(\sigma)}}\right)_{T^{(\sigma)},\gamma,\omega_{(B)}^{(\sigma)}(B \neq A,N)}$$

and

$$\hat{\Phi}^{(\sigma)} = \sum_{A=1}^{N} \overline{\overline{\Phi}}_{(A)}^{(\sigma)} \omega_{(A)}$$

Determine that

$$\sum_{A=1}^{N} \left(\frac{\partial \overline{\overline{\Phi}}_{(A)}^{(\sigma)}}{\partial \omega_{(B)}^{(\sigma)}}\right)_{T^{(\sigma)},\gamma,\omega_{(C)}^{(\sigma)}(C \neq B,N)} \omega_{(A)}^{(\sigma)} = 0$$

5.8.6-6 *partial molal surface variables* Let $\tilde{\Phi}^{(\sigma)}$ be any intensive (per unit mole) variable

$$\tilde{\Phi}^{(\sigma)} = \tilde{\Phi}^{(\sigma)}\left(T^{(\sigma)}, \gamma, x_{(1)}^{(\sigma)}, x_{(2)}^{(\sigma)}, \ldots, x_{(N-1)}^{(\sigma)} \right)$$

We will define the partial molal surface variable $\overline{\overline{\Phi}}_{(A)}^{(\sigma,m)}$ by requiring (see Exercise 5.8.3-3 and Slattery 1981, p. 289)

$$\overline{\overline{\Phi}}_{(A)}^{(\sigma,m)} - \overline{\overline{\Phi}}_{(N)}^{(\sigma,m)} \equiv \left(\frac{\partial \tilde{\Phi}^{(\sigma)}}{\partial x_{(A)}^{(\sigma)}}\right)_{T^{(\sigma)},\gamma,x_{(B)}^{(\sigma)}(B \neq A,N)}$$

and

$$\tilde{\Phi}^{(\sigma)} = \sum_{A=1}^{N} \overline{\overline{\Phi}}_{(A)}^{(\sigma,m)} x_{(A)}^{(\sigma)}$$

Determine that

$$\sum_{A=1}^{N} \left(\frac{\partial \overline{\overline{\Phi}}_{(A)}^{(\sigma,m)}}{\partial x_{(B)}^{(\sigma)}} \right)_{T^{(\sigma)}, \gamma, x_{(C)}^{(\sigma)} (C \neq B, N)} x_{(A)}^{(\sigma)} = 0$$

5.8.6-7 *relation between partial mass and partial molal surface variables* Prove that

$$\overline{\overline{\Phi}}_{(A)}^{(\sigma,m)} = M_{(A)} \overline{\overline{\Phi}}_{(A)}^{(\sigma)}$$

5.8.6-8 Determine that

i) $$\overline{\overline{S}}_{(A)}^{(\sigma)} = - \left(\frac{\partial \mu_{(A)}^{(\sigma)}}{\partial T^{(\sigma)}} \right)_{\gamma, \omega_{(B)}^{(\sigma)}}$$

ii) $$\overline{\overline{\mathcal{A}}}_{(A)} = - \left(\frac{\partial \mu_{(A)}^{(\sigma)}}{\partial \gamma} \right)_{T^{(\sigma)}, \omega_{(B)}^{(\sigma)}}$$

(Hint: See Exercise 5.8.6-2 and the surface Gibbs-Duhem equation.)

5.8.7 Surface behavior: adsorption isotherms and equations of state

Relationships between interfacial chemical potentials, interfacial concentrations, interfacial tension, and bulk concentrations will be of particular interest in considering the effects of adsorption.

Gibbs adsorption isotherm From the surface Gibbs-Duhem equation (6-13), we conclude that at constant $T^{(\sigma)}$

$$d\gamma + \sum_{A=1}^{N} c_{(A)}^{(\sigma)} \, d\mu_{(A)}^{(\sigma,m)} = 0 \tag{7-1}$$

or

$$c_{(A)}^{(\sigma)} = -\left(\frac{\partial\gamma}{\partial\mu_{(A)}^{(\sigma,m)}}\right)_{T^{(\sigma)},\mu_{(C)}^{(\sigma,m)}(C\neq A)} \tag{7-2}$$

When we assume that chemical potential is continuous across the dividing surface (see Sec. 5.10.3),

$$\mu_{(A)}^{(\sigma,m)} = \mu_{(A)}^{(i,m)} \tag{7-3}$$

and (7-2) becomes

$$c_{(A)}^{(\sigma)} = -\left(\frac{\partial\gamma}{\partial\mu_{(A)}^{(i,m)}}\right)_{T^{(\sigma)},\mu_{(C)}^{(i,m)}(C\neq A)} \tag{7-4}$$

Here $\mu_{(A)}^{(i,m)}$ is the chemical potential (on a molar basis) of species A in an adjacent phase i. When it is possible to vary one chemical potential while holding all of the others fixed, (7-4) can be the basis for an experiment to determine $c_{(A)}^{(\sigma)}$. Note in particular that in arriving at (7-4) we have not had to choose a convention for locating the dividing surface (see Sec. 5.2.3).

Often it is not possible to use (7-4). For example, the chemical potentials can not be varied independently, when the temperature and pressure are both fixed in this phase (see the Gibbs-Duhem equation in Sec. 5.8.3). When the temperature is fixed and total molar density is assumed to be independent of concentration (dilute solutions; see caloric equation of state in Sec. 5.8.3), we have the same problem. In such cases, it is convenient to adopt the Gibbs convention for locating the dividing surface (see Sec. 5.2.3), in which case the Gibbs-Duhem equation becomes

$$d\gamma + \sum_{\substack{A=1\\A\neq B}}^{N} c_{(A)}^{(\sigma,B)}\, d\mu_{(A)}^{(\sigma,m)} = 0 \tag{7-5}$$

and by the same reasoning that led to (7-4) we have (Gibbs 1948, p. 235)

$$c_{(A)}^{(\sigma,B)} = -\left(\frac{\partial\gamma}{\partial\mu_{(A)}^{(i,m)}}\right)_{T^{(\sigma)},\mu_{(C)}^{(i,m)}(C\neq A,B)} \tag{7-6}$$

When we are concerned with ideal solutions which are also dilute, we are probably justified in identifying $c_{(A)}^{(\sigma,B)}$ with the surface molar density corresponding to another convention for locating the dividing surface (see Sec. 5.2.3). An isotherm constructed using (7-6) is often referred to as a **Gibbs adsorption isotherm**.

For at least some purposes, this last can be expressed in a more convenient form. If $f_{(A)}^{(i)}$ is the **fugacity** of species A in an adjacent phase i evaluated at the dividing surface,

$$\mu^{(i,m)}_{(A)} = \mu^{(i,m)\circ}_{(A)} + RT^{(\sigma)}\ln\left(\frac{f^{(i)}_{(A)}}{f^{(i)\circ}_{(A)}}\right) \tag{7-7}$$

Here $\mu^{(i,m)\circ}_{(A)}$ is the value of $\mu^{(i,m)}_{(A)}$ in the standard state for species A, and $f^{(i)\circ}_{(A)}$ is the corresponding fugacity. If we define

$$b^{(i)}_{(A)} \equiv \frac{1}{x^{(i)}_{(A)}}\frac{f^{(i)}_{(A)}}{f^{(i)\circ}_{(A)}} \tag{7-8}$$

as the **activity coefficient**, we conclude from (7-6) through (7-8) that

$$c^{(\sigma,B)}_{(A)} = -\frac{b^{(i)}_{(A)}x^{(i)}_{(A)}}{RT^{(\sigma)}}\left[\frac{\partial\gamma}{\partial\left(b^{(i)}_{(A)}x^{(i)}_{(A)}\right)}\right]_{b^{(i)}_{(C)}x^{(i)}_{(C)}(C\neq A,B)} \tag{7-9}$$

For ideal solutions, $b^{(i)}_{(C)} = 1$ for C = 1,...,N and (7-9) reduces to

$$c^{(\sigma,B)}_{(A)} = -\frac{x^{(i)}_{(A)}}{RT^{(\sigma)}}\left(\frac{\partial\gamma}{\partial x^{(i)}_{(A)}}\right)_{x^{(i)}_{(C)}(C\neq A,B)} \tag{7-10}$$

surface chemical potential Now let us consider an N component system. For the moment, we will not choose any particular convention for locating our dividing surface.

Regarding

$$\mu^{(\sigma,m)}_{(A)} = \mu^{(\sigma,m)}_{(A)}\left(T^{(\sigma)}, \gamma, x^{(\sigma)}_{(1)}, \ldots, x^{(\sigma)}_{(N-1)}\right) \tag{7-11}$$

we can write the change in chemical potential as we move away from the standard state for species A in the interface as

$$\begin{aligned}
\mu^{(\sigma,m)}_{(A)} &- \mu^{(\sigma,m)\circ}_{(A)} \\
&= \sum_{B=1}^{N-1}\int_{x^{(\sigma)\circ}_{(B)}}^{x^{(\sigma)}_{(B)}}\left(\frac{\partial\mu^{(\sigma,m)}_{(A)}}{\partial x^{(\sigma)}_{(B)}}\right)_{T^{(\sigma)},\gamma,x^{(\sigma)}_{(C)}(C\neq B,N)} dx^{(\sigma)}_{(B)} \\
&\quad + \int_{\gamma^{(\sigma)\circ}}^{\gamma}\left(\frac{\partial\mu^{(\sigma,m)}_{(A)}}{\partial\gamma}\right)_{T^{(\sigma)},x^{(\sigma)}_{(C)}(C\neq N)} d\gamma
\end{aligned} \tag{7-12}$$

Here $\gamma^{(A)\circ}$ is the interfacial tension in the standard state for species A in the interface. For the moment, we will not define this standard state. Equation (7-12) can be simplified somewhat, if we introduce the **surface**

fugacity of species A $f^{(\sigma)}_{(A)}$

$$\left(\frac{\partial \mu^{(\sigma,m)}_{(A)}}{\partial x^{(\sigma)}_{(B)}} \right)_{T^{(\sigma)}, \gamma, x^{(\sigma)}_{(C)} (C \neq B, N)}$$

$$= RT^{(\sigma)} \left(\frac{\partial \ln f^{(\sigma)}_{(A)}}{\partial x^{(\sigma)}_{(B)}} \right)_{T^{(\sigma)}, \gamma, x^{(\sigma)}_{(C)} (C \neq B, N)} \tag{7-13}$$

and if we observe from Exercise 5.8.6-8 that

$$\left(\frac{\partial \mu^{(\sigma,m)}_{(A)}}{\partial \gamma} \right)_{T^{(\sigma)}, x^{(\sigma)}_{(C)}} = - \overline{\overline{\mathcal{A}}}^{(m)}_{(A)} \tag{7-14}$$

Here $\overline{\overline{\mathcal{A}}}^{(m)}_{(A)}$ is the partial molal area defined in Exercise 5.8.6-6. Combining (7-12) through (7-14), we have

$$\mu^{(\sigma,m)}_{(A)} - \mu^{(\sigma,m)\circ}_{(A)} = RT^{(\sigma)} \ln \left(\frac{f^{(\sigma,\gamma)}_{(A)}}{f^{(\sigma)\circ}_{(A)}} \right) - \int_{\gamma^{(\sigma)\circ}}^{\gamma} \overline{\overline{\mathcal{A}}}^{(m)}_{(A)} \, d\gamma \tag{7-15}$$

in which we have introduced $f^{(\sigma,\gamma)}_{(A)}$ as the surface fugacity at the current temperature and composition of the interface but at the interfacial tension of the standard state. In terms of the constant-interfacial-tension surface activity coefficient

$$b^{(\sigma)}_{(A)} \equiv \frac{1}{x^{(\sigma)}_{(A)}} \frac{f^{(\sigma,\gamma)}_{(A)}}{f^{(\sigma)\circ}_{(A)}} \tag{7-16}$$

this becomes

$$\mu^{(\sigma,m)}_{(A)} - \mu^{(\sigma,m)\circ}_{(A)} = RT^{(\sigma)} \ln \left(b^{(\sigma)}_{(A)} \, x^{(\sigma)}_{(A)} \right) - \int_{\gamma^{(\sigma)\circ}}^{\gamma} \overline{\overline{\mathcal{A}}}^{(m)}_{(A)} \, d\gamma \tag{7-17}$$

For the sake of simplicity, we will assume in what follows that $\overline{\overline{\mathcal{A}}}^{(m)}_{(A)}$ is nearly independent of γ at constant temperature in order that we can write (7-17) as

$$\mu^{(\sigma,m)}_{(A)} - \mu^{(\sigma,m)\circ}_{(A)} = RT^{(\sigma)} \ln \left(b^{(\sigma)}_{(A)} \, x^{(\sigma)}_{(A)} \right) + \overline{\overline{\mathcal{A}}}^{(m)}_{(A)} \left(\gamma^{(A)\circ} - \gamma \right) \tag{7-18}$$

We will see shortly that, when the dividing surface is located according to the Lucassen-Reynders and van den Tempel convention, $\overline{\overline{\mathcal{A}}}^{(m)}_{(A)}$ is a constant and (7-18) is exact [see (7-33)].

general relationships assuming local equilibrium Sometimes we will be willing to assume that we have local equilibrium at the phase interface in the sense that chemical potential is continuous across the dividing surface (5.10.3):

$$\mu_{(A)}^{(i,m)} = \mu_{(A)}^{(\sigma,m)}$$

$$= \mu_{(A)}^{(\sigma,m)\circ} + RT^{(\sigma)} \ln\left(b_{(A)}^{(\sigma)}\, x_{(A)}^{(\sigma)} \right) + \overline{\overline{\mathcal{A}}}_{(A)}^{(m)} \left(\gamma^{(\sigma)\circ} - \gamma \right) \tag{7-19}$$

This relationship was derived in a different manner by Butler (1932) for a single layer of surfactant molecules at the phase interface. Without further derivation, it was applied by Lucassen-Reynders and van den Tempel (1967)[a] to a dividing surface.

The chemical potential of species A in phase i in the limit as the dividing surface is approached can also be expressed in terms of its corresponding activity coefficient $b_{(A)}^{(i)}$:

$$\mu_{(A)}^{(i,m)} - \mu_{(A)}^{(i,m)\circ} = RT \ln\left(b_{(A)}^{(i)}\, x_{(A)}^{(i)} \right) \tag{7-20}$$

Eliminating $\mu_{(A)}^{(i,m)}$ between (7-19) and (7-20) and rearranging, we find

$$\frac{b_{(A)}^{(\sigma)} x_{(A)}^{(\sigma)}}{b_{(A)}^{(i)} x_{(A)}^{(i)}}$$

$$= \exp\left[\frac{\mu_{(A)}^{(i,m)\circ} - \mu_{(A)}^{(\sigma,m)\circ}}{RT} \right] \exp\left[- \frac{\overline{\overline{\mathcal{A}}}_{(A)}^{(m)} \left(\gamma^{(A)\circ} - \gamma \right)}{RT} \right] \tag{7-21}$$

or (Lucassen-Reynders 1976)

$$x_{(A)}^{(\sigma)} = \frac{b_{(A)}^{(i)} x_{(A)}^{(i)}}{a_{(A)}^{(i)\circ}\, b_{(A)}^{(\sigma)}} \exp\left[- \frac{\overline{\overline{\mathcal{A}}}_{(A)}^{(m)} \left(\gamma^{(A)\circ} - \gamma \right)}{RT} \right] \tag{7-22}$$

in which we have introduced

$$a_{(A)}^{(i)\circ} \equiv \exp\left[\frac{\mu_{(A)}^{(\sigma,m)\circ} - \mu_{(A)}^{(i,m)\circ}}{RT} \right]$$

---

a) The derivation of (7-19) offered by Lucassen-Reynders (1976) appears to be unsatisfactory, since it assumes that the interfacial area is a function of temperature, *pressure*, and composition (her equation 11).

$$= \text{limit } \gamma \to \gamma^{(A)\circ}: \ \frac{b_{A}^{(i)}\ x_{A}^{(i)}}{b_{A}^{(\sigma)}\ x_{A}^{(\sigma)}} \tag{7-23}$$

Equation (7-22) tells us how γ varies as a function of $x_{A}^{(\sigma)}$ and $x_{A}^{(i)}$ with the assumption that chemical potential and temperature are continuous across the dividing surface.

Note that, because

$$\sum_{A=1}^{N} x_{A}^{(\sigma)} = 1 \tag{7-24}$$

(7-22) implies (Lucassen-Reynders 1976)

$$\sum_{A=1}^{N} \left(\frac{b_{A}^{(i)}\ x_{A}^{(i)}}{a_{A}^{(i)\circ} b_{A}^{(\sigma)}} \right) \exp\left[- \frac{\overline{\overline{\mathcal{A}}}_{A}^{(m)}(\gamma^{(A)\circ} - \gamma)}{RT} \right] = 1 \tag{7-25}$$

Employing

$$\sum_{A=1}^{N} x_{A}^{(i)} = 1 \tag{7-26}$$

we can eliminate $x_{1}^{(i)}$ in (7-25) to find

$$\sum_{A=2}^{N} \left\{ \frac{b_{A}^{(i)}}{a_{A}^{(i)\circ} b_{A}^{(\sigma)}} \exp\left[- \frac{\overline{\overline{\mathcal{A}}}_{A}^{(m)}(\gamma^{(A)\circ} - \gamma)}{RT} \right] - \frac{b_{1}^{(i)}}{a_{1}^{(i)\circ} b_{1}^{(\sigma)}} \exp\left[- \frac{\overline{\overline{\mathcal{A}}}_{1}^{(m)}(\gamma^{(1)\circ} - \gamma)}{RT} \right] \right\} x_{A}^{(i)}$$
$$+ \frac{b_{1}^{(i)}}{a_{1}^{(i)\circ} b_{1}^{(\sigma)}} \exp\left[- \frac{\overline{\overline{\mathcal{A}}}_{1}^{(m)}(\gamma^{(1)\circ} - \gamma)}{RT} \right] = 1 \tag{7-27}$$

Equation (7-27) describes how γ varies as a function of $x_{1}^{(i)}, \ldots, x_{N-1}^{(i)}$ with the understanding that chemical potential and temperature are continuous across the dividing

In a similar fashion (7-22) and (7-26) imply

$$\sum_{A=1}^{N} \frac{a_{A}^{(i)\circ} b_{A}^{(\sigma)}\ x_{A}^{(\sigma)}}{b_{A}^{(i)}} \exp\left[\frac{\overline{\overline{\mathcal{A}}}_{A}^{(m)}(\gamma^{(A)\circ} - \gamma)}{RT} \right] = 1 \tag{7-28}$$

Using (7-24), we can remove $x_{1}^{(\sigma)}$ in (7-28) to conclude

$$\sum_{A=2}^{N} \left\{ \frac{a^{(1)}_{(A)}{}^{\circ} b^{(\sigma)}_{(A)}}{b^{(i)}_{(A)}} \exp\left[\frac{\overline{\overline{\mathcal{A}}}^{(m)}_{(A)} \left(\gamma^{(A)\circ} - \gamma \right)}{RT} \right] \right.$$

$$\left. - \left\{ \frac{a^{(i)}_{(1)}{}^{\circ} b^{(\sigma)}_{(1)}}{b^{(i)}_{(1)}} \exp\left[\frac{\overline{\overline{\mathcal{A}}}^{(m)}_{(1)} \left(\gamma^{(1)\circ} - \gamma \right)}{RT} \right] \right\} \right\} x^{(\sigma)}_{(A)}$$

$$+ \frac{a^{(i)}_{(1)}{}^{\circ} b^{(\sigma)}_{(1)}}{b^{(i)}_{(1)}} \exp\left[\frac{\overline{\overline{\mathcal{A}}}^{(m)}_{(1)} \left(\gamma^{(1)\circ} - \gamma \right)}{RT} \right] = 1 \tag{7-29}$$

Equation (7-29) describes how γ varies as a function of $x^{(\sigma)}_{(1)}, \ldots, x^{(\sigma)}_{(N-1)}$. Although the chemical potential and temperature have been assumed to be continuous across the dividing surface in deriving (7-29), it is easily extended to non-equilibrium systems as suggested in Exercise 5.8.7-1.

Lucassen-Reynders and van den Tempel convention It is clear from (7-17) that we do not want $x^{(\sigma)}_{(A)}$ to go to zero except in the limit as species A disappears from the system. And we certainly do not want $x^{(\sigma)}_{(A)}$ to become negative. The Lucassen-Reynders and van den Tempel convention (see Sec. 5.2.3) avoids these difficulties. In this convention, the dividing surface is located such that

$$c^{(\sigma)} \equiv \sum_{A=1}^{N} c^{(\sigma, L)}_{(A)} = c^{(\sigma)\infty} \tag{7-30}$$

The total number of moles adsorbed in the dividing surface is a constant $c^{(\sigma)\infty}$, independent of composition. From the definition of the partial molal areas in Exercise 5.8.6-6, we have for A = 1,..., N–1

$$\overline{\overline{\mathcal{A}}}^{(m)}_{(A)} - \overline{\overline{\mathcal{A}}}^{(m)}_{(N)} = \left(\frac{\partial c^{(\sigma)\,-1}}{\partial x^{(\sigma)}_{(A)}} \right)_{T^{(\sigma)}, \gamma, x^{(\sigma)}_{(B)} (B \neq A, N)}$$

$$= 0 \tag{7-31}$$

and

$$\frac{1}{c^{(\sigma)}} = \sum_{A=1}^{N} \overline{\overline{\mathcal{A}}}^{(m)}_{(A)} x^{(\sigma)}_{(A)}$$

$$= \sum_{A=1}^{N-1} \left(\overline{\overline{\mathcal{A}}}^{(m)}_{(A)} - \overline{\overline{\mathcal{A}}}^{(m)}_{(N)} \right) x^{(\sigma)}_{(A)} + \overline{\overline{\mathcal{A}}}^{(m)}_{(N)}$$

$$= \overline{\overline{\mathcal{A}}}^{(m)}_{(N)}$$

$$= \frac{1}{c^{(\sigma)\infty}} \tag{7-32}$$

This means that for all A = 1,..., N

$$\overline{\overline{\mathcal{A}}}^{(m)}_{(A)} = \frac{1}{c^{(\sigma)\infty}} \tag{7-33}$$

pure components as standard states Although it would not be strictly necessary, I will choose the same standard state to apply to both a component within a bulk phase and the same component in an interface.

For the moment, let us confine our attention to components that are liquids at the system temperature. It will be convenient to choose the standard state for each component A to be the pure material at the system temperature and its corresponding vapor pressure. With this interpretation

$$\text{as } x^{(i)}_{(A)} \to 1 : b^{(i)}_{(A)} \to 1 \tag{7-34}$$

$$\text{as } x^{(\sigma)}_{(A)} \to 1 : b^{(\sigma)}_{(A)} \to 1 \tag{7-35}$$

and

$$\gamma^{(A)\circ} = \gamma^{\circ}_{(A)} \tag{7-36}$$

where $\gamma^{\circ}_{(A)}$ should be interpreted as the surface tension for pure species A at the system temperature.

Since the chemical potentials of each species in the surface and in the substrate are equal in the standard state (see Sec. 5.10.3), it follows from (7-18) that for all A = 1,..., N

$$a^{(i)\circ}_{(A)} = 1 \tag{7-37}$$

In view of (7-31) and (7-32), we can express (7-17), (7-22), and (7-24) as

$$x^{(\sigma)}_{(A)} = \frac{b^{(i)}_{(A)}\, x^{(i)}_{(A)}}{b^{(\sigma)}_{(A)}} \exp\left[- \frac{\overline{\overline{\mathcal{A}}}^{(m)}_{(A)}\left(\gamma^{\circ}_{(A)} - \gamma \right)}{RT} \right] \tag{7-38}$$

$$\sum_{A=2}^{N} \left\{ \frac{b_{(A)}^{(i)}}{b_{(A)}^{(\sigma)}} \exp\left[- \frac{\overline{\overline{\mathcal{A}}}_{(A)}^{(m)}(\gamma_{(A)}^{\circ} - \gamma)}{RT} \right] - \frac{b_{(1)}^{(i)}}{b_{(1)}^{(\sigma)}} \exp\left[- \frac{\overline{\overline{\mathcal{A}}}_{(1)}^{(m)}(\gamma_{(1)}^{\circ} - \gamma)}{RT} \right] \right\} x_{(A)}^{(i)} + \frac{b_{(1)}^{(i)}}{b_{(1)}^{(\sigma)}} \exp\left[- \frac{\overline{\overline{\mathcal{A}}}_{(1)}^{(m)}(\gamma_{(1)}^{\circ} - \gamma)}{RT} \right] = 1 \tag{7-39}$$

$$\sum_{A=2}^{N} \left\{ \frac{b_{(A)}^{(\sigma)}}{b_{(A)}^{(i)}} \exp\left[\frac{\overline{\overline{\mathcal{A}}}_{(A)}^{(m)}(\gamma_{(A)}^{\circ} - \gamma)}{RT} \right] - \frac{b_{(1)}^{(\sigma)}}{b_{(1)}^{(i)}} \exp\left[\frac{\overline{\overline{\mathcal{A}}}_{(1)}^{(m)}(\gamma_{(1)}^{\circ} - \gamma)}{RT} \right] \right\} x_{(A)}^{(\sigma)} + \frac{b_{(1)}^{(\sigma)}}{b_{(1)}^{(i)}} \exp\left[\frac{\overline{\overline{\mathcal{A}}}_{(1)}^{(m)}(\gamma_{(1)}^{\circ} - \gamma)}{RT} \right] = 1 \tag{7-40}$$

If we adopt the Lucassen-Reynders and van den Tempel convention for locating the dividing surface, (7-38) through (7-40) further reduce to respectively

$$\frac{c_{(A)}^{(\sigma,L)}}{c^{(\sigma)\infty}} = \frac{b_{(A)}^{(i)}\ x_{(A)}^{(i)}}{b_{(A)}^{(\sigma)}} \exp\left(- \frac{\gamma_{(A)}^{\circ} - \gamma}{c^{(\sigma)\infty} RT} \right) \tag{7-41}$$

$$\frac{b_{(1)}^{(i)}}{b_{(1)}^{(\sigma)}} + \sum_{A=2}^{N} \left[\frac{b_{(A)}^{(i)}}{b_{(A)}^{(\sigma)}} \exp\left(- \frac{\gamma_{(A)}^{\circ} - \gamma_{(1)}^{\circ}}{c^{(\sigma)\infty} RT} \right) - \frac{b_{(1)}^{(i)}}{b_{(1)}^{(\sigma)}} \right] x_{(A)}^{(i)} = \exp\left(\frac{\gamma_{(1)}^{\circ} - \gamma}{c^{(\sigma)\infty} RT} \right) \tag{7-42}$$

$$\frac{b_{(1)}^{(\sigma)}}{b_{(1)}^{(i)}} - \sum_{A=2}^{N} \left[\frac{b_{(1)}^{(\sigma)}}{b_{(1)}^{(i)}} - \frac{b_{(A)}^{(\sigma)}}{b_{(A)}^{(i)}} \exp\left(\frac{\gamma_{(A)}^{\circ} - \gamma_{(1)}^{\circ}}{c^{(\sigma)\infty} RT} \right) \right] \frac{c_{(A)}^{(\sigma,L)}}{c^{(\sigma)\infty}}$$

$$= \exp\left(- \frac{\gamma^{\circ}_{(1)} - \gamma}{c^{(\sigma)\infty} RT} \right) \tag{7-43}$$

standard states referred to dilute solutions In many applications, we may be willing to say that either one or both phases adjoining a dividing surface are dilute solutions. It may be more convenient to define our standard states with respect to solution behavior at infinite dilution.

In particular, the standard states for all solutes within a phase are chosen such that the activity coefficients go to unity in the limit of infinite dilution at the system temperature and the vapor pressure of the solvent. If species 1 is the solvent (principal component) of phase i, then for all A ≠ 1

$$\text{as } x^{(i)}_{(A)} \to 0 \text{ and } x^{(i)}_{(1)} \to 1 : \quad b^{(i)*}_{(A)} \to 1, \quad b^{(i)}_{(1)} \to 1 \tag{7-44}$$

A star ...* will be used to denote an activity coefficient that approaches unity as its mole fraction goes to zero. The standard states for all solutes in the dividing surface are defined such that the surface activity coefficients go to unity in the limit of infinite surface dilution at the system temperature and the corresponding surface tension for pure solvent. For all A ≠ 1,

$$\text{as } x^{(\sigma)}_{(A)} \to 0 \text{ and } x^{(\sigma)}_{(1)} \to 1 : \quad b^{(\sigma)*}_{(A)} \to 1, \quad b^{(\sigma)}_{(1)} \to 1 \tag{7-45}$$

and

$$\gamma^{(A)\circ} = \gamma^{(1)\circ} = \gamma^{\circ}_{(1)} \tag{7-46}$$

Note that the chemical potential for a solute in its surface standard state is not normally equal to the chemical potential for this species in its substrate standard state. For this reason (7-37) does not hold for the solutes, although we can say

$$a^{(i)\circ}_{(1)} = 1 \tag{7-47}$$

Let us immediately adopt the Lucassen-Reynders and van den Tempel convention for locating the dividing surface. In view of (7-33), (7-46) and (7-47), we can express (7-22), (7-27) and (7-29) as

$$\frac{c^{(\sigma,L)}_{(A)}}{c^{(\sigma)\infty}} = \frac{b^{(i)*}_{(A)} x^{(i)}_{(A)}}{a^{(i)\circ}_{(A)} \, b^{(\sigma)}_{(A)}} \exp\left(- \frac{\gamma^{\circ}_{(1)} - \gamma}{c^{(\sigma)\infty} RT} \right) \tag{7-48}$$

$$\frac{b^{(i)}_{(1)}}{b^{(\sigma)}_{(1)}} + \sum_{A=2}^{N} \left(\frac{b^{(i)*}_{(A)}}{a^{(i)\circ}_{(A)}\, b^{(\sigma)*}_{(A)}} - \frac{b^{(i)}_{(1)}}{b^{(\sigma)}_{(1)}} \right) x^{(i)}_{(A)} = \exp\left(\frac{\gamma^{\circ}_{(1)} - \gamma}{c^{(\sigma)\infty} RT} \right) \tag{7-49}$$

$$\frac{b^{(\sigma)}_{(1)}}{b^{(i)}_{(1)}} - \sum_{A=2}^{N} \left(\frac{b^{(\sigma)}_{(1)}}{b^{(i)}_{(1)}} - \frac{a^{(i)\circ}_{(A)}\, b^{(\sigma)*}_{(A)}}{b^{(i)*}_{(A)}} \right) \frac{c^{(\sigma,L)}_{(A)}}{c^{(\sigma)\infty}} = \exp\left(- \frac{\gamma^{\circ}_{(1)} - \gamma}{c^{(\sigma)\infty} RT} \right) \tag{7-50}$$

dilute solutions Phase i may often be sufficiently dilute that $b^{(i)}_{(1)} x^{(i)}_{(1)} \doteq 1$, in which case both (7-41) and (7-48) require for the solvent

$$1 - \sum_{A=2}^{N} \frac{c^{(\sigma,L)}_{(A)}}{c^{(\sigma)\infty}} = \frac{1}{b^{(\sigma)}_{(1)}} \exp\left(- \frac{\gamma^{\circ}_{(1)} - \gamma}{c^{(\sigma)\infty} RT} \right) \tag{7-51}$$

In starting from (7-41), we have employed (7-24); with (7-48), we have used both (7-24) and (7-47). For a two-component solution so dilute that $b^{(\sigma)}_{(1)} \doteq 1$, (7-51) reduces to the **Frumkin equation of state** (Frumkin 1925; Lucassen-Reynders and van den Tempel 1967; Lucassen-Reynders 1976):

$$1 - \frac{c^{(\sigma,L)}_{(2)}}{c^{(\sigma)\infty}} = \exp\left(- \frac{\gamma^{\circ}_{(1)} - \gamma}{c^{(\sigma)\infty} RT} \right) \tag{7-52}$$

A major limitation to (7-51) and the Frumkin equation of state (7-52) is that local equilibrium is assumed between the dividing surface and the substrate [see (7-22), (7-41), and (7-48)]. Equations (7-29), (7-40), (7-43), and (7-50) avoid this restriction (see Exercise 5.8.7-1).

If phase i is sufficiently dilute that

$$b^{(i)}_{(1)} \left(1 - \sum_{A=2}^{N} x^{(i)}_{(A)} \right) \doteq 1 \tag{7-53}$$

(7-49) becomes

$$\frac{1}{b^{(\sigma)}_{(1)}} + \sum_{A=2}^{N} \frac{b^{(i)*}_{(A)}}{a^{(i)\circ}_{(A)}\, b^{(\sigma)*}_{(A)}} x^{(i)}_{(A)} = \exp\left(\frac{\gamma^{\circ}_{(1)} - \gamma}{c^{(\sigma)\infty} RT} \right) \tag{7-54}$$

For a two-component solution sufficiently dilute that $b^{(\sigma)}_{(1)} \doteq 1$, this simplifies to the **von Szyszkowski equation of state** (von Szyszkowski 1908; Lucassen-Reynders and van den Tempel 1967; Lucassen-Reynders 1976):

$$1 + \frac{b\left\{{}^{i}_{(2)}\right\}^{*}}{a\left\{{}^{i}_{(2)}\right\}^{\circ} b\left\{{}^{\sigma}_{(2)}\right\}^{*}} x\left\{{}^{i}_{(2)}\right\} = \exp\left(\frac{\gamma^{\circ}_{(1)} - \gamma}{c^{(\sigma)\infty} RT}\right) \tag{7-55}$$

Alternatively, (7-53) allows us to write (7-42) as

$$\frac{1}{b\left\{{}^{\sigma}_{(1)}\right\}} + \sum_{A=2}^{N} \left[\frac{b\left\{{}^{i}_{(A)}\right\}}{b\left\{{}^{\sigma}_{(A)}\right\}} \exp\left(-\frac{\gamma^{\circ}_{(A)} - \gamma^{\circ}_{(1)}}{c^{(\sigma)\infty} RT}\right)\right] x\left\{{}^{i}_{(A)}\right\}$$

$$= \exp\left(\frac{\gamma^{\circ}_{(1)} - \gamma}{c^{(\sigma)\infty} RT}\right) \tag{7-56}$$

For a two-component solution sufficiently dilute that $b\left\{{}^{\sigma}_{(1)}\right\} \doteq 1$, this becomes

$$1 + \left[\frac{b\left\{{}^{i}_{(2)}\right\}}{b\left\{{}^{\sigma}_{(2)}\right\}} \exp\left(-\frac{\gamma^{\circ}_{(2)} - \gamma^{\circ}_{(1)}}{c^{(\sigma)\infty} RT}\right)\right] x\left\{{}^{i}_{(2)}\right\} = \exp\left(\frac{\gamma^{\circ}_{(1)} - \gamma}{c^{(\sigma)\infty} RT}\right) \tag{7-57}$$

which has the same form as the von Szyszkowski equation of state (7-50), although it involves one less parameter $a\left\{{}^{i}_{(2)}\right\}^{\circ}$.

ionized surfactants with or without excess counterions Although there are many nonionic surfactants, many more ionize in solution.

We will treat each ion as an individual species. As a result, we quickly conclude that (7-22) holds for every species in the solution, whether it be an ion or an electrically neutral component.[b]

The major difference from the preceding discussion for uncharged species is that in what follows we will be concerned to preserve electrical neutrality at each point in the solution. In particular, for the surface we can compactly write the condition for electrical neutrality as

---

b) Lucassen-Reynders (1976) chooses to apply (7-22) to electrically neutral combinations of ions. In this way, she requires only that the chemical potentials of pairs of ions are continuous across the dividing surface; the chemical potentials of individual ions are not necessarily the same both in the dividing surface and in the substrate. We require that the chemical potentials of the individual ions are continuous across the dividing surface in addition to requiring electrical neutrality both in the dividing surface and in the bulk solution.

$$\sum_{A=1}^{N} \alpha_{(A)} c^{(\sigma)}_{(A)} = 0 \tag{7-58}$$

where $\alpha_{(A)}$ is the valence or charge number for each species ($\alpha_{(B)} = 0$), if species B happens to be uncharged). This together with (7-24) implies that

$$\sum_{A=1}^{N-1} \left(1 - \frac{\alpha_{(A)}}{\alpha_{(N)}} \right) x^{(\sigma)}_{(A)} = 1 \tag{7-59}$$

Similarly, for the adjoining bulk phases

$$\sum_{A=1}^{N-1} \left(1 - \frac{\alpha_{(A)}}{\alpha_{(N)}} \right) x^{(i)}_{(A)} = 1 \tag{7-60}$$

We will adopt the Lucassen-Reynders and van den Tempel convention for locating the dividing surface. We will again choose the standard states within a phase such that, for all solutes A except the undissociated electrolytes, (7-44) describes the limiting behavior of the activity coefficients. The standard states within the dividing surface are fixed in such a manner that, for all solutes A except the undissociated electrolytes, (7-45) and (7-46) apply.

Our primary concern here will be with relatively strong electrolytes. If X dissociates as

$$X \rightarrow \beta_{(+)} X^{\alpha(+)} + \beta_{(-)} X^{\alpha(-)} \tag{7-61}$$

with

$$\alpha_{(+)}\beta_{(+)} + \alpha_{(-)}\beta_{(-)} = 0 \tag{7-62}$$

that we can identify the molar density of the cation $X^{\alpha(+)}$ as $\beta_{(+)}$ times the stoichiometric molar density of X. The molar density of the anion $X^{\alpha(-)}$ is similarly $\beta_{(-)}$ times the stoichiometric molar density of X.

In the absence of reliable information regarding the concentration of the undissociated salt X, we will choose its standard states in phases i and j such that the equilibrium constant for reaction (7-61) at the system temperature and the vapor pressure of the solvent is unity (Lewis *et al.* 1961, p. 309):

$$\frac{\left(b^{(i)}_{(+)} x^{(i)}_{(+)} \right)^{\beta(+)} \left(b^{(i)}_{(-)} x^{(i)}_{(-)} \right)^{\beta(-)}}{a^{(i)}_{(X)}}$$

$$= \frac{\left(b^{(j)}_{(+)} \, x^{(j)}_{(+)} \right)^{\beta(+)} \left(b^{(j)}_{(-)} \, x^{(j)}_{(-)} \right)^{\beta(-)}}{a^{(i)}_{(X)}}$$

$$= 1 \tag{7-63}$$

The standard state for X in the dividing surface is defined such that the equilibrium constant for reaction (7-61) at the system temperature and the surface tension of the solvent is again unity

$$\frac{\left(b^{(\sigma)}_{(+)} \, x^{(\sigma)}_{(+)} \right)^{\beta(+)} \left(b^{(\sigma)}_{(-)} \, x^{(\sigma)}_{(-)} \right)^{\beta(-)}}{a^{(\sigma)}_{(X)}} = 1 \tag{7-64}$$

and the interfacial tension in this standard state is $\gamma^{\circ}_{(1)}$ (see Exercise 5.8.7-1). Here $a^{(i)}_{(X)}$ and $a^{(j)}_{(X)}$ are the activities of the undissociated salt X in phases i and j respectively; $a^{(\sigma)}_{(X)}$ is the corresponding activity in the dividing surface, the Lucassen-Reynders and van den Tempel convention having been adopted; the subscript (+) refers to the cation $X^{\alpha(+)}$ and the subscript (–) to the anion $X^{\alpha(-)}$.

For the solvent 1 in phase i, (7-22), (7-33), (7-47) and (7-59) require

$$1 - \sum_{A=2}^{N-1} \left(1 - \frac{\alpha_{(A)}}{\alpha_{(N)}} \right) \frac{c^{(\sigma, L)}_{(A)}}{c^{(\sigma)\infty}} = \frac{1}{b^{(\sigma)}_{(1)}} \exp\left(- \frac{\gamma^{\circ}_{(1)} - \gamma}{c^{(\sigma)\infty} RT} \right) \tag{7-65}$$

with the assumption phase i is sufficiently dilute that $b^{(i)}_{(1)} x^{(i)}_{(1)} \doteq 1$.

In applying (7-65) to solutions of strong electrolytes, it should be satisfactory to take the surface concentrations of undissociated species as negligible with respect to their corresponding anions and cations. For a specific example, see Exercise 5.8.7-3.

We can also derive an expression for interfacial tension as a function of the composition of the substrate. Because of (7-33) and (7-59), (7-22) implies

$$\sum_{A=1}^{N-1} \left(1 - \frac{\alpha_{(A)}}{\alpha_{(N)}} \right) \left(\frac{b^{(i)}_{(A)} \, x^{(i)}_{(A)}}{a^{(i)\circ}_{(A)} \, b^{(\sigma)}_{(A)}} \right) = \exp\left(\frac{\gamma^{\circ}_{(1)} - \gamma}{c^{(\sigma)\infty} RT} \right) \tag{7-66}$$

We can eliminate $x^{(i)}_{(A)}$ from this using (7-60) to obtain

$$\frac{b^{(i)}_{(1)}}{b^{(\sigma)}_{(1)}} + \sum_{A=2}^{N-1} \left(1 - \frac{\alpha_{(A)}}{\alpha_{(N)}} \right) \left(\frac{b^{(i)}_{(A)}}{a^{(i)\circ}_{(A)} \, b^{(\sigma)}_{(A)}} - \frac{b^{(i)}_{(1)}}{b^{(\sigma)}_{(1)}} \right) x^{(i)}_{(A)}$$

$$= \exp\left(\frac{\gamma^{\circ}_{(1)} - \gamma}{c^{(\sigma)\infty} RT} \right) \tag{7-67}$$

in which we have observed (7-47). When phase i is so dilute that we are willing to approximate

$$b^{(i)}_{(1)}\left[1-\sum_{A=2}^{N-1}\left(1-\frac{\alpha_{(A)}}{\alpha_{(N)}}\right)x^{(i)}_{(A)}\right] \doteq 1 \tag{7-68}$$

(7-67) tells us

$$\frac{1}{b^{(\sigma)}_{(1)}}+\sum_{A=2}^{N-1}\left(1-\frac{\alpha_{(A)}}{\alpha_{(N)}}\right)\left(\frac{b^{(i)}_{(A)}\,x^{(i)}_{(A)}}{a^{(i)}_{(A)}{}^{\circ}\,b^{(\sigma)}_{(A)}}\right)=\exp\left(\frac{\gamma^{\circ}_{(1)}-\gamma}{c^{(\sigma)\infty}RT}\right) \tag{7-69}$$

in using (7-69) to describe solutions of strong electrolytes, $a^{(i)}_{(X)} = b^{(i)}_{(X)}x^{(i)}_{(X)}$ should be expressed in terms of the concentrations of its cation and anion using (7-63). For a specific application of (7-69), see Exercise 5.8.7-4.

In a similar manner, we can arrive at an expression for γ as a function of the surface concentrations that is somewhat less restrictive than (7-65). Because of (7-33) and (7-60), (7-22) demands

$$\sum_{A=1}^{N-1}\left(1-\frac{\alpha_{(A)}}{\alpha_{(N)}}\right)\left(\frac{a^{(i)}_{(A)}{}^{\circ}\,b^{(\sigma)}_{(A)}}{b^{(i)}_{(A)}}\right)\frac{c^{(\sigma,L)}}{c^{(\sigma)\infty}}=\exp\left(-\frac{\gamma^{\circ}_{(1)}-\gamma}{c^{(\sigma)\infty}RT}\right) \tag{7-70}$$

Eliminating $x^{(\sigma)}_{(1)}$ by means of (7-59), we find

$$\frac{b^{(\sigma)}_{(1)}}{b^{(i)}_{(1)}}-\sum_{A=2}^{N-1}\left(1-\frac{\alpha_{(A)}}{\alpha_{(N)}}\right)\left(\frac{b^{(\sigma)}_{(1)}}{b^{(i)}_{(1)}}-\frac{a^{(i)}_{(A)}{}^{\circ}\,b^{(\sigma)}_{(A)}}{b^{(i)}_{(A)}}\right)\frac{c^{(\sigma,L)}_{(A)}}{c^{(\sigma)\infty}}$$

$$=\exp\left(-\frac{\gamma^{\circ}_{(1)}-\gamma}{c^{(\sigma)\infty}RT}\right) \tag{7-71}$$

in which we have employed (7-47). In applying (7-71) to a solution of strong electrolytes, $a^{(\sigma)}_{(X)} = b^{(\sigma)}_{(X)}x^{(\sigma)}_{(X)}$ should be expressed in terms of the concentrations of its cation and anion using (7-64). Although the chemical potential and temperature have been assumed to be continuous across the dividing surface in deriving (7-22), (7-71) is easily extended to non-equilibrium systems as suggested in Exercise 5.8.7-1. For an example of the use of (7-71), see Exercise 5.8.7-5.

For very dilute solutions, we may be tempted to say $b^{(\sigma)}_{(1)} \doteq 1$ in (7-69) and $b^{(\sigma)}_{(1)}/b^{(i)}_{(1)} \doteq 1$ in (7-71). But in view of the well established behavior of simple brine solutions (see Exercise 5.8.7-6), caution is advised.

regular solution behavior There has been modest success in extending regular solution theory to the thermodynamic behavior of dividing surfaces (Defay *et al.* 1966; Sprow and Prausnitz 1966, 1967; Schay 1969; Joos 1971; Lucassen-Reynders 1973, 1976). In particular, Reid *et al.*

(1977, p. 621) recommended the Sprow and Prausnitz (1966, 1967) method for predicting the surface tension of non-aqueous, non-polar mixtures, which is based upon (7-39).

References

Butler, J. A. V., *Proc. R. Soc. London A* **135**, 348 (1932).

Defay, R., I. Prigogine, A. Bellemans, and D. H. Everett, "Surface Tension and Adsorption," John Wiley, New York (1966).

Dole, M., and J. A. Swartout, *J. Am. Chem. Soc.* **62**, 3039 (1940).

Frumkin, A., *Z. Phys. Chem.* **116**, 466 (1925).

Gibbs, J. W., "The Collected Works," vol. 1, Yale University Press, New Haven, Conn. (1948).

Jones, G., and W. A. Ray, *J. Am. Chem. Soc.* **59**, 187 (1937).

Joos, P., *J. Colloid Interface Sci.* **35**, 215 (1971).

Langmuir, I., *J. Am. Chem. Soc.* **39**, 1848 (1917).

Lewis, G. N., M. Randall, K. S. Pitzer, and L. Brewer, "Thermodynamics," second edition, McGraw-Hill, New York (1961).

Lin, C. Y., personal communication, June 1978.

Lucassen-Reynders, E. H., and M. van den Tempel, "Proceedings of the Fourth International Congress on Surface Active Substances," vol. 2, p. 779, Gordon and Breach, New York (1967).

Lucassen-Reynders, E. H., *J. Colloid Interface Sci.* **42**, 554 (1973).

Lucassen-Reynders, E. H., *Progress in Surface and Membrane Science*, **10**, 253 (1976).

Possoth, G., *Z. Phys. Chem.* **211**, 129 (1959).

Reid, R. C., J. M. Prausnitz, and T. K. Sherwood, "The Properties of Gases and Liquids," third edition, McGraw-Hill, New York (1977).

Schay, G., *Surface and Colloid Science* **2**, 155 (1969).

Sprow, F. B., and J. M. Prausnitz, *Trans. Faraday Soc.* **62**, 1105 (1966).

Sprow, F. B., and J. M. Prausnitz, *Can. J. Chem. Eng.* **45**, 25 (1967).

von Szyszkowski, B., *Z. Phys. Chem.* **64**, 385 (1908).

Exercises

5.8.7-1 *when dividing surface is not in equilibrium with substrate (Lin 1978)* Consider a dividing surface that is not in equilibrium with its substrate (an adjoining bulk phase in the limit as the dividing surface is approached). Derive an equation similar to (7-17) in which properties of the bulk phase refer to a fictitious substrate that would be in equilibrium with the dividing surface. In this way, extend to nonequilibrium systems (7-29), (7-40), (7-43), (7-50), and (7-71).

5.8.7-2 *Langmuir adsorption isotherm* Eliminate interfacial tension between the Frumkin equation of state and the von Szyszkowski equation of state to arrive at the **Langmuir adsorption isotherm** (Langmuir 1917; Lucassen-Reynders and van den Tempel 1967; Lucassen-Reynders 1976):

$$\frac{c_{(2)}^{(\sigma,L)}}{c^{(\sigma)\infty}} = \frac{x_{(2)}^{(i)}}{x_{(2)}^{(i)} + a_{(2)}^{(i)\circ} b_{(2)}^{(\sigma)*} / b_{(2)}^{(i)*}}$$

5.8.7-3 *single ionized surfactant in brine solution* Consider a dilute aqueous solution of NaCl and a single ionized surfactant NaR against its vapor. Let species N refer to the cation Na^+. Determine that (7-65) requires

$$1 - 2\,\frac{c_{(Cl)}^{(\sigma,L)}}{c^{(\sigma)\infty}} - 2\,\frac{c_{(R)}^{(\sigma,L)}}{c^{(\sigma)\infty}} = \frac{1}{b_{(1)}^{(\sigma)}} \exp\left(- \frac{\gamma_{(1)}^{\circ} - \gamma}{c^{(\sigma)\infty} RT} \right)$$

so long as the surface concentrations of the undissociated NaCl and NaR can be neglected.

5.8.7-4 *more about single ionized surfactant in brine solution* Consider the same dilute aqueous solution of NaCl and a single ionized surfactant NaR, again letting species N refer to the cation Na^+.

i) Show that (7-69) together with (7-64) require

$$\frac{1}{b_{(1)}^{(\sigma)}} + 2\,\frac{b_{(Cl)}^{(i)*}}{a_{(Cl)}^{(i)\circ} b_{(Cl)}^{(\sigma)*}}\, x_{(Cl)}^{(i)} + \frac{b_{(Na)}^{(i)*} b_{(Cl)}^{(i)*}}{a_{(NaCl)}^{(i)\circ} b_{(NaCl)}^{(\sigma)}}\, x_{(Na)}^{(i)} x_{(Cl)}^{(i)}$$

$$+ 2 \frac{b^{(i)}_{(R)}{}^{*}}{a^{(i)}_{(R)}\, {}^{\circ}b^{(\sigma)}_{(R)}{}^{*}} x^{(i)}_{(R)} + \frac{b^{(i)}_{(Na)}{}^{*} b^{(i)}_{(R)}{}^{*}}{a^{(i)}_{(Na\,R)}\, {}^{\circ}b^{(\sigma)}_{(NaR)}} x^{(i)}_{(Na)} x^{(i)}_{(R)}$$

$$= \exp\left(\frac{\gamma^{\circ}_{(1)} - \gamma}{c^{(\sigma)\infty} RT} \right)$$

ii) Local electrical neutrality requires

$$\sum_{A=1}^{N} \alpha_{(A)} x^{(i)}_{(A)} = 0$$

Show as a result

$$\frac{1}{b^{(\sigma)}_{(1)}} + 2 \frac{b^{(i)}_{(Cl)}{}^{*}}{a^{(i)}_{(Cl)}\, {}^{\circ}b^{(\sigma)}_{(Cl)}{}^{*}} x^{(i)}_{(Cl)}$$

$$+ \frac{b^{(i)}_{(Na)}{}^{*} b^{(i)}_{(Cl)}{}^{*}}{a^{(i)}_{(NaCl)}\, {}^{\circ}b^{(\sigma)}_{(NaCl)}} \left(x^{(i)}_{(Cl)} + x^{(i)}_{(R)} \right) x^{(i)}_{(Cl)}$$

$$+ 2 \frac{b^{(i)}_{(R)}{}^{*}}{a^{(i)}_{(R)}\, {}^{\circ}b^{(\sigma)}_{(R)}{}^{*}} x^{(i)}_{(R)}$$

$$+ \frac{b^{(i)}_{(Na)}{}^{*} b^{(i)}_{(R)}{}^{*}}{a^{(i)}_{(Na\,R)}\, {}^{\circ}b^{(\sigma)}_{(NaR)}} \left(x^{(i)}_{(Cl)} + x^{(i)}_{(R)} \right) x^{(i)}_{(R)}$$

$$= \exp\left(\frac{\gamma^{\circ}_{(1)} - \gamma}{c^{(\sigma)\infty} RT} \right)$$

iii) Rewrite this in terms of the *stoichiometric* mole fractions of NaCl and NaR, $x^{(i)}_{(NaCl)}$ and $x^{(i)}_{(NaR)}$, to find

$$\frac{1}{b^{(\sigma)}_{(1)}} + 2 \frac{b^{(i)}_{(Cl)}{}^{*}}{a^{(i)}_{(Cl)}\, {}^{\circ}b^{(\sigma)}_{(Cl)}{}^{*}} x^{(i)}_{(NaCl)}$$

$$+ \frac{b^{(i)}_{(Na)}{}^{*} b^{(i)}_{(Cl)}{}^{*}}{a^{(i)}_{(NaCl)}\, {}^{\circ}b^{(\sigma)}_{(NaCl)}} \left(x^{(i)}_{(NaCl)} + x^{(i)}_{(NaR)} \right) x^{(i)}_{(NaCl)}$$

$$+ 2 \frac{b^{(i)\,*}_{(R)}}{a^{(i)\,\circ}_{(R)}\, b^{(\sigma)\,*}_{(R)}}\, x^{(i)}_{(NaR)}$$

$$+ \frac{b^{(i)\,*}_{(Na)}\, b^{(i)\,*}_{(R)}}{a^{(i)\,\circ}_{(Na\,R)}\, b^{(\sigma)}_{(NaR)}} \left(x^{(i)}_{(NaCl)} + x^{(i)}_{(NaR)} \right) x^{(i)}_{(NaR)}$$

$$= \exp\left(\frac{\gamma^{\circ}_{(1)} - \gamma}{c^{(\sigma)\infty} RT} \right)$$

5.8.7-5 *still more about single ionized surfactant in brine solution* Consider the same dilute aqueous solution of NaCl and a single ionized surfactant NaR used in Exercises 5.8.7-3 and 5.8.7-4. Show that (7-71) together with (7-64) and the requirement of electrical neutrality implies

$$\frac{b^{(\sigma)}_{(1)}}{b^{(i)}_{(1)}} - 2 \left(\frac{b^{(\sigma)}_{(1)}}{b^{(i)}_{(1)}} - \frac{a^{(i)\,\circ}_{(Cl)}\, b^{(\sigma)}_{(Cl)}}{b^{(i)}_{(Cl)}} \right) \left(\frac{1}{c^{(\sigma)\infty}} \right) c^{(\sigma,L)}_{(Cl)}$$

$$- \left(\frac{b^{(\sigma)}_{(1)}}{b^{(i)}_{(1)}} - \frac{a^{(i)\,\circ}_{(NaCl)}\, b^{(\sigma)}_{(NaCl)}}{b^{(i)}_{(NaCl)}} \right) \left(\frac{b^{(\sigma)\,*}_{(Na)}\, b^{(\sigma)\,*}_{(Cl)}}{\left(c^{(\sigma)\infty}\right)^2\, b^{(\sigma)}_{(NaCl)}} \right)$$

$$\times \left(c^{(\sigma,L)}_{(Cl)} + c^{(\sigma,L)}_{(R)} \right) c^{(\sigma,L)}_{(Cl)}$$

$$- 2 \left(\frac{b^{(\sigma)}_{(1)}}{b^{(i)}_{(1)}} - \frac{a^{(i)\,\circ}_{(R)}\, b^{(\sigma)}_{(R)}}{b^{(i)}_{(R)}} \right) \left(\frac{1}{c^{(\sigma)\infty}} \right) c^{(\sigma,L)}_{(R)}$$

$$- \left(\frac{b^{(\sigma)}_{(1)}}{b^{(i)}_{(1)}} - \frac{a^{(i)\,\circ}_{(NaR)}\, b^{(\sigma)}_{(NaR)}}{b^{(i)}_{(NaR)}} \right) \left(\frac{b^{(\sigma)\,*}_{(Na)}\, b^{(\sigma)\,*}_{(R)}}{\left(c^{(\sigma)\infty}\right)^2\, b^{(\sigma)}_{(NaR)}} \right)$$

$$\times \left(c^{(\sigma,L)}_{(Cl)} + c^{(\sigma,L)}_{(R)} \right) c^{(\sigma,L)}_{(R)}$$

$$= \exp\left(- \frac{\gamma^{\circ}_{(1)} - \gamma}{c^{(\sigma)\infty} RT} \right)$$

5.8.7-6 *brine solutions* For a solution of NaCl, Exercise 5.8.7-4 requires

$$\frac{1}{b^{(\sigma)}_{(1)}} + 2\,\frac{b^{(i)\star}_{(Cl)}}{a^{(i)\circ}_{(Cl)}\, b^{(\sigma)\star}_{(Cl)}}\, x^{(i)}_{(NaCl)} + \frac{b^{(i)\star}_{(Na)}\, b^{(i)\star}_{(Cl)}}{a^{(i)\circ}_{(NaCl)}\, b^{(\sigma)}_{(NaCl)}} \left(x^{(i)}_{(NaCl)}\right)^2$$

$$= \exp\left(\frac{\gamma^{\circ}_{(1)} - \gamma}{c^{(\sigma)\infty} RT}\right)$$

Argue that for sufficiently dilute solutions of NaCl, $b^{(\sigma)}_{(1)} \to 1$ and $(\gamma^{\circ}_{(1)} - \gamma) > 0$.

This is in agreement with experimental observation for strong electrolytes (Jones and Ray 1937; Dole and Swartout 1940; Passoth 1959). But at very low concentrations this trend is reversed and the surface tension increases with increasing concentration to become larger than the surface tension of water. This suggests that at very low salt concentrations $b^{(\sigma)}_{(1)}$ may become an exponential function of concentration which dominates the concentration dependence of γ.

5.8.7-7 *when there are two solvents* When there are two solvents, it will normally be more convenient to determine the interfacial tension in the presence of surfactant relative to the interfacial tension in the absence of surfactant.

As an example, consider a three component system: solvent 1, solvent 2, and surfactant 3. You may consider both phases to be dilute.

i) In the absence of surfactant, use (7-54) to show that, for phase i in which solvent 1 is the principal component,

$$\frac{1}{b^{(\sigma)}_{(1)}} + \frac{b^{(i)\star}_{(2)}}{a^{(i)\circ}_{(2)}\, b^{(\sigma)\star}_{(2)}}\, x^{(i)}_{(2)eq} = \exp\left(\frac{\gamma^{\circ}_{(1)} - \gamma_{(12)}}{c^{(\sigma)\infty} RT}\right)$$

Here $\gamma_{(12)}$ is the equilibrium interfacial tension between solvents 1 and 2 and $x^{(i)}_{(2)eq}$ is the corresponding mole fraction of solvent 2 in phase i.

ii) Show that in the presence of surfactant

$$A + B\, x^{(i)}_{(3)} = \exp\left(\frac{\gamma_{(12)} - \gamma}{c^{(\sigma)\infty} RT}\right)$$

where

$$A \equiv \left[\frac{1}{b^{(\sigma)}_{(1)}} + \frac{b^{(i)\star}_{(2)}}{a^{(i)\circ}_{(2)}\, b^{(\sigma)\star}_{(2)}}\, x^{(i)}_{(2)}\right]\left[\frac{1}{b^{(\sigma)}_{(1)}} + \frac{b^{(i)\star}_{(2)}}{a^{(i)\circ}_{(2)}\, b^{(\sigma)\star}_{(2)}}\, x^{(i)}_{(2)eq}\right]^{-1}$$

$$B \equiv \frac{b_{(3)}^{(i)*}}{a_{(3)}^{(i)\circ} b_{(3)}^{(\sigma)*}} \left[\frac{1}{b_{(1)}^{(\sigma)}} + \frac{B_{(2)}^{(i)*}}{a_{(2)}^{(i)\circ} b_{(2)}^{(\sigma)*}} x_{(2)eq}^{(i)} \right]^{-1}$$

Only if $x_{(2)}^{(i)}$ is independent of $x_{(3)}^{(i)}$ (perhaps zero), can we say A = 1.

5.8.7-8 *single ionized surfactant in brine solution against organic solvent* Let us now consider a dilute aqueous solution of NaCl and a single ionized surfactant NaR against an organic solvent 2.

i) Extend Exercise 5.8.7-3 to this system. Following the suggestion given in Exercise 5.8.7-7, determine that

$$A_1 - A_2 \frac{c_{(R)}^{(\sigma,L)}}{c^{(\sigma)\infty}} = \exp\left(- \frac{\gamma_{(2,brine)} - \gamma}{c^{(\sigma)\infty} RT} \right)$$

where

$$A_1 \equiv \left[1 - \frac{c_{(2)}^{(\sigma,L)}}{c^{(\sigma)\infty}} - 2 \frac{c_{(Cl)}^{(\sigma,L)}}{c^{(\sigma)\infty}} \right] \left[1 - \frac{c_{(2)eq}^{(\sigma,L)}}{c^{(\sigma)\infty}} - 2 \frac{c_{(Cl)eq}^{(\sigma,L)}}{c^{(\sigma)\infty}} \right]^{-1}$$

$$A_2 \equiv 2 \left[1 - \frac{c_{(2)eq}^{(\sigma,L)}}{c^{(\sigma)\infty}} - 2 \frac{c_{(Cl)eq}^{(\sigma,L)}}{c^{(\sigma)\infty}} \right]^{-1}$$

By $\gamma_{(2,brine)}$ we mean the interfacial tension of the organic solvent 2 against brine in the absence of NaR; $c_{(2)eq}^{(\sigma,L)}$ and $c_{(Cl)eq}^{(\sigma,L)}$ are the corresponding interfacial concentrations of solvent 2 and chloride ion. Only if we are willing to say $c_{(2)}^{(\sigma,L)} = c_{(2)eq}^{(\sigma,L)}$ and $c_{(Cl)}^{(\sigma,L)} = c_{(Cl)eq}^{(\sigma,L)}$, can we set $A_1 = 1$.

ii) In a similar manner, extend Exercise 5.8.7-4 (iii) to this system to conclude

$$B_1 + B_2 \, x_{(NaR)}^{(i)} + B_3 \left(x_{(NaR)}^{(i)}\right)^2 = \exp\left(\frac{\gamma_{(2,brine)} - \gamma}{c^{(\sigma)\infty} RT} \right)$$

in which

$$B_1 \equiv \left[\frac{1}{b_{(1)}^{(\sigma)}} + \frac{b_{(2)}^{(i)*}}{a_{(2)}^{(i)\circ} b_{(2)}^{(\sigma)*}} x_{(2)}^{(i)} + 2 \frac{b_{(Cl)}^{(i)*}}{a_{(Cl)}^{(i)\circ} b_{(Cl)}^{(\sigma)*}} x_{(NaCl)}^{(i)} \right.$$

$$+ \frac{b^{(i)*}_{(Na)} b^{(i)*}_{(Cl)}}{a^{(i)}_{(NaCl)} b^{(\sigma)}_{(NaCl)}} \left(x^{(i)}_{(NaCl)} \right)^2 \Bigg] E$$

$$B_2 \equiv \left[2 \frac{b^{(i)*}_{(R)}}{a^{(i)}_{(R)} b^{(\sigma)*}_{(R)}} + \left(\frac{b^{(i)*}_{(Na)} b^{(i)*}_{(Cl)}}{a^{(i)}_{(NaCl)} b^{(\sigma)}_{(NaCl)}} \right. \right.$$

$$\left. \left. + \frac{b^{(i)*}_{(Na)} b^{(i)*}_{(R)}}{a^{(i)}_{(NaR)} b^{(\sigma)}_{(NaR)}} \right) x^{(i)}_{(NaCl)} \right] E$$

$$B_3 \equiv \frac{b^{(i)*}_{(Na)} b^{(i)*}_{(R)}}{a^{(i)}_{(NaR)} b^{(\sigma)}_{(NaR)}} E$$

$$E \equiv \left[\frac{1}{b^{(\sigma)}} + \frac{b^{(i)*}_{(2)}}{a^{(i)}_{(2)} b^{(\sigma)*}_{(2)}} x^{(i)}_{(2)eq} + 2 \frac{b^{(i)*}_{(Cl)}}{a^{(i)}_{(Cl)} b^{(\sigma)*}_{(Cl)}} x^{(i)}_{(NaCl)eq} \right.$$

$$\left. + \frac{b^{(i)*}_{(Na)} b^{(i)*}_{(Cl)}}{a^{(i)}_{(NaCl)} b^{(\sigma)}_{(NaCl)}} \left(x^{(i)}_{(NaCl)eq} \right)^2 \right]^{-1}$$

If $x^{(i)}_{(2)} = x^{(i)}_{(2)eq}$ and $x^{(i)}_{(NaCl)} = x^{(i)}_{(NaCl)eq}$, we have $B_1 = 1$.

iii) Extend Exercise 5.8.7-5 to this system to find a relationship of the form

$$C_1 - C_2 \frac{c^{(\sigma,L)}_{(R)}}{c^{(\sigma)\infty}} - C_3 \left(\frac{c^{(\sigma,L)}_{(R)}}{c^{(\sigma)\infty}} \right)^2 = \exp\left(- \frac{\gamma_{(2,\,brine)} - \gamma}{c^{(\sigma)\infty} RT} \right)$$

If we assume $c^{(\sigma,L)}_{(2)} = c^{(\sigma,L)}_{(2)eq}$ and $c^{(\sigma,L)}_{(Cl)} = c^{(\sigma,L)}_{(Cl)eq}$, then $C_1 = 1$. While this result is somewhat similar to that derived in (i), remember that Exercise 5.8.7-3 assumes equilibrium between the dividing surface and the substrate while Exercise 5.8.7-5 does not.

iv) The adsorption isotherm for this system may be derived by eliminating γ between the results of (i) and (ii):

$$\frac{c^{(\sigma,L)}_{(R)}}{c^{(\sigma)\infty}} = \frac{A_1 \left[B_1 + B_2 x^{(i)}_{(NaR)} + B_3 \left(x^{(i)}_{(NaR)} \right)^2 \right] - 1}{A_2 \left[B_1 + B_2 x^{(i)}_{(NaR)} + B_3 \left(x^{(i)}_{(NaR)} \right)^2 \right]}$$

This reduces to the form of the Langmuir isotherm (see Exercise 5.8.7-2), when $A_1 = B_1 = 1$ and $B_3 = 0$.

5.8.8 Surface behavior: more on implications of jump Clausius -Duhem inequality In Sec. 5.8.5, we found that, as the result of the jump Clausius-Duhem inequality, the surface thermal energy flux vector $\mathbf{e}^{(\sigma)}$ is related to the surface energy flux vector $\mathbf{q}^{(\sigma)}$ by

$$\mathbf{e}^{(\sigma)} = \mathbf{q}^{(\sigma)} - \sum_{B=1}^{N-1} \left(\frac{\partial \hat{A}^{(\sigma)}}{\partial \omega_{(B)}} \right)_{\hat{\mathcal{A}}, T^{(\sigma)}, \omega_{(C)}(C \neq B, N)} \mathbf{j}_{(B)} \tag{8-1}$$

In the last section, we introduced the surface chemical potential and saw

$$\left(\frac{\partial \hat{A}^{(\sigma)}}{\partial \omega_{(B)}} \right)_{\hat{\mathcal{A}}, T^{(\sigma)}, \omega_{(C)}(C \neq B, N)} = \mu_{(B)} - \mu_{(N)} \tag{8-2}$$

Since (Table 5.2.2-4)

$$\sum_{B=1}^{N} \mathbf{j}_{(B)} = 0 \tag{8-3}$$

we have

$$\mathbf{e}^{(\sigma)} = \mathbf{q}^{(\sigma)} - \sum_{B=1}^{N} \mu_{(B)} \mathbf{j}_{(B)} \tag{8-4}$$

Observing that (Sec. 5.4.1)

$$\sum_{B=1}^{N} r_{(B)} = 0 \tag{8-5}$$

we may express (5-26) in terms of the thermodynamic surface tension (Sec. 5.8.6)

$$\gamma \equiv \left(\frac{\partial \hat{A}^{(\sigma)}}{\partial \mathcal{A}} \right)_{T^{(\sigma)}, \omega_{(B)}} \tag{8-6}$$

and the chemical potentials (8-2):

$$
\begin{aligned}
&\frac{1}{T^{(\sigma)}}\,\mathbf{e}^{(\sigma)} \cdot \nabla_{(\sigma)} T^{(\sigma)} - \text{tr}[(\,\mathbf{T}^{(\sigma)} - \gamma \mathbf{P}\,) \cdot \mathbf{D}^{(\sigma)}\,] \\
&\quad + \sum_{B=1}^{N} \mathbf{j}^{(\sigma)}_{(B)} \cdot (\,\nabla_{(\sigma)} \mu^{(\sigma)}_{(B)} - \mathbf{b}^{(\sigma)}_{(B)}\,) + \sum_{B=1}^{N} \mu^{(\sigma)}_{(B)}\, r^{(\sigma)}_{(B)} \\
&\quad + \gamma \hat{\mathcal{A}}\,[\,\rho(\,\mathbf{v} \cdot \boldsymbol{\xi} - v^{(\sigma)}_{(\xi)}\,)] \\
&\quad - \sum_{B=1}^{N} \mu^{(\sigma)}_{(B)} [\,\mathbf{j}_{(B)} \cdot \boldsymbol{\xi} + \rho(\,\omega_{(B)} - \omega^{(\sigma)}_{(B)}\,)(\,\mathbf{v} \cdot \boldsymbol{\xi} - v^{(\sigma)}_{(\xi)}\,)] \\
&\quad + \Big[\rho \left(\hat{U} - \hat{U}^{(\sigma)} + \frac{1}{2}\left|\mathbf{v} - \mathbf{v}^{(\sigma)}\right|^2\right)(\mathbf{v} \cdot \boldsymbol{\xi} - v^{(\sigma)}_{(\xi)}) \\
&\qquad - (\,\mathbf{v} - \mathbf{v}^{(\sigma)}\,) \cdot \mathbf{T} \cdot \boldsymbol{\xi} + \mathbf{q} \cdot \boldsymbol{\xi} \\
&\qquad - \rho T^{(\sigma)}(\,\hat{S} - \hat{S}^{(\sigma)}\,)(\,\mathbf{v} \cdot \boldsymbol{\xi} - v^{(\sigma)}_{(\xi)}\,) - \frac{T^{(\sigma)}}{T}\,\mathbf{e} \cdot \boldsymbol{\xi}\Big] \\
&\leq 0
\end{aligned}
\tag{8-7}
$$

In view of the surface Euler equation (6-11), this reduces to

$$
\begin{aligned}
&\frac{1}{T^{(\sigma)}}\,\mathbf{e}^{(\sigma)} \cdot \nabla_{(\sigma)} T^{(\sigma)} - \text{tr}[(\,\mathbf{T}^{(\sigma)} - \gamma \mathbf{P}\,) \cdot \mathbf{D}^{(\sigma)}\,] \\
&\quad + \sum_{B=1}^{N} \mathbf{j}^{(\sigma)}_{(B)} \cdot (\,\nabla_{(\sigma)} \mu^{(\sigma)}_{(B)} - \mathbf{b}^{(\sigma)}_{(B)}\,) + \sum_{B=1}^{N} \mu^{(\sigma)}_{(B)}\, r^{(\sigma)}_{(B)} \\
&\quad - \sum_{B=1}^{N} \mu^{(\sigma)}_{(B)} [\,\mathbf{j}_{(B)} \cdot \boldsymbol{\xi} + \rho \omega_{(B)}(\,\mathbf{v} \cdot \boldsymbol{\xi} - v^{(\sigma)}_{(\xi)}\,)] \\
&\quad + \Big[\rho \left(\,\hat{U} + \frac{1}{2}\left|\mathbf{v} - \mathbf{v}^{(\sigma)}\right|^2\right)(\,\mathbf{v} \cdot \boldsymbol{\xi} - v^{(\sigma)}_{(\xi)}\,) \\
&\qquad - (\,\mathbf{v} - \mathbf{v}^{(\sigma)}\,) \cdot \mathbf{T} \cdot \boldsymbol{\xi} + \mathbf{q} \cdot \boldsymbol{\xi}
\end{aligned}
$$

$$- \rho T^{(\sigma)} \hat{S} (\mathbf{v} \cdot \boldsymbol{\xi} - v^{(\sigma)}_{(\xi)}) - \frac{T^{(\sigma)}}{T} \mathbf{e} \cdot \boldsymbol{\xi} \Big]$$

$$\leq 0 \tag{8-8}$$

As I point out in my introduction to Chapter 6, we will normally assume that chemical potential, temperature, and the tangential components of the mass-averaged velocity are continuous across phase interfaces,

$$\text{at } \Sigma : \quad \mu_{(B)} = \mu^{(\sigma)}_{(B)} \tag{8-9}$$

$$\text{at } \Sigma : \quad T = T^{(\sigma)} \tag{8-10}$$

$$\text{at } \Sigma : \quad \mathbf{P} \cdot \mathbf{v} = \mathbf{P} \cdot \mathbf{v}^{(\sigma)} \tag{8-11}$$

With these assumptions, (8-8) further simplifies to

$$\frac{1}{T^{(\sigma)}} \mathbf{e}^{(\sigma)} \cdot \nabla_{(\sigma)} T^{(\sigma)} - \text{tr}[(\mathbf{T}^{(\sigma)} - \gamma \mathbf{P}) \cdot \mathbf{D}^{(\sigma)}]$$

$$+ \sum_{B=1}^{N} \mathbf{j}^{(\sigma)}_{(B)} \cdot (\nabla_{(\sigma)} \mu^{(\sigma)}_{(B)} - \mathbf{b}^{(\sigma)}_{(B)}) + \sum_{B=1}^{N} \mu^{(\sigma)}_{(B)} r^{(\sigma)}_{(B)}$$

$$+ \Big[\frac{1}{2} \rho (\mathbf{v} \cdot \boldsymbol{\xi} - v^{(\sigma)}_{(\xi)})^3 - (\mathbf{v} \cdot \boldsymbol{\xi} - v^{(\sigma)}_{(\xi)}) \boldsymbol{\xi} \cdot (\mathbf{T} + P \mathbf{I}) \cdot \boldsymbol{\xi} \Big]$$

$$\leq 0 \tag{8-12}$$

To arrive at this result, we have employed both the Euler equation (3-11) and the expression for the thermal energy flux in terms of the energy flux (4-4). This is a useful alternative form of the jump Clausius-Duhem inequality.

Where there is no mass transfer between the interface and the adjoining phases, we see that this alternative form of the jump Clausius-Duhem inequality requires

$$\frac{1}{T^{(\sigma)}} \mathbf{e}^{(\sigma)} \cdot \nabla_{(\sigma)} T^{(\sigma)} - \text{tr}[(\mathbf{T}^{(\sigma)} - \gamma \mathbf{P}) \cdot \mathbf{D}^{(\sigma)}]$$

$$+ \sum_{B=1}^{N} \mathbf{j}^{(\sigma)}_{(B)} \cdot (\nabla_{(\sigma)} \mu^{(\sigma)}_{(B)} - \mathbf{b}^{(\sigma)}_{(B)}) + \sum_{B=1}^{N} \mu^{(\sigma)}_{(B)} r^{(\sigma)}_{(B)}$$

$$\leq 0 \tag{8-13}$$

Exercises

5.8.8-1 *alternative expression for* $\mathbf{e}^{(\sigma)}$ The **surface Gibbs free energy** per unit mass $\hat{G}^{(\sigma)}$ is defined in terms of the **surface enthalpy** per unit mass

$$\hat{H}^{(\sigma)} \equiv \hat{U}^{(\sigma)} - \gamma \hat{\mathcal{A}}$$

by

$$\begin{aligned}\hat{G}^{(\sigma)} &\equiv \hat{H}^{(\sigma)} - T^{(\sigma)} \hat{S}^{(\sigma)} \\ &= \hat{U}^{(\sigma)} - \gamma \hat{\mathcal{A}} - T^{(\sigma)} \hat{S}^{(\sigma)} \\ &= \hat{A}^{(\sigma)} - \gamma \hat{\mathcal{A}}\end{aligned}$$

i) Use the surface Gibbs equation to conclude

$$\left(\frac{\partial \hat{G}^{(\sigma)}}{\partial \omega_{(B)}^{(\sigma)}} \right)_{T^{(\sigma)}, \gamma, \omega_{(C)}^{(\sigma)} (C \neq B, N)} = \mu_{(B)}^{(\sigma)} - \mu_{(N)}^{(\sigma)}$$

ii) The **partial mass surface Gibbs free energies** are defined by (Exercise 5.8.6- 5)

$$\overline{\overline{G}}_{(A)}^{(\sigma)} - \overline{\overline{G}}_{(N)}^{(\sigma)} \equiv \left(\frac{\partial \hat{G}^{(\sigma)}}{\partial \omega_{(A)}^{(\sigma)}} \right)_{T^{(\sigma)}, \gamma, \omega_{(B)}^{(\sigma)} (B \neq A, N)}$$

and the requirement that

$$\hat{G} = \sum_{A=1}^{N} \overline{\overline{G}}_{(A)}^{(\sigma)} \, \omega_{(A)}^{(\sigma)}$$

Prove that

$$\mathbf{e}^{(\sigma)} = \mathbf{q}^{(\sigma)} - \sum_{A=1}^{N} \overline{\overline{G}}_{(A)}^{(\sigma)} \, \mathbf{j}_{(A)}^{(\sigma)}$$

5.8.8-2 *alternative form for (8-12)* The partial mass surface entropies are defined by (Exercise 5.8.6-5)

$$\overline{\overline{S}}^{(\sigma)}_{(A)} - \overline{\overline{S}}^{(\sigma)}_{(N)} \equiv \left(\frac{\partial \hat{S}^{(\sigma)}}{\partial \omega^{(\sigma)}_{(A)}} \right)_{T^{(\sigma)}, \gamma, \omega^{(\sigma)}_{(B)} (B \neq A, N)}$$

and the requirement that

$$\hat{S}^{(\sigma)} = \sum_{A=1}^{N} \overline{\overline{S}}^{(\sigma)}_{(A)} \, \omega^{(\sigma)}_{(A)}$$

Let us define

$$\mathbf{d}^{(\sigma)}_{(A)} \equiv \frac{\rho^{(\sigma)}_{(A)}}{c^{(\sigma)} R T^{(\sigma)}} \Big(\nabla_{(\sigma)} \mu^{(\sigma)}_{(A)} - \mathbf{b}^{(\sigma)}_{(A)} + \overline{\overline{S}}^{(\sigma)}_{(A)} \nabla_{(\sigma)} T^{(\sigma)}$$

$$+ \hat{\mathcal{A}} \nabla_{(\sigma)} \gamma + \sum_{B=1}^{N} \omega^{(\sigma)}_{(B)} \, \mathbf{b}^{(\sigma)}_{(B)} \Big)$$

i) Prove that (8-12) may be expressed as

$$\frac{1}{T^{(\sigma)}} \Big(\mathbf{e}^{(\sigma)} - \sum_{B=1}^{N} T^{(\sigma)} \overline{\overline{S}}^{(\sigma)}_{(B)} \, \mathbf{j}^{(\sigma)}_{(B)} \Big) \cdot \nabla_{(\sigma)} T^{(\sigma)}$$

$$- \operatorname{tr}[(\mathbf{T}^{(\sigma)} - \gamma \mathbf{P}) \cdot \mathbf{D}^{(\sigma)}]$$

$$+ \sum_{B=1}^{N} \frac{c^{(\sigma)} R T^{(\sigma)}}{\rho^{(\sigma)}_{(B)}} \mathbf{j}^{(\sigma)}_{(B)} \cdot \mathbf{d}^{(\sigma)}_{(B)} + \sum_{B=1}^{N} \mu^{(\sigma)}_{(B)} \, r^{(\sigma)}_{(B)}$$

$$+ \Big[\frac{1}{2} \rho (\mathbf{v} \cdot \boldsymbol{\xi} - v^{(\xi)})^3$$

$$- (\mathbf{v} \cdot \boldsymbol{\xi} - v^{(\xi)}) \, \boldsymbol{\xi} \cdot (\mathbf{T} + P \mathbf{I}) \cdot \boldsymbol{\xi} \Big]$$

$$\leq 0$$

ii) Noting (Exercise 5.8.6-8)

$$\overline{\overline{S}}^{(\sigma)}_{(A)} = -\left(\frac{\partial \mu^{(\sigma)}_{(A)}}{\partial T^{(\sigma)}}\right)_{\gamma,\omega^{(\sigma)}_{(B)}}$$

$$\overline{\overline{\mathcal{A}}}_{(A)} = -\left(\frac{\partial \mu^{(\sigma)}_{(A)}}{\partial \gamma}\right)_{T^{(\sigma)},\omega^{(\sigma)}_{(B)}}$$

determine that

$$\mathbf{d}^{(\sigma)}_{(A)} = \frac{\rho^{(\sigma)}_{(A)}}{c^{(\sigma)}RT^{(\sigma)}}\left[\sum_{\substack{B=1\\ B\neq A}}^{N}\left(\frac{\partial \mu^{(\sigma)}_{(A)}}{\partial \omega^{(\sigma)}_{(B)}}\right)_{T^{(\sigma)},\gamma,\omega^{(\sigma)}_{(C)}(C\neq A,B)}\nabla_{(\sigma)}\omega^{(\sigma)}_{(B)}\right.$$

$$\left. -\left(\overline{\overline{\mathcal{A}}}_{(A)} - \hat{\mathcal{A}}\right)\nabla_{(\sigma)}\gamma - \left(\mathbf{b}^{(\sigma)}_{(A)} - \sum_{B=1}^{N}\omega^{(\sigma)}_{(B)}\,\mathbf{b}^{(\sigma)}_{(B)}\right)\right]$$

iii) From Exercise 5.8.6-1, we see that

$$\mu^{(\sigma,m)}_{(A)} = M_{(A)}\mu^{(\sigma)}_{(A)}$$

where $M_{(A)}$ is the molecular weight of species A. We also know that (Exercise 5.8.6-7)

$$\overline{\overline{\mathcal{A}}}^{(m)}_{(A)} = M_{(A)}\overline{\overline{\mathcal{A}}}_{(A)}$$

Here $\overline{\overline{\mathcal{A}}}^{(m)}_{(A)}$ is the partial molal interfacial area defined by (Exercise 5.8.6-6)

$$\overline{\overline{\mathcal{A}}}^{(m)}_{(A)} - \overline{\overline{\mathcal{A}}}^{(m)}_{(N)} \equiv \left(\frac{\partial \tilde{\mathcal{A}}}{\partial x^{(\sigma)}_{(A)}}\right)_{T^{(\sigma)},\gamma,x^{(\sigma)}_{(B)}(B\neq A,N)}$$

and

$$\tilde{\mathcal{A}} = \sum_{A=1}^{N}\overline{\overline{\mathcal{A}}}^{(m)}_{(A)}\,x^{(\sigma)}_{(A)}$$

Reason that we can also write

$$\mathbf{d}^{(\sigma)}_{(A)} = \frac{x^{(\sigma)}_{(A)}}{RT^{(\sigma)}}\left[\sum_{\substack{B=1\\ B\neq A}}^{N}\left(\frac{\partial \mu^{(\sigma,m)}_{(A)}}{\partial x^{(\sigma)}_{(B)}}\right)_{T^{(\sigma)},\gamma,x^{(\sigma)}_{(C)}(C\neq A,B)}\nabla_{(\sigma)}x^{(\sigma)}_{(B)}\right.$$

$$- M_{(A)} \left(\frac{\bar{\bar{\mathcal{A}}}_{(A)}^{(m)}}{M_{(A)}} - \hat{\mathcal{A}} \right) \nabla_{(\sigma)}\gamma$$

$$- M_{(A)} \left(\mathbf{b}_{(A)}^{(\sigma)} - \sum_{B=1}^{N} \omega_{(B)}^{(\sigma)} \mathbf{b}_{(B)}^{(\sigma)} \right) \Bigg] \Bigg]$$

iv) It will be useful to observe that, as the result of the Gibbs-Duhem equation,

$$\sum_{A=1}^{N} \mathbf{d}_{(A)}^{(\sigma)} = 0$$

5.8.9 Alternative forms for the energy balances and the Clausius-Duhem inequalities In Sec. 5.6.4, we derived two forms of the jump energy balance: equations (A) and (B) of Table 5.8.9-1. We are usually more interested in determining the temperature distribution within a material than the distribution of internal energy.

From the definition of the surface enthalpy per unit mass $\hat{H}^{(\sigma)}$ (see Exercise 5.8.6-2)

$$\hat{H}^{(\sigma)} \equiv \hat{U}^{(\sigma)} - \gamma\hat{\mathcal{A}} \tag{9-1}$$

it follows that

$$\frac{d_{(\mathbf{v}^{(\sigma)})}\hat{U}^{(\sigma)}}{dt} = \frac{d_{(\mathbf{v}^{(\sigma)})}\hat{H}^{(\sigma)}}{dt} + \hat{\mathcal{A}}\frac{d_{(\mathbf{v}^{(\sigma)})}\gamma}{dt} + \gamma\frac{d_{(\mathbf{v}^{(\sigma)})}\hat{\mathcal{A}}}{dt} \tag{9-2}$$

With the observation that the overall jump mass balance for a multicomponent dividing surface requires (Sec. 5.4.1)

$$\rho^{(\sigma)}\frac{d_{(\mathbf{v}^{(\sigma)})}\hat{\mathcal{A}}}{dt} = -\frac{1}{\rho^{(\sigma)}}\frac{d_{(\mathbf{v}^{(\sigma)})}\rho^{(\sigma)}}{dt}$$

$$= \frac{1}{\rho^{(\sigma)}}\left\{ \rho^{(\sigma)}\text{div}_{(\sigma)}\mathbf{v}^{(\sigma)} + [\,\rho(\,\mathbf{v}\cdot\boldsymbol{\xi} - v_{(\xi)}^{(\sigma)}\,)]\right\} \tag{9-3}$$

we can use (9-2) to express equation (B) of Table 5.8.9-1 as

$$\rho^{(\sigma)} \frac{d_{(\mathbf{v}^{(\sigma)})}\hat{H}^{(\sigma)}}{dt} = -\operatorname{div}_{(\sigma)}\mathbf{q}^{(\sigma)} - \frac{d_{(\mathbf{v}^{(\sigma)})}\gamma}{dt} + \operatorname{tr}(\, \mathbf{S}^{(\sigma)} \cdot \nabla_{(\sigma)}\mathbf{v}^{(\sigma)}\,)$$

$$+ \sum_{A=1}^{N} \mathbf{j}_{(A)}^{(\sigma)} \cdot \mathbf{b}_{(A)}^{(\sigma)} + \rho^{(\sigma)}Q^{(\sigma)}$$

$$+ \Big[-\rho(\, \hat{U} - \hat{H}^{(\sigma)}\,)(\, \mathbf{v}\cdot\boldsymbol{\xi} - v_{(\xi)}^{(\sigma)}\,) - \frac{1}{2}\rho(\, \mathbf{v}\cdot\boldsymbol{\xi} - v_{(\xi)}^{(\sigma)}\,)^3$$

$$+ (\, \mathbf{v}\cdot\boldsymbol{\xi} - v_{(\xi)}^{(\sigma)}\,)\boldsymbol{\xi}\cdot\mathbf{T}\cdot\boldsymbol{\xi} - \mathbf{q}\cdot\boldsymbol{\xi} \Big] \tag{9-4}$$

This is the enthalpy form for the jump energy balance, equation (C) of Table 5.8.9-1. Starting with the surface Gibbs-Duhem equation (Sec. 5.8.6), we find

$$d\hat{H}^{(\sigma)} = T^{(\sigma)}d\hat{S}^{(\sigma)} - \hat{\mathcal{A}}\,d\gamma + \sum_{A=1}^{N-1} (\, \mu_{(A)}^{(\sigma)} - \mu_{(N)}^{(\sigma)}\,)d\omega_{(B)}^{(\sigma)}$$

$$= T^{(\sigma)}\left(\frac{\partial \hat{S}^{(\sigma)}}{\partial T^{(\sigma)}}\right)_{\gamma,\omega_{(B)}^{(\sigma)}} dT^{(\sigma)}$$

$$+ \left[T^{(\sigma)}\left(\frac{\partial \hat{S}^{(\sigma)}}{\partial \gamma}\right)_{T^{(\sigma)},\omega_{(B)}^{(\sigma)}} - \hat{\mathcal{A}} \right] d\gamma$$

$$+ \sum_{A=1}^{N-1} \left[T^{(\sigma)}\left(\frac{\partial \hat{S}^{(\sigma)}}{\partial \omega_{(A)}^{(\sigma)}}\right)_{T^{(\sigma)},\gamma,\omega_{(B)}^{(\sigma)}(B \neq A,N)} \right.$$

$$\left. + \mu_{(A)}^{(\sigma)} - \mu_{(N)}^{(\sigma)} \right] d\omega_{(A)}^{(\sigma)}$$

$$= \hat{c}_{\gamma}^{(\sigma)}dT^{(\sigma)} + \left[T^{(\sigma)}\left(\frac{\partial \hat{\mathcal{A}}}{\partial T^{(\sigma)}}\right)_{\gamma,\omega_{(B)}^{(\sigma)}} - \hat{\mathcal{A}} \right] d\gamma$$

$$+ \sum_{A=1}^{N-1} \left[-\, T^{(\sigma)}\left(\frac{\partial(\, \mu_{(A)}^{(\sigma)} - \mu_{(N)}^{(\sigma)}\,)}{\partial T^{(\sigma)}}\right)_{\gamma,\omega(B)} \right.$$

$$\left. + \mu_{(A)}^{(\sigma)} - \mu_{(N)}^{(\sigma)} \right] d\omega_{(A)}^{(\sigma)} \tag{9-5}$$

In the third line, I introduce the surface heat capacity per unit mass at constant surface tension (Exercise 5.8.6-4)

$$\hat{c}_{\gamma}^{(\sigma)} \equiv T^{(\sigma)} \left(\frac{\partial \hat{S}^{(\sigma)}}{\partial T^{(\sigma)}} \right)_{\gamma, \omega(B)} \tag{9-6}$$

and one of the Maxwell relations for the interface (Exercise 5.8.6-2)

$$\left(\frac{\partial \hat{S}^{(\sigma)}}{\partial \omega_{(A)}^{(\sigma)}} \right)_{T^{(\sigma)}, \gamma, \omega_{(B)}^{(\sigma)} (B \neq A, N)} = - \left(\frac{\partial (\mu_{(A)}^{(\sigma)} - \mu_{(N)}^{(\sigma)})}{\partial T^{(\sigma)}} \right)_{\gamma, \omega_{(B)}^{(\sigma)}} \tag{9-7}$$

Equation (9-5) implies that

$$\begin{aligned} \rho^{(\sigma)} \frac{d_{(v^{(\sigma)})} \hat{H}^{(\sigma)}}{dt} &= \rho^{(\sigma)} \hat{c}_{\gamma}^{(\sigma)} \frac{d_{(v^{(\sigma)})} T^{(\sigma)}}{dt} \\ &\quad + \left[\left(\frac{\partial \ln \hat{A}}{\partial \ln T^{(\sigma)}} \right)_{\gamma, \omega_{(B)}^{(\sigma)}} - 1 \right] \frac{d_{(v^{(\sigma)})} \omega_{(A)}^{(\sigma)}}{dt} \end{aligned} \tag{9-8}$$

which allows us to express (9-4) in terms of the time rate of change of the surface temperature:

$$\begin{aligned} \rho^{(\sigma)} \hat{c}_{\gamma}^{(\sigma)} \frac{d_{(v^{(\sigma)})} T^{(\sigma)}}{dt} &= - \operatorname{div}_{(\sigma)} \mathbf{q}^{(\sigma)} - \left(\frac{\partial \ln \hat{A}}{\partial \ln T^{(\sigma)}} \right)_{\gamma, \omega_{(B)}^{(\sigma)}} \frac{d_{(v^{(\sigma)})} \gamma}{dt} \\ &\quad + \operatorname{tr}(\mathbf{S}^{(\sigma)} \cdot \nabla_{(\sigma)} \mathbf{v}^{(\sigma)}) + \sum_{A=1}^{N} \mathbf{j}_{(A)}^{(\sigma)} \cdot \mathbf{b}_{(A)}^{(\sigma)} + \rho^{(\sigma)} Q^{(\sigma)} \\ &\quad - \sum_{A=1}^{N} \left[\mu_{(A)}^{(\sigma)} - T^{(\sigma)} \left(\frac{\partial \mu_{(A)}^{(\sigma)}}{\partial T^{(\sigma)}} \right)_{\gamma, \omega_{(B)}^{(\sigma)}} \right] \frac{d_{(v^{(\sigma)})} \omega_{(A)}^{(\sigma)}}{dt} \\ &\quad + \Big[- \rho (\hat{U} - \hat{H}^{(\sigma)})(\mathbf{v} \cdot \boldsymbol{\xi} - v_{(\xi)}^{(\sigma)}) - \frac{1}{2} \rho (\mathbf{v} \cdot \boldsymbol{\xi} - v_{(\xi)}^{(\sigma)})^3 \\ &\quad + (\mathbf{v} \cdot \boldsymbol{\xi} - v_{(\xi)}^{(\sigma)}) \boldsymbol{\xi} \cdot \mathbf{T} \cdot \boldsymbol{\xi} - \mathbf{q} \cdot \boldsymbol{\xi} \Big] \end{aligned} \tag{9-9}$$

This form of the jump energy balance, equation (F) of Table 5.8.9-1, is particularly useful for those situations where one is willing to say that the interfacial tension is independent of time and position within the dividing surface.

Derivations for the other two forms of the jump energy balance listed in Table 5.8.9-1 are outlined in Exercise 5.8.9-1.

A previously given list of alternative forms for the differential energy balance (Slattery 1981, p. 464) is consistent with the treatment given here after a minor change.[a]

With respect to the jump Clausius-Duhem inequality, the three useful forms that we have encountered in the preceding sections are listed in Table 5.8.9-2. Equation (A) is derived in Sec. 5.7.3; equation (B) in Sec. 5.8.8; equation (C) in Exercise 5.8.8-2.

For comparison, Table 5.8.9-3 summarizes a similar list of alternative forms for the differential Clausius-Duhem inequality from Secs. 5.7.3 and 5.8.4. They are in agreement with expressions derived previously[a] (Slattery 1981, pp. 466-470; see footnote a of Sec. 5.8.4).

Reference

Slattery, J. C., "Momentum, Energy, and Mass Transfer in Continua," McGraw-Hill, New York (1972); second edition, Robert E. Krieger, Malabar, FL 32950 (1981).

Exercises

5.8.9-1 *alternative forms of the jump energy balance* Two forms of the jump energy balance in Table 5.8.9-1 are not derived in the text: equations (D) and (E).

i) Use the surface Gibbs equation of Sec. 5.8.6 to say

$$\rho^{(\sigma)} \frac{d_{(v^{(\sigma)})}\hat{H}^{(\sigma)}}{dt} = \rho^{(\sigma)} T^{(\sigma)} \frac{d_{(v^{(\sigma)})}\hat{S}^{(\sigma)}}{dt} - \frac{d_{(v^{(\sigma)})}\gamma}{dt}$$

$$+ \rho^{(\sigma)} \sum_{B=1}^{N} \mu_{(B)}^{(\sigma)} \frac{d_{(v^{(\sigma)})}\omega_{(A)}^{(\sigma)}}{dt}$$

---

a) The terms involving the relative kinetic energies of the individual species do not appear in my discussion here. See footnote a, Sec. 5.6.4.

This expression together with equation (C) of Table 5.8.9-1 leads immediately to equation (D).

ii) Starting again with the surface Gibbs equation, employ a train of reasoning very similar to that shown in (9-5) to determine that

$$\rho^{(\sigma)} \frac{d_{(\mathbf{v}^{(\sigma)})} \hat{U}^{(\sigma)}}{dt} = \rho^{(\sigma)} \hat{c}_{\mathcal{A}}^{(\sigma)} \frac{d_{(\mathbf{v}^{(\sigma)})} T^{(\sigma)}}{dt}$$

$$+ \left\{ \gamma - T^{(\sigma)} \left(\frac{\partial \gamma}{\partial T^{(\sigma)}} \right)_{\hat{\mathcal{A}}, \omega_{(B)}^{(\sigma)}} \right\}$$

$$\times \left\{ \text{div}_{(\sigma)} \mathbf{v}^{(\sigma)} + \left[\frac{\rho}{\rho^{(\sigma)}} \left(\mathbf{v} \cdot \boldsymbol{\xi} - v_{(\xi)}^{(\sigma)} \right) \right] \right\}$$

$$+ \sum_{A=1}^{N} \left[\mu_{(A)}^{(\sigma)} - T^{(\sigma)} \left(\frac{\partial \mu_{(A)}^{(\sigma)}}{\partial T^{(\sigma)}} \right)_{\hat{\mathcal{A}}, \omega_{(B)}^{(\sigma)}} \right] \rho^{(\sigma)} \frac{d_{(\mathbf{v}^{(\sigma)})} \omega_{(A)}^{(\sigma)}}{dt}$$

where $\hat{c}_{\mathcal{A}}^{(\sigma)}$ is the surface heat capacity at constant specific area introduced in Exercise 5.8.6-4. Use this to rearrange equation (B) of Table 5.8.9-1 in the form of equation (E).

Table 5.8.9-1: *Alternative forms of the jump energy balance*

$$\rho^{(\sigma)} \frac{d_{(\mathbf{v}^{(\sigma)})}}{dt}\left(\hat{U}^{(\sigma)} + \frac{1}{2} v^{(\sigma)2} \right) = - \operatorname{div}_{(\sigma)}\mathbf{q}^{(\sigma)} + \operatorname{div}_{(\sigma)}\left(\mathbf{T}^{(\sigma)} \cdot \mathbf{v}^{(\sigma)} \right)$$

$$+ \sum_{A=1}^{N} \mathbf{n}^{(\sigma)}_{(A)} \cdot \mathbf{b}^{(\sigma)}_{(A)} + \rho^{(\sigma)} Q^{(\sigma)}$$

$$+ \Big[- \rho \left(\hat{U} + \frac{1}{2} v^{2} - \hat{U}^{(\sigma)} - \frac{1}{2} v^{(\sigma)2} \right)$$

$$\times \left(\mathbf{v} \cdot \boldsymbol{\xi} - v^{(\sigma)}_{(\xi)} \right) + \mathbf{v} \cdot \mathbf{T} \cdot \boldsymbol{\xi} - \mathbf{q} \cdot \boldsymbol{\xi} \Big] \qquad \text{(A)}$$

$$\rho^{(\sigma)} \frac{d_{(\mathbf{v}^{(\sigma)})} \hat{U}^{(\sigma)}}{dt} = - \operatorname{div}_{(\sigma)}\mathbf{q}^{(\sigma)} + \gamma \operatorname{div}_{(\sigma)}\mathbf{v}^{(\sigma)} + \operatorname{tr}\left(\mathbf{S}^{(\sigma)} \cdot \nabla_{(\sigma)}\mathbf{v}^{(\sigma)} \right)$$

$$+ \sum_{A=1}^{N} \mathbf{j}^{(\sigma)}_{(A)} \cdot \mathbf{b}^{(\sigma)}_{(A)} + \rho^{(\sigma)} Q^{(\sigma)}$$

$$+ \Big[- \rho \left(\hat{U} - \hat{U}^{(\sigma)} \right)\left(\mathbf{v} \cdot \boldsymbol{\xi} - v^{(\sigma)}_{(\xi)} \right) - \frac{1}{2} \rho \left(\mathbf{v} \cdot \boldsymbol{\xi} - v^{(\sigma)}_{(\xi)} \right)^{3}$$

$$+ \left(\mathbf{v} \cdot \boldsymbol{\xi} - v^{(\sigma)}_{(\xi)} \right) \boldsymbol{\xi} \cdot \mathbf{T} \cdot \boldsymbol{\xi} - \mathbf{q} \cdot \boldsymbol{\xi} \Big] \qquad \text{(B)}$$

$$\rho^{(\sigma)} \frac{d_{(\mathbf{v}^{(\sigma)})} \hat{H}^{(\sigma)}}{dt} = - \operatorname{div}_{(\sigma)}\mathbf{q}^{(\sigma)} - \frac{d_{(\mathbf{v}^{(\sigma)})} \gamma}{dt} + \operatorname{tr}\left(\mathbf{S}^{(\sigma)} \cdot \nabla_{(\sigma)}\mathbf{v}^{(\sigma)} \right)$$

$$+ \sum_{A=1}^{N} \mathbf{j}^{(\sigma)}_{(A)} \cdot \mathbf{b}^{(\sigma)}_{(A)} + \rho^{(\sigma)} Q^{(\sigma)}$$

$$+ \Big[- \rho \left(\hat{U} - \hat{H}^{(\sigma)} \right)\left(\mathbf{v} \cdot \boldsymbol{\xi} - v^{(\sigma)}_{(\xi)} \right) - \frac{1}{2} \rho \left(\mathbf{v} \cdot \boldsymbol{\xi} - v^{(\sigma)}_{(\xi)} \right)^{3}$$

$$+ \left(\mathbf{v} \cdot \boldsymbol{\xi} - v^{(\sigma)}_{(\xi)} \right) \boldsymbol{\xi} \cdot \mathbf{T} \cdot \boldsymbol{\xi} - \mathbf{q} \cdot \boldsymbol{\xi} \Big] \qquad \text{(C)}$$

Table 5.8.9-1: *Alternative forms of the jump energy balance (cont.)*

$$
\rho^{(\sigma)} T^{(\sigma)} \frac{d_{(\mathbf{v}^{(\sigma)})} \hat{S}^{(\sigma)}}{dt} = - \operatorname{div}_{(\sigma)} \mathbf{q}^{(\sigma)} + \operatorname{tr}\left(\mathbf{S}^{(\sigma)} \cdot \nabla_{(\sigma)} \mathbf{v}^{(\sigma)} \right)
$$

$$
+ \sum_{A=1}^{N} \mathbf{j}_{(A)}^{(\sigma)} \cdot \mathbf{b}_{(A)}^{(\sigma)} + \rho^{(\sigma)} Q^{(\sigma)} - \rho^{(\sigma)} \sum_{B=1}^{N} \mu_{(B)}^{(\sigma)} \frac{d_{(\mathbf{v}^{(\sigma)})} \omega_{(B)}^{(\sigma)}}{dt}
$$

$$
+ \Big[- \rho (\hat{U} - \hat{H}^{(\sigma)}) (\mathbf{v} \cdot \boldsymbol{\xi} - v_{(\xi)}^{(\sigma)}) - \frac{1}{2} \rho (\mathbf{v} \cdot \boldsymbol{\xi} - v_{(\xi)}^{(\sigma)})^3
$$

$$
+ (\mathbf{v} \cdot \boldsymbol{\xi} - v_{(\xi)}^{(\sigma)}) \boldsymbol{\xi} \cdot \mathbf{T} \cdot \boldsymbol{\xi} - \mathbf{q} \cdot \boldsymbol{\xi} \Big] \tag{D}
$$

$$
\rho^{(\sigma)} \hat{c}_{\mathcal{A}}^{(\sigma)} \frac{d_{(\mathbf{v}^{(\sigma)})} T^{(\sigma)}}{dt} = - \operatorname{div}_{(\sigma)} \mathbf{q}^{(\sigma)}
$$

$$
+ T^{(\sigma)} \left(\frac{\partial \gamma}{\partial T^{(\sigma)}} \right)_{\hat{\mathcal{A}}, \omega_{(B)}^{(\sigma)}} \rho^{(\sigma)} \frac{d_{(\mathbf{v}^{(\sigma)})} \hat{\mathcal{A}}}{dt} + \operatorname{tr}\left(\mathbf{S}^{(\sigma)} \cdot \nabla_{(\sigma)} \mathbf{v}^{(\sigma)} \right)
$$

$$
+ \sum_{A=1}^{N} \mathbf{j}_{(A)}^{(\sigma)} \cdot \mathbf{b}_{(A)}^{(\sigma)} + \rho^{(\sigma)} Q^{(\sigma)}
$$

$$
- \sum_{A=1}^{N} \left[\mu_{(A)}^{(\sigma)} - T^{(\sigma)} \left(\frac{\partial \mu_{(A)}^{(\sigma)}}{\partial T^{(\sigma)}} \right)_{\hat{\mathcal{A}}, \omega_{(B)}^{(\sigma)}} \right] \rho^{(\sigma)} \frac{d_{(\mathbf{v}^{(\sigma)})} \omega_{(A)}^{(\sigma)}}{dt}
$$

$$
+ \Big[- \rho (\hat{U} - \hat{H}^{(\sigma)}) (\mathbf{v} \cdot \boldsymbol{\xi} - v_{(\xi)}^{(\sigma)}) - \frac{1}{2} \rho (\mathbf{v} \cdot \boldsymbol{\xi} - v_{(\xi)}^{(\sigma)})^3
$$

$$
+ (\mathbf{v} \cdot \boldsymbol{\xi} - v_{(\xi)}^{(\sigma)}) \boldsymbol{\xi} \cdot \mathbf{T} \cdot \boldsymbol{\xi} - \mathbf{q} \cdot \boldsymbol{\xi} \Big] \tag{E}
$$

Table 5.8.9-1: *Alternative forms of the jump energy balance (cont.)*

$$\rho^{(\sigma)}\hat{c}^{(\sigma)}_{\gamma}\frac{d_{(v^{(\sigma)})}T^{(\sigma)}}{dt} = -\operatorname{div}_{(\sigma)}\mathbf{q}^{(\sigma)} - \left(\frac{\partial \ln \hat{A}}{\partial \ln T^{(\sigma)}}\right)_{\gamma,\omega_{(B)}^{(\sigma)}}\frac{d_{(v^{(\sigma)})}\gamma}{dt}$$

$$+ \operatorname{tr}(\mathbf{S}^{(\sigma)} \cdot \nabla_{(\sigma)}\mathbf{v}^{(\sigma)}) + \sum_{A=1}^{N} \mathbf{j}_{(A)}^{(\sigma)} \cdot \mathbf{b}_{(A)}^{(\sigma)} + \rho^{(\sigma)}Q^{(\sigma)}$$

$$- \sum_{A=1}^{N}\left[\mu_{(A)}^{(\sigma)} - T^{(\sigma)}\left(\frac{\partial \mu_{(A)}^{(\sigma)}}{\partial T^{(\sigma)}}\right)_{\gamma,\omega_{(B)}^{(\sigma)}}\right]\frac{d_{(v^{(\sigma)})}\omega_{(A)}^{(\sigma)}}{dt}$$

$$+ \Big[-\rho(\hat{U} - \hat{H}^{(\sigma)})(\mathbf{v}\cdot\boldsymbol{\xi} - v_{(\xi)}^{(\sigma)}) - \frac{1}{2}\rho(\mathbf{v}\cdot\boldsymbol{\xi} - v_{(\xi)}^{(\sigma)})^{3}$$

$$+ (\mathbf{v}\cdot\boldsymbol{\xi} - v_{(\xi)}^{(\sigma)})\boldsymbol{\xi}\cdot\mathbf{T}\cdot\boldsymbol{\xi} - \mathbf{q}\cdot\boldsymbol{\xi}\Big] \qquad \text{(F)}$$

Table 5.8.9-2: *Alternative forms of jump Clausius-Duhem inequality*

$$\rho^{(\sigma)} \frac{d_{(\mathbf{v}^{(\sigma)})} \hat{S}^{(\sigma)}}{dt} + \text{div}_{(\sigma)} \left(\frac{\mathbf{e}^{(\sigma)}}{T^{(\sigma)}} \right) - \rho^{(\sigma)} \frac{Q^{(\sigma)}}{T^{(\sigma)}}$$

$$+ \left[\rho (\hat{S} - \hat{S}^{(\sigma)})(\mathbf{v} \cdot \boldsymbol{\xi} - v_{(\xi)}) + \frac{1}{T} \mathbf{e} \cdot \boldsymbol{\xi} \right] \geq 0 \qquad \text{(A)}$$

$$\frac{1}{T^{(\sigma)}} \mathbf{e}^{(\sigma)} \cdot \nabla_{(\sigma)} T^{(\sigma)} - \text{tr}[(\mathbf{T}^{(\sigma)} - \gamma \mathbf{P}) \cdot \mathbf{D}^{(\sigma)}]$$

$$+ \sum_{B=1}^{N} \mathbf{j}_{(B)} \cdot (\nabla_{(\sigma)} \mu_{(B)} - \mathbf{b}_{(B)}) + \sum_{B=1}^{N} \mu_{(B)} r_{(B)}$$

$$+ \left[\frac{1}{2} \rho (\mathbf{v} \cdot \boldsymbol{\xi} - v_{(\xi)})^3 \right.$$

$$\left. - (\mathbf{v} \cdot \boldsymbol{\xi} - v_{(\xi)}) \boldsymbol{\xi} \cdot (\mathbf{T} + P \mathbf{I}) \cdot \boldsymbol{\xi} \right] \leq 0 \qquad \text{(B)}$$

$$\frac{1}{T^{(\sigma)}} \left(\mathbf{e}^{(\sigma)} - \sum_{B=1}^{N} T^{(\sigma)} \overline{\overline{S}}_{(B)} \mathbf{j}_{(B)} \right) \cdot \nabla_{(\sigma)} T^{(\sigma)}$$

$$- \text{tr}[(\mathbf{T}^{(\sigma)} - \gamma \mathbf{P}) \cdot \mathbf{D}^{(\sigma)}] - \text{tr}[(\mathbf{T}^{(\sigma)} - \gamma \mathbf{P}) \cdot \mathbf{D}^{(\sigma)}]$$

$$+ \sum_{B=1}^{N} \frac{c^{(\sigma)} R T^{(\sigma)}}{\rho_{(B)}} \mathbf{j}_{(B)} \cdot \mathbf{d}_{(B)} + \sum_{B=1}^{N} \mu_{(B)} r_{(B)}$$

$$+ \left[\frac{1}{2} \rho (\mathbf{v} \cdot \boldsymbol{\xi} - v_{(\xi)})^3 \right.$$

$$\left. - (\mathbf{v} \cdot \boldsymbol{\xi} - v_{(\xi)}) \boldsymbol{\xi} \cdot (\mathbf{T} + P \mathbf{I}) \cdot \boldsymbol{\xi} \right] \leq 0 \qquad \text{(C)}$$

Table 5.8.9-2: *Alternative forms of jump Clausius-Duhem inequality (cont.)*

where

$$\mathbf{d}^{(\sigma)}_{(A)} = \frac{\rho^{(\sigma)}_{(A)}}{c^{(\sigma)}RT^{(\sigma)}}\left[\sum_{\substack{B=1\\ B\neq A}}^{N}\left(\frac{\partial\mu^{(\sigma)}_{(A)}}{\partial\omega^{(\sigma)}_{(B)}}\right)_{T^{(\sigma)},\gamma,\omega^{(\sigma)}_{(C)}(C\neq A,B)}\nabla_{(\sigma)}\omega^{(\sigma)}_{(B)}\right.$$

$$\left.-\left(\overline{\overline{\mathcal{A}}}_{(A)}-\hat{\mathcal{A}}\right)\nabla_{(\sigma)}\gamma-\left(\mathbf{b}^{(\sigma)}_{(A)}-\sum_{B=1}^{N}\omega^{(\sigma)}_{(B)}\,\mathbf{b}^{(\sigma)}_{(B)}\right)\right]$$

and

$$\sum_{A=1}^{N}\mathbf{d}^{(\sigma)}_{(A)}=0$$

Table 5.8.9-3: *Alternative forms of differential Clausius-Duhem inequality*

$$\rho\frac{d_{(v)}\hat{S}}{dt}+\operatorname{div}\left(\frac{\mathbf{e}}{T}\right)-\rho\frac{Q}{T}\geq 0 \qquad \text{(A)}$$

$$\frac{1}{T}\mathbf{e}\cdot\nabla T-\operatorname{tr}[(\mathbf{T}+P\mathbf{I})\cdot\mathbf{D}]+\sum_{B=1}^{N}\mathbf{j}_{(B)}\cdot\left(\nabla\mu_{(B)}-\mathbf{b}_{(B)}\right)$$

$$+\sum_{B=1}^{N}\mu_{(B)}\,r_{(B)}\leq 0 \qquad \text{(B)}$$

$$\frac{1}{T}\left(\mathbf{e}-\sum_{B=1}^{N}T\,\overline{\overline{S}}_{(B)}\,\mathbf{j}_{(B)}\right)\cdot\nabla T-\operatorname{tr}[(\mathbf{T}+P\mathbf{I})\cdot\mathbf{D}]$$

$$+\sum_{B=1}^{N}\frac{cRT}{\rho_{(B)}}\mathbf{j}_{(B)}\cdot\mathbf{d}_{(B)}+\sum_{B=1}^{N}\mu_{(B)}\,r_{(B)}\leq 0 \qquad \text{(C)}$$

Table 5.8.9-3: *Alternative forms of differential Clausius-Duhem inequality (cont.)*

where

$$\mathbf{d}_{(A)} = \frac{\rho_{(A)}}{cRT}\left[\sum_{\substack{B=1\\ B\neq A}}^{N} \left(\frac{\partial \mu_{(A)}}{\partial \omega_{(B)}}\right)_{T,P,\omega_{(C)}(C\neq A,B)} \nabla \omega_{(B)} + \left(\overline{\overline{V}}_{(A)} - \frac{1}{\rho}\right)\nabla P - \left(\mathbf{b}_{(A)} - \sum_{B=1}^{N} \omega_{(B)}\mathbf{b}_{(B)}\right)\right]$$

and

$$\sum_{A=1}^{N} \mathbf{d}_{(A)} = 0$$

5.9 Behavior as restricted by frame indifference

5.9.1 Other principles to be considered In Secs. 5.8.2 and 5.8.5, we made some assumptions about the independent variables appearing in constitutive equations for $\mathbf{T}$, $\mathbf{q}$, $\mathbf{j}_{(A)}$, $\mathbf{T}^{(\sigma)}$, $\mathbf{q}^{(\sigma)}$, $\mathbf{j}^{(\sigma)}_{(A)}$. No further restrictions were placed upon the forms of these equations by the differential Clausius-Duhem inequality or by the jump Clausius-Duhem inequality. But these constitutive equations are subject to further restrictions.

It seems obvious that what happens to a body in the future can have no influence upon its present behavior. Generalizing a statement made in Sec. 2.2.1, we can lay down the **principle of determinism**:

The behavior of a body is determined by its (temperature, concentration, deformation,...) history.

The constitutive equations assumed in Secs. 5.8.2 and 5.8.5 already satisfy the principle of determinism.

A candle is a multiphase body. The behavior of the wax at its base is not necessarily affected by lighting its wick. In the displacement of residual oil by water in an oil-bearing rock, the momentum, energy, and mass transfer at any one of the phase interfaces is not an immediate or direct function of the polymers and surfactants introduced with the water at one of the wells in the pattern. This suggests that we can generalize the **principle of local action** introduced in Sec. 2.2.1:

The (temperature, concentration, deformation,...) history of the material outside an arbitrarily small neighborhood of a mass- averaged material particle may be ignored in determining the behavior at this particle.

The principle of local action is already satisfied by the constitutive equations assumed in Secs. 5.8.2 and 5.8.5.

However, the principle of frame indifference introduced in Sec. 1.4.4 does restrict the forms of these constitutive equations. It is this point that we shall explore in the sections that follow.

But before we do so, let us reconsider the independent variables chosen in Secs. 5.8.2 and 5.8.5.

5.9.2 Alternative independent variables in constitutive equations In Secs. 5.8.2 and 5.8.5, I made assumptions about the set of independent variables appearing in constitutive equations for the bulk material and for the interface. Very little motivation was given for the particular sets chosen. Perhaps we could have picked smaller or more concise sets of variables.

Let us consider for a moment what we know to be useful in

describing the behavior of single-component materials. Newton's law of viscosity says that the extra stress $\mathbf{S} \equiv \mathbf{T} + P\mathbf{I}$ is a linear function of the rate of deformation $\mathbf{D}$ (Slattery 1981, p. 49). Fourier's law of heat conduction gives the energy flux $\mathbf{q}$ as a linear function of the temperature gradient ∇T (Slattery 1981, p. 297). Notice that these same terms appear as products describing the rate of production of entropy beyond that supplied by the outside world in the differential Clausius-Duhem in equality expressed as equation (B) of Table 5.8.9-3.

This indicates that our intuition may be assisted in formulating bulk constitutive equations for multicomponent materials by examining the *fluxes* and corresponding *affinities* as they appear in the Clausius-Duhem inequality. Unfortunately, there is not a unique arrangement. Influenced by relationships developed for dilute gases in the context of statistical mechanics, I have listed in Table 5.9.2-1 only the fluxes and affinities suggested by equation (C) of Table 5.8.9-3.

With much the same thought in mind, we can now look at the products appearing in the jump Clausius-Duhem inequality as fluxes and their corresponding affinities. Somewhat arbitrarily, I have chosen to develop relationships that are analogous to those used in describing bulk behavior. For that reason, I have listed in Table 5.9.2-2 only the fluxes and affinities suggested by equation (C) of Table 5.8.9-2.

The linear theory of irreversible processes (de Groot 1951, p. 34). which received considerable attention for a time, states that each flux is a linear homogeneous function of all affinities with the restriction that quantities whose tensorial order differ by an odd integer cannot interact in nonoriented media. (A nonoriented or isotropic material is one whose behavior is unaffected by orthogonal transformations, rotations or reflections, of some reference configuration of the material (Truesdell 1966a, p. 60). It may also be thought of as a substance that has no natural direction when it assumes its reference configuration.) There are also symmetries attributed to the coefficients appearing in these linear constitutive equations. This theory is without firm foundations (Coleman and Truesdell 1960; Truesdell and Toupin 1960, pp. 643 and 646; Truesdell 1966b, p. 49) and will not be used here.

However, the affinities indicated in Tables 5.9.2-1 and 5.9.2-2 can suggest useful arrangements of the independent variables assumed in Secs. 5.8.2 and 5.8.5. It is only in this sense that we shall employ them.

But, as I note in the next section, we will have to be cautious.

References

Coleman, B., and C. Truesdell, *J. Chem. Phys.* **33**, 28 (1960).

de Groot, S. R., "Thermodynamics of Irreversible Processes," Interscience, New York (1951).

Slattery, J. C., "Momentum, Energy, and Mass Transfer in Continua,"

McGraw-Hill, New York (1972); second edition, Robert E. Krieger, Malabar, FL 32950 (1981).

Truesdell, C., and R. A. Toupin, "Handbuch der Physik," vol. 3/1, edited by S. Flügge, Springer-Verlag, Berlin (1960).

Truesdell, C., "The Elements of Continuum Mechanics," Springer-Verlag, New York (1966).

Truesdell, C., "Six Lectures on Modern Natural Philosophy," Springer-Verlag, New York (1966).

Table 5.9.2-1: *flux-affinity relations for bulk behavior from Table 5.8.9-3, equation (C)*

flux	**affinity**
$\mathbf{S} \equiv \mathbf{T} + P\mathbf{I}$	$\mathbf{D}$
$\mathbf{e} - \sum_{B=1}^{N} T\overline{\overline{S}}_{(B)} \mathbf{j}_{(B)} = \mathbf{q} - \sum_{B=1}^{N} \overline{\overline{H}}_{(B)} \mathbf{j}_{(B)}$	$\nabla \ln T$
$\mathbf{j}_{(A)}$	$\frac{cRT}{\rho_{(A)}} \mathbf{d}_{(A)} = \sum_{\substack{B=1 \\ B \neq A}}^{N} \left(\frac{\partial \mu_{(A)}}{\partial \omega_{(B)}} \right)_{T,P,\omega_{(C)}(C \neq A,B)} \nabla \omega_{(B)} + \left(\overline{\overline{V}}_{(A)} - \frac{1}{\rho} \right) \nabla P - \left(\mathbf{b}_{(A)} - \sum_{B=1}^{N} \omega_{(B)} \mathbf{b}_{(B)} \right)$
$r_{(A)}$	$\mu_{(A)}$

Table 5.9.2-2: *flux-affinity relations for surface behavior from Table 5.8.9-2, equation (C)*

flux	**affinity**
$\mathbf{S}^{(\sigma)} \equiv \mathbf{T}^{(\sigma)} - \gamma \mathbf{P}$	$\mathbf{D}^{(\sigma)}$
$\mathbf{e}^{(\sigma)} - \sum_{B=1}^{N} T^{(\sigma)} \overline{\overline{S}}_{(B)}^{(\sigma)} \mathbf{j}_{(B)}^{(\sigma)}$ $= \mathbf{q} - \sum_{B=1}^{N} \overline{\overline{H}}_{(B)}^{(\sigma)} \mathbf{j}_{(B)}^{(\sigma)}$	$\nabla_{(A)} \ln T^{(\sigma)}$
$\mathbf{j}_{(A)}^{(\sigma)}$	$\frac{c^{(\sigma)} R T^{(\sigma)}}{\rho_{(A)}^{(\sigma)}} \mathbf{d}_{(A)}^{(\sigma)}$ $= \sum_{\substack{B=1 \\ B \neq A}}^{N} \left(\frac{\partial \mu_{(A)}^{(\sigma)}}{\partial \omega_{(B)}^{(\sigma)}} \right)_{T^{(\sigma)}, \gamma, \omega_{(C)}^{(\sigma)} (C \neq A, B)}$ $\times \nabla_{(\sigma)} \omega_{(B)}^{(\sigma)}$ $- (\overline{\overline{\mathcal{A}}}_{(A)} - \hat{\mathcal{A}}) \nabla_{(\sigma)} \gamma$ $- \left(\mathbf{b}_{(A)}^{(\sigma)} - \sum_{B=1}^{N} \omega_{(B)}^{(\sigma)} \mathbf{b}_{(B)}^{(\sigma)} \right)$
$r_{(A)}^{(\sigma)}$	$\mu_{(A)}^{(\sigma)}$

5.9.3 Bulk behavior: constitutive equations for stress tensor We already know that the stress tensor for single-component materials can be a very complicated function of the history of the motion of the material (Coleman *et al.* 1966). The extra stress tensor $\mathbf{S} \equiv \mathbf{T} + P\mathbf{I}$ can realistically be considered to be a function of the rate of deformation tensor $\mathbf{D}$ only for low molecular weight materials for which the newtonian fluid applies (Slattery 1981, p. 49). This is a good illustration of how one can be misled in choosing the independent variables to appear in a constitutive equation based upon the affinities suggested by the Clausius-Duhem inequality (see Table 5.9.2-1).

In general, we should expect the stress tensor in a multicomponent material to be a function of the temperature and concentration gradients of all the species present (Bowen 1967; Müller 1968). Since there is little or no experimental evidence available to guide us, I recommend current practice in engineering, which is to use constitutive equations for **T** developed for pure materials (Slattery 1981, p. 45; Coleman *et al.* 1966), recognizing that all parameters should be functions of the local thermodynamic state variables $\hat{V}$, T, $\omega_{(1)}$, $\omega_{(2)}$, ..., $\omega_{(N-1)}$.

References

Bowen, R. M., *Arch. Ration. Mech. Anal.* **24,** 370 (1967).

Coleman, B. D., H. Markovitz, and W. Noll, "Viscometric Flows of Non-Newtonian Fluids," Springer-Verlag, New York (1966).

Müller, I., *Arch. Ration. Mech. Anal.* **28,** 1 (1968).

Slattery, J. C., "Momentum, Energy, and Mass Transfer in Continua," McGraw-Hill, New York (1972); second edition, Robert E. Krieger, Malabar, FL 32950 (1981).

5.9.4 Bulk behavior: constitutive equations for energy flux vector

Our assumptions about material behavior in Sec. 5.8.2, our discussion in Sec. 5.9.2 (see Table 5.9.2-1), and kinetic theory (Hirschfelder *et al.* 1964, pp. 483 and 715) suggest that the functional dependence of the thermal energy flux **e** may be as simple as

$$\boldsymbol{\varepsilon} \equiv \mathbf{e} - \sum_{B=1}^{N} T\overline{\overline{S}}_{(B)}\mathbf{j}_{(B)} = \mathbf{q} - \sum_{B=1}^{N} \overline{\overline{H}}_{(B)}\mathbf{j}_{(B)}$$

$$= \boldsymbol{\varepsilon}(\boldsymbol{\Pi}) \tag{4-1}$$

where

$$\boldsymbol{\Pi} \equiv \left(\hat{V}, T, \omega_{(1)}, \ldots, \omega_{(N-1)}, \nabla \ln T, \frac{cRT}{\rho_{(1)}}\mathbf{d}_{(1)}, \ldots, \frac{cRT}{\rho_{(N)}}\mathbf{d}_{(B)}\right) \tag{4-2}$$

There is no experimental evidence indicating that **e** should be a function of the history of the motion of the material.

Since **e**, **q**, and $\mathbf{j}_{(A)}$ are all frame-indifferent vectors (Exercises 5.2.2-1, 5.6.4-3, and 5.7.3-8), the principle of frame indifference requires that $\boldsymbol{\varepsilon}$ be an isotropic function of $\boldsymbol{\Pi}$

$$\mathbf{Q} \cdot \boldsymbol{\varepsilon}(\boldsymbol{\Pi}) = \boldsymbol{\varepsilon}(\mathbf{Q} \cdot \boldsymbol{\Pi}) \tag{4-3}$$

where I have defined

$$\mathbf{Q} \cdot \boldsymbol{\Pi} \equiv \left(\hat{V}, T, \omega_{(1)}, \ldots, \omega_{(N-1)}, \mathbf{Q} \cdot \nabla \ln T, \frac{cRT}{\rho_{(1)}} \mathbf{Q} \cdot \mathbf{d}_{(1)}, \ldots, \frac{cRT}{\rho_{(N)}} \mathbf{Q} \cdot \mathbf{d}_{(N)} \right) \tag{4-4}$$

Using the representation theorems of Spencer and Rivlin (1960), and of Smith (1965), we find that the most general polynomial vector function of this form is[a]

$$\boldsymbol{\varepsilon} = \alpha \, \nabla \ln T + cRT \sum_{A=1}^{N} \frac{\beta_{(A)}}{\rho_{(A)}} \mathbf{d}_{(A)} \tag{4-5}$$

It is to be understood here that the coefficients α and $\beta_{(A)}$ (A = 1, 2, ..., N) are all functions of the local thermodynamic state variables as well as of all the scalar products involving $\nabla \ln T$ and $\mathbf{d}_{(A)}$ (A = 1, 2, ..., N) (including the magnitudes of these vectors).

---

a) In applying the theorem of Spencer and Rivlin (1960), we identify the vector **b** that has covariant components b_i with the skew-symmetric tensor that has contravariant components $\varepsilon^{ijk} b_i$. Their theorem requires additional terms in (4-5) of the form

$$\sum_{A=1}^{N} \gamma_{(OA)} \nabla \ln T \wedge \mathbf{d}_{(A)} + \sum_{A=1}^{N} \sum_{B=A+1}^{N} \gamma_{(AB)} \mathbf{d}_{(A)} \wedge \mathbf{d}_{(B)}$$

The terms are not consistent with the requirement that $\boldsymbol{\varepsilon}$ be isotropic (Truesdell and Noll 1965, p. 24) and are consequently dropped.

The kinetic theory result (Hirschfelder *et al.* 1964, pp. 483 and 715)

$$\boldsymbol{\varepsilon} = -\lambda\nabla T - cRT\sum_{A=1}^{N}\frac{D^T_{(A)}}{\rho_{(A)}}\mathbf{d}_{(A)} \tag{4-6}$$

is a special case of (4-5), in the sense that the coefficients λ and $D^T_{(A)}$ (A = 1, 2,..., N) are functions only of the thermodynamic state variables.

The direct dependence of $\boldsymbol{\varepsilon}$ upon concentration and pressure gradients through the $\mathbf{d}_{(A)}$ is usually referred to as the **Dufour effect**. It is generally believed to be small (Hirschfelder *et al.* 1964, pp. 717) and is often neglected along with the dependence upon the external forces. In this case, (4-6) reduces to

$$\boldsymbol{\varepsilon} = \mathbf{q} - \sum_{A=1}^{N}\overline{\overline{H}}_{(A)}\mathbf{j}_{(A)} = -k\nabla T \tag{4-7}$$

If $\boldsymbol{\varepsilon}$ is a linear function of the temperature gradient, the scalar $k = k(\hat{V}, T, \omega_{(1)}, \omega_{(2)}, \ldots, \omega_{(N-1)})$ is referred to as the **thermal conductivity**.

References

Hirschfelder, J. O., C. F. Curtiss, and R. B. Bird, "Molecular Theory of Gases and Liquids," Wiley, New York (1954); corrected with notes added (1964).

Smith, G. F., *Arch. Ration. Mech. Anal.* **18**, 282 (1965).

Spencer, A. J., and R. S. Rivlin, *Arch. Ration. Mech. Anal.* **4**, 214 (1960).

Truesdell, C., and W. Noll, "Handbuch der Physik," vol. 3/3, edited by S. Flügge, Springer-Verlag, Berlin (1965).

5.9.5 Bulk behavior: constitutive equations for mass flux vector In view of our assumptions about material behavior in Sec. 5.8.5, our discussion in Sec. 5.9.2 (see Table 5.9.2-1), and kinetic theory (Hirschfelder *et al.* 1964, pp. 479 and 715), the functional dependence of the mass flux $\mathbf{j}_{(A)}$ may be as simple as

$$\mathbf{j}_{(A)} = \mathbf{j}_{(A)}(\boldsymbol{\Pi}) \tag{5-1}$$

where $\boldsymbol{\Pi}$ is defined by (4-2). I am not aware of any experimental evidence suggesting that $\mathbf{j}_{(A)}$ should be a function of the history of the motion of the material.

Since $\mathbf{j}_{(A)}$ is a frame-indifferent vector (see Exercise 5.2.2-1), the principle of frame indifference requires that $\mathbf{j}_{(A)}$ be an isotropic function of $\mathbf{\Pi}$:

$$\mathbf{Q} \cdot \mathbf{j}_{(A)}(\mathbf{\Pi}) = \mathbf{j}_{(A)}(\mathbf{Q} \cdot \mathbf{\Pi}) \tag{5-2}$$

The set $\mathbf{Q} \cdot \mathbf{\Pi}$ is defined by (4-4). Using the representation theorems of Spencer and Rivlin (1960) and of Smith (1965), we find that the most general polynomial vector function of this form is (see Sec. 5.9.4, footnote a)

$$\mathbf{j}_{(A)} = \alpha_{(A0)} \nabla \ln T + cRT \sum_{B=1}^{N} \frac{\beta_{(AB)}}{\rho_{(B)}} \mathbf{d}_{(B)} \tag{5-3}$$

It is understood here that the coefficients $\alpha_{(A0)}$ and $\beta_{(AB)}$ (A, B = 1, 2,..., N) are all functions of the local thermodynamic state variables as well as of all the scalar products involving $\nabla \ln T$ and $\mathbf{d}_{(C)}$ (C = 1, 2,..., N) (including the magnitudes of these vectors).

The kinetic theory result (Hirschfelder *et al.* 1964, pp. 479 and 715)

$$\mathbf{j}_{(A)} = -D^T_{(A)} \nabla \ln T + \frac{c^2}{\rho} \sum_{B=1}^{N} M_{(A)} M_{(B)} D_{(AB)} \mathbf{d}_{(B)} \tag{5-4}$$

is a special case of (5-3), in the sense that the coefficients $D^T_{(A)}$ and $D_{(AB)}$ (B = 1, 2,..., N) are functions only of the thermodynamic state variables. Note that in Exercise 5.8.4-2 we find

$$\sum_{A=1}^{N} \mathbf{d}_{(A)} = 0 \tag{5-5}$$

The resulting indeterminateness in the coefficients $D_{(AB)}$ is removed by requiring (A = 1, 2..., N)

$$D_{(AA)} = 0 \tag{5-6}$$

Because (see Table 5.2.2-4)

$$\sum_{A=1}^{N} \mathbf{j}_{(A)} = 0 \tag{5-7}$$

we must also require

$$\sum_{A=1}^{N} D^{T}_{(A)} = 0 \tag{5-8}$$

and

$$\sum_{A=1}^{N} \left(M_{(A)} M_{(B)} D_{(AB)} - M_{(A)} M_{(C)} D_{(AC)} \right) = 0 \tag{5-9}$$

Alternative forms and special cases of (5-4) are discussed in more detail elsewhere (Slattery 1981, pp. 475-485).

References

Hirschfelder, J. O., C. F. Curtiss, and R. B. Bird, "Molecular Theory of Gases and Liquids," Wiley, New York (1954); corrected with notes added (1964).

Slattery, J. C., "Momentum, Energy, and Mass Transfer in Continua," McGraw-Hill, New York (1972); second edition, Robert E. Krieger, Malabar, FL 32950 (1981).

Smith, G. F., *Arch. Ration. Mech. Anal.* **18**, 282 (1965).

Spencer, A. J., and R. S. Rivlin, *Arch. Ration. Mech. Anal.* **4**, 214 (1960).

5.9.6 Surface behavior: constitutive equations for surface stress tensor In Chapter 2, I indicated that we might expect the surface stress tensor in general to be a very complicated function of the history of the motion of the surface. The discussion in Sec. 5.9.2 suggests that $\mathbf{T}^{(\sigma)}$ could be a function of the surface gradients of surface temperature and of surface chemical potential as well. But, since there are no relevant experimental studies of either surface or bulk stress-deformation behavior to guide us, I suggest that we use the same constitutive equations for $\mathbf{T}^{(\sigma)}$ developed in Chapter 2, recognizing that all parameters may now be functions of the local thermodynamic state variables for the surface: $\hat{\mathcal{A}}$, $T^{(\sigma)}$, $\omega^{(\sigma)}_{(1)}$, $\omega^{(\sigma)}_{(2)}$, ..., $\omega^{(\sigma)}_{(N-1)}$.

Exercise

5.9.6-1 *implications of jump Clausius-Duhem inequality* In Sec. 2.2.2, we saw that the most general linear relation between the surface stress tensor and the surface rate of deformation tensor which is consistent with the principle of frame indifference is

$$\mathbf{T}^{(\sigma)} = (\ \alpha + \lambda\ \text{div}_{(\sigma)}\mathbf{v}^{(\sigma)}\ )\mathbf{P} + 2\varepsilon\ \mathbf{D}^{(\sigma)}$$

i) In the absence of thermal energy transfer and mass transfer, show that the jump Clausius-Duhem inequality implies (see Sec. 5.8.9)

$$(\ \alpha - \gamma\ )\text{tr}\ \mathbf{D}^{(\sigma)} + \lambda(\ \text{tr}\ \mathbf{D}^{(\sigma)}\ )^2 + 2\varepsilon\ \text{tr}(\ \mathbf{D}^{(\sigma)} \cdot \mathbf{D}^{(\sigma)}\ ) \geq 0$$

ii) Determine that this implies

$$\alpha = \gamma$$

$$\kappa \equiv \lambda + \varepsilon \geq 0$$

$$\varepsilon \geq 0$$

Hint: In order to show that $\alpha = \gamma$ and $\lambda + \varepsilon \geq 0$, realize that the six components of the rate of deformation tensor may be chosen independently. Set the nondiagonal components of $\mathbf{D}^{(\sigma)}$ equal to zero, require that $D_{11}^{(\sigma)} = D_{22}^{(\sigma)}$, and let $D_{11}^{(\sigma)}$ assume both positive and negative values.

(See also Exercises 5.9.7-1 and 5.9.8-1.)

5.9.7 Surface behavior: constitutive equations for surface energy flux vector Our assumptions about surface material behavior in Sec. 5.8.5 and our discussion in Sec. 5.9.2 (see Table 5.9.2-2) suggest that the functional dependence of the surface thermal energy flux $\mathbf{e}^{(\sigma)}$ might be no more complicated than

$$\boldsymbol{\varepsilon}^{(\sigma)} \equiv \mathbf{e}^{(\sigma)} - \sum_{B=1}^{N} T^{(\sigma)}\ \overline{\overline{S}}_{(B)}^{(\sigma)}\ \mathbf{j}_{(B)}^{(\sigma)}$$

$$= \mathbf{q} - \sum_{B=1}^{N} \overline{\overline{H}}_{(B)}^{(\sigma)}\ \mathbf{j}_{(B)}^{(\sigma)}$$

$$= \boldsymbol{\varepsilon}^{(\sigma)}(\ \boldsymbol{\Pi}^{(\sigma)}\ ) \tag{7-1}$$

where

$$\Pi^{(\sigma)} \equiv \left(\hat{A}, T^{(\sigma)}, \omega_{(1)}^{(\sigma)}, \omega_{(2)}^{(\sigma)}, \ldots, \omega_{(N-1)}^{(\sigma)}, \nabla_{(\sigma)} \ln T^{(\sigma)}, \right.$$

$$\left. \frac{c^{(\sigma)} R T^{(\sigma)}}{\rho_{(1)}^{(\sigma)}} \mathbf{d}_{(1)}^{(\sigma)}, \frac{c^{(\sigma)} R T^{(\sigma)}}{\rho_{(2)}^{(\sigma)}} \mathbf{d}_{(2)}^{(\sigma)}, \ldots, \frac{c^{(\sigma)} R T^{(\sigma)}}{\rho_{(N)}^{(\sigma)}} \mathbf{d}_{(N)}^{(\sigma)} \right) \tag{7-2}$$

Since there is no experimental evidence that $\mathbf{e}$ should be a function of the history of the motion of the material, it seems reasonable that we can neglect any explicit dependence of $\mathbf{e}^{(\sigma)}$ upon the motion of the interface.

Because $\mathbf{e}^{(\sigma)}$, $\mathbf{q}^{(\sigma)}$, and $\mathbf{j}_{(A)}^{(\sigma)}$ are all frame-indifferent vectors (Exercises 5.2.2-1, 5.6.4-3, and 5.7.3-8), the principle of frame indifference requires that $\boldsymbol{\varepsilon}^{(\sigma)}$ be an isotropic function of $\Pi^{(\sigma)}$

$$\mathbf{Q} \cdot \boldsymbol{\varepsilon}^{(\sigma)}(\Pi^{(\sigma)}) = \boldsymbol{\varepsilon}^{(\sigma)}(\mathbf{Q} \cdot \Pi^{(\sigma)}) \tag{7-3}$$

where I have defined

$$\mathbf{Q} \cdot \Pi^{(\sigma)} \equiv \left(\hat{A}, T^{(\sigma)}, \omega_{(1)}^{(\sigma)}, \omega_{(2)}^{(\sigma)}, \ldots, \omega_{(N-1)}^{(\sigma)}, \nabla_{(\sigma)} \ln T^{(\sigma)}, \right.$$

$$\frac{c^{(\sigma)} R T^{(\sigma)}}{\rho_{(1)}^{(\sigma)}} \mathbf{Q} \cdot \mathbf{d}_{(1)}^{(\sigma)}, \frac{c^{(\sigma)} R T^{(\sigma)}}{\rho_{(2)}^{(\sigma)}} \mathbf{Q} \cdot \mathbf{d}_{(2)}^{(\sigma)}, \ldots,$$

$$\left. \frac{c^{(\sigma)} R T^{(\sigma)}}{\rho_{(N)}^{(\sigma)}} \mathbf{Q} \cdot \mathbf{d}_{(N)}^{(\sigma)} \right) \tag{7-4}$$

Using the representation theorems of Spencer and Rivlin (1960) and of Smith (1965), we find that the most general polynomial vector function of this form is (see Sec. 5.9.4, footnote a)

$$\boldsymbol{\varepsilon}^{(\sigma)} = \alpha^{(\sigma)} \nabla_{(\sigma)} \ln T^{(\sigma)} + c^{(\sigma)} R T^{(\sigma)} \sum_{A=1}^{N} \frac{\beta_{(A)}^{(\sigma)}}{\rho_{(A)}^{(\sigma)}} \mathbf{d}_{(A)}^{(\sigma)} \tag{7-5}$$

The coefficients $\alpha^{(\sigma)}$ and $\beta_{(A)}^{(\sigma)}$ ($A = 1, 2, \ldots, N$) may be functions of the local thermodynamic state variables for the surface ($\hat{A}$, $T^{(\sigma)}$, $\omega_{(1)}^{(\sigma)}$, $\omega_{(2)}^{(\sigma)}$, ..., $\omega_{(N-1)}^{(\sigma)}$) as well as of all the scalar products involving $\nabla_{(\sigma)} \ln T^{(\sigma)}$ and $\mathbf{d}_{(B)}^{(\sigma)}$ ($B = 1, 2, \ldots, N$) (including the magnitudes of these vectors).

We can expect to normally neglect the direct dependence of $\boldsymbol{\varepsilon}^{(\sigma)}$ upon the surface gradients of the surface chemical potentials and upon the external forces. Equation (7-5) reduces under these circumstances to

$$\boldsymbol{\varepsilon}^{(\sigma)} = \mathbf{q}^{(\sigma)} - \sum_{A=1}^{N} \overline{\overline{H}}_{(A)}^{(\sigma)} \mathbf{j}_{(A)}^{(\sigma)} = -k^{(\sigma)} \nabla_{(\sigma)} T^{(\sigma)} \tag{7-6}$$

If $\boldsymbol{\varepsilon}^{(\sigma)}$ is a linear function of $\nabla_{(\sigma)} T^{(\sigma)}$, then

$$k^{(\sigma)} = k^{(\sigma)}(\hat{\mathcal{A}}, T^{(\sigma)}, \omega_{(1)}^{(\sigma)}, \omega_{(2)}^{(\sigma)}, \ldots, \omega_{(N-1)}^{(\sigma)}) \tag{7-7}$$

In Exercise 5.9.7-1, we find that

$$k^{(\sigma)} \geq 0 \tag{7-8}$$

References

Smith, G. F., *Arch. Ration. Mech. Anal.* **18**, 282 (1965).

Spencer, A. J., and R. S. Rivlin, *Arch. Ration. Mech. Anal.* **4**, 214 (1960).

Exercise

5.9.7-1 *implications of the jump Clausius-Duhem inequality* Let us assume the surface thermal energy flux vector $\boldsymbol{\varepsilon}^{(\sigma)}$ is described by (7-6). Extend the reasoning outlined in Exercise 5.9.6-1 to conclude that

$$k^{(\sigma)} \geq 0$$

5.9.8 Surface behavior: constitutive equations for surface mass flux vector In view of our assumptions about surface material behavior in Sec. 5.8.5 and our discussion in Sec. 5.9.2 (see Table 5.9.2-2), the functional dependence of the surface mass flux $\mathbf{j}_{(A)}^{(\sigma)}$ may be as simple as

$$\mathbf{j}_{(A)}^{(\sigma)} = \mathbf{j}_{(A)}^{(\sigma)}(\boldsymbol{\Pi}^{(\sigma)}) \tag{8-1}$$

where $\boldsymbol{\Pi}^{(\sigma)}$ is defined by (7-2). There appears to be no experimental evidence that $\mathbf{j}_{(A)}$ should be a function of the history of the motion of the material. This suggests that we neglect any explicit dependence of $\mathbf{j}_{(A)}^{(\sigma)}$ upon the motion of the interface.

Since $\mathbf{j}_{(A)}^{(\sigma)}$ is a frame-indifferent vector (see Exercise 5.2.2-1), the principle of frame indifference requires that $\mathbf{j}_{(A)}^{(\sigma)}$ be an isotropic function of $\boldsymbol{\Pi}^{(\sigma)}$:

$$\mathbf{Q} \cdot \mathbf{j}_{(A)}(\,\boldsymbol{\Pi}^{(\sigma)}\,) = \mathbf{j}_{(A)}(\,\mathbf{Q} \cdot \boldsymbol{\Pi}^{(\sigma)}\,) \tag{8-2}$$

The set $\mathbf{Q} \cdot \boldsymbol{\Pi}^{(\sigma)}$ is defined by (7-4). Using the representation theorems of Spencer and Rivlin (1960) and of Smith (1965), we find that the most general polynomial vector function of this form is (see Sec. 5.9.4, footnote a)

$$\mathbf{j}_{(A)} = \alpha_{(A0)} \nabla_{(\sigma)} \ln T^{(\sigma)} + c^{(\sigma)} R T^{(\sigma)} \sum_{B=1}^{N} \frac{\beta_{(AB)}}{\rho_{(B)}} \mathbf{d}_{(B)} \tag{8-3}$$

The coefficients $\alpha_{(A0)}$ and $\beta_{(AB)}$ (A, B = 1, 2,..., N) are all functions of the thermodynamic state variables for the surface ($\hat{A}$, $T^{(\sigma)}$, $\omega_{(1)}$, $\omega_{(2)}$, ..., $\omega_{(N-1)}$) as well as of all the scalar products involving $\nabla_{(\sigma)} \ln T^{(\sigma)}$ and $\mathbf{d}_{(C)}$ (C = 1, 2,..., N) (including the magnitudes of these vectors).

Note that in Exercise 5.8.7-2 we find

$$\sum_{A=1}^{N} \mathbf{d}_{(A)} = 0 \tag{8-4}$$

The resulting indeterminateness in the coefficients $\beta_{(AB)}$ is removed by requiring (A = 1, 2,..., N)

$$\beta_{(AA)} = 0 \tag{8-5}$$

Because (see Table 5.2.2-4)

$$\sum_{A=1}^{N} \mathbf{j}_{(A)} = 0 \tag{8-6}$$

we must also require

$$\sum_{A=1}^{N} \alpha_{(A0)} = 0 \tag{8-7}$$

and

$$\sum_{A=1}^{N} \left(\frac{\beta_{(AB)}}{\rho_{(B)}} - \frac{\beta_{(AC)}}{\rho_{(C)}} \right) = 0 \tag{8-8}$$

For a material composed of only two species, (8-3) reduces to

$$\mathbf{j}_{(A)} = c^{(\sigma)} R T^{(\sigma)} \frac{\beta_{(AB)}}{\rho_{(B)}} \mathbf{d}_{(B)} + \alpha_{(A0)} \nabla_{(\sigma)} \ln T^{(\sigma)} \tag{8-9}$$

where (8-8) demands

$$\frac{\beta^{(\sigma)}_{(AB)}}{\rho^{(\sigma)}_{(B)}} = \frac{\beta^{(\sigma)}_{(BA)}}{\rho^{(\sigma)}_{(A)}} \tag{8-10}$$

In view of (8-4), (8-10), and the expression for $\mathbf{d}^{(\sigma)}_{(A)}$ given in Table 5.9.2-2, (8-9) may be expressed as

$$\mathbf{j}^{(\sigma)}_{(A)} = -\, c^{(\sigma)} R T^{(\sigma)} \frac{\beta^{(\sigma)}_{(BA)}}{\rho^{(\sigma)}_{(A)}} \mathbf{d}^{(\sigma)}_{(A)} + \alpha^{(\sigma)}_{(A0)} \nabla_{(\sigma)} \ln T^{(\sigma)}$$

$$= -\, \beta^{(\sigma)}_{(BA)} \left[\left(\frac{\partial \mu^{(\sigma)}_{(A)}}{\partial \omega^{(\sigma)}_{(A)}} \right)_{T^{(\sigma)},\gamma} \nabla \omega^{(\sigma)}_{(A)} - \left(\overline{\overline{\mathcal{A}}}_{(A)} - \hat{\mathcal{A}} \right) \nabla_{(\sigma)} \gamma \right.$$

$$\left. - \, \omega^{(\sigma)}_{(B)} \left(\mathbf{b}^{(\sigma)}_{(A)} - \mathbf{b}^{(\sigma)}_{(B)} \right) \right] + \alpha^{(\sigma)}_{(A0)} \nabla_{(\sigma)} \ln T^{(\sigma)} \tag{8-11}$$

Using the jump Clausius-Duhem inequality, we can show that (Exercise 5.9.8-1)

$$\beta^{(\sigma)}_{(BA)} \geq 0 \tag{8-12}$$

Normally we will not be given the concentration dependence of the surface chemical potentials and we will find it more convenient to express (8-11) as

$$\mathbf{j}^{(\sigma)}_{(A)} = -\, \rho^{(\sigma)} \mathcal{D}^{(\sigma)}_{(AB)} \nabla_{(\sigma)} \omega^{(\sigma)}_{(A)} + \beta^{(\sigma)}_{(AB)} \left[\left(\overline{\overline{\mathcal{A}}}_{(A)} - \hat{\mathcal{A}} \right) \nabla_{(\sigma)} \gamma \right.$$

$$\left. + \, \omega^{(\sigma)}_{(B)} \left(\mathbf{b}^{(\sigma)}_{(A)} - \mathbf{b}^{(\sigma)}_{(B)} \right) \right] + \alpha^{(\sigma)}_{(A0)} \nabla_{(\sigma)} \ln T^{(\sigma)} \tag{8-13}$$

where the **surface diffusion coefficient** $\mathcal{D}^{(\sigma)}_{(AB)}$ is symmetric (see Exercise 5.9.8-2):

$$\mathcal{D}^{(\sigma)}_{(AB)} = \mathcal{D}^{(\sigma)}_{(BA)} \tag{8-14}$$

If $\mathbf{j}^{(\sigma)}_{(A)}$ is a linear function of $\nabla_{(\sigma)} \omega^{(\sigma)}_{(A)}$, then

$$\mathcal{D}^{(\sigma)}_{(AB)} = \mathcal{D}^{(\sigma)}_{(AB)} \left(\hat{\mathcal{A}}, T^{(\sigma)}, \omega^{(\sigma)}_{(1)}, \omega^{(\sigma)}_{(2)}, \ldots, \omega^{(\sigma)}_{(N-1)} \right) \tag{8-15}$$

When we are willing to neglect the direct dependence of $\mathbf{j}^{(\sigma)}_{(A)}$ upon $\nabla_{(\sigma)} T^{(\sigma)}$, $\nabla_{(\sigma)} \gamma$, and the external forces, (8-9) takes the form for

$$\mathbf{j}^{(\sigma)}_{(A)} = -\rho^{(\sigma)} \mathcal{D}^{(\sigma)}_{(AB)} \nabla_{(\sigma)} \omega^{(\sigma)}_{(A)} \tag{8-16}$$

Table 5.9.8-1 presents several alternative forms of (8-16) that follow immediately using the relationships developed in Sec. 5.2.2.

Table 5.9.8-1 *equivalent forms of (8-16) for binary surface diffusion*

$$\mathbf{n}^{(\sigma)}_{(A)} = \omega^{(\sigma)}_{(A)}\left(\mathbf{n}^{(\sigma)}_{(A)} + \mathbf{n}^{(\sigma)}_{(B)} \right) - \rho^{(\sigma)} \mathcal{D}^{(\sigma)}_{(AB)} \nabla_{(\sigma)} \omega^{(\sigma)}_{(A)}$$

$$\mathbf{j}^{(\sigma)}_{(A)} = -\rho^{(\sigma)} \mathcal{D}^{(\sigma)}_{(AB)} \nabla_{(\sigma)} \omega^{(\sigma)}_{(A)}$$

$$\mathbf{j}^{(\sigma)}_{(A)} = -\left(\frac{c^{(\sigma)}}{\rho^{(\sigma)}} \right)^2 M_{(A)} M_{(B)} \mathcal{D}^{(\sigma)}_{(AB)} \nabla_{(\sigma)} x^{(\sigma)}_{(A)}$$

$$\mathbf{N}^{(\sigma)}_{(A)} = x_{(A)}\left(\mathbf{N}^{(\sigma)}_{(A)} + \mathbf{N}^{(\sigma)}_{(B)} \right) - c^{(\sigma)} \mathcal{D}^{(\sigma)}_{(AB)} \nabla_{(\sigma)} x^{(\sigma)}_{(A)}$$

$$\mathbf{J}^{(\sigma)*}_{(A)} = -c^{(\sigma)} \mathcal{D}^{(\sigma)}_{(AB)} \nabla_{(\sigma)} x^{(\sigma)}_{(A)}$$

$$\mathbf{J}^{(\sigma)*}_{(A)} = -\left(\frac{\rho^{(\sigma)2}}{c^{(\sigma)} M_{(A)} M_{(B)}} \right) \mathcal{D}^{(\sigma)}_{(AB)} \nabla_{(\sigma)} \omega^{(\sigma)}_{(A)}$$

References

Smith, G. F., *Arch. Ration. Mech. Anal.* **18**, 282 (1965).

Spencer, A. J., and R. S. Rivlin, *Arch. Ration. Mech. Anal.* **4**, 214 (1960).

Exercises

5.9.8-1 *implications of the jump Clausius-Duhem inequality* Let us restrict ourselves to a two-component system and let us assume that the surface mass flux $\mathbf{j}^{(\sigma)}_{(A)}$ is represented by (8-9). Extend the reasoning outlined in Exercises 5.9.6-1 and 5.9.7-1 to conclude that (see Exercise 5.8.7-1)

$$\beta^{(\sigma)}_{(AB)} \geq 0$$

This implies that, if we neglect the direct dependence of $\mathbf{j}_{(A)}^{(\sigma)}$ upon the surface gradient of surface temperature and upon the external forces,

$$\mathcal{D}_{(AB)}^{(\sigma)} \geq 0$$

in (8-11).

5.9.8-2 *symmetry of* $\mathcal{D}_{(AB)}^{(\sigma)}$

i) Starting with the surface Gibbs-Duhem equation, prove that

$$\omega_{(A)}^{(\sigma)} \left(\frac{\partial \mu_{(A)}^{(\sigma)}}{\partial \omega_{(B)}^{(\sigma)}} \right)_{T^{(\sigma)},\gamma} = \omega_{(B)}^{(\sigma)} \left(\frac{\partial \mu_{(B)}^{(\sigma)}}{\partial \omega_{(A)}^{(\sigma)}} \right)_{T^{(\sigma)},\gamma}$$

ii) Use (8-10) to conclude that

$$\beta_{(BA)}^{(\sigma)} \left(\frac{\partial \mu_{(A)}^{(\sigma)}}{\partial \omega_{(B)}^{(\sigma)}} \right)_{T^{(\sigma)},\gamma} = \beta_{(AB)}^{(\sigma)} \left(\frac{\partial \mu_{(B)}^{(\sigma)}}{\partial \omega_{(A)}^{(\sigma)}} \right)_{T^{(\sigma)},\gamma}$$

and that

$$\mathcal{D}_{(AB)}^{(\sigma)} = \mathcal{D}_{(BA)}^{(\sigma)}$$

5.10 Intrinsically stable equilibrium (Slattery 1977)

5.10.1 Stable equilibrium The temperature at which water vaporizes at atmospheric pressure depends upon the circumstances. In the presence of a powder or boiling chips that favor the formation of large bubbles, we find that it boils at very nearly 100°C. In a very clean, smooth glass container, it is possible to superheat water somewhat above 100°C before boiling occurs. As an extreme example, Dufour (1861) has super heated water suspended in oil to 178°C.

In each of these experiments, we have a relatively isolated body whose form, below some limiting temperature, remains stable when subjected to the random disturbances that are always present in real experiments. But at this limiting temperature (or above), the original form of the body is unstable to all disturbances, no matter how small.

I define **equilibrium** to be achieved by a body, when the Clausius-Duhem inequality becomes an equality. Let us assume that a body initially at equilibrium is momentarily disturbed. If after the disturbance ceases the body returns to very nearly its original state, we say that the equilibrium is **stable**. If instead the body evolves into a distinctly different state, we refer to the equilibrium as being **unstable**. An equilibrium may be unstable to relatively large disturbances, but stable when subjected to small disturbances. We will say that an equilibrium is **intrinsically stable**, if the conditions for equilibrium are not immediately violated by the imposition of small disturbances, so small that mass transfer to, from and across the phase interfaces are unaffected. A supersaturated salt solution is intrinsically stable, although it may be unstable to macroscopic transients such as those imposed by scraping a glass stirring rod across the wall of the beaker containing the solution.

Necessary and sufficient conditions for an intrinsically stable equilibrium of a single-phase, multicomponent body have been discussed previously[a] (Gibbs 1948, chapter 3; Tisza 1951; Prigogine and Defay 1954, chapters 15 and 16; Haase 1956, p. 78; Denbigh 1955, p. 137; Callen 1960, chapter 8; Rowlinson 1969, chapter 5; Münster 1970, chapters 2 and 6; Beegle *et al.* 1974; Modell and Reid 1974, chapter 7). Slattery (1981, p. 485) considered a multiphase, multicomponent body, but neglected all interfacial effects.

---

a) With the exception of Slattery (1981, p. 485), all of these writers have discussed the necessary and sufficient conditions for the achievement of a stable state of rest (stable equilibrium state). A stable state of rest is defined to have a corresponding entropy that is greater than the entropy associated with any other possible state of rest. For example, Gibbs (1948, p. 100; see also Coleman and Noll 1960) restricted his comparisons to all states of rest that a body might achieve with a fixed volume and internal energy.

Beegle *et al.* (1974) have pointed out that for a single phase, multicomponent body a particular one of these conditions will fail before all the others. This is the **limiting criterion for intrinsic stability**. The failure of this condition defines the **limits of intrinsic stability** for the body.

In the following sections, we will develop necessary and sufficient conditions for an intrinsically stable equilibrium in a multiphase, multicomponent body totally enclosed by fixed, impermeable, adiabatic walls. Interfacial effects are taken fully into account by including the mass, momentum, energy, and entropy associated with all of the phase interfaces. We will find that there is a limiting criterion for intrinsic stability which must be satisfied on every phase interface. The limits of intrinsic stability for a multiphase, multicomponent body are defined by the failure either of the first criterion within one of the phases or of the second criterion on one of the phase interfaces.

References

Beegle, B. L., M. Modell, and R. C. Reid, *AIChE J* **20**, 1200 (1974). See also R. A. Heidemann, *AIChE J* **21**, 824 (1975) as well as B. L. Beegle, M. Modell, and R. C. Reid, *AIChE J* **21**, 826 (1975).

Callen, H. B., "Thermodynamics," Wiley, New York (1960).

Coleman, B. D., and W. Noll, *Arch. Ration. Mech. Anal.* **4**, 97 (1960).

Denbigh, K. G., "The Principles of Chemical Equilibrium," Cambridge, London (1955).

Dufour, L., Supplément à la Bibliothèque Universelle et Revue Suisse, Archives des Sciences Physiques et Naturelles, Genève (2nd series) **12**, 210 (1961); referred to by R. Defay, I. Prigogine, A. Bellemans, and D. H. Everett, "Surface Tension and Adsorption," p. 243, Wiley, New York (1966).

Gibbs, J. W., "The Collected Works," vol. 1, Yale University Press, New Haven, Conn. (1948).

Haase, R., "Thermodynamik der Mischphasen," Springer-Verlag, Berlin (1956).

Modell, M., and R. C. Reid, "Thermodynamics and Its Applications," Prentice-Hall, Englewood Cliffs, New Jersey (1974).

Münster, A., "Classical Thermodynamics," Wiley-Interscience, New York (1970).

Prigogine, I., and R. Defay, "Chemical Thermodynamics," Wiley, New York

(1954).

Rowlinson, J. S., "Liquids and Liquid Mixtures," second edition, Butterworth, London (1969).

Slattery, J. C., "Momentum, Energy, and Mass Transfer in Continua," McGraw-Hill, New York (1972); second edition, Robert E. Krieger, Malabar, FL 32950 (1981).

Slattery, J. C., *AIChE J* **23**, 275 (1977).

Tisza, L., in "Phase Transformations in Solids," edited by R. Smoluchowski, J. C. Mayer, and W. A. Weyl, Wiley, New York (1951); see also "Generalized Thermodynamics," p. 194, MIT Press, Cambridge, MA (1966).

5.10.2 Constraints on isolated systems Let us begin by examining the constraints imposed upon an isolated, multiphase, multicomponent body by the mass balances for the individual species, by Euler's first law, by the energy balance, and by the Clausius-Duhem inequality.

Species mass balance In Sec. 5.2.1, we found that each species is constrained by a mass balance which requires the time rate of change of the mass of each component A to be equal to the rate at which the mass of that component is produced by chemical reactions:

$$\frac{d}{dt}\left(\int_R \rho_{(A)}\, dV + \int_\Sigma \rho^{(\sigma)}_{(A)}\, dA \right) = \int_R \sum_{j=1}^{J} r_{(A,j)}\, dV + \int_\Sigma \sum_{k=1}^{K} r^{(\sigma)}_{(A,k)}\, dA \tag{2-1}$$

Here R denotes the region occupied by the body, Σ represents its internal phase interfaces, $r_{(A,j)}$ is the rate at which mass of species A is produced per unit volume as the result of the homogeneous chemical reaction j (j = 1, 2, ..., J), and $r^{(\sigma)}_{(A,k)}$ is the rate at which mass of species A is produced per unit area by the heterogeneous chemical reaction k (k = 1, 2, ..., K). After an application of the transport theorem of Exercise 5.2.1-2, (2-1) takes the form

$$\int_R \left\{ \frac{\partial \rho_{(A)}}{\partial t} + \text{div}(\rho_{(A)} \mathbf{v}_{(A)}) - \sum_{j=1}^{J} r_{(A,j)} \right\} dV$$

$$+ \int_{\Sigma} \left\{ \frac{\partial \rho^{(\sigma)}_{(A)}}{\partial t} - \nabla_{(\sigma)} \rho^{(\sigma)}_{(A)} \cdot \mathbf{u} + \text{div}_{(\sigma)}(\rho^{(\sigma)}_{(A)} \mathbf{v}^{(\sigma)}_{(A)}) - \sum_{k=1}^{K} r^{(\sigma)}_{(A,k)} + [\, \rho_{(A)} (\mathbf{v}_{(A)} \cdot \boldsymbol{\xi} - \mathrm{v}^{(\sigma)}_{(\xi)})] \right\} dA = 0 \tag{2-2}$$

Since this is an isolated body, there is no mass transfer at the bounding surfaces of R. In other words, at the fixed bounding surfaces of R the normal component of the velocity of an individual species is zero. Under these conditions, the divergence theorem requires

$$\int_{R} \text{div}(\rho_{(A)} \mathbf{v}_{(A)}) \, dV + \int_{\Sigma} [\, \rho_{(A)} \mathbf{v}_{(A)} \cdot \boldsymbol{\xi} \,] \, dA = 0 \tag{2-3}$$

Similarly, since there is no mass transfer between Σ and the fixed bounding surfaces of R, the surface divergence theorem tells us that

$$\int_{\Sigma} \text{div}_{(\sigma)}(\rho^{(\sigma)}_{(A)} \mathbf{v}^{(\sigma)}_{(A)}) \, dA = - \int_{\Sigma} 2H \rho^{(\sigma)}_{(A)} \mathbf{v}^{(\sigma)}_{(A)} \cdot \boldsymbol{\xi} \, dA \tag{2-4}$$

Let us introduce the *jth* homogeneous reaction coordinate $\Psi_{(j)}$ by defining for all A,B = 1, 2, ..., N

$$\frac{\partial \psi_{(j)}}{\partial t} \equiv \frac{r_{(A,j)}}{M_{(A)} \nu_{(A,j)}} = \frac{r_{(B,j)}}{M_{(B)} \nu_{(B,j)}} \tag{2-5}$$

The right side of this equation represents a normalized rate of production of moles of species A by the homogeneous chemical reaction j at a fixed point in space; $M_{(A)}$ is the molecular weight of species A and $\nu_{(A,j)}$ is the stoichiometric coefficient for species A in the homogeneous chemical reaction j. The stoichiometric coefficient is taken to be a positive number for a species consumed in the reaction. As we saw in Sec. 5.4.1, overall mass conservation requires that for each chemical reaction

$$\sum_{A=1}^{N} r_{(A,j)} = \frac{\partial \psi_{(j)}}{\partial t} \sum_{A=1}^{N} M_{(A)} \nu_{(A,j)} = 0 \tag{2-6}$$

Our assumption is that

$$\frac{\partial \psi_{(j)}}{\partial t} \neq 0 \tag{2-7}$$

which implies (j = 1, 2, ..., J)

$$\sum_{A=1}^{N} M_{(A)} \nu_{(A,j)} = 0 \tag{2-8}$$

In much the same manner, we can introduce the k*th* heterogeneous reaction coordinate $\psi^{(\sigma)}_{(k)}$ by defining for all A,B = 1, 2, ..., N

$$\frac{\partial \psi^{(\sigma)}_{(k)}}{\partial t} - \nabla_{(\sigma)} \psi^{(\sigma)}_{(k)} \cdot \mathbf{u} \equiv \frac{r^{(\sigma)}_{(A,k)}}{M_{(A)} \nu^{(\sigma)}_{(A,k)}} = \frac{r^{(\sigma)}_{(B,k)}}{M_{(B)} \nu^{(\sigma)}_{(B,k)}} \tag{2-9}$$

On the right side of this equation, we have a normalized rate of production of moles of species A by the heterogeneous chemical reaction k at a point fixed on the surface in such a way that its only motion is normal to the interface; $\nu^{(\sigma)}_{(A,k)}$ is the stoichiometric coefficient for species A in the heterogeneous chemical reaction k. Again, overall mass conservation requries that for each chemical rection (see Sec. 5.4.1)

$$\sum_{A=1}^{N} r^{(\sigma)}_{(A,k)} = \left(\frac{\partial \psi^{(\sigma)}_{(k)}}{\partial t} - \nabla_{(\sigma)} \psi^{(\sigma)}_{(k)} \cdot \mathbf{u} \right) \sum_{A=1}^{N} M_{(A)} \nu^{(\sigma)}_{(A,k)} = 0 \tag{2-10}$$

Since

$$\frac{\partial \Psi^{(\sigma)}_{(k)}}{\partial t} - \nabla_{(\sigma)} \psi^{(\sigma)}_{(k)} \cdot \mathbf{u} \neq 0 \tag{2-11}$$

we have (k = 1, 2, ..., K)

$$\sum_{A=1}^{N} M_{(A)} \nu^{(\sigma)}_{(A,k)} = 0 \tag{2-12}$$

In view of (2-3) through (2-5) and of (2-9), we may express (2-2) as

$$Z_{(A)} \equiv \int_R \frac{\partial}{\partial t} \left(\rho_{(A)} - \sum_{j=1}^{J} M_{(A)} \nu_{(A,j)} \Psi_{(j)} \right) dV$$

$$+ \int_\Sigma \left\{ \frac{\partial}{\partial t} \left(\rho^{(\sigma)}_{(A)} - \sum_{k=1}^{K} M_{(A)} \, \nu^{(\sigma)}_{(A,k)} \, \psi^{(\sigma)}_{(k)} \right) \right.$$

$$- \nabla_{(\sigma)} \left(\rho^{(\sigma)}_{(A)} - \sum_{k=1}^{K} M_{(A)} \nu^{(\sigma)}_{(A,k)} \psi^{(\sigma)}_{(k)} \right) \cdot \mathbf{u}$$

$$- 2H\rho^{(\sigma)}_{(A)} \mathbf{v}^{(\sigma)}_{(A)} \cdot \boldsymbol{\xi} - [\, \rho_{(A)} v^{(\sigma)}_{(\xi)} \,] \Big\} \, dA$$

$$= 0 \tag{2-13}$$

Euler's first law Since this body is isolated, the sum of the forces exerted upon the body is zero. Referring to Sec. 5.5.2, we find that Euler's first law requires the time rate of change of the momentum of the body to be zero:

$$\frac{d}{dt} \left(\int_R \rho \mathbf{v} \, dV + \int_\Sigma \rho^{(\sigma)} \mathbf{v}^{(\sigma)} \, dA \right) = 0 \tag{2-14}$$

An application of the transport theorem of Exercise 5.3.1-2 yields

$$\mathbf{Z}_m \equiv \int_R \left\{ \frac{\partial(\rho \mathbf{v})}{\partial t} + \operatorname{div}(\rho \mathbf{v}\mathbf{v}) \right\} dV + \int_\Sigma \left\{ \frac{\partial(\rho^{(\sigma)} \mathbf{v}^{(\sigma)})}{\partial t} \right.$$

$$- \nabla_{(\sigma)}(\rho^{(\sigma)} \mathbf{v}^{(\sigma)}) \cdot \mathbf{u} + \operatorname{div}_{(\sigma)}(\rho^{(\sigma)} \mathbf{v}^{(\sigma)} \mathbf{v}^{(\sigma)})$$

$$\left. + [\, \rho \mathbf{v}(\mathbf{v} \cdot \boldsymbol{\xi} - v^{(\sigma)}_{(\xi)})] \right\} dA$$

$$= \int_R \frac{\partial(\rho \mathbf{v})}{\partial t} dV + \int_\Sigma \left\{ \frac{\partial(\rho^{(\sigma)} \mathbf{v}^{(\sigma)})}{\partial t} - \nabla_{(\sigma)}(\rho^{(\sigma)} \mathbf{v}^{(\sigma)}) \cdot \mathbf{u} \right.$$

$$\left. - \rho^{(\sigma)}(\mathbf{B} \cdot \mathbf{v}^{(\sigma)}) v^{(\sigma)}_{(\xi)} - [\, \rho \mathbf{v} v^{(\sigma)}_{(\xi)} \,] \right\} dA$$

$$= 0 \tag{2-15}$$

In arriving at this result, I have used both Green's transformation (Slattery 1981, p. 661) and the surface divergence theorem (see Exercise A.6.3-2). I have also recognized that $\mathbf{v}$ and $\mathbf{v}^{(\sigma)}$ must be zero at the surface bounding R.

Energy balance For this isolated body, the energy balance of Sec. 5.6.3 states that

$$\frac{d}{dt}\left[\int_R \rho\left(\hat{U}+\frac{1}{2}v^2\right) dV + \int_\Sigma \rho^{(\sigma)}\left(\hat{U}^{(\sigma)}+\frac{1}{2}v^{(\sigma)2}\right) dA\right]$$

$$= \int_R \sum_{A=1}^{N} \rho_{(\sigma)}\mathbf{v}_{(\sigma)} \cdot \mathbf{b}_{(A)}\, dV + \int_\Sigma \sum_{A=1}^{N} \rho^{(\sigma)}_{(A)}\, \mathbf{v}^{(\sigma)}_{(A)} \cdot \mathbf{b}^{(\sigma)}_{(A)}\, dA \qquad (2\text{-}16)$$

The contact forces do no work, since the boundary of the body is fixed in space. The boundary is adiabatic, which we interpret here as meaning that there is neither contact energy transmission with the surroundings nor external radiant energy transmission. We neglect the possibility of mutual radiant energy transmission. The time rate of change of the internal and kinetic energy of the body is the result only of work done by the body forces.

An application of the transport theorem of Exercise 5.3.1-2 allows us to express the left side of (2-16) as

$$\frac{d}{dt}\left[\int_R \rho\left(\hat{U}+\frac{1}{2}v^2\right) dV + \int_\Sigma \rho^{(\sigma)}\left(\hat{U}^{(\sigma)}+\frac{1}{2}v^{(\sigma)2}\right) dA\right]$$

$$= \int_R \left\{\frac{\partial}{\partial t}\left[\rho\left(\hat{U}+\frac{1}{2}v^2\right)\right] + \text{div}\left[\rho\left(\hat{U}+\frac{1}{2}v^2\right)\mathbf{v}\right]\right\} dV$$

$$+ \int_\Sigma \left\{\frac{\partial}{\partial t}\left[\rho^{(\sigma)}\left(\hat{U}^{(\sigma)}+\frac{1}{2}v^{(\sigma)2}\right)\right]\right.$$

$$- \nabla_{(\sigma)}\left[\rho^{(\sigma)}\left(\hat{U}^{(\sigma)}+\frac{1}{2}v^{(\sigma)2}\right)\right] \cdot \mathbf{u}$$

$$+ \text{div}_{(\sigma)}\left[\rho^{(\sigma)}\left(\hat{U}^{(\sigma)}+\frac{1}{2}v^{(\sigma)2}\right)\mathbf{v}^{(\sigma)}\right]$$

$$\left. + \left[\rho\left(\hat{U}+\frac{1}{2}v^2\right)(\mathbf{v}\cdot\boldsymbol{\xi} - v_{(\xi)}^{(g)})\right]\right\} dA \qquad (2\text{-}17)$$

The divergence theorem permits us to write

$$\int_R \text{div}\left[\rho\left(\hat{U}+\frac{1}{2}v^2\right)\mathbf{v}\right] dV$$

$$= -\int_{\Sigma} \left[\rho \left(\hat{U} + \frac{1}{2} v^2 \right) (\mathbf{v} \cdot \boldsymbol{\xi}) \right] dA \tag{2-18}$$

recognizing that $\mathbf{v}$ must be zero at the bounding surface of R. Using the surface divergence theorem, we find

$$\int_{\Sigma} \mathrm{div}_{(\sigma)} \left[\rho^{(\sigma)} \left(\hat{U}^{(\sigma)} + \frac{1}{2} v^{(\sigma)2} \right) \mathbf{v}^{(\sigma)} \right] dA$$

$$= -\int_{\Sigma} 2H\rho^{(\sigma)} \left(\hat{U}^{(\sigma)} + \frac{1}{2} v^{(\sigma)2} \right) v_{\{\xi\}}^{\{\sigma\}} \, dA \tag{2-19}$$

again observing that $\mathbf{v}^{(\sigma)}$ must be zero at the bounding surface of R. In view of (2-17) through (2-19), (2-16) takes the form

$$\int_{R} \left\{ \frac{\partial}{\partial t} \left[\rho \left(\hat{U} + \frac{1}{2} v^2 \right) \right] - \sum_{A=1}^{N} \rho_{(A)} \mathbf{v}_{(A)} \cdot \mathbf{b}_{(A)} \right\} dV$$

$$+ \int_{\Sigma} \left\{ \frac{\partial}{\partial t} \left[\rho^{(\sigma)} \left(\hat{U}^{(\sigma)} + \frac{1}{2} v^{(\sigma)2} \right) \right] \right.$$

$$- \nabla_{(\sigma)} \left[\rho^{(\sigma)} \left(\hat{U}^{(\sigma)} + \frac{1}{2} v^{(\sigma)2} \right) \right] \cdot \mathbf{u}$$

$$- \sum_{A=1}^{N} \rho_{\{A\}}^{\{\sigma\}} \mathbf{v}_{\{A\}}^{\{\sigma\}} \cdot \mathbf{b}_{\{A\}}^{\{\sigma\}} - 2H\rho^{(\sigma)} \left(\hat{U}^{(\sigma)} + \frac{1}{2} v^{(\sigma)2} \right) v_{\{\xi\}}^{\{\sigma\}}$$

$$\left. - \left[\rho \left(\hat{U} + \frac{1}{2} v^2 \right) v_{\{\xi\}}^{\{\sigma\}} \right] \right\} dA$$

$$= 0 \tag{2-20}$$

If we assume that the external force for each species A is representable by a potential

$$\mathbf{b}_{(A)} = - \nabla \phi_{(A)} \tag{2-21}$$

and if we assume that this potential is independent of time for a fixed position in space, we may rearrange

$$\rho_{(A)}\mathbf{v}_{(A)} \cdot \mathbf{b}_{(A)} = -\operatorname{div}(\rho_{(A)}\phi_{(A)}\mathbf{v}_{(A)}) + \phi_{(A)}\operatorname{div}(\rho_{(A)}\mathbf{v}_{(A)})$$

$$= -\operatorname{div}(\rho_{(A)}\phi_{(A)}\mathbf{v}_{(A)})$$

$$+ \phi_{(A)}\left[\frac{\partial\rho_{(A)}}{\partial t} + \operatorname{div}(\rho_{(A)}\mathbf{v}_{(A)})\right]$$

$$- \frac{\partial(\rho_{(A)}\phi_{(A)})}{\partial t} \tag{2-22}$$

The differential mass balance for species A together with (2-5) says that

$$\frac{\partial\rho_{(A)}}{\partial t} + \operatorname{div}(\rho_{(A)}\mathbf{v}_{(A)}) = \sum_{j=1}^{J} r_{(A,j)}$$

$$= \sum_{j=1}^{J} \frac{\partial}{\partial t}(M_{(A)}v_{(A,j)}\Psi_{(j)}) \tag{2-23}$$

This allows us to express (2-22) as

$$\rho_{(A)}\mathbf{v}_{(A)} \cdot \mathbf{b}_{(A)} = -\operatorname{div}(\rho_{(A)}\phi_{(A)}\mathbf{v}_{(A)})$$

$$+ \frac{\partial}{\partial t}\left(\sum_{j=1}^{J} \phi_{(A)}M_{(A)}v_{(A,j)}\Psi_{(j)} - \rho_{(A)}\phi_{(A)}\right) \tag{2-24}$$

This together with the divergence theorem permit us to reason that

$$\int_R \sum_{A=1}^{N} \rho_{(A)}\mathbf{v}_{(A)} \cdot \mathbf{b}_{(A)}\, dV = \int_R \frac{\partial}{\partial t}\sum_{A=1}^{N}\sum_{j=1}^{J} (\phi_{(A)}M_{(A)}v_{(A,j)}\Psi_{(j)}$$

$$- \rho_{(A)}\phi_{(A)})\, dV + \int_\Sigma \left[\sum_{A=1}^{N} \rho_{(A)}\phi_{(A)}\mathbf{v}_{(A)} \cdot \boldsymbol{\xi}\right] dA \tag{2-25}$$

In much the same way, let us also assume that the external force acting on each species A in the dividing surface may also be represented in terms of a potential (see Secs. A.3.2 and A.5.2):

$$\mathbf{b}^{(\sigma)}_{(A)} = -\nabla\phi^{(\sigma)}_{(A)}$$

$$= -\mathbf{I} \cdot \nabla\phi^{(\sigma)}_{(A)}$$

$$= - \mathbf{P} \cdot \nabla \phi^{(\sigma)}_{(A)} - (\nabla \phi^{(\sigma)}_{(A)} \cdot \boldsymbol{\xi}) \boldsymbol{\xi}$$

$$= - \nabla_{(\sigma)} \phi^{(\sigma)}_{(A)} - (\nabla \phi^{(\sigma)}_{(A)} \cdot \boldsymbol{\xi}) \boldsymbol{\xi} \qquad (2\text{-}26)$$

Employing this expression, we can say

$$\begin{aligned}
&\rho^{(\sigma)}_{(A)} \mathbf{v}^{(\sigma)}_{(A)} \cdot \mathbf{b}^{(\sigma)}_{(A)} \\
&\quad = - \operatorname{div}_{(\sigma)} (\rho^{(\sigma)}_{(A)} \phi^{(\sigma)}_{(A)} \mathbf{v}^{(\sigma)}_{(A)}) + \phi^{(\sigma)}_{(A)} \operatorname{div}_{(\sigma)} (\rho^{(\sigma)}_{(A)} \mathbf{v}^{(\sigma)}_{(A)}) \\
&\qquad - \rho^{(\sigma)}_{(A)} (\nabla \phi^{(\sigma)}_{(A)} \cdot \boldsymbol{\xi})(\mathbf{v}^{(\sigma)}_{(A)} \cdot \boldsymbol{\xi}) \\
&\quad = - \operatorname{div}_{(\sigma)} (\rho^{(\sigma)}_{(A)} \phi^{(\sigma)}_{(A)} \mathbf{v}^{(\sigma)}_{(A)}) \\
&\qquad + \phi^{(\sigma)}_{(A)} \left[\frac{\partial \rho^{(\sigma)}_{(A)}}{\partial t} + \operatorname{div}_{(\sigma)} (\rho^{(\sigma)}_{(A)} \mathbf{v}^{(\sigma)}_{(A)}) \right] \\
&\qquad - \phi^{(\sigma)}_{(A)} \frac{\partial \rho^{(\sigma)}_{(A)}}{\partial t} - \rho^{(\sigma)}_{(A)} (\nabla \phi^{(\sigma)}_{(A)} \cdot \boldsymbol{\xi})(\mathbf{v}^{(\sigma)}_{(A)} \cdot \boldsymbol{\xi})
\end{aligned} \qquad (2\text{-}27)$$

Because of (2-9), the jump mass balance for species A may be written as

$$\begin{aligned}
&\frac{\partial \rho^{(\sigma)}_{(A)}}{\partial t} - \nabla_{(\sigma)} \rho^{(\sigma)}_{(A)} \cdot \mathbf{u} + \operatorname{div}_{(\sigma)} (\rho^{(\sigma)}_{(A)} \mathbf{v}^{(\sigma)}_{(A)}) + [\rho_{(A)} (\mathbf{v}_{(A)} \cdot \boldsymbol{\xi} - v^{(\sigma)}_{(\xi)})] \\
&\quad = \sum_{k=1}^{K} r^{(\sigma)}_{(A,k)} \\
&\quad = \sum_{k=1}^{K} \left\{ \frac{\partial}{\partial t} (M_{(A)} \nu^{(\sigma)}_{(A,k)} \psi^{(\sigma)}_{(k)}) \right. \\
&\qquad \left. - \nabla_{(\sigma)} (M_{(A)} \nu^{(\sigma)}_{(A,k)} \psi^{(\sigma)}_{(k)}) \cdot \mathbf{u} \right\}
\end{aligned} \qquad (2\text{-}28)$$

This may be used to rearrange (2-27) in the form

$$\begin{aligned}
&\rho^{(\sigma)}_{(A)} \mathbf{v}^{(\sigma)}_{(A)} \cdot \mathbf{b}^{(\sigma)}_{(A)} = - \operatorname{div}^{(\sigma)} \left(\rho^{(\sigma)}_{(A)} \phi^{(\sigma)}_{(A)} \mathbf{v}^{(\sigma)}_{(A)} \right) \\
&\quad + \frac{\partial}{\partial t} \left(\sum_{k=1}^{K} \phi^{(\sigma)}_{(A)} M_{(A)} \nu^{(\sigma)}_{(A,k)} \psi^{(\sigma)}_{(k)} - \rho^{(\sigma)}_{(A)} \phi^{(\sigma)}_{(A)} \right) \\
&\qquad - \nabla_{(\sigma)} \left(\sum_{k=1}^{K} \phi^{(\sigma)}_{(A)} M_{(A)} \nu^{(\sigma)}_{(A,k)} \psi^{(\sigma)}_{(k)} - \rho^{(\sigma)}_{(A)} \phi^{(\sigma)}_{(A)} \right) \cdot \mathbf{u}
\end{aligned}$$

$$+ \left(\sum_{k=1}^{K} M_{(A)} \, v^{(\sigma)}_{(A,k)} \, \psi^{(\sigma)}_{(k)} - \rho^{(\sigma)}_{(A)} \right) \left(- \frac{\partial \phi^{(\sigma)}_{(A)}}{\partial t} \right.$$

$$\left. + \nabla_{(\sigma)} \phi^{(\sigma)}_{(A)} \cdot \mathbf{u} \right) - [\, \rho_{(A)} \phi^{(\sigma)}_{(A)} (\mathbf{v}_{(A)} \cdot \boldsymbol{\xi} - v^{(\sigma)}_{(\xi)})]$$

$$- \rho^{(\sigma)}_{(A)} (\nabla \phi^{(\sigma)}_{(A)} \cdot \boldsymbol{\xi})(\mathbf{v}^{(\sigma)}_{(A)} - \boldsymbol{\xi}) \qquad (2\text{-}29)$$

Using the surface divergence theorem, we find

$$\int_{\Sigma} \sum_{A=1}^{N} \rho^{(\sigma)}_{(A)} \, \mathbf{v}^{(\sigma)}_{(A)} \cdot \mathbf{b}^{(\sigma)}_{(A)} \, dA = \int_{\Sigma} \left\{ 2H \sum_{A=1}^{N} \rho^{(\sigma)}_{(A)} \, \phi^{(\sigma)}_{(A)} \, \mathbf{v}^{(\sigma)}_{(A)} \cdot \boldsymbol{\xi} \right.$$

$$+ \frac{\partial}{\partial t} \sum_{A=1}^{N} \sum_{k=1}^{K} (\phi^{(\sigma)}_{(A)} M_{(A)} \, v^{(\sigma)}_{(A,k)} \, \psi^{(\sigma)}_{(k)} - \rho^{(\sigma)}_{(A)} \, \phi^{(\sigma)}_{(A)})$$

$$- \nabla_{(\sigma)} \sum_{A=1}^{N} \sum_{k=1}^{K} (\phi^{(\sigma)}_{(A)} \, M_{(A)} \, v^{(\sigma)}_{(A,k)} \, \psi^{(\sigma)}_{(k)}$$

$$- \rho^{(\sigma)}_{(A)} \, \phi^{(\sigma)}_{(A)}) \cdot \mathbf{u}$$

$$+ \sum_{A=1}^{N} \sum_{k=1}^{K} (M_{(A)} \, v^{(\sigma)}_{(A,k)} \, \psi^{(\sigma)}_{(k)} - \rho^{(\sigma)}_{(A)})$$

$$\times \left(- \frac{\partial \phi^{(\sigma)}_{(A)}}{\partial t} + \nabla_{(\sigma)} \phi^{(\sigma)}_{(A)} \cdot \mathbf{u} \right)$$

$$- \left[\sum_{A=1}^{N} \rho_{(A)} \phi^{(\sigma)}_{(A)} (\mathbf{v}_{(A)} \cdot \boldsymbol{\xi} - v^{(\sigma)}_{(\xi)}) \right]$$

$$\left. - \sum_{A=1}^{N} \rho^{(\sigma)}_{(A)} (\nabla \phi^{(\sigma)}_{(A)} \cdot \boldsymbol{\xi})(\mathbf{v}^{(\sigma)}_{(A)} \cdot \boldsymbol{\xi}) \right\} dA \qquad (2\text{-}30)$$

We can now use (2-25) and (2-30) to express (2-20) in a form more convenient for our purposes:

$$
\begin{aligned}
Z_e \equiv & \int_R \frac{\partial}{\partial t}\left(\check{E} - \sum_{A=1}^{N}\sum_{j=1}^{J} \phi_{(A)} M_{(A)} \nu_{(A,j)} \psi_{(j)} \right) dV \\
& + \int_\Sigma \Bigg\{ \frac{\partial}{\partial t}\left(\overline{E}^{(\sigma)} - \sum_{A=1}^{N}\sum_{k=1}^{K} \phi_{(A)}^{(\sigma)} M_{(A)} \nu_{(A,k)}^{(\sigma)} \psi_{(k)}^{(\sigma)} \right) \\
& \quad - \nabla_{(\sigma)}\left(\overline{E}^{(\sigma)} - \sum_{A=1}^{N}\sum_{k=1}^{K} \phi_{(A)}^{(\sigma)} M_{(A)} \nu_{(A,k)}^{(\sigma)} \psi_{(k)}^{(\sigma)} \right) \cdot \mathbf{u} \\
& \quad + \sum_{A=1}^{N}\sum_{k=1}^{K} \left(\rho_{(A)}^{(\sigma)} - M_{(A)} \nu_{(A,k)}^{(\sigma)} \psi_{(k)}^{(\sigma)} \right) \\
& \qquad \times \left(-\frac{\partial \phi_{(A)}^{(\sigma)}}{\partial t} + \nabla_{(\sigma)} \phi_{(A)}^{(\sigma)} \cdot \mathbf{u} \right) \\
& \quad - 2H \left[\overline{E}^{(\sigma)} v_{(\xi)}^{(\sigma)} + \sum_{A=1}^{N} \rho_{(A)}^{(\sigma)} \phi_{(A)}^{(\sigma)} \left(\mathbf{v}_{(A)}^{(\sigma)} \cdot \boldsymbol{\xi} - v_{(\xi)}^{(\sigma)} \right) \right] \\
& \quad - \Bigg[\check{E}\, v_{(\xi)}^{(\sigma)} + \sum_{A=1}^{N} \rho_{(A)} \left(\phi_{(A)} - \phi_{(A)}^{(\sigma)} \right) \\
& \qquad \times \left(\mathbf{v}_{(A)} \cdot \boldsymbol{\xi} - v_{(\xi)}^{(\sigma)} \right) \Bigg] \Bigg\} dA \\
= & \; 0
\end{aligned}
\tag{2-31}
$$

Here we have introduced the total energy per unit volume associated with each phase

$$
\check{E} \equiv \rho \left(\hat{U} + \frac{1}{2} v^2 \right) + \sum_{A=1}^{N} \rho_{(A)} \phi_{(A)}
\tag{2-32}
$$

and the total energy per unit area associated with the dividing surface

$$
\overline{E}^{(\sigma)} \equiv \rho^{(\sigma)} \left(\hat{U}^{(\sigma)} + \frac{1}{2} v^{(\sigma)2} \right) + \sum_{A=1}^{N} \rho_{(A)}^{(\sigma)} \phi_{(A)}^{(\sigma)}
\tag{2-33}
$$

Clausius-Duhem inequality For the isolated body under consideration here, the Clausius-Duhem inequality of Sec. 5.7.3 says that the time rate of change of the body's entropy must be greater than or equal to zero:

$$\frac{d}{dt}\left(\int_R \check{S}\, dV + \int_\Sigma \overline{S}^{(\sigma)}\, dA\right) \geq 0 \tag{2-34}$$

In arriving at this result, I have used expressions for the thermal energy flux vector **e** and the surface thermal energy flux vector $\mathbf{e}^{(\sigma)}$ in Secs. 5.8.4 and 5.8.7. Applying the transport theorem of Exercise 5.3.1-2 as well as the divergence theorem, we find that this may also be written as

$$\int_R \frac{\partial \check{S}}{\partial t}\, dV + \int_\Sigma \left\{ \frac{\partial \overline{S}^{(\sigma)}}{\partial t} - \nabla_{(\sigma)}\overline{S}^{(\sigma)} \cdot \mathbf{u} - 2H\overline{S}^{(\sigma)} v_{\{\xi\}} - [\, \check{S} v_{\{\xi\}}\,] \right\} dA$$

$$\geq 0 \tag{2-35}$$

As indicated in Sec. 5.10.1, we define **equilibrium** to be attained, when the entropy inequality becomes an equality. If equilibrium is reached by our isolated body at some time t_e, (2-34) requires

$$\text{at } t = t_e: \quad \frac{d}{dt}\left(\int_R \check{S}\, dV + \int_\Sigma \overline{S}^{(\sigma)}\, dA\right) = 0 \tag{2-36}$$

This together with (2-34) implies that we should also expect

$$\text{at } t = t_e: \quad \frac{d^2}{dt^2}\left(\int_R \check{S}\, dV + \int_\Sigma \overline{S}^{(\sigma)}\, dA\right)$$

$$\equiv \text{limit } t \to t_e: \quad \frac{1}{t_e - t}\left\{ \frac{d}{dt}\left(\int_R \check{S}\, dV + \int_\Sigma \overline{S}^{(\sigma)}\, dA\right)\Bigg|_{t_e}\right.$$

$$\left. - \frac{d}{dt}\left(\int_R \check{S}\, dV + \int_\Sigma \overline{S}^{(\sigma)}\, dA\right)\Bigg|_{t}\right\}$$

$$< 0 \tag{2-37}$$

Equation (2-36) and inequality (2-37) tell us that, as an isolated body approaches equilibrium, its entropy approaches a local maximum as a function of time.

Reference

Slattery, J. C., "Momentum, Energy, and Mass Transfer in Continua," McGraw-Hill, New York (1972); second edition, Robert E. Krieger, Malabar, FL 32950 (1981).

5.10.3 Implications of (2-36) for intrinsically stable equilibrium If equilibrium is to be achieved by the isolated multicomponent body considered here, (2-36) must be satisfied within the constraints imposed by the mass balance for each species (2-13), by Euler's first law (2-15), and by the energy balance (2-31). Let us recognize these constraints with lagrangian multipliers:

$$\begin{aligned}\frac{d}{dt}\left(\int_R \check{S}\, dV + \int_\Sigma \bar{S}^{(\sigma)}\, dA\right) &= \int_R \frac{\partial \check{S}}{\partial t}\, dV + \int_\Sigma \left(\frac{\partial \bar{S}^{(\sigma)}}{\partial t} - \nabla_{(\sigma)}\bar{S}^{(\sigma)} \cdot \mathbf{u} - 2H\bar{S}^{(\sigma)} v_{(\xi)}\right. \\ &\qquad \left. - [\, \check{S} v_{(\xi)} \,]\right) dA + \sum_{A=1}^{N} \lambda_A Z_A + \boldsymbol{\lambda}_m \cdot \mathbf{Z}_m + \lambda_e Z_e \\ &= 0 \end{aligned} \tag{3-1}$$

Here $\lambda_{(A)}$ (A = 1, 2, ..., N) and λ_e are constants or lagrangian multipliers; $\boldsymbol{\lambda}_m$ is a constant spatial vector, the components of which are lagrangian multipliers; Z_A, $\mathbf{Z}_m$, Z_e are defined by (2-13), (2-15) and (2-31). From Secs. 5.8.3 and 5.8.6, we have

$$\frac{\partial \check{S}}{\partial t} = \frac{1}{T}\frac{\partial \check{U}}{\partial t} - \sum_{A=1}^{N} \frac{\mu_{(A)}}{T}\frac{\partial \rho_{(A)}}{\partial t} \tag{3-2}$$

and

$$\frac{\partial \bar{S}^{(\sigma)}}{\partial t} = \frac{1}{T^{(\sigma)}}\frac{\partial \bar{U}^{(\sigma)}}{\partial t} - \sum_{A=1}^{N} \frac{\mu^{(\sigma)}_{(A)}}{T^{(\sigma)}}\frac{\partial \rho^{(\sigma)}_{(A)}}{\partial t} \tag{3-3}$$

Differentiating (2-32) and (2-33), we find

$$\frac{\partial \check{E}}{\partial t} = \frac{\partial \check{U}}{\partial t} + \rho \mathbf{v} \cdot \frac{\partial \mathbf{v}}{\partial t} + \frac{1}{2} v^2 \frac{\partial \rho}{\partial t} + \sum_{A=1}^{N} \phi_{(A)} \frac{\partial \rho_{(A)}}{\partial t} \tag{3-4}$$

and

$$\frac{\partial \bar{E}^{(\sigma)}}{\partial t} = \frac{\partial \bar{U}^{(\sigma)}}{\partial t} + \rho^{(\sigma)} \mathbf{v}^{(\sigma)} \cdot \frac{\partial \mathbf{v}^{(\sigma)}}{\partial t} + \frac{1}{2} v^{(\sigma)2} \frac{\partial \rho^{(\sigma)}}{\partial t}$$

$$+ \sum_{A=1}^{N} \phi_{\{ {\sigma \atop A} \}} \frac{\partial \rho_{\{ {\sigma \atop A} \}}}{\partial t} + \sum_{A=1}^{N} \rho_{\{ {\sigma \atop A} \}} \frac{\partial \phi_{\{ {\sigma \atop A} \}}}{\partial t} \tag{3-5}$$

where I have assumed that the potentials $\phi_{(A)}$ and $\phi_{\{ {\sigma \atop A} \}}$ are independent of time. After rearranging (3-1) with the help of (3-2) through (3-5), we can say that at equilibrium

$$\int_R \Bigg\{ \left(\frac{1}{T} + \lambda_e \right) \frac{\partial \check{U}}{\partial t}$$

$$+ \sum_{A=1}^{N} \left(- \frac{\mu_{(A)}}{T} + \lambda_{(A)} + \boldsymbol{\lambda}_m \cdot \mathbf{v} + \lambda_e \frac{1}{2} v^2 + \lambda_e \phi_{(A)} \right) \frac{\partial \rho_{(A)}}{\partial t}$$

$$- \sum_{j=1}^{J} \sum_{A=1}^{N} \left(\lambda_{(A)} M_{(A)} \nu_{(A,j)} + \lambda_e \phi_{(A)} M_{(A)} \nu_{(A,j)} \right) \frac{\partial \psi_{(j)}}{\partial t}$$

$$+ \rho (\boldsymbol{\lambda}_m + \lambda_e \mathbf{v}) \cdot \frac{\partial \mathbf{v}}{\partial t} \Bigg\} dV$$

$$+ \int_\Sigma \Bigg\{ \left(\frac{1}{T^{(\sigma)}} + \lambda_e \right) \left(\frac{\partial \bar{U}^{(\sigma)}}{\partial t} - \nabla_{(\sigma)} \bar{U}^{(\sigma)} \cdot \mathbf{u} \right)$$

$$+ \sum_{A=1}^{N} \left(- \frac{\mu_{\{ {\sigma \atop A} \}}}{T^{(\sigma)}} + \lambda_{(A)} + \boldsymbol{\lambda}_m \cdot \mathbf{v}^{(\sigma)} + \lambda_e \frac{1}{2} v^{(\sigma)2} + \lambda_e \phi_{\{ {\sigma \atop A} \}} \right)$$

$$\times \left(\frac{\partial \rho_{\{ {\sigma \atop A} \}}}{\partial t} - \nabla_{(\sigma)} \rho_{\{ {\sigma \atop A} \}} \cdot \mathbf{u} \right)$$

$$- \sum_{k=1}^{K} \sum_{A=1}^{N} \left(\lambda_{(A)} M_{(A)} \nu_{\{ {\sigma \atop A},k \}} + \lambda_e \phi^{(\sigma)}_{(A)} M_{(A)} \nu_{\{ {\sigma \atop A},k \}} \right)$$

$$\times \left(\frac{\partial \psi^{(\sigma)}_{(k)}}{\partial t} - \nabla_{(\sigma)} \psi^{(\sigma)}_{(k)} \cdot \mathbf{u} \right)$$

$$+ \rho^{(\sigma)} (\boldsymbol{\lambda}_m + \lambda_e \mathbf{v}^{(\sigma)}) \left(\frac{\partial \mathbf{v}^{(\sigma)}}{\partial T} - \nabla_{(\sigma)} \mathbf{v}^{(\sigma)} \cdot \mathbf{u} \right)$$

$$- 2H v^{(\sigma)}_{(\xi)} \left[\left(\frac{1}{T^{(\sigma)}} + \lambda_e \right) \bar{U}^{(\sigma)} - \frac{\gamma}{T^{(\sigma)}} \right.$$

$$\left. - \sum_{A=1}^{N} \frac{\mu^{(\sigma)}_{(A)} \rho^{(\sigma)}_{(A)}}{T^{(\sigma)}} + \lambda_e \frac{1}{2} \rho^{(\sigma)} v^{(\sigma)2} \right]$$

$$- 2H \sum_{A=1}^{N} \rho^{(\sigma)}_{(A)} (\lambda_{(A)} + \lambda_e \phi^{(\sigma)}_{(A)}) \mathbf{v}^{(\sigma)}_{(A)} \cdot \boldsymbol{\xi}$$

$$- (\boldsymbol{\lambda}_m \cdot \mathbf{B} \cdot \mathbf{v}^{(\sigma)}) \rho^{(\sigma)} v^{(\sigma)}_{(\xi)}$$

$$+ \left[\left(- \frac{\check{U}}{T} - \frac{P}{T} + \sum_{A=1}^{N} \frac{\mu_{(A)} \rho_{(A)}}{T} - \lambda_e \check{U} - \lambda_e \frac{1}{2} \rho v^2 \right. \right.$$

$$\left. - \lambda_e \sum_{A=1}^{N} \rho_{(A)} \phi_{(A)} - \sum_{A=1}^{N} \lambda_{(A)} \rho_{(A)} - \rho \boldsymbol{\lambda}_m \cdot \mathbf{v} \right) v_{(\xi)}$$

$$\left. \left. - \lambda_e \sum_{A=1}^{N} \rho_{(A)} (\phi_{(A)} - \phi^{(\sigma)}_{(A)}) (\mathbf{v}_{(A)} \cdot \boldsymbol{\xi} - v^{(\sigma)}_{(\xi)}) \right] \right\} dA$$

$$= 0 \tag{3-6}$$

In order that (3-6) be satisfied at equilibrium, it is sufficient that all velocities be zero and that all intensive quantities be independent of time. But this is not necessary.

If this equilibrium is intrinsically stable, (3-6) will not be immediately violated by the imposition of arbitrary or random transients so small that mass transfer across the phase interfaces is unaffected. Equivalently, if this equilibrium is intrinsically stable, (3-6) must be satisfied to the first perturbation of an arbitrary disturbance. This implies that the integrands for the integrations of R and Σ in (3-6) (or their first perturbations) are zero everywhere. For an intrinsically stable equilibrium,

$$\frac{1}{T} + \lambda_e = 0 \tag{3-7}$$

$$-\frac{\mu_{(A)}}{T} + \lambda_{(A)} + \boldsymbol{\lambda}_m \cdot \mathbf{v} + \lambda_e \frac{1}{2} v^2 + \lambda_e \phi_{(A)} = 0 \tag{3-8}$$

$$\sum_{A=1}^{N} \left(\lambda_{(A)} M_{(A)} \nu_{(A,j)} + \lambda_e \phi_{(A)} M_{(A)} \nu_{(A,j)} \right) = 0 \tag{3-9}$$

$$\boldsymbol{\lambda}_m + \lambda_e \mathbf{v} = 0 \tag{3-10}$$

$$\frac{1}{T^{(\sigma)}} + \lambda_e = 0 \tag{3-11}$$

$$-\frac{\mu^{(\sigma)}_{(A)}}{T^{(\sigma)}} + \lambda_{(A)} + \boldsymbol{\lambda}_m \cdot \mathbf{v}^{(\sigma)} + \lambda_e \frac{1}{2} v^{(\sigma)2} + \lambda_e \phi^{(\sigma)}_{(A)} = 0 \tag{3-12}$$

$$\sum_{A=1}^{N} \left(\lambda_{(A)} M_{(A)} \nu^{(\sigma)}_{(A,k)} + \lambda_e \phi^{(\sigma)}_{(A)} M_{(A)} \nu^{(\sigma)}_{(A,k)} \right) = 0 \tag{3-13}$$

$$\boldsymbol{\lambda}_m + \lambda_e \mathbf{v}^{(\sigma)} = 0 \tag{3-14}$$

and the integrand for the second surface integral on the left of (3-6) is zero. Equations (3-7) and (3-11) say

$$T = T^{(\sigma)} = -\frac{1}{\lambda_e} = \text{a constant} \tag{3-15}$$

From (3-10) and (3-14) we have

$$\mathbf{v} = \mathbf{v}^{(\sigma)} = -\frac{\boldsymbol{\lambda}_m}{\lambda_e} = 0 \tag{3-16}$$

when we remember that the boundaries of the system are stationary and impermeable. Equations (3-8) and (3-12) tell us that

$$\mu_{(A)} + \phi_{(A)} = \mu^{(\sigma)}_{(A)} + \phi^{(\sigma)}_{(A)} = \lambda_{(A)} T = \text{a constant} \tag{3-17}$$

We see from (3-9) and (3-13) that

$$\sum_{A=1}^{N} \mu_{(A)} M_{(A)} v_{(A,j)} = 0 \tag{3-18}$$

and

$$\sum_{A=1}^{N} \mu^{(\sigma)}_{(A)} M_{(A)} v^{(\sigma)}_{(A,k)} = 0 \tag{3-19}$$

In view of (3-15) through (3-17), the integrand for the second surface integral on the left of (3-6) reduces to

$$-2H \sum_{A=1}^{N} \frac{\mu^{(\sigma)}_{(A)} \rho^{(\sigma)}_{(A)}}{T^{(\sigma)}} \left(\mathbf{v}^{(\sigma)}_{(A)} \cdot \boldsymbol{\xi} - v_{(\xi)} \right)$$

$$+ \sum_{A=1}^{N} \left[\frac{\rho_{(A)}}{T} \left(\phi_{(A)} - \phi^{(\sigma)}_{(A)} \right) \left(\mathbf{v}_{(A)} \cdot \boldsymbol{\xi} - v_{(\xi)} \right) \right] = 0 \tag{3-20}$$

Since the surface chemical potentials $\mu^{(\sigma)}_{(A)}$ and the potential energies $\phi_{(A)}$ and $\phi^{(\sigma)}_{(A)}$ are arbitrary, this in turn implies that for an intrinsically stable equilibrium there can be no mass transfer to, from, or across the phase interfaces:

$$\text{at } \Sigma: \ \mathbf{v}_{(A)} \cdot \boldsymbol{\xi} - v_{(\xi)} = \mathbf{v}^{(\sigma)}_{(A)} \cdot \boldsymbol{\xi} - v_{(\xi)} = 0 \tag{3-21}$$

This is consistent with but not implied by our original statement that the random transients are too small to immediately affect mass transfer at the phase interfaces.

To summarize, (3-15) through (3-19) and (3-21) define necessary conditions for an intrinsically stable equilibrium in an isolated body. We can not say that they are sufficient, since we have not as yet examined the implications of (2-37).

Exercises

5.10.3-1 *At equilibrium, the Helmholtz free energy of an isothermal body is minimized for an isothermal system having a constant volume, so long as the effects of inertial forces and body forces can be neglected.* Consider an isothermal, multiphase, multicomponent body totally enclosed by fixed, impermeable walls.

i) If we neglect the effects of inertial forces and body forces, argue that the energy balance for this body requires (Table 7.1.3-1, Equation A)

$$\frac{d}{dt}\left(\int_R \rho\hat{U}\,dV + \int_\Sigma \rho^{(\sigma)}\hat{U}^{(\sigma)}\,dA\right) = -\int_S \mathbf{q}\cdot\mathbf{n}\,dA$$

$$-\int_C \mathbf{q}^{(\sigma)}\cdot\boldsymbol{\mu}\,ds + \int_R \rho Q\,dV + \int_\Sigma \rho^{(\sigma)}Q^{(\sigma)}\,dA$$

Here S is the closed surface bounding the body, C represents the collection of all closed curves formed by the intersection of S with any phase interfaces present, $\mathbf{n}$ is the outwardly directed unit vector normal to S, and $\boldsymbol{\mu}$ is the outwardly directed unit vector normal to C and tangent to the phase interface.

ii) Determine that the Clausius-Duhem inequality demands (Sec. 7.1.4)

$$\frac{d}{dt}\left(\int_R \rho\hat{S}\,dV + \int_\Sigma \rho^{(\sigma)}\hat{S}^{(\sigma)}\,dA\right) \geq -\frac{1}{T}\int_S \mathbf{q}\cdot\mathbf{n}\,dA$$

$$-\frac{1}{T}\int_C \mathbf{q}^{(\sigma)}\cdot\boldsymbol{\mu}\,ds + \frac{1}{T}\int_R \rho Q\,dV + \frac{1}{T}\int_\Sigma \rho^{(\sigma)}Q^{(\sigma)}\,dA$$

We have recognized here that, since the bounding surfaces of the system are impermeable, they do not permit adsorption.

iii) Prove that at equilibrium the Helmholtz free energy

$$\mathcal{A} \equiv \int_R \rho\hat{A}\,dV + \int_\Sigma \rho^{(\sigma)}\hat{A}^{(\sigma)}\,dA$$

is minimized.

iv) Use Euler's equation (Sec. 5.8.3) and the surface Euler equation (Sec. 5.8.6) in writing the Helmholtz free energy for this body as

$$\mathcal{A} = \int_R \left(-P + \sum_{B=1}^{N} \mu_{(B)}\rho_{(B)}\right) dV + \int_\Sigma \left(\gamma + \sum_{B=1}^{N} \mu^{(\sigma)}_{(B)}\,\rho^{(\sigma)}_{(B)}\right) dA$$

In arriving at this result, we have neglected the Helmholtz free energy associated with the fixed impermeable boundaries. Here and in what follows, Σ should be identified as only the internal phase interfaces, not including the system boundaries.

v) In the limit of equilibrium, mass transfer across interfaces can be neglected. Use the transport theorem for a multicomponent body containing intersecting dividing surfaces (Exercise 5.4.1-5) to conclude that

$$\frac{dA}{dt} = \frac{d}{dt}\left(- \int_{R} P \, dV + \int_{\Sigma} \gamma \, dA \right) + \int_{R} \rho \frac{d_{(v)}}{dt}\left(\sum_{B=1}^{N} \mu_{(B)} \omega_{(B)} \right) dV$$

$$+ \int_{\Sigma} \rho^{(\sigma)} \frac{d_{(v^{(\sigma)})}}{dt}\left(\sum_{B=1}^{N} \mu_{\{B\}} \, \omega_{\{B\}} \right) dA$$

vi) In the limit of equilibrium, concentrations and chemical potentials are independent of time and

$$\frac{dJ}{dt} = 0$$

where

$$J \equiv - \int_{R} P \, dV + \int_{\Sigma} \gamma \, dA$$

It should be understood here that Σ represents the internal phase interfaces and does not include the fixed impermeable boundaries of the system.

vii) Argue in a similar fashion that at equilibrium

$$\frac{d^2 J}{dt^2} > 0$$

5.10.3-2 *At equilibrium, the sum of the Helmholtz free energy and the potential energy of an isothermal body is minimized for an isothermal system having a constant volume, so long as the effects of inertial forces can be neglected.* Let us extend the results of Exercise 5.10.3-1 by assuming that all species are subject to the same body force, which can be represented in terms of a scalar potential energy per unit mass ϕ:

$$\mathbf{b} = \mathbf{b}^{(\sigma)} = - \nabla \phi$$

i) If we neglect the effects of inertial forces, argue that the energy balance for this body requires (Table 7.1.3-1, Equation A)

$$\frac{d}{dt}\left[\int_R \rho(\hat{U}+\phi)\,dV+\int_\Sigma \rho^{(\sigma)}(\hat{U}^{(\sigma)}+\phi)\,dA\right]=-\int_S \mathbf{q}\cdot\mathbf{n}\,dA$$

$$-\int_C \mathbf{q}^{(\sigma)}\cdot\boldsymbol{\mu}\,ds+\int_R \rho Q\,dV+\int_\Sigma \rho^{(\sigma)}Q^{(\sigma)}\,dA$$

ii) Prove that at equilibrium the sum of the Helmholtz free energy and of the potential energy

$$A+\Phi\equiv\int_R \rho(\hat{A}+\phi)\,dV+\int_\Sigma \rho^{(\sigma)}(\hat{A}^{(\sigma)}+\phi)\,dA$$

is minimized.

iii) Follow the argument in Exercise 5.10.3-1 to conclude that at equilibrium

$$\frac{dJ}{dt}=0 \text{ and } \frac{d^2J}{dt^2}>0$$

where now

$$J\equiv-\int_R (P+\rho\phi)\,dV+\int_\Sigma (\gamma+\rho^{(\sigma)}\phi)\,dA$$

As in Exercise 5.10.3-1, Σ represents only the internal interfaces and does not include the system boundaries.

5.10.3-3 *At equilibrium, the* Gibbs *free energy of an isothermal body composed of isobaric fluids is minimized.* Consider an isolated, isothermal, multiphase, multicomponent body composed of isobaric fluids.

i) Argue that the energy balance for this body requires (Table 7.1.3-1, Equation E)

$$\frac{d}{dt}\left(\int_R \rho\hat{H}\,dV+\int_\Sigma \rho^{(\sigma)}\hat{U}^{(\sigma)}\,dA\right)=-\int_S \mathbf{q}\cdot\mathbf{n}\,dA-\int_C \mathbf{q}^{(\sigma)}\cdot\boldsymbol{\mu}\,ds$$

$$+ \int_R \rho Q \, dV + \int_\Sigma \left(\gamma \operatorname{div}_{(\sigma)} \mathbf{v}^{(\sigma)} + \rho^{(\sigma)} Q^{(\sigma)} \right) dA$$

Here S is the closed surface bounding the body, C represents the collection of all closed curves formed by the intersection of S with any phase interfaces present, $\mathbf{n}$ is the outwardly directed unit vector normal to S, and $\boldsymbol{\mu}$ is the outwardly directed unit vector normal to C and tangent to the phase interface. To reach this result, we have assumed that in the limit of equilibrium we can neglect the effects of viscous dissipation, of mass transfer, and of work done at common lines.

ii) In the limit of equilibrium, γ is independent of time and position in each interface and mass transfer across interfaces can be neglected. Use the transport theorem for intersecting dividing surfaces (Exercise 5.4.1-5) and the overall jump mass balance (Sec. 5.4.1) to conclude

$$\frac{d}{dt}\left(\int_R \rho \hat{H} \, dV + \int_\Sigma \rho^{(\sigma)} \hat{H}^{(\sigma)} \, dA \right) = - \int_S \mathbf{q} \cdot \mathbf{n} \, dA - \int_C \mathbf{q}^{(\sigma)} \cdot \boldsymbol{\mu} \, ds$$

$$+ \int_R \rho Q \, dV + \int_\Sigma \rho^{(\sigma)} Q^{(\sigma)} \, dA$$

iii) Follow the argument in Exercise 5.10.3-1 to prove that at equilibrium the Gibbs free energy

$$G \equiv \int_R \rho \hat{G} \, dV + \int_\Sigma \rho^{(\sigma)} \hat{G}^{(\sigma)} \, dA$$

is minimized.

5.10.4 Implications of (2-37) for instrinsically stable equilibrium For equilibrium to be achieved at $t = t_e$ by the isolated multicomponent body considered here, (2-37) must also be satisfied within the constraints imposed by the mass balance for each species (2-13), by Euler's first law (2-15), and by the energy balance (2-31). A convenient way of recognizing these constraints is to evaluate the second derivative on the left of (2-37) by differentiating the left side of (3-6). Using the transport theorem of Exercise 5.3.1-2 as well as (3-15) through (3-17), we find that for an intrinsically stable equilibrium (2-37) requires (see also footnote a of Sec. 5.8.2)

$$
\begin{aligned}
&\frac{d^2}{dt^2}\left(\int_R \check{S}\, dV + \int_\Sigma \bar{S}^{(\sigma)}\, dA \right) \\
&= \int_R \Bigg\{ -\frac{1}{T^2} \left[\left(\frac{\partial \check{U}}{\partial \check{S}} \right)_{\rho_{(C)}} \frac{\partial \check{S}}{\partial t} + \sum_{A=1}^{N} \left(\frac{\partial \check{U}}{\partial \rho_{(A)}} \right)_{\check{S},\rho_{(C)}(C \neq A)} \frac{\partial \rho_{(A)}}{\partial t} \right] \\
&\qquad \times \left[\left(\frac{\partial T}{\partial \check{S}} \right)_{\rho_{(C)}} \frac{\partial \check{S}}{\partial t} + \sum_{B=1}^{N} \left(\frac{\partial T}{\partial \rho_{(B)}} \right)_{\check{S},\rho_{(C)}(C \neq B)} \frac{\partial \rho_{(B)}}{\partial t} \right] \\
&\quad - \sum_{A=1}^{N} \left[\left(\frac{\partial}{\partial \check{S}} \left[\mu_{(A)} T^{-1} \right] \right)_{\rho_{(C)}} \frac{\partial \check{S}}{\partial t} \right. \\
&\qquad \left. + \sum_{B=1}^{N} \left(\frac{\partial}{\partial \rho_{(B)}} \left[\mu_{(A)} T^{-1} \right] \right)_{\hat{S},\rho_{(C)}(C \neq B)} \frac{\partial \rho_{(B)}}{\partial t} \right] \frac{\partial \rho_{(A)}}{\partial t} \\
&\quad - \frac{\rho}{T} \frac{\partial \mathbf{v}}{\partial t} \cdot \frac{\partial \mathbf{v}}{\partial t} \Bigg\} dV \\
&+ \int_\Sigma \Bigg\{ -\frac{1}{T^{(\sigma)2}} \left[\left(\frac{\partial \bar{U}^{(\sigma)}}{\partial \bar{S}^{(\sigma)}} \right)_{\rho\left\{{}^{\sigma}_{C}\right\}} \left(\frac{\partial \bar{S}^{(\sigma)}}{\partial t} - \nabla_{(\sigma)} \bar{S}^{(\sigma)} \cdot \mathbf{u} \right) \right. \\
&\qquad + \sum_{A=1}^{N} \left(\frac{\partial \bar{U}^{(\sigma)}}{\partial \rho\left\{{}^{\sigma}_{A}\right\}} \right)_{\bar{S}^{(\sigma)},\rho\left\{{}^{\sigma}_{A}\right\}(C \neq A)} \\
&\qquad\qquad \left. \times \left(\frac{\partial \rho\left\{{}^{\sigma}_{A}\right\}}{\partial t} - \nabla_{(\sigma)} \rho\left\{{}^{\sigma}_{A}\right\} \cdot \mathbf{u} \right) \right] \\
&\qquad \times \left[\left(\frac{\partial T^{(\sigma)}}{\partial \bar{S}^{(\sigma)}} \right)_{\rho\left\{{}^{\sigma}_{C}\right\}} \left(\frac{\partial \bar{S}^{(\sigma)}}{\partial t} - \nabla_{(\sigma)} \bar{S}^{(\sigma)} \cdot \mathbf{u} \right) \right. \\
&\qquad \left. + \sum_{B=1}^{N} \left(\frac{\partial T^{(\sigma)}}{\partial \rho\left\{{}^{\sigma}_{B}\right\}} \right)_{\bar{S}^{(\sigma)},\rho\left\{{}^{\sigma}_{C}\right\}(C \neq B)} \left(\frac{\partial \rho\left\{{}^{\sigma}_{B}\right\}}{\partial t} - \nabla_{(\sigma)} \rho\left\{{}^{\sigma}_{B}\right\} \cdot \mathbf{u} \right) \right]
\end{aligned}
$$

$$- \sum_{A=1}^{N} \left[\left(\frac{\partial}{\partial \overline{S}^{(\sigma)}} [\mu_{(A)}^{(\sigma)} T^{(\sigma) -1}] \right)_{\rho_{(B)}^{(\sigma)}} \right.$$

$$\times \left(\frac{\partial \overline{S}^{(\sigma)}}{\partial t} - \nabla_{(\sigma)} \overline{S}^{(\sigma)} \cdot \mathbf{u} \right)$$

$$+ \sum_{B=1}^{N} \left(\frac{\partial}{\partial \rho_{(B)}^{(\sigma)}} [\mu_{(A)}^{(\sigma)} T^{(\sigma) -1}] \right)_{\overline{S}^{(\sigma)}, \rho_{(C)}^{(\sigma)} (C \neq B)}$$

$$\left. \times \left(\frac{\partial \rho_{(B)}^{(\sigma)}}{\partial t} - \nabla_{(\sigma)} \rho_{(B)}^{(\sigma)} \cdot \mathbf{u} \right) \right] \left[\frac{\partial \rho_{(A)}^{(\sigma)}}{\partial t} - \nabla_{(\sigma)} \rho_{(A)}^{(\sigma)} \cdot \mathbf{u} \right]$$

$$- \sum_{A=1}^{N} \frac{1}{T^{(\sigma)}} \frac{\partial \phi_{(A)}^{(\sigma)}}{\partial \mathbf{p}^{(\sigma)}} \cdot (\mathbf{u} - \nabla_{(\sigma)} \mathbf{p}^{(\sigma)} \cdot \mathbf{u})$$

$$\times \left(\frac{\partial \rho_{(A)}^{(\sigma)}}{\partial t} - \nabla_{(\sigma)} \rho_{(A)}^{(\sigma)} \cdot \mathbf{u} \right)$$

$$+ \sum_{k=1}^{K} \sum_{A=1}^{N} \frac{1}{T^{(\sigma)}} M_{(A)} \nu_{(A,k)}^{(\sigma)} \frac{\partial \phi_{(A)}^{(\sigma)}}{\partial \mathbf{p}^{(\sigma)}} \cdot (\mathbf{u} - \nabla_{(\sigma)} \mathbf{p}^{(\sigma)} \cdot \mathbf{u})$$

$$\times \left(\frac{\partial \psi_{(k)}^{(\sigma)}}{\partial t} - \nabla_{(\sigma)} \psi_{(k)}^{(\sigma)} \cdot \mathbf{u} \right)$$

$$- \frac{\rho^{(\sigma)}}{T^{(\sigma)}} \frac{\partial \mathbf{v}^{(\sigma)}}{\partial t} \cdot \frac{\partial \mathbf{v}^{(\sigma)}}{\partial t} + 2H\gamma \frac{1}{T^{(\sigma)}} \frac{\partial v_{(\xi)}^{(\sigma)}}{\partial t}$$

$$- \sum_{A=1}^{N} 2H \rho_{(A)}^{(\sigma)} \mu_{(A)}^{(\sigma)} \frac{1}{T^{(\sigma)}} \frac{\partial}{\partial t} (\mathbf{v}_{(A)}^{(\sigma)} \cdot \boldsymbol{\xi} - v_{(\xi)}^{(\sigma)})$$

$$+ \left[- \frac{P}{T} \frac{\partial v_{(\xi)}^{(\sigma)}}{\partial t} \right.$$

$$\left. \left. + \sum_{A=1}^{N} \frac{1}{T} \rho_{(\sigma)} (\phi_{(A)} - \phi_{(A)}^{(\sigma)}) \frac{\partial}{\partial t} (\mathbf{v}_{(A)} \cdot \boldsymbol{\xi} - v_{(\xi)}^{(\sigma)}) \right] \right\} dA$$

$$< 0 \tag{4-1}$$

We can see that

$$
\begin{aligned}
\mathbf{u} - \nabla_{(\sigma)}\mathbf{p}^{(\sigma)} \cdot \mathbf{u} &= \mathbf{u} - \mathbf{P} \cdot \mathbf{u} \\
&= (\mathbf{I} - \mathbf{P}) \cdot \mathbf{u} \\
&= (\mathbf{u} \cdot \boldsymbol{\xi})\boldsymbol{\xi} \\
&= v_{(\xi)}^{(\sigma)}\boldsymbol{\xi} \\
&= 0
\end{aligned}
\tag{4-2}
$$

where in the first line we have used a result from Sec. A.5.1 and in the second one from Sec. A.3.2. We also can reason that

$$
\begin{aligned}
&\int_{\Sigma} \Bigg\{ - \frac{\rho^{(\sigma)}}{T^{(\sigma)}} \frac{\partial \mathbf{v}^{(\sigma)}}{\partial t} \cdot \frac{\partial \mathbf{v}^{(\sigma)}}{\partial t} + 2H\gamma \frac{1}{T^{(\sigma)}} \frac{\partial v_{(\xi)}^{(\sigma)}}{\partial t} \\
&\quad - \sum_{A=1}^{N} 2H\rho_{(A)}^{(\sigma)} \mu_{(A)}^{(\sigma)} \frac{1}{T^{(\sigma)}} \frac{\partial}{\partial t} (\mathbf{v}_{(A)}^{(\sigma)} \cdot \boldsymbol{\xi} - v_{(\xi)}^{(\sigma)}) \\
&\quad + \Bigg[- \frac{P}{T} \frac{\partial v_{(\xi)}^{(\sigma)}}{\partial t} + \sum_{A=1}^{N} \frac{1}{T} \rho_{(\sigma)} (\phi_{(\sigma)} - \phi_{(A)}^{(\sigma)}) \frac{\partial}{\partial t} (\mathbf{v}_{(\sigma)} \cdot \boldsymbol{\xi} - v_{(\xi)}^{(\sigma)}) \Bigg] \Bigg\} \, dA \\
&= \int_{\Sigma} \Bigg\{ \frac{1}{T^{(\sigma)}} \left(- \rho^{(\sigma)} \frac{\partial \mathbf{v}^{(\sigma)}}{\partial t} + 2H\gamma \boldsymbol{\xi} - [P\boldsymbol{\xi}] \right) \cdot \frac{\partial \mathbf{v}^{(\sigma)}}{\partial t} \\
&\quad - \sum_{A=1}^{N} 2H\rho_{(A)}^{(\sigma)} \mu_{(A)}^{(\sigma)} \frac{1}{T^{(\sigma)}} \frac{\partial}{\partial t} (\mathbf{v}_{(A)}^{(\sigma)} \cdot \boldsymbol{\xi} - v_{(\xi)}^{(\sigma)}) \\
&\quad + \Bigg[\sum_{A=1}^{N} \frac{1}{T} \rho_{(A)} (\phi_{(\sigma)} - \phi_{(A)}^{(\sigma)}) \frac{\partial}{\partial t} (\mathbf{v}_{(A)} \cdot \boldsymbol{\xi} - v_{(\xi)}^{(\sigma)}) \Bigg] \Bigg\} \, dA \\
&= 0
\end{aligned}
\tag{4-3}
$$

Here we have observed that for an intrinsically stable equilibrium the jump momentum balance of Sec. 2.1.7 reduces to

$$
\rho^{(\sigma)} \frac{\partial \mathbf{v}^{(\sigma)}}{\partial t} = 2H\gamma \boldsymbol{\xi} - [P\boldsymbol{\xi}]
\tag{4-4}
$$

We have also said that at Σ

$$\frac{\partial}{\partial t}\left(\mathbf{v}_{(A)} \cdot \boldsymbol{\xi} - \mathbf{v}^{(\sigma)} \right) = \frac{\partial}{\partial t}\left(\mathbf{v}^{(\sigma)}_{(A)} \cdot \boldsymbol{\xi} - \mathbf{v}^{(\sigma)} \right) = 0 \tag{4-5}$$

for all A = 1, 2, ..., N, since the random transients are so small that mass transfer to, from, and across the phase interfaces is not immediately affected. Finally, we note that

$$\frac{1}{T}\left(\frac{\partial T}{\partial \rho_{(A)}}\right)_{\check{S},\rho_{(C)}(C\neq A)} + \frac{1}{T^2}\mu_{(A)}\left(\frac{\partial T}{\partial \check{S}}\right)_{\rho_{(C)}} + \left[\frac{\partial}{\partial \check{S}}\left(\mu_{(A)}T^{-1}\right)\right]_{\rho_{(C)}}$$

$$= \frac{2}{T}\left(\frac{\partial \mu_{(A)}}{\partial \check{S}}\right)_{\rho_{(C)}} \tag{4-6}$$

$$\frac{1}{T^2}\mu_{(A)}\left(\frac{\partial T}{\partial \rho_{(B)}}\right)_{\check{S},\rho_{(C)}(C\neq B)} + \left(\frac{\partial}{\partial \rho_{(B)}}\left[\mu_{(A)}T^{-1}\right]\right)_{\check{S},\rho_{(C)}(C\neq B)}$$

$$= \frac{1}{T}\left(\frac{\partial \mu_{(A)}}{\partial \rho_{(B)}}\right)_{\check{S},\rho_{(C)}(C\neq B)} \tag{4-7}$$

and in much the same manner

$$\frac{1}{T^{(\sigma)}}\left(\frac{\partial T^{(\sigma)}}{\partial \rho^{(\sigma)}_{(A)}}\right)_{\bar{S}^{(\sigma)},\rho^{(\sigma)}_{(C)}(C\neq A)} + \frac{1}{T^{(\sigma)2}}\mu^{(\sigma)}_{(A)}\left(\frac{\partial T^{(\sigma)}}{\partial \bar{S}^{(\sigma)}}\right)_{\rho^{(\sigma)}_{(C)}}$$

$$+ \left[\frac{\partial}{\partial \bar{S}^{(\sigma)}}\left(\mu^{(\sigma)}_{(A)}\, T^{(\sigma)\,-1}\right)\right]_{\rho^{(\sigma)}_{(C)}}$$

$$= \frac{2}{T^{(\sigma)}}\left(\frac{\partial \mu^{(\sigma)}_{(A)}}{\partial \bar{S}^{(\sigma)}}\right)_{\rho^{(\sigma)}_{(C)}} \tag{4-8}$$

$$\frac{1}{T^{(\sigma)2}}\mu^{(\sigma)}_{(A)}\left(\frac{\partial T^{(\sigma)}}{\partial \rho^{(\sigma)}_{(B)}}\right)_{\bar{S}^{(\sigma)},\rho^{(\sigma)}_{(C)}(C\neq B)}$$

$$+ \left[\frac{\partial}{\partial \rho^{(\sigma)}_{(B)}}\left(\mu^{(\sigma)}_{(A)}\, T^{(\sigma)\,-1}\right)\right]_{\bar{S}^{(\sigma)},\rho^{(\sigma)}_{(C)}(C\neq B)}$$

$$= \frac{1}{T^{(\sigma)}} \left(\frac{\partial \mu^{(\sigma)}_{(A)}}{\partial \rho^{(\sigma)}_{(B)}} \right)_{\bar{S}^{(\sigma)}, \rho^{(\sigma)}_{(C)} (C \neq B)} \tag{4-9}$$

In view of (4-2), (4-3), and (4-6) through (4-9), the second surface integral on the left of (4-1) is zero and (4-1) may be rearranged to read

$$\begin{aligned}
&\int_R \Bigg\{ \frac{1}{T} \left(\frac{\partial T}{\partial \check{S}} \right)_{\rho_{(C)}} \left(\frac{\partial \check{S}}{\partial t} \right)^2 + \sum_{A=1}^{N} \frac{2}{T} \left(\frac{\partial \mu_{(A)}}{\partial \check{S}} \right)_{\rho_{(C)}} \frac{\partial \check{S}}{\partial t} \frac{\partial \rho_{(A)}}{\partial t} \\
&\quad + \sum_{A=1}^{N} \sum_{B=1}^{N} \frac{1}{T} \left(\frac{\partial \mu_{(A)}}{\partial \rho_{(B)}} \right)_{\check{S}, \rho_{(C)} (C \neq B)} \frac{\partial \rho_{(A)}}{\partial t} \frac{\partial \rho_{(B)}}{\partial t} + \frac{\rho}{T} \frac{\partial \mathbf{v}}{\partial t} \cdot \frac{\partial \mathbf{v}}{\partial t} \Bigg\} dV \\
&+ \int_\Sigma \Bigg\{ \frac{1}{T^{(\sigma)}} \left(\frac{\partial T^{(\sigma)}}{\partial \bar{S}^{(\sigma)}} \right)_{\rho^{(\sigma)}_{(C)}} \left(\frac{\partial \bar{S}^{(\sigma)}}{\partial t} - \nabla_{(\sigma)} \bar{S}^{(\sigma)} \cdot \mathbf{u} \right)^2 \\
&\quad + \sum_{A=1}^{N} \frac{2}{T^{(\sigma)}} \left(\frac{\partial \mu^{(\sigma)}_{(A)}}{\partial \bar{S}^{(\sigma)}} \right)_{\rho^{(\sigma)}_{(C)}} \left(\frac{\partial \bar{S}^{(\sigma)}}{\partial t} - \nabla_{(\sigma)} \bar{S}^{(\sigma)} \cdot \mathbf{u} \right) \\
&\quad \times \left(\frac{\partial \rho^{(\sigma)}_{(A)}}{\partial t} - \nabla_{(\sigma)} \rho^{(\sigma)}_{(A)} \cdot \mathbf{u} \right) \\
&\quad + \sum_{A=1}^{N} \sum_{B=1}^{N} \frac{1}{T^{(\sigma)}} \left(\frac{\partial \mu^{(\sigma)}_{(A)}}{\partial \rho^{(\sigma)}_{(B)}} \right)_{\bar{S}^{(\sigma)}, \rho^{(\sigma)}_{(C)} (C \neq B)} \\
&\quad \times \left(\frac{\partial \rho^{(\sigma)}_{(A)}}{\partial t} - \nabla_{(\sigma)} \rho^{(\sigma)}_{(A)} \cdot \mathbf{u} \right) \left(\frac{\partial \rho^{(\sigma)}_{(B)}}{\partial t} - \nabla_{(\sigma)} \rho^{(\sigma)}_{(B)} \cdot \mathbf{u} \right) \Bigg\} dA \\
&> 0
\end{aligned} \tag{4-10}$$

If an equilibrium is intrinsically stable, inequality (4-10) will not be immediately violated by the imposition of arbitrary or random transients so small that mass transfer across the phase interfaces is unaffected. Equivalently, if this equilibrium is intrinsically stable, (4-10) must be satisfied to the second perturbation of an arbitrary disturbance.

For a single phase system, the first integral on the left of (4-10) must be greater than zero. Since the dimensions of the system are arbitrary, this implies (see argument developed in Exercise 5.7.3-3)

$$\frac{1}{T}\left(\frac{\partial T}{\partial \check{S}}\right)_{\rho_{(C)}}\left(\frac{\partial \check{S}}{\partial t}\right)^2 + \sum_{A=1}^{N}\frac{2}{T}\left(\frac{\partial \mu_{(A)}}{\partial \check{S}}\right)_{\rho_{(C)}}\frac{\partial \check{S}}{\partial t}\frac{\partial \rho_{(A)}}{\partial t}$$

$$+\sum_{A=1}^{N}\sum_{B=1}^{N}\frac{1}{T}\left(\frac{\partial \mu_{(A)}}{\partial \rho_{(B)}}\right)_{\check{S},\rho_{(C)}(C\neq B)}\frac{\partial \rho_{(A)}}{\partial t}\frac{\partial \rho_{(B)}}{\partial t}$$

$$+\frac{\rho}{T}\left|\frac{\partial \mathbf{v}}{\partial t}\right|^2$$

$$> 0 \tag{4-11}$$

Consider now the vector space whose elements are ordered sets of N+2, real-valued functions of time and position (as the field of scalars, we take all real-valued functions of time and position with the usual rules for addition and multiplication of functions); this is a generalization of the vector space of N+2-tuples of real numbers (Halmos 1958, p. 1 and 5). Let $\mathbf{x}$ be an element of this vector space:

$$\mathbf{x} \equiv \left(\frac{\partial \check{S}}{\partial t}, \frac{\partial \rho_{(1)}}{\partial t}, \ldots, \frac{\partial \rho_{(N)}}{\partial t}, \left|\frac{\partial \mathbf{v}}{\partial t}\right|\right) \tag{4-12}$$

In these terms, inequality (4-11) may be written as

$$(\mathbf{x}, \mathbf{Sx}) > 0 \tag{4-13}$$

where $(\mathbf{x}, \mathbf{Sx})$ represents an inner product of the vectors $\mathbf{x}$ and $\mathbf{Sx}$ of this vector space. Here $\mathbf{S}$ is a transformation of the vector space into itself. If we take

$$\boldsymbol{\gamma}_1 \equiv (1, 0, \ldots, 0)$$

$$\boldsymbol{\gamma}_2 \equiv (0, 1, 0, \ldots, 0)$$

$$\vdots$$

$$\boldsymbol{\gamma}_{N+2} \equiv (0, \ldots, 0, 1) \tag{4-14}$$

as a basis for this vector space, then the elements of the matrix of the transformation $\mathbf{S}$ with respect to this basis are (Halmos 1958, p. 65)

$$S_{11} \equiv \frac{1}{T}\left(\frac{\partial T}{\partial \check{S}}\right)_{\rho_{(C)}} \tag{4-15}$$

$$S_{1,1+A} = S_{1+A,1} \equiv \frac{1}{T}\left(\frac{\partial \mu_{(A)}}{\partial \check{S}}\right)_{\rho_{(C)}} \quad \text{for } A = 1, 2, \ldots, N \tag{4-16}$$

$$S_{1+A,1+B} = S_{1+B,1+A} \equiv \frac{1}{T}\left(\frac{\partial \mu_{(A)}}{\partial \rho_{(B)}}\right)_{\check{S},\rho_{(C)}(C \neq B)}$$

$$\text{for } A, B = 1, 2, \ldots, N \tag{4-17}$$

$$S_{N+2,R} = S_{R,N+2} \equiv 0 \quad \text{for } R = 1, 2, \ldots, N+1 \tag{4-18}$$

$$S_{N+2,N+2} \equiv \frac{\rho}{T} \tag{4-19}$$

In writing (4-17), I have observed that (Slattery 1981, p. 289)

$$\left(\frac{\partial \mu_{(A)}}{\partial \rho_{(B)}}\right)_{T,\rho_{(C)}(C \neq B)} = \left(\frac{\partial \mu_{(B)}}{\partial \rho_{(A)}}\right)_{T,\rho_{(C)}(C \neq A)} \tag{4-20}$$

Notice that **S** is symmetric.

Since (4-10) applies to every multiphase body, we can use the arguments indicated in Exercises 5.7.3-4 and 5.7.3-7 to conclude not only that the surface integral on the left of (4-10) is greater than zero but also that its integrand is greater than zero everywhere on Σ:

$$\begin{aligned}
&\frac{1}{T^{(\sigma)}}\left(\frac{\partial T^{(\sigma)}}{\partial \bar{S}^{(\sigma)}}\right)_{\rho^{(\sigma)}_{(C)}} \left(\frac{\partial \bar{S}^{(\sigma)}}{\partial t} - \nabla_{(\sigma)}\bar{S}^{(\sigma)} \cdot \mathbf{u}\right)^2 \\
&\quad + \sum_{A=1}^{N} \frac{2}{T^{(\sigma)}}\left(\frac{\partial \mu^{(\sigma)}_{(A)}}{\partial \bar{S}^{(\sigma)}}\right)_{\rho^{(\sigma)}_{(C)}} \left(\frac{\partial \bar{S}^{(\sigma)}}{\partial t} - \nabla_{(\sigma)}\bar{S}^{(\sigma)} \cdot \mathbf{u}\right) \\
&\qquad \times \left(\frac{\partial \rho^{(\sigma)}_{(A)}}{\partial t} - \nabla_{(\sigma)}\rho^{(\sigma)}_{(A)} \cdot \mathbf{u}\right) \\
&\quad + \sum_{A=1}^{N}\sum_{B=1}^{N} \frac{1}{T^{(\sigma)}}\left(\frac{\partial \mu^{(\sigma)}_{(A)}}{\partial \rho^{(\sigma)}_{(B)}}\right)_{\bar{S}^{(\sigma)},\rho^{(\sigma)}_{(C)}(C \neq B)} \\
&\qquad \times \left(\frac{\partial \rho^{(\sigma)}_{(A)}}{\partial t} - \nabla_{(\sigma)}\rho^{(\sigma)}_{(A)} \cdot \mathbf{u}\right)\left(\frac{\partial \rho^{(\sigma)}_{(B)}}{\partial t} - \nabla_{(\sigma)}\rho^{(\sigma)}_{(B)} \cdot \mathbf{u}\right) \\
&> 0
\end{aligned} \tag{4-21}$$

The vector space whose elements are ordered sets of N+1, real-valued functions of time and position on the dividing surface Σ (as the field of scalars, we take all real-valued functions of time and position on Σ with the usual rules for addition and multiplication of functions) is a generalization of the vector space of N+1-tuples of real numbers (Halmos 1958, p. 1 and 5). If

$$\mathbf{x}^{(\sigma)} \equiv \left(\left\{ \frac{\partial \bar{S}^{(\sigma)}}{\partial t} - \nabla_{(\sigma)} \bar{S}^{(\sigma)} \cdot \mathbf{u} \right\}, \left\{ \frac{\partial \rho^{(\sigma)}_{(1)}}{\partial t} - \nabla_{(\sigma)} \rho^{(\sigma)}_{(1)} \cdot \mathbf{u} \right\}, \dots, \right.$$

$$\left. \left\{ \frac{\partial \rho^{(\sigma)}_{(N)}}{\partial t} - \nabla_{(\sigma)} \rho^{(\sigma)}_{(N)} \cdot \mathbf{u} \right\} \right) \tag{4-22}$$

is an element of this vector space, inequality (4-21) may be written as

$$(\mathbf{x}^{(\sigma)}, \mathbf{S}^{(\sigma)} \mathbf{x}^{(\sigma)}) > 0 \tag{4-23}$$

where ($\mathbf{x}^{(\sigma)}$, $\mathbf{S}^{(\sigma)}\mathbf{x}^{(\sigma)}$) represents an inner product of the vectors $\mathbf{x}^{(\sigma)}$ and $\mathbf{S}^{(\sigma)}\mathbf{x}^{(\sigma)}$ of this vector space. Here $\mathbf{S}^{(\sigma)}$ is a transformation of the vector space into itself. If we take

$$\gamma_1^{(\sigma)} \equiv (1, 0, \dots, 0)$$

$$\gamma_2^{(\sigma)} \equiv (0, 1, 0, \dots, 0)$$

$$\vdots$$

$$\gamma_{N+1}^{(\sigma)} \equiv (0, \dots, 0, 1) \tag{4-24}$$

as a basis for this vector space, then the elements of the matrix of the transformation $\mathbf{S}^{(\sigma)}$ with respect to this basis are (Halmos 1958, p. 65)

$$S^{(\sigma)}_{11} \equiv \frac{1}{T^{(\sigma)}} \left(\frac{\partial T^{(\sigma)}}{\partial \bar{S}^{(\sigma)}} \right)_{\rho^{(\sigma)}_{(C)}} \tag{4-25}$$

$$S^{(\sigma)}_{1,1+A} \equiv S^{(\sigma)}_{1+A,1} \equiv \frac{1}{T^{(\sigma)}} \left(\frac{\partial \rho^{(\sigma)}_{(A)}}{\partial \bar{S}^{(\sigma)}} \right)_{\rho^{(\sigma)}_{(C)}} \quad \text{for } A = 1, 2, \dots, N \tag{4-26}$$

$$S^{(\sigma)}_{1+A,1+B} \equiv S^{(\sigma)}_{1+B,1+A} \equiv \frac{1}{T^{(\sigma)}} \left(\frac{\partial \mu^{(\sigma)}_{(A)}}{\partial \rho^{(\sigma)}_{(B)}} \right)_{\bar{S}^{(\sigma)}, \rho^{(\sigma)}_{(C)} (C \neq B)}$$

$$\text{for } A, B = 1, 2, \dots, N \tag{4-27}$$

Like $\mathbf{S}$, $\mathbf{S}^{(\sigma)}$ is also symmetric. In writing (4-27), I have observed that

(Exercise 5.8.6-3)

$$\left(\frac{\partial\mu_{(A)}}{\partial\rho_{(B)}}\right)_{\bar{S}^{(\sigma)},\rho_{(C)}(C\neq B)} = \left(\frac{\partial\mu_{(B)}}{\partial\rho_{(A)}}\right)_{\bar{S}^{(\sigma)},\rho_{(C)}(C\neq A)} \tag{4-28}$$

In summary, there are two conclusions that should be drawn from this discussion.

1. For an intrinsically stable equilibrium, (4-13) must hold at each point within a phase, where S is the self-adjoint transformation whose matrix with respect to the basis (4-14) is defined by (4-15) through (4-19). This means that S is a positive transformation (Halmos 1958, p. 140). In order that S be positive, it is necessary and sufficient that the principal minors of the determinant of S all be greater than zero (Halmos 1958, p. 167). (If we cross out rows and columns bearing the same number in the matrix of S, the remaining small matrix is still positive and so is its determinant. The determinants found in this manner are the principal minors of the determinant of S.) These are constraints that must be satisfied by any caloric equation of state (see Sec. 5.8.3) at those conditions capable of supporting an intrinsically stable equilibrium.

2. For an intrinsically stable equilibrium, (4-23) must be satisfied at each point on the dividing surface Σ, where $S^{(\sigma)}$ is the self-adjoint transformation whose matrix with respect to the basis (4-24) is defined by (4-25) through (4-27). We conclude that $S^{(\sigma)}$ is also positive. The necessary and sufficient conditions for $S^{(\sigma)}$ to be positive are that the principal minors of the deterinant of $S^{(\sigma)}$ all be greater than zero. These are constraints that must be satisified by any surface caloric equation of state (see Sec. 5.8.6) at those conditions capable of supporting an intrisically stable equilibrium.

References

Halmos, P. R., "Finite-dimensional Vector Spaces," Van Nostrand, New York (1958).

Slattery, J. C., "Momentum, Energy, and Mass Transfer in Continua," McGraw-Hill, New York (1972); second edition, Robert E. Krieger, Malabar, FL 32950 (1981).

5.10.5 Limiting criteria for intrinsically stable equilibrium Both S and $\mathbf{S}^{(\sigma)}$ are symmetric tensors. Consequently, there is a basis for the vector space whose elements are ordered sets of N+2, real-valued functions of time and position with respect to which the matrix of **S** assumes a diagonal form (Birkhoff and MacLane 1953, p. 277; Halmos 1958, p. 156). For the same reason, there is a basis for the vector space whose elements are ordered sets of N+1, real-valued functions of time and position on the dividing surface Σ with respect to which the matrix of $\mathbf{S}^{(\sigma)}$ assumes a diagonal form.

I think it is helpful to construct the diagonal matrix for the transformation **S**. In the last section, we defined the matrix of **S**

$$[\,\mathbf{S}\,] = \begin{bmatrix} S_{11} & S_{12} & \cdots & S_{1,N+1} & 0 \\ S_{21} & S_{22} & \cdots & S_{2,N+1} & 0 \\ \vdots & \vdots & & \vdots & \vdots \\ S_{N+1,1} & S_{N+1,2} & \cdots & S_{N+1,N+1} & 0 \\ 0 & 0 & \cdots & 0 & \dfrac{\rho}{T} \end{bmatrix} \tag{5-1}$$

with respect to a particular basis $\{\, \boldsymbol{\gamma}_{(1)}, \boldsymbol{\gamma}_{(2)}, \ldots, \boldsymbol{\gamma}_{(N+2)} \,\}$. If we introduce a new basis (Hoffman and Kunze 1961, p. 249)

$$\boldsymbol{\gamma}'_{(1)} \equiv \boldsymbol{\gamma}_{(1)}$$

$$\boldsymbol{\gamma}'_{(k)} \equiv \boldsymbol{\gamma}_{(k)} - \frac{S_{1k}}{S_{11}}\,\boldsymbol{\gamma}_{(1)} \qquad \text{for } 2 \le k \le N+1$$

$$\boldsymbol{\gamma}'_{(N+2)} \equiv \boldsymbol{\gamma}_{(N+2)} \tag{5-2}$$

we find

$$[\,S\,] \equiv \begin{bmatrix} S_{11} & 0 & 0 & \dots & 0 & 0 \\ 0 & S'_{22} & S'_{23} & \dots & S'_{2,N+1} & 0 \\ 0 & S'_{32} & S'_{33} & \dots & S'_{3,N+1} & 0 \\ \cdot & \cdot & \cdot & & \cdot & \cdot \\ \cdot & \cdot & \cdot & & \cdot & \cdot \\ \cdot & \cdot & \cdot & & \cdot & \cdot \\ S'_{N+1,1} & S'_{N+1,2} & S'_{N+1,3} & \dots & S_{N+1,N+1} & 0 \\ 0 & 0 & 0 & \dots & 0 & \frac{\rho}{T} \end{bmatrix} \tag{5-3}$$

where

$$S'_{jk} = S_{jk} - \frac{S_{1k}}{S_{11}} S_{j1} \qquad \text{for } 2 \le j \le N+1,\ 2 \le k \le N+1 \tag{5-4}$$

We can repeat this process with still another basis

$$\boldsymbol{\gamma}''_{(1)} \equiv \boldsymbol{\gamma}_{(1)}$$

$$\boldsymbol{\gamma}''_{(2)} \equiv \boldsymbol{\gamma}'_{(2)}$$

$$\boldsymbol{\gamma}''_{(k)} \equiv \boldsymbol{\gamma}'_{(k)} - \frac{S'_{2k}}{S'_{22}} \boldsymbol{\gamma}'_{(2)} \qquad \text{for } 3 \le k \le N+1$$

$$\boldsymbol{\gamma}''_{(N+2)} \equiv \boldsymbol{\gamma}_{(N+2)} \tag{5-5}$$

in terms of which

$$[\,S\,] \equiv \begin{bmatrix} S_{11} & 0 & 0 & \ldots & 0 & 0 \\ 0 & S'_{22} & 0 & \ldots & 0 & 0 \\ 0 & 0 & S''_{33} & \ldots & S''_{3,N+1} & 0 \\ . & . & . & & . & . \\ . & . & . & & . & . \\ . & . & . & & . & . \\ 0 & 0 & S''_{N+1,3} & \ldots & S''_{N+1,N+1} & 0 \\ 0 & 0 & 0 & \ldots & 0 & \frac{\rho}{T} \end{bmatrix} \tag{5-6}$$

where

$$S''_{jk} = S'_{jk} - \frac{S'_{2k}}{S'_{22}} S'_{j2} \qquad \text{for } 3 \leq j \leq N+1,\ 3 \leq k \leq N+1 \tag{5-7}$$

We can construct in this manner a basis with respect to which the matrix of **S** is diagonal. The entries on the diagonal are known as the **proper** or characteristic values of the transformation. Without constructing the specific forms for these proper values, there are three interesting conclusions that we can draw.

1) At those conditions capable of supporting an intrinsically stable equilibrium, the caloric equation of state must be such that the determinant of **S** as well as the proper values of **S** must all be greater than zero.

2) At those conditions capable of supporting an intrinsically stable equilibrium, there are no more than N + 2 independent constraints upon the caloric equation of state (for example, the N + 2 proper values of **S**).

As soon as any one of the proper values of **S** goes to zero, det **S** goes to zero. Therefore, the limiting criterion for an intrinsically stable equilibrium is det **S** > 0. But we can be more specific. In constructing the diagonal form for the matrix of **S**, observe that S'_{22} goes to zero before S_{11} and that S''_{33} goes to zero before either S_{11} or S'_{22}. Each succeeding proper value is more critical in defining the limits of intrinsic stability for a particular caloric equation of state. This leads us to our third conclusion.

3) The limiting criterion for an intrinsically stable equilibrium with a particular caloric equation of state is

$$\frac{1}{D_N} \det \mathbf{S} = \frac{\rho}{T} \frac{D_{N+1}}{D_N} > 0 \tag{5-8}$$

with the determinants D_K defined by

$$D_K \equiv \begin{vmatrix} S_{11} & \cdot & \cdot & \cdot & S_{1K} \\ \cdot & & & & \\ \cdot & & & & \\ \cdot & & & & \\ S_{K1} & \cdot & \cdot & \cdot & S_{KK} \end{vmatrix} \tag{5-9}$$

This has been recognized previously by Modell and Reid (1974, p. 211; see also Beegle *et al.* 1974b).

By analogy, we can make three similar statements about the surface caloric equation of state..

4) At those conditions capable of supporting an intrinsically stable equilibrium, the surface caloric equation of state must be such that the determinant of $\mathbf{S}^{(\sigma)}$ as well as the proper values of $\mathbf{S}^{(\sigma)}$ must be greater than zero.

5) At those conditions capable of supporting an intrinsically stable equilibrium, there are no more than N + 1 independent constraints upon the surface caloric equation of state (for example, the N + 1 proper values of $\mathbf{S}^{(\sigma)}$).

6) The limiting criterion for an intrinsically stable equilibrium with a particular surface caloric equation of state is

$$\frac{1}{D_N^{(\sigma)}} \det \mathbf{S}^{(\sigma)} = \frac{D_{N+1}^{(\sigma)}}{D_N^{(\sigma)}} > 0 \tag{5-10}$$

with the determinants $D_K^{(\sigma)}$ defined by

$$D_K^{(\sigma)} \equiv \begin{vmatrix} S_{11}^{(\sigma)} & . \; . \; . & S_{1K}^{(\sigma)} \\ . & & \\ . & & \\ . & & \\ S_{K1}^{(\sigma)} & . \; . \; . & S_{KK}^{(\sigma)} \end{vmatrix} \tag{5-11}$$

The implications of inequalities (5-8) and (5-10) for one- and two-component systems are detailed in Exercises 5.10.5-1 and 5.10.5-2. More generally, in a multicomponent system, inequality (5-8) implies that the limiting criterion for an intrinsically stable equilibrium with a particular caloric equation of state is (Exercise 5.10.5-3)

$$\begin{aligned} \frac{1}{D_N} \det \mathbf{S} &= \frac{\rho}{T} \frac{D_{N+1}}{D_N} \\ &= \frac{\rho}{T^2} \left(\frac{\partial \mu_{(N)}}{\partial \rho_{(N)}} \right)_{T, \mu_{(K)} (K \neq N)} \\ &= \frac{1}{T^2 \omega_{(N)}} \left(\frac{\partial P}{\partial \rho_{(N)}} \right)_{T, \mu_{(K)} (K \neq N)} \\ &> 0 \end{aligned} \tag{5-12}$$

where N may refer to any one of the N species, since the ordering of terms in defining **S** is abritrary. For this same reason, an equivalent limiting stability criterion is

$$\frac{\rho}{T^2} \left(\frac{\partial T}{\partial \check{S}} \right)_{\mu_{(K)}} = \frac{1}{T^2 \hat{S}} \left(\frac{\partial P}{\partial \check{S}} \right)_{\mu_{(K)}} > 0 \tag{5-13}$$

Inequality (5-10) requires that, in a multicomponent system, the limiting criterion for an intrinsically stable equilibrium with a particular surface caloric equation of state is (Exercise 5.10.5-3)

$$\frac{1}{D_N^{(\sigma)}} \det \mathbf{S}^{(\sigma)} = \frac{D_{N+1}^{(\sigma)}}{D_N^{(\sigma)}}$$

$$= \frac{1}{T^{(\sigma)}} \left(\frac{\partial \mu^{(\sigma)}_{(N)}}{\partial \rho^{(\sigma)}_{(N)}} \right)_{T^{(\sigma)}, \mu^{(\sigma)}_{(K)} (K \neq N)}$$

$$= - \frac{1}{T^{(\sigma)} \rho^{(\sigma)}_{(N)}} \left(\frac{\partial \gamma}{\partial \rho^{(\sigma)}_{(N)}} \right)_{T^{(\sigma)}, \mu^{(\sigma)}_{(K)} (K \neq N)}$$

$$> 0 \tag{5-14}$$

where again N may refer to any one of the N species, since the ordering of terms in defining $S^{(\sigma)}$ is also arbitrary. An equivalent limiting stability criterion is

$$\frac{1}{T^{(\sigma)}} \left(\frac{\partial T^{(\sigma)}}{\partial \overline{S}^{(\sigma)}} \right)_{\mu^{(\sigma)}_{(K)}} = - \frac{1}{T^{(\sigma)} \, \overline{S}^{(\sigma)}} \left(\frac{\partial \gamma}{\partial \overline{S}^{(\sigma)}} \right)_{\mu^{(\sigma)}_{(K)}} > 0 \tag{5-15}$$

In order for a multiphase, multicomponent system to exist in an intrinsically stable equilibrium state, inequality (5-12) [or equivalently (5-13)] must be satisfied within every phase and inequality (5-14) [or [5-15)] must be satisfied on every phase interface. The limits of intrinsically stable equilibrium are defined by the failure of inequality (5-12) within one of the phases or by the failure of inequality (5-14) on one of the phase interfaces.

Consider a relatively insoluble material spread as a monolayer over a portion of a liquid-gas interface bounded by moveable barriers. As the spread film is slowly compressed, the interfacial tension γ decreases as the surface mass density $\rho^{(\sigma)}_{(N)}$ of the insoluble material increases in accord with inequality (5-14). Note that, if the compression takes place sufficiently slowly, we would expect the surface chemical potentials $\mu^{(\sigma)}_{(K)}$ (K = 1, 2, ..., N–1) of the other species present to remain fixed at the values of their chemical potentials in the adjoining phases. But there is a minimum value of the interfacial tension that can be achieved in this way by compression of the monolayer. If futher compression is attempted, the monolayer collapses and the spread material is forced out of the monolayer into the adjacent phase (Gaines 1966, p. 144). This collapse point appears to correspond to a limit of intrinsic stability for the multiphase system at which inequality (5-14) is violated on that portion of the liquid-gas interface within the barriers.

The interpretation of experimental observations in terms of the limiting criteria for intrinsic stability involves a certain amount of speculation. Although it seems likely, we can not be certain that, in the experiment described above, inequality (5-14) is violated on the interface within the barriers before inequality (5-12) fails within the adjacent liquid phase. It is perhaps safe to say only that the experimentally observed collapse point occurs within the neighborhood of the limit of intrinsic stability.

It is more appropriate to look forward to using inequalities (5-12) and (5-14) in predicting the limits of intrinsic stability in multiphase, multicomponent systems. In addition to the collapse point of a monolayer, one might consider the limiting conditions at which phase transitions occur in monolayers (Gaines 1966, p. 156), the limiting conditions under which a macroemulsion is transformed into a micellar solution (or microemulsion) (Gerbacia and Rosano 1973), and the limiting conditions for the multiple phase transitions that can be observed upon the addition of a mixed surfactant system to a two-phase mixture of crude oil and brine (Robbins 1976). In studies of organic-brine-mixed surfactant systems, interfacial tension displays a distinct minimum as a function of concentration (Cayias *et al.* 1976, Wilson *et al.* 1976). In the absence of contrary evidence, one might speculate that the minimum interfacial tension occurs within the immediate neighborhood of a limit of intrinsic stability. This speculation may be supported by the observation of Healy *et al.* (1976) that very low interfacial tensions were accompanied by a phase transition.

References

Beegle, B. L., M. Modell, and R. C. Reid, *AIChE J.* **20**, 1194 (1974a).

Beegle, B. L., M. Modell, and R. C. Reid, *AIChE J.* **20**, 1200 (1974b). See also R. A. Heidemann, *AIChE J.* **21**, 824 (1975) as well as B. L. Beegle, M. Modell, and R. C. Reid, *AIChE J.* **21**, 826 (1975).

Birkhoff, G., and S. MacLane, "A Survey of Modern Algebra," Macmillan, New York (1953).

Cayias, J. L., R. S. Schechter, and W. H. Wade, *Soc. Petrol. Eng. J.* **16**, 351 (1976).

Gaines, G. L., Jr., "Insoluble Monolayers at Liquid-Gas Interfaces," Interscience, New York (1966).

Gerbacia, W., and H. L. Rosano, *J. Colloid Interface Sci.* **44**, 242 (1973).

Halmos, P. R., "Finite-dimensional Vector Spaces," Van Nostrand, New York (1958).

Healy, R. N., R. L. Reed, and D. G. Stenmark, *Soc. Petrol. Eng. J.* **16**, 147 (1976).

Hoffman, K., and R. Kunze, "Linear Algebra," Prentice-Hall, Englewood Cliffs, New Jersey (1961).

Modell, M., and R. C. Reid, "Thermodynamics and Its Applications," Prentice-Hall, Englewood Cliffs, New Jersey (1974).

Robbins, M. L., "Theory for the Phase Behavior of Microemulsions," paper 5839, *Society of Petroleum Engineers Symposium on Improved Oil Recovery*, Tulsa, OK (March 22-24, 1976).

Slattery, J. C., "Momentum, Energy, and Mass Transfer in Continua," McGraw-Hill, New York (1972); second edition, Robert E. Krieger, Malabar, FL 32950 (1981).

Wilson, P. M., C. L. Murphy, and W. R. Foster, "The Effects of Sulfonate Molecular Weight and Salt Concentration on the Interfacial Tension of Oil-Brine-Surfactant Systems," paper 5812, *Society of Petroleum Engineers Symposium on Improved Oil Recovery*, Tulsa, OK (March 22-24, 1976).

Exercises

5.10.5-1 *limiting stability criteria for single component system* Prove the following statements.

i) For intrinsic stability, it is necessary and sufficient that any caloric equation of state for a single component system satisfy

$$S_{11} \equiv \frac{1}{T}\left(\frac{\partial T}{\partial \check{S}}\right)_{\rho}$$

$$= \frac{1}{\rho \hat{c}_V} > 0 \tag{1}$$

$$S_{22} \equiv \frac{1}{T}\left(\frac{\partial \mu}{\partial \rho}\right)_{\check{S}} > 0 \tag{2}$$

$$S_{33} \equiv \frac{\rho}{T} > 0 \tag{3}$$

Here $\hat{c}_V$ is the heat capacity per unit mass at constant specific volume (Slattery 1981, p. 290).

ii) For intrinsic stability, it is necessary and sufficient that any surface caloric equation of state for a single component system satisfy

$$S_{11}^{(\sigma)} \equiv \frac{1}{T^{(\sigma)}} \left(\frac{\partial T^{(\sigma)}}{\partial \bar{S}^{(\sigma)}} \right)_{\rho^{(\sigma)}}$$

$$= \frac{1}{\rho^{(\sigma)} \, \hat{c}_{\mathcal{A}}^{(\sigma)}} > 0 \tag{4}$$

$$S_{22}^{(\sigma)} \equiv \frac{1}{T^{(\sigma)}} \left(\frac{\partial \mu^{(\sigma)}}{\partial \rho^{(\sigma)}} \right)_{\bar{S}^{(\sigma)}} > 0 \tag{5}$$

Here $\hat{c}_{\mathcal{A}}^{(\sigma)}$ is the surface heat capacity per unit mass at constant specific area (see Exercise 5.8.6-4).

iii) The limiting criterion for an intrinsically stable equilibrium with a particular caloric equation of state is

$$\frac{1}{D_1} \det \mathbf{S} = \frac{\rho}{T} \frac{D_2}{D_1}$$

$$= \frac{\rho}{T} \frac{\begin{vmatrix} \frac{1}{T} \left(\frac{\partial T}{\partial \check{S}} \right)_\rho & \frac{1}{T} \left(\frac{\partial \mu}{\partial \check{S}} \right)_\rho \\ \frac{1}{T} \left(\frac{\partial \mu}{\partial \check{S}} \right)_\rho & \frac{1}{T} \left(\frac{\partial \mu}{\partial \rho} \right)_{\check{S}} \end{vmatrix}}{\frac{1}{T} \left(\frac{\partial T}{\partial \check{S}} \right)_\rho}$$

$$= \frac{\rho}{T^2} \frac{\begin{vmatrix} \frac{\partial^2 \check{U}}{\partial \check{S}^2} & \frac{\partial^2 \check{U}}{\partial \check{S} \partial \rho} \\ \frac{\partial^2 \check{U}}{\partial \check{S} \partial \rho} & \frac{\partial^2 \check{U}}{\partial \rho^2} \end{vmatrix}}{\frac{\partial^2 \check{U}}{\partial \check{S}^2}}$$

$$= \frac{\rho}{T^2} \left[\frac{\partial^2}{\partial \rho^2} (\check{U} - T\check{S}) \right]_T$$

$$= \frac{\rho}{T^2} \left(\frac{\partial^2 \check{A}}{\partial \rho^2} \right)_T$$

$$= \frac{\rho}{T^2} \left(\frac{\partial \mu}{\partial \rho} \right)_T$$

$$= \frac{1}{T^2} \left(\frac{\partial P}{\partial \rho} \right)_T$$

$$= - \frac{1}{T^2 \rho^2} \left(\frac{\partial P}{\partial \hat{V}} \right)_T$$

$$> 0 \tag{6}$$

Our ordering of variables in defining **S** is arbitrary. An equivalent limiting stability criterion is

$$\frac{\rho}{T} \frac{\begin{vmatrix} \frac{1}{T}\left(\frac{\partial \mu}{\partial \rho}\right)_{\check{S}} & \frac{1}{T}\left(\frac{\partial \mu}{\partial \check{S}}\right)_{\rho} \\ \frac{1}{T}\left(\frac{\partial \mu}{\partial \check{S}}\right)_{\rho} & \frac{1}{T}\left(\frac{\partial T}{\partial \check{S}}\right)_{\rho} \end{vmatrix}}{\frac{1}{T}\left(\frac{\partial \mu}{\partial \rho}\right)_{\check{S}}}$$

$$= \frac{\rho}{T^2} \left[\frac{\partial^2}{\partial \check{S}^2} (\check{U} - \mu\rho) \right]_{\mu}$$

$$= \frac{\rho}{T^2} \left(\frac{\partial T}{\partial \check{S}} \right)_{\mu}$$

$$= \frac{1}{T^2 \hat{S}} \left(\frac{\partial P}{\partial \hat{S}} \right)_{\mu}$$

$$> 0 \tag{7}$$

These alternative stability criteria are reported by Beegle *et al.* (1974b).

iv) The limiting criterion for an intrinsically stable equilibrium with a particular surface caloric equation of state is

$$\frac{1}{D_1^{(\sigma)}} \det \mathbf{S}^{(\sigma)} = \frac{D_2^{(\sigma)}}{D_1^{(\sigma)}}$$

$$= \frac{\begin{vmatrix} \frac{1}{T^{(\sigma)}}\left(\frac{\partial T^{(\sigma)}}{\partial \bar{S}^{(\sigma)}}\right)_{\rho^{(\sigma)}} & \frac{1}{T^{(\sigma)}}\left(\frac{\partial \mu^{(\sigma)}}{\partial \bar{S}^{(\sigma)}}\right)_{\rho^{(\sigma)}} \\ \frac{1}{T^{(\sigma)}}\left(\frac{\partial \mu^{(\sigma)}}{\partial \bar{S}^{(\sigma)}}\right)_{\rho^{(\sigma)}} & \frac{1}{T^{(\sigma)}}\left(\frac{\partial \mu^{(\sigma)}}{\partial \rho^{(\sigma)}}\right)_{\bar{S}^{(\sigma)}} \end{vmatrix}}{\frac{1}{T^{(\sigma)}}\left(\frac{\partial T^{(\sigma)}}{\partial \bar{S}^{(\sigma)}}\right)_{\rho^{(\sigma)}}}$$

$$= \frac{1}{T^{(\sigma)}} \frac{\begin{vmatrix} \frac{\partial^2 \bar{U}^{(\sigma)}}{\partial \bar{S}^{(\sigma)2}} & \frac{\partial^2 \bar{U}^{(\sigma)}}{\partial \bar{S}^{(\sigma)}\,\partial \rho^{\sigma)}} \\ \frac{\partial^2 \bar{U}^{(\sigma)}}{\partial \bar{S}^{(\sigma)}\,\partial \rho^{(\sigma)}} & \frac{\partial^2 \bar{U}^{(\sigma)}}{\partial \rho^{(\sigma)2}} \end{vmatrix}}{\frac{\partial^2 \bar{U}^{(\sigma)}}{\partial \bar{S}^{(\sigma)2}}}$$

$$= \frac{1}{T^{(\sigma)}} \left[\frac{\partial^2}{\partial \rho^{(\sigma)2}} \left(\bar{U}^{(\sigma)} - T^{(\sigma)} \bar{S}^{(\sigma)} \right) \right]_{T^{(\sigma)}}$$

$$= \frac{1}{T^{(\sigma)}} \left(\frac{\partial^2 \bar{A}^{(\sigma)}}{\partial \rho^{(\sigma)2}} \right)_{T^{(\sigma)}}$$

$$= \frac{1}{T^{(\sigma)}} \left(\frac{\partial \mu^{(\sigma)}}{\partial \rho^{(\sigma)}} \right)_{T^{(\sigma)}}$$

$$= - \frac{1}{\rho^{(\sigma)} T^{(\sigma)}} \left(\frac{\partial \gamma}{\partial \rho^{(\sigma)}} \right)_{T^{(\sigma)}}$$

$$= \frac{1}{\rho^{(\sigma)3} T^{(\sigma)}} \left(\frac{\partial \gamma}{\partial \hat{\mathcal{A}}} \right)_{T^{(\sigma)}}$$

$$> 0 \tag{8}$$

Our ordering of variables in defining $\mathbf{S}^{(\sigma)}$ is arbitrary. An equivalent limiting stability criterion is

$$\frac{\begin{vmatrix} \dfrac{1}{T^{(\sigma)}} \left(\dfrac{\partial \mu^{(\sigma)}}{\partial \rho^{(\sigma)}} \right)_{\overline{S}^{(\sigma)}} & \dfrac{1}{T^{(\sigma)}} \left(\dfrac{\partial \mu^{(\sigma)}}{\partial \overline{S}^{(\sigma)}} \right)_{\rho^{(\sigma)}} \\ \dfrac{1}{T^{(\sigma)}} \left(\dfrac{\partial \mu^{(\sigma)}}{\partial \overline{S}^{(\sigma)}} \right)_{\rho^{(\sigma)}} & \dfrac{1}{T^{(\sigma)}} \left(\dfrac{\partial T^{(\sigma)}}{\partial \overline{S}^{(\sigma)}} \right)_{\rho^{(\sigma)}} \end{vmatrix}}{\dfrac{1}{T^{(\sigma)}} \left(\dfrac{\partial \mu^{(\sigma)}}{\partial \rho^{(\sigma)}} \right)_{\overline{S}^{(\sigma)}}}$$

$$= \frac{1}{T^{(\sigma)}} \left[\frac{\partial^2}{\partial \overline{S}^{(\sigma)2}} \left(\overline{U}^{(\sigma)} - \mu^{(\sigma)} \rho^{(\sigma)} \right) \right]_{\mu^{(\sigma)}}$$

$$= \frac{1}{T^{(\sigma)}} \left(\frac{\partial T^{(\sigma)}}{\partial \overline{S}^{(\sigma)}} \right)_{\mu^{(\sigma)}}$$

$$> 0 \tag{9}$$

(Hint: In going from the third to the fourth lines of (6) and (8), I have employed an expression for the second derivative of a first Legendre transform (Beegle *et al.* 1974a, equations 31 and 35).

5.10.5-2 *limiting stability criteria for binary system* Prove the following statements.

i) The limiting criterion for an intrinsically stable equilibrium with a particular caloric equation of state is (Beegle *et al.* 1974b)

$$\frac{1}{D_2}\det \mathbf{S} = \frac{\rho}{T}\frac{D_3}{D_2}$$

$$= \frac{\rho}{T^2}\left[\frac{\partial^2}{\partial \rho_{(2)}^2}(\check{U} - T\check{S} - \mu_{(1)}\rho_{(1)})\right]_{T,\mu_{(1)}}$$

$$= \frac{\rho}{T^2}\left(\frac{\partial \mu_{(2)}}{\partial \rho_{(2)}}\right)_{T,\mu_{(1)}}$$

$$= \frac{1}{T^2\omega_{(2)}}\left(\frac{\partial P}{\partial \rho_{(2)}}\right)_{T,\mu_{(1)}}$$

$$> 0 \tag{1}$$

The ordering of terms in defining **S** is arbitrary. Equivalent limiting stability criteria are

$$\frac{\rho}{T^2}\left[\frac{\partial^2}{\partial \rho_{(1)}^2}(\check{U} - T\check{S} - \mu_{(2)}\rho_{(2)})\right]_{T,\mu_{(2)}}$$

$$= \frac{\rho}{T^2}\left(\frac{\partial \mu_{(1)}}{\partial \rho_{(1)}}\right)_{T,\mu_{(2)}}$$

$$= \frac{1}{T^2\omega_{(1)}}\left(\frac{\partial P}{\partial \rho_{(1)}}\right)_{T,\mu_{(2)}}$$

$$> 0 \tag{2}$$

and

$$\frac{\rho}{T^2}\left[\frac{\partial^2}{\partial \check{S}^2}(\check{U} - \mu_{(1)}\rho_{(1)} - \mu_{(2)}\rho_{(2)})\right]_{\mu_{(1)},\mu_{(2)}}$$

$$= \frac{\rho}{T^2}\left(\frac{\partial T}{\partial \check{S}}\right)_{\mu_{(1)},\mu_{(2)}}$$

$$= \frac{1}{T^2\hat{S}}\left(\frac{\partial P}{\partial \check{S}}\right)_{\mu_{(1)},\mu_{(2)}}$$

$$> 0 \tag{3}$$

ii) The limiting criterion for an intrinsically stable equilibrium with a particular surface caloric equation of state is

$$
\begin{aligned}
\frac{1}{D_2^{(\sigma)}} \det \mathbf{S}^{(\sigma)} &= \frac{D_3^{(\sigma)}}{D_2^{(\sigma)}} \\
&= \frac{1}{T^{(\sigma)}} \left[\frac{\partial^2}{\partial \rho_{(2)}^{(\sigma)\,2}} \left(\overline{U}^{(\sigma)} - T^{(\sigma)} \overline{S}^{(\sigma)} \right.\right. \\
&\qquad \left.\left. - \mu_{(1)}^{(\sigma)} \rho_{(1)}^{(\sigma)} \right) \right]_{T^{(\sigma)}, \mu_{(1)}^{(\sigma)}} \\
&= \frac{1}{T^{(\sigma)}} \left(\frac{\partial \mu_{(2)}^{(\sigma)}}{\partial \rho_{(2)}^{(\sigma)}} \right)_{T^{(\sigma)}, \mu_{(1)}^{(\sigma)}} \\
&= - \frac{1}{T^{(\sigma)} \rho_{(2)}^{(\sigma)}} \left(\frac{\partial \gamma}{\partial \rho_{(2)}^{(\sigma)}} \right)_{T^{(\sigma)}, \mu_{(1)}^{(\sigma)}} \\
&> 0
\end{aligned}
\tag{4}
$$

The ordering of terms in defining $\mathbf{S}^{(\sigma)}$ is arbitrary. Equivalent limiting stability criteria are

$$
\begin{aligned}
& \frac{1}{T^{(\sigma)}} \left[\frac{\partial^2}{\partial \rho_{(1)}^{(\sigma)\,2}} \left(\overline{U}^{(\sigma)} - T^{(\sigma)} \overline{S}^{(\sigma)} - \mu_{(2)}^{(\sigma)} \rho_{(2)}^{(\sigma)} \right) \right]_{T^{(\sigma)}, \mu_{(2)}^{(\sigma)}} \\
&= \frac{1}{T^{(\sigma)}} \left(\frac{\partial \mu_{(1)}^{(\sigma)}}{\partial \rho_{(1)}^{(\sigma)}} \right)_{T^{(\sigma)}, \mu_{(2)}^{(\sigma)}} \\
&= - \frac{1}{T^{(\sigma)} \rho_{(1)}^{(\sigma)}} \left(\frac{\partial \gamma}{\partial \rho_{(1)}^{(\sigma)}} \right)_{T^{(\sigma)}, \mu_{(2)}^{(\sigma)}} \\
&> 0
\end{aligned}
\tag{5}
$$

and

$$
\frac{1}{T^{(\sigma)}} \left[\frac{\partial^2}{\partial \overline{S}^{(\sigma)\,2}} \left(\overline{U}^{(\sigma)} - \mu_{(1)}^{(\sigma)} \rho_{(1)}^{(\sigma)} - \mu_{(2)}^{(\sigma)} \rho_{(2)}^{(\sigma)} \right) \right]_{\mu_{(1)}^{(\sigma)}, \mu_{(2)}^{(\sigma)}}
$$

$$= \frac{1}{T^{(\sigma)}} \left(\frac{\partial T^{(\sigma)}}{\partial \overline{S}^{(\sigma)}} \right)_{\mu \left\{ \substack{\sigma \\ 1} \right\}, \mu \left\{ \substack{\sigma \\ 2} \right\}}$$

$$= - \frac{1}{T^{(\sigma)} \overline{S}^{(\sigma)}} \left(\frac{\partial \gamma}{\partial \overline{S}^{(\sigma)}} \right)_{\mu \left\{ \substack{\sigma \\ 1} \right\}, \mu \left\{ \substack{\sigma \\ 2} \right\}}$$

$$> 0 \tag{6}$$

(Hint: In going from the first to the second lines of (1) and (4), I have employed an expression for the second derivative of a second Legendre transform (Beegle *et al.* 1974a, equations 31 and 35).)

5.10.5-3 *limiting stability criteria for multicomponent systems*

i) Prove that the limiting criterion for an intrinsically stable equilibrium with a particular caloric equation of state is inequality (5-12) or equivalently inequality (5-13).

ii) Prove that the limiting criterion for an intrinsically stable equilibrium with a particular surface caloric equation of state is inequality (5-14) or ineqality (5-15).

(Hint: In going from the first to the second lines of inequality (5-12) and of inequality (5-14), I have employed an expression for the second derivative of a second Legendre transform (Beegle *et al.* (1974a, equations 31 and 35).)

5.10.5-4 *limiting stability criteria for multicomponent systems with Lucassen-Reynders and van den Tempel convention* With the Lucassen-Reynders and van den Tempel convention for locating the reference dividing surface, the total surface molar density $c^{(\sigma)}$ is a constant $c^{(\sigma)\infty}$ (see Sec. 5.2.3). This means that there are only N-1 independent surface molar concentrations.

This has no effect upon the limiting criterion for an intrinsically stable equilibrium with a particular caloric equation of state: inequality (5-12) or equivalently (5-13).

Prove that the limiting criterion for an intrinsically stable equilibrium with a particular surface caloric equation of state is now

$$\frac{1}{T^{(\sigma)}} \left(\frac{\partial \mu \left(\substack{\sigma \\ A} \right)^{m)}}{\partial c \left(\substack{\sigma \\ A} \right)^{L)}} \right)_{T^{(\sigma)} \mu \left(\substack{\sigma \\ B} \right)^{m)} (B \neq A, N)}$$

$$= -\frac{1}{T^{(\sigma)} c_{(A)}^{(L)}} \left(\frac{\partial \gamma}{\partial c_{(A)}^{(L)}} \right)_{T^{(\sigma)}, \mu_{(B)}^{(m)} (B \neq A, N)}$$

$$> 0$$

where A and N may refer to any one of the N species present in the dividing surface. Also prove that an equivalent limiting stability criterion is inequality (5-15).

5.10.5-5 *Frumkin equation of state* For the Frumkin equation (see Sec. 5.8.7), prove that

$$\left(\frac{\partial \gamma}{\partial c_{(2)}^{(L)}} \right)_{T^{(\sigma)}} = -\frac{RT^{(\sigma)}}{1 - c_{(2)}^{(L)}/c^{(\sigma)\infty}}$$

The Frumkin equation of state does not admit instabilities and should not be used in the neighborhood of surface phase transitions.

5.10.5-6 *limiting stability criteria for multicomponent electrolyte solutions with Lucassen-Reynders and van den Tempel convention* Let N now refer to the total number of species present, whether they be cations, anions, undissociated species or solvent.

We see in Exercise 5.10.5-4 that, as the result of adopting the Lucassen-Reynders and van den Tempel convention, there are only N-1 independent surface molar densities.

For an electrolyte solution, we must impose local electrical neutrality, which implies that there are only N-1 independent molar densities in each bulk phase and N-2 independent surface molar densities in the dividing surface.

Write the limiting stability criteria for multicomponent electrolyte solutions by analogy with (5-12) through (5-15) and Exercise 5.10.5-4.

5.10.6 Equilibrium conditions for nucleation (with R. W. Flumerfelt and W. Ruengphrathuengsuka) Nucleation occurs in most boiling, condensation, and crystallization processes. The gas evolution observed upon opening a champagne bottle is one of the most dramatic examples. Others are encountered in meteorology (formation of clouds, fog, rain, sleet, and snow), in materials processing (polymer foam manufacturing, polymer foam devolatalization), and in many heat exchange processes involving phase transitions.

Consider a pan of cool water which is gently stirred as it is slowly heated on a stove at sea level. If we were to record the temperature of the water as a function of time, we would find that it does not boil at 100°C. The temperature would rise somewhat above 100°C before small bubbles would appear and boiling commenced. Liquid water and water vapor are at equilibrium at 100°C and 1 atm. only when the interface separating the two phases is a plane.

We commonly observe that the air in bathroom becomes foggy after a hot shower. Alternatively, consider slowly dropping the temperature of a container of superheated water vapor that is suitably stirred and maintained at a constant pressure of 1 atm. Condensation would not begin at 100°C. The temperature would have to drop below 100°C before droplets would begin to appear on the inside surface of the container.

There are three related phenomena that we must consider: homogeneous nucleation, heterogeneous nucleation, and bubble or droplet growth and detachment.

By *homogeneous nucleation*, we mean the formation of bubbles or droplets within a phase.

By *heterogeneous nucleation*, we mean the formation of bubbles or droplets at bounding surfaces. Often, but not always, heterogeneous nucleation will occur at small imperfections, defects, or fissures in these surfaces that are filled with the continuous phase.

After heterogeneous nucleation has occurred in a fissure, it is partially or completely filled with the discontinuous phase, which can grow as the result of mass transfer from the adjacent continuous phase. A fissure can also be filled with a discontinuous phase as the result of its past treatment. For example, a champagne glass is purposely made with a fissure which is filled with air following washing and a thorough drying and from which the air is likely never to be completely displaced by the wine. Long after the initial burst of bubbles is formed, likely by homogeneous nucleation, this air bubble grows by mass transfer of CO_2 from the champagne, periodically detaching a portion of itself when it becomes too large. We will not consider here the rate of *bubble or droplet growth and detachment.*

In these developments, we will consider homogeneous nucleation and heterogeneous nucleation separately. For general references in this area, the reader is referred to Frenkel (1946), Feder *et al.* (1966), Ward *et al.* (1970), Cole (1974), Blander and Katz (1975), Hodgson (1984), Wilt (1986), and Narsimhan and Ruckenstein (1989).

Homogeneous nucleation For the moment, let us ignore the possibility of heterogeneous nucleation.

Consider a container of liquid or vapor that is sealed with a piston which maintains a constant pressure. The container is submerged in an isothermal bath, the temperature of which is gradually changed. Given the temperature of the system under conditions such that liquid would be in equilibrium across a plane interface with its vapor at a specified pressure, how much must the temperature be increased to produce a small bubble in equilibrium with the liquid or a small droplet in equilibrium with the vapor? In the context of equilibrium, at what temperature is the Gibbs free energy minimized by the formation of a second phase?

We will find it convenient to make the following assumptions.

i) At all times, the system is nearly isothermal. The system is sufficiently close to equilibrium that temperature may assumed to be continuous across the interface. Any effects attributable to the time rate of change of temperature will be neglected in the limit as equilibrium is approached.

ii) The pressure $P^{(C)}$ in the continuous liquid phase is a constant, independent of position and of time.

iii) The pressure $P^{(D)}$ in the discontinuous vapor phase is independent of position.

iv) The effects of inertia are neglected.

v) At equilibrium, the interface is spherical.

vi) The effects of mass transfer are neglected in the limit as equilibrium is approached.

Let us choose as our system all of the liquid and vapor in the container, excluding the solid boundaries. In view of assumptions iv and v, the energy balance requires (Table 7.1.3-1, Equation A)

$$\frac{dU}{dt} = - \int_S \mathbf{q} \cdot \mathbf{n} \, dA - \int_C \mathbf{q}^{(\sigma)} \cdot \boldsymbol{\mu} \, ds + \int_R \rho Q \, dV + \int_\Sigma \rho^{(\sigma)} Q^{(\sigma)} \, dA - P^{(C)} \frac{dV}{dt} \tag{6-1}$$

Here U is the internal energy of the system; V is the volume of the system. The Clausius-Duhem inequality demands (Sec. 7.1.4)

$$\frac{dS}{dt} \geq - \frac{1}{T} \int_S \mathbf{q} \cdot \mathbf{n} \, dA - \frac{1}{T} \int_C \mathbf{q}^{(\sigma)} \cdot \boldsymbol{\mu} \, ds$$

$$+ \frac{1}{T} \int_R \rho Q \, dV + \frac{1}{T} \int_\Sigma \rho^{(\sigma)} Q^{(\sigma)} \, dA \tag{6-2}$$

in which we have recognized assumptions i and vi. Here S denotes the entropy of the system. Equations (6-1) and (6-2) imply

$$\frac{dA}{dt} + P^{(C)} \frac{dV}{dt} \leq 0 \tag{6-3}$$

or

$$\frac{dG}{dt} + P^{(C)} \frac{dV}{dt} \leq \frac{d}{dt} \left(P^{(C)} V^{(C)} + P^{(D)} V^{(D)} - \gamma \mathcal{A} \right) \tag{6-4}$$

where A is the Helmholtz free energy of the system, G the Gibbs free energy of the system, $\mathcal{A}$ the area of the bubble surface, $...^{(C)}$ a property of the continuous phase, and $...^{(D)}$ a property of the discontinuous (bubble or droplet) phase.

Equation (6-4) can be integrated to find

$$\Delta G + P^{(C)} \left(\Delta V^{(C)} + \Delta V^{(D)} \right)$$

$$\leq P^{(C)} \Delta V^{(C)} + \Delta \left(P^{(D)} V^{(D)} \right) - \gamma \, \Delta \mathcal{A} \tag{6-5}$$

or

$$\begin{aligned} \Delta G &\leq \Delta \left(P^{(D)} V^{(D)} \right) - P^{(C)} \, \Delta V^{(D)} - \gamma \, \Delta \mathcal{A} \\ &= P^{(D)} V^{(D)} - P^{(C)} \, V^{(D)} - \gamma \mathcal{A} \\ &= \left(P^{(D)} - P^{(C)} \right) V^{(D)} - \gamma \mathcal{A} \end{aligned} \tag{6-6}$$

Here Δ represents the difference between the variable in the vapor-liquid system and that in the original system (liquid in equilibrium with a plane interface). If R^* is the bubble radius for which the jump momentum balance is satisfied

$$P^{(D)} - P^{(C)} = \frac{2\gamma}{R^*} \tag{6-7}$$

we can express (6-6) as

$$\Delta G \leq \frac{2\gamma}{R^*} V^{(D)} - \gamma \mathcal{A}$$

$$= \frac{8}{3} \pi\gamma \frac{R^3}{R^*} - 4\pi R^2 \gamma \tag{6-8}$$

In writing this, we have not assumed that equilibrium necessarily occurs when the jump momentum balance is satisfied.

With the restrictions of assumptions i through iii, the Gibbs free energy of the system is minimized as a function of R at equilibrium (Exercise 5.10.3-3). Equivalently, (6-8) requires

$$\frac{d}{dR}\left(\frac{8}{3} \pi\gamma \frac{R^3}{R^*} - 4\pi R^2 \gamma \right) = 8\pi\gamma \frac{R^2}{R^*} - 8\pi R\gamma = 0 \tag{6-9}$$

or as expected

$$R = R^* \tag{6-10}$$

It follows immediately that (Gibbs 1948, p. 254)

$$\Delta G = - \frac{4}{3} \pi R^{*2} \gamma$$

$$= - \frac{16\pi\gamma^3}{3\left(P^{(D)} - P^{(C)} \right)^2} \tag{6-11}$$

The Gibbs free energy of the system in equilibrium with a bubble is smaller than that for the system in equilibrium with a plane interface. This is in agreement with our expectation that the Gibbs free energy of the system is minimized.

At this point, it is helpful to introduce an additional assumption:

vii) In relating R^* to the pressure in the continuous phase and the vapor pressure P_v observed over a plane liquid-vapor interface at temperature T, we will assume that the bubble is sufficiently large that chemical potential is continuous across the phase interface and that the effects of London-van der Waals forces and of electrostatic double-layer forces can be neglected.

Recognizing that the system is composed of a single species, we observe

$$\left(\frac{\partial \mu^{(i)}}{\partial P} \right)_T = \hat{V}^{(i)} \tag{6-12}$$

Integrating, we find

$$\mu^{(G)}(P^{(G)}, T) = \mu^{(G)}(P_v , T) + \int_{P_v}^{P^{(G)}} \hat{V}^{(G)} \, dP \tag{6-13}$$

$$\mu^{(L)}(P^{(L)}, T) = \mu^{(L)}(P_v , T) + \int_{P_v}^{P^{(L)}} \hat{V}^{(L)} \, dP \tag{6-14}$$

In view of assumption vii, for equilibrium at the surface of the bubble or droplet

$$\mu^{(L)}(P^{(L)}, T) = \mu^{(G)}(P^{(G)}, T) \tag{6-15}$$

For equilibrium at a plane interface at the same temperature T

$$\mu^{(L)}(P_v, T) = \mu^{(G)}(P_v , T) \tag{6-16}$$

From (6-13) through (6-16),

$$\int_{P_v}^{P^{(G)}} \hat{V}^{(G)} \, dP = \int_{P_v}^{P^{(L)}} \hat{V}^{(L)} \, dP \tag{6-17}$$

For an incompressible liquid, this reduces to

$$\int_{P_v}^{P^{(G)}} \hat{V}^{(G)} \, dP = - \hat{V}^{(L)}(P_v - P^{(L)}) \tag{6-18}$$

For an ideal gas

$$\hat{V}^{(G)} = \frac{\mathcal{R}T}{MP} \tag{6-19}$$

and (6-18) simplifies to

$$\frac{\mathcal{R}T}{M} \ln\left(\frac{P^{(G)}}{P_v} \right) = - \hat{V}^{(L)}(P_v - P^{(L)}) \tag{6-20}$$

By $\mathcal{R}$ we mean the gas law constant; M is the molecular weight. For a discontinuous gas phase and therefore a given pressure $P^{(L)}$ in the continuous liquid phase, (6-20) implies in view of (6-7)

$$R^* = 2\gamma \left[P_v \exp\left(- \frac{M\hat{V}^{(L)}}{\mathcal{R}T} (P_v - P^{(L)}) \right) - P^{(L)} \right]^{-1} \tag{6-21}$$

For a discontinuous liquid phase and a specified pressure $P^{(G)}$ in the continuous gas phase, we have instead

$$R^* = 2\gamma \left[\frac{\mathcal{R}T}{M\hat{V}^{(L)}} \ln\left(\frac{P^{(G)}}{P_v} \right) + P_v - P^{(G)} \right]^{-1} \tag{6-22}$$

In the limit

$$\frac{P^{(G)}}{P_v} \to 1: \quad \frac{M\hat{V}^{(L)}}{\mathcal{R}T} (P_v - P^{(L)}) \to 0 \tag{6-23}$$

(6-21) for a discontinuous gas phase reduces to

$$R^* = 2\gamma [P_v - P^{(L)}]^{-1} \tag{6-24}$$

and (6-22) for a discontinuous liquid phase to

$$R^* = 2\gamma \left[\frac{\mathcal{R}T}{M\hat{V}^{(L)}} \left(\frac{P^{(G)}}{P_v} - 1 \right) \right]^{-1} \tag{6-25}$$

Homogeneous bubble nucleation occurs as $P^{(L)}$ is reduced below P_v; homogeneous droplet nucleation takes place as $P^{(G)}$ is raised above P_v.

Heterogeneous bubble nucleation Now let us consider heterogeneous bubble nucleation in liquid-filled defects. We will again choose all of the liquid and vapor in the container, excluding the fluid-solid boundaries, as the system to which the integral energy balance and the entropy inequality are to be applied. We will invoke assumptions i through iv and vi as well as

v') At equilibrium, the liquid-vapor interface is spherical. In effect, this assumes that the bubble is sufficiently large that the effects of London-van der Waals forces and electrostatic double-layer forces can be neglected.

viii) A defect can be approximated as the conical pit depicted in Figure 6-1a.

ix) In order for the defect to be initially filled with liquid, we will assume that $\Theta < \pi/2$, where Θ is the contact angle measured through the liquid phase.

a)

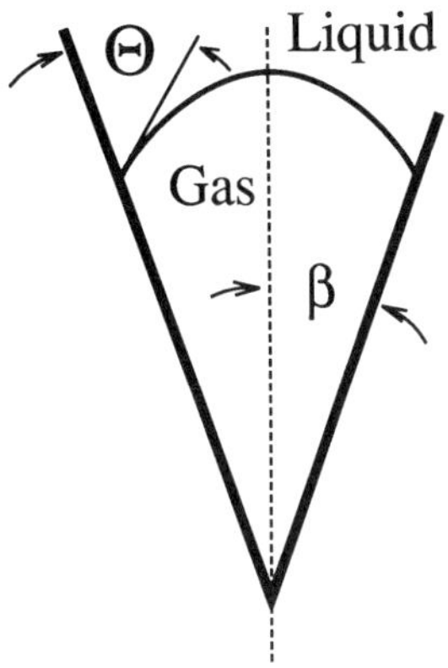

b)

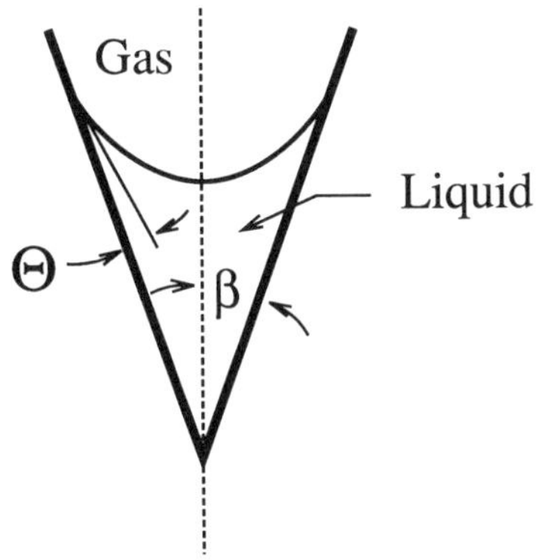

Figure 5.10.6-1: (a) Heterogeneous bubble nucleation with the contact angle Θ measured through the liquid phase. (b) Heterogeneous droplet nucleation.

With these assumptions, we find

$$\Delta(\, P^{(G)} V^{(G)} \,) = P^{(G)} \left[\frac{1}{3} \pi R^3 (\, 2 - 3m + m^3 \,) + \frac{1}{3} \pi R^3 \frac{\cos^3(\, \Theta - \beta \,)}{\tan \beta} \right] \tag{6-26}$$

$$\Delta\mathcal{A} = 2\pi R^2 (1 - m) \tag{6-27}$$

$$\Delta V^{(G)} = \frac{1}{3}\pi R^3 (2 - 3m + m^3) + \frac{1}{3}\pi R^3 \frac{\cos^3(\Theta - \beta)}{\tan\beta} \tag{6-28}$$

where

$$m \equiv \sin(\Theta - \beta) \tag{6-29}$$

At equilibrium, (6-6) and (6-7) specify

$$\Delta G = -\frac{2}{3}\pi R^{*2}\gamma \left(1 - m^3 - \frac{\cos^3(\Theta - \beta)}{\tan\beta}\right) \tag{6-30}$$

with the recognition that (6-10) must be satisfied and that Θ is determined by the properties of the system.

Small surface defects can be characterized by $\beta \to 0$. In the limit $\beta \to 0$ for $\Theta < \pi/2$, we have $\Delta G \to \infty$, and homogeneous bubble nucleation would be the dominant mode. For $\Theta > \pi/2$, we find $\Delta G \to -\infty$ and heterogeneous bubble nucleation would be dominant.

A very smooth surface can be characterized by $\beta \to \pi/2$. In the limit $\beta \to \pi/2$ for $\Theta = 0$, we find ΔG is the same as that for homogeneous bubble nucleation [see (6-11)], which suggests that both modes of nucleation should be observed. In the limit $\beta \to \pi/2$ for $\Theta = \pi/2$,

$$\Delta G = -\frac{2}{3}\pi R^{*2}\gamma \tag{6-31}$$

and homogeneous bubble nucleation would be the dominant mode. In the limit $\beta \to \pi/2$ for $\Theta = \pi$, $\Delta G = 0$ and again homogeneous bubble nucleation would be dominant.

In view of assumptions v' and ix and with the additional assumption vii, R^* is determined in this case by (6-21).

It is interesting to examine an alternative development based upon choosing the system to include the fluid-solid boundaries. With the observation that

$$\Delta(\gamma\mathcal{A}) = 2\pi R^2 \gamma (1 - m) - (\gamma^{(LS)} - \gamma^{(GS)})\Delta\mathcal{A}^{(GS)} \tag{6-32}$$

a development which parallels that for (6-30) yields at equilibrium

$$\Delta G \le \Delta(\, P^{(G)}V^{(G)}\,) - P^{(L)}\,\Delta V^{(G)} - \Delta(\,\gamma\mathcal{A}\,)$$

$$\le -\frac{2}{3}\pi R^{*2}\gamma\left(1 - m^3 - \frac{\cos^3(\,\Theta - \beta\,)}{\tan\,\beta}\right)$$

$$+ (\,\gamma^{(LS)} - \gamma^{(GS)}\,)\Delta\mathcal{A}^{(GS)} \tag{6-33}$$

Here we have defined $\gamma^{(LS)}$ to be the interfacial tension acting in the liquid solid interface, $\gamma^{(GS)}$ the interfacial tension acting in the gas-solid interface, and $\Delta\mathcal{A}^{(GS)}$ the change in the area of the gas-solid interface as the result of the bubble expanding beyond the defect:

$$\Delta\mathcal{A}^{(GS)} = \pi R^{*2}\,\frac{\cos^2(\,\Theta - \beta\,)}{\sin\,\beta} \tag{6-34}$$

Young's equation (see critique in Sec. 2.1.10) requires

$$\gamma^{(LS)} - \gamma^{(GS)} = -\,\gamma\cos\Theta_0 \tag{6-35}$$

Θ_0 is the contact angle measured at the common line, in contrast with Θ which is measured at some distance from the common line. With (6-34) and (6-35), (6-33) reduces at equilibrium to

$$\Delta G = -\frac{2}{3}\pi R^{*2}\gamma\left(1 - m^3 - \frac{\cos^3(\,\Theta - \beta\,)}{\tan\,\beta}\right.$$

$$\left. + \frac{3}{2}\cos\Theta_0\,\frac{\cos^2(\,\Theta - \beta\,)}{\sin\,\beta}\right) \tag{6-36}$$

In the limit $\beta \to 0$ for $\Theta < \pi/2$, we have $\Theta_0 < \Theta$, $\Delta G \to -\infty$, and heterogeneous bubble nucleation would be the dominant mode. In the limit $\beta \to 0$ for $\Theta > \pi/2$, we find $\Theta_0 > \Theta$, $\Delta G \to \infty$, and homogeneous bubble nucleation would be dominant. In the limit $\beta \to \pi/2$ for $\Theta = 0$, ΔG is the same as that for homogeneous bubble nucleation [see (6-11)], and both modes of nucleation should be observed. In the limit $\beta \to \pi/2$ for $\Theta < \pi/2$ and $\Theta_0 < 0.268\pi$, there is a solution Θ_{c1} of

$$-\,1 - m^3 + \frac{3}{2}\cos\Theta_0\,(\,1 - m^2\,) = 0 \tag{6-37}$$

at which homogeneous and heterogeneous nucleation also would be equally likely. For $0 < \Theta < \Theta_{c1}$, homogeneous bubble nucleation would be dominant. For $\Theta_{c1} < \Theta < \pi/2$, heterogeneous bubble nucleation would be observed. For $\Theta_0 > 0.268\pi$, only homogeneous nucleation would be seen. For $\pi/2 < \Theta \leq \pi$, there are no solutions of (6-37) for which $\Theta_0 > \Theta$, and homogeneous nucleation would be dominant.

Why are the conclusions based upon (6-36) different from those derived from (6-30)? As noted above we question the validity of Young's equation (6-35). As an alternative explanation of the same objection, we assume that interfacial tension is independent of time and position in all interfaces in proving that the Gibbs free energy of an isothermal body consisting of isobaric phases is minimized (see Exercise 5.10.3-3). The surface stress distributions in the fluid-solid interfaces are dependent upon the stress distribution in the solid (see Sec. 2.1.10), and both are altered by the presence of the three-phase line of contact. The surface stress in the solid-fluid interface would not be independent of position as assumed. For this reason, we believe that the development is flawed and that (6-30) is the preferred result. Homogeneous bubble nucleation is more likely to be observed than heterogeneous bubble nucleation.

Let us consider what actually happens when we uncap a bottle of beer. If we were to say that the glass was perfectly smooth and that $0 < \Theta < 0.268\pi$, both (6-30) and (6-36) predict that only homogeneous nucleation would be observed. If we assumed that $\Theta = 0$, (6-30) predicts that both homogeneous nucleation and heterogeneous nucleation are equally likely. If we said that $\Theta < \pi/2$ and surface defects were present, (6-36) indicates that heterogeneous nucleation would be dominant. Our experimental observations suggest that nucleation is homogeneous. [Nucleation should be distinguished from simple bubble growth as the result of mass transfer of CO_2 from the adjoining supersaturated solution. When we pour beer into a glass, bubbles are generated at isolated surface defects that were likely filled with air before the beer was introduced. This phenomena results in massive foaming, when we use a polystyrene-foam cup (which would have many surface defects filled with air). This is similar to the rapid foaming that occurs on opening a bottle of beer after vigorously shaking it. Shaking disperses in the beer many small bubbles that grow rapidly after the pressure drops.]

Heterogeneous droplet nucleation In a similar manner, we can consider heterogeneous droplet nucleation within a vapor-filled defect. Referring to Figure 6-1b, we recognize that in the initial state of the system the defect is filled with vapor. Employing assumptions i through iv, v', vi, and viii, we can develop an argument similar to that used in arriving at (6-30). We find that for heterogeneous droplet nucleation at vapor-filled defects

$$\Delta\left(P^{(L)} V^{(L)} \right) = P^{(L)} \left[-\frac{1}{3} \pi R^3 \left(2 - 3m' + m'^3 \right) + \frac{1}{3} \pi R^3 \frac{\cos^3 (\Theta + \beta)}{\tan \beta} \right] \tag{6-38}$$

$$\Delta\mathcal{A} = 2\pi R^2 (1 - m') \tag{6-39}$$

$$\Delta V^{(L)} = -\frac{1}{3}\pi R^3 (2 - 3m' + m'^3) + \frac{1}{3}\pi R^3 \frac{\cos^3(\Theta + \beta)}{\tan \beta} \tag{6-40}$$

where

$$m' \equiv \sin(\Theta + \beta) \tag{6-41}$$

Here the contact angle Θ continues to be measured through the liquid phase. At equilibrium, (6-6) and (6-7) require

$$\Delta G = -\frac{2}{3}\pi R^{*2}\gamma \left(5 - 6m' + m'^3 - \frac{\cos^3(\Theta + \beta)}{\tan \beta} \right) \tag{6-42}$$

Small surface defects can be characterized by $\beta \to 0$. In the limit $\beta \to 0$ for $\Theta < \pi/2$, we have $\Delta G \to \infty$ and homogeneous droplet nucleation would be the dominant mode. For $\Theta > \pi/2$, we find $\Delta G \to -\infty$ and heterogeneous droplet nucleation would be dominant.

A very smooth surface can be characterized by $\beta \to \pi/2$. In the limit $\beta \to \pi/2$ for $\Theta = 0$, we find $\Delta G = 0$ and only homogeneous droplet nucleation would be observed. For $\Theta \simeq 0.324\pi$, we observe that ΔG is the same as that for homogeneous nucleation [see (6-11)], which suggests that both modes of nucleation should be observed. For $\Theta < 0.324\pi$, only homogeneous nucleation is likely to be observed; for $\Theta > 0.324\pi$, heterogeneous nucleation will be dominant.

Employing assumptions v' and vii, we find that, for $\Theta < \pi/2$, the equilibrium radius R* is given by

$$R^* = 2\gamma \left[\frac{\mathcal{R}T}{\hat{V}^{(L)}M} \ln\left(\frac{P_v}{P^{(G)}} \right) \right]^{-1} \tag{6-43}$$

For $\Theta > \pi/2$, R* is determined by (6-22).

As we did in the case of heterogeneous bubble nucleation, let us examine an alternative development based upon choosing the system to include the fluid-solid boundaries. With the observation that

$$\Delta(\gamma\mathcal{A}) = 2\pi R^2\gamma (1 - m') + (\gamma^{(LS)} - \gamma^{(GS)})\Delta\mathcal{A}^{(LS)} \tag{6-44}$$

a development which parallels that for (6-42) yields at equilibrium

$$\Delta G \leq \Delta(\, P^{(L)} V^{(L)}\,) - P^{(G)}\, \Delta V^{(L)} - \Delta(\, \gamma \mathcal{A}\,)$$

$$\leq -\frac{2}{3}\, \pi R^{*2} \gamma \left(5 - 6m' + m'^3 - \frac{\cos^3(\, \Theta + \beta\,)}{\tan\, \beta} \right)$$

$$- (\, \gamma^{(LS)} - \gamma^{(GS)}\,) \Delta \mathcal{A}^{(LS)} \tag{6-45}$$

Here $\Delta \mathcal{A}^{(LS)}$ is the change in the area of the gas-solid interface as the result of the bubble expanding beyond the defect:

$$\Delta \mathcal{A}^{(GS)} = \pi R^{*2} \frac{\cos^2(\, \Theta + \beta\,)}{\sin\, \beta} \tag{6-46}$$

With (6-35) and (6-46), (6-45) reduces at equilibrium to

$$\Delta G = -\frac{2}{3}\, \pi R^{*2} \gamma \left(5 - 6m' + m'^3 - \frac{\cos^3(\, \Theta + \beta\,)}{\tan\, \beta} \right.$$

$$\left. - \frac{3}{2} \cos \Theta_0 \frac{\cos^2(\, \Theta + \beta\,)}{\sin\, \beta} \right) \tag{6-47}$$

In the limit $\beta \to 0$ for $\Theta < \pi/2$, we have $\Theta_0 < \Theta$, $\Delta G \to \infty$, and homogeneous droplet nucleation would be the dominant mode. In the limit $\beta \to 0$ for $\Theta > \pi/2$, we find $\Theta_0 > \Theta$, $\Delta G \to -\infty$, and heterogeneous droplet nucleation would be dominant. In the limit $\beta \to \pi/2$ for $\Theta = 0$, $\Delta G = 0$ and only homogeneous droplet nucleation would be observed. At some $\Theta_{c2} < \pi/2$ satisfying

$$3 - 6m' + m'^3 - \frac{3}{2} \cos \Theta_0\, (\, 1 - m'^2\,) = 0 \tag{6-48}$$

both homogeneous and heterogeneous droplet nucleation are equally likely. In the limit $\beta \to \pi/2$, for $0 \leq \Theta < \Theta_{c2}$ only homogeneous droplet nucleation would be observed. For $\Theta_{c2} < \Theta \leq \pi$, heterogeneous droplet nucleation would be the dominant mode.

The conclusions based upon (6-47) differ from those derived upon (6-42) for the same reason that those based upon (6-36) disagree with those derived from (6-30) (see last paragraph of ***Heterogeneous bubble nucleation***). Equations (6-36) and (6-47) employ Young's equation; (6-30) and (6-42) do not.

Commonly we experience foggy air (resulting from homogeneous nucleation) in the bathroom after a hot shower. Usually the walls will have been designed such that $\Theta < 0.324\pi$. Many small surface defects can be expected to be present in the walls, but relatively few on the mirror. Such observations can be explained either using the conclusions of (6-42) or those of (6-47). The relatively small amount of condensation on the walls and the mirror are likely attributable to condensation driven by their somewhat lower temperatures, analogous to the condensation which collects on a cold bottle of beer. This is not taken into account here, since we have assumed that the system is isothermal system.

Summary Several conclusions can be drawn from this discussion.

1) Let us assume that very small ($\beta \to 0$) surface defects are present. For $\Theta < \pi/2$ (where Θ is measured through the liquid phase), homogeneous nucleation is the dominant mode in boiling and in condensation. For $\Theta > \pi/2$, heterogeneous nucleation is dominant.

2) For boiling on very smooth ($\beta \to \pi/2$) surfaces, homogeneous bubble nucleation would be the dominant mode, except in the limit $\Theta \to 0$ where both homogeneous and heterogeneous nucleation appear to be equally likely.

3) For condensation on very smooth ($\beta \to \pi/2$), homogeneous droplet nucleation will be the dominant mode for $\Theta < 0.324\pi$; for $\Theta > 0.324\pi$, heterogeneous droplet nucleation will be observed.

But note how the acceptance of Young's equation (see last paragraph of ***Heterogeneous bubble nucleation***) alters these conclusions:

1') If very small ($\beta \to 0$) surface defects are present, for $\Theta < \pi/2$ heterogeneous bubble nucleation would be the dominant mode in boiling and homogeneous droplet nucleation in condensation. For $\Theta > \pi/2$, homogeneous nucleation would be the dominant mode in boiling and heterogeneous nucleation in condensation.

2') For boiling on very smooth ($\beta \to \pi/2$) surfaces, in the limit $\Theta \to 0$, both modes of nucleation would be equally likely. For $\Theta < \pi/2$ and $\Theta_0 < 0.268\pi$, there is a solution Θ_{c1} of (6-37) at which homogeneous and heterogeneous nucleation also would be equally likely. For $0 < \Theta < \Theta_{c1}$, homogeneous bubble nucleation would be dominant. For $\Theta_{c1} < \Theta < \pi/2$, heterogeneous bubble nucleation would be observed. For $\Theta_0 > 0.268\pi$, only homogeneous nucleation would be seen. For $\pi/2 < \Theta \leq \pi$, there are no solutions of (6-37) for which $\Theta_0 > \Theta$, and homogeneous nucleation would be dominant.

3') For condensation on very smooth ($\beta \to \pi/2$) surfaces, with $0 \leq \Theta < \Theta_{c2}$ only homogeneous droplet nucleation would be observed. For $\Theta_{c2} < \Theta \leq \pi$, heterogeneous droplet nucleation would be the dominant mode.

References

Blander, M., and J. L. Katz, *AIChE J.* **21**, 833 (1975).

Cole, R., *Adv. Heat Transfer*, edited by J. P. Hartnett and T. F. Irvine Jr., **10**, 85 (1974).

Feder, J., K. C. Russell, J. Lothe, and G. M. Pound, *Adv. Phys.* **15**, 111 (1966).

Frenkel, J., "Kinetic Theory of Liquids," chapter VII, Oxford University Press, Oxford (1946).

Gibbs, J. W., "The Collected Works," vol. 1, Yale University Press, New Haven, CT (1948).

Hodgson, A. W., *Adv. Colloid Interface Sci.* **21**, 303 (1984).

Narsimhan, G., and E. Ruckenstein, *J. Colloid Interface Sci.* **128**, 549 (1989).

Ward, C. A., A. Balakrishnan, and F. C. Hooper, "Symposium on the Role of Nucleation in Boiling and Cavitation," Paper No. 70-FE-26, The American Society of Mechanical Engineers, New York (May, 1970).

Wilt, P. M., *J. Colloid Interface Sci.* **112**, 530 (1986).

Exercise

5.10.6-1 *multicomponent systems* The development in the text leading to (6-22) and (6-23) assumes that the system is composed of only one component. For a multicomponent system, the argument must be revised.

We can begin by observing that for species A

$$\left(\frac{\partial \mu_{(A)}^{(i)}}{\partial P} \right)_T = \overline{\overline{V}}_{(A)}^{(i)} \tag{1}$$

where $\overline{\overline{V}}_{(A)}^{(i)}$ is the partial mass volume of species A in phase i (Exercise 5.8.3-2). Integrating, we find

$$\mu_{(A)}^{(G)}(P^{(G)}, T) = \mu_{(A)}^{(G)}(P_v , T) + \int_{P_v}^{P^{(G)}} \overline{\overline{V}}_{(A)}^{(G)} \, dP \tag{2}$$

$$\mu_{\{A\}}^{(L)}(P^{(L)}, T) = \mu_{\{A\}}^{(L)}(P_v, T) + \int_{P_v}^{P^{(L)}} \overline{\overline{V}}_{\{A\}}^{(L)} \, dP \tag{3}$$

For equilibrium at the surface of the bubble or droplet

$$\mu_{\{A\}}^{(L)}(P^{(L)}, T) = \mu_{\{A\}}^{(G)}(P^{(G)}, T) \tag{4}$$

and for equilibrium at a plane interface at the same temperature T

$$\mu_{\{A\}}^{(L)}(P_v, T) = \mu_{\{A\}}^{(G)}(P_v, T) \tag{5}$$

From (6-13) through (6-16),

$$\int_{P_v}^{P^{(G)}} \overline{\overline{V}}_{\{A\}}^{(G)} \, dP = \int_{P_v}^{P^{(G)}} \overline{\overline{V}}_{\{A\}}^{(L)} \, dP - \int_{P^{(L)}}^{P^{(G)}} \overline{\overline{V}}_{\{A\}}^{(L)} \, dP \tag{6}$$

If we are willing to assume that $\overline{\overline{V}}_{\{A\}}^{(L)}$ is nearly constant between $P^{(L)}$ and $P^{(G)}$, this reduces in view of (6-7) to

$$\int_{P_v}^{P^{(G)}} \left(\overline{\overline{V}}_{\{A\}}^{(G)} - \overline{\overline{V}}_{\{A\}}^{(L)} \right) dP = - \overline{\overline{V}}_{\{A\}}^{(L)} \left(P^{(G)} - P^{(L)} \right)$$

$$= \pm \frac{2\gamma}{R^*} \overline{\overline{V}}^{(L)} \tag{7}$$

Here the + sign assumes that the liquid is the discontinuous phase; the – sign that the gas is the discontinuous phase. For a discontinuous gas phase, we must recognize from (6-7) that

$$P^{(G)} = P^{(L)} + \frac{2\gamma}{R^*} \tag{8}$$

in order to eliminate $P^{(G)}$ from (7) before solving for R^*. For a discontinuous liquid phase, (7) can be rearranged to give

$$R^* = 2\gamma \, \overline{\overline{V}}^{(L)} \left[\int_{P_v}^{P^{(G)}} \left(\overline{\overline{V}}_{\{A\}}^{(G)} - \overline{\overline{V}}_{\{A\}}^{(L)} \right) dP \right]^{-1} \tag{9}$$

5.11 Summary

5.11.1 Summary of useful equations This is actually the third summary of useful relationships and equations presented in this chapter. In Tables 5.2.2-1 through 5.2.2-5 I examined the various concentrations, velocities, and mass fluxes that one might wish to use as well as a number of alternative forms for the jump mass balance for species A. Tables 5.8.9-1 through 5.8.9-3 list alternative forms for the jump energy balance, the jump Clausius-Duhem inequality, and the differential Clausius-Duhem inequality.

Here we concentrate on the forms that the jump mass balance for species A and the jump energy balance take for a few simple interfacial configurations. These should be thought of as extensions of Tables 2.3.1-1 through 2.3.1-8. In fact, you may wish to refer back to those tables for further details.

It is impractical to list even for this restricted set of configurations all possible forms of these equations that may be useful in specific contexts. Those given here are representative in the sense that the reader may use them as guides in immediately writing the various other forms of the jump mass balance for species A in Table 5.2.2-5 and of the jump energy balance in Table 5.8.9-1.

In order to use the jump species mass balance and the jump energy balance, we must adopt specific constitutive equations for the surface mass flux vector and for the surface energy flux vector. For purposes of illustration, I have chosen to use the simplest such relationships in these tables. I have limited myself to two component materials and I have neglected the direct dependence of $\mathbf{j}_{(A)}^{(\sigma)}$ upon the surface gradient of the surface temperature and upon the external forces in stating from Sec. 5.9.8 that

$$\mathbf{j}_{(A)}^{(\sigma)} = - \mathcal{D}_{(AB)}^{(\sigma)} \nabla_{(\sigma)} \mu_{(A)}^{(\sigma)} \tag{1-1}$$

Similarly, I have neglected the direct dependence of $\mathbf{e}^{(\sigma)}$ upon the surface gradients of the surface chemical potentials and upon the external forces in saying from Sec. 5.9.7 that

$$\mathbf{e}^{(\sigma)} = \mathbf{q}^{(\sigma)} - \sum_{B=1}^{N} \mu_{(B)}^{(\sigma)} \mathbf{j}_{(B)}^{(\sigma)} = - k^{(\sigma)} \nabla T^{(\sigma)} \tag{1-2}$$

If any of these assumptions prove to be inappropriate for situations that you are considering, it should be straightforward to use the tabulated components of (1-1) and (1-2) as guides in writing the components of more general constitutive equations.

Table 5.11.1-1: Stationary plane dividing surface viewed in a rectangular cartesian coordinate system (refer to Exercises A.2.1-4 and A.5.3-5 and to Table 2.4.2-1)

dividing surface

z_3 = a constant

surface coordinates

$y^1 \equiv z_1 \qquad y^2 \equiv z_2$

jump mass balance for species A (Table 5.2.2-5, equation C)

$$\rho^{(\sigma)}\left(\frac{\partial \omega^{(\sigma)}_{(A)}}{\partial t} + v_1^{(\sigma)}\frac{\partial \omega^{(\sigma)}_{(A)}}{\partial z_1} + v_2^{(\sigma)}\frac{\partial \omega^{(\sigma)}_{(A)}}{\partial z_2}\right)$$

$$+ \frac{\partial j^{(\sigma)}_{(A)1}}{\partial z_1} + \frac{\partial j^{(\sigma)}_{(A)2}}{\partial z_2} - r^{(\sigma)}_{(A)}$$

$$+ \left[\, j_{(A)3}\xi_3 + \rho(\,\omega_{(A)} - \omega^{(\sigma)}_{(A)}\,)v_3\xi_3\,\right] = 0$$

components of $j^{(\sigma)}_{(A)}$ ***when* (1-1) *applies for two-component materials***

$$j^{(\sigma)}_{(A)1} = -\,\mathcal{D}^{(\sigma)}_{(AB)}\frac{\partial \mu^{(\sigma)}_{(A)}}{\partial z_1}$$

$$j^{(\sigma)}_{(A)2} = -\,\mathcal{D}^{(\sigma)}_{(AB)}\frac{\partial \mu^{(\sigma)}_{(A)}}{\partial z_2}$$

jump energy balance (Table 5.8.9-1, equation F)

$$\rho^{(\sigma)}\hat{c}_{\gamma}^{(\sigma)}\frac{d_{(v(\sigma))}T^{(\sigma)}}{dt} = -\left(\frac{\partial q_1^{(\sigma)}}{\partial z_1} + \frac{\partial q_2^{(\sigma)}}{\partial z_2}\right)$$

$$-\left(\frac{\partial \ln \hat{\mathcal{A}}}{\partial \ln T^{(\sigma)}}\right)_{\gamma,\,\omega^{(\sigma)}_{(B)}}\frac{d_{(v(\sigma))}\gamma}{dt}$$

$$+ S_{11}^{(\sigma)} D_{11}^{(\sigma)} + 2 S_{12}^{(\sigma)} D_{21}^{(\sigma)} + S_{22}^{(\sigma)} D_{22}^{(\sigma)}$$

$$+ \sum_{A=1}^{N} \left(j_{(A)1}^{(\sigma)} b_{(A)1}^{(\sigma)} + j_{(A)2}^{(\sigma)} b_{(A)2}^{(\sigma)} \right) + \rho^{(\sigma)} Q^{(\sigma)}$$

$$- \sum_{A=1}^{N} \left[\mu_{(A)}^{(\sigma)} - T^{(\sigma)} \left(\frac{\partial \mu_{(A)}^{(\sigma)}}{\partial T^{(\sigma)}} \right)_{\gamma, \omega_{(B)}^{(\sigma)}} \right] \frac{d_{(v^{(\sigma)})} \omega_{(A)}^{(\sigma)}}{dt}$$

$$+ \left[- \rho (\hat{U} - \hat{H}^{(\sigma)}) v_3 \xi_3 - \frac{1}{2} \rho (v_3 \xi_3)^3 + v_3 T_{33} \xi_3 - q_3 \xi_3 \right]$$

Here

$$\frac{d_{(v^{(\sigma)})} T^{(\sigma)}}{dt}, \quad \frac{d_{(v^{(\sigma)})} \gamma}{dt}, \quad \text{and} \quad \frac{d_{(v^{(\sigma)})} \omega_{(A)}^{(\sigma)}}{dt}$$

have similar forms, for example

$$\frac{d_{(v^{(\sigma)})} T^{(\sigma)}}{dt} = \frac{\partial T^{(\sigma)}}{\partial t} + v_1^{(\sigma)} \frac{\partial T^{(\sigma)}}{\partial z_1} + v_2^{(\sigma)} \frac{\partial T^{(\sigma)}}{\partial z_2}$$

components of* $\mathbf{q}^{(\sigma)}$ *when* (1-2) *applies

$$q_1^{(\sigma)} = - k^{(\sigma)} \frac{\partial T^{(\sigma)}}{\partial z_1} + \sum_{B=1}^{N} \mu_{(B)}^{(\sigma)} j_{(B)1}^{(\sigma)}$$

$$q_2^{(\sigma)} = - k^{(\sigma)} \frac{\partial T^{(\sigma)}}{\partial z_2} + \sum_{B=1}^{N} \mu_{(B)}^{(\sigma)} j_{(B)2}^{(\sigma)}$$

Table 5.11.1-2: Stationary plane dividing surface viewed in a cylindrical coordinate system (refer to Exercises A.2.1-5 and A.5.3-6 and to Table 2.4.2-2)

dividing surface

z = a constant

surface coordinates

$y^1 \equiv r \qquad y^2 \equiv \theta$

jump mass balance for species A (Table 5.2.2-5, equation C)

$$\rho^{(\sigma)} \left(\frac{\partial \omega^{(\sigma)}_{(A)}}{\partial t} + v_r^{(\sigma)} \frac{\partial \omega^{(\sigma)}_{(A)}}{\partial r} + \frac{v_\theta^{(\sigma)}}{r} \frac{\partial \omega^{(\sigma)}_{(A)}}{\partial \theta} \right) + \frac{1}{r} \frac{\partial}{\partial r} \left(r \, j^{(\sigma)}_{(A)r} \right)$$

$$+ \frac{1}{r} \frac{\partial j^{(\sigma)}_{(A)\theta}}{\partial \theta} - r^{(\sigma)}_{(A)} + [\, j_{(A)z} \xi_z + \rho (\omega_{(A)} - \omega^{(\sigma)}_{(A)}) v_z \xi_z \,] = 0$$

components of $j^{(\sigma)}_{(A)}$ ***when*** **(1-1)** ***applies for two-component materials***

$$j^{(\sigma)}_{(A)r} = - \mathcal{D}^{(\sigma)}_{(AB)} \frac{\partial \mu^{(\sigma)}_{(A)}}{\partial r}$$

$$j^{(\sigma)}_{(A)\theta} = - \mathcal{D}^{(\sigma)}_{(AB)} \frac{1}{r} \frac{\partial \mu^{(\sigma)}_{(A)}}{\partial \theta}$$

jump energy balance (Table 5.8.9-1, equation F)

$$\rho^{(\sigma)} \hat{c}^{(\sigma)}_{\gamma} \frac{d_{(v^{(\sigma)})} T^{(\sigma)}}{dt} = - \left[\frac{1}{r} \frac{\partial}{\partial r} (r \, q_r^{(\sigma)}) + \frac{1}{r} \frac{\partial q_\theta^{(\sigma)}}{\partial \theta} \right]$$

$$- \left(\frac{\partial \ln \hat{\mathcal{A}}}{\partial \ln T^{(\sigma)}} \right)_{\gamma, \omega^{(\sigma)}_{(B)}} \frac{d_{(v^{(\sigma)})} \gamma}{dt}$$

$$+ S_{rr}^{(\sigma)} D_{rr}^{(\sigma)} + 2 S_{r\theta}^{(\sigma)} D_{\theta r}^{(\sigma)} + S_{\theta\theta}^{(\sigma)} D_{\theta\theta}^{(\sigma)}$$

$$+ \sum_{A=1}^{N} \left(j_{(A)r}^{(\sigma)} b_r^{(\sigma)} + j_{(A)\theta}^{(\sigma)} b_{(A)\theta}^{(\sigma)} \right) + \rho^{(\sigma)} Q^{(\sigma)}$$

$$- \sum_{A=1}^{N} \left[\mu_{(A)}^{(\sigma)} - T^{(\sigma)} \left(\frac{\partial \mu_{(A)}^{(\sigma)}}{\partial T^{(\sigma)}} \right)_{\gamma, \omega_{(B)}^{(\sigma)}} \right] \frac{d_{(v^{(\sigma)})} \omega_{(A)}^{(\sigma)}}{dt}$$

$$+ \left[- \rho (\hat{U} - \hat{H}^{(\sigma)}) v_z \xi_z - \frac{1}{2} \rho (v_z \xi_z)^3 + v_z T_{zz} \xi_z - q_z \xi_z \right]$$

Here

$$\frac{d_{(v^{(\sigma)})} T^{(\sigma)}}{dt}, \quad \frac{d_{(v^{(\sigma)})} \gamma}{dt}, \quad \text{and} \quad \frac{d_{(v^{(\sigma)})} \omega_{(A)}^{(\sigma)}}{dt}$$

have similar forms, for example

$$\frac{d_{(v^{(\sigma)})} T^{(\sigma)}}{dt} = \frac{\partial T^{(\sigma)}}{\partial t} + v_r^{(\sigma)} \frac{\partial T^{(\sigma)}}{\partial r} + \frac{v_\theta^{(\sigma)}}{r} \frac{\partial T^{(\sigma)}}{\partial \theta}$$

components of* q$^{(\sigma)}$ *when* (1-2) *applies

$$q_r^{(\sigma)} = - k^{(\sigma)} \frac{\partial T^{(\sigma)}}{\partial r} + \sum_{B=1}^{N} \mu_{(B)}^{(\sigma)} j_{(B)r}^{(\sigma)}$$

$$q_\theta^{(\sigma)} = - k^{(\sigma)} \frac{1}{r} \frac{\partial T^{(\sigma)}}{\partial \theta} + \sum_{B=1}^{N} \mu_{(B)}^{(\sigma)} j_{(B)\theta}^{(\sigma)}$$

5.11.1-3: Alternative form for stationary plane dividing surface viewed in a cylindrical coordinate system (refer to Exercises A.2.1-6 and A.5.3-7 and to Table 2.4.2-3)

dividing surface

$$\theta = \text{a constant}$$

surface coordinates

$$y^1 = r \qquad y^2 = z$$

jump mass balance for species A (Table 5.2.2-5, equation C)

$$\rho^{(\sigma)} \left(\frac{\partial \omega_{(A)}^{(\sigma)}}{\partial t} + v_r^{(\sigma)} \frac{\partial \omega_{(A)}^{(\sigma)}}{\partial r} + v_z^{(\sigma)} \frac{\partial \omega_{(A)}^{(\sigma)}}{\partial z} \right) + \frac{\partial j_{(A)r}^{(\sigma)}}{\partial r} + \frac{\partial j_{(A)z}^{(\sigma)}}{\partial z}$$

$$- r_{(A)}^{(\sigma)} + [\, j_{(A)\theta} \xi_\theta + \rho (\omega_{(A)} - \omega_{(A)}^{(\sigma)}) v_\theta \xi_\theta \,] = 0$$

components of $j_{(A)}^{(\sigma)}$ ***when*** **(1-1)** ***applies for two-component materials***

$$j_{(A)r}^{(\sigma)} = - \mathcal{D}_{(AB)}^{(\sigma)} \frac{\partial \mu_{(A)}^{(\sigma)}}{\partial r}$$

$$j_{(A)z}^{(\sigma)} = - \mathcal{D}_{(AB)}^{(\sigma)} \frac{\partial \mu_{(A)}^{(\sigma)}}{\partial z}$$

jump energy balance (Table 5.8.9-1, equation F)

$$\rho^{(\sigma)} \hat{c}_\gamma^{(\sigma)} \frac{d_{(v^{(\sigma)})} T^{(\sigma)}}{dt} = - \left(\frac{\partial q_r^{(\sigma)}}{\partial r} + \frac{\partial q_z^{(\sigma)}}{\partial z} \right)$$

$$- \left(\frac{\partial \ln \hat{\mathcal{A}}}{\partial \ln T^{(\sigma)}} \right)_{\gamma, \omega_{(B)}^{(\sigma)}} \frac{d_{(v^{(\sigma)})} \gamma}{dt}$$

$$+ S_{rr}^{(\sigma)} D_{rr}^{(\sigma)} + 2 S_{rz}^{(\sigma)} D_{zr}^{(\sigma)} + S_{zz}^{(\sigma)} D_{zz}^{(\sigma)}$$

$$+ \sum_{A=1}^{N} \left(j_{(A)r}^{(\sigma)} \, b_{(A)r}^{(\sigma)} + j_{(A)z}^{(\sigma)} \, b_{(A)z}^{(\sigma)} \right) + \rho^{(\sigma)} Q^{(\sigma)}$$

$$- \sum_{A=1}^{N} \left[\mu_{(A)}^{(\sigma)} - T^{(\sigma)} \left(\frac{\partial \mu_{(A)}^{(\sigma)}}{\partial T^{(\sigma)}} \right)_{\gamma, \omega_{(B)}^{(\sigma)}} \right] \frac{d_{(v^{(\sigma)})} \omega_{(A)}^{(\sigma)}}{dt}$$

$$+ \left[- \rho (\hat{U} - \hat{H}^{(\sigma)}) v_\theta \xi_\theta - \frac{1}{2} \rho (v_\theta \xi_\theta)^3 + v_\theta T_{\theta\theta} \xi_\theta - q_\theta \xi_\theta \right]$$

Here

$$\frac{d_{(v^{(\sigma)})} T^{(\sigma)}}{dt}, \quad \frac{d_{(v^{(\sigma)})} \gamma}{dt}, \quad \text{and} \quad \frac{d_{(v^{(\sigma)})} \omega_{(A)}^{(\sigma)}}{dt}$$

have similar forms, for example

$$\frac{d_{(v^{(\sigma)})} T^{(\sigma)}}{dt} = \frac{\partial T^{(\sigma)}}{\partial t} + v_r^{(\sigma)} \frac{\partial T^{(\sigma)}}{\partial r} + v_z^{(\sigma)} \frac{\partial T^{(\sigma)}}{\partial z}$$

components of* $\mathbf{q}^{(\sigma)}$ *when* (1-2) *applies

$$q_r^{(\sigma)} = - k^{(\sigma)} \frac{\partial T^{(\sigma)}}{\partial r} + \sum_{B=1}^{N} \mu_{(B)}^{(\sigma)} \, j_{(B)r}^{(\sigma)}$$

$$q_z^{(\sigma)} = - k^{(\sigma)} \frac{\partial T^{(\sigma)}}{\partial z} + \sum_{B=1}^{N} \mu_{(B)}^{(\sigma)} \, j_{(B)z}^{(\sigma)}$$

Table 5.11.1-4: Stationary cylindrical dividing surface viewed in a cylindrical coordinate system (refer to Exercises A.2.1-7 and A.5.3-8 and to Table 2.4.2-4)

dividing surface

$$r = R$$

surface coordinates

$$y^1 \equiv \theta \qquad y^2 \equiv z$$

jump mass balance for species A (Table 5.2.2-5, equation C)

$$\rho^{(\sigma)} \left(\frac{\partial \omega_{(A)}^{(\sigma)}}{\partial t} + \frac{v_\theta^{(\sigma)}}{R} \frac{\partial \omega_{(A)}^{(\sigma)}}{\partial \theta} + v_z^{(\sigma)} \frac{\partial \omega_{(A)}^{(\sigma)}}{\partial z} \right) + \frac{1}{R} \frac{\partial j_{(A)\theta}^{(\sigma)}}{\partial \theta}$$

$$+ \frac{\partial j_{(A)z}^{(\sigma)}}{\partial z} - r_{(A)}^{(\sigma)} + [\, j_{(A)r} \xi_r + \rho (\omega_{(A)} - \omega_{(A)}^{(\sigma)}) v_r \xi_r \,] = 0$$

components of $j_{(A)}^{(\sigma)}$ ***when* (1-1) *applies for two-component materials***

$$j_{(A)\theta}^{(\sigma)} = - \mathcal{D}_{(AB)}^{(\sigma)} \frac{1}{R} \frac{\partial \mu_{(A)}^{(\sigma)}}{\partial \theta}$$

$$j_{(A)z}^{(\sigma)} = - \mathcal{D}_{(AB)}^{(\sigma)} \frac{\partial \mu_{(A)}^{(\sigma)}}{\partial z}$$

jump energy balance (Table 5.8.9-1, equation F)

$$\rho^{(\sigma)} \hat{c}_\gamma^{(\sigma)} \frac{d_{(v^{(\sigma)})} T^{(\sigma)}}{dt} = - \left(\frac{1}{R} \frac{\partial q_\theta^{(\sigma)}}{\partial \theta} + \frac{\partial q_z^{(\sigma)}}{\partial z} \right)$$

$$- \left(\frac{\partial \ln \hat{\mathcal{A}}}{\partial \ln T^{(\sigma)}} \right)_{\gamma, \omega_{(B)}^{(\sigma)}} \frac{d_{(v^{(\sigma)})} \gamma}{dt}$$

$$+ S_{\theta\theta}^{(\sigma)} D_{\theta\theta}^{(\sigma)} + 2 S_{\theta z}^{(\sigma)} D_{z\theta}^{(\sigma)} + S_{zz}^{(\sigma)} D_{zz}^{(\sigma)}$$

$$+\sum_{A=1}^{N}\left(j_{(A)\theta}^{(\sigma)}\, b_{(A)\theta}^{(\sigma)} + j_{(A)z}^{(\sigma)}\, b_{(A)z}^{(\sigma)} \right) + \rho^{(\sigma)} Q^{(\sigma)}$$

$$-\sum_{A=1}^{N}\left[\mu_{(A)}^{(\sigma)} - T^{(\sigma)} \left(\frac{\partial \mu_{(A)}^{(\sigma)}}{\partial T^{(\sigma)}} \right)_{\gamma,\omega_{(B)}^{(\sigma)}} \right] \frac{d_{(v^{(\sigma)})}\omega_{(A)}^{(\sigma)}}{dt}$$

$$+\left[-\rho(\hat{U} - \hat{H}^{(\sigma)})v_r\xi_r - \frac{1}{2}\rho(v_r\xi_r)^3 + v_r T_{rr}\xi_r - q_r\xi_r \right]$$

Here

$$\frac{d_{(v^{(\sigma)})}T^{(\sigma)}}{dt}, \quad \frac{d_{(v^{(\sigma)})}\gamma}{dt}, \quad \text{and} \quad \frac{d_{(v^{(\sigma)})}\omega_{(A)}^{(\sigma)}}{dt}$$

have similar forms, for example

$$\frac{d_{(v^{(\sigma)})}T^{(\sigma)}}{dt} = \frac{\partial T^{(\sigma)}}{\partial t} + \frac{v_\theta^{(\sigma)}}{R}\frac{\partial T^{(\sigma)}}{\partial \theta} + v_z^{(\sigma)}\frac{\partial T^{(\sigma)}}{\partial z}$$

components of $q^{(\sigma)}$ ***when* (1-2) *applies***

$$q_\theta^{(\sigma)} = -k^{(\sigma)}\frac{1}{R}\frac{\partial T^{(\sigma)}}{\partial \theta} + \sum_{B=1}^{N}\mu_{(B)}^{(\sigma)}\, j_{(B)\theta}^{(\sigma)}$$

$$q_z^{(\sigma)} = -k^{(\sigma)}\frac{\partial T^{(\sigma)}}{\partial z} + \sum_{B=1}^{N}\mu_{(B)}^{(\sigma)}\, j_{(B)z}^{(\sigma)}$$

Table 5.11.1-5: Stationary spherical dividing surface viewed in a spherical coordinate system (refer to Exercises A.2.1-8 and A.5.3-9 and to Table 2.4.2-5)

dividing surface

$$r = R$$

surface coordinates

$$y^1 \equiv \theta \qquad y^2 \equiv \phi$$

jump mass balance for species A (Table 5.2.2-5, equation C)

$$\rho^{(\sigma)} \left(\frac{\partial \omega_{(A)}^{(\sigma)}}{\partial t} + \frac{v_\theta^{(\sigma)}}{R} \frac{\partial \omega_{(A)}^{(\sigma)}}{\partial \theta} + \frac{v_\phi^{(\sigma)}}{R \sin \theta} \frac{\partial \omega_{(A)}^{(\sigma)}}{\partial \phi} \right)$$

$$+ \frac{1}{R \sin \theta} \frac{\partial}{\partial \theta} (j_{(A)\theta}^{(\sigma)} \sin \theta) + \frac{1}{R \sin \theta} \frac{\partial j_{(A)\phi}^{(\sigma)}}{\partial \phi}$$

$$- r_{(A)}^{(\sigma)} + [j_{(A)r} \xi_r + \rho (\omega_{(A)} - \omega_{(A)}^{(\sigma)}) v_r \xi_r] = 0$$

components of $\mathbf{j}_{(A)}^{(\sigma)}$ ***when*** **(1-1)** ***applies for two-component materials***

$$j_{(A)\theta}^{(\sigma)} = - \mathcal{D}_{(AB)}^{(\sigma)} \frac{1}{R} \frac{\partial \mu_{(A)}^{(\sigma)}}{\partial \theta}$$

$$j_{(A)\phi}^{(\sigma)} = - \mathcal{D}_{(AB)}^{(\sigma)} \frac{1}{R \sin \theta} \frac{\partial \mu_{(A)}^{(\sigma)}}{\partial \phi}$$

jump energy balance (Table 5.8.9-1, equation F)

$$\rho^{(\sigma)} \hat{c}_\gamma^{(\sigma)} \frac{d_{(v^{(\sigma)})} T^{(\sigma)}}{dt} = - \left[\frac{1}{R \sin \theta} \frac{\partial}{\partial \theta} (q_\theta^{(\sigma)} \sin \theta) + \frac{1}{R \sin \theta} \frac{\partial q_\phi^{(\sigma)}}{\partial \phi} \right]$$

$$- \left(\frac{\partial \ln \hat{A}}{\partial \ln T^{(\sigma)}} \right)_{\gamma, \omega_{(B)}^{(\sigma)}} \frac{d_{(v^{(\sigma)})} \gamma}{dt}$$

$$+ S_{\theta\theta}^{(\sigma)} D_{\theta\theta}^{(\sigma)} + 2S_{\theta\phi}^{(\sigma)} D_{\theta\phi}^{(\sigma)} + S_{\phi\phi}^{(\sigma)} D_{\phi\phi}^{(\sigma)}$$

$$+ \sum_{A=1}^{N} \left(j_{(A)\theta}^{(\sigma)} b_{(A)\theta}^{(\sigma)} + j_{(A)\phi}^{(\sigma)} b_{(A)\phi}^{(\sigma)} \right) + \rho^{(\sigma)} Q^{(\sigma)}$$

$$- \sum_{A=1}^{N} \left[\mu_{(A)}^{(\sigma)} - T^{(\sigma)} \left(\frac{\partial \mu_{(A)}^{(\sigma)}}{\partial T^{(\sigma)}} \right)_{\gamma, \omega_{(B)}^{(\sigma)}} \right] \frac{d_{(v(\sigma))} \omega_{(A)}^{(\sigma)}}{dt}$$

$$+ \left[- \rho(\hat{U} - \hat{H}^{(\sigma)}) v_r \xi_r - \frac{1}{2} \rho(v_r \xi_r)^3 + v_r T_{rr} \xi_r - q_r \xi_r \right]$$

Here

$$\frac{d_{(v(\sigma))} T^{(\sigma)}}{dt}, \quad \frac{d_{(v(\sigma))} \gamma}{dt}, \quad \text{and} \quad \frac{d_{(v(\sigma))} \omega_{(A)}^{(\sigma)}}{dt}$$

have similar forms, for example

$$\frac{d_{(v(\sigma))} T^{(\sigma)}}{dt} = \frac{\partial T^{(\sigma)}}{\partial t} + \frac{v_\theta^{(\sigma)}}{R} \frac{\partial T^{(\sigma)}}{\partial \theta} + \frac{v_\phi^{(\sigma)}}{R \sin \theta} \frac{\partial T^{(\sigma)}}{\partial \phi}$$

components of $\mathbf{q}^{(\sigma)}$ ***when* (1-2) *applies***

$$q_\theta^{(\sigma)} = - k^{(\sigma)} \frac{1}{R} \frac{\partial T^{(\sigma)}}{\partial \theta} + \sum_{B=1}^{N} \mu_{(B)}^{(\sigma)} j_{(B)\theta}^{(\sigma)}$$

$$q_\phi^{(\sigma)} = - k^{(\sigma)} \frac{1}{R \sin \theta} \frac{\partial T^{(\sigma)}}{\partial \phi} + \sum_{B=1}^{N} \mu_{(B)}^{(\sigma)} j_{(B)\phi}^{(\sigma)}$$

Table 5.11.1-6: Deforming two-dimensional surface viewed in a rectangular cartesian coordinate system (refer to Exercises A.2.1-9, A.2.6-1, and A.5.3-10 and to Table 2.4.2-6)

dividing surface

$$z_3 = h(\, z_1, t\,)$$

assumptions (see also Table 2.4.2-6)

$$\omega_{(A)}^{(\sigma)} = \omega_{(A)}^{(\sigma)}(\, z_1, t\,) \qquad A = 1, \ldots, N$$

$$T^{(\sigma)} = T^{(\sigma)}(\, z_1, t\,)$$

$$\mathbf{j}_{(A)}^{(\sigma)} = \bar{j}_{(A)\alpha}^{(\sigma)}\, \mathbf{a}^{\alpha} = \bar{j}_{(A)<1>}^{(\sigma)}\, \mathbf{a}_{<1>} \qquad A = 1, \ldots, N$$

$$\mathbf{q}^{(\sigma)} = \bar{q}_{\alpha}^{(\sigma)}\, \mathbf{a}^{\alpha} = \bar{q}_{<1>}^{(\sigma)}\, \mathbf{a}_{<1>}$$

surface coordinates

$$y^1 \equiv z_1 \qquad y^2 \equiv z_2$$

jump mass balance for species A (Table 5.2.2-5, equation C)

$$\rho^{(\sigma)} \left(\frac{\partial \omega_{(A)}^{(\sigma)}}{\partial t} + \frac{\partial \omega_{(A)}^{(\sigma)}}{\partial z_1} v_1^{(\sigma)} \right) + \frac{1}{(\, a_{11}\,)^{1/2}} \frac{\partial}{\partial z_1} \bar{j}_{(A)<1>}^{(\sigma)} - r_{(A)}^{(\sigma)}$$

$$+ \Bigg[j_{(A)1}\xi_1 + j_{(A)3}\xi_3 + \rho(\, \omega_{(A)} - \omega_{(A)}^{(\sigma)}\,)$$

$$\times \left\{ v_1 \xi_1 + v_3 \xi_3 - \frac{\partial h}{\partial t} \left[1 + \left(\frac{\partial h}{\partial z_1} \right)^2 \right]^{-1/2} \right\} \Bigg]$$

$$= 0$$

components of $\mathbf{j}_{(A)}^{(\sigma)}$ ***when* (1-1) *applies for two-component materials***

$$\bar{j}_{(A)<1>}^{(\sigma)} = - \frac{\mathcal{D}_{(AB)}^{(\sigma)}}{(\, a_{11}\,)^{1/2}} \frac{\partial \mu_{(A)}^{(\sigma)}}{\partial z_1}$$

$$\overline{j}_{(A)<2>}^{(\sigma)} = 0$$

jump energy balance (Table 5.8.9-1, equation F)

$$\rho^{(\sigma)}\hat{c}_{\gamma}^{(\sigma)} \frac{d_{(v(\sigma))}T^{(\sigma)}}{dt}$$

$$= -\frac{1}{(a_{11})^{1/2}}\frac{\partial}{\partial z_1}\overline{q}_{<1>}^{(\sigma)} - \left(\frac{\partial \ln \hat{A}}{\partial \ln T^{(\sigma)}}\right)_{\gamma,\omega_{(B)}^{(\sigma)}} \frac{d_{(v(\sigma))}\gamma}{dt}$$

$$+ \overline{S}_{<11>}^{(\sigma)}\overline{D}_{<11>}^{(\sigma)} + 2\overline{S}_{<12>}^{(\sigma)}\overline{D}_{<21>}^{(\sigma)} + \overline{S}_{<22>}^{(\sigma)}\overline{D}_{<22>}^{(\sigma)}$$

$$+ \sum_{A=1}^{N} \frac{1}{(a_{11})^{1/2}}\overline{j}_{(A)<1>}^{(\sigma)} \left(b_{(A)1}^{(\sigma)} + \frac{\partial h}{\partial z_1} b_{(A)3}^{(\sigma)}\right) + \rho^{(\sigma)}Q^{(\sigma)}$$

$$- \sum_{A=1}^{N} \left[\mu_{(A)}^{(\sigma)} - T^{(\sigma)}\left(\frac{\partial \mu_{(A)}^{(\sigma)}}{\partial T^{(\sigma)}}\right)_{\gamma,\omega_{(B)}^{(\sigma)}}\right] \frac{d_{(v(\sigma))}\omega_{(A)}^{(\sigma)}}{dt}$$

$$+ \Big[-\rho(\hat{U} - \hat{H}^{(\sigma)})(\mathbf{v}\cdot\boldsymbol{\xi} - v_{(\xi)}^{(\sigma)}) - \frac{1}{2}\rho(\mathbf{v}\cdot\boldsymbol{\xi} - v_{(\xi)}^{(\sigma)})^3$$

$$+ (\mathbf{v}\cdot\boldsymbol{\xi} - v_{(\xi)}^{(\sigma)})(\xi_1 T_{11}\xi_1 + 2\xi_1 T_{13}\xi_3 + \xi_3 T_{33}\xi_3)$$

$$- q_1\xi_1 - q_3\xi_3 \Big]$$

Here

$$v_{(\xi)}^{(\sigma)} = \frac{\partial h}{\partial t}\left[1 + \left(\frac{\partial h}{\partial z_1}\right)^2\right]^{-1/2}$$

$$\mathbf{v}\cdot\boldsymbol{\xi} - v_{(\xi)}^{(\sigma)} = v_1\xi_1 + v_3\xi_3 - \frac{\partial h}{\partial t}\left[1 + \left(\frac{\partial h}{\partial z_1}\right)^2\right]^{-1/2}$$

and

$$\frac{d_{(v(\sigma))}T^{(\sigma)}}{dt}, \quad \frac{d_{(v(\sigma))}\gamma}{dt}, \quad \text{and} \quad \frac{d_{(v(\sigma))}\omega_{(A)}^{(\sigma)}}{dt}$$

have similar forms, for example

$$\frac{d_{(v^{(\sigma)})} T^{(\sigma)}}{dt} = \frac{\partial T^{(\sigma)}}{\partial t} + \frac{\partial T^{(\sigma)}}{\partial z_1} v_1^{(\sigma)}$$

components of $\mathbf{q}^{(\sigma)}$ ***when* (1-2) *applies***

$$\bar{q}^{(\sigma)}_{<1>} = - \frac{k^{(\sigma)}}{(a_{11})^{1/2}} \frac{\partial T^{(\sigma)}}{\partial z_1} + \sum_{B=1}^{N} \mu^{\{\sigma\}}_{(B)} \bar{j}^{\{\sigma\}}_{(B)<1>}$$

$$\bar{q}^{(\sigma)}_{<2>} = 0$$

Table 5.11.1-7: Rotating and deforming axially symmetric surface viewed in a cylindrical coordinate system (refer to Exercises A.2.1-10, A.2.6-2, and A.5.3-11 and to Table 2.4.2-7)

dividing surface

$$z = h(r, t)$$

assumptions (see also Table 2.4.2-7)

$$\omega^{(\sigma)}_{(A)} = \omega^{(\sigma)}_{(A)}(r, t) \qquad A = 1, \ldots, N$$

$$T^{(\sigma)} = T^{(\sigma)}(r, t)$$

$$\mathbf{j}^{(\sigma)}_{(A)} = \bar{j}^{(\sigma)}_{(A)\alpha}\mathbf{a}^{\alpha} = \bar{j}^{(\sigma)}_{(A)<1>}\mathbf{a}_{<1>} \qquad A = 1, \ldots, N$$

$$\mathbf{q}^{(\sigma)} = \bar{q}^{(\sigma)}_{\alpha}\mathbf{a}^{\alpha} = \bar{q}^{(\sigma)}_{<1>}\mathbf{a}_{<1>}$$

surface coordinates

$$y^1 \equiv r \qquad y^2 \equiv \theta$$

jump mass balance for species A (Table 5.2.2-5, equation C)

$$\rho^{(\sigma)}\left(\frac{\partial \omega^{(\sigma)}_{(A)}}{\partial t} + \frac{\partial \omega^{(\sigma)}_{(A)}}{\partial r} v^{(\sigma)}_r\right) + \frac{1}{r\,(a_{11})^{1/2}}\frac{\partial}{\partial r}\left(r\,\bar{j}^{(\sigma)}_{(A)<1>}\right) - r^{(\sigma)}_{(A)}$$

$$+ \Bigg[j_{(A)r}\xi_r + j_{(A)z}\xi_z + \rho\left(\omega_{(A)} - \omega^{(\sigma)}_{(A)}\right)$$

$$\times \left\{ v_r\xi_r + v_z\xi_z - \frac{\partial h}{\partial t}\left[1 + \left(\frac{\partial h}{\partial r}\right)^2\right]^{-1/2} \right\}\Bigg]$$

$$= 0$$

components of $\mathbf{j}^{(\sigma)}_{(A)}$ ***when* (1-1) *applies for two-component materials***

$$\bar{j}^{(\sigma)}_{(A)<1>} = -\frac{\mathcal{D}^{(\sigma)}_{(AB)}}{(a_{11})^{1/2}}\frac{\partial \mu^{(\sigma)}_{(A)}}{\partial r}$$

$$\overline{j}_{(A)<2>}^{(\sigma)} = 0$$

jump energy balance (Table 5.8.9-1, equation F)

$$\rho^{(\sigma)} \hat{c}_{\gamma}^{(\sigma)} \frac{d_{(v(\sigma))} T^{(\sigma)}}{\partial t}$$

$$= - \frac{1}{r \, (\, a_{11} \,)^{1/2}} \frac{\partial}{\partial r} \left(r \, \overline{q}_{<1>}^{(\sigma)} \right)$$

$$- \left(\frac{\partial \ln \hat{A}}{\partial \ln T} \right)_{\gamma, \omega_{(B)}^{(\sigma)}} \frac{d_{(v(\sigma))} \gamma}{dt}$$

$$+ \overline{S}_{<11>}^{(\sigma)} \overline{D}_{<11>}^{(\sigma)} + 2 \overline{S}_{<12>}^{(\sigma)} \overline{D}_{<21>}^{(\sigma)} + \overline{S}_{<22>}^{(\sigma)} \overline{D}_{<22>}^{(\sigma)}$$

$$+ \sum_{A=1}^{N} \frac{1}{(\, a_{11} \,)^{1/2}} \overline{j}_{(A)<1>}^{(\sigma)} \left(v_r^{(\sigma)} + \frac{\partial h}{\partial r} v_z^{(\sigma)} \right) + \rho^{(\sigma)} Q^{(\sigma)}$$

$$- \sum_{A=1}^{N} \left[\mu_{(A)}^{(\sigma)} - T^{(\sigma)} \left(\frac{\partial \mu_{(A)}^{(\sigma)}}{\partial T^{(\sigma)}} \right)_{\gamma, \omega_{(B)}^{(\sigma)}} \right] \frac{d_{(v(\sigma))} \omega_{(A)}^{(\sigma)}}{dt}$$

$$+ \Big[- \rho (\, \hat{U} - \hat{H}^{(\sigma)} \,)(\, \mathbf{v} \cdot \boldsymbol{\xi} - v_{(\xi)}^{(\sigma)} \,) - \frac{1}{2} \rho (\, \mathbf{v} \cdot \boldsymbol{\xi} - v_{(\xi)}^{(\sigma)} \,)^3$$

$$+ (\, \mathbf{v} \cdot \boldsymbol{\xi} - v_{(\xi)}^{(\sigma)} \,)(\, \xi_r T_{rr} \xi_r + 2 \xi_r T_{rz} \xi_z + \xi_z T_{zz} \xi_z \,)$$

$$- q_r \xi_r - q_z \xi_z \Big]$$

Here

$$v_{(\xi)}^{(\sigma)} = \frac{\partial h}{\partial t} \left[1 + \left(\frac{\partial h}{\partial r} \right)^2 \right]^{-1/2}$$

$$\mathbf{v} \cdot \boldsymbol{\xi} - v_{(\xi)}^{(\sigma)} = v_r \xi_r + v_z \xi_z - \frac{\partial h}{\partial t} \left[1 + \left(\frac{\partial h}{\partial r} \right)^2 \right]^{-1/2}$$

and

$$\frac{d_{(v^{(\sigma)})} T^{(\sigma)}}{dt}, \quad \frac{d_{(v^{(\sigma)})} \gamma}{dt}, \quad \text{and} \quad \frac{d_{(v^{(\sigma)})} \omega^{(\sigma)}_{(A)}}{dt}$$

have similar forms, for example

$$\frac{d_{(v^{(\sigma)})} T^{(\sigma)}}{dt} = \frac{\partial T^{(\sigma)}}{\partial t} + \frac{\partial T^{(\sigma)}}{\partial r} v_r^{(\sigma)}$$

components of $\mathbf{q}^{(\sigma)}$ ***when* (1-2) *applies***

$$\bar{q}^{(\sigma)}_{<1>} = - \frac{k^{(\sigma)}}{(a_{11})^{1/2}} \frac{\partial T^{(\sigma)}}{\partial r} + \sum_{B=1}^{N} \mu^{(\sigma)}_{(B)} \bar{j}^{(\sigma)}_{(B)<1>}$$

$$\bar{q}^{(\sigma)}_{<2>} = 0$$

Table 5.11.1-8: Alternative form for rotating and deforming axially symmetric surface viewed in a cylindrical coordinate system (refer to Exercises A.2.1-11, A.2.6-3, and A.5.3-12 and to Table 2.4.2-8)

dividing surface

$$r = c(z, t)$$

assumptions (see also Table 2.4.2-8)

$$\omega^{(\sigma)}_{(A)} = \omega^{(\sigma)}_{(A)}(z, t) \qquad A = 1, \ldots, N$$

$$T^{(\sigma)} = T^{(\sigma)}(z, t)$$

$$\mathbf{j}^{(\sigma)}_{(A)} = \overline{j}^{(\sigma)}_{(A)\alpha}\mathbf{a}^{\alpha} = \overline{j}^{(\sigma)}_{(A)<1>}\mathbf{a}_{<1>} \qquad A = 1, \ldots, N$$

$$\mathbf{q}^{(\sigma)} = \overline{q}^{(\sigma)}_{\alpha}\mathbf{a}^{\alpha} = \overline{q}^{(\sigma)}_{<1>}\mathbf{a}_{<1>}$$

surface coordinates

$$y^1 \equiv z \qquad y^2 \equiv \theta$$

jump mass balance for species A (Table 5.2.2-5, equation C)

$$\rho^{(\sigma)}\left(\frac{\partial \omega^{(\sigma)}_{(A)}}{\partial t} + \frac{\partial \omega^{(\sigma)}_{(A)}}{\partial z} v^{(\sigma)}_{z}\right) + \frac{1}{c(a_{11})^{1/2}}\frac{\partial}{\partial z}\left(c\,\overline{j}^{(\sigma)}_{(A)<1>}\right) - r^{(\sigma)}_{(A)}$$

$$+ \left[j_{(A)r}\xi_r + j_{(A)z}\xi_z + \rho\left(\omega_{(A)} - \omega^{(\sigma)}_{(A)}\right)\right.$$

$$\left. \times \left\{ v_r\xi_r + v_z\xi_z - \frac{\partial c}{\partial t}\left[1 + \left(\frac{\partial c}{\partial z}\right)^2\right]^{-1/2}\right\}\right]$$

$$= 0$$

components of $\mathbf{j}^{(\sigma)}_{(A)}$ ***when*** **(1-1)** ***applies for two-component materials***

$$\overline{j}^{(\sigma)}_{(A)<1>} = -\frac{\mathcal{D}^{(\sigma)}_{(AB)}}{(a_{11})^{1/2}}\frac{\partial \mu^{(\sigma)}_{(A)}}{\partial z}$$

$$\overline{j}^{(\sigma)}_{(A)<2>} = 0$$

jump energy balance (Table 5.8.9-1, equation F)

$$\rho^{(\sigma)}\hat{c}^{(\sigma)}_{\gamma} \frac{d_{(v^{(\sigma)})}T^{(\sigma)}}{dt}$$

$$= - \frac{1}{c(\ a_{11}\ )^{1/2}} \frac{\partial}{\partial z} \left(c\ \overline{q}^{(\sigma)}_{<1>} \right)$$

$$- \left(\frac{\partial \ln \hat{\mathcal{A}}}{\partial \ln T} \right)_{\gamma, \omega^{(\sigma)}_{(B)}} \frac{d_{(v^{(\sigma)})}\gamma}{dt}$$

$$+ \overline{S}^{(\sigma)}_{<11>}\overline{D}^{(\sigma)}_{<11>} + 2\overline{S}^{(\sigma)}_{<12>}\overline{D}^{(\sigma)}_{<21>} + \overline{S}^{(\sigma)}_{<22>}\overline{D}^{(\sigma)}_{<22>}$$

$$+ \sum_{A=1}^{N} \frac{1}{(\ a_{11}\ )^{1/2}} \overline{j}^{(\sigma)}_{(A)<1>} \left(\frac{\partial c}{\partial z} v^{(\sigma)}_{r} + v^{(\sigma)}_{z} \right) + \rho^{(\sigma)}Q^{(\sigma)}$$

$$- \sum_{A=1}^{N} \left[\mu^{(\sigma)}_{(A)} - T^{(\sigma)} \left(\frac{\partial \mu^{(\sigma)}_{(A)}}{\partial T^{(\sigma)}} \right)_{\gamma, \omega^{(\sigma)}_{(B)}} \right] \frac{d_{(v^{(\sigma)})}\omega^{(\sigma)}_{(A)}}{dt}$$

$$+ \Big[- \rho(\ \hat{U} - \hat{H}^{(\sigma)}\ )(\ \mathbf{v} \cdot \boldsymbol{\xi} - v^{(\xi)}\ ) - \frac{1}{2}\rho(\ \mathbf{v} \cdot \boldsymbol{\xi} - v^{(\xi)}\ )^{3}$$

$$+ (\ \mathbf{v} \cdot \boldsymbol{\xi} - v^{(\xi)}\ )(\ \xi_{r}T_{rr}\xi_{r} + 2\xi_{r}T_{rz}\xi_{z} + \xi_{z}T_{zz}\xi_{z}\ )$$

$$- q_{r}\xi_{r} - q_{z}\xi_{z} \Big]$$

Here

$$v^{(\xi)} = \frac{\partial c}{\partial t} \left[1 + \left(\frac{\partial c}{\partial z} \right)^{2} \right]^{-1/2}$$

$$\mathbf{v} \cdot \boldsymbol{\xi} - v^{(\xi)} = v_{r}\xi_{r} + v_{z}\xi_{z} - \frac{\partial c}{\partial t} \left[1 + \left(\frac{\partial c}{\partial z} \right)^{2} \right]^{-1/2}$$

and

$$\frac{d_{(v(\sigma))} T^{(\sigma)}}{dt}, \quad \frac{d_{(v(\sigma))} \gamma}{dt}, \quad \text{and} \quad \frac{d_{(v(\sigma))} \omega^{\{\sigma\}}_{(A)}}{dt}$$

have similar forms, for example

$$\frac{d_{(v(\sigma))} T^{(\sigma)}}{dt} = \frac{\partial T^{(\sigma)}}{\partial t} + \frac{\partial T^{(\sigma)}}{\partial z} v_z^{(\sigma)}$$

components of $\mathbf{q}^{(\sigma)}$ ***when* (1-2) *applies***

$$\bar{q}^{\{\sigma\}}_{<1>} = - \frac{k^{(\sigma)}}{(a_{11})^{1/2}} \frac{\partial T^{(\sigma)}}{\partial z} + \sum_{B=1}^{N} \mu^{\{\sigma\}}_{(B)} \bar{j}^{\{\sigma\}}_{(B)<1>}$$

$$\bar{q}^{\{\sigma\}}_{<2>} = 0$$

Notation for chapter 5

$\mathbf{a}_\alpha$	natural basis vectors, introduced in Sec. A.2.1
$\hat{A}$	Helmholtz free energy per unit mass, introduced in Sec. 5.8.2
$\hat{A}^{(\sigma)}$	surface Helmholtz free energy per unit mass, introduced in Exercise 5.8.6-2
$\overline{\mathcal{A}}^{(m)}_{(A)}$	partial molal area, introduced in Exercise 5.8.6-7
$b^{(i)}_{(A)}$	activity coefficient, introduced in Sec. 5.8.7
$b^{(i)*}_{(A)}$	activity coefficient that approaches unity as its mole fraction go to zero, introduced in Sec. 5.8.7
$b^{(\sigma)}_{(A)}$	constant-interfacial-tension surface activity coefficient, introduced in Sec. 5.8.7
$\mathbf{b}_{(A)}$	body force per unit mass acting upon species A within a phase, introduced in Sec. 5.5.2
$\mathbf{b}^{(\sigma)}_{(A)}$	body force per unit mass acting upon species A in a dividing surface, introduced in Sec. 5.5.2
B^e	exterior or surroundings of the body B
c	total molar density
$c_{(A)}$	molar density of species A
$c^{(\sigma)}$	total surface molar density
$c^{(\sigma)}_{(A)}$	surface molar density of species A
$\hat{c}^{(\sigma)}_{\mathcal{A}}$	surface heat capacity per unit mass at constant specific area, introduced in Sec. 5.8.6
$\hat{c}^{(\sigma)}_{\gamma}$	surface heat capacity per unit mass at constant surface tension, introduced in Sec. 5.8.6
$C_{(v)}$	curve of intersection between the multicomponent dividing surface Σ and $S_{(v)}$
$C^{(cl)}$	common line
$c^{(\sigma)\infty}$	total number of moles adsorbed in dividing surface, which is a constant in Lucassen-Reynders and van den Tempel

	convention, introducted in Sec. 5.2.3
dA	denotes that an area integration is to be performed
ds	denotes that a line integration with respect to arc length is to be performed
dV	denotes that a volume integration is to be performed
$\mathbf{D}$	rate of deformation tensor (Slattery 1981, p. 28)
$\mathbf{D}^{(\sigma)}$	surface rate of deformation tensor, introduced in Sec. 1.2.8
$\mathcal{D}^{(\sigma)}_{(AB)}$	surface diffusion coefficient, introduced in Sec. 5.9.8
$\mathbf{e}$	thermal energy flux vector, introduced in Sec. 5.7.3
$\mathbf{e}_j$	basis vectors for rectangular cartesian coordinate system
$\mathbf{e}^{(\sigma)}$	surface thermal energy flux vector, introduced in Sec. 5.7.3
E_c	rate of contact entropy transmission per unit area to a body at its surrounding surface $S_{(v)}$
$E_c^{(\sigma)}$	rate of contact entropy transmission per unit length at the curve $C_{(v)}$
$\mathcal{E}$	rate of entropy transmission to a body, introduced in Sec. 5.7.2
$\mathcal{E}_c$	rate of contact entropy transmission to a body, introduced in Sec. 5.7.2
$\mathcal{E}_r$	rate of radiant entropy transmission to a body, introduced in Sec. 5.7.2
$f^{(i)}_{(A)}$	fugacity of species A in phase i, introduced in Sec. 5.8.7
$f^{(i)\circ}_{(A)}$	fugacity of species A in phase i in the standard state, introduced in Sec. 5.8.7
$f^{(\sigma)}_{(A)}$	surface fugacity of species A, introduced in Sec. 5.8.7
$f^{(\sigma,\gamma)}_{(A)}$	surface fugacity at the current temperature and composition of the interface but at the interfacial tension of the standard state, introduced in Sec. 5.8.7
$\mathbf{g}_j$	natural basis vectors for curvilinear coordinate system, introduced in Sec. A.1.2

$\hat{G}$ Gibbs free energy per unit mass within a phase, introduced in Exercise 5.8.4-1

$\hat{H}$ enthalpy per unit mass within a phase, introduced in Exercise 5.8.4-1

$\hat{H}^{(\sigma)}$ surface enthalpy per unit mass, introduced in Exercise 5.8.6-2

$\mathbf{I}$ identity tensor that transform every vector into itself

$\mathbf{j}_{(A)}$ mass flux of species A with respect to $\mathbf{v}$, introduced in Sec. 5.2.2

$\mathbf{j}^{\star}_{(A)}$ mass flux of species A with respect to $\mathbf{v}^{\star}$, introduced in Sec. 5.2.2

$\mathbf{j}^{(\sigma)}_{(A)}$ surface mass flux of species A with respect to $\mathbf{v}^{(\sigma)}$, introduced in Sec. 5.2.2

$\mathbf{j}^{(\sigma)\star}_{(A)}$ surface mass flux of species A with respect to $\mathbf{v}^{(\sigma)\star}$, introduced in Sec. 5.2.2

$\mathbf{J}_{(A)}$ molar flux of species A with respect to $\mathbf{v}$, introduced in Sec. 5.2.2

$\mathbf{J}^{\star}_{(A)}$ molar flux of species A with respect to $\mathbf{v}^{\star}$, introduced in Sec. 5.2.2

$\mathbf{J}^{(\sigma)}_{(A)}$ surface molar flux of species A with respect to $\mathbf{v}^{(\sigma)}$

$\mathbf{J}^{(\sigma)\star}_{(A)}$ surface molar flux of species A with respect to $\mathbf{v}^{(\sigma)\star}$, introduced in Sec. 5.2.2

$\mathcal{K}^{M}_{(A)}$ intrinsic configuration of surface particles of A, introduced in Sec. 5.1.3

$\mathcal{K}^{M}_{X\,(A)}$ intrinsic deformation of surface particles of A in the reference configuration from the current configuration, introduced in Sec. 5.1.3

$M_{(A)}$ molecular weight of species A

$M^{(\sigma)}$ molar-averaged molecular weight of surface, introduced in Sec. 5.2.2

$\mathcal{M}$ mass of a multicomponent body

$\mathbf{n}_{(A)}$	mass flux of species A with respect to the frame of reference, introduced in Sec. 5.2.2
$\mathbf{n}_{(A)}^{(\sigma)}$	surface mass flux of species A with respect to the frame of reference, introduced in Sec. 5.2.2.
$\mathbf{N}_{(A)}$	molar flux of species A with respect to the frame of reference, introduced in Sec. 5.2.2
$\mathbf{N}_{(A)}^{(\sigma)}$	surface molar flux of species A with respect to the frame of reference, introduced in Sec. 5.2.2
$\mathbf{p}^{(\sigma)}$	position vector field on Σ, introduced in Sec. 1.2.5
P	thermodynamic pressure, defined in Sec. 5.8.3
$\mathbf{P}$	projection tensor, defined in Sec. A.3.2
$\mathbf{q}$	energy flux vector, introduced in Sec. 5.6.4
$\mathbf{q}^{(\sigma)}$	surface energy flux vector, introduced in Sec. 5.6.4
Q	rate of radiant energy transmission per unit mass to the material within a phase, introduced in Sec. 5.6.3
Q_c	rate of contact energy transmission per unit area to a body at its bounding surface $S_{(v)}$, introduced in Sec. 5.6.3
$Q^{(\sigma)}$	rate of radiant energy transmission per unit mass to the material in the dividing surface, introduced in Sec. 5.6.3
$Q_c^{(\sigma)}$	rate of contact energy transmission per unit length at the curve $C_{(v)}$, introduced in Sec. 5.6.3
$\mathbf{Q}$	orthogonal transformation that describes the rotation and (possibly) reflection that takes one frame into another, introduced in Sec. 1.4.1
$\mathcal{Q}$	rate of energy transmission to a body from the surroundings, introduced in Sec. 5.6.1
$\mathcal{Q}_c$	rate of contact energy transmission, introduced in Sec. 5.6.3
$\mathcal{Q}_r$	rate of radiant energy transmission, introduced in Sec. 5.6.3
$r_{(A)}$	rate of production of mass of species A per unit volume by homogeneous chemical reactions, introduced in Sec. 5.2.1

$r_{(A)}^{(\sigma)}$	rate of production of mass of species A per unit area by heterogeneous chemical reactions, introduced in Sec. 5.2.1
R	gas law constant
$R_{(A)}$	region occupied by body of species A
$R_{(v)}$	region of space occupied by a set of mass-averaged material particles, introduced in Sec. 5.3.1
$S_{(v)}$	closed bounding surface of $R_{(v)}$, introduced in Sec. 5.3.1
$\hat{S}$	entropy per unit mass within a phase, introduced in Sec. 5.7.3
$\hat{S}^{(\sigma)}$	surface entropy per unit mass, introduced in Sec. 5.7.3
t	time
t_κ	time at which reference configuration is defined
$\mathbf{t}$	stress vector
$\mathbf{t}^{(\sigma)}$	surface stress vector
T	temperature, introduced in Sec. 5.7.2
$T^{(\sigma)}$	surface temperature, introduced in Sec. 5.7.2
$\mathbf{T}$	stress tensor, introduced in Sec. 2.1.4
$\mathbf{T}^{(\sigma)}$	surface stress tensor, introduced in Sec. 2.1.5
$\mathbf{u}$	time rate of change of spatial position following a surface point, introduced in Sec. 1.2.7
$u_{(v)}^{(cl)}$	speed of displacment of common line, defined in Sec. 1.3.7
$\hat{U}$	internal energy per unit mass within a phase, introduced in Sec. 5.6.2
$\hat{U}^{(\sigma)}$	surface internal energy per unit mass, introduced in Sec. 5.6.2
$\mathbf{v}$	mass-averaged velocity, introduced in Sec. 5.2.2
$\mathbf{v}_{(A)}$	velocity of species A, introduced in Sec. 5.2.2

$\mathbf{v}^{(\sigma)}$	mass-averaged surface velocity, introduced in Sec. 5.2.2
$\mathbf{v}^{*}$	molar-averaged velocity, introduced in Sec. 5.2.2
$\mathbf{v}^{(\sigma)*}$	molar-averaged surface velocity, introduced in Sec. 5.2.2
$\mathbf{v}^{(\sigma)}_{(A)}$	surface velocity of species A, introduced in Sec. 5.2.2
$v^{(\sigma)}_{(\xi)}$	speed of displacment of the surface, introduced in Sec. 1.2.7
$\hat{V}$	volume per unit mass, introduced in Sec. 5.8.2
$x_{(A)}$	mole fraction of A, introduced in Sec. 5.2.2
$x^{(\sigma)}_{(A)}$	surface mole fraction of species A, introduced in Sec. 5.2.2
$X^{\alpha}_{(A)}$	intrinsic configuration or motion for the surface particles of species A on Σ, introduced in Sec. 5.1.3
$X^{\alpha}_{K(A)}$	intrinsic deformation of the surface particles of A from reference intrinsic configuration, introduced in Sec. 5.1.3
y^{α}	surface coordinates
y^{M}_{κ}	surface coordinates used in reference configuration, introduced in Sec. 5.1.3
$\dot{y}_{(A)}$	intrinsic surface velocity of species A, introduced in Sec. 5.1.4
$\mathbf{z}$	position vector
$\mathbf{z}_{\kappa}$	position vector denoting a place in the reference configuration, introduced in Sec. 5.1.3
$\mathbf{z}_{\kappa(A)}$	position vector of a material particle of A in the reference configuration, introduced in Sec. 5.1.2

greek letters

$\alpha_{(i)}$	valence or charge number for phase i, introduced in Sec. 5.8.7
γ	thermodynamic surface tension, defined in Sec. 5.8.6
$\gamma^{\circ}_{(A)}$	interfacial tension for pure species A at the system temperature, introduced in Sec. 5.8.7

$\gamma^{(A)\circ}$	interfacial tension in the standard state for species A, introduced in Sec. 5.8.7
$\zeta_{(A)}$	material particle of species A, introduced in Sec. 5.1.2
$\zeta^{(\sigma)}_{(A)}$	surface particle of species A on Σ, introduced in Sec. 5.1.3
$\boldsymbol{\kappa}_{(A)}$	reference configuration of body of species A, introduced in Sec. 5.1.2
$\boldsymbol{\kappa}^{(\sigma)}$	reference dividing surface, introduced in Sec. 5.1.3
$\boldsymbol{\kappa}^{(\sigma)}_{K(A)}$	reference configuration of the surface particles of A, introduced in Sec. 5.1.3
$\boldsymbol{\Lambda}$	indicates that a vector cross product is to be performed (Slattery 1981, p. 653), except where noted
$\boldsymbol{\mu}$	unit vector tangent to dividing surface, normal to closed bounding curve, and outwardly directed with respect to this curve
$\mu_{(A)}$	chemical potential of species A, introduced in Sec. 5.8.3
$\mu^{(\sigma)}_{(A)}$	surface chemical potential of species A, introduced in Sec. 5.8.6
$\mu^{(i,m)}_{(A)}$	chemical potential of species A in phase i on a molar basis, introduced in Sec. 5.8.7
$\mu^{(i,m)\circ}_{(A)}$	chemical potential of species A in phase i on a molar basis in the standand state, introduced in Sec. 5.8.7
$\mu^{(\sigma,m)}_{(A)}$	surface chemical potential of species A on a molar basis, introduced in Sec. 5.8.7
$\mu^{(\sigma,m)\circ}_{(A)}$	surface chemical potential of species A on a molar basis in the standard state, introduced in Sec. 5.8.7
$\boldsymbol{\nu}$	unit vector normal to common line and tangent to dividing surface
$\boldsymbol{\xi}$	unit normal to Σ, introduced in Sec. 1.2.7
ρ	total mass density
$\rho_{(A)}$	mass density of species A

$\rho^{(\sigma)}$	total surface mass density
$\rho^{(\sigma)}_{(A)}$	surface mass density of species A
Σ	dividing surface
$\chi_{(A)}$	configuration of material body of species A, introduced in Sec. 5.1.2
χ^{t}	history of the motion of the multicomponent material, introduced in Sec. 5.8.2
$\chi^{(\sigma)t}$	history of the motion of the multicomponent dividing surface, introduced in Sec. 5.8.5
$\chi^{(\sigma)}_{(A)}$	configuration of the surface particles of A, introduced in Sec. 5.1.3
$\chi^{(\sigma)}_{K(A)}$	deformation of speciesA on Σ from the reference configuration, introduced in Sec. 5.1.3
$\chi^{(\sigma)}_{X(A)}$	deformation of speciesA on Σ from the reference configuration, introduced in Sec. 5.1.3
$\omega_{(A)}$	mass fraction of species A, introduced in Sec. 5.2.2
$\omega^{(\sigma)}_{(A)}$	surface mass fraction of species A, introduced in Sec. 5.2.2

others

div	divergence operator
$\text{div}_{(\sigma)}$	surface divergence operator, introduced in Secs. A.5.1 through A.5.6
tr	trace operator
$\ldots^{(\sigma,c)}$	denotes a reference dividing surface located such that the total surface molar density is zero, introduced in Sec. 5.2.3
$\ldots^{(\sigma,B)}$	denotes a reference dividing surface located such that the surface molar density of one of the components B is zero, introduced in Sec. 5.2.3
$\ldots^{(\sigma,L)}$	denotes a reference dividing surface located according to the Lucassen-Reynders and van den Tempel convention in which the total surface molar density $c^{(\sigma)\infty}$ is a constant, introduced in Sec. 5.2.3

$\ldots^{(\sigma,\rho)}$	denotes a reference dividing surface located such that the total surrface mass density is zero, introduced in Sec. 5.2.3
$\ldots^{T}$	transpose
$\ldots^{-1}$	inverse
$\ldots^{*}$	denotes an alternative frame of reference
$\ldots^{\star}$	denotes an activity coefficient that approaches unity as its mole fraction goes to zero, introduced in Sec. 5.8.7
$\overline{\ldots}$	denotes per unit area
$\overline{\overline{\ldots}}$	denotes partial mass variable, defined in Exercise 5.8.3-2
$\overline{\overline{\ldots}}^{(m)}$	denotes partial molal variable, defined in Exercise 5.8.3-3
$\overline{\overline{\ldots}}^{(\sigma)}$	denotes partial mass surface variable, defined in Exercise 5.8.6-5
$\overline{\overline{\ldots}}^{(\sigma,m)}$	denotes partial molal surface variable, defined in Exercise 5.8.6-6
$\overset{\vee}{\ldots}$	denotes per unit volume
$\overset{\wedge}{\ldots}$	denotes per unit mass
(...)	jump for common line, defined in Sec. 1.3.7
[...]	jump for dividing surface, defined in Sec. 1.3.4
∇	gradient with respect to current configuration
$\nabla_{(\sigma)}$	surface gradient operation, introduced in Secs. A.5.1 through A.5.6
$\frac{d_{(A)}}{dt}$	material derivative following a particle of species A, introduced in Sec. 5.1.2
$\frac{d_{(A,s)}}{dt}$	material derivative following a particle of species A in a dividing surface, introduced in Sec. 5.1.3
$\frac{d_{(v)}}{dt}$	material derivative following a mass-averaged material particle, introduced in Sec. 5.3.1

$\frac{d_{(\mathbf{v}^{(\sigma)})}}{dt}$	material derivative following a mass-averaged surface particle, introduced in Sec. 5.3.1

6
Applications of the differential balances to energy and mass transfer

In Chapter 3, we discussed detailed descriptions of flows in which the effects of any temperature or concentration gradients could be neglected. In what follows, we offer a brief guide to the solution of problems in which these effects can not be dismissed.

I assume in what follows that you are already familiar with the more standard problems of convective energy and mass transfer involving a single phase (Slattery 1981, chapters 6 and 9; Bird *et al.* 1960, chapters 9 through 11 and 17 through 19). While there are many practical multiphase flows involving energy and mass transfer, we will focus our attention only on those in which the interfacial effects are decidedly different from those considered in Chapter 3.

References

Bird, R. B., W. E. Stewart, and E. N. Lightfoot, "Transport Phenomena," John Wiley, New York (1960).

Slattery, J. C., "Momentum, Energy, and Mass Transfer in Continua," McGraw-Hill, New York (1972); second edition, Robert E. Krieger, Malabar, FL 32950 (1981).

6.1 Philosophy

6.1.1 Structure of problems involving energy transfer To make things simple, let us begin by restricting our attention to those problems in which there are no concentration gradients. Our objective in these problems is to describe the configurations and motions of several adjoining phases as well as the temperature distributions within these phases and the rate at which energy is transferred between the various phases.

Since these are problems involving simultaneous momentum and energy transfer, the eleven differential equations and boundary conditions for momentum transfer outlined in Sec. 3.1.1 apply equally well here. The energy transfer is described by four additional conditions.

xii) *Within each phase,* we must satisfy the differential energy balance (Sec. 5.11.1; Slattery 1981, p. 293).

xiii) *On each dividing surface,* we must satisfy the jump energy balance (Sec. 5.11.2).

With few exceptions (see Sec. 6.3.2), our concern is with interphase energy transfer rather than with the energy associated with the dividing surface. Under these circumstances, we will normally neglect all interfacial effects and write the jump energy balance as

$$\Big[-\rho\hat{U}(\mathbf{v}\cdot\boldsymbol{\xi} - v_{(\xi)}) - \frac{1}{2}\rho(\mathbf{v}\cdot\boldsymbol{\xi} - v_{(\xi)})^3 + (\mathbf{v}\cdot\boldsymbol{\xi} - v_{(\xi)})\,\boldsymbol{\xi}\cdot\mathbf{T}\cdot\boldsymbol{\xi} - \mathbf{q}\cdot\boldsymbol{\xi} \Big] = 0 \tag{1-1}$$

While there is little experimental evidence to suggest that interfacial effects in the jump energy balance significantly affect interphase energy transfer, it is clear that these effects may alter the velocity distribution within the immediate neighborhood of the interface and in this way indirectly influence the interphase energy transfer. It is also obvious that, even if the interfacial energy has no practical effect upon interphase energy transfer, it is not zero, since interfacial tension is a derivative of $U^{(\sigma)}$ (see Sec. 5.8.6).

Normally we will be willing to neglect the effect of mass transfer upon the interchange of kinetic energy and upon the work done at the phase interface with respect to its effect upon the interchange of internal energy at the interface. In this case, (1-1) simplifies to

$$[-\rho\hat{U}(\mathbf{v}\cdot\boldsymbol{\xi} - v_{(\xi)}) - \mathbf{q}\cdot\boldsymbol{\xi}] = 0 \tag{1-2}$$

If there is no mass transfer, (1-1) further reduces to

$$[\mathbf{q}\cdot\boldsymbol{\xi}] = 0 \tag{1-3}$$

xiv) *At each common line*, there is an energy balance to be obeyed (Sec. 5.11.3).

I am not aware of any problem in which it would be necessary to apply the energy balance at the common line.

xv) *At each dividing surface*, temperature is continuous.

This is a consequence of assuming nearly local equilibrium at the phase interface (see Sec. 5.10.3), analogous with our assumptions that the tangential components of velocity (see Sec. 3.1.1) and chemical potentials (see Sec. 6.1.2) are continuous across an interface. Lacking contrary experimental evidence, this condition is applied even when there is a phase change, the normal component of velocity is not continuous at the interface, and the adjacent phases are clearly not in local equilibrium. This condition is known to fail either when one of the phases is a rarefied gas or when radiation plays a significant role (Siegel and Howell 1981, p. 689).

As I indicated above, problems in which it is important to account for interfacial effects in the jump energy balance or problems in which we must employ the energy balance at the common line are not common. In the context of energy transfer, the most common interfacial phenomena arise either from the temperature dependence of the interfacial tension (and any parameters required to describe the interfacial stress-deformation behavior, such as the interfacial viscosities) or from phase changes.

References

Siegel, R., and J. R. Howell, "Thermal Radiation Heat Transfer," second edition, Hemisphere, Washington, DC (1981).

Slattery, J. C., "Momentum, Energy, and Mass Transfer in Continua," McGraw-Hill, New York (1972); second edition, Robert E. Krieger, Malabar, FL 32950 (1981).

6.1.2 Structure of problems involving mass transfer Now let us restrict our view to problems in which there are no temperature gradients. Once we understand problems involving only energy transfer and those involving only mass transfer, it is relatively easy to pose problems involving simultaneous energy and mass transfer (although the solution of the resultant boundary value problem may be another matter).

These are problems involving simultaneous momentum and mass transfer and the eleven differential equations and boundary conditions for momentum transfer outlined in Sec. 3.1.1 are again applicable. The mass transfer is described by four additional conditions.

xii) *Within each phase*, we must satisfy the differential mass balance or equation of continuity for each species (Sec. 5.11.1; Slattery 1981, p. 450). Alternatively, the differential mass balance for one of these species can be replaced by the overall equation of continuity.

xiii) *On each dividing surface*, we must satisfy the jump mass balance (Sec. 5.11.2). In contrast with energy transfer, it is relatively common to include the interfacial effects in describing mass transfer problems.

xiv) *At each common line*, there is a mass balance to be obeyed for each species (Sec. 5.11.3). I am not currently aware of any problem in which it would be necessary to apply the mass balance at the common line.

xv) *At each dividing surface*, we may or may not have local equilibrium. There are several cases that we can consider.

If there are no surface active species present, it is reasonable to assume that all chemical potentials (plus body force potentials; see Sec. 5.10.3) are continuous at the phase interface, which is equivalent to assuming local equilibrium at the interface.

If there is a surface active species present, it will normally have a preferred orientation in the phase interface. This surface active species will arrive at the substrate adjacent to the interface with a random orientation and it must rotate into the preferred orientation before it can enter the interface.

If the rate of diffusion of this species from the interior of the phase to the substrate is sufficiently slow, this orientation process can be carried out without affecting the rate of mass transfer to the interface. In this limit, we will again be willing to assume that all chemical potentials are continuous at the interface.

More generally, this orientation process will affect the rate of mass transfer and the chemical potentials will not be continuous at the interface. In this case, it is common to assume that the mass flux $\mathbf{j}_{(A)} \cdot \boldsymbol{\xi}$ of any species A at the interface will be proportional to the difference between its surface mass density $\rho^{(\sigma)}_{(A)}$ and the surface mass density $\rho^{(\sigma)}_{(A)eq}$ that would exist if the surface were in equilibrium with the adjacent substrate into which the unit normal $\boldsymbol{\xi}$ is pointing:

$$\mathbf{j}_{(A)} \cdot \boldsymbol{\xi} = k\left(\rho^{(\sigma)}_{(A)} - \rho^{(\sigma)}_{(A)eq} \right) \tag{2-1}$$

The adsorption rate constant k characterizes the rate at which this orientation process takes place. As $k \to \infty$, diffusion from the interior of the phase to the substrate becomes the rate controlling step and $\rho^{(\sigma)}_{(A)} \to \rho^{(\sigma)}_{(A)eq}$. As $k \to 0$, the rate of adsorption from the substrate to the interface becomes the rate controlling step and $\nabla \rho_{(A)} \cdot \boldsymbol{\xi} \to 0$ in the limit as the interface is approached.

References

Slattery, J. C., "Momentum, Energy, and Mass Transfer in Continua," McGraw-Hill, New York (1972); second edition, Robert E. Krieger, Malabar, FL 32950 (1981).

6.2 Complete solutions

6.2.1 There are no complete solutions To my knowledge, there are no complete solutions for problems involving either energy or mass transfer at moving and deforming phase interfaces that also include significant interfacial effects. As you will see in the next series of sections, all of the available solutions are for limiting cases (zeroth perturbations) in which one or more dimensionless parameters approach zero.

6.3 Limiting cases of energy transfer

6.3.1 Motion of a drop or bubble (LeVan 1981; with D. Li) A small drop or bubble moves through an immiscible, unbounded fluid in which there is a uniform vertical temperature gradient T'. We will find it convenient to examine this problem in a frame of reference in which the drop is stationary and the fluid very far away from the drop moves with a uniform velocity v_∞ in the positive z_3 direction as shown in Figure 1-1. The radius of a sphere having the same volume as the drop is R. Our objective is to determine the effects of the axially-symmetric interfacial tension gradient induced by this uniform temperature gradient.

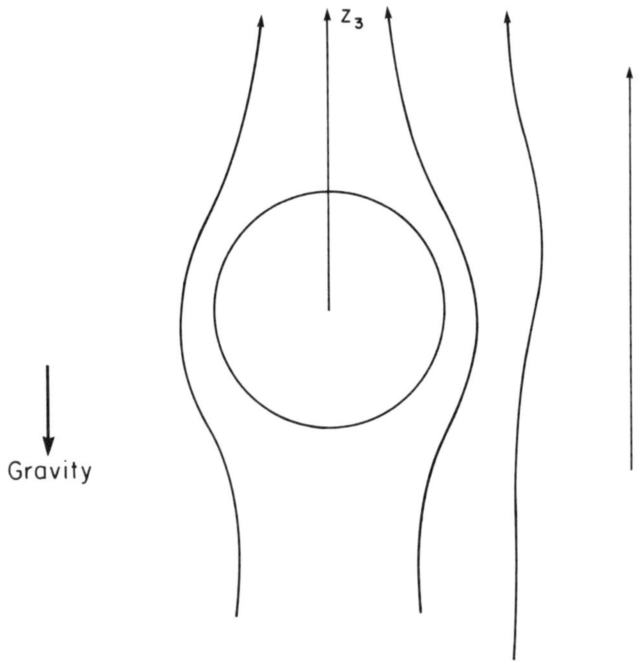

Figure 6.3.1-1: Flow past drop in a frame of reference in which drop is stationary.

We shall make the following assumptions in this analysis.

i) The Reynolds number

$$N_{Re} \equiv \frac{R v_\infty \rho}{\mu} << 1 \tag{1-1}$$

and all inertial effects are neglected in Cauchy's first law.

ii) The Peclet number

$$N_{Pe} \equiv \frac{\rho \hat{c} R v_\infty}{k} << 1 \tag{1-2}$$

and convection is neglected with respect to conduction in the differential energy balance.

iii) The Bond number

$$N_{Bo} \equiv \frac{|\rho - \hat{\rho}| g R^2}{\gamma_0} << 1 \tag{1-3}$$

which means that we neglect the effect of gravity upon the configuration of the interface. Under static conditions, the drop would be spherical.

iv) Both phases are incompressible Newtonian fluids.

v) The two fluids are immiscible and there is no mass transfer across the phase interface.

vi) The location of the dividing surface is chosen such that

$$\frac{\rho^{(\sigma)}}{\rho R} << 1 \tag{1-4}$$

Defining $\rho^{(\sigma)} = 0$ should be avoided in this case, because the energy balance used (Table 5.8.9-1, equation E) involves $\rho^{(\sigma)-1}$ [see (1-15)].

vii) We will neglect the effects of any viscous stress-deformation behavior, conduction, and radiation in the fluid-fluid interface.

viii) All physical properties other than the interfacial tension are constants. In particular, we neglect any effects attributable to natural convection.

ix) We will assume that there are no surfactants present and that the interfacial tension gradient which is developed is attributable only to a temperature gradient in the interface.

x) In a frame of reference in which the drop is stationary, the flow is independent of time.

We seek velocity and temperature distributions such that very far away from the drop

$$\text{as } r \to \infty : \quad \mathbf{v} \to v_\infty \mathbf{e}_3 \tag{1-5}$$

$$\text{as } r \to \infty : \quad T \to T_\infty + T' \, r \cos \theta \tag{1-6}$$

We will begin by asking whether there is a solution such that the interface is a sphere. We will assume that in spherical coordinates these velocity and temperature distributions take the forms

$$v_r = v_r(r, \theta)$$

$$v_\theta = v_\theta(r, \theta)$$

$$v_\phi = 0$$

$$T = T(r, \theta) \tag{1-7}$$

$$\hat{v}_r = \hat{v}_r(r, \theta)$$

$$\hat{v}_\theta = \hat{v}_\theta(r, \theta)$$

$$\hat{v}_\phi = 0$$

$$\hat{T} = \hat{T}(r, \theta) \tag{1-8}$$

where the carat ^ denotes a variable associated with the drop phase. For the moment, we will also assume that the drop remains a sphere in the temperature gradient. With these assumptions, the absence of mass transfer, the continuity of the tangential components of velocity, and the continuity of temperature require

$$\text{at } r = R : \quad v_r = \hat{v}_r = 0 \tag{1-9}$$

$$\text{at } r = R : \quad v_\theta = \hat{v}_\theta \tag{1-10}$$

$$\text{at } r = R : \quad T = \hat{T} \tag{1-11}$$

In view of assumption iv, the jump mass balance reduces to

$$\frac{\partial \rho^{(\sigma)}}{\partial t} + \rho^{(\sigma)} \text{div}_{(\sigma)} \mathbf{v}^{(\sigma)} = 0 \tag{1-12}$$

The normal and tangential components of the jump momentum balance reduce to (Table 2.4.2-5)

$$\text{at } r = R: \ \hat{p} - p + 2\mu \frac{\partial v_r}{\partial r} - 2\hat{\mu} \frac{\partial \hat{v}_r}{\partial r} = \frac{2\gamma}{R} \tag{1-13}$$

$$\text{at } r = R: \ \mu \left[r \frac{\partial}{\partial r}\left(\frac{v_\theta}{r} \right) + \frac{1}{r}\frac{\partial v_r}{\partial \theta} \right] - \hat{\mu} \left[r \frac{\partial}{\partial r}\left(\frac{\hat{v}_\theta}{r} \right) + \frac{1}{r}\frac{\partial \hat{v}_r}{\partial \theta} \right]$$

$$= - \frac{1}{R}\frac{\partial \gamma}{\partial \theta} \tag{1-14}$$

and, with (1-12), the jump energy balance takes the form (Table 5.8.9-1, equation E)

$$\text{at } r = R: \ k \frac{\partial T}{\partial r} - \hat{k} \frac{\partial \hat{T}}{\partial r}$$

$$= - T^{(\sigma)} \frac{d\gamma}{dT^{(\sigma)}} \operatorname{div}_{(\sigma)} \mathbf{v}^{(\sigma)}$$

$$= - T^{(\sigma)} \frac{d\gamma}{dT^{(\sigma)}} \frac{1}{R \sin\theta} \frac{d}{d\theta} \left(v_\theta^{(\sigma)} \sin\theta \right) \tag{1-15}$$

On the right side of (1-15), we have that portion of the convection terms that remain in the limit (1-4).

In solving this problem it is convenient to introduce stream functions for the exterior and interior phases, since in this way the associated equations of continuity can be automatically satisfied:

$$v_r = \frac{1}{r^2 \sin\theta} \frac{\partial \psi}{\partial \theta}$$

$$v_\theta = - \frac{1}{r \sin\theta} \frac{\partial \psi}{\partial r} \tag{1-16}$$

$$\hat{v}_r = \frac{1}{r^2 \sin\theta} \frac{\partial \hat{\psi}}{\partial r}$$

$$\hat{v}_\theta = - \frac{1}{r \sin\theta} \frac{\partial \hat{\psi}}{\partial r} \tag{1-17}$$

In terms of these stream functions, Cauchy's first law for the creeping flow of an incompressible Newtonian fluid requires for the inner and outer phases (Slattery 1981, p. 66)

$$E^4 \psi = 0 \tag{1-18}$$

$$E^4 \hat{\psi} = 0 \tag{1-19}$$

where

$$E^2 = \frac{\partial^2}{\partial r^2} + \frac{\sin \theta}{r^2} \frac{\partial}{\partial \theta} \left(\frac{1}{\sin \theta} \frac{\partial}{\partial \theta} \right) \tag{1-20}$$

With the interfacial tension gradient left unspecified, the stream functions consistent with (1-5), (1-9), (1-10), (1-14), (1-18), and (1-19) are

$$\psi = \frac{v_\infty}{2} \left(1 - \frac{R^3}{r^3} \right) r^2 \sin^2 \theta - (r^2 - R^2) \sum_{n=2}^{\infty} D_n \, r^{-n+1} \, C_n^{-1/2} (\cos \theta) \tag{1-21}$$

$$\hat{\psi} = - \frac{3}{4} v_\infty \left(1 - \frac{r^2}{R^2} \right) r^2 \sin^2 \theta + (R^2 - r^2) \sum_{n=2}^{\infty} D_n \, R^{-2n+1} \, r^n \, C_n^{-1/2} (\cos \theta) \tag{1-22}$$

in which $C_n^{-1/2}$ denotes the Gegenbauer polynomial of order n and degree −1/2 and

$$D_2 = - \frac{v_\infty R(6\mu + 9\hat{\mu}) - 2Rf_2}{6\mu + 6 \hat{\mu}} \tag{1-23}$$

$$\text{for } n > 2 : \ D_n = \frac{n(n - 1) R^{n-1} f_n}{2(2n - 1) (\mu + \hat{\mu})} \tag{1-24}$$

$$f_n \equiv - \frac{(2n - 1)}{2} \int_0^\pi C_n^{-1/2} (\cos \theta) \frac{d\gamma}{d\theta} \, d\theta \tag{1-25}$$

The corresponding interfacial velocity is

$$v_\theta^{(\sigma)} = v_0^{(\sigma)} \sin\theta - \sum_{n=2}^{\infty} \frac{n(n-1)f_n}{(2n-1)(\mu+\hat{\mu})} \frac{C_n^{-1/2}(\cos\theta)}{\sin\theta} \tag{1-26}$$

where

$$v_0^{(\sigma)} \equiv -\frac{\mu\, v_\infty}{2(\mu+\hat{\mu})} \tag{1-27}$$

is the interfacial velocity at the equator.

The differential energy balances for the inner and outer phases appropriate in the limit (1-2) reduce to

$$\nabla^2 T = 0 \tag{1-28}$$

$$\nabla^2 \hat{T} = 0 \tag{1-29}$$

These can be solved consistent with (1-6), (1-11), and (1-15). The important portion of this result is the temperature distribution in the interface

$$T^{(\sigma)} = T_0^{(\sigma)} + \left(\frac{3kR}{2k+\hat{k}} T' + \frac{2T_\infty v_0^{(\sigma)}}{2k+\hat{k}} \frac{d\gamma}{dT^{(\sigma)}} \right) \cos\theta \tag{1-30}$$

since this permits us to write the interfacial tension distribution as

$$\gamma = \gamma_0 + a_1 \cos\theta \tag{1-31}$$

with

$$a_1 \equiv \left(\frac{3kR}{2k+\hat{k}} T' + \frac{2T_\infty v_0^{(\sigma)}}{2k+\hat{k}} \frac{d\gamma}{dT^{(\sigma)}} \right) \frac{d\gamma}{dT^{(\sigma)}} \tag{1-32}$$

Given the interfacial tension distribution, we can now go back to the normal component of the jump momentum balance and show that it is consistent with our assumption that the interface is a sphere.

Finally, the integral momentum balance for the drop allows us to determine the terminal velocity of the drop as

$$v_\infty = \frac{6(\mu+\hat{\mu})\, D + 2G(1-F)}{6\mu + 9\hat{\mu} + 3F\mu} \tag{1-33}$$

in which we have introduced

$$D \equiv \frac{R^2 g(\hat{\rho} - \rho)}{3\mu} \tag{1-34}$$

$$F \equiv \left[1 + \frac{3(\mu + \hat{\mu})(2k + \hat{k})}{2T_\infty (d\gamma/dT^{(\sigma)})^2} \right]^{-1} \tag{1-35}$$

$$G \equiv \frac{3kR}{2k + \hat{k}} \frac{d\gamma}{dT^{(\sigma)}} T' \tag{1-36}$$

The results presented here reduce to those of Hadamard (1911) and Rybczynski (1911) in the absence of a temperature gradient very far away from the drop ($T' = 0$), when the interfacial tension is independent of temperature ($d\gamma/dT^{(\sigma)} = 0$) and there is no surfactant present. As summarized by Levich (1962), the Hadamard-Rybczynski expression for the terminal velocity of the drop normally does not give a good representation of available experimental data.

Harper *et al.* (1967) and of Kenning (1969) examined the effect of temperature gradients generated by the local expansion and contraction of the interface. There was a question whether this effect might be sufficient to explain the deviations between experimental data and the Hadamard-Rybczynski result for the terminal velocity. Harper *et al.* (1967) demonstrated that this effect is too small to be of practical significance. The results presented here are consistent with theirs for the case where $T' = 0$.

The original treatment of this problem by LeVan (1981) also took into account the effect of the interfacial viscosities. To the present, interfacial viscosities have been observed only in those systems containing surfactants. LeVan (1981) did not take into account the surfactant distribution in the interface.

When surfactants are present, their effects may be more significant than those due either to convection of energy within the interface or to the interfacial viscosities (Levich 1962). For a discussion of the effect of a surfactant upon the terminal velocity of a drop, see Sec. 6.4.1.

References

Hadamard, J. S., *Comp. Rend. Acad. Sci.* (Paris) **152**, 1735 (1911); **154**, 109 (1912). As quoted by Happel and Brenner (1973).

Happel, J., and H. Brenner, "Low Reynolds Number Hydrodynamics," second edition, p. 157, Noordhoff, Leyden (1973).

Harper, J. F., D. W. Moore, and J. R. A. Pearson, *J. Fluid Mech.* **27**, 361

(1967).

Kenning, D. B. R., *Chem. Eng. Sci.* **24**, 1385 (1969).

LeVan, M. D., *J. Colloid Interface Sci.* **83**, 11 (1981).

Levich, V. G., "Physicochemical Hydrodynamics," Prentice-Hall, Englewood Cliffs, NJ (1962).

Rybczynski, W., *Bull. Acad. Sci. Cracovie (ser. A)*, 40 (1911). As quoted by Happel and Brenner (1973).

Slattery, J. C., "Momentum, Energy, and Mass Transfer in Continua," McGraw-Hill, New York (1972); second edition, Robert E. Krieger, Malabar, FL 32950 (1981).

6.4 Limiting cases of mass transfer

6.4.1 Motion of a drop or bubble (LeVan and Newman 1976; with D. Li) A small drop or bubble moves through an immiscible, unbounded fluid that contains a trace of surfactant S. The surfactant S is confined to the outer phase. The transfer of this species to and from the interface is assumed to be controlled by its rate of diffusion. We will find it convenient to examine this problem in a frame of reference in which the drop is stationary and the fluid very far away from the drop moves with a uniform velocity v_∞ in the positive z_3 direction as shown in Figure 6.3.1-1. The radius of a sphere having the same volume as the drop is R. Our objective is to determine the effect of the axially-symmetric interfacial tension gradient attributable to the interfacial concentration gradient that develops in the interface.

We shall make the following assumptions in this analysis.

i) The Reynolds number

$$N_{Re} \equiv \frac{R v_\infty \rho}{\mu} << 1 \tag{1-1}$$

and all inertial effects are neglected in Cauchy's first law.

ii) The Bond number

$$N_{Bo} \equiv \frac{|\rho - \hat{\rho}| g R^2}{\gamma_\infty} << 1 \tag{1-2}$$

which means that we neglect the effect of gravity upon the configuration of the interface. Under static conditions, the drop would be spherical.

iii) Both phases are incompressible Newtonian fluids.

iv) The two fluids are immiscible.

v) The location of the dividing surface is chosen such that

$$\rho^{(\sigma)} = 0 \tag{1-3}$$

vi) The outer phase contains a trace of surfactant S; the inner phase contains none.

vii) The transfer of S to and from the interface is limited by its rate of diffusion.

viii) We will neglect the effects of any viscous stress-deformation behavior in the fluid-fluid interface.

ix) We will neglect the effects of any surface diffusion.

x) All physical properties other than the interfacial tension are constants.

xi) The temperature of the system is uniform.

xii) In a frame of reference in which the drop is stationary, the flow is independent of time.

We seek velocity and concentration distributions such that very far away from the drop

$$\text{as } r \to \infty : \quad \mathbf{v} \to v_\infty \mathbf{e}_3 \tag{1-4}$$

$$\text{as } r \to \infty : \quad \rho_{(S)} \to \rho_{(S)\infty} \tag{1-5}$$

We will begin by asking whether there is a solution such that the interface is a sphere. We will assume that in spherical coordinates these velocity and concentration distributions take the forms

$$v_r = v_r(r, \theta)$$

$$v_\theta = v_\theta(r, \theta)$$

$$v_\phi = 0$$

$$\rho_{(S)} = \rho_{(S)}(r, \theta) \tag{1-6}$$

$$\hat{v}_r = \hat{v}_r(r, \theta)$$

$$\hat{v}_\theta = \hat{v}_\theta(r, \theta)$$

$$\hat{v}_\phi = 0 \tag{1-7}$$

where the carat ^ denotes a variable associated with the drop phase. In view of assumption v, the overall jump mass balance (Sec. 5.4.1) and the continuity of the tangential components of velocity require

$$\text{at } r = R : \quad v_r = \hat{v}_r = 0 \tag{1-8}$$

$$\text{at } r = R : \quad v_\theta = \hat{v}_\theta \tag{1-9}$$

The tangential component of the jump momentum balance reduces to (Table 2.4.2-5)

$$\text{at } r = R: \quad \mu\left[r\frac{\partial}{\partial r}\left(\frac{v_\theta}{r}\right) + \frac{1}{r}\frac{\partial v_r}{\partial \theta}\right] - \hat{\mu}\left[r\frac{\partial}{\partial r}\left(\frac{\hat{v}_\theta}{r}\right) + \frac{1}{r}\frac{\partial \hat{v}_r}{\partial \theta}\right] = -\frac{1}{R}\frac{\partial \gamma}{\partial \theta} \tag{1-10}$$

and the jump mass balance for surfactant becomes (Table 5.2.2-5, Eq. E)

$$\text{at } r = R: \quad \frac{1}{R\sin\theta}\frac{\partial}{\partial \theta}\left(\rho_{(S)}^{(\sigma)} v_\theta^{(\sigma)} \sin\theta\right) - \mathcal{D}_{(Sm)}\frac{\partial \rho_{(S)}}{\partial r} = 0 \tag{1-11}$$

By $\mathcal{D}_{(Sm)}$ we mean the diffusion coefficient for S in the mixture (Slattery 1981, p. 484). Since the transfer of surfactant to and from the interface is controlled by its rate of diffusion, the surfactant distribution in the interface is in equilibrium with that in the substrate. Since there is only a trace of surfactant in the system, we will write

$$\text{at } r = R: \quad \rho_{(S)}^{(\sigma)} = \beta\rho_{(S)} \tag{1-12}$$

In solving this problem, it is convenient to introduce stream functions for the exterior and interior phases, since in this way the associated equations of continuity can be automatically satisfied:

$$v_r = \frac{1}{r^2\sin\theta}\frac{\partial \psi}{\partial \theta}$$

$$v_\theta = -\frac{1}{r\sin\theta}\frac{\partial \psi}{\partial r} \tag{1-13}$$

$$\hat{v}_r = \frac{1}{r^2\sin\theta}\frac{\partial \hat{\psi}}{\partial \theta}$$

$$\hat{v}_\theta = -\frac{1}{r\sin\theta}\frac{\partial \hat{\psi}}{\partial r} \tag{1-14}$$

Since the outer phase contains only a trace of surfactant, we can write

$$\gamma = \gamma_\infty + \gamma'\left(\rho_{(S)}^{(\sigma)} - \rho_{(S)\infty}^{(\sigma)}\right) \tag{1-15}$$

where

$$\rho_{(S)\infty}^{(\sigma)} = \beta\,\rho_{(S)\infty} \tag{1-16}$$

is the equilibrium surface concentration of surfactant and γ_∞ is the corresponding equilibrium interfacial tension. Assuming

$$\eta \equiv \frac{-\gamma' \rho^{\{g\}}_\infty}{\gamma_\infty} = \frac{-\gamma' \beta \rho_{(s)\infty}}{\gamma_\infty} << 1 \tag{1-17}$$

we will seek a perturbation solution of the form

$$\mathbf{v} = \mathbf{v}_0 + \eta\, \mathbf{v}_1 + \ldots$$

$$\rho_{(s)} = \rho_{(s)0} + \eta\, \rho_{(s)1} + \ldots$$

$$\rho^{\{g\}} = \rho^{\{g\}}_0 + \eta\, \rho^{\{g\}}_1 + \ldots \tag{1-18}$$

At the zeroth perturbation, the stream functions for the outer and inner phases are (see Sec. 6.3.1 in the limit where the interfacial tension is a constant)

$$\psi_0 = \frac{v_\infty}{4}\left(2 - \frac{2\mu + 3\hat{\mu}}{\mu + \hat{\mu}}\frac{R}{r} + \frac{\hat{\mu}}{\mu + \hat{\mu}}\frac{R^3}{r^3}\right) r^2 \sin^2\theta \tag{1-19}$$

$$\hat{\psi}_0 = -\frac{v_\infty}{4}\frac{\mu}{\mu + \hat{\mu}}\left(1 - \frac{r^2}{R^2}\right) r^2 \sin^2\theta \tag{1-20}$$

LeVan and Newman (1976) have solved numerically the zeroth perturbation of the equation of continuity for the surfactant S (Slattery 1981, p. 482)

$$\frac{\partial \rho_{(s)0}}{\partial r} v_{0r} + \frac{\partial \rho_{(s)0}}{\partial \theta}\frac{v_{0\theta}}{r} = \mathcal{D}_{(sm)}\left[\frac{1}{r^2}\frac{\partial}{\partial r}\left(r^2 \frac{\partial \rho_{(s)0}}{\partial r}\right) + \frac{1}{r^2 \sin\theta}\frac{\partial}{\partial \theta}\left(\sin\theta \frac{\partial \rho_{(s)0}}{\partial \theta}\right)\right] \tag{1-21}$$

consistent with (1-5), (1-11), (1-12), and

$$\text{at } \theta = 0, \pi : \quad \frac{\partial \rho_{(s)0}}{\partial \theta} = 0 \tag{1-22}$$

for the case

$$N_{Pe} \equiv \frac{2Rv_{\infty}}{\mathcal{D}_{(Sm)}} = 60$$

$$\beta^* = \frac{\beta}{2R} = 0.02$$

$$\frac{\hat{\mu}}{\mu} = 0 \tag{1-23}$$

Their results for

$$\rho_{\{g\}}^* \equiv \frac{\rho_{\{g\}0}}{\rho_{\{g\}\infty}} \tag{1-24}$$

are shown in Figure 1-1.

At the first perturbation, the stream functions for the inner and outer phases are (These are rearrangements of the stream functions found in Sec. 6.3.1.)

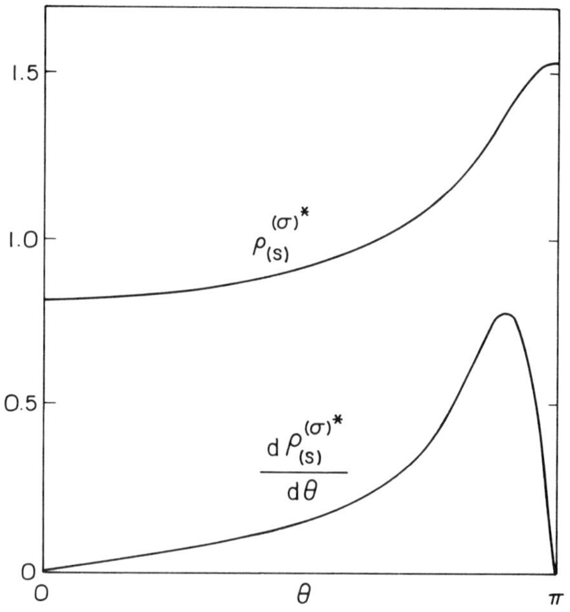

Figure 6.4.1-1: For the special case (1-22), $\rho_{\{g\}}^*$ and $d\rho_{\{g\}}^*/d\theta$ as functions of θ (LeVan and Newman 1976).

$$\psi = \frac{v_\infty}{4}\left(2 - \frac{2\mu + 3\hat{\mu}}{\mu + \hat{\mu}}\frac{R}{r} + \frac{\hat{\mu}}{\mu + \hat{\mu}}\frac{R^3}{r^3}\right) r^2 \sin^2\theta$$
$$+ \eta \frac{\gamma_\infty}{4}\frac{r^2 - R^2}{\mu + \hat{\mu}} \sum_{n=2}^{\infty} n(n-1)\left[\int_0^\pi C_n^{-1/2}(\cos\theta)\right.$$
$$\left.\times \frac{d\rho_s^{(\sigma)*}}{d\theta} d\theta\right] \frac{R^{n-1}}{r^{n-1}} C_n^{-1/2}(\cos\theta) + \ldots \tag{1-25}$$

$$\hat{\psi} = -\frac{v_\infty}{4}\frac{\mu}{\mu + \hat{\mu}}\left(1 - \frac{r^2}{R^2}\right) r^2 \sin^2\theta - \eta \frac{\gamma_\infty}{4}\frac{R^2 - r^2}{\mu + \hat{\mu}}$$
$$\times \sum_{n=2}^{\infty} n(n-1)\left[\int_0^\pi C_n^{-1/2}(\cos\theta)\frac{d\rho_s^{(\sigma)*}}{d\theta} d\theta\right]$$
$$\times \frac{r^n}{R^n} C_n^{-1/2}(\cos\theta) + \ldots \tag{1-26}$$

The corresponding interfacial velocity is

$$v_\theta^{(\sigma)} = -\frac{v_\infty}{2}\frac{\mu}{\mu + \hat{\mu}}\sin\theta - \eta\frac{\gamma_\infty}{2(\mu + \hat{\mu})}\sum_{n=2}^{\infty} n(n-1)$$
$$\times \left[\int_0^\pi C_n^{-1/2}(\cos\theta)\frac{d\rho_s^{(\sigma)*}}{d\theta} d\theta\right]\frac{C_n^{-1/2}(\cos\theta)}{\sin\theta}$$
$$+ \ldots \tag{1-27}$$

If we wished to do so, we could now use the first perturbation of the normal component of the jump momentum balance to determine the first perturbation of the configuration of the interface. While the result would not be a sphere, it would be only slight deformation from a sphere (Wasserman and Slattery 1969; LeVan and Newman 1976), since it would be a first perturbation result.

Finally the integral momentum balance for the drop permits us to determine the terminal speed of the drop as

$$v_\infty = \left[\frac{2\mu + 2\hat{\mu}}{2\mu + 3\hat{\mu}} \right] \left[\frac{R^2 g(\hat{\rho} - \rho)}{3\mu} \right]$$

$$+ \eta \frac{\gamma_\infty}{4\mu + 6\hat{\mu}} \int_0^\pi \sin^2\theta \frac{d\rho\{g\}^*}{d\theta} d\theta \tag{1-28}$$

For the special case (1-23), LeVan and Newman (1976) compute (see also Figure 1-1)

$$\int_0^\pi \sin^2\theta \frac{d\rho\{g\}^*}{d\theta} d\theta = 0.284 \tag{1-29}$$

If we say more specifically that

$$R = 1 \times 10^{-3} \text{ cm}$$

$$\rho = 1 \text{ gm/cm}^3$$

$$\mu = 1 \text{ mPa s}$$

$$\gamma_\infty = 70 \text{ mN/m}$$

$$\rho_{(S)\infty} = 1 \times 10^{-8} \text{ gm/cm}^3 \tag{1-30}$$

(1-28) indicates

$$v_\infty = 3.27 \times 10^{-2} + 497\, \eta \text{ cm/s} \tag{1-31}$$

From Lucassen and Hansen (1966, 1967), we can estimate

$$c^{(\sigma)\infty} RT \sim 0(10) \text{ mN/m}$$

$$M \sim 0(10)$$

$$M_{(S)} \sim 0(100)$$

$$a \sim 0(10^{-4}) \tag{1-32}$$

This together with Exercise 6.4.1-1 suggest

$$\eta \sim 0(10^{-6}) \tag{1-33}$$

Equations (1-31) and (1-33) indicate that, in this example computation, the concentration of surfactant must be very small ($\rho_{(S)\infty} = 1 \times 10^{-8}$ gm/cm^3), in order that the correction for the effect of the surface concentration gradient be on the order of 1%.

Wasserman and Slattery (1969) were the first to consider this problem in such detail. They recognized that a perturbation solution was appropriate and that the interface is a slightly deformed sphere at the first perturbation. Although their computations appear to be correct (LeVan and Newman 1976), they did not take full advantage of the simplicity of their perturbation analysis and their results are difficult to generalize. LeVan and Newman (1976) gave a more lucid analysis of the problem. [Holbrook and LeVan (1983a, b) and Sadhal and Johnson (1983) did not explain their treatments as perturbation analyses, and they did not investigate the deformation of the drop or bubble.]

This discussion focuses on only one limiting case of the motion of a drop or bubble in which the mass transfer of surfactant to and from the drop surface is controlled by its rate of diffusion. More generally,

any surfactant present may be isolated in the interface as a stagnant cap covering the trailing surface of the drop,

mass transfer of surfactant to and from the drop surface may be controlled by its rate of diffusion within the adjacent phase (the case considered above),

mass transfer of surfactant to and from the drop surface may be controlled by its rate of adsorption from the substrate,

or mass transfer of surfactant to and from the drop surface may be controlled by both its rate of diffusion and its rate of adsorption.

In a manner similar to that illustrated above, it may be possible to develop first-order perturbation analyses of these problems based upon the discussions given by Holbrook and LeVan (1983a, b) and by Sadhal and Johnson (1983). [These authors assume as an approximation that the drop or bubble would be spherical. It would not be, as can be confirmed by checking the normal component of the jump momentum balance.]

References

Holbrook, J.A., and M. D. LeVan, *Chem. Eng. Commun.* **20**, 191 (1983a).

Holbrook, J.A., and M. D. LeVan, *Chem. Eng. Commun.* **20**, 273 (1983b).

LeVan, M. D., and J. Newman, *AIChE J.* **22**, 695 (1976).

Lucassen, J. and R. S. Hansen, *J. Colloid Interface Sci.* **22**, 32 (1966).

Lucassen, J. and R. S. Hansen, *J. Colloid Interface Sci.* **22**, 319 (1967).

Sadhal, S. S., and R. E. Johnson, *J. Fluid Mech.* **126**, 237 (1983).

Slattery, J. C., "Momentum, Energy, and Mass Transfer in Continua," McGraw-Hill, New York (1972); second edition, Robert E. Krieger, Malabar, FL 32950 (1981).

Wasserman, M. L., and J. C. Slattery, *AIChE J.* **15**, 533 (1969).

Exercise

6.4.1-1 *estimate of* γ' Start with the Frumkin equation of state (Sec. 5.8.7) and the Langmuir adsorption isotherm (Exercise 5.8.7-2) to estimate

$$-\gamma' = \frac{RTM \quad c^{(\sigma)\infty}}{a\,\rho\,M_{(S)}} \frac{\rho_{(S)\infty}}{\rho_{(S)\infty}^{(g)}} \tag{1}$$

Here R is the gas law constant, T temperature, M the molar-averaged molecular weight of the solution, $M_{(S)}$ the molecular weight of the surfactant, and a is a parameter in the Langmuir adsorption isotherm. In the Lucassen-Reynders and van den Tempel convention, the total number of moles adsorbed in the dividing surface is a constant $c^{(\sigma)\infty}$.

6.4.2 **Longitudinal and transverse waves** (Hajiloo and Slattery 1986)

Let us consider the role of diffusion and adsorption of surfactant upon the formation of longitudinal and transverse waves at a liquid-gas phase interface.

Introduction: longitudinal waves Longitudinal waves have been studied intensively, both theoretically (Lucassen 1968a; Lucassen and Barnes 1972; Lucassen and van den Tempel 1972a; Lucassen and Giles 1975; Maru *et al.* 1979) and experimentally (Lucassen 1968a, b; Lucassen and van den Tempel 1972a, b; Crone *et al.* 1979; Maru and Wasan 1979; Graham and Phillips 1980; Miyano *et al.* 1983).

Lucassen (1968a) suggested that longitudinal waves could be used to measure the Gibbs surface elasticity.

Lucassen-Reynders and Lucassen (1969; see also Crone *et al.* 1979) discuss the possibility of measuring a "surface viscosity," but this had nothing to do with one of the rheological coefficients appearing in the Boussinesq surface fluid. Maru *et al.* (1979) considered linear surface viscoelasticity.

Lucassen and van den Tempel (1972a) and Lucassen and Giles (1975) consider the limiting case of mass transfer controlled by diffusion. Maru *et al.* (1979) in addition consider the limiting case in which mass transfer is controlled by the rate of adsorption at the surface.

Introduction: transverse waves There have been a number of theoretical (Lamb 1945; Goodrich 1961a, b; Hansen and Mann 1964; van den Tempel and van de Riet 1964; Hansen *et al.* 1968; Hegde and Slattery 1971; Mann and Du 1971; Mayer and Eliassen 1971) and experimental (Goodrich 1961c, 1962; Mann and Hansen 1963; Davies and Vose 1965; Lucassen and Hansen 1966, 1967; Lucassen and Lucassen-Reynders 1967; Bendure and Hansen 1967; Mayer and Eliassen 1971; Sohl *et al.* 1978) studies of transverse waves created by forced oscillations.

Lamb (1945) neglected all viscous effects in the adjoining bulk phases as well as all interfacial effects other than a uniform interfacial tension.

Goodrich (1961a, b) used a description of surface stress-deformation behavior that can be classified as a simple surface material (Sec. 2.2.3). But it appears neither to be a simple surface fluid (Secs. 2.2.7 and 2.2.8) nor to include as a special case the Boussinesq surface fluid (Sec. 2.2.2) in terms of which the surface viscosities are defined. It is not clear that the models of surface stress-deformation behavior employed by Hansen and Mann (1964) and by Mayer and Eliassen (1971) are simple surface materials. Hegde and Slattery (1971) were apparently the first to use the Boussinesq surface fluid directly.

Goodrich (1961a, b) studied insoluble monolayers. Hansen and Mann (1964) and van den Temple and van de Riet (1964) assumed that mass transfer was controlled by diffusion within the adjacent liquid phase. Hegde and Slattery (1971) incorrectly assumed that surface tension was a function of the concentration of surfactant in the liquid phase immediately adjacent to the interface, while at the same time accounting for the rate of adsorption of surfactant in the interface. In the limit of diffusion controlled mass transfer, their results are correct and do include the effect of the surface viscosities.

Of particular note in the available experimental studies is the novel means of generating and detecting transverse waves at liquid-gas interfaces developed by Sohl *et al.* (1978).

For a limiting case, Hansen *et al.* (1968) have extended their analysis to a liquid-liquid interface. But it is not clear that such an experiment could be carried out without the wave generator directly disturbing one of the liquid phases.

Statement of problem Our objective here is to describe both for longitudinal waves and for transverse waves at a liquid-gas interface the effects of the surface viscosities, the effects of diffusion within the liquid phase, and the effects of the rate of adsorption of surfactant at the surface. We will place particular emphasis upon the ranges of wave length and of damping coefficient that are sensitive to these effects.

Consider a liquid-gas system that, when it is at rest, is in equilibrium. With reference to a rectangular, Cartesian coordinate system (x, y, z), the interface at rest is located in the plane $z = 0$ and gravity acts in the negative z direction. Small amplitude sinusoidal oscillations of frequency α are imposed by a knife-edge touching the interface. Two-dimensional waves are created that move away from the knife-edge in both the positive and negative x directions. If the waves are created by vertical oscillations in the plane $x = 0$, they will be referred to as transverse (or *capillary*) waves. If they are generated by horizontal oscillations in the plane $z = 0$, they will be designated as longitudinal waves.

We will require the following assumptions about the physical problem.

i) The liquid is incompressible and Newtonian.

ii) Viscous effects can be neglected in the gas phase, where the constant ambient pressure is p_o.

iii) The liquid-gas interfacial stress-deformation behavior can be represented by the linear Boussinesq surface fluid model (Sec. 2.2.2).

iv) All species including the surfactant S are present in vanishingly small amounts in the liquid phase with the exception of the solvent A. This means that diffusion of S in the multicomponent mixture can be described as binary diffusion of S with respect to A (Slattery 1981, p. 483). The solution is sufficiently dilute that all physical properties other than interfacial tension and the two interfacial viscosities will be considered to be constants.

v) Mass transfer from the interface to the gas phase is neglected.

vi) The dividing surface is located in such a manner that

$$\rho^{(\sigma)} = 0 \tag{2-1}$$

vii) Surface diffusion is neglected with respect to surface convection.

viii) The liquid phase is unbounded.

ix) We will assume that the waves are generated remotely. We will describe the velocity distribution and the configuration of the interface only very far to one side of the oscillating knife-edge. Because the oscillating knife-edge is outside the solution domain, we do not satisfy the boundary conditions associated with it.

Governing equations Our first objective is to characterize the surface waves generated.

For the liquid phase, the overall equation of continuity requires (Sec. 5.4.1)

$$\operatorname{div} \mathbf{v}^* = 0 \tag{2-2}$$

Cauchy's first law demands (Sec. 5.5.2)

$$\frac{\partial \mathbf{v}^*}{\partial t^*} + \nabla \mathbf{v}^* \cdot \mathbf{v}^* = -\nabla \mathcal{P}^* + \frac{1}{N_{Re}} \operatorname{div} \mathbf{S}^* \tag{2-3}$$

and the equation of continuity for surfactant S takes the form

$$\frac{\partial \rho^*_{(S)}}{\partial t^*} + \nabla \rho^*_{(S)} \cdot \mathbf{v}^* = \frac{1}{N_{Re} N_{Sc}} \operatorname{div}(\nabla \rho^*_{(S)}) \tag{2-4}$$

Here

$$\mathcal{P} \equiv p + \rho \phi \tag{2-5}$$

with the understanding that the force per unit mass of gravity is representable as

$$\mathbf{b} = -\nabla \phi \tag{2-6}$$

In (2-2) through (2-6), we have introduced as dimensionless variables

$$\mathbf{v}^* \equiv \frac{\mu \mathbf{v}}{\gamma_0} \qquad \mathcal{P}^* \equiv \frac{\mu^2 (\mathcal{P} - p_0)}{\rho \gamma_0^2}$$

$$\mathbf{S}^* \equiv \frac{\mathbf{S}}{\mu \alpha} \qquad \mathbf{S}^{(\sigma)*} \equiv \frac{\mathbf{S}^{(\sigma)}}{\varepsilon \alpha}$$

$$x^* \equiv \frac{\alpha \mu x}{\gamma_0} \qquad z^* \equiv \frac{\alpha \mu z}{\gamma_0}$$

$$t^* \equiv \alpha t \qquad \rho^*_{(S)} \equiv \frac{\rho_{(S)}}{\rho_{(S)0}} \tag{2-7}$$

and the dimensionless parameters

$$N_{Re} \equiv \frac{\rho \gamma_0^2}{\alpha \mu^3} \qquad N_{Sc} \equiv \frac{\mu}{\rho \mathcal{D}_{(SA)}} \tag{2-8}$$

Here $\rho_{(S)}$ is the mass density of species S in the liquid mixture; $\rho_{(S)0}$ the mass density of species S in the liquid mixture in the limit of an undisturbed (plane) interface; $\mathcal{D}_{(SA)}$ the binary diffusion coefficient of S with respect to A, the principal constituent of the liquid mixture. In making (2-2) through (2-6) dimensionless, we have defined our characteristic time

$$t_0 \equiv \frac{1}{\alpha} \tag{2-9}$$

and we have introduced a characteristic length L_0 by requiring

$$N_{ca} \equiv \frac{\mu L_0}{\gamma_0 \, t_0} = 1 \tag{2-10}$$

As a result of assumptions v through vii, the overall jump mass balance requires (Sec. 5.4.1)

$$(\mathbf{v}^\star - \mathbf{v}^{(\sigma)\star}) \cdot \boldsymbol{\xi} = 0 \tag{2-11}$$

the jump momentum balance becomes (Sec. 5.5.2)

$$\nabla_{(\sigma)}\gamma^\star + 2H^\star\gamma^\star\boldsymbol{\xi} + N_\varepsilon \, \text{div}_{(\sigma)}\mathbf{S}^{(\sigma)\star} + N_{Re} \, \mathcal{P}^\star \, \boldsymbol{\xi} - \mathbf{S}^\star \cdot \boldsymbol{\xi}$$

$$- N_{Re}N_g \, \phi^\star\boldsymbol{\xi} = 0 \tag{2-12}$$

and the jump mass balance for surfactant S takes the form (Sec. 5.2.2)

$$\frac{\partial \rho_{(S)}^{(\sigma)\star}}{\partial t^\star} + \nabla_{(\sigma)}\rho_{(S)}^{(\sigma)\star} \cdot (\mathbf{v}^{(\sigma)\star} - \mathbf{u}^\star) + \rho_{(S)}^{(\sigma)\star} \, \text{div}_{(\sigma)}\mathbf{v}^{(\sigma)\star}$$

$$- \mathbf{j}^\star_{(S)} \cdot \boldsymbol{\xi} = 0 \tag{2-13}$$

Here we have introduced as additional dimensionless variables

$$\gamma^\star \equiv \frac{\gamma}{\gamma_0} \qquad H^\star \equiv \frac{\gamma_0 H}{\alpha\mu}$$

$$\phi^\star \equiv \frac{\alpha\mu\phi}{\gamma_0 \, g} \qquad \rho_{(S)}^{(\sigma)\star} \equiv \frac{\rho_{(S)}^{(\sigma)}}{\rho_{(S)0}^{(\sigma)}}$$

$$u^* \equiv \frac{\mu u}{\gamma_0} \qquad \mathbf{j}^*_{(S)} \equiv \frac{\mathbf{j}_{(S)}}{\alpha \rho^{\{S\}}_0} \tag{2-14}$$

and as additional dimensionless parameters

$$N_\varepsilon \equiv \frac{\alpha \varepsilon}{\gamma_0} \qquad N_\kappa \equiv \frac{\alpha \kappa}{\gamma_0}$$

$$N_g \equiv \frac{\mu g}{\alpha \gamma_0} \tag{2-15}$$

By $\boldsymbol{\xi}$ I mean the unit normal to the dividing surface pointing into the gas phase, g the magnitude of the acceleration of gravity, $\rho^{\{S\}}_0$ the equilibrium surface concentration in the limit of an undisturbed (plane) interface, γ_0 the interfacial tension corresponding to $\rho^{\{S\}}_0$, and $\mathbf{j}_{(S)}$ the mass flux of surfactant S within the liquid phase measured with respect to the mass-averaged velocity. Note that in arriving at (2-12) and (2-13) we have employed (2-11).

The rate ($\mathbf{j}^*_{(S)} \cdot \boldsymbol{\xi}$) at which mass of surfactant S is transferred to the interface in (2-13) is determined by two competing physical phenomena: the rate of diffusion of S from the interior of the liquid to that portion of the liquid adjacent to the interface

$$\mathbf{j}^*_{(S)} \cdot \boldsymbol{\xi} = - N_{\mathcal{D}} \nabla \rho^*_{(S)} \cdot \boldsymbol{\xi} \tag{2-16}$$

and the rate of adsorption of S from the adjacent liquid onto the interface

$$\mathbf{j}^*_{(S)} \cdot \boldsymbol{\xi} = - N_k (\rho^{\{S\}*} - \rho^{\{S\}*}_{eq}) \tag{2-17}$$

Here $\rho^{\{S\}*}_{eq}$ is the dimensionless mass density of surfactant that would exist in the interface if it were in equilibrium with the adjacent liquid phase,

$$N_{\mathcal{D}} \equiv \frac{\rho_{(S)0} \mathcal{D}_{(SA)} \mu}{\gamma_0 \rho^{\{S\}}_0}$$

$$N_k \equiv \frac{k}{\alpha} \tag{2-18}$$

and k is the adsorption rate coefficient.

The configuration of the interface is

$$z^* = h^*(x^*, t^*) \tag{2-19}$$

Equations (2-2) through (2-4), (2-11) through (2-13), (2-16), and (2-17) are to be solved simultaneously. This solution must be consistent with the requirements that very far away from the dividing surface both phases are static, that the bulk phases are Newtonian fluids (assumption i), that the interfacial stress-deformation behavior can be represented by the linear Boussinesq surface fluid model (assumption iii), and that the interfacial tension can be described as a function of the surface concentration of surfactant by a series expansion of $\rho^{\{S\}}$ with respect to $\rho^{\{S\}}_0$:

$$\gamma^* = 1 + \gamma'^* (\rho^{\{S\}*} - 1) + \dots \tag{2-20}$$

in which γ'^* is the dimensionless Gibbs surface elasticity evaluated at the equilibrium surface concentration of surfactant. Similar expansions for the two interfacial viscosities are also assumed, but we find that only the first terms will appear in the final result.

Solution We seek a perturbation solution to this problem, taking as our perturbation parameter

$$f^* \equiv \frac{f}{L_0} \tag{2-21}$$

where f is the amplitude of the oscillation of the knife-edge.

The zeroth perturbation corresponds to the static problem with a stationary knife-edge. The zeroth perturbations of all variables other than the concentrations are identically zero. The zeroth perturbations of the concentrations are unity.

The first perturbation of the equation of continuity (2-2) takes the same form and is satisfied identically by the introduction of the dimensionless stream function ψ^*:

$$v^*_{x1} = \frac{\partial \psi^*}{\partial z^*} \qquad v^*_{z1} = -\frac{\partial \psi^*}{\partial x^*} \tag{2-22}$$

In terms of this stream function, the first perturbation of Cauchy's first law (2-3) for an incompressible Newtonian fluid requires (Slattery 1981, p. 66)

$$N_{Re} \frac{\partial (E^2 \psi^*)}{\partial t^*} = E^4 \psi^* \tag{2-23}$$

in which we have introduced

$$E^2 \equiv \frac{\partial^2}{\partial x^{*2}} + \frac{\partial^2}{\partial z^{*2}} \tag{2-24}$$

The first perturbation of (2-4), the equation of continuity for surfactant S, reduces to

$$N_{Re} N_{Sc} \frac{\partial \rho^*_{(S)1}}{\partial t^*} = E^2 \rho^*_{(S)1} \tag{2-25}$$

The first perturbation of the overall jump mass balance (2-11) demands at $z^* = 0$

$$\frac{\partial h^*_1}{\partial t^*} = - \frac{\partial \psi^*}{\partial x^*} \tag{2-26}$$

in which we have recognized (2-19). The first perturbation of the tangential component of the jump momentum balance (2-12) takes the form at $z^* = 0$

$$\gamma'^*_1 \frac{\partial \rho^{\{S\}*}_1}{\partial x^*} + (N_\kappa + N_\varepsilon) \frac{\partial^3 \psi^*}{\partial z^* \partial x^{*2}} = \frac{\partial^2 \psi^*}{\partial z^{*2}} - \frac{\partial^2 \psi^*}{\partial x^{*2}} \tag{2-27}$$

where we have employed (2-20). The first perturbation of the normal component of the jump momentum balance (2-12) is at $z^* = 0$

$$\frac{\partial^2 h^*_1}{\partial x^{*2}} + 2 \frac{\partial^2 \psi^*}{\partial x^* \partial z^*} + N_{Re} \mathcal{P}^*_1 - N_{Re} N_g h^*_1 = 0 \tag{2-28}$$

Taking the first perturbation of the jump mass balance for surfactant S (2-13), we find at $z^* = 0$

$$\frac{\partial \rho^{\{S\}*}_1}{\partial t^*} + \frac{\partial^2 \psi^*}{\partial x^* \partial z^*} + N_{\mathcal{D}} \frac{\partial \rho^*_{(S)1}}{\partial z^*} = 0 \tag{2-29}$$

when we recognized (2-16). Finally, the first perturbations of (2-16) and (2-17) together require at $z^* = 0$

$$N_{\mathcal{D}} \frac{\partial \rho^*_{(S)1}}{\partial z^*} = N_k (\rho^{\{S\}*}_1 - \beta^* \rho^*_{(S)1}) \tag{2-30}$$

In arriving at (2-30), we have recognized that $\rho^{\{S\}}_{eq}$ is a function of $\rho_{(S)}$, which in the neighborhood of $\rho_{(S)0}$ takes the form

$$\rho^{\{g\}*}_{(S)eq} = 1 + \beta^*(\rho^*_{(S)} - 1) + \dots \tag{2-31}$$

We will seek a first-order perturbation solution such that, very far away from the oscillating knife-edge, the fluid is undisturbed,

$$\text{as } x^* \to \infty \quad : \frac{\partial \psi^*}{\partial x^*} = \frac{\partial \psi^*}{\partial z^*} = 0 \tag{2-32}$$

$$\text{as } z^* \to -\infty : \frac{\partial \psi^*}{\partial x^*} = \frac{\partial \psi^*}{\partial z^*} = 0 \tag{2-33}$$

$$\text{as } z^* \to -\infty : \rho^*_{(S)1} = 0 \tag{2-34}$$

A solution of the form (see assumption ix)

$$\psi^* = A \exp(- it^* + imx^* + \ell z^*) + B \exp(- it^* + imx^* + mz^*) \tag{2-35}$$

with

$$\ell \equiv (m^2 - iN_{Re})^{1/2} \tag{2-36}$$

satisfies the first perturbation of Cauchy's first law for an incompressible Newtonian fluid (2-23). In order that (2-32) and (2-33) be satisfied, both the real and imaginary parts of m must be positive and the real part of ℓ must be positive.

A solution to the first perturbation of the equation of continuity for surfactant S (2-25) is

$$\rho^*_{(S)1} = D \exp(- it^* + imx^* + nz^*) \tag{2-37}$$

where the real part of

$$n \equiv (m^2 - iN_{Re}N_{Sc})^{1/2} \tag{2-38}$$

must be positive in order to satisfy (2-34).

Integrating the first perturbation of the overall jump mass balance (2-26), we find

$$h_1^* = m(A + B)\exp(i[- t^* + mx^*]) \tag{2-39}$$

Equation (2-39) represents a wave travelling to the right (see assumption ix).

The partial derivatives of the first perturbation of the pressure distribution are determined from Cauchy's first law (2-3) for an incompressible Newtonian fluid. Carrying out a line integration from the point $x^\star = 0$ on the phase interface, we find

$$\mathcal{P}_1^\star = B \exp(- it^\star + imx^\star + mz^\star) \tag{2-40}$$

Here we have recognized that, in order for (2-28), the first perturbation of the normal component of the jump momentum balance, to be satisfied, we must require

$$\frac{A}{B} = \frac{m^3 - 2im^2 - N_{Re} + N_{Re} N_g m}{2im\ell - m^3 - N_{Re} N_g m} \tag{2-41}$$

Eliminating $\rho_{(S)1}^{(\sigma)\star}$ by (2-30), we find that (2-29), the first perturbation of the jump mass balance for surfactant S, fixes

$$D = im(\ell A + mB) \left(in \frac{N_{\mathcal{D}}}{N_k} + i\beta^\star - nN_{\mathcal{D}} \right)^{-1} \tag{2-42}$$

This, together with (2-37), permits us to rearrange (2-30) as

$$\rho_{(S)1}^{(\sigma)\star} = \frac{N_{\mathcal{D}}}{N_k} \frac{\partial \rho_{(S)1}^\star}{\partial z^\star} + \beta^\star \rho_{(S)1}^\star$$

$$= \left(n \frac{N_{\mathcal{D}}}{N_k} + \beta^\star \right) D \exp(- it^\star + imx^\star) \tag{2-43}$$

Finally, we find that (2-27), the first perturbation of the tangential component of the jump momentum balance, demands

$$(nN_{\mathcal{D}} + \beta^\star N_k)(nN_{\mathcal{D}} N_k - inN_{\mathcal{D}} - i\beta^\star N_k)^{-1} - N_{\kappa+\varepsilon} \, \gamma'^{\star -1} = C \tag{2-44}$$

in which we have introduced

$$C \equiv \frac{(\ell^2 + m^2)A/B + 2m^2}{(\ell A/B + m) m^2 \gamma'^\star} \tag{2-45}$$

$$N_{\kappa+\varepsilon} \equiv N_\kappa + N_\varepsilon \tag{2-46}$$

Equation (2-44) specifies m, if we are given a priori all pertinent physical parameters.

Alternatively, the real and imaginary portions of (2-44) can be used to determine N_k and $N_{\kappa+\varepsilon}$, assuming that we are given all of the other physical parameters as well as the real and imaginary portions of m:

$$m = m_1 + i m_2 \tag{2-47}$$

In order to determine m_1 and m_2, the experimentalist must measure the

$$\text{wave length} \equiv \frac{2\pi\gamma_0}{m_1 \alpha\mu} \tag{2-48}$$

and the

$$\text{damping coefficient} \equiv m_2 \frac{\alpha\mu}{\gamma_0} \tag{2-49}$$

of the waves. In what follows, we assume that this is the context in which this analysis will be used.

With the recognition that (Hegde and Slattery 1971)

$$| m^2 | < N_{Re} N_{Sc} \tag{2-50}$$

and therefore

$$n = \left[\frac{N_{Re} N_{Sc}}{2} \right]^{1/2} (1 - i) \tag{2-51}$$

the imaginary portion of (2-44) reduces to

$$X N_k^2 + Y N_k + Z = 0 \tag{2-52}$$

For convenience, we have introduced here

$$X \equiv (2u^2 + \beta^{*2} + 2\beta^* u) \mathrm{Im}(C) - \beta^* u - \beta^{*2} \tag{2-53}$$

$$Y \equiv 2\beta^* u [\mathrm{Im}(C) - 1] \tag{2-54}$$

$$Z \equiv 2u^2 [\mathrm{Im}(C) - 1] \tag{2-55}$$

$$u \equiv N_{\mathcal{D}} \left[\frac{N_{Re} N_{Sc}}{2} \right]^{1/2} \tag{2-56}$$

where Im(C) denotes the imaginary portion of C. Since N_k must be positive, the wave length and damping coefficient or, equivalently, m_1 and m_2 [see (2-48) and (2-49)] must be such that

$$Y^2 - 4XZ \geq 0 \tag{2-57}$$

and either

$$\frac{Z}{X} < 0 \tag{2-58}$$

in the case of one positive root to (2-52) or

$$\frac{Z}{X} > 0 \,, \; -\frac{Y}{X} > 0 \tag{2-59}$$

in the case of two positive roots to (2-52). It is obvious that the signs of Z/X and Y/X are always the same and (2-59) can never be satisfied. For one positive root to exist, (2-57) and (2-58) require

$$\frac{\beta^* u + \beta^{*2}}{2u^2 + 2\beta^* u + \beta^{*2}} < \mathrm{Im}(\,C\,) < 1 \tag{2-60}$$

Having solved for N_k using (2-52), we can compute $N_{\kappa+\varepsilon}$ from the real portion of (2-44). Since $\gamma'^* < 0$ (surface tension decreases with increasing surface concentration),

$$-\,N_{\kappa+\varepsilon}\,\gamma'^{*-1} = \mathrm{Re}(C) - \frac{2u^2 N_k + \beta u N_k^2}{u^2(\,N_k - 1\,)^2 + (\,uN_k + u + \beta^* N_k\,)^2}$$

$$\geq 0 \tag{2-61}$$

in order for $N_{\kappa+\varepsilon}$ to be positive. Here Re(C) indicates the real portion of C.

Inequalities (2-60) and (2-61), together with (2-45) and (2-47) through (2-49), place constraints on the experimentally measured values of the wave length and damping coefficient. Should these constraints be violated, it would indicate that a physical property had been estimated incorrectly, that an error had been made in observing either the wave length or the damping coefficient, or that the analysis developed here did not represent the physical system being observed.

Limiting cases In the preceding analysis, we have assumed that both diffusion and adsorption are important. This is not always the case. There are various limits that can be considered corresponding to rapid adsorption, slow adsorption, rapid diffusion, slow diffusion, or some combination of these cases.

The characteristic length introduced in (2-10) is convenient from the point of view of the experimentalist, but it is not helpful in discussing these limiting cases. The problem is that with this scaling the magnitude of $\partial \rho^{\star}_{(S)1}/\partial z^{\star}$ is not $O(1)$ in (2-29) and (2-30). As a result, $N_{\mathcal{D}}$ does not characterize the order of magnitude of the effects of diffusion in the mass transfer process at the interface.

For the purpose of characterizing the effects of diffusion at the interface, a more appropriate choice for the characteristic length is

$$L_m \equiv \frac{\rho^{(\sigma)}_{(S)0}}{\rho_{(S)0}} \tag{2-62}$$

If we use L_m rather than L_0, (2-29) and (2-30) take the same form, but $N_{\mathcal{D}}$ is replaced by

$$N'_{\mathcal{D}} \equiv \frac{\mathcal{D}_{(SA)}}{\alpha}\left(\frac{\rho_{(S)0}}{\rho^{(\sigma)}_{(S)0}}\right)^2 \tag{2-63}$$

Our approach in discussing the limiting cases that follow is to use L_m and $N'_{\mathcal{D}}$ when discussing the relative importance of terms in (2-29) and (2-30). Having chosen perhaps to discard a term, we return to using L_0 and $N_{\mathcal{D}}$ when analyzing experimental data.

Limiting cases: no exchange If $N'_{\mathcal{D}} << 1$, diffusion will be too slow to respond to the oscillations of the interface. If $N_k << 1$, adsorption will be slow to respond. In either limit, (2-16) and (2-17) indicate that the rate of mass transfer between the bulk phase and the interface vanishes.

In terms of our original dimensionless variables, (2-44) reduces in this limit to

$$i\gamma'^{\star} - N_{\kappa+\varepsilon} = \gamma'^{\star} C \tag{2-64}$$

In order that this limiting case be applicable, it is necessary that the measured values of wave length and damping coefficient be such that [see (2-47) through (2-49)]

$$-\gamma'^{\star} = \operatorname{Im}(G) > 0 \tag{2-65}$$

and

$$N_{\kappa+\varepsilon} = \operatorname{Re}(G) \geq 0 \tag{2-66}$$

where

$$G \equiv -\frac{(\ell^2 + m^2)A/B + 2m^2}{(\ell A/B + m)m^2} \tag{2-67}$$

Should these constraints be violated, it would indicate that a physical property had been estimated incorrectly, that an error had been made in observing either the wave length or the damping coefficient, or that this limiting case did not represent the physical system being observed.

Limiting cases: diffusion controlled If $N'_{\mathcal{D}} \sim 0(1)$ and $N_k >> 1$, diffusion occurs much more slowly than adsorption, and it controls the mass transfer process. Equation (2-44) reduces in this limit to

$$\frac{\beta^*}{nN_{\mathcal{D}} - i\beta^*} - N_{\kappa+\varepsilon}\,\gamma'^{*-1} = C \tag{2-68}$$

In order that this limiting case be applicable, it is necessary that the measured values of wave length and damping coefficient be such that [see (2-47) through (2-49)]

$$0 < \mathrm{Im}(C) < 1 \tag{2-69}$$

$$\frac{4N_{\mathcal{D}}}{\beta^*}\left[\frac{N_{Re}N_{Sc}}{2}\right]^{1/2}$$

$$= \frac{4u}{\beta^*}$$

$$= [\,\mathrm{Im}(C)]^{-1}\left[1 - 2\mathrm{Im}(C) + \{1 + 4\mathrm{Im}(C) - 4\mathrm{Im}^2(C)\}^{1/2}\right]$$

$$> 0 \tag{2-70}$$

and

$$-N_{\kappa+\varepsilon}\,\gamma'^{*-1}$$

$$= \mathrm{Re}(C)$$

$$- \mathrm{Im}(C)[\,1 - 2\mathrm{Im}(C) + \{(1 + 4\mathrm{Im}(C) - 4\mathrm{Im}^2(C)\}^{1/2}\,]$$

$$\times [\,1 + 2\mathrm{Im}(C) + \{1 + \mathrm{Im}(C) - 4\mathrm{Im}^2(C)\}^{1/2}\,]^{-1}$$

$$\geq 0 \tag{2-71}$$

Should these constraints be violated, it would indicate that a physical property had been estimated incorrectly, that an error had been made in observing either the wave length or the damping coefficient, or that this limiting case did not represent the physical system being observed.

Limiting cases: adsorption controlled If $N_k \sim 0(1)$ and $N'_{\mathcal{D}} >> 1$, adsorption occurs much more slowly than diffusion, and it controls the exchange process. For this limit, (2-44) simplifies to

$$\frac{1}{N_k - i} - N_{\kappa+\varepsilon}\,\gamma'^{\star-1} = C \tag{2-72}$$

In order that this limiting case be applicable, it is necessary that the measured values of wave length and damping coefficient be such that [see (2-47) through (2-49)]

$$N_k = \left\{ \frac{[\ 1 - \mathrm{Im}(\ C\ )]}{\mathrm{Im}(\ C\ )} \right\}^{1/2} > 0 \tag{2-73}$$

and

$$-\ N_{\kappa+\varepsilon}\,\gamma'^{\star-1} = \mathrm{Re}(\ C\ ) - \{\ \mathrm{Im}(\ C\ )[\ 1 - \mathrm{Im}(\ C\ )]\}^{1/2} \geq 0 \tag{2-74}$$

as well as inequality (2-69). Should these constraints be violated, it would indicate that a physical property had been estimated incorrectly, that an error had been made in observing either the wave length or the damping coefficient, or that this limiting case did not represent the physical system being observed.

Limiting cases: instantaneous diffusion and adsorption If both $N'_{\mathcal{D}} >> 1$ and $N_k >> 1$, diffusion and adsorption occur nearly instantaneously. This means that at all times the interface will be nearly in equilibrium with the adjacent bulk phase and there will be no interfacial tension gradient. In this limit, (2-44) requires

$$-\ N_{\kappa-\varepsilon}\,\gamma'^{\star-1} = \mathrm{Re}(\ C\ ) \geq 0 \tag{2-75}$$

and

$$\mathrm{Im}(\ C\ ) = 0 \tag{2-76}$$

Should these constraints be violated, it would indicate that a physical property had been estimated incorrectly, that an error had been made in observing either the wave length or the damping coefficient, or that this limiting case did not represent the physical system being observed.

Discussion We would like to emphasize that this analysis applies both to transverse (or *capillary*) waves and to longitudinal waves, so long as the wave length and the damping coefficient are measured at a point sufficiently far away from the oscillating knife-edge generating the waves. But it may help in evaluating our results to note that Lucassen-Reynders and Lucassen (1969) have observed

$$m_1 \sim 0(\, m_2 \,) \tag{2-77}$$

for longitudinal waves whereas

$$m_1 >> m_2 \tag{2-78}$$

for transverse waves.

The results of van den Tempel and van de Riet (1964) for an insoluble monolayer are in agreement with our limiting case of no exchange of mass between the interface and the adjacent liquid phase, when we neglect the effects of the interfacial viscosities in our results. However, they did not recognize the constraints imposed by (2-65) and (2-66).

Their analysis for diffusion controlled mass transfer also agrees with that obtained here, when we neglect the effects of the interfacial viscosities in our results. They did not obtain the limits of applicability described by (2-69) through (2-71).

Hegde and Slattery (1971) incorrectly assume that surface tension is a function of the concentration of surfactant in the liquid phase immediately adjacent to the interface, while at the same time accounting for the rate of adsorption of surfactant in the interface. In the limit of diffusion controlled mass transfer, their results are correct and do include the effect of the surface viscosities. Like van den Tempel and van de Riet (1964), they do not derive the limits of applicability described by (2-69) through (2-71).

Neither van den Tempel and van de Riet (1964) nor Hegde and Slattery (1971) realized that their analyses applied to both longitudinal waves and to transverse waves.

Unfortunately, there are currently no experimental studies that are sufficiently complete and accurate to permit evaluation by our analysis. From Exercises 2-1 and 2-2, we see that, in order to estimate $\gamma'^{\star}$ and $\beta^{\star}$, we must have measurements of the equilibrium surface tension as a function of the concentration of surfactant. While there are a number of studies that provide measurements of the wave length and damping coefficient, none provide independent measurements of the equilibrium surface tension as a function of concentration for their systems. We attempted to evaluate the data of Lucassen and Hansen (1966, 1967) using equilibrium surface tension data from another source. Over most of the concentration range, either Re(C) or Im(C) or both were negative, contradicting (2-60) and (2-61). Either our surface tension data did not accurately represent the behavior of their system or the experimental measurements of Lucassen and Hansen (1966, 1967) were not sufficiently accurate.

As an alternative to an evaluation of experimental data, we present a parametric study to illustrate the effects of surface properties upon m_1 and

m_2 or, equivalently, upon the wave length and damping coefficient [see (2-48) and (2-49)].

We can typically expect $10^4 < N_{Re} < 10^8$ [which follows from (2-8) and our estimates that $\rho \sim 0(1)$ gm/cm^3, $\gamma_0 \sim 0(10)$ mN/m, $\mu \sim 0(1)$ mN s/m^2, $1 < \alpha < 10^4$ s^{-1}]. In what follows, we will restrict our attention to $N_{Re} = 10^5$. The results for $N_{Re} = 10^4$ and $N_{Re} = 10^8$ are qualitatively similar.

For simplicity, we have also set $N_g = 0$. [For $\alpha = 10^3$ s^{-1} corresponding to $N_{Re} = 10^5$, it follows from (2-15) that $N_g = 10^{-3}$.] Our result for capillary waves driven by thermal noise (see next section) suggest that the effects of N_g are generally small.

Let us begin by examining the various limiting cases.

Discussion: no exchange For the limiting case of no exchange of mass between the interface and the adjacent liquid phase, Figure 2-1 shows how

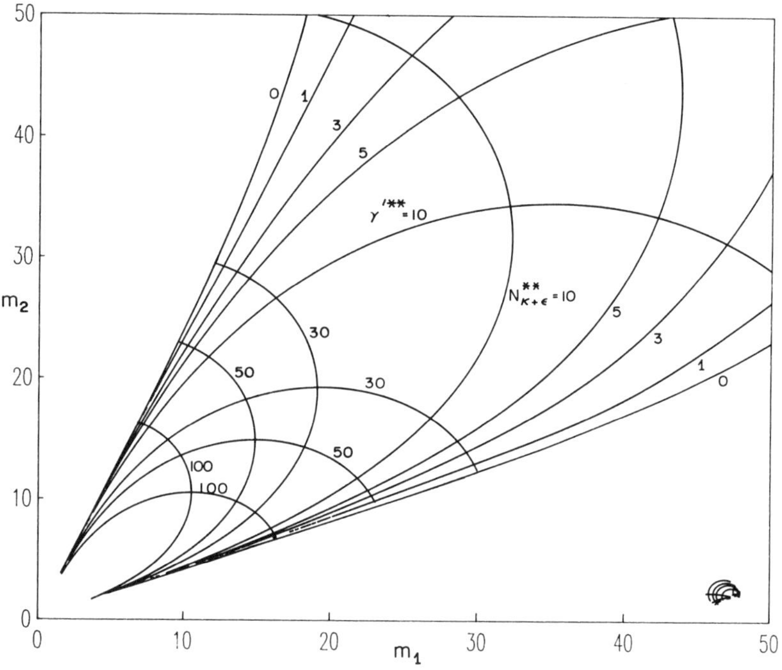

Figure 6.4.2-1: No exchange of mass between the interface and the adjacent liquid phase: range of applicability of longitudinal wave solution for $N_{Re} = 10^5$.

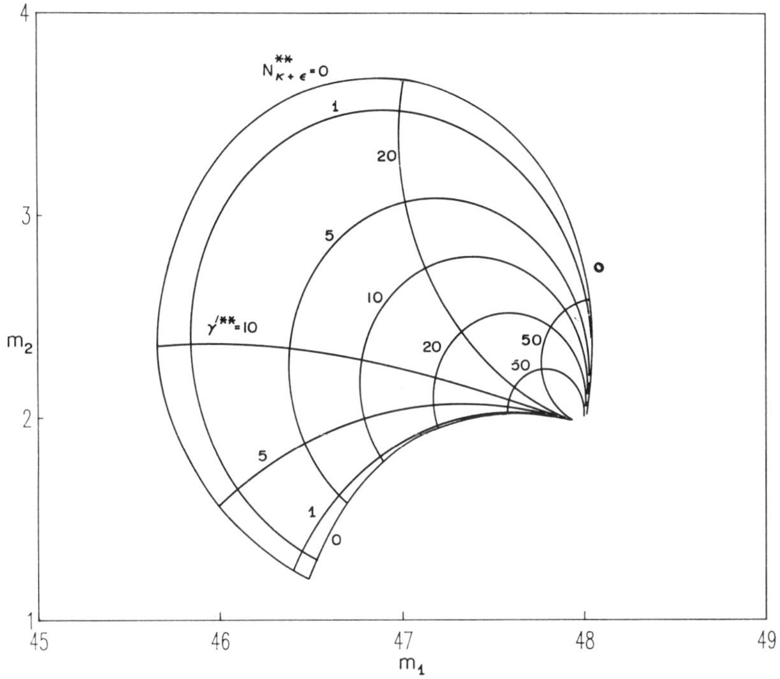

Figure 6.4.2-2: No exchange of mass between the interface and the adjacent liquid phase: range of applicability of transverse wave solution for $N_{Re} = 10^5$.

$$\gamma'^{**} \equiv -100\,\gamma'^{*} \tag{2-79}$$

$$N^{**}_{\kappa+\varepsilon} \equiv 100\,N_{\kappa+\varepsilon} \tag{2-80}$$

change as functions of m_1 and m_2. The region that is empty corresponds to the values of m_1 and m_2 for which (2-65) and (2-66) are not satisfied.

Note that there are two regions for which (2-65) and (2-66) are satisfied. The region where $m_1 \sim O(m_2)$ corresponds to longitudinal waves and the region where $m_1 >> m_2$ to transverse waves (Lucassen-Reynders and Lucassen 1969). This latter region is enlarged in Figure 2-2.

Discussion: diffusion controlled For the limiting case of diffusion controlled mass transfer between the interface and the adjacent liquid phase, Figures 2-3 through 2-6 indicate how

$$N_{\mathcal{D}}^{**} \equiv \frac{4u}{\beta^*}$$

$$= \frac{4N_{\mathcal{D}}}{\beta}\left(\frac{N_{Re}N_{Sc}}{2}\right)^{1/2}$$

$$= 894\left(\frac{N_{\mathcal{D}}(N_{Sc})^{1/2}}{\beta^*}\right) \tag{2-81}$$

and $N_{\kappa+\varepsilon}^{**}$ change as functions of m_1 and m_2 for two typical values of $-\gamma'^*$: 0.1 and 1. [From Lucassen and Hansen (1966, 1967), $c^{(\sigma)\infty}RT \sim 0(10)$ mN/m, $M \sim 0(10)$, $M_{(S)} \sim 0(100)$, $\rho_{(S)} \sim 0(10^{-5}$ or $10^{-6})$ gm/cm^3, $a \sim 0(10^{-4})$. Our estimate of γ'^* follows from Exercise 2-1.] Figures 2-3 and 2-4 apply to longitudinal waves; Figures 2-5 and 2-6 to transverse waves. The open regions are outside the range of validity of (2-69) through (2-71).

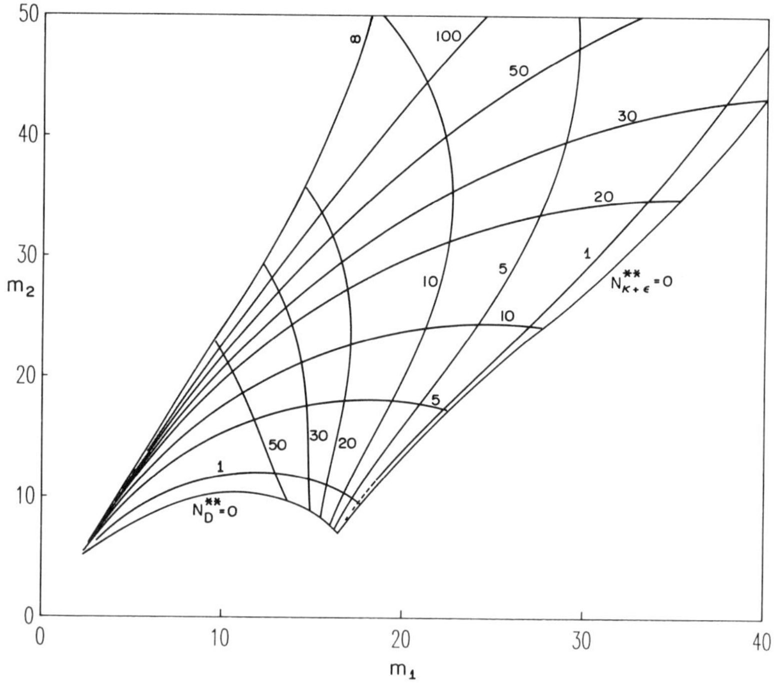

Figure 6.4.2-3: Diffusion controlled mass transfer between the interface and the adjacent liquid phase: range of applicability of longitudinal wave solution for $N_{Re} = 10^5$ and $-\gamma'^* = 1$.

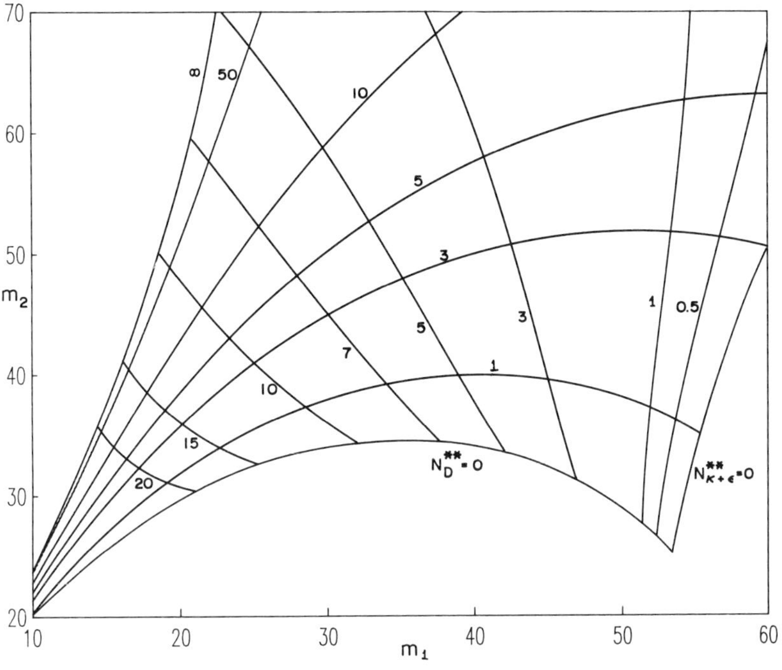

Figure 6.4.2-4: Diffusion controlled mass transfer between the interface and the adjacent liquid phase: range of applicability of longitudinal wave solution for $N_{Re} = 10^5$ and $-\gamma'^{*} = 0.1$.

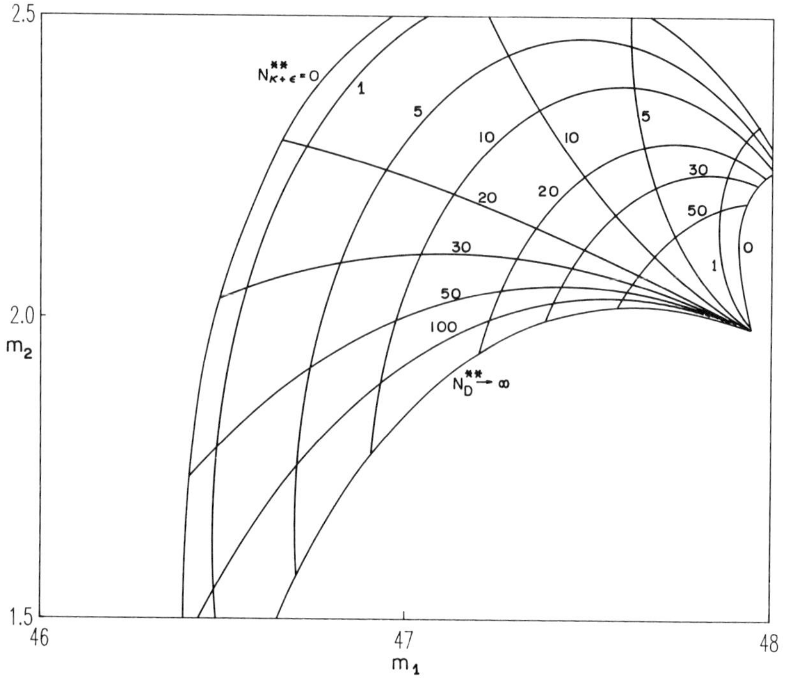

Figure 6.4.2-5: Diffusion controlled mass transfer between the interface and the adjacent liquid phase: range of applicability of transverse wave solution for $N_{Re} = 10^5$ and $-\gamma'^{*} = 1$.

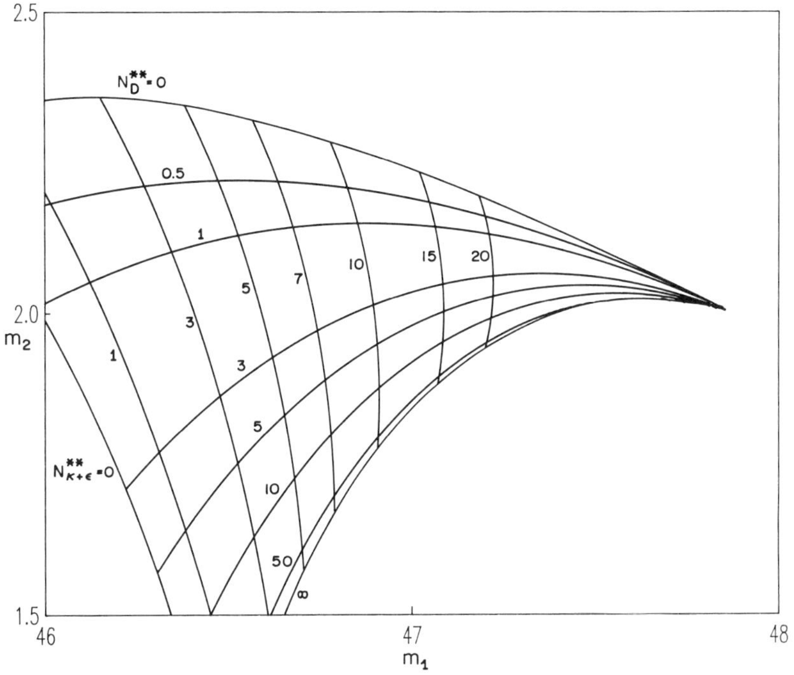

Figure 6.4.2-6: Diffusion controlled mass transfer between the interface and the adjacent liquid phase: range of applicability of transverse wave solution for $N_{Re} = 10^5$ and $-\gamma'^{\star} = 0.1$.

Discussion: adsorption controlled For the limiting case of adsorption controlled mass transfer between the interface and the adjacent liquid phase, Figures 2-7 through 2-10 illustrate how N_k and $N_{\kappa+\varepsilon}^{\star\star}$ vary as functions of m_1 and m_2 for $-\gamma'^{\star} = 0.1$ and 1. Figures 2-7 and 2-8 apply to longitudinal waves; Figures 2-9 and 2-10 to transverse waves. The open regions are outside the range of validity of (2-69), (2-73), and (2-74).

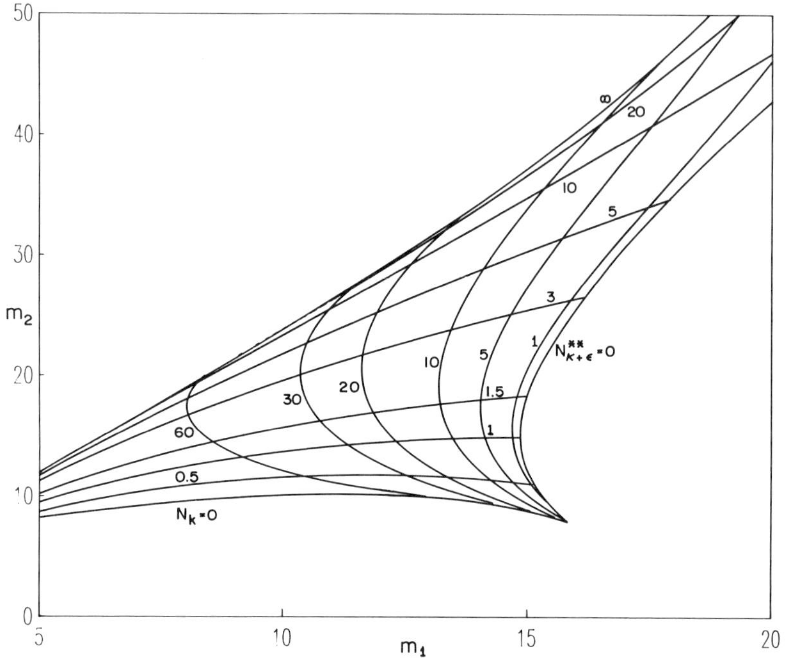

Figure 6.4.2-7: Adsorption controlled mass transfer between the interface and the adjacent liquid phase: range of applicability of longitudinal wave solution for $N_{Re} = 10^5$ and $-\gamma'^* = 1$.

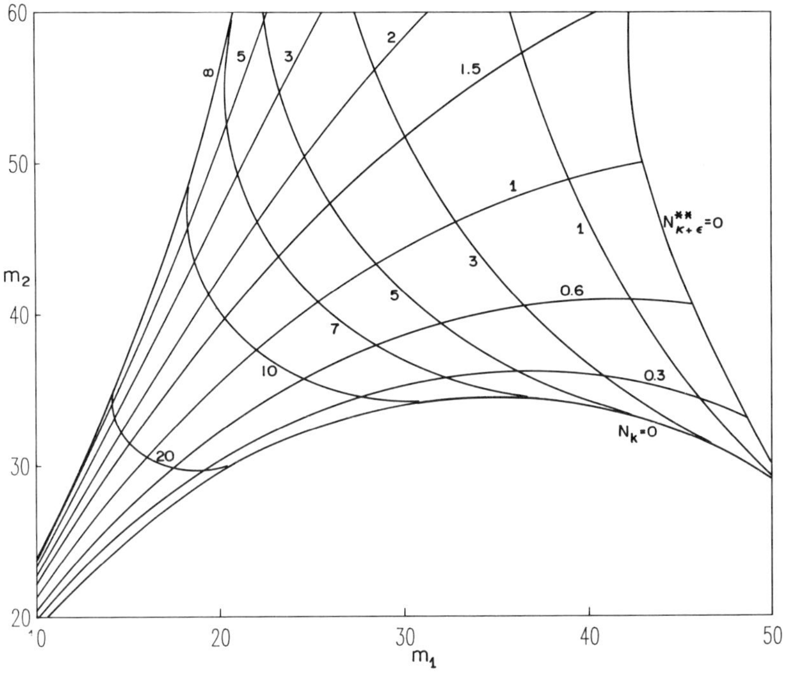

Figure 6.4.2-8: Adsorption controlled mass transfer between the interface and the adjacent liquid phase: range of applicability of longitudinal wave solution for $N_{Re} = 10^5$ and $-\gamma'^* = 0.1$.

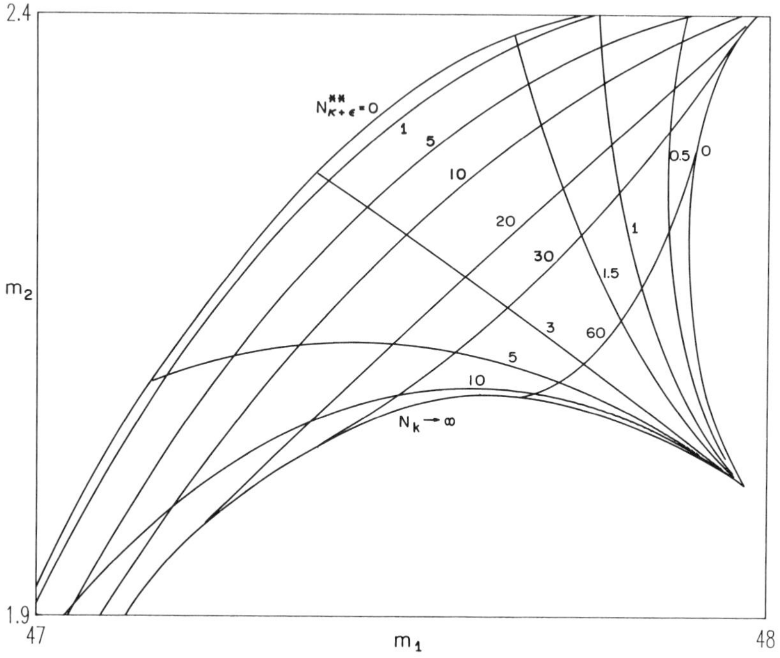

Figure 6.4.2-9: Adsorption controlled mass transfer between the interface and the adjacent liquid phase: range of applicability of transverse wave solution for $N_{Re} = 10^5$ and $-\gamma'^* = 1$.

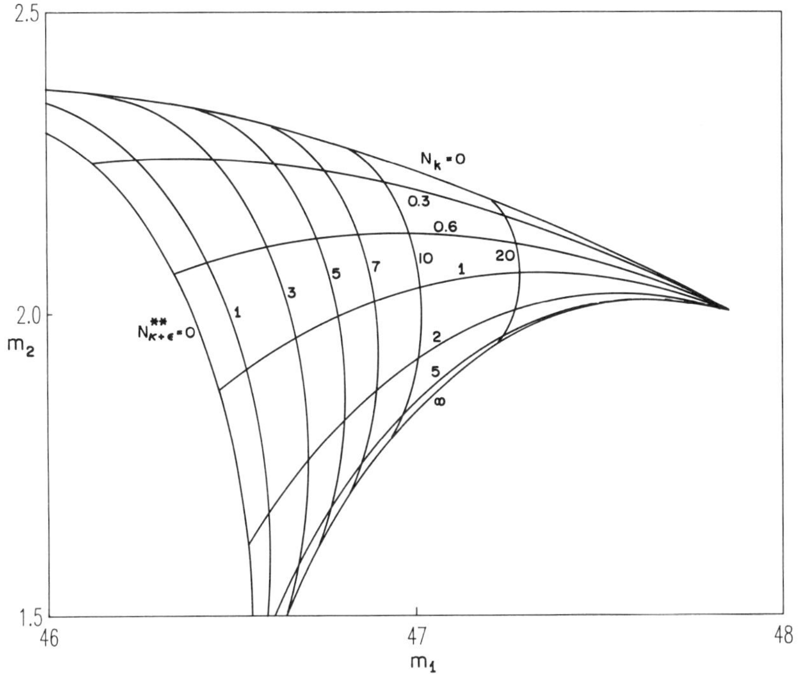

Figure 6.4.2-10: Adsorption controlled mass transfer between the interface and the adjacent liquid phase: range of applicability of transverse wave solution for $N_{Re} = 10^5$ and $-\gamma'^* = 0.1$.

Discussion: instantaneous diffusion and adsorption For the limiting case of instantaneous diffusion and adsorption or instantaneous mass transfer between the interface and the adjacent liquid phase, the interface is always in equilibrium with the liquid phase. This correspond to the limiting case of no exchange shown in Figures 2-1 and 2-2 with $\gamma'^* = 0$.

Discussion: general case For the general case in which both diffusion and adsorption are important, Figures 2-11 through 2-14 illustrate how N_k and $N^{**}_{\kappa+\varepsilon}$ vary as functions of m_1 and m_2 for $-\gamma'^* = 0.1$ and 1 and $\beta^* = 1$. [The argument (see ***Discussion: diffusion controlled***) leading to an estimate of γ'^* can be used together with Exercise 2-2 to estimate β^*.] Figures 2-11 and 2-12 apply to longitudinal waves; Figures 2-13 and 2-14 to transverse waves. The open regions are outside the range of validity of (2-60) and (2-61).

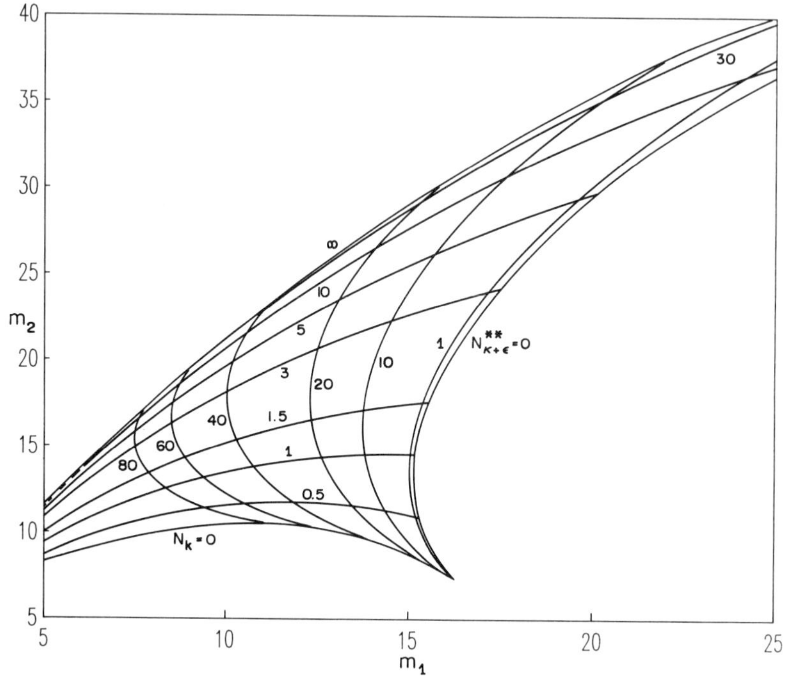

Figure 6.4.2-11: General case in which both diffusion and adsorption are important: range of applicability of longitudinal wave solution for $N_{Re} = 10^5$, $-\gamma'^{*} = 1$, $u = 10$, and $\beta^{*} = 1$.

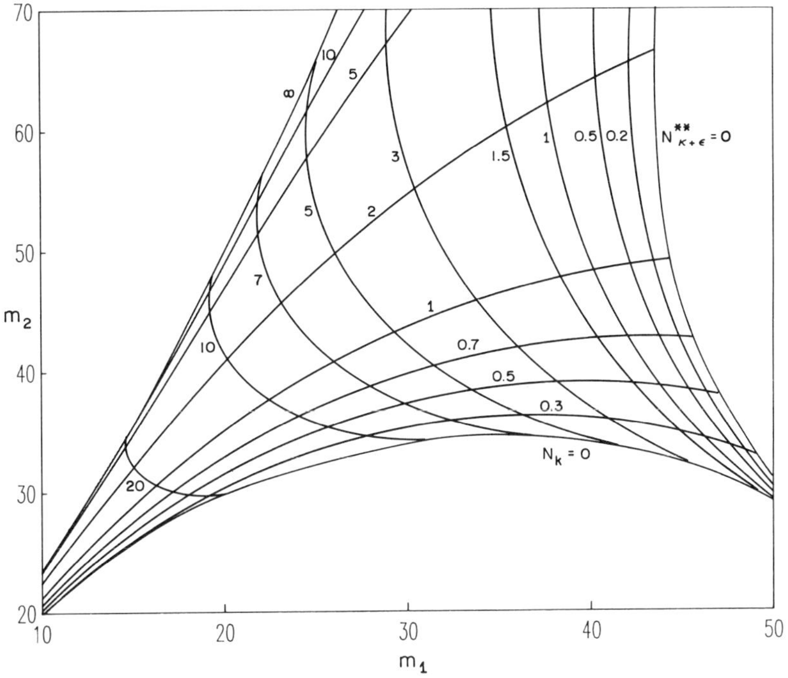

Figure 6.4.2-12: General case in which both diffusion and adsorption are important: range of applicability of longitudinal wave solution for $N_{Re} = 10^5$, $-\gamma'^{*} = 0.1$, $u = 10$, and $\beta^{*} = 1$.

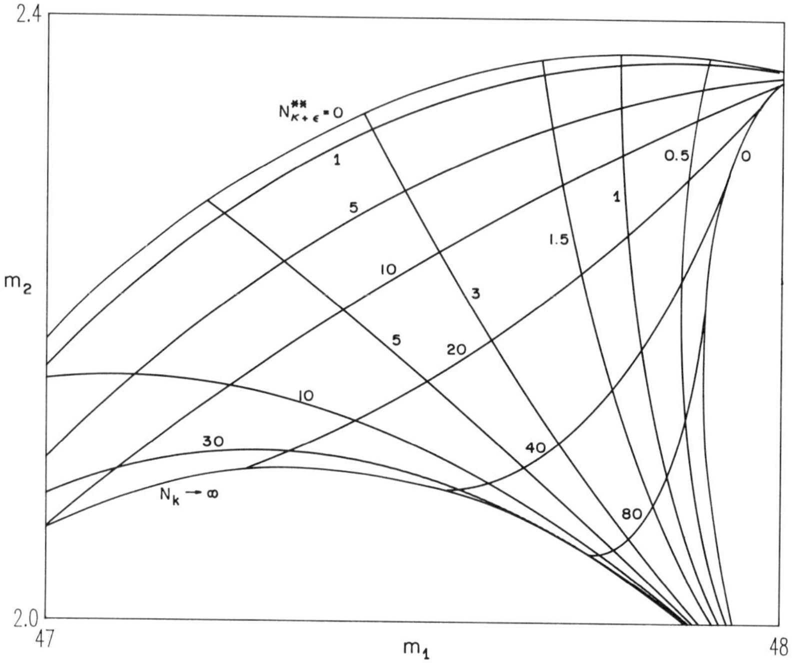

Figure 6.4.2-13: General case in which both diffusion and adsorption are important: range of applicability of transverse wave solution for $N_{Re} = 10^5$, $-\gamma'^* = 1$, $u = 10$, and $\beta^* = 1$.

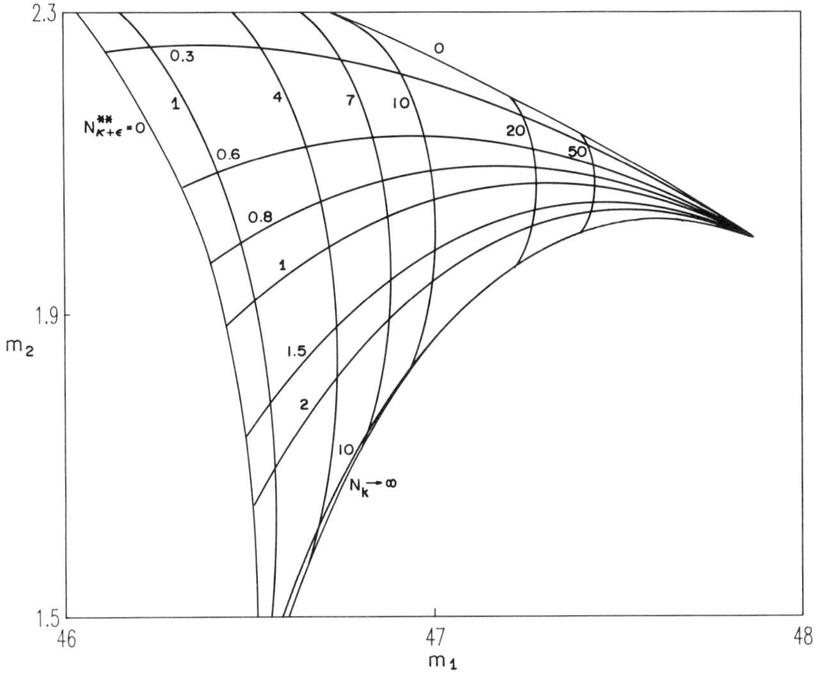

Figure 6.4.2-14: General case in which both diffusion and adsorption are important: range of applicability of transverse wave solution for $N_{Re} = 10^5$, $-\gamma'^* = 0.1$, u = 10, and $\beta^* = 1$.

General observations Whether one is studying longitudinal waves or transverse waves, we can not overemphasize the importance of choosing the correct model to represent the surfactant mass exchange between the surface and the adjacent phase. As an illustration, assume that we have measured $m_1 = 45$ and $m_2 = 42$ with a system for which $N_{Re} = 10^5$ and $-\gamma'^* = 0.1$. From Figures 2-1, 2-4, 2-8, and 2-12, we see that, depending upon the mechanism chosen to represent the mass exchange, the following result will be obtained for $N^{**}_{\kappa+\varepsilon}$.

a) no exchange: $N^{**}_{\kappa+\varepsilon} = 4.64$

b) diffusion controlled: $N^{**}_{\kappa+\varepsilon} = 2.80$

c) adsorption controlled: $N^{**}_{\kappa+\varepsilon} = 0.06$

d) general case: $N^{**}_{\kappa+\varepsilon} = 0.22$ (assuming in addition that u = 10 and $\beta^* = 1$).

In order to choose among these results, it is necessary that we know a priori only $N'_{\mathcal{D}}$ [defined by (2-63)]. If $N'_{\mathcal{D}} << 1$, alternatives b through d are eliminated and we are left with only one choice. If $N'_{\mathcal{D}} >> 1$, only alternative c is appropriate. If $N'_{\mathcal{D}} \sim 0(1)$, alternatives b and d are possible. In order to choose between them, assume that alternative b is correct and compute the corresponding value of $\mathcal{D}_{(SA)}$. If it agrees with the value given a priori, alternative b is confirmed as being correct. If it does not, alternative d should be considered.

In this solution, we have not considered the effects of any reflections from the walls of the container (assumption viii). Yet this is clearly an important consideration for the experimentalist. Let L denote the length of the container as measured in the x direction from the oscillating knife-edge to the boundary. From (2-39), we must require that

$$m_2 \frac{L\alpha\mu}{\gamma_0} >> 1 \tag{2-82}$$

in order to ensure that the waves are fully damped before they reach this boundary. It is only in this way that we can eliminate the effects of reflections.

References

Bendure, R. L., and R. S. Hansen, *J. Phys. Chem.* **71**, 2889 (1967).

Crone, A. H. M., A. F. M. Snik, J. A. Poulis, A. J. Kruger, and M. van den Tempel, *J. Colloid Interface Sci.* **74**, 1 (1979).

Davies, J. T., and R. W. Vose, *Proc. Roy. Soc. London Ser. A* **286**, 218 (1965).

Goodrich, F. C., *Proc. Roy. Soc. London Ser. A* **260**, 481 (1961a).

Goodrich, F. C., *Proc. Roy. Soc. London Ser. A* **260**, 490 (1961b).

Goodrich, F. C., *Proc. Roy. Soc. London Ser. A* **260**, 503 (1961c).

Goodrich, F. C., *J. Phys. Chem.* **66**, 1858 (1962).

Graham, D. E., and M. C. Phillips, *J. Colloid Interface Sci.* **76**, 227 (1980).

Hajiloo, A., and J. C. Slattery, *J. Colloid Interface Sci.* **111**, 169 (1986).

Hansen, R. S., J. Lucassen, R. L. Bendure, and G. P. Bierwagen, *J. Colloid Interface Sci.* **26**, 198 (1968).

Hansen, R. S., and J. A. Mann Jr., *J. App. Phys.* **35**, 152 (1964).

Hegde, M. G., and J. C. Slattery, *J. Colloid Interface Sci.* **35**, 183 (1971).

Lamb, H., "Hydrodynamics," p. 456, Dover, New York (1945).

Lucassen, J., *Trans. Faraday Soc.* **64**, 2221 (1968a).

Lucassen, J., *Trans. Faraday Soc.* **64**, 2230 (1968b).

Lucassen, J., and G. T. Barnes, *J. Chem. Soc. Faraday Trans. I* **68**, 2129 (1972).

Lucassen, J., and D. Giles, *J. Chem. Soc. Faraday Trans. I* **71**, 217 (1975).

Lucassen, J., and R. S. Hansen, *J. Colloid Interface Sci.* **22**, 32 (1966).

Lucassen, J., and R. S. Hansen, *J. Colloid Interface Sci.* **23**, 319 (1967).

Lucassen, J., and E. H. Lucassen-Reynders, *J. Colloid Interface Sci.* **25**, 496 (1967).

Lucassen, J., and M. van den Tempel, *Chem. Eng. Sci.* **27**, 1283 (1972a).

Lucassen, J., and M. van den Tempel, *J. Colloid Interface Sci.* **41**, 491 (1972b).

Lucassen-Reynders, E. H., and J. Lucassen, *Adv. Colloid Interface Sci.* **2**, 347 (1969).

Mann, J. A. Jr., and G. Du, *J. Colloid Interface Sci.* **37**, 2 (1971).

Mann, J. A. Jr., and R. S. Hansen, *J. Colloid Sci.* **18**, 757 (1963).

Maru, H. C., V. Mohan, and D. T. Wasan, *Chem. Eng. Sci.* **34**, 1283 (1979).

Maru, H. C., and D. T. Wasan, *Chem. Eng. Sci.* **34**, 1295 (1979).

Mayer, E., and J. D. Eliassen, *J. Colloid Interface Sci.* **37**, 228 (1971).

Miyano, K., B. M. Abraham, L. Ting, and D. T. Wasan, *J. Colloid Interface Sci.* **92**, 297 (1983).

Slattery, J. C., "Momentum, Energy, and Mass Transfer in Continua," McGraw-Hill, New York (1972); second edition, Robert E. Krieger, Malabar, FL (1981).

Sohl, C. H., K. Miyano, and J. B. Ketterson, *Rev. Sci. Instrum.* **49**, 1464

(1978).

van den Tempel, M., and R. P. van de Riet, *J. Chem. Phys.* **42**, 2769 (1964).

Exercises

6.4.2-1 *estimate of* γ'^{*} Start with Exercise 6.4.1-1 to estimate

$$-\gamma'^{*} = \frac{RTM\ c^{(\sigma)\infty}\rho_{(S)0}}{a\,\rho M_{(S)}\gamma_0}$$

6.4.2-2 *estimate of* β^{*} Start with the Langmuir adsorption isotherm (Exercise 5.8.7-2) to estimate

$$\beta^{*} = \frac{aM_{(S)}\rho}{M\rho_{(S)0} + aM_{(S)}\rho}$$

Here M is the molar-averaged molecular weight of the solution, $M_{(S)}$ the molecular weight of the surfactant, and a is a parameter in the Langmuir adsorption isotherm. In the Lucassen-Reynders and van den Tempel convention, the total number of moles adsorbed in the dividing surface is a constant $c^{(\sigma)\infty}$.

6.4.3 Stochastic interfacial disturbances created by thermal noise and the importance of the interfacial viscosities (Hajiloo and Slattery 1986) Smoluchowski (1908) and Mandelstam (1913) observed the scattering of the light reflected from an interface between two static fluids and identified this phenomena with the existence of small amplitude random waves at the interface. Mandelstam (1913) related the mean value of the intensity of light scattered in a particular direction to the mean square of the amplitude of a Fourier component of these waves, which is inversely proportional to the interfacial tension but independent of the bulk viscosities and of the interfacial viscosities. (See below for details.) Lachaise *et al.* (1981) have shown that this can be employed as the basis for an experiment to measure the interfacial tension.

The appearance of lasers and the development of spectroscopic techniques made it possible to measure statistical properties of the scattered light that could be simply related to the statistical properties of the capillary waves. Papoular (1968), Meunier *et al.* (1969), and Bouchiat and Meunier (1971) developed expressions for the spectral density of the amplitude of any given Fourier component of these waves. Langevin and Bouchiat (1971; see also Langevin 1981) computed the spectral density assuming a form of surface viscoelasticity. Critiques of these developments are given below.

After an initial series of experiments that demonstrated the viability of the technique (Katyl and Ingard 1968; Bouchiat *et al.* 1968; Bouchiat and Meunier 1968, 1969; Huang and Webb 1969), Pouchelon *et al.* (1980) illustrated its potential for measuring ultra-low interfacial tensions. Kim et al. (1982) not only measured ultra-low interfacial tensions, but they also showed agreement between these measurements and those obtained using the sessile drop technique. There also have been several studies of the parametetrs in a model for surface viscoelasticity (Langevin 1981; Hard and Löfgren 1977; Byrne and Earnshaw 1979; Hard and Newman 1981).

More recently, Dorshow *et al.* (1988) presented the optical design of an adjustable resolution surface laser-light scattering spectrometer that enabled them to accurately measure the viscosity and the surface tension against air of several liquids. Dorshow and Swofford (1989a,b) have extended these measurements to crude oil both at ambient conditions and at elevated temperatures and pressures. The viscosities measured in this way at ambient conditions were in good agreement with those obtained using a low-shear viscometer; the interfacial tensions agreed well with those obtained using the Wilhelmy plate technique at ambient conditions and a pendant drop tensiometer at elevated temperatures and pressures (Dorshow and Pallas 1990).

Statement of problem Our objective here is to present a new analysis of this technique in which it is explicitly recognized that these small amplitude capillary waves are the result of thermal noise, a stochastic mutual force.

Consider two fluids at equilibrium in a closed, rigid, perfectly insulated container. Small amplitude capillary waves are created by thermal noise at the interface between these otherwise static fluids. The light from a laser beam reflected from this interface is dispersed by these capillary waves. We intend to relate some characteristics of this light dispersion to the interfacial tension and the two interfacial viscosities.

We will require the following assumptions about the physical problem.

i) The mutual force per unit mass $\mathbf{b}_{(I)}^{(j)}$ or thermal noise in phase j is representable as the gradient of a scalar potential $\phi^{(j)}$ that is a random function of time at any given position in space:

$$\mathbf{b}_{(I)}^{(j)} = - \nabla\Phi^{(j)} \tag{3-1}$$

ii) The fluids are incompressible and Newtonian.

iii) The fluid-fluid interfacial stress-deformation behavior can be represented by the linear Boussinesq surface fluid model (Sec. 2.2.2).

iv) The tangential components of velocity are continuous across the dividing surface.

v) The dividing surface is located in such a manner that

$$\rho^{(\sigma)} = 0 \tag{3-2}$$

vi) Any mass transfer to or from the dividing surface is neglected. The adjacent phases are in equilibrium. The characteristic time for the interfacial disturbances is so small that we can neglect any transfer of surfactant between the interface and the adjacent phases either by diffusion or by adsorption.

vii) The characteristic time for the interfacial disturbances is so small that we can neglect the effect of surface diffusion with respect to surface convection.

viii) The two adjacent phases are unbounded in the z direction but

$$-L \leq x \leq L$$

$$-L \leq y \leq L \tag{3-3}$$

The cross section of the container is square, measuring 2L on a side outside the immediate neighborhood and influence of the walls. We do not require the velocity vector to go to zero at these walls nor do we impose a fixed contact angle at the three-phase line of contact.

We will assume here that the positive direction along the coordinate z points into the lighter phase 2 in the opposite direction to gravity.

Governing equations and their solution Our first objective is to characterize the capillary waves generated by thermal noise.

In terms of dimensionless variables, the equation of continuity for each phase j = 1, 2 requires (Sec. 5.4.1)

$$\text{div } \mathbf{v}^{(j)\star} = 0 \tag{3-4}$$

and Cauchy's first law demands (Sec. 5.5.2)

$$\frac{\partial \mathbf{v}^{(1)\star}}{\partial t^\star} + \nabla \mathbf{v}^{(1)\star} \cdot \mathbf{v}^{(1)\star} = -\nabla \mathcal{P}^{(1)\star} + \frac{1}{N_{Re}} \text{div } \mathbf{S}^{(1)\star} \tag{3-5}$$

$$\frac{\partial \mathbf{v}^{(2)*}}{\partial t^*} + \nabla \mathbf{v}^{(2)*} \cdot \mathbf{v}^{(2)*} = - \nabla \mathcal{P}^{(2)*} + \frac{N_\mu}{N_\rho \, N_{Re}} \operatorname{div} \mathbf{S}^{(2)*} \tag{3-6}$$

Here

$$\mathcal{P}^{(j)} \equiv p^{(j)} + \rho^{(j)}\phi + \rho^{(j)}\Phi^{(j)} \tag{3-7}$$

with the understanding that the force per unit mass of gravity is representable as

$$\mathbf{b}_{(g)} = - \nabla \phi \tag{3-8}$$

In (3-4) through (3-6), we have introduced as dimensionless variables

$$\mathbf{v}^{(j)*} \equiv \frac{t_0 \mathbf{v}^{(j)}}{L_0} \qquad \mathcal{P}^{(j)*} \equiv \frac{t_0^2 (p^{(j)} - p_0)}{\rho^{(j)} L_0^2}$$

$$\mathbf{S}^{(j)*} \equiv \frac{t_0 \mathbf{S}^{(j)}}{\mu^{(j)}}$$

$$x^* \equiv \frac{x}{L_0} \qquad y^* \equiv \frac{y}{L_0}$$

$$z^* \equiv \frac{z}{L_0} \qquad t^* \equiv \frac{t}{t_0} \tag{3-9}$$

and as dimensionless parameters

$$N_\rho \equiv \frac{\rho^{(2)}}{\rho^{(1)}} \qquad N_\mu \equiv \frac{\mu^{(2)}}{\mu^{(1)}}$$

$$N_{Re} \equiv \frac{\rho^{(1)} L_0^2}{\mu^{(1)} \, t_0} \tag{3-10}$$

By t_0, we mean a characteristic time to be defined later; L_0 is a characteristic wavelength to be later; p_0 is the pressure that would exist at the dividing surface in the absence of thermal noise.

The overall jump mass balance or mass balance at the dividing surface (Sec. 5.4.1) is satisfied identically as the result of assumptions v and vi. The jump momentum balance or force balance at the dividing surface becomes in terms of dimensionless variables (Sec. 5.5.2)

$$\nabla_{(\sigma)}\gamma^* + 2H^*\gamma^*\boldsymbol{\xi} + N_\varepsilon N_{ca}\text{div}_{(\sigma)}\mathbf{S}^{(\sigma)*} + N_{Re}N_{ca}\left(\mathcal{P}^{(1)*} - \frac{1}{N_\rho}\mathcal{P}^{(2)*} \right)\boldsymbol{\xi}$$

$$+ N_\mu N_{ca}\mathbf{S}^{(2)*}\cdot\boldsymbol{\xi} - N_{ca}\mathbf{S}^{(1)*}\cdot\boldsymbol{\xi} - N_{Re}N_{ca}N_g(1 - N_\rho)\phi^*\boldsymbol{\xi}$$

$$- N_{Re}N_{ca}N_\Phi\Phi^*\boldsymbol{\xi} = 0 \qquad (3\text{-}11)$$

We define as dimensionless variables

$$H^* \equiv L_0 H \qquad \gamma^* \equiv \frac{\gamma}{\gamma_0}$$

$$\phi^* \equiv \frac{\phi}{gL_0} \qquad \Phi^* \equiv \frac{\rho^{(1)}\Phi^{(1)} - \rho^{(2)}\Phi^{(2)}}{\rho^{(1)}\Phi_0^{(1)} - \rho^{(2)}\Phi_0^{(2)}} \qquad (3\text{-}12)$$

and as dimensionless parameters

$$N_\varepsilon \equiv \frac{\varepsilon}{\mu^{(1)}L_0} \qquad N_\kappa \equiv \frac{\kappa}{\mu^{(1)}L_0}$$

$$N_{ca} \equiv \frac{\mu^{(1)}L_0}{\gamma t_0} \qquad N_g \equiv \frac{g t_0^2}{L_0}$$

$$N_\Phi \equiv \frac{(\rho^{(1)}\Phi_0^{(1)} - \rho^{(2)}\Phi_0^{(2)})t_0^2}{\rho^{(1)}L_0^2} \qquad (3\text{-}13)$$

Here $\Phi_0^{(j)}$ is a characteristic value of the thermal noise in phase j, g the magnitude of the acceleration of gravity, and γ_0 the equilibrium interfacial tension or the interfacial tension at the concentration of surfactant corresponding to a planar interface.

In view of assumptions vi and vii, the jump mass balance for surfactant S reduces to (Sec. 5.2.2)

$$\frac{\partial\rho_{(S)}^{\sigma*}}{\partial t^*} + \nabla_{(\sigma)}\rho_{(S)}^{\sigma*}\cdot(\mathbf{v}^{(\sigma)*} - \mathbf{u}^*) + \rho_{(S)}^{\sigma*}\,\text{div}_{(\sigma)}\mathbf{v}^{(\sigma)*} = 0 \qquad (3\text{-}14)$$

in which we have introduced as dimensionless variables

$$\rho_{(S)}^{\sigma*} \equiv \frac{\rho_{(S)}^{\sigma}}{\rho_{(S)0}^{\sigma}} \qquad \mathbf{u}^* \equiv \frac{t_0\mathbf{u}}{L_0}$$

$$v^{(\sigma)*} \equiv \frac{t_0 v^{(\sigma)}}{L_0} \tag{3-15}$$

By $\rho_{(s)0}$ we mean the surface mass density of surfactant corresponding to a planar interface.

Equations (3-4) through (3-6), (3-11) and (3-14) are to be solved simultaneously consistent with the requirements that velocity be continuous at the dividing surface (assumptions iv and vi), that very far away from the dividing surface both phases are static, that the bulk phases are Newtonian fluids (assumption ii), and that the interfacial stress-deformation behavior can be represented by the linear Boussinesq surface fluid model (assumption iii). In solving this set of equations, we will recognize that the interfacial tension can be described as a function of the surface concentration of surfactant by a series expansion of $\gamma(\rho_{(s)})$ with respect to $\rho_{(s)0}$:

$$\gamma^* = 1 + \gamma'^* (\rho_{(s)}^* - 1) + \ldots \tag{3-16}$$

in which γ'^* is the dimensionless Gibbs surface elasticity evaluated at the equilibrium surface concentration of surfactant.

We seek a perturbation solution to this problem, taking as our perturbation parameter N_Φ, a measure of the thermal noise generating the capillary waves.

The zeroth perturbations of all variables, corresponding to the static problem in the absence of thermal noise, are identically zero.

The first perturbations of (3-4) through (3-6) reduce to

$$\text{div } v_1^{(j)*} = 0 \tag{3-17}$$

$$\frac{\partial v_1^{(1)*}}{\partial t^*} = - \nabla \mathcal{P}_1^{(1)*} + \frac{1}{N_{Re}} \text{div}(\nabla v_1^{(1)*}) \tag{3-18}$$

$$\frac{\partial v_1^{(2)*}}{\partial t^*} = - \nabla \mathcal{P}_1^{(2)*} + \frac{N_\mu}{N_\rho N_{Re}} \text{div}(\nabla v_1^{(2)*}) \tag{3-19}$$

Let u_1^*, v_1^*, and w_1^* be the x^*, y^*, and z^* components of v_1^*, the first perturbation of velocity; let $\rho_{(s)1}^*$ be the first perturbation of $\rho_{(s)}^*$. In view of (3-16), the x^*, y^*, and z^* components of the first perturbation of (3-11) demand at $z^* = 0$

$$\frac{\gamma'^*}{N_{ca}} \frac{\partial \rho_{(s)1}^*}{\partial x^*} + (N_\kappa + N_\varepsilon) \frac{\partial^2 u_1^{(\sigma)*}}{\partial x^{*2}} + N_\varepsilon \frac{\partial^2 u_1^{(\sigma)*}}{\partial y^{*2}} + N_\kappa \frac{\partial^2 v_1^{(\sigma)*}}{\partial x^* \partial y^*}$$
$$= \frac{\partial u_1^{(1)*}}{\partial z^*} + \frac{\partial w_1^{(1)*}}{\partial x^*} - N_\mu \left(\frac{\partial u_1^{(2)*}}{\partial z^*} + \frac{\partial w_1^{(2)*}}{\partial x^*} \right) \tag{3-20}$$

$$\frac{\gamma'^{*}}{N_{ca}} \frac{\partial \rho\{g\}_1^*}{\partial y^*} + (N_\kappa + N_\varepsilon) \frac{\partial^2 v_1^{(\sigma)*}}{\partial y^{*2}} + N_\varepsilon \frac{\partial^2 v_1^{(\sigma)*}}{\partial x^{*2}} + N_\kappa \frac{\partial^2 u_1^{(\sigma)*}}{\partial x^* \partial y^*}$$

$$= \frac{\partial v_1^{(1)*}}{\partial z^*} + \frac{\partial w_1^{(1)*}}{\partial x^*} - N_\mu \left(\frac{\partial v_1^{(2)*}}{\partial z^*} + \frac{\partial w_1^{(2)*}}{\partial y^*} \right) \tag{3-21}$$

$$\frac{\partial^2 h_1^*}{\partial x^{*2}} + \frac{\partial^2 h_1^*}{\partial y^{*2}} - 2N_{ca} \left(\frac{\partial w_1^{(1)*}}{\partial z^*} - N_\mu \frac{\partial w_1^{(2)*}}{\partial z^*} \right)$$

$$+ N_{ca} N_{Re} (\mathcal{P}_1^{(1)*} - N_\rho \mathcal{P}_1^{(2)*}) - N_{ca} N_{Re} N_g (1 - N_\rho) h_1^*$$

$$- N_{ca} N_{Re} \Phi^*$$

$$= 0 \tag{3-22}$$

Our understanding here is that the configuration of the interface takes the form

$$z = h(x, y, t) = h_1 (x, y, t) N_\Phi + \ldots \tag{3-23}$$

with

$$h_1^* \equiv \frac{h_1}{L_0} \tag{3-24}$$

The first perturbation of (3-14) reduces to

$$\frac{\partial \rho\{g\}_1^*}{\partial t^*} + \frac{\partial u_1^{(\sigma)*}}{\partial x^*} + \frac{\partial v_1^{(\sigma)*}}{\partial y^*} = 0 \tag{3-25}$$

From (3-17) through (3-19), we can establish that

$$\left(N_{Re} \frac{\partial}{\partial t^*} - D^2 \right) \left(D^2 w_1^{(1)*} \right) = 0 \tag{3-26}$$

$$\left(N_{Re} N_\rho \frac{\partial}{\partial t^*} - N_\mu D^2 \right) \left(D^2 w_1^{(2)*} \right) = 0 \tag{3-27}$$

where

$$D^2 \equiv \frac{\partial^2}{\partial x^{*2}} + \frac{\partial^2}{\partial y^{*2}} + \frac{\partial^2}{\partial z^{*2}} \tag{3-28}$$

In arriving at (3-26) for example, we have differentiated the x^* component of (3-18) with respect to x^* and the y^* component of (3-18) with respect to y^*, we have added the results, and we have eliminated u_1^* and v_1^* using (3-17). We subsequently subtracted the derivative of this with respect to z^* from the equation obtained by operating $(\partial^2/\partial x^{*2} + \partial^2/\partial y^{*2})$ on the z^* component of (3-18).

In a similar fashion, we can take the derivative of (3-20) with respect to x^*, the derivative of (3-21) with respect to y^*, add the results, and take advantage of (3-17) to find that at $z^* = 0$

$$\frac{\gamma'^*}{N_{ca}} \left(\frac{\partial^2 \rho_{(s)1}^{(g)*}}{\partial x^{*2}} + \frac{\partial^2 \rho_{(s)1}^{(g)*}}{\partial y^{*2}} \right) - (N_\kappa + N_\varepsilon) \frac{\partial}{\partial z^*} \left(\frac{\partial^2 w_1^{(1)*}}{\partial x^{*2}} + \frac{\partial^2 w_1^{(1)*}}{\partial y^{*2}} \right)$$
$$+ \left(\frac{\partial^2}{\partial z^{*2}} - \frac{\partial^2}{\partial x^{*2}} - \frac{\partial^2}{\partial y^{*2}} \right) (w_1^{(1)*} - N_\mu w_1^{(2)*})$$
$$= 0 \tag{3-29}$$

In order to determine h_1^*, $w_1^{(j)*}$, $w_1^{(\sigma)*}$, and $\mathcal{P}_1^{(j)*}$ as functions of position and time, we must solve simultaneously (3-26), (3-27) as well as the x^* and y^* components of (3-18) and (3-19) consistent with (3-22), (3-25), (3-29) and the following boundary conditions. Since there is no mass transfer at the dividing surface (assumption vi), we have

$$\text{at } z^* = 0 : \quad w_1^{(1)*} = w_1^{(2)*} = w_1^{(\sigma)*} = \frac{\partial h^*}{\partial t^*} \tag{3-30}$$

Because the tangential components of velocity are continuous across the dividing surface (assumption iv), we can use the equation of continuity (3-17) to say

$$\text{at } z^* = 0 : \quad \frac{\partial w_1^{(1)*}}{\partial z^*} = \frac{\partial w_1^{(2)*}}{\partial z^*} \tag{3-31}$$

Finally, very far away from the dividing surface,

$$\text{as } z^* \to \infty : \quad w_1^{(2)*} \to 0 \tag{3-32}$$

and

$$\text{as } z^* \to -\infty : \quad w_1^{(1)*} \to 0 \tag{3-33}$$

Seeking a solution of the form

$$h_1^*, \quad w_1^{(j)*}, \quad w_1^{(\sigma)*}, \quad \mathcal{P}_1^{(j)*}, \quad \rho_{(S)1}^{\{S\}*} = \sum_{m,n=-\infty}^{\infty} [\, h_{mn}(t^*),$$

$$w_{mn}^{(j)}(t^*, z^*), \quad w_{mn}^{(\sigma)}(t^*), \quad \mathcal{P}_{mn}^{(j)}(t^*, z^*),$$

$$\rho_{(S)mn}^{\{S\}}(t^*)] \exp\left[i\pi(mx^* + ny^*)\frac{L_0}{L} \right] \tag{3-34}$$

we conclude from (3-26) and (3-27) that

$$\left(N_{Re}\frac{\partial}{\partial t^*} - D_1^2 \right)(D_1^2 w_{mn}^{(1)}) = 0 \tag{3-35}$$

$$\left(N_{Re}N_\rho \frac{\partial}{\partial t^*} - N_\mu D_1^2 \right)(D_1^2 w_{mn}^{(2)}) = 0 \tag{3-36}$$

in which

$$D_1^2 \equiv \frac{\partial^2}{\partial z^{*2}} - a^2 \tag{3-37}$$

$$a^2 \equiv \pi^2(m^2 + n^2)\frac{L_0^2}{L^2} \tag{3-38}$$

Here we have taken advantage of the orthogonality of the functions $\exp[\, i\pi(mx^* + ny^*)L_0/L \,]$. Using the x^* and y^* components of (3-18) and (3-19), we can establish

$$a^2 N_{Re}\mathcal{P}_{mn}^{(1)} = \left(\frac{\partial^2}{\partial z^{*2}} - a^2 - N_{Re}\frac{\partial}{\partial t^*} \right)\left(\frac{\partial w_{mn}^{(1)}}{\partial z^*} \right) \tag{3-39}$$

$$a^2 N_\rho N_{Re}\mathcal{P}_{mn}^{(2)} = \left(N_\mu \frac{\partial^2}{\partial z^{*2}} - N_\mu a^2 - N_\rho N_{Re}\frac{\partial}{\partial t^*} \right)\left(\frac{\partial w_{mn}^{(2)}}{\partial z^*} \right) \tag{3-40}$$

In arriving at (3-39) for example, we have differentiated the x^* component of (3-18) with respect to x^* and the y^* component of (3-18) with respect to y^*, we have added the results, and we have eliminated u_1^* and v_1^* using the equation of continuity. With the assumption that we can represent

$$\Phi^* = \sum_{m,n=-\infty}^{\infty} \Phi_{mn}(t) \exp\left[i\pi(mx^* + ny^*) \frac{L_0}{L} \right] \tag{3-41}$$

we find that (3-22) and (3-29) through (3-31) require at $z^* = 0$

$$[a^2 + N_{ca}N_{Re}N_g(1 - N_\rho)]h_{mn} + 2N_{ca}\left(\frac{\partial w_{mn}^{(1)}}{\partial z^*} - N_\mu \frac{\partial w_{mn}^{(2)}}{\partial z^*} \right)$$

$$- N_{ca}N_{Re}(\mathcal{P}_{mn}^{(1)} - N_\rho \mathcal{P}_{mn}^{(2)}) + N_{ca}N_{Re}\Phi_{mn} = 0 \tag{3-42}$$

$$-\frac{a^2\gamma'^*}{N_{ca}}\rho^{\{g\}}_{mn} + a^2(N_\kappa + N_\varepsilon)\frac{\partial w_{mn}^{(1)}}{\partial z^*}$$

$$+ \left(\frac{\partial^2}{\partial z^{*2}} + a^2 \right)\left(w_{mn}^{(1)} - N_\mu w_{mn}^{(2)} \right) = 0 \tag{3-43}$$

$$w_{mn}^{(1)} = w_{mn}^{(2)} = w_{mn}^{(\sigma)} = \frac{\partial h_{mn}}{\partial t^*} \tag{3-44}$$

$$\frac{\partial w_{mn}^{(1)}}{\partial z^*} = \frac{\partial w_{mn}^{(2)}}{\partial z^*} \tag{3-45}$$

Similarly, (3-32) and (3-33) demand

$$\text{as } z^* \to \infty : \quad w_{mn}^{(2)} \to 0 \tag{3-46}$$

$$\text{as } z^* \to -\infty : \quad w_{mn}^{(1)} \to 0 \tag{3-47}$$

Equation (3-25) together with the equation of continuity (3-17) and the fact that the tangential components of velocity are continuous across the interface require at $z^* = 0$

$$\frac{\partial \rho^{\{g\}}_{mn}}{\partial t^*} - \frac{\partial w_{mn}^{(1)}}{\partial z^*} = 0 \tag{3-48}$$

Since $\Phi_{mn}(t^*)$ is a random function of time, all dependent variables in this problem are also random functions of time. Our objective will be to determine certain statistical properties of these variables such as $B_{mn}(t^*, \tau^*)$, the correlation function in time of $h_{mn}(t^*)$,

$$B_{mn}(t^*, \tau^*) \equiv < h_{mn}(t^*) h_{mn}(t^* + \tau^*) > N_\Phi^2 \tag{3-49}$$

By <x>, we mean the expected or mean value of x. The system that we are discussing here is stationary in the sense that $B_{mn}(t^*, \tau^*)$ is independent of t^* and is only a function of τ^*. For such a stationary process, a good estimate for the correlation function can be obtained by using the ergodic theorem [or a less general result, the law of large numbers (Yaglom 1973)]:

$$B_{mn}(\tau^*) = \text{limit } T^* \to \infty: \frac{N_\Phi^2}{T^*} \int_0^{T^*} h_{mn}(t^*) h_{mn}(t^* + \tau^*) dt^* \tag{3-50}$$

An expression for $B_{mn}(\tau^*)$ that is more convenient for our present purposes than either (3-49) or (3-50) is (Lindgren 1976)

$$B_{mn}(\tau^*) = N_\Phi^2 \int_{-\infty}^{\infty} \dots \int_{-\infty}^{\infty} h'_{11} \dots h'_{mn} f(h'_{11}, \dots, h'_{mn} \dots)$$

$$\times < h_{mn}(\tau^*) \mid h_{11}(0) = h'_{11}, \dots, h_{mn}(0) = h'_{mn}, \dots >$$

$$\times dh'_{11} \dots dh'_{mn} \dots \tag{3-51}$$

Here $f(h_{11}, \dots, h_{mn}, \dots)$ is the joint probability density and

$$< h_{mn}(\tau^*) \mid h_{11}(0) = h'_{11}, \dots, h_{mn}(0) = h'_{mn}, \dots >$$

is the conditional expectation or the expected (mean) value of the random variable $h_{mn}(\tau^*)$ assuming $h_{11}(0) = h'_{11}, \dots, h_{mn}(0) = h'_{mn}, \dots$. Our understanding in writing (3-51) is that the series for h_1^* in (3-34) has been truncated with a finite number of terms.

Before computing $B_{mn}(\tau^*)$, we will determine in the next two sections expressions for the conditional expectation and the joint probability density appearing in (3-51).

Conditional expectation In computing the conditional expectation of $h_{mn}(\tau^*)$, we will require two additional assumptions:

ix) The conditional expectation of $h_{mn}(\tau^*)$ is independent of the velocity distribution that existed at $t^* = 0$.

x) As in the case of Brownian motion, the scalar potentials for the thermal noise can be represented as white noise, which implies among other things that (Schuss 1980)

$$< \Phi_{mn}(t^*) > = 0 \tag{3-52}$$

In view of assumption ix, we will seek solutions to (3-35), (3-36), (3-39), and (3-40) consistent with boundary conditions (3-42) through (3-48) and initial conditions

$$\text{at } t^* = 0 : \; h_{mn} = h'_{mn} \quad \text{for all m, n} \tag{3-53}$$

and

$$\text{at } t^* = 0 : \; w_{mn}^{(j)} = 0 \quad \text{for all m, n and j = 1, 2} \tag{3-54}$$

The Laplace transforms in time of (3-35), (3-36), (3-39), and (3-40) are

$$\left(sN_{Re} + a^2 - \frac{\partial^2}{\partial z^{*2}} \right)\left(a^2 - \frac{\partial^2}{\partial z^{*2}} \right) \tilde{w}_{mn}^{(1)} = 0 \tag{3-55}$$

$$\left(sN_{Re}N_\rho + N_\mu a^2 - \frac{\partial^2}{\partial z^{*2}} \right)\left(a^2 - \frac{\partial^2}{\partial z^{*2}} \right) \tilde{w}_{mn}^{(2)} = 0 \tag{3-56}$$

$$a^2 N_{Re} \tilde{\mathcal{P}}_{mn}^{(1)} = \left(\frac{\partial^2}{\partial z^{*2}} - a^2 - sN_{Re} \right)\left(\frac{\partial \tilde{w}_{mn}^{(1)}}{\partial z^*} \right) \tag{3-57}$$

$$a^2 N_\rho N_{Re} \tilde{\mathcal{P}}_{mn}^{(2)} = \left(N_\mu \frac{\partial^2}{\partial z^{*2}} - N_\mu a^2 - sN_\rho N_{Re} \right)\left(\frac{\partial \tilde{w}_{mn}^{(2)}}{\partial z^*} \right) \tag{3-58}$$

in which the Laplace transform of a variable is denoted by $\tilde{\cdot}$.. The Laplace transforms of (3-42) through (3-48) are at $z^* = 0$

$$[a^2 + N_{ca}N_{Re}N_g(1 - N_\rho)]\tilde{h}_{mn} + 2N_{ca}(1 - N_\mu) \frac{\partial \tilde{w}_{mn}^{(1)}}{\partial z^*}$$

$$- N_{ca}N_{Re}(\tilde{\mathcal{P}}_{mn}^{(1)} - N_\rho \tilde{\mathcal{P}}_{mn}^{(2)}) + N_{ca}N_{Re}\tilde{\phi}_{mn} = 0 \tag{3-59}$$

$$-\frac{a^2\gamma'^*}{N_{ca}}\tilde{\rho}\{g\}_{mn}+a^2(N_\kappa+N_\varepsilon)\frac{\partial\tilde{w}^{(1)}_{mn}}{\partial z^*}$$

$$+\left(\frac{\partial^2}{\partial z^{*2}}+a^2\right)\left(\tilde{w}^{(1)}_{mn}-N_\mu\tilde{w}^{(2)}_{mn}\right)=0 \quad (3\text{-}60)$$

$$\tilde{w}^{(1)}_{mn}=\tilde{w}^{(2)}_{mn}=\tilde{w}^{(\sigma)}_{mn}=s\tilde{h}_{mn}-h'_{mn} \quad (3\text{-}61)$$

$$\frac{\partial\tilde{w}^{(1)}_{mn}}{\partial z^*}=\frac{\partial\tilde{w}^{(2)}_{mn}}{\partial z^*} \quad (3\text{-}62)$$

$$\text{as } z^*\to\infty:\ \tilde{w}^{(2)}_{mn}\to 0 \quad (3\text{-}63)$$

$$\text{as } z^*\to-\infty:\ \tilde{w}^{(1)}_{mn}\to 0 \quad (3\text{-}64)$$

$$s\tilde{\rho}\{g\}_{mn}-\frac{\partial\tilde{w}^{(1)}_{mn}}{\partial z^*}=0 \quad (3\text{-}65)$$

Solutions to (3-55) and (3-56) consistent with (3-63) and (3-64) are

$$\tilde{w}^{(1)}_{mn}=A\exp(az^*)+C\exp(bz^*) \quad (3\text{-}66)$$

$$\tilde{w}^{(2)}_{mn}=B\exp(-az^*)+D\exp(cz^*) \quad (3\text{-}67)$$

with

$$b=\sqrt{a^2+sN_{Re}}\quad\text{(root with positive real part)} \quad (3\text{-}68)$$

$$c=-\sqrt{a^2+sN_{Re}N_\rho/N_\mu}\quad\text{(root with negative real part)} \quad (3\text{-}69)$$

Boundary conditions (3-61) and (3-62) demand

$$A+C=B+D \quad (3\text{-}70)$$

$$aA+bC=-aB+cD \quad (3\text{-}71)$$

Equations (3-60) and (3-65) require

$$\left[a^3\left(N_\kappa+N_\varepsilon+\frac{\gamma'^*}{N_{ca}}\frac{1}{s}\right)+2a^2\right]A+\left[a^2b\left(N_\kappa+N_\varepsilon+\frac{\gamma'^*}{N_{ca}}\frac{1}{s}\right)\right.$$

$$+ a^2 + b^2 \Big] C - 2a^2 N_\mu B - N_\mu (a^2 + c^2) D = 0 \qquad (3\text{-}72)$$

Finally, (3-59) tells us

$$\begin{aligned}
&[a^3 + aN_{ca}N_{Re}N_g(1 - N_\rho) + 2a^2N_{ca}(1 - N_\mu) s + N_{ca}N_{Re}s^2]A \\
&\quad + N_\rho N_{ca}N_{Re}s^2 B + [a^3 + aN_{ca}N_{Re}N_g(1 - N_\rho) \\
&\qquad + 2ab(1 - N_\mu)N_{ca}s]C + asN_{ca}N_{Re}\tilde{\phi}_{mn} \\
&\quad + [a^3 + aN_{ca}N_{Re}N_g(1 - N_\rho)]h'_{mn} \\
&= 0 \qquad (3\text{-}73)
\end{aligned}$$

Equations (3-70) through (3-73) may be solved for A, B, C, and D. Using these results as well as (3-61) and (3-66), we can calculate

$$\begin{aligned}
\tilde{h}_{mn} &= \frac{1}{s}(\tilde{w}^{(1)}_{mn} + h'_{mn}) \\
&= \frac{1}{s}(A + C + h'_{mn}) \\
&= \frac{h'_{mn}}{s}\{ 1 + [a^3 + aN_{ca}N_{Re}N_g(1 - N_\rho)] \tilde{J} \} \\
&\quad + aN_{ca}N_{Re}\tilde{J}\tilde{\phi}_{mn} \qquad (3\text{-}74)
\end{aligned}$$

where

$$\begin{aligned}
\tilde{J} = \tilde{J}(s) &\equiv (a + c)(b - a)[a^2(N_\kappa + N_\varepsilon) + a + b \\
&\quad - N_\mu(c - a)][\det(K_{ij})]^{-1} \qquad (3\text{-}75)
\end{aligned}$$

and $\det(K_{ij})$ is the 3×3 determinant whose entries are

$$K_{11} \equiv a - c$$

$$K_{12} \equiv b - c$$

$$K_{13} \equiv a + c$$

$$K_{21} \equiv a^3\left(N_\kappa + N_\varepsilon + \frac{\gamma'^*}{N_{ca}}\frac{1}{s}\right) + 2a^2 - N_\mu(a^2 + c^2)$$

$$K_{22} \equiv a^2 b\left(N_\kappa + N_\varepsilon + \frac{\gamma'^*}{N_{ca}}\frac{1}{s}\right) + a^2 + b^2 - N_\mu(a^2 + c^2)$$

$$K_{23} \equiv N_\mu(c^2 - a^2)$$

$$K_{31} \equiv a^3 + aN_{ca}N_{Re}N_g(1 - N_\rho) + 2a^2 N_{ca}(1 - N_\mu)s + N_{ca}N_{Re}s^2$$

$$K_{32} \equiv a^3 + aN_{ca}N_{Re}N_g(1 - N_\rho) + 2ab(1 - N_\mu)N_{ca}s$$

$$K_{33} \equiv N_\rho N_{ca}N_{Re}s^2 \tag{3-76}$$

Taking the inverse Laplace transform, we find

$$h_{mn}(t^*) = \frac{1}{2\pi i}\int_{\nu-i\infty}^{\nu+i\infty} e^{st^*}\,\tilde{h}_{mn}\,ds \qquad \text{for } t^* > 0 \tag{3-77}$$

The real number ν is arbitrary, so long as the line along which the integration is to be performed lies on the right of all singularities of the argument in the complex plane.

From (3-74) and (3-77), we see that the stochastic portion of $h_{mn}(t^*)$

$$S\left(h_{mn}(t^*)\right) \equiv \frac{1}{2\pi i}aN_{ca}N_{Re}\int_{\nu-i\infty}^{\nu+i\infty} e^{st^*}\,\tilde{J}\tilde{\Phi}_{mn}\,ds \tag{3-78}$$

can be expressed as

$$S\left(h_{mn}(t^*)\right) \equiv \frac{1}{2\pi i}aN_{ca}N_{Re}$$

$$\times\int_{\nu-i\infty}^{\nu+i\infty}\int_0^\infty\int_0^\infty e^{s(t^*-u-v)}J(u)\,\Phi_{mn}(v)\,du\,dv\,ds \tag{3-79}$$

where $J(u)$ and $\Phi_{mn}(v)$ are the variables whose Laplace transforms are $\tilde{J}$ and $\tilde{\Phi}_{mn}$. Noting that [Proof follows using the Fourier integral theorem (Spiegel 1965).]

$$\int_{-\infty}^{\infty} e^{iy(t^*-u-v)}\,dy = 2\pi\delta(t^* - u - v) \tag{3-80}$$

in which $\delta(x)$ is the Dirac delta function, we can integrate (3-79) twice to find

$$S\left(h_{mn}(t^*)\right) = aN_{ca}N_{Re} \int_0^{\infty} J(t^* - v)\, \Phi_{mn}(v)\, dv \tag{3-81}$$

Since the future beyond time t will have no influence on $S\left(h_{mn}(t^*)\right)$, (3-81) reduces to

$$S\left(h_{mn}(t^*)\right) = aN_{ca}N_{Re} \int_0^{t^*} J(t^* - v)\, \Phi_{mn}(v)\, dv \tag{3-82}$$

or alternatively

$$S\left(h_{mn}(t^*)\right) = aN_{ca}N_{Re} \left\{ \text{limit } N \to \infty: \sum_{n=0}^{\infty} J\left(t^*\left[1 - \frac{n}{N}\right]\right) \times \Phi_{mn}\left(\frac{n}{N} t^* \right) \frac{t^*}{N} \right\} \tag{3-83}$$

In view of (3-52), we observe that

$$< S\left(h_{mn}(t^*)\right) > = 0 \tag{3-84}$$

Given (3-84), we conclude from (3-74) that for $\tau^* > 0$

$$< h_{mn}(\tau^*) > = \frac{1}{2\pi i} \int_{\nu - i\infty}^{\nu + i\infty} e^{s\tau^*} \frac{h'_{mn}}{s} \times \{ 1 + [a^3 + aN_{ca}N_{Re}N_g(1 - N_\rho)]\tilde{J} \}ds \tag{3-85}$$

Because $B_{mn}(\tau^*)$, defined by (3-49), is independent of t^* for this stationary random process, it must be a symmetric function of τ^*. By (3-51), the conditional expectation must also be a symmetric function of τ^* and for $t^* < 0$

$$< h_{mn}(\tau^*) > = \frac{1}{2\pi i} \int_{\nu - i\infty}^{\nu + i\infty} e^{-s\tau^*} \frac{h'_{mn}}{s} \times \{ 1 + [a^3 + aN_{ca}N_{Re}N_g(1 - N_\rho)]\tilde{J} \}ds \tag{3-86}$$

An expression valid for all τ^* can be obtained by adding (3-85) and (3-86) (Spiegel 1965):

$$< h_{mn}(\tau^*) > = \frac{h'_{mn}}{2\pi i} \int_{\nu - i\infty}^{\nu + i\infty} \frac{1}{s} (e^{s\tau^*} + e^{-s\tau^*})$$

$$\times \{ 1 + [a^3 + aN_{ca}N_{Re}N_g(1 - N_\rho)]\tilde{J} \}ds \qquad (3\text{-}87)$$

For all cases tested, the two singularities of the argument of this integral lie to the left of the imaginary axis, either on the negative real axis or symmetric with respect to the negative real axis. For this reason, we are free to take $\nu = 0$ and write (3-87) as

$$< h_{mn}(\tau^*) > = \frac{h'_{mn}}{2\pi} \int_{\infty}^{\infty} \frac{1}{i\omega} (e^{i\omega\tau^*} + e^{-i\omega\tau^*})$$

$$\times \{ 1 + [a^3 + aN_{ca}N_{Re}N_g(1 - N_\rho)]\tilde{J} \}d\omega \qquad (3\text{-}88)$$

Note that from (3-75)

$$\overline{\tilde{J}}(i\omega) = \tilde{J}(i\omega) \qquad (3\text{-}89)$$

where the overbar denotes the complex conjugate. In view of (3-89), equation (3-88) yields the desired conditional expectation

$$< h_{mn}(\tau^*) \mid h_{11}(0) = h'_{11}, ..., h_{mn}(0) = h'_{mn}, ..., >$$

$$= \frac{1}{\pi} h'_{mn}[a^3 + aN_{ca}N_{Re}N_g(1 - N_\rho)]$$

$$\times \int_{-\infty}^{\infty} \frac{1}{\omega} \operatorname{Im}\left(\tilde{J}(i\omega)\right) \exp(i\omega\tau^*)\, d\omega \qquad (3\text{-}90)$$

Joint probability density By Boltzmann's principle (Kestin and Dorfman 1971), the joint probability density for our isolated, isothermal system takes the form

$$f(h'_{11}, ..., h'_{mn}, ...) \sim \exp\left(- \frac{\Delta A + \Delta P}{kT} \right) \qquad (3\text{-}91)$$

Here ΔA is the difference between the Helmholtz free energy corresponding to this deformed state at time $t = 0$ and that corresponding to the equilibrium plane interface; ΔP the difference between the potential energies corresponding to this deformed state and to the equilibrium plane interface; T temperature; k the Boltzmann constant. The proportionality constant is determined by requiring

$$\int_{-\infty}^{\infty} \ldots \int_{-\infty}^{\infty} \ldots f(h'_{1\,1}, \ldots, h'_{m\,n}, \ldots)dh'_{1\,1} \ldots dh'_{m\,n} \ldots = 1 \qquad (3\text{-}92)$$

We can compute

$$\Delta A = \gamma \int_{-L}^{L} \int_{-L}^{L} \left\{ \left[1 + \left(\frac{\partial h}{\partial x} \right)^2 + \left(\frac{\partial h}{\partial y} \right)^2 \right]^{1/2} - 1 \right\} dxdy$$

$$= \frac{\gamma L_0^2}{2} \int_{-L^*}^{L^*} \int_{-L^*}^{L^*} \left[\left(\frac{\partial h^*}{\partial x^*} \right)^2 + \left(\frac{\partial h^*}{\partial y^*} \right)^2 \right] dx^* dy^*$$

$$= 2\,\gamma L^2\, N_\Phi^2 \sum_{m,n=-\infty}^{\infty} a^2 (h_{m\,n})^2 \qquad (3\text{-}93)$$

and

$$\Delta P = \int_{-L}^{L} \int_{-L}^{L} \int_{0}^{h} (\rho^{(1)} - \rho^{(2)})g\, z\, dzdxdy$$

$$= \frac{1}{2} (\rho^{(1)} - \rho^{(2)})gL_0^4 \int_{-L^*}^{L^*} \int_{-L^*}^{L^*} h^{*2}\, dx^* dy^*$$

$$= 2(\rho^{(1)} - \rho^{(2)})gL_0^2 L^2\, N_\Phi^2 \sum_{m,n=-\infty}^{\infty} (h_{m\,n})^2$$

$$= 2\,\gamma L^2\, N_\Phi^2\, N_{ca} N_{Re} N_g (1 - N_\rho) \sum_{m,n=-\infty}^{\infty} (h_{m\,n})^2 \qquad (3\text{-}94)$$

Substituting (3-93) and (3-94) in (3-91), we see that the joint probability density can be rearranged in the form

$f(h'_{1\,1}, \ldots, h'_{m\,n}, \ldots)$

$$\sim \exp\left\{ -\frac{2\gamma L^2\ N_\Phi^2}{kT} \sum_{m,n=-N}^{N} [\ a^2 + N_{ca}N_{Re}N_g(\ 1 - N_\rho\ )](\ h_{mn}\ )^2 \right\}$$

$$= f_{11}(\ h'_{11}\ ) \ldots f_{mn}(\ h'_{mn}\ ) \ldots \tag{3-95}$$

in which

$$f_{mn}(\ h'_{mn}\ ) \equiv D_{mn}\exp\left\{ -\frac{2\gamma L^2\ N_\Phi^2}{kT} \right.$$

$$\left. \times [\ a^2 + N_{ca}N_{Re}N_g(\ 1 - N_\rho\ )](\ h'_{mn}\ )^2 \right\} \tag{3-96}$$

Remember that in arriving at (3-51), the series for h_1^* in (3-34) was truncated with a finite number of terms. Since the joint probability density has taken the form of the product of the probability densities of independent random variables in (3-96), we can require

$$\int_{-\infty}^{\infty} f_{mn}(\ h'_{mn}\ )dh'_{mn} = 1 \tag{3-97}$$

and in this way determine

$$D_{mn} = \frac{1}{\sqrt{\pi}}\left\{ \frac{2\gamma L^2\ N_\Phi^2}{kT} [\ a^2 + N_{ca}N_{Re}N_g(\ 1 - N_\rho\ )] \right\}^{1/2} \tag{3-98}$$

Spectral density of $h_{mn}(t^*)$ Rather than $B_{mn}(\tau^*)$, the correlation function in terms of $h_{mn}(t^*)$ defined by (3-49), we will find it more convenient to speak in terms of the spectral density of $h_{mn}(t^*)$, the Fourier transform of $B_{mn}(\tau^*)$:

$$C_{mn}(\omega) \equiv \int_{-\infty}^{\infty} B_{mn}(\ \tau^*\ ) \exp(\ -\ i\omega\tau^*\ )\ d\tau^* \tag{3-99}$$

In view of (3-51), (3-90), (3-95), (3-96), and (3-98), this takes the explicit form

$$C_{mn}(\omega) = 2\ N_\Phi^2\ [\ a^3 + aN_{ca}N_{Re}N_g(\ 1 - N_\rho\ )]\ \frac{1}{\omega}\ \mathrm{Im}\left(\ \tilde{J}(\ i\omega\ )\right)$$

$$\times \int_{-\infty}^{\infty} \dots \int_{-\infty}^{\infty} \dots (h'_{mn})^2 f(h'_{11}, \dots, h'_{mn}, \dots)\, dh'_{11} \dots dh'_{mn} \dots$$

$$= 2 N_\Phi^2 [\, a^3 + aN_{ca}N_{Re}N_g(1 - N_\rho)] \frac{1}{\omega} \mathrm{Im}\left(\tilde{J}(i\omega) \right)$$

$$\times \int_{-\infty}^{\infty} (h'_{mn})^2 f_{mn}(h'_{mn})\, dh'_{mn}$$

$$= \frac{akT}{2\gamma L^2} \frac{1}{\omega} \mathrm{Im}\left(\tilde{J}(i\omega) \right) \tag{3-100}$$

Experimentally, we can measure $G_{mn}(\tau)$, the correlation function in time of the intensity of the scattered light $I_{mn}(t)$ (Bouchiat *et al.* 1968; Bouchiat and Meunier 1968, 1969; Huang and Webb 1969; Kim *et al.* 1982); Pouchelon *et al.* 1980, 1981)

$$G_{mn}(\tau^*) \equiv \langle I_{mn}(t^*) I_{mn}(t^* + \tau^*) \rangle \tag{3-101}$$

Arguing by analogy with the scattering of light by a diffraction grating, Langevin and Meunier (1976) have suggested that the spectral density of the scattered light intensity

$$P_{mn}(\omega) \equiv \int_{-\infty}^{\infty} G_{mn}(\tau^*) \exp(-i\omega\tau^*)\, d\tau^* \tag{3-102}$$

and $C_{mn}(\omega)$ are proportional to one another,

$$P_{mn}(\omega) \propto C_{mn}(\omega) \tag{3-103}$$

The proportionality factor is assumed to depend only on the wave length of the incident beam, the angle of incidence, and the scattering angle.

The interfacial tension, the viscosity, and in some cases the sum of the interfacial viscosities can be determined through measurements of $P_{mn}(\omega)$ by fitting the plot of experimental data to

$$P_{mn}(\omega) = AC_{mn}(\omega) + B \tag{3-104}$$

where B compensates for noise.

The parameters m and n are defined to the extent that (Mandelstam 1913; Chu 1974; Pike 1976)

$$(m^2 + n^2)^{1/2} = \frac{2Ln_i}{\lambda}(\sin\theta_s - \sin\theta_i) \tag{3-105}$$

Here λ is the wave length of the incident light, n_i the index of refraction in the incident medium, θ_s the angle of scattering measured with respect to t the normal to a plane interface and θ_i the angle of incidence measured with respect to this same normal. For convenience, we define L_0 by requiring

$$a = 1 \tag{3-106}$$

or

$$L_0 \equiv \frac{L}{\pi}(m^2 + n^2)^{-1/2} \tag{3-107}$$

The characteristic time t_0 is fixed by demanding

$$N_{Re} = 1 \tag{3-108}$$

or

$$t_0 \equiv \frac{\rho^{(1)} L_0^2}{\mu^{(1)}} \tag{3-109}$$

As a result of these definitions, we will find it convenient to observe

$$N_{ca} = \frac{\mu^{(1)2}}{\rho^{(1)}\gamma L_0}$$

$$N_{\kappa+\varepsilon} = N_\kappa + N_\varepsilon = \frac{\kappa + \varepsilon}{\mu^{(1)} L_0}$$

$$N_g = \frac{g\rho^{(1)} L_0^3}{\mu^{(1)2}} \tag{3-110}$$

Discussion Callen and Greene (1952; see also Greene and Callen 1952) have proposed an alternative derivation of the conditional expectation required here. Consider an artificial system similar to the one described above. But rather than a stochastic, dimensionless difference ϕ^* in force potentials per unit volume generating capillary waves at the interface, a constant $\phi^{*\prime}$ maintains the interface in a stationary configuration whose first perturbation in N_Φ is

$$h_1^{*\prime} = \sum_{m,n=-\infty}^{\infty} h'_{mn} \exp\left[i\pi(mx^* + ny^*)\frac{L_0}{L} \right] \tag{3-111}$$

At time t = 0, this force potential is removed and the system decays to equilibrium. Callen and Greene (1952) argue without proof that the value of the function observed in this system is equal to the conditional expectation for the real problem with which we are concerned for positive values of τ^*. This can be proved by observing that their initial value problem is nearly identical with that described by (3-35), (3-36), (3-39), (3-40), (3-42) through (3-48), (3-53), and (3-54). The only difference is that Φ_{mn} does not appear in (3-42) for their initial value problem. Equation (3-90) for the conditional expectation follows immediately.

It appears that there is an alternative solution to this same initial value problem

$$\begin{aligned} & h_{mn}(t^*),\quad w_{mn}^{(j)}(t^*, z^*),\quad w_{mn}^{(\sigma)}(t^*),\quad \mathcal{P}_{mn}^{(j)}(t^*) \\ &= (\hat{h}_1,\quad \hat{w}_1^{(j)}(z),\quad \hat{w}_1^{(\sigma)},\quad \hat{\mathcal{P}}_1^{(j)}) \exp(s_1 t^*) \\ &\quad + (\hat{h}_1,\quad \hat{w}_2^{(j)}(z),\quad \hat{w}_2^{(\sigma)},\quad \hat{\mathcal{P}}_2^{(j)}) \exp(s_2 t^*) \end{aligned} \tag{3-112}$$

but it leads to a trivial solution when the initial condition (3-54) is imposed. In order to avoid this trivial solution, Papoular (1968) and Meunier *et al.* (1969) replace (3-54) by

$$\text{at } t^* = 0:\ \frac{\partial h_{mn}}{\partial t^*} = 0 \tag{3-113}$$

in discussing the case where the effects of the Gibbs elasticity, of the interfacial tension gradient, and of all surface stress-deformation behavior can be neglected. The resulting expression for the spectral density of $h_{mn}(t^*)$ is

$$\begin{aligned} C_{mn}(\omega) = & -\frac{kT}{2\gamma L^2} [1 + N_{ca} N_g (1 - N_\rho)]^{-1} \\ & \times s_1 s_2 (s_1 + s_2)(s_1^2 + \omega^2)^{-1} (s_2^2 + \omega^2)^{-1} \end{aligned} \tag{3-114}$$

where s_1 and s_2 are solutions of

$$\det(K_{ij}) = 0 \tag{3-115}$$

In the range of the parametric study described below, s_1 and s_2 are either complex conjugates

$$s_1 = - s_r + is_i$$

$$s_2 = - s_r - is_i \tag{3-116}$$

or they are both real and negative. When (3-116) applies, (3-114) reduces to

$$C_{mn}(\omega) = \frac{kT}{\gamma L^2} [1 + N_{ca} N_g (1 - N_\rho)]^{-1}$$

$$\times s_r(s_r^2 + s_i^2)[\omega^4 + s(s_r^2 - s_i^2)\omega^2 + (s_r^2 + s_i^2)^2]^{-1} \tag{3-117}$$

This is similar to the spectral density of the Brownian motion of a harmonic oscillator (Yaglom 1973).

Instead of (3-112), we look for a solution of the form

$$h_{mn}(t^*), \quad w_{mn}^{(j)}(t^*, z^*), \quad w_{mn}^{(\sigma)}(t^*), \quad \mathcal{P}_{mn}^{(j)}(t^*)$$

$$= (\hat{h}_1, \quad \hat{w}_1^{(j)}(z^*), \quad \hat{w}_1^{(\sigma)}, \quad \hat{\mathcal{P}}_1^{(j)}) \exp [(- s_r + is_i)t^*] \tag{3-118}$$

we can satisfy neither (3-54) nor (3-113). The corresponding Lorentzian (Edwards *et al.* 1982) spectral density of $h_{mn}(t^*)$

$$C_{mn}(\omega) = \frac{kT}{4\gamma L^2} [1 + N_{ca} N_g (1 - N_\rho)]^{-1} s_r$$

$$\times \left(\frac{1}{s_r^2 + (\omega - s_i)^2} + \frac{1}{s_r^2 + (\omega + s_i)^2} \right) \tag{3-119}$$

is commonly employed in discussing the case where the effects of the Gibbs elasticity, of the interfacial tension gradient, and of all surface stress-deformation behavior can be neglected (Hard and Lofgren 1977; Edwards et al. 1982). The quality of this approximation is examined below.

Bouchiat and Meunier (1971) have presented three different solutions for the case where the effects of the Gibbs elasticity, of the interfacial tension gradient, and of all surface stress-deformation behavior can be neglected. In the first, they follow Callen and Greene (1952) in computing the conditional expectation, but they fail to satisfy all of the initial conditions [see (3-112)]. In the second, they again follow Callen and Greene (1952) in computing the conditional expectation, but they do not recognize that the solution of their initial value problem is equal to the conditional expectation only for positive values of t. (Their equation 42 is correct, but it does not follow from their equation 41. A factor of 2 is missing.) In their third solution, they used an expression proposed by Callen and Greene (1952; see also Greene and Callen 1952) that incorporates a correct result for the conditional expectation. This approach is identical with that used in the second solution, if their second solution had been carried out correctly.

Langevin and Bouchiat (1971; see also Langevin 1981) follow the second solution of Bouchiat and Meunier (1971) in computing the spectral density for a form of surface viscoelasticity (Goodrich 1961) (including the same error described above). They however do not include the effects of the Gibbs elasticity or of the interfacial tension gradient (Langevin 1981, her first equation on page 415). In addition, although their description of surface stress-deformation behavior is a simple surface material (Sec. 2.2.3), it is not a simple surface fluid (Sec. 2.2.7), and it does not include as a special case the Boussinesq surface fluid (Sec. 2.2.2). In this sense, their work does not consider the effects of the surface viscosities. However, the linearized jump momentum balance of Langevin and Bouchiat (1971, their equation 2) reduces to equations similar to (3-20) and (3-21) (for two-dimensional liquid-gas systems, not including the effects of thermal noise, of the Gibbs elasticity, or the interfacial tension gradient), if their surface elasticity and transverse viscosity are set to zero.

In contrast, Mandelstam (1913) was concerned with the mean square of the amplitude of a Fourier component. By (3-49), (3-51), (3-95) through (3-98), (3-106) and (3-108), this may be expressed as

$$
\begin{aligned}
< [\, h_m (t^*)]^2 > &= B_{m\,n}(0) \\
&= N_\Phi^2 \int_{-\infty}^{\infty} \ldots \int_{-\infty}^{\infty} \ldots (h'_{m\,n})^2 \\
&\qquad f(h'_{1\,1}, \ldots, h'_{m\,n}, \ldots)\, dh'_{1\,1} \ldots dh'_{m\,n} \ldots \\
&= N_\Phi^2 \int_{-\infty}^{\infty} (h'_{m\,n})^2 f(h'_{m\,n})\, dh'_{m\,n} \\
&= \frac{kT}{4\gamma L^2} [\, 1 + N_{c\,a} N_g (1 - N_\rho)]^{-1}
\end{aligned}
\qquad (3\text{-}120)
$$

Notice that by (3-106) through (3-110) this expression depends upon neither the bulk viscosities nor the interfacial viscosities. Equation (3-120) can be employed as the basis for an experiment to measure the interfacial tension (Lachaise *et al.* 1981).

Results: liquid-gas interfaces We consider here only the limiting case of liquid-gas systems: $N_\mu \to 0$, $N_\rho \to 0$. For simplicity, we plot

$$
C^{*\,*}_{m\,n}(\omega) \equiv \frac{2\gamma L^2}{kT} C_{m\,n}(\omega) \qquad (3\text{-}121)
$$

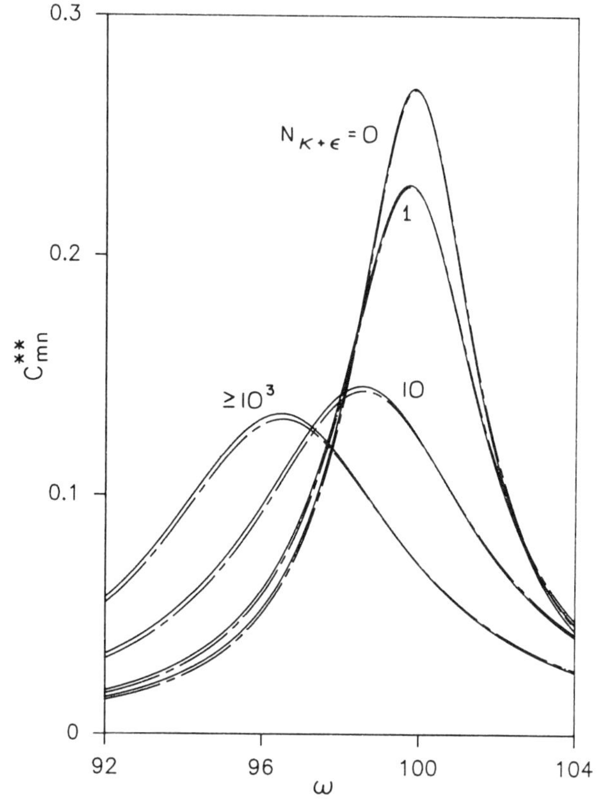

Figure 6.4.3-1: C^{**}_{mn} as a function of ω for a liquid-gas system with $N_{ca} = 10^{-4}$, $N_g \leq 10$, $\gamma'^{*} = 0$, and various values of $N_{\kappa+\varepsilon}$. The solid lines represent (3-100) and the dashed lines (3-119).

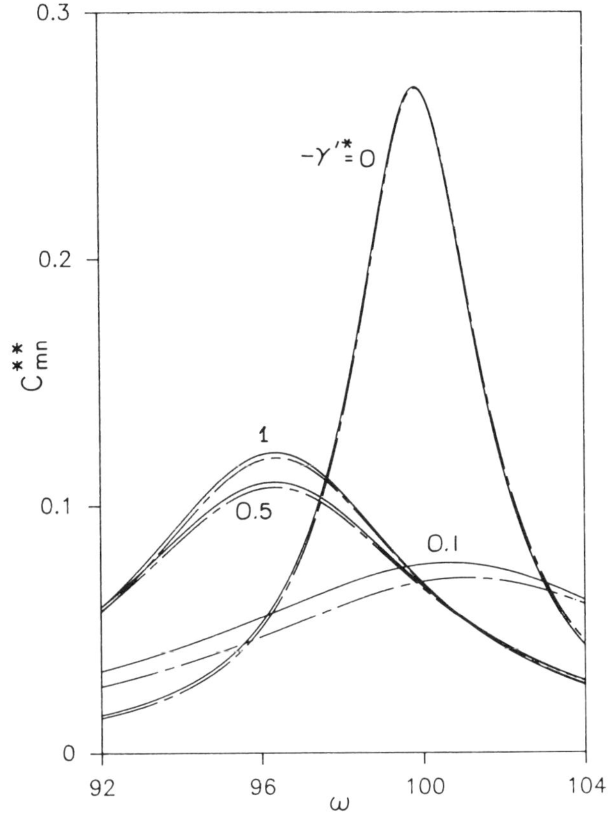

Figure 6.4.3-2: C^{**}_{mn} as a function of ω for a liquid-gas system with $N_{ca} = 10^{-4}$, $N_g \leq 10$, $N_{\kappa+\varepsilon} = 0$, and various values of $-\gamma'^{*}$. The solid lines represent (3-100) and the dashed lines (3-119).

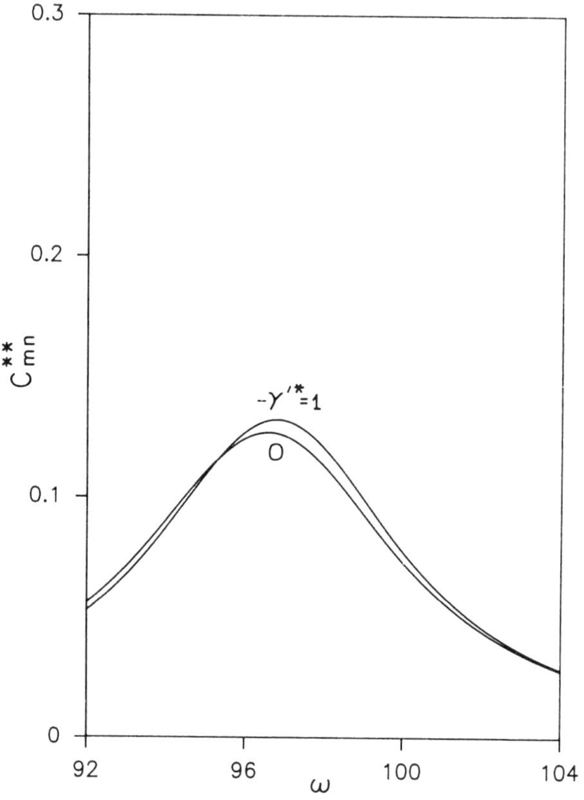

Figure 6.4.3-3: C^{**}_{mn} as a function of ω for a liquid-gas system with $N_{ca} = 10^{-4}$, $N_g \leq 10$, $N_{\kappa+\varepsilon} = 100$, and $-\gamma'^{*} \leq 1$, computed from (3-100).

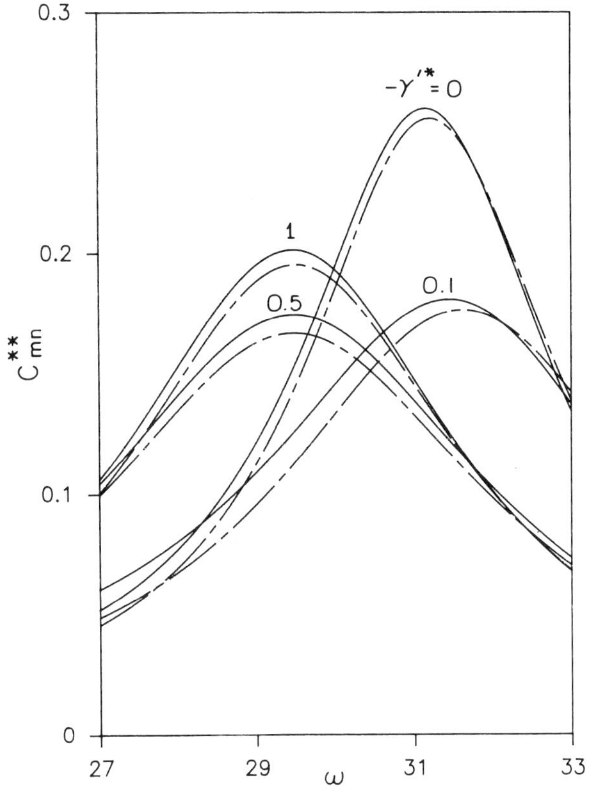

Figure 6.4.3-4: C^{**}_{mn} as a function of ω for a liquid-gas system with $N_{ca} = 10^{-3}$, $N_g \leq 10$, $N_{\kappa+\varepsilon} = 1$, and various values of $-\gamma'^{*}$. The solid lines represent (3-100) and the dashed lines (3-119).

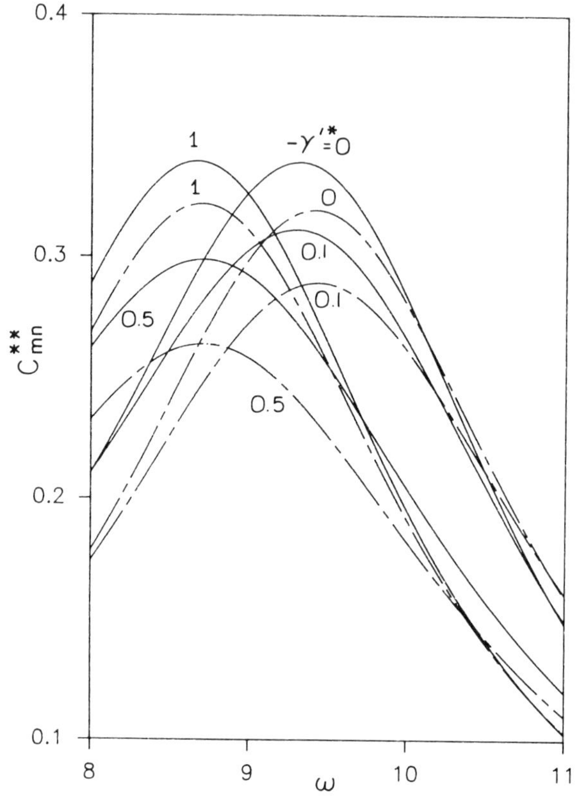

Figure 6.4.3-5: C^{**}_{mn} as a function of ω for a liquid-gas system with $N_{ca} = 10^{-2}$, $N_g \leq 10$, $N_{\kappa+\varepsilon} = 1$, and various values of $-\gamma'^{*}$. The solid lines represent (3-100) and the dashed lines (3-119).

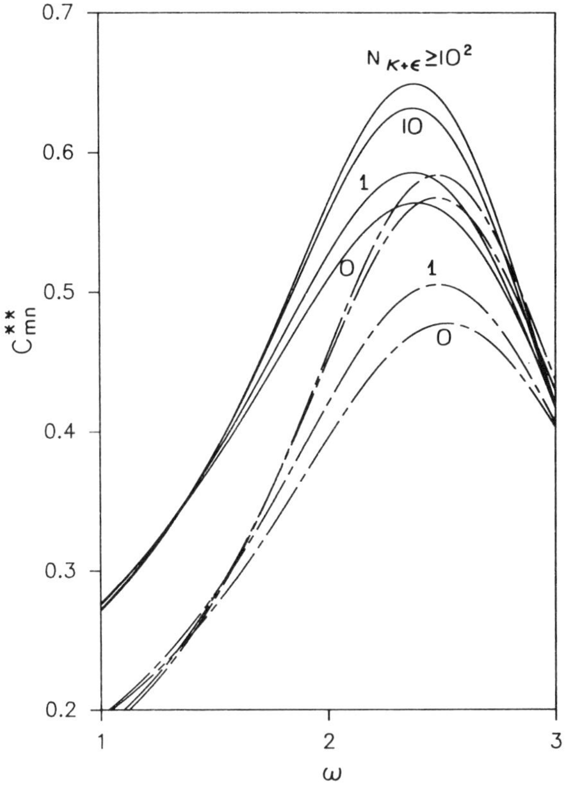

Figure 6.4.3-6: C^{**}_{mn} as a function of ω for a liquid-gas system with $N_{ca} = 0.1$, $N_g \leq 10$, $-\gamma'^{*} \leq 0.1$ and various values of $N_{\kappa+\varepsilon}$. The solid lines represent (3-100) and the dashed lines (3-119).

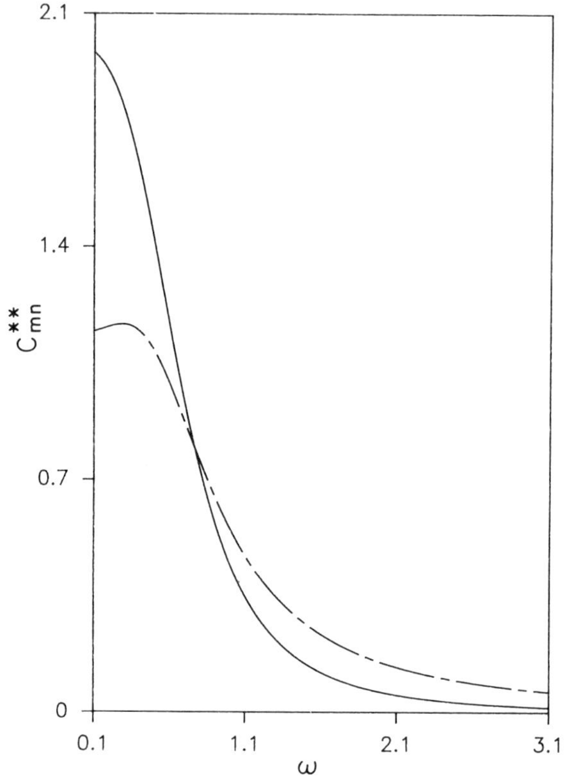

Figure 6.4.3-7: C^{**}_{mn} as a function of ω for a liquid-gas system with $N_{ca} = 1$, $N_g = 0$, and all values of $N_{\kappa+\varepsilon}$ and γ'^{*}. The solid line represents (3-100) and the dashed line (3-119).

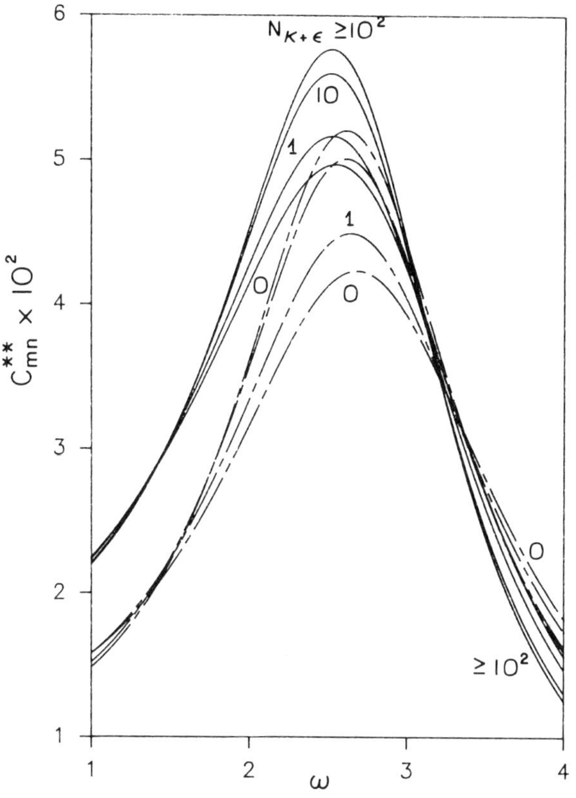

Figure 6.4.3-8: C^{**}_{mn} as a function of ω for a liquid-gas system with $N_{ca} = 1$, $N_g = 10$, $-\gamma'^{*} \leq 1$, and various values of $N_{\kappa+\varepsilon}$. The solid lines represent (3-100) and the dashed lines (3-119).

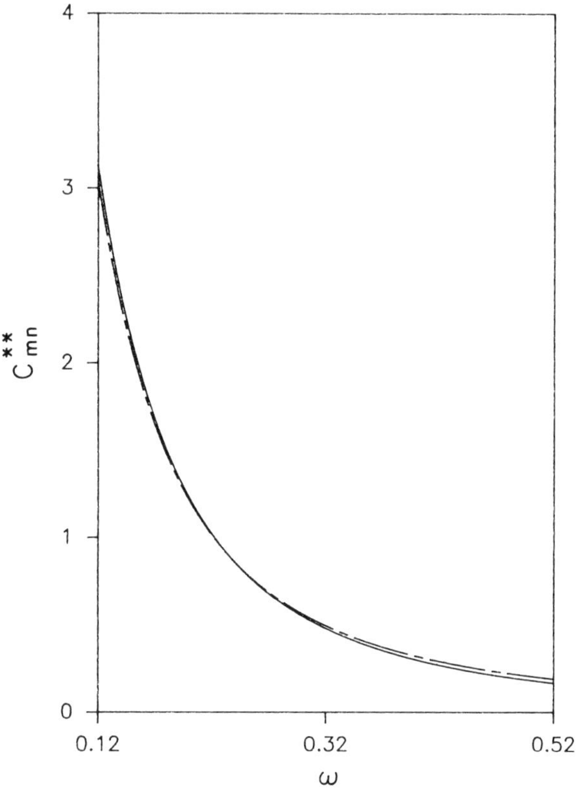

Figure 6.4.3-9: C^{**}_{mn} as a function of ω for a liquid-gas system with $N_{ca} = 10$, $N_g = 0$, and all values of $N_{\kappa+\epsilon}$ and γ'^*. The solid line represents (3-100) and the dashed line (3-119).

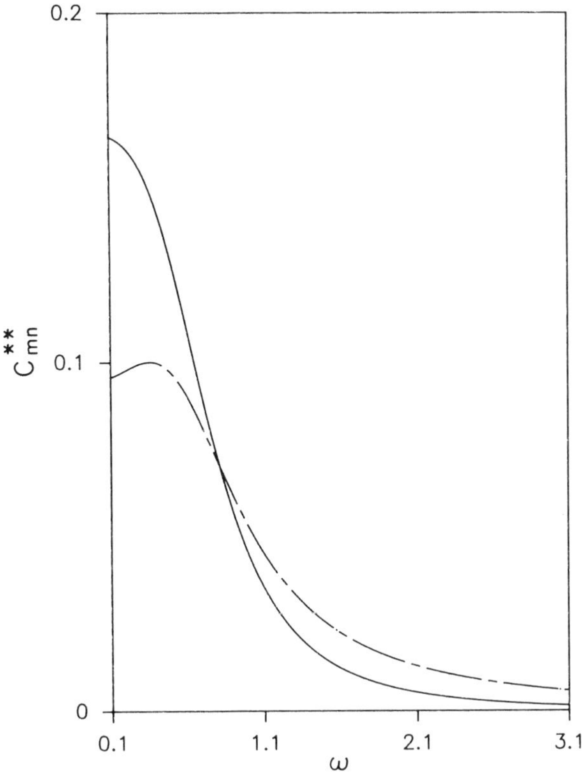

Figure 6.4.3-10: C^{**}_{mn} as a function of ω for a liquid-gas system with $N_{ca} = 10$, $N_g = 1$, and all values of $N_{\kappa+\varepsilon}$ and γ'^{*}. The solid line represents (3-100) and the dashed line (3-119).

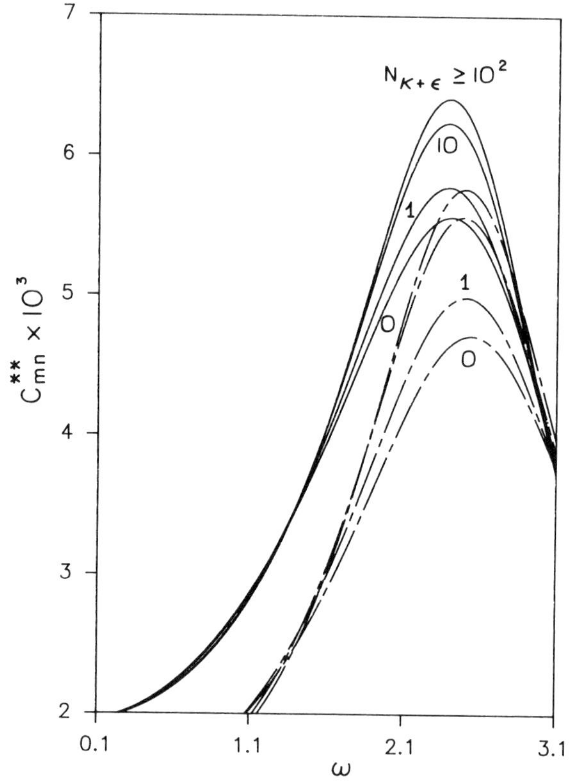

Figure 6.4.3-11: C^{**}_{mn} as a function of ω for a liquid-gas system with $N_{ca} = 10$, $N_g = 10$, $-\gamma'^{*} \leq 1$, and various values of $N_{\kappa+\varepsilon}$. The solid lines represent (3-100) and the dashed line (3-119).

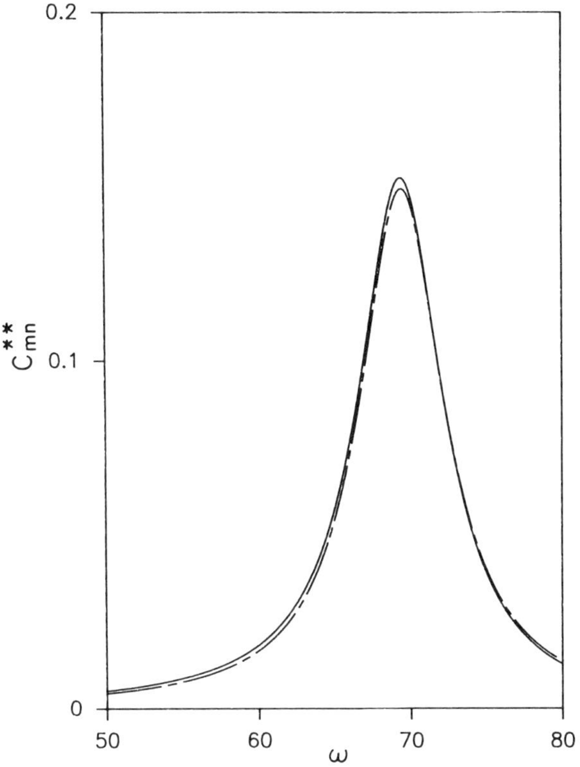

Figure 6.4.3-12: C^{**}_{mn} as a function of ω for a liquid-liquid system with $N_{ca} = 10^{-4}$, $N_{\mu} = 1$, $N_{\rho} = 0.9$, $N_g \leq 10$, and all values of γ'^{*} and $N_{\kappa+\varepsilon}$. The solid line represents (3-100) and the dashed line (3-119).

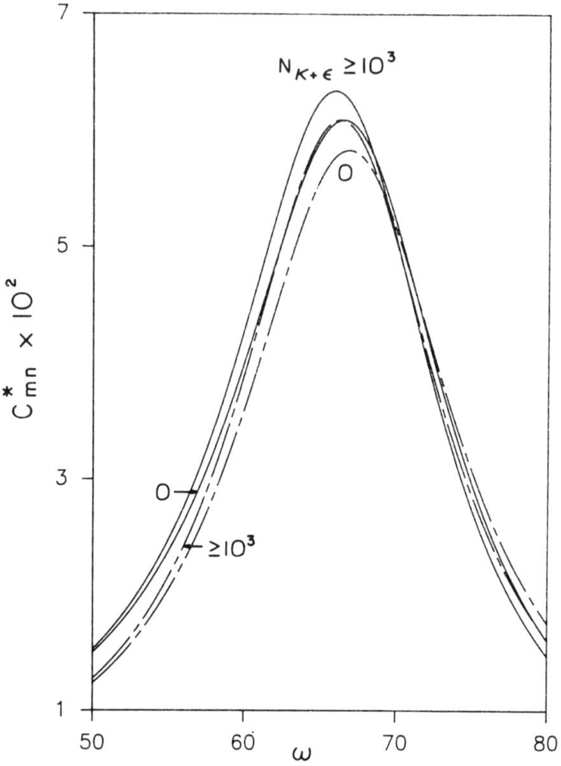

Figure 6.4.3-13: $C^{\star\star}_{mn}$ as a function of ω for a liquid-liquid system with $N_{ca} = 10^{-4}$, $N_{\mu} = 10$, $N_{\rho} = 0.9$, $N_{g} \le 10$, $\gamma'^{\star} = 0$, and various values of and $N_{\kappa+\varepsilon}$. The solid lines represent (3-100) and the dashed line (3-119).

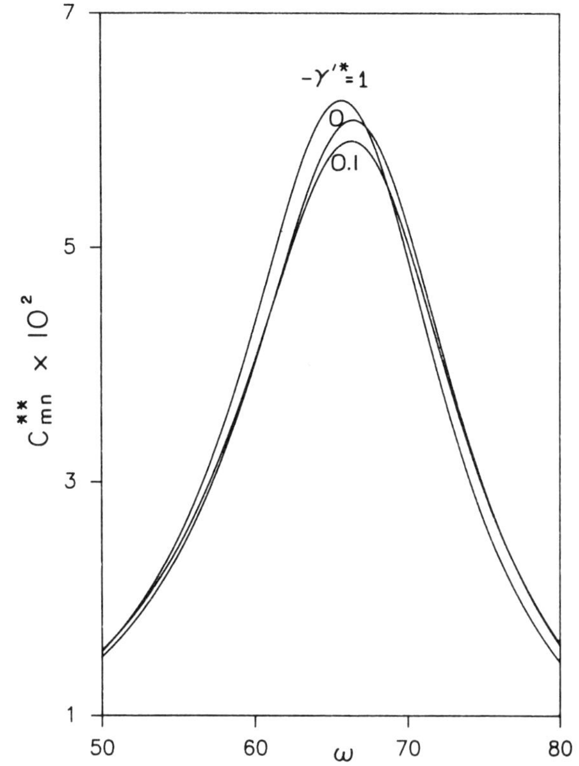

Figure 6.4.3-14: C^{**}_{mn} as a function of ω for a liquid-liquid system with $N_{ca} = 10^{-4}$, $N_{\mu} = 10$, $N_{\rho} = 0.9$, $N_g \leq 10$, $N_{\kappa+\varepsilon} = 0$, and various values of $-\gamma'^*$. The solid lines represent both (3-100) and (3-119).

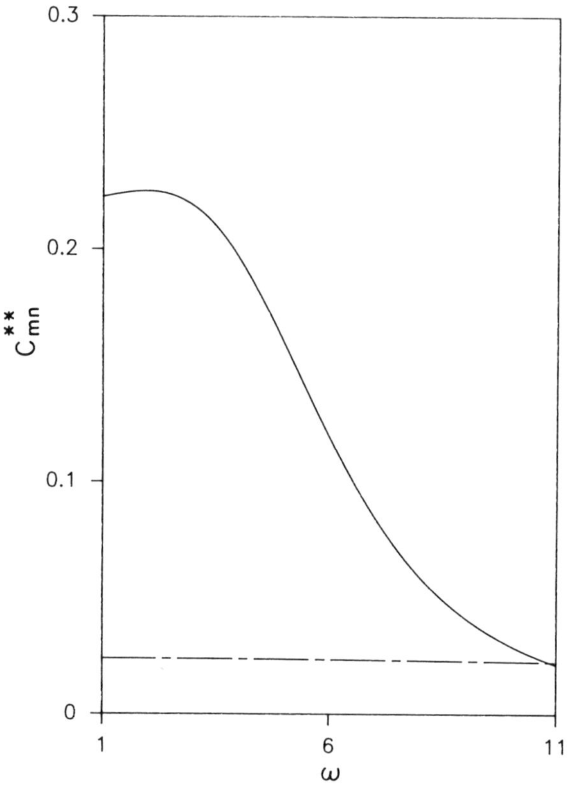

Figure 6.4.3-15: C^{**}_{mn} as a function of ω for a liquid-liquid system with $N_{ca} = 10^{-2}$, $N_{\mu} = 10$, $N_{\rho} = 0.9$, $N_g \leq 10$, and all values of γ'^{*} and $N_{\kappa+\varepsilon}$. The solid line represents (3-100) and the dashed line (3-119).

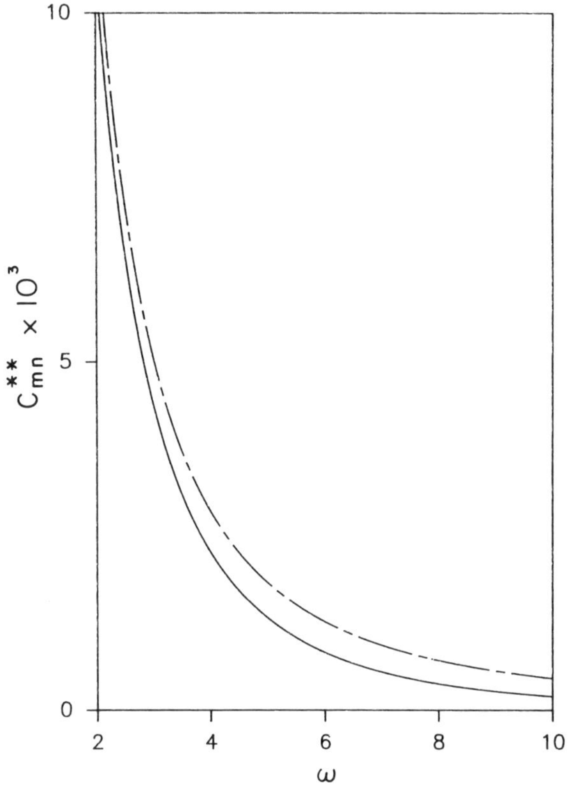

Figure 6.4.3-16: C^{**}_{mn} as a function of ω for a liquid-liquid system with $N_{ca} = 1$, $N_{\mu} = 10$, $N_{\rho} = 0.9$, $N_{g} \leq 10$, and all values of γ'^{*} and $N_{\kappa+\varepsilon}$. The solid line represents (3-100) and the dashed line (3-119).

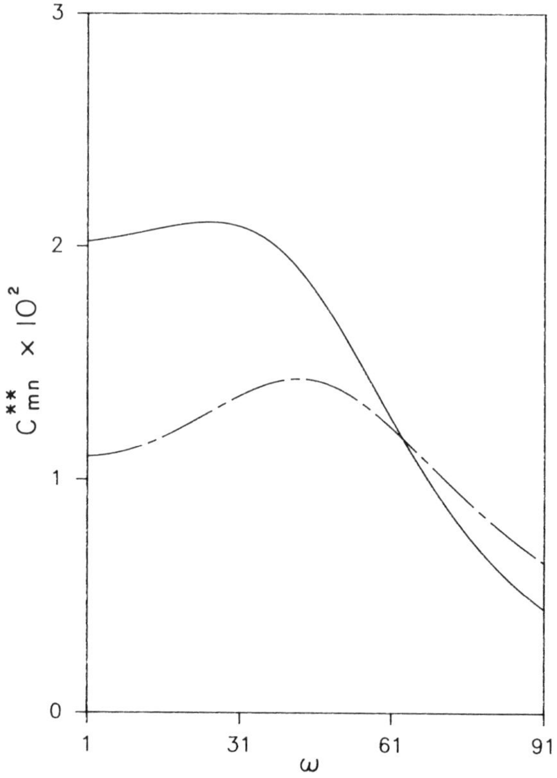

Figure 6.4.3-17: C^{**}_{mn} as a function of ω for a liquid-liquid system with $N_{ca} = 10^{-4}$, $N_\mu = 100$, $N_\rho = 0.9$, $N_g \leq 10$, and all values of γ'^* and $N_{\kappa+\varepsilon}$. The solid line represents (3-100) and the dashed line (3-119).

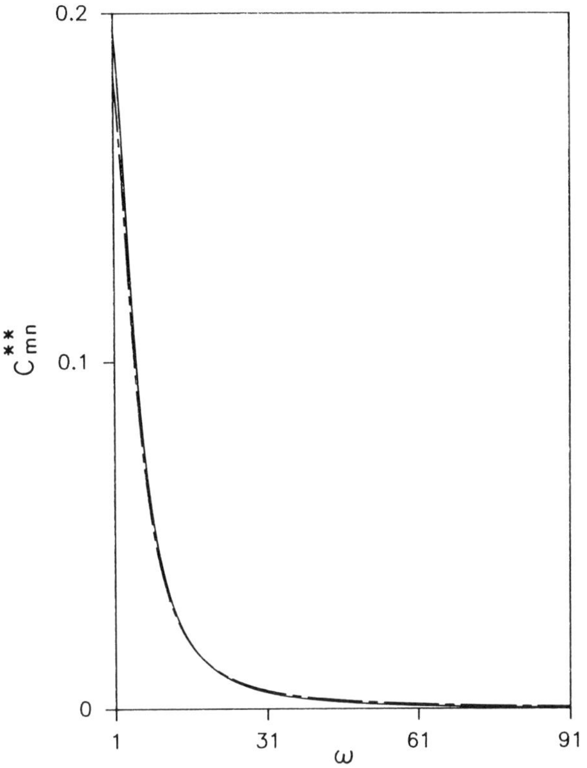

Figure 6.4.3-18: C^{**}_{mn} as a function of ω for a liquid-liquid system with $N_{ca} = 10^{-3}$, $N_{\mu} = 100$, $N_{\rho} = 0.9$, $N_g \leq 10$, and all values of γ'^* and $N_{\kappa+\varepsilon}$. The solid line represents (3-100) and the dashed line (3-119).

as functions of N_{ca}, $N_{\kappa+\varepsilon}$, γ'^*, and N_g. We will restrict our illustrations to $-\gamma'^* \leq 1$ [see discussion following (2-81)].

Figures 3-1, 3-2, 3-4, and 3-5 indicate that the spectral density is a significant function of the interfacial viscosities and of γ'^* for $N_{ca} \leq 10^{-2}$. Figure 3-3 shows that, with increasing N_{ca}, the effect of γ'^* on the spectral density disappears. With systems for which $N_{ca} \leq 10^{-2}$, it may be possible to use this technique to determine the interfacial tension, $N_{\kappa+\varepsilon}$, as well as γ'^*. As N_{ca} increases beyond 10^{-1}, first the effect of γ'^* and then that of $N_{\kappa+\varepsilon}$ disappear.

There is no effect of $N_g \leq 10$ for $N_{ca} \leq 10^{-1}$. Figures 3-7 through 3-11 indicate that there is a significant effect of gravity for $N_{ca} \geq 1$. As N_g increases, the effects of the interfacial viscosities begin to reappear.

Figures 3-1 through 3-11 suggest that the Lorentzian form (3-119) is usually a poor approximation to the actual spectrum unless $N_{ca} < 10^{-4}$.

Results: liquid-liquid systems Figures 3-12 through 3-18 illustrate the effects of N_{ca} and N_μ for only one density ratio: $N_\rho = 0.9$.

The effects of the dimensionless Gibbs surface elasticity γ'^* and of the dimensionless sum of the interfacial viscosities $N_{\kappa+\varepsilon}$ virtually disappear for these systems. They reappear only when N_{ca} and N_μ are sufficiently small. Both N_{ca} and N_μ are important parameters; N_g is of little significance.

Similar to the liquid-gas case, the Lorentzian spectrum (3-119) is generally a poor approximation.

Results: propagation of errors Errors in the experimental spectral density data propagate in computing $N_{\kappa+\varepsilon}$ and γ'^*. In order to examine the effect of such experimental errors upon the calculated values of these parameters, we have carried out the following computer experiments.

For each spectral density curve represented in Figures 3-1 through 3-6, 3-8, and 3-11, we chose 100 equally spaced points. (The effects of $N_{\kappa+\varepsilon}$ and γ'^* essentially disappear in the other figures.) We then generated a random error bar for each of these points, the mean value of which was zero and the standard deviation of which was either 0.5% or 1% of the maximum spectral density on the curve. New sets of noisy data were generated by adding the old values to these generated error bars. New values of either $N_{\kappa+\varepsilon}$ or γ'^* and the corresponding standard deviations then were computed by fitting (3-100) to these noisy data.

Since this computer experiment was for illustrative purposes only, we determined only one parameter, either $N_{\kappa+\varepsilon}$ or γ'^*, in using (3-100) to fit these data, assuming the other was known. In reality, the experimentalist may wish to determine both of these parameters as well as N_{ca}, in which case the standard deviation for each parameter must be estimated.

Tables 3-1 through 3-8 represent the result of this computer experiment. The first column of each table represents the parameter of the original curve that was distorted to represent the noisy data. Columns 2 and 3 show the mean value and the standard deviation of the fitted parameter corresponding to the specified standard deviation on the error bar of the spectral density data. The rows in Tables 3-1 through 3-8 that are marked by a * correspond to a large standard deviation, indicating that under these conditions meaningful values of the parameter can not be estimated in this manner.

The results have the same character that we observed above in examining the figures. Both Tables and Figures 3-1, 3-2, 3-4, and 3-5 indicate that the spectral density is a significant function of $N_{\kappa+\varepsilon}$ and of γ'^* for $N_{ca} \leq 10^{-2}$. Table and Figure 3-3 show that, with increasing N_{ca}, the effect of γ'^* on the spectral density disappears. With systems for which $N_{ca} \leq 10^{-2}$, it may be possible to use this technique to determine N_{ca} as well as $N_{\kappa+\varepsilon}$ and γ'^*, depending upon the quality of the spectral density data. As N_{ca} increases beyond 10^{-1}, first the effect of γ'^* and then that of $N_{\kappa+\varepsilon}$ disappear.

In what has been presented here, it has been assumed that the experimentalist, by fixing the angle of the incident light with respect to that of the measured scattered light, will measure only the correlation function $G_{mn}(\tau)$ or the spectral density $P_{mn}(\omega)$ corresponding to a single value of

$$q \equiv (m^2 + n^2)^{1/2} \tag{3-122}$$

In reality, the sample surface is illuminated by a Gaussian laser beam having a finite width (Dorshow *et al.* 1988). The light intensity registered at the detector is contributed from a range Δq, which can be determined experimentally. The measured power spectrum at the detector is a convolution of P_{mn} defined in (3-102) and a Gaussian distribution

$$P(\omega, q, \Delta q) = \int_{-\infty}^{\infty} P_{mn}(\omega, q + \delta) \exp\left(- \frac{2\delta^2}{(\Delta q)^2} \right) d\delta \tag{3-123}$$

and (3-104), used to fit the experimental data, should be replaced by

$$P(\omega, q, \Delta q) = A \int_{-\infty}^{\infty} C_{mn}(\omega, q + \delta) \exp\left(- \frac{2\delta^2}{(\Delta q)^2} \right) d\delta + B \tag{3-124}$$

Table 6.4.3-1: Mean and standard deviation for $N_{\kappa+\varepsilon}$ as the fitted parameter in the computer experiment for the results shown in Figure 3-1 with $N_{ca} = 10^{-4}$, $N_g \leq 10$, and $\gamma'^{*} = 0$.

	$\sigma = 0.5\%$		$\sigma = 1\%$	
$N_{\kappa+\varepsilon}$	$< N_{\kappa+\varepsilon} >$	$\sigma_{\kappa+\varepsilon}$	$< N_{\kappa+\varepsilon} >$	$\sigma_{\kappa+\varepsilon}$
0	5.3×10^{-3}	1.6×10^{-2}	1.8×10^{-2}	3.8×10^{-2}
1	0.992	0.034	0.985	0.068
10	9.98	0.773	9.96	1.50
*1000	1005	$*8 \times 10^{6}$	1010	$*1.6 \times 10^{7}$

Table 6.4.3-2: Mean and standard deviation for γ'^{*} as the fitted parameter in the computer experiment for the results shown in Figure 3-2 with $N_{ca} = 10^{-4}$, $N_g \leq 10$, and $N_{\kappa+\varepsilon} = 0$.

	$\sigma = 0.5\%$		$\sigma = 1\%$	
$-\gamma'^{*}$	$< -\gamma'^{*} >$	σ_{γ}	$< -\gamma^{*} >$	σ_{γ}
0	6.7×10^{-5}	0.019	1.3×10^{-4}	0.038
0.1	0.099	0.134	0.099	*0.269
*0.5	0.499	*40.6	0.498	*80.6
*1	0.995	$*0.71 \times 10^{3}$	0.989	$*1.4 \times 10^{3}$

Table 6.4.3-3: Mean and standard deviation for γ'^* as the fitted parameter in the computer experiment for the results shown in Figure 3-3 with $N_{ca} = 10^{-4}$, $N_g \leq 10$, and $N_{\kappa+\varepsilon} = 100$.

	$\sigma = 0.5\%$		$\sigma = 1\%$	
$-\gamma'^*$	$< -\gamma'^* >$	σ_γ	$< -\gamma'^* >$	σ_γ
0	0.007	0.85×10^3	0.005	1.7×10^3
*1	1.004	$*3.2 \times 10^3$	1.006	$*0.64 \times 10^4$

Table 6.4.3-4: Mean and standard deviation for γ'^* as the fitted parameter in the computer experiment for the results shown in Figure 3-4 with $N_{ca} = 10^{-3}$, $N_g \leq 10$, and $N_{\kappa+\varepsilon} = 1$.

	$\sigma = 0.5\%$		$\sigma = 1\%$	
$-\gamma'^*$	$< -\gamma'^* >$	σ_γ	$< -\gamma'^* >$	σ_γ
0	9.9×10^{-6}	0.025	1.9×10^{-5}	0.054
0.1	0.099	0.029	0.099	0.058
*0.5	0.49	*1.34	0.49	*2.67
*1	0.99	*18	0.99	*37

Table 6.4.3-5: Mean and standard deviation for γ'^* as the fitted parameter in the computer experiment for the results shown in Figure 3-5 with $N_{ca} = 10^{-2}$, $N_g \leq 10$, and $N_{\kappa+\varepsilon} = 1$.

	$\sigma = 0.5\%$		$\sigma = 1\%$	
$-\gamma'^*$	$<-\gamma'^*>$	σ_γ	$<-\gamma'^*>$	σ_γ
0	1.9×10^{-4}	0.026	3.9×10^{-4}	0.053
0.1	0.100	0.028	0.1	0.057
0.5	0.50	0.11	0.50	0.22
1	0.99	0.89	0.99	1.77

Table 6.4.3-6: Mean and standard deviation for $N_{\kappa+\varepsilon}$ as the fitted parameter in the computer experiment for the results shown in Figure 3-6 with $N_{ca} = 0.1$, $N_g \leq 10$, and $-\gamma'^* \leq 0.1$.

	$\sigma = 0.5\%$		$\sigma = 1\%$	
$N_{\kappa+\varepsilon}$	$< N_{\kappa+\varepsilon} >$	$\sigma_{\kappa+\varepsilon}$	$< N_{\kappa+\varepsilon} >$	$\sigma_{\kappa+\varepsilon}$
0	3.2×10^{-3}	4.6×10^{-2}	6.48×10^{-3}	9.3×10^{-2}
1	1.009	0.150	1.018	0.305
*10	10.20	22.3	10.41	*47.8
*100	117	$*1.6 \times 10^5$	120	$*3 \times 10^6$

Table 6.4.3-7: Mean and standard deviation for $N_{\kappa+\varepsilon}$ as the fitted parameter in the computer experiment for the results shown in Figure 3-8 with $N_{ca} = 1$, $N_g = 10$, and $-\gamma'^{*} \leq 1$.

	$\sigma = 0.5\%$		$\sigma = 1\%$	
$N_{\kappa+\varepsilon}$	< $N_{\kappa+\varepsilon}$ >	$\sigma_{\kappa+\varepsilon}$	< $N_{\kappa+\varepsilon}$)	$\sigma_{\kappa+\varepsilon}$
0	3.1×10^{-2}	0.57	6.4×10^{-2}	1.19
*1	1.06	*1.97	1.11	*4.2
*10	10.6	*335	11.4	*830
*100	162	*7×10^6	410	*6×10^8

Table 6.4.3-8: Mean and standard deviation for $N_{\kappa+\varepsilon}$ as the fitted parameter in the computer experiment for the results shown in Figure 3-11 with $N_{ca} = 10$, $N_g = 10$, and $-\gamma'^{*} \leq 1$.

	$\sigma = 0.5\%$		$\sigma = 1\%$	
$N_{\kappa+\varepsilon}$	< $N_{\kappa+\varepsilon}$ >	$\sigma_{\kappa+\varepsilon}$	< $N_{\kappa+\varepsilon}$ >	$\sigma_{\kappa+\varepsilon}$
*0	1.3×10^{-2}	*6.4	2.7×10^{-2}	*13.1
*1	1.03	*24.1	1.06	*50
*10	10.46	*3650	10.96	*8450
*100	230	*10^8	236	*8×10^8

References

Bouchiat, M. A., and J. Meunier, *C. R. Hebd. Seances Acad. Sci. Ser. B* **266**, 301 (1968).

Bouchiat, M. A., and J. Meunier, *Phys. Rev. Lett.* **23**, 752 (1969).

Bouchiat, M. A., and J. Meunier, *J. Phys.* **32**, 561 (1971).

Bouchiat, M. A., J. Meunier, and J. Brossel, *C. R. Hebd. Seances Acad. Sci. Ser. B* **266**, 255 (1968).

Byrne, D., and J. C. Earnshaw, *J. Phys.* **D12**, 1145 (1979).

Callen, H. B., and R. F. Greene, *Phys. Rev.* **86**, 702 (1952).

Chu, B., "Laser Light Scattering," Academic Press, New York (1974).

Dorshow, R. B., A. Hajiloo, and R. L. Swofford, *J. Appl. Phys.* **63**, 1265 (1988).

Dorshow, R. B., and R. L. Swofford, *J. Appl. Phys.* **65**, 3756 (1989).

Dorshow, R. B., and R. L. Swofford, *Colloids & Surfaces* (1989).

Dorshow, R. B., and N. R. Pallas, *currently being reviewed* (1990).

Edwards, R. V., R. S. Sirohi, J. A. Mann, L. B. Shih, and L. A. Lading, *Appl. Optics* **21**, 3555 (1982).

Goodrich, F. C., *Proc. Roy. Soc. London* **A260**, 490 (1961).

Greene, R. F., and H. B. Callen, *Phys. Rev.* **88**, 1387 (1952).

Hajiloo, A., and J. C. Slattery, *J. Colloid Interface Sci.* **112**, 325 (1986).

Hard, S., and H. Löfgren, *J. Colloid Interface Sci.* **60**, 529 (1977).

Hard, S., and R. D. Neuman, *J. Colloid Interface Sci.* **83**, 315 (1981).

Huang, J. S., and W. W. Webb, *Phys. Rev. Lett.* **23**, 160 (1969).

Katyl, R. H., and U. Ingard, *Phys. Rev. Lett.* **20**, 248 (1968).

Kestin, J., and J. R. Dorfman, "A Course in Statistical Thermodynamics," p. 184, Academic Press, New York (1971).

Kim, M. W., J. S. Huang, and J. Bock, "Interfacial Light Scattering Study in Microemulsions," SPE/DOE 10788, Third Joint Symposium on Enhanced Oil Recovery of the Society of Petroleum Engineers, Tulsa, OK, April 4-7, 1982.

Lachaise, J., A. Graciaa, A. Martinez, and A. Roussei, *Thin Solid Films* **82**, 55 (1981).

Langevin, D., *J. Colloid Interface Sci.* **80**, 412 (1981).

Langevin, D., and M. A. Bouchiat, *C. R. Hebd. Seances Acad. Sci. Ser. B* **272**, 1422 (1971).

Langevin, D., and J. Meunier, in "Photon Correlation Spectroscopy and Velocimetry," edited by H. Z. Cummins and E. R. Pike, Plenum Press (1976).

Lindgren, B. W., "Statistical Theory," third edition, Macmillan, New York (1976).

Mandelstam, L., *Ann. Phys.* **41**, 609 (1913).

Meunier, J., D. Cruchon, and M. Bouchiat, *C. R. Hebd. Seances Acad. Sci. Ser. B* **268**, 92 (1969).

Papoular, M., *J. Phys.* **29**, 81 (1968).

Pike, E. R., in "Photon Correlation Spectroscopy and Velocimetry," edited by H. Z. Cummins and E. R. Pike, Plenum Press (1976).

Pouchelon, A., D. Chatenay, J. Meunier, and D. Langevin, *J. Colloid Interface Sci.* **82**, 418 (1981).

Pouchelon, A., J. Meunier, D. Langevin, and A. M. Cazabat, *J. Phys. Lett.* **41**, 239 (1980).

Schuss, Z., "Theory and Application of Stochastic Differential Equations," Wiley, New York (1980).

Smoluchowski, M. V., *Ann. Phys.* **25**, 225 (1908).

Spiegel, M. R., "Laplace Transforms," p. 176, 201, McGraw-Hill, New York (1965).

Yaglom, A. M., "Stationary Random Functions," Dover, New York (1973).

Exercise

6.4.3-1 *limiting expression for* $C_{m\,n}(\omega)$ For a vapor-liquid interface in a single component system, it is reasonable to consider the limiting case

$$N_\rho \to 0, \quad N_\mu \to 0, \quad N_\kappa \to 0, \quad N_\varepsilon \to 0, \quad \gamma'^* \to 0$$

If in addition

$$N_g \to 0$$

such that the effects of gravity can be neglected, show that (3-100) reduces to

$$C_{mn}(\omega) = \frac{kT}{2\gamma L^2 \omega} \, \mathrm{Im}\left(\left[-1 - 4N_{ca} + \omega^2 N_{ca} - 4N_{ca}\, i\omega + 4N_{ca}\sqrt{1 + i\omega} \right]^{-1} \right)$$

Notation for chapter 6

$\mathbf{b}$	external force per unit mass, introduced in Sec. 2.1.3
$\mathcal{D}_{(jm)}$	diffusion coefficient for species j in the mixture
$\mathbf{e}_j$	basis vectors for rectangular cartesian coordinate system (Slattery 1981, pp. 607 and 614)
g	magnitude of the acceleration of gravity
H	mean curature of dividing surface, introduced in Exercise A.5.3-3
$j_{(A)}$	mass flux of species A with respect to mass-averaged velocity, defined in Table 5.2.2-4
M	molar-averaged molecular weight of the solution, defined in Table 5.2.2-1
N_{Bo}	dimensionless Bond number, defined in Sec. 6.3.1
N_{ca}	dimensionless capillary number, defined in Sec. 6.4.2
N_{Pe}	dimensionless Pelect number, defined in Sec. 6.3.1
N_{Re}	dimensionless Reynolds number, defined in Sec. 6.3.1
N_{Sc}	dimensionless Schmidt number, defined in Sec. 6.4.2
$N_{\kappa+\varepsilon}$	dimensionless sum of the characteristic interfacial viscosities, defined in Sec. 6.4.3
p	mean pressure, defined in Sec. 2.4.1
$\mathcal{P}$	modified pressure, defined in Sec. 2.4.1
$\mathbf{q}$	energy flux vector, introduced in Sec. 5.6.4
r	spherical coordinate
$\mathbf{S}$	viscous portion of stress tensor, defined in Sec. 2.4.1
$\mathbf{S}^{(\sigma)}$	viscous portion of surface stress tensor, defined in Sec. 2.4.2
t	time

T Temperature, introduced in Sec. 5.7.2

$T^{(\sigma)}$ surface temperature, introduced in Sec. 5.7.2

$\mathbf{T}$ stress tensor, introduced in Sec. 2.1.4

$\mathbf{u}$ time rate of change of spatial position following a surface point, introduced in Sec. 1.2.7

$\hat{U}$ internal energy per unit mass within a phase, introduced in Sec. 5.6.2

$\mathbf{v}$ velocity vector, defined in Sec. 1.1.1

$\mathbf{v}^{(\sigma)}$ surface velocity vector, defined in Sec. 1.2.5

$v_{(\xi)}^{\{g\}}$ speed of displacement, introduced in Sec. 1.2.7

z_j rectangular cartesian coordinates

greek letters

γ thermodynamic interfacial tension, defined in Sec. 5.8.6

γ^* dimensionless Gibbs surface elasticity, defined in Sec. 6.4.2

θ spherical coordinate

μ viscosity

$\mu^{(i)}$ chemical potential, defined in Sec. 5.8.3

ξ unit normal to Σ, introduced in Sec. 1.2.7

π 3.14159...

ρ mass density

$\rho_{(j)}^{\{\sigma\}}$ surface mass density of species j, introduced in Sec. 5.2.1

ϕ spherical coordinate; potential energy per unit mass, defined in Secs. 2.1.3 and 2.3.4

others

div divergence operator

$\text{div}_{(\sigma)}$	surface divergence operator, introduced in Secs. A.5.1 through A.5.6
$\ldots^*$	denotes dimensionless variable
$\ldots^{(i)}$	variable associated with phase i
$[\ldots]$	jump for dividing surface, defined in Sec. 1.3.4
∇	gradient operator
$\nabla_{(\sigma)}$	surface gradient operation, introduced in Secs. A.5.1 through A.5.6

7
Applications of integral averaging to energy and mass transfer

The discussion of integral averaging techniques begun in Chapter 4 is continued here, but the emphasis is now upon energy and mass transfer. If you have not already done so, I recommend that you read Chapter 4 before beginning this material. The derivations are more detailed there and more attention is paid to motivation.

7.1 More integral balances

7.1.1 Introduction We began our discussion of integral balances in Secs. 4.1.1 through 4.1.9. If you have not already done so, I recommend that you read those sections before proceeding here. The general approach taken there is the same as that to be used here. But the integral balances developed there are by their very nature somewhat easier to understand than those to be discussed here.

The integral balances derived in Secs. 4.1.1 through 4.1.5 literally apply only to single-component, multiphase systems. Fortunately, the overall mass and momentum balances derived in Chapter 5 have essentially the same form as those we found in Chapters 1 and 2. Besides the straightforward difference in interpretation of variables, one need recognize only that a different external force may operate on each of the species present. As a result, the integral balances derived in Secs. 4.1.1 through 4.1.5 can be applied immediately to multicomponent, multiphase systems with little or no change. For this reason, I will say no more about the integral overall mass balance, the integral momentum balance, the integral mechanical energy balance, and the integral moment of momentum balance.

Nor do I wish to repeat what I have said previously about the integral mass balance for an individual species, the integral energy balance, and the integral Clausius-Duhem inequality (Slattery 1981). That discussion includes the effects of turbulence, the construction of needed empiricisms, and several simple examples. But it does not include interfacial effects. In what follows, our primary interest is these interfacial effects.

References

Slattery, J. C., "Momentum, Energy, and Mass Transfer in Continua," McGraw-Hill, New York (1972); second edition, Robert E. Krieger, Malabar, FL 32950 (1981).

7.1.2 The integral mass balance for species A The derivation of the integral mass balance for species A parallels that given for the integral mass balance in Sec. 4.1.2. We can use the generalized transport theorem (Exercise 1.3.7-1) to write an expression for the time rate of change of mass of species A within an arbitrary system. From this, we subtract the volume integral of the equation of continuity for species A (Sec. 5.2.1) over $R_{(sys)}$, the area integral of the jump mass balance for species A (Sec. 5.2.1) over $\Sigma_{(sys)}$, and the line integral of the mass balance for species A at a multicomponent common line (Sec. 5.2.1) over $C^{(cl)}_{(sys)}$.

We conclude that:

$$\begin{aligned}
\frac{d}{dt}\left(\int_{R_{(sys)}} \rho_{(A)}\, dV + \int_{\Sigma_{(sys)}} \rho^{(\sigma)}_{(A)}\, dA \right) \\
= \int_{S_{(ent\ ex)}} \rho_{(A)}(\mathbf{v} - \mathbf{v}_{(sys)}) \cdot (- \mathbf{n})\, dA \\
+ \int_{C_{(ent\ ex)}} \rho^{(\sigma)}_{(A)} (\mathbf{v}^{(\sigma)} - \mathbf{v}_{(sys)}) \cdot (- \boldsymbol{\mu})\, ds \\
+ \int_{S_{(ent\ ex)}} \mathbf{j}_{(A)} \cdot (- \mathbf{n})\, dA + \int_{C_{(ent\ ex)}} \mathbf{j}^{(\sigma)}_{(A)} \cdot (- \boldsymbol{\mu})\, dA \\
+ \int_{R_{(sys)}} r_{(A)}\, dV + \int_{\Sigma_{(sys)}} r^{(\sigma)}_{(A)}\, dA
\end{aligned} \tag{2-1}$$

We will refer to this as the **integral mass balance for species A**. It says that the time rate of change of the mass of species A in an arbitrary system is equal to the net rate at which the mass of species A is brought into the system through the entrances and exits by convection, the net rate at which the mass of species A diffuses into the system (relative to the mass-averaged velocity), and the rate at which the mass of species A is produced in the system by homogeneous and heterogeneous chemical reactions.

For more on the integral mass balance excluding interfacial effects, see Slattery (1981, p. 584) and Bird *et al.* (1960, p. 686).

References

Bird, R. B., W. E. Stewart, and E. N. Lightfoot, "Transport Phenomena," John Wiley, New York (1960).

Slattery, J. C., "Momentum, Energy, and Mass Transport in Continua," McGraw-Hill, New York (1972); second edition, Robert E. Krieger, Malabar, FL 32950 (1981).

7.1.3 The integral energy balance Various forms of the integral energy balance are given in Table 7.1.3. They are all derived in the same manner. The generalized transport theorem (Exercise 1.3.7-1) is used to write an expression for the time rate of change of the energy within an arbitrary system. From this, we subtract the volume integral of the differential energy balance (Sec. 5.6.4) over $S_{(sys)}$, the area integral of the jump energy balance (Sec. 5.6.4) over $\Sigma_{(sys)}$, and the line integral of the energy balance at a multicomponent common line over $C^{(cl)}_{(sys)}$ (Exercise 5.6.4-4).

In Table 7.1.3-1,

$$\mathcal{Q} \equiv \int_{S_{(sys)}} \mathbf{q} \cdot (- \mathbf{n}) \, dA + \int_{C_{(sys)}} \mathbf{q}^{(\sigma)} \cdot (- \boldsymbol{\mu}) \, ds \tag{3-1}$$

is the **rate of (contact) energy transfer to the system** from the surroundings across the bounding surfaces of the system and

$$\mathcal{W} \equiv \int_{S_{(sys)} - S_{(ent\ ex)}} \mathbf{v} \cdot [(\mathbf{T} + p_0 \mathbf{I}) \cdot (- \mathbf{n})] \, dA + \int_{C_{(sys)} - C_{ent\ ex)}} \mathbf{v}^{(\sigma)} \cdot \mathbf{T}^{(\sigma)} \cdot (- \boldsymbol{\mu}) \, ds \tag{3-2}$$

is the **rate at which work is done by the system on the surroundings** at the moving impermeable surfaces of the system (beyond any work done on these surfaces by the ambient pressure p_0).

For more on the integral energy balance excluding interfacial effects, see Slattery (1981, p. 420 and 595) and Bird *et al.* (1960, p. 457 and 689).

References

Bird, R. B., W. E. Stewart, and E. N. Lightfoot, "Transport Phenomena," John Wiley, New York (1960).

Israelachvili, J. N., "Intermolecular and Surface Forces," Academic Press, London (1985).

Slattery, J. C., "Momentum, Energy, and Mass Transport in Continua," McGraw-Hill, New York (1972); second edition, Robert E. Krieger, Malabar, FL 32950 (1981).

Table 7.1.3-1: General forms of the integral overall energy balance

$$\frac{d}{dt}\left[\int_{R_{(sys)}} \rho \left(\hat{U} + \frac{1}{2} v^2 + \phi + \frac{p_0}{\rho} \right) dV + \int_{\Sigma_{(sys)}} \rho^{(\sigma)} \left(\hat{U}^{(\sigma)} + \frac{1}{2} v^{(\sigma)2} + \phi \right) dA \right]$$

$$= \int_{S_{(ent\ ex)}} \rho \left(\hat{H} + \frac{1}{2} v^2 + \phi \right) (\mathbf{v} - \mathbf{v}_{(sys)}) \cdot (- \mathbf{n}) \, dA$$

$$+ \int_{C_{(ent\ ex)}} \rho^{(\sigma)} \left(\hat{H}^{(\sigma)} + \frac{1}{2} v^{(\sigma)2} + \phi \right)$$

$$\times (\mathbf{v}^{(\sigma)} - \mathbf{v}_{(sys)}) \cdot (- \boldsymbol{\mu}) \, ds + \mathcal{Q} - \mathcal{W}$$

$$+ \int_{R_{(sys)}} \left(\sum_{A=1}^{N} \mathbf{j}_{(A)} \cdot \mathbf{b}_{(A)} + \rho Q \right) dV$$

$$+ \int_{\Sigma_{(sys)}} \left(\sum_{A=1}^{N} \mathbf{j}_{(A)}^{(\sigma)} \cdot \mathbf{b}_{(A)}^{(\sigma)} + \rho^{(\sigma)} Q^{(\sigma)} \right) dA$$

$$+ \int_{S_{(ent\ ex)}} [- (P - p_0)(\mathbf{v}_{(sys)} \cdot \mathbf{n}) + \mathbf{v} \cdot \mathbf{S} \cdot \mathbf{n}] \, dA$$

$$+ \int_{C_{(ent\ ex)}} (\gamma \, \mathbf{v}_{(sys)} \cdot \boldsymbol{\mu} + \mathbf{v}^{(\sigma)} \cdot \mathbf{S}^{(\sigma)} \cdot \boldsymbol{\mu}) \, ds \qquad \text{(A)}^{a}$$

---

a) We assume here that

$$\sum_{A=1}^{N} \omega_{(A)} \mathbf{b}_{(A)} = - \nabla \phi$$

(cont.)

Table 7.1.3-1: (cont.)

$$\frac{d}{dt}\left[\int_{R_{(sys)}} \rho\left(\hat{U}+\frac{1}{2}v^2+\frac{p_0}{\rho}\right)dV + \int_{\Sigma_{(sys)}} \rho^{(\sigma)}\left(\hat{U}^{(\sigma)}+\frac{1}{2}v^{(\sigma)2}\right)dA\right]$$

$$= \int_{S_{(ent\ ex)}} \rho\left(\hat{H}+\frac{1}{2}v^2\right)(\mathbf{v}-\mathbf{v}_{(sys)})\cdot(-\mathbf{n})\,dA$$

$$+ \int_{C_{(ent\ ex)}} \rho^{(\sigma)}\left(\hat{H}^{(\sigma)}+\frac{1}{2}v^{(\sigma)2}\right)(\mathbf{v}^{(\sigma)}-\mathbf{v}_{(sys)})\cdot(-\boldsymbol{\mu})\,ds$$

$$+ \mathcal{Q} - \mathcal{W} + \int_{R_{(sys)}} \left[\sum_{A=1}^{N} \mathbf{n}_{(A)}\cdot\mathbf{b}_{(A)} + \rho Q\right] dV$$

$$+ \int_{\Sigma_{(sys)}} \left[\sum_{A=1}^{N} \mathbf{n}^{(\sigma)}_{(A)}\cdot\mathbf{b}^{(\sigma)}_{(A)} + \rho^{(\sigma)}Q^{(\sigma)}\right] dA$$

$$+ \int_{S_{(ent\ ex)}} \left[-(P-p_0)(\mathbf{v}_{(sys)}\cdot\mathbf{n}) + \mathbf{v}\cdot\mathbf{S}\cdot\mathbf{n}\right] dA$$

$$+ \int_{C_{(ent\ ex)}} (\gamma\,\mathbf{v}_{(sys)}\cdot\boldsymbol{\mu} + \mathbf{v}^{(\sigma)}\cdot\mathbf{S}^{(\sigma)}\cdot\boldsymbol{\mu})\,ds \qquad \text{(B)}$$

---

a) (continued from previous page) and that on every divid[illegible]rface

$$\sum_{A=1}^{N} \omega^{(\sigma)}_{(A)}\mathbf{b}^{(\sigma)}_{(A)} = -\nabla\phi$$

where ϕ is not an explicit function of time. [illegible] cise 4.1.4-2 for more details.

Table 7.1.3-1: (cont.)

$$\frac{d}{dt}\left[\int_{R_{(sys)}} \rho\left(\hat{U} + \frac{p_0}{\rho}\right) dV + \int_{\Sigma_{(sys)}} \rho^{(\sigma)}\hat{U}^{(\sigma)}\, dA\right]$$

$$= \int_{S_{(ent\ ex)}} \rho\left(\hat{U} + \frac{p_0}{\rho}\right)(\mathbf{v} - \mathbf{v}_{(sys)}) \cdot (-\mathbf{n})\, dA$$

$$+ \int_{C_{(ent\ ex)}} \rho^{(\sigma)}\hat{U}^{(\sigma)}(\mathbf{v}^{(\sigma)} - \mathbf{v}_{(sys)}) \cdot (-\boldsymbol{\mu})\, ds + Q$$

$$+ \int_{R_{(sys)}} \left[-(P - p_0)\,\text{div}\,\mathbf{v} + \text{tr}(\mathbf{S} \cdot \nabla\mathbf{v})\right.$$

$$\left. + \sum_{A=1}^{N} \mathbf{j}_{(A)} \cdot \mathbf{b}_{(A)} + \rho Q\right] dV$$

$$+ \int_{\Sigma_{(sys)}} \left\{\gamma\,\text{div}_{(\sigma)}\mathbf{v}^{(\sigma)} + \text{tr}(\mathbf{S}^{(\sigma)} \cdot \nabla_{(\sigma)}\mathbf{v}^{(\sigma)})\right.$$

$$+ \sum_{A=1}^{N} \mathbf{j}^{(\sigma)}_{(A)} \cdot \mathbf{b}^{(\sigma)}_{(A)} + \rho^{(\sigma)}Q^{(\sigma)}$$

$$+ \left[-(P - p_0)(\mathbf{v} \cdot \boldsymbol{\xi} - v_{(\xi)}) - \frac{1}{2}\rho(\mathbf{v} \cdot \boldsymbol{\xi} - v_{(\xi)})^3\right.$$

$$\left.\left. + (\boldsymbol{\xi} \cdot \mathbf{S} \cdot \boldsymbol{\xi})(\mathbf{v} \cdot \boldsymbol{\xi} - v_{(\xi)})\right]\right\} dA$$

$$\left(-\frac{1}{2}\rho^{(\sigma)}\, v^{(\sigma)2}\left[\mathbf{v}^{(\sigma)} \cdot \mathbf{v} - u^{(cl)}_{(\nu)}\right]\right.$$

$$\left. \mathbf{T}^{(\sigma)} \cdot \mathbf{v}\right) ds \qquad \text{(C)}$$

Table 7.1.3-1: (cont.)

For an incompressible fluid:

$$\frac{d}{dt}\left[\int_{R_{(sys)}} \rho \hat{U} \, dV + \int_{\Sigma_{(sys)}} \rho^{(\sigma)} \hat{U}^{(\sigma)} \, dA \right]$$

$$= \int_{S_{(ent\ ex)}} \rho \hat{U} (\mathbf{v} - \mathbf{v}_{(sys)}) \cdot (- \mathbf{n}) \, dA$$

$$+ \int_{C_{(ent\ ex)}} \rho^{(\sigma)} \hat{U}^{(\sigma)} (\mathbf{v}^{(\sigma)} - \mathbf{v}_{(sys)}) \cdot (- \boldsymbol{\mu}) \, ds + \mathcal{Q}$$

$$+ \int_{R_{(sys)}} \left[\text{tr}(\mathbf{S} \cdot \nabla \mathbf{v}) + \sum_{A=1}^{N} \mathbf{j}_{(A)} \cdot \mathbf{b}_{(A)} + \rho Q \right] dV$$

$$+ \int_{\Sigma_{(sys)}} \Big\{ \gamma \, \text{div}_{(\sigma)} \mathbf{v}^{(\sigma)} + \text{tr}(\mathbf{S}^{(\sigma)} \cdot \nabla_{(\sigma)} \mathbf{v}^{(\sigma)})$$

$$+ \sum_{A=1}^{N} \mathbf{j}^{(\sigma)}_{(A)} \cdot \mathbf{b}^{(\sigma)}_{(A)} + \rho^{(\sigma)} Q^{(\sigma)}$$

$$+ \Big[- p(\mathbf{v} \cdot \boldsymbol{\xi} - v^{(\sigma)}_{(\xi)}) - \frac{1}{2} \rho (\mathbf{v} \cdot \boldsymbol{\xi} - v^{(\sigma)}_{(\xi)})^3$$

$$+ (\boldsymbol{\xi} \cdot \mathbf{S} \cdot \boldsymbol{\xi})(\mathbf{v} \cdot \boldsymbol{\xi} - v^{(\sigma)}_{(\xi)}) \Big] \Big\} \, dA$$

$$+ \int_{C^{(cl)}_{(sys)}} \Big(- \frac{1}{2} \rho^{(\sigma)} v^{(\sigma)2} [\mathbf{v}^{(\sigma)} \cdot \boldsymbol{\nu} - u^{(cl)}_{(\nu)}]$$

$$+ \mathbf{v}^{(\sigma)} \cdot \mathbf{T}^{(\sigma)} \cdot \boldsymbol{\nu} \Big) \, ds \qquad \text{(D)}$$

Table 7.1.3-1: (cont.)

For an isobaric fluid:

$$\frac{d}{dt}\left[\int_{R_{(sys)}} \rho\hat{H}\, dV + \int_{\Sigma_{(sys)}} \rho^{(\sigma)}\hat{U}^{(\sigma)}\, dA\right]$$

$$= \int_{S_{(ent\ ex)}} \rho\hat{H}(\mathbf{v} - \mathbf{v}_{(sys)}) \cdot (-\mathbf{n})\, dA$$

$$+ \int_{C_{(ent\ ex)}} \rho^{(\sigma)}\hat{U}^{(\sigma)} (\mathbf{v}^{(\sigma)} - \mathbf{v}_{(sys)}) \cdot (-\boldsymbol{\mu})\, ds + \mathcal{Q}$$

$$+ \int_{R_{(sys)}} \left[\operatorname{tr}(\mathbf{S} \cdot \nabla\mathbf{v}) + \sum_{A=1}^{N} \mathbf{j}_{(A)} \cdot \mathbf{b}_{(A)} + \rho Q\right] dV$$

$$+ \int_{\Sigma_{(sys)}} \left\{\gamma \operatorname{div}_{(\sigma)}\mathbf{v}^{(\sigma)} + \operatorname{tr}(\mathbf{S}^{(\sigma)} \cdot \nabla_{(\sigma)} \mathbf{v}^{(\sigma)})\right.$$

$$+ \sum_{A=1}^{N} \mathbf{j}^{(\sigma)}_{(A)} \cdot \mathbf{b}^{(\sigma)}_{(A)} + \rho^{(\sigma)}Q^{(\sigma)} + \left[-\frac{1}{2}\rho(\mathbf{v} \cdot \boldsymbol{\xi} - v^{(\sigma)}_{(\xi)})^3\right.$$

$$\left.\left. + (\boldsymbol{\xi} \cdot \mathbf{S} \cdot \boldsymbol{\xi})(\mathbf{v} \cdot \boldsymbol{\xi} - v^{(\sigma)}_{(\xi)})\right]\right\} dA$$

$$+ \int_{C^{(cl)}_{(sys)}} \left(-\frac{1}{2}\rho^{(\sigma)}\, v^{(\sigma)2}[\, \mathbf{v}^{(\sigma)} \cdot \boldsymbol{\nu} - u^{(cl)}_{(\nu)}\,]\right.$$

$$\left. + \mathbf{v}^{(\sigma)} \cdot \mathbf{T}^{(\sigma)} \cdot \boldsymbol{\nu}\right) ds \qquad \text{(E)}$$

Exercise

7.1.3-1 *measurement of force potential in thin films* We saw in Exercises 2.1.3-1 and 2.1.3-2 that it is common to express the mutual force per unit mass **f** (London-van der Waals, electrostatic) which acts in thin films in terms of a force potential ϕ,

$$\mathbf{f} \equiv - \nabla \phi \tag{1}$$

How can ϕ be studied experimentally?

Let us assume that we have two parallel plates submerged in a closed bath of fluid to be studied. The plate at $z_3 = 0$ is stationary; the other at $z_3 = L$ is moveable. The force potential ϕ in this fluid can be studied by measuring the work required to slowly bring the two planes into close proximity and in this way to create a thin film. Alternatively, one can measure the force required to hold the moveable plate in a stationary position. [In practice, spheres or crossed cylinders are normally employed, in order to minimize alignment problems. For references to work in this area, see Israelachvili (1985, p. 159).]

In what follows, we will show that these experiments can be analyzed starting with the energy balance, the mechanical energy balance, or the momentum balance. Although the assumptions required vary somewhat, the results of all three analyses are the same.

We will take as our system all of the fluid in the bath.

integral overall energy balance Starting with the integral overall energy balance in the form of Table 7.1.3-1 (A), let us make the following observations and assumptions:

a) Since the bath is closed, there are no entrances or exits.

b) The effects of gravity can be neglected with respect to those of the mutual forces.

c) Any potential energy associated with the interfaces can be neglected.

d) Since ϕ is arbitrary to a constant, it is taken to be zero outside the immediate neighborhood of the fluid-solid interfaces.

e) Inertial effects can be neglected.

f) The internal energy of the system does not change with time. Pressure and temperature are nearly uniform throughout the fluid.

g) The effect of mass transfer can be neglected: either only one species is present or the same forces act on all species.

h) The effects of radiation will be neglected.

i) The system is well-insulated, and the rate of energy transfer to the system from the surroundings is zero.

With these assumptions, determine that the integral overall energy balance reduces to

$$\frac{d\Phi}{dt} = - \mathcal{W} \tag{2}$$

where

$$\Phi \equiv \int_{R_{(sys)}} \rho\phi \, dV = \int_0^L \int_S \rho\phi \, dA \, dz_3 \tag{3}$$

is the potential energy of the film and S is a surface parallel to the plates having the same area as each of the plates. Alternatively conclude that (2) can be expressed as

$$\frac{d\Phi}{dL} = \int_{S_{(mov)}} [\, \rho\phi \,]_{z_3=L} \, dA = [\, \rho\phi \,]_{z_3=L} \, \mathcal{A}_{(mov)} = - \mathcal{F}_3 \tag{4}$$

in which $\mathcal{F}_3$ is the z_3 component of the force which the liquid film exerts on the moving plate, $S_{(mov)}$ is the liquid-solid interface of the moving plate, and $\mathcal{A}_{(mov)}$ is the area of $S_{(mov)}$.

integral mechanical energy balance Starting with the integral mechanical energy balance (Sec. 4.1.4), let us again make assumptions (a) through (e). In addition, we will say:

f) The effects of viscous dissipation can be neglected.

Under these circumstances, show that the mechanical energy balance again reduces to (2) or (4).

integral momentum balance Let us now restrict our attention to the alternative experiment in which one measures the force required to hold the moveable plate in a stationary position. Determine that, as the result of assumptions (a) through (d), the integral momentum balance (Sec. 4.1.3) reduces to (4).

Any concern that you might have about the assumptions required in using the integral overall energy balance should be relieved by observing that fewer assumptions are required to obtain the same result with either the integral mechanical energy balance or the integral momentum balance.

7.1.4 The integral Clausius-Duhem inequality The integral Clausius-Duhem inequality is derived in the same manner used to derive the integral mass balance for species A in Sec. 7.1.2 and the integral energy balance in Sec. 7.1.3. By means of the generalized transport theorem (Exercise 1.3.7-1), we write an expression for the time rate of change of the entropy within an arbitrary system. From this, we subtract the volume integral of the differential Clausius-Duhem inequality (Sec. 5.7.3) over $R_{(sys)}$, the area integral of the jump Clausius-Duhem inequality (Sec. 5.7.3) over $\Sigma_{(sys)}$, and the line integral of the Clausius-Duhem inequality at a multicomponent common line over $C^{(cl)}_{(sys)}$ (Exercise 5.7.3-9).

We conclude that

$$\frac{d}{dt}\left[\int_{R_{(sys)}} \rho \hat{S}\, dV + \int_{\Sigma_{(sys)}} \rho^{(\sigma)} \hat{S}^{(\sigma)}\, dA\right]$$

$$\geq \int_{S_{(ent\ ex)}} \rho \hat{S} \left(\mathbf{v} - \mathbf{v}_{(sys)} \right) \cdot (-\mathbf{n})\, dA$$

$$+ \int_{C_{(ent\ ex)}} \rho^{(\sigma)} \hat{S}^{(\sigma)} \left(\mathbf{v}^{(\sigma)} - \mathbf{v}_{(sys)} \right) \cdot (-\boldsymbol{\mu})\, ds$$

$$+ \int_{S_{(sys)}} \frac{1}{T}\left[\mathbf{q} - \sum_{A=1}^{N} \mu_{(A)} \mathbf{j}_{(A)}\right] \cdot (-\mathbf{n})\, dA$$

$$+ \int_{C_{(sys)}} \frac{1}{T^{(\sigma)}}\left[\mathbf{q}^{(\sigma)} - \sum_{A=1}^{N} \mu^{(\sigma)}_{(A)} \mathbf{j}^{(\sigma)}_{(A)}\right] \cdot (-\boldsymbol{\mu})\, ds$$

$$+ \int_{R_{(sys)}} \rho \frac{Q}{T}\, dV + \int_{\Sigma_{(sys)}} \frac{\rho^{(\sigma)} Q^{(\sigma)}}{T^{(\sigma)}}\, dA \qquad (4\text{-}1)$$

We will refer to this as the **integral Clausius-Duhem inequality**. It says that the time rate of change of the entropy of the system is greater than or equal to the net rate at which entropy is brought into the system by convection, the net rate of contact entropy transmission to the system, and the net rate of radiant entropy transmission to the system.

For more on the integral Clausius-Duhem inequality excluding interfacial effects, see Slattery (1981, p. 441 and 600).

References

Giordano, R. M., and J. C. Slattery *J. Colloid Interface Sci.* **92**, 13 (1983).

Slattery, J. C., "Momentum, Energy, and Mass Transfer in Continua," McGraw-Hill, New York (1972); second edition, Robert E. Krieger, Malabar, FL 32950 (1981).

Exercises

7.1.4-1 *integral inequality for Helmholtz free energy* Let us confine our attention to an isothermal system. Multiply the integral Clausius-Duhem inequality (4-1) by temperature and subtract the result from the integral energy balance in the form of either equation A or equation B in Table 7.1.3-1. Arrange the result in the form of an integral inequality for the Helmholtz free energy.

For the special case of a single-phase system, see Slattery (1981, p. 601).

7.1.4-2 *more on the integral inequality for Helmholtz free energy* In addition to saying that the system is isothermal, we will often be willing to make the following assumptions.

a) The boundaries of the system are fixed in space.

b) The effects of kinetic energy can be neglected.

c) The effects of external and mutual forces (such as gravity) can be neglected.

d) The rate of energy transfer as the result of diffusion is neglected at the boundaries of the system:

$$\int_{S_{(sys)}} \sum_{B=1}^{N} \mu_{(B)} \, \mathbf{j}_{(B)} \cdot \mathbf{n} \, dA + \int_{C_{(sys)}} \sum_{B=1}^{N} \mu_{\{\sigma, B\}} \, \mathbf{j}_{\{\sigma, B\}} \cdot \boldsymbol{\mu} \, ds \doteq 0$$

This assumes that there are no concentration gradients in the immediate neighborhoods of the entrances and exits.

e) The rate at which work is done by viscous forces at the entrances and exits of the system are neglected:

$$\int_{S_{(ent\ ex)}} \mathbf{v} \cdot \mathbf{S} \cdot \mathbf{n} \, dA + \int_{C_{(ent\ ex)}} \mathbf{v}^{(\sigma)} \cdot \mathbf{S}^{(\sigma)} \cdot \boldsymbol{\mu} \, ds \doteq 0$$

This condition will be satisfied identically to the extent we are concerned with single-phase flow through uniform diameter tubes (Poiseuille flow) at the entrance and exit surfaces.

With these restrictions, the integral inequality for Helmholtz free energy reduces to

$$\frac{d}{dt}\left(\int_{R_{(sys)}} \rho \hat{A} \, dV + \int_{\Sigma_{(sys)}} \rho^{(\sigma)} \hat{A}^{(\sigma)} \, dA \right)$$

$$\leq \int_{S_{(ent\ ex)}} (\rho \hat{A} + P) \mathbf{v} \cdot (- \mathbf{n}) \, dA$$

$$+ \int_{C_{(ent\ ex)}} (\rho^{(\sigma)} \hat{A}^{(\sigma)} - \gamma) \mathbf{v}^{(\sigma)} \cdot (- \boldsymbol{\mu}) \, ds$$

7.1.4-3 *still more on the integral inequality for Helmholtz free energy* In addition to the assumptions made in Exercises 7.1.4-1 and 7.1.4-2, Giordano and Slattery (1983) further restricted the system by saying:

f) There is only one entrance and one exit.

g) There are no intersections between $S_{(ent\ ex)}$ and $\Sigma_{(sys)}$, which means $C_{(ent\ ex)} = 0$.

h) Pressure is independent of position on $S_{(ent)}$ and on $S_{(ex)}$, which means

$$\int_{S_{(ent\ ex)}} P \, \mathbf{v} \cdot \mathbf{n} \, dA = - F \, \Delta P$$

Here F is the volume rate of flow through the capillary,

$$\Delta P \equiv P_{(ent)} - P_{(ex)}$$

and $P_{(ent)}$ and $P_{(ex)}$ are the pressures at $S_{(ent)}$ and $S_{(ex)}$ respectively.

i) The interchange of Helmholtz free energy between the various interfaces and the adjacent bulk phases can be neglected so as to say

$$\frac{d}{dt}\int_{R_{(sys)}} \rho\hat{A}\, dV = -\int_{S_{(ent\ ex)}} \rho\hat{A}\, \mathbf{v}\cdot\mathbf{n}\, dA$$

With these additional restrictions, the integral inequality for Helmholtz free energy found in Exercise 7.1.4-2 further reduces to

$$\frac{d}{dt}\int_{\Sigma_{(sys)}} \rho^{(\sigma)}\hat{A}^{(\sigma)}\, dA \le F\,\Delta P$$

7.1.5 Stability of static interfaces in a sinusoidal capillary (Giordano and Slattery 1983) Production of oil from an underground reservoir or percolation of water through soil involves two-phase displacement through the microscopic pores of which the reservoir rock or soil is composed. (For more on oil displacement, see Secs. 3.2.2 and 4.1.6 through 4.1.8.)

Haines (1930) observed that, when a water-air interface moves through a porous medium, it does so in an episodic manner; it will slowly creep forward until an instability develops somewhere in the system, jump ahead a short distance, and begin another period of creeping motion. Since these Haines jumps take place very rapidly (Heller 1968), the local displacement process appears to be controlled by the period of slow advancement between jumps.

When an instability develops, the displaced and displacing phases need not remain intact as suggested above. Roof (1970) and Wardlaw (1982) have observed that, when one of the phases is strongly wetting, an instability may lead to the snap-off or pinch-off and ultimate entrapment of a portion of the non-wetting phase. The probability of snap-off is heightened by the existence of sharp grooves or irregularities in the pore throat (Roof 1970).

The occurrence of jumps or capillary instabilities is a characteristic of flow through periodically-constricted pores, since static configurations of the interface are not possible at all positions within the pore.

Referring to Figures 5-1 and 3.2.2-1, let us consider as a system all of phases 1 and 2 in the capillary between $S_{(ent)}$ and $S_{(ex)}$, including all of the fluid-fluid and fluid-solid interfaces (and, for clarity in the derivation, a thin layer of the solid wall). The radius r_w of the capillary is a sinusoidal function of axial position

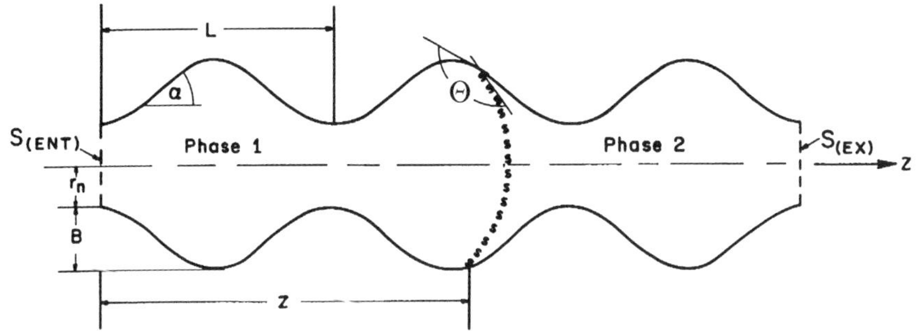

Figure 7.1.5-1: Static configuration of phases 1 and 2 in a sinusoidal pore.

$$r_w = r_n + \frac{B}{2}\left[1 - \cos\left(\frac{2\pi z}{L}\right)\right] \tag{5-1}$$

In what follows, we use the integral inequality for Helmholtz free energy developed in Exercise 7.1.4-3 to test the stability of two-phase systems in these periodically constricted pores, when they are subjected to small perturbations from a static state. These limits of static stability may be approximately identified with the limits of the creeping motion between Haines jumps discussed in Secs. 4.1.6 and 4.1.7.

The surface Euler equation (Sec. 5.8.6) implies that

$$\frac{d}{dt}\int_{\Sigma_{(sys)}} \rho^{(\sigma)}\hat{A}^{(\sigma)}\, dA = \frac{d}{dt}\int_{\Sigma_{(sys)}} \left(\gamma + \sum_{B=1}^{N} \mu_{(B)}^{(\sigma)}\, \rho_{(B)}^{(\sigma)}\right) dA \tag{5-2}$$

For an isothermal system, the surface Gibbs-Duhem equation (Sec. 5.8.6) demands

$$\frac{d\gamma}{dt} + \sum_{B=1}^{N} \rho_{(B)}^{(\sigma)}\, \frac{d\mu_{(B)}^{(\sigma)}}{dt} = 0 \tag{5-3}$$

Continuing the list of assumptions begun in Exercises 7.1.4-2 and 7.1.4-3, we shall say

j) Interfacial tension γ and the various interfacial chemical potentials $\mu_{(B)}^{(\sigma)}$ (B=1,...,N) are independent of position within the dividing surface, although they may be functions of time. The more basic assumption is that the interfacial concentrations of the various species present are independent of position.

k) Any effects attributable to changes in the mass associated with the dividing surface can be neglected and

$$\sum_{\text{All }\Sigma} \sum_{B=1}^{N} \mu^{(B)} \frac{d}{dt} \int_{\Sigma} \rho^{(B)} \, dA = 0 \tag{5-4}$$

By a summation over all Σ, we mean a summation over all fluid-fluid and fluid-solid interfaces within the system.

With assumptions j and k and (5-3), equation (5-2) becomes

$$\frac{d}{dt} \int_{\Sigma_{(sys)}} \rho^{(\sigma)} \hat{A}^{(\sigma)} \, dA = \sum_{\text{All }\Sigma} \gamma \frac{d}{dt} \int_{\Sigma} dA \tag{5-5}$$

By the generalized transport theorem (Exercise 1.3.7-1) and (5-5),

$$\begin{aligned} \frac{d}{dt} \int_{\Sigma_{(sys)}} \rho^{(\sigma)} \hat{A}^{(\sigma)} \, dA &= - \int_{\Sigma_{(sys)}} 2H \, \gamma \, \mathbf{v}^{(\sigma)} \cdot \boldsymbol{\xi} \, dA \\ &\quad - \int_{C^{(cl)}} (\, \gamma \boldsymbol{\nu} \,) \cdot \mathbf{u}^{(cl)} \, ds \\ &= - \gamma^{(12)} \int_{\Sigma^{(12)}} 2H \, \mathbf{v}^{(\sigma)} \cdot \boldsymbol{\xi} \, dA \end{aligned} \tag{5-6}$$

Here we have recognized that the fluid-solid interfaces are stationary and we have assumed

l) Momentum transfer at the common line as the result of mass transfer and the action of surface viscous stresses can be neglected. In this static limit, the momentum balance at the multicomponent common line reduces to

$$(\, \gamma \boldsymbol{\nu} \,) = 0 \tag{5-7}$$

By $\Sigma^{(12)}$ we mean the dividing surface between phases 1 and 2; $\gamma^{(12)}$ is the interfacial tension in this dividing surface.

We will assume that

m) During the small perturbations to be considered, the interface maintains its static configuration, a segment of a sphere.

For the single interface shown in Figure 5-1,

$$\int_{\Sigma^{(12)}} 2H \, \mathbf{v}^{(\sigma)} \cdot \boldsymbol{\xi} \, dA = \frac{2F}{r_w} \cos(\Theta + \alpha) \tag{5-8}$$

In view of (5-1), the angle α is defined by

$$\tan \alpha \equiv \frac{dr_w}{dz} = \frac{\pi B}{L} \sin\left(\frac{2\pi z}{L}\right) \tag{5-9}$$

and is illustrated in Figure 5-1.

With (5-6) and (5-8), the integral inequality for Helmholtz free energy developed in Exercise 7.1.4-3 can be rearranged for the single interface shown in Figure 5-1 as

$$-F\left[\frac{r_n \Delta P}{\gamma^{(12)}} + \frac{2 r_n}{r_w} \cos(\Theta + \alpha)\right] \leq 0 \tag{5-10}$$

For the two interfaces shown in Figure 3.2.2-1, we find in a similar manner

$$-F\left\{\frac{r_n \Delta P}{\gamma^{(12)}} + 2r_n\left[\frac{\cos(\Theta_{(L)} + \alpha_{(L)})}{r_{w(L)}} - \frac{\cos(\Theta_{(R)} - \alpha_{(R)})}{r_{w(R)}}\right]\right\}$$
$$\leq 0 \tag{5-11}$$

The subscript $\ldots_{(L)}$ refers to the left interface in Figure 3.2.2-1; the subscript $\ldots_{(R)}$ to the right interface.

Equations (5-10) and (5-11) place a constraint on the motion of phases 1 and 2 that permit the stability of a static state to be investigated.

For the single interface shown in Figure 5-1, any static state satisfies (see Sec. 3.2.2)

$$\frac{r_n \Delta P}{\gamma^{(12)}} + \frac{2 r_n}{r_w} \cos(\Theta + \alpha) = 0 \tag{5-12}$$

We have tested the stability of such static states to small perturbations in the position of the common line under conditions such that the pressure drop remains fixed and there is no contact angle hysteresis (the contact angle is independent of the direction of flow). The configuration of the interface is assumed to remain a spherical segment during a perturbation (assumption m). Since the pressure drop and the axial position of the common line are known in the perturbed state, the sign of F can be calculated from (5-10). If F is negative for a forward perturbation and positive for a backward perturbation, the static state is stable. If F is positive for a forward perturbation or negative for a backward perturbation, the static state is unstable. Since any random disturbance will cause both forward and backward motion of the common line, a static interface need be unstable to a perturbation in only one direction to make the interface as a whole unstable.

It is sometimes possible to pose a physically different stability problem appropriate for investigation in the sphere of thermostatics that results in the same conclusions as the original problem solved in the context of nonequilibrium thermodynamics (Slattery 1977, 1981). This problem is such a case. Replace our open system, which is exchanging mass with the surroundings, with a similar system that has been closed. In place of the entrance and exit, the new system has pistons riding in cylindrical tubes. The pistons are connected externally in such a manner that they are forced to move together to maintain a constant volume in the system. By examining a sequence of static states of this system corresponding to different positions of the pistons, we can arrive at relations similar to (5-10) and (5-11). The only difference is that F in (5-10) and (5-11) would now be replaced by the displacement of one of the pistons from a reference position. According to the sign of this displacement, we draw exactly the same conclusions regarding stability as those described above. But there are at least two disadvantages to this approach. Instead of solving the original problem, one is forced to solve a physically different problem that is intuitively identified with the original. The physical constraints on the original problem necessary to derive (5-10) and (5-11) are not immediately obvious in the thermostatic development.

Usually, the advancing and receding contact angles are not the same. When the contact angle measured through phase 1 exceeds a critical value $\Theta_{(a)}$ known as the critical advancing contact angle, the common line moves and phase 1 advances. When the contact angle measured through phase 1 is less than a critical value $\Theta_{(r)}$ known as the critical receding contact angle, phase 1 recedes. For intermediate contact angles, the common line is stationary. This phenomenon, known as contact angle hysteresis, can significantly affect displacement in porous media (see Sec. 3.2.2).

We will assume:

n) For the small perturbations in the position of the common line assumed here, $\Theta_{(a)}$ and $\Theta_{(r)}$ are independent of the magnitude of the perturbation. We know that $\Theta_{(a)}$ and $\Theta_{(r)}$ are independent of the speed of displacement of the common line, when the speed of displacement is sufficiently small (see Sec. 2.1.11).

Because of contact angle hysteresis, a perturbation of the system will lead to common line motion only when the contact angle is at one of its critical values. If the static contact angle is between these critical values, a perturbation will cause the contact angle only to increase or decrease without movement of the common line. One must be careful to distinguish perturbations at critical angles from perturbations at non-critical contact angles.

The static positions of two interfaces in a sinusoidal pore, illustrated in Figure 3.2.2-1, must satisfied (see Sec. 3.2.2)

$$\frac{r_n \Delta P}{\gamma^{(12)}} + 2r_n \left[\frac{\cos(\Theta_{(L)} + \alpha_{(L)})}{r_{w(L)}} - \frac{\cos(\Theta_{(R)} - \alpha_{(R)})}{r_{w(R)}} \right] = 0 \quad (5\text{-}13)$$

in addition to a constraint on the volume of phase 2. The stability of any such static state to small perturbations in the positions of the common lines can be tested in the manner described above using (5-11).

In order to facilitate comparisons, we use the model pore assumed in the illustrations of Sec. 3.2.2: $B/r_n = 1.5$ and $L/r_n = 6$.

In our numerical tests for stability, 0.02L is considered to be a small perturbation in the position of a common line.

Our results are as follows.

Single interface, no contact angle hysteresis Consider a single interface in a static configuration as shown in Figure 5-1. We will assume that there is no contact angle hysteresis. For illustrative purposes, we will assume that the contact angle Θ is 140°.

Figure 5-2 gives the pressure drop ΔP as a function of the position of the static interface as determined by (5-12). The unstable static positions of the interface are labeled U; along the curve U the common line moves spontaneously either forward or backward, depending upon the direction of the perturbation. The stable static positions of the interface are designated S; along the curve S the common line is stable to any small perturbation of the system. If the pressure drop is gradually changed, an interface located along curve S will assume a sequence of quasi-static states. The interface creeps forward or backward until the U curve is met, at which point the interface moves spontaneously.

Single interface contact angle hysteresis Figure 5-3 shows pressure drop ΔP as a function of the position of the static interface as determined by (5-12) for a system in which $\Theta_{(a)} = 140°$ and $\Theta_{(r)} = 38°$. The unstable static positions of the interface are labeled U; any perturbation leads to spontaneous movement of the common line. The segments marked S_n indicate the neutrally stable, static positions of the interface. A perturbation along S_n will not lead to spontaneous motion. But, because of contact angle hysteresis, the interface may assume a new stable, static state that neighbors but is not identical to the original static state. Within the region bounded by the critical contact angle curves, all static states are stable and the common line remains stationary during any small perturbations of the system. In comparing Figures 5-2 and 5-3, note how contact angle hysteresis alters the critical pressure drop at which spontaneous movements occur to the left in Figure 5-1.

To illustrate the effect of contact angle hysteresis, consider an interface initially at $z/r_n = 7$ with $\Theta = 38°$. Let us examine the sequence of static positions assumed by the interface as we slowly lower and then raise ΔP. The arrows in Figure 5-3 trace out the movement of the common line and the variation of contact angle for this situation. The interface creeps to the left in Figure 5-1 as ΔP is lowered. The pressure drop is increased at $z/r_n = 6$, which causes the common line to remain stationary and the contact angle to increase to 140°. As the pressure drop is further increased, the interface creeps to the right in Figure 5-1 along the segment S_n in Figure 5-3, until the U curve is met. At this point, the interface moves spontaneously to the right in Figure 5-1. Note that movement along S_n is limited to only one direction, because of contact angle hysteresis.

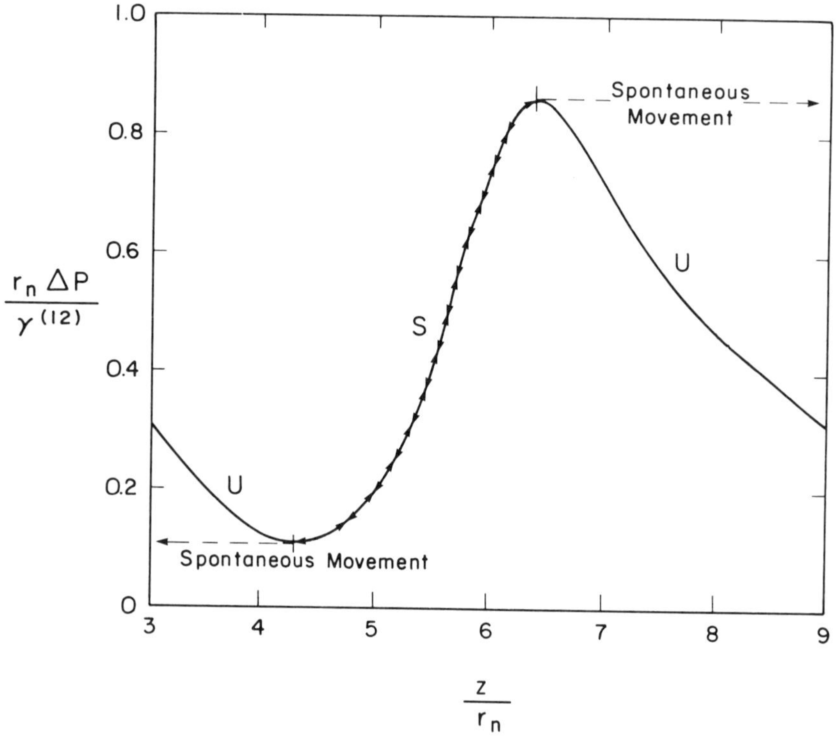

Figure 7.1.5-2: Stability of a single, static interface as a function of position in a sinusoidal pore, when contact angle hysteresis is absent. Here $\Theta = 140°$, $B/r_n = 1.5$, and $L/r_n = 6$.

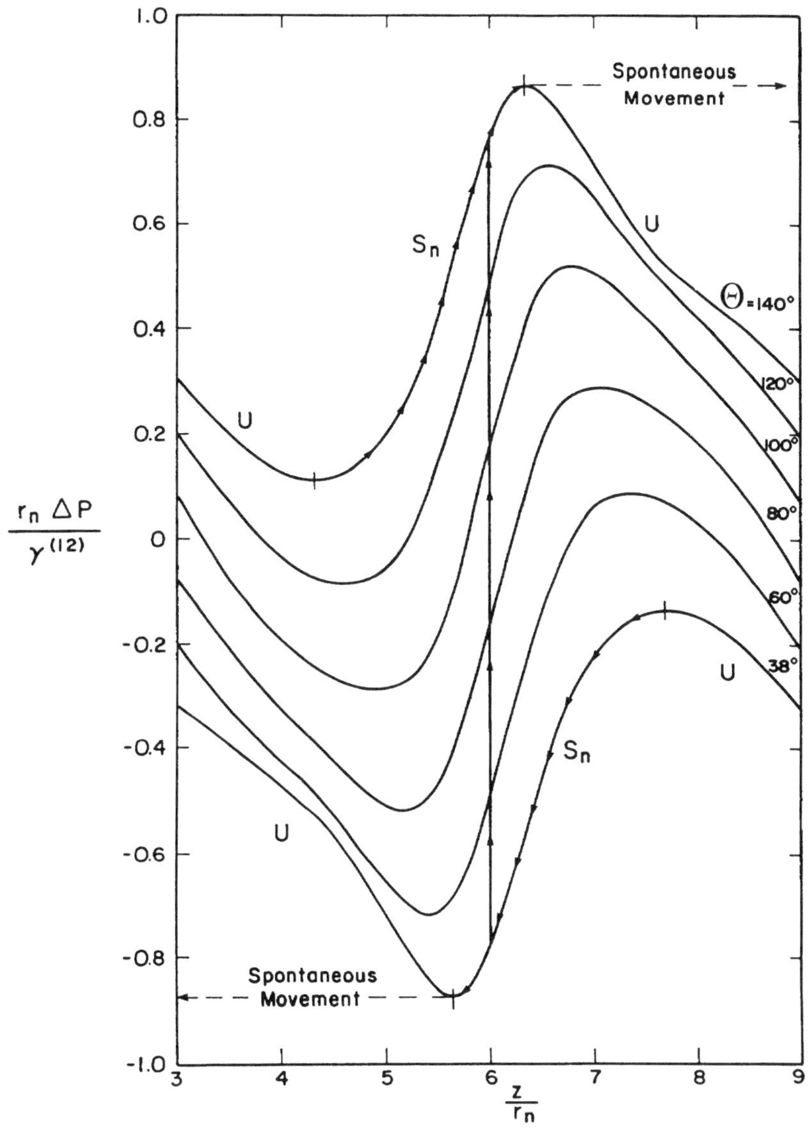

Figure 7.1.5-3: Effect of contact angle hysteresis upon the stability of a single, static interface in a sinusoidal pore. Here $\Theta_{(a)} = 140°$, $\Theta_{(r)} = 38°$, $B/r_n = 1.5$, and $L/r_n = 6$.

Figure 5-4 shows pressure drop ΔP as a function of the position of the static interface as determined by (5-12) for a system in which $\Theta_{(a)} = 40°$ and $\Theta_{(r)} = 4°$. Here we see a region of *contact angle instability*. Upon approaching the U_θ curve from the stationary common line region by reducing ΔP, the contact angle decreases spontaneously to 4° and the interface moves to the left in Figure 5-1. In the region where U and U_θ coincide, the common line moves spontaneously forward or backward, depending upon the initial direction of the perturbation.

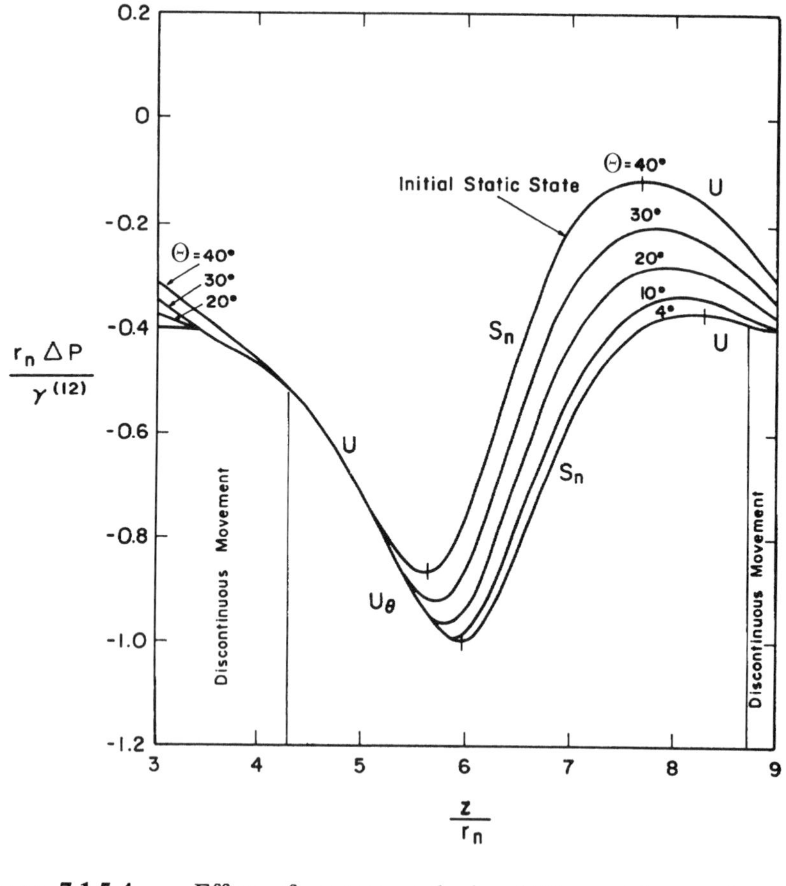

Figure 7.1.5-4: Effect of contact angle instability and hysteresis upon the stability of a single, static interface in a sinusoidal pore. Here $\Theta_{(a)}$ = 40°, $\Theta_{(r)}$ = 4°, B/r_n = 1.5, and L/r_n = 6.

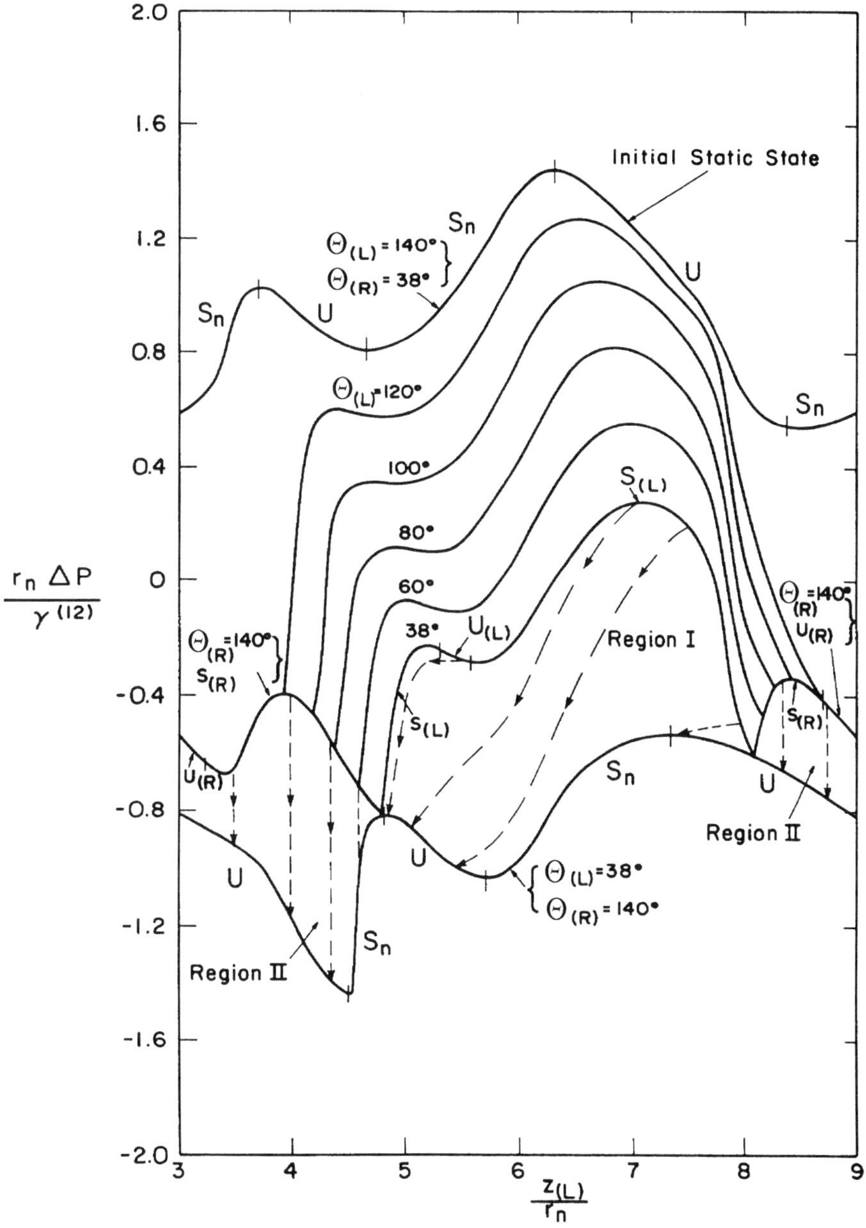

Figure 7.1.5-5: Stability of a segment of phase 2 trapped by phase 1 as a function of the position of the left interface. Here $\Theta_{(a)} = 140°$, $\Theta_{(r)} = 38°$, $B/r_n = 1.5$, $L/r_n = 6$, and $V^{(2)} = 80\ r_n^3$.

Because of contact angle instability, a unique representation of the quasi-static states assumed by the interfaces as ΔP is slowly varied demands that we specify the process path. Figure 5-4 applies to initial static states along $\Theta = 40°$ and other static states reached by decreasing ΔP. In general, other process paths or initial conditions require separate figures.

In Sec. 3.2.2, we recognize that, for very large or small values of the contact angle, one phase will be trapped on the wall of a sinusoidal pore during displacement. The term *discontinuous movement* is used to describe the apparent jump of the common line over this ring of trapped fluid. Figure 5-4 shows the location of these discontinuous movements. Note that the discontinuity occurs only in movements to the left in Figure 5-1, since the contact angle in the initial static state ($\Theta = 40°$) is not sufficiently large to cause trapping.

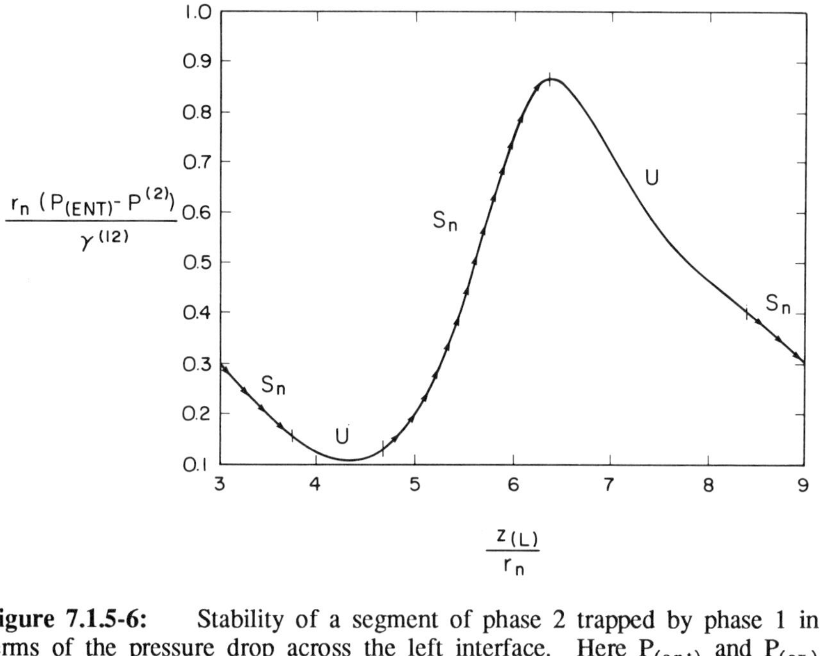

Figure 7.1.5-6: Stability of a segment of phase 2 trapped by phase 1 in terms of the pressure drop across the left interface. Here $P_{(ent)}$ and $P_{(ex)}$ are constants, $\Theta_{(L)} = \Theta_{(a)} = 140°$, $\Theta_{(R)} = \Theta_{(r)} = 38°$, $B/r_n = 1.5$, $L/r_n = 6$, and $V^{(2)} = 80\ r_n^3$.

Two interfaces, contact angle hysteresis Let us now consider two interfaces in a sinusoidal pore as illustrated in Figure 3.2.2-1. In order to facilitate comparisons with the preceding cases as well as with the example computations of Sec. 3.2.2, we will take $\Theta_{(a)} = 140°$, $\Theta_{(r)} = 38°$, and $V^{(2)} = 80 r_n^3$ as the volume of phase 2. Any perturbation along U causes both common lines to move; a perturbation along $U_{(L)}$ causes spontaneous movement of only the left common line in Figure 2; a perturbation along

$U_{(R)}$ causes spontaneous movement of only the right common line. The curves marked S_n indicate neutrally stable positions for both interfaces. Those marked $S_{(L)}$ indicate that the left interface is neutrally stable; the right common line is stationary. Those curves marked $S_{(R)}$ indicate that the right interface is neutrally stable; the left common line is stationary. In constructing this figure, we have focused our attention on the left interface in Figure 3.2.2-1; the contact angle and position of the right interface can be found from (5-13) and the volume of phase 2.

We can use Figure 5-5 to determine the sequence of static states that an interface will assume as ΔP is slowly decreased. In region I where $\Theta_{(L)} = \Theta_{(r)} = 38°$, small perturbations to the system can cause only the left common line to move. Upon decreasing ΔP, an interface located along $S_{(L)}$ will creep backward along the arrows shown. Similarly, in region II where $\Theta_{(R)} = \Theta_{(a)} = 140°$, a decrease in ΔP causes movement of only the right interface; $z_{(L)}$ remains fixed.

Because of contact angle hysteresis, the paths in regions I and II are not reversible. Figure 5-5 has been constructed assuming that all initial static states lie along the curve for which $\Theta_{(L)} = 140°$ and $\Theta_{(R)} = 38°$. All other static states are reached by decreasing ΔP from one of these initial states. Other process paths or initial conditions require separate figures, at least within regions I and II.

Stability of interacting interfaces In practice, we are often faced with more complex problems than the preceding examples. For example, in the tertiary recovery of residual oil, we must displace a blob or ganglion of oil that may occupy thousands of neighboring pores in a rock structure. The oil ganglion may be bounded by hundreds of interacting crude oil-brine interfaces.

The temptation is to ignore the influence that other interfaces in the system will have upon the stability of a particular interface.

The magnitude of the problem can be appreciated by considering the limiting case of an oil ganglion that occupies a single pore as shown in Figure 3.2.2-1. It is bounded by two interacting phase interfaces. Let us assume that there is incipient motion from left to right; let $\Theta_{(L)} = \Theta_{(a)} = 140°$ and $\Theta_{(R)} = \Theta_{(r)} = 38°$. If we assume that the pressure drop across the left crude oil-brine interface in Figure 3.2.2-1 is a constant, independent of the positions of the two interfaces, the top curve in Figure 5-3 describes the stability of the left interface. If we assume that the pressure drop across the oil ganglion in Figure 3.2.2-1 is a constant, then the computations leading to Figure 5-5 give the substantially different result shown in Figure 5-6. Note that, in order to compare the result with the top curve in Figure 5-3, Figure 5-6 is a plot of the pressure drop across the left interface.

References

Giordano, R. M., and J. C. Slattery *J. Colloid Interface Sci.* **92**, 13 (1983).

Haines, W. B., *J. Agric. Sci.* **20**, 97 (1930).

Heller, J. P., *Soil Sci. Soc. Am. Proc.* **32**, 778 (1968).

Roof, J. G., *Soc. Pet. Eng. J.* **5**, 259 (1965).

Slattery, J. C., *AIChE J.* **23**, 275 (1977).

Slattery, J. C., "Momentum, Energy, and Mass Transfer in Continua," McGraw-Hill, New York (1972); second edition, Robert E. Krieger, Malabar, FL 32950 (1981).

Wardlaw, N. C., *J. Can Pet. Technol.* **21**, 21 (May-June 1982).

Notation for chapter 7

$\hat{A}$	Helmholtz free energy per unit mass, introduced in Sec. 5.8.2
$\hat{A}^{(\sigma)}$	surface Helmholtz free energy per unit mass, introduced in Sec. 5.8.4
$\mathbf{b}_{(A)}$	body force per unit mass acting upon species A, introduced in Sec. 5.5.2
$\mathbf{b}^{(\sigma)}_{(A)}$	body force per unit mass acting upon species A in a dividing surface, introduced in Sec. 5.5.2
$C_{(ent\ ex)}$	entrance and exit portions of $C_{(sys)}$
$C_{(sys)}$	lines formed by the intersection of the dividing surfaces with $S_{(sys)}$
$C^{(cl)}_{(sys)}$	common lines contained within the system system, introduced in Sec. 7.1.2
dA	denotes that an area integration is to be performed
ds	denotes that a line integration with respect to arc length is to be performed
dV	denotes that a volume integration is to be performed
H	mean curvature of dividing surface, introduced in Exercise A.5.3-3
$\hat{H}$	enthalpy per unit mass, introduced in Exercise 5.8.4-1
$\hat{H}^{(\sigma)}$	surface enthalpy per unit mass, introduced in Exercise 5.8.6-2
$\mathbf{I}$	identity tensor that transforms every vector into itself
$\mathbf{j}_{(A)}$	mass flux of species A with respect $\mathbf{v}$, introduced in Sec. 5.2.2
$\mathbf{j}^{(\sigma)}_{(A)}$	surface mass flux of species A with respect to $\mathbf{v}^{(\sigma)}$, introduced in Sec. 5.2.2
$\mathbf{n}$	outwardly directed unit normal to $S_{(sys)}$

$\mathbf{n}_{(A)}$	mass flux of species A with respect to the frame of reference, introduced in Sec. 5.2.2
p	mean pressure, defined in Sec. 2.4.1
p_0	ambient pressure
P	thermodynamic pressure
$\mathbf{q}$	energy flux vector, introduced in Sec. 5.6.4
$\mathbf{q}^{(\sigma)}$	surface energy flux vector, introduced in Sec. 5.6.4
Q	rate of radiant energy transmission per unit mass to the material within a phase, introduced in Sec. 5.6.3
$Q^{(\sigma)}$	rate of radiant energy transmission per unit mass to the material in the dividing surface, introduced in Sec. 5.6.3
$\mathcal{Q}$	rate of (contact) energy transfer to the system, introduced in Sec. 7.1.3
$r_{(A)}$	rate of production of mass of species A per unit volume by homogeneous chemical reactions, introduced in Sec. 5.2.1
$r_{(A)}^{(\sigma)}$	rate of production of mass of species A per unit area by heterogeneous chemical reactions, introduced in Sec. 5.2.1
$R_{(sys)}$	region occupied by the system
$\mathbf{S}$	viscous portion of stress tensor, defined in Sec. 2.4.1
$\hat{S}$	entropy per unit mass within a phase, introduced in Sec. 5.7.3
$S_{(ent\ ex)}$	entrance and exit portions of $S_{(sys)}$
$S_{(sys)}$	closed surface bounding system
$\hat{S}^{(\sigma)}$	surface entropy per unit mass, introduced in Sec. 5.7.3
$\mathbf{S}^{(\sigma)}$	viscous portion of surface stress tensor, defined in Sec. 2.4.2
t	time
T	temperature, introduced in Sec. 5.7.2

$T^{(\sigma)}$	surface temperature, introduced in Sec. 5.7.2
$\mathbf{T}$	stress tensor, introduced in Sec. 2.1.4
$\mathbf{T}^{(\sigma)}$	surface stress tensor, introduced in Sec. 2.1.5
$u_{(v)}^{(cl)}$	speed of displacement of common line, defined in Sec. 1.3.7
$\hat{U}$	internal energy per unit mass within a phase, introduced in Sec. 5.6.2
$\hat{U}^{(\sigma)}$	surface internal energy per unit mass, introduced in Sec. 5.6.2
$\mathbf{v}$	mass-averaged velocity, introduced in Sec. 5.2.2
$\mathbf{v}_{(sys)}$	local velocity of the boundary of the system
$\mathbf{v}^{(\sigma)}$	mass-averaged surface velocity vector, introduced in Sec. 5.2.2
$v_{(\xi)}^{(\sigma)}$	speed of displacment of surface, introduced in Sec. 1.2.7
$\mathcal{W}$	rate at which work is done by the system on the surroundings at the moving impermeable surface of the system (beyond any work done on these surfaces by the ambient pressure p_0), defined in Sec. 4.1.4
z	cylindrical coordinate
	greek letters
γ	thermodynamic surface tension, defined in Sec. 5.8.6
$\gamma^{(12)}$	interfacial tension between phases 1 and 2, introduced in Sec. 7.1.5
$\Theta_{(a)}$	critical advancing contact angle, introduced in Sec. 7.1.5
$\Theta_{(r)}$	critical receding contact angle, introduced in Sec. 7.1.5
$\boldsymbol{\mu}$	unit vector normal to $C_{(sys)}$ that is both tangent and outwardly directed with respect to the dividing surface
$\mu_{(A)}^{(\sigma)}$	surface chemical potential of species A, introduced in Sec. 5.8.6

$\mathbf{v}$ unit vector normal to common line and tangent to dividing surface

$\boldsymbol{\xi}$ unit normal to Σ, introduced in Sec. 1.2.7

π 3.14159...

ρ total mass density

$\rho_{(A)}$ mass density of species A

$\rho^{(\sigma)}$ total surface mass density

$\rho_{(A)}^{(\sigma)}$ surface mass density of species A

$\Sigma_{(sys)}$ dividing surface within system

$\Sigma^{(12)}$ dividing surface between phases 1 and 2

ϕ potential energy per unit mass, defined in Sec. 2.3.4

$\omega_{(A)}$ mass fraction of species A, introduced in Sec. 5.2.2

$\omega_{(A)}^{(\sigma)}$ surface mass fraction of species A, introduced in Sec. 5.2.2

others

div divergence operator

$\text{div}_{(\sigma)}$ surface divergence operator, introduced in Secs. A.5.1 through A.5.6

tr trace operator

∇ gradient operator with respect to the current configuration

$\nabla_{(\sigma)}$ surface gradient operation, introduced in Secs. A.5.1 through A.5.6

Appendix A
Differential geometry[a]

My objective in the text is to describe momentum, energy, and mass transfer in real multiphase materials. Upon a little thought, it quickly becomes apparent that one of our first requirements is a mathematical representation for the space in which the physical world around us is situated. There are at least two features of our physical space that we wish to retain: length and relative direction. A euclidean space incorporates both of these properties.

I assume here that the reader is familiar with tensor analysis. If he is not, I suggest that he first stop and read the introduction given by Slattery (1981), by Leigh (1968), or by Coleman *et al.* (1966). I also have found helpful the discussions by Ericksen (1960) and by McConnell (1957).

In this introduction to differential geometry, we shall confine our attention to scalar, vector, and tensor fields defined on or evaluated at surfaces in a euclidean space. Much of this material is drawn from McConnell (1957). My primary modification has been to explicitly recognize the role of various bases in the vector space. I have employed wherever possible a coordinate-free notation, which I feel is a little easier on the eyes.

References

Coleman, B. D., H. Markovitz, and W. Noll, "Viscometric Flows of Non-Newtonian Fluids," Springer-Verlag, New York (1966).

---

a) Based in part on work by Norman Lifshutz (1970).

Ericksen, J. L., "Handbuch der Physik," vol. 3/1, edited by S. Flügge, Springer-Verlag, Berlin (1960).

Leigh, D. C., "Nonlinear Continuum Mechanics," McGraw-Hill, New York (1968).

Lifshutz, N., Ph.D. dissertation, Northwestern University, Evanston, IL (1970).

McConnell, A. J., "Applications of Tensor Analysis," Dover, New York (1957).

Slattery, J. C., "Momentum, Energy, and Mass Transfer in Continua," McGraw-Hill, New York (1972); second edition, Robert E. Krieger, Malabar, FL 32950 (1981).

A.1 Physical space

A.1.1 Euclidean space Let E^n be a set of elements a, b, ... which we will refer to as **points**. Let V^n be a real, n-dimensional inner product space (linear vector space upon which an inner product is defined). We can define a relation between points and vectors in the following way.

1. To every ordered pair (a, b) of points in E^n there is assigned a vector of V^n called the **difference vector** and denoted by

$$\overrightarrow{ab}$$

2. If o is an arbitrary point in E^n, then to every vector **a** of V^n there corresponds a unique point such that

$$\overrightarrow{oa} = \mathbf{a}$$

3. If a, b and c are three arbitrary points in E^n, then

$$\overrightarrow{ab} + \overrightarrow{bc} = \overrightarrow{ac} \tag{1-1}$$

We refer to V^n as the **translation space** corresponding to E^n; the couple (E^n, V^n) is a **euclidean space.**

The distance between any two points a and b is defined by

$$d(\,a, b\,) \equiv |\,\overrightarrow{ab}\,| \equiv \left(\overrightarrow{ab} \cdot \overrightarrow{ab}\right)^{1/2} \tag{1-2}$$

We use the notation ($\mathbf{a} \cdot \mathbf{b}$) for the inner product of two vectors **a** and **b** that belong to V^n.

We represent the physical space in which we find ourselves by the three-dimensional euclidean space (E^3, V^3). This is the only euclidean space with which we shall be concerned.

Our primary concern in the text is with surfaces in this euclidean space. Many of our discussions will focus on $V^2(\mathbf{z})$, the subspace of V^3 composed of the vectors in V^3 that are tangent to a surface at some point **z**. I would like to emphasize that V^2 will in general be a function of position on the surface. If the surface in question is moving and deforming as a function of time, V^2 will be a function of both time and position.

One's first thought might be that these surfaces can themselves be thought of as euclidean spaces of the form (E^2, V^3), where E^2 is a subset of E^3 corresponding to points on the surface. This certainly works well for planes; planes are euclidean spaces. But curved surfaces are noneuclidean, because the difference vectors are not tangent to the surface and do not belong to V^2 evaluated at either of the points defining the difference. The distance between two points measured along a curved surface is in general not equal to the distance between these same two points measured in the three-dimensional euclidean space (E^3, V^3).

We could associate a two-dimensional euclidean space (plane) with a curved surface in much the same manner as we prepare maps of the world. Such a space has the unpleasant feature that distance measured in the plane does not correspond to distance measured either along the curved surface or to distance measured in (E^3, V^3). This objection outweighs the potential usefulness of such a map in our discussion.

This viewpoint of a euclidean space is largely drawn from Greub (1967, p. 282) and Lichnerowicz (1962, p. 24), both of which I recommend for further reading.

References

Greub, W. H., "Linear Algebra," third edition, Springer-Verlag, New York (1967).

Lichnerowicz, A., "Elements of Tensor Calculus," John Wiley, New York (1962).

A.1.2 Notation in (E^2, V^3) This discussion of differential geometry is based upon a previous treatment of tensor analysis (Slattery 1972, appendix A). The definitions and notation introduced there for (E^3, V^3) are certainly necessary here. While I must refer the reader to the original development for many points, I thought it might be helpful to review some of the most basic notation before going any further.

The inner product space V^3 will be referred to as the space of **spatial vectors.**

We find it convenient to choose some point O in E^3 as a reference point or **origin** and to locate all points in E^3 relative to O. Instead of referring to the point z, we will use the **position vector**

$$\mathbf{z} \equiv \overrightarrow{\mathrm{Oz}}$$

Temperature, concentration, and pressure are examples of real numerically valued functions of position. We call any real numerically valued function of position a **real scalar field.**

When we think of water flowing through a pipe or in a river, we recognize that the velocity of water is a function of position. At the wall of the pipe, the velocity of the water is zero; at the center, it is a maximum. The velocity of the water in the pipe is an example of a spatial vector-valued function of position. We shall term any spatial vector-valued function a **spatial vector field.**

As another example, consider the **position vector field** $\mathbf{p}(z)$. It maps every point z of E^3 into the corresponding position vector $\mathbf{z}$ measured with respect to a previously chosen origin O:

$$\mathbf{z} = \mathbf{p}(z) \tag{2-1}$$

Every spatial vector field $\mathbf{u}$ may be written as a linear combination of rectangular cartesian basis fields $\{ \mathbf{e}_1, \mathbf{e}_2, \mathbf{e}_3 \}$ (Slattery 1981, p. 607):

$$\begin{aligned} \mathbf{u} &= u_1 \mathbf{e}_1 + u_2 \mathbf{e}_2 + u_3 \mathbf{e}_3 \\ &= \sum_{i=1}^{3} u_i \mathbf{e}_i \\ &= u_i \mathbf{e}_i \end{aligned} \tag{2-2}$$

A special case is the position vector field

$$\begin{aligned} \mathbf{p} &= z_1 \mathbf{e}_1 + z_2 \mathbf{e}_2 + z_3 \mathbf{e}_3 \\ &= \sum_{i=1}^{3} z_i \mathbf{e}_i \\ &= z_i \mathbf{e}_i \end{aligned} \tag{2-3}$$

The rectangular cartesian components $\{ z_1, z_2, z_3 \}$ of the position vector field $\mathbf{p}$ are called the **rectangular cartesian coordinates** with respect to the previously chosen origin O. They are one-to-one functions of position z in E^3:

$$z_i = z_i(z) \qquad \text{for } i = 1, 2, 3 \tag{2-4}$$

For this reason we will often find it convenient to think of $\mathbf{p}$ as being a function of the rectangular cartesian coordinates:

$$\mathbf{z} = \mathbf{p}(z_1, z_2, z_3) \tag{2-5}$$

Note that in (2-2_3) and (2-3_3) we have employed the summation convention (Slattery 1981, p. 610). We find it convenient to use this convention hereafter.

Let us assume that each z_i (i = 1, 2, 3) may be regarded as a function of three parameters $\{ x^1, x^2, x^3 \}$ called **curvilinear coordinates**:

$$z_i = z_i(x^1, x^2, x^3) \quad \text{for } i = 1, 2, 3 \tag{2-6}$$

Here we use the common notation preserving device of employing the same symbol for both the function z_i and its value $z_i(x^1, x^2, x^3)$. So long as

$$\det\left(\frac{\partial z_i}{\partial x^j}\right) = \begin{vmatrix} \frac{\partial z_1}{\partial x^1} & \frac{\partial z_1}{\partial x^2} & \frac{\partial z_1}{\partial x^3} \\ \frac{\partial z_2}{\partial x^1} & \frac{\partial z_2}{\partial x^2} & \frac{\partial z_2}{\partial x^3} \\ \frac{\partial z_3}{\partial x^1} & \frac{\partial z_3}{\partial x^2} & \frac{\partial z_3}{\partial x^3} \end{vmatrix} \neq 0 \tag{2-7}$$

we can solve (2-6) for the x^i to find

$$x^i = x^i(z_1, z_2, z_3) \quad \text{for } i = 1, 2, 3 \tag{2-8}$$

This means that for each set $\{ x^1, x^2, x^3 \}$ there is a unique set $\{ z_1, z_2, z_3 \}$ and vice versa. Consequently, each set $\{ x^1, x^2, x^3 \}$ determines a point in space.

Setting x^1 = constant in (2-6) gives us a family of surfaces, one member corresponding to each value of the constant. Similarly, x^2 = constant and x^3 = constant define two other families of surfaces. We will refer to all of these surfaces as coordinate surfaces. The line of intersection of two coordinate surfaces defines a coordinate curve. Because of (2-7), three coordinate surfaces obtained by taking a member from each family intersect in one and only one point.

Every spatial vector field $\mathbf{u}$ may be written as

$$\mathbf{u} = u^i \mathbf{g}_i \tag{2-9}$$

$$\mathbf{u} = u_i \mathbf{g}^i \tag{2-10}$$

or in the case of an orthogonal coordinate system

$$\mathbf{u} = u_{<i>} \mathbf{g}_{<i>} \tag{2-11}$$

Here $\{ \mathbf{g}_1, \mathbf{g}_2, \mathbf{g}_3 \}$ are the **natural** basis fields

$$\mathbf{g}_i \equiv \frac{\partial \mathbf{p}}{\partial x^i} \tag{2-12}$$

We refer to $\mathbf{g}^1, \mathbf{g}^2, \mathbf{g}^3$ as the **dual** basis fields,

$$\mathbf{g}^i \equiv \nabla x^i \tag{2-13}$$

For orthogonal coordinate systems, we define the **physical** basis fields $\{ \mathbf{g}_{<1>}, \mathbf{g}_{<2>}, \mathbf{g}_{<3>} \}$ in this way:

$$\mathbf{g}_{<i>} \equiv \frac{\mathbf{g}_i}{\sqrt{g_{i\,i}}} = \sqrt{g_{ii}}\; \mathbf{g}^i \text{ (no sum on i)} \tag{2-14}$$

We say that the u^i (i = 1, 2, 3) are the **contravariant** components of **u**; u_i the **covariant** components of **u**; $u_{<i>}$ the **physical** components of **u**. The scalar fields

$$g_{ij} \equiv \mathbf{g}_i \cdot \mathbf{g}_j \tag{2-15}$$

are the covariant components of the identity tensor (Slattery 1981, p. 628); $\sqrt{g_{ii}}$ (no sum on i) is the magnitude of $\mathbf{g}_i$.

A **second-order spatial tensor** field **T** is a transformation (or mapping or rule) that assigns to each given spatial vector field **v** another spatial vector field **T** · **v** such that the rules

$$\mathbf{T} \cdot (\mathbf{v} + \mathbf{w}) = \mathbf{T} \cdot \mathbf{v} + \mathbf{T} \cdot \mathbf{w} \tag{2-16}$$

$$\mathbf{T} \cdot (\alpha \mathbf{v}) = \alpha(\mathbf{T} \cdot \mathbf{v}) \tag{2-17}$$

hold. By α we mean here a real scalar field. [Here the dot denotes that the tensor **T** operates on or transforms the vector **v**. It does not indicate a scalar product. Our choice of notation is suggestive, however, of the rules for transformation in (2-18) and (2-21).] If two spatial vector fields **a** and **b** are given, we can define a second-order tensor field **ab** by the requirement that it transform every vector field **v** into another field (**ab**) · **v** according to the rule

$$(\mathbf{ab}) \cdot \mathbf{v} \equiv \mathbf{a} (\mathbf{b} \cdot \mathbf{v}) \tag{2-18}$$

(On the left side, the dot indicates the transformation of a vector by a tensor; on the right, the dot denotes the scalar product of **b** with **v**.) This tensor field **ab** is called the **tensor product** or dyadic product of the spatial vector fields **a** and **b**. Every second-order spatial tensor field **T** can be written as a linear combination of tensor products:

$$\begin{aligned} \mathbf{T} &= T^{ij}\mathbf{g}_i\mathbf{g}_j = T_{ij}\mathbf{g}^i\mathbf{g}^j \\ &= T^i{}_j\mathbf{g}_i\mathbf{g}^j = T_i{}^j\mathbf{g}^i\mathbf{g}_j \end{aligned} \tag{2-19}$$

For an orthogonal coordinate system, we can write

$$\mathbf{T} = T_{<ij>}\mathbf{g}_{<i>}\mathbf{g}_{<j>} \tag{2-20}$$

We refer to the T^{ij} (i, j = 1, 2, 3) as the **contravariant** components of **T**; T_{ij} the **covariant** components of **T**; $T^i{}_j$ and $T_i{}^j$ the **mixed** components of

T; $T_{<ij>}$ the **physical** components of **T**. In terms of these components, the vector **T** · **v** takes the form

$$\begin{aligned}\mathbf{T}\cdot\mathbf{v} &= (\,T^{ij}\mathbf{g}_i\mathbf{g}_j\,)\cdot\mathbf{v}\\ &= T^{ij}\mathbf{g}_i(\,\mathbf{g}_j\cdot\mathbf{v}\,)\\ &= T^{ij}v_j\mathbf{g}_i\\ &= T_{ij}v^j\mathbf{g}^i\\ &= T^i{}_jv^j\mathbf{g}_i\\ &= T_i{}^jv_j\mathbf{g}^i \end{aligned} \qquad (2\text{-}21)$$

A **third-order tensor** field $\boldsymbol{\beta}$ is a transformation (or mapping or rule) that assigns to each given spatial vector field **v** a second-order tensor field $\boldsymbol{\beta}\cdot\mathbf{v}$ such that the rules

$$\boldsymbol{\beta}\cdot(\,\mathbf{v}+\mathbf{w}\,) = \boldsymbol{\beta}\cdot\mathbf{v} + \boldsymbol{\beta}\cdot\mathbf{w} \qquad (2\text{-}22)$$

$$\boldsymbol{\beta}\cdot(\,\alpha\mathbf{v}\,) = \alpha(\,\boldsymbol{\beta}\cdot\mathbf{v}\,) \qquad (2\text{-}23)$$

hold. Again α is a real scalar field. (The dot denotes that the tensor $\boldsymbol{\beta}$ operates on, or transforms, the vector **v**. This is not a scalar product between two vectors.) In this text, we deal with only one third-order spatial tensor field

$$\boldsymbol{\varepsilon} = \varepsilon^{ijk}\mathbf{g}_i\mathbf{g}_j\mathbf{g}_k = \varepsilon_{ijk}\mathbf{g}^i\mathbf{g}^j\mathbf{g}^k \qquad (2\text{-}24)$$

where

$$\varepsilon^{ijk} \equiv \frac{1}{\sqrt{g}}\,e^{ijk} \qquad (2\text{-}25)$$

$$\varepsilon_{ijk} \equiv \sqrt{g}\;e_{ijk} \qquad (2\text{-}26)$$

$$g \equiv \det(\,g_{ij}\,) \qquad (2\text{-}27)$$

and $e^{ijk} = e_{ijk}$ have only three distinct values:

0, when any two of the indices are equal

+1, when ijk is an even permutation of 1 2 3

−1, when ijk is an odd permutation of 1 2 3

This tensor is used in forming the **vector product** of two spatial vector fields **a** and **b**:

$$\mathbf{a} \wedge \mathbf{b} \equiv \boldsymbol{\varepsilon} : \mathbf{ba} \equiv (\boldsymbol{\varepsilon} \cdot \mathbf{b}) \cdot \mathbf{a} = \varepsilon^{ijk} a_j b_k \mathbf{g}_i \tag{2-28}$$

It is also employed in the definition for the **curl** of a spatial vector field **v**,

$$\text{curl } \mathbf{v} \equiv \boldsymbol{\varepsilon} : \nabla \mathbf{v} = \varepsilon^{ijk} v_{k,j} \mathbf{g}_i \tag{2-29}$$

where $v_{k,j}$ is referred to as the **covariant derivative** of v_k, the covariant component of the spatial vector field **v**.

For more details, I suggest the treatments by Slattery (1981), Leigh (1968), Coleman *et al.* (1966), Ericksen (1960), and McConnell (1957).

References

Coleman, B. D., H. Markovitz, and W. Noll, "Viscometric Flows of Non-Newtonian Fluids," Springer-Verlag, New York (1966).

Ericksen, J. L., "Handbuch der Physik," vol. 3/1, edited by S. Flügge, Springer-Verlag, Berlin (1960).

Leigh, D. C., "Nonlinear Continuum Mechanics," McGraw-Hill, New York (1968).

McConnell, A. J., "Applications of Tensor Analysis," Dover, New York (1957).

Slattery, J. C., "Momentum, Energy, and Mass Transfer in Continua," McGraw-Hill, New York (1972); second edition, Robert E. Krieger, Malabar, FL 32950 (1981).

A.1.3 Surface in (E^3, V^3) As illustrated in Figure 3-1, a surface in (E^3, V^3) is the locus of a point whose position is a function of two parameters y^1 and y^2:

$$\mathbf{z} = \mathbf{p}^{(\sigma)}(y^1, y^2) \tag{3-1}$$

Since the two numbers y^1 and y^2 uniquely determine a point on the surface, we call them the **surface coordinates**. A y^1 coordinate curve is a line in the surface along which y^1 varies while y^2 takes a fixed value.

Similarly, a y^2 coordinate curve is one along which y^2 varies while y^1 assumes a constant value.

For any surface, there is an infinite number of surface coordinate systems that might be used. Any two families of lines may be chosen as coordinate curves, so long as each member of one family intersects each member of the other in one and only one point.

Equation (3-1) represents three scalar equations. If we eliminate the surface coordinates y^1 and y^2 among the three components of (3-1), we are left with one scalar equation of the form

$$f(\mathbf{z}) = 0 \tag{3-2}$$

It is sometimes more convenient to think in terms of this single scalar equation for the surface.

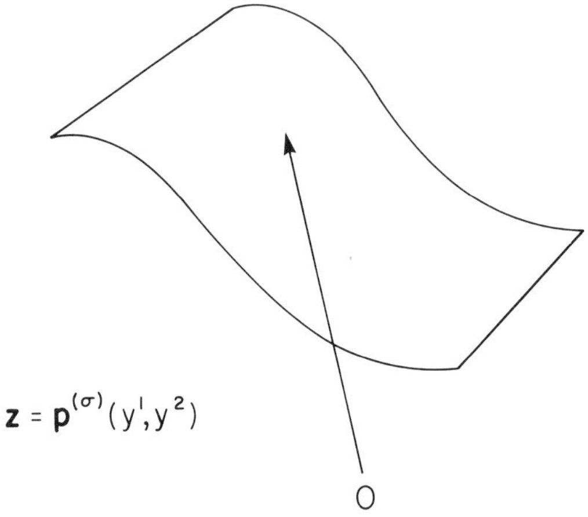

Figure A.1.3-1: A surface in (E^3, V^3) is the locus of a point whose position is a function of two parameters y^1 and y^2.

A.2 Vector fields

A.2.1 Natural basis A spatial vector field has been previously defined as any spatial vector-valued function of position (Slattery 1981, p. 605). In the present context, we are concerned with spatial vector-valued functions of position on a surface.

Referring to the parametric representation for a surface in Sec. A.1.3, we define

$$\mathbf{a}_\alpha \equiv \frac{\partial \mathbf{p}^{(\sigma)}}{\partial y^\alpha} = \frac{\partial x^i}{\partial y^\alpha} \mathbf{g}_i \tag{1-1}$$

At every point on the surface, the values of these spatial vector fields are tangent to the y^α coordinate curves and therefore tangent to the surface. These two spatial vector fields are also linearly independent; at no point is a member of one family of surface coordinate curves allowed to be tangent to a member of the other family. Note that the definition for $\mathbf{a}_\alpha$ parallels that for the natural basis vector field $\mathbf{g}_k$ for V^3 (Slattery 1981, p. 617; see also Sec. A.1.2).

Let $\boldsymbol{\xi}$ be the unit normal to the surface (Slattery 1981, p. 619). Since the vector fields $\mathbf{a}_1$, $\mathbf{a}_2$, and $\boldsymbol{\xi}$ are linearly independent, they form a basis for the spatial vector fields on the surface. We are particularly interested in the **tangential vector fields**, tangent vector-valued functions of position on a surface. The tangential vector fields are a two-dimensional subspace of the spatial vector fields on a surface in the sense that every tangential vector field $\mathbf{c}$ can be expressed as a linear combination of $\mathbf{a}_1$ and $\mathbf{a}_2$:

$$\begin{aligned} \mathbf{c} &= c^1 \mathbf{a}_1 + c^2 \mathbf{a}_2 \\ &= \sum_{\alpha=1}^{2} c^\alpha \mathbf{a}_\alpha \\ &= c^\alpha \mathbf{a}_\alpha \end{aligned} \tag{1-2}$$

(Note that we have introduced here a modification of the summation convention. A repeated Greek index in any term is to be summed from 1 to 2. A repeated italic index in any term will continue to indicate a summation from 1 to 3.) In what follows, $\{ \mathbf{a}_1, \mathbf{a}_2 \}$ are known as the **natural basis** for the surface coordinate system $\{ y^1, y^2 \}$.

Since it occurs quite frequently, let us define

$$\begin{aligned} a_{\alpha\beta} &\equiv \mathbf{a}_\alpha \cdot \mathbf{a}_\beta \\ &= \frac{\partial \mathbf{p}^{(\sigma)}}{\partial y^\alpha} \cdot \frac{\partial \mathbf{p}^{(\sigma)}}{\partial y^\beta} \end{aligned}$$

$$= g_{ij} \frac{\partial x^i}{\partial y^\alpha} \frac{\partial x^j}{\partial y^\beta} \tag{1-3}$$

We also will be concerned with the determinant

$$a \equiv \det(a_{\alpha\beta}) \tag{1-4}$$

which has as its typical entry $a_{\alpha\beta}$. Expanding by columns, we find (McConnell 1957, p. 10; Slattery 1981, p. 611)

$$e^{\alpha\beta} a_{\alpha\mu} a_{\beta\nu} = a e_{\mu\nu} \tag{1-5}$$

An expansion by rows gives

$$e^{\alpha\beta} a_{\mu\alpha} a_{\nu\beta} = a e_{\mu\nu} \tag{1-6}$$

The symbols $e_{\alpha\beta}$ and $e^{\alpha\beta}$ have only three distinct values:

0, when the indices are equal,
+1, when $\alpha\beta$ is 12
–1, when $\alpha\beta$ is 21

If we define the normalized cofactor $a^{\alpha\beta}$ as

$$a^{\alpha\beta} \equiv \frac{1}{a} e^{\alpha\nu} e^{\beta\gamma} a_{\gamma\nu} \tag{1-7}$$

we see

$$a^{\alpha\beta} a_{\beta\gamma} = a_{\gamma\beta} a^{\beta\alpha} = \delta^\alpha_\gamma \tag{1-8}$$

Here δ^α_γ is the Kronecker delta.

References

Giordano, R. M., personal communication, October, 1979.

Halmos, P. R., "Finite-Dimensional Vector Spaces," second edition, Van Nostrand, New York (1958).

Kaplan, W., "Advanced Calculus," second edition, Addison-Wesley, Cambridge, MA (1973).

McConnell, A. J., "Applications of Tensor Analysis," Dover, New York

(1957).

Slattery, J. C., "Momentum, Energy, and Mass Transfer in Continua," McGraw-Hill, New York (1972); second edition, Robert E. Krieger, Malabar, FL 32950 (1981).

Exercises

A.2.1-1 *transformation of surface coordinates* Let the $\mathbf{a}_\alpha$ ($\alpha = 1, 2$) be the natural basis vector fields associated with one surface coordinate system $\{y^1, y^2\}$. A change of surface coordinates is suggested:

$$y^\alpha = y^\alpha(\overline{y}^1, \overline{y}^2) \qquad \alpha = 1, 2$$

i) Prove that

$$\overline{\mathbf{a}}_\alpha = \frac{\partial y^\beta}{\partial \overline{y}^\alpha} \mathbf{a}_\beta$$

ii) Given one surface coordinate system $\{ y^1, y^2 \}$, we will admit $\{ \overline{y}^1, \overline{y}^2 \}$ as a new surface coordinate system, only if the corresponding natural basis vector fields $\{ \overline{\mathbf{a}}_1, \overline{\mathbf{a}}_2 \}$ actually form a basis for the tangential vector fields. They can be said to form a basis, if every tangential vector field can be represented as a linear combination of them (Halmos 1958, p. 14). In particular, $\{ \overline{\mathbf{a}}_1, \overline{\mathbf{a}}_2 \}$ form a basis for the tangential vector fields, if the natural basis vector fields $\{ \mathbf{a}_1, \mathbf{a}_2 \}$ can be represented as a linear combination of them:

$$\mathbf{a}_\alpha = \frac{\partial \overline{y}^\beta}{\partial y^\alpha} \overline{\mathbf{a}}_\beta$$

Conclude that $\{ \overline{\mathbf{a}}_1, \overline{\mathbf{a}}_2 \}$ form a basis for the tangential vector fields, if everywhere

$$\det\left(\frac{\partial y^\alpha}{\partial \overline{y}^\beta}\right) \neq 0$$

For another point of view leading to the same conclusion, see a discussion of the implicit function theorem (Kaplan 1973, p. 150).

A.2.1-2 If **c** is a tangential vector field, determine that

$$\mathbf{c} = c^{\alpha} \frac{\partial x^i}{\partial y^{\alpha}} \mathbf{g}_i$$

and that the contravariant spatial components are

$$c^i = \frac{\partial x^i}{\partial y^{\alpha}} c^{\alpha}$$

A.2.1-3 If the $\mathbf{a}_{\alpha}$ (α = 1, 2) are the natural basis vector fields associated with one surface coordinate system $\{y^1, y^2\}$ and if the $\overline{\mathbf{a}}_{\beta}$ (β = 1, 2) are the natural basis vector fields associated with another surface coordinate system $\{\overline{y}^1, \overline{y}^2\}$, prove that

$$\det(\overline{a}_{\alpha\beta}) = \det(a_{\mu\nu})\left[\det\left(\frac{\partial y^{\gamma}}{\partial \overline{y}^{\delta}} \right)\right]^2$$

A.2.1-4 *plane surface* Given a rectangular cartesian coordinate system and a plane surface z_3 = a constant, let us choose

$$y^1 \equiv z_1$$

$$y^2 \equiv z_2$$

Prove that

$$a_{11} = \frac{1}{a^{11}} = a_{22} = \frac{1}{a^{22}} = 1$$

$$a_{12} = a_{21} = 0$$

and

$$a = 1$$

(See also Exercise A.5.3-5.)

A.2.1-5 *plane surface in polar coordinates* Given a cylindrical coordinate system

$$z_1 = x^1 \cos x^2 = r \cos \theta$$

$$z_2 = x^1 \sin x^2 = r \sin \theta$$

$$z_3 = x^3 \equiv z$$

and a plane surface z_3 = a constant, let us choose

$$y^1 \equiv x^1 \equiv r$$

$$y^2 \equiv x^2 \equiv \theta$$

Prove that

$$a_{11} = \frac{1}{a^{11}} = 1$$

$$a_{22} = \frac{1}{a^{22}} = r^2$$

$$a_{12} = a_{21} = 0$$

and

$$a = r^2$$

(See also Exercise A.5.3-6.)

A.2.1-6 *alternative form for plane surface in polar coordinates* (Giordano 1979) Given the cylindrical coordinate system described in Exercise A.2.1-5 and a plane surface θ = a constant, let us choose

$$y^1 \equiv x^1 \equiv r$$

$$y^2 \equiv x^3 \equiv z$$

Prove that

$$a_{11} = \frac{1}{a^{11}} = 1$$

$$a_{22} = \frac{1}{a^{22}} = 1$$

$$a_{12} = a_{21} = 0$$

and

$$a = 1$$

(See also Exercise A.5.3-7.)

A.2.1-7 *cylindrical surface* Given the above cylindrical coordinate system described in Exercise A.2.1-5 and a cylindrical surface of radius R, let us choose

$$y^1 \equiv x^2 \equiv \theta$$

$$y^2 \equiv x^3 \equiv z$$

Prove that

$$a_{11} = \frac{1}{a^{11}} = R^2$$

$$a_{22} = \frac{1}{a^{22}} = 1$$

$$a_{12} = a_{21} = 0$$

and

$$a = R^2$$

(See also Exercise A.5.3-8.)

A.2.1-8 *spherical surface* Given a spherical coordinate system

$$z_1 = x^1 \sin x^2 \cos x^3 = r \sin \theta \cos \phi$$

$$z_2 = x^1 \sin x^2 \sin x^3 = r \sin \theta \sin \phi$$

$$z_3 = x^1 \cos x^2 = r \cos \theta$$

and a spherical surface of radius R, let us choose

$$y^1 \equiv x^2 \equiv \theta$$

$$y^2 \equiv x^3 \equiv \phi$$

Prove that

$$a_{11} = \frac{1}{a^{11}} = R^2$$

$$a_{22} = \frac{1}{a^{22}} = R^2 \sin^2\theta$$

$$a_{12} = a_{21} = 0$$

and

$$a = R^4 \sin^2\theta$$

(See also Exercise A.5.3-9.)

A.2.1-9 *two-dimensional waves* Given a rectangular cartesian coordinate system and a surface

$$z_3 = h(z_1, t)$$

let us choose

$$y_1 \equiv z_1$$

$$y^2 \equiv z_2$$

Prove that

$$a_{11} = 1 + \left(\frac{\partial h}{\partial z_1}\right)^2$$

$$a_{22} = 1$$

$$a_{12} = a_{21} = 0$$

and

$$a = 1 + \left(\frac{\partial h}{\partial z_1}\right)^2$$

(See also Exercises A.2.6-1 and A.5.3-10.)

A.2.1-10 *axially symmetric surface in cylindrical coordinates* Given the cylindrical coordinate system described in Exercise A.2.1-5 and an axially symmetric surface

$$z = h(r)$$

let us choose

$$y^1 \equiv r$$

$$y^2 \equiv \theta$$

Prove that

$$a_{11} = 1 + \left(\frac{\partial h}{\partial r}\right)^2$$

$$a_{22} = r^2$$

$$a_{12} = a_{21} = 0$$

and

$$a = r^2 \left[1 + \left(\frac{\partial h}{\partial r}\right)^2 \right]$$

(See also Exercises A.2.6-2 and A.5.3-11.)

A.2.1-11 *alternative form for axially symmetric surface in cylindrical coordinates* Given the cylindrical coordinate system described in Exercise A.2.1-5 and an axially symmetric surface

$$r = c(z)$$

let us choose

$$y^1 = x^3 = z$$

$$y^2 = x^2 = \theta$$

Prove that

$$a_{11} = 1 + \left(\frac{dc}{dz}\right)^2$$

$$a_{22} = c^2$$

$$a_{12} = a_{21} = 0$$

and

$$a = c^2 \left[1 + \left(\frac{dc}{dz} \right)^2 \right]$$

(See also Exercises A.2.6-3 and A.5.3-12.)

A.2.2 Surface gradient of scalar field We will frequently be concerned with scalar fields on a surface. It is easy, for example, to visualize temperature as a function of position on a phase interface. It is for this reason that we must be concerned with derivatives on a surface.

By analogy with a spatial gradient (Slattery 1981, p. 613), the **surface gradient** of a scalar field ϕ is a tangential vector field denoted by $\nabla_{(\sigma)}\phi$ and specified by defining its inner product with an arbitrary tangential vector field **c**:

$$\nabla_{(\sigma)}\phi(y^1, y^2) \cdot \mathbf{c}$$

$$\equiv \text{limit } s \to 0: \frac{1}{s} \{ \phi(y^1 + sc^1, y^2 + sc^2) - \phi(y^1, y^2)\} \qquad (2\text{-}1)$$

Equation (2-1) may be rearranged into a more easily applied expression,

$$\nabla_{(\sigma)}\phi(y^1, y^2) \cdot \mathbf{c}$$

$$= \text{limit } s \to 0: \frac{1}{s} \{ \phi(y^1 + sc^1, y^2 + sc^2) - \phi(y^1, y^2 + sc^2)\}$$

$$+ \text{limit } s \to 0: \frac{1}{s} \{ \phi(y^1, y^2 + sc^2) - \phi(y^1, y^2)\}$$

$$= c^1 \text{ limit } sc^1 \to 0: \frac{1}{sc^1} \{ \phi(y^1 + sc^1, y^2 + sc^2) - \phi(y^1, y^2 + sc^2)\}$$

$$+ c^2 \text{ limit } sc^2 \to 0: \frac{1}{sc^2} \{ \phi(y^1, y^1 + sc^2) - \phi(y^1, y^2)\}$$

$$= c^1 \frac{\partial \phi}{\partial y^1} (y^1 , y^2) + c^2 \frac{\partial \phi}{\partial y^2} (y^1 , y^2)$$

$$= c^\alpha \frac{\partial \phi}{\partial y^\alpha} (y^1 , y^2) \tag{2-2}$$

Since $\mathbf{c}$ is an arbitrary tangential vector field, take $\mathbf{c} = \mathbf{a}_\beta$:

$$\nabla_{(\sigma)}\phi \cdot \mathbf{a}_\beta = \frac{\partial \phi}{\partial y^\beta} \tag{2-3}$$

Since $\nabla_{(\sigma)}\phi$ is defined to be a tangential vector field, we conclude that

$$\nabla_{(\sigma)}\phi = \frac{\partial \phi}{\partial y^\alpha} a^{\alpha\beta} \mathbf{a}_\beta \tag{2-4}$$

In arriving at this result, you will find (1-8) helpful.

Reference

Slattery, J. C., "Momentum, Energy, and Mass Transfer in Continua," McGraw-Hill, New York (1972); second edition, Robert E. Krieger, Malabar, FL 32950 (1981).

A.2.3 Dual basis The **dual** tangential vector fields $\mathbf{a}^\alpha$ ($\alpha = 1, 2$) are defined as the surface gradients of the surface coordinates,

$$\mathbf{a}^\alpha \equiv \nabla_{(\sigma)} y^\alpha \tag{3-1}$$

Applying the expression for the surface gradient developed in Sec. A.2.2, we find

$$\mathbf{a}^\alpha = \frac{\partial y^\alpha}{\partial y^\beta} a^{\beta\gamma} \mathbf{a}_\gamma$$

$$= a^{\alpha\gamma} \mathbf{a}_\gamma \tag{3-2}$$

Observe that

$$a_{\beta\alpha}\,\mathbf{a}^{\alpha} = a_{\beta\alpha}\,a^{\alpha\gamma}\mathbf{a}_{\gamma}$$

$$= \delta^{\gamma}_{\beta}\,\mathbf{a}_{\gamma}$$

$$= \mathbf{a}_{\beta} \tag{3-3}$$

or that each natural basis vector may be written as a linear combination of the dual vectors:

$$\mathbf{a}_{\alpha} = a_{\alpha\beta}\,\mathbf{a}^{\beta} \tag{3-4}$$

Since $\mathbf{a}_1$ and $\mathbf{a}_2$ form a basis for the tangential vector fields, it follows that the two dual vectors also comprise a basis.

Exercises

A.2.3-1 If the $\mathbf{a}^{\alpha}$ ($\alpha = 1, 2$) are the dual basis vector fields associated with one surface coordinate system $\{ y^1, y^2 \}$ and if the $\overline{\mathbf{a}}^{\alpha}$ ($\alpha = 1, 2$) are the dual basis vector fields associated with another surface coordinate system $\{ \overline{y}^1, \overline{y}^2 \}$, prove that

$$\overline{\mathbf{a}}^{\alpha} = \frac{\partial \overline{y}^{\alpha}}{\partial y^{\beta}}\,\mathbf{a}^{\beta}$$

A.2.3-2 Prove that

$$\mathbf{a}^{\alpha} \cdot \mathbf{a}^{\beta} = a^{\alpha\beta}$$

A.2.3-3 Prove that

$$\mathbf{a}^{\alpha} \cdot \mathbf{a}_{\beta} = \delta^{\alpha}_{\beta}$$

A.2.4 Covariant and contravariant components By definition, every tangential vector field may be written as a linear combination of the two natural basis vectors,

$$\mathbf{c} = c^{\alpha} \mathbf{a}_{\alpha} \tag{4-1}$$

In the last section, we concluded that every tangential field may also be written as a linear combination of the dual basis fields,

$$\mathbf{c} = c_{\alpha} \mathbf{a}^{\alpha} \tag{4-2}$$

The c^{α} and c_{α} are referred to respectively as the **contravariant** and **covariant** surface components of the tangential vector field **c**.

Since for any tangential vector field

$$\mathbf{c} = c_{\alpha} \mathbf{a}^{\alpha} = c_{\alpha} a^{\alpha\beta} \mathbf{a}_{\beta} = c^{\beta} \mathbf{a}_{\beta} \tag{4-3}$$

we may write

$$(c^{\beta} - c_{\alpha} a^{\alpha\beta}) \mathbf{a}_{\beta} = 0 \tag{4-4}$$

The dual basis fields are linearly independent and (4-4) implies that

$$c^{\beta} = a^{\alpha\beta} c_{\alpha} \tag{4-5}$$

In the same way,

$$\mathbf{c} = c^{\alpha} \mathbf{a}_{\alpha} = c^{\alpha} a_{\alpha\beta} \mathbf{a}^{\beta} = c_{\beta} \mathbf{a}^{\beta} \tag{4-6}$$

so that we may identify

$$c_{\beta} = a_{\alpha\beta} c^{\alpha} \tag{4-7}$$

We find in this way that the $a_{\alpha\beta}$ and the $a^{\alpha\beta}$ may be used to *raise and lower* indices.

Exercise

A.2.4-1 i)4 Let **c** be some tangential vector field. If the c^{α} ($\alpha = 1, 2$) are the contravariant components of **c** with respect to one surface coordinate system $\{y^1, y^2\}$ and if the $\overline{c}^{\beta}$ ($\beta = 1, 2$) are the contravariant components of **c** with respect to another surface coordinate system $\{\overline{y}^1, \overline{y}^2\}$, determine that

$$c^{\alpha} = \frac{\partial y^{\alpha}}{\partial \overline{y}^{\beta}} \overline{c}^{\beta}$$

ii) Similarly, show that

$$c_\alpha = \frac{\partial \bar{y}^\beta}{\partial y^\alpha} \bar{c}_\beta$$

A.2.5 Physical components The natural basis fields are **orthogonal,** if

$$\mathbf{a}_\alpha \cdot \mathbf{a}_\beta = g_{ij} \frac{\partial x^i}{\partial y^\alpha} \frac{\partial x^j}{\partial y^\beta} = 0 \qquad \text{for } \alpha \neq \beta \tag{5-1}$$

When the natural basis fields are orthogonal to one another, we say that they correspond to an orthogonal surface coordinate system. Four examples of orthogonal surface coordinate systems are given in Exercises A.2.1-4 through A.2.1-7.

When possible, it is usually more convenient to work in terms of an orthonormal basis, one consisting of orthogonal unit vectors. For an orthogonal surface coordinate system, the natural basis fields defined in (1-1) may be normalized to form an orthonormal basis $\{\mathbf{a}_{<1>}, \mathbf{a}_{<2>}\}$

$$\mathbf{a}_{<\alpha>} \equiv \frac{\mathbf{a}_\alpha}{\sqrt{a_{\alpha\alpha}}} \qquad \text{(no summation on } \alpha) \tag{5-2}$$

This basis is referred to as the **physical basis** for the surface coordinate system.

In this text, we will employ a normalized natural basis only for orthogonal surface coordinate systems. For these coordinate systems, the physical basis has a particularly convenient relation to the normalized dual basis:

$$\mathbf{a}_{<\alpha>} = \frac{\mathbf{a}^\alpha}{\sqrt{a^{\alpha\alpha}}}$$

$$= \sqrt{a_{\alpha\alpha}}\, \mathbf{a}^\alpha \qquad \text{(no summation on } \alpha) \tag{5-3}$$

Any tangential vector field **c** may consequently be expressed as a linear combination of the two physical basis vector fields associated with an orthogonal surface coordinate system:

$$\mathbf{c} = c_{<\alpha>} \mathbf{a}_{<\alpha>} \tag{5-4}$$

The two coefficients $\{c_{<1>}, c_{<2>}\}$

$$c_{<\alpha>} \equiv \sqrt{a_{\alpha\alpha}}\, c^{\alpha}$$

$$= \frac{c_{\alpha}}{\sqrt{a_{\alpha\alpha}}} \quad \text{(no summation on } \alpha\text{)} \tag{5-5}$$

are known as the **physical** surface components of **c** with respect to this particular surface coordinate system.

A.2.6 Tangential and normal components In the text, spatial vector fields defined on a surface play an important role. It is often convenient to think in terms of their tangential and normal components.

If as suggested in Sec. A.1.3 we describe our surface by the single scalar equation

$$f(\mathbf{z}) = 0 \tag{6-1}$$

then the unit normal to the surface is given by (Slattery 1981, p. 619)

$$\boldsymbol{\xi} \equiv \frac{\nabla f}{|\nabla f|} \tag{6-2}$$

We require that the sign of the function f in (6-1) is such that $\boldsymbol{\xi} \cdot (\mathbf{a}_1 \wedge \mathbf{a}_2)$ is positive. This means that, with respect to "right-handed" spatial coordinate systems, the spatial vector fields $\mathbf{a}_1$, $\mathbf{a}_2$ and $\boldsymbol{\xi}$ have the same orientation as the index finger, middle finger and thumb on the right hand (Slattery 1981, p. 656).

Since $\mathbf{a}_1$, $\mathbf{a}_2$ and $\boldsymbol{\xi}$ are linearly independent, they form a basis for the spatial vector fields on the surface. If **v** is any spatial vector field on the surface, we can write

$$\mathbf{v} = v^{\alpha}\mathbf{a}_{\alpha} + v_{(\xi)}\boldsymbol{\xi} \tag{6-3}$$

By the same argument, **v** can also be expressed as

$$\mathbf{v} = v_{\alpha}\mathbf{a}^{\alpha} + v_{(\xi)}\boldsymbol{\xi} \tag{6-4}$$

Here $v_{(\xi)}$ is known as the normal component of the spatial vector field **v**.

Reference

Slattery, J. C., "Momentum, Energy, and Mass Transfer in Continua," McGraw-Hill, New York (1972); second edition, Robert E. Krieger, Malabar, FL 32950 (1981).

Exercises

A.2.6-1 *two-dimensional waves* Given a rectangular cartesian coordinate system and a surface

$$z_3 = h(z_1, t)$$

determine that the rectangular cartesian components of $\boldsymbol{\xi}$ are

$$\xi_1 = -\frac{\partial h}{\partial z_1}\left[1 + \left(\frac{\partial h}{\partial z_1}\right)^2\right]^{-1/2}$$

$$\xi_2 = 0$$

$$\xi_3 = \left[1 + \left(\frac{\partial h}{\partial z_1}\right)^2\right]^{-1/2}$$

(See also Exercises A.2.1-9 and A.5.3-10.)

A.2.6-2 *axially symmetric surface in cylindrical coordinates* Given the cylindrical coordinate system described in Exercise A.2.1-5 and an axially symmetric surface

$$z = h(r)$$

determine that the cylindrical components of $\boldsymbol{\xi}$ are

$$\xi_r = -\frac{dh}{dr}\left[1 + \left(\frac{dh}{dr}\right)^2\right]^{-1/2}$$

$$\xi_\theta = 0$$

$$\xi_z = \left[1 + \left(\frac{dh}{dr}\right)^2\right]^{-1/2}$$

(See also Exercises A.2.1-10 and A.5.3-11.)

A.2.6-3 *alternative form for axially symmetric surface in cylindrical coordinates* Given the cylindrical coordinate system described in Exercise A.2.1-5 and an axially symmetric surface

$$r = c(z)$$

determine that the cylindrical components of $\boldsymbol{\xi}$ are

$$\xi_r = \left[1 + \left(\frac{dc}{dr} \right)^2 \right]^{-1/2}$$

$$\xi_\theta = 0$$

$$\xi_z = -\frac{dc}{dr} \left[1 + \left(\frac{dc}{dr} \right)^2 \right]^{-1/2}$$

(See also Exercises A.2.1-11 and A.5.3-12.)

A.3 Second-order tensor fields

A.3.1 Tangential transformations and surface tensors A second-order tensor field **A** is a transformation (or mapping or rule) that assigns to each given spatial vector field **v** another spatial vector field $\mathbf{A} \cdot \mathbf{v}$ such that the rules

$$\mathbf{A} \cdot (\mathbf{v} + \mathbf{w}) = \mathbf{A} \cdot \mathbf{v} + \mathbf{A} \cdot \mathbf{w}$$

$$\mathbf{A} \cdot (\alpha \mathbf{v}) = \alpha (\mathbf{A} \cdot \mathbf{v}) \tag{1-1}$$

hold. By α we mean a real scalar field.

A **tangential transformation** $\boldsymbol{\mathcal{T}}$ is a particular type of second-order tensor field that is defined only on the surface, that assigns to each given tangential vector field **c** on a surface Σ another tangential vector field $\boldsymbol{\mathcal{T}} \cdot \mathbf{c}$ on a surface $\overline{\Sigma}$, and that transforms every spatial vector field normal to Σ into the zero vector. Let $\{ \mathbf{a}_1, \mathbf{a}_2 \}$ and $\{ \mathbf{a}^1, \mathbf{a}^2 \}$ be the natural and dual basis fields on Σ; $\{ \overline{\mathbf{a}}_1, \overline{\mathbf{a}}_2 \}$ and $\{ \overline{\mathbf{a}}^1, \overline{\mathbf{a}}^2 \}$ denote the natural and dual basis fields on $\overline{\Sigma}$. A tangential transformation $\boldsymbol{\mathcal{T}}$ can be defined by the way in which it transforms the natural basis fields $\{ \mathbf{a}_1, \mathbf{a}_2 \}$:

$$\boldsymbol{\mathcal{T}} \cdot \mathbf{a}_\alpha = \mathcal{T}^{A}_{\cdot\ \alpha} \overline{\mathbf{a}}_A \tag{1-2}$$

By an argument which is similar to that given for general second-order tensor fields (Slattery 1981, p. 627), we conclude[a]

$$\boldsymbol{\mathcal{T}} = \mathcal{T}^{A}_{\cdot\ \alpha} \overline{\mathbf{a}}_A \mathbf{a}^\alpha \tag{1-3}$$

Alternatively we may write

$$\boldsymbol{\mathcal{T}} = \mathcal{T}^{A\,\alpha} \overline{\mathbf{a}}_A \mathbf{a}_\alpha = \mathcal{T}_{A\,\alpha} \overline{\mathbf{a}}^A \mathbf{a}^\alpha \tag{1-4}$$

where we have defined

$$\mathcal{T}^{A\,\alpha} \equiv a^{\alpha\beta} \mathcal{T}^{A}_{\cdot\ \beta} \tag{1-5}$$

and

$$\mathcal{T}_{A\,\alpha} \equiv \overline{a}_{A\,B} \mathcal{T}^{B}_{\cdot\ \alpha} \tag{1-6}$$

A very important special case of the tangential transformation is the **tangential tensor** for which $\overline{\Sigma} = \Sigma$. A tangential tensor **T** sends every

---

a) As individual indices are raised, their order will be preserved with a dot in the proper subscript position.

tangential vector field $\mathbf{c}$ on a surface Σ into another tangential vector field $\mathbf{T} \cdot \mathbf{c}$ on Σ; it transforms every spatial vector field normal to Σ into the zero vector. We see from (1-3) and (1-4) that a tangential tensor will have the form

$$\mathbf{T} = T^{\alpha\beta}\mathbf{a}_\alpha \mathbf{a}_\beta = T_{\alpha\beta}\mathbf{a}^\alpha \mathbf{a}^\beta \tag{1-7}$$

where $T^{\alpha\beta}$ and $T_{\alpha\beta}$ are known respectively as the **contravariant** and **covariant** surface components of $\mathbf{T}$.

We will usually find it convenient to introduce an orthogonal surface coordinate system. If $\{\mathbf{a}_{<1>}, \mathbf{a}_{<2>}\}$ are the associated physical basis fields (see Sec. A.2.5), we can again construct an argument similar to that given above to conclude

$$\mathbf{T} = T_{<\alpha\beta>}\mathbf{a}_{<\alpha>}\mathbf{a}_{<\beta>} \tag{1-8}$$

Here $T_{<\alpha\beta>}$ are the **physical** surface components of $\mathbf{T}$:

$$T_{<\alpha\beta>} \equiv \sqrt{a_{\alpha\alpha}}\ \sqrt{a_{\beta\beta}}\ T^{\alpha\beta}$$

$$= \frac{T_{\alpha\beta}}{\sqrt{a_{\alpha\alpha}}\ \sqrt{a_{\beta\beta}}} \quad \text{(no summation on } \alpha \text{ and } \beta) \tag{1-9}$$

The tangential tensors are special cases of what we can refer to as surface tensors. A second-order **surface tensor S** is a particular type of second-order tensor field that is defined only on the surface, that assigns to each given tangential vector field $\mathbf{c}$ on a surface another spatial vector field $\mathbf{S} \cdot \mathbf{c}$, and that transforms every spatial vector field normal to the surface into the zero vector. A surface tensor can be defined by the way in which it transforms the natural basis fields $\{ \mathbf{a}_1, \mathbf{a}_2 \}$:

$$\mathbf{S} \cdot \mathbf{a}_\alpha = S^i_{.\alpha}\mathbf{g}_i \tag{1-10}$$

Arguing as we did in writing (1-3), we can express

$$\mathbf{S} = S^i_{.\alpha}\mathbf{g}_i\mathbf{a}^\alpha = S^{i\alpha}\mathbf{g}_i\mathbf{a}_\alpha = S_{i\alpha}\mathbf{g}^i\mathbf{a}^\alpha \tag{1-11}$$

where

$$S^{i\alpha} \equiv a^{\alpha\beta}S^i_{.\beta} \tag{1-12}$$

and

$$S_{i\alpha} \equiv g_{ij}S^j_{.\alpha} \tag{1-13}$$

It is easy to think of more general second-order tensor fields that could be defined on a surface. However, we shall stop with these three classes, since they are sufficient for the physical applications which we discuss in the text.

Reference

Slattery, J. C., "Momentum, Energy, and Mass Transfer in Continua," McGraw-Hill, New York (1972); second edition, Robert E. Krieger, Malabar, FL 32950 (1981).

Exercise

A.3.1-1 i) Let **T** be some tangential tensor field. If the $T^{\alpha\beta}$ ($\alpha = 1, 2$; $\beta = 1, 2$) are the contravariant components of **T** with respect to one surface coordinate system $\{y^1, y^2\}$ and if the $\overline{T}^{\mu\nu}$ ($\mu = 1, 2$; $\nu = 1, 2$) are the contravariant components of **T** with respect to another surface coordinate system $\{\overline{y}^1, \overline{y}^2\}$, determine that

$$T^{\alpha\beta} = \frac{\partial y^\alpha}{\partial \overline{y}^\mu} \frac{\partial y^\beta}{\partial \overline{y}^\nu} \overline{T}^{\mu\nu}$$

ii) Similarly, show that

$$T_{\alpha\beta} = \frac{\partial \overline{y}^\mu}{\partial y^\alpha} \frac{\partial \overline{y}^\nu}{\partial y^\beta} \overline{T}_{\mu\nu}$$

A.3.2 Projection tensor The **projection tensor P** is a tangential second-order tensor field that transforms every tangential vector field into itself,

$$\mathbf{P} \cdot \mathbf{a}_\beta = P^{\alpha}_{\cdot\,\beta}\, \mathbf{a}_\alpha = \mathbf{a}_\beta = \delta^\alpha_\beta\, \mathbf{a}_\alpha \tag{2-1}$$

This indicates that

$$(P^{\alpha}_{\cdot\,\beta} - \delta^\alpha_\beta)\mathbf{a}_\alpha = 0 \tag{2-2}$$

Since the natural basis fields $\mathbf{a}_\alpha$ ($\alpha = 1, 2$) are linearly independent, we conclude that

$$P^{\alpha}_{.\beta} = \delta^{\alpha}_{\beta} \tag{2-3}$$

or

$$\mathbf{P} = \mathbf{a}_{\alpha}\mathbf{a}^{\alpha} = \mathbf{a}^{\alpha}\mathbf{a}_{\alpha} = a_{\alpha\beta}\mathbf{a}^{\alpha}\mathbf{a}^{\beta} = a^{\alpha\beta}\mathbf{a}_{\alpha}\mathbf{a}_{\beta} \tag{2-4}$$

The quantities $a_{\alpha\beta}$ and $a^{\alpha\beta}$ are often referred to as respectively the covariant and contravariant components of the surface metric tensor (McConnell 1957, p. 167). We prefer to think of them either in terms of their definitions in Sec. A.2.1 or as respectively the covariant and contravariant components of the projection tensor.

In Sec. A.2.6, we found that every spatial vector field in the surface can be written in terms of its tangential and normal components,

$$\mathbf{v} = v^{\alpha}\mathbf{a}_{\alpha} + v_{(\xi)}\boldsymbol{\xi} \tag{2-5}$$

This can also be expressed as

$$\mathbf{v} = (\mathbf{v} \cdot \mathbf{a}^{\alpha})\mathbf{a}_{\alpha} + v_{(\xi)}\boldsymbol{\xi} \tag{2-6}$$

or

$$\mathbf{I} \cdot \mathbf{v} = (\mathbf{P} + \boldsymbol{\xi}\boldsymbol{\xi}) \cdot \mathbf{v} \tag{2-7}$$

where $\mathbf{I}$ is the identity transformation (Slattery 1981, p. 628). Since $\mathbf{v}$ is any spatial vector field on the surface, this implies that

$$\mathbf{I} = \mathbf{P} + \boldsymbol{\xi}\boldsymbol{\xi} \tag{2-8}$$

This provides us with a convenient alternate interpretation for the projection tensor,

$$\mathbf{P} = \mathbf{I} - \boldsymbol{\xi}\boldsymbol{\xi} \tag{2-9}$$

The contravariant components of this relationship are well known (McConnell 1957, p. 197):

$$a^{\alpha\beta}\frac{\partial x^i}{\partial y^{\alpha}}\frac{\partial x^j}{\partial y^{\beta}} = g^{ij} - \xi^i\xi^j \tag{2-10}$$

It is important to keep in mind that the projection tensor plays the role of the identity tensor for the set of all tangential vector fields.

References

McConnell, A. J., "Applications of Tensor Analysis," Dover, New York (1957).

Slattery, J. C., "Momentum, Energy, and Mass Transfer in Continua," McGraw-Hill, New York (1972); second edition, Robert E. Krieger, Malabar, FL 32950 (1981).

Exercises

A.3.2-1 *relations between spatial, tangential, and normal components* Let **v** be a spatial vector field defined on a surface.

i) Prove that

$$v^{\alpha} = a^{\alpha\beta} \frac{\partial x^i}{\partial y^{\beta}} v_i$$

ii) Prove that

$$v^i = v^{\alpha} \frac{\partial x^i}{\partial y^{\alpha}} + v_{(\xi)} \xi^i$$

A.3.2-2 *alternative interpretation for surface gradient of scalar field* Let ψ be an explicit function of position in space. Determine that

$$\nabla_{(\sigma)} \psi = \mathbf{P} \cdot \nabla \psi$$

The surface gradient of ψ is the projection of the spatial gradient of ψ.

A.3.3 Tangential cross tensor Let us investigate the properties of the tangential second-order tensor[a]

$$\boldsymbol{\varepsilon}^{(\sigma)} \equiv \varepsilon^{\alpha\beta} \mathbf{a}_{\alpha} \mathbf{a}_{\beta} = \varepsilon_{\alpha\beta} \mathbf{a}^{\alpha} \mathbf{a}^{\beta} \tag{3-1}$$

where we introduce

$$\varepsilon^{\alpha\beta} \equiv \frac{1}{\sqrt{a}} e^{\alpha\beta} \tag{3-2}$$

and

$$\varepsilon_{\alpha\beta} \equiv \sqrt{a}\, e_{\alpha\beta} \tag{3-3}$$

The symbols $e^{\alpha\beta}$ and $e_{\alpha\beta}$ are defined in Sec. A.2.1.

If $\boldsymbol{\lambda}$ is any tangential vector field

$$\boldsymbol{\lambda} = \lambda^{\alpha} \mathbf{a}_{\alpha} \tag{3-4}$$

we shall define

$$\mathbf{v} \equiv - \boldsymbol{\varepsilon}^{(\sigma)} \cdot \boldsymbol{\lambda} \tag{3-5}$$

We can observe that the tangential vector field $\mathbf{v}$ has the same length as the tangential vector field $\boldsymbol{\lambda}$:

$$\begin{aligned} \mathbf{v} \cdot \mathbf{v} &= (\varepsilon^{\alpha\beta} \lambda_{\beta} \mathbf{a}_{\alpha}) \cdot (\varepsilon^{\gamma\zeta} \lambda_{\zeta} \mathbf{a}_{\gamma}) \\ &= \varepsilon^{\alpha\beta} \varepsilon^{\gamma\zeta} a_{\alpha\gamma} \lambda_{\beta} \lambda_{\zeta} \\ &= a^{\beta\zeta} \lambda_{\beta} \lambda_{\zeta} \\ &= \boldsymbol{\lambda} \cdot \boldsymbol{\lambda} \end{aligned} \tag{3-6}$$

More important, $\mathbf{v}$ is orthogonal to $\boldsymbol{\lambda}$,

$$\begin{aligned} \mathbf{v} \cdot \boldsymbol{\lambda} &= (- \varepsilon^{\alpha\beta} \lambda_{\beta} \mathbf{a}_{\alpha}) \cdot (\lambda^{\gamma} \mathbf{a}_{\gamma}) \\ &= - \varepsilon^{\alpha\beta} \lambda_{\alpha} \lambda_{\beta} \\ &= - \frac{1}{2} (\varepsilon^{\alpha\beta} + \varepsilon^{\beta\alpha}) \lambda_{\alpha} \lambda_{\beta} \\ &= 0 \end{aligned} \tag{3-7}$$

---

a) The superscript $^{(\sigma)}$ is used in order to distinguish this from the closely related third-order tensor (see Sec. A.1.2 or Slattery 1981, p. 648)

$$\boldsymbol{\varepsilon} = \varepsilon^{ijk} \mathbf{g}_i \mathbf{g}_j \mathbf{g}_k = \varepsilon_{ijk} \mathbf{g}^i \mathbf{g}^j \mathbf{g}^k$$

Because of this last property, we shall refer to $\boldsymbol{\varepsilon}^{(\sigma)}$ as the **tangential cross tensor**.

Notice that

$$(-\boldsymbol{\varepsilon}^{(\sigma)} \cdot \mathbf{a}_1) \cdot \mathbf{a}_2 = \varepsilon^{\alpha\beta} a_{\alpha 1} a_{\beta 2}$$
$$= \sqrt{a} \tag{3-8}$$

This operation is illustrated in Figure A.3.3-1.

If $\boldsymbol{\lambda}$ and $\boldsymbol{\mu}$ are any two tangential vector fields, we define the **rotation** from $\boldsymbol{\lambda}$ to $\boldsymbol{\mu}$ to be **positive**, if $(-\boldsymbol{\varepsilon}^{(\sigma)} \cdot \boldsymbol{\lambda}) \cdot \boldsymbol{\mu}$ is positive. In Figs. A.3.3-2 and A.3.3-3, consider the angle between $\boldsymbol{\lambda}$ and $\boldsymbol{\mu}$ that is measured in the same direction as the angle between $\mathbf{a}_1$ and $\mathbf{a}_2$. When this angle is less than 180°, the rotation is referred to as positive. When it is greater than 180° or negative, we say that the rotation is negative.

Note that the rotation from $\boldsymbol{\lambda}$ to $\mathbf{v}$ must always be positive:

$$(-\boldsymbol{\varepsilon}^{(\sigma)} \cdot \boldsymbol{\lambda}) \cdot \mathbf{v} = \mathbf{v} \cdot \mathbf{v} \tag{3-9}$$

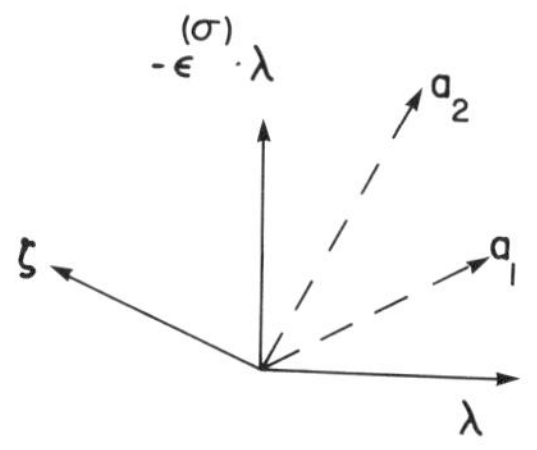

Figure A.3.3-1: Rotation from $\mathbf{a}_1$ to $\mathbf{a}_2$ is positive.

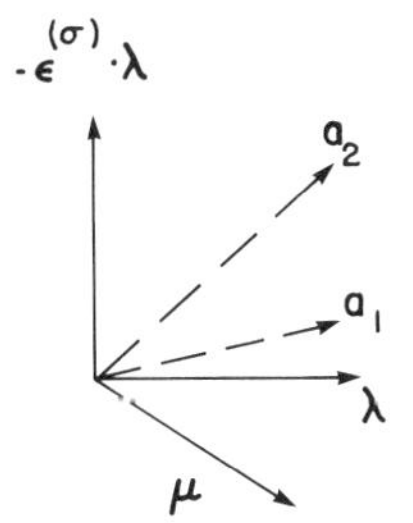

Figure A.3.3-2: Rotation from $\boldsymbol{\lambda}$ to $\boldsymbol{\mu}$ is negative.

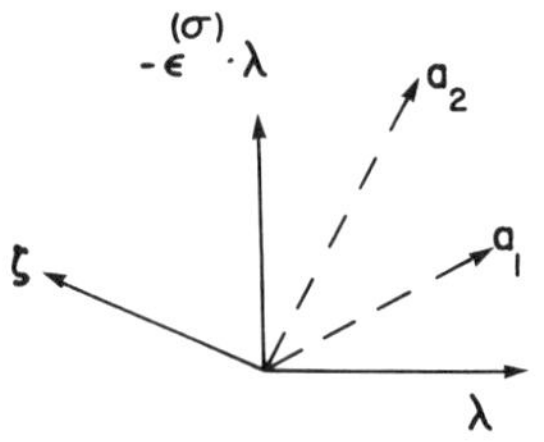

Figure A.3.3-3: Rotation from $\boldsymbol{\lambda}$ to $\boldsymbol{\zeta}$ is positive.

References

McConnell, A. J., "Applications of Tensor Analysis" Dover, New York (1957).

Slattery, J. C., "Momentum, Energy, and Mass Transfer in Continua," McGraw-Hill, New York (1972); second edition, Robert E. Krieger, Malabar, FL 32950 (1981).

Exercises

A.3.3-1 Prove that the two expressions for $\boldsymbol{\varepsilon}^{(\sigma)}$ given in (3-1) are equivalent.

A.3.3-2 At any point (y^1, y^2) on a surface, two tangential vector fields $\boldsymbol{\lambda}$ and $\boldsymbol{\mu}$ may be viewed as forming two edges of a parallelogram. If the rotation from $\boldsymbol{\lambda}$ to $\boldsymbol{\mu}$ is positive, show that $\boldsymbol{\lambda} \cdot (\boldsymbol{\varepsilon}^{(\sigma)} \cdot \boldsymbol{\mu})$ determines the area of the corresponding parallelogram.

A.3.3-3 Let $\boldsymbol{\lambda}$ and $\boldsymbol{\mu}$ be unit tangent vector fields.

a) If at any point (y^1, y^2) on a surface the rotation from $\boldsymbol{\lambda}$ to $\boldsymbol{\mu}$ is positive, prove that

$$\boldsymbol{\lambda} \cdot (\boldsymbol{\varepsilon}^{(\sigma)} \cdot \boldsymbol{\mu}) = \sin \theta$$

where θ is the angle measured between the two directions.

b) Conclude that at this point

$$\boldsymbol{\xi} \, [\boldsymbol{\lambda} \cdot (\boldsymbol{\varepsilon}^{(\sigma)} \cdot \boldsymbol{\mu})] = \boldsymbol{\lambda} \wedge \boldsymbol{\mu}$$

or

$$\xi_i \varepsilon_{\alpha\beta} = \varepsilon_{ijk} \frac{\partial x^j}{\partial y^\alpha} \frac{\partial x^k}{\partial y^\beta}$$

where $\boldsymbol{\xi}$ is the unit vector normal to the surface.

c) Prove that

$$\varepsilon^{\alpha\beta} \frac{\partial x^j}{\partial y^\alpha} \frac{\partial x^k}{\partial y^\beta} = \varepsilon^{ijk} \xi_i$$

A.3.3-4 Prove that, upon a change of surface coordinate systems, $\varepsilon^{\alpha\beta}$ and $\varepsilon_{\alpha\beta}$ transform according to the rules appropriate to the contravariant and covariant components of a tangential second-order tensor field (see Exercise A.3.3-1).

Hint: Start with

$$e^{\alpha\beta} \det\left(\frac{\partial \overline{y}^\phi}{\partial y^\gamma}\right) = e^{\mu\nu} \frac{\partial \overline{y}^\alpha}{\partial y^\mu} \frac{\partial \overline{y}^\beta}{\partial y^\nu}$$

$$e_{\alpha\beta} \det\left(\frac{\partial y^\gamma}{\partial \overline{y}^\phi}\right) = e_{\mu\nu} \frac{\partial y^\mu}{\partial \overline{y}^\alpha} \frac{\partial y^\nu}{\partial \overline{y}^\beta}$$

and with Exercise A.2.1-3. You may also find it helpful to review the rules for determinants (Slattery 1981, p. 611; McConnell 1957, p. 10).

A.3.3-5 Prove that

$$-\boldsymbol{\varepsilon}^{(\sigma)} \cdot \boldsymbol{\varepsilon}^{(\sigma)} = \mathbf{P}$$

A.3.3-6 *relations between natural and dual basis fields* Prove that

$$-\boldsymbol{\varepsilon}^{(\sigma)} \cdot \mathbf{a}_1 = \sqrt{a}\, \mathbf{a}^2$$

and

$$-\boldsymbol{\varepsilon}^{(\sigma)} \cdot \mathbf{a}_2 = -\sqrt{a}\, \mathbf{a}^1$$

This implies that

$$| \mathbf{a}^2 | = \frac{1}{\sqrt{a}} | \mathbf{a}_1 | \tag{1}$$

that $\mathbf{a}^2$ is orthogonal to $\mathbf{a}_1$, and that the rotation from $\mathbf{a}_1$ to $\mathbf{a}^2$ is positive. In a similar manner, we conclude that

$$| \mathbf{a}^1 | = \frac{1}{\sqrt{a}} | \mathbf{a}_2 | \tag{2}$$

that $\mathbf{a}^1$ is orthogonal to $\mathbf{a}_2$, and that the rotation from $\mathbf{a}_2$ to $\mathbf{a}^1$ is negative. It is also interesting to observe that (1) and (2) imply

$$a^{22} = \frac{a_{11}}{a}$$

$$a^{11} = \frac{a_{22}}{a}$$

A.3.4 Transpose The **transpose** of any second-order tensor field **A** is defined to be that second-order tensor field $\mathbf{A}^T$ such that, if **u** and **v** are any two spatial vector fields (Slattery 1981, p. 632)

$$(\mathbf{A} \cdot \mathbf{u}) \cdot \mathbf{v} = \mathbf{u} \cdot (\mathbf{A}^T \cdot \mathbf{v}) \tag{4-1}$$

Let us now consider the particular case of a tangential transformation $\mathcal{T}$:

$$\boldsymbol{\mathcal{T}} = \mathcal{T}_{A\beta}\, \bar{\mathbf{a}}^A\, \mathbf{a}^\beta \tag{4-2}$$

To determine the relation of $\boldsymbol{\mathcal{T}}^T$ to $\boldsymbol{\mathcal{T}}$, let

$$\mathbf{u} = \mathbf{a}_\gamma$$

$$\mathbf{v} = \bar{\mathbf{a}}_C$$

$$\boldsymbol{\mathcal{T}}^T = \mathcal{T}^T_{\mu N}\, \mathbf{a}^\mu\, \bar{\mathbf{a}}^N \tag{4-3}$$

From (4-1), it follows that

$$(\mathcal{T} \cdot \mathbf{a}_\gamma) \cdot \bar{\mathbf{a}}_C = \mathbf{a}_\gamma \cdot (\mathcal{T}^T \cdot \bar{\mathbf{a}}_C)$$

$$(\mathcal{T}_{A\beta} \bar{\mathbf{a}}^A \mathbf{a}^\beta \cdot \mathbf{a}_\gamma) \cdot \bar{\mathbf{a}}_C = \mathbf{a}_\gamma \cdot (\mathcal{T}^T_{\mu N} \mathbf{a}^\mu \bar{\mathbf{a}}^N \cdot \bar{\mathbf{a}}_C)$$

$$\mathcal{T}_{C\gamma} = \mathcal{T}^T_{\gamma C} \tag{4-4}$$

If $\mathcal{T}$ is represented by (4-2), then its transpose takes the form

$$\mathcal{T}^T = \mathcal{T}_{A\beta} \mathbf{a}^\beta \bar{\mathbf{a}}^A \tag{4-5}$$

Reference

Slattery, J. C., "Momentum, Energy, and Mass Transfer in Continua," McGraw-Hill, New York (1972); second edition, Robert E. Krieger, Malabar, FL 32950 (1981).

A.3.5 Inverse A second-order tensor field **A** is **invertible**, when the following conditions are satisfied.

(1) If $\mathbf{u}_1$ and $\mathbf{u}_2$ are spatial vector fields such that $\mathbf{A} \cdot \mathbf{u}_1 = \mathbf{A} \cdot \mathbf{u}_2$, then $\mathbf{u}_1 = \mathbf{u}_2$.

(2) There corresponds to every spatial vector field **v** at least one spatial vector field **u** such that $\mathbf{A} \cdot \mathbf{u} = \mathbf{v}$.

If **A** is invertible, we define as follows a second-order tensor field $\mathbf{A}^{-1}$, called the inverse of **A**. If $\mathbf{v}_1$ is any spatial vector field, by property (2) we may find a spatial vector field $\mathbf{u}_1$ for which $\mathbf{A} \cdot \mathbf{u}_1 = \mathbf{v}_1$. Say that $\mathbf{u}_1$ is not uniquely determined, such that $\mathbf{v}_1 = \mathbf{A} \cdot \mathbf{u}_1 = \mathbf{A} \cdot \mathbf{u}_2$. By property (1), $\mathbf{u}_1 = \mathbf{u}_2$ and we have a contradiction. The spatial vector field $\mathbf{u}_1$ is uniquely determined. We define $\mathbf{A}^{-1} \cdot \mathbf{v}_1$ to be $\mathbf{u}_1$.

It can easily be shown (Slattery 1981, p. 636) that $\mathbf{A}^{-1}$ satisfies the linearity rules (1-1) for a second-order tensor and that

$$\mathbf{A}^{-1} \cdot \mathbf{A} = \mathbf{A} \cdot \mathbf{A}^{-1} = \mathbf{I} \tag{5-1}$$

for some second-order tensor field $\mathbf{A}^{-1}$, if and only if **A** is invertible. Here **I** is the identity tensor.

Note that a tangential transformation is not invertible in this sense, since it transforms every spatial vector field normal to a surface into the zero vector. However, the concept of an inverse remains valid so long as we restrict our attention to tangential vector fields.

Let $\mathcal{T}$ be a tangential transformation from the space of tangential vector fields on Σ to the space of tangential vector fields $\bar{\Sigma}$. We will say that $\mathcal{T}$ is **invertible** or **nonsingular**, when the following two conditions are met.

(a) If $\mathbf{c}_1$ and $\mathbf{c}_2$ are tangential vector fields on Σ such that $\mathcal{T} \cdot \mathbf{c}_1 = \mathcal{T} \cdot \mathbf{c}_2$, then $\mathbf{c}_1 = \mathbf{c}_2$.

(b) There corresponds to every tangential vector field $\mathbf{d}$ on $\bar{\Sigma}$ at least one tangential vector field $\mathbf{c}$ on Σ such that $\mathcal{T} \cdot \mathbf{c} = \mathbf{d}$.

We define the tangential transformation $\mathcal{T}^{-1}$, the **inverse** of $\mathcal{T}$, in the following way. If $\mathbf{d}_1$ is any tangential vector field on $\bar{\Sigma}$, by property (b) we may find a tangential vector field $\mathbf{c}_1$ on Σ for which $\mathcal{T} \cdot \mathbf{c}_1 = \mathbf{d}_1$. Say that $\mathbf{c}_1$ is not uniquely determined, such that $\mathbf{d}_1 = \mathcal{T} \cdot \mathbf{c}_1 = \mathcal{T} \cdot \mathbf{c}_2$. By property (a), $\mathbf{c}_1 = \mathbf{c}_2$ and we have a contradiction. The tangential vector field $\mathbf{c}_1$ is uniquely determined. We define $\mathcal{T}^{-1} \cdot \mathbf{d}_1$ to be $\mathbf{c}_1$. Observe that $\mathcal{T}^{-1}$ is a tangential transformation from the space of tangential vector fields on $\bar{\Sigma}$ to the space of tangential vector fields on Σ.

In order to prove that $\mathcal{T}^{-1}$ satisfies the linearity rules for a second-order tensor field, we may evaluate $\mathcal{T}^{-1} \cdot (\alpha_1 \mathbf{c}_1 + \alpha_2 \mathbf{c}_2)$ where α_1 and α_2 are real scalar fields. If $\mathcal{T} \cdot \mathbf{c}_1 = \mathbf{d}_1$ and $\mathcal{T} \cdot \mathbf{c}_2 = \mathbf{d}_2$, we have

$$
\begin{aligned}
\mathcal{T} \cdot (\alpha_1 \mathbf{c}_1 + \alpha_2 \mathbf{c}_2) &= \alpha_1 \mathcal{T} \cdot \mathbf{c}_1 + \alpha_2 \mathcal{T} \cdot \mathbf{c}_2 \\
&= \alpha_1 \mathbf{d}_1 + \alpha_2 \mathbf{d}_2
\end{aligned}
\tag{5-2}
$$

This means that

$$
\begin{aligned}
\mathcal{T}^{-1} \cdot (\alpha_1 \mathbf{d}_1 + \alpha_2 \mathbf{d}_2) &= \alpha_1 \mathbf{c}_1 + \alpha_2 \mathbf{c}_2 \\
&= \alpha_1 \mathcal{T}^{-1} \cdot \mathbf{d}_1 + \alpha_2 \mathcal{T}^{-1} \cdot \mathbf{d}_2
\end{aligned}
\tag{5-3}
$$

It follows immediately from the definition that, for any invertible tangential transformation $\mathcal{T}$,

$$
\begin{aligned}
&\mathcal{T}^{-1} \cdot \mathcal{T} = \mathbf{P} \\
&\mathcal{T} \cdot \mathcal{T}^{-1} = \bar{\mathbf{P}}
\end{aligned}
\tag{5-4}
$$

Here $\mathbf{P}$ is the projection tensor for the space of tangential vector fields on Σ; $\bar{\mathbf{P}}$ is the projection tensor for the space of tangential vector fields on $\bar{\Sigma}$.

Finally, we can show that (5-4) is valid for some tangential transformation $\mathcal{T}^{-1}$, if and only if $\mathcal{T}$ is invertible. If $\mathcal{T}$, $\mathcal{B}$, and $\mathcal{C}$ are tangential transformations such that

$$\mathcal{B} \cdot \mathcal{T} = \mathbf{P}$$

$$\mathcal{T} \cdot \mathcal{C} = \overline{\mathbf{P}} \tag{5-5}$$

we will show that $\mathcal{T}$ is invertible and that $\mathcal{T}^{-1} = \mathcal{B} = \mathcal{C}$. If $\mathcal{T} \cdot \mathbf{c}_1 = \mathcal{T} \cdot \mathbf{c}_2$, we have from (5-5)

$$(\mathcal{B} \cdot \mathcal{T}) \cdot \mathbf{c}_1 = (\mathcal{B} \cdot \mathcal{T}) \cdot \mathbf{c}_2$$

$$\mathbf{c}_1 = \mathbf{c}_2 \tag{5-6}$$

This fulfills property (a) of an invertible tangential transformation. If $\mathbf{d}$ is any tangential vector field on Σ and if $\mathbf{c} = \mathcal{C} \cdot \mathbf{d}$, then by (5-5)

$$\mathcal{T} \cdot \mathbf{c} = (\mathcal{T} \cdot \mathcal{C}) \cdot \mathbf{d}$$

$$= \mathbf{d} \tag{5-7}$$

and property (b) is also satisfied. If $\mathcal{T}$ is invertible, from (5-5)

$$\mathcal{B} \cdot \mathcal{T} \cdot \mathcal{T}^{-1} = \mathcal{T}^{-1}$$

$$\mathcal{B} = \mathcal{T}^{-1} \tag{5-8}$$

and

$$\mathcal{T}^{-1} \cdot \mathcal{T} \cdot \mathcal{C} = \mathcal{T}^{-1}$$

$$\mathcal{C} = \mathcal{T}^{-1} \tag{5-9}$$

References

Slattery, J. C., "Momentum, Energy, and Mass Transfer in Continua," McGraw-Hill, New York (1972); second edition, Robert E. Krieger, Malabar, FL 32950 (1981).

A.3.6 Orthogonal tangential transformation An **orthogonal tensor** field is one that preserves lengths and angles in V^3. If $\mathbf{u}$ and $\mathbf{v}$ are any two spatial vector fields and $\mathbf{Q}$ is an orthogonal tensor field, we require

$$(\mathbf{Q} \cdot \mathbf{u}) \cdot (\mathbf{Q} \cdot \mathbf{v}) = \mathbf{u} \cdot \mathbf{v} \tag{6-1}$$

But this means that

$$\begin{aligned} [\mathbf{Q}^T \cdot (\mathbf{Q} \cdot \mathbf{u})] \cdot \mathbf{v} &= [(\mathbf{Q}^T \cdot \mathbf{Q}) \cdot \mathbf{u}] \cdot \mathbf{v} \\ &= \mathbf{u} \cdot \mathbf{v} \end{aligned} \tag{6-2}$$

or

$$[(\mathbf{Q}^T \cdot \mathbf{Q}) \cdot \mathbf{u} - \mathbf{u}] \cdot \mathbf{v} = 0 \tag{6-3}$$

Since **u** and **v** are arbitrary spatial vector fields, we conclude that

$$(\mathbf{Q}^T \cdot \mathbf{Q}) \cdot \mathbf{u} = \mathbf{u} \tag{6-4}$$

and

$$\mathbf{Q}^T \cdot \mathbf{Q} = \mathbf{I} \tag{6-5}$$

where **I** is the identity tensor (Slattery 1981, p. 628). By taking the transpose of (6-5), we have

$$\mathbf{Q} \cdot \mathbf{Q}^T = \mathbf{Q}^T \cdot \mathbf{Q} = \mathbf{I} \tag{6-6}$$

It follows easily that **Q** is an orthogonal tensor, if and only if (6-6) is satisfied.

In much the same way, we can define an **orthogonal tangential transformation** to be one that preserves lengths and angles in sending every tangential vector field on a surface Σ into another tangential vector field on a surface $\overline{\Sigma}$. Like every tangential transformation, it sends every spatial vector field normal to the surface Σ into the zero vector. If **c** and **d** are any two tangential vector fields on Σ and $\mathfrak{R}$ is an orthogonal tangential transformation, we demand

$$(\mathfrak{R} \cdot \mathbf{c}) \cdot (\mathfrak{R} \cdot \mathbf{d}) = \mathbf{c} \cdot \mathbf{d} \tag{6-7}$$

We have consequently

$$\begin{aligned} [\mathfrak{R}^T \cdot (\mathfrak{R} \cdot \mathbf{c})] \cdot \mathbf{d} &= [(\mathfrak{R}^T \cdot \mathfrak{R}) \cdot \mathbf{c}] \cdot \mathbf{d} \\ &= \mathbf{c} \cdot \mathbf{d} \end{aligned} \tag{6-8}$$

or

$$[(\mathfrak{R}^T \cdot \mathfrak{R}) \cdot \mathbf{c} - \mathbf{c}] \cdot \mathbf{d} = 0 \tag{6-9}$$

Since **c** and **d** are arbitrary tangential vector fields on Σ, we find that

$$(\mathfrak{R}^T \cdot \mathfrak{R}) \cdot \mathbf{c} = \mathbf{c} \tag{6-10}$$

and

$$\mathfrak{R}^T \cdot \mathfrak{R} = \mathbf{P} \tag{6-11}$$

where **P** is the projection tensor for tangential vector fields on Σ.

It is easy to conclude that $\mathfrak{R}$ is an orthogonal tangential transformation, if and only if (6-11) is satisfied.

A similar argument can be constructed to find that $\mathfrak{R}^T$ is an orthogonal tangential transformation, if and only if

$$\mathfrak{R} \cdot \mathfrak{R}^T = \overline{\mathbf{P}} \tag{6-12}$$

Here $\overline{\mathbf{P}}$ is the projection tensor for tangential vector fields on $\overline{\Sigma}$.

From Sec. A.3.5, we see that an orthogonal tangential transformation $\mathfrak{R}$ is invertible, if and only if both (6-11) and (6-12) are satisfied.

Reference

Slattery, J. C., "Momentum, Energy, and Mass Transfer in Continua," McGraw-Hill, New York (1972); second edition, Robert E. Krieger, Malabar, FL 32950 (1981).

Exercise

A.3.6-1 *isotropic tangential tensors* Let **A** be a tangential tensor field and $\mathcal{Q}$ an orthogonal tangential transformation. If

$$\mathcal{Q} \cdot \mathbf{A} \cdot \mathcal{Q}^T = \mathbf{A}$$

for all orthogonal tangential transformations $\mathcal{Q}$ we will refer to **A** as an isotropic tangential tensor. Prove that

$$\mathbf{A} = \alpha \, \mathbf{P}$$

where α is a scalar field.

Hint: Let [$\mathcal{Q}$] denote the matrix (array) of the components of $\mathcal{Q}$ with respect to an appropriate basis.

i) Let

$$[\mathcal{Q}] = \begin{bmatrix} 1 & 0 \\ 0 & -1 \end{bmatrix}$$

to conclude that $A^1_{.2} = A^2_{.1} = 0$.

ii) Let

$$[\mathcal{Q}] = \begin{bmatrix} 0 & 1 \\ 1 & 0 \end{bmatrix}$$

to find that $A^1_{.1} = A^2_{.2}$.

A.3.7 Surface determinant of tangential transformation Let $\mathcal{T}$ be a tangential transformation that assigns to every tangential vector field $\mathbf{c}$ on Σ another tangential field $\mathcal{T} \cdot \mathbf{c}$ on $\overline{\Sigma}$. Let $\mathbf{a}_\alpha$ ($\alpha = 1, 2$) be the natural basis fields corresponding to an arbitrary surface coordinate system on Σ. At any point on Σ, the basis vectors may be thought of as two edges of a parallelogram as shown in Fig. A.3.7-1a. The vectors $\mathcal{T} \cdot \mathbf{a}_1$ and $\mathcal{T} \cdot \mathbf{a}_2$ also form two edges of a parallelogram at the corresponding point on $\overline{\Sigma}$ as indicated in Fig. A.3.7-1b. We define the magnitude of the surface determinant of a tangential transformation $\mathcal{T}$ at any point on Σ as the ratio of the area of the parallelogram spanned by $\mathcal{T} \cdot \mathbf{a}_1$ and $\mathcal{T} \cdot \mathbf{a}_2$ to the area of the parallelogram spanned by $\mathbf{a}_1$ and $\mathbf{a}_2$. We choose the sign of the surface determinant of $\mathcal{T}$ to be positive, if the rotation from $\mathcal{T} \cdot \mathbf{a}_1$ to $\mathcal{T} \cdot \mathbf{a}_2$ is positive; the sign is negative, if the rotation from $\mathcal{T} \cdot \mathbf{a}_1$ to $\mathcal{T} \cdot \mathbf{a}_2$ is negative (see Sec. A.3.3).

This together with Exercise A.3.3-2 suggests that we take as our formal definition of the **surface determinant** $\det_{(\sigma)}\mathcal{T}$:

$$\det_{(\sigma)}\mathcal{T} \equiv \frac{(\mathcal{T} \cdot \mathbf{a}_1) \cdot [\overline{\boldsymbol{\varepsilon}}^{(\sigma)} \cdot (\mathcal{T} \cdot \mathbf{a}_2)]}{\mathbf{a}_1 \cdot (\boldsymbol{\varepsilon}^{(\sigma)} \cdot \mathbf{a}_2)} \tag{7-1}$$

Here $\overline{\boldsymbol{\varepsilon}}^{(\sigma)}$ and $\boldsymbol{\varepsilon}^{(\sigma)}$ are the tangential cross tensors appropriate to $\overline{\Sigma}$ and Σ respectively. Observe that $\det_{(\sigma)}\mathcal{T}$ is a scalar field on Σ.

Expressing $\mathcal{T}$ in terms of its covariant components, we can rewrite (7-1) as

$$\det_{(\sigma)}\mathcal{T} = \frac{(\mathcal{T}_{A1}\overline{\mathbf{a}}^A) \cdot [\overline{\varepsilon}^{BC}\overline{\mathbf{a}}_B\overline{\mathbf{a}}_C \cdot (\mathcal{T}_{M2}\overline{\mathbf{a}}^M)]}{\mathbf{a}_1 \cdot (\varepsilon_{\mu\nu}\mathbf{a}^\mu\mathbf{a}^\nu \cdot \mathbf{a}_2)}$$

$$= \frac{\mathcal{T}_{B\,1}\,\overline{\varepsilon}^{B\,C}\,\mathcal{T}_{C\,2}}{\varepsilon_{1\,2}} \tag{7-2}$$

or

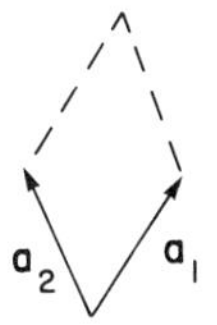

Figure A.3.7-1a: At any point on Σ, the natural basis vectors $\mathbf{a}_1$ and $\mathbf{a}_2$ form two edges of a parallelogram.

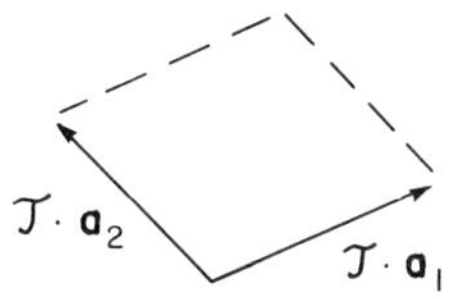

Figure A.3.7-1b: At the corresponding point on $\overline{\Sigma}$, $\mathcal{T} \cdot \mathbf{a}_1$ and $\mathcal{T} \cdot \mathbf{a}_2$ form two edges of a parallelogram.

$$\overline{\varepsilon}^{A\,B}\,\mathcal{T}_{A\,1}\,\mathcal{T}_{B\,2} = \varepsilon_{1\,2}\ \det_{(\sigma)}\boldsymbol{\mathcal{T}} \tag{7-3}$$

This implies (Slattery 1981, p. 611)

$$\overline{\varepsilon}^{A\,B}\,\mathcal{T}_{A\,\mu}\,\mathcal{T}_{B\,\nu} = \varepsilon_{\mu\,\nu}\det_{(\sigma)}\boldsymbol{\mathcal{T}} \tag{7-4}$$

A similar result can be obtained in terms of the contravariant components of $\boldsymbol{\mathcal{T}}$,

$$\overline{\varepsilon}_{A\,B}\,\mathcal{T}^{A\,\mu}\,\mathcal{T}^{B\,\nu} = \varepsilon^{\mu\,\nu}\det_{(\sigma)}\boldsymbol{\mathcal{T}} \tag{7-5}$$

For computations, it will often be more convenient to use one of these expressions:

$$\det_{(\sigma)}\boldsymbol{\mathcal{T}} = \frac{1}{2}\,\overline{\varepsilon}^{A\,B}\,\varepsilon^{\mu\,\nu}\,\mathcal{T}_{A\,\mu}\,\mathcal{T}_{B\,\nu} \tag{7-6}$$

$$\det_{(\sigma)}\boldsymbol{\mathcal{T}} = \frac{1}{2}\,\overline{\varepsilon}_{A\,B}\,\varepsilon_{\mu\,\nu}\,\mathcal{T}^{A\,\mu}\,\mathcal{T}^{B\,\nu} \tag{7-7}$$

$$\det_{(\sigma)}\mathcal{T} = \frac{1}{2}\,(\,\mathcal{T}\cdot \mathbf{a}_\alpha\,)\cdot[\,\boldsymbol{\varepsilon}^{(\sigma)}\cdot(\,\mathcal{T}\cdot \mathbf{a}_\beta\,)]\,\varepsilon^{\alpha\beta} \tag{7-8}$$

It follows that

$$\det_{(\sigma)}(\,\mathcal{S}\cdot\mathcal{T}^{\mathrm{T}}\,) = (\,\det_{(\sigma)}\mathcal{S}\,)(\,\det_{(\sigma)}\mathcal{T}\,) \tag{7-9}$$

$$\det_{(\sigma)}(\,\mathcal{T}^{-1}\,) = \frac{1}{\det_{(\sigma)}\mathcal{T}} \tag{7-10}$$

and

$$\det_{(\sigma)}(\,\mathcal{T}^{\mathrm{T}}\,) = \det_{(\sigma)}\mathcal{T} \tag{7-11}$$

Here $\mathcal{T}^{-1}$ and $\mathcal{T}^{\mathrm{T}}$ are respectively the inverse and transpose of $\mathcal{T}$.

For the special case of a tangential tensor field $\mathbf{T}$, $\overline{\mathrm{V}}^2 = \mathrm{V}^2$ and (7-1) becomes

$$\det_{(\sigma)}\mathbf{T} \equiv \frac{(\,\mathbf{T}\cdot\mathbf{a}_1\,)\cdot[\,\boldsymbol{\varepsilon}^{(\sigma)}\cdot(\,\mathbf{T}\cdot\mathbf{a}_2\,)]}{\mathbf{a}_1\cdot(\,\boldsymbol{\varepsilon}^{(\sigma)}\cdot\mathbf{a}_2\,)} \tag{7-12}$$

Reference

Slattery, J. C., "Momentum, Energy, and Mass Transfer in Continua," McGraw-Hill, New York (1972); second edition, Robert E. Krieger, Malabar, FL 32950 (1981).

Exercises

A.3.7-1 Prove (7-9), (7-10), and (7-11).

A.3.7-2 *projection tensor*

i) Prove that

$$\det_{(\sigma)}\mathbf{P} = 1$$

ii) Derive (A.2.1-5) and (A.2.1-6).

A.3.7-3 *determinant of the components of a tangential transformation* Let $\boldsymbol{\mathcal{T}}$ be a tangential transformation from the space of tangential vector fields on Σ to the space of tangential vector fields on $\overline{\Sigma}$. Prove that

$$\det(\mathcal{T}_{A\beta}) = \sqrt{\overline{a}}\,\sqrt{a}\,\det_{(\sigma)}\boldsymbol{\mathcal{T}}$$

and

$$\det(\mathcal{T}^{A}_{.\beta}) = \frac{\sqrt{a}}{\sqrt{\overline{a}}}\det_{(\sigma)}\boldsymbol{\mathcal{T}}$$

Here $\det(\mathcal{T}_{A\beta})$ is the determinant whose typical element is $\mathcal{T}_{A\beta}$. Recall from Sec. A.2.1 that on Σ

$$a \equiv \det(a_{\alpha\beta})$$

Similarly, on $\overline{\Sigma}$ we have

$$\overline{a} \equiv \det(\overline{a}_{AB})$$

A.3.7-4 *surface determinant of a sum of tangential tensors* Let the $S^{\alpha}_{.\beta}$ and the $T^{\alpha}_{.\beta}$ (α, β = 1, 2) be the mixed components of two tangential tensor fields **S** and **T**. We denote by $\det(S^{\alpha}_{.\beta} + T^{\alpha}_{.\beta})$ the determinant whose typical element is $S^{\alpha}_{.\beta} + T^{\alpha}_{.\beta}$.

i) Prove that

$$\det(S^{\alpha}_{.\beta} + T^{\alpha}_{.\beta}) = \det(S^{\alpha}_{.\beta}) + \det(T^{\alpha}_{.\beta}) - \operatorname{tr}(\mathbf{S} \cdot \mathbf{T}) + (\operatorname{tr} \mathbf{S})(\operatorname{tr} \mathbf{T})$$

ii) Conclude that

$$\det_{(\sigma)}(\mathbf{S} + \mathbf{T}) = \det_{(\sigma)}\mathbf{S} + \det_{(\sigma)}\mathbf{T} - \operatorname{tr}(\mathbf{S} \cdot \mathbf{T}) + (\operatorname{tr} \mathbf{S})(\operatorname{tr} \mathbf{T})$$

and

$$\det(S_{\alpha\beta} + T_{\alpha\beta}) = \det (S_{\alpha\beta}) + \det(T_{\alpha\beta}) - a \operatorname{tr}(\mathbf{S} \cdot \mathbf{T}) + a(\operatorname{tr} \mathbf{S})(\operatorname{tr} \mathbf{T})$$

A.3.8 Polar decomposition The polar decomposition theorem is generally stated and proved for a linear transformation on a finite-dimensional inner product space (Halmos 1958, p. 169; Ericksen 1960, p. 841). Because of the importance of this theorem to the development of the simple surface fluid model in the text, I will give a proof explicitly for tangential transformations.

Let $\mathcal{T}$ be a tangential transformation from the space of tangential vector fields on Σ to the space of tangential vector fields on $\bar{\Sigma}$. The polar decomposition theorem requires that any invertible, tangential transformation $\mathcal{T}$ may be written in two different forms

$$\begin{aligned}\mathcal{T} &= \mathcal{R} \cdot \mathbf{U} \\ &= \bar{\mathbf{V}} \cdot \mathcal{R}\end{aligned} \tag{8-1}$$

Here $\mathcal{R}$ is a unique orthogonal tangential transformation; $\mathbf{U}$ is a unique, positive-definite (Halmos 1958, p. 140), symmetric tangential tensor on Σ; $\bar{\mathbf{V}}$ is a unique, positive-definite, symmetric tangential tensor on $\bar{\Sigma}$.

Since $\mathcal{T}^T \cdot \mathcal{T}$ is a positive tangential tensor on Σ, we may find its unique positive square root (Halmos 1958, p. 166)

$$\mathbf{U} \equiv \sqrt{\mathcal{T}^T \cdot \mathcal{T}} \tag{8-2}$$

Let us define

$$\mathcal{M} \equiv \mathbf{U} \cdot \mathcal{T}^{-1} \tag{8-3}$$

Taking its transpose

$$\mathcal{M}^T = (\mathcal{T}^{-1})^T \cdot \mathbf{u}^T = (\mathcal{T}^T)^{-1} \cdot \mathbf{U} \tag{8-4}$$

and calculating

$$\begin{aligned}\mathcal{M}^T \cdot \mathcal{M} &= (\mathcal{T}^T)^{-1} \cdot \mathbf{U} \cdot \mathbf{U} \cdot \mathcal{T}^{-1} \\ &= (\mathcal{T}^T)^{-1} \cdot \mathcal{T}^T \cdot \mathcal{T} \cdot \mathcal{T}^{-1} \\ &= \bar{\mathbf{P}}\end{aligned} \tag{8-5}$$

we see that $\mathcal{M}$ is an orthogonal tangential transformation. Here $\bar{\mathbf{P}}$ is the projection tensor for the tangential vector fields on $\bar{\Sigma}$. We can take advantage of the fact that both $\mathcal{T}$ and $\mathcal{M}$ are invertible in (8-3) and write

$$\mathcal{T} = \mathcal{M}^{-1} \cdot \mathbf{U} \tag{8-6}$$

On comparing this statement with ($8\text{-}1_1$), we can identify $\mathcal{R}$ as the orthogonal tangential transformation

$$\mathcal{R} \equiv \mathcal{M}^{-1} = \mathcal{T} \cdot \mathrm{U}^{-1} \tag{8-7}$$

Let us now prove that $\mathcal{R}$ and **U** are unique. Observe that

$$\mathcal{R} \cdot \mathrm{U} = \mathrm{R}_0 \cdot \mathrm{U}_0 \tag{8-8}$$

implies

$$\mathrm{U} \cdot \mathcal{R}^{\mathrm{T}} = \mathrm{U}_0 \cdot \mathcal{R}_0^{\mathrm{T}} \tag{8-9}$$

and therefore

$$\begin{aligned} \mathrm{U}^2 &= \mathrm{U} \cdot \mathcal{R}^{\mathrm{T}} \cdot \mathcal{R} \cdot \mathrm{U} \\ &= \mathrm{U}_0 \cdot \mathcal{R}_0^{\mathrm{T}} \cdot \mathcal{R}_0 \cdot \mathrm{U}_0 \\ &= \mathrm{U}_0^2 \end{aligned} \tag{8-10}$$

Since the positive tangential tensor $\mathbf{U}^2$ has only one positive square root, it follows that

$$\mathrm{U} = \mathrm{U}_0 \tag{8-11}$$

Returning to (8-1_1), we can write

$$\mathrm{U} = \mathcal{R}^{-1} \cdot \mathcal{T} \tag{8-12}$$

and reason

$$\begin{aligned} \mathcal{R} \cdot \mathrm{U} &= \mathcal{R}_0 \cdot \mathrm{U}_0 \\ \mathcal{R} \cdot \mathrm{U} \cdot \mathrm{U}^{-1} &= \mathcal{R}_0 \cdot \mathrm{U}_0 \cdot \mathrm{U}_0^{-1} \\ \mathcal{R} &= \mathcal{R}_0 \end{aligned} \tag{8-13}$$

In the same way as above, we can obtain a unique decomposition

$$\mathcal{T} = \overline{\mathrm{V}} \cdot \mathcal{R}_1 \tag{8-14}$$

with

$$\overline{\mathrm{V}} \equiv \sqrt{\mathcal{T} \cdot \mathcal{T}^{\mathrm{T}}} \tag{8-15}$$

and

$$\mathcal{R}_1 \equiv \mathrm{U}^{-1} \cdot \mathcal{T} \tag{8-16}$$

Observe that

$$\mathcal{T} = \mathcal{R} \cdot \mathrm{U} \cdot \mathcal{R}^{\mathrm{T}} \cdot \mathcal{R} \tag{8-17}$$

Since both (8-14) and (8-17) are unique decompositions, we conclude

$$\overline{\mathrm{V}} = \mathcal{R} \cdot \mathrm{U} \cdot \mathcal{R}^{\mathrm{T}} \tag{8-18}$$

and

$$\mathcal{R}_1 = \mathcal{R} \tag{8-19}$$

Our proof of the polar decomposition theorem for invertible, tangential transformations is complete. I think that it is probably obvious to you what minor modifications in the statement and proof of this theorem would be necessary, if we were talking about second-order tensors on V^3.

References

Ericksen, J. L., "Handbuch der Physik," vol. 3/1, edited by S. Flügge, Springer-Verlag, Berlin (1960).

Halmos, P. R., "Finite-Dimensional Vector Spaces," second edition, Van Nostrand, Princeton, NJ (1958).

A.4 Third-order tensor fields

A.4.1 Surface tensors A third-order tensor field $\mathbf{B}$ is a transformation (or mapping or rule) that assigns to each given spatial vector field $\mathbf{v}$ a second-order tensor field $\mathbf{B} \cdot \mathbf{v}$ such that the rules

$$\mathbf{B} \cdot (\mathbf{v} + \mathbf{w}) = \mathbf{B} \cdot \mathbf{v} + \mathbf{B} \cdot \mathbf{w}$$

$$\mathbf{B} \cdot (\alpha \mathbf{v}) = \alpha (\mathbf{B} \cdot \mathbf{v}) \tag{1-1}$$

hold. By α we mean a real scalar field.

A third-order surface tensor field $\boldsymbol{\beta}$ is a particular type of third-order tensor field that is defined only on the surface, that assigns to each given tangential vector field $\mathbf{c}$ some second-order tensor $\boldsymbol{\beta} \cdot \mathbf{c}$, and that transforms every spatial vector field normal to the surface into the zero second-order tensor. If the $\mathbf{a}_\alpha$ ($\alpha = 1, 2$) are the natural basis fields associated with some surface coordinate system, and if the $\mathbf{g}_i$ (i = 1, 2, 3) are the natural basis fields associated with some spatial coordinate system, then we may write

$$\boldsymbol{\beta} \cdot \mathbf{a}_\alpha = \beta^{ij}_{\cdot\cdot\alpha} \mathbf{g}_i \mathbf{g}_j \tag{1-2}$$

Using an argument that is similar to that given for third-order tensor fields (Slattery 1981, p. 646), we conclude

$$\boldsymbol{\beta} = \beta^{ij}_{\cdot\cdot\alpha} \mathbf{g}_i \mathbf{g}_j \mathbf{a}^\alpha \tag{1-3}$$

Alternatively, we may say

$$\boldsymbol{\beta} = \beta^{ij\alpha} \mathbf{g}_i \mathbf{g}_j \mathbf{a}_\alpha = \beta_{ij\alpha} \mathbf{g}^i \mathbf{g}^j \mathbf{a}^\alpha \tag{1-4}$$

where we have defined

$$\beta^{ij\alpha} \equiv a^{\alpha\gamma} \beta^{ij}_{\cdot\cdot\gamma} \tag{1-5}$$

and

$$\beta_{ij\alpha} \equiv g_{im}\, g_{jn}\, \beta^{mn}_{\cdot\;\cdot\;\alpha} \tag{1-6}$$

Reference

Slattery, J. C., "Momentum, Energy, and Mass Transfer in Continua," McGraw-Hill, New York (1972); second edition, Robert E. Krieger, Malabar, FL 32950 (1981).

Exercise

A.4.1-1 *tangential tensor* A third-order **tangential** tensor field $\boldsymbol{\tau}$ is a particular type of third-order tensor field that assigns to each given tangential vector field $\mathbf{c}$ some second-order tangential tensor field $\boldsymbol{\tau} \cdot \mathbf{c}$ and that transforms every spatial vector field normal to the surface into the zero second-order tensor. Using an argument similar to that given in the text for third-order surface tensor fields, conclude that $\boldsymbol{\tau}$ takes the form

$$\boldsymbol{\tau} = \tau^{\alpha\beta\gamma}\mathbf{a}_\alpha \mathbf{a}_\beta \mathbf{a}_\gamma = \tau_{\alpha\beta\gamma}\mathbf{a}^\alpha \mathbf{a}^\beta \mathbf{a}^\gamma$$

A.5 Surface gradient

A.5.1 Spatial vector field Since the text deals with moving and deforming phase interfaces, you can appreciate that we will usually be concerned with situations in which velocity is a function of position. In attempting to describe the motion and deformation of a surface, derivatives of velocity with respect to position on the surface will arise quite naturally. The surface gradient, introduced in Sec. A.2.2 for scalar fields, is extended here to spatial vector fields.

The **surface gradient** of a spatial vector field $\mathbf{v}$ is a second-order surface tensor field denoted by $\nabla_{(\sigma)}\mathbf{v}$ and defined by how it transforms an arbitrary tangential vector field $\mathbf{c}$ at all points on the surface:

$$\nabla_{(\sigma)}\mathbf{v}(\, y^1 , y^2 \,) \cdot \mathbf{c}$$

$$\equiv \text{limit } s \to 0: \ \frac{1}{s} [\ \mathbf{v}(\, y^1 + sc^1 , y^2 + sc^2 \,) - \mathbf{v}(\, y^1 , y^2 \,)] \tag{1-1}$$

In a manner similar to that indicated in Sec. A.2.2, (1-1) may be rearranged in the form

$$\nabla_{(\sigma)}\mathbf{v} \cdot \mathbf{c} = \frac{\partial \mathbf{v}}{\partial y^\alpha} c^\alpha \tag{1-2}$$

Since $\mathbf{c}$ is an arbitrary tangential vector field, we can take $\mathbf{c} = \mathbf{a}_\beta$,

$$\nabla_{(\sigma)}\mathbf{v} \cdot \mathbf{a}_\beta = \frac{\partial \mathbf{v}}{\partial y^\beta} \tag{1-3}$$

We conclude that

$$\nabla_{(\sigma)}\mathbf{v} = \frac{\partial \mathbf{v}}{\partial y^\alpha} \mathbf{a}^\alpha \tag{1-4}$$

It is worthwhile to consider as an example the surface gradient of the position vector field $\mathbf{p}^{(\sigma)}$ introduced in Sec. A.1.3:

$$\nabla_{(\sigma)}\mathbf{p}^{(\sigma)} = \frac{\partial \mathbf{p}^{(\sigma)}}{\partial y^\alpha} \mathbf{a}^\alpha = \mathbf{a}_\alpha \mathbf{a}^\alpha = \mathbf{P} \tag{1-5}$$

This provides us with a useful alternative expression for the projection tensor.

The usual divergence operation defined for spatial vector fields (Slattery 1981, p. 639) suggests that we define the **surface divergence** as

$$\operatorname{div}_{(\sigma)}\mathbf{v} \equiv \operatorname{tr}(\nabla_{(\sigma)}\mathbf{v}) = \frac{\partial \mathbf{v}}{\partial y^{\alpha}} \cdot \mathbf{a}^{\alpha} \tag{1-6}$$

We have not as yet expressed the surface gradient and surface divergence in terms of the components of **v**. The spatial vector field **v** may be an explicit function of position in space or it may be an explicit function of position on a surface. We will develop these two cases separately in the next sections.

Reference

Slattery, J. C., "Momentum, Energy, and Mass Transfer in Continua," McGraw-Hill, New York (1972); second edition, Robert E. Krieger, Malabar, FL 32950 (1981).

A.5.2 Vector field is explicit function of position in space In Sec. A.5.1 we introduced the surface gradient of a spatial vector field **v**. We now wish to determine how this operation is to be expressed in terms of the components of **v**, with the assumption that it is an explicit function of position in space.

If **v** is an explicit function of the curvilinear coordinates (x^1, x^2, x^3)

$$\mathbf{v} = \mathbf{v}(x^1, x^2, x^3) \tag{2-1}$$

we see from (1-4) that the surface gradient becomes

$$\begin{aligned}\nabla_{(\sigma)}\mathbf{v} &= \frac{\partial \mathbf{v}}{\partial x^i}\frac{\partial x^i}{\partial y^{\alpha}}\,\mathbf{a}^{\alpha} \\ &= \frac{\partial \mathbf{v}}{\partial x^i}(\mathbf{g}^i \cdot \mathbf{a}_{\alpha})\mathbf{a}^{\alpha} \\ &= \nabla\mathbf{v} \cdot \mathbf{P} \end{aligned} \tag{2-2}$$

where **P** is the projection tensor (see Sec. A.3.2). This may be expressed either in terms of the covariant components of **v**

$$\nabla_{(\sigma)}\mathbf{v} = v_{j,i}\,\frac{\partial x^i}{\partial y^\alpha}\,\mathbf{g}^j\mathbf{a}^\alpha \tag{2-3}$$

or in terms of its contravariant components

$$\nabla_{(\sigma)}\mathbf{v} = v^j{}_{,i}\,\frac{\partial x^i}{\partial y^\alpha}\,\mathbf{g}_j\mathbf{a}^\alpha \tag{2-4}$$

Here $v_{j,i}$ and $v^j{}_{,i}$ are the covariant derivatives of the covariant and contravariant components of **v** respectively (Slattery 1981, p. 640).

From (1-6) and (2-2), the surface divergence of **v** reduces for this case to

$$\begin{aligned} \operatorname{div}_{(\sigma)}\mathbf{v} &= \operatorname{tr}(\ \nabla_{(\sigma)}\mathbf{v}\ ) \\ &= \operatorname{tr}(\ \nabla\mathbf{v}\cdot\mathbf{P}\ ) \end{aligned} \tag{2-5}$$

This may also be written in terms of the covariant and contravariant components of **v** as

$$\operatorname{div}_{(\sigma)}\mathbf{v} = v_{j,i}\,\frac{\partial x^i}{\partial y^\alpha}\,\frac{\partial x^j}{\partial y^\beta}\,a^{\alpha\beta} \tag{2-6}$$

and

$$\operatorname{div}_{(\sigma)}\mathbf{v} = v^j{}_{,i}\,\frac{\partial x^i}{\partial y^\alpha}\,\frac{\partial x^k}{\partial y^\beta}\,a^{\alpha\beta}g_{jk} \tag{2-7}$$

Reference

Slattery, J. C., "Momentum, Energy, and Mass Transfer in Continua," McGraw-Hill, New York (1972); second edition, Robert E. Krieger, Malabar, FL 32950 (1981).

A.5.3 Vector field is explicit function of position on surface Let us now determine how the surface gradient is to be expressed in terms of the components of **w**, when **w** is an explicit function of position (y^1, y^2) on the surface:

$$\mathbf{w} = \mathbf{w}(y^1, y^2) \tag{3-1}$$

If **w** is given in terms of the natural basis fields for some curvilinear coordinate system

$$\mathbf{w} = w^i \mathbf{g}_i \tag{3-2}$$

the surface gradient takes the form

$$\nabla_{(\sigma)}\mathbf{w} = \frac{\partial}{\partial y^\alpha}(w^i \mathbf{g}_i)\,\mathbf{a}^\alpha$$

$$= \frac{\partial w^i}{\partial y^\alpha}\mathbf{g}_i\mathbf{a}^\alpha + w^i \frac{\partial \mathbf{g}_i}{\partial x^j}\frac{\partial x^j}{\partial y^\alpha}\mathbf{a}^\alpha \tag{3-3}$$

This can be simplified by expressing (Slattery 1981, p. 640)

$$\frac{\partial \mathbf{g}_i}{\partial x^j} = \left\{ {m \atop j\ i} \right\} \mathbf{g}_m \tag{3-4}$$

where

$$\left\{ {m \atop j\ i} \right\}$$

is known as the Christoffel symbol of the second kind. Substituting (3-4) into (3-3), we have

$$\nabla_{(\sigma)}\mathbf{w} = \frac{\partial w^i}{\partial y^\alpha}\mathbf{g}_i\mathbf{a}^\alpha + w^i \left\{ {m \atop j\ i} \right\} \frac{\partial x^j}{\partial y^\alpha}\mathbf{g}_m\mathbf{a}^\alpha$$

$$= \left[\frac{\partial w^i}{\partial y^\alpha} + \left\{ {i \atop j\ m} \right\} w^m \frac{\partial x^j}{\partial y^\alpha} \right] \mathbf{g}_i\mathbf{a}^\alpha$$

$$= w^i{}_{,\alpha}\,\mathbf{g}_i\mathbf{a}^\alpha \tag{3-5}$$

I have defined here

$$w^i{}_{,\alpha} \equiv \frac{\partial w^i}{\partial y^\alpha} + \left\{ {i \atop j\,m} \right\} w^m \frac{\partial x^j}{\partial y^\alpha} \tag{3-6}$$

which is known as the **surface covariant derivative** of w^i.

If instead $\mathbf{w}$ is written as a linear combination of the dual basis fields

$$\mathbf{w} = w_i \mathbf{g}^i \tag{3-7}$$

the surface gradient becomes

$$\nabla_{(\sigma)} \mathbf{w} = w_{i,\alpha}\, \mathbf{g}^i \mathbf{a}^\alpha \tag{3-8}$$

where

$$w_{i,\alpha} \equiv \frac{\partial w_i}{\partial y^\alpha} - \left\{ {m \atop j\;\,i} \right\} w_m \frac{\partial x^j}{\partial y^\alpha} \tag{3-9}$$

is referred to as the **surface covariant derivative** of w_i. In deriving (3-8), we have noted (Slattery 1981, p. 643)

$$\frac{\partial \mathbf{g}^i}{\partial x^j} = - \left\{ {i \atop j\,m} \right\} \mathbf{g}^m \tag{3-10}$$

We might more commonly think of $\mathbf{w}$ in terms of its tangential and normal components (see Secs. A.2.6 and A.3.2), either

$$\begin{aligned} \mathbf{w} &= \mathbf{w} \cdot \mathbf{I} \\ &= \mathbf{w} \cdot \mathbf{P} + (\mathbf{w} \cdot \boldsymbol{\xi})\, \boldsymbol{\xi} \\ &= w^\alpha \mathbf{a}_\alpha + w_{(\xi)} \boldsymbol{\xi} \end{aligned} \tag{3-11}$$

or

$$\mathbf{w} = w_\alpha \mathbf{a}^\alpha + w_{(\xi)} \boldsymbol{\xi} \tag{3-12}$$

The surface gradient consequently becomes either

$$\nabla_{(\sigma)} \mathbf{w} = \frac{\partial w^\alpha}{\partial y^\beta} \mathbf{a}_\alpha \mathbf{a}^\beta + w^\alpha \frac{\partial \mathbf{a}_\alpha}{\partial y^\beta} \mathbf{a}^\beta + \frac{\partial w_{(\xi)}}{\partial y^\beta} \boldsymbol{\xi} \mathbf{a}^\beta + w_{(\xi)} \frac{\partial \boldsymbol{\xi}}{\partial y^\beta} \mathbf{a}^\beta \tag{3-13}$$

or

$$\nabla_{(\sigma)}\mathbf{w} = \frac{\partial w_\alpha}{\partial y^\beta}\,\mathbf{a}^\alpha\,\mathbf{a}^\beta + w_\alpha\,\frac{\partial \mathbf{a}^\alpha}{\partial y^\beta}\,\mathbf{a}^\beta + \frac{\partial w_{(\xi)}}{\partial y^\beta}\,\boldsymbol{\xi}\mathbf{a}^\beta + w_{(\xi)}\,\frac{\partial \boldsymbol{\xi}}{\partial y^\beta}\,\mathbf{a}^\beta \qquad (3\text{-}14)$$

Let us rearrange (3-13) and (3-14) into forms that are more convenient for computations. We will begin by examining individual terms in these equations.

Let us start with the second term on the right of (3-13). We can say

$$\begin{aligned}
\frac{\partial \mathbf{a}_\alpha}{\partial y^\beta} &= \frac{\partial}{\partial y^\beta}\left(\frac{\partial x^i}{\partial y^\alpha}\,\mathbf{g}_i\right) \\
&= \frac{\partial^2 x^i}{\partial y^\beta\,\partial y^\alpha}\,\mathbf{g}_i + \frac{\partial x^i}{\partial y^\alpha}\,\frac{\partial \mathbf{g}_i}{\partial y^\beta} \\
&= \frac{\partial^2 x^i}{\partial y^\beta\,\partial y^\alpha}\,\mathbf{g}_i + \frac{\partial x^i}{\partial y^\alpha}\,\frac{\partial x^j}{\partial y^\beta}\left\{\begin{matrix} m \\ j\ \ i \end{matrix}\right\}\mathbf{g}_m \\
&= \left(\frac{\partial^2 x^i}{\partial y^\beta\,\partial y^\alpha} + \frac{\partial x^j}{\partial y^\beta}\,\frac{\partial x^m}{\partial y^\alpha}\left\{\begin{matrix} i \\ j\ m \end{matrix}\right\}\right)\mathbf{g}_i
\end{aligned} \qquad (3\text{-}15)$$

where in the third line we have again employed (3-4). We may also observe that

$$\begin{aligned}
\mathbf{g}_i &= \mathbf{I}\cdot\mathbf{g}_i \\
&= (\,\mathbf{P} + \boldsymbol{\xi}\boldsymbol{\xi}\,)\cdot\mathbf{g}_i \\
&= (\,\mathbf{a}_\mu\,\mathbf{a}^\mu + \boldsymbol{\xi}\boldsymbol{\xi}\,)\cdot\mathbf{g}_i \\
&= \left(\mathbf{a}_\mu\,a^{\mu\nu}\,\frac{\partial x^k}{\partial y^\nu}\,\mathbf{g}_k + \boldsymbol{\xi}\boldsymbol{\xi}\right)\cdot\mathbf{g}_i \\
&= a^{\mu\nu}\,g_{ik}\,\frac{\partial x^k}{\partial y^\nu}\,\mathbf{a}_\mu + \xi_i\,\boldsymbol{\xi}
\end{aligned} \qquad (3\text{-}16)$$

In the second line of this development, we have expressed the identity tensor in terms of the projection tensor as indicated in Sec. A.3.2. Equation (3-16) may be used to learn from (3-15)

$$\frac{\partial \mathbf{a}_\alpha}{\partial y^\beta} = \left(\frac{\partial^2 x^i}{\partial y^\beta \partial y^\alpha} + \frac{\partial x^j}{\partial y^\beta} \frac{\partial x^m}{\partial y^\alpha} \left\{ {i \atop j\,m} \right\} \right) \cdot \left(a^{\mu\nu} g_{ik} \frac{\partial x^k}{\partial y^\nu} \mathbf{a}_\mu + \xi_i \boldsymbol{\xi} \right)$$

$$= \left\{ {\mu \atop \beta\,\alpha} \right\}_a \mathbf{a}_\mu + B_{\beta\alpha} \boldsymbol{\xi} \tag{3-17}$$

Here

$$\left\{ {\mu \atop \beta\,\alpha} \right\}_a \equiv \left(\frac{\partial^2 x^i}{\partial y^\beta \partial y^\alpha} + \frac{\partial x^j}{\partial y^\beta} \frac{\partial x^m}{\partial y^\alpha} \left\{ {i \atop j\,m} \right\} \right) a^{\mu\nu} g_{ik} \frac{\partial x^k}{\partial y^\nu} \tag{3-18}$$

is the **surface Christoffel symbol** of the **second kind** (see also Exercise A.5.3-1) and

$$B_{\beta\alpha} \equiv \left(\frac{\partial^2 x^i}{\partial y^\beta \partial y^\alpha} + \frac{\partial x^j}{\partial y^\beta} \frac{\partial x^m}{\partial y^\alpha} \left\{ {i \atop j\,m} \right\} \right) \xi_i$$

$$= B_{\alpha\beta} \tag{3-19}$$

are the components of the symmetric **second groundform** tangential tensor field (McConnell 1957, p. 201; see also Exercises A.5.3-2 and A.5.6-2):

$$\mathbf{B} \equiv B_{\alpha\beta} \mathbf{a}^\alpha \mathbf{a}^\beta \tag{3-20}$$

The physical significance of **B** will become obvious shortly.

Let us next determine the components of $\partial \boldsymbol{\xi} / \partial y^\beta$ in the fourth term on the right of (3-13). We can start by observing that

$$\mathbf{a}_\gamma \cdot \boldsymbol{\xi} = 0 \tag{3-21}$$

and

$$\boldsymbol{\xi} \cdot \boldsymbol{\xi} = 1 \tag{3-22}$$

Differentiating these equations with respect to the surface coordinate y^β, we have

$$\mathbf{a}_\gamma \cdot \frac{\partial \boldsymbol{\xi}}{\partial y^\beta} = - B_{\gamma\beta} \tag{3-23}$$

and

$$\boldsymbol{\xi} \cdot \frac{\partial \boldsymbol{\xi}}{\partial y^{\beta}} = 0 \tag{3-24}$$

From (3-23) and (3-24), we conclude that $\partial\boldsymbol{\xi}/\partial y^{\beta}$ is a tangential vector field:

$$\frac{\partial \boldsymbol{\xi}}{\partial y^{\beta}} = - B_{\gamma\beta}\, \mathbf{a}^{\gamma} \tag{3-25}$$

We can find the components of $\partial \mathbf{a}^{\alpha}/\partial y^{\beta}$ in the second term on the right of (3-14) in a similar manner, noting that

$$\mathbf{a}_{\mu} \cdot \mathbf{a}^{\alpha} = \delta^{\alpha}_{\mu} \tag{3-26}$$

and

$$\boldsymbol{\xi} \cdot \mathbf{a}^{\alpha} = 0 \tag{3-27}$$

Differentiating these equations with respect to y^{β} and rearranging, we see

$$\begin{aligned} \mathbf{a}_{\mu} \cdot \frac{\partial \mathbf{a}^{\alpha}}{\partial y^{\beta}} &= - \frac{\partial \mathbf{a}_{\mu}}{\partial y^{\beta}} \cdot \mathbf{a}^{\alpha} \\ &= - \left\{ {\nu \atop \beta\ \mu} \right\}_{a} \mathbf{a}_{\nu} \cdot \mathbf{a}^{\alpha} \\ &= - \left\{ {\alpha \atop \beta\ \mu} \right\}_{a} \end{aligned} \tag{3-28}$$

and

$$\begin{aligned} \boldsymbol{\xi} \cdot \frac{\partial \mathbf{a}^{\alpha}}{\partial y^{\beta}} &= - \frac{\partial \boldsymbol{\xi}}{\partial y^{\beta}} \cdot \mathbf{a}^{\alpha} \\ &= B_{\gamma\beta}\, a^{\gamma\alpha} \end{aligned} \tag{3-29}$$

In the second line of (3-28), we have used (3-17); in (3-29), (3-25) has been employed. Equations (3-28) and (3-29) give us the tangential and normal components of $\partial \mathbf{a}^{\alpha}/\partial y^{\beta}$ and we conclude

$$\frac{\partial \mathbf{a}^{\alpha}}{\partial y^{\beta}} = -\left\{ \begin{matrix} \alpha \\ \beta \ \gamma \end{matrix} \right\}_{a} \mathbf{a}^{\gamma} + a^{\alpha\gamma} B_{\beta\gamma} \boldsymbol{\xi} \tag{3-30}$$

With (3-17) and (3-25), we are now in a position to write (3-13) in a more useful form:

$$\begin{aligned} \nabla_{(\sigma)} \mathbf{w} &= \frac{\partial w^{\alpha}}{\partial y^{\beta}} \mathbf{a}_{\alpha} \mathbf{a}^{\beta} + \left\{ \begin{matrix} \mu \\ \beta \ \alpha \end{matrix} \right\}_{a} w^{\alpha} \mathbf{a}_{\mu} \mathbf{a}^{\beta} + B_{\beta\alpha} w^{\alpha} \boldsymbol{\xi} \mathbf{a}^{\beta} \\ &\quad + \frac{\partial w_{(\xi)}}{\partial y^{\beta}} \boldsymbol{\xi} \mathbf{a}^{\beta} - w_{(\xi)} B_{\gamma\beta} \mathbf{a}^{\gamma} \mathbf{a}^{\beta} \\ &= w^{\alpha}{}_{,\beta} \mathbf{a}_{\alpha} \mathbf{a}^{\beta} + \boldsymbol{\xi} (\mathbf{B} \cdot \mathbf{w}) + \boldsymbol{\xi} \nabla_{(\sigma)} w_{(\xi)} - w_{(\xi)} \mathbf{B} \\ &= \mathbf{P} \cdot \nabla_{(\sigma)} (\mathbf{P} \cdot \mathbf{w}) + \boldsymbol{\xi} (\mathbf{B} \cdot \mathbf{w}) + \boldsymbol{\xi} \nabla_{(\sigma)} w_{(\xi)} - w_{(\xi)} \mathbf{B} \end{aligned} \tag{3-31}$$

Here we have introduced the **surface covariant derivative** of w^{α},

$$w^{\alpha}{}_{,\beta} \equiv \frac{\partial w^{\alpha}}{\partial y^{\beta}} + \left\{ \begin{matrix} \alpha \\ \beta \ \mu \end{matrix} \right\}_{a} w^{\mu} \tag{3-32}$$

and we have observed that

$$\mathbf{P} \cdot \nabla_{(\sigma)} (\mathbf{P} \cdot \mathbf{w}) = w^{\alpha}{}_{,\beta} \mathbf{a}_{\alpha} \mathbf{a}^{\beta} \tag{3-33}$$

Equations (3-25) and (3-30) can be used to rearrange (3-14) in a similar manner

$$\nabla_{(\sigma)} \mathbf{w} = w_{\alpha,\beta} \mathbf{a}^{\alpha} \mathbf{a}^{\beta} + \boldsymbol{\xi} (\mathbf{B} \cdot \mathbf{w}) + \boldsymbol{\xi} \nabla_{(\sigma)} w_{(\xi)} - w_{(\xi)} \mathbf{B} \tag{3-34}$$

where

$$w_{\alpha,\beta} \equiv \frac{\partial w_{\alpha}}{\partial y^{\beta}} - \left\{ \begin{matrix} \mu \\ \beta \ \alpha \end{matrix} \right\}_{a} w_{\mu} \tag{3-35}$$

is the **surface covariant derivative** of w_{α}.

Useful expressions for the **surface divergence** of **w**

$$\text{div}_{(\sigma)} \mathbf{w} \equiv \text{tr}(\nabla_{(\sigma)} \mathbf{w}) \tag{3-36}$$

may be obtained either from (3-5) and (3-8)

$$\text{div}_{(\sigma)}\mathbf{w} = w_{i,\alpha} \frac{\partial x^i}{\partial y^\beta} a^{\alpha\beta} \tag{3-37}$$

or from (3-31) and (3-34)

$$\begin{aligned}\text{div}_{(\sigma)}\mathbf{w} &= w^\alpha{}_{,\alpha} - w_{(\xi)}\text{tr}\mathbf{B} \\ &= \text{div}_{(\sigma)}(\mathbf{P} \cdot \mathbf{w}) - 2Hw_{(\xi)}\end{aligned} \tag{3-38}$$

The **mean curvature** of the surface (McConnell 1957, p. 205; see also Exercise A.5.3-3) is defined

$$H \equiv \frac{1}{2}\,\text{tr}\mathbf{B} = \frac{1}{2}\,B_{\alpha\beta}a^{\alpha\beta} \tag{3-39}$$

It follows immediately from (3-38) that

$$H = \frac{1}{2}\,\text{div}_{(\sigma)}\boldsymbol{\xi} \tag{3-40}$$

Finally, the physical significance of **B** may be seen from (3-31):

$$\mathbf{B} = -\,\nabla_{(\sigma)}\boldsymbol{\xi} \tag{3-41}$$

References

Giordano, R. M., personal communication, October 1979.

McConnell, A. J., "Applications of Tensor Analysis," Dover, New York (1957).

Slattery, J. C., "Momentum, Energy, and Mass Transfer in Continua," McGraw-Hill, New York (1972); second edition, Robert E. Krieger, Malabar, FL 32950 (1981).

Exercises

A.5.3-1 *alternative expression for surface Christoffel symbol of the second kind* While (3-18) is sufficient to define the surface Christoffel symbols of the second kind, it is rarely used in practice. A more convenient expression may be developed by looking first at the **surface**

Christoffel symbol of the **first kind**:

$$[\,\beta\alpha, \gamma\,]_a \equiv a_{\gamma\mu} \left\{ {\mu \atop \beta\ \alpha} \right\}_a$$

$$= \left(\frac{\partial^2 x^i}{\partial y^\beta \partial y^\alpha} + \frac{\partial x^j}{\partial y^\beta} \frac{\partial x^m}{\partial y^\alpha} \left\{ {i \atop j\ m} \right\} \right) g_{ik} \frac{\partial x^k}{\partial y^\gamma} \tag{1}$$

i) Prove that

$$\frac{\partial a_{\gamma\beta}}{\partial y^\alpha} = g_{ij} \frac{\partial^2 x^i}{\partial y^\alpha \partial y^\gamma} \frac{\partial x^j}{\partial y^\beta} + g_{ij} \frac{\partial x^i}{\partial y^\gamma} \frac{\partial^2 x^j}{\partial y^\alpha \partial y^\beta} + \left(\left\{ {m \atop k\ i} \right\} g_{mj}\right.$$

$$\left. + \left(\left\{ {m \atop k\ i} \right\} g_{mj} + \left\{ {m \atop k\ j} \right\} g_{im} \right) \frac{\partial x^i}{\partial y^\gamma} \frac{\partial x^j}{\partial y^\beta} \frac{\partial x^k}{\partial y^\alpha} \tag{2}$$

Hint: Observe that

$$g_{ij,k} \equiv \frac{\partial g_{ij}}{\partial x^k} - \left\{ {m \atop k\ i} \right\} g_{mj} - \left\{ {m \atop k\ j} \right\} g_{im} = 0$$

ii) Two similar expressions may be obtained from (2) by rotating the indices α, β, and γ. Combine these to learn

$$\frac{1}{2}\left(\frac{\partial a_{\gamma\beta}}{\partial y^\alpha} + \frac{\partial a_{\alpha\gamma}}{\partial y^\beta} - \frac{\partial a_{\alpha\beta}}{\partial y^\gamma} \right)$$

$$= g_{ij} \frac{\partial^2 x^i}{\partial y^\alpha \partial y^\beta} \frac{\partial x^j}{\partial y^\gamma} + \left\{ {m \atop i\ j} \right\} g_{mk} \frac{\partial x^i}{\partial y^\alpha} \frac{\partial x^j}{\partial y^\beta} \frac{\partial x^k}{\partial y^\gamma} \tag{3}$$

iii) From (1) and (3), conclude that

$$[\,\alpha\beta, \gamma\,]_a = \frac{1}{2}\left(\frac{\partial a_{\gamma\beta}}{\partial y^\alpha} + \frac{\partial a_{\alpha\gamma}}{\partial y^\beta} - \frac{\partial a_{\alpha\beta}}{\partial y^\gamma} \right) \tag{4}$$

and

$$\left\{ {\mu \atop \alpha \ \beta} \right\}_{a} = \frac{a^{\mu\gamma}}{2} \left(\frac{\partial a_{\gamma\beta}}{\partial y^{\alpha}} + \frac{\partial a_{\alpha\gamma}}{\partial y^{\beta}} - \frac{\partial a_{\alpha\beta}}{\partial y^{\gamma}} \right) \tag{5}$$

A.5.3-2 *alternative expression for the components of the second groundform* It is more common to express the components of the second groundform tangential tensor field in terms of

$$\left(\frac{\partial x^{m}}{\partial y^{\alpha}} \right)_{,\beta} \equiv \frac{\partial^{2} x^{m}}{\partial y^{\beta} \partial y^{\alpha}} + \left\{ {m \atop j \ p} \right\} \frac{\partial x^{j}}{\partial y^{\beta}} \frac{\partial x^{p}}{\partial y^{\alpha}} - \left\{ {\mu \atop \beta \ \alpha} \right\}_{a} \frac{\partial x^{m}}{\partial y^{\mu}}$$

Starting with (3-19), prove that

$$B_{\alpha\beta} = \left(\frac{\partial x^{i}}{\partial y^{\alpha}} \right)_{,\beta} \xi_{i}$$

For a different point of view, refer to Exercise A.5.6-2.

A.5.3-3 *curvature of the surface* (McConnell 1957, p. 210) As I have introduced the mean curvature H of a surface in the text, it is a measure of the second groundform tensor

$$\mathbf{B} = - \nabla_{(\sigma)} \boldsymbol{\xi}$$

There are several alternative measures of the curvature of a surface:

mean curvature

$$H \equiv \frac{1}{2} \operatorname{tr} \mathbf{B} = \frac{1}{2} a^{\alpha\beta} B_{\alpha\beta}$$

total curvature

$$K \equiv \det_{(\sigma)} \mathbf{B} = \frac{B_{11} B_{22} - (B_{12})^{2}}{a}$$

normal curvature in the direction $\boldsymbol{\lambda}$

$$\kappa_{n} \equiv \boldsymbol{\lambda} \cdot \mathbf{B} \cdot \boldsymbol{\lambda}$$

Here $\boldsymbol{\lambda}$ is a unit vector tangent to the surface at the point where the normal curvature is being measured. Alternatively, it can be shown that κ_n is the magnitude of the curvature of a normal plane section of the surface in the direction $\boldsymbol{\lambda}$ (McConnell 1957, p. 208).

principal curvatures κ_1 and κ_2 The principal curvatures are the maximum and minimum values of the normal curvature κ_n.

i) Prove that a direction $\boldsymbol{\lambda}$ which corresponds to one of the principal curvatures satisfies

$$(\mathbf{B} - \kappa_n \mathbf{P}) \cdot \boldsymbol{\lambda} = 0$$

ii) How does one argue to conclude that, if a non-zero $\boldsymbol{\lambda}$ exists, we must require

$$\det_{(\sigma)}(\mathbf{B} - \kappa_n \mathbf{P}) = 0$$

or

$$(\kappa_n)^2 - \text{tr}\mathbf{B}\ \kappa_n + \det_{(\sigma)}\mathbf{B} = 0$$

iii) Conclude that the principal curvatures κ_1 and κ_2 may be expressed in terms of the mean and total curvatures by

$$\kappa_1 + \kappa_2 = 2H$$

$$\kappa_1 \kappa_2 = K$$

A.5.3-4 *magnitude of* **B** Prove that

$$| \mathbf{B} |^2 \equiv \text{tr}(\mathbf{B} \cdot \mathbf{B}) = 4H^2 - 2K$$

A.5.3-5 *plane surface* Given a rectangular cartesian coordinate system and a plane surface z_3 = a constant, let us choose

$$y^1 \equiv z_1$$

$$y^2 \equiv z_2$$

i) Prove that all of the surface Christoffel symbols of the second kind are zero (see Exercise A.2.1-4).

ii) Determine that

$$\mathbf{B} = 0$$

iii) Conclude that

$$H = 0$$

$$K = 0$$

A.5.3-6 *plane surface in polar coordinates* Given a cylindrical coordinate system

$$z_1 = x^1 \cos x^2 = r \cos \theta$$

$$z_2 = x^1 \sin x^2 = r \sin \theta$$

$$z_3 = x^3 \equiv z$$

and a plane surface z_3 = a constant, let us choose

$$y^1 \equiv x^1 \equiv r$$

$$y^2 \equiv x^2 \equiv \theta$$

i) Prove that the only non-zero surface Christoffel symbols of the second kind are (see Exercise A.2.1-5)

$$\left\{ {1 \atop 2\;2} \right\}_a = -r$$

$$\left\{ {2 \atop 1\;2} \right\}_a = \left\{ {2 \atop 2\;1} \right\}_a = \frac{1}{r}$$

ii) Determine that

$$\mathbf{B} = 0$$

iii) Conclude that

$$H = 0$$

$$K = 0$$

A.5.3-7 *alternative form for plane surface in polar coordinates* (Giordano 1979) Given the cylindrical coordinate system described in Exercise A.5.3-6 and a plane surface θ = a constant, let us choose

$$y^1 \equiv x^1 \equiv r$$

$$y^2 \equiv x^3 \equiv z$$

i) Prove that all surface Christoffel symbols of the second kind are zero (see Exercise A.2.1-6).

ii) Determine that

$$\mathbf{B} = 0$$

iii) Conclude that

$$H = 0$$

$$K = 0$$

A.5.3-8 *cylindrical surface* Given the above cylindrical coordinate system described in Exercise A.5.3-6 and a cylindrical surface of radius R, let us choose

$$y^1 \equiv x^2 \equiv \theta$$

$$y^2 \equiv x^3 \equiv z$$

i) Prove that all of the surface Christoffel symbols of the second kind are zero (see Exercise A.2.1-7).

ii) Determine that

$$B_{11} = - R$$

$$B_{22} = B_{12} = B_{21} = 0$$

iii) Conclude that

$$H = -\frac{1}{2R}$$

$$K = 0$$

A.5.3-9 *spherical surface* Given a spherical coordinate system

$$z_1 = x^1 \sin x^2 \cos x^3 = r \sin\theta \cos\phi$$

$$z_2 = x^1 \sin x^2 \sin x^3 = r \sin\theta \sin\phi$$

$$z_3 = x^1 \cos x^2 = r \cos\theta$$

and a spherical surface of radius R, let us choose

$$y^1 \equiv x^2 \equiv \theta$$

$$y^2 \equiv x^3 \equiv \phi$$

i) Prove that the only non-zero surface Christoffel symbols of the second kind are (see Exercise A.2.1-8)

$$\left\{ {1 \atop 2\;2} \right\}_a = -\sin\theta \cos\theta$$

and

$$\left\{ {2 \atop 1\;2} \right\}_a = \left\{ {2 \atop 2\;1} \right\}_a = \cot\theta$$

ii) Determine that

$$B_{11} = -R$$

$$B_{22} = -R \sin^2\theta$$

$$B_{12} = B_{21} = 0$$

iii) Conclude that

$$H = -\frac{1}{R}$$

$$K = \frac{1}{R^2}$$

A.5.3-10 *two-dimensional waves* Given a rectangular cartesian coordinate system and a surface

$$z_3 = h(z_1, t)$$

let us choose

$$y^1 \equiv z_1$$

$$y^2 \equiv z_2$$

i) Prove that the only non-zero Christoffel symbol of the second kind is (see Exercise A.2.1-9)

$$\left\{ {1 \atop 1\ 1} \right\}_a = \frac{\partial h}{\partial z_1} \frac{\partial^2 h}{\partial z_1^2} \left[1 + \left(\frac{\partial h}{\partial z_1} \right)^2 \right]^{-1}$$

ii) Determine that

$$B_{11} = \frac{\partial^2 h}{\partial z_1^2} \left[1 + \left(\frac{\partial h}{\partial z_1} \right)^2 \right]^{-1/2}$$

$$B_{22} = B_{12} = B_{21} = 0$$

iii) Conclude that

$$H = \frac{1}{2} \frac{\partial^2 h}{\partial z_1^2} \left[1 + \left(\frac{\partial h}{\partial z_1} \right)^2 \right]^{-3/2}$$

$$K = 0$$

A.5.3-11 *axially symmetric surface in cylindrical coordinates* Given the cylindrical coordinate system described in Exercise A.5.3-6 and an axially symmetric surface

$$z = h(r)$$

let us choose

$$y^1 \equiv r$$

$$y^2 \equiv \theta$$

i) Prove that the only non-zero Christoffel symbols of the second kind are (see Exercise A.2.1-10)

$$\left\{ {1 \atop 1\ 1} \right\}_a = h'h'' \left[1 + (h')^2 \right]^{-1}$$

$$\left\{ {1 \atop 2\ 2} \right\}_a = -r \left[1 + (h')^2 \right]^{-1}$$

$$\left\{ {2 \atop 1\ 2} \right\}_a = \left\{ {2 \atop 2\ 1} \right\}_a = \frac{1}{r}$$

where the primes denote differentiation with respect to r.

ii) Determine that

$$B_{11} = h'' \left[1 + (h')^2 \right]^{-1/2}$$

$$B_{22} = rh' \left[1 + (h')^2 \right]^{-1/2}$$

$$B_{12} = B_{21} = 0$$

iii) Conclude that

$$H = \frac{1}{2r} \left[rh'' + h' + (h')^3 \right] \left[1 + (h')^2 \right]^{-3/2}$$

$$K = \frac{1}{r} h'h'' \left[1 + (h')^2 \right]^{-2}$$

A.5.3-12 *alternative form for rotationally symmetric surface in cylindrical coordinates* Given the cylindrical coordinate system described in Exercise A.2.1-11 and an axially symmetric surface

$$r = c(z)$$

let us choose

$$y^1 \equiv x^3 = z$$

$$y^2 \equiv x^2 = \theta$$

i) Prove that the only non-zero Christoffel symbols of the second kind are (see Exercise A.2.1-11)

$$\left\{ {1 \atop 1\ 1} \right\}_a = c'c'' \left[1 + (c')^2 \right]^{-1}$$

$$\left\{ {1 \atop 2\ 2} \right\}_a = - cc' \left[1 + (c')^2 \right]^{-1}$$

$$\left\{ {2 \atop 1\ 2} \right\}_a = \left\{ {2 \atop 2\ 1} \right\}_a = \frac{c'}{c}$$

where the primes denote differentiation with respect to z.

ii) Determine that

$$B_{11} = c''\left[1 + (c')^2 \right]^{-1/2}$$

$$B_{22} = - c\left[1 + (c')^2 \right]^{-1/2}$$

$$B_{12} = B_{21} = 0$$

iii) Conclude that

$$H = \frac{1}{2c} [cc'' - (c')^2 - 1][1 + (c')^2]^{-3/2}$$

$$K = -\frac{c''}{c} [1 + (c')^2]^{-2}$$

A.5.3-13 Show that

$$(a_{\alpha\beta} w^{\beta})_{,\gamma} = a_{\alpha\beta} w^{\beta}{}_{,\gamma}$$

implying that the $a_{\alpha\beta}$ may be treated as constants with respect to surface covariant differentiation. See also Exercise A.5.6-2.

A.5.3-14 Starting with

$$(a_{\alpha\beta} w^{\beta})_{,\gamma} = \frac{\partial(a_{\alpha\beta} w^{\beta})}{\partial y^{\gamma}} - \left\{ {\mu \atop \gamma\ \alpha} \right\}_a a_{\mu\beta} w^{\beta}$$

rework Exercise A.5.3-13.

A.5.4 Second-order tensor field In a dynamic system, the stress tensor will normally be a function of position on any phase interface. In formulating force balances, consider derivatives of the stress tensor with respect to position on such a surface. With this application in mind, we introduce the surface gradient of a second-order tensor field.

The surface gradient of a second-order tensor field **A** is a third-order tensor field denoted by $\nabla_{(\sigma)}\mathbf{A}$. The gradient is specified by defining how it transforms an arbitrary tangential vector field **c** at all points on the surface:

$$\nabla_{(\sigma)}\mathbf{A}(y^1, y^2) \cdot \mathbf{c} \equiv \text{limit } s \to 0: \frac{1}{s} [\mathbf{A}(y^1 + sc^1, y^2 + sc^2) - \mathbf{A}(y^1, y^2)] \tag{4-1}$$

Equation (4-1) may be rearranged in the manner similar to that suggested in Sec. A.2.2:

$$\nabla_{(\sigma)}\mathbf{A} \cdot \mathbf{c} = \frac{\partial \mathbf{A}}{\partial y^{\alpha}} c^{\alpha} \tag{4-2}$$

Since $\mathbf{c}$ is an arbitrary tangential vector field, we can take $\mathbf{c} = \mathbf{a}_{\beta}$ to conclude that

$$\nabla_{(\sigma)}\mathbf{A} = \frac{\partial \mathbf{A}}{\partial y^{\alpha}} \mathbf{a}^{\alpha} \tag{4-3}$$

The usual divergence operation defined for second-order tensor fields (Slattery 1981, p. 650) suggests that we define the **surface divergence** as

$$\text{div}_{(\sigma)}\mathbf{A} \equiv \frac{\partial \mathbf{A}}{\partial y^{\alpha}} \cdot \mathbf{a}^{\alpha} \tag{4-4}$$

We have not expressed the surface gradient and surface divergence in terms of the components of **A**. In the next sections, we consider two cases: either **A** is an explicit function of position in space or it is an explicit function of position on a surface.

Reference

Slattery, J. C., "Momentum, Energy, and Mass Transfer in Continua," McGraw-Hill, New York (1972); second edition, Robert E. Krieger, Malabar, FL 32950 (1981).

A.5.5 Tensor field is explicit function of position in space In Sec. A.5.4 we defined the surface gradient of a second-order tensor field **A**. Let us see how this operation is to be expressed in terms of the components of **A**, assuming that it is an explicit function of position in space.

If **A** is an explicit function of the curvilinear coordinates (x^1, x^2, x^3)

$$\mathbf{A} = \mathbf{A}(x^1, x^2, x^3) \tag{5-1}$$

we find from (4-3) that the surface gradient becomes

$$\nabla_{(\sigma)}\mathbf{A} = \frac{\partial \mathbf{A}}{\partial x^i}\frac{\partial x^i}{\partial y^\alpha}\,\mathbf{a}^\alpha$$

$$= \frac{\partial \mathbf{A}}{\partial x^i}(\,\mathbf{g}^i \cdot \mathbf{a}_\alpha\,)\mathbf{a}^\alpha$$

$$= \nabla\mathbf{A} \cdot \mathbf{P} \tag{5-2}$$

where **P** is the projection tensor. We may write this either in terms of the covariant components of **A**

$$\nabla_{(\sigma)}\mathbf{A} = A_{jk,i}\,\frac{\partial x^i}{\partial y^\alpha}\,\mathbf{g}^j\mathbf{g}^k\,\mathbf{a}^\alpha \tag{5-3}$$

or its terms of its contravariant components

$$\nabla_{(\sigma)}\mathbf{A} = A^{jk}{}_{,i}\,\frac{\partial x^i}{\partial y^\alpha}\,\mathbf{g}_j\mathbf{g}_k\,\mathbf{a}^\alpha \tag{5-4}$$

Here $A_{jk,i}$ and $A^{jk}{}_{,i}$ are the covariant derivatives of the covariant and contravariant components of **A** respectively (Slattery 1981, p. 651).

From (4-4), (5-3), and (5-4), the surface divergence of **A** may be expressed in terms of its covariant and contravariant components as

$$\text{div}_{(\sigma)}\mathbf{A} = A_{jk,i}\,\frac{\partial x^k}{\partial y^\alpha}\,\frac{\partial x^i}{\partial y^\beta}\,a^{\alpha\beta}\,\mathbf{g}^j \tag{5-5}$$

and

$$\text{div}_{(\sigma)}\mathbf{A} = A^{jk}{}_{,i}\,\frac{\partial x^m}{\partial y^\alpha}\,\frac{\partial x^i}{\partial y^\beta}\,a^{\alpha\beta}\,g_{km}\,\mathbf{g}_j \tag{5-6}$$

Reference

Slattery, J. C., "Momentum, Energy, and Mass Transfer in Continua," McGraw-Hill, New York (1972); second edition, Robert E. Krieger, Malabar, FL 32950 (1981).

A.5.6 Tensor field is explicit function of position on surface Our objective here is to express in terms of its components the surface gradient of a tensor that is an explicit function of position on a surface. We will consider two cases in detail: a surface tensor field and a tangential tensor field.

Let us begin with a surface tensor field **S**

$$\mathbf{S} = S^{i\alpha}\mathbf{g}_i\mathbf{a}_\alpha$$
$$= S_{i\alpha}\mathbf{g}^i\mathbf{a}^\alpha \qquad (6\text{-}1)$$

that is an explicit function of position (y^1, y^2) on a surface:

$$\mathbf{S} = \mathbf{S}(y^1, y^2) \qquad (6\text{-}2)$$

Using (4-3) and ($6\text{-}1_1$), we may write the surface gradient **S** in terms of its contravariant components as

$$\nabla_{(\sigma)}\mathbf{S} = \frac{\partial S^{i\alpha}}{\partial y^\beta}\mathbf{g}_i\mathbf{a}_\alpha\mathbf{a}^\beta + S^{i\alpha}\frac{\partial \mathbf{g}_i}{\partial y^\beta}\mathbf{a}_\alpha\mathbf{a}^\beta + S^{i\alpha}\mathbf{g}_i\frac{\partial \mathbf{a}_\alpha}{\partial y^\beta}\mathbf{a}^\beta$$

$$= \frac{\partial S^{i\alpha}}{\partial y^\beta}\mathbf{g}_i\mathbf{a}_\alpha\mathbf{a}^\beta + S^{i\alpha}\left\{ {n \atop m\ i} \right\}\mathbf{g}_n\frac{\partial x^m}{\partial y^\beta}\mathbf{a}_\alpha\mathbf{a}^\beta$$

$$+ S^{i\alpha}\mathbf{g}_i\left(\left\{ {\mu \atop \beta\ \alpha} \right\}_a\mathbf{a}_\mu + B_{\beta\alpha}\boldsymbol{\xi}\right)\mathbf{a}^\beta$$

$$= S^{i\alpha}{}_{,\beta}\,\mathbf{g}_i\mathbf{a}_\alpha\mathbf{a}^\beta + S^{i\alpha}B_{\alpha\beta}\mathbf{g}_i\boldsymbol{\xi}\mathbf{a}^\beta \qquad (6\text{-}3)$$

where we have introduced the **surface covariant derivative** of $S^{i\alpha}$

$$S^{i\alpha}{}_{,\beta} \equiv \frac{\partial S^{i\alpha}}{\partial y^\beta} + \left\{ {i \atop m\ n} \right\}S^{n\alpha}\frac{\partial x^m}{\partial y^\beta} + \left\{ {\alpha \atop \beta\ \mu} \right\}_a S^{i\mu} \qquad (6\text{-}4)$$

In the second line of (6-3) we have employed (Slattery 1981, p. 640)

$$\frac{\partial \mathbf{g}_i}{\partial x^m} = \left\{ {n \atop m\ i} \right\}\mathbf{g}_n \qquad (6\text{-}5)$$

Here

$$\left\{ {i \atop m\ n} \right\}$$

is the Christoffel symbol of the second kind (Slattery 1981, p. 640) and

$$\left\{ {\alpha \atop \beta\ \mu} \right\}_a$$

is the **surface** Christoffel symbol of the second kind [see (3-18) and Exercise A.5.3-1]. We find in a similar manner that we may also write the surface gradient of **S** in terms of its covariant components:

$$\nabla_{(\sigma)}\mathbf{S} = S_{i\alpha,\beta}\,\mathbf{g}^i\mathbf{a}^\alpha\mathbf{a}^\beta + S_{i\alpha}B_{\gamma\beta}\,a^{\alpha\gamma}\mathbf{g}^i\boldsymbol{\xi}\mathbf{a}^\beta \tag{6-6}$$

where we have defined the **surface covariant derivative** of $S_{i\alpha}$ as

$$S_{i\alpha,\beta} \equiv \frac{\partial S_{i\alpha}}{\partial y^\beta} - \left\{ {n \atop m\ i} \right\} S_{n\alpha}\frac{\partial x^m}{\partial y^\beta} - \left\{ {\mu \atop \beta\ \alpha} \right\}_a S_{i\mu} \tag{6-7}$$

In deriving (6-7), we have noted that (Slattery 1981, p. 643)

$$\frac{\partial \mathbf{g}^i}{\partial x^m} = -\left\{ {i \atop m\ n} \right\}\mathbf{g}^n \tag{6-8}$$

From (4-4), (6-3), and (6-6), the corresponding expressions for the surface divergence of **S** are

$$\mathrm{div}_{(\sigma)}\mathbf{S} = S^{i\alpha}{}_{,\alpha}\,\mathbf{g}_i = S_{i\alpha,\beta}\,a^{\alpha\beta}\mathbf{g}^i \tag{6-9}$$

We find in a completely analogous fashion for a tangential second-order tensor field **T**

$$\mathbf{T} = T^{\alpha\beta}\mathbf{a}_\alpha\mathbf{a}_\beta = T_{\alpha\beta}\,\mathbf{a}^\alpha\mathbf{a}^\beta \tag{6-10}$$

the surface gradient may be developed either in terms of the contravariant components of **T**

$$\nabla_{(\sigma)}\mathbf{T} = T^{\alpha\beta}{}_{,\gamma}\,\mathbf{a}_\alpha\mathbf{a}_\beta\mathbf{a}^\gamma + T^{\mu\beta}B_{\mu\gamma}\boldsymbol{\xi}\mathbf{a}_\beta\mathbf{a}^\gamma + T^{\alpha\mu}B_{\mu\gamma}\mathbf{a}_\alpha\boldsymbol{\xi}\mathbf{a}^\gamma \tag{6-11}$$

or in terms of its covariant components

$$\nabla_{(\sigma)}\mathbf{T} = T_{\alpha\beta,\gamma}\,\mathbf{a}^\alpha\mathbf{a}^\beta\mathbf{a}^\gamma + T_{\mu\beta}B_{\nu\gamma}a^{\mu\nu}\boldsymbol{\xi}\mathbf{a}^\beta\mathbf{a}^\gamma + T_{\alpha\mu}B_{\nu\gamma}a^{\mu\nu}\mathbf{a}^\alpha\boldsymbol{\xi}\mathbf{a}^\gamma \tag{6-12}$$

We have defined here the **surface covariant derivatives** of $T^{\alpha\beta}$ and $T_{\alpha\beta}$:

$$T^{\alpha\beta}{}_{,\gamma} \equiv \frac{\partial T^{\alpha\beta}}{\partial y^{\gamma}} + \left\{ {\alpha \atop \gamma\ \mu} \right\}_{a} T^{\mu\beta} + \left\{ {\beta \atop \gamma\ \mu} \right\}_{a} T^{\alpha\mu} \tag{6-13}$$

$$T_{\alpha\beta,\gamma} \equiv \frac{\partial T_{\alpha\beta}}{\partial y^{\gamma}} - \left\{ {\mu \atop \gamma\ \alpha} \right\}_{a} T_{\mu\beta} - \left\{ {\mu \atop \gamma\ \beta} \right\}_{a} T_{\alpha\mu} \tag{6-14}$$

From (4-4), (6-11), and (6-12), we observe that the surface divergence of **T** takes the forms

$$\begin{aligned} \mathrm{div}_{(\sigma)}\mathbf{T} &= T^{\alpha\beta}{}_{,\beta}\mathbf{a}_{\alpha} + T^{\mu\beta}B_{\mu\beta}\boldsymbol{\xi} \\ &= T_{\alpha\beta,\gamma}a^{\beta\gamma}\mathbf{a}^{\alpha} + T_{\mu\beta}B_{\nu\gamma}a^{\mu\nu}a^{\beta\gamma}\boldsymbol{\xi} \end{aligned} \tag{6-15}$$

References

McConnell, A. J., "Applications of Tensor Analysis," Dover, New York (1957).

Slattery, J. C., "Momentum, Energy, and Mass Transfer in Continua," McGraw-Hill, New York (1972); second edition, Robert E. Krieger, Malabar, FL 32950 (1981).

Exercises

A.5.6-1 Starting with (6-1) and (6-10), derive (6-6), (6-11) and (6-12).

A.5.6-2 *surface gradient of the projection tensor*

i) Prove that

$$\frac{\partial a_{\alpha\beta}}{\partial y^{\gamma}} = a_{\alpha\delta}\left\{ {\delta \atop \gamma\ \beta} \right\}_{a} + a_{\beta\delta}\left\{ {\delta \atop \gamma\ \alpha} \right\}_{a}$$

ii) Deduce that

$$a_{\alpha\beta,\gamma} = 0$$

$$\nabla_{(\sigma)}\mathbf{P} = B_{\alpha\beta}[\, \boldsymbol{\xi}\mathbf{a}^{\alpha}\mathbf{a}^{\beta} + \mathbf{a}^{\alpha}\boldsymbol{\xi}\mathbf{a}^{\beta}\,]$$

and

$$\text{div}_{(\sigma)}\mathbf{P} = \text{tr}\mathbf{B}\boldsymbol{\xi} = 2H\boldsymbol{\xi}$$

Here

$$H \equiv \frac{1}{2}\,\text{tr}\mathbf{B}$$

is the mean curvature.

iii) Prove that

$$\nabla_{(\sigma)}\mathbf{P} = \left(\frac{\partial x^i}{\partial y^{\alpha}}\right)_{,\beta}\mathbf{g}_i\mathbf{a}^{\alpha}\mathbf{a}^{\beta} + B_{\alpha\beta}\mathbf{a}^{\alpha}\boldsymbol{\xi}\mathbf{a}^{\beta}$$

iv) Conclude that (McConnell 1957, p. 200)

$$\left(\frac{\partial x^i}{\partial y^{\alpha}}\right)_{,\beta} = B_{\alpha\beta}\xi^i$$

or

$$B_{\alpha\beta} = \left(\frac{\partial x^i}{\partial y^{\alpha}}\right)_{,\beta}\xi_i$$

Compare this derivation with that of Exercise A.5.3-2.

A.5.6-3 *geodesic coordinates and the surface gradient of the tangential cross tensor* From Exercise A.4.1-1, we recognize

$$\nabla_{(\sigma)}\boldsymbol{\varepsilon}^{(\sigma)} - \varepsilon^{\mu\alpha}B_{\beta\mu}[\, \boldsymbol{\xi}\mathbf{a}_{\alpha}\mathbf{a}^{\beta} - \mathbf{a}_{\alpha}\boldsymbol{\xi}\mathbf{a}^{\beta}\,] = \varepsilon_{\alpha\beta,\gamma}\mathbf{a}^{\alpha}\mathbf{a}^{\beta}\mathbf{a}^{\gamma}$$

as a tangential third-order tensor field, the properties of which are independent of the surface coordinate system chosen.

We can always choose a surface coordinate system in which all of the surface Christoffel symbols vanish at a given point 0. Such coordinates are known as the **geodesic coordinates** for the point 0 (McConnell 1957, p. 175).

Use a geodesic coordinate system to prove that in general (McConnell 1957, p. 181)

$$\varepsilon^{\alpha\beta}{}_{,\gamma} = \varepsilon_{\alpha\beta,\gamma} = 0$$

and

$$\nabla_{(\sigma)}\boldsymbol{\varepsilon}^{(\sigma)} = \varepsilon^{\mu\alpha} B_{\beta\mu} \left[\boldsymbol{\xi}\mathbf{a}_{\alpha}\mathbf{a}^{\beta} - \mathbf{a}_{\alpha}\boldsymbol{\xi}\mathbf{a}^{\beta} \right]$$

A.5.6-4 Prove that (McConnell 1957, p. 182)

$$\frac{\partial \ln\sqrt{a}}{\partial y^{\gamma}} = \left\{ \begin{matrix} \beta \\ \gamma\ \beta \end{matrix} \right\}_{a}$$

(Hint: Write out in full $\varepsilon_{\alpha\beta,\gamma} = 0$ and set $\alpha, \beta = 1, 2$)

A.5.6-5 *surface divergence of a tangential vector field* Let $\mathbf{w}$ be a tangential vector field. Use Exercise A.5.6-4 to prove

$$\begin{aligned} \text{div}_{(\sigma)}\mathbf{w} &= w^{\alpha}{}_{,\alpha} \\ &= \frac{1}{\sqrt{a}} \frac{\partial}{\partial y^{\alpha}} \left(\sqrt{a}\, w^{\alpha} \right) \end{aligned}$$

A.5.6-6 *alternative expression for the surface divergence of a tangential second-order tensor field* Let $\mathbf{T}$ be a tangential second-order tensor field. Prove that

$$\text{div}_{(\sigma)}\mathbf{T} = \left(\frac{\partial x^{i}}{\partial y^{\alpha}} T^{\alpha\beta} \right)_{,\beta} \mathbf{g}_{i}$$

A.5.6-7 *second covariant surface derivative* $v^{\alpha}{}_{,\beta\gamma}$ *is not symmetric in* β *and* γ It is common to use a comma followed by a double index as a shorthand for the **second covariant derivative**. We will write for example

$$v^{i}{}_{,\alpha\beta} \equiv \left(v^{i}{}_{,\alpha} \right)_{,\beta}$$

i) Prove that

$$v^i_{,\alpha\beta} = v^i_{,\beta\alpha}$$

ii) Calculate that for a tangential vector field **v**

$$0 = v^i_{,\alpha\beta} - v^i_{,\beta\alpha}$$

$$= \xi^i_{,\beta} B_{\gamma\alpha} v^\gamma + \xi^i B_{\gamma\alpha,\beta} v^\gamma + \frac{\partial x^i}{\partial y^\gamma} v^\gamma_{,\alpha\beta}$$

$$- \xi^i_{,\alpha} B_{\gamma\beta} v^\gamma - \xi^i B_{\gamma\beta,\alpha} v^\gamma - \frac{\partial x^i}{\partial y^\gamma} v^\gamma_{,\beta\alpha}$$

iii) Observe from Sec. A.5.3 that

$$\xi^i_{,\beta} = - B^\mu_\beta \frac{\partial x^i}{\partial y^\mu}$$

Starting with the result of (ii) and

$$g_{ij} \frac{\partial x^j}{\partial y^\nu} (v^i_{,\alpha\beta} - v^i_{,\beta\alpha}) = 0$$

conclude that

$$v_{\alpha,\beta\gamma} - v_{\alpha,\gamma\beta} = R_{\rho\alpha\beta\gamma} v^\rho$$

which means that $v_{\alpha,\beta\gamma}$ is not symmetric in β and γ. Here

$$R_{\rho\alpha\beta\gamma} \equiv K \varepsilon_{\rho\alpha} \varepsilon_{\beta\gamma}$$

are known as the components of the **surface Riemann-Christoffel tensor** (McConnell 1957, p. 182 and 204). The total curvature K is defined in Exercise A.5.3-3.

A.5.6-8 *Codazzi equations of the surface* Starting with the result of Exercise A.5.6-7 (ii), prove that

$$B_{\alpha\beta,\gamma} = B_{\alpha\gamma,\beta}$$

These are known as the Codazzi equations of the surface (McConnell 1957, p. 204). Hint: Multiply by ξ_i.

A.6 Integration

A.6.1 Line integration On a number of occasions in the text, we use line integrations. These are in the form of integrations with respect to arc length s along a curve C in space:

$$\int_C F \, ds$$

Usually the integrand will not be an explicit function of arc length. We wish to consider here what form the integral will take when expressed in terms of some other parameter t.

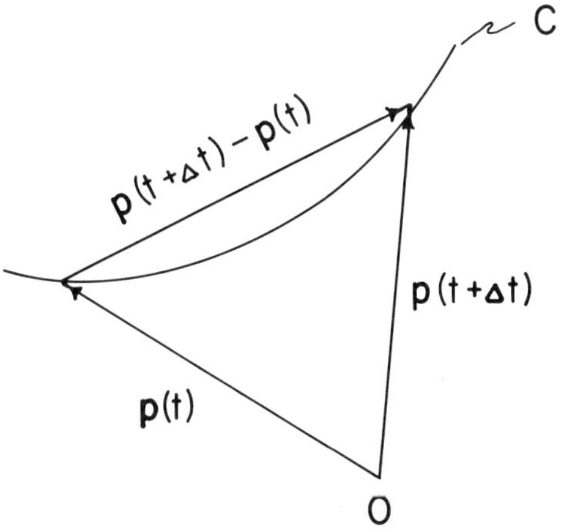

Figure A.6.1-1: The parameter t is measured along the curve C.

If t is a parameter measured along the curve C in Figure A.6.1-1, we define

$$\frac{d\mathbf{p}}{dt}(\,t\,) \equiv \text{limit } \Delta t \to 0: \quad \frac{\mathbf{p}(\,t + \Delta t\,) - \mathbf{p}(\,t\,)}{\Delta t} \tag{1-1}$$

Figure A.6.1-1 suggests that $d\mathbf{p}(t)/dt$ is a tangent vector to the curve at point C. **Arc length** s is that parameter such that the length of the vector [$\mathbf{p}(\,s + \Delta s\,) - \mathbf{p}(s)$] approaches Δs in the limit as $\Delta s \to 0$. This means that $d\mathbf{p}/ds$ is a unit vector or

$$\frac{d\mathbf{p}}{ds} \cdot \frac{d\mathbf{p}}{ds} = 1 \tag{1-2}$$

In terms of any parameter t, it follows from (1-2) that

$$ds = \left(\frac{d\mathbf{p}}{ds} \cdot \frac{d\mathbf{p}}{ds}\right)^{1/2} ds$$

$$= \left[\left(\frac{d\mathbf{p}}{dt}\frac{dt}{ds}\right) \cdot \left(\frac{d\mathbf{p}}{dt}\frac{dt}{ds}\right)\right]^{1/2} ds$$

$$= \left(\frac{d\mathbf{p}}{dt} \cdot \frac{d\mathbf{p}}{dt}\right)^{1/2} dt \qquad (1\text{-}3)$$

and the line integration may be expressed as

$$\int_C F\, ds = \int_C F(t) \left(\frac{d\mathbf{p}}{dt} \cdot \frac{d\mathbf{p}}{dt}\right)^{1/2} dt \qquad (1\text{-}4)$$

Let us consider a particular example in which we have a line integration over a segment $C_{(2)}$ of the y^2 coordinate curve. The integrand with which we are concerned may be an explicit function of y^2. From (1-3), we find that

$$ds_{(2)} = \left(\frac{\partial \mathbf{p}}{\partial y^2} \cdot \frac{\partial \mathbf{p}}{\partial y^2}\right)^{1/2} dy^2$$

$$= \sqrt{\mathbf{a}_2 \cdot \mathbf{a}_2}\, dy^2$$

$$= \sqrt{a_{22}}\, dy^2 \qquad (1\text{-}5)$$

and the line integration becomes

$$\int_{C_{(2)}} F\, ds_{(2)} = \int_{C_{(2)}} F(y^2)\sqrt{a_{22}}\, dy^2 \qquad (1\text{-}6)$$

A.6.2 Surface integration We repeatedly use surface integrations in the text. These are in the form of integrations with respect to area over some surface Σ in space, for which we employ the notation

$$\int_{\Sigma} F \, dA$$

Let us see how this integral is expressed as a double integration over the surface coordinates y^1 and y^2.

At any point on Σ, let us denote the unit tangent vectors to the two surface coordinate curves by $\partial \mathbf{p}/\partial s_{(1)}$ and $\partial \mathbf{p}/\partial s_{(2)}$. Here $s_{(\alpha)}$ denotes the arc length measured along the y^α coordinate curve. Exercise A.3.3-2 indicates that the area dA of the parallelogram formed from $\partial \mathbf{p}/\partial s_{(1)}$ and $\partial \mathbf{p}/\partial s_{(2)}$ with sides of length $ds_{(1)}$ and $ds_{(2)}$ is

$$dA \equiv \frac{\partial \mathbf{p}}{\partial s_{(1)}} \cdot \left(\boldsymbol{\varepsilon}^{(\sigma)} \cdot \frac{\partial \mathbf{p}}{\partial s_{(2)}} \right) ds_{(1)} ds_{(2)} \tag{2-1}$$

or

$$\begin{aligned} dA &= \frac{\partial \mathbf{p}}{\partial y^1} \cdot \left(\boldsymbol{\varepsilon}^{(\sigma)} \cdot \frac{\partial \mathbf{p}}{\partial y^2} \right) dy^1 dy^2 \\ &= \mathbf{a}_1 \cdot (\boldsymbol{\varepsilon}^{(\sigma)} \cdot \mathbf{a}_2) \, dy^1 dy^2 \\ &= \varepsilon^{\alpha\beta} a_{\alpha 1} a_{\beta 2} dy^1 dy^2 \\ &= \sqrt{a} \, dy^1 dy^2 \end{aligned} \tag{2-2}$$

where (see Sec. A.2.1)

$$a \equiv \det(a_{\alpha\beta}) \tag{2-3}$$

Consider a polyhedron

$$\sum_{k=1}^{N} \Sigma_k$$

each planar element Σ_k (k = 1,...,N) of which is tangent to Σ. Identify the maximum diagonal D_N of any element of this polyhedron. Form a sequence of such polyhedra, ordered with decreasing maximum diagonals. The areas of these polyhedra approach the area of Σ as $N \to \infty$ and $D_N \to 0$, which leads us to express an area integration over Σ as

$$\int_{\Sigma} F\, dA = \text{limit } N \to \infty \text{ and } D_N \to 0: \sum_{k=1}^{N} \int_{\Sigma_k} F\, dA$$

$$= \iint_{\Sigma} F(y^1, y^2)\, \sqrt{a}\, dy^1 dy^2 \tag{2-4}$$

A.6.3 Surface divergence theorem Green's transformation (Slattery 1981, p. 661; Ericksen 1960, p. 815) is an invaluable tool. Our objective here is to develop a similar theorem for surfaces.

If $\mathbf{w}$ is a tangential vector field

$$\mathbf{w} = w^{\alpha} \mathbf{a}_{\alpha} \tag{3-1}$$

then

$$\int_{\Sigma} \text{div}_{(\sigma)} \mathbf{w}\, dA = \int\int_{\Sigma} w^{\alpha}{}_{,\alpha}\, \sqrt{a}\, dy^1 dy^2$$

$$= \int\int_{\Sigma} \frac{\partial(\sqrt{a}\, w^{\alpha})}{\partial y^{\alpha}}\, dy^1 dy^2$$

$$= \int\int_{\Sigma} \left[\frac{\partial(\sqrt{a}\, w^1)}{\partial y^1} + \frac{\partial(\sqrt{a}\, w^2)}{\partial y^2} \right] dy^1 dy^2 \tag{3-2}$$

In the second line, we have employed a result from Exercise A.5.6-5.

Green's theorem (Kaplan 1952, p. 239) for a surface tells us that, if $P(y^1, y^2)$ and $Q(y^1, y^2)$ are continuous functions having continuous first partial derivatives on the surface, then

$$\int\int_{\Sigma} \left(\frac{\partial P}{\partial y^1} + \frac{\partial Q}{\partial y^2} \right) dy^1 dy^2 = \int_{C} \left(P \frac{dy^2}{ds} - Q \frac{dy^1}{ds} \right) ds \tag{3-3}$$

Here C is the piecewise smooth simple closed curve bounding Σ; s indicates the arc length measured along this curve in the positive sense. A sense of direction along a closed curve on the surface is defined by referring to the simple circuit shown in Figure A.6.3-1: $y^1 = 0$, $y^2 = 0$, $y^1 = 1$, $y^2 = 1$. The **positive sense** of direction along this circuit is that in which the curves are named; the opposite sense is **negative**. The sense of direction for every other circuit is obtained by comparison with this simple one. Equation (3-3) may now be used to arrange (3-2) in the form

$$\int_{\Sigma} \text{div}_{(\sigma)}\mathbf{w}\, dA = \int_{C} \sqrt{a} \left(w^1 \frac{dy^2}{ds} - w^2 \frac{dy^1}{ds} \right) ds$$

$$= \int_{C} \varepsilon_{\alpha\beta} w^{\alpha} \frac{dy^{\beta}}{ds} ds \tag{3-4}$$

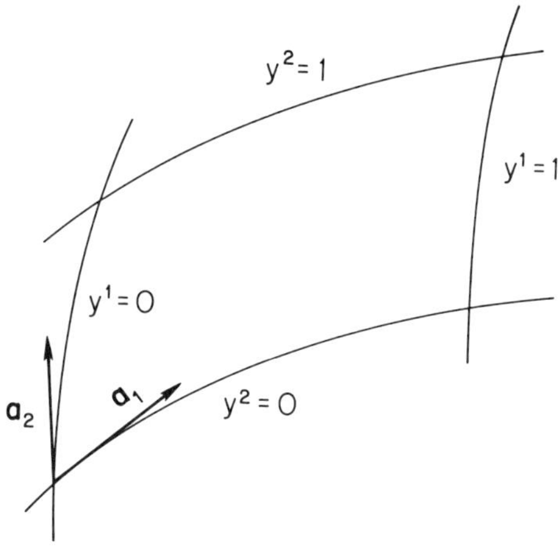

Figure A.6.3-1: The positive sense of direction along the circuit $y^1 = 0$, $y^2 = 0$, $y^1 = 1$, $y^2 = 1$ is defined to be that in which the curves are named; the opposite sense is negative. The sense of direction for every other circuit is obtained by comparison with this simple one.

Clearly

$$\boldsymbol{\lambda} \equiv \frac{d\mathbf{p}}{ds} = \frac{\partial \mathbf{p}}{\partial y^{\beta}} \frac{dy^{\beta}}{ds} = \frac{dy^{\beta}}{ds} \mathbf{a}_{\beta} \tag{3-5}$$

is the unit tangent vector to the curve C. From Sec. A.3.3, it follows that

$$\boldsymbol{\mu} \equiv \boldsymbol{\varepsilon}^{(\sigma)} \cdot \boldsymbol{\lambda} = \varepsilon_{\alpha\beta} \frac{dy^{\beta}}{ds} \mathbf{a}^{\alpha} \tag{3-6}$$

is the tangent vector field normal to the curve C; it is oriented with respect to C in such a manner that the rotation from $\boldsymbol{\lambda}$ to $\boldsymbol{\mu}$ is negative. Because of our requirement that s is measured along C in the positive sense, it is clear that $\boldsymbol{\mu}$ is outwardly directed with respect to the closed curve. In terms of $\boldsymbol{\mu}$, (3-4) becomes (McConnell 1957, p. 188)

$$\int_{\Sigma} \text{div}_{(\sigma)}\mathbf{w} \, dA = \int_{C} \mathbf{w} \cdot \boldsymbol{\mu} \, ds \tag{3-7}$$

More generally, if $\mathbf{v}$ is a spatial vector field defined on the surface, we found in Sec. A.5.3 that

$$\text{div}_{(\sigma)} \mathbf{v} = \text{div}_{(\sigma)}(\mathbf{P} \cdot \mathbf{v}) - 2H \, \mathbf{v} \cdot \boldsymbol{\xi} \tag{3-8}$$

We may consequently use (3-7) to conclude that

$$\int_{\Sigma} \text{div}_{(\sigma)}\mathbf{v} \, dA = \int_{C} \mathbf{v} \cdot \boldsymbol{\mu} \, ds - \int_{\Sigma} 2H \, \mathbf{v} \cdot \boldsymbol{\xi} \, dA \tag{3-9}$$

We will refer to this hereafter as the **surface divergence theorem**.

In Exercise A.6.3-1, the surface divergence theorem is extended to second-order surface tensors.

References

Ericksen, J. L., "Handbuch der Physik," vol. 3/1, edited by S. Flügge, Springer-Verlag, Berlin (1960).

Kaplan, Wilfred, "Advanced Calculus," Addison-Wesley, Cambridge, Massachusetts (1952).

McConnell, A. J., "Applications of Tensor Analysis," Dover, New York (1957).

Slattery, J. C., "Momentum, Energy and Mass Transfer in Continua," McGraw-Hill, New York (1972); second edition, Robert E. Krieger, Malabar, FL 32950 (1981).

Exercises

A.6.3-1 *surface divergence theorem for surface tensors* If $\mathbf{S}$ is a second-order surface tensor, prove that

$$\int_{\Sigma} \text{div}_{(\sigma)}\mathbf{S}\, dA = \int_{C} \mathbf{S} \cdot \boldsymbol{\mu}\, ds$$

A.6.3-2 *surface divergence theorem for second-order tensors* Let **A** be any second-order tensor that is an explicit function of position on the surface Σ. Starting with

$$\mathbf{A} = \mathbf{A} \cdot \mathbf{I}$$
$$= \mathbf{A} \cdot \mathbf{P} + \mathbf{A} \cdot \boldsymbol{\xi}\boldsymbol{\xi}$$

prove that

$$\nabla_{(\sigma)}\mathbf{A} = \nabla_{(\sigma)}(\mathbf{A} \cdot \mathbf{P}) + \frac{\partial(\mathbf{A} \cdot \boldsymbol{\xi})}{\partial y^{\alpha}}\boldsymbol{\xi}\mathbf{a}^{\alpha} - (\mathbf{A} \cdot \boldsymbol{\xi})\mathbf{B}$$

and

$$\text{div}_{(\sigma)}\mathbf{A} = \text{div}_{(\sigma)}(\mathbf{A} \cdot \mathbf{P}) - 2H(\mathbf{A} \cdot \boldsymbol{\xi})$$

Conclude that

$$\int_{\Sigma} \text{div}_{(\sigma)}\mathbf{A}\, dA = \int_{C} \mathbf{A} \cdot \boldsymbol{\mu}\, ds - \int_{\Sigma} 2H(\mathbf{A} \cdot \boldsymbol{\xi})\, dA$$

A.6.3-3 *another consequence of the surface divergence theorem* Let ϕ be a scalar field on Σ. Prove that

$$\int_{\Sigma} (\nabla_{(\sigma)}\phi + 2H\phi\boldsymbol{\xi})dA = \int_{\Sigma} [\nabla_{(\sigma)}\phi - (\text{div}_{(\sigma)}\boldsymbol{\xi})\phi\boldsymbol{\xi}]dA$$
$$= \int_{C} \phi\boldsymbol{\mu}\, ds$$

Hint: Apply the surface divergence theorem to $\phi\mathbf{P}$.

A.6.3-4 *Stokes' theorem* Use Exercises A.3.3-3 and A.5.6-2 in order to prove

$$\varepsilon^{ijk} v_{k,j} \xi_i = (\varepsilon^{\alpha\beta} v_\beta)_{,\alpha}$$

Employ this expression to prove Stokes' theorem:

$$\int_\Sigma \boldsymbol{\xi} \cdot \operatorname{curl} \mathbf{v} \, dA = \int_C \mathbf{v} \cdot \boldsymbol{\lambda} \, ds$$

Here $\boldsymbol{\lambda}$ is the unit tangent vector to the curve C bounding Σ. The understanding here is that $\boldsymbol{\xi} \cdot (\mathbf{a}_1 \wedge \mathbf{a}_2)$ is positive (Sec. A.2.6) and that the rotation from $\boldsymbol{\lambda}$ to $\boldsymbol{\mu}$ is positive (Sec. A.3.3).

Name index[a,b]

---

a) References to "Momentum, Energy, and Mass Transfer in Continua" [J. C. Slattery, McGraw-Hill, New York (1972); second edition, Robert E. Krieger, Malabar, FL 32950 (1981)] have been left out.

b) When there are multiple references in a sub-section, only the first is listed.

Subject Index[a]

---

a) When there are multiple references in a sub-section, only the first is listed.